THIRD EDITION

FOOD LIPIDS
Chemistry, Nutrition, and Biotechnology

Edited by
Casimir C. Akoh • David B. Min

CRC Press
Taylor & Francis Group
Boca Raton   London   New York

CRC Press is an imprint of the
Taylor & Francis Group, an **informa** business

CRC Press
Taylor & Francis Group
6000 Broken Sound Parkway NW, Suite 300
Boca Raton, FL 33487-2742

© 2008 by Taylor & Francis Group, LLC
CRC Press is an imprint of Taylor & Francis Group, an Informa business

No claim to original U.S. Government works
Printed in the United States of America on acid-free paper
10 9 8 7 6 5 4 3 2 1

International Standard Book Number-13: 978-1-4200-4663-2 (Hardcover)

Library of Congress Cataloging-in-Publication Data

Food lipids : chemistry, nutrition, and biotechnology / edited by Casimir C. Akoh and David B. Min.
   -- 3rd ed.
      p. ; cm.
   Includes bibliographical references and index.
   ISBN-13: 978-1-4200-4663-2 (hardcover : alk. paper)
   ISBN-10: 1-4200-4663-2 (hardcover : alk. paper)
   1. Lipids. 2. Lipids in human nutrition. 3. Lipids--Biotechnology. 4. Lipids--Metabolism. I. Akoh, Casimir C., 1955- II. Min, David B.
   [DNLM: 1. Lipids--chemistry. 2. Lipids--physiology. 3. Biotechnology--methods. 4. Food. 5. Nutrition Physiology. QU 85 F6865 2008] I. Title.

QP751.F647 2008
612'.01577--dc22   2007031989

Visit the Taylor & Francis Web site at
http://www.taylorandfrancis.com

and the CRC Press Web site at
http://www.crcpress.com

Contents

PART I Chemistry and Properties

PART II Processing

PART III   Oxidation and Antioxidants

PART IV   Nutrition

PART V   Biotechnology and Biochemistry

Preface to the Third Edition

The first edition of *Food Lipids* was published in 1998 and the second edition in 2002 by Marcel Dekker, Inc. Taylor & Francis Group, LLC, acquired Marcel Dekker and the rights to publish the third edition. We firmly believe that this book has been of interest and will help those involved in lipid research and instruction. Many have bought the previous editions and we thank you for your support. The need to update the information in the second edition cannot be overstated, as more data and new technologies are constantly becoming available. We have received good comments and suggestions on how to improve the second edition. The response reassured us that there was indeed a great need for a textbook suitable for teaching food lipids, nutritional aspects of lipids, and lipid chemistry courses to food science and nutrition majors. The aim of the first and second editions remains unchanged: to provide a modern, easy-to-read textbook for students and instructors. The book is also suitable for upper-level undergraduate, graduate, and postgraduate instruction. Scientists who have left the university and are engaged in research and development in the industry, government, or academics will find this book a useful reference. In this edition, we have expanded on lipid oxidation and antioxidants, as these continue to be topics of great interest to the modern consumer. The title of Part III has also been changed to reflect the recent interest on the importance of antioxidants and health. Again, we have made every effort to select contributors who are internationally recognized experts. We thank them for their exceptional attention to details and timely submissions of their chapters.

Overall, the text has been updated with new and available information. We removed some chapters and added new ones. Chapter 2 includes a brief discussion of sphingolipids, and Chapter 31 includes one on diacylglycerols. The new additions are Chapters 13, 16, 17, and 25. Although it is not possible to cover all aspects of lipids, we feel we have added and covered most topics that are of interest to our readers. The book still is divided into five main parts: Chemistry and Properties; Processing; Oxidation and Antioxidants; Nutrition; and Biotechnology and Biochemistry.

We are grateful to the readers and users of the previous editions and can only hope that we have improved and updated the latest edition to your satisfaction. We welcome comments on the third edition to help us continue to provide our readers with factual information on the science of lipids. Based on the comments of readers and reviewers of the past editions, we have improved the third edition—we hope, without creating new errors, which are sometimes unavoidable for a book this size and complexity. We apologize for any errors in advance and urge you to contact us if you find mistakes or have suggestions to improve the readability and comprehension of this text.

Special thanks to our readers and students, and to the editorial staff of Taylor & Francis Group, LLC, for their helpful suggestions toward improving the quality of this edition.

Casimir C. Akoh
David B. Min

Editors

Casimir C. Akoh is a distinguished research professor of food science and technology and an adjunct professor of foods and nutrition at the University of Georgia, Athens. He is the coeditor of the book *Carbohydrates as Fat Substitutes* (Marcel Dekker, Inc.), coeditor of *Healthful Lipids* (AOCS Press), editor of *Handbook of Functional Lipids* (CRC Press), the author or coauthor of over 162 referenced SCI publications, more than 30 book chapters, and the holder of three U.S. patents. He is a fellow of the Institute of Food Technologists (2005), American Oil Chemists' Society (2006), and the American Chemical Society (2006). He serves on the editorial boards of five journals and is a member of the Institute of Food Technologists, the American Oil Chemists' Society, and the American Chemical Society. He has received numerous international professional awards for his work on lipids including the 1998 IFT Samuel Cate Prescott Award, the 2003 D.W. Brooks Award, and the 2004 AOCS Stephen S. Chang Award. He received his PhD (1988) in food science from Washington State University, Pullman. He holds MS and BS degrees in biochemistry from Washington State University and the University of Nigeria, Nsukka, respectively.

David B. Min's major research objective is to improve the oxidative and flavor stability of foods by understanding and controlling the chemical mechanisms for the flavor compound formation by a combination of GC, HPLC, IR, NMR, ESR, and MS. Dr. Min's group painstakingly, conclusively, and scientifically developed the novel chemical mechanisms for the formation of sunlight flavor in milk, reversion flavor in soybean oil, and light sensitivity of riboflavin. He is a pioneer for the formation, reaction mechanisms and kinetics, quenching mechanisms and kinetics singlet oxygen in foods. He has published 6 books and more than 200 publications.

He has been scientific editor of *Journal of Food Science* and *Journal of the American Oil Chemists' Society* and has been on the editorial board of *Journal of Critical Reviews on Food Science and Nutrition, Journal of Food Quality, Food Chemistry, International News on Fats and Oils, Food Science and Biochemistry*, and Marcel Dekker Publications.

He has received more than 30 national and international awards including the 1995 IFT Achievement Award of Lipid and Flavor Chemistry, the 1999 Distinguished Senior Faculty Research Award, the 2001 IFT Food Chemistry Lectureship Award, the 2002 Professor of the Year Award, and the 2004 Outstanding Teaching Award. He has been an elected member of the Korean National Academy of Science, and a fellow of the Institute of Food Technologists, the American Oil Chemists' Society, the American Institute of Chemists, and the International Academy of Food Science and Technology.

Contributors

Casimir C. Akoh
Department of Food Science and Technology
University of Georgia
Athens, Georgia

Angela D. Bell
Department of Chemistry
Auburn University
Auburn, Alabama

Eunok Choe
Department of Food and Nutrition
Inha University
Incheon, Korea

Eric A. Decker
Department of Food Science
University of Massachusetts
Amherst, Massachusetts

Anthony J. Del Vecchio
Monsanto, Inc.
Davis, California

Paul S. Dimick
Department of Food Science
Pennsylvania State University
University Park, Pennsylvania

Ronald R. Eitenmiller
Department of Food Science and Technology
University of Georgia
Athens, Georgia

Marilyn C. Erickson
Center for Food Safety
Department of Food Science and Technology
University of Georgia
Griffin, Georgia

J. Bruce German
Department of Food Science and Technology
University of California
Davis, California

Howard Perry Glauert
Department of Nutrition and Food Science
University of Kentucky
Lexington, Kentucky

Barbara Mullen Grossman
Department of Foods and Nutrition
University of Georgia
Athens, Georgia

Frank D. Gunstone
Scottish Crop Research Institute
Invergowrie, Dundee, Scotland

Dorothy B. Hausman
Department of Foods and Nutrition
University of Georgia
Athens, Georgia

Lawrence A. Johnson
Center for Crops Utilization Research
Department of Food Science and Human
  Nutrition
Iowa State University
Ames, Iowa

Byung Hee Kim
Department of Food Science and Technology
University of Georgia
Athens, Georgia

Hyun Jung Kim
Department of Food Science and Technology
Ohio State University
Columbus, Ohio

David M. Klurfeld
United States Department of Agriculture
Agricultural Research Service
Beltsville, Maryland

Vic C. Knauf
Monsanto, Inc.
Davis, California

David Kritchevsky (Late)
Wistar Institute
Philadelphia, Pennsylvania

Oi-Ming Lai
Department of Bioprocess Technology
Universiti Putra Malaysia
Serdang Selangor, Malaysia

Patrick J. Lawler
McCormick and Company, Inc.
Cockeysville, Maryland

Shengrong Li
Department of Chemistry
Auburn University
Auburn, Alabama

Yong Li
Department of Food Science
Purdue University
West Lafayette, Indiana

Dorris A. Lillard
Department of Food Science and Technology
University of Georgia
Athens, Georgia

Alejandro G. Marangoni
Department of Food Science
University of Guelph
Guelph, Ontario, Canada

D. Julian McClements
Department of Food Science
University of Massachusetts
Amherst, Massachusetts

Richard E. McDonald
Food and Drug Administration
National Center for Food Safety and
  Technology
Summit-Argo, Illinois

Ronald P. Mensink
Department of Human Biology
Maastricht University
Maastricht, The Netherlands

David B. Min
Department of Food Science and Technology
Ohio State University
Columbus, Ohio

Kazuo Miyashita
Graduate School of Fisheries Sciences
Hokkaido University
Hakodate, Japan

Magdi M. Mossoba
Food and Drug Administration
Center for Food Safety and Applied Nutrition
College Park, Maryland

Kumar D. Mukherjee
Institute for Lipid Research
Federal Research Centre for Nutrition and Food
Munster, Germany

Sean Francis O'Keefe
Department of Food Science and Technology
Virginia Polytechnic Institute and State
  University
Blacksburg, Virginia

Edward J. Parish
Department of Chemistry
Auburn University
Auburn, Alabama

Jogchum Plat
Department of Human Biology
Maastricht University
Maastricht, The Netherlands

David W. Reische
Dannon Company, Inc.
Fort Worth, Texas

Dérick Rousseau
School of Nutrition
Ryerson Polytechnic University
Toronto, Ontario, Canada

Fereidoon Shahidi
Department of Biochemistry
Memorial University of Newfoundland
St. John's, Newfoundland, Canada

P.K.J.P.D. Wanasundara
Agriculture Agri-Food Canada
Saskatoon Research Centre
Saskatoon, Saskatchewan, Canada

Udaya N. Wanasundara
POS Pilot Plant Corporation
Saskatoon, Canada

Kathleen Warner
National Center for Agricultural Utilization
  Research
Agricultural Research Service
U.S. Department of Agriculture
Peoria, Illinois

Bruce A. Watkins
Department of Food Science
Purdue University
West Lafayette, Indiana

Steven M. Watkins
FAME Analytics
West Sacramento, California

Nikolaus Weber
Institute for Lipid Research
Federal Research Centre for Nutrition and Food
Munster, Germany

John D. Weete
West Virginia University
Morgantown, West Virginia

Wendy M. Willis
Yves Veggie Cuisine
Vancouver, British Columbia, Canada

Part I

Chemistry and Properties

1 Nomenclature and Classification of Lipids

Sean Francis O'Keefe

CONTENTS

I. DEFINITIONS OF LIPIDS

No exact definition of lipids exists. Christie [1] defines lipids as "a wide variety of natural products including fatty acids and their derivatives, steroids, terpenes, carotenoids, and bile acids, which have in common a ready solubility in organic solvents such as diethyl ether, hexane, benzene, chloroform, or methanol."

Kates [2] says that lipids are "those substances which are (a) insoluble in water; (b) soluble in organic solvents such as chloroform, ether or benzene; (c) contain long-chain hydrocarbon groups in their molecules; and (d) are present in or derived from living organisms."

Gurr and James [3] point out that the standard definition includes "a chemically heterogeneous group of substances, having in common the property of insolubility in water, but solubility in nonpolar solvents such as chloroform, hydrocarbons or alcohols."

Despite common usage, definitions based on solubility have obvious problems. Some compounds that are considered lipids, such as C1–C4 very short-chain fatty acids (VSCFAs), are completely miscible with water and insoluble in nonpolar solvents. Some researchers have accepted this solubility definition strictly and exclude C1–C3 fatty acids in a definition of lipids, keeping C4 (butyric acid) only because of its presence in dairy fats. Additionally, some compounds that are considered lipids, such as some *trans* fatty acids (those not derived from bacterial hydrogenation), are not derived directly from living organisms. The development of synthetic acaloric and reduced calorie lipids complicates the issue because they may fit into solubility-based definitions but are not derived from living organisms, may be acaloric, and may contain esters of VSCFAs.

The traditional definition of total fat of foods used by the U.S. Food and Drug Administration (FDA) has been the "sum of the components with lipid characteristics that are extracted by Association of Official Analytical Chemists (AOAC) methods or by reliable and appropriate procedures." The FDA has changed from a solubility-based definition to "total lipid fatty acids expressed as triglycerides" [4], with the intent to measure caloric fatty acids. Solubility and size of fatty acids affect their caloric values. This is important for products that take advantage of this, such as Benefat/Salatrim, so these products would be examined on a case-by-case basis. Food products containing sucrose polyesters would require special methodology to calculate caloric fatty acids. Foods containing vinegar (~4.5% acetic acid) present a problem because they will be considered to have 4.5% fat unless the definition is modified to exclude water-soluble fatty acids or the caloric weighting for acetic acid is lowered [4].

Despite the problems with accepted definitions, a more precise working definition is difficult, given the complexity and heterogeneity of lipids. This chapter introduces the main lipid structures and their nomenclature.

II. LIPID CLASSIFICATIONS

Classification of lipid structures is possible based on physical properties at room temperature (oils are liquid and fats are solid), their polarity (polar and neutral lipids), their essentiality for humans (essential and nonessential fatty acids), or their structure (simple or complex). Neutral lipids include fatty acids, alcohols, glycerides, and sterols, whereas polar lipids include glycerophospholipids and glyceroglycolipids. The separation into polarity classes is rather arbitrary, as some short-chain fatty acids are very polar. A classification based on structure is, therefore, preferable.

Based on structure, lipids can be classified as derived, simple, or complex. The derived lipids include fatty acids and alcohols, which are the building blocks for the simple and complex lipids. Simple lipids, composed of fatty acids and alcohol components, include acylglycerols, ether acylglycerols, sterols, and their esters and wax esters. In general terms, simple lipids can be hydrolyzed to two different components, usually an alcohol and an acid. Complex lipids include glycerophospholipids (phospholipids), glyceroglycolipids (glycolipids), and sphingolipids. These structures yield three or more different compounds on hydrolysis.

The fatty acids constitute the obvious starting point in lipid structures. However, a short review of standard nomenclature is appropriate. Over the years, a large number of different nomenclature systems have been proposed [5]. The resulting confusion has led to a need for nomenclature standardization. The International Union of Pure and Applied Chemists (IUPAC) and International Union of Biochemistry (IUB) collaborative efforts have resulted in comprehensive nomenclature standards [6], and the nomenclature for lipids has been reported [7–9]. Only the main aspects of the standardized IUPAC nomenclature relating to lipid structures will be presented; greater detail is available elsewhere [7–9].

Standard rules for nomenclature must take into consideration the difficulty in maintaining strict adherence to structure-based nomenclature and elimination of common terminology [5]. For example, the compound known as vitamin K_1 can be described as 2-methyl-3-phytyl-1,4-naphtho-quinone. Vitamin K_1 and many other trivial names have been included into standardized nomenclature to avoid confusion arising from long chemical names. Standard nomenclature rules will be discussed in separate sections relating to various lipid compounds.

Fatty acid terminology is complicated by the existence of several different nomenclature systems. The IUPAC nomenclature, common (trivial) names, and shorthand (n- or ω) terminology will be discussed. As a lipid class, the fatty acids are often called free fatty acids (FFAs) or nonesterified fatty acids (NEFAs). IUPAC has recommended that fatty acids as a class be called fatty acids and the terms FFA and NEFA eliminated [6].

A. STANDARD IUPAC NOMENCLATURE OF FATTY ACIDS

In standard IUPAC terminology [6], the fatty acid is named after the parent hydrocarbon. Table 1.1 lists common hydrocarbon names. For example, an 18-carbon carboxylic acid is called octadecanoic acid, from octadecane, the 18-carbon aliphatic hydrocarbon. The name octadecanecarboxylic acid may also be used, but it is more cumbersome and less common. Table 1.2 summarizes the rules for hydrocarbon nomenclature.

Double bonds are designated using the Δ configuration, which represents the distance from the carboxyl carbon, naming the carboxyl carbon number 1. A double bond between the ninth and tenth carbons from the carboxylic acid group is a $\Delta 9$ bond. The hydrocarbon name is changed to indicate the presence of the double bond; an 18-carbon fatty acid with one double bond is called octadece-noic acid, one with two double bonds octadecadienoic acid, etc. The double-bond positions are designated with numbers before the fatty acid name ($\Delta 9$-octadecenoic acid or simply 9-octadecenoic acid). The Δ is assumed and often not placed explicitly in structures.

TABLE 1.1
Systematic Names of Hydrocarbons

Carbon Number	Name	Carbon Number	Name
1n	Methane	19	Nonadecane
2	Ethane	20	Eicosane
3	Propane	21	Henicosane
4	Butane	22	Docosane
5	Pentane	23	Tricosane
6	Hexane	24	Tetracosane
7	Heptane	25	Pentacosane
8	Octane	26	Hexacosane
9	Nonane	27	Heptacosane
10	Decane	28	Octacosane
11	Hendecane	29	Nonacosane
12	Dodecane	30	Triacontane
13	Tridecane	40	Tetracontane
14	Tetradecane	50	Pentacontane
15	Pentadecane	60	Hexacontane
16	Hexadecane	70	Heptacontane
17	Heptadecane	80	Octacontane
18	Octadecane		

TABLE 1.2

IUPAC Rules for Hydrocarbon Nomenclature

1. Saturated unbranched acyclic hydrocarbons are named with a numerical prefix and the termination "ane." The first four in this series use trivial prefix names (methane, ethane, propane, and butane), whereas the rest use prefixes that represent the number of carbon atoms.
2. Saturated branched acyclic hydrocarbons are named by prefixing the side chain designation to the name of the longest chain present in the structure.
3. The longest chain is numbered to give the lowest number possible to the side chains, irrespective of the substituents.
4. If more than two side chains are present, they can be cited either in alphabetical order or in order of increasing complexity.
5. If two or more side chains are present in equivalent positions, the one assigned the lowest number is cited first in the name. Order can be based on alphabetical order or complexity.
6. Unsaturated unbranched acyclic hydrocarbons with one double bond have the "ane" replaced with "ene." If there is more than one double bond, the "ane" is replaced with "diene," "triene," "tetraene," etc. The chain is numbered to give the lowest possible number to the double bonds.

Source: From IUPAC in *Nomenclature of Organic Chemistry, Sections A, B, C, D, E, F, and H,* Pergamon Press, London, 1979, 182.

Double-bond geometry is designated with the *cis–trans* or *E/Z* nomenclature systems [6]. The *cis/trans* terms are used to describe the positions of atoms or groups connected to doubly bonded atoms. They can also be used to indicate relative positions in ring structures. Atoms/groups are *cis* or *trans* if they lie on same (*cis*) or opposite (*trans*) sides of a reference plane in the molecule. Some examples are shown in Figure 1.1. The prefixes *cis* and *trans* can be abbreviated as *c* and *t* in structural formulas.

The *cis/trans* configuration rules are not applicable to double bonds that are terminal in a structure or to double bonds that join rings to chains. For these conditions, a sequence preference ordering must be conducted. Since *cis/trans* nomenclature is applicable only in some cases, a new nomenclature system was introduced by the Chemical Abstracts Service and subsequently adopted by IUPAC (the *E/Z* nomenclature). This system was developed as a more applicable system to describe isomers by using sequence ordering rules, as is done using the *R/S* system (rules to decide which ligand has priority). The sequence rule-preferred atom/group attached to one of a pair of doubly bonded carbon atoms is compared with the sequence rule-preferred atom/group of the other of the doubly bonded carbon atoms. If the preferred atom/groups are on the same side of the reference plane, it is the *Z* configuration. If they are on the opposite sides of the plane, it is the *E* configuration. Table 1.3 summarizes some of the rules for sequence preference [10]. Although *cis* and *Z* (or *trans* and *E*) do not always refer to the same configurations, for most fatty acids *E* and *trans* are equivalent, as are *Z* and *cis*.

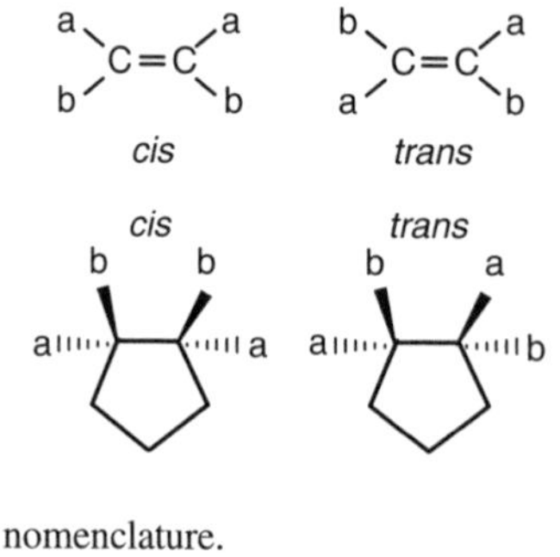

FIGURE 1.1 Examples of *cis/trans* nomenclature.

TABLE 1.3

Summary of Sequence Priority Rules for *E/Z* Nomenclature

1. Higher atomic number precedes lower.
2. For isotopes, higher atomic mass precedes lower.
3. If the atoms attached to one of the double-bonded carbons are the same, proceed outward concurrently until a point of difference is reached considering atomic mass and atomic number.
4. Double bonds are treated as if each bonded atom is duplicated.

Source: From Streitwieser Jr., A. and Heathcock, C.H. in *Introduction to Organic Chemistry*, Macmillan, New York, 1976, 111.

B. Common (Trivial) Nomenclature of Fatty Acids

Common names have been introduced throughout the years and, for certain fatty acids, are a great deal more common than standard (IUPAC) terminology. For example, oleic acid is much more common than *cis*-9-octadecenoic acid. Common names for saturated and unsaturated fatty acids are illustrated in Tables 1.4 and 1.5. Many of the common names originate from the first identified

TABLE 1.4

**Systematic, Common, and Shorthand Names
of Saturated Fatty Acids**

Systematic Name	Common Name	Shorthand
Methanoic	Formic	1:0
Ethanoic	Acetic	2:0
Propanoic	Propionic	3:0
Butanoic	Butyric	4:0
Pentanoic	Valeric	5:0
Hexanoic	Caproic	6:0
Heptanoic	Enanthic	7:0
Octanoic	Caprylic	8:0
Nonanoic	Pelargonic	9:0
Decanoic	Capric	10:0
Undecanoic	—	11:0
Dodecanoic	Lauric	12:0
Tridecanoic	—	13:0
Tetradecanoic	Myristic	14:0
Pentadecanoic	—	15:0
Hexadecanoic	Palmitic	16:0
Heptadecanoic	Margaric	17:0
Octadecanoic	Stearic	18:0
Nonadecanoic	—	19:0
Eicosanoic	Arachidic	20:0
Docosanoic	Behenic	22:0
Tetracosanoic	Lignoceric	24:0
Hexacosanoic	Cerotic	26:0
Octacosanoic	Montanic	28:0
Tricontanoic	Melissic	30:0
Dotriacontanoic	Lacceroic	32:0

TABLE 1.5

Systematic, Common, and Shorthand Names of Unsaturated Fatty Acids

Systematic Name	Common Name	Shorthand
c-9-Dodecenoic	Lauroleic	12:1ω3
c-5-Tetradecenoic	Physeteric	14:1ω9
c-9-Tetradecenoic	Myristoleic	14:1ω5
c-9-Hexadecenoic	Palmitoleic	16:1ω7
c-7,*c*-10,*c*-13-Hexadecatrienoic	—	16:3ω3
c-4,*c*-7,*c*-10,*c*-13-Hexadecatetraenoic	—	16:4ω3
c-9-Octadecenoic	Oleic	18:1ω9
c-11-Octadecenoic	*cis*-Vaccenic (Asclepic)	18:1ω7
t-11-Octadecenoic	Vaccenic	[a]
t-9-Octadecenoic	Elaidic	[a]
c-9,*c*-12-Octadecadienoic	Linoleic	18:2ω6
c-9-*t*-11-Octadecadienoic acid	Rumenic[b]	[a]
c-9,*c*-12,*c*-15-Octadecatrienoic	Linolenic	18:3ω3
c-6,*c*-9,*c*-12-Octadecatrienoic	γ-Linolenic	18:3ω6
c-6,*c*-9,*c*-12,*c*-15-Octadecatetraenoic	Stearidonic	18:4ω3
c-11-Eicosenoic	Gondoic	20:1ω9
c-9-Eicosenoic	Gadoleic	20:1ω11
c-8,*c*-11,*c*-14-Eicosatrienoic	Dihomo-γ-linolenic	20:3ω6
c-5,*c*-8,*c*-11-Eicosatrienoic	Mead's	20:3ω9
c-5,*c*-8,*c*-11,*c*-14-Eicosatetraenoic	Arachidonic	20:4ω6
c-5,*c*-8,*c*-11,*c*-14,*c*-17-Eicosapentaenoic	Eicosapentaenoic	20:5ω3
c-13-Docosenoic	Erucic	22:1ω9
c-11-Docosenoic	Cetoleic	22:1ω11
c-7,*c*-10,*c*-13,*c*-16,*c*-19-Docosapentaenoic	DPA, Clupanodonic	22:5ω3
c-4,*c*-7,*c*-10,*c*-13,*c*-16,*c*-19-Docosahexaenoic	DHA, Cervonic	22:6ω3
c-15-Tetracosenoic	Nervonic (Selacholeic)	24:1ω9

[a] Shorthand nomenclature cannot be used to name *trans* fatty acids.
[b] One of the conjugated linoleic acid (CLA) isomers.

botanical or zoological origins for those fatty acids. Myristic acid is found in seed oils from the Myristicaceae family. Mistakes have been memorialized into fatty acid common names; margaric acid (heptadecanoic acid) was once incorrectly thought to be present in margarine. Some of the common names can pose memorization difficulties, such as the following combinations: caproic, caprylic, and capric; arachidic and arachidonic; linoleic, linolenic, γ-linolenic, and dihomo-γ-linolenic. Even more complicated is the naming of EPA, or eicosapentaenoic acid, usually meant to refer to *c*-5,*c*-8,*c*-11,*c*-14,*c*-17-eicosapentaenoic acid, a fatty acid found in fish oils. However, a different isomer *c*-2,*c*-5,*c*-8,*c*-11,*c*-14-eicosapentaenoic acid is also found in nature. Both can be referred to as eicosapentaenoic acids using standard nomenclature. Nevertheless, in common nomenclature, EPA refers to the *c*-5,*c*-8,*c*-11,*c*-14,*c*-17 isomer. Docosahexaenoic acid (DHA) refers to all-*cis* 4,7,10,13,16,19-docosahexaenoic acid.

C. SHORTHAND (ω) NOMENCLATURE OF FATTY ACIDS

Shorthand (*n*- or ω) identifications of fatty acids are found in common usage. The shorthand designation is the carbon number in the fatty acid chain followed by a colon, then the number of double bonds and the position of the double bond closest to the methyl side of the fatty acid

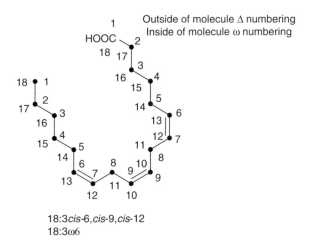

Outside of molecule Δ numbering
Inside of molecule ω numbering

18:3*cis*-6,*cis*-9,*cis*-12
18:3ω6

FIGURE 1.2 IUPAC Δ and common ω numbering systems.

molecule. The methyl group is number 1 (the last character in the Greek alphabet is ω, hence the end). In shorthand notation, the unsaturated fatty acids are assumed to have *cis* bonding and, if the fatty acid is polyunsaturated, double bonds are in the methylene-interrupted positions (Figure 1.2). In this example, CH₂ (methylene) groups at Δ8 and Δ11 interrupt what would otherwise be a conjugated bond system.

Shorthand terminology cannot be used for fatty acids with *trans* or acetylene bonds, for those with additional functional groups (branched, hydroxy, etc.), or for double-bond systems (≥2 double bonds) that are not methylene interrupted (isolated or conjugated). Despite the limitations, shorthand terminology is very popular because of its simplicity and because most of the fatty acids of nutritional importance can be named. Sometimes the ω is replaced by *n*- (18:2*n*-6 instead of 18:2ω6). Although there have been recommendations to eliminate ω and use *n*- exclusively [6], both *n*- and ω are commonly used in the literature and are equivalent.

Shorthand designations for polyunsaturated fatty acids (PUFAs) are sometimes reported without the ω term (18:3). However, this notation is ambiguous, since 18:3 could represent 18:3ω1, 18:3ω3, 18:3ω6, or 18:3ω9 fatty acids, which are completely different in their origins and nutritional significances. Two or more fatty acids with the same carbon and double-bond numbers are possible in many common oils. Therefore, the ω terminology should always be used with the ω term specified.

III. LIPID CLASSES

A. Fatty Acids

1. Saturated Fatty Acids

The saturated fatty acids begin with methanoic (formic) acid. Methanoic, ethanoic, and propanoic acids are uncommon in natural fats and are often omitted from definitions of lipids. However, they are found nonesterified in many food products. Omitting these fatty acids because they are water soluble would argue for also eliminating butyric acid, which would be difficult given its importance in dairy fats. The simplest solution is to accept the very short-chain carboxylic acids as fatty acids while acknowledging the rarity in natural fats of these water-soluble compounds. The systematic, common, and shorthand designations of some saturated fatty acids are given in Table 1.4.

2. Unsaturated Fatty Acids

By far, the most common monounsaturated fatty acid is oleic acid (18:1ω9), although more than 100 monounsaturated fatty acids have been identified in nature. The most common double-bond position for monoenes is Δ9. However, certain families of plants have been shown to accumulate what would be considered unusual fatty acid patterns. For example, *Eranthis* seed oil contains Δ5 monoenes and nonmethylene-interrupted PUFAs containing Δ5 bonds [11]. Erucic acid (22:1ω9) is found at high levels (40%–50%) in Cruciferae such as rapeseed and mustard seed. Canola is a rapeseed oil that is low in erucic acid (<2% 22:1ω9).

PUFAs are best described in terms of families because of the metabolism that allows interconversion within, but not among, families of PUFA. The essentiality of ω6 fatty acids has been known since the late 1920s. Signs of ω6 fatty acid deficiency include decreased growth, increased epidermal water loss, impaired wound healing, and impaired reproduction [12,13]. Early studies did not provide clear evidence that ω3 fatty acids are essential. However, since the 1970s, evidence has accumulated illustrating the essentiality of the ω3 PUFA.

Not all PUFAs are essential fatty acids (EFAs). Plants are able to synthesize de novo and interconvert ω3 and ω6 fatty acid families via desaturases with specificity in the Δ12 and Δ15 positions. Animals have Δ5, Δ6, and Δ9 desaturase enzymes and are unable to synthesize the ω3 and ω6 PUFAs de novo. However, extensive elongation and desaturation of EFA occurs (primarily in the liver). The elongation and desaturation of 18:2ω6 is illustrated in Figure 1.3. The most common of the ω6 fatty acids in our diets is 18:2ω6. Often considered the parent of the ω6 family, 18:2ω6 is first desaturated to 18:3ω6. The rate of this first desaturation is thought to be limiting in premature infants, in the elderly, and under certain disease states. Thus, a great deal of interest has been placed in the few oils that contain 18:3ω6, γ-linolenic acid (GLA). Relatively rich sources of GLA include black currant, evening primrose, and borage oils. GLA is elongated to 20:3ω6, dihomo-γ-linolenic acid (DHGLA). DHGLA is the precursor molecule to the 1-series prostaglandins. DHGLA is further desaturated to 20:4ω6, precursor to the 2-series prostaglandins. Further elongation and desaturation to 22:4ω6 and 22:5ω6 can occur, although the exact function of these fatty acids remains obscure. Relatively high levels of these fatty acids are found in caviar from wild but not cultured sturgeon.

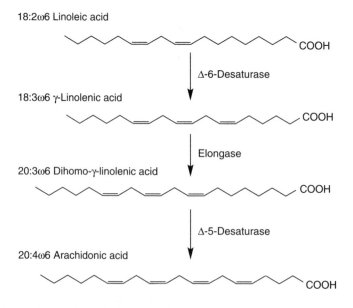

FIGURE 1.3 Pathway of 18:2ω6 metabolism to 20:4ω6.

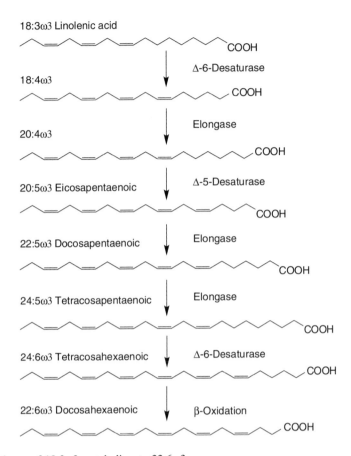

FIGURE 1.4 Pathway of 18:3ω3 metabolism to 22:6ω3.

Figure 1.4 illustrates analogous elongation and desaturation of 18:3ω3. The elongation of 20:5ω3 to 22:5ω3 was thought for many years to be via Δ4 desaturase. The inexplicable difficulty in identifying and isolating the putative Δ4 desaturase led to the conclusion that it did not exist, and the pathway from 20:5ω3 to 22:6ω3 was elucidated as a double elongation, desaturation, and β-oxidation.

One of the main functions of the EFAs is their conversion to metabolically active prostaglandins and leukotrienes [14,15]. Examples of some of the possible conversions from 20:4ω6 are shown in Figures 1.5 and 1.6 [15]. The prostaglandins are called eicosanoids as a class and originate from the action of cyclooxygenase on 20:4ω6 to produce PGG_2. The standard nomenclature of prostaglandins allows usage of the names presented in Figure 1.5. For a name such as PGG_2, the PG represents prostaglandin, the next letter (G) refers to its structure (Figure 1.7), and the subscript number refers to the number of double bonds in the molecule.

The parent structure for most of the prostaglandins is prostanoic acid (Figure 1.7) [14]. Thus, the prostaglandins can be named based on this parent structure. In addition, they can be named using standard nomenclature rules. For example, prostaglandin E_2 (PGE$_2$) is named (5Z,11α,13E,15S)-11,15-dihydroxy-9-oxoprosta-5,13-dienoic acid using the prostanoic acid template. It can also be named using standard nomenclature as 7-[3-hydroxy-2-(3-hydroxy-1-octenyl)-5-oxocyclopentyl]-*cis*-5-heptenoic acid.

The leukotrienes are produced from 20:4ω6 via 5-, 12-, or 15-lipoxygenases to a wide range of metabolically active molecules. The nomenclature is shown in Figure 1.6.

It is important to realize that there are 1-, 2-, and 3-series prostaglandins originating from 20:3ω6, 20:4ω6, and 20:5ω3, respectively. The structures of the 1-, 2-, and 3-prostaglandins differ

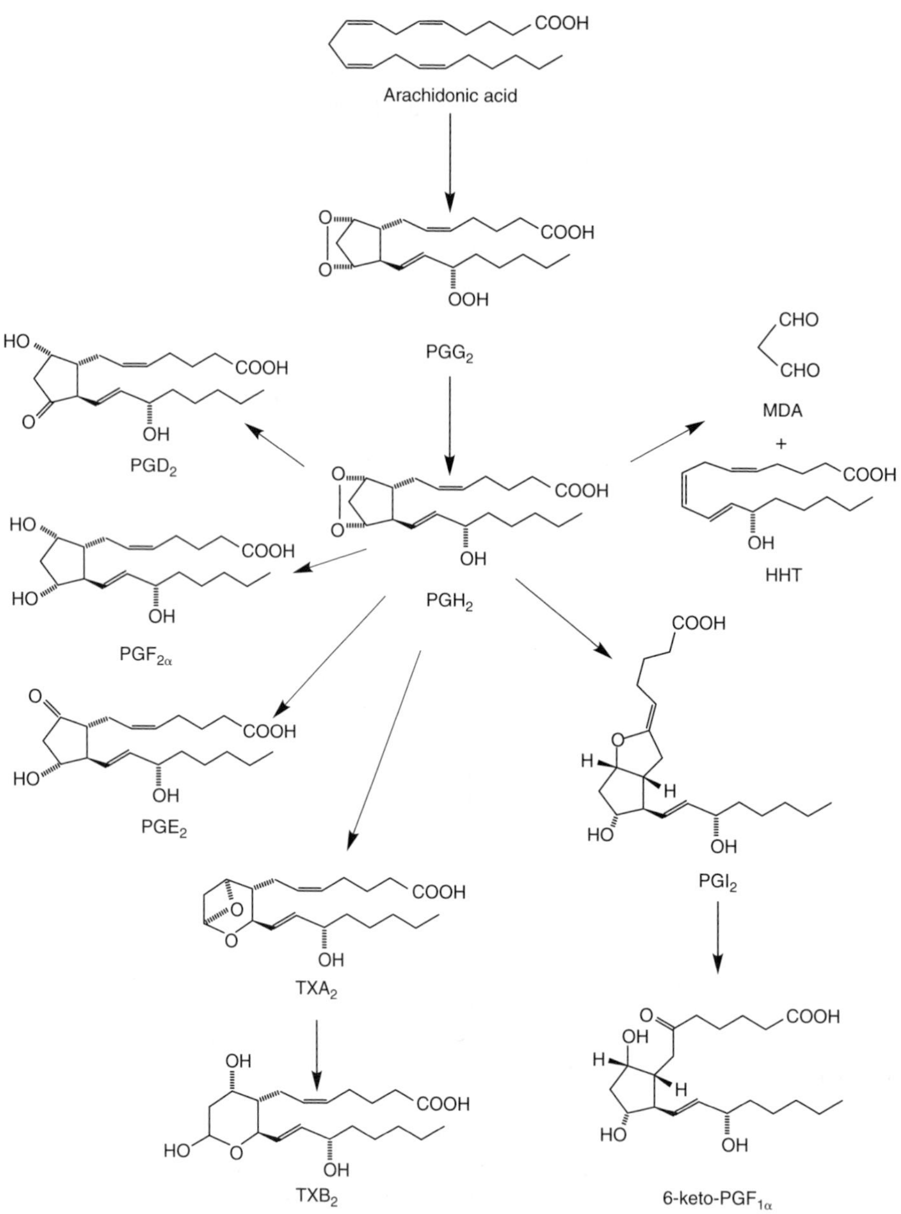

FIGURE 1.5 Prostaglandin metabolites of 20:4ω6.

by the removal or addition of the appropriate double bonds. Leukotrienes of the 3-, 4-, and 5-series are formed via lipoxygenase activity on 20:3ω6, 20:4ω6, and 20:5ω3. A great deal of interest has been focused on changing proportions of the prostaglandins and leukotrienes of the various series by diet to modulate various diseases.

3. Acetylenic Fatty Acids

A number of different fatty acids have been identified having triple bonds [16]. The nomenclature is similar to double bonds, except that the -ane ending of the parent alkane is replaced with -ynoic acid, -diynoic acid, etc.

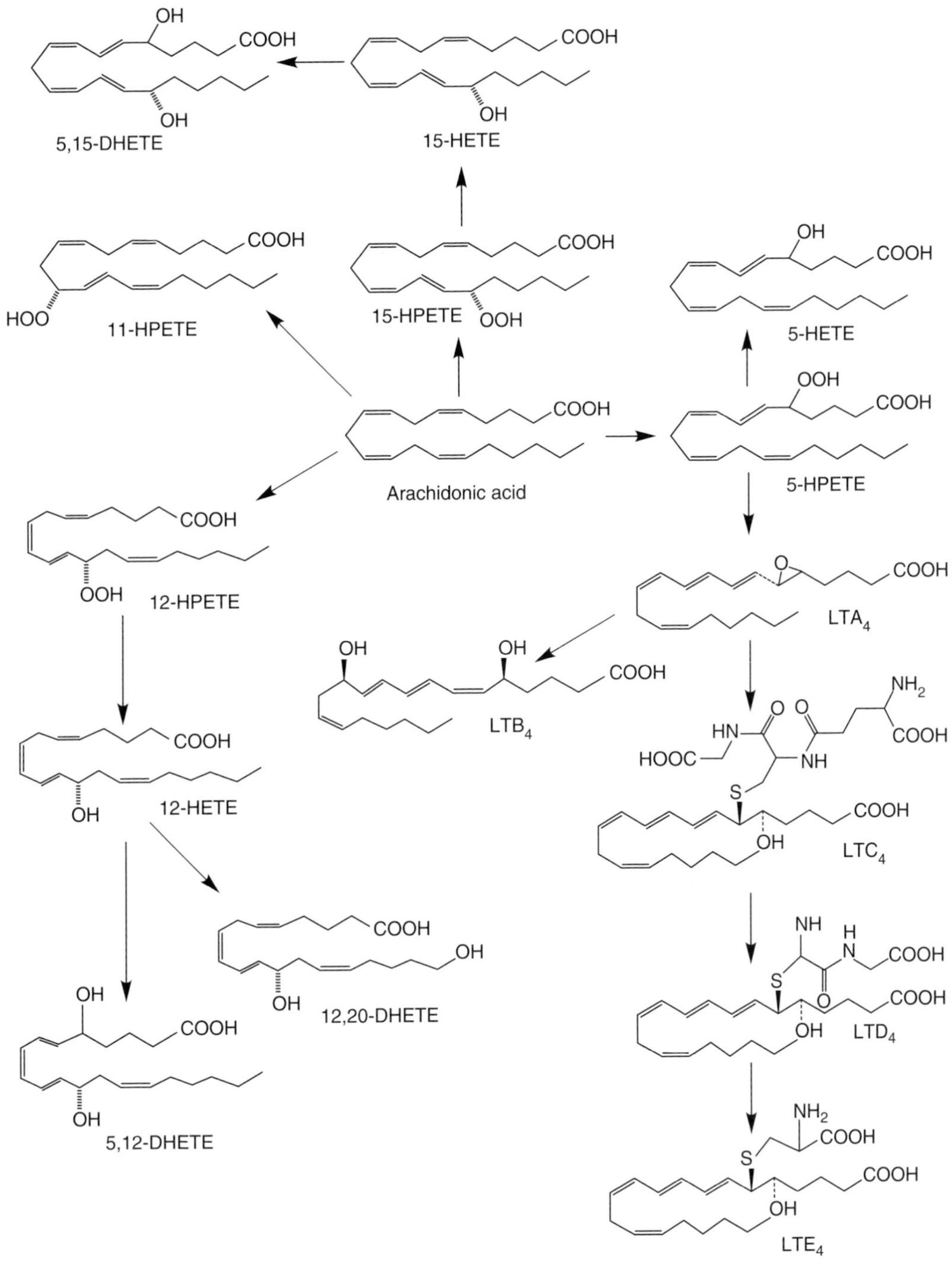

FIGURE 1.6 Leucotriene metabolites of 20:4ω6.

Shorthand nomenclature uses a lowercase "a" to represent the acetylenic bond; 9c,12a-18:2 is an octadecynoic acid with a double bond in position 9 and the triple bond in position 12. Figure 1.8 shows the common names and standard nomenclature for some acetylenic fatty acids. Since the ligands attached to triple-bonded carbons are 180° from one another (the structure through the bond is linear), the second representation in Figure 1.8 is more accurate.

PGA
PGB
PGC
PGD
PGE
PGFα
COOH
CH3
Prostanoic acid

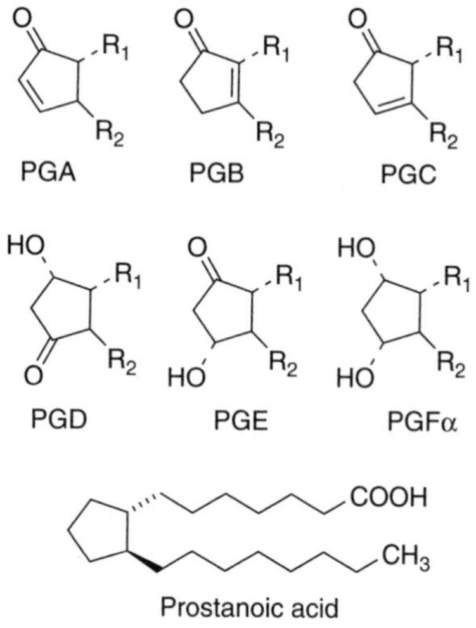

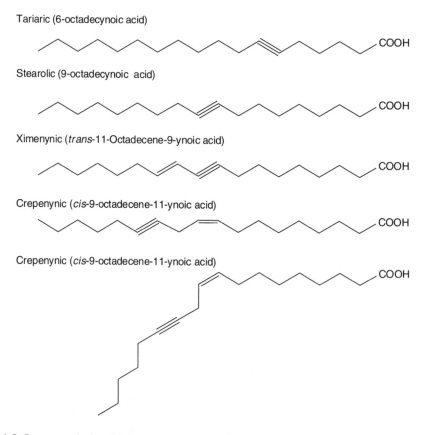

FIGURE 1.7 Prostanoic acid and prostaglandin ring nomenclature.

Tariaric (6-octadecynoic acid)
Stearolic (9-octadecynoic acid)
Ximenynic (trans-11-Octadecene-9-ynoic acid)
Crepenynic (cis-9-octadecene-11-ynoic acid)
Crepenynic (cis-9-octadecene-11-ynoic acid)
COOH

FIGURE 1.8 Some acetylenic acid structures and nomenclature.

The acetylenic fatty acids found in nature are usually 18-carbon molecules with unsaturation starting at Δ9 consisting of conjugated double–triple bonds [9,16]. Acetylenic fatty acids are rare.

4. *Trans* Fatty Acids

Trans fatty acids include any unsaturated fatty acid that contains double-bond geometry in the *E* (*trans*) configuration. Nomenclature differs from normal *cis* fatty acids only in the configuration of the double bonds.

The three main origins of *trans* fatty acids in our diet are bacteria, deodorized oils, and partially hydrogenated oils. The preponderance of *trans* fatty acids in our diets is derived from the hydrogenation process.

Hydrogenation is used to stabilize and improve oxidative stability of oils and to create plastic fats from oils [17]. The isomers that are formed during hydrogenation depend on the nature and amount of catalyst, the extent of hydrogenation, and other factors. The identification of the exact composition of a partially hydrogenated oil is extremely complicated and time consuming. The partial hydrogenation process produces a mixture of positional and geometrical isomers. Identification of the fatty acid isomers in a hydrogenated menhaden oil has been described [18]. The 20:1 isomers originally present in the unhydrogenated oil were predominantly *cis*-Δ11 (73% of total 20:1) and *cis*-Δ13 (15% of total 20:1). After hydrogenation from an initial iodine value of 159 to 96.5, the 20:1 isomers were distributed broadly across the molecules from Δ3 to Δ17 (Figure 1.9). The major *trans* isomers were Δ11 and Δ13, whereas the main *cis* isomers were Δ6, Δ9, and Δ11. Similar broad ranges of isomers are produced in hydrogenated vegetable oils [17].

Geometrical isomers of essential fatty acids linoleic and linolenic were first reported in deodorized rapeseed oils [19]. The geometrical isomers that result from deodorization are found

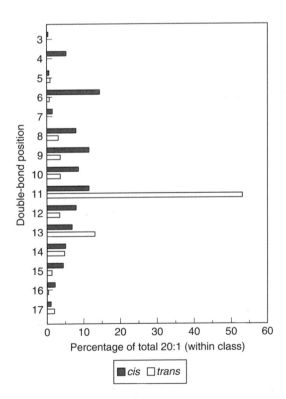

FIGURE 1.9 Eicosenoid isomers in partially hydrogenated menhaden oil. (From Sebedio, J.L. and Ackman, R.G., *J. Am. Oil Chem. Soc.*, 60, 1986, 1983.)

in vegetable oils and products made from vegetable oils (infant formulas) and include 9c,12t-18:2; 9t,12c-18:2; and 9t,12t-18:2, as well as 9c,12c,15t-18:3; 9t,12c,15c-18:3; 9c,12t,15c-18:3; and 9t,12t,15t-18:3 [19–22]. These *trans*-EFA isomers have been shown to have altered biological effects and are incorporated into nervous tissue membranes [23,24], although the importance of these findings has not been elucidated. Geometrical isomers of long-chain ω3 fatty acids have been identified in deodorized fish oils.

Trans fatty acids are formed by some bacteria, primarily under anaerobic conditions [25]. It is believed that the formation of *trans* fatty acids in bacterial cell membranes is an adaptation response to decrease membrane fluidity, perhaps as a reaction to elevated temperature or stress from solvents or other lipophilic compounds that affects membrane fluidity.

Not all bacteria produce appreciable levels of *trans* fatty acids. The *trans*-producing bacteria are predominantly gram negative and produce *trans* fatty acids under anaerobic conditions. The predominant formation of *trans* is via double-bond migration and isomerization, although some bacteria appear to be capable of isomerization without bond migration. The action of bacteria in the anaerobic rumen results in biohydrogenation of fatty acids and results in *trans* fatty acid formation in dairy fats (2%–6% of total fatty acids). The double-bond positions of the *trans* acids in dairy fats are predominantly in the Δ11 position, with smaller amounts in Δ9, Δ10, Δ13, and Δ14 positions [26].

5. Branched Fatty Acids

A large number of branched fatty acids have been identified [16]. The fatty acids can be named according to rules for branching in hydrocarbons (Table 1.2). Besides standard nomenclature, several common terms have been retained, including iso-, with a methyl branch on the penultimate (ω2) carbon, and anteiso, with a methyl branch on the antepenultimate (ω3) carbon. The iso and anteiso fatty acids are thought to originate from a modification of the normal de novo biosynthesis, with acetate replaced by 2-methyl propanoate and 2-methylbutanoate, respectively [16]. Other branched fatty acids are derived from isoprenoid biosynthesis including pristanic acid (2,6,10,14-tetramethylpentadecanoic acid) and phytanic acid (3,7,11,15-tetramethylhexadecanoic acid).

6. Cyclic Fatty Acids

Many fatty acids that exist in nature contain cyclic carbon rings [27]. Ring structures contain either three (cyclopropyl and cyclopropenyl), five (cyclopentenyl), or six (cyclohexenyl) carbon atoms and may be saturated or unsaturated. As well, cyclic fatty acid structures resulting from heating the vegetable oils have been identified [27–29].

In nomenclature of cyclic fatty acids, the parent fatty acid is the chain from the carboxyl group to the ring structure. The ring structure and additional ligands are considered a substituent of the parent fatty acid. An example is given in Figure 1.10. The parent in this example is nonanoic acid (not pentadecanoic acid, which would result if the chain were extended through the ring structure). The substituted group is a cyclopentyl group with a 2-butyl ligand (2-butylcyclopentyl). Thus, the correct standard nomenclature is 9-(2-butylcyclopentyl)nonanoic acid. The 2 is sometimes expressed as 2′ to indicate that the numbering is for the ring, and not the parent chain. The C-1 and C-2 carbons of the cyclopentyl ring are chiral, and two possible configurations are possible. Both the carboxyl and the longest hydrocarbon substituents can be on the same side of the ring, or they can be on opposite sides. These are referred to as *cis* and *trans*, respectively.

The cyclopropene and cyclopropane fatty acids can be named by means of the standard nomenclature noted in the example above. They are also commonly named using the parent structure that carries through the ring structure. In the example in Figure 1.11, the fatty acid (commonly named lactobacillic acid or phycomonic acid) is named 10-(2-hexylcyclopropyl) decanonic acid in standard nomenclature. An older naming system would refer to this fatty acid as *cis*-11,12-methyleneoctadecanoic acid, where *cis* designates the configuration of the ring

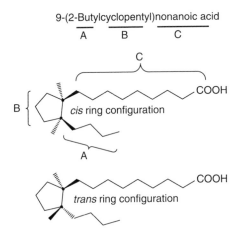

FIGURE 1.10 Nomenclature of cyclic fatty acids.

structure. If the fatty acid is unsaturated, the term methylene is retained but the double-bond position is noted in the parent fatty acid structure (*cis*-11,12-methylene-*cis*-octadec-9-enoic acid) (Figure 1.12).

7. Hydroxy and Epoxy Fatty Acids

Saturated and unsaturated fatty acids containing hydroxy and epoxy functional groups have been identified [1,16]. Hydroxy fatty acids are named by means of the parent fatty acid and the hydroxy group numbered with its Δ location. For example, the fatty acid with the trivial name ricinoleic (Figure 1.13) is named *R*-12-hydroxy-*cis*-9-octadecenoic acid. Ricinoleic acid is found in the seeds of *Ricinus* species and accounts for about 90% of the fatty acids in castor bean oil.

Because the hydroxy group is chiral, stereoisomers are possible. The *R/S* system is used to identify the exact structure of the fatty acid. Table 1.6 reviews the rules for *R/S* nomenclature. The *R/S* system can be used instead of the α/β and *cis/trans* nomenclature systems. A fatty acid with a hydroxy substituent in the $\Delta 2$ position is commonly called an α-hydroxy acid; fatty acids with hydroxy substituents in the $\Delta 3$ and $\Delta 4$ positions are called β-hydroxy acids and γ-hydroxy acids, respectively. Some common hydroxy acids are shown in Figure 1.13. Cutins, which are found in the outer layer of fruit skins, are composed of hydroxy acid polymers, which also may contain epoxy groups [16].

Epoxy acids, found in some seed oils, are formed on prolonged storage of seeds [16]. They are named similarly to cyclopropane fatty acids, with the parent acid considered to have a substituted oxirane substituent. An example of epoxy fatty acids and their nomenclature is shown in Figure 1.14. The fatty acid with the common name vernolic acid is named (using standard nomenclature) 11-(3-pentyloxyranyl)-9-undecenoic acid. In older nomenclature, where the carbon chain is carried through the oxirane ring, vernolic acid would be called 12,13-epoxyoleic acid or 12-13-epoxy-9-octadecenoic acid. The configuration of the oxirane ring substituents can be named in the *cis/trans*, *E/Z*, or *R/S* configuration systems.

cis-11,12-Methyleneoctadecanoic acid

cis-10-(2-Hexylcyclopropyl)decanoic acid

Lactobacillic acid/phytomonic acid

FIGURE 1.11 Nomenclature for a cyclopropenoid fatty acid.

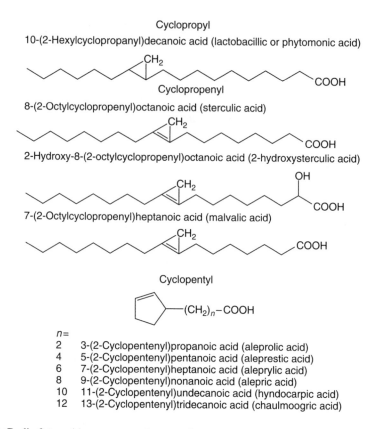

Cyclopropyl
10-(2-Hexylcyclopropanyl)decanoic acid (lactobacillic or phytomonic acid)
CH2
COOH
Cyclopropenyl
8-(2-Octylcyclopropenyl)octanoic acid (sterculic acid)
CH2
COOH
2-Hydroxy-8-(2-octylcyclopropenyl)octanoic acid (2-hydroxysterculic acid)
OH
CH2
COOH
7-(2-Octylcyclopropenyl)heptanoic acid (malvalic acid)
CH2
COOH
Cyclopentyl
(CH2)n–COOH
n=
2 3-(2-Cyclopentenyl)propanoic acid (aleprolic acid)
4 5-(2-Cyclopentenyl)pentanoic acid (aleprestic acid)
6 7-(2-Cyclopentenyl)heptanoic acid (aleprylic acid)
8 9-(2-Cyclopentenyl)nonanoic acid (alepric acid)
10 11-(2-Cyclopentenyl)undecanoic acid (hyndocarpic acid)
12 13-(2-Cyclopentenyl)tridecanoic acid (chaulmoogric acid)

FIGURE 1.12 Cyclic fatty acid structures and nomenclature.

8. Furanoid Fatty Acids

Some fatty acids contain an unsaturated oxolane heterocyclic group. There are more commonly called furanoid fatty acids because a furan structure (diunsaturated oxolane) is present in the molecule. Furanoid fatty acids have been identified in *Exocarpus* seed oils. They have also been identified in plants, algae, and bacteria and are a major component in triacylglycerols (TAGs) from

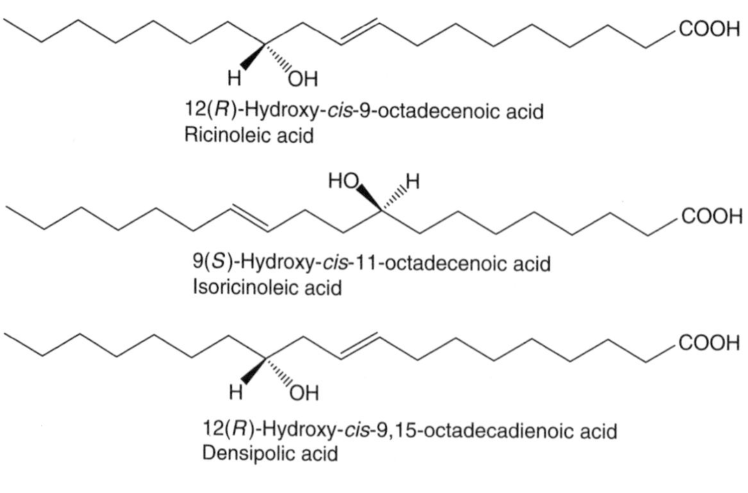

COOH
H OH
12(R)-Hydroxy-cis-9-octadecenoic acid
Ricinoleic acid
HO H
COOH
9(S)-Hydroxy-cis-11-octadecenoic acid
Isoricinoleic acid
COOH
H OH
12(R)-Hydroxy-cis-9,15-octadecadienoic acid
Densipolic acid

FIGURE 1.13 Hydroxy fatty acid structures and nomenclature.

TABLE 1.6

Summary of Rules for *R/S* Nomenclature

1. The sequence priority rules (Table 1.3) are used to prioritize the ligands attached to the chiral center ($a > b > c > d$).
2. The molecule is viewed with the *d* substituent facing away from the viewer.
3. The remaining three ligands (*a, b, c*) will be oriented with the order *a–b–c* in a clockwise or counterclockwise direction.
4. Clockwise describes the *R* (*rectus*, right) conformation, and counterclockwise describes the *S* (*sinister*, left) conformation.

Source: From Streitwieser Jr., A. and Heathcock, C.H. in *Introduction to Organic Chemistry*, Macmillan, New York, 1976, 111.

latex rubber [1,16]. They are important in marine oils and may total several percentage points of the total fatty acids or more in liver and testes [1,30].

Furanoid fatty acids have a general structure as shown in Figure 1.15. A common nomenclature describing the furanoid fatty acids (as F_1, F_2, etc.) is used [30]. The naming of the fatty acids in this nomenclature is arbitrary and originated from elution order in gas chromatography. A shorthand notation that is more descriptive gives the methyl substitution followed by F, and then the carbon lengths of the carboxyl and terminal chains in parentheses: MeF(9,5). Standard nomenclature follows the same principles outlined in Section IV.A.6. The parent fatty acid chain extends only to the furan structure, which is named as a ligand attached to the parent molecule. For example, the fatty acid named F5 in Figure 1.15 is named 11-(3,4-dimethyl-5-pentyl-2-furyl)undecanoic acid. Shorthand notation for this fatty acid would be F_5 or MeF(11,5). Numbering for the furan ring starts at the oxygen and proceeds clockwise.

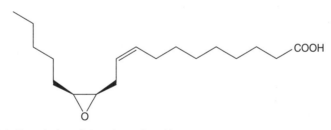

11-(3-Pentyloxiranyl)-9-undecanoic acid
cis-12-13-Epoxy-*cis*-9-octadecenoic acid
Vernolic acid
Both the (+) 12*S*, 13*R* and (−) 12*R*, 13*S* forms are found in nature

8-(3-*cis*-2'-Nonenyloxiranyl)-octanoic acid
cis-9,10-Epoxy-*cis*-12-octadecenoic acid
Coronaric acid

FIGURE 1.14 Epoxy fatty acid structures and nomenclature.

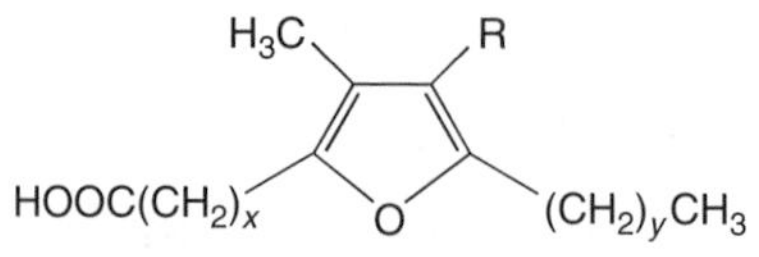

Name	x	y	R
F1	8	2	CH_3
F2	8	4	H
F3	8	4	CH_3
F4	10	2	CH_3
F5	10	4	H
F6	10	4	CH_3
F7	12	4	H
F8	12	4	CH_3

FIGURE 1.15 Furanoid fatty acid structure and shorthand nomenclature.

B. ACYLGLYCEROLS

Acylglycerols are the predominant constituent in oils and fats of commercial importance. Glycerol can be esterified with one, two, or three fatty acids, and the individual fatty acids can be located on different carbons of glycerol. The terms monoacylglycerol, diacylglycerol, and TAG are preferred for these compounds over the older and confusing names mono-, di-, and triglycerides [6,7].

Fatty acids can be esterified on the primary or secondary hydroxyl groups of glycerol. Although glycerol itself has no chiral center, it becomes chiral if different fatty acids are esterified to the primary hydroxyls or if one of the primary hydroxyls is esterified. Thus, terminology must differentiate between the two possible configurations (Figure 1.16). The most common convention to differentiate these stereoisomers is the *sn* convention of Hirshmann (see Ref. [31]). In the numbering that describes the hydroxyl groups on the glycerol molecule in Fisher projection, *sn*1, *sn*2, and *sn*3 designations are used for the top (C1), middle (C2), and bottom (C3) OH groups (Figure 1.17). The *sn* term indicates stereospecific numbering [1].

In common nomenclature, esters are called α on primary and β on secondary OH groups. If the two primary-bonded fatty acids are present, the primary carbons are called α and α′. If one or two

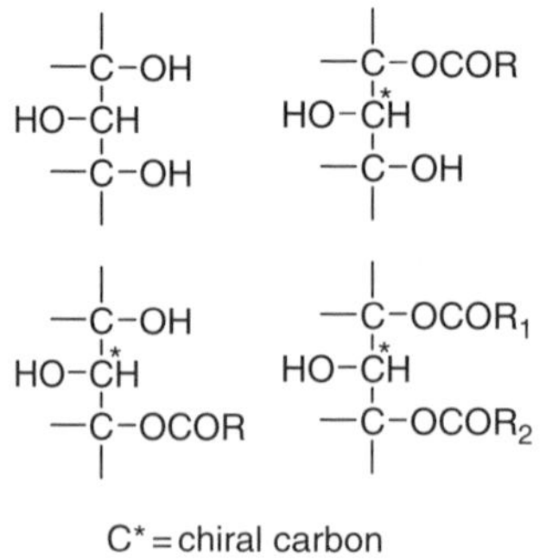

C* = chiral carbon

FIGURE 1.16 Chiral carbons in acylglycerols.

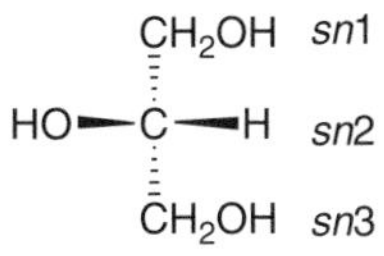

FIGURE 1.17 Stereospecific numbering (*sn*) of triacylglycerols.

acyl groups are present, the term partial glyceride is sometimes used. Nomenclature of the common partial glycerides is shown in Figure 1.18.

Standard nomenclature allows several different names for each TAG [6]. A TAG with three stearic acid esters can be named as glycerol tristearate, tristearoyl glycerol, or tri-*O*-stearoyl glycerol. The *O* locant can be omitted if the fatty acid is esterified to the hydroxyl group. More commonly, TAG nomenclature uses the designation -in to indicate the molecule in a TAG (e.g., tristearin). If different fatty acids are esterified to the TAG—for example, the TAG with *sn*-1 palmitic acid, *sn*-2 oleic acid, and *sn*-3 stearic acid—the name replaces the -ic in the fatty acid name with -oyl, and fatty acids are named in *sn*1, *sn*2, and *sn*3 order (1-palmitoyl-2-oleoyl-3-stearoyl-*sn*-glycerol). This TAG also can be named as *sn*-1-palmito-2-oleo-3-stearin or *sn*-glycerol-1-palmitate-2-oleate-3-stearate. If two of the fatty acids are identical, the name incorporates the designation di- (e.g., 1,2-dipalmitoyl-3-oleoyl-*sn*-glycerol, 1-stearoyl-2,3-dilinolenoyl-*sn*-glycerol, etc.).

To facilitate TAG descriptions, fatty acids are abbreviated using one or two letters (Table 1.7). The TAGs can be named after the EFAs using shorthand nomenclature. For example, *sn*-POSt is shorthand description for the molecule 1-palmitoyl-2-oleoyl-3-stearoyl-*sn*-glycerol. If the *sn*- is omitted, the stereospecific positions of the fatty acids are unknown. POSt could be a mixture of *sn*-POSt, *sn*-StOP, *sn*-PStO, *sn*-OStP, *sn*-OPSt, or *sn*-StPO in any proportion. An equal mixture of both stereoisomers (the racemate) is designated as *rac*. Thus, *rac*-OPP represents equal amounts of *sn*-OPP and *sn*-PPO. If only the *sn*-2 substituent is known with certainty in a TAG, the designation β- is used. For example, β-POSt is a mixture (unknown amounts) of *sn*-POSt and *sn*-StOP.

TAGs are also sometimes described by means of the ω nomenclature. For example, *sn*-18: 0–18:2ω6–16:0 represents 1-stearoyl-2-linoleoyl-3-palmitoyl-*sn*-glycerol.

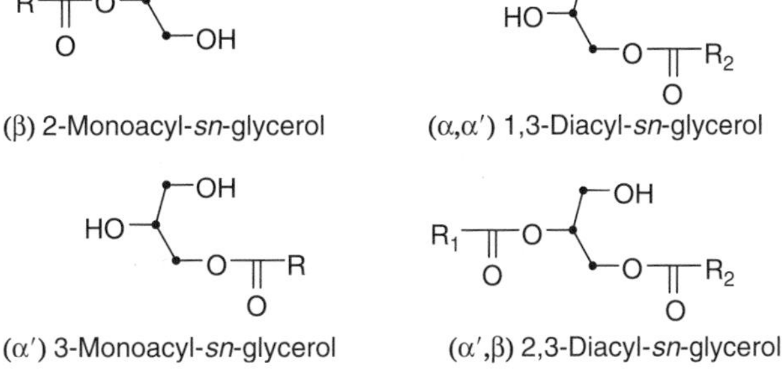

(α) 1-Monoacyl-*sn*-glycerol   (α,β) 1,2-Diacyl-*sn*-glycerol

(β) 2-Monoacyl-*sn*-glycerol   (α,α′) 1,3-Diacyl-*sn*-glycerol

(α′) 3-Monoacyl-*sn*-glycerol   (α′,β) 2,3-Diacyl-*sn*-glycerol

FIGURE 1.18 Mono- and diacylglycerol structures.

TABLE 1.7
Short Abbreviations for Some Common Fatty Acids

AC	Acetic	Ln	Linolenic
Ad	Arachidic	M	Myristic
An	Arachidonic	N	Nervonic
B	Butyric	O	Oleic
Be	Behenic	Oc	Octanoic
D	Decanoic	P	Palmitic
E	Erucic	Po	Palmitoleic
El	Elaidic	R	Ricinoleic
G	Eicosenoic	S	Saturated (any)
H	Hexanoic	St	Stearic
L	Linoleic	U	Unsaturated (any)
La	Lauric	V	Vaccenic
Lg	Lingnoceric	X	Unknown

Source: From Litchfield, C. in *Analysis of Triglycerides*, Academic Press, New York, 1972, 355.

C. STEROLS AND STEROL ESTERS

The steroid class of organic compounds includes sterols of importance in lipid chemistry. Although the term sterol is widely used, it has never been formally defined. The following working definition was proposed some years ago: "Any hydroxylated steroid that retains some or all of the carbon atoms of squalene in its side chain and partitions nearly completely into the ether layer when it is shaken with equal volumes of ether and water" [32]. Thus, for this definition, sterols are a subset of steroids and exclude the steroid hormones and bile acids. The importance of bile acids and their intimate origin from cholesterol make this definition difficult. In addition, nonhydroxylated structures such as cholestane, which retain the steroid structure, are sometimes considered sterols.

The sterols may be derived from plant (phytosterols) or animal (zoosterols) sources. They are widely distributed and are important in cell membranes. The predominant zoosterol is cholesterol. Although a few phytosterols predominate, the sterol composition of plants can be very complex. For example, as many as 65 different sterols have been identified in corn (*Zea mays*) [33].

In the standard ring and carbon numbering (Figure 1.19) [33], the actual three-dimensional configuration of the tetra ring structure is almost flat, so the ring substituents are either in the same

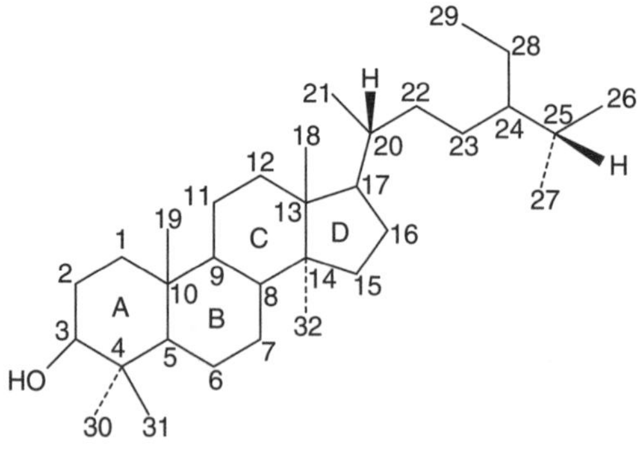

FIGURE 1.19 Carbon numbering in cholesterol structure.

plane as the rings or in front or behind the rings. If the structure in Figure 1.19 lacks one or more of the carbon atoms, the numbering of the remainder will not be changed.

The methyl group at position 10 is axial and lies in front of the general plane of the molecule. This is the β configuration and is designated by connection using a solid or thickened line. Atoms or groups behind the molecule plane are joined to the ring structure by a dotted or broken line and are given the α configuration. If the stereochemical configuration is not known, a wavy line is used and the configuration is referred to as ε. Unfortunately, actual three-dimensional position of the substituents may be in plane, in front of, or behind the plane of the molecule. The difficulties with this nomenclature have been discussed elsewhere [32,33].

The nomenclature of the steroids is based on parent ring structures. Some of the basic steroid structures are presented in Figure 1.20 [6]. Because cholesterol is a derivative of the cholestane structure (with the H at C-5 eliminated because of the double bond), the correct standard nomenclature for cholesterol is 3β-cholest-5-en-3-ol. The complexity of standardized nomenclature has led to the retention of trivial names for some of the common structures (e.g., cholesterol). However,

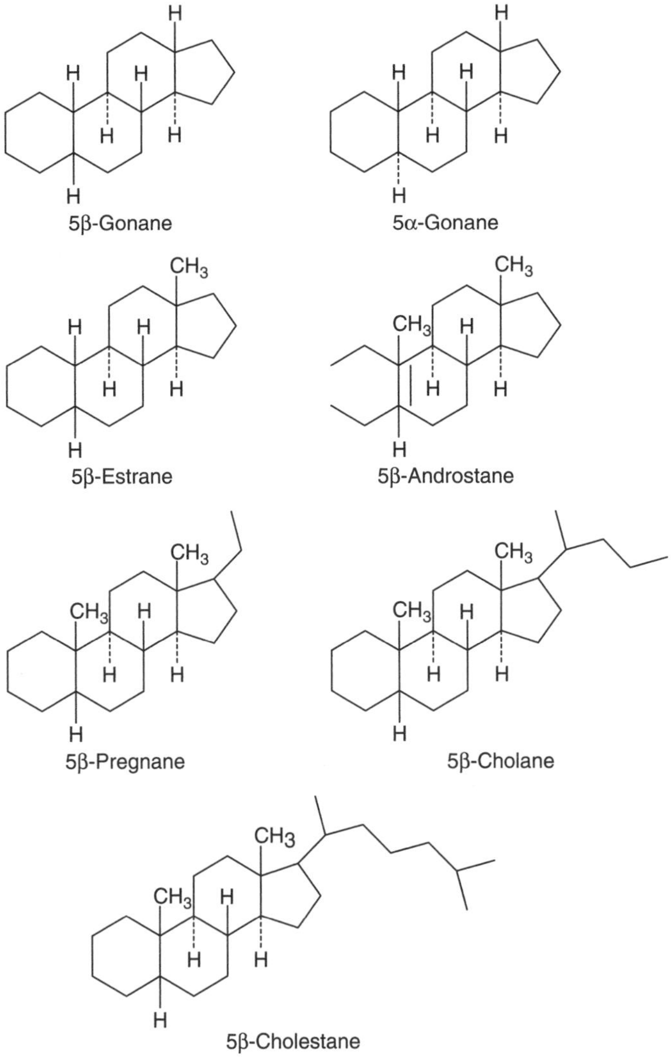

FIGURE 1.20 Steroid nomenclature.

when the structure is changed—for example, with the addition of a ketone group to cholesterol at the 7 position—the proper name is 3β-hydroxycholest-5-en-7-one, although this molecule is also called 7-ketocholesterol in common usage.

A number of other sterols of importance in foods are shown in Figure 1.21. The trivial names are retained for these compounds, but based on the nomenclature system discussed for sterols, stigmasterol can be named 3β-hydroxy-24-ethylcholesta-5,22-diene. Recent studies have suggested that plant sterols and stanols (saturated derivatives of sterols) have cholesterol-lowering properties in humans [34].

Cholesterol has been reported to oxidize in vivo and during food processing [35–38]. These cholesterol oxides have come under intense scrutiny because they have been implicated in development of atherosclerosis. Some of the more commonly reported oxidation products are shown in Figures 1.22 and 1.23. Nomenclature in common usage in this field often refers to the oxides as derivatives of the cholesterol parent molecule: 7-β-hydroxycholesterol, 7-ketocholesterol, 5,6β-epoxycholesterol, etc. The standard nomenclature follows described rules and is shown in Figures 1.22 and 1.23.

Sterol esters exist commonly and are named using standard rules for esters. For example, the ester of cholesterol with palmitic acid would be named cholesterol palmitate. The standard nomenclature would also allow this molecule to be named 3-O-palmitoyl-3β-cholest-5-en-3-ol or 3-palmitoyl-3β-cholest-5-en-3-ol.

Cholesterol

Ergosterol

Stigmasterol

β-Sitosterol

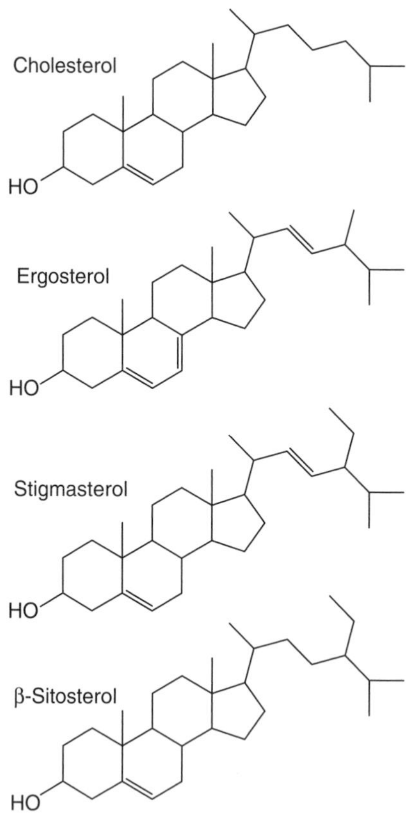

FIGURE 1.21 Common steroid structures.

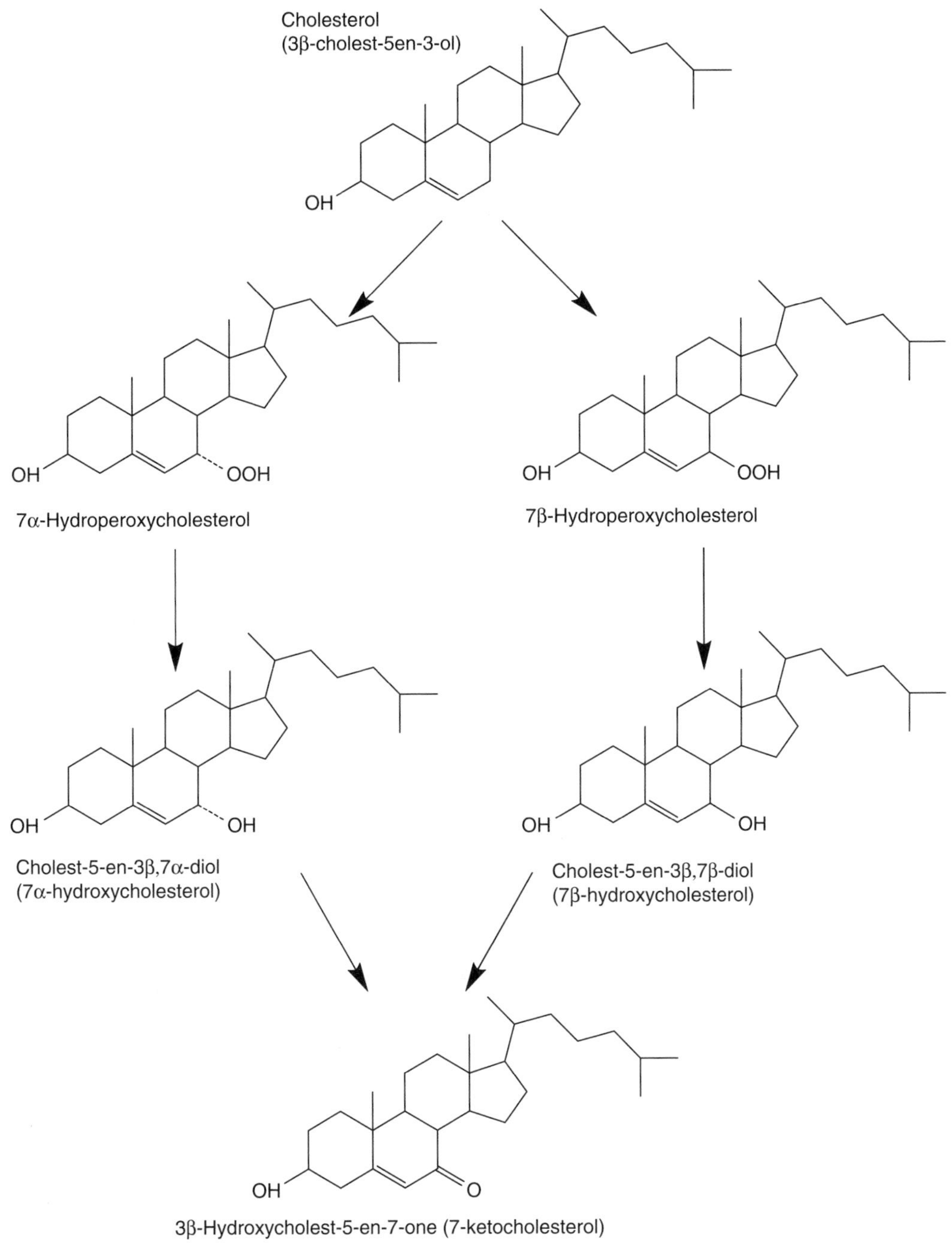

FIGURE 1.22 Cholesterol oxidation products and nomenclature I. (From Smith, L.L., *Chem. Phys. Lipids*, 44, 87, 1987.)

D. WAXES

Waxes (commonly called wax esters) are esters of fatty acids and long-chain alcohols. Simple waxes are esters of medium-chain fatty acids (16:0, 18:0, 18:1ω9) and long-chain aliphatic alcohols. The alcohols range in size from C8 to C18. Simple waxes are found on the surfaces of animals, plants,

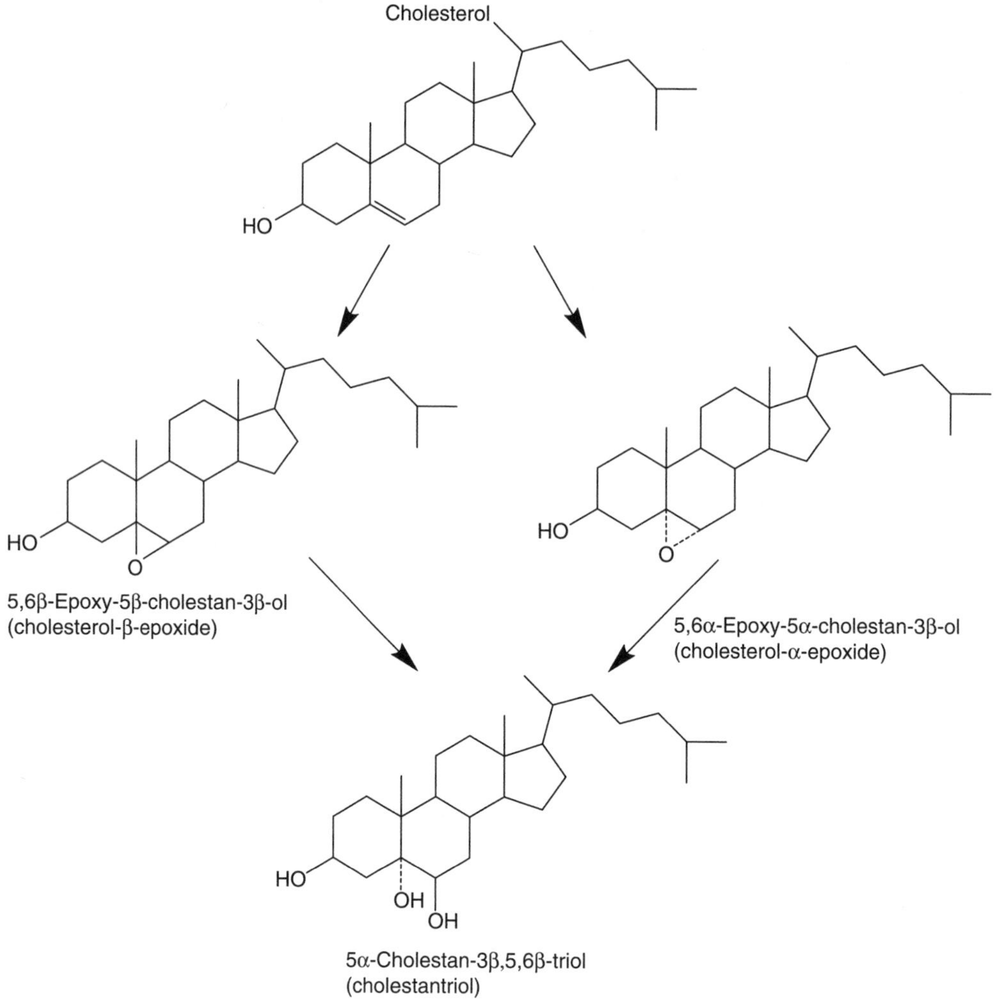

Cholesterol

5,6β-Epoxy-5β-cholestan-3β-ol
(cholesterol-β-epoxide)

5,6α-Epoxy-5α-cholestan-3β-ol
(cholesterol-α-epoxide)

5α-Cholestan-3β,5,6β-triol
(cholestantriol)

FIGURE 1.23 Cholesterol oxidation products and nomenclature II. (From Smith, L.L., *Chem. Phys. Lipids*, 44, 87, 1987.)

and insects and play a role in prevention of water loss. Complex waxes are formed from diols or from alcohol acids. Di- and triesters as well as acid and alcohol esters have been described.

Simple waxes can be named by removing the -ol from the alcohol and replacing it with -yl, and replacing the -ic from the acid with -oate. For example, the wax ester from hexadecanol and oleic acid would be named hexadecyl oleate or hexadecyl-*cis*-9-octadecenoate. Some of the long-chain alcohols have common names derived from the fatty acid parent (e.g., lauryl alcohol, stearyl alcohol). The C16 alcohol (1-hexadecanol) is commonly called cetyl alcohol. Thus, cetyl oleate is another acceptable name for this compound.

Waxes are found in animal, insect, and plant secretions as protective coatings. Waxes of importance in foods as additives include beeswax, carnauba wax, and candelilla wax.

E. PHOSPHOGLYCERIDES (PHOSPHOLIPIDS)

Phosphoglycerides (PLs) are composed of glycerol, fatty acids, phosphate, and (usually) an organic base or polyhydroxy compound. The phosphate is almost always linked to the *sn*-3 position of glycerol molecule.

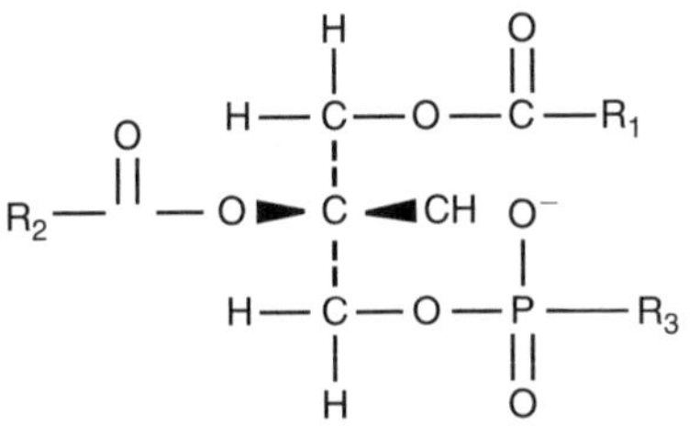

R_3	= Phospholipid class/nomenclature
–OH	Phosphatidic acid (PA)
$–O–CH_2–CH(NH_2)–COOH$	(3-*sn*-phosphatidyl)-L-serine (PS) 1,2-diacyl-*sn*-glycero(3)phospho-L-serine
$–O–CH_2–CH_2–NH_2$	(3-*sn*-phosphatidyl)ethanolamine (PE) 1,2-diacyl-*sn*-glycero(3)phosphoethanolamine
$–O–CH_2–CH_2–N^+(CH_3)_3$	(3-*sn*-phosphatidyl)choline (PC) 1,2-diacyl-*sn*-glycero(3)phosphocholine
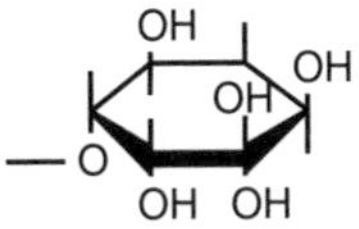	(3-*sn*-phosphatidyl)-L-*myo*inositol (PI) 1,2-diacyl-*sn*-glycero(3)phospho-L-*myo*-inositol

FIGURE 1.24 Nomenclature for glycerophospholipids.

The parent structure of the PLs is phosphatidic acid (*sn*-1,2-diacylglycerol-3-phosphate). The terminology for PLs is analogous to that of acylglycerols with the exception of the no acyl group at *sn*-3. The prefix lyso-, when used for PLs, indicates that the *sn*-2 position has been hydrolyzed, and a fatty acid is esterified to the *sn*-1 position only.

Some common PL structures and nomenclature are presented in Figure 1.24. Phospholipid classes are denoted using shorthand designation (PC = phosphatidylcholine, etc.). The standard nomenclature is based on the PL type. For example, a PC with an oleic acid on *sn*-1 and linolenic acid on *sn*-2 would be named 1-oleoyl-2-linolenoyl-*sn*-glycerol-3-phosphocholine. The name phosphorycholine is sometimes used but is not recommended [8]. The terms lecithin and cephalin, sometimes used for PC and PE, respectively, are not recommended [8].

Cardiolipin is a PL that is present in heart muscle mitochondria and bacterial membranes. Its structure and nomenclature are shown in Figure 1.25. Some cardiolipins contain the maximum possible number of 18:2ω6 molecules (4 mol/mol).

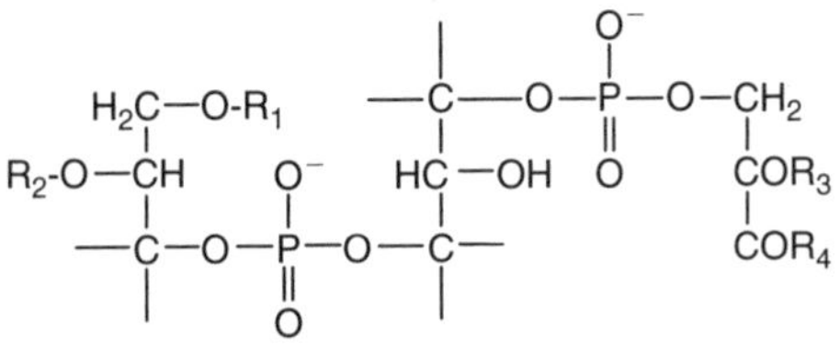

Cardiolipin
1′,3′-di-*O*-(3-*sn*-phosphatidyl)-*sn*-glycerol
R1–R4 are fatty acids

FIGURE 1.25 Cardiolipin structure and nomenclature.

F. Ether(Phospho)Glycerides (Plasmalogens)

Plasmalogens are formed when a vinyl (1-alkenyl) ether bond is found in a phospholipid
or acylglycerol. The 1-alkenyl-2,3-diacylglycerols are termed neutral plasmalogens. A 2-acyl-1-
(1-alkenyl)-*sn*-glycerophosphocholine is named a plasmalogen or plasmenylcholine. The related
1-alkyl compound is named plasmanylcholine.

G. Glyceroglycolipids (Glycosylglycolipids)

The glyceroglycolipids or glycolipids are formed when a 1,2-diacyl-*sn*-3-glycerol is linked via the
sn-3 position to a carbohydrate molecule. The carbohydrate is usually a mono- or a disaccharide,
less commonly a tri- or tetrasaccharide. Galactose is the most common carbohydrate molecule in
plant glyceroglycolipids.

Structures and nomenclature for some glyceroglycolipids are shown in Figure 1.26. The
names monogalactosyldiacylglycerol (MGDG) and digalactosyldiacylglycerol (DGDG) are used

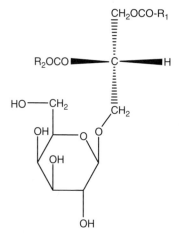

Monogalactosyldiacylglycerol (MGDG)
1,2-diacyl-3β-D-galactopyranosyl-L-glycerol

Digalactosyldiacylglycerol (DGDG)
1,2-diacyl-3-(α-D-galactopyranosyl-1,6-β-D-galactopryanosyl)-L-glycerol

FIGURE 1.26 Glyceroglycolipid structures and nomenclature.

in common nomenclature. The standard nomenclature identifies the ring structure and bonding of the carbohydrate groups (Figure 1.26).

H. SPHINGOLIPIDS

The glycosphingolipids are a class of lipids containing a long-chain base, fatty acids, and various other compounds, such as phosphate and monosaccharides. The base is commonly sphingosine, although more than 50 bases have been identified. The ceramides are composed of sphingosine and a fatty acid (Figure 1.27). Sphingomyelin is one example of a sphingophospholipid. It is a ceramide with a phosphocholine group connected to the primary hydroxyl of sphingosine. The ceramides can also be attached to carbohydrate molecules (sphingoglycolipids or cerebrosides) via the primary

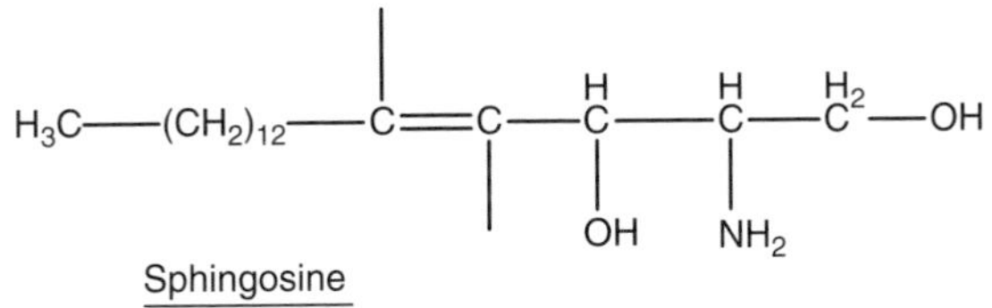

Sphingosine
D-*erythro*-1,3-dihydroxy-2-amino-*trans*-4-octadecene

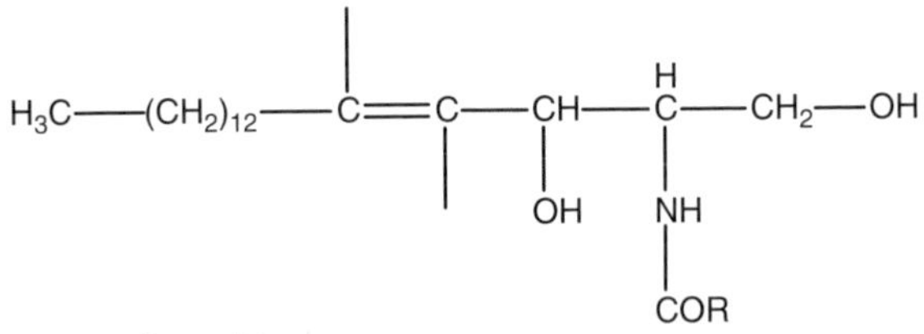

Ceramide
D-*erythro*-1,3-dihydroxy-2(*N*-acyl)-amino-*trans*-4-octadecene
(*N*-acyl-sphingosine)

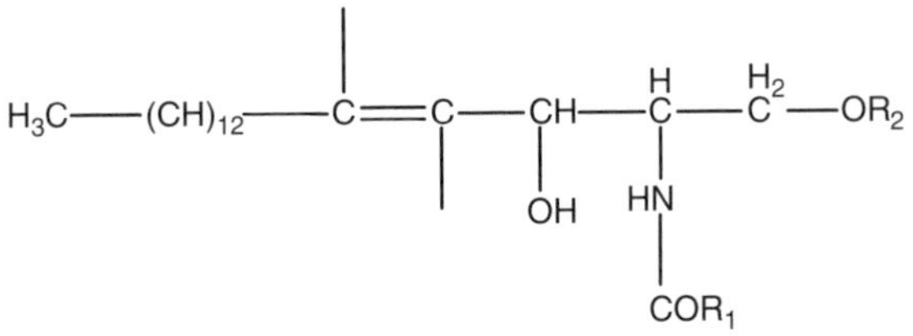

Cerebroside
1-*O*-β-D-galactopyranosyl-*N*-acyl-sphingosine
R1 = fatty acid
R2 = galactopyranose

Ganglioside

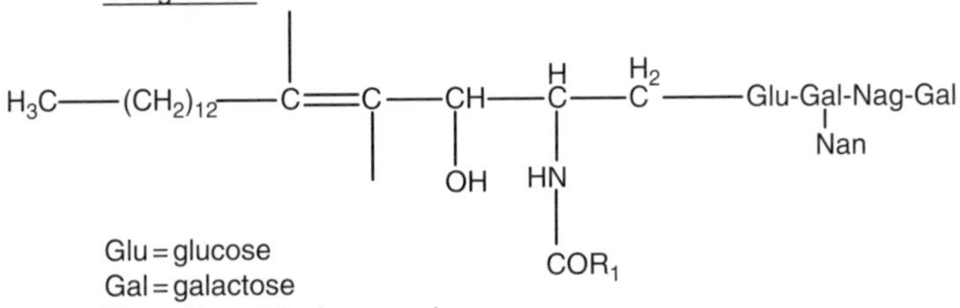

Glu = glucose
Gal = galactose
Nag = *N*-acetylgalactosamine
Nan = *N*-acetylneuraminic acid

FIGURE 1.27 Sphingolipid structures and nomenclature.

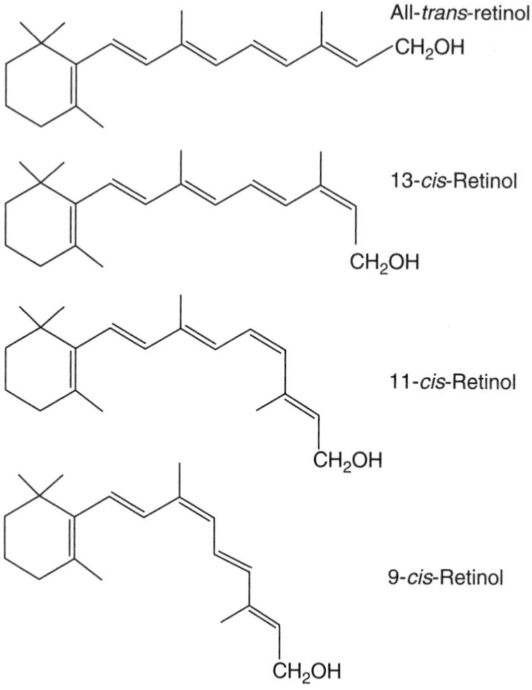

FIGURE 1.28 Structures of some vitamin A compounds.

hydroxyl group of sphingosine. Gangliosides are complex cerebrosides with the ceramide residue connected to a carbohydrate-containing glucose-galactosamine-N-acetylneuraminic acid. These lipids are important in cell membranes and the brain, and they act as antigenic sites on cell surfaces. Nomenclature and structures of some cerebrosides are shown in Figure 1.27.

I. FAT-SOLUBLE VITAMINS

1. Vitamin A

Vitamin A exists in the diet in many forms (Figure 1.28). The most bioactive form is the all-*trans* retinol, and *cis* forms are created via light-induced isomerization (Table 1.8). The 13-*cis* isomer is

TABLE 1.8
Approximate Biological Activity
Relationships of Vitamin A Compounds

Compound	Activity of All-*trans* Retinol (%)
All-*trans* retinol	100
9-*cis* Retinol	21
11-*cis* Retinol	24
13-*cis* Retinol	75
9,13-Di-*cis* retinol	24
11,13-Di-*cis* retinol	15
α-Carotene	8.4
β-Carotene	16.7

the most biopotent of the mono- and di-*cis* isomers. The α- and β-carotenes have biopotencies of about 8.7% and 16.7% of the all-*trans* retinol activity, respectively. The daily value (DV) for vitamin A is 1000 retinol equivalents (RE), which represents 1000 μg of all-*trans* retinol or 6000 μg of β-carotene. Vitamin A can be toxic when taken in levels exceeding the %DV. Some reports suggest that levels of 15,000 RE/day can be toxic [39].

Toxic symptoms of hypervitaminosis A include drowsiness, headache, vomiting, and muscle pain. Vitamin A can be teratogenic at high doses [39]. Vitamin A deficiency results in night blindness and ultimately total blindness, abnormal bone growth, increased cerebrospinal pressure, reproductive defects, abnormal cornification, loss of mucus secretion cells in the intestine, and decreased growth. The importance of beef liver, an excellent source of vitamin A, in cure of night blindness was known to the ancient Egyptians about 1500 bc [40].

2. Vitamin D

Although as many as five vitamin D compounds have been described (Figure 1.29), only two of these are biologically active: ergocalciferol (vitamin D_2) and cholecalciferol (vitamin D_3). Vitamin

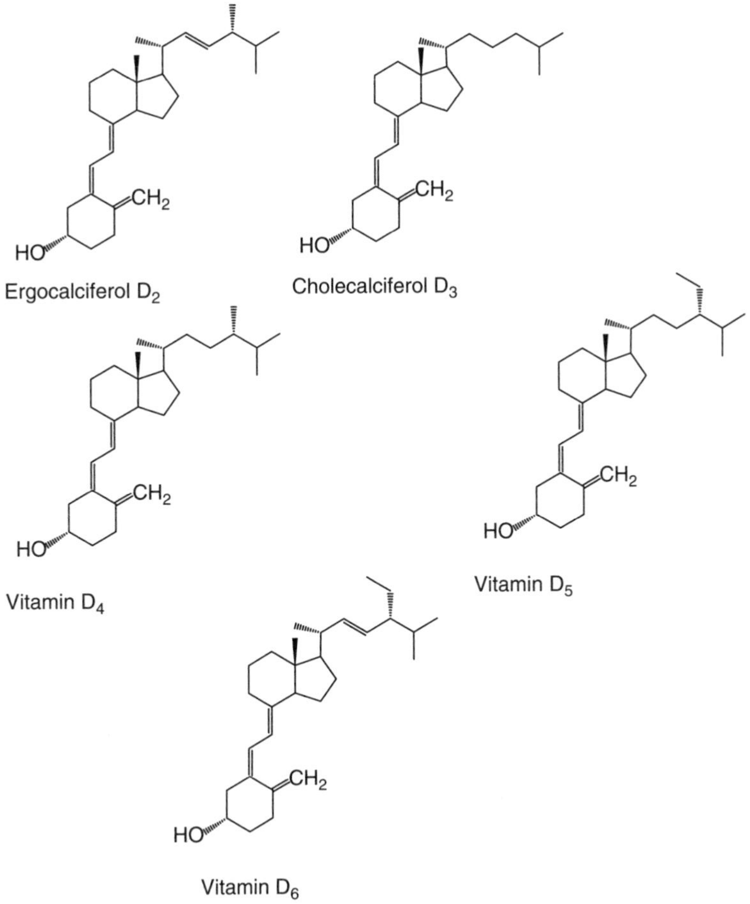

FIGURE 1.29 Structures of some vitamin D compounds.

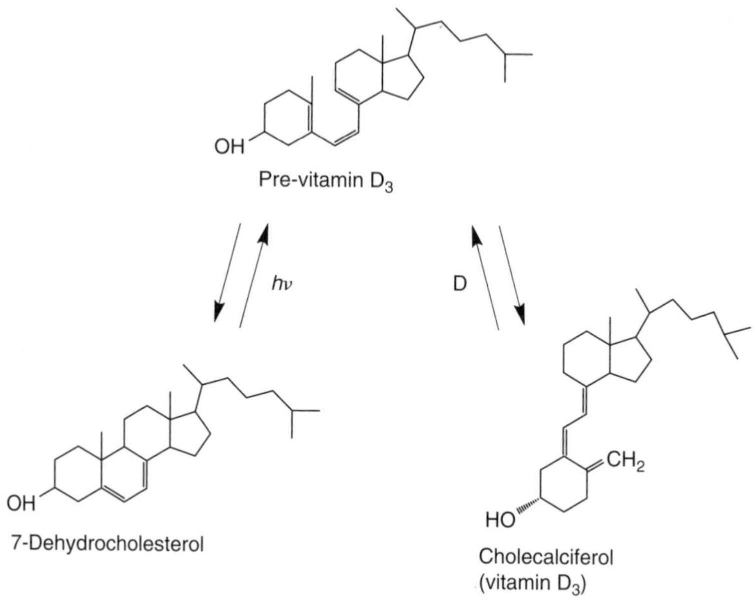

FIGURE 1.30 Formation of vitamin D in vivo.

D_3 can be synthesized in humans from 7-dehydrocholesterol, which occurs naturally in the skin, via light irradiation (Figure 1.30).

The actual hormonal forms of the D vitamins are the hydroxylated derivatives. Vitamin D is converted to 25-OH-D in the kidney and further hydroxylated to 1,25-diOH-D in the liver. The dihydroxy form is the most biologically active form in humans.

3. Vitamin E

Vitamin E compounds include the tocopherols and tocotrienols. Tocotrienols have a conjugated triene double-bond system in the phytyl side chain, whereas tocopherols do not. The basic nomenclature is shown in Figure 1.31. The bioactivity of the various vitamin E compounds is shown in Table 1.9. Methyl substitution affects the bioactivity of vitamin E, as well as its in vitro antioxidant activity.

4. Vitamin K

Several forms of vitamin K have been described (Figure 1.32). Vitamin K (phylloquinone) is found in green leaves, and vitamin K_2 (menaquinone) is synthesized by intestinal bacteria. Vitamin K is involved in blood clotting as an essential cofactor in the synthesis of γ-carboxyglutamate necessary for active prothrombin. Vitamin K deficiency is rare due to intestinal microflora synthesis. Warfarin and dicoumerol prevent vitamin K regeneration and may result in fatal hemorrhaging.

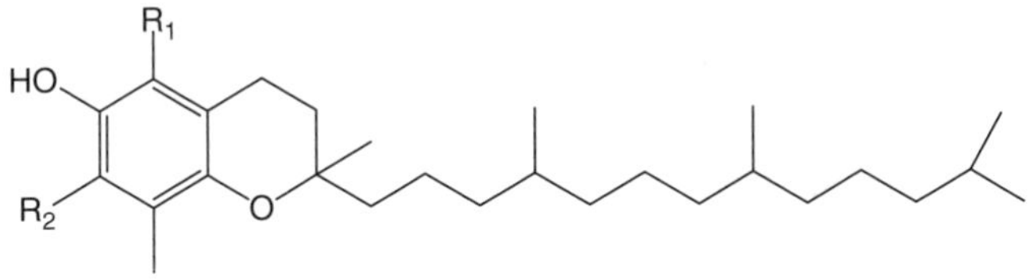

Tocopherol	R_1	R_2
α	CH_3	CH_3
β	CH_3	H
γ	H	CH_3
δ	H	H

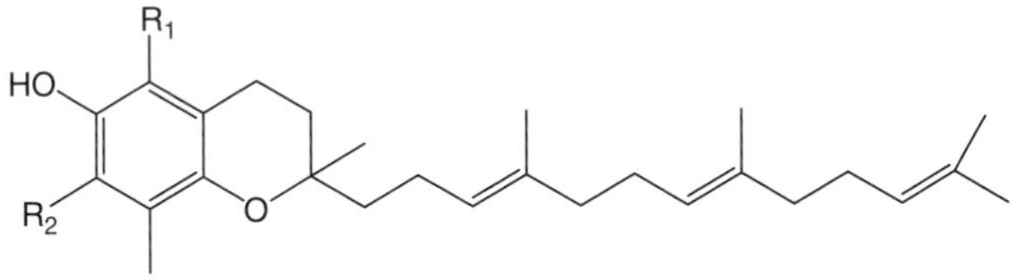

Tocotrienol	R_1	R_2
α	CH_3	CH_3
β	CH_3	H
γ	H	CH_3
δ	H	H

FIGURE 1.31 Structures of some vitamin E compounds.

TABLE 1.9

Approximate Biological Activity Relationships of Vitamin E Compounds

Compound	Activity of d-α-Tocopherol (%)
d-α-Tocopherol	100
l-α-Tocopherol	26
dl-α-Tocopherol	74
dl-α-Tocopheryl acetate	68
d-β-Tocopherol	8
d-γ-Tocopherol	3
d-δ-Tocopherol	—
d-α-Tocotrienol	22
d-β-Tocotrienol	3
d-γ-Tocotrienol	—
d-δ-Tocotrienol	—

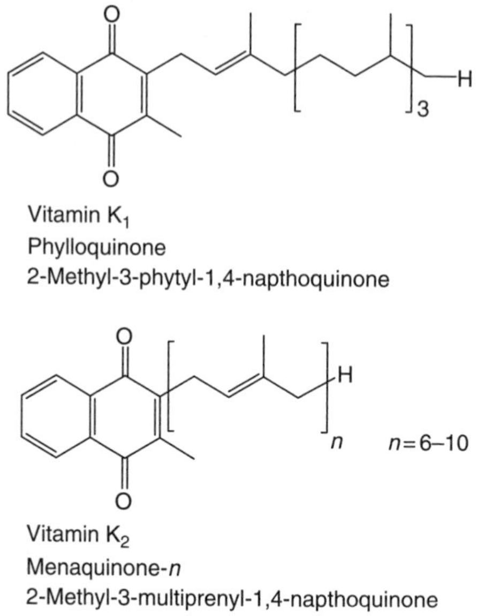

Vitamin K$_1$
Phylloquinone
2-Methyl-3-phytyl-1,4-napthoquinone

Vitamin K$_2$
Menaquinone-n
2-Methyl-3-multiprenyl-1,4-napthoquinone

FIGURE 1.32 Structures of some vitamin K compounds.

J. HYDROCARBONS

The hydrocarbons include normal, branched, saturated, and unsaturated compounds of varying chain lengths. The nomenclature for hydrocarbons has already been discussed. The hydrocarbons of most interest to lipid chemists are the isoprenoids and their oxygenated derivatives.

The basic isoprene unit (2-methyl-1,3-butadiene) is the building block for a large number of interesting compounds, including carotenoids (Figure 1.33), oxygenated carotenoids or xanthophylls (Figure 1.34), sterols, and unsaturated and saturated isoprenoids (isopranes). Recently, it has been discovered that 15-carbon and 20-carbon isoprenoids are covalently attached to some proteins and may be involved in control of cell growth [41]. Members of this class of protein-isoprenoid molecules are called prenylated proteins.

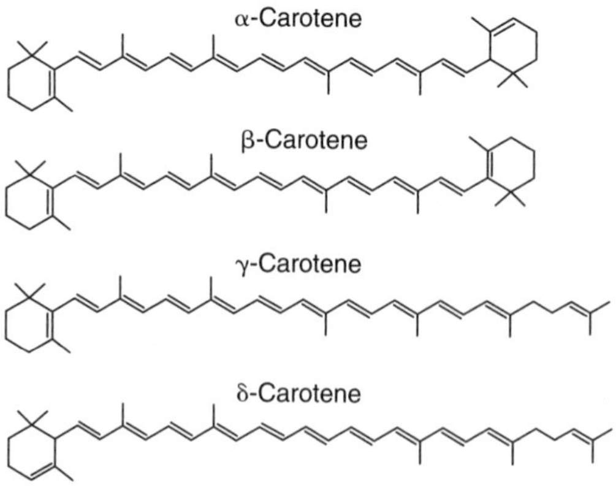

α-Carotene

β-Carotene

γ-Carotene

δ-Carotene

FIGURE 1.33 Structures and nomenclature of carotenoids.

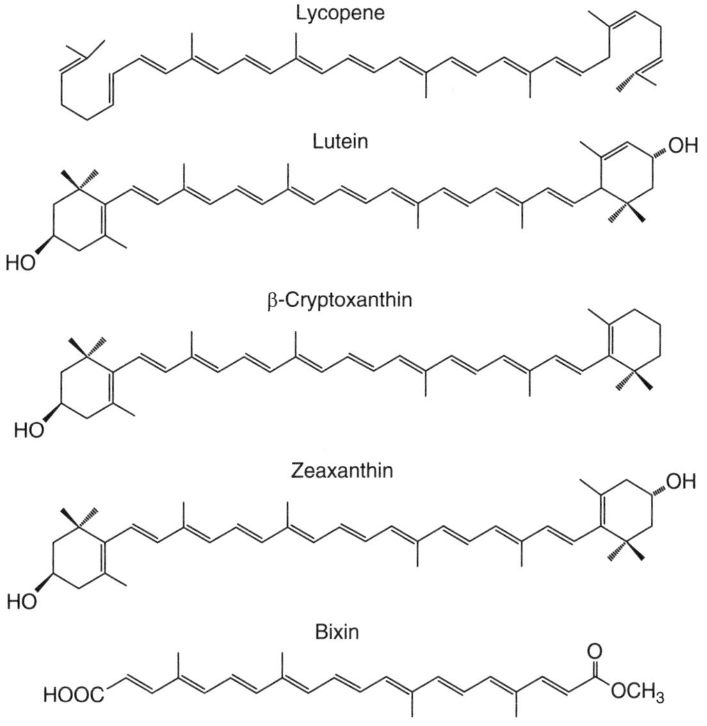

FIGURE 1.34 Structures and nomenclature of some oxygenated carotenoids.

IV. SUMMARY

It would be impossible to describe the structures and nomenclature of all known lipids even in one entire book. The information presented in this chapter is a brief overview of the complex and interesting compounds we call lipids.

REFERENCES

1. W.W. Christie. *Lipid Analysis*. Pergamon Press, New York, NY, 1982, p. 1.
2. M. Kates. *Techniques of Lipidology: Isolation, Analysis and Identification of Lipids*. Elsevier, New York, NY, 1986, p. 1.
3. M.I. Gurr and A.T. James. *Lipid Biochemistry and Introduction*. Cornell University Press, Ithaca, NY, 1971, p. 1.
4. R.H. Schmidt, M.R. Marshall, and S.F. O'Keefe. Total fat. In: *Analyzing Food for Nutrition Labeling and Hazardous Contaminants* (I.J. Jeon and W.G. Ikins, eds.). Dekker, New York, NY, 1995, pp. 29–56.
5. P.E. Verkade. *A History of the Nomenclature of Organic Chemistry*. Reidel, Boston, 1985, 507 pp.
6. IUPAC. *Nomenclature of Organic Chemistry, Sections A, B, C, D, E, F, and H*. Pergamon Press, London, 1979, p. 182.
7. IUPAC-IUB Commission on Biochemical Nomenclature. The nomenclature of lipids. *Lipids* 12:455–468 (1977).
8. IUPAC-IUB Commission on Biochemical Nomenclature. Nomenclature of phosphorus-containing compounds of biological importance. *Chem. Phys. Lipids* 21:141–158 (1978).
9. IUPAC-IUB Commission on Biochemical Nomenclature. The nomenclature of lipids. *Chem. Phys. Lipids* 21:159–173 (1978).

10. A. Streitwieser Jr. and C.H. Heathcock. *Introduction to Organic Chemistry*. Macmillan, New York, NY, 1976, p. 111.

11. K. Aitzetmuller. An unusual fatty acid pattern in *Eranthis* seed oil. *Lipids* 31:201–205 (1996).

12. J.F. Mead, R.B. Alfin-Slater, D.R. Howton, and G. Popjak. *Lipids: Chemistry, Biochemistry and Nutrition*. Plenum Press, New York, NY, 1986, 486 pp.

13. R.S. Chapkin. Reappraisal of the essential fatty acids. In: *Fatty Acids in Foods and Their Health Implications* (C.K. Chow, ed.). Dekker, New York, NY, 1992, pp. 429–436.

14. E. Granstrom and M. Kumlin. Metabolism of prostaglandins and lipoxygenase products: Relevance for eicosanoid assay. In: *Prostaglandins and Related Substances* (C. Benedetto, R.G. McDonald-Gibson, S. Nigram, and T.F. Slater, eds.). IRL Press, Oxford, 1987, pp. 5–27.

15. T.F. Slater and R.G. McDonald-Gibson. Introduction to the eicosanoids. In: *Prostaglandins and Related Substances* (C. Benedetto, R.G. McDonald-Gibson, S. Nigram, and T.F. Slater, eds.). IRL Press, Oxford, 1987, pp. 1–4.

16. F.D. Gunstone, J.L. Harwood, and F.D. Padley. *The Lipid Handbook*. Chapman and Hall, New York, NY, 1994, p. 9.

17. H.J. Dutton. Hydrogenation of fats and its significance. In: *Geometrical and Positional Fatty Acid Homers* (E.A. Emken and H.J. Dutton, eds.). Association of Official Analytical Chemists, Champaign, IL, 1979, pp. 1–16.

18. J.L. Sebedio and R.G. Ackman. Hydrogenation of menhaden oil: Fatty acid and C20 monoethylenic isomer compositions as a function of the degree of hydrogenation. *J. Am. Oil Chem. Soc.* 60:1986–1991 (1983).

19. R.G. Ackman, S.N. Hooper, and D.L. Hooper. Linolenic acid artifacts from the deodorization of oils. *J. Am. Oil Chem. Soc.* 51:42–49 (1974).

20. S. O'Keefe, S. Gaskins-Wright, V. Wiley, and I.-C. Chen. Levels of *trans* geometrical isomers of essential fatty acids in some unhydrogenated U.S. vegetable oils. *J. Food Lipids* 1:165–176 (1994).

21. S.F. O'Keefe, S. Gaskins, and V. Wiley. Levels of *trans* geometrical isomers of essential fatty acids in liquid infant formulas. *Food Res. Intern.* 27:7–13 (1994).

22. J.M. Chardigny, R.L. Wolff, E. Mager, C.C. Bayard, J.L. Sebedio, L. Martine, and W.M.N. Ratnayake. Fatty acid composition of French infant formulas with emphasis on the content and detailed profile of *trans* fatty acids. *J. Am. Oil Chem. Soc.* 73:1595–1601 (1996).

23. A. Grandgirard, J.M. Bourre, F. Juilliard, P. Homayoun, O. Dumont, M. Piciotti, and J.L. Sebedio. Incorporation of *trans* long-chain *n*-3 polyunsaturated fatty acids in rat brain structures and retina. *Lipids* 29:251–258 (1994).

24. J.M. Chardigny, J.L. Sebedio, P. Juaneda, J.-M. Vatele, and A. Grandgirard. Effects of *trans n-3* polyunsaturated fatty acids on human platelet aggregation. *Nutr. Res.* 15:1463–1471 (1995).

25. H. Keweloh and H.J. Heipieper. *trans* Unsaturated fatty acids in bacteria. *Lipids* 31:129–137 (1996).

26. R.G. Jensen. Fatty acids in milk and dairy products. In: *Fatty Acids in Foods and Their Health Implications* (C.K. Chow, ed.). Dekker, New York, NY, 1992, pp. 95–135.

27. J.-L. Sebedio and A. Grandgirard. Cyclic fatty acids: Natural sources, formation during heat treatment, synthesis and biological properties. *Prog. Lipid Res.* 28:303–336 (1989).

28. G. Dobson, W.W. Christie, and J.-L. Sebedio. Gas chromatographic properties of cyclic dienoic fatty acids formed in heated linseed oil. *J. Chromatogr. A* 723:349–354 (1996).

29. J.-L. LeQuere, J.L. Sebedio, R. Henry, F. Coudere, N. Dumont, and J.C. Prome. Gas chromatography–mass spectrometry and gas chromatography–tandem mass spectrometry of cyclic fatty acid monomers isolated from heated fats. *J. Chromatogr.* 562:659–672 (1991).

30. M.E. Stansby, H. Schlenk, and E.H. Gruger, Jr. Fatty acid composition of fish. In: *Fish Oils in Nutrition* (M. Stansby, ed.). Van Nostrand Reinhold, New York, NY, 1990, pp. 6–39.

31. C. Litchfield. *Analysis of Triglycerides*. Academic Press, New York, NY, 1972, 355 pp.

32. W.R. Nes and M.L. McKean. *Biochemistry of Steroids and Other Isopentenoids*. University Park Press, Baltimore, 1977, p. 37.

33. D.A. Guo, M. Venkatramesh, and W.D. Nes. Development regulation of sterol biosynthesis in *Zea mays*. *Lipids* 30:203–219 (1995).

34. M. Law. Plant sterol and stanol margarines and health. *Br. Med. J.* 320:861–864 (2000).

35. K.T. Hwang and G. Maerker. Quantification of cholesterol oxidation products in unirradiated and irradiated meats. *J. Am. Oil Chem. Soc.* 70:371–375 (1993).

36. S.K. Kim and W.W. Nawar. Parameters affecting cholesterol oxidation. *J. Am. Oil Chem. Soc.* 28:917–922 (1993).

37. N. Li, T. Oshima, K.-I. Shozen, H. Ushio, and C. Koizumi. Effects of the degree of unsaturation of coexisting triacylglycerols on cholesterol oxidation. *J. Am. Oil Chem. Soc.* 71:623–627 (1994).

38. L.L. Smith. Cholesterol oxidation. *Chem. Phys. Lipids* 44:87–125 (1987).

39. G. Wolf. *Vitamin A,* Vol. 3B. In: *Human Nutrition Series* (R.B. Alfin-Slater and D. Kritchevsky, eds.). Plenum Press, New York, NY.

40. L.M. DeLuca. Vitamin A. In: *The Fat Soluble Vitamins* (H.F. DeLuca, ed.). Plenum Press, New York, NY, 1978, p. 1.

41. M. Sinensky and R.J. Lutz. The prenylation of proteins. *Bioessays* 14:25–31 (1992).

2 Chemistry and Function of Phospholipids

Marilyn C. Erickson

CONTENTS

I. INTRODUCTION

Phospholipids can generally be regarded as fatty acyl-containing lipids with a phosphoric acid residue. Although hydrolysis is inherent to their ester and phosphoester bonds, other physical and chemical reactions associated with phospholipids are dictated by the kind of head group and by the chain length and degree of unsaturation of the constituent aliphatic groups. These activities constitute the focus of this chapter. In addition, the ramifications of the amphiphilic nature of phospholipids and their propensity to aggregate as bilayers and segregate into specific domains will be discussed in relation to their functional role in biological systems and foods.

II. PHOSPHOLIPID CLASSIFICATION

Phospholipids are divided into two main classes depending on whether they contain a glycerol or a sphingosyl backbone (Figure 2.1). These differences in base structure affect their chemical reactivity.

Glycerophospholipids are named after and contain structures that are based on phosphatidic acid (PA), with the group attached to the phosphate being choline, ethanolamine, serine, inositol, or glycerol. In most tissues, the diacyl forms of the glycerophospholipids predominate, but small amounts of plasmalogens, monoacyl monoalk-1-enyl ether derivatives, are also found. Choline and ethanolamine plasmalogens are the most common forms, although serine plasmalogen has also been found. The phosphonolipids that contain a covalent bond between the phosphorus atom and the carbon of the nitrogenous base comprise another glycerophospholipid variant [1]. Phosphonolipids are major constituents in three phyla and are synthesized by phytoplanktons, the base of the food chains of the ocean.

The second major group of phospholipids are sphingomyelins (SM) that consist of a long-chain base sphingosine (*trans*-D-erythro-1,3-dihydroxy-2-amino-4-octadecene), amidated to a long-chain fatty acyl chain and attached to a phosphocholine group at the primary alcohol group. Structurally, SM resembles the glycerophospholipid phosphatidylcholine (PC), with both phospholipids containing a phosphocholine hydrophilic headgroup and two long hydrophobic hydrocarbon chains. SPH, however, is only of minor importance in plants and probably is absent from bacteria.

III. PHOSPHOLIPID PHYSICAL STRUCTURES

Phospholipids are characterized by the presence of a polar or hydrophilic head group and a nonpolar or hydrophobic fatty acid region. Dissolution of the phospholipid in water is therefore limited by these structural features to a critical concentration, typically in the range of 10^{-5} to 10^{-10} mol/L. Above this concentration, the amphipathic character of phospholipids drives its assembly to form a variety of macromolecular structures in the presence of water, the chief structure being a bilayer in which the polar regions tend to orient toward the aqueous phase and the hydrophobic regions are sequestered from water (Figure 2.2A). Another macromolecular structure commonly adopted by phospholipids and compatible with their amphipathic constraints is the hexagonal (H_{II}) phase (Figure 2.2B). This phase consists of a hydrocarbon matrix penetrated by hexagonally packed aqueous cylinders with diameters of about 20Å. Table 2.1 lists less common macromolecular structures that may be adopted by phospholipids in a solid or liquid state. Note that SM and PC exist primarily as a bilayer, and this aggregation state prevails in suspensions characterized by wide ranges in temperatures, pHs, and ionic strengths. Other phospholipids, in contrast, adopt a variety of structures, and this capability is known as lipid polymorphism. Additional information on the properties of these phospholipid structures may be found in the review of Seddon and Cevc [2].

IV. BIOLOGICAL MEMBRANES

Phospholipids, along with proteins, are major components of biological membranes, which, in turn, are an integral part of prokaryotes (bacteria) and eukaryotes (plants and animals). The predominant structures assumed by phospholipids in membranes are the bilayer and H_{II} structure, which is dictated by the phase preference of the individual phospholipids (Table 2.2). It is immediately apparent that a significant proportion of membrane lipids adopt or promote H_{II} phase structure under appropriate conditions. The most striking example is phosphatidylethanolamine (PE), which may compose up to 30% of membrane phospholipids. Under such conditions, portions of the membrane that adopt an H_{II} phase would be expected to be incompatible with maintenance of a permeability barrier between external and internal compartments in those areas. Consequently, alternative roles for those structures must exist.

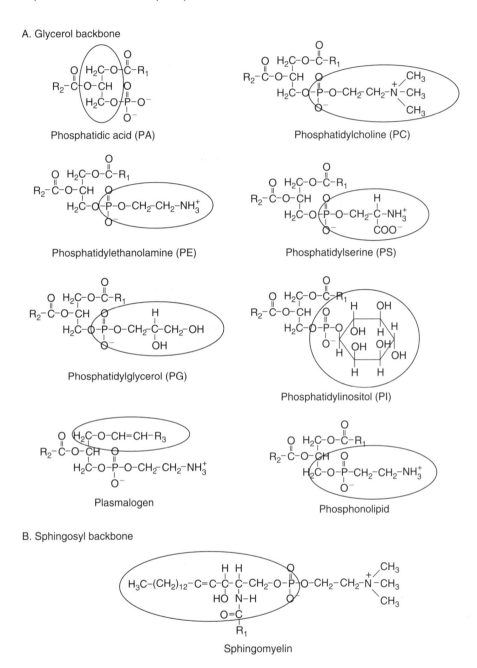

FIGURE 2.1 Structure of phospholipids. Circled areas show distinguishing features of each phospholipid.

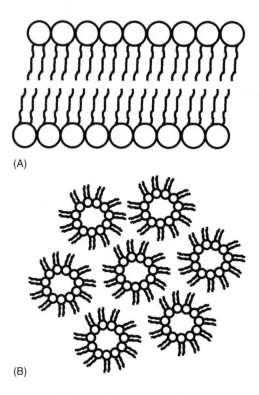

(A)

(B)

FIGURE 2.2 Mesomorphic structures of phospholipids: (A) lamellar and (B) hexagonal II.

TABLE 2.1
Macromolecular Structures
Adopted by Phospholipids

Phase	Phase Structure
Liquid	Fluid lamellar
	Hexagonal
	Complex hexagonal
	Rectangular
	Oblique
	Cubic
	Tetragonal
	Rhombohedral
Solid	Three-dimensional crystal
	Two-dimensional crystal
	Rippled gel
	Ordered ribbon phase
	Untilted gel
	Tilted gel
	Interdigitated gel
	Partial gel

TABLE 2.2

Phase Preference of Membrane Phospholipids

Bilayer	Hexagonal H$_{II}$
Phosphatidylcholine	
Sphingomyelin	
	Phosphatidylethanolamine
Phosphatidylserine	Phosphatidylserine (pH < 3)
Phosphatidylglycerol	
Phosphatidylinositol	
Phosphatidic acid	Phosphatidic acid (+Ca^{2+})
	Phosphatidic acid (pH < 3)

A. MEMBRANE PROPERTIES

1. Membrane Permeability

The ability of lipids to provide a bilayer permeability barrier between external and internal environments constitutes one of their most important functions in a biological membrane. In the case of water, permeability coefficients are determined through light scattering measurements of swelling membranes [3] and are typically high in the range of 10^{-2}–10^{-4} cm/s [4]. Factors that increase water permeability include increased unsaturation of the fatty acids of the membrane, whereas the presence of ether-linked phospholipids (absence of a carbonyl group) and branched-chain fatty acids have led to reduced rates of water permeation [5,6]. Moreover, since cholesterol reduces water permeability, the general conclusion has been made that factors contributing to increased order in the hydrocarbon region reduce water permeability.

Since oxygen transport is fundamental to all aerobic organisms, the permeability of this gaseous component in membranes is also of interest and has been estimated from the paramagnetic enhancements in relaxation of lipid-soluble spin labels in the presence of oxygen [7]. Using such a tool, oxygen permeability coefficients of 210 cm/s have been measured in lipid bilayers of dimyristoylPC with rates appearing to be dictated by the penetration of water [8].

The diffusion properties of nonelectrolytes (uncharged polar solutes) also appear to depend on the properties of the lipid matrix in much the same manner as does the diffusion of water. That is, decreased unsaturation of phospholipids or increased cholesterol content results in lower permeability coefficients. In the case of nonelectrolytes, however, the permeability coefficients are at least two orders of magnitude smaller than those of water. Furthermore, for a given homologous series of compounds, the permeability increases as the solubility in a hydrocarbon environment increases, indicating that the rate-limiting step in diffusion is the initial partitioning of the molecule into the lipid bilayer [9].

Measures of the permeability of membranes to small ions are complicated, since for free permeation to proceed, a counterflow of other ions of equivalent charge is required. In the absence of such a counterflow, a membrane potential is established that is equal and opposite to the chemical potential of the diffusing species. A remarkable impermeability of lipid bilayers exists for small ions with permeability coefficients of less than 10^{-10} cm/s commonly observed. Although permeability coefficients for Na$^+$ and K$^+$ may be as small as 10^{-14} cm/s, lipid bilayers appear to be much more permeable to H$^+$ or OH$^-$ ions, which have been reported to have permeability coefficients in the range of 10^{-4} cm/s [10]. One of the hypotheses put forth to explain this anomaly involves hydrogen-bonded wires across membranes. Such water wires could have transient existence in lipid membranes, and when such structures connect the two aqueous phases, proton flux could result as a consequence of H–O–H $\cdots$ O–H bond rearrangements. Such a mechanism does not involve physical movement of a proton all the way across the membrane; hence, proton flux occurring by this mechanism is expected to be significantly faster when compared with that of other monovalent ions that lack such a mechanism. As support for the existence of this mechanism, an increase in the level of cholesterol decreased the rate of proton transport that correlated to the decrease in the membrane's water content [11].

Two alternative mechanisms are frequently used to describe ionic permeation of lipid bilayers. In the first, the solubility-diffusion mechanism, ions partition and diffuse across the hydrophobic phase. In the second, the pore mechanism, ions traverse the bilayer through transient hydrophilic defects caused by thermal fluctuations. Based on the dependence of halide permeability coefficients on bilayer thickness and on ionic size, a solubility-diffusion mechanism was ascribed to these ions [12]. In contrast, permeation by monovalent cations, such as potassium, has been accounted for by a combination of both mechanisms. In terms of the relationship between lipid composition and membrane permeability, ion permeability appears to be related to the order in the hydrocarbon region, where increased order leads to a decrease in permeability. The charge on the phospholipid polar head group can also strongly influence permeability by virtue of the resulting surface potential. Depending on whether the surface potential is positive or negative, anions and cations could be attracted or repelled to the lipid–water interface.

2. Membrane Fluidity

The current concept of biological membranes is a dynamic molecular assembly characterized by the coexistence of structures with highly restricted mobility and components having great rotational freedom. These membrane lipids and proteins comprising domains of highly restricted mobility appear to exist on a micrometer scale in a number of cell types [13,14]. Despite this heterogeneity, membrane fluidity is still considered as a bulk, uniform property of the lipid phase that is governed by a complex pattern of the components' mobilities. Individual lipid molecules can display diffusion of three different types: lateral, rotational, and transversal [15]. Lateral diffusion of lipids in biological membranes refers to the two-dimensional translocation of the molecules in the plane of the membrane. Rotational diffusion of lipid molecules is restricted to the plane of biological membranes, whereas transverse diffusion (flip-flop) is the out-of-plane rotation or redistribution of lipid molecules between the two leaflets of the bilayer. Although the presence of docosahexaenoic acid (22:6) in the phospholipid supports faster flip-flop [16], transverse diffusion is very low in lipid bilayers, and flippase enzymes are required to mediate the process [17,18].

There are two major components of membrane fluidity. The first component is the order parameter (S), also called the structural, static, or range component of membrane fluidity. This is a measure of angular range of rotational motion, with more tightly packed chains resulting in a more ordered or less fluid bilayer. The second component of membrane fluidity is microviscosity and is the dynamic component of membrane fluidity. This component measures the rate of rotational motion and is a more accurate reflection of membrane microviscosity.

There are many physical and chemical factors that regulate the fluidity properties of biological membranes, including temperature, pressure, membrane potential, fatty acid composition, protein incorporation, and Ca^{2+} concentration. For example, calcium influenced the structure of membranes containing acidic phospholipids by nonspecifically cross-linking the negative charges. Consequently, increasing the calcium concentration in systems induced structural rearrangements and a decrease in membrane fluidity [19]. Similarly, changes in microfluidity and lateral diffusion fluidity were exhibited when polyunsaturated fatty acids oxidized [20].

Fluidity is an important property of membranes because of its role in various cellular functions. Activities of integral membrane-bound enzymes, such as Na^+, K^+-ATPase, can be regulated to some extent by changes in the lipid portions of biological membranes. In turn, changes in enzyme activities tightly connected to ion transport processes could affect translocations of ions.

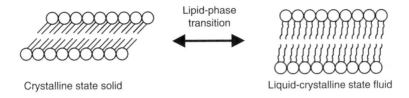

Lipid-phase
transition

Crystalline state solid   Liquid-crystalline state fluid

FIGURE 2.3 Phospholipids gel–liquid crystalline phase transition.

3. Phase Transitions

As is the case for triacylglycerols, phospholipids can exist in a frozen gel state or in a fluid liquid crystalline state depending on the temperature [21] as illustrated in Figure 2.3. Transitions between the gel and liquid crystalline phases can be monitored by a variety of techniques, including nuclear magnetic resonance (NMR), electron spin resonance, fluorescence, and differential scanning calorimetry (DSC). With DSC, both enthalpy and cooperativity of the transition may be determined, enthalpy being the energy required to melt the acyl chains and cooperativity reflecting the number of molecules that undergo a transition simultaneously. However, difficulties in determining membrane transitions have been attributed to entropy/enthalpy compensations in that enthalpy lost by lipids undergoing transition is absorbed by membrane proteins as they partition into the more fluid phase of the bilayer [21].

For complex mixtures of lipids found in biological membranes, at temperatures above the phase transition, all component lipids are liquid crystalline, exhibiting characteristics consistent with complete mixing of the various lipids. At temperatures below the phase transition of the phospholipid with the highest melting temperature, separation of the component into crystalline domains (lateral phase separation) can occur. This ability of individual lipid components to adopt gel or liquid crystalline arrangements has led to the suggestion that particular lipids in a biological membrane may become segregated into a local gel state. This segregation could affect protein function by restricting protein mobility in the bilayer matrix, or it could provide packing defects, resulting in permeability changes. Exposure of plant food tissues to refrigerator temperatures could thus induce localized membrane phase transitions, upset metabolic activity, and create an environment that serves to reduce the quality of the product [22].

Several compositional factors play a role in determining transition temperatures of membranes. Membranes whose phospholipids contain more saturated fatty acids have a higher transition temperature than membranes containing unsaturated fatty acids as the presence of *cis* double bonds inhibits hydrocarbon chain packing in the gel state. Fatty acids whose chain length is longer will also have higher transition temperature than shorter fatty acids. Hence, naturally occurring SPH whose fatty acids are more saturated and longer (50% of SM have fatty acids >20 carbons) than naturally occurring PC has a transition temperature in the physiological range, whereas PC has a transition temperature below the physiological range [23]. In the case of PS membranes, cation binding decreases the phase transition temperature. On the other hand, the presence of the free fatty acid, oleic acid, had negligible effects on the bilayer phase transition, whereas the free fatty acid, palmitic acid, increased the bilayer phase transition temperature [24]. Differential effects on bilayer properties were also seen by the incorporation of cholesterol, and these effects were dependent on the cholesterol concentration [25]. In small amounts (≤ 3 mol %), a softening of the bilayers in the transition region occurred. However, higher cholesterol concentrations led to a rigidification of the bilayer that was characterized as a liquid-ordered phase. This phase is liquid in the sense that the molecules diffuse laterally as in a fluid, but at the same time the lipid-acyl chains have a high degree of conformational order.

4. Heat Capacity

Another characteristic of the phospholipid bilayer system is its heat capacity that is dependent on the chain-melting temperatures of its component fatty acids. Using calorimetric methods, it has been found that relaxation times of lipid systems close to the chain-melting transition are proportional to the excess heat capacity. Hence, in multilamellar phospholipid systems, relaxation times were indicative of a very pronounced and narrow heat capacity maximum, whereas relaxation times of extruded vesicles were indicative of a much broader melting profile and a lower maximum heat capacity [26].

Currently, calorimetric experiments have been unable to measure heat capacity maxima of biomembranes because of the inability to distinguish heat capacity events originating from lipids and from proteins. Furthermore, the diverse phospholipid composition within biomembranes gives rise to a continuous process of melting events. Hence, it is presumed that chain melting does not occur on a global level but rather occurs locally at domain interfaces or in the lipid interface of proteins. Such local events may modify protein activity as has been noted with melting and phospholipase A_2 activity [27].

5. Curvature/Stress/Surface Tension

As individual phospholipid molecules associate, mono- and bilayers exhibit a characteristic curvature to their structures. To determine curvature elasticity of lipid bilayer membranes, several experimental methods have been used including amplitude and frequency of thermal fluctuations in the membrane contour [28], tether formation of large vesicular membranes [29], and x-ray diffraction of osmotically stressed systems [30]. Measurements of spontaneous curvatures are useful in providing some indication of phospholipid phase preference with the zero curvatures of PC forming flat lamellar L_α phases, negative curvatures of PE forming reverse hexagonal H_{II} phases, and positive curvatures of lysolipids with large polar groups forming either micelles or the hexagonal H_I phases [31]. Phospholipid curvatures are thought to affect both membrane-associated enzymes [32] and bilayer/membrane fusion [33].

B. Fundamentals of Phospholipid Interactions

1. Complexation of Phospholipid to Ions

To comprehend ion binding to phospholipid molecules or to phospholipid membranes, it is necessary to understand the behavior of ions in bulk solution and in the vicinity of a membrane–solution interface. If ion–solvent interactions are stronger than the intermolecular interactions in the solvent, ions are prone to be positively hydrating or structure-making (cosmotropic) entities. The entropy of water is decreased for such ions, whereas it is increased near other ion types with a low charge density. The latter ions are thus considered to be negatively hydrating or structure-breaking (chaotropic) entities.

When an ion approaches a phospholipid membrane it experiences several forces, the best known of which is the long-range electrostatic, Coulombic force. This force is proportional to the product of all involved charges (on both ions and phospholipids) and inversely proportional to the local dielectric constant. Since phospholipid polar head groups in an aqueous medium are typically hydrated, ion–phospholipid interactions are mediated by dehydration on binding. Similarly, dehydration of the binding ion may occur. For instance, a strong dehydration effect is observed upon cation binding to the acidic phospholipids, where up to eight water molecules are expelled from the interface once cation–phospholipid association has taken place [34–36].

Various degrees of binding exist between phospholipids and ions. When several water molecules are intercalated between the ion and its binding site, there is actually an association between the ion and phospholipid rather than binding. Outer-sphere complex formation between ion and phospholipid exists when only one water molecule is shared between the ion and its ligand. On the other hand, complete displacement of the water molecules from the region between an ion and its binding site corresponds to an inner-sphere complex. Forces involved in the inner-sphere complex formation include ion–dipole, ion–induced dipole, induced dipole–induced dipole, and ion–quadrupole forces, in addition to the Coulombic interaction. Hydrogen bonding can also participate in inner-sphere complex formation. Under appropriate circumstances, the outer-sphere complexes may also be stabilized by through-water hydrogen bonding.

Phospholipid affinity for cations appears to follow the sequence lanthanides > transition metals > alkaline earths > alkali metals, thus documenting the significance of electrostatic interactions in the process of ion–membrane binding. Electrostatic forces also play a strong role in lipid–anion binding with affinity for anions by PC, following the sequence $ClO_4^- > I^- \geq SCN^- > NO_3^- \geq Br^- > Cl^- > SO_4^{2-}$. Here anion size also has an important role in the process of binding, partly as a result of the transfer of the local excess charges from the anion to the phospholipid head groups and vice versa. However, strength of anion binding to phospholipid membranes decreases with increasing net negative charge density of the membrane [37].

Results of NMR, infrared spectroscopy, and neutron diffraction studies strongly imply that the inorganic cations interact predominantly with the phosphodiester groups of the phospholipid head groups [35,36,38–40]. On the other hand, inorganic anions may interact specifically with the trimethylammonium residues of the PC head groups [41,42].

Temperature may influence binding of ions to phospholipids. Under conditions of phase transitions, phospholipid chain melting results in a lateral expansion of the lipid bilayers, which for charged systems is also associated with the decrease in the net surface charge density. In the case of negatively charged membranes, this transition leads to lowering of the interfacial proton concentration and decreases the apparent pK value of the anion phospholipids [43].

2. Phospholipid–Lipid Interactions

Interactions between different phospholipids may be discerned from studying temperature–composition phase diagrams, and many of these are found in the review of Koynova and Caffrey [44] as well as the LIPIDAG database (http://www.lipidat.chemistry.ohio-state.edu/). Using mixtures of cholesterol and either dipalmitoyl-PC or dilauroyl-PC, complexes preferentially existed in the 2:1 and 1:1 stochiometries, respectively, and were attributed to differences in packing geometries and phospholipid conformations possible with the differing tail lengths of the two PC lipids [45]. When exposed to different acyl chain phospholipids concurrently, sterols preferentially associate with C18-acyl chain phospholipids over C14-acyl chain phospholipids especially when the sterol concentration in the bilayer is high [46]. Sphingolipids, however, are favored over phospholipids for association with cholesterol since the fully saturated acyl chains of sphingolipids can interact by their complete length with the steroid ring [47].

In cases where two phospholipid components differ in their fluid-phase/ordered-phase transition temperatures by several degrees, fluid-phase and ordered-phase domains can emerge in the temperature range between the two transition temperatures [48]. Such immiscibility contributes to the establishment of domains that are associated with many important cellular processes (i.e., signal transduction, membrane fusion, and membrane trafficking) as well as diseased states. Rafts are an example of one type of domain that is established when long-chain, saturated PC or SM and physiological amounts of cholesterol are present, with SM being the preferred partner of cholesterol [49]. Complexes formed between SM and cholesterol can, in turn, have a repulsive interaction with other phospholipids, leading to immiscibility [50]. For those phospholipids (unsaturated SM and unsaturated glycerolphospholipids) that do not segregate with cholesterol and instead have an affinity with each other [51], these associations are described as nonraft domains [52]. One specific type of nonraft domain is the ripple, a corrugated structure with defined periodicity ranging from 100Å to 300Å, depending on the lipid [53]. Ripples appear in a temperature range below the main phase transition and above a low enthalpy transition called the pretransition. They have been found to exist in the fluid-phase/ordered-phase coexistence temperature range of a binary phospholipid mixture where the fluid-phase domains elongate parallel to the ripples [54].

3. Phospholipid–Protein Interactions

Complete functioning of a biomembrane is controlled by both the protein and the lipid, mainly phospholipid, components. In a bilayer membrane that contains a heterogeneous distribution of both peripheral and integral proteins, there will be a certain proportion of the phospholipids interacting with the protein component to give the membrane its integrity at both the structural and the functional levels. Thus, the proportion of phospholipids in the bilayer interacting with protein at any one time is dictated by protein density, protein type, protein size, and aggregation state of the proteins.

The major structural element of the transmembrane part of many integral proteins is the α-helix bundle, and the disposition and packing of such helices determine the degree of protein–lipid interactions. A single α-helix passing through a bilayer membrane has a diameter of about 0.8–1 nm, depending on side-chain extension, which is similar to the long dimension of the cross-section of a diacyl phospholipid (~0.9–1.0 nm) [55]. In the absence of any significant lateral restriction of such an individual peptide helix, the lateral and rotational motion of the peptide will be similar to that for the lipids. As the protein mass in or on the membrane increases, however, the motional restriction of the adjacent lipid also increases.

Phospholipids may interact with protein interfaces in selective or nonselective ways. In the absence of selectivity, the lipids act as solvating species, maintaining the membrane protein in a suitable form for activity and mobility. Under these conditions, bilayer fluidity may alter the activity of the protein, with rigid bilayers reducing or inhibiting protein function and fluid bilayers permitting or enhancing protein activity. Nonspecific binding shows relatively little structural specificity, although the presence of a charged or polar headgroup is required to provide good localization of the molecule at the phospholipid–water interface and to interact with charged residues on the protein flanking the transmembrane region. Through high-resolution structural studies of membranes, details of these nonspecific phospholipid–protein interactions are emerging [56]. Such studies have revealed that a shell of disordered lipids surrounds the hydrophobic surface of a membrane protein. This shell of lipids, referred to as boundary or annular lipids, is equivalent to the solvent layer around a water-soluble protein. Strong evidence for the presence of these annular lipids is the close relationship between the number of lipid molecules estimated to surround a membrane protein and the circumference of the protein. Lipid molecules within the annular shell typically exchange with bulk lipids at a rate of approximately $1–2 \times 10^7$ per s at 30°C [57]. Distinct from these annular lipids, however, are tightly bound phospholipid molecules found in deep clefts of protein transmembrane α-helices. Such associations are likely to show much more specificity than binding to the annular sites. Examples of membrane proteins that require specific phospholipids for optimal activity include electron transfer complexes I and III, cytochrome c oxidase, ion channels, and transporter proteins [58]. Selective binding of cytosolic proteins to specific lipids or lipid domains may also occur. For example, the earthworm toxin, lysenin, recognizes SM-rich membrane domains in membranes and initiates oligomerization of the toxin on binding and formation of pores in this membrane section [59]. PA has also been recognized as a lipid ligand for a cytosolic protein, Opi1p, in yeast. On PA breakdown, the protein is released, translocated to the nucleus, and represses the target genes [60].

Phospholipid–protein interactions have important functional consequences. As one example, most ion gradients are set up by active transport proteins, which subsequently are used to drive secondary transport processes. If the ion gradients are lost too quickly by nonspecific leakage (e.g., through membrane regions at the protein–lipid interface), energy will be lost unnecessarily and thus will not be converted to useful work. Another consequence of protein–lipid interactions may be a result of the mutual dynamic influence of one component on the other. It is possible that for biochemical activity to take place, a fluidity window is required within the bilayer part of a membrane for the proteins to undergo the requisite rates and degrees of molecular motion around the active site. When proteins, such as ion-translocating ATPases, undergo significant conformational changes, these rearrangements may not be possible in a solid matrix. Since it is the lipid component of such bilayers that provides this fluidity window, changes in this component can alter activities of the proteins.

C. MEMBRANE DEGRADATION

Quality losses in both plant and animal tissues may be attributed to membrane breakdown following slaughter or harvest. However, postmortem changes in animal tissues occur more rapidly than those in plant tissues. In animals, cessation of circulation in the organism leads to lack of oxygen and accumulation of waste products, whereas in plants, respiratory gases can still diffuse across cell membranes, and waste products are removed by accumulation in vacuoles.

Two different membrane breakdown pathways predominate in food tissues: free radical lipid oxidation, and loss of plasma and organelle membrane integrity. Some representative modifications that occur in membranes in response to lipid peroxidation include uncoupling of oxidative phosphorylation in mitochondria; alteration of endoplasmic reticulum function; increased permeability; altered activity; inactivation of membrane-bound enzymes; and polymerization, cross-linking, and covalent binding of proteins [61]. Another consequence of lipid peroxidation is formation of the volatile aldehydes that contribute to the aroma characteristics of many vegetables. With regard to loss of plasma and organelle membrane integrity, influx and efflux of solutes may occur, leading to intimate contact among formerly separated catalytic molecules. Thus, in plants in which small changes in calcium flux bring about a wide range of physiological responses, catastrophic changes may proceed in the event of loss of membrane integrity. Specific examples of membrane deterioration in both animal and plant tissues are listed in Table 2.3. A more detailed discussion on these types of membrane deterioration may be found in the review by Stanley [22].

V. EMULSIFYING PROPERTIES OF PHOSPHOLIPIDS

When one of two immiscible liquid phases is dispersed in the other as droplets, the resulting mixture is referred to as an emulsion. To aid in the stabilization of mainly oil/water emulsions, phospholipids may act as an emulsifier by adsorbing at the interface of the two phases, their amphipathic character contributing to the lowering of interfacial tension. To characterize this process more specifically, a sequence of phases or pseudophase transitions was described near the phase boundary between immiscible liquids on hydration of an adsorbed phospholipid in n-decane [62]. These transitions were spherical reverse micelles $\rightarrow$ three-dimensional network from entangled wormlike

TABLE 2.3
Membrane Deterioration in Animal and Plant Tissues

Tissue	Description of Deterioration	Manifestations of Deterioration
Animal	Loss of membrane integrity	Drip
	Oxidative degradation of membrane lipids	Generation of off-flavors: rancid, warmed-over
Plant	Loss of membrane integrity	Loss of crispness
	Chilling injury	Surface pitting; discoloration
	Senescence/aging	Premature yellowing
	Dehydration	Failure to rehydrate

micelles → organogel separation into a diluted solution and a compact gel or solid mass precipitating on the interfacial boundary. When prepared in the presence of electrolytes, however, these phospholipid emulsions have a poor stability due to the ability of electrolytes to enhance the vibration of the phospholipid groups at the interface [63]. To circumvent the destabilizing effect of electrolytes, steric surfactants at low concentrations (0.025%–0.05%) may be added [64].

Both soybean lecithin and egg yolk are used commercially as emulsifying agents. Egg yolk contains 10% phospholipid and has been used to help form and stabilize emulsions in mayonnaise, salad dressing, and cakes. Commercial soybean lecithin, containing equal amounts of PC and inositol, has also been used as an emulsifying agent in ice creams, cakes, candies, and margarines. To expand the range of food grade emulsifiers having different hydrophilic and lipophilic properties, lecithins have been modified physically and enzymatically.

VI. HYDROLYSIS OF PHOSPHOLIPIDS

Several types of ester functionality, all capable of hydrolysis, are present in the component parts of phospholipids. These may be hydrolyzed totally by chemical methods or selectively by either chemical or enzymatic methods.

A. CHEMICAL HYDROLYSIS

Mild acid hydrolysis (trichloroacetic acid, acetic acid, HCl, and a little $HgCl_2$) results in the complete cleavage of alk-1-enyl bonds of plasmalogens, producing long-chain aldehydes. With increasing strength of acid and heating (e.g., 2N HCl or glacial acetic acid at 100°C), diacylglycerol and inositol phosphate are formed from phosphatidylinositol (PI), and diacylglycerol and glyceroldiphosphate are formed from diphosphatidylglycerol. Total hydrolysis into each of the component parts of all phospholipids can be accomplished by strong acid (HCl, H_2SO_4) catalysis in 6N aqueous or 5%–10% methanolic solutions [65]. Kinetics and mechanism for hydrolysis in 2N HCl at 120°C have been described by DeKoning and McMullan [66]. Deacylation occurs first, followed by formation of a cyclic phosphate triester as an intermediate to cyclic glycerophosphate and choline. Eventually an equilibrium mixture of α- and β-glycerophosphates is formed.

Mild alkaline hydrolysis of ester bonds in phospholipids at 37°C (0.025–0.1 M NaOH in methanolic or ethanolic solutions) leads to fatty acids and glycerophosphates. In contrast, phosphosphingolipids are not affected unless subjected to strong alkaline conditions. Some selectivity is seen in the susceptibility of phosphoglycerides to hydrolyze with diacyl > alk-1-enyl, acyl > alkyl, acyl. With more vigorous alkaline hydrolysis, the glycerophosphates are apt to undergo further hydrolysis because the phosphoester bond linking the hydrophilic component to the phospholipid moiety is not stable enough under alkaline conditions and splits, yielding a cyclic phosphate. When the cycle opens up, it gives a 1:1 mixture of 2- and 3-glycerophosphates. Both state of aggregation and specific polar group have been shown to affect the reaction rates for alkaline hydrolysis of glycerophospholipids [67]. Higher activation energies were observed for hydrolysis of glycerophospholipids in membrane vesicles than when glycerophospholipids were present as monomers or Triton X-100 micelles. Alkaline hydrolysis of PC, on the other hand, was three times faster than hydrolysis of PE.

Hydrolysis of glycerophospholipids has also been demonstrated through the application of hypochlorous acid, with unsaturated acyl lipids being a requirement. Formation of chlorohydrin groups on the unsaturated fatty acids has been conjectured as the trigger for rendering the ester group more accessible to hydrolyzing agents. Given the levels of hypochlorous acid generated with acute inflammatory responses, lysophospholipid formation by this mechanism could be relevant in vivo [68].

B. ENZYMATIC HYDROLYSIS

Selective hydrolysis of glycerophospholipids can be achieved by the application of phospholipases. One beneficial aspect to application of phospholipase is improved emulsifying properties to a PC mixture [69]. Unfortunately, although these enzymes may be isolated from a variety of sources, in general they are expensive. Interest in phospholipid hydrolysis, however, has not been lacking because of the role phospholipases have in vivo. In particular, hydrolysis products of phospholipase serve as secondary messengers in cell-signaling pathways in both plants and animals [70,71].

Several phospholipases exist differing in their preferential site of attack. The ester linkage between the glycerol backbone and the phosphoryl group is hydrolyzed by phospholipase C whereas the ester linkage on the other side of the phosphoryl group is hydrolyzed by phospholipase D. Hydrolysis of the acyl groups at the sn-1 and sn-2 positions of phospholipids is carried out by phospholipases A_1 and A_2, respectively.

Although binding of phospholipase A_2 to membrane phospholipids has been enhanced 10-fold by the presence of calcium [72], in the absence of calcium, electrostatic interactions play a major role in interactions between enzyme and substrate [73,74]. A highly cationic enzyme ($pI > 10.5$), phospholipase A_2, has a marked preference for anionic phospholipid interfaces. Thus, PA and palmitic acid promoted the binding of phospholipase A_2 to the bilayer surface [73,75]. Perturbations and a loosening of the structure associated with the presence of these hydrolysis products and indicative of phase separation were suggested as the properties contributing to enhanced binding and increases in activity [76,77]. Through a similar mechanism, the presence of phospholipid hydroperoxides has also facilitated enhanced binding of phospholipases [78]. Localized changes at the interfacial water activity at these binding sites have been suggested as the mechanism of control of phospholipase A_2 [79].

Another enzymatic hydrolysis reaction of significant merit in biological systems that has come to light in recent years is catalyzed by sphingomyelinase. Targeting raft lipid domains of SM and cholesterol, both acid and neutral sphingomyelinase hydrolyze SM to ceramide and phosphocholine [80]. Subsequent to this activity, secretory vesicle fusion associated with exocytosis at these microdomains, consisting of SM and cholesterol, is inhibited [81]. Ceramides, in turn, are believed to act as a second messenger in processes such as apoptosis, cell growth and differentiation, and cellular responses to stress through the coalescence of raised small lipid domains that serve as large signaling platforms for the oligomerization of cell surface receptors [82].

VII. HYDROGENATION OF PHOSPHOLIPIDS

Hydrogenation of fats involves the addition of hydrogen to double bonds in the chains of fatty acids. Although hydrogenation is more typically applied to triacylglycerols to generate semisolid or plastic fats more suitable for specific applications, it may also be applied to phospholipid fractions. Hydrogenated lecithins are more stable and more easily bleached to a light color, and therefore are more useful as emulsifiers than the natural, highly unsaturated lecithin from soybean oil. These advantages are exemplified by a report that hydrogenated lecithin functions well as an emulsifier and as an inhibitor of fat bloom in chocolate [83].

In practice, hydrogenation involves the mixing of the lipid with a suitable catalyst (usually nickel), heating, and then exposing the mixture to hydrogen at high pressures during agitation. Phospholipids are not as easily hydrogenated as triacylglycerols; as a result, their presence decreases the catalyst activity toward triacylglycerols [84]. In this situation, PA was the most potent poisoning agent; however, fine-grained nickel catalyst was more resistant to the poisoning effect of phospholipids than moderate-grained catalyst. In any event, hydrogenation of phospholipids requires higher temperatures and higher hydrogenation pressures. For example, hydrogenation of lecithin is carried out at 75°C–80°C in at least 70 atm pressure and in the presence of a flaked nickel catalyst [85]. In chlorinated solvents or in mixtures of these solvents with alcohol, much lower temperatures and pressures can be used for hydrogenation, particularly when a palladium catalyst is used [86].

VIII. HYDROXYLATION

Hydroxylation of the double bonds in the unsaturated fatty acids of lecithin improves the stability of the lecithin and its dispersibility in water and aqueous media. Total hydroxylating agents for lecithin include hydrogen peroxide in glacial acetic acid and sulfuric acid [87]. Such products have been advocated as useful in candy manufacture in which sharp moldings can be obtained when the hydroxylated product is used with starch molds.

IX. HYDRATION

The interaction of phospholipids with water is critical to the formation, maintenance, and function of membranes and organelles. It is the low solubility of the acyl chains in water combined with the strong hydrogen bonding between the water molecules that furnishes the attractive force that holds together polar lipids as supramolecular complexes (the hydrophobic bond). These ordered structures are generated when the phospholipid concentration exceeds its critical micelle concentration (cmc), which is dependent on the free energy gained when an isolated amphiphile in solution enters an aggregate [88]. For diacyl phospholipids in water, the cmc in general is quite low, but it depends on both the chain length and the head group. For a given chain length, the solubility of charged phospholipids is higher, whereas the cmc of a single-chain phospholipid is higher than that of a diacyl phospholipid with the same head group and the same chain length [88]. In terms of its role in membrane function, phospholipid hydration affects many membrane processes like membrane transport, ion conductance, and insertion of proteins and other molecules into membranes, and their translocation across the membrane. Dehydration of phospholipids is also advocated to play a role in membrane fusion events.

The amount of water absorbed by phospholipids has been measured by a number of different methods, including gravimetry, x-ray diffraction, neutron diffraction, NMR, and DSC [89]. For any measurement, however, Klose et al. [90] cautioned that the morphology and method of sample preparation can induce the formation of defects in and between the bilayers, and therefore will influence the water content of lamellar phospholipids.

The electrical charge on the phospholipid head group does not in itself determine the nature of the water binding [91]. However, it does affect the amount of water bound, with the amount of hydration increasing as the distance between adjacent headgroups is increased. For example, PI or PS imbibed water without limit [92,93] whereas PC imbibed up to 34 water molecules when directly mixed with bulk water [94,95]. Considerably less water was imbibed by PE, with a maximum of about 18 water molecules per lipid [96]. Method of sample preparation, however, influences the number of imbibed water molecules. For example, the amount of water absorbed by PC from the vapor phase increased monotonically from 0 water molecules per PC molecule at 0% humidity to only 14 [97] and 20 [98] water molecules per lipid molecule at 100% relative humidity. Moreover, PE only absorbed about 10 water molecules per lipid molecule from the saturated vapor phase [97]. Observed differences between absorption from bulk water and saturated vapor have been ascribed to the difficulty of exerting accurate control over relative humidities near 100%. Other factors also determining the number of water molecules in the hydration shell include the lipid phase, acyl chain composition, the presence of double bonds, and the presence of sterols [97,99–101]. For example, inclusion of cholesterol in PC membranes increased the number of water molecules in the gel state but had no influence on PS membranes [102].

Hydration of a phospholipid appears to be cooperative. A water molecule that initiated hydration of a site facilitated access of additional water molecules, until the hydration of the whole site composed of many different interacting polar residues was completed [103]. Incorporation of the first 3–4 water molecules on each phospholipid occurs on the phosphate of the lipid head group and is exothermic [104]. The remaining water molecules are incorporated endothermically.

Neutron diffraction experiments on multilayers containing PC [105,106], PE [107], and PI [108] have revealed that water distributions are centered between adjacent bilayers and overlap the head group peaks in the neutron scattering profile of the bilayer. These results imply that water penetrates into the bilayer head group region, but appreciable quantities of water do not reach the hydrocarbon core. By combining x-ray diffraction and dilatometry data, McIntosh and Simon [109,110] were able to calculate the number of water molecules in the interbilayer space and in the head group region for dilauroyl-PE bilayers. They found that there are about 7 and 10 water molecules in the gel and liquid crystalline phases, respectively, with about half of these water molecules located between adjacent bilayers and the other half in the head group region.

The amount of water taken up by a given phospholipid depends on interactions between the lipid molecules, including interbilayer forces (those perpendicular to the plane of the bilayer) and intrabilayer forces (those in the plane of the bilayer). For interbilayer forces, at least four repulsive interactions have been shown to operate between bilayer surfaces. These are the electrostatic, undulation, hydration (solvation), and steric pressures. Attractive pressures include the relatively long-range van der Waals pressure and short-range bonds between the molecules in apposing bilayers, such as hydrogen bonds or bridges formed by divalent salts. Several of the same repulsive and attractive interactions act in the plane of the bilayer, including electrostatic repulsion, hydration repulsion, steric repulsion, and van der Waals attraction. In addition, interfacial tension plays an important role in determining the area per lipid molecule [111]. Thus, as the area per molecule increases, more water can be incorporated into the head group region of the bilayer. Such a situation is found with bilayers having an interdigitated gel phase compared with the normal gel phase and with bilayers having unsaturated fatty acids in the phospholipid compared with saturated fatty acids [112,113].

The presence of monovalent and divalent cations in the fluid phase changes the hydration properties of the phospholipids. For example, the partial fluid thickness between dipalmitoyl PC bilayers increased from about 20Å in water to more than 90Å in 1 mM $CaCl_2$ [114]. In contrast, monovalent cations such as Na^+, K^+, or Cs^+ decrease the fluid spaces between adjacent charged PS or PG bilayers as a result of screening of the charge [37,115]. Divalent cations have a dehydrating effect on glycerophospholipids. For example, Ca^{2+}, the most extensively studied divalent cation, binds to the phosphate group of PS [116], liberates water between bilayers and from the lipid polar groups [88], crystallizes the lipid hydrocarbon chains [115,116], and raises the gel to the liquid crystalline melting temperature of dipalmitoyl PS by more than 100°C [115]. Cu^{2+} and Zn^{2+}, on the other hand, caused considerable dehydration of the phosphate and carbonyl groups [117]. In any event, hydration alterations by these ions would likely alter membrane permeability.

X. OXIDATION

Unsaturated fatty acids of phospholipids are susceptible to oxidation through both enzymatically controlled processes and random autoxidation processes. The mechanism of autoxidation is basically similar to the oxidative mechanism of fatty acids or esters in the bulk phase or in inert organic solvents. This mechanism is characterized by three main phases: initiation, propagation, and termination. Initiation occurs as hydrogen is abstracted from an unsaturated fatty acid of a phospholipid, resulting in a lipid-free radical. The lipid-free radical, in turn, reacts with molecular oxygen to form a lipid peroxyl radical. Although irradiation can directly abstract hydrogen from phospholipids, initiation is frequently attributed to reaction of the fatty acids with active oxygen species, such as the hydroxyl-free radical and the protonated form of superoxide. These active oxygen species are produced when a metal ion, particularly iron, interacts with triplet oxygen, hydrogen peroxide, and superoxide anion. On the other hand, enzymatic abstraction of hydrogen from an unsaturated fatty acid occurs when Fe^{3+} at the active site of lipoxygenase is reduced to Fe^{2+}. While the majority of lipoxygenases require free fatty acids, there have been reports of lipoxygenase acting directly on fatty acids in phospholipids [118,119]. Hence, enzymatic hydrolysis may not always be required before lipoxygenase activity.

During propagation, lipid–lipid interactions foster propagation of free radicals produced during initiation by abstracting hydrogen from adjacent molecules; the result is a lipid hydroperoxide and a new lipid-free radical. Magnification of initiation by a factor of 10 [120] to 100 [121] may occur through free radical chain propagation. Further magnification may occur through branching reactions (also known as secondary initiation) in which Fe^{2+} interacts with a hydroperoxide to form a lipid alkoxyl radical and hydroxyl radical, which will then abstract hydrogens from unsaturated fatty acids.

There are many consequences to phospholipid peroxidation in biological and membrane systems. On a molecular level, lipid peroxidation has been manifested in a decreased hydrocarbon core width and molecular volume [122]. In food, the decomposition of hydroperoxides to aldehydes and ketones is responsible for the characteristic flavors and aromas that collectively are often described by the terms rancid and warmed-over. Numerous studies, on the other hand, have shown that specific oxidation products may be desirable flavor components [123–126], particularly when formed in more precise (i.e., less random) reactions by the action of lipoxygenase enzymes [127–132] and by the modifying influence of tocopherol on autoxidation reactions [133].

Through in vitro studies, membrane phospholipids have been shown to oxidize faster than emulsified triacylglycerols [134], apparently because propagation is facilitated by the arrangement of phospholipid fatty acids in the membrane. However, when phospholipids are in an oil state, they are more resistant to oxidation than triacylglycerols or free fatty acids [135]. Evidence that phospholipids are the major contributors to the development of warmed-over flavor in meat from different animal species has been described in several sources [136–139]. Similarly, during frozen storage of salmon fillets, hydrolysis followed by oxidation of the n-3 fatty acids in phospholipids was noted [140]. The relative importance of phospholipids in these food samples has been attributed to the high degree of polyunsaturation in this lipid fraction and the proximity of the phospholipids to catalytic sites of oxidation (enzymic lipid peroxidation, heme-containing compounds) [141]. However, the importance of phospholipids has not been restricted to animal and fish tissues. In an accelerated storage test of potato granules, both the amount of phospholipids and their unsaturation decreased [142]. Moreover, with pecans, a much stronger negative correlation was found between headspace hexanal and its precursor fatty acid (18:2) from the phospholipid fraction ($R = -0.98$) than from the triacylglycerol fraction ($R = -0.66$) or free fatty acid fraction ($R = -0.79$) [143]. These results suggest that despite the fact that membrane lipid constitutes a small percentage of the total lipid (0.5%), early stages of oxidation may actually occur primarily within the phospholipids.

The presence of phospholipids does not preclude acceleration of lipid oxidation. When present as a minor component of oil systems, solubilized phospholipids have limited the oxidation of the triacylglycerols [144–146]. Order of effectiveness of individual phospholipids was as follows: SPH = LPC = PC = PE > PS > PI > PG [147], with both the amino and hydroxy groups in the side chain participating in the antioxidant activity [148]. It was postulated that antioxidant Maillard reaction products were formed when aldehydes reacted with the amino group of the nitrogen-containing phospholipid. Alternatively, antioxidant activity occurred when complexes between peroxyl-free radicals and the amino group were formed [149]. The latter activity is supported by an extended induction period when both tocopherol and phospholipids were present.

Fatty acid composition is a major factor affecting the susceptibility of a phospholipid to assume an oxidized state, with carbon–hydrogen dissociation energies decreasing as the number of bisallylic methylene positions increases [150,151]. However, lipid unsaturation also physically affects oxidation. In model membrane bilayers made from single unilamellar vesicles, lipid unsaturation resulted in smaller vesicles and therefore a larger curvature of the outer bilayer leaflet. The increased lipid–lipid spacing of these highly curved bilayers, in turn, facilitated penetration by oxidants [152,153]. Other functional groups on the phospholipid will also impact their oxidative stability. For example, the presence of an enol ether bond at position 1 of the glycerol backbone in plasmalogen phospholipids has led to inhibition of lipid oxidation, possibly through the binding of the enol ether double bond to initiating peroxyl radicals [154]. Apparently, products of enol ether oxidation do not readily propagate oxidation of polyunsaturated fatty acids. Alternatively, inhibition of lipid oxidation by plasmalogens has been attributed to the iron-binding properties of these compounds [155]. Variation within the phospholipid classes toward oxidation has also been ascribed to the iron-trapping ability of the polar head group [156]. For example, PS was shown to inhibit lipid peroxidation induced by a ferrous–ascorbate system in the presence of PC hydroperoxides [157]. However, stimulation of phospholipid oxidation by trivalent metal ions (Al^{3+}, Sc^{3+}, Ga^{3+}, In^{3+}, Be^{2+}, Y^{3+}, and La^{3+}) has been attributed to the capacity of the ions to increase lipid packing and promote the formation of rigid clusters or displacement to the gel state—processes that bring phospholipid acyl chains closer together to favor propagation steps [158–160].

XI. SUMMARY

This chapter has attempted to highlight the major chemical activities associated with phospholipids and the relevance of these activities to the function of phospholipids in foods and biological systems. When present in oils or formulated floods, phospholipids may have either detrimental or beneficial effects. As a major component of membranes, phospholipids may also impact the quality of food tissues to a significant extent. Consequently, their modifying presence should not be overlooked, even when they represent a small proportion of the total lipid of a given food tissue.

REFERENCES

1. M.C. Moschidis. Phosphonolipids. *Prog. Lipid Res.* 23:223–246 (1985).
2. J.M. Seddon and G. Cevc. Lipid polymorphism: structure and stability of lyotropic mesophases of phospholipids. In: *Phospholipids Handbook* (G. Cevc, ed.). Dekker, New York, NY, 1993, pp. 403–454.
3. M.C. Blok, L.L.M. van Deenen, and J. De Gier. Effect of the gel to liquid crystalline phase transition on the osmotic behavior of phosphatidylcholine liposomes. *Biochim. Biophys. Acta* 433:1–12 (1976).
4. R. Fettiplace, I.G.H. Gordon, S.B. Hladky, J. Requens, H.B. Zingshen, and D.A. Haydon. Techniques in the formation and examination of black lipid bilayer membranes. In: *Methods in Membrane Biology*, Vol. 4 (E.D. Korn, ed.). Plenum Press, New York, NY, 1974, pp. 1–75.
5. K. Shinoda, W. Shinoda, T. Baba, and M. Mikami. Comparative molecular dynamics study of ether and ester-linked phospholipid bilayers. *J. Chem. Phys.* 121:9648–9654 (2004).
6. W. Shinoda, M. Mikami, T. Baba, and M. Hato. Dynamics of a highly branched lipid bilayer: a molecular dynamics study. *Chem. Phys. Lipids* 390:35–40 (2004).
7. J.S. Hyde, and W.K. Subczynski. Simulation of electron spin resonance spectra of the oxygen-sensitive spin label probe CTPO. *J. Magn. Reson.* 56:125–130 (1984).
8. B.G. Dzikovski, V.A. Livshits, and D. Marsh. Oxygen permeation profile in lipid membranes: comparison with transmembrane polarity profile. *Biophys. J.* 85:1005–1012 (2003).
9. M. Poznansky, S. Tang, P.C. White, J.M. Milgram, and M. Selenen. Nonelectrolyte diffusion across lipid bilayer systems. *J. Gen. Physiol.* 67:45–66 (1976).
10. D.W. Deamer. Proton permeability in biological and model membranes. In: *Intracellular pH: Its Measurement, Regulation and Utilization in Cellular Functions* (R. Nuccitelli and D.W. Deamer, eds.). Liss, New York, NY, 1982, pp. 173–187.
11. I. Krishnamoorthy and G. Krishnamoorthy. Probing the link between proton transport and water content in lipid membranes. *J. Phys. Chem. B* 105:1484–1488 (2001).
12. S. Paula, A.G. Volkov, and D.W. Deamer. Permeation of halide anions through phospholipid bilayers occurs by the solubility-diffusion mechanism. *Biophys. J.* 74:319–327 (1998).
13. E. Yechiel and M. Edidin. Micrometer-scale domains in fibroblast plasma membranes. *J. Cell. Biol.* 105:755–760 (1987).
14. M. Edidin and I. Stroynowski. Differences between the lateral organization of conventional and inositol phospholipid-anchored membrane proteins. A further definition of micrometer scale membrane domains. *J. Cell Biol.* 112:1143–1150 (1991).
15. R.L. Smith and E. Oldfield. Dynamic structure of membranes by deuterium NMR. *Science* 225:280–288 (1984).

16. V.T. Armstrong, M.R. Brzustowicz, S.R. Wassall, L.J. Jenski, and W. Stillwell. Rapid flip-flop in polyunsaturated (docosahexaenoate) phospholipid membranes. *Arch. Biochem. Biophys.* 414:74–82 (2003).

17. A. Hermann, A. Zachowski, and P.F. Devaux. Protein-mediated phospholipid translocation in the endoplasmic reticulum with a low lipid specificity. *Biochemistry* 29:2023–2027 (1990).

18. J. Kubelt, A.K. Menon, P. Müller, and A. Herrmann. Transbilayer movement of fluorescent phospholipid analogs in the cytoplasmic membrane of *E. coli. Biochemistry* 41:5605–5612 (2002).

19. M. Shinitzky. Membrane fluidity and cellular functions. In: *Physiology of Membrane Fluidity* (M. Shinitzky, ed.). CRC Press, Boca Raton, FL, 1984, Chap. 1.

20. J.W. Borst, N.V. Visser, O. Kouptsova, and A.J.W.G. Visser. Oxidation of unsaturated phospholipids in membrane bilayer mixtures is accompanied by membrane fluidity changes. *Biochim. Biophys. Acta* 1487:61–73 (2000).

21. J.R. Silvius. Thermotropic phase transitions of pure lipids in model membranes and their modification by membrane proteins. In: *Lipid–Protein Interactions*, Vol. 2 (P.C. Jost and O.H. Griffith, eds.). Wiley, New York, NY, 1982, Chap. 7.

22. D.W. Stanley. Biological membrane deterioration and associated quality losses in food tissues. *Crit. Rev. Food Sci. Nutr.* 30:487–553 (1991).

23. Y. Barenholz and T.E. Thompson. Sphingomyelin: biophysical aspects. *Chem. Phys. Lipids* 102:29–34 (1999).

24. T. Inoue, S.-I. Yanagihara, Y. Misono, and M. Suzuki. Effect of fatty acids on phase behavior of hydrated dipalmitoylphosphatidylcholine bilayer: saturated versus unsaturated fatty acids. *Chem. Phys. Lipids* 109:117–133 (2001).

25. J. Lemmich, K. Mortensen, J.H. Ipsen, T. Hønger, R. Bauer, and O.B. Mouritsen. The effect of cholesterol in small amounts on lipid-bilayer softness in the region of the main phase transition. *Eur. Biophys. J.* 25:293–304 (1997).

26. P. Grabitz, V.P. Ivanova, and T. Heimburg. Relaxation kinetics of lipid membranes and its relation to the heat capacity. *Biophys. J.* 82:299–309 (2002).

27. W. Burack, Q. Yuan, and R. Biltonen. Role of lateral phase separation in the modulation of phospholipase A_2 activity. *Biochemistry* 32:583–589 (1993).

28. R.M. Servuss, W. Harbich, and W. Helfrich. Measurement of the curvature-elastic modulus of egg lecithin bilayers. *Biochim. Biophys. Acta* 436:900–903 (1976).

29. R.E. Waugh, J. Song, S. Svetina, and B. Zeks. Local and nonlocal curvature elasticity in bilayer membranes by tether formation from lecithin vesicles. *Biophys. J.* 61:974–982 (1992).

30. N. Fuller, and R.P. Rand. The influence of lysolipids on the spontaneous curvature and bending elasticity of phospholipid membranes. *Biophys. J.* 81:243–254 (2001).

31. J.A. Szule, N.L. Fuller, and R.P. Rand. The effects of acyl chain length and saturation of diacylglycerols and phosphatidylcholines on membrane monolayer curvature. *Biophys. J.* 83:977–984 (2002).

32. G.S. Attard, R.H. Templer, W.S. Smith, A.N. Hunt, and S. Jackowski. Modulation of CTP: phosphocholine cytidylyltransferase by membrane curvature elastic stress. *Proc. Natl. Acad. Sci. U.S.A.* 97:9032–9036 (2000).

33. L.V. Chernomordik and M.M. Kozlov. Protein–lipid interplay in fusion and fission of biological membranes: tipping the balance of membrane stability. *Annu. Rev. Biochem.* 72:175–207 (2003).

34. R.A. Dluhy, D.G. Cameron, H.H. Mantsch, and R. Mendelsohn. Fourier transform infrared spectroscopic studies of the effect of calcium ions on phosphatidylserine. *Biochemistry* 22:6318–6325 (1983).

35. H.L. Casal, H.H. Mantsch, and H. Hauser. Infrared studies of fully hydrated saturated and phosphatidylserine bilayers. Effect of Li^+ and Ca^{2+}. *Biochemistry* 26:4408–4416 (1987).

36. H.L. Casal, A. Martin, H.H. Mantsch, F. Paltauf, and H. Hauser. Infrared studies of fully hydrated unsaturated phosphatidylserine bilayers. Effect of Li^+ and Ca^{2+}. *Biochemistry* 26:7395–7401 (1987).

37. M.E. Loosley-Millman, R.P. Rand, and V.A. Parsegian. Effects of monovalent ion binding and screening on measured electrostatic forces between charged phospholipid bilayers. *Biophys. J.* 40:221–232 (1982).

38. H. Hauser, M.C. Phillips, B.A. Levine, and R.J.P. Williams. Ion-binding to phospholipids. Interaction of calcium and lanthanide ions with phosphatidylcholine. *Eur. J. Biochem.* 58:133–144 (1975).

39. P.W. Nolden and T. Ackermann. A high-resolution NMR study (^{1}H, ^{13}C, ^{31}P) of the interaction of paramagnetic ions with phospholipids in aqueous dispersions. *Biophys. Chem.* 4:297–304 (1976).

40. L. Herbette, C. Napolitano, and R.V. McDaniel. Direct determination of the calcium profile structure for dipalmitoyllecithin multilayers using neutron diffraction. *Biophys. J.* 46:677–685 (1984).

41. G.L. Jendrasiak. Halide interaction with phospholipids: proton magnetic resonance studies. *Chem. Phys. Lipids* 9:133–146 (1972).

42. P.M. MacDonald and J. Seelig. Anion binding to neutral and positively charged lipid membranes. *Biochemistry* 27:6769–6775 (1988).

43. H. Träuble. Membrane electrostatics. In: *Structure of Biological Membranes* (S. Abrahamsson and I. Pascher, eds.). Plenum Press, New York, NY, 1977, pp. 509–550.

44. R. Koynova and M. Caffrey. An index of lipid phase diagrams. *Chem. Phys. Lipids* 115:107–219 (2002).

45. S.A. Pandit, D. Bostick, and M.L. Berkowitz. Complexation of phosphatidylcholine lipids with cholesterol. *Biophys. J.* 86:1345–1356 (2004).

46. M. Sugahara, M. Uragami, and S.L. Regen. Selective sterol–phospholipid associations in fluid bilayers. *J. Am. Chem. Soc.* 124:4253–4256 (2002).

47. J.P. Slotte. Sphingomyelin–cholesterol interactions in biological and model membranes. *Chem. Phys. Lipids* 102:13–27 (1999).

48. L.A. Bagatolli and E. Gratton. Two photon fluorescence microscopy of coexisting lipid domains in giant unilamellar vesicles of binary phospholipid mixtures. *Biophys. J.* 78:290–305 (2000).

49. D. Scherfeld, N. Kahya, and P. Schwille. Lipid dynamics and domain formation in model membranes composed of ternary mixtures of unsaturated and saturated phosphatidylcholines and cholesterol. *Biophys. J.* 85:3758–3768 (2003).

50. A. Radhakrishnan and H. McConnell. Condensed complexes in vesicles containing cholesterol and phospholipids. *Proc. Natl. Acad. Sci. U.S.A.* 102:12662–12666 (2005).

51. R.M. Epand and R.F. Epand. Non-raft forming sphingomyelin-cholesterol mixtures. *Chem. Phys. Lipids* 132:37–46 (2004).

52. S.R. Shaikh and M.A. Edidin. Membranes are not just rafts. *Chem. Phys. Lipids* 144:1–3 (2006).

53. M.J. Janiak, D.M. Small, and G.G. Shipley. Temperature and compositional dependence of the structure of hydrated dimyristoyl lecithin. *J. Biol. Chem.* 254:6068–6078 (1979).

54. C. Leidy, T. Kaasgaard, J.H. Crowe, O.G. Mouritsen, and K. Jørgensen. Ripples and the formation of anisotropic lipid domains: imaging two-component supported double bilayers by atomic force microscopy. *Biophys. J.* 83:2625–2633 (2002).

55. A. Watts. Magnetic resonance studies of phospholipid–protein interactions in bilayers. In: *Phospholipids Handbook* (G. Cevc, ed.). Dekker, New York, NY, 1993, pp. 687–741.

56. A.G. Lee. Lipid–protein interactions in biological membranes: a structural perspective. *Biochim. Biophys. Acta* 1612:1–40 (2003).

57. D. Marsh and L.I. Horváth. Structure, dynamics and composition of the lipid–protein interface. Perspectives from spin-labelling. *Biochim. Biophys. Acta* 1376:267–296 (1998).

58. M.M. Sperotto, S. May, and A. Baumgaertner. Modelling of proteins in membranes. *Chem. Phys. Lipids* 141:2–29 (2006).

59. R. Ishitsuka, A. Yamaji-Hasegawa, A. Makino, Y. Hirbayashi, and T. Kobayashi. A lipid-specific toxin reveals heterogeneity of sphingomyelin-containing membranes. *Biophys. J.* 86:296–307 (2004).

60. X. Wang, S.P. Devaiah, W. Zhang, and R. Welti. Signaling functions of phosphatidic acid. *Prog. Lipid Res.* 45:250–278 (2006).

61. P.J. O'Brien. Oxidation of lipids in biological membranes and intracellular consequences. In: *Autoxidation of Unsaturated Lipids* (H.-W. Chan, ed.). Academic Press, San Diego, CA, 1987, pp. 233–280.

62. Y.A. Shchipunov and P. Schmiedel. Phase behavior of lecithin at the oil/water interface. *Langmuir* 12:6443–6445 (1996).

63. J.M. Whittinghill, J. Norton, and A. Proctor. Stability determination of soy lecithin based emulsions by Fourier transform infrared spectroscopy. *J. Am. Oil Chem. Soc.* 77:37–42 (2000).

64. D. De Vleeschauwer and P. Van der Meeren. Colloid chemical stability and interfacial properties of mixed phospholipid-non-ionic surfactant stabilised oil-in-water emulsions. *Colloids Surf. A* 152:59–66 (1999).

65. D.J. Hanahan, J. Ekholm, and C.M. Jackson. The structure of glyceryl ethers and the glyceryl ether phospholipids of bovine erythrocytes. *Biochemistry* 2:630–641 (1963).

66. A.J. DeKoning and K.B. McMullan. Hydrolysis of phospholipids with hydrochloric acid. *Biochim. Biophys. Acta* 106:519–526 (1965).

67. C.R. Kensil and E.A. Dennis. Alkaline hydrolysis in model membranes and the dependence on their state of aggregation. *Biochemistry* 20:6079–6085 (1981).

68. J. Arnhold, A.N. Osipov, H. Spalteholz, O.M. Panasenko, and J. Schiller. Formation of lysophospholipids from unsaturated phosphatidylcholines under the influence of hypochlorous acid. *Biochim. Biophys. Acta* 1572:91–100 (2002).

69. T. Yamane. Enzyme technology for the lipids industry: an engineering overview. In: *Proceedings World Conference on Biotechnology for the Fats and Oils Industry* (T.H. Applewhite, ed.). American Oil Chemists' Society, Champaign, IL, 1988, pp. 17–22.

70. J.H. Hurley, Y. Tsujishita, and M.A. Pearson. Floundering about at cell membranes: a structural view of phospholipid signaling. *Curr. Opin. Struct. Biol.* 10:737–743 (2000).

71. S.B. Ryu. Phospholipid-derived signaling mediated by phospholipase A in plants. *Trends Plant Sci.* 9:229–235 (2004).

72. M.S. Hixon, A. Ball, and M.H. Gelb. Calcium-dependent and independent interfacial binding and catalysis of cytosolic group IV phospholipase A_2. *Biochemistry* 37:8516–8526 (1998).

73. A.R. Kinkaid, R. Othman, J. Voysey, and D.C. Wilton. Phospholipase D and phosphatidic acid enhance the hydrolysis of phospholipids in vesicles and in cell membranes by human secreted phospholipase A_2. *Biochim. Biophys. Acta* 1390:173–185 (1998).

74. S.A. Tatulian. Toward understanding interfacial activation of secretory phospholipase A_2 (PLA_2): membrane surface properties and membrane-induced structural changes in the enzyme contribute synergistically to PLA_2 activation. *Biophys. J.* 80:789–800 (2001).

75. J.B. Henshaw, C.A. Olsen, A.R. Farnback, K.H. Nielson, and J.D. Bell. Definition of the specific roles of lysolecithin and palmitic acid in altering the susceptibility of dipalmitoylphosphatidylcholine bilayers to phospholipase A_2. *Biochemistry* 37:10709–10721 (1998).

76. M.T. Hyvonen, K. Oorni, P.T. Kovanen, and M. Ala-Korpela. Changes in a phospholipid bilayer induced by the hydrolysis of a phospholipase A_2 enzyme: a molecular dynamics simulation study. *Biophys. J.* 80:565–578 (2001).

77. L.K. Nielsen, K. Balashev, T.H. Callisen, and T. Bjørnholm. Influence of product phase separation on phospholipase A_2 hydrolysis of supported phospholipid bilayers studied by force microscopy. *Biophys. J.* 83:2617–2624 (2002).

78. J. RashbaStep, A. Tatoyan, R. Duncan, D. Ann, T.R. PushpaRehka, and A. Sevanian. Phospholipid peroxidation induces cytosolic phospholipase A_2 activity: membrane effects versus enzyme phosphorylation. *Arch. Biochem. Biophys.* 343:44–54 (1997).

79. C.S. Rao and S. Damodaran. Surface pressure dependence of phospholipase A_2 activity in lipid monolayers is linked to interfacial water activity. *Colloids Surf. B Biointerfaces* 34:197–204 (2004).

80. A.E. Cremesti, F.M. Goni, and R. Kolesnick. Role of sphingomyelinase and ceramide in modulating rafts: do biophysical properties determine biologic outcome? *FEBS Lett.* 531:47–53 (2002).

81. T. Rogasevskaia and J.R. Coorssen. Sphingomyelin-enriched microdomains define the efficiency of native Ca^{2+}-triggered membrane fusion. *J. Cell Sci.* 119:2688–2694 (2006).

82. I. Johnston and L.J. Johnston. Ceramide promotes restructuring of model raft membranes. *Langmuir* 22:11284–11289 (2006).

83. P.L. Julian. *Treating Phosphatides*. U.S. Patent 2,629,662 (1953).

84. E. Szukalska. Effect of phospholipid structure on kinetics and chemistry of soybean oil hydrogenation with nickel catalysts. *Eur. J. Lipid Sci. Technol.* 102:739–745 (2000).

85. G. Jacini. *Hydrogenation of Phosphatides*. U.S. Patent 2,870,179 (1959).

86. R.D. Cole. *Hydrogenated Lecithin*. U.S. Patent 2,907,777 (1959).

87. H. Wittcoff. *Hydroxyphosphatides*. U.S. Patent 2,445,948 (1948).

88. C. Tanford. *The Hydrophobic Effect*, 2nd ed. Wiley, New York, NY, 1980.

89. T.J. McIntosh and A.D. Magid. Phospholipid hydration. In: *Phospholipids Handbook* (G. Cevc, ed.). Dekker, New York, NY, 1993, pp. 553–577.

90. G. Klose, B. Konig, H.W. Meyer, G. Schulze, and G. Degovics. Small-angle x-ray scattering and electron microscopy of crude dispersions of swelling lipids and the influence of the morphology on the repeat distance. *Chem. Phys. Lipids* 47:225–234 (1988).

91. G.L. Jendrasiak and R.L. Smith. The interaction of water with the phospholipid head group and its relationship to the lipid electrical conductivity. *Chem. Phys. Lipids* 131:183–195 (2004).

92. H. Hauser, F. Paltauf, and G.G. Shipley. Structure and thermotropic behavior of phosphatidylserine bilayer membranes. *Biochemistry* 21:1061–1067 (1982).

93. R.V. McDaniel. Neutron diffraction studies of digalactosyldiacylglycerol. *Biochim. Biophys. Acta* 940:158–161 (1988).

94. D.M. Small. Phase equilibria and structure of dry and hydrated egg lecithin. *J. Lipid Res.* 8:551–557 (1967).

95. D.M. LeNeveu, R.P. Rand, V.A. Parsegian, and D. Gingell. Measurement and modification of forces between lecithin bilayers. *Biophys. J.* 18:209–230 (1977).

96. S.M. Gruner, M.W. Tate, G.L. Kirk, P.T.C. So, D.C. Turner, D.T. Keane, C.P.S. Tilcock, and P.R. Cullis. x-ray diffraction study of the polymorphic behavior of N-methylated dioleoylphosphatidylethanolamine. *Biochemistry* 27:2853–2866 (1988).

97. G.L. Jendrasiak and J.H. Hasty. The hydration of phospholipids. *Biochim. Biophys. Acta* 337:79–91 (1974).

98. P.H. Elworthy. The adsorption of water vapor by lecithin and lysolecithin and the hydration of lysolecithin micelles. *J. Chem. Soc.* 5385–5389 (1961).

99. T.J. McIntosh. Hydration properties of lamellar and non-lamellar phases of phosphatidylcholine and phosphatidylethanolamine. *Chem. Phys. Lipids* 81:117–131 (1996).

100. K. Murzyn, T. Róg, G. Jezierski, Y. Takaoka, and M. Pasenkiewicz-Gierula. Effects of phospholipid unsaturation on the membrane/water interface: a molecular simulation study. *Biophys. J.* 81:170–183 (2001).

101. A.S. Klymchenko, Y. Mély, A.P. Demchenko, and G. Duportail. Simultaneous probing of hydration and polarity of lipid bilayers with 3-hydroxyflavone fluorescent dyes. *Biochim. Biophys. Acta* 1665:6–19 (2004).

102. D. Bach and I.R. Miller. Hydration of phospholipid bilayers in the presence and absence of cholesterol. *Chem. Phys. Lipids* 136:67–72 (2005).

103. I.R. Miller and D. Bach. Hydration of phosphatidyl serine multilayers and its modulation by conformational change induced by correlated electrostatic interaction. *Bioelectrochem. Bioenerg.* 48:361–367 (1999).

104. N. Markova, E. Sparr, L. Wadso, and H. Wennerstrom. A calorimetric study of phospholipid hydration. Simultaneous monitoring of enthalpy and free energy. *J. Phys. Chem. B* 104:8053–8060 (2000).

105. G. Zaccai, J.K. Blasie, and B.P. Schoenborn. Neutron diffraction studies on the location of water in lecithin bilayer model membranes. *Proc. Natl. Acad. Sci. U.S.A.* 72:376–380 (1975).

106. D.L. Worcester and N.P. Franks. Structural analysis of hydrated egg lecithin and cholesterol bilayers. II. Neutron diffraction. *J. Mol. Biol.* 100:359–378 (1976).

107. S.A. Simon and T.J. McIntosh. Depth of water penetration into lipid bilayers. *Meth. Enzymol.* 127:511–521 (1986).

108. R.V. McDaniel and T.J. McIntosh. Neutron and x-ray diffraction structural analysis of phosphatidylinositol bilayers. *Biochim. Biophys. Acta* 983:241–246 (1989).

109. T.J. McIntosh and S.A. Simon. Area per molecule and distribution of water in fully hydrated dilauroylphosphatidylethanolamine bilayers. *Biochemistry* 25:4948–4952 (1986).

110. T.J. McIntosh and S.A. Simon. Area per molecule and distribution of water in fully hydrated dilauroylphosphatidylethanolamine bilayers. Corrections. *Biochemistry* 25:8474 (1986).

111. J.N. Israelachivili. *Intermolecular and Surface Forces.* Academic Press, London, 1985.

112. C. Selle, W. Pohle, and H. Fritzsche. Monitoring the stepwise hydration of phospholipids in films by FT–IR spectroscopy. *Mikrochim. Acta* 14:449–450 (1997).

113. W. Pohle, C. Selle, H. Fritzsche, and H. Binder. Fourier transform infrared spectroscopy as a probe for the study of the hydration of lipid self-assemblies. I. Methodology and general phenomena. *Biospectroscopy* 4:267–280 (1998).

114. L.J. Lis, V.A. Parsegian, and R.P. Rand. Binding of divalent cations to dipalmitoylphosphatidylcholine bilayers and its effects on bilayer interaction. *Biochemistry* 20:1761–1770 (1981).

115. H. Hauser and G.G. Shipley. Interaction of divalent cations with phosphatidylserine bilayer membranes. *Biochemistry* 23:34–41 (1984).

116. H. Hauser, E.G. Finer, and A. Darke. Crystalline anhydrous Ca-phosphatidylserine bilayers. *Biochem. Biophys. Res. Commun.* 76:267–274 (1977).

117. H. Binder and O. Zschörnig. The effect of metal cations on the phase behavior and hydration characteristics of phospholipid membranes. *Chem. Phys. Lipids* 115:39–61 (2002).

118. H. Kuhn, J. Belkner, R. Wiesner, and A.R. Brash. Oxygenation of biological membranes by the pure reticulocyte lipoxygenase. *J. Biol. Chem.* 265:18351–18361 (1990).

119. M. Maccarrone, P.G.M. van Aarle, G.A. Veldink, and J.F.G. Vliegenthart. In vitro oxygenation of soybean biomembranes by lipoxygenase-2. *Biochim. Biophys. Acta* 1190:164–169 (1994).

120. D.C. Borg and K.M. Schaich. Iron and hydroxyl radicals in lipid oxidation: Fenton reactions in lipid and nucleic acids co-oxidized with lipid. In: *Oxy-Radicals in Molecular Biology and Pathology* (P.A. Cerutti, I. Fridovich, and J.M. McCord, eds.). Liss, New York, NY, 1988, pp. 427–441.

121. J.M.C. Gutteridge and B. Halliwell. The measurement and mechanism of lipid peroxidation in biological systems. *Trends Biochem. Sci.* 15:129–135 (1990).

122. R.P. Mason, M.F. Walter, and P.E. Mason. Effect of oxidative stress on membrane structure: small-angle x-ray diffraction analysis. *Free Radical Biol. Med.* 23:419–425 (1997).

123. D.B. Josephson, R.C. Lindsay, and D.A. Stuiber. Identification of compounds characterizing the aroma of fresh whitefish (*Coregonus clupeaformis*). *J. Agric. Food Chem.* 31:326–330 (1983).

124. D.B. Josephson, R.C. Lindsay, and G. Olafsdottir. Measurement of volatile aroma constituents as a means for following sensory deterioration of fresh fish and fishery products. In: *Seafood Quality Determination* (D.E. Kramer and J. Liston, eds.). Elsevier Science, Amsterdam, 1986, pp. 27–47.

125. C. Karahadian and R.D. Lindsay. Role of oxidative processes in the formation and stability of fish flavors. In: *Flavor Chemistry: Trends and Developments* (R. Teranishi, R.G. Buttery, and F. Shahidi, eds.). American Chemical Society, Washington, DC, 1989, pp. 60–75.

126. R.C. Lindsay. Fish flavors. *Food Rev. Int.* 6:437–455 (1990).

127. D.B. Josephson, R.C. Lindsay, and D.A. Stuiber. Variations in the occurrences of enzymically derived volatile aroma compounds in salt- and freshwater fresh. *J. Agric. Food Chem.* 32:1344–1347 (1984).

128. D.B. Josephson, R.C. Lindsay, and D.A. Stuiber. Biogenesis of lipid-derived volatile aroma compounds in the emerald shiner (*Notropis atherinoides*). *J. Agric. Food Chem.* 32:1347–1352 (1984).

129. R.J. Hsieh and J.E. Kinsella. Lipoxygenase-catalyzed oxidation of N-6 and N-3 polyunsaturated fatty acids: relevance to and activity in fish tissue. *J. Food Sci.* 51: 940–945, 996 (1986).

130. R.J. Hsieh and J.E. Kinsella. Lipoxygenase generation of specific volatile flavor carbonyl compounds in fish tissues. *J. Agric. Food Chem.* 37:279–286 (1989).

131. J.B. German, H. Zhang, and R. Berger. Role of lipoxygenases in lipid oxidation in foods. In: *Lipid Oxidation in Food* (A.J. St. Angelo, ed.). American Chemical Society, Washington, DC, 1992, pp. 74–92.

132. R.J. Hsieh. Contribution of lipoxygenase pathway to food flavors. In: *Lipids in Food Flavors* (C.-T. Ho and T.G. Hartman, eds.). American Chemical Society, Washington, DC, 1994, pp. 30–48.

133. C. Karahadian and R.C. Lindsay. Action of tocopherol-type compounds in directing reactions forming flavor compounds in autoxidizing fish oils. *J. Am. Oil Chem. Soc.* 66:1302–1308 (1989).

134. B.M. Slabyj and H.O. Hultin. Oxidation of a lipid emulsion by a peroxidizing microsomal fraction from herring muscle. *J. Food Sci.* 49:1392–1393 (1984).

135. J.H. Song, Y. Inoue, and T. Miyazawa. Oxidative stability of docosahexaenoic acid containing oils in the form of phospholipids, triacylglycerols, and ethyl esters. *Biosci. Biotechnol. Biochem.* 61:2085–2088.

136. B.R. Wilson, A.M. Pearson, and F.B. Shorland. Effect of total lipids and phospholipids on warmed-over flavor in red and white muscle from several species as measured by thiobarbituric acid analysis. *J. Agric. Food Chem.* 24:7–11 (1976).

137. J.O. Igene, A.M. Pearson, and J.I. Gray. Effects of length of frozen storage, cooking and holding temperatures upon component phospholipids and the fatty acid composition of meat triglycerides and phospholipids. *Food Chem.* 7:289–303 (1981).

138. C. Willemot, L.M. Poste, J. Salvador, and D.F. Wood. Lipid degradation in pork during warmed-over flavor development. *Can. Inst. Food Sci. Technol.* 18:316–322 (1985).

139. T.C. Wu and B.W. Sheldon. Influence of phospholipids on the development of oxidized off flavors in cooked turkey rolls. *J. Food Sci.* 53:55–61 (1988).

140. S.M. Polvi, R.G. Ackman, S.P. Lall, and R.L. Saunders. Stability of lipids and omega-3 fatty acids during frozen storage of Atlantic salmon. *J. Food Proc. Preserv.* 15:167–181 (1991).

141. H.O. Hultin, E.A. Decker, S.D. Kelleher, and J.E. Osinchak. Control of lipid oxidation processes in minced fatty fish. In: *Seafood Science and Technology* (E.G. Bligh, ed.). Fishing News Books, Oxford, 1990, pp. 93–100.

142. M.L. Hallberg and H. Lingnert. The relationship between lipid composition and oxidative stability of potato granules. *Food Chem.* 38:201–210 (1990).

143. M.C. Erickson. Contribution of phospholipids to headspace volatiles during storage of pecans. *J. Food Qual.* 16:13–24 (1993).

144. S.Z. Dziedzic and B.J.F. Hudson. Phosphatidyl ethanolamine as a synergist for primary antioxidants in edible oils. *J. Am. Oil Chem. Soc.* 61:1042–1045 (1984).

145. D.H. Hildebrand, J. Terao, and M. Kito. Phospholipids plus tocopherols increase soybean oil stability. *J. Am. Oil Chem. Soc.* 61:552–555 (1984).

146. M.C. King, L.C. Boyd, and B.W. Sheldon. Effect of phospholipids on lipid oxidation of a salmon oil model system. *J. Am. Oil Chem. Soc.* 69:237–242 (1992).

147. M.C. King, L.C. Boyd, and B.W. Sheldon. Antioxidant properties of individual phospholipids in a salmon oil model system. *J. Am. Oil Chem. Soc.* 69:545–551 (1992).

148. H. Saito and K. Ishihara. Antioxidant activity and active sites of phospholipids as antioxidants. *J. Am. Oil Chem. Soc.* 74:1531–1536 (1997).

149. B. Saadan, B. Le Tutour, and F. Quemeneur. Oxidation properties of phospholipids: mechanistic studies. *New J. Chem.* 22:801–807 (1998).

150. J.O. Igene, A.M. Pearson, L.R. Dugan, Jr., and J.F. Price. Role of triglycerides and phospholipids on development of rancidity in model meat systems during frozen storage. *Food Chem.* 5:263–276 (1980).

151. S. Adachi, T. Ishiguro, and R. Matsuno. Autoxidation kinetics for fatty acids and their esters. *J. Am. Oil Chem. Soc.* 72:547–551 (1995).

152. Q.T. Li, M.H. Yee, and B.K. Tan. Lipid peroxidation in small and large phospholipid unilamellar vesicles induced by water-soluble free radical sources. *Biochem. Biophys. Res. Commun.* 273:72–76 (2000).

153. J.W. Borst, N.V. Visser, O. Kouptsova, and A.J.W.G. Visser. Oxidation of unsaturated phospholipids in membrane bilayer mixtures is accompanied by membrane fluidity changes. *Biochim. Biophys. Acta* 1487:61–73 (2000).

154. D. Reiss, K. Beyer, and B. Engelmann. Delayed oxidative degradation of polyunsaturated diacyl phospholipids in the presence of plasmalogen phospholipids in vitro. *Biochem. J.* 323:807–814 (1997).

155. M. Zommara, N. Tachibana, K. Mitsui, N. Nakatani, M. Sakono, I. Ikeda, and K. Imaizumi. Inhibitory effect of ethanolamine plasmalogen on iron- and copper-dependent lipid peroxidation. *Free Radic. Biol. Med.* 18:599–602 (1995).

156. B. Tadolini, P. Motta, and C.A. Rossi. Iron binding to liposomes of different phospholipid composition. *Biochem. Mol. Biol. Int.* 29:299–305 (1993).

157. K. Yoshida, J. Terao, T. Suzuki, and K. Takama. Inhibitory effect of phosphatidylserine on iron-dependent lipid peroxidation. *Biochem. Biophys. Res. Commun.* 179:1077–1081 (1991).

158. P.I. Oteiza. A mechanism for the stimulatory effect of aluminum on iron-induced lipid peroxidation. *Arch. Biochem. Biophys.* 308:374–379 (1994).

159. S.V. Verstraeten, L.V. Nogueira, S. Schreier, and P.I. Oteiza. Effect of trivalent metal ions on phase separation and membrane lipid packing: role in lipid peroxidation. *Arch. Biochem. Biophys.* 338:121–127 (1997).

160. S.V. Verstraeten and P.I. Oteiza. Effects of Al^{3+} and related metals on membrane phase state and hydration: correlation with lipid oxidation. *Arch. Biochem. Biophys.* 375:340–346 (2000).

3 Lipid-Based Emulsions and Emulsifiers

D. Julian McClements

CONTENTS

I. INTRODUCTION

Many natural and processed foods exist either partly or wholly as emulsions, or have been in an emulsified state at some time during their existence [1–7]. Milk is the most common example of a naturally occurring food emulsion [8]. Mayonnaise, salad dressing, cream, ice cream, butter, and margarine are all examples of manufactured food emulsions. Powdered coffee whiteners, sauces, and many desserts are examples of foods that were emulsions at one stage during their production but subsequently were converted into another form. The bulk physicochemical properties of food emulsions, such as appearance, texture, and stability, depend ultimately on the type of molecules the food contains and their interactions with one another. Food emulsions contain a variety of ingredients, including water, lipids, surfactants, proteins, carbohydrates, minerals, preservatives, colors, and flavors [5]. By a combination of covalent and physical interactions, these ingredients form the individual phases and structural components that give the final product its characteristic physicochemical properties [9]. It is the role of food scientists to untangle the complex relationship between the molecular, structural, and bulk properties of foods, so that foods with improved properties can be created in a more systematic fashion.

II. EMULSIONS

An emulsion is a dispersion of droplets of one liquid in another liquid with which it is incompletely miscible [1,10]. In foods, the two immiscible liquids are oil and water. The diameter of the droplets in food emulsions is typically within the range 0.1–50 μm [4,5]. A system that consists of oil droplets dispersed in an aqueous phase is called an oil-in-water (O/W) emulsion. A system that consists of water droplets dispersed in an oil phase is called a water-in-oil (W/O) emulsion. The material that makes up the droplets in an emulsion is referred to as the dispersed or internal phase, whereas the material that makes up the surrounding liquid is called the continuous or external phase. Multiple emulsions can be prepared that consist of oil droplets contained in larger water droplets,

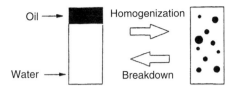

FIGURE 3.1 Emulsions are thermodynamically unstable systems that tend to revert back to the individual oil and water phases with time. To produce an emulsion, energy must be supplied.

which are themselves dispersed in an oil phase (O/W/O), or vice versa (W/O/W). Multiple emulsions can be used for protecting certain ingredients, for controlling the release of ingredients, or for creating low-fat products [11].

Emulsions are thermodynamically unstable systems due to the positive free energy required to increase the surface area between the oil and water phases [5]. The origin of this energy is the unfavorable interaction between oil and water, which exists because water molecules are capable of forming strong hydrogen bonds with other water molecules but not with oil molecules [10,11]. Thus, emulsions tend to reduce the surface area between the two immiscible liquids by separating into a system that consists of a layer of oil (lower density) on top of a layer of water (higher density). This is clearly seen if one tries to homogenize pure oil and pure water together: initially an emulsion is formed, but after a few minutes phase separation occurs (Figure 3.1).

Emulsion instability can manifest itself through a variety of physicochemical mechanisms, including creaming, flocculation, coalescence, partial coalescence, Ostwald ripening, and phase inversion (Section VI). To form emulsions that are kinetically stable for a reasonable period (a few weeks, months, or even years), chemical substances known as emulsifiers must be added before homogenization. Emulsifiers are surface-active molecules that adsorb to the surface of freshly formed droplets during homogenization, forming a protective membrane that prevents the droplets from coming close enough together to aggregate [5]. Most food emulsifiers are amphiphilic molecules, that is, they have both polar and nonpolar regions on the same molecule. The most common types used in the food industry are lipid-based emulsifiers (small-molecule surfactants and phospholipids) and amphiphilic biopolymers (proteins and polysaccharides) [4,5]. In addition, some types of small solid particles are also surface active and can act as emulsifiers in foods, for example, granules from egg or mustard.

Most food emulsions are more complex than the simple three-component (oil, water, and emulsifier) system described earlier [5,7,11]. The aqueous phase may contain water-soluble ingredients of many different kinds, including sugars, salts, acids, bases, surfactants, proteins, polysaccharides, flavors, and preservatives [1]. The oil phase may contain a variety of lipid-soluble components, such as triacylglycerols, diacylglycerols, monoacylglycerols, fatty acids, vitamins, cholesterol, and flavors [1]. The interfacial region may be composed of surface-active components of a variety of types, including small-molecule surfactants, phospholipids, polysaccharides, and proteins. It should be noted that the composition of the interfacial region may evolve over time after an emulsion is produced, due to competitive adsorption with other surface-active substances or due to adsorption of oppositely charged substances, for example, polysaccharides [1]. Some of the ingredients in food emulsions are not located exclusively in one phase but are distributed between the oil, water, and interfacial phases according to their partition coefficients. Despite having low concentrations, many of the minor components present in an emulsion can have a pronounced influence on its bulk physicochemical properties. For example, addition of small amounts (about a few millimolar) of multivalent mineral ions can destabilize an electrostatically stabilized emulsion [1]. Food emulsions may consist of oil droplets dispersed in an aqueous phase (e.g., mayonnaise, milk, cream, soups) or water droplets dispersed in an oil phase (e.g., margarine, butter, spreads). The droplets and the continuous phase may be fluid, gelled, crystalline, or glassy. The size of the droplets may vary from less than a micrometer to a few hundred micrometers, and the droplets themselves may be more or less polydispersive. In addition, many emulsions may contain air bubbles that have a pronounced influence on the sensory and physicochemical properties of the system, for example, ice cream, whipped cream, and desserts [1].

To complicate matters further, the properties of food emulsions are constantly changing with time because of the action of various chemical (e.g., lipid oxidation, biopolymer hydrolysis), physical (e.g., creaming, flocculation, coalescence), and biological (e.g., bacterial growth) processes. In addition, during their processing, storage, transport, and handling, food emulsions are subjected to variations in their temperature (e.g., via sterilization, cooking, chilling, freezing) and to various mechanical forces (e.g., stirring, mixing, whipping, flow-through pipes, centrifugation high pressure) that alter their physicochemical properties. Despite the compositional, structural, and dynamic complexity of food emulsions, considerable progress has been made in understanding the major factors that determine their bulk physicochemical properties.

III. LIPID-BASED EMULSIFIERS

A. Molecular Characteristics

The most important types of lipid-based emulsifiers used in the food industry are small-molecule surfactants (e.g., Tweens, Spans, and salts of fatty acids) and phospholipids (e.g., lecithin). The principal role of lipid-based emulsifiers in food emulsions is to enhance the formation and stability of the product; however, they may also alter the bulk physicochemical properties by interacting with proteins or polysaccharides or by modifying the structure of fat crystals [1,11]. All lipid-based emulsifiers are amphiphilic molecules that have a hydrophilic head group with a high affinity for water and lipophilic tail group with a high affinity for oil [1,12,13]. These emulsifiers can be represented by the formula RX, where X represents the hydrophilic head and R the lipophilic tail. Lipid-based emulsifiers differ with respect to type of head group and tail group. The head group may be anionic, cationic, zwitterionic, or nonionic. The lipid-based emulsifiers used in the food industry are mainly nonionic (e.g., monoacylglycerols, sucrose esters, Tweens, and Spans), anionic (e.g., fatty acids), or zwitterionic (e.g., lecithin). The tail group usually consists of one or more hydrocarbon chains, having between 10 and 20 carbon atoms per chain. The chains may be saturated or unsaturated, linear or branched, aliphatic or aromatic. Most lipid-based emulsifiers used in foods have either one or two linear aliphatic chains, which may be saturated or unsaturated. Each type of emulsifier has unique functional properties that depend on its chemical structure.

Lipid-based emulsifiers aggregate spontaneously in solution to form a variety of thermodynamically stable structures known as association colloids (e.g., micelles, bilayers, vesicles, reversed micelles) (Figure 3.2). These structural types are adopted because they minimize the unfavorable contact area between the nonpolar tails of the emulsifier molecules and water [12]. The type of association colloid formed depends principally on the polarity and molecular geometry of the emulsifier molecules (Section III.C.3). The forces holding association colloids together are relatively weak, and so they have highly dynamic and flexible structures [10]. Their size and shape is continually fluctuating, and individual emulsifier molecules rapidly exchange between the micelle and the surrounding liquid. The relative weakness of the forces holding association colloids together also means that their structures are particularly sensitive to changes in environmental conditions, such as temperature, pH, ionic strength, and ion type. Surfactant micelles are the most important type of association colloid formed in many food emulsions, and we focus principally on their properties.

B. Functional Properties

1. Critical Micelle Concentration

A surfactant forms micelles in an aqueous solution when its concentration exceeds some critical level, known as the critical micelle concentration (cmc). Below the cmc, surfactant molecules are

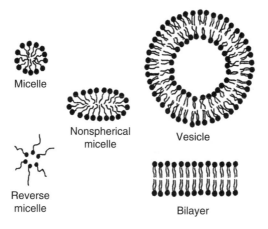

FIGURE 3.2 Association colloids formed by surfactant molecules.

dispersed predominantly as monomers, but once the cmc has been exceeded, any additional surfactant molecules form micelles, and the monomer concentration remains constant. Despite the highly dynamic nature of their structure, surfactant micelles do form particles that have a well-defined average size. Thus, when surfactant is added to a solution above the cmc, the number of micelles increases, rather than their size. When the cmc is exceeded, there is an abrupt change in the physicochemical properties of a surfactant solution (e.g., surface tension, electrical conductivity, turbidity, osmotic pressure) [14]. This is because the properties of surfactant molecules dispersed as monomers are different from those in micelles. For example, surfactant monomers are amphiphilic and have a high surface activity, whereas micelles have little surface activity because their surface is covered with hydrophilic head groups. Consequently, the surface tension of a solution decreases with increasing surfactant concentration below the cmc but remains fairly constant above it.

2. Cloud Point

When a surfactant solution is heated above a certain temperature, known as the cloud point, it becomes turbid. As the temperature is raised, the hydrophilic head groups become increasingly dehydrated, which causes the emulsifier molecules to aggregate. These aggregates are large enough to scatter light, and so the solution appears turbid. At temperatures above the cloud point, the aggregates grow so large that they sediment under the influence of gravity and form a separate phase. The cloud point increases as the hydrophobicity of a surfactant molecule increases; that is, the length of its hydrocarbon tail increases or the size of its hydrophilic head group decreases [15,16].

3. Solubilization

Nonpolar molecules, which are normally insoluble or only sparingly soluble in water, can be solubilized in an aqueous surfactant solution by incorporation into micelles or other types of association colloids [11]. The resulting system is thermodynamically stable; however, equilibrium may take an appreciable time to achieve because of the activation energy associated with transferring a nonpolar molecule from a bulk phase to a micelle. Micelles containing solubilized materials are referred to as swollen micelles or microemulsions, whereas the material solubilized within the micelle is referred to as the solubilizate. The ability of micellar solutions to solubilize nonpolar molecules has a number of potentially important applications in the food industry, including selective extraction of nonpolar molecules from oils, controlled ingredient release, incorporation of nonpolar substances into aqueous solutions, transport of nonpolar molecules across aqueous membranes, and modification of chemical reactions [11]. Three important factors determine the functional properties of swollen micellar solutions: the location of the solubilizate within the micelles, the maximum amount of material that can be solubilized per unit mass of surfactant, and the rate at which solubilization proceeds [11].

4. Surface Activity and Droplet Stabilization

Lipid-based emulsifiers are used widely in the food industry to enhance the formation and stability of food emulsions. To do this they must adsorb to the surface of emulsion droplets during homogenization and form a protective membrane that prevents the droplets from aggregating with each other [1]. Emulsifier molecules adsorb to oil–water interfaces because they can adopt an orientation in which the hydrophilic part of the molecule is located in the water while the hydrophobic part is located in the oil. This minimizes the unfavorable free energy associated with the contact of hydrophilic and hydrophobic regions, and therefore reduces the interfacial tension. This reduction in interfacial tension is important because it facilitates the further disruption of emulsion droplets; that is, less energy is needed to break up a droplet when the interfacial tension is lowered.

Once adsorbed to the surface of a droplet, the emulsifier must provide a repulsive force that is strong enough to prevent the droplet from aggregating with its neighbors. Ionic surfactants provide stability by causing all the emulsion droplets to have the same electric charge, hence to repel each other electrostatically. Nonionic surfactants provide stability by generating a number of short-range repulsive forces (e.g., steric overlap, hydration, thermal fluctuation interactions) that prevent the droplets from getting too close together [1,13]. Some emulsifiers form multilayers (rather than monolayers) at the surface of an emulsion droplet, which greatly enhances the stability of the droplets against aggregation.

In summary, emulsifiers must have three characteristics to be effective. First, they must rapidly adsorb to the surface of the freshly formed emulsion droplets during homogenization. Second, they must reduce the interfacial tension by a significant amount. Third, they must form a membrane that prevents the droplets from aggregating.

C. INGREDIENT SELECTION

A large number of different types of lipid-based emulsifiers can be used as food ingredients, and a manufacturer must select the one that is most suitable for each particular product. Suitability, in turn, depends on factors such as an emulsifier's legal status as a food ingredient, its cost and availability, the consistency in its properties from batch to batch, its ease of handling and dispersion, its shelf life, its compatibility with other ingredients, and the processing, storage, and handling conditions it will experience, as well as the expected shelf life and physicochemical properties of the final product.

How does a food manufacturer decide which emulsifier is most suitable for a product? There have been various attempts to develop classification systems that can be used to select the most appropriate emulsifier for a particular application. Classification schemes have been developed that are based on an emulsifier's solubility in oil and water (Bancroft's rule), its ratio of hydrophilic to lipophilic groups (HLB number) [17,18], and its molecular geometry [19]. Ultimately, all of these properties depend on the chemical structure of the emulsifier, and so all the different classification schemes are closely related.

1. Bancroft's Rule

One of the first empirical rules developed to describe the type of emulsion that could be stabilized by a given emulsifier was proposed by Bancroft. Bancroft's rule states that the phase in which the emulsifier is most soluble forms the continuous phase of an emulsion. Hence, a water-soluble emulsifier stabilizes O/W emulsions, whereas an oil-soluble emulsifier stabilizes W/O emulsions.

2. Hydrophile–Lipophile Balance

The hydrophile–lipophile balance (HLB) concept underlies a semiempirical method for selecting an appropriate emulsifier or combination of emulsifiers to stabilize an emulsion. The HLB is described by a number, which gives an indication of the overall affinity of an emulsifier for the oil and aqueous phases [14]. Each emulsifier is assigned an HLB number according to its chemical structure. A molecule with a high HLB number has a high ratio of hydrophilic groups to lipophilic groups, and vice versa. The HLB number of an emulsifier can be calculated from knowledge of the number and type of hydrophilic and lipophilic groups it contains, or it can be estimated from experimental measurements of its cloud point. The HLB numbers of many emulsifiers have been tabulated in the literature [17,18]. A widely used semiempirical method of calculating the HLB number of a lipid-based emulsifier is as follows:

$$\text{HLB} = 7 + \sum (\text{hydrophilic group numbers}) - (\text{lipophilic group numbers}). \qquad (3.1)$$

As indicated in Table 3.1 [20], group numbers have been assigned to hydrophilic and lipophilic groups of many types. The sums of the group numbers of all the lipophilic groups and of all the hydrophilic groups are substituted into Equation 3.1, and the HLB number is calculated. The semiempirical equation above has been found to have a firm thermodynamic basis, with the sums corresponding to the free energy changes in the hydrophilic and lipophilic parts of the molecule when micelles are formed.

The HLB number of an emulsifier gives a useful indication of its solubility in the oil and water phases, and it can be used to predict the type of emulsion that will be formed. An emulsifier with a low HLB number (4–6) is predominantly hydrophobic, dissolves preferentially in oil, stabilizes W/O emulsions, and forms reversed micelles in oil. An emulsifier with a high HLB number (8–18) is predominantly hydrophilic, dissolves preferentially in water, stabilizes O/W emulsions, and forms micelles in water. An emulsifier with an intermediate HLB number (6–8) has no particular preference for either oil or water. Nonionic molecules with HLB numbers below 4 and above 18 are less surface active and are therefore less likely to preferentially accumulate at an oil–water interface.

Emulsion droplets are particularly prone to coalescence when they are stabilized by emulsifiers that have extreme or intermediate HLB numbers. At very high or very low HLB numbers, a nonionic emulsifier has such a low surface activity that it does not accumulate appreciably at the droplet surface and therefore does not provide protection against coalescence. At intermediate HLB numbers (6–8), emulsions are unstable to coalescence because the interfacial tension is so low that very little energy is required to disrupt the membrane. Maximum stability of emulsions is obtained for O/W emulsions using an emulsifier with an HLB number around 10–12, and for W/O emulsions

TABLE 3.1
Selected HLB Group Numbers

Hydrophilic Group	Group Number	Lipophilic Group	Group Number
$-SO_4NA^+$	38.7	$-CH-$	0.475
$-COO^-H^+$	21.2	$-CH_2-$	0.475
Tertiary amine	9.4	$-CH_3-$	0.475
Sorbitan ring	6.8		
$-COOH$	2.1		
$-O-$	1.3		

Source: Adapted from Davis, H.T., *Colloids Surf. A*, 91, 9, 1994.

around 3–5. This is because the emulsifiers are sufficiently surface active but do not lower the interfacial tension so much that the droplets are easily disrupted. It is possible to adjust the effective HLB number by using a combination of two or more emulsifiers with different HLB numbers.

One of the major drawbacks of the HLB concept is its failure to account for the significant alterations in the functional properties of an emulsifier molecule that result from changes in temperature or solution conditions, even though the chemical structure of the molecule does not change. Thus, an emulsifier may be capable of stabilizing O/W emulsions at one temperature but W/O emulsions at another temperature.

3.   Molecular Geometry and Phase Inversion Temperature

The molecular geometry of an emulsifier molecule is described by a packing parameter p (see Figure 3.3) as follows:

$$p = \frac{v}{la_0},$$ (3.2)

where
   v and l are the volume and length of the hydrophobic tail, respectively
   a_0 is the cross-sectional area of the hydrophilic head group

When surfactant molecules associate with each other, they tend to form monolayers having a curvature that allows the most efficient packing of the molecules. At this optimum curvature, the monolayer has its lowest free energy, and any deviation from this curvature requires the expenditure of energy [10,13]. The optimum curvature of a monolayer depends on the packing parameter of the emulsifier: for $p = 1$, monolayers with zero curvature are preferred; for $p < 1$, the optimum curvature is convex; and for $p > 1$, the optimum curvature is concave (Figure 3.3). Simple geometrical considerations indicate that spherical micelles are formed when p is less than 0.33, nonspherical micelles when p is between 0.33 and 0.5, and bilayers when p is between 0.5 and 1 [13]. Above a certain concentration, bilayers join up to form vesicles because energetically unfavorable end effects are eliminated. At values of p greater than 1, reversed micelles are formed, in which the hydrophilic head groups are located in the interior (away from the oil), and the hydrophobic tail groups are located at the exterior (in contact with the oil) (Figure 3.2). The packing parameter therefore gives a useful indication of the type of association colloid that is formed by an emulsifier molecule in solution.

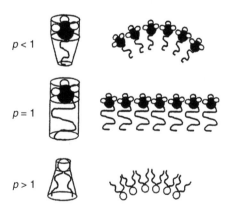

FIGURE 3.3 Relationship between the molecular geometry of surfactant molecules and their optimum curvature.

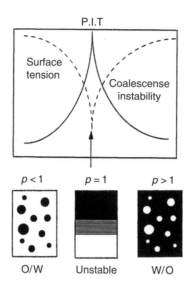

FIGURE 3.4 Phase inversion temperature in emulsions.

The packing parameter is also useful because it accounts for the temperature dependence of the physicochemical properties of surfactant solutions and emulsions. The temperature at which an emulsifier solution converts from a micellar to a reversed micellar system or an O/W emulsion converts to a W/O emulsion is known as the phase inversion temperature (PIT). Consider what happens when an emulsion that is stabilized by a lipid-based emulsifier is heated (Figure 3.4). At temperatures well below the PIT ($\approx 20°C$), the packing parameter is significantly less than unity, and so a system that consists of O/W emulsion in equilibrium with a swollen micellar solution is favored. As the temperature is raised, the hydrophilic head groups of the emulsifier molecules become increasingly dehydrated, which causes p to increase toward unity. Thus, the emulsion droplets become more prone to coalescence and the swollen micelles grow in size. At the PIT, $p \approx 1$, and the emulsion breaks down because the droplets have an ultralow interfacial tension and therefore readily coalesce with each other. The resulting system consists of excess oil and excess water (containing some emulsifier monomers), separated by a third phase that contains emulsifier molecules aggregated into bilayer structures. At temperatures sufficiently greater than the PIT, the packing parameter is much larger than unity, and the formation of a system that consists of a W/O emulsion in equilibrium with swollen reversed micelles is favored. A further increase in temperature leads to a decrease in the size of the reversed micelles and in the amount of water solubilized within them. The method of categorizing emulsifier molecules according to their molecular geometry is now widely accepted as the most useful means of determining the types of emulsions they tend to stabilize [19].

4. Other Factors

The classification schemes mentioned above provide information about the type of emulsion an emulsifier tends to stabilize (i.e., O/W or W/O), but they do not provide much insight into the size of the droplets that form during homogenization or the stability of the emulsion droplets once formed [1]. In choosing a suitable emulsifier for a particular application, these factors must also be considered. The speed at which an emulsifier adsorbs to the surface of the emulsion droplets produced during homogenization determines the minimum droplet size that can be produced: the faster the adsorption rate, the smaller the size. The magnitude and range of the repulsive forces generated by a membrane, and its viscoelasticity, determine the stability of the droplets to aggregation.

IV. BIOPOLYMERS

Proteins and polysaccharides are the two most important biopolymers used as functional ingredients in food emulsions. These biopolymers are used principally for their ability to stabilize emulsions, enhance viscosity, and form gels.

A. MOLECULAR CHARACTERISTICS

Molecular characteristics of biopolymers, such as molecular weight, conformation, flexibility, and polarity, ultimately determine the properties of biopolymer solutions. These characteristics are determined by the type, number, and sequence of monomers that make up the polymer. Proteins are polymers of amino acids [21], whereas polysaccharides are polymers of monosaccharides [22]. The three-dimensional structures of biopolymers in aqueous solution can be categorized as globular, fibrous, or random coil (Figure 3.5). Globular biopolymers have fairly rigid compact structures; fibrous biopolymers have fairly rigid, rodlike structures; and random-coil biopolymers have highly dynamic and flexible structures. Biopolymers can also be classified according to the degree of branching of the chain. Most proteins have linear chains, whereas polysaccharides can have either linear (e.g., amylose) or branched (e.g., amylopectin) chains.

The conformation of a biopolymer in solution depends on the relative magnitude of the various types of attractive and repulsive interactions that occur within and between molecules, as well as the configurational entropy of the molecule. Biopolymers that have substantial proportions of nonpolar groups tend to fold into globular structures, in which the nonpolar groups are located in the interior (away from the water) and the polar groups are located at the exterior (in contact with the water) because this arrangement minimizes the number of unfavorable contacts between hydrophobic regions and water. However, since stereochemical constraints and the influence of other types of molecular interactions usually make it impossible for all the nonpolar groups to be located in the interior, the surfaces of globular biopolymers have some hydrophobic character. Many kinds of food proteins have compact globular structures, including β-lactoglobulin, α-lactalbumin, and bovine serum albumin [8]. Biopolymers that contain a high proportion of polar monomers, distributed fairly evenly along their backbone, often have rodlike conformations with substantial amounts of helical structure stabilized by hydrogen bonding. Such biopolymers (e.g., collagen, cellulose) usually have low water solubilities because they tend to associate strongly with each other rather than with water; consequently, they often have poor functional properties. However, if the chains are branched, the molecules may be prevented from getting close enough together to aggregate, and so they may exist in solution as individual molecules. Predominantly polar biopolymers containing monomers that are incompatible with helix formation (e.g., β-casein) tend to form random-coil structures.

In practice, biopolymers may have some regions along their backbone that have one type of conformation and others that have a different conformation. Biopolymers may also exist as isolated molecules or as aggregates in solution, depending on the relative magnitude of the biopolymer–biopolymer, biopolymer–solvent, and solvent–solvent interactions. Biopolymers are also capable of undergoing transitions from one type of conformation to another in response to environmental changes such as alterations in their pH, ionic strength, solvent composition, and temperature. Examples include helix ⇔ random coil and globular ⇔ random coil. In many food biopolymers, this type of transition plays an important role in determining the functional properties (e.g., gelation).

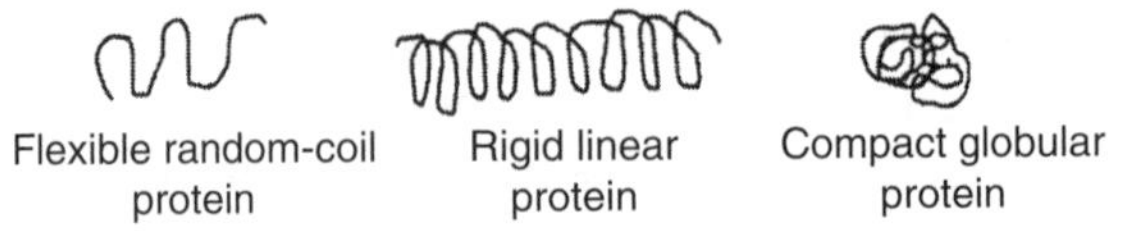

Flexible random-coil protein   Rigid linear protein   Compact globular protein

FIGURE 3.5 Typical molecular conformations adopted by biopolymers in aqueous solution.

B. FUNCTIONAL PROPERTIES

1. Emulsification

Biopolymers that have a high proportion of nonpolar groups tend to be surface active, that is, they can accumulate at oil–water interfaces [1–6]. The major driving force for adsorption is the hydrophobic effect. When the biopolymer is dispersed in an aqueous phase, some of the nonpolar groups are in contact with water, which is a thermodynamically unfavorable condition. By adsorbing to an interface, the biopolymer can adopt a conformation of nonpolar groups in contact with the oil phase (away from the water) and hydrophilic groups located in the aqueous phase (in contact with the water). In addition, adsorption reduces the number of contacts between the oil and water molecules at the interface, thereby reducing the interfacial tension. The conformation a biopolymer adopts at an oil–water interface and the physicochemical properties of the membrane formed depend on its molecular structure. Flexible random-coil biopolymers adopt an arrangement in which the predominantly nonpolar segments protrude into the oil phase, the predominantly polar segments protrude into the aqueous phase, and the neutral regions lie flat against the interface (Figure 3.6, left). The membranes formed by molecules of these types tend to have relatively open structures, to be relatively thick, and to have low viscoelasticities. Globular biopolymers (usually proteins) adsorb to an interface so that the predominantly nonpolar regions on their surface face the oil phase; thus, they tend to have a definite orientation at an interface (Figure 3.6, right). Once they have adsorbed to an interface, biopolymers often undergo structural rearrangements that permit them to maximize the number of contacts between nonpolar groups and oil [6].

Random-coil biopolymers have flexible conformations and therefore rearrange their structures rapidly, whereas globular biopolymers are more rigid and therefore unfold more slowly. The unfolding of a globular protein at an interface often exposes amino acids that were originally located in the hydrophobic interior of the molecule, which can lead to enhanced interactions with neighboring protein molecules through hydrophobic attraction or disulfide bond formation. Consequently, globular proteins tend to form relatively thin and compact membranes, high in viscoelasticity. Thus, membranes formed from globular proteins tend to be more resistant to rupture than those formed from random-coil proteins [5].

To be effective emulsifiers, biopolymers must rapidly adsorb to the surface of the emulsion droplets formed during homogenization and provide a membrane that prevents the droplets from aggregating. Biopolymer membranes can stabilize emulsion droplets against aggregation by a number of different physical mechanisms [1]. All biopolymers are capable of providing short-range steric repulsive forces that are usually strong enough to prevent droplets from getting sufficiently close together to coalesce. If the membrane is sufficiently thick, it can also prevent droplets from flocculating. Otherwise, it must be electrically charged so that it can prevent flocculation by electrostatic repulsion. The properties of emulsions stabilized by charged biopolymers are particularly sensitive to

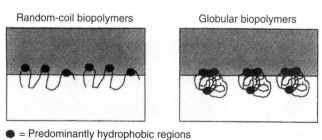

Random-coil biopolymers   Globular biopolymers

● = Predominantly hydrophobic regions

FIGURE 3.6 Conformation and unfolding of biopolymers at oil–water interfaces depend on their molecular structure.

the pH and ionic strength of aqueous solutions [1]. At pH values near the isoelectric point of proteins, or at high ionic strengths, the electrostatic repulsion between droplets may not be large enough to prevent the droplets from aggregating (see Section VI. A.5).

Proteins are commonly used as emulsifiers in foods because many of them naturally have a high proportion of nonpolar groups. Most polysaccharides are so hydrophilic that they are not surface active. However, a small number of naturally occurring polysaccharides have some hydrophobic character (e.g., gum arabic) or have been chemically modified to introduce nonpolar groups (e.g., some hydrophobically modified starches), and these biopolymers can be used as emulsifiers.

2. Thickening and Stabilization

The second major role of biopolymers in food emulsions is to increase the viscosity of the aqueous phase [1]. This modifies the texture and mouthfeel of the food product (thickening), as well as reducing the rate at which particles sediment or cream (stabilization). Both proteins and polysaccharides can be used as thickening agents, but polysaccharides are usually preferred because they can be used at much lower concentrations. The biopolymers used to increase the viscosity of aqueous solutions are usually highly hydrated and extended molecules or molecular aggregates. Their ability to increase the viscosity depends principally on their molecular weight, degree of branching, conformation, and flexibility. The viscosity of a dilute solution of particles increases as the concentration of particles increases [5]:

$$\eta = \eta_0(1 + 2.5\phi), \tag{3.3}$$

where
    η is the viscosity of the solution
    η_0 is the viscosity of the pure solvent
    ϕ is the volume fraction of particles in solution

Biopolymers are able to enhance the viscosity of aqueous solutions at low concentrations because they have an effective volume fraction that is much greater than their actual volume fraction [1]. A biopolymer rapidly rotates in solution because of its thermal energy, and so it sweeps out a spherical volume of water that has a diameter approximately equal to the end-to-end length of the molecule (Figure 3.7). The

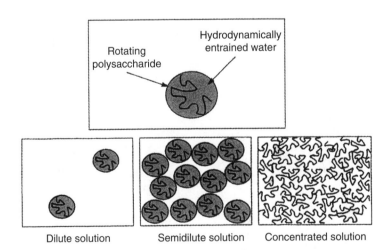

FIGURE 3.7 Extended biopolymers in aqueous solutions sweep out a large volume of water as they rotate, which increases their effective volume fraction and therefore their viscosity.

volume of the biopolymer molecule is only a small fraction of the total volume of the sphere swept out, and so the effective volume fraction of a biopolymer is much greater than its actual volume fraction. Consequently, small concentrations of biopolymer can dramatically increase the viscosity of a solution (Equation 3.3). The effectiveness of a biopolymer at increasing the viscosity increases as the volume fraction it occupies within the sphere it sweeps out decreases. Thus, large, highly extended linear biopolymers increase the viscosity more effectively than small compact or branched biopolymers.

In a dilute biopolymer solution, the individual molecules (or aggregates) do not interact with each other. When the concentration of biopolymer increases above some critical value c^*, the viscosity increases rapidly because the spheres swept out by the biopolymers overlap with each another. This type of solution is known as a semidilute solution, because even though the molecules are interacting with one another, each individual biopolymer is still largely surrounded by solvent molecules. At still higher polymer concentrations, the molecules pack so close together that they become entangled, and the system has more gel-like characteristics. Biopolymers that are used to thicken the aqueous phase of emulsions are often used in the semidilute concentration range [5].

Solutions containing extended biopolymers often exhibit strong shear-thinning behavior; that is, their apparent viscosity decreases with increasing shear stress. Some biopolymer solutions even have a characteristic yield stress. When a stress is applied below the yield stress, the solution acts like an elastic solid, but when it exceeds the yield stress the solution acts like a liquid. Shear thinning tends to occur because the biopolymer molecules become aligned with the shear field, or because the weak physical interactions responsible for biopolymer–biopolymer interactions are disrupted. The characteristic rheological behavior of biopolymer solutions plays an important role in determining their functional properties in food emulsions. For example, a salad dressing must be able to flow when it is poured from a container, but must maintain its shape under its own weight after it has been poured onto a salad. The amount and type of biopolymer used must therefore be carefully selected to provide a low viscosity when the salad dressing is poured (high applied stress), but a high viscosity when the salad dressing is allowed to sit under its own weight (low applied stress).

The viscosity of biopolymer solutions is also related to the mouthfeel of a food product. Liquids that do not exhibit extensive shear-thinning behavior at the shear stresses experienced in the mouth are perceived as being slimy. On the other hand, a certain amount of viscosity is needed to contribute to the creaminess of a product.

The shear-thinning behavior of biopolymer solutions is also important for determining the stability of food emulsions to creaming [1]. As oil droplets move through an emulsion, they exert very small shear stresses on the surrounding liquid. Consequently, they experience a very large viscosity, which greatly slows down the rate at which they cream and therefore enhances stability. Many biopolymer solutions also exhibit thixotropic behavior (i.e., their viscosity decreases with time when they are sheared at a constant rate) as a result of disruption of the weak physical interactions that cause biopolymer molecules to aggregate. A food manufacturer must therefore select an appropriate biopolymer or combination of biopolymers to produce a final product that has a desirable mouthfeel and texture.

3. Gelation

Biopolymers are used as functional ingredients in many food emulsions (e.g., yogurts, cheeses, desserts, eggs, and meat products) because of their ability to cause the aqueous phase to gel [1]. Gel formation imparts desirable textural and sensory attributes, as well as preventing the droplets from creaming. A biopolymer gel consists of a three-dimensional network of aggregated or entangled biopolymers that entraps a large volume of water, giving the whole structure some solid-like characteristics. The appearance, texture, water-holding capacity, reversibility, and gelation temperature of biopolymer gels depend on the type, structure, and interactions of the molecules they contain.

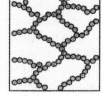

Gel containing   Gel containing
particulate aggregates   filamentous aggregates

FIGURE 3.8 Biopolymer molecules or aggregates can form various types of gel structures, such as particulate or filamentous.

Gels may be transparent or opaque, hard or soft, brittle or rubbery, homogeneous or heterogeneous; they may exhibit syneresis or have good water-holding capacity. Gelation may be induced by a variety of different methods, including altering the temperature, pH, ionic strength, or solvent quality; adding enzymes; and increasing the biopolymer concentration. Biopolymers may be cross-linked by covalent and noncovalent bonds.

It is convenient to distinguish between two types of gels: particulate and filamentous (Figure 3.8). Particulate gels consist of biopolymer aggregates (particles or clumps) that are assembled together to form a three-dimensional network. This type of gel tends to be formed when there are strong attractive forces over the whole surface of the individual biopolymer molecules. Particulate gels are optically opaque because the particles scatter light, and they are prone to syneresis because the large interparticle pore sizes mean that the water is not held tightly in the gel network by capillary forces. Filamentous gels consist of filaments of individual or aggregated biopolymer molecules that are relatively thin and tend to be formed by biopolymers that can form junction zones only at a limited number of sites on the surface of a molecule, or when the attractive forces between the molecules are so strong that they stick firmly together and do not undergo subsequent rearrangement [5]. Filamentous gels tend to be optically transparent because the filaments are so thin that they do not scatter light significantly, and they tend to have good water-holding capacity because the small pore size of the gel network means that the water molecules are held tightly by capillary forces.

In some foods a gel is formed on heating (heat-setting gels), whereas in others it is formed on cooling (cold-setting gels). Gels may also be either thermoreversible or thermoirreversible, depending on whether gelation is or is not reversible. Gelatin is an example of a cold-setting thermoreversible gel: when a solution of gelatin molecules is cooled below a certain temperature, a gel is formed, but when it is reheated the gel melts. Egg white is an example of a heat-setting thermoirreversible gel: when an egg is heated above a temperature at which gelation occurs, a characteristic white gel is formed; when the egg is cooled back to room temperature, however, the gel remains white (i.e., it does not revert back to its earlier liquid form). Whether a gel is reversible or irreversible depends on the changes in the molecular structure and organization of the molecules during gelation. Biopolymer gels that are stabilized by noncovalent interactions and do not involve large changes in the structure of the individual molecules before gelation tend to be reversible. On the other hand, gels that are held together by covalent bonds or involve large changes in the structure of the individual molecules before gelation tend to form irreversible gels.

The type of force holding the molecules together in gels varies from biopolymer to biopolymer. Some proteins and polysaccharides (e.g., gelatin, starch) form helical junction zones through extensive hydrogen bond formation. This type of junction zone tends to form when a gel is cooled, becoming disrupted when it is heated, and thus it is responsible for cold-setting gels. Below the gelatin temperature, the attractive hydrogen bonds favor junction zone formation, but above this temperature the configurational entropy favors a random-coil type of structure. Biopolymers with extensive nonpolar groups (e.g., caseins, denatured whey proteins) tend to associate via hydrophobic interactions. Electrostatic interactions play an important role in determining the gelation behavior of many biopolymers, and so gelation is particularly sensitive to the pH and ionic strength of the solution containing the biopolymers. For example, at pH values sufficiently far from their isoelectric point, proteins may be prevented from gelling because of the electrostatic repulsion between the molecules. However, if the pH of the same solution is adjusted near to the isoelectric point, or salt is added, the proteins gel.

The addition of multivalent ions, such as Ca^{2+}, can promote gelation of charged biopolymer molecules by forming salt bridges between the molecules. Proteins with thiol groups are capable of forming covalent linkages through thiol–disulfide interchanges, which help to strengthen and enhance the stability of gels. The tendency for a biopolymer to form a gel under certain conditions and the physical properties of the gel formed depend on a delicate balance of biopolymer–biopolymer, biopolymer–solvent, and solvent–solvent interactions of various kinds.

C. Ingredient Selection

A wide variety of proteins and polysaccharides are available as ingredients in foods, each with its own unique functional properties and optimum range of applications. Food manufacturers must decide which biopolymer is the most suitable for each type of food product. The selection of the most appropriate ingredient is often the key to success of a particular product. The factors a manufacturer must consider include the desired properties of the final product (appearance, rheology, mouthfeel, stability), the composition of the product, and the processing, storage, and handling conditions the food experiences during its lifetime, as well as the cost, availability, consistency from batch to batch, ease of handling, dispersibility, and functional properties of the biopolymer ingredient.

V. EMULSION FORMATION

The formation of an emulsion may involve a single step or a number of consecutive steps, depending on the nature of the starting material, the desired properties of the end product, and the instrument used to create it [1]. Before separate oil and aqueous phases are converted to an emulsion, it is usually necessary to disperse the various ingredients into the phase in which they are most soluble. Oil-soluble ingredients, such as certain vitamins, coloring agents, antioxidants, and surfactants, are mixed with the oil, whereas water-soluble ingredients, such as proteins, polysaccharides, sugars, salts, and some vitamins, coloring agents, antioxidants, and surfactants, are mixed with the water. The intensity and duration of the mixing process depend on the time required to solvate and uniformly distribute the ingredients. Adequate solvation is important for the functionality of a number of food components. If the lipid phase contains any crystalline material, it is usually necessary to warm it before homogenization to a temperature at which all the fat melts; otherwise it is difficult, if not impossible, to efficiently create a stable emulsion.

The process of converting two immiscible liquids to an emulsion is known as homogenization, and a mechanical device designed to carry out this process is called a homogenizer. To distinguish the nature of the starting material, it is convenient to divide homogenization into two categories. The creation of an emulsion directly from two separate liquids will be referred to as primary homogenization, whereas the reduction in size of droplets in an existing emulsion will be referred to as secondary homogenization (Figure 3.9). The creation of a food emulsion may involve the use of one or the other form of homogenization, or a combination of both. For example, salad dressing is formed by direct homogenization of the aqueous and oil phases and is therefore an example of primary homogenization, whereas homogenized milk is manufactured by reducing the size of the fat globules in natural milk and hence is an example of secondary homogenization.

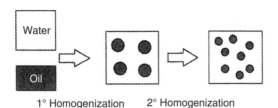

1° Homogenization   2° Homogenization

FIGURE 3.9 Homogenization process can be divided into two steps: primary homogenization (creating an emulsion from two separate phases) and secondary homogenization (reducing the size of the droplets in a preexisting emulsion).

In many food-processing operations and laboratory studies it is more effective to prepare an emulsion in two steps. The separate oil and water phases are converted to a coarse emulsion, with fairly large droplets, using one type of homogenizer (e.g., high-speed blender). Then the droplet size is reduced by means of another type of homogenizer (e.g., colloid mill, high-pressure valve homogenizer). In reality, many of the same physical processes that occur during primary homogenization also occur during secondary homogenization, and there is no clear distinction between them. Emulsions that have undergone secondary homogenization usually contain smaller droplets than those that have undergone primary homogenization, although this is not always the case. Some homogenizers (e.g., ultrasound, microfluidizers, membrane homogenizers) are capable of producing emulsions with small droplet sizes directly from separate oil and water phases (see Section V.C).

To highlight the important physical mechanisms that occur during homogenization, it is useful to consider the formation of an emulsion from pure oil and pure water. When the two liquids are placed in a container, they tend to adopt their thermodynamically most stable state, which consists of a layer of oil on top of the water (Figure 3.1). This arrangement is adopted because it minimizes the contact area between the two immiscible liquids and because oil has a lower density than water. To create an emulsion, it is necessary to mechanically agitate the system, to disrupt and intermingle the oil and water phases. The type of emulsion formed in the absence of an emulsifier depends primarily on the initial concentration of the two liquids. At high oil concentrations a W/O emulsion tends to form, but at low oil concentrations an O/W emulsion tends to form. In this example, it is assumed that the oil concentration is so low that an O/W emulsion is formed. Mechanical agitation can be applied in a variety of ways, the simplest being to vigorously shake the oil and water together in a sealed container. An emulsion is formed immediately after shaking, and it appears optically opaque (because light is scattered from the emulsion droplets). With time, the system rapidly reverts back to its initial state—a layer of oil sitting on top of the water. This is because the droplets formed during the application of the mechanical agitation are constantly moving around and frequently collide and coalesce with neighboring droplets. As this process continues, the large droplets formed rise to the top of the container and merge together to form a separate layer.

To form a stable emulsion, one must prevent the droplets from merging after they have been formed. This is achieved by having a sufficiently high concentration of a surface-active substance, known as an emulsifier, present during the homogenization process. The emulsifier rapidly adsorbs to the droplet surfaces during homogenization, forming a protective membrane that prevents the droplets from coming close enough together to coalesce. One of the major objectives of homogenization is to produce droplets as small as possible because this usually increases the shelf life of the final product. It is therefore important for the food scientist to understand the factors that determine the size of the droplets produced during homogenization. It should be noted that homogenization is only one step in the formation of a food emulsion, and many of the other unit operations (e.g., pasteurization, cooking, drying, freezing, whipping) also affect the final quality of the product.

A. Physical Principles of Emulsion Formation

The size of the emulsion droplets produced by a homogenizer depends on a balance between two opposing mechanisms: droplet disruption and droplet coalescence (Figure 3.10). The tendency for emulsion droplets to break up during homogenization depends on the strength of the interfacial forces that hold the droplets together, compared with the strength of the disruptive forces in the homogenizer. In the absence of any applied external forces, emulsion droplets tend to be spherical because this shape minimizes the contact area between oil and water phases. Changing the shape of a droplet, or breaking it into smaller droplets, increases this contact area and therefore requires the input of energy. The interfacial force holding a droplet together is given by the Laplace pressure (ΔP_1):

$$\Delta P_1 = \frac{2\gamma}{r}, \tag{3.4}$$

where
  γ is the interfacial tension between oil and water
  r is the droplet radius

This equation indicates that it is easier to disrupt large droplets than small ones and that the lower the interfacial tension, the easier it is to disrupt a droplet. The nature of the disruptive forces that act on a droplet during homogenization depends on the flow conditions (i.e., laminar, turbulent, or cavitational) the droplet experiences and therefore on the type of homogenizer used to create the emulsion. To deform and disrupt a droplet during homogenization, it is necessary to generate a stress that is greater than the Laplace pressure and to ensure that this stress is applied to the droplet long enough to enable it to become disrupted [23–25].

Emulsions are highly dynamic systems in which the droplets continuously move around and frequently collide with each other. Droplet–droplet collisions are particularly rapid during homogenization due to the intense mechanical agitation of the emulsion. If droplets are not protected by a sufficiently strong emulsifier membrane, they tend to coalesce during collision. Immediately after the disruption of an emulsion droplet during homogenization, there is insufficient emulsifier present to completely cover the newly formed surface, and therefore the new droplets are more likely to coalesce with their neighbors. To prevent coalescence from occurring, it is necessary to form a sufficiently concentrated emulsifier membrane around a droplet before it has time to collide with its neighbors. The size of droplets produced during homogenization therefore depends on the time

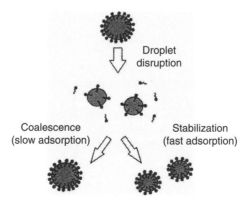

Droplet
disruption

Coalescence
(slow adsorption)

Stabilization
(fast adsorption)

FIGURE 3.10 Size of the droplets produced in an emulsion is a balance between droplet disruption and droplet coalescence.

taken for the emulsifier to be adsorbed to the surface of the droplets ($\tau_{adsorption}$) compared with the time between droplet–droplet collisions ($\tau_{collision}$). If $\tau_{adsorption} \ll \tau_{collision}$, the droplets are rapidly coated with emulsifier as soon as they are formed and are stable; but if $\tau_{adsorption} \gg \tau_{collision}$, the droplets tend to rapidly coalesce because they are not completely coated with emulsifier before colliding with one of their neighbors. The values of these two times depend on the flow profile the droplets experience during homogenization, as well as the physicochemical properties of the bulk phases and the emulsifier [1,25].

B. ROLE OF EMULSIFIERS

The preceding discussion has highlighted two of the most important roles of emulsifiers during the homogenization process:

1. Their ability to decrease the interfacial tension between oil and water phases and thus reduce the amount of energy required to deform and disrupt a droplet (Equation 3.4). It has been demonstrated experimentally that when the movement of an emulsifier to the surface of a droplet is not rate limiting ($\tau_{adsorption} \ll \tau_{collision}$), there is a decrease in the droplet size produced during homogenization with a decrease in the equilibrium interfacial tension [26].
2. Their ability to form a protective membrane that prevents droplets from coalescing with their neighbors during a collision.

The effectiveness of emulsifiers at creating emulsions containing small droplets depends on a number of factors: (1) the concentration of emulsifier present relative to the dispersed phase; (2) the time required for the emulsifier to move from the bulk phase to the droplet surface; (3) the probability that an emulsifier molecule will be adsorbed to the surface of a droplet during a droplet–emulsifier encounter (i.e., the adsorption efficiency); (4) the amount by which the emulsifier reduces the interfacial tension; and (5) the effectiveness of the emulsifier membrane in protecting the droplets against coalescence.

It is often assumed that small emulsifier molecules adsorb to the surface of emulsion droplets during homogenization more rapidly than larger ones. This assumption is based on the observation that small molecules diffuse to the interface more rapidly than larger ones under quiescent conditions [5]. It has been demonstrated that under turbulent conditions large surface-active molecules tend to accumulate at the droplet surface during homogenization preferentially to smaller ones [25].

C. HOMOGENIZATION DEVICES

There are a wide variety of food emulsions, and each one is created from different ingredients and must have different final characteristic properties. Consequently, a number of homogenization devices have been developed for the chemical production of food emulsions, each with its own particular advantages and disadvantages, and each having a range of foods to which it is most suitably applied [1]. The choice of a particular homogenizer depends on many factors, including the equipment available, the site of the process (i.e., a factory or a laboratory), the physicochemical properties of the starting materials and final product, the volume of material to be homogenized, the throughput, the desired droplet size of the final product, and the cost of purchasing and running the equipment. The most important types of homogenizers used in the food industry are discussed in the subsections that follow.

1. High-Speed Blenders

High-speed blenders are the most commonly used means of directly homogenizing bulk oil and aqueous phases. The oil and aqueous phases are placed in a suitable container, which may contain as little as a few milliliters or as much as several liters of liquid, and agitated by a stirrer that rotates at high speeds. The rapid rotation of the blade generates intense velocity gradients that cause disruption of the interface between the oil and water, intermingling of the two immiscible liquids, and breakdown of larger droplets to smaller ones [27]. Baffles are often fixed to the inside of the container to increase the efficiency of the blending process by disrupting the flow profile. High-speed blenders are particularly useful for preparing emulsions with low or intermediate viscosities. Typically, they produce droplets that are between 1 and 10 μm in diameter.

2. Colloid Mills

The separate oil and water phases are usually blended together to form a coarse emulsion premix before their introduction into a colloid mill because this increases the efficiency of the homogenization process. The premix is fed into the homogenizer, where it passes between two disks separated by a narrow gap. One of the disks is usually stationary, whereas the other rotates at a high speed, thus generating intense shear stresses in the premix. These shear stresses are large enough to cause the droplets in the coarse emulsion to be broken down. The efficiency of the homogenization process can be improved by increasing the rotation speed, decreasing the flow rate, decreasing the size of the gap between the disks, and increasing the surface roughness of the disks. Colloid mills are more suitable than most other types of homogenizers for homogenizing intermediate- or high-viscosity fluids (e.g., peanut butter, fish or meat pastes), and they typically produce emulsions with droplet diameters between 1 and 5 μm.

3. High-Pressure Value Homogenizers

Like colloid mills, high-pressure valve homogenizers are more efficient at reducing the size of the droplets in a coarse emulsion premix than at directly homogenizing two separate phases [28]. The coarse emulsion premix is forced through a narrow orifice under high pressure, which causes the droplets to be broken down because of the intense disruptive stresses (e.g., impact forces, shear forces, cavitation, turbulence) generated inside the homogenizer [29]. Decreasing the size of the orifice increases the pressure the emulsion experiences, which causes a greater degree of droplet disruption and therefore the production of smaller droplets. Nevertheless, the throughput is reduced and more energy must be expended. A food manufacturer must therefore select the most appropriate homogenization conditions for each particular application, depending on the compromise between droplet size, throughput, and energy expenditure. High-pressure valve homogenizers can be used to homogenize a wide variety of food products, ranging from low viscosity liquids to viscoelastic pastes, and can produce emulsions with droplet sizes as small as 0.1 μm.

4. Ultrasonic Homogenizers

A fourth type of homogenizer uses high-intensity ultrasonic waves that generate intense shear and pressure gradients. When applied to a sample containing oil and water, these waves cause the two liquids to intermingle and the large droplets formed to be broken down to smaller ones. There are two types of ultrasonic homogenizers commonly used in the food industry: piezoelectric transducers and liquid jet generators [30]. Piezoelectric transducers are most commonly found in the small benchtop ultrasonic homogenizers used in many laboratories. They are ideal for preparing small volumes of emulsion (a few milliliters to a few hundred milliliters), a property that is often important in fundamental research when expensive components are used. The ultrasonic transducer consists of a piezoelectric crystal contained in some form of protective metal casing, which is tapered at the end. A high-intensity electrical wave is applied to the transducer, which causes the piezoelectric crystal inside to oscillate and generate an ultrasonic wave. The ultrasonic wave is directed toward the tip of the transducer, where it radiates into the surrounding liquids, generating intense pressure and shear gradients (mainly due to cavitational affects) that cause the liquids to be broken up into smaller fragments and intermingled with one another. It is usually necessary to irradiate a sample with ultrasound for a few seconds to a few minutes to create a stable emulsion. Continuous application of ultrasound to a sample can cause appreciable heating, and so it is often advantageous to apply the ultrasound in a number of short bursts.

Ultrasonic jet homogenizers are used mainly for industrial applications. A stream of fluid is made to impinge on a sharp-edged blade, which causes the blade to rapidly vibrate, thus generating an intense ultrasonic field that breaks up any droplets in its immediate vicinity though a combination of cavitation, shear, and turbulence [30]. This device has three major advantages: it can be used for continuous production of emulsions; it can generate very small droplets; and it is more energy efficient than high-pressure valve homogenizers (since less energy is needed to form droplets of the same size).

5. Microfluidization

Microfluidization is a technique that is capable of creating an emulsion with small droplet sizes directly from the individual oil and aqueous phases [31]. Separate streams of an oil and an aqueous phase are accelerated to a high velocity and then made to simultaneously impinge on a surface, which causes them to be intermingled and leads to effective homogenization. Microfluidizers can be used to produce emulsions that contain droplets as small as 0.1 μm.

6. Membrane Homogenizers

Membrane homogenizers form emulsions by forcing one immiscible liquid into another through a glass membrane that is uniform in pore size. The size of the droplets formed depends on the diameter of the pores in the membrane and on the interfacial tension between the oil and water phases [32]. Membranes can be manufactured with different pore diameters, with the result that emulsions with different droplet sizes can be produced [32]. The membrane technique can be used either as a batch or a continuous process, depending on the design of the homogenizer. Increasing numbers of applications for membrane homogenizers are being identified, and the technique can now be purchased for preparing emulsions in the laboratory or commercially. These instruments can be used to produce O/W, W/O, and multiple emulsions. Membrane homogenizers have the ability to produce emulsions with very narrow droplet size distributions, and they are highly energy efficient since there is much less energy loss due to viscous dissipation.

7. Energy Efficiency of Homogenization

The efficiency of the homogenization process can be calculated by comparing the energy required to increase the surface area between the oil and water phases with the actual amount of energy required to create an emulsion. The difference in free energy between the two separate immiscible liquids and an emulsion can be estimated by calculating the amount of energy needed to increase the interfacial area between the oil and aqueous phases ($\Delta G = \gamma \Delta A$). Typically, this is less than 0.1% of the total energy input into the system during the homogenization process because most of the energy supplied to the system is dissipated as heat, owing to frictional losses associated with the movement of molecules past one another [25]. This heat exchange accounts for the significant increase in temperature of emulsions during homogenization.

8. Choosing a Homogenizer

The choice of a homogenizer for a given application depends on a number of factors, including volume of sample to be homogenized, desired throughput, energy requirements, nature of the sample, final droplet size distribution required, equipment available, and initial and running costs. Even after the most suitable homogenization technique has been chosen, the operator must select the optimum processing conditions, such as temperature, time, flow rate, pressure, valve gaps, rotation rates, and sample composition. If an application does not require that the droplets in an emulsion be particularly small, it is usually easiest to use a high-speed blender. High-speed blenders are also used frequently to produce the coarse emulsion premix that is fed into other devices.

To create an emulsion that contains small droplets (<1 μm), either industrially or in the laboratory, it is necessary to use one of the other methods. Colloid mills are the most efficient type of homogenizer for high-viscosity fluids, whereas high-pressure valve, ultrasonic, or micro-fluidization homogenizers are more efficient for liquids that are low or intermediate in viscosity. In fundamental studies one often uses small volumes of sample, and therefore a number of laboratory homogenizers have been developed that are either scaled-down versions of industrial equipment or instruments specifically designed for use in the laboratory. For studies involving ingredients that are limited in availability or expensive, an ultrasonic piezoelectric transducer can be used because it requires only small sample volumes. When it is important to have monodisperse emulsions, the use of a membrane homogenizer would be advantageous.

D. Factors That Determine Droplet Size

The food manufacturer is often interested in producing emulsion droplets that are as small as possible, using the minimum amount of energy input and the shortest amount of time. The size of the droplets produced in an emulsion depends on many different factors, some of which are summarized below [29–32].

Emulsifier concentration. Up to a certain level, the size of the droplets usually decreases as the emulsifier concentration increases; above this level, droplet size remains constant. When the emulsifier concentration exceeds the critical level, the size of the droplets is governed primarily by the energy input of the homogenization device.

Emulsifier type. At the same concentration, different types of emulsifiers produce different sized droplets, depending on their surface load, the speed at which they reach the oil–water interface, and the ability of the emulsifier membrane to prevent droplet coalescence.

Homogenization conditions. The size of the emulsion droplets usually decreases as the energy input or homogenization time increases.

Physicochemical properties of bulk liquids. The homogenization efficiency depends on the physicochemical properties of the lipids that comprise an emulsion (e.g., their viscosity, interfacial tension, density, or physical state).

VI. EMULSION STABILITY

Emulsions are thermodynamically unstable systems that tend, with time, to separate back into individual oil and water phases (Figure 3.1). The term emulsion stability refers to the ability of an emulsion to resist changes in its properties with time: the greater the emulsion stability, the longer the time taken for the emulsion to alter its properties [1]. Changes in the properties of emulsions may be the result of physical processes that cause alterations in the spatial distribution of the ingredients (e.g., creaming, flocculation, coalescence, phase inversion) or chemical processes that cause alterations in the chemical structure of the ingredients (e.g., oxidation, hydrolysis). It is important for food scientists to elucidate the relative importance of each of these mechanisms, the relationship between them, and the factors that affect them, so that effective means of controlling the properties of food emulsions can be established.

A. Droplet–Droplet Interactions

The bulk properties of food emulsions are largely determined by the interaction of the droplets with each other. If the droplets exert a strong mutual attraction, they tend to aggregate, but if they are strongly repelled they tend to remain as separate entities. The overall interaction between droplets depends on the magnitude and range of a number of different types of attractive and repulsive interactions. Knowledge of the origin and nature of these interactions are important because it enables food scientists to predict and control the stability and physicochemical properties of food emulsions.

Droplet–droplet interactions are characterized by an interaction potential $\Delta G(s)$, which describes the variation of the free energy with droplet separation. The overall interaction potential between emulsion droplets is the sum of various attractive and repulsive contributions [5]:

$$\Delta G(s) = \Delta G_{\text{VDW}}(s) + \Delta G_{\text{electrostatic}}(s) + \Delta G_{\text{hydrophobic}}(s) + \Delta G_{\text{short range}}(s), \qquad (3.5)$$

where ΔG_{VDW}, $\Delta G_{\text{electrostatic}}$, $\Delta G_{\text{hydrophobic}}$, and $\Delta G_{\text{short range}}$ refer to the free energies associated with van der Waals, electrostatic, hydrophobic, and various short-range forces, respectively. In certain systems, there are additional contributions to the overall interaction potential from other types of mechanisms, such as depletion or bridging [1,2]. The stability of food emulsions to aggregation depends on the shape of the free energy versus separation curve, which is governed by the relative contributions of the different types of interactions [1–5].

1. van der Waals Interactions

The van der Waals interactions act between emulsion droplets of all types and are always attractive. At close separations, the van der Waals interaction potential between two emulsion droplets of equal radius r separated by a distance s is given by the following equation 3.12:

$$\Delta G_{\text{VDW}}(s) = -\frac{Ar}{12s}, \qquad (3.6)$$

where A is the Hamaker parameter, which depends on the physical properties of the oil and water phases. This equation provides a useful insight into the nature of the van der Waals interaction. The strength of the interaction decreases with the reciprocal of droplet separation, and so van der Waals interactions are fairly long range compared with other types of interactions. In addition, the strength of the interaction increases as the size of the emulsion droplets increases. In practice, Equation 3.6 tends to overestimate the attractive forces because it ignores the effects of electrostatic screening, radiation, and the presence of the droplet membrane on the Hamaker parameter [13].

2. Electrostatic Interactions

Electrostatic interactions occur only between emulsion droplets that have electrically charged surfaces (e.g., those established by ionic surfactants or biopolymers). The electrostatic interaction between two droplets at close separation is given by the following relationship [7]:

$$\Delta G_{\text{electrostatic}}(s) = 4.3 \times 10^{-9} r\psi_0^2 \ln(1 + e^{-4.5}), \qquad (3.7)$$

where

$$\kappa^{-1} = \left(\frac{\varepsilon_0 \varepsilon_r kT}{e^2 \Sigma c_i z_i^2}\right)^{1/2}.$$

Here κ^{-1} is the thickness of the electric double layer, c_i and z_i are the molar concentration and valency of ions of species i, ε_0 is the dielectric constant of a vacuum, ε_r is the relative dielectric constant of the medium surrounding the droplet, e is the electrical charge, ψ_0 is the surface potential, k is the Boltzmann constant, and T is the temperature. These equations provide a useful insight into the nature of the electrostatic interactions between emulsion droplets. Usually all the droplets in food emulsions have the same electrical charge, hence repel each other. Electrostatic interactions are therefore important for preventing droplets from aggregating. The strength of the interactions increases as the magnitude of the surface potential increases; thus, greater the number of charges per unit area at a surface, greater the protection against aggregation. The strength of the repulsive interaction decreases as the concentration of valency of ions in the aqueous phase increases because counterions screen the charges between droplets, which causes a decrease in the thickness of the electrical double layer. Emulsions stabilized by proteins are particularly sensitive to the pH and ionic strength of the aqueous solution, since altering pH changes ψ_0 and altering ionic strength changes κ^{-1}. The strength of the electrostatic interaction also increases as the size of the emulsion droplets increases.

3. Hydrophobic Interactions

The surfaces of emulsion droplets may not be completely covered by emulsifier molecules, or the droplet membrane may have some nonpolar groups exposed to the aqueous phase [1]. Consequently, there may be attractive hydrophobic interactions between nonpolar groups and water. The interaction potential energy per unit area between two hydrophobic surfaces separated by water is given by

$$\Delta G_{\text{hydrophobic}}(s) = -0.69 \times 10^{-10} r\phi \exp\left(-\frac{s}{\lambda_0}\right), \tag{3.8}$$

where
  ϕ is the fraction of the droplet surface (which is hydrophobic)
  λ_0, the decay length, is of the order of 1–2 nm [13]

The hydrophobic attraction between droplets with nonpolar surfaces is fairly strong and relatively long range [13]. Hydrophobic interactions therefore play an important role in determining the stability of a number of food emulsions. Protein-stabilized emulsions often have nonpolar groups on the protein molecules exposed to the aqueous phase, and therefore hydrophobic interactions are important. They are also important during homogenization because the droplets are not covered by emulsifier molecules.

4. Short-Range Forces

When two emulsion droplets come sufficiently close together, their interfacial layers start to interact. A number of short-range forces result from these interactions, including steric (osmotic and elastic components), hydration, protrusion, and undulation forces [13,14]. Some progress has been made in developing theories to predict the magnitude and range of short-range forces associated with interfacial layers of fairly simple geometry. Nevertheless, both magnitude and range of these forces are particularly sensitive to the size, shape, conformation, packing, interactions, mobility, and hydration of the molecules in the adsorbed layer, and so it is difficult to predict their contribution to the overall interaction potential with any certainty. Even so, they are usually repulsive and tend to increase strongly as the interfacial layers overlap.

5. Overall Interaction Potential

It is often difficult to accurately calculate the contribution of each type of interaction to the overall interdroplet pair potential because information about the relevant physicochemical properties of the system is lacking. Nevertheless, it is informative to examine the characteristics of certain combinations of interactions that are particularly important in food emulsions, for this provides a valuable insight into the factors that affect the tendency of droplets to aggregate. Consider an emulsion in which the only important types of droplet–droplet interactions are van der Waals attraction, electrostatic repulsion, and steric repulsion (e.g., an emulsion stabilized by a charged biopolymer).

The van der Waals interaction potential is fairly long range and always negative (attractive), the electrostatic interaction potential is fairly long range and always positive (repulsive), whereas the steric interaction is short range and highly positive (strongly repulsive). The overall interdroplet pair potential has a complex dependence on separation because it is the sum of these three different interactions, and it may be attractive at some separations and repulsive at others. Figure 3.11 shows a typical profile of interdroplet pair potential versus separation for an emulsion stabilized by a charged biopolymer. When the two droplets are separated by a large distance, there is no effective interaction between them. As they move closer together, the van der Waals attraction dominates initially and there is a shallow minimum in the profile, which is referred to as the secondary minimum. If the depth of this minimum is large compared with the thermal energy ($|\Delta G(s_{min2})| > kT$), the droplets tend to be flocculated. However, if it is small compared with the thermal energy, the droplets tend to remain unaggregated. At closer separations, the repulsive electrostatic interactions dominate, and there is an energy barrier $\Delta G(s_{max})$ that must be overcome before the droplets can come any closer. If this energy barrier is sufficiently large compared with the thermal energy $\Delta G(s_{max}) \gg kT$, it prevents the droplets from falling into the deep primary minimum at close separations. On the other hand, if it is not large compared with the thermal energy, the droplets tend to fall into the primary minimum, leading to strong flocculation of the droplets. In this situation, the droplets would be prevented from coalescing because of the domination of the strong steric repulsion at close separations.

Emulsions that are stabilized by repulsive electrostatic interactions are particularly sensitive to the ionic strength and pH of the aqueous phase [1,2]. At low ion concentrations there may be a sufficiently high energy barrier to prevent the droplets from getting close enough together to aggregate into the primary minimum. As the ion concentration is increased, the screening of the electrostatic interactions becomes more effective, which reduces the height of the energy barrier. Above a certain ion concentration, the energy barrier is not high enough to prevent the droplets from falling into the primary minimum, and so the droplets become strongly flocculated. This

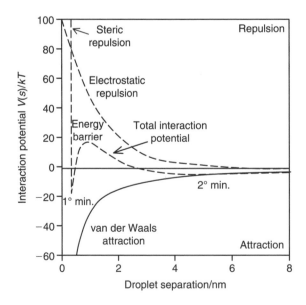

FIGURE 3.11 Overall interaction potential for an emulsion stabilized by a charged biopolymer.

phenomenon accounts for the tendency of droplets to flocculate when salt is added to emulsions stabilized by ionic emulsifiers. The surface charge density of protein-stabilized emulsions decreases as the pH tends toward the isoelectric point, which reduces the magnitude of the repulsive electrostatic interactions between the droplets and also leads to droplet flocculation.

B. Mechanisms of Emulsion Instability

As mentioned earlier, emulsions are thermodynamically unstable systems that tend with time to revert back to the separate oil and water phases of which they were made. The rate at which this process occurs and the route that is taken depend on the physicochemical properties of the emulsion and the prevailing environmental conditions. The most important mechanisms of physical instability are creaming, flocculation, coalescence, Ostwald ripening, and phase inversion. In practice, all these mechanisms act in concert and can influence one another. However, one mechanism often dominates the others, facilitating the identification of the most effective method of controlling emulsion stability.

The length of time an emulsion must remain stable depends on the nature of the food product. Some food emulsions (e.g., cake batters, ice cream mix, margarine premix) are formed as intermediate steps during a manufacturing process and need remain stable for only a few seconds, minutes, or hours. Other emulsions (e.g., mayonnaise, creme liqueurs) must persist in a stable state for days, months, or even years before sale and consumption. Some food-processing operations (e.g., the production of butter, margarine, whipped cream, and ice cream) rely on controlled destabilization of an emulsion. We now turn to a discussion of the origin of the major destabilization mechanisms, the factors that influence them, and methods of controlling them. This type of information is useful for food scientists because it facilitates the selection of the most appropriate ingredients and processing conditions required to produce a food emulsion with particular properties.

1. Creaming and Sedimentation

The droplets in an emulsion have a density different from that of the liquid that surrounds them, and so a net gravitational force acts on them [1,2]. If the droplets have lower density than the surrounding liquid, they tend to move up, that is, to cream. Conversely, if they have a higher density they tend to move down, resulting in what is referred to as sedimentation. Most liquid oils have densities lower than that of water, and so there is a tendency for oil to accumulate at the top of an emulsion and water at the bottom. Thus, droplets in an O/W emulsion tend to cream, whereas those in a W/O emulsion tend to sediment. The creaming rate of a single isolated spherical droplet in a viscous liquid is given by the Stokes equation:

$$v = -\frac{2gr^2(\rho_2 - \rho_1)}{9\eta_1},\tag{3.9}$$

where
  v is the creaming rate
  g is the acceleration due to gravity
  ρ is the density
  η is the shear viscosity
  the subscripts 1 and 2 refer to the continuous phase and droplet, respectively
  The sign of v determines whether the droplet moves up $(+)$ or down $(-)$

Equation 3.9 can be used to estimate the stability of an emulsion to creaming. For example, an oil droplet $(\rho_2 = 910 \text{ kg/m}^3)$ with a radius of 1 μm suspended in water $(\eta_1 = 1 \text{ mPa} \cdot \text{s},$

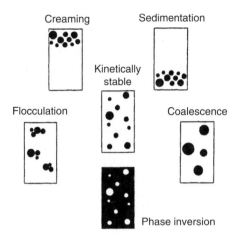

FIGURE 3.12 Mechanisms of emulsion instability.

$\rho_1 = 1000$ kg/m^3) will cream at a rate of about 5 mm/day. Thus, one would not expect an emulsion containing droplets of this size to have a particularly long shelf life. As a useful rule of thumb, an emulsion in which the creaming rate is less than about 1 mm/day can be considered to be stable toward creaming [5].

In the initial stages of creaming (Figure 3.12), the droplets move upward and a droplet-depleted layer is observed at the bottom of the container. When the droplets reach the top of the emulsion, they cannot move up any further and so they pack together to form the creamed layer. The thickness of the final creamed layer depends on the packing of the droplets in it. Droplets may pack very tightly together, or they may pack loosely, depending on their polydispersity and the magnitude of the forces between them. Close-packed droplets tend to form a thin creamed layer, whereas loosely packed droplets form a thick creamed layer. The same factors that affect the packing of the droplets in a creamed layer determine the nature of the flocs formed (see Section VIB.2). If the attractive forces between the droplets are fairly weak, the creamed emulsion can be redispersed by slightly agitating the system. On the other hand, if an emulsion is centrifuged, or if the droplets in a creamed layer are allowed to remain in contact for extended periods, significant coalescence of the droplets may occur, with the result that the emulsion droplets can no longer be redispersed by mild agitation.

Creaming of emulsion droplets is usually an undesirable process, which food manufacturers try to avoid. Equation 3.9 indicates that creaming can be retarded by minimizing the density difference $(\rho_2 - \rho_1)$ between the droplets and the surrounding liquid, reducing the droplet size, or increasing the viscosity of the continuous phase. The Stokes equation is strictly applicable only to isolated rigid spheres suspended in an infinite viscous liquid. Since these assumptions are not valid for food emulsions, the equation must be modified to take into account hydrodynamic interactions, droplet fluidity, droplet aggregation, non-Newtonian aqueous phases, droplet crystallization, the adsorbed layer, and Brownian motion [1,4].

2. Flocculation and Coalescence

The droplets in emulsions are in continual motion because of their thermal energy, gravitational forces, or applied mechanical forces, and as they move about they collide with their neighbors. After a collision, emulsion droplets may either move apart or remain aggregated, depending on the relative magnitude of the attractive and repulsive forces between them. If the net force acting between the droplets is strongly attractive, they aggregate, but if it is strongly repulsive they remain unaggregated. Two types of aggregations are commonly observed in emulsions: flocculation and coalescence. In flocculations (Figure 3.12), two or more droplets come together to form an aggregate in which the emulsion droplets retain their individual integrity. Coalescence is the process whereby two or more droplets merge together to form a single larger droplet (Figure 3.12). Improvements in the quality of emulsion-based food products largely depend on an understanding of the factors that cause droplets to aggregate. The rate at which droplet aggregation occurs in an emulsion depends on two factors: collision frequency and collision efficiency [1,2].

The collision frequency is the number of encounters between droplets per unit time per unit volume. Any factor that increases the collision frequency is likely to increase the aggregation rate. The frequency of collisions between droplets depends on whether the emulsion is subjected to mechanical agitation. For dilute emulsions containing identical spherical particles, the collision frequency N has been calculated for both quiescent and stirred systems [5]:

$$N = \frac{4kTn_0^2}{3\eta},$$ (3.10)

$$N = \frac{16}{3}Gr^3n_0^2,$$ (3.11)

where
  n_0 is the initial number of particles per unit volume
  G is the shear rate

The collision efficiency, E, is the fraction of encounters between droplets that lead to aggregation. Its value ranges from 0 (no flocculation) to 1 (fast flocculation) and depends on the interaction potential. The equations for the collision frequency must therefore be modified to take into account droplet–droplet interactions:

$$N = \frac{4kTn_0^2}{3\eta}E,$$ (3.12)

where

$$E = \int_{2r}^{x} \left\{ \exp\left|\frac{\Delta G(x)}{kT}\right| x^{-2}dx \right\}^{-1}$$

with x the distance between the centers of the droplets ($x = 2r + s$) and $\Delta G(x)$ the droplet–droplet interaction potential (Section VI.A). Emulsion droplets may remain unaggregated, or they may aggregate into the primary or secondary minima depending on $\Delta G(x)$.

The equations above are applicable only to the initial stages of aggregation in dilute emulsions containing identical spherical particles. In practice, most food emulsions are fairly concentrated systems, and interactions between flocs as well as between individual droplets are important. The equations above must therefore be modified to take into account the interactions and properties of flocculated droplets.

The nature of the droplet–droplet interaction potential also determines the structure of the flocs formed, and the rheology and stability of the resulting emulsion [1]. When the attractive force between them is relatively strong, two droplets tend to become locked together as soon as they encounter each other. This leads to the formation of flocs that have quite open structures [5]. When the attractive forces are not particularly strong, the droplets may roll around each other after a collision, which allows them to pack more efficiently to form denser flocs. These two extremes of floc structure are similar to those formed by filamentous and particulate gels, respectively (Figure 3.8).

The structure of the flocs formed in an emulsion has a pronounced influence on its bulk physicochemical properties. An emulsion containing flocculated droplets has a higher viscosity than one containing unflocculated droplets, since the water trapped between the flocculated droplets increases the effective diameter (and therefore volume fraction) of the particles (Equation 3.3). Flocculated particles also exhibit strong shear-thinning behavior: as the shear rate is increased, the viscosity of the emulsion decreases because the flocs are disrupted and so their effective volume fraction decreases. If flocculation is extensive, a three-dimensional network of aggregated particles extends throughout the system and the emulsion has a yield stress that must be overcome before the system flows. The creaming rate of droplets is also strongly dependent on flocculation. At low droplet concentrations, flocculation increases the creaming rate because the effective size of the particles is increased (Equation 3.9), but at high droplet concentrations, it retards creaming because the droplets are trapped within the three-dimensional network of aggregated emulsion droplets.

In coalescence (Figure 3.12), two or more liquid droplets collide and merge into a single larger droplet. Extensive coalescence eventually leads to oiling off, that is, formation of free oil on the top of an emulsion. Because coalescence involves a decrease in the surface area of oil exposed to the continuous phase, it is one of the principal mechanisms by which an emulsion reverts to its most thermodynamically stable state (Figure 3.1). Coalescence occurs rapidly between droplets that are not protected by emulsifier molecules; for example, if one homogenizes oil and water in the absence of an emulsifier, the droplets readily coalesce. When droplets are stabilized by an emulsifier membrane, the tendency for coalescence to occur is governed by the droplet–droplet interaction potential and the stability of the film to rupture. If there is a strong repulsive force between the droplets at close separations, or if the film is highly resistant to rupture, the droplets tend not to coalesce. Most food emulsions are stable to coalescence, but they become unstable when subjected to high shear forces that cause the droplets to frequently collide with each other or when the droplets remain in contact with each other for extended periods (e.g., droplets in flocs, creamed layers, or highly concentrated emulsions).

3. Partial Coalescence

Normal coalescence involves the aggregation of two or more liquid droplets to form a single larger spherical droplet, but partial coalescence occurs when two or more partially crystalline droplets encounter each other and form a single irregularly shaped aggregate (Figure 3.13). The aggregate is irregular in shape because some of the structure of the fat crystal network contained in the original droplets is maintained within it. It has been proposed that partial coalescence occurs when two partially crystalline droplets collide and a crystal from one of them penetrates the intervening membranes and protrudes into the liquid region of the other droplet [1]. Normally, the crystal would stick out into the aqueous phase, thus becoming surrounded by water; however, when it penetrates another droplet, it is surrounded by oil, and because this arrangement is energetically favorable the droplets remain aggregated. With time, the droplets slowly fuse more closely together,

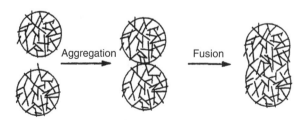

FIGURE 3.13 Partial coalescence occurs when two partly crystalline emulsion droplets collide and aggregate because a crystal in one droplet penetrates the other droplet.

with the result that the total surface area of oil exposed to the aqueous phase is reduced. Partial coalescence occurs only when the droplets have a certain ratio of solid fat and liquid oil. If the solid fat content of the droplets is either too low or too high, the droplets tend not to undergo partial coalescence [7].

Partial coalescence is particularly important in dairy products because milk fat globules are partially crystalline at temperatures commonly found in foods. The application of shear forces or temperature cycling to cream containing partly crystalline milk fat globules can cause extensive aggregation of the droplets, leading to a marked increase in viscosity (thickening) and subsequent phase separation [11]. Partial coalescence is an essential process in the production of ice cream, whipped toppings, butter, and margarine. O/W emulsions are cooled to a temperature at which the droplets are partly crystalline, and a shear force is then applied that causes droplet aggregation via partial coalescence. In butter and margarine, aggregation results in phase inversion, whereas in ice cream and whipped cream the aggregated fat droplets form a network that surrounds air cells and provides the mechanical strength needed to produce good stability and texture.

4. Ostwald Ripening

Ostwald ripening is the growth of large droplets at the expense of smaller ones [1]. This process occurs because the solubility of the material in a spherical droplet increases as the size of the droplet decreases:

$$S(r) = S(\infty) \exp\left(\frac{2\gamma V_{\mathrm{m}}}{RTr}\right). \tag{3.13}$$

Here V_{m} is the molar volume of the solute, γ is the interfacial tension, R is the gas constant, $S(\infty)$ is the solubility of the solute in the continuous phase for a droplet with infinite curvature (i.e., a planar interface), and $S(r)$ is the solubility of the solute when contained in a spherical droplet of radius r. The greater solubility of the material in smaller droplets means that there is a higher concentration of solubilized material around a small droplet than around a larger one. Consequently, solubilized molecules move from small droplets to large droplets because of this concentration gradient, which causes the larger droplets to grow at the expense of the smaller ones. Once steady-state conditions have been achieved, the growth in droplet radius with time due to Ostwald ripening is given by

$$\frac{d\langle r\rangle^3}{dt} = \frac{8\gamma V_{\mathrm{m}}S(\infty)D}{9RT}, \tag{3.14}$$

where D is the diffusion coefficient of the material through the continuous phase. This equation assumes that the emulsion is dilute and that the rate-limiting step is the diffusion of the solute molecules across the continuous phase. In practice, most food emulsions are concentrated systems, and so the effects of the neighboring droplets on the growth rate have to be considered. Some droplets are surrounded by interfacial membranes that retard the diffusion of solute molecules in and out of droplets, and in such cases the equation must be modified accordingly. Ostwald ripening is negligible in many foods because triacylglyercols have extremely low water solubilities, and therefore the mass transport rate is insignificant (Equation 3.14). Nevertheless, in emulsions that contain more water-soluble lipids, such as flavor oils, Ostwald ripening may be important.

5. Phase Inversion

In phase inversion (Figure 3.12), a system changes from an O/W emulsion to a W/O emulsion or vice versa. This process usually occurs as a result of some alteration in the system's composition or environmental conditions, such as dispersed phase volume fraction, emulsifier type, emulsifier concentration, temperature, or application of mechanical forces. Phase inversion is believed to occur by means of a complex mechanism that involves a combination of the processes that occur during flocculation, coalescence, and emulsion formation. At the point where phase inversion occurs, the system may briefly contain regions of O/W emulsion, W/O emulsion, multiple emulsions, and bicontinuous phases, before converting to its final state.

6. Chemical and Biochemical Stability

Chemical and biochemical reactions of various types (e.g., oxidation, reduction, or hydrolysis of lipids, polysaccharides, and proteins) can cause detrimental changes in the quality of food emulsions. Many of these reactions are catalyzed by specific enzymes that may be present in the food. The reactions that are important in a given food emulsion depend on the concentration, type, and distribution of ingredients, and the thermal and shear history of the food. Chemical and biochemical reactions can alter the stability, texture, flavor, odor, color, and toxicity of food emulsions. Thus, it is important to identify the most critical reactions that occur in each type of food so that they can be controlled in a systematic fashion.

VII. CHARACTERIZATION OF EMULSION PROPERTIES

Ultimately, food manufacturers want to produce a high-quality product at the lowest possible cost. To achieve this goal they must have a good appreciation of the factors that determine the properties of the final product. This knowledge, in turn, is used to formulate and manufacture a product with the desired characteristics (e.g., appearance, texture, mouthfeel, taste, shelf life). These bulk physicochemical and sensory properties are determined by such molecular and colloidal properties of emulsions as dispersed volume fraction, droplet size distribution, droplet–droplet interactions, and interfacial properties. Consequently, a wide variety of experimental techniques have been developed to characterize the molecular, colloidal, microscopic, and macroscopic properties of food emulsions [1]. Analytical techniques are needed to characterize the properties of food emulsions in the laboratory, where they are used to improve our understanding of the factors that determine emulsion properties, and in the factory, where they are used to monitor the properties of foods during processing to ensure that the manufacturing process is operating in an appropriate manner. The subsections that follow highlight some of the most important properties of food emulsions and outline experimental techniques for their measurement.

A. DISPERSED PHASE VOLUME FRACTION

The dispersed phase volume fraction or φ is the volume of emulsion droplets (V_D) divided by the total volume of the emulsion (V_E): $\phi = V_D/V_E$. The dispersed phase volume fraction determines the relative proportion of oil and water in a product, as well as influencing many of the bulk physicochemical and sensory properties of emulsions, such as appearance, rheology, taste, and stability. For example, an emulsion tends to become more turbid and to have a higher viscosity when the concentration of droplets is increased [1]. Methods for measuring the dispersed phase volume fraction of emulsions are outlined in Table 3.2. Traditional proximate analysis techniques, such as solvent extraction to determine oil content and oven drying to determine moisture content, can be used to analyze the dispersed phase volume fraction of emulsions. Nevertheless, proximate analysis techniques are often destructive and quite time consuming to carry out, and are therefore unsuitable for rapid quality control or online measurements. If the densities of the separate oil and aqueous phases are known, the dispersed phase volume fraction of an emulsion can simply be determined from a measurement of its density:

$$\phi = (\rho_{\text{emulsion}} - \rho_{\text{continuous phase}})(\rho_{\text{droplet}} - \rho_{\text{continuous phase}}). \qquad (3.15)$$

TABLE 3.2

Experimental Techniques for Characterizing the Physicochemical Properties of Food Emulsions

Dispersed phase volume fraction	Proximate analysis, density, electrical conductivity, light scattering, NMR, ultrasound
Droplet size distribution	Light scattering (static and dynamic), electrical conductivity, optical microscopy, electron microscopy, ultrasound, NMR
Microstructure	Optical microscopy, electron microscopy, atomic force microscopy
Creaming and sedimentation	Light scattering, ultrasound, NMR, visual observation
Droplet charge	Electrokinetic techniques, electroacoustic techniques
Droplet crystallization	Density, NMR, ultrasound, differential scanning calorimetry, polarized optical microscopy
Emulsion rheology	Viscometers, dynamic shear rheometers
Interfacial tension	Interfacial tensiometers (static and dynamic)
Interfacial thickness	Ellipsometry, neutron reflection, neutron scattering, light scattering, surface force apparatus

Source: From McClements, D.J. in *Food Emulsions: Principles, Practice and Techniques,* 2nd edn., CRC, Boca Raton, FL, 2005.

The electrical conductivity of an emulsion decreases as the concentration of oil within it increases, and so instruments based on electrical conductivity can also be used to determine φ. Light-scattering techniques can be used to measure the dispersed phase volume fraction of dilute emulsions ($\varphi < 0.001$), whereas NMR and ultrasound spectroscopy can be used to rapidly and nondestructively determine φ of concentrated and optically opaque emulsions. A number of these experimental techniques (e.g., ultrasound, NMR, electrical conductivity, density measurements) are particularly suitable for online determination of the composition of food emulsions during processing.

B. DROPLET SIZE DISTRIBUTION

The size of the droplets in an emulsion influences many of their sensory and bulk physicochemical properties, including rheology, appearance, mouthfeel, and stability [5,7]. It is therefore important for food manufacturers to carefully control the size of the droplets in a food product and to have analytical techniques to measure droplet size. Typically, the droplets in a food emulsion are somewhere in the size range of 0.1–50 μm in diameter.

Food emulsions always contain droplets that have a range of sizes, and so it is usually important to characterize both the average size and the size distribution of the droplets. The droplet size distribution is usually represented by a plot of droplet frequency (number or volume) versus droplet size (radius or diameter). Some of the most important experimental techniques for measuring droplet size distributions are included in Table 3.2.*

Light scattering and electrical conductivity techniques are capable of providing a full particle size distribution of a sample in a few minutes. Since, however, these techniques usually require that the droplet concentration be very low ($\varphi < 0.001$), samples must be diluted considerably before analysis. Optical and electron microscopy techniques, which provide the most direct measurement of droplet size distribution, are often time consuming and laborious to operate, and sample preparation can cause considerable artifacts in the results. In contrast, recently developed techniques based on NMR and ultrasonic spectroscopy can be used to rapidly and nondestructively measure the droplet size distribution of concentrated and optically opaque emulsions [1]. These techniques are particularly useful for online characterization of emulsion properties.

---

* A comprehensive review of analytical methods for measuring particle size in emulsions has recently been published [33].

C. Microstructure

The structural organization and interactions of the droplets in an emulsion often play an important role in determining the properties of a food. For example, two emulsions may have the same droplet concentration and size distribution, but very different properties, due to differences in the degree of droplet flocculation. Various forms of microscopy are available for providing information about the microstructure of food emulsions. The unaided human eye can resolve objects that are farther apart than about 0.1 mm (100 μm). Most of the structural components in food emulsions (e.g., emulsion droplets, surfactant micelles, fat crystals, ice crystals, small air cells, protein aggregates) are much smaller than this lower limit and cannot therefore be observed directly by the eye.

Optical microscopy can be used to study components of size between about 0.5 and 100 μm. The characteristics of specific components can be highlighted by selectively staining certain ingredients or by using special lenses. Electron microscopy can be used to study components that have sizes down to about 0.5 nm. Atomic force microscopy can be used to provide information about the arrangements and interactions of single atoms or molecules. All these techniques are burdened by sample preparation steps that often are laborious and time consuming, and subject to alter the properties of the material being examined. Nevertheless, when carried out correctly, the advanced microscopic techniques provide extremely valuable information about the arrangement and interactions of emulsion droplets with each other and with the other structural entities found in food emulsions.

D. Physical State

The physical state of the components in a food emulsion often has a pronounced influence on its overall properties [1]. For example, O/W emulsions are particularly prone to partial coalescence when the droplets contain a certain percentage of crystalline fat (Section VI.B). Partial coalescence leads to extensive droplet aggregation, which decreases the stability of emulsions to creaming and greatly increases their viscosity. In W/O emulsions, such as margarine or butter, the formation of a network of aggregated fat crystals provides the characteristic rheological properties. The most important data for food scientists are the temperature at which melting or crystallization begins, the temperature range over which the phase transition occurs, and the value of the solid fat content at any particular temperature. Phase transitions can be monitored by measuring changes in any property (e.g., density, compressibility, heat capacity, absorption, or scattering of radiation) that is altered on conversion of an ingredient from a solid to a liquid (Table 3.2). The density of a component often changes when it undergoes a phase transition, and so melting or crystallization can be monitored by measuring changes in the density of a sample with temperature or time.

Phase transitions can also be monitored by measuring the amount of heat absorbed or released when a solid melts or a liquid crystallizes, respectively. This type of measurement can be carried out by means of differential thermal analysis or differential scanning calorimetry. These techniques also provide valuable information about the polymorphic form of the fat crystals in an emulsion. More recently, rapid instrumental methods based on NMR and ultrasound have been developed to measure solid fat contents [1]. These instruments are capable of nondestructively determining the solid fat content of a sample in a few seconds and are extremely valuable analytical tools for rapid quality control and online procedures. Phase transitions can be observed in a more direct manner by means of polarized optical microscopy.

E. Creaming and Sedimentation Profiles

Over the past decade, a number of instruments have been developed to quantify the creaming or sedimentation of the droplets in emulsions. Basically the same light scattering, NMR, and ultrasound techniques used to measure the dispersed phase volume fraction or droplet size distributions of emulsions are applied to creaming or sedimentation, but the measurements are carried out as a function of sample height to permit the acquisition of a profile of droplet concentrations or sizes. Techniques based on the scattering of light can be used to study creaming and sedimentation in fairly dilute emulsions. A light beam is passed through a sample at a number of different heights, and the reflection and transmission coefficients are measured and related to the droplet concentration and size. By measuring the ultrasonic velocity or attenuation as a function of sample height and time, it is possible to quantify the rate and extent of creaming in concentrated and optically opaque emulsions. This technique can be fully automated and has the two additional advantages: creaming can be detected before it is visible to the eye and a detailed creaming profile can be determined rather than a single boundary. By measuring the ultrasound properties as a function of frequency, it is possible to determine both the concentration and size of the droplets as a function of sample height. Thus, a detailed analysis of creaming and sedimentation in complex food systems can be monitored noninvasively. Recently developed NMR imaging techniques can also measure the concentration and size of droplets in any region in an emulsion [11]. These ultrasound and NMR techniques will prove particularly useful for understanding the kinetics of creaming and sedimentation in emulsions and for predicting the long-term stability of food emulsions.

F. Emulsion Rheology

The rheology of an emulsion is one of its most important overall physical attributes because it largely determines the mouthfeel, flowability, and stability of emulsions [5]. A variety of experimental techniques are available for measuring the rheological properties of food emulsions. The rheology of emulsions that have low viscosities and act like ideal liquids can be characterized by capillary viscometers. For nonideal liquids or viscoelastic emulsions, more sophisticated instrumental techniques called dynamic shear rheometers are available to measure the relationship between the stress applied to an emulsion and the resulting strain, or vice versa. As well as providing valuable information about the bulk physicochemical properties of emulsions (e.g., texture, flow-through pipes), rheological measurements can provide information about droplet–droplet interactions and the properties of any flocs formed in an emulsion.

G. Interfacial Properties

Despite comprising only a small fraction of the total volume of an emulsion, the interfacial region that separates the oil from the aqueous phase plays a major role in determining stability, rheology, chemical reactivity, flavor release, and other overall physicochemical properties of emulsions. The most important properties of the interface are the concentration of emulsifier molecules present (the surface load), the packing of the emulsifier molecules, and the thickness, viscoelasticity, electrical charge, and (interfacial) tension of the interface.

A variety of experimental techniques are available for characterizing the properties of oil–water interfaces (Table 3.2). The surface load is determined by measuring the amount of emulsifier that adsorbs per unit area of oil–water interface. The thickness of an interfacial membrane can be determined by light scattering, neutron scattering, neutron reflection, surface force, and ellipsometry techniques. The rheological properties of the interfacial membrane can be determined by means of the two-dimensional analog of normal rheological techniques. The electrical charge of the droplets in an emulsion determines their susceptibility to aggregation. Experimental techniques based on electrokinetic and electroacoustic techniques are available for determining the charge on emulsion droplets. The dynamic or equilibrium interfacial tension of an oil–water interface can be determined by means of a number of interfacial tension meters, including the Wilhelmy plate, Du Nouy ring, maximum bubble pressure, and pendant drop methods.

REFERENCES

1. D.J. McClements. *Food Emulsions: Principles, Practice and Techniques*. 2nd edn., CRC, Boca Raton, FL, 2005.
2. S. Friberg and K. Larsson. *Food Emulsions*. 3rd edn., Dekker, New York, NY, 1997.
3. S. Friberg, K. Larsson, and J. Sjoblom. *Food Emulsions*. 4th edn., Dekker, New York, NY, 2004.
4. E. Dickinson and G. Stainsby. *Colloids in Foods*. Applied Science, London, 1982.
5. E. Dickinson. *Introduction to Food Colloids*. Oxford University Press, Oxford, 1992.
6. D.G. Dalgleish. Food emulsions. In: *Emulsions and Emulsion Stability* (J. Sjoblom, ed.). Dekker, New York, NY, 1996.
7. P. Walstra. Disperse systems: basic considerations. In: *Food Chemistry* (O.R. Fennema, ed.). Dekker, New York, NY, 1996, p. 85.
8. H.E. Swaisgood. Characteristics of milk. In: *Food Chemistry* (O.R. Fennema, ed.). Dekker, New York, NY, 1996, p. 841.
9. T.M. Eads. Molecular origins of structure and functionality in foods. *Trends Food Sci. Technol.* 5:147 (1994).
10. J. Israelachvili. The science and applications of emulsions—an overview. *Colloids Surf. A* 91:1 (1994).
11. E. Dickinson and D.J. McClements. *Advances in Food Colloids*. Blackie Academic and Professional, Glasgow, 1995.
12. D.F. Evans and H. Wennerstrom. *The Colloid Domain: Where Physics, Chemistry, Biology and Technology Meet*. VCH Publishers, New York, NY, 1994.
13. J. Israelachvili. *Intermolecular and Surface Forces*. Academic Press, London, 1992.
14. P.C. Hiemenz and R. Rejogopolan. *Principles of Colloid and Surface Science*. 3rd edn., Dekker, New York, NY, 1997.
15. R. Aveyard, B.P. Binks, S. Clark, and P.D.I. Fletcher. Cloud points, solubilization and interfacial tensions in systems containing nonionic surfactants. *Chem. Tech. Biotechnol.* 48:161 (1990).
16. R. Aveyard, B.P. Binks, P. Cooper, and P.D.I. Fletcher. Mixing of oils with surfactant monolayers. *Prog. Colloid Polym. Sci.* 81:36 (1990).
17. P. Becher. Hydrophile–lipophile balance: an updated bibliography. In: *Encyclopedia of Emulsion Technology*, Vol. 2 (P. Becher, ed.). Dekker, New York, 1985, p. 425.
18. P. Becher. HLB: update III. In: *Encyclopedia of Emulsion Technology*, Vol. 4 (P. Becher, ed.). Dekker, New York, NY, 1996.
19. A. Kabalnov and H. Wennerstrom. Macroemulsion stability: the oriented wedge theory revisited. *Langmuir* 12:276 (1996).
20. H.T. Davis. Factors determining emulsion type: hydrophile–lipophile balance and beyond. *Colloids Surf. A* 91:9 (1994).
21. S. Damodaran. Amino acids, peptides and proteins. In: *Food Chemistry* (O.R. Fennema, ed.). Dekker, New York, NY, 1996, p. 321.
22. J.N. BeMiller and R.L. Whistler. Carbohydrates. In: *Food Chemistry* (O.R. Fennema, ed.). Dekker, New York, NY, 1996, p. 157.
23. H. Schubert and H. Armbruster. Principles of formation and stability of emulsions. *Int. Chem. Eng.* 32:14 (1992).
24. H. Karbstein and H. Schubert. Developments in the continuous mechanical production of oil-in-water macroemulsions. *Chem. Eng. Process.* 34:205 (1995).
25. P. Walstra. Formation of emulsions. In: *Encyclopedia of Emulsion Technology*, Vol. 1 (P. Becher, ed.). Dekker, New York, NY, 1983.
26. M. Stang, H. Karbstein, and H. Schubert. Adsorption kinetics of emulsifiers at oil–water interfaces and their effect on mechanical emulsification. *Chem. Eng. Process.* 33:307 (1994).
27. P.J. Fellows. *Food Processing Technology: Principles and Practice*. Ellis Horwood, New York, NY, 1988.
28. W.D. Pandolfe. Effect of premix condition, surfactant concentration, and oil level on the formation of oil-in-water emulsions by homogenization. *J. Dispers. Sci. Technol.* 16:633 (1995).
29. L.W. Phipps. *The High Pressure Dairy Homogenizer*. Technical Bulletin 6, National Institute of Research in Dairying. NIRD, Reading, England, 1985.

30. E.S.R. Gopal. Principles of emulsion formation. In: *Emulsion Science* (P. Sherman, ed.). Academic Press, London and New York, 1968, p. 1.

31. E. Dickinson and G. Stainsby. Emulsion stability. In: *Advances in Food Emulsions and Foams* (E. Dickinson and G. Stainsby, eds.). Elsevier Applied Science, London, 1988, p. 1.

32. K. Kandori. Applications of microporous glass membranes: membrane emulsification. In: *Food Processing: Recent Developments* (A.G. Gaonkar, ed.). Elsevier Science Publishers, Amsterdam, 1995.

33. R.A. Meyers. *Encyclopedia of Analytical Chemistry: Applications, Theory and Instrumentation*. Vol. 6. Wiley, Chichester, UK, 2000.

4 Chemistry of Waxes and Sterols

Edward J. Parish, Shengrong Li, and Angela D. Bell

CONTENTS

I. CHEMISTRY OF WAXES

A. INTRODUCTION

The term waxes commonly refer to the mixtures of long-chain apolar compounds found on the surface of plants and animals. By a strict chemical definition, a wax is the ester of a long-chain acid and a long-chain alcohol. However, this academic definition is much too narrow both for the wax chemist and for the requirements of the industry. The following description from the German Society for Fat Technology [1] better fits the reality:

> Wax is the collective term for a series of natural or synthetically produced substances that normally possess the following properties: kneadable at 20°C, brittle to solid, coarse to finely crystalline,

translucent to opaque, relatively low viscosity even slightly above the melting point, not tending to stinginess, consistency and solubility depending on the temperature and capable of being polished by slight pressure.

The collective properties of wax as just defined clearly distinguish waxes from other articles of commerce. Chemically, waxes constitute a large array of different chemical classes, including hydrocarbons, wax esters, sterol esters, ketones, aldehydes, alcohols, and sterols. The chain length of these compounds may vary from C_2, as in the acetate of a long-chain ester, to C_{62}, as in the case of some hydrocarbons [2,3].

Waxes can be classified according to their origins as naturally occurring or synthetic. The naturally occurring waxes can be subclassified into animal, vegetable, and mineral waxes. Beeswax, spermaceti, wool grease, and lanolin are important animal waxes. Beeswax, wool grease, and lanolin are by-products of other industries. The vegetable waxes include carnauba wax, the so-called queen of waxes, ouricouri (another palm wax), and candelilla. These three waxes account for the major proportion of the consumption of vegetable waxes. The mineral waxes are further classified into the petroleum waxes, ozokerite, and montan. Based on their chemical structure, waxes represent a very broad spectrum of chemical types from polyethylene, polymers of ethylene oxide, derivatives of montan wax, alkyl esters of monocarboxylic acids, alkyl esters of hydroxy acids, polyhydric alcohol esters of hydroxy acids, Fisher–Tropsch waxes, and hydrogenated waxes, to long-chain amide waxes.

We begin with an overview of the diverse class of lipids known as waxes. The discussion presented that follows, which touches on source, structure, function, and biosynthesis, is intended to serve as an entry to the literature, enabling the reader to pursue this topic in greater detail.

B. Properties and Characteristics of Waxes

The ancient Egyptians used beeswax to make writing tablets and models, and waxes are now described as man's first plastic. Indeed, the plastic property of waxes and cold-flow yield values allow manual working at room temperature, corresponding to the practices of the Egyptians. The melting points of waxes usually vary within the range 40°C–120°C.

Waxes dissolve in fat solvents, and their solubility is dependent on temperature. They can also wet and disperse pigments, and can be emulsified with water, which makes them useful in the furniture, pharmaceutical, and food industries. Their combustibility, associated with low ash content, is important in candle manufacture and solid fuel preparation. Waxes also find application in industry as lubricants and insulators, where their properties as natural plastics, their high flash points, and their high dielectric constants are advantageous.

The physical and technical properties of waxes depend more on molecular structure than on molecular size and chemical constitution. The chemical components of waxes range from hydrocarbons, esters, ketones, aldehydes, and alcohols to acids, mostly as aliphatic long-chain molecules. The hydrocarbons in petroleum waxes are mainly alkanes, though some unsaturated and branched chain compounds are found. The common esters are those of saturated acids with 12–28 carbon atoms combining with saturated alcohols of similar chain length. Primary alcohols, acids, and esters have been characterized and have been found to contain an even straight chain of carbon atoms. By contrast, most ketones, secondary alcohols, and hydrocarbons have odd number of carbon atoms. The chemical constitution of waxes varies in great degree depending on the origin of the material. A high proportion of cholesterol and lanosterol is found in wool wax. Commercial waxes are characterized by a number of properties. These properties are used in wax grading [4].

1. Physical Properties of Waxes

Color and odor are determined by comparison with standard samples in a molten state. In the National Petroleum Association scale, the palest color is rated 0, whereas amber colors are rated 8.

Refined waxes are usually free from taste, this property being especially important in products such as candelilla when it is used in chewing gum. Melting and softening points are important physical properties. The melting points can be determined by the capillary tube method or the drop point method. The softening point of a wax is the temperature at which the solid wax begins to soften. The penetration property measures the depth to which a needle with a definite top load penetrates the wax sample.

Shrinkage and flash point are two frequently measured physical properties of waxes. The flash point is the temperature at which a flash occurs if a small flame is passed over the surface of the sample. In the liquid state, a molten wax shrinks uniformly until the temperature approaches the solidification point. This property is measured as the percentage shrinkage of the volume.

2. Chemical Properties of Waxes

a. Acid Value
The acid value is the number of milligrams of potassium hydroxide required to neutralize a gram of the wax. It is determined by the titration of the wax solution in ethanol–toluene with 0.5 M potassium hydroxide. Phenolphthalein is normally used as the titration indicator.

$$\text{Acid value} = \frac{V_w \times 56.104}{w},$$

where
  V_w is the number of milliliters (mL) of potassium hydroxide used in the titration
  w is the mass of wax

b. Saponification Number
The saponification number is the number of milligrams of potassium hydroxide required to hydrolyze 1 g of wax:

$$\text{Saponification number} = \frac{(V_b - V_w) \times 56.105}{w},$$

where
  w is the weight of wax samples
  V_b the volume (mL) of hydrochloric acid used in the blank
  V_w the volume (mL) of hydrochloric acid used in the actual analysis

The wax (2 g) is dissolved in hot toluene (910 mL). Alcoholic potassium hydroxide (25 mL of 0.5 M KOH) is added, and the solution is refluxed for 2 h. A few drops of phenolphthalein are added and the residual potassium hydroxide is titrated with 0.5 M hydrochloric acid. A blank titration is also performed with 25 mL of 0.5 M alcoholic potassium hydroxide plus toluene.

c. Ester Value
Ester value, the difference between the saponification number and the acid value, shows the amount of potassium hydroxide consumed in the saponification of esters.

d. Iodine Number
The iodine number expresses the amount of iodine that is absorbed by the wax. It is a measure of the degree of unsaturation.

e. Acetyl Number
The acetyl number indicates the milligrams of potassium hydroxide required for the saponification of the acetyl group assimilated in 1 g of wax on acetylation. The difference of this number and the ester value reflects the amount of free hydroxy groups (or alcohol composition) in a wax. The wax

sample is first acetylated by acetic anhydride. A certain amount of acetylated wax (about 2 g) is taken out to be saponified with the standard procedure in the measurement of the saponification number. The acetyl number is the saponification number of the acetylated wax.

3. Properties of Important Naturally Occurring Waxes

a. Beeswax

Beeswax is a hard amorphous solid, usually light yellow to amber depending on the source and manufacturing process. It has a high solubility in warm benzene, toluene, chloroform, and other polar organic solvents. Typically, beeswax has an acid value of 17–36, a saponification number of 90–147, melting point of 60°C–67°C, an ester number of 64–84, a specific gravity of 0.927–0.970, and an iodine number of 7–16. Pure beeswax consists of about 70%–80% of long-chain esters, 12%–15% of free acids, 10%–15% of hydrocarbon, and small amounts of diols and cholesterol esters. Beeswax is one of the most useful and valuable of waxes. Its consumption is not limited to the candle industry, the oldest field of wax consumption. It is also used in electrical insulation and in the food, paper, and rubber industries.

b. Wool Grease and Lanolin

Wool grease is a by-product of the wool industry, and the finest wool grease yields lanolin. Pharmaceutical grade lanolin accounts for about 80% of all wool grease consumption. Wool grease has a melting point of 35°C–42°C, an acid value of 7–15, a saponification value of 100–110, an ester value of 85–100, a specific gravity of 0.932–0.945, and an iodine value of 22–30.

c. Carnauba Wax

Carnauba wax, queen of waxes, is a vegetable wax produced in Brazil. Carnauba wax is hard, amorphous, and tough, with a pleasant smell. It is usually used in cosmetics and by the food industry, in paper coatings, and in making inks. In the food industry, it is a minor component in glazes for candies, gums, and fruit coatings. Carnauba wax is soluble in most polar organic solvents. It contains esters (84%–85%), free acids (3%–3.5%), resins (4%–6%), alcohols (2%–3%), and hydrocarbons (1.5%–3.0%). Typically, carnauba has an acid value of 2.9–9.7, an ester value of 39–55, a saponification value of 79–95, an iodine value of 7–14, and a melting range of 78°C–85°C.

d. Candelilla Wax

Candelilla wax is a vegetable wax produced mainly in Mexico. It is used chiefly in the manufacturing of chewing gum and cosmetics, which represent about 40% of the market. It is also used in furniture polish, in the production of lubricants, and in paper coating. Candelilla wax has a specific gravity of 0.98, an acid value of 12–22, a saponification value of 43–65, a melting point of 66°C–71°C, an ester value of 65–75, and an iodine value of 12–22. The chemical composition of candelilla wax is 28%–29% esters, 50%–51% hydrocarbons, 7%–9% free acids, and small amounts of alcohols and cholesterols.

e. Ozocerite

Ozocerite is a mineral wax found in Galicia, Russia, Iran, and the United States. Most ozocerite consists of hydrocarbons, but the chemical composition varies with the source. Typically, ozocerite has an ester value of 56–66, an acid value of 31–38, a saponification value of 87–104, a melting point of 93°C–89°C, and an iodine value of 14–18. Ozocerite is graded as unbleached (black), single bleached (yellow), and double bleached (white). It is mainly used in making lubricants, lipsticks, polishes, and adhesives.

C. Isolation, Separation, and Analysis of Natural Waxes

Knowledge of the chemical analysis of natural waxes is essential for understanding wax biosynthesis, manufacture, and application. Although the chemical compositions of synthetic waxes are constant and depend on the manufacturing process, the natural waxes are much more complicated in chemical composition. In general, natural waxes are isolated by chemical extraction, separated by chromatographic methods, and analyzed by means of mass spectrometry (MS); both gas chromatography (GC) and high-performance liquid chromatography (HPLC) techniques are used. The following discussion on chemical analysis is based on an understanding of the general principles of chemical extraction, chromatography, and MS. There are numerous textbooks detailing these principles [5–7].

1. Isolation

Natural waxes are mixtures of long-chain apolar compounds found on the surface of plants and animals. However, internal lipids also exist in most organisms. In earlier times, the plant or animal tissue was dried, whereupon the total lipid material could be extracted with hexane or chloroform by means of a Soxhlet extractor. The time of exposure to the organic solvent, particularly chloroform, is kept short to minimize or avoid the extraction of internal lipids. Because processors are interested in surface waxes, it became routine to harvest them by a dipping procedure. For plants this was usually done in the cold, but occasionally at the boiling point of light petroleum or by swabbing to remove surface lipids. Chloroform, which has been widely used, is now known to be toxic; dichloromethane can be substituted. After removal of the solvent under vacuum, the residue can be weighed. Alternatively, the efficiency of the extraction can be determined by adding a known quantity of a standard wax component (not present naturally in the sample) and performing a quantification based on this component following column chromatography.

2. Separation

The extract of surface lipids contains hydrocarbons, as well as long-chain alcohols, aldehydes and ketones, short-chain acid esters of the long-chain alcohols, fatty acids, sterols and sterol esters, and oxygenated forms of these compounds. In most cases, it is necessary to separate the lipid extract into lipid classes before the identification of components. Separation of waxes into their component classes is first achieved by column chromatography. The extract residue is redissolved in the least polar solvent possible, usually hexane or light petroleum, and transferred to the chromatographic column. When the residue is not soluble in hexane or light petroleum, a hot solution or a more polar solvent, like chloroform of dichloromethane, may be used to load the column. By gradually increasing the polarity of the eluting solvent, it is possible to obtain hydrocarbons, esters, aldehydes and ketones, triglycerides, alcohols, hydroxydiketones, sterols, and fatty acids separately from the column. Most separations have been achieved on alumina or silica gel. However, Sephadex LH-20 was used to separate the alkanes from Green River Shale. Linde 5Å sieve can remove the *n*-alkanes to provide concentrated branched and alicyclic hydrocarbons. Additionally, silver nitrate can be impregnated into alumina or silica gel columns or thin-layer chromatography (TLC) plates for separating components according to the degree of unsaturation.

As the means of further identifying lipids become more sophisticated, it is possible to obtain a sufficient quantity of the separated wax components by TLC. One of the major advantages of TLC is that it can be modified very easily, and minor changes to the system have allowed major changes in separation to be achieved. Most components of wax esters can be partially or completely separated by TLC on 25 μm silica gel G plates developed in hexane–diethyl ether or benzene–hexane. The retardation factor (R_f) values of most wax components are listed in Table 4.1 [8].

If TLC is used, the components must be visualized, and the methods employed can be either destructive or nondestructive. The commonly used destructive method is to spray TLC plates with sulfuric or molybdic acid in ethanol and heat them. This technique is very sensitive, but it destroys the compounds and does not work well with free acids. Iodine vapors will cause a colored band to appear, particularly with unsaturated compounds, and are widely used to both locate and quantify the lipids. Since the iodine can evaporate from the plate readily after removal from iodine chamber, the components usually remain unchanged. Iodine vapor is one of the ideal visualization media in

TABLE 4.1

TLC Separation of Wax Components on Silica Gel: R_f Values for Common Wax Components

Component	\multicolumn{8}{c}{Solvent Systems}
	A	B	C	D	E	F	G	H
Hydrocarbon	0.95	0.96	0.95	0.85	0.83	0.95	0.85	
Squalene							0.80	
Trialkylglyceryl ethers	0.90							
Steryl esters	0.90					0.95	0.57	
Wax esters	0.90	0.82	0.84	0.71	0.65	0.91	0.75	
β-Diketones		0.75	0.54					
Monoketones					0.53			
Fatty acid methyl esters	0.65				0.47	0.75		
Aldehydes	0.55	0.65		0.47				0.66
Triterpenyl acetates							0.53	
Secondary alcohols				0.36				
Triacylglycerols	0.35				0.61	0.37		
Free fatty acids	0.18	0.00	0.00			0.35	0.20	
Triterphenols								0.22
Primary alcohols	0.15	0.14	0.16	0.09	0.15	0.21		0.19
Sterols	0.10				0.16	0.10	0.12	
Hydroxy-β-diketones	0.09	0.04						
Triterpenoid acid							0.05	

Note: A, petroleum ether (b.p. 60°C–70°C)–diethyl ether–glacial acetic acid (90:10:1, v/v); B, benzene; C, chloroform containing 1% ethanol; D, petroleum ether (b.p. 40°C–60°C)–diethyl ether (80:20, v/v); E, chloroform containing 1% ethanol; F, hexane–heptane–diethyl ether–glacial acetic acid (63:18.5:18.5, v/v) to 2 cm from top, then full development with carbon tetrachloride; G (1) petroleum ether–diethyl ether–glacial acetic acid (80:20:1, v/v); (2) petroleum ether; H, benzene–chloroform (70:30 v/v).

the isolation of lipid classes from TLC plates. Commercial TLC plates with fluorescent indicators are available as well, and bands can be visualized under UV light. However, if it is necessary to use solvents more polar than diethyl ether to extract polar components from the matrix, the fluorescent indicators may also be extracted, and these additives interfere with subsequent analyses.

To isolate lipid classes from TLC plates after a nondestructive method of visualization, the silica gel can be scraped into a champagne funnel and eluted with an appropriate solvent. Or, the gel can be scraped into a test tube and the apolar lipid extracted with diethyl ether by vortexing, centrifuging, and decanting off the ether. Polar lipids are extracted in the same manner, using a more polar solvent such as chloroform and methanol. HPLC has been used in the separation and analysis of natural waxes, but its application was halted by the lack of a suitable detector, since most wax components have no useful UV chromophore. Application of UV detection was limited to wavelength around 210 nm. Some components with isolated double bonds and carbonyl group (e.g., esters, aldehydes, ketones) can be detected in this wavelength. Hamilton and coworkers have examined an alternative detection system, infrared detection at 5.74 μm, which allowed the hydrocarbon components to be detected [9]. Although the sensitivity of this method of detection could not match that of UV detection, it has merit for use in the preparative mode, where it is feasible to allow the whole output from the column to flow through the detector. The third useful mode for HPLC is MS. The coupling of HPLC and MS makes this form of chromatography a very important analytical technique.

3. Analysis

When individual classes of waxes have been isolated, the identity of each must be determined. Due to the complex composition of these materials, combined analytical approaches (e.g., GC–MS) have been used to analyze individual wax classes. MS is a major analytical method for the analysis of this class of compounds. With the electron impact–mass spectrometry (EI–MS), the wax molecules tend to give cleavage fragments rather than parent ions. Thus, soft (chemical) ionization (CI) and fast atom bombardment (FAB) have been frequently used to give additional information for wax analysis.

In GC–MS analysis, the hydrocarbon fraction and many components of the wax ester fraction can be analyzed directly, whereas long-chain alcohols, the aldehydes, and fatty acids are often analyzed as their acetate esters of alcohols, dimethylhydrazones of aldehydes, and methyl ethers of fatty acids. The analysis of wax esters after hydrolysis and derivatization will provide additional information on high molecular weight esters. For example, the chain branching of a certain component might be primarily examined with respect to its unusual retention time on GC analysis, then determined by converting to the corresponding hydrocarbon through the reduction of its iodide intermediate with $LiAID_4$ (the functional group end is labeled by the deuterium atom). A similar approach is to convert the alcohol of the target component to an alkyl chloride via methanesulfonyl chloride. This method labels the functional end with a chlorine atom, and its mass spectra are easily interpreted because of the chlorine isotopes. As mentioned earlier, unsaturated hydrocarbons can be separated from saturated hydrocarbons and unsaturated isomers by column chromatography or TLC with silver nitrate silica gel or alumina gel media. The position and number of double bonds affect the volatility of the hydrocarbons, thereby altering their retention in GC and HPLC analysis. The location of a double bond is based on the mass spectra of their derivatives, using either positive or negative CI.

D. BIOSYNTHESIS OF NATURAL WAXES

Epicuticular waxes (from the outermost layer of plant and insect cuticles) comprise very long chain nonpolar lipid molecules that are soluble in organic solvents. In many cases, this lipid layer may contain proteins and pigments, and great variability in molecular architecture is possible, depending on the chemical composition of the wax and on environmental factors [10,11].

A variety of waxes can be found in the cuticle. On the outer surface of plants these intracuticular waxes entrap cutin, which is an insoluble lipid polymer of hydroxy and epoxy fatty acids. In underlying layers, associated with the suberin matrix, another cutin-like lipid polymer containing aliphatic and aromatic components is found [12]. In some instances, internal nonsuberin waxes, which are stored in plant seeds, are the major energy reserves rather than triacylglycerols. In insects, intracuticular waxes are the major constituents of the inner epicuticular layer [13–15].

A variety of aliphatic lipid classes occur in epicuticular waxes. These include hydrocarbons, alcohols, esters, ketones, aldehydes, and free fatty acids of numerous types [16,17]. Frequently, a series of 10 carbon atom homologs occurs, while chain lengths of 10–35 carbon atoms are most often found. However, fatty acids and hydrocarbons with fewer than 20 carbon atoms are known, as are esters with more than 60 carbon atoms. Other minor lipids such as terpenoids, flavonoids, and sterols also occur in epicuticular waxes. The composition and quantity of epicuticular wax vary widely from one species to another and from one organ, tissue, or cell type to another [16]. In insects, wax composition depends on stage of life cycle, age, sex, and external environment [17].

In waxes, the biosynthesis of long-chain carbon skeletons is accomplished by a basic condensation–elongation mechanism. Elongases are enzyme complexes that repetitively condense short activated carbon chains to an activated primer and prepare the growing chain for the next addition. The coordinated action of two such soluble complexes in plastid results in the synthesis of the 16- and 18-carbon acyl chains characterizing plant membranes [18–20]. Each condensation introduces a β-keto group into the elongating chain. This keto group is normally removed by a series of three reactions: a β-keto reduction, a β-hydroxy dehydration, and an enol reduction.

Variations of the foregoing basic biosynthetic mechanism occur, giving rise to compounds classified as polyketides. Their modified acyl chains can be recognized by the presence of keto groups, hydroxy groups, or double bonds that were not removed before the next condensation took place. It is well established that the very long carbon skeletons of the wax lipids are synthesized by a condensation–elongation mechanism. The primary elongated products in the form of free fatty acids are often minor components of epicuticular waxes. Most of them, however, serve as substrates for the associated enzyme systems discussed. The total length attained during elongation is reflected by the chain lengths of the members of the various wax classes [15–21]. Normal, branched, and unsaturated hydrocarbons and fatty acids are prominent components of plant waxes, whereas insect waxes usually lack long-chain free fatty acids [22–26].

II. CHEMISTRY OF STEROLS

A. INTRODUCTION

Sterols constitute a large group of compounds with a broad range of biological activities and physical properties. The natural occurring sterols usually possess the 1,2-cyclopentano-phenanthrene skeleton with a stereochemistry similar to the *trans-syn-trans-anti-trans-anti* configuration at their ring junctions, and have 27–30 carbon atoms with an hydroxy group at C-3 and a side chain of at least seven carbons at C-17 (Figure 4.1). Sterols can exhibit both nuclear variations (differences within the ring system) and side chain variations. The examples of the three subclasses of sterols in Figure 4.1 represent the major variations of sterols. Sterols have been defined as hydroxylated steroids that retain some or all of the carbon atoms of squalene in the side chain and partition almost completely into an ether layer when shaken with equal volumes of water and ether [27].

Sterols are common in eukaryotic cells but rare in prokaryotes. Without exception, vertebrates confine their sterol biosynthetic activity to producing cholesterol. Most invertebrates do not have the enzymatic machinery for sterol biosynthesis and must rely on an outside supply. Sterols of invertebrates have been found to comprise most complex mixtures arising through food chains. In plants, cholesterol exists only as a minor component. Sitosterol and stigmasterol are the most abundant and widely distributed plant sterols, whereas ergosterol is the major occurring sterol in fungus and yeast. The plant sterols are characterized by an additional alkyl group at C-24 on the cholesterol nucleus with either α or β chirality. Sterols with methylene and ethylidene substitutes are also found in plants (e.g., 24-methylene cholesterol, fucosterol). The other major characteristics

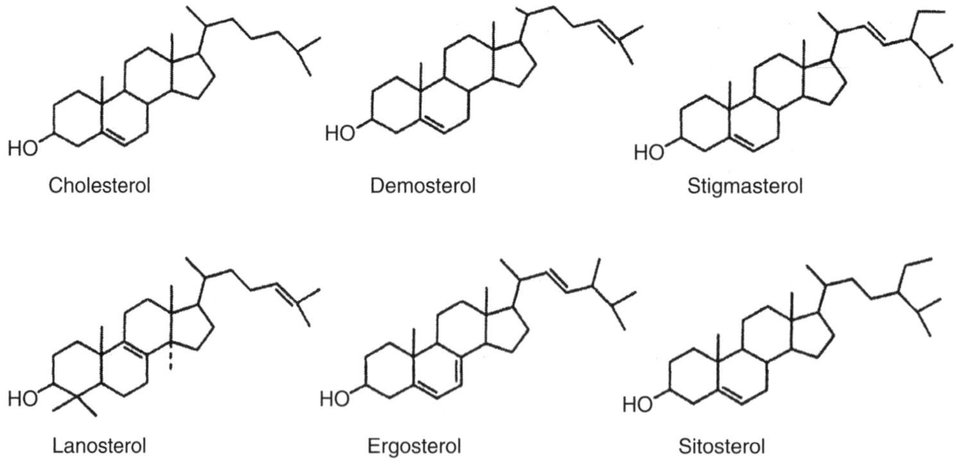

| Cholesterol | Demosterol | Stigmasterol |
| Lanosterol | Ergosterol | Sitosterol |

FIGURE 4.1 Examples of naturally occurring sterols.

of plant sterols are the presence of additional double bonds in the side chain, as in porifeasterol, cyclosadol, and closterol.

Despite the diversity of plant sterols and sterols of invertebrates, cholesterol is considered the most important sterol. Cholesterol is an important structural component of cell membranes and is also the precursor of bile acids and steroid hormones [28]. Cholesterol and its metabolism are of importance in human disease. Abnormalities in the biosynthesis or metabolism of cholesterol and bile acid are associated with cardiovascular disease and gallstone formation [29,30]. Our discussion will mainly focus on cholesterol and its metabolites, with a brief comparison of the biosynthesis of cholesterol and plant sterols (see Section II.B.2). The biosynthesis of plant sterols and sterols of invertebrates was reviewed by Goodwin [31] and Ikekawa [32].

The chemistry of sterols encompasses a large amount of knowledge relating to the chemical properties, chemical synthesis, and analysis of sterols. A detailed discussion on all these topics is impossible in one chapter. We consider the analysis of sterols to be of primary interest, and therefore our treatment of the chemistry of sterols is confined to the isolation, purification, and characterization of sterols from various sources. Readers interested in the chemical reactions and total syntheses of sterols may refer to the monographs in these areas [33–35].

B. BIOSYNTHETIC ORIGINS OF STEROLS

1. Cholesterol Biosynthesis

Cholesterol is the principal mammalian sterol and the steroid that modulates the fluidity of eukaryotic membranes. Cholesterol is also the precursor of steroid hormones such as progesterone, testosterone, estradiol, cortisol, and vitamin D. The elucidation of the cholesterol biosynthesis pathway has challenged the ingenuity of chemists for many years. The early work of Konrad Bloch in the 1940s showed that cholesterol is synthesized from acetyl coenzyme A (acetyl CoA) [36]. Acetate isotopically labeled in its carbon atoms was prepared and fed to rats. The cholesterol that was synthesized by these rats contained the isotopic label, which showed that acetate is a precursor of cholesterol. In fact, all 27 carbon atoms of cholesterol are derived from acetyl CoA. Since then, many chemists have put forward enormous efforts to elucidate this biosynthetic pathway, and this work has yielded our present detailed knowledge of sterol biosynthesis. This outstanding scientific endeavor was recognized by the awarding of several Nobel prizes to investigators in research areas related to sterol [1].

The cholesterol biosynthetic pathway can be generally divided into four stages: (1) the formation of mevalonic acid from three molecules of acetyl CoA, (2) the biosynthesis of squalene from six molecules and mevalonic acid through a series of phosphorylated intermediates, (3) the biosynthesis of lanosterol from squalene via cyclization of 2,3-epoxysqualene, and (4) the modification of lanosterol to produce cholesterol.

The first stage in the synthesis of cholesterol is the formation of mevalonic acid and isopentyl pyrophosphate from acetyl CoA. Three molecules of acetyl CoA are combined to produce mevalonic acid, as shown in Scheme 4.1. The first step of this synthesis is catalyzed by a thiolase enzyme and results in the production of acetoacetyl CoA, which is then combined with the third molecule of acetyl CoA by the action of 3-hydroxy-3-methylglutaryl CoA (HMG-CoA) is cleaved to acetyl CoA and acetoacetate. Acetoacetate is further reduced to d-3-hydroxybutyrate in the mitochondrial matrix. Since it is a β-keto acid, acetoacetate also undergoes a slow, spontaneous decarboxylation to acetone. Acetoacetate, d-3-hydroxybutyrate, and acetone, sometime referred to as ketone bodies, occur in fasting or diabetic individuals. Alternatively, HMG-CoA can be reduced to mevalonate and is present in both the cytosol and the mitochondria of liver cells. The mitochondrial pool of this intermediate is mainly a precursor of ketone bodies, whereas the cytoplasmic pool gives rise to mevalonate for the biosynthesis of cholesterol.

The reduction of HMG-CoA to give the mevalonic acid is catalyzed by a microsomal enzyme, HMG-CoA reductase, which is of prime importance in the control of cholesterol biosynthesis.

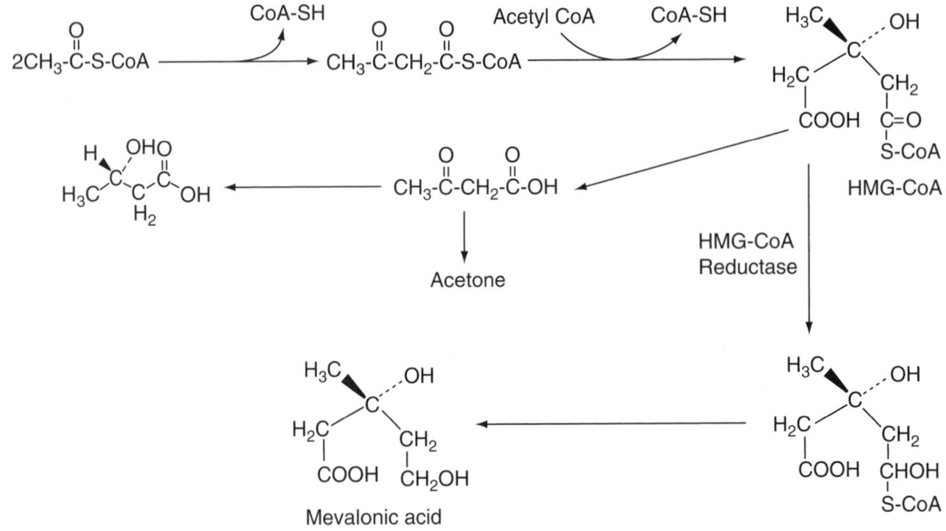

SCHEME 4.1 Synthesis of mevalonic acid from acetyl CoA.

The biomedical reduction of HMG-CoA is an essential step in cholesterol biosynthesis. The reduction of HMG-CoA is irreversible and proceeds in two steps, each requiring NADPH as the reducing reagent. A hemithioacetal derivative of mevalonic acid is considered to be an intermediate. The concentration of HMG-CoA reductase is determined by rates of its synthesis and degradation, which are, in turn, regulated by the amount of cholesterol in the cell. Cholesterol content is influenced by the rate of biosynthesis, dietary uptake, and a lipoprotein system that traffics in the intercellular movement of cholesterol. During growth, cholesterol is mainly incorporated into the cell membrane. However, in homeostasis, cholesterol is mainly converted to bile acids and is transported to other tissues via low density lipoprotein (LDL). High density lipoprotein (HDL) also serves as a cholesterol carrier, which carries cholesterol from peripheral tissues to the liver. The major metabolic route of cholesterol is its conversion to bile acids and neutral sterols, which are excreted from the liver via the bile. Kandutsch and Chen have shown that oxysterols regulate the biosynthesis of HMG-CoA reductase as well as its digression, which controls cholesterol bio-synthesis [37]; the regulation of HMG-CoA reductase by oxysterols is discussed in more detail in a later section. A number of substrate analogs have been tested for their inhibition of HMG-CoA reductase. Some of them (e.g., compactin and melinolin) were found to be very effective in treating hypocholesterol diseases [38,39].

    The coupling of six molecules of mevalonic acid to produce squalene proceeds through a series of phosphorylated compounds. Mevalonate is first phosphorylated by mevalonic kinase to form a 5-phosphomevalonate, which serves as the substrate for the second phosphorylation to form 5-pyrophosphomevalonate (Scheme 4.2). There is then a concerted decarboxylation and loss of a tertiary hydroxy group from 5-pyrophosphomevalonate to form 3-isopentyl pyrophosphate, and in each step one molecule of ATP must be consumed. 3-Isopentyl pyrophosphate is regarded as the basic biological isoprene unit from which all isoprenoid compounds are elaborated. Squalene is synthesized from isopentyl pyrophosphate by sequence coupling reactions. This stage in the cholesterol biosynthesis starts with the isomerization of isopentyl pyrophosphate to dimethylallyl pyrophosphate. The coupling reaction shown in Scheme 4.2 is catalyzed by a soluble sulfydryl enzyme, isopentyl pyrophosphate–dimethylallyl pyrophosphate isomerase. The coupling of these two isomeric C5 units yields geranyl pyrophosphate, which is catalyzed by geranyl pyrophosphate synthetase (Scheme 4.3). This reaction proceeds by the head-to-tail joining of isopentyl pyrophosphate to

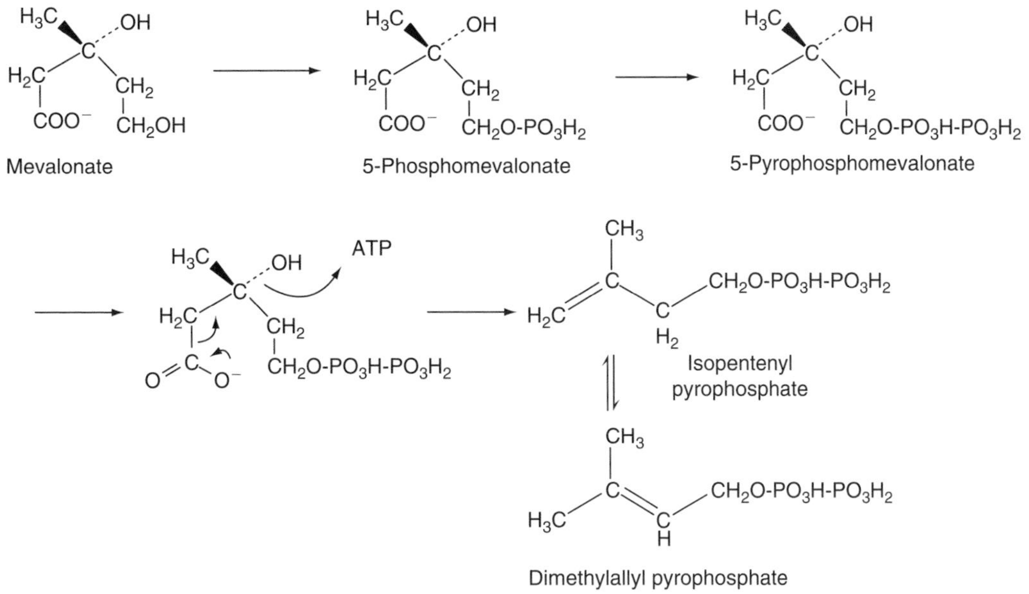

SCHEME 4.2 Synthesis of isopentenyl pyrophosphate, the biological isoprene unit, and dimethylallyl pyrophosphate.

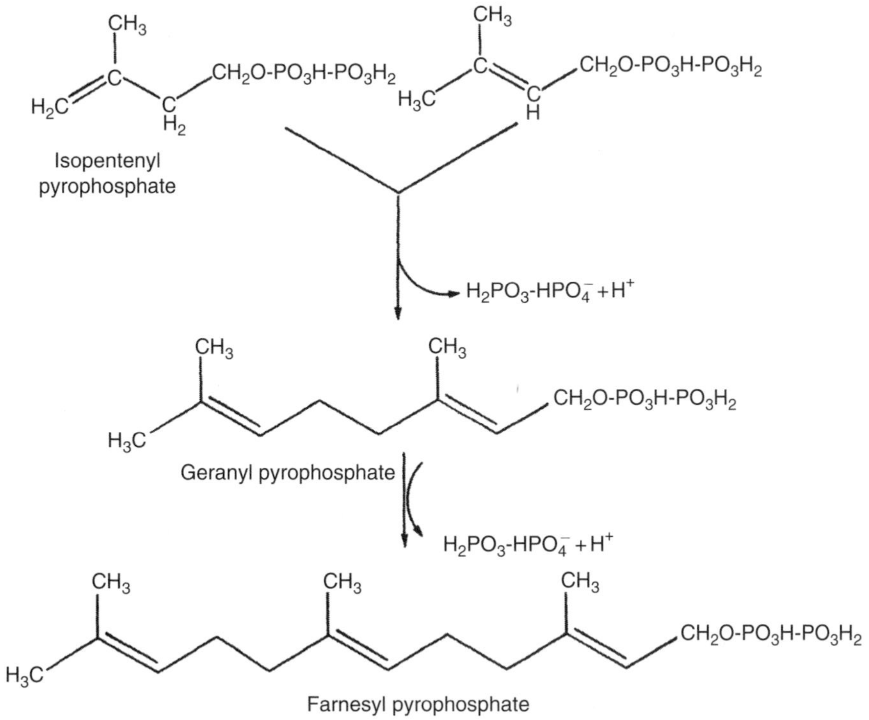

SCHEME 4.3 Synthesis of farnesyl pyrophosphate from the biological isoprenyl unit.

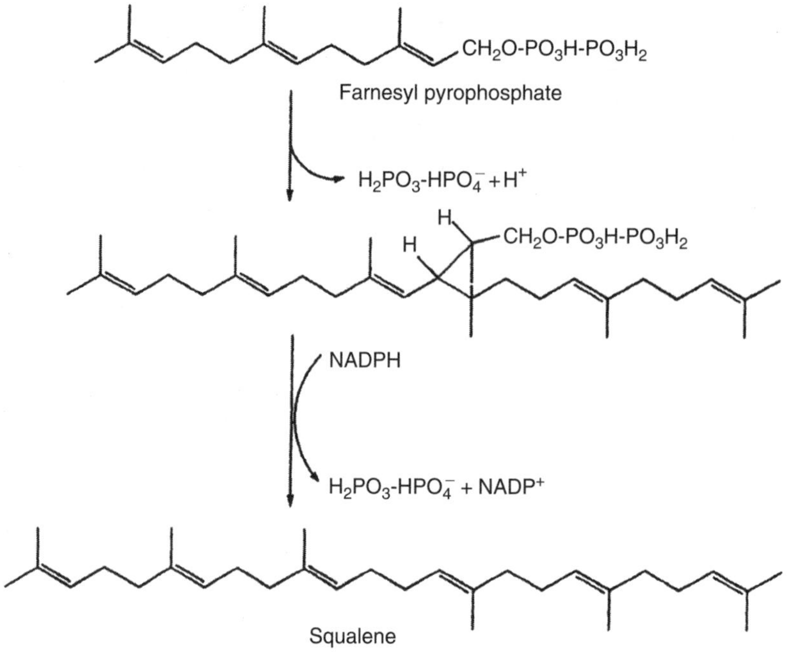

SCHEME 4.4 Synthesis of squalene from the coupling of two molecules of farnesyl pyrophosphate.

dimethylallyl pyrophosphate. A new carbon–carbon bond is formed between the C-1 of dimethy-lallyl pyrophosphate and C-4 of isopentyl pyrophosphate. Consequently, geranyl pyrophosphate can couple in a similar manner with a second molecule of isopentyl pyrophosphate to produce farnesyl pyrosphate (C15 structure). The last step in the synthesis of squalene is a reductive condensation of two molecules of farnesyl pyrophosphate (Scheme 4.4). This step is actually a two-step sequence, catalyzed by squalene synthetase. In the first reaction, presqualene pyrophosphate is produced by a tail-to-tail coupling of two farnesyl pyrophosphate molecules. In the following conversion of presqualene pyrophosphate to squalene, the cyclopropane ring of presqualene pyrophosphate is opened with a loss of the pyrophosphate moiety. A molecule of NADPH is required in the second conversion.

The third stage of cholesterol biosynthesis is the cyclization of squalene to lanosterol (Scheme 4.5). Squalene cyclization proceeds in two steps requiring molecular oxygen, NADPH, squalene epoxidase, and 2,3-oxidosqualene–sterol cyclase. The first step is the epoxidation of squalene to form 2,3-oxidosqualene–sterol cyclase. The 2,3-oxidosqualene is oriented as a chair–boat–chair–boat conformation in the enzyme active center. The acid-catalyzed epoxide ring opening initiates the cyclization to produce a tetracyclic protosterol cation. This is followed by a series of concerted 1,2-*trans* migrations of hydrogen and methyl groups to produce lanosterol.

The last stage of cholesterol biosynthesis is the metabolism of lanosterol to cholesterol. Scheme 4.6 gives the general biosynthetic pathway from lanosterol to cholesterol. The C-14 methyl group is first oxidized to an aldehyde, and removed as formic acid. The oxidation of the C-4α methyl group leads to an intermediate, 3-oxo-4α-carboxylic acid, which undergoes a decarboxylation to form 3-oxo-4β-3-methylsterol. This compound is then reduced by an NADPH-dependent microsomal 3-oxosteroid reductase to produce 3β-hydroxy-4α-methyl sterol, which undergoes a similar series of reactions to produce a 4,4-dimethylsterol. In animal tissues, C-14 demethylation and the subsequent double-bond modification are independent of the reduction of the Δ^{24} double bond. Desmosterol (cholesta-5,24-dien-3β-ol) is found in animal tissues and can serve as a cholesterol precursor. The double-bond isomerization of 8 to 5 involves the pathway $\Delta^8 \rightarrow \Delta^7 \rightarrow \Delta^{5.7} \rightarrow \Delta^5$.

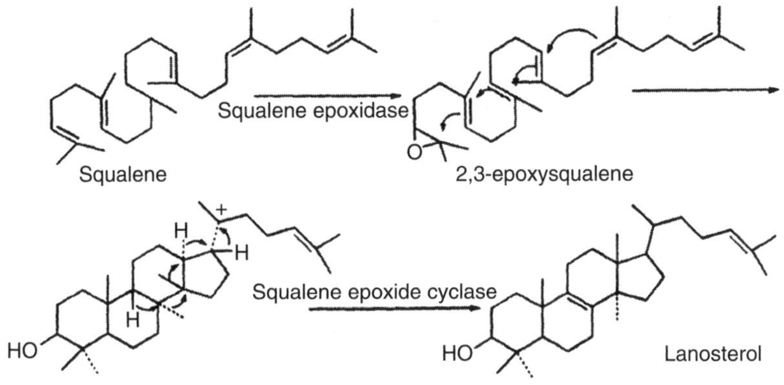

SCHEME 4.5 Cyclization of squalene.

2. Biosynthesis of Plant Sterols

In animals, 2,3-oxidosqualene is first converted to lanosterol through a concerted cyclization reaction. This reaction also occurs in yeast. However, in higher plants and algae the first cyclic product is cycloartenol (Scheme 4.7). The cyclization intermediate, tetracyclic protosterol cation,

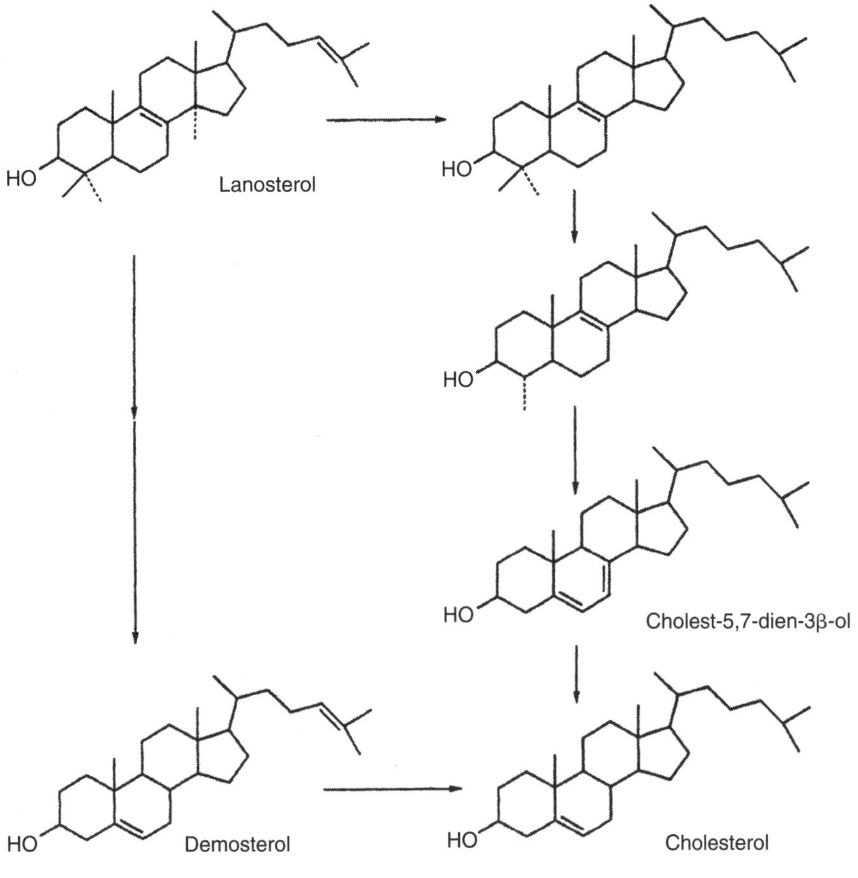

SCHEME 4.6 Biosynthesis of cholesterol from lanosterol.

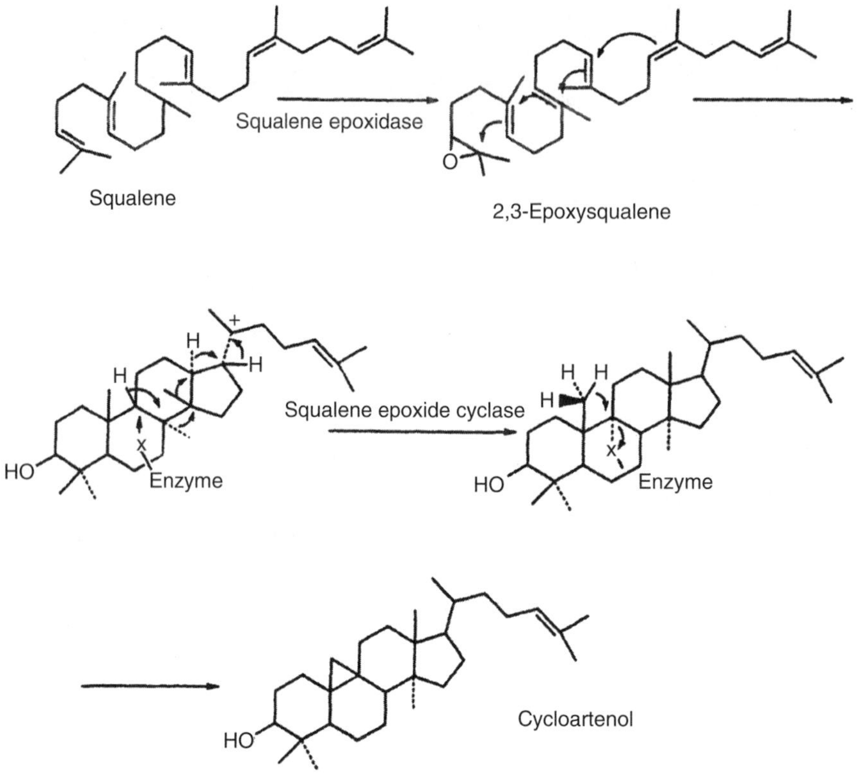

SCHEME 4.7 Cyclization of squalene to cycloartenol.

undergoes a different series of concerted 1,2-*trans* migrations of hydrogen and methyl groups. Instead of the 8,9 double bond, a stabilized C-9 cation intermediate is formed. Following a *trans* elimination of enzyme-X⁻ and H⁺ from C-19, with the concomitant formation of the 9,19-cyclopropane ring, cycloartenol is formed. A nearby α-face nucleophile from the enzyme is necessary to stabilize the C-9 cation and allow the final step to be a *trans* elimination according to the isoprene rule. The biosynthesis pathway from acetyl CoA to 2,3-oxisqualene in plants is the same as that in animals (see detailed discussion of the biosynthesis of cholesterol in Section II.B.1).

The conversion of cycloartenol to other plant sterols can be generally divided into three steps, which are the alkylation of the side chain at C-24, demethylation of the C-4 and C-14 methyl groups, and double-bond manipulation. Alkylation in the formation of plant sterols involves methylation at C-24 with *S*-adenosylmethionine to produce C28 sterols. The further methylation of a C-24 methylene substrate yields C-24 ethyl sterols. The details of the mechanism of demethylation and double-bond manipulation in plants are not clear, but it is highly likely to be very similar to that in animals. In plants, C-4 methyl groups are removed before the methyl group at C-14, whereas in animals it is the other way around. Sterols found in plants are very diversified. The structural features of major plant sterols are depicted in Figure 4.2.

C. REGULATION OF STEROL BIOSYNTHESIS IN ANIMALS

Sterol biosynthesis in mammalian systems has been intensely studied for several decades. Interest in the cholesterol biosynthesis pathway increased following clinical observations that the incidence of cardiovascular disease is greater in individuals with levels of serum cholesterol higher than average. More recently, the results of numerous clinical studies have indicated that lowering serum

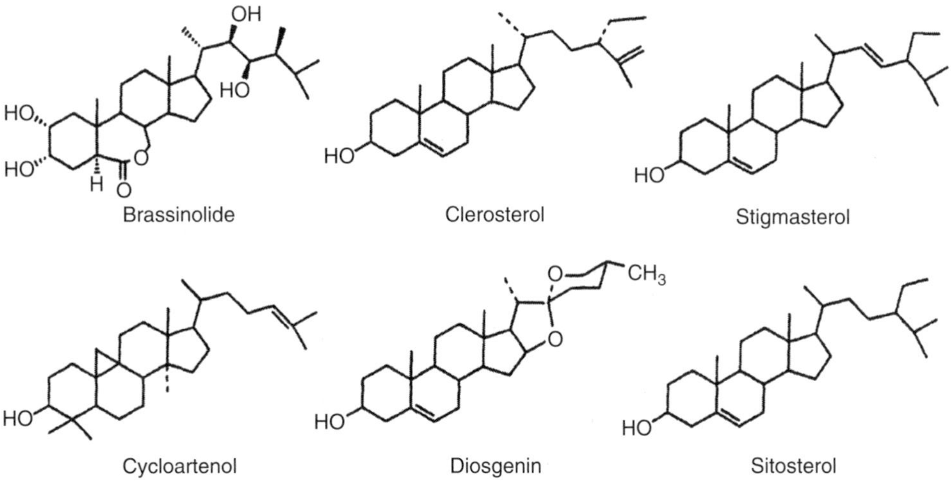

FIGURE 4.2 Examples of plant sterols.

cholesterol levels may reduce the risk of coronary heart disease and promote the regression of atherosclerotic lesions [40,41]. The total exchangeable cholesterol of the human body is estimated at 60 g for a 60 kg man. The cholesterol turnover rate is in the order of 0.8–1.4 g/day. Both the dietary uptake and the biosynthesis contribute significantly to body cholesterol. In Western countries, the daily ingestion of cholesterol ranges from 0.5 to 3.0 g, although only a portion of this sterol is absorbed from the intestine. The absorption of dietary cholesterol ranges from 25% (high dietary sterol intake) to 50% (low dietary sterol intake). The total body cholesterol level is determined by the interaction of dietary cholesterol, the excretion of cholesterol and bile acid, and the biosynthesis of cholesterol in tissue.

The major site of cholesterol synthesis in mammals is the liver. Appreciable amounts of cholesterol are also formed by the intestine. The rate of cholesterol formation by these organs is highly responsive to the amount of cholesterol absorbed from dietary sources. This feedback regulation is mediated by changes in the activity of HMG-CoA reductase. As discussed in connection with pathway for the biosynthesis of cholesterol, this enzyme catalyzes the formation of mevalonate, which is the committed step in cholesterol biosynthesis. Dietary cholesterol suppresses cholesterol biosynthesis in these organs through the regulation of HMC-CoA reductase activity. In 1974, Kandutsch and Chen observed that highly purified cholesterol (in contrast to crude cholesterol) is rather ineffective in lowering HMG-CoA reductase activity in culture cells [37]. This perception led to the recognition that oxidized derivatives of sterols (oxysterols), rather than cholesterol, may function as the natural regulators of HMG-CoA reductase activity. Furthermore, oxysterols display a high degree of versatility ranging from substrates in sterol biosynthesis to regulators of gene expression to cellular transporters.

Cholesterol, triacylglycerols, and other lipids are transported in body fluids to specific targets by lipoproteins. A lipoprotein is a particle consisting of a core of hydrophobic lipids surrounded by a shell of polar lipids and apoproteins. Lipoproteins are classified according to their densities. LDL, the major carrier of cholesterol in blood, has a diameter of 22 nm and a mass of about 3×10^6 Da. LDL is composed of globular particles, with lipid constituting about 75% of the weight and protein (apoprotein B) the remainder. Cholesterol esters (about 1500 molecules) are located at the core, which is surrounded by a more polar layer of phospholipids and free cholesterol. The shell of LDL contains a single copy of apoprotein B-100, a very large protein (514 kDa). The major functions of LDL are to transport cholesterol to peripheral tissues and to regulate de novo cholesterol synthesis at these sites.

As we discussed earlier, the major site of cholesterol biosynthesis is the liver. The mode of control in the liver has also been discussed: dietary cholesterol (possibly oxysterols) reduces the activity and amount of HMG-CoA reductase, the enzyme catalyzing the committed step of cholesterol biosynthesis. In some tissues, such as adrenal gland, spleen, lung, and kidney, biosynthesis contributes only a relatively small proportion of the total tissue cholesterol, with the bulk derived by uptake from LDL in the blood. Investigation on the interaction of plasma LDL with specific receptors on the surface of some nonhepatic cells has led to a new understanding of the mechanisms of cellular regulation of cholesterol uptake, storage, and biosynthesis in peripheral tissues.

Michael Brown and Joseph Goldstein did pioneering work concerning the control of cholesterol metabolism in nonhepatic cells based on studies of cultured human fibroblasts [42,43]. In general, cells outside the liver and intestine obtain cholesterol from the plasma rather than by synthesizing them de novo. LDL, the primary source of cholesterol, is first bound to a specific high-affinity receptor on the cell surface; endocytosis then transfers it to internal lysosomes, where the LDL cholesteryl ester and protein are hydrolyzed. The released cholesterol suppresses the transcription of the gene from HMG-CoA reductase, hence blocking de novo synthesis of cholesterol. In the meantime, the LDL receptor itself is subject to feedback regulation. The raised cholesterol concentration also suppresses new LDL receptor synthesis. So the uptake of additional cholesterol from plasma LDL is blocked. After the drop of HMG-CoA reductase activity, there is a reciprocal increase in the microsomal acyl CoA-cholesterol acyltransferase (ACAT), with the result that the excess free cholesterol is reesterified for storage. In addition, the reduction in the rate of cholesterol biosynthesis, which is attributed to uptake of LDL cholesterol by cells, may in fact be due to the presence of a small amount of oxygenated sterol in the LDL [44]. Hydroxylated sterols are known to be far more potent inhibitors of cholesterol biosynthesis and microsomal HMG-CoA reductase activity than is pure cholesterol [45,46]. The development of the hypothesis that oxysterols are regulators of cholesterol biosynthesis has attracted much attention. A comprehensive review has been published by George Schroepfer Jr. [47]. This work could lead to the development of new drugs for the treatment of hypocholesterol diseases [48].

D. Cholesterol Metabolism

In mammals, cholesterol is metabolized into three major classes of metabolic products: (1) the C18, C19, and C21 steroid hormones and vitamin D; (2) the fecal neutral sterols such as 5α-cholestan-3β-ol and 5β-cholestan-3β-ol; and (3) the C_{24} bile acids. Only small amounts of cholesterol are metabolized to steroid hormones and vitamin D. These metabolites are very important physiologically. A detailed discussion of steroid hormones is beyond the scope of this chapter. Vitamin D, also considered a steroid hormone, is discussed individually (see Section II.E). The neutral sterols and bile acids are quantitatively the most important excretory metabolites of cholesterol.

The fecal excretion of neutral sterols in humans is estimated to range from 0.5 to 0.7 g/day. These sterols are complex mixtures of cholesterol, 5α-cholestan-3β-ol, 5β-cholestan-3β-ol, cholest-4-en-3-one, and a number of cholesterol precursor sterols. The major sterol, 5β-cholestan-3β-ol, is found in the feces as a microbial transformation product of cholesterol.

The principal C24 bile acids are cholic acid and chenodeoxycholic acid. The conversion of cholesterol to bile acids takes place in the liver. These bile acids are conjugated with either glycine or taurine to produce bile salts. The bile salts produced in the liver are secreted into the bile and enter the small intestine, where they facilitate lipid and fat absorption. Most bile acids are reabsorbed from the intestine and pass back to the liver and the enterohepatic circulation. The excretion of bile acids in the feces is estimated to range from 0.4 to 0.6 g/day.

The metabolic pathway of cholesterol to bile acids has been studied for many years. Recent advances in oxysterol syntheses have aided the study of this metabolic pathway [49–51]. Several reviews are available describing the formation of bile acids from cholesterol [52,53]. There are three general stages in the biotransformation of cholesterol to bile acids (Scheme 4.8). The first stage

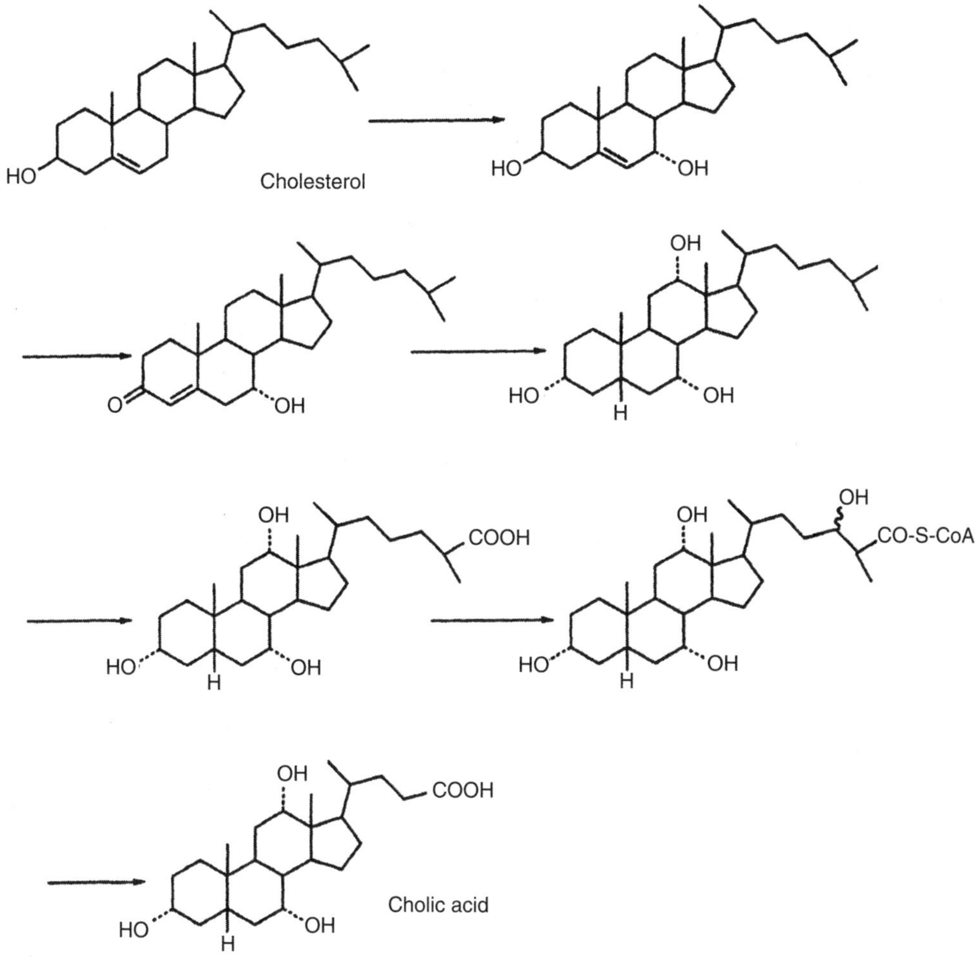

SCHEME 4.8 Biosynthesis of cholic acid from cholesterol.

is the hydroxylation of cholesterol at the 7α position to form cholest-5-ene-3β,7α-diol. Elucidation of the role of LXR receptors has furthered our knowledge of 7α-hydroxylase and the role of oxysterols in sterol metabolism [54]. In the second stage, cholest-5-ene-3β,7α-diol is first oxidized to 7α-hydroxycholest-5-en-3-one, which is isomerized to 7α-hy-droxycholest-4-en-3-one. Further enzymatic transformation leads to 5β-cholesta-3β,7α-diol and 5β-cholesta-3β,7α,12α-triol. The third stage is the degradation of the hydrocarbon side chain, which is less well understood. However, in cholic acid formation it is generally considered to commence when the steroid ring modifications have been completed. The side chain oxidation begins at the C-26 position; 3β,7α,12α-trihydroxy-5β-cholestan-26-oic acid is an important intermediate. The removal of the three terminal atoms is believed to proceed by a β-oxidation mechanism analogous to that occurring in fatty acid catabolism.

E. CHEMISTRY OF VITAMIN D AND RELATED STEROLS

The discovery of vitamin D dates back to the 1930s, following studies of rickets, a well-known disease resulting from deficiency of vitamin D [55]. Vitamin D has basically two functions in mammals: to stimulate the intestinal absorption of calcium and to metabolize bone calcium. A deficiency of vitamin D results in rickets in young growing animals and osteomalacia in adult

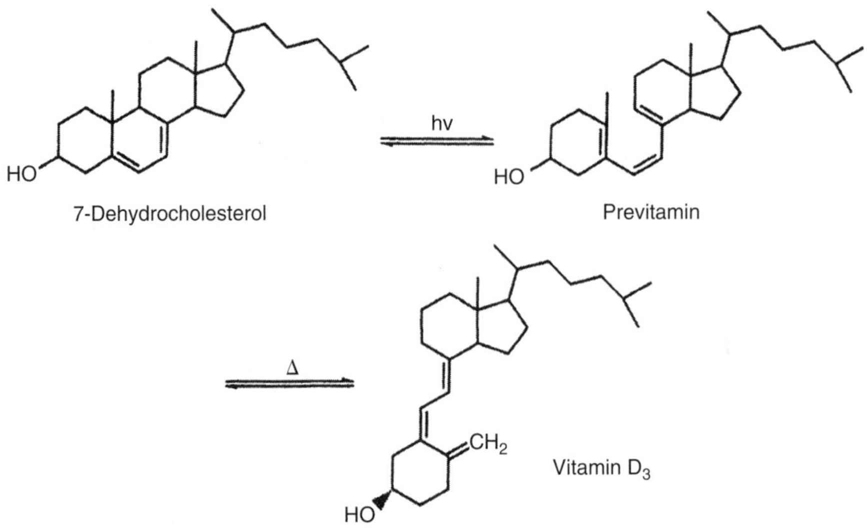

SCHEME 4.9 Photochemical synthesis of vitamin D_3.

animals. In both cases, the collagen fibrils are soft and pliable and are unable to carry out the structural role of the skeleton. As a result, bones become bent and twisted under the stress of the body's weight and muscle function.

Vitamin D is obtained from dietary uptake or via biosynthesis in the skin by means of the UV irradiation of 7-dehydrocholesterol. The UV irradiation of 7-dehydrocholesterol first produces provitamin D_3, which results from a rupture in the 9–10 bond followed by a 5,7-sigmatropic shift (Scheme 4.9). Provitamin then undergoes the thermally dependent isomerization to vitamin D_3 in liver, and further is metabolized to 1,25-dihydroxyvitamin D_3, which is 10 times more active than vitamin D_3, whereas 25-hydroxyvitamin D_3 (Scheme 4.10) is approximately twice as active as vitamin D_3.

The most important nutritional forms of vitamin D are shown in Figure 4.3. Of these structures, the two most important are vitamin D_2 and vitamin D_3. These two forms of vitamin D are prepared from their respective 5,7-diene sterols. Vitamins D_4, D_5, and D_6 have also been prepared chemically, but they have much lower biological activity than vitamins D_2 and D_3. In addition, many analogs of vitamin D metabolites have been synthesized. Some of these compounds exhibit similar vitamin D hormone responses and have found use in the treatment of vitamin D deficiency diseases (Figure 4.4). Recently, vitamin D metabolites were found to be potent inducers of cancer cells,

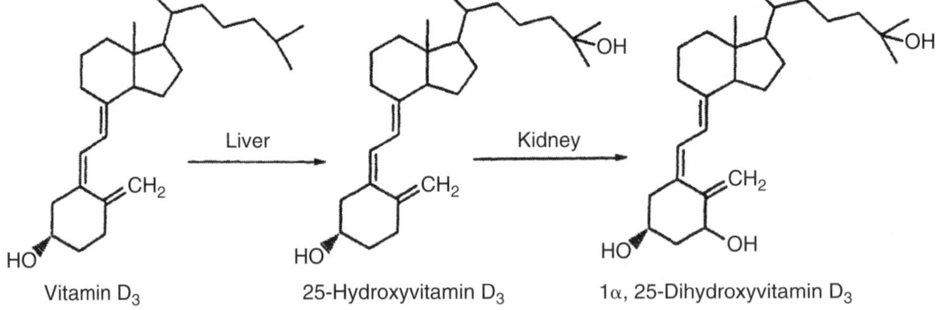

SCHEME 4.10 Metabolic alterations of vitamin D_3.

Vitamin D_2

Vitamin D_3

Vitamin D_4

Vitamin D_5

Vitamin D_6

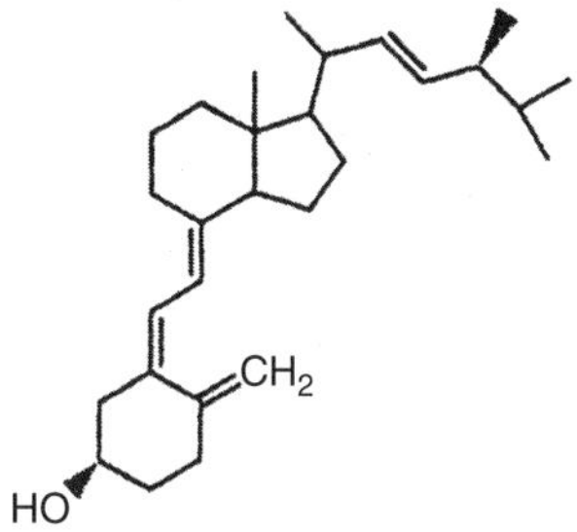

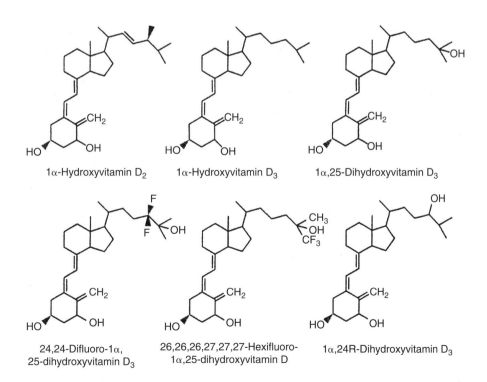

FIGURE 4.3 Structures of known nutritional forms of vitamin D.

1α-Hydroxyvitamin D_2

1α-Hydroxyvitamin D_3

1α,25-Dihydroxyvitamin D_3

24,24-Difluoro-1α,
25-dihydroxyvitamin D_3

26,26,26,27,27,27-Hexifluoro-
1α,25-dihydroxyvitamin D

1α,24R-Dihydroxyvitamin D_3

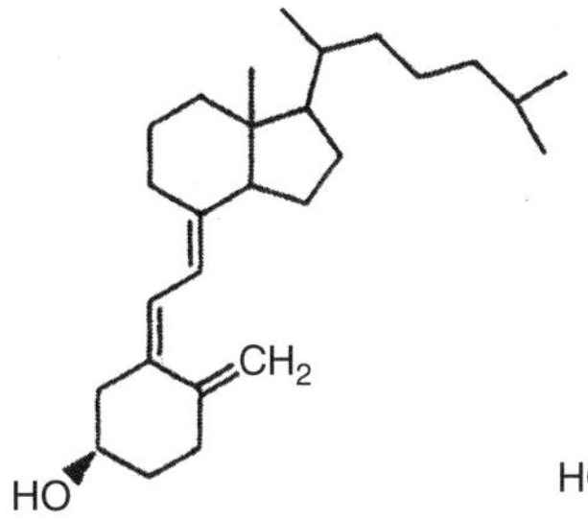

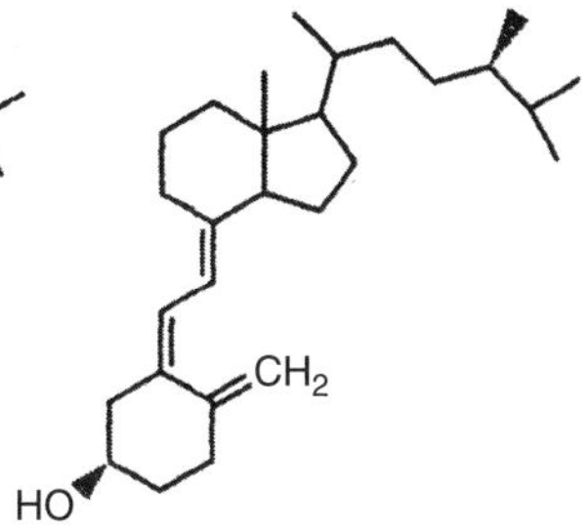

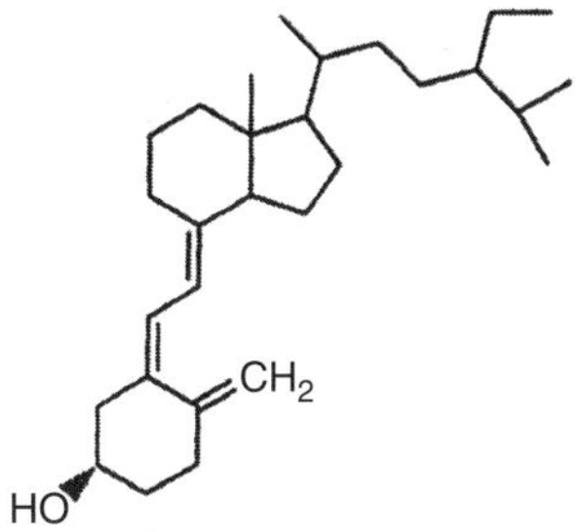

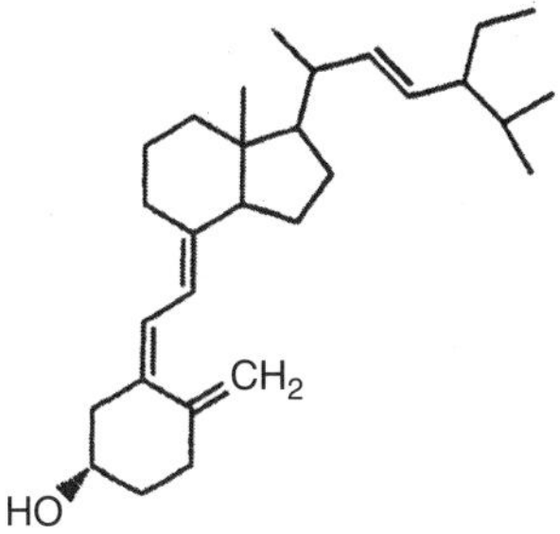

FIGURE 4.4 Examples of vitamin D_3 analogs.

which make this steroid hormone and its analogs (biosynthetic inhibitors) potential candidates for the treatment of cancers and other diseases [56]. The most interesting analogs possessing hormonal activity are the 26,26,26,27,27,27-hexafluoro-1,25-$(OH)_2D_3$ and 24,24-F2–1,25$(OH)_2D_3$, which possess all responses to the vitamin D hormone but are 10–100 times more active than the native hormone. The fluoro groups on the side chain block the metabolism of these compounds.

All vitamin D compounds possess a common triene structure. Thus, it is not surprising that they have the same UV maximum at 265 nm, a minimum of 228 nm, and a molar extinction coefficient of 18,200. Since they possess intense UV absorption, they are labile to light-induced isomerization. In addition, the triene system is easily protonated, resulting in isomerization that produces isotachysterol, which is essentially devoid of biological activity. The lability of the triene structure has markedly limited the chemical approaches to modification of this molecule. There is a great deal of chemistry relating to the chemical properties and syntheses of vitamin D compounds. The detailed chemistry and chemical synthesis of the D vitamins are beyond the scope of this chapter, and interested readers are referred to reviews in this important area [57,58].

F. ANALYSIS OF STEROLS

1. Extraction of Sterols

To analyze the sterols in specific biological tissues, sterols are first extracted from these tissues with organic solvents. The choice of an extraction technique is often determined by the nature of the source and the amount of information the investigator chooses to obtain concerning the forms of sterols present as free, glycosylated, or esterified through the 3-hydroxy group. Different extraction procedures may vary dramatically depending on the extraction efficiency required for different classes of sterols [59,60]. Irrespective of the nature of the source, one of the four methods of sample preparation is usually employed. Samples can be extracted directly, with little or no preparation; after drying and powdering; after homogenization of the fresh materials; or after freeze drying or fresh freezing, followed by powdering, sonication, or homogenization. Extraction procedures vary. The analyst may simply mix the prepared material with the extraction solvent (the most frequently used solvents include mixtures of chloroform and methanol, or dichloromethane and acetone) for a short time (0.5–1 h) and separate the organic solvent phase from the aqueous phases and debris by centrifugation. Another common procedure is extraction from a homogenized material by means of a refluxing solvent in a Soxhlet apparatus for 18 h or with a boiling solvent for 1 h. To obtain a total lipid extraction, saponification under basic (i.e., in 10% KOH in 95% ethanol) or acidic conditions is usually conducted before the organic solvent extraction. Most of the glycosylated sterols and some esterified sterols cannot be easily extracted into organic solvents without the hydrolysis step.

Many oxysterols contain functional groups (e.g., epoxides and ketones) that may be sensitive to high concentration of acids or bases. Epoxides may undergo nucleophilic attack by strong bases (e.g., NaOH and KOH), followed by ring opening. Moreover, treatment with strong acids can result in ring opening to form alcohols, alkenes, and ketones. The hydrolysis of cholesterol epoxides under mild acidic conditions has been studied [61]. Modified procedures are available for the isolation of steroidal epoxides from tissues and cultured cells by saponification or extraction [62–64]. Ketones are known to form enolates under the influence of strong bases, which may then form condensation products of higher molecular weight [65]. To circumvent these potential problems, procedures using different extraction techniques are sometimes preferred to a saponification followed by extraction. Mild methods for the removal of the ester function without ketone enolization include extraction by means of sodium or potassium carbonate in heated aqueous solutions of methanol or ethanol. The addition of tetrahydrofuran to these mixtures has been found to significantly increase the solubility of the more polar oxysterols [66].

There are not many studies on the efficiency of various extraction methods. Most extraction procedures were designed to compare the extraction of lipids from cells or tissues of a single source and have been applied subsequently to plants and animals of various types. The errors in the quantitative analysis of sterols, which are probably introduced in the extraction steps, could be eliminated by using in situ labeling of key sterols. Sterols labeled with deuterium and ^{14}C have been used to monitor the extraction recovery in human plasma oxysterol analysis.

2. Isolation of Sterols

Conventional column chromatography, with ordinary phase (silica gel or alumina oxide), reversed phase, and argentation stationary phase, is still the most important method for the isolation and purification of sterols, especially if the total lipid extraction is complex and high in weight (>200 mg) [67]. Chromatographic methods with an organic phase involve the binding of a substrate to the surface of a stationary polar phase through hydrogen bonding and dipole–dipole interaction. A solvent gradient with increasing polarity is used to elute the substrate from the stationary phase. The order of substrate movement will be alkyl > ketone > hindered alcohol > unhindered alcohol. The elution profile is routinely monitored by GC or TLC. Reversed phase column chromatography involves the use of lipophilic dextran (Sephadex LH-20 or Lipidex 5000) and elution with nonpolar solvents like a mixture of methanol and hexane. The substrate could be separated in order of increasing polarity.

Argentation column chromatography is a very powerful chromatographic method in the separation of different alkene isomers. The argentation stationary phase is typically made by mixing AgNO$_3$ solution (10 g of AgNO$_3$ in 10 mL of water) and silica gel or aluminum oxide (90 g of stationary phase in 200 mL of acetone). After acetone has been evaporated at a moderate temperature (<35°C) under vacuum, the resulting argentation stationary phase is dried under vacuum at room temperature until a constant weight is obtained. Usually argentation chromatography is suitable only for the separation of sterol acetates. For free sterol, on-column decomposition of substrate was observed. Compared with the conditions of normal adsorption column chromatography, a slightly polar solvent system is usually needed to elute the substrate from the column. For a preliminary cleanup step, good normal silica gel column chromatography running under the gravity gives good results. Column chromatography with fine silica gel and aluminum oxide (230–400 mesh) running under medium pressure (10–100 psi, MPLC) could be used in the separation of individual sterols and sterol subclasses.

TLC is primarily used as an analytical method to detect compounds. In addition, small samples (<20 mg) can be separated by means of preparative TLC. For an analytical TLC plate, the thickness of the stationary phase (usually silica gel) is less than 0.25 mm; for preparative TLC a much thicker stationary phase (>0.5 mm) is required. No more than 1 mg of substrate should be loaded on 1 cm analytical plates and 2 mg on preparative plates. Otherwise overloading prevents the achievement of a good separation. Nondestructive detection methods (such as iodine and UV detection) are usually used to locate individual compounds. The silica gel containing pure individual compounds is scraped from the plates and eluted with polar solvents.

HPLC, adsorption or reversed phase, is becoming the most commonly used technique for the separation of individual sterols from subclass fractions [67,68]. In adsorption HPLC, microspheres of silica (silicic acid such as μ-Poracil) are employed. As in adsorption, open column, and TLC, this method is characterized by hydrogen bonding and other electronic attractions between the sterol and stationary phase. The smaller diameter (3–5 μm) and high porosity of the microspheres create a very large effective surface area and allow the simultaneous analysis of many more theoretical plates than the other systems can handle. TLC solvent systems can be used directly for adsorption HPLC. In general, the solvents that give best separations are binary (or trinary) systems of mostly low to moderate polarity solvents with a small amount of a strongly polar solvent such as hexane/benzene (9:1) or dichloromethane-n-hexane-ethyl acetate (94:5:1).

In reversed phase HPLC (RP-HPLC) systems, alkyl groups at C-8 or C-18 have been chemically bonded to microspheres of silica. These highly porous particles with small diameters

(2–5 μm) give the stationary phases very large effective surface areas for the interaction of the sterols with the bonded alkyl groups. This large area, in turn, makes possible rapid equilibration of the sterol between the stationary phase and an appropriate mobile phase, yielding columns with as many as 10^4 theoretical plates. A variety of polar mobile phases have been employed in RP-HPLC (e.g., 100% acetonitrile, acetonitrile–water mixtures (0%–20%), and methanol–water mixtures (2%–20% water)). Samples are usually injected onto the analytical column at levels of 1–100 μg/component (up to 1.0 mg total) for routine separations. Retention volumes of eluted sterols are usually expressed relative to cholesterol.

GLC is most frequently used as an analytical technique to monitor fractions during the isolation and separation of sterols; however, this technique can also be used in preparative separations. The use of GLC as an analytical and preparative technique for sterols and related steroids has been discussed in numerous reviews [69,70]. The separation of sterols in gas–liquid systems depends on the polarity and molecular weight (frequently correlating with size and volume) of the molecule.

3. Characterization of Sterols

GLC is most frequently used to quantitate sterols in extracts, subtraction mixtures, and isolated sterol fractions. Selectivity is provided by gas–liquid partitioning. In general, three different methods of detection have been employed for quantitation: flame ionization detection (FID), electron capture detection (ECD), and mass detection (MD) [71,72]. However, FID is by far the most commonly employed method because it is relatively insensitive to temperature changes during analysis and to minor structural differences in sterols and related steroids, and because it has a large linear mass range of response. For quantitative analysis, it is necessary to determine the linear range or response to a sterol standard (e.g., cholesterol or β-cholestanol) for each FID and accompanying chromatographic system. For sample analysis, dried fractions are routinely weighed and dissolved in a known amount of solvent to give a mass-to-volume ratio within the linear range of the detector, whereupon the samples are injected onto the column.

Quantification of sterols in extracts, subclass fractions, or isolated sterol fractions by HPLC is somewhat limited. The selectivity provided by either adsorption or reversed phase chromatography is in many cases greater than that provided by GLC. Most frequently, UV detectors are employed and are set at end-adsorption wavelengths [73]. These detectors have limited sensitivity, with monochromatic detectors being most sensitive. Additionally, UV detectors cannot be considered universal sterol detectors. Most sterols differ significantly in their UV absorption properties, even in end-absorption regions. Thus, it would not be possible to select a single wavelength for the quantitation of complex sterol mixtures. Therefore, HPLC coupled to variable-wavelength detectors or multidiode detectors can be used to quantitate specific sterols in a mixture if the compound in question has a unique absorption spectrum relative to other members of the mixture.

For a full discussion of the identification of sterols with NMR [74,75], mass [76], UV [77], infrared, and x-ray [78] spectrographic techniques, the reader is referred to recent comprehensive reviews in this area.

The early analysis of sterols relied on color tests. The Liebermann–Burchard color test is one of the most common tests. It has been extensively used to detect the presence of sterols in biological samples [79–81]. A chloroform solution of a sterol is treated with a few drops of acetic anhydride and concentrated sulfuric acid while the temperature is held at 15°C–20°C. Changing from rose-red to blue and finally green, a pigment develops in the solution. The time of color development is characteristic of the specific sterol. In most cases, the saturated sterols give this color sequence much more slowly than the unsaturated. When the solid sterol is used instead of the chloroform solution, the test is then known as the Liebermann reaction.

The digitonin precipitation is another common method for the detection of 3β-hydroxy sterols. Cholesterol forms a very stable and almost insoluble complex with digitonin, a glycosidic saponin, which is available commercially [79–81]. It was found that saponins react only with free sterols, not with sterol esters. In addition, the epi (3α-ol) forms of sterols will not precipitate with digitonin. Free sterol is determined by direct precipitation, whereas the full amount of sterol is determined after saponification of the extract, ester sterol being obtained by difference. This method is commonly used to estimate 3β-sterols in tissues. The recovery of the sterol may be obtained from the digitonide.

REFERENCES

1. R.J. Hamilton. Commercial waxes: Their composition and application. In: *Waxes: Chemistry, Molecular Biology and Functions* (R.J. Hamilton, ed.). The Oily Press, Ayr, Scotland, 1995, p. 257.
2. A.H. Warth. *Chemistry and Technology of Waxes*. Reinhold, New York, NY, 1956.
3. R. Sayers. *Wax—An Introduction*. European Wax Federation and Gentry Books, London, 1983.
4. *Kirk-Othmer Encyclopedia of Chemical Technology*, Vol. 24. Wiley, New York, NY, 1984, pp. 446–481.
5. P.E. Kolattukudy. *Chemistry and Biochemistry of Natural Waxes*. Elsevier-North Holland, Amsterdam, 1976.
6. J.C. Touchstone and M.F. Dobbins. *Practice of Thin Layer Chromatography*, 2nd edn. Wiley, New York, NY, 1983.
7. W.W. Christie. *Gas Chromatography and Lipids, A Practical Guide*. The Oily Press, Ayr, Scotland, 1989.
8. R.J. Hamilton. Commercial waxes: Their composition and application. In: *Waxes: Chemistry, Molecular Biology and Functions* (R.J. Hamilton, ed.). The Oily Press, Ayr, Scotland, 1995, p. 315.
9. S. Atkins, R.J.N. Hamilton, S.F. Mitchell, and P.A. Sewell. Analysis of lipids and hydrocarbons by HPLC. *Chromatographia* 15:97–100 (1982).
10. N.C. Hadley. Wax secretion of the desert tenebrionid beetle. *Science* 203:367–369 (1979).
11. S.A. Hanrahan, E. McClain, and S.J. Warner. Protein component of the surface wax of a desert tenebrionid. *S. Afr. J. Sci.* 83:495–497 (1987).
12. P.E. Kolattukudy and K.E. Espelie. Biosynthesis of cutin, suberin, and associated waxes. In: *Biosynthesis and Biodegradation of Wood Components* (T. Higuchi, ed.). Academic Press, New York, NY, 1985, pp. 161–207.
13. A.R. Kartha and S.P. Singh. The in vivo synthesis of reserve seed waxes. *Chem. Ind.* 38:1342–1343 (1969).
14. T. Miwa. Jojoba oil wax esters and fatty acids and alcohols: GLC analysis. *J. Am. Oil Chem. Soc.* 48:259–264 (1971).
15. G.F. Spencer, R.D. Planner, and T. Miwa. Jojoba oil analysis by HPLC and GC–MS. *J. Am. Oil Chem. Soc.* 54:187–189 (1977).
16. A.D. Bary. Composition and occurrence of epicuticular wax in plants. *Bot. Z.* 29:129–619 (1871).
17. G.J. Blomquist, D.R. Nelson, and M. de Renolbales. Chemistry, biochemistry and physiology of insect cuticular lipids. *Arch. Insect Biochem. Physiol.* 6:227–265 (1987).
18. J.L. Harwood and P.K. Stumpf. Fat metabolism in higher plants. *Arch. Biochem. Biophys.* 142:281–291 (1971).
19. J.G. Jaworski, E.E. Goldsmith, and P.K. Stumpf. Fat metabolism in higher plants. *Arch. Biochem. Biophys.* 163:769–776 (1974).
20. T. Shimakata and P.K. Stumpf. Isolation and function of spinach leaf ketoacyl synthases. *Proc. Natl. Acad. Sci. USA* 79:5808–5812 (1982).
21. P.E. Kolattukudy. Biosynthesis of wax in *Brassica oleracea*. *Biochemistry* 5:2265–2275 (1966).
22. P.E. Kolattukudy. Biosynthesis of plant waxes. *Plant Physiol.* 43:375–383 (1968).
23. P.E. Kolattukudy. Biosynthesis of plant waxes. *Plant Physiol.* 43:1466–1470 (1988).
24. P.E. Kolattukudy. Biosynthesis of surface lipids. *Science* 159:498–505 (1968).
25. C. Cassagne and R. Lessire. Studies on alkane biosynthesis in plant leaves. *Arch. Biochem. Biophys.* 165:274–280 (1974).
26. P. von Wettstein-Knowles. Biosynthesis and genetics of waxes. In: *Waxes: Chemistry, Molecular Biology and Functions* (R.J. Hamilton, ed.). The Oily Press, Ayr, Scotland, 1995, pp. 91–129.
27. W.R. Nes and M.L. McKean. *Biochemistry of Steroids and Other Isopentenoids*. University Park Press, Baltimore, 1977, pp. 1–37.
28. M.E. Dempsey. Regulation of steroid biosynthesis. *Annu. Rev. Biochem.* 43:967 (1974).
29. P. Leren. The effect of a plasma cholesterol lowering diet. *Acta Med. Scand. Suppl.* 466:1–92 (1966).
30. H.W. Chen, A.A. Kandutsch, and H.W. Heiniger. Stimulation of sterol synthesis. *Cancer Res.* 34:1034–1037 (1974).
31. T.W. Goodwin. Biosynthesis of plant sterols. In: *Sterols and Bile Acids* (H. Danielsson and J. Sijovall, eds.). Elsevier, New York, NY, 1985, p. 175.

32. N. Ikekawa. Structures, biosynthesis and function of sterols in invertebrates. In: *Sterols and Bile Acids* (H. Danielsson and J. Sijovall, eds.). Elsevier, New York, NY, 1985, p. 199.
33. J. Fried and J.A. Edwards. *Organic Reaction in Steroid Chemistry*. Van Nostrand Reinhold, New York, NY, 1972.
34. R.T. Blickenstaff, A.C. Ghosh, and G.C. Wolf. *Total Synthesis of Steroids*. Academic Press, New York, NY, 1974.
35. J.B. Dence. *Steroids and Peptides*. Wiley, New York, NY, 1976.
36. K. Bloch. The biological synthesis of cholesterol. *Science* 150:19, 23 (1965).
37. A.A. Kandutsch and H.W. Chen. Inhibition of sterol biosynthesis in cultured mouse cells by oxysterols. *J. Biol. Chem.* 249:6057, 6062 (1974).
38. Y. Tanzawa and A. Endo. Kinetic analysis of HMG-CoA reductase inhibition. *Eur. J. Biochem.* 98:195, 203 (1979).
39. A.W. Alberts, J. Chen, G. Kurton, V. Hunt, J. Huff, C. Hoffman, J. Rothrock, M. Lopez, H. Joshua, E. Harris, A. Parchett, R. Monaghan, S. Currie, E. Stapley, G. Albetschonbert, O. Hensens, J. Hirschfield, K. Hoogsteen, J. Liesch, and J. Springer. Mevinolin: A potent inhibitor of HMG-CoA reductase. *Proc. Natl. Acad. Sci. USA* 77:3957–3961 (1980).
40. Lipid Research Clinics Program. *JAMA* 251:351 (1984).
41. E.J. Parish, U.B.B. Nanduri, J.M. Seikel, H.H. Kohl, and K.E. Nusbaum. Hypocholesterolemia. *Steroids* 48:407–412 (1986).
42. M.S. Brown and J.L. Goldstein. How LDL receptors influence cholesterol and atherosclerosis. *Science* 232:34–38 (1984).
43. M.S. Brown and J.L. Goldstein. A receptor-mediated pathway for cholesterol homostasis. *Sci. Am.* 251(6):58–66 (1984).
44. G.J. Schroepfer. Sterol biosynthesis. *Annu. Rev. Biochem.* 51:555–569 (1982).
45. A.A. Kandutsch, H.W. Chen, and H.W. Heiniger. Biological activity of some oxygenated sterols. *Science* 201:498–501 (1978).
46. E.J. Parish, S.C. Parish, and S. Li. Side-chain oxysterol regulation of 3-hydroxy-3-methylglutaryl coenzyme A reductase activity. *Lipids* 30:247–254 (1995).
47. G.J. Schroepfer Jr. Oxysterols: Modulators of cholesterol metabolism and other processes. *Physiol. Rev.* 80:361–554 (2000).
48. G.J. Schroepfer Jr. Design of new oxysterols for regulation of cholesterol metabolism. *Curr. Pharm. Des.* 2:103–112 (1996).
49. E.J. Corey and M.J. Grogan. Stereocontrolled synthesis of 24(S),25-epoxycholesterol and related oxysterols for studies on the activation of LXR receptors. *Tetrahedron Lett.* 39:9351–9354 (1998).
50. J.S. Russel and T.A. Spencer. Efficient, stereoselective synthesis of 24(S),25-epoxy-cholesterol. *J. Org. Chem.* 63:9919–9923 (1998).
51. E.J. Parish, N. Aksara, and T.L. Boos. Remote functionalization of the steroidal side chain. *Lipids* 32:1325–1330 (1997).
52. H. Danielsson and J. Sijovall. Bile acid metabolism. *Annu. Rev. Biochem.* 44:233–264 (1977).
53. D.W. Gibson, R.A. Parker, C.S. Stewart, and K.J. Evenson. Short-term regulations of HMG-CoA reductase by phosphorylation. *Adv. Enzyme Regul.* 20:263–269 (1982).
54. B.A. Janowski, M.J. Grogan, S.A. Jones, G.B. Wisely, S.A. Kliewer, E.J. Corey, and D.J. Mangelsdorf. Structural requirements of ligands for the oxysterol liver X receptors LXRα and LXRβ. *Proc. Natl. Acad. Sci. USA* 96:266–271 (1999).
55. H.F. Deluca. The metabolism and function of vitamin D. In: *Biochemistry of Steroid Hormones* (H.L.J. Makin, ed.). Blackwell Scientific Publications, London, 1984, p. 71.
56. K. Silvia, H. Sebastian, P.V. Jan, and R. Wolfgang. A novel class of vitamin D analogs, synthesis and preliminary biological evaluation. *Bioorg. Med. Chem. Lett.* 6:1865–1873 (1996).
57. P.A. Bell. The chemistry of vitamin D. In: *Vitamin D* (D.E.M. Lawson, ed.). Academic Press, New York, NY, 1978, pp. 1–46.
58. P.E. Georghiou. The chemical synthesis of vitamin D. *Chem. Soc. Rev.* 6:83–98 (1977).
59. J. Folch, M. Lees, and G.H. Stanley. A simple method for the isolation and purification of total lipids from animal tissues. *J. Biol. Chem.* 226:497–506 (1957).
60. A.U. Osagie and M. Kates. Lipid composition of millet seeds. *Lipids.* 19:958–961 (1984).
61. G. Maerker and F.J. Bunick. Cholesterol oxides III. *J. Am. Oil Chem. Soc.* 63:771–774 (1986).

62. S.E. Saucier, A.A. Kandutsch, F.R. Taylor, T.A. Spencer, S. Phirwa, and A.K. Gayen. Identification of regulatory oxysterols. *J. Biol. Chem.* 260:14571–14579 (1985).

63. T.A. Spencer, A.K. Gayen, S. Phirwa, J.A. Nelson, F.R. Taylor, A.A. Kandutsch, and S. Erickson. 24(S), 25-Epoxy cholesterol. *J. Biol. Chem.* 260:13391–13399 (1985).

64. S.R. Panini, R.C. Sexton, A.K. Gupta, E.J. Parish, S. Chitrakorn, and H. Rudney. Regulation of 3-hydroxy-3-methylglutaryl coenzyme A reductase activity and cholesterol biosynthesis by oxylanosterols. *J. Lipid Res.* 27:1190–1196 (1986).

65. G. Maerker and J. Unrun. Cholesterol oxides, II. *J. Am. Chem. Soc.* 63:767–771 (1986).

66. E.J. Parish, H. Honda, S. Chitrakorn, and F.R. Taylor. A facile synthesis of 24-ketolanosterol. *Chem. Phys. Lipids* 48:255–261 (1988).

67. R.C. Heupel. Isolation and primary characterization of sterols. In: *Analysis of Sterols and Other Biologically Significant Steroids* (W.D. Nes and E.J. Parish, eds.). Academic Press, San Diego, CA, 1989, pp. 1–32.

68. M.P. Kautsky. *Steroid Analysis by HPLC.* Dekker, New York, NY, 1981.

69. G.W. Patterson. Chemical and physical methods in the analysis of plant sterols. In: *Isopentenoids in Plants* (W.D. Nes, G. Fuller, and L.S. Tsai, eds.). Dekker, New York, NY, 1984, pp. 293–312.

70. W.R. Nes. A comparison of methods for the identification of sterols. In: *Methods in Enzymology,* Vol. III (J.H. Laio and H.C. Rilling, eds.). Academic Press, New York, NY, 1985, pp. 3–29.

71. F. Bernini, G. Sangiovanni, S. Pezzotta, G. Galli, R. Fumagalli, and P. Paoletti. Selected ion monitoring technique for the evaluation of sterols in cerebrospinal fluid. *J. Neuro. Oncol.* 4:31–40 (1986).

72. B.A. Knight. Quantitative analysis of plant sterols. *Lipids* 17:204–211 (1982).

73. R.C. Heupel. Varietal similarities and differences in the polycyclic isopentenoid composition of sorghum. *Phytochemistry* 24:2929–2936 (1985).

74. W.K. Wilson and G.J. Schroepfer Jr. ^{1}H and ^{13}C NMR spectroscopy of sterols. In: *Molecular Structure and Biological Activity of Steroids* (M. Bohl and W.L. Duax, eds.). CRC Press, Boca Raton, FL, 1992, pp. 33–42.

75. T. Akihisa.^{13}C-NMR identification of sterols. In: *Analysis of Sterols and Other Biologically Significant Steroids* (W.D. New and E.J. Parish, eds.). Academic Press, San Diego, CA, 1989, pp. 251–262.

76. A. Rahier and P. Beneveniste. Mass spectral identification of phytosterols. In: *Analysis of Sterols and Other Biologically Significant Steroids* (W.D. Nes and E.J. Parish, eds.). Academic Press, San Diego, CA, 1989, pp. 223–236.

77. P. Acuna-Johnson and A.C. Oehlschlager. Identification of sterols and biologically significant steroids by UV and IR spectroscopy. In: *Analysis of Sterols and Other Biologically Significant Steroids* (W.D. Nes and E.J. Parish, eds.). Academic Press, San Diego, CA, 1989, pp. 267–278.

78. W.L. Duax, J.F. Griffin, and G. Cheer. Steroid conformational analysis based on x-ray crystal structure determination. In: *Analysis of Sterols and Other Biologically Significant Steroids* (W.D. Nes and E.J. Parish, eds.). Academic Press, San Diego, CA, 1989, pp. 207–229.

79. L.F. Fieser and M. Fieser. *Natural Products Related to Phenanthrene.* Reinhold Publishing Corp., New York, NY, 1949.

80. L.F. Fieser and M. Fieser. *Steroids.* Reinhold Publishing Corp., New York, NY, 1959.

81. R.P. Cook. *Cholesterol: Chemistry, Biochemistry, and Pathology.* Academic Press Inc., New York, NY, 1958.

5 Extraction and Analysis of Lipids

Fereidoon Shahidi and P.K.J.P.D. Wanasundara

CONTENTS

I. INTRODUCTION

Lipids are among the major components of food of plant and animal origin. There is no precise definition available for the term lipid; however, it usually includes a broad category of compounds that have some common properties and compositional similarities. Lipids are materials that are sparingly soluble or insoluble in water, but soluble in selected organic solvents such as benzene, chloroform, diethyl ether, hexane, and methanol. Together with carbohydrates and proteins, lipids constitute the principal structural components of tissues. However, the common and unique features of lipids relate to their solubility rather than their structural characteristics [1]. Many classification systems have been proposed for lipids. From the nutrition point of view, according to the National Academy of Sciences' report on nutrition labeling, fats and oils are defined as the complex organic molecules that are formed by combining three fatty acid molecules with one molecule of glycerol [2]. As indicated in Table 5.1 [3–5], lipids are generally classified as simple and compound (complex) or derived lipids according to the Bloor [3] classification.

Foods contain any or all of these lipid compounds; however, triacylglycerols (TAGs) and phospholipids (PLs) are the most abundant and important ones. Liquid TAGs at room temperature are referred to as oils, and are generally of plant or marine origin (e.g., vegetable and marine oils). Solid TAGs at room temperature are termed fats, which are generally of animal origin (e.g., lard and tallow).

Accurate and precise analysis of lipids in foods is important for determining constituting components and nutritive value, standardizing identity and uniformity, preparing nutritional labeling material, as well as for promoting and understanding the effects of fats and oils on food functionality. At the same time, knowledge about the structural characteristics of lipids may allow development of tailor-made products designed for a particular function or application.

II. EXTRACTION OF LIPIDS FROM FOODS AND BIOLOGICAL MATERIALS

Lipids in nature are associated with other molecules via (1) van der Waals interaction, for example, interaction of several lipid molecules with proteins; (2) electrostatic and hydrogen bonding, mainly between lipids and proteins; and (3) covalent bonding among lipids, carbohydrates, and proteins. Therefore, to separate and isolate lipids from a complex cellular matrix, different chemical and physical treatments must be administered. Water insolubility is the general property used for the separation of lipids from other cellular components. Complete extraction may require longer extraction time or a series or combination of solvents so that lipids can be solubilized from the matrix.

The existing procedures of lipid extraction from animal or plant tissues usually include several steps: (1) pretreatment of the sample, which includes drying, size reduction, or hydrolysis;

TABLE 5.1
General Classification of Lipids

Simple lipids: Compounds with two types of structural moieties

Glyceryl esters	Esters of glycerol and fatty acids (e.g., triacylglycerols, partial acylglycerols)
Cholesteryl esters	Esters of cholesterol and fatty acids
Waxes	True waxes are esters of long-chain alcohols and fatty acids; esters of vitamins A and D are also included
Ceramides	Amides of fatty acids with long-chain di- or trihydroxy bases containing 12–22 carbon atoms in the aliphatic chain (e.g., sphingosine)

Complex lipids: Compounds with more than two types of structural moieties

Phospholipids	Glycerol esters of fatty acids, phosphoric acid, and other groups containing nitrogen
Phosphatidic acid	Diacylglycerol esterified to phosphoric acid
Phosphatidylcholine	Phosphatidic acid linked to choline, known also as lecithin
Phosphatidylethanolamine	
Phosphatidylserine	
Phosphatidylinositol	
Phosphatidyl acylglycerol	More than one glycerol molecule is esterified to phosphoric acid (e.g., cardiolipin, diphosphatidyl acylglycerol)
Glycoglycerolipids	1,2-Diacylglycerol joined by a glycosidic linkage through position *sn*-3 with a carbohydrate moiety [e.g., monogalactosyl diacylglycerol, digalactosyl monoacylglycerol, sulfoquinovosyl diacylglycerol (monoglycosyl diacylglycerol at position 6 of the disaccharide moiety is linked by carbon–sulfur bond to a sulfonic acid)]
Gangliosides	Glycolipids that are structurally similar to ceramide polyhexoside and also contain 1–3 sialic acid residues; most contain an amino sugar in addition to the other sugars
Sphingolipids	Derivatives of ceramides
Sphingomyelin	Ceramide phosphorylcholine
Cerebroside	Ceramide monohexoside (i.e., ceramide linked to a single sugar moiety at the terminal hydroxyl group of the base)
Ceramide dihexoside	Linked to a disaccharide
Ceramide polyhexoside	Linked to a tri- or oligosaccharide
Cerebroside sulfate	Ceramide monohexoside esterified to a sulfate group

Derived lipids: Compounds that occur as such or released from simple or complex lipids because of hydrolysis (e.g., fatty acids; fatty alcohols; fat-soluble vitamins A, D, E, and K; hydrocarbons; sterols)

Source: Adapted from Bloor, W.R., *Proc. Soc. Exp. Biol. Med.*, 17, 138, 1920; Christie, W.W. in *Lipid Analysis*, Pergamon Press, Oxford, 1982; Pomeranz, Y. and Meloan, C.L. in *Food Analysis; Theory and Practice, 4th edn.*, AVI, Westport, Connecticut, 1994.

(2) homogenization of the tissue in the presence of a solvent; (3) separation of liquid (organic and aqueous) and solid phases; (4) removal of nonlipid contaminants; and (5) removal of solvent and drying of the extract. Standard methods for lipid extraction have been established by the Association of Official Analytical Chemists (AOAC) International for different types of materials/tissues. However, when it comes to practical situations, each case might require modification of the method.

A. SAMPLE PREPARATION

As with any form of chemical analysis, proper sampling and storage of the samples are essential for obtaining valid results. According to Pomeranz and Meloan [5], an ideal sample should be identical in all of its intrinsic properties to the bulk of the material from which it is taken. In practice, a sample is satisfactory if its properties under investigation correspond to those of the bulk material within the limits set by the nature of the test. Sample preparation for lipid analysis depends on the type of food and the nature of its lipids. Effective analysis calls for a knowledge of the structure, chemistry, and occurrence of principal lipid classes and their constituents. Therefore, it is not possible to devise a single standard method for extraction of all kinds of lipids in different foods.

Extraction of lipids should be performed as soon as possible after the removal of tissues from the living organism so as to minimize any subsequent changes. Immediate extraction is not always possible; however, the samples usually are stored at very low temperatures in sealed containers, under an inert (nitrogen) atmosphere or on dry ice. Yet the freezing process itself may permanently damage the tissues as a result of osmotic shock, which alters the original environment of the tissue lipids and brings them into contact with enzymes from which they are normally protected. Thawing the sample taken from frozen storage before extraction may enhance this deterioration. Therefore, tissue samples should be homogenized and extracted with solvents without being allowed to thaw [4]. Lipolytic enzymes of animal and plant tissues are usually deactivated irreversibly by homogenization with polar solvents. Use of high temperatures should be avoided; it is also advisable, when possible, to maintain an inert atmosphere during sample preparation and extraction, which may minimize oxidation reactions of unsaturated lipids.

B. PRETREATMENTS

1. Drying

Sometimes nonpolar solvents, such as diethyl ether and hexane, do not easily penetrate the moist tissues (>8% moisture); therefore, effective lipid extraction does not occur. Diethyl ether is hygroscopic and becomes saturated with water and thus inefficient for lipid extraction. Therefore, reducing moisture content of the samples may facilitate lipid extraction. Vacuum oven drying at low temperatures or lyophilization is usually recommended. Predrying facilitates the grinding of the sample, enhances extraction, and may break fat–water emulsions to make fat dissolve easily in the organic solvent and helps to free tissue lipids. Drying the samples at elevated temperatures is undesirable because lipids become bound to proteins and carbohydrates, and such bound lipids are not easily extracted with organic solvents [5].

2. Particle Size Reduction

The extraction efficiency of lipids from a dried sample also depends on the size of the particles. Therefore, particle size reduction increases surface area, allowing more intimate contact of the solvent, and enhances lipid extraction (e.g., grinding of oilseeds before lipid extraction). In some cases, homogenizing the sample together with the extracting solvent (or solvent system) is carried out instead of performing these operations separately. Ultasonication, together with homogenization in excess amount of extracting solvent, has been successfully used to recover lipids of microalgae [6] and organ tissues [7].

3. Acid/Alkali Hydrolysis

To make lipids more available for the extracting solvent, food matrices are often treated with acid or alkali before extraction. Acid or alkali hydrolysis is required to release covalently and ionically bound lipids to proteins and carbohydrates as well as to break emulsified fats. Digestion of the sample with acid (usually 3–6 M HCl) under reflux conditions converts such bound lipids to an easily extractable form. Many dairy products, including butter, cheese, milk, and milk-based products, require alkali pretreatment with ammonia to break emulsified fat, neutralize any acid, and solubilize proteins before solvent extraction [8]. Enzymes are also employed to hydrolyze food carbohydrates and proteins (e.g., use of Clarase, a mixture of α-amylase and protease) [2].

C. LIPID EXTRACTION WITH SOLVENTS

The insolubility of lipids in water makes possible their separation from proteins, carbohydrates, and water in the tissues. Lipids have a wide range of relative hydrophobicity depending on their molecular constituents. In routine food analysis, fat content (sometimes called the ether extract, neutral fat, or crude fat) refers to free lipid constituents that can be extracted into less polar solvents, such as light petroleum ether or diethyl ether. The bound lipid constituents require more polar solvents, such as alkanols, for their extraction. Therefore, use of a single universal solvent for extraction of lipids from tissues is not possible. During solvent extraction, van der Waals and electrostatic interactions as well as hydrogen bonds are broken to different extents; however, covalent bonds remain intact.

Neutral lipids are hydrophobically bound and can be extracted from tissues by nonpolar solvents, whereas polar lipids, which are bound predominantly by electrostatic forces and hydrogen bonding, require polar solvents capable of breaking such bonds. However, less polar neutral lipids, such as TAGs and cholesterol esters, may also be extracted incompletely with nonpolar solvents, probably due to inaccessibility of a significant part of these lipids to the solvents. Lipids that are covalently bound to polypeptide and polysaccharide groups will not be extracted at all by organic solvents and will remain in the nonlipid residue. Therefore, a hydrolysis step may be required to release covalently bound lipids to render them fully extractable.

1. Properties of Solvents and Their Mode of Extraction

The type of solvent and the actual method of lipid extraction depend on both the chemical nature of the sample and the type of lipid extract (e.g., total lipids, surface lipids of leaves) desired. The most important characteristic of the ideal solvent for lipid extraction is the high solubility of lipids coupled with low or no solubility of proteins, amino acids, and carbohydrates. The extracting solvent may also prevent enzymatic hydrolysis of lipids, thus ensuring the absence of side reactions. The solvent should readily penetrate sample particles and should have a relatively low boiling point to evaporate readily without leaving any residues when recovering lipids. The solvents mostly used for isolation of lipids are alcohols (methanol, ethanol, isopropanol, *n*-butanol), acetone, acetonitrile, ethers (diethyl ether, isopropyl ether, dioxane, tetrahydrofuran), halocarbons (chloroform, dichloromethane), hydrocarbons (hexane, benzene, cyclohexane, isooctane), or their mixtures. Although solvents such as benzene are useful in lipid extraction, it is advisable to look for alternative solvents because of the potential carcinogenicity of such products. Flammability and toxicity of the solvent are also important considerations to minimize potential hazards as well as cost and nonhygroscopicity.

Solubility of lipids in organic solvents is dictated by the proportion of the nonpolar hydrocarbon chain of the fatty acids or other aliphatic moieties and polar functional groups, such as phosphate or sugar moieties, in their molecules. Lipids containing no distinguishable polar groups (e.g., TAGs or cholesterol esters) are highly soluble in hydrocarbon solvents such as hexane, benzene, or cyclohexane and in more polar solvents such as chloroform or diethyl ether, but remain insoluble in polar solvents such as methanol. The solubility of such lipids in alcoholic solvents increases with the chain length of the hydrocarbon moiety of the alcohol; therefore, they are more soluble in ethanol and completely soluble in *n*-butanol. Similarly, the shorter-chain fatty acid residues in the lipids have greater solubility in more polar solvents (e.g., tributyrin is completely soluble in methanol, whereas tripalmitin is insoluble). Polar lipids are only sparingly soluble in hydrocarbon solvents unless solubilized by association with other lipids; however, they dissolve readily in more polar solvents such as methanol, ethanol, or chloroform [4].

2. Extraction Methods with Single Organic Solvent

Diethyl ether and petroleum ether are the most commonly used solvents for extraction of lipids. In addition, hexane and sometimes pentane are preferred to obtain lipids from oilseeds. Diethyl ether (bp 34.6°C) has a better solvation ability for lipids compared with petroleum ether. Petroleum ether is the low boiling point fraction (bp 35°C–38°C) of petroleum and mainly contains hexanes and pentanes. It is more hydrophobic than diethyl ether and therefore selective for more hydrophobic lipids [5,9]. The main component (>95%) of dietary lipids are TAGs, whereas the remaining lipids are mono- and diacylglycerols, phospho- and glycolipids, and sterols. Therefore, nonpolar solvent extractions have been widely employed to extract and determine lipid content of foods. However, oil-soluble flavors, vitamins, and color compounds may also be extracted and determined as lipids when less polar solvents are used.

In determining total lipid content, several equipments and methods have been developed that utilize single-solvent extraction. Among them the gravimetric methods are most commonly used for routine analysis purposes. In gravimetric methods, lipids of the sample are extracted with a suitable solvent continuously, semicontinuously, or discontinuously. The fat content is quantified as weight loss of the sample or by weight of the fat removed. The continuous solvent extraction (e.g., Goldfisch and Foss-Let) gives a continuous flow of boiling solvent to flow over the sample (held in a ceramic thimble) for a long period. This gives a faster and more efficient extraction than semicontinuous methods but may result in incomplete extraction due to channeling. In the semicontinuous solvent extraction (e.g., Soxhlet, Soxtec), the solvent accumulates in the extraction chamber (sample is held in a filter paper thimble) for 5–10 min and then siphons back to the boiling flasks. This method requires a longer time than the continuous method, provides a soaking effect for the sample, and does not result in channeling. In the direct or discontinuous solvent extraction, there is no continuous flow of solvent and the sample is extracted with a fixed volume of solvent. After a certain period of time the solvent layer is recovered, and the dissolved fat is isolated by evaporating the organic solvent. Rose-Gottlieb, modified Mojonnier, and Schmid–Boudzynski–Ratzlaff (SBR) methods are examples, and these always include acid or base dissolution of proteins to release lipids [8]. Such procedures sometimes employ a combination extraction with diethyl and petroleum ethers to obtain lipids from dairy products. Use of these solvents may allow extraction of mono-, di-, and triacylglycerols, most of the sterols and glycolipids, but may not remove PLs and free fatty acids (FFAs).

3. Methods Using Organic Solvent Combination

A single nonpolar solvent may not extract the polar lipids from tissues under most circumstances. To ensure a complete and quantitative recovery of tissue lipids, a solvent system composed of varying proportions of polar and nonpolar components may be used. Such a mixture extracts total lipids more exhaustively and the extract is suitable for further lipid characterization. The methods of Folch et al. [10] and Bligh and Dyer [11] are most widely used for total lipid extraction. Use of a polar solvent alone may leave nonpolar lipids in the residue; when lipid-free apoproteins are to be isolated, tissues are defatted with polar solvents only [12]. It is also accepted that the water in tissues or water used to wash lipid extracts markedly alters the properties of organic solvents used for lipid extraction.

Commonly the chloroform–methanol (2:1, v/v) solvent system [10] provides an efficient medium for complete extraction of lipids from animal, plant, or bacterial tissues. The initial solvent system is binary; during the extraction process, it becomes a ternary system consisting of chloroform, methanol, and water in various proportions, depending on the moisture content of the sample [11]. The method of Bligh and Dyer [11] specifically recognizes the importance of water in the extraction of lipids from most tissues and also plays an important role in purifying the resulting lipid extract. A typical Folch procedure uses a solvent-to-sample ratio of 2:1 (v/w) with a mixture of chloroform and methanol (2:1, v/v) in a two-step extraction. The sample is homogenized with the solvent and the resultant mixture is filtered to recover the lipid mixture from the residue. Repeated extractions are usually carried out, separated by washings with fresh solvent mixtures of a similar composition. It is usually accepted that about 95% of tissue lipids are extracted during the first step. In this method, if the initial sample contains a significant amount of water, it may be necessary to perform a preliminary extraction with 1:2 (v/v) chloroform–methanol in order to obtain a one-phase solution. This extract is then diluted with water or a salt solution (0.08% KCl, w/v) until the phases separate and the lower phase containing lipids is collected. Bligh and Dyer [11] use 1:1 (v/v) chloroform–methanol for the first step extraction and the ratio is adjusted to 2:1 (v/v) in the alternate step of extraction and washing. The original procedure of Folch or of Bligh and Dyer uses large amounts of sample (40–100 g) and solvents; therefore, the amounts may be scaled down when a small amount of sample is present or for routine analysis in the laboratory. Hence, Lee and coworkers [13] have described a method that uses the same solvent combination, but in different proportions, based on the anticipated lipid content of the sample. According to this method, chloroform–methanol ratios of 2:1 (v/v) for fatty tissues (>10% lipid) and 1:2 (v/v) for lean (<2%) tissues are recommended. A modified Bligh and Dyer extraction that replaces methanol with propan-2-ol and chloroform with cyclohexane has been described by Smedes [14]. This procedure eliminates the use of chlorinated organic solvents. In this extraction, a mixture of water:propan-2-ol:cyclohexane (11:8:10, v/v/v) is employed, and subsequent separation step brings all lipids to the upper most layer containing cyclohexane [14].

Folch extraction recovers neutral lipids, diacylglycerophospholipids, and most of the sphingolipids. Lysophospholipids are only partly recovered, and more polar acidic PLs and glycolipids may be lost during washing with water. However, both Folch and Bligh–Dyer procedures may fail to transfer all of the lipids to the organic phase. Lysophospholipids, phosphoinositides, and other highly polar lipid substances are selectively lost. According to Christie [4], tissues rich in phosphoinositides should be stored in such a manner as to minimize their enzymatic degradation, and solvent extraction should be performed initially in the presence of $CaCl_2$. When lysophosphatides are the major component of the tissue extract, it is recommended that acids or inorganic salts be added during extraction with chloroform–methanol or *n*-butanol saturated with water. Therefore, specific applications and modifications of the method are required to ensure complete recovery of tissue lipids.

Due to the potential health hazards of chloroform, solvent mixtures containing alkane–alcohol–water mixtures such as hexane and isopropanol, with or without water, have been successfully used to extract tissue [15,16] and fish meal lipids [17]. Hexane–isopropanol (3:2, v/v) [15,17,18], heptane–ethanol–water–sodium dodecylsulfate (1:1:1, 0.05, v/v/v/w) [19], methylene chloride–methanol (2:1, v/v) [20,21], and hexane–acetone (1:1, v/v) [22] are such solvent combinations employed to extract lipids from biological materials. Azeotropes of isopropanol have also been used to extract lipids from oilseeds as substitutes for hexane [23–25]. Water-saturated *n*-butanol [26] has been most effective in extracting lipids from cereals that are rich in starch. This solvent mixture is used extensively for extracting lipids from starchy foods; however, acid hydrolysis might be needed to release bound lipids or inclusion complexes before their extraction.

Pressurized fluid extraction (PFE), pressurized liquid extraction (PLE), or accelerated solvent extraction (ASE; trade name used by Dionex) techniques have been developed to enhance the capabilities of conventional solvent extraction. These techniques use classical solvent systems to extract lipids, but under varying extraction parameters such as temperature, pressure, and volume. The extraction media are organic solvents or their aqueous mixtures in which lipids are soluble. The system is operated under high pressure and the solvent is kept at much higher temperatures above its atmospheric boiling point. The elevated temperature at which the extraction is conducted increases the capacity of the solvent to solubilize the analyte. Elevated temperature is also known to weaken the bonds between the analyte and the matrix, thus decrease the viscosity of the solvent with improved penetration into the matrix, resulting in an increased extraction yield. PFE can cut down the solvent consumption by 50% when compared with conventional methods such as Folch extraction procedure [27].

ASE system or ASE is the automated PFE/PLE [28,29]. The ASE process consumes a much lower solvent volume and time as lipid is extracted at temperatures well above the boiling point of the solvent because of the elevated pressure used in the process. This enhances solubilization and diffusion of lipids from samples into the solvent, significantly shortening the extraction time and solvent consumption. The fat could be extracted with no outflow of solvent (static mode) or allowing fresh solvent to flow continuously through the sample (dynamic mode) during extraction. Under elevated temperature and pressure, dissolved lipids diffuse from the core to the surface of the sample particles and then are transferred to the extraction solvent. Compressed gas then purges the solubilized fat into a collection vessel and can then be quantified gravimetrically [9]. According to Shafer [29], the content of fatty acids of the lipids extracted from muscle matrices using ASE (Dionex 200 or 300 System, chloroform–methanol solvent system) was similar or better in comparison with the conventional Folch extraction. The automated solvent extractors contain microwave moisture analyzer to dry the sample before extraction, redry to remove solvent and moisture, and to determine the percentage of fat as weight loss due to the extraction process [2]. Possibility of using ASE system for sequential extraction of lipid classes has been described by Poerschmann and Carlson [30]. Under optimum extraction conditions n-hexane/acetone (9:1, v/v) at 50°C (two cycles, 10 min each) to obtain neutral lipids followed by chloroform/methanol (1:4, v/v) extraction at 110°C (two cycles, 10 min) has been used for phytoflankton lipid extraction with ASE.

4. Methods Using Nonorganic Solvents

a. Microwave-Assisted Extraction

Due to environmental concerns and potential health hazards of organic solvents, nonorganic solvents have become popular. The use of microwave digestion for isolating lipids has recently been reported [31]. It is suggested that microwave energy, by increasing the rotational force on bonds connecting dipolar moieties to adjacent molecules, reduces the energy required to disrupt hydrophobic associations, hydrogen bonding, and electrostatic forces, thus helping to dissolve all kinds of lipids [31]. A solvent with sufficient dielectric constant is a requirement to absorb microwave energy. Closed-vessel microwave-assisted extraction is performed at high pressure and therefore, allows extraction at temperatures above boiling point of the solvent. Open-vessel microwave-assisted extraction system works at ambient pressure and refluxing of the solvent. Since extraction time is also a function of the temperature, a high boiling point of the solvent accelerates the extraction. However, evaporation of sample water is a possibility when open-vessel microwave-assisted extraction is used, which facilitates decomposition of cell structure and removal of lipids from their association with cell membrane and lipoprotein. However, low lipid yield is a concern due to increased polarity of solvent when the water content is high; thus, sample drying is recommended [32].

Microwave technology has allowed the development of rapid, safe, and cost-effective methods for extracting lipids and does not require that samples be devoid of water [33]. Performance of microwave lipid extraction was qualitatively (all lipid classes) and quantitatively comparable with that of the conventional Folch method for various biological samples [31].

b. Supercritical Fluid Extraction

When carbon dioxide is compressed at a temperature (31.1°C) and pressure (72.9 atm) above its critical point, it does not liquify but attains a dense gaseous state that behaves like a solvent. Thus, it is called supercritical CO_2 (SC-CO_2). Use of SC-CO_2 for lipid extraction significantly reduces the use of organic solvents, avoids waste disposal problems, eliminates the use of potentially toxic and flammable solvents, and reduces the extraction time. Lipids so extracted are not subjected to high temperatures during the extraction process.

Extraction using SC-CO_2 yields a good recovery of nonpolar lipids including esterified fatty acids, acylglycerols, and unsaponifiable matter. Complex polar lipids are only sparingly soluble in SC-CO_2. The polarity of SC-CO_2 can be varied by using an entrainer or modifier such as methanol, ethanol, or even water to improve the extraction of polar lipids [34–37]. This technique has been used for the extraction of lipids from various matrices, including dehydrated foods [38,39], meats [40–42], oilseeds [43], and fried foods [44]. Particle size also affects lipid recovery because it influences the surface area exposed to SC-CO_2. High moisture content decreases contact between sample and SC-CO_2 as well as the diffusion lipids outside the sample [45]. The extracted lipids from meat or hydrolytic products from the acid hydrolysis step are allowed to absorb onto a solid phase extraction (SPE) matrix and SC-CO_2 can be used to extract the adsorbed lipids [41]. An increased lipid recovery with decreased moisture content has been demonstrated in wet samples, such as meat [40,46–49]. Therefore, lyophilization is suggested to improve the extraction efficiency of lipids from samples with a high moisture content. The SC-CO_2 extraction is able to recover 97%–100% of lipids when compared with the conventional solvent extraction methods [50,51]; no significant differences between fatty acids extracted were observed. Several researchers have shown that supercritical fluid extraction (SFE) could replace solvent extraction methods in a large variety of samples. In fact, SFE has recently been included in the recommended methods of the AOAC to extract lipids from oilseeds [52]. According to Barthet and Daun [43], for canola and flaxseed a modifier ethanol (15%, v/v) and multiple extractions are needed to obtain similar values of oil content on exhaustive solvent extraction with petroleum ether. They noted that mustard always rendered 10% lower values when SC-CO_2 extraction was used. The main drawback of SC-CO_2 is equipment cost and the extraction of nonfat materials, such as water [53].

D. LIPID EXTRACTION WITHOUT SOLVENTS

Lipid extraction methods are mostly wet extraction procedures that do not use solvents, and lipid content is quantified by volumetric means. Such procedures are well utilized in determining fat content of dairy foods, especially fresh milk, and require the use of specifically designed glasswares and equipment.

1. Acid Digestion Methods

Babcock and Gerber methods are classical examples for acid digestion methods. The basic principle of these methods is destabilization and release of fat from the emulsion with a strong acid (e.g., sulfuric). The less dense fat rises in the calibrated neck of the Babcock bottle, and the centrifugation step helps the separation. Added sulfuric acid digests proteins, generates heat, and releases fat. The content of fat is measured volumetrically and expressed as weight percent. The modified Babcock method uses an acetic–perchloric acid mixture rather than sulfuric acid and is employed to determine essential oil in flavor extracts and products containing sugar and chocolate. The Gerber method uses a principle similar to that of the Babcock method but utilizes sulfuric acid and pentanol. Pentanol prevents charring of sugar, which can occur with the Babcock method; therefore, the Gerber method could be applied to a wide variety of dairy-based foods [2,8].

2. Detergent Method

The detergent method uses a detergent to form a protein–detergent complex to break up emulsion and release fat. For milk, the anionic detergent dioctyl sodium phosphate is added to disperse the protein layer that stabilizes and liberates fat. Then a strong hydrophilic nonionic polyoxyethylene detergent, sorbitan monolaurate, is added to separate the fat from other food components [5].

3. Physical Methods

External compression forces may be used to release tissue contents and extract lipids, especially from the dry matter. Oilseeds (moisture <5%, oil >30%) are generally subjected to expeller pressing to obtain lipids without using solvents, however, may not afford complete recovery of oils.

E. REMOVAL OF NONLIPID CONTAMINANTS FROM LIPID EXTRACTS AND OTHER PRACTICAL CONSIDERATIONS

Removal of nonlipid contaminants from the lipid extract is necessary since most of the solvents employed also dissolve significant amounts of oil-soluble flavors, pigments, sugars, amino acids, short-chain peptides, inorganic salts, and urea. The nonlipid matter must be removed before gravimetric determination of total lipids in order to prevent contamination during subsequent fractionation of the total extract. In chloroform–methanol extract, the commonly used method for removing nonlipid contaminants includes washing with water or a diluted KCl solution (0.88%, w/v). Use of salt solution has the advantage of preventing or minimizing the formation of an intermediate phase. When chloroform–methanol (2:1, v/v) is used for extraction of the sample, addition of water or diluted salt solution results in the formation of a two-phase system, that is, a lower phase consisting of chloroform–methanol–water (86:14:1, v/v/v) and an upper phase consisting of the same, but in the ratio of 3:48:47 (v/v/v). The lower phase composes of about two-thirds of the total volume and contains the lipid components, but the upper phase retains the nonlipid contaminants. However, more polar lipids, such as some PLs and glycolipids and all gangliosides, may remain in the upper phase [4,54]. Nonlipid contaminants may also be removed partly or completely by evaporation of the lipid extract to dryness in vacuo or under nitrogen and then reextracted with a nonpolar solvent, such as hexane.

In the Bligh and Dyer [11] method, the sample is homogenized with chloroform and methanol in such proportions that a miscible system is formed with water in the sample. Dilution of the homogenate with chloroform and water separates it into two layers; the chloroform layer contains all the lipids and the methanol–water layer contains all the nonlipid matter. A purified lipid extract could be obtained by isolating the chloroform layer. Traces of moisture can be removed by passing the chloroform extract through a bed of anhydrous sodium sulfate.

Removal of nonlipid contaminants by liquid–liquid partition chromatography on a dextran gel was introduced by Wells and Dittmer [55]. This is done by passing the crude lipid extract through a Sephadex G-25 column (packed in the upper phase of chloroform–methanol–water, 8:4:3, v/v/v or Folch wash). Lipids free of contaminants would be eluted rapidly from the column by employing the lower phase of the Folch wash. Gangliosides and nonlipids are retained in the column and can be recovered by washing with the upper phase of Folch wash and at the same time regenerating the column [56].

Use of predistilled solvents for lipid extraction is advisable since all solvents contain small amounts of lipid contaminants. Use of plastic containers and non-Teflon apparatus should also be avoided as plasticizers may leach out to the lipid extract. To prevent autoxidation of unsaturated lipids it is advisable to add an antioxidant (e.g., BHT) to the solvent (at a level of 50–100 mg/L). Furthermore, extraction should be performed under an inert nitrogen atmosphere, and both tissue and tissue extracts should be stored at −20°C under nitrogen, if possible. Most of the methods described earlier are suitable for quantifying total lipid content of the sample of interest. When high temperatures are involved in extraction, the resulting lipid extract is not suitable for further composition analysis. Folch method-based extraction is usually the preferred procedure to obtain total lipids for further analysis. Lipids are recovered from the chloroform layer by removal of solvent at low temperature under vacuum. Acid hydrolysis also results in decomposition of PLs and possibly TAGs to a certain extent [8].

Quantification of lipids from the extracts is mostly carried out as the weight difference of an aliquot after solvent removal. Removal of the solvent from lipid extracts should be conducted under vacuum in a rotary evaporator at or near room temperature. When a large amount of solvent must be evaporated, the solution must be concentrated and transferred to a small vessel so that the lipids do not dry out as a thin film over a large area of glass. Lipids should be stored immediately in an inert nonalcoholic solvent such as chloroform, rather than being allowed to remain in dry state for long [4].

III. INDIRECT METHODS OF TOTAL LIPID DETERMINATION

Several techniques and instruments have been developed and applied to the indirect and rapid determination of total lipid content of samples. These methods are really not lipid extraction methods, but they are gaining popularity because they are rapid and largely nondestructive. Most of these methods rely on a standard reference procedure and must be calibrated against a methodology to be validated.

A. Density Measurement

It has been reported that the density of flaxseed is highly correlated ($r = 0.96$) with its oil content [57]. Thus, measurement of seed density may be used as a means of screening flax genetic lines for high oil content.

B. Dielectric Method

The dielectric constant of a selected solvent changes when fat is dissolved in it. After an oilseed sample has been ground with a solvent and the dielectric constant of the mixture measured, the lipid content is determined from standard charts that show variation of the dielectric constants of different amounts of lipid in the same solvent [5]. According to Hunt et al. [58], the amount of induced current and oil content determined by solvent extraction of soybean are linearly related ($r = 0.98$).

C. Near-Infrared Spectroscopy

The near-infrared (NIR) reflectance is in the range of 14,300–4,000 cm^{-1} (700–2500 nm) due to overtone and combination bands of C–H, O–H, and N–H. NIR spectrometry can be used for determination of contents of oil, protein, and moisture and serves as a very useful tool in the routine analysis of oilseeds.

Rodrigneuz-Otero et al. [59] have used NIR spectroscopy for measurement of fat, protein, and total solids of cheese. Lee et al. [60] have used short-wavelength (700–1100 nm) NIR with a bifurcated fiberoptic probe to estimate the crude lipid content in the muscle of whole rainbow trout. A very good correlation was observed between fat content determined by chemical analysis and NIR reflectance spectroscopic values obtained for farmed Atlantic salmon fillets [61]. The use of mid-IR spectroscopy to determine lipid content of milk and dairy products has been described by Biggs [62]. Lipids absorb IR energy at the wavelength of 5730 nm, and the energy absorbed depends on the lipid content of the sample. Quantification is carried out by the standard curve of the IR absorption and lipid content determined by standard analytical methods [63]. Details about the use of IR spectroscopy for lipid analysis is provided in a later section of this chapter.

D. Low-Resolution Nuclear Magnetic Resonance Spectroscopy

Time domain low-resolution nuclear magnetic resonance (NMR) (referred to as wide-line NMR) and frequency domain NMR could be used to determine the total lipid content of foods. In time domain NMR, signals from the hydrogen nuclei (^{1}H or protons) of different food components are distinguished by their different rates of decay or nuclear relaxation. Protons of solid phases relax (signal disappear) quickly, whereas protons in the liquid phase relax very slowly. Protons of water in the sample relax faster than protons of the lipid. The intensity of the signal is proportional to the number of protons and, therefore, to the hydrogen content. Thus, the intensity of the NMR signal can be converted to oil content of the sample using calibration curves or tables [64–67]. This method can be used to determine the contents of water, oil, and solid-fat and solid-to-liquid ratio of the sample. Time domain NMR has been used to analyze the fat content of foods, including butter, margarine, shortening, chocolate, oilseed, meat, milk and milk powder, and cheese [68–70].

Frequency domain NMR distinguishes food components by resonance frequency (chemical shift) of the peaks in the spectrum. The pattern of oil resonances reflects the degree of unsaturation and other chemical properties [63,68].

E. Turbidimetric/Colorimetric Methods

Haugaard and Pettinati [71] have described a turbidimetric method for rapid determination of lipid in milk. The milk fat is homogenized to obtain uniform globules, and the milk proteins are retained with chelating agents such as EDTA. Light transmission of the sample is measured and then converted to the lipid content with the use of a conversion chart.

The lipid content of milk can also be determined using a colorimetric method [72]. The lipids of milk are allowed to react with an alkaline solution of hydroxamic acid for a specified period. On acidification and addition of ferric chloride, a relatively stable chromophore with a maximal absorbance at 540 nm is formed [73]. A colorimetric method suitable for plasma PL quantitation has been described by Hojjati and Jiang [74]. First suitable enzymes are employed to release choline from plasma PLs; that is, PC-specific phospolipase D is used for phosphotidylcholine, and spingo-myelinase and alkaline phosphatase combination can be used for spingomyelin. The resulting choline is directed to generate H_2O_2 in a reaction catalyzed by choline oxidase. Generated H_2O_2 reacts with N-ethyl-n-(2-hydroxy-3-sulfopropyl)-3,5-dimethoxyanilline sodium salt (DAOS, Trinders reagent) and 4-aminoantipyrine in a reaction that generates blue chromaphore (maximum absorbance at 595 nm) and peroxidase functions as a catalyst [74].

F. ULTRASONIC METHOD

Fitzgerald et al. [75] have described an ultrasonic method to determine the amount of fat and nonfat solids of liquid milk. The velocity of sound increases or decreases directly with the lipid content above or below a certain critical temperature. This method of fat determination is based on the speed of sound passing through the milk at various temperatures.

G. X-RAY ABSORPTION

It is known that lean meat absorbs more x-ray than high-fat meat [76]. This fact has been used to determine lipid content in meat and meat products using a standard curve of the relationship between x-ray absorption and the lipid content determined by usual solvent extraction methods [5].

IV. ANALYSIS OF LIPID EXTRACTS

Lipid analysis is usually required to determine the composition and structure of the lipid extracted from the sample. Foods must be analyzed to reveal the content and type of saturated and unsaturated lipids as well as their cholesterol content. Such characterization provides information about the caloric value, as well as other properties, including nutritional quality and safety of lipids with respect to their cholesterol and saturated fatty acid contents. In addition, quantification of quality characteristics such as degree of unsaturation, saponification value, refractive index (RI), FFA content, solid-fat index (SFI), and oxidative stability are required to determine the market value and potential application of fats and oils. Analysis of individual components of lipids is beyond the scope of this chapter and hence is not discussed here.

A. BULK OIL PROPERTIES

The analysis of bulk properties of lipids is primarily important for defining quality characteristics of oils and fats. Therefore, methods applicable to bulk vegetable oils, confectionary fats (e.g., cocoa butter), and table fats (e.g., butter, margarine) are discussed in the following sections.

1. Degree of Unsaturation

Iodine value (IV) measures the degree of unsaturation of a lipid and is defined as the number of grams of iodine absorbed by 100 g of lipid. The source of iodine (or other halogen, usually Br_2 and Cl_2) for the reaction is Wijs or Hanus reagent; the reaction involved is essentially a volumetric titration.

In a microanalytical method for determination of the IV of lipids reported by Iskander [77], ethylenic double bonds of the lipid are saturated with bromine vapors, after which the amount of absorbed bromine is determined by neutron activation analysis (NAA).

Determination of IV gives a reasonable quantitation of lipid unsaturation if the double bonds are not conjugated with each other or with a carbonyl oxygen. Furthermore, the determination should be carried out in the absence of light for a given period and with an excess of halogen reagent used [78].

Use of hydrogenation methods to determine degree of unsaturation overcomes the limitations of halogenation methods. Hydrogenation is used to measure the degree of unsaturation of acetylenic or conjugated double bonds. Such fats do not absorb halogen readily; however, the addition of hydrogen to them is considered to be quantitative. This method is essentially a catalytic reaction of heated lipid; the amount of hydrogen absorbed is determined under standard conditions. The results are expressed on a mole basis or on the basis of IV [5].

At the low-frequency end of the fingerprint region of IR (1500–900 cm^{-1}) a band due to the CH=CH bending absorption of isolated *trans* double bonds would be observed. Beyond the isolated *trans* bond is another group of CH absorption, in this case bending vibrations, including a very strong *cis* absorption. The combination of *cis* and *trans* absorption provides a measure of the total unsaturation or IV [79–81].

2. Free Fatty Acid (FFA) Content

The presence of FFAs in an oil is an indication of insufficient processing, lipase activity, or other hydrolytic actions. Classically, the acid value, which is defined as the number of milligrams of KOH required to neutralize the free acids in 1 g of sample, is a measure of FFA content. FFAs of oils can be determined colorimetrically by dissolving oil in chloroform (or benzene), then allowing the FFAs to react with a cupric acetate solution. The organic solvent turns to a blue color because of the FFA–cupric ion complex, which has a maximum absorbance between 640 and 690 nm [82].

As there is a band attributed to the carboxyl group (COOH) in the center region of the mid-IR spectrum, Fourier transform IR (FTIR) spectroscopy can be used to determine the content of FFAs [83,84].

3. Oxidative Stability and Oxidation Products

Owing to their degree of unsaturation, lipids are very susceptible to autoxidation. Autoxidation occurs via a self-sustaining free radical mechanism that produces hydroperoxides (primary products), which in turn undergo scission to form various aldehydes, ketones, alcohols, and hydrocarbons (secondary products). The presence of secondary lipid oxidation products influences the overall quality of a lipid. Methods of determination of oxidative stability and oxidation products are discussed in detail in another chapter.

4. Refractive Index

The RI of an oil is defined as the ratio of the speed of light in vacuum (practically in air) to the speed of light in oil at a specified temperature. This ratio also provides a measure of purity of oils and may be used as a means of identifying them. The RI is measured with a refractometer, usually at 20°C–25°C for oils and 40°C for solid fats, which generally liquify at 40°C. The RI declines linearly with decreasing IV; thus, it is also used as an index for reporting the degree of hydrogenation of the oil [5].

5. Saponification Value

The saponification value provides an indication of the average molecular weight of lipids. It is defined as the amount of KOH, in milligrams, required to saponify 1 g of fat, that is, to neutralize the existing FFAs and those liberated from TAG [5].

6. Solid-Fat Index

The SFI, an empirical expression of the ratio of liquids in fat at a given temperature, is measured as the change in specific volume with respect to temperature. As the solid fat melts, the volume of the sample increases, and this change is measured by dilatometry. Detection of analysis of phase transformation of fat may also be performed, because lipids expand on melting and contract on polymorphic change to a more stable fat [5]. Use of low-resolution pulse NMR and FTIR [85–89] for determination of solid-fat content has been detailed in the literature.

B. Chromatographic Procedures for Lipid Characterization

Lipid extracts are complex mixtures of individual classes of compounds and require further separation to pure components if needed. Analysis of chemical components of lipid (e.g., lipid classes, fatty acids, *trans* fatty acids, sterols, tocopherols, pigments, etc.) primarily involves chromatographic and spectroscopic methods. Usually a combination of separation techniques is used to achieve a high degree of purity of respective lipid components and this could be analytical (for quantitation) or preparative.

The first step in the analyses involves separation of lipids into their various polarity components. It may simply separate the lipid into its polar and nonpolar fractions or may entail analysis of TAG, FFAs, sterols, steryl esters, glycolipids, and PLs. Traditionally, liquid–liquid extraction, thin-layer chromatography (TLC), or liquid–solid column chromatography have been used for fractionation, cleaning, and concentration of lipid extracts. The most commonly used chromatographic techniques for lipid analysis include column chromatography, gas chromatography (GC), high-performance liquid chromatography (HPLC), supercritical fluid chromatography (SFC), and TLC. Applications of these techniques for the analysis of food lipids are discussed in the following sections.

1. Column Chromatography

Lipid extracts are usually fractionated by column chromatography on a preparative scale before subjecting them to detailed analysis. Solid–liquid (adsorption), liquid–liquid (partition), and ion exchange chromatography are among the widely used methods of lipid fractionation. In solid–liquid chromatography, separation is based on partitioning and adsorption of the lipid components between solid and liquid (mobile) phases. Elution of the desired lipid class is achieved by varying the polarity and strength of the mobile phase. Common stationary phases for column chromatography are silica, alumina, and ion exchange resins, whereas the preferred column materials for lipid analysis are silicic acid as well as florisil (magnesium silicate).

Low-pressure column chromatography using 50–500 mesh adsorbents has been used commonly for separation of different lipid classes. The main parameters involved in column chromatography include weight of the adsorbent, conditioning of the adsorbent (moisture content), and column size. It is generally accepted that long narrow columns give the best resolution, but large-diameter columns increase sample capacity. For convenience, diameters over 5 cm and heights over 45 cm are not recommended for typical laboratory use [4,90].

In adsorption chromatography, compounds are bound to the solid adsorbent by polar, ionic, and, to a lesser extent, nonpolar or van der Waals forces. Therefore, separation of lipid components takes place according to the relative polarities of the individual components, which are determined by the number and type of nonpolar hydrophobic groups. In general, elution of the column with solvents with increasing polarity separates the lipid mixture according to increasing polarity of its components in the following order: saturated hydrocarbons, unsaturated hydrocarbons, wax esters, steryl esters, long-chain aldehydes, TAGs, long-chain alcohols, FFAs, quinones, sterols, diacylglycerols, monoacylglycerols, cerebrosides, glycosyl diacylglycerols, sulfolipids, acidic glycerophosphatides, phosphatidylethanolamine, lysophosphatidylethanolamine, phosphatidylcholine, sphingomyelin, and lysophosphatidylcholine [91]. The procedure applicable to most lipid mixtures is stepwise elution on a silicic acid column with the solvent sequence of *n*-hexane, acetone, and methanol to separate into neutral lipids, glycolipids, and PLs, respectively [4,91]. The shortcomings of this SPE are incomplete elution from silicic acid, potential contamination of other lipid classes which requires further purification and arbitrary assignment of solvents in predefined lipid classes without proof.

Complexing the adsorbent material with silver nitrate enables separation of lipid mixtures according to the number, position, and *cis* and *trans* isomerism of double bonds in unsaturated fatty acids and their derivatives. Use of borate treatment of the column material complexes the compounds containing hydroxyl groups on adjacent carbon atoms and assists the separation of glycolipids [92]. Complexing of adsorbent materials is discussed in detail in the section on TLC.

Nowadays commercial columns are prepacked with a variety of solid stationary phases, which are available for separation of lipid classes and may be referred to as SPE columns. SPE requires less time, solvent, and packing material than classical column chromatography [93]. SPE can be used for isolation, concentration, purification, and fractionation of analytes from complex mixtures [94,95]. Aminopropyl-bonded phase has been used for separation of total lipids in lipid classes obtained from different sources [96–99].

The ion exchange columns carry ionic groups that bind to the opposite charge of the ionic groups of lipids. Thus, a polymer with fixed cations binds anionic lipids from a mixture, provided that the pH of the solvent mixture allows ionization of the anionic groups. At the same time, the concentration of nonlipid anions in the solvent mixture should not compete for all of the fixed ions [92]. Some of the ion exchange chromatographic materials commonly used for lipid analysis are given in Table 5.2. Diethylaminoethyl (DEAE) cellulose is used for separation of lipid classes, and triethylaminoethyl (TEAE) cellulose is useful for separation of lipids having only ionic carboxyl groups (e.g., fatty acids, bile acids, gangliosides) or phosphatidylethanolamine from ceramide polyhexosides [91]. In polar lipid analysis, DEAE cellulose in the acetate form is the most frequently used anion exchange material. It is most effective in the pH range 3–6, and often separation of polar lipid is achieved by stepwise elution with ammonium acetate buffer in water–ethanol. The cation exchanger carboxymethylcellulose (CMC) as its sodium salt has been used occasionally over the same pH range for separation of PLs [92].

Immunoaffinity column chromatography has been used for isolation and purification of apolipoproteins. Ligand molecules (antibody, antibody to apolipoprotein or antigen, purified apolipoprotein) are immobilized on the solid support (matrix) and bind to corresponding target molecules (antigen or antibody) in a mixture of macromolecules. The bound target molecule (e.g., apolipoprotein) can then be desorbed from the ligands and eluted in the purified form using appropriate buffers [100].

2. Gas Chromatography

The GC (or GLC) analysis of lipids has been much studied in the literature. This method involves partitioning of the components of the lipid mixture in the vapor state between a mobile gas phase and a stationary nonvolatile liquid phase dispersed on an inert support.

TABLE 5.2
Ion Exchange Chromatography Materials Used in Lipid Analysis

Ionizing Group	Commercial Classification	Analytical Use
$-(CH_2)_2N + H(C_2H_5)_2$ (diethylaminoethyl)	DEAE (anionic exchanger)	Anionic lipids (phospholipids, sulfolipids, sialoglycolipids, fatty acids)
$-(CH_2)_2N + (C_2H_5)_3$ (triethylaminoethyl)	TEAE (anionic exchanger)	Anionic lipids
$-CH_2COO-$ (carboxymethyl)	CM (cationic exchanger)	Phospholipid mixtures

Source: Adapted from Hemming, F.W. and Hawthrone, J.N. in *Lipid Analysis*, BIOS Scientific, Oxford, 1996.

Analysis of fatty acid composition by GC usually requires derivatization of fatty acids to increase their volatility. Fatty acid methyl esters (FAME) may be prepared by different transmethylation techniques and then separated on GC columns and detected by flame ionization detection (FID). The gas phase for GC is usually nitrogen or helium for packed columns and helium or hydrogen for capillary columns. The identification of chromatographic peaks is based on comparison of their retention times with those of authentic samples. GC analysis of TAG of food lipids may also provide information about positional distribution of fatty acids in the molecules. Naturally occurring TAGs that are purified by TLC can then be resolved without derivatization on the basis of their carbon number or molecular weight using capillary GC equipped with 8–15 m long columns coated with methylphenyl-, methyl-, or dimethylsilicone (nonpolar capillary). Use of helium or hydrogen as a carrier gas for separation of TAG on such columns requires higher temperatures than those employed for separation of methyl esters. Mono- and diacylglycerols have to be converted to trimethylsilyl (TMS) or *tert*-butyldimethylsilyl ethers (TBDMS) for complete resolution [101,102]. A combined GC and mass spectrometric technique has been applied for determining molecular species in the glycerol esters. TMS or TBDMS derivatives of glycerol esters separated on GC may be subjected to mass spectrometric analysis in order to obtain information on their molecular structure [103]. Identification of cholesterol oxidation products has great clinical importance. Recently, several authors have described GC separation combined with mass spectroscopic identification of cholesterol oxidation products [104,105].

3. High-Performance Liquid Chromatography

HPLC is a highly efficient form of adsorption, partition, or ion exchange LC that uses a very uniform, finely divided, microspherical (5–10 μm) support of controlled porosity and degree of hydration. The adsorbent is tightly packed into a stainless steel column (10–30 cm long, 2–4 mm diameter) and requires a high-pressure pump to obtain an adequate and constant flow of solvent through the column. Elution of the column may be carried out either isocratically with a solvent mixture of constant composition, or by gradient elution in which the solvent composition may be varied linearly or in a stepwise fashion with both binary and ternary solvent systems. The column eluates are continuously monitored by means of a flow-through detector, which should be insensitive to solvent flow rate, temperature, and composition [91].

Sample derivatization is employed to facilitate the separation and to enhance the limit of detection for the HPLC analysis. Hydrolysis or saponification is done to cleave ester linkages and to obtain FFAs for their subsequent analysis. Although lipids do not possess specific UV absorption peaks, they could be detected in the region of 203–210 nm because of the presence of double bonds in the fatty acyl groups, or the functionalities such as carbonyl, carboxyl, phosphate, amino, or quaternary ammonium groups. However, this low UV range greatly restricts the choice of solvent, and it is advisable to avoid chloroform–methanol mixtures because they display a strong absorption below 245 nm. Frequently, diacylglycerols require preparation of UV-absorbing derivatives (e.g., benzoate, dinitrobenzoates, pentafluorobenzoates, and TBDMS ethers) for detection. Fatty acids can be analyzed by forming 9-anthryl-diazomethane (ADAM) derivatives and employing a fluorescence detector. Fatty acids increase their hydrophobicity when ADAM is bound to the carboxyl group; thus, the derivatives are retained longer on a reversed-phase column than fatty acids and good separation is obtained [106]. Fluorescence detectors are specific for detection and quantification of tocopherols and fluorescent derivatives of fatty acids. Evaporative light-scattering detection (ELSD) and FID have been in wide use for detection of all types of lipids following HPLC separation [101,107]. These are sometimes referred to as universal detectors. The principle of ELSD involves evaporation of mobile phase of the separated lipid fraction by a nebulizer (spray the eluent stream with a large volume of nitrogen or air) to obtain droplets of solute (lipids). These solute droplets are directed through a light source (may be a laser light source); the degree of scattering of the light is proportional to the mass of the solute [108]. The RI detection may also be used for lipid analysis [109].

Normal phase HPLC also allows the separation of normal chain and hydroxy fatty acid-containing TAG. Normal phase HPLC on silver ion-loaded anion exchange columns is currently used to resolve TAG based on their degree of unsaturation [101]. Reversed-phase HPLC (using C18 columns) is also widely applied for separation and quantification of tri-, di-, and monoacylglycerols. Sehat et al. [110] have described that silver-ion HPLC can be employed to separate and identify conjugated linoleic acid isomers.

Since glycerol possesses a prochiral carbon, asymmetrical esterification of the primary position leads to the formation of enantiomers. Although enantiomeric TAGs cannot be resolved by normal HPLC, their diastereomeric naphthylethylurethane derivatives can be separated by HPLC on silica gel [111,112]. Use of a stationary phase with chiral moieties to separate enantiomers of mono- and diacyl *sn*-glycerols after derivatization with 3,5-dinitrophenylurethane (DNPU) by HPLC separation of enantiomers has been reported [113–115].

Separation of PLs is very laborious when TLC is used; however, HPLC provides a better means of separation and quantification. At present, use of gradient (binary or tertiary) elution on silica columns is frequently used for separation of different classes of PLs. Several polar solvent systems that are suitable for such separations are available [116,117]. UV detection, FID, or ELSD is suitable for PL identification and quantification using silica or normal phase chromatography [116–118]. Separation of glycolipids can be achieved using a silica column with a binary gradient (hexane–IPA–2.8 mM ammonium acetate) [119] or reversed-phase C18 (ODS) column [108].

Isocratic, normal phase separation of cholesterol esters, FFAs, and free sterols is widely used. Simultaneous analysis of nonpolar and polar lipids using HPLC silica gel column has also been reported. In addition to normal- and reversed-phase methods, size exclusion HPLC has been used to separate TAG and other nonpolar lipids. This has been specifically employed for the analysis of polymerized lipids such as those generated during deep-fat frying [67]. Christopoulou and Perkins [120] have described separation of monomers, dimers, and trimers of fatty acids in oxidized lipids using size exclusion column with an RI detector, whereas Burkov and Henderson [121] have reported the use of a similar column with ELSD to analyze polymers in autoxidized marine oils.

Micro-HPLC columns have the volume of one-hundredth of that of a conventional column and benefit from low consumption of material used as stationary phase, sample and mobile phase, operation at low flow rate, and temperature programming. It can also be easily coupled with a mass spectrometer and an FTIR detector [106].

4. Supercritical Fluid Chromatography

Use of SFC for lipid extraction was discussed in a previous section. When CO_2 is compressed at a temperature and pressure above its critical point it does not liquify but forms a dense gas; thus, as a mobile phase SFC is gaseous and solvating. Such a dense gas has a number of properties (e.g., relatively high densities and diffusivities) that make it attractive for use as a mobile phase for LC. SFC with open tubular columns acts as a substitute for GC, but with the analysis temperature much lower than that employed in GC.

The temperature and pressure required for SC-CO_2 are much lower than that for HPLC. As CO_2 is nonpolar its SFC can dissolve less polar compounds and is suitable for the analysis of less polar species. To analyze polar components, polar solvents such as methanol or ethanol may be added to the SFC to cover active sites ($-Si-OH$) on the surface of the supporting material and to increase the dissolving power of the mobile phase. Both packed and capillary columns are used for SFC. Packing materials developed for HPLC are suitable for SFC-packed columns [122]. Similarly, fused silica capillary tubes used for GC are suitable for SFC, and stationary phases may employ dimethylpolysiloxane, methylphenylpolysiloxane, diphenylpolysiloxane, and cyanopropylpolysiloxane [123].

Capillary SFC with FID or UV has been used for analysis of TAG, FFAs, and their derivatives [124–126]. SFC argentation chromatography has been used for the separation of TAG according to the number of double bonds, chain length, and the nature of the double bonds [127,128]. It is necessary to add acetonitrile as a polar modifier to CO_2 to facilitate elution of TAGs. For improved analytical efficiency of lipid having a narrow range of unsaturation, mobile phase gradient (e.g., temperature, pressure and density, velocity) can be employed. Detectors for ELSD, FID, UV, mass spectrometry (MS), and FTIR developed for GC and HPLC are applicable to SFC. Combination of SFC with SFE has been successful for analyzing lipids of different food samples. SFE may replace any conventional lipid extraction method, and the quantification of extracted lipids (instead of gravimetric analysis) can be performed with a detector (ELSD) that has been directly connected to the extraction cell. SFE–SFC has also been used to characterize TAG patterns of seed oils [122,129]. SFC is a viable alternative for reducing any solvent use in lipid extraction and analysis, and has a great potential for further development.

5. Thin-Layer Chromatography

TLC is one of the main analytical tools used for lipid analysis. TLC can be used for fractionation of complex lipid mixtures, assay of purity, identification, information on the structure, as well as for monitoring extraction and separation of components via preparative column chromatography for routine and experimental purposes. The principles and theory of TLC are based on the difference in the affinity of a component toward a stationary and a mobile phase. The important components of TLC are the stationary phase, mobile phase, detection, and quantification [5]. The adsorbent generally used in TLC for lipid analysis is very fine grade silica gel and may contain calcium sulfate as a binder to ensure adhesion to the plate [4]. Alumina and kieselguhr are also used as stationary phases. These adsorbents can also be modified by impregnation with other substances so as to achieve the desired separations. The most popular impregnations are with boric acid or silver nitrate. Silver nitrate-impregnated (argentation) TLC may be used to separate TAGs or FAMEs according to the number of double bonds and also by virtue of the geometry (e.g., *cis* or *trans*) and position of the double bonds in the alkyl chain. The silver ion forms a reversible complex with the π electrons of the double bond of unsaturated fatty acids, thereby decreasing their mobility [130]. Structural isomers of TAGs (due to their fatty acid constituents) may also be separated on this type of TLC plate [107]. Impregnation of the TLC plates with boric acid (3%, w/v) prevents isomerization of mono- and diacylglycerols while separating neutral lipids [91]. Boric acid complexes with vicinal hydroxyl groups and leads to slower migration of these compounds [130].

Samples of lipid extracts are applied as discrete spots or as narrow streaks, 1.5–2 cm from the bottom of the plate. The plate is then developed in a chamber containing the developing solvent or a solvent mixture. The solvent moves up the plate by capillary action, taking the various components with it at different rates, depending on their polarity and how tightly they might be held by the adsorbent. The plate is removed from the developing chamber when the solvent approaches the top of it and then dried in the air or under a flow of nitrogen. Solvents with low boiling point, viscosity, and toxicity are suitable for TLC application. A low boiling point helps in the quick evaporation of the solvent from the surface layer, and low viscosity facilitates faster movement of the solvent during development. The selection of a suitable solvent is very important for good separation of the lipid classes. Several solvent systems may be used to resolve individual lipid classes, as exemplified in Table 5.3.

The location of the corresponding lipid spots on a developed plate has to be detected before their isolation. Detection of the spots may be done using a reagent directly on the plate. This reagent could be specific to certain functional groups of the lipid molecules or may be a nonspecific reagent that renders all lipids visible. There are nondestructive chemical reagents, such as 2′,7′-dichlorofluorescin in 95% methanol (1%, w/v), iodine, rhodamine 6G, and water, which allow recovery of lipids after detection. Lipids exhibit a yellow color and in the presence of rhodamine 6G (0.01%, w/v) produce pink spots under UV light. These chemicals may also be

TABLE 5.3

Solvent Systems That Could Be Used for Separation of Lipids by TLC

Lipid Component	TLC Adsorbent	Solvent System	References
Complex lipids (animal tissues)	Silica gel G	Chloroform–methanol–water, 25:10:1 (v/v/v)	[131]
	Silica gel H	Chloroform–methanol–acetic acid–water, 25:15:4:2 (v/v/v/v)	[131]
	Silica gel H	First developing system, pyridine–hexane, 3:1 (v/v) and second developing system, chloroform–methanol–pyridine–2 M ammonia, 35:12:65:1 (v/v/v/v)	[131]
Complex lipids (plant tissues)	Silica gel G	Acetone–acetic acid–water, 100:2:1 (v/v/v)	[132]
	Silica gel G	Diisobutyl ketone–acetic acid, 40:25:3.7 (v/v/v)	[4]
Simple lipids	Silica gel G	Hexane–diethyl ether–formic acid, 80:20:2 (v/v/v)	[4,133]
	Silica gel G	Benzene–diethyl ether–ethyl acetate–acetic acid, 80:10:10:0.2 (v/v/v/v)	[4,133]
Partial acylglycerol	Silica gel G containing 5% (w/v) boric acid	Chloroform–acetone, 96:4 (v/v)	[4]
Neutral plasmalogens	Silica gel G	Hexane–diethyl ether, 95:5 (v/v) in first direction, hexane–diethyl ether, 80:20 (v/v) in second direction	[4]

used as a nondestructive spray for preparative TLC. When water is used separated lipids may appear as white spots in a translucent background and can easily be distinguished. Developed plates may also be subjected to saturated iodine vapor in a chamber and this may produce brown spots because of the reaction of iodine with unsaturated bonds of the lipid molecules. However, unsaturated lipids may form artifacts with iodine, if sufficient time is allowed. The destructive methods include spraying of the plate with sulfuric acid (50%, v/v) and drying at 180°C for 1 h to make lipids visible as black deposits of carbon [4]. Potassium dichromate (5%, w/v) in 40% (v/v) sulfuric acid also works in a similar manner to sulfuric acid spray. Molybdophosphoric acid (5%, w/v) in ethanol turns lipids into blue then black when heated at 120°C for 1 h. Coomassie blue (0.03% in 20% methanol) turns lipid into blue spots on white background. Examples of specific reagents that react selectively with specific functional groups include $FeCl_3$ to detect cholesterol and cholesterol esters, ninhydrin for choline-containing PLs, and orcinol or naphthol/H_2SO_4 for glycolipids [134]. Some lipids contain chromophores and can be visualized directly under UV or visible light without staining.

Lipids detected by nondestructive methods can be recovered by scraping the bands of interest and dissolving them in an appropriate solvent. Complex lipid mixtures cannot always be separated by one-dimensional TLC; however, two-dimensional TLC may resolve such mixtures. Use of high-performance TLC (HP-TLC) plates with spot-focusing slits to minimize spot diffusion on the plate can be used for quantitative determination of analytes. This technique has been demonstrated by Kozutsumi and group [135] for determining indigenous DHA levels of cows' milk after converting to DHA methyl esters.

The separation efficiency of TLC plates is affected by the degree of hydration of the adsorbent. Therefore, activation of the adsorbent, which depends on both the time and the temperature as well as the storage conditions of the plates and the relative humidity of the atmosphere, must be considered. Use of authentic lipid standards would allow direct comparison of the resolved lipids in an unknown mixture. HP-TLC plates have high resolving power and speed of separation. They are available commercially as precoated plates with fine (5 μm) and uniform particle size silica gel. However, the amount of sample that can be applied to such HP-TLC plates is very small. TLC is a preferred method of purification and separation of lipid classes before subjecting them to further separation of individual components.

For quantification of TLC-separated lipids, traditional scraping followed by quantitation and in situ determination are available. The separated lipid classes on silica gel can be scraped off, extracted by means of suitable solvents, and quantitated gravimetrically, spectrophotometrically, or by GC. Determination of the phosphorus content of the eluted PLs is a classic example for spectrophotometric quantitation. GC can be employed to quantify separated neutral and PL classes by derivatizing the constituent fatty acids into their methyl esters. Densitometric methods provide an in situ quantification method for lipids. Lipids are sprayed with reagents, and their absorption or fluorescence can be measured under UV or visible light by densitometry.

The spots detected by charring can be measured by scanning photodensitometer areas of peaks, which are proportional to the amount of original lipid present. Scintillation counting is also possible after introducing a correct scintillator (e.g., mixture of 2,5-diphenyloxazole and 1,4-bis-2-(5-phenyloxazoyl)benzene, PPO, and POPOP in toluene) into the lipid. Scanning fluorometry allows resolution of lipids on an adsorbent containing a fluorescent dye or by spraying with such reagents.

TLC could be coupled with other methods to facilitate detection, quantitative identification, or quantitation of separated lipids. These include coupling with HPLC (HPLC/TLC), Fourier transform infrared (TLC/FTIR), nuclear magnetic resonance (TLC/NMR), and Raman spectroscopy (TLC/RS) [130]. The Iatroscan (Iatron Laboratories of Japan) applies the same principles of TLC to separate lipid mixtures followed by their detection using FID. The TLC medium is silica (7.5 μm thick) sintered onto quartz rods (0.9 mm in ID, 15 cm long, called Chromarods). Chromatographic resolution of lipid classes on Chromarods and the composition of solvent systems used are similar to those employed in classical TLC with modifications. Copper sulfate, silver nitrate, boric acid, or oxalic acid impregnation has been reported to produce better resolution and increased responses in the determination [136]. Parrish and Ackman [137] have shown that stepwise scanning and developing in solvent systems of increasing polarity resolve individual lipid components in neutral lipids of marine origin. FID gives a linear response that is proportional to the number of nonoxidized carbon atoms in the material entering the flame. The Chromarods are reusable up to 100–150 times and require thorough cleaning after each use. The main advantage of TLC-FID is the short analysis time, with the possibility of determining all components in a single analysis. The partial scanning and redevelopment [138–140] that can be done with TLC-FID give a unique advantage over any other lipid analysis technique. TLC-FID system has been used to analyze different types of lipids as detailed by Kaitaranta and Ke [141], Sebedio and Ackman [142], Tanaka et al. [143], Walton et al. [144], Indrasena et al. [145], Kramer et al. [146], Parrish et al. [147], and Pryzbylski and Eskin [148].

C. SPECTROSCOPIC METHODS OF LIPID ANALYSIS

UV and visible spectroscopy are less frequently used but have specific applications for the identification and quantification of lipids. IR spectroscopy was the first spectroscopic method applied for the analysis of lipids. NMR and MS have been widely used for lipid structure determination; however, new applications other than these have been developed. IR, UV, and NMR are nondestructive spectroscopic methodologies.

1. UV-Visible Spectroscopy

UV and visible spectra of organic compounds are attributable to electronic excitations or transitions. Functional groups with high electron density, such as carbonyl and nitro groups with double, triple, or conjugated double bonds, absorb strongly in the ultraviolet or visible range at characteristic wavelengths (λ_{max}) and molar extinction coefficients (ε_{max}). Table 5.4 provides some of the diagnostic UV absorption bands for lipid analysis. It should be noted that the λ_{max} for a compound may vary, depending on the solvent used [91,92].

TABLE 5.4

UV Absorption Characteristics of Some Chromophoric Groups

Chromophore	Example	λ_{max} (nm)	ε_{max}	Solvent
–C=C–	Octene	177	12,600	Heptane
–C≡C–	Octyne	178	10,000	Heptane
		196	2,100	
–C=C–C=C–	Butadiene	217	20,900	Hexane
–(C=C)$_n$–	Conjugated polyenes	217 + 30 (n–2)	20,000–100,000	Hexane
C$_6$H$_6$	Benzene	184	47,000	Cyclohexane
		202	7,000	
		255	230	
–(C=C–C=C)$_n$–	β–Carotene	452	139,000	Hexane
		478	12,200	
HC=O	Acetaldehyde	290	17	Hexane
C=O	Acetone	275	17	Ethanol
–COOH	Acetic to palmitic acid	208–210	32–50	Ethanol

Source: Adapted from Kates, M. in *Techniques of Lipidology, 2nd edn.*, Elsevier, New York, 1986.

2. Infrared Absorption Spectroscopy

The IR spectrum of an oil provides substantial information on the structure and functional groups of the lipid and also about the impurities associated with it. These information are represented as peaks or shoulders of the spectrum as illustrated in Figure 5.1. At the high-frequency end of the spectrum (3700–3400 cm^{-1}), there is a region where compounds containing hydroxyl groups are absorbed,

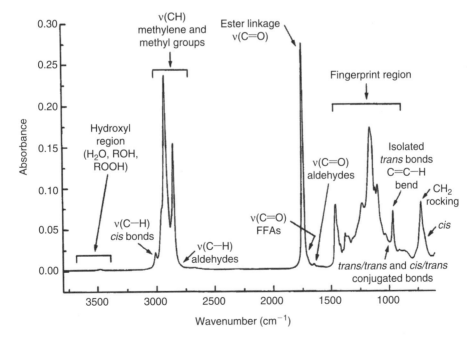

FIGURE 5.1 Mid-infrared spectrum of an edible oil collected on an attenuated total reflectance crystal. Labels indicate absorption bands or regions associated with triacylglycerols or other constituents that may be present in an oil or free fatty acid. (From van de Voort, F.R. and Sedman, J., *Inform*, 11, 614, 2000. With permission.)

including water (H–OH) and hydroperoxides (RO–H) and their breakdown products. At some lower frequencies (3025–2850 cm^{-1}) is the CH stretching region where three bands are visible: a weak *cis* double bond CH absorption (CH=CH) and strong bonds due to the CH_2 groups of the aliphatic chains of TAGs as well as the terminal methyl groups. Just beyond this region are secondary oxidation products of lipids such as aldehydes and ketones that absorb energy albeit weakly. Toward the center of the spectrum is a very strong band due to the C=O stretching absorption due to the ester linkage attaching the fatty acids to the glycerol backbone of TAGs. Next to it is a band due to COOH group of FFAs if the lipid is hydrolyzed. In the same region there would be carbonyl absorption bands of aldehydes (R–CHO) and ketones (R–CO–R) if the oil is oxidized. Continuing to lower frequencies area is the fingerprinting region (1500–900 cm^{-1}) as the pattern of bands in this region is very characteristic of molecular composition of the lipid and could be used to identify different components. At the low-frequency end of the fingerprinting region, a band due to the CH=CH bonding absorption of isolated *trans* double bonds could be seen when *trans*-containing TAGs are present in the oil. The corresponding absorption of conjugated dienes containing *trans–trans* and *cis–trans* double bonds appears at slightly higher frequencies in oxidized polyunsaturated fatty acid-containing oils. Beyond the isolated *trans* bond is another group of CH absorption bending vibrations including a very strong *cis* absorption [149].

The IR absorption signal could be employed to analyze and obtain information about qualitative structural and functional groups of lipids. In principle, since IR band intensities are linearly related to the concentration of the absorbing molecular species, quantitative information about the lipid can also be obtained [149]. IR spectroscopy has been applied to solid lipids to obtain information about polymorphism, crystal structure, conformation, and chain length. In oils, IR is commonly used to determine the presence and the content of *trans* unsaturation. Single *trans* double bonds show a characteristic absorption band at 968 cm^{-1}, and the frequency does not change for additional double bonds unless these are conjugated. There is no similar diagnostic IR absorption band for *cis* unsaturation; however, Raman spectra show strong absorption bands at 1665 ± 1 cm^{-1} (*cis*-olefin), 1670 ± 1 cm^{-1} (*trans*-olefin), and 2230 ± 1 and 2291 ± 2 cm^{-1} (acetylene) for the type of unsaturation shown [102]. Kates [91] has provided the characteristic IR absorption frequencies that have diagnostic values for identification of major classes of lipids. It has also been reported that ionic forms of PLs influence the absorption bands associated with phosphate groups that influence the interpretation of the spectra [91]. The FTIR spectrometer finds its uses in measuring IV, saponification value, and FFAs [150]. Oxidative stability of lipids as reflected in the formation of peroxides and secondary oxidation products may also be determined by FTIR [149,151].

3. Nuclear Magnetic Resonance Spectroscopy

Low-resolution pulsed ^{1}H NMR spectroscopy is employed to determine solid-fat content of lipids as well as the oil content of seeds, as discussed earlier in this chapter. High-resolution ^{1}H NMR applied to vegetable oils gives several signals with designated chemical shifts, coupling constant, splitting pattern, and area. This information can be used to obtain structural and quantitative information about lipids. Methyl stearate (saturated fatty acid ester) may give five distinct ^{1}H NMR signals as summarized in Table 5.5. Similar signals appear in methyl oleate and linoleate, but methyl oleate also gives signals for olefinic (5.35 ppm, 2H) and allylic (2.05 ppm, 4H) hydrogen atoms, and for linoleate these are at 5.35 (4H), 2.05 (4H, C8, and C14), and 2.77 ppm (2H, C11). When a double bond gets close to the methyl group, as in α-linolenate and other ω-3 esters, the CH_3 signal is shifted to 0.98 ppm; oils containing other esters (ω-6, ω-9, and saturated) exhibit a triplet at 0.89 ppm. The area associated with these various signals can be used to obtain semiquantitative information in terms of ω3 fatty acids (α-linolenate), other polyenic, monoenic (oleate), and saturated acids [102,152].

Acylglycerols show signals associated with the five hydrogen atoms on the glycerol moiety. There is a one-proton signal at 5.25 ppm (CHOCOR), which overlaps with the olefinic signals and a

TABLE 5.5
Chemical Shift (ΔH) for Methyl Alkanoates Observed for ^{1}H NMR

ΔH (ppm)	Splitting	H	Group[a]
0.90	Triplet	3	CH_3
1.31	Broad	$2n$	$(CH_2)_n$
1.58	Quintet	2	$-CH_2CH_2COOCH_3$
2.30	Triplet	2	$-CH_2CH_2COOCH_3$
3.65	Singlet	3	$-CH_2CH_2COOCH_3$

Source: Adapted from Gunstone, F.D. in *Fatty Acid and Lipid Chemistry*, Blackie, London, UK, 1996.

[a] Assigned hydrogen is designated as H.

four-proton signal located between 4.12 and 4.28 ppm (CH_2OCOR). PLs display characteristic signals for phosphatidylcholine and phosphatidylethanolamine [102].

High-resolution ^{13}C NMR spectra are more complex than ^{1}H spectra and provide more structural information than quantitative ones. It is also possible to locate functional groups such as hydroxy, epoxy, acetylenic, and branched chains in the molecules. The application of ^{13}C NMR to TAG allows determination of the positional distribution of fatty acids on the glycerol backbone [153,154]. The ^{13}C resonance of the carbonyl group of fatty acids in the *sn*-1 and *sn*-3 positions is well resolved from those esterified at the *sn*-2 position. Most unsaturated fatty acids in the *sn*-2 position are nondegenerative and could be easily differentiated.^{13}C NMR has been successfully applied for determination of positional distribution of TAG fatty acids in vegetable oils [155–157] and marine oils [158,159].

The NMR imaging is based on manipulation of magnetic field gradients oriented at right angles to each other to provide spacial encoding of signals from an object, which are converted by FT techniques to three-dimensional NMR images [160]. It produces three-dimensional data by selecting two-dimensional cross-sections in all directions. Application of NMR imaging or magnetic resonance imaging (MRI) to foods has been of interest as it is a noninvasive technique that can be applied to track the dynamic changes in foods during storage, processing, packaging, and distribution.

Most magnetic resonance images of foods are based on proton resonances from either water or lipids. Simoneau et al. [161] have applied MRI to the study of fat crystallization in bulk or dispersed systems. Halloin et al. [162] described two MRI techniques, spin-echo imaging (SEI) and chemical shift imagining (CSI), for the study of lipid distribution in pecan embryos. Insect- or fungus-damaged embryos gave images that were less intense than those of normal embryos, reflecting lower oil content. When MRI and NMR was employed to determine the oil content of French-style salad dressings, results were within ±2% of expected values and were in agreement with oil content determined by traditional methods [163]. Pilhofer et al. [164] have studied the use of MRI to investigate the formation and stability of oil-in-water emulsions formed with vegetable oil, milk fat, and milk fat fractions. Distribution of lean and fat in retail meat as a means of quality can be measured [165] using MRI and also to visualize oil and water concentration gradient during deep-fat frying food [166].

4. Mass Spectrometry

In conventional MS, compounds in their gaseous state are ionized by bombardment with electrons (electron impact) in an ionization chamber. The resulting mass spectrum consists of a characteristic pattern of peaks representing molecular fragments with different mass-to-charge (m/z) ratios. Some of these peaks or patterns of peaks are structurally diagnostic. The parent ion peak that arises from the unfragmented ionized molecule has the highest m/z ratio, but may not be always present, depending on the volatility and thermal stability of the compound. Thus, lipids containing polar groups, such as PLs, with low thermal stability and volatility and high molecular weight cannot be analyzed by conventional electron impact (EI) MS. Therefore, fast atom bombardment (FAB), chemical ionization (CI), field desorption (FD), or secondary ion (SI) MS are required for such lipid analysis [103].

MS is very useful in identifying structural modification of chain length such as branching or the presence of rings for saturated species. In this regard, matrix-assisted laser desorption ionization and time-of-flight mass spectrometry (MALDI-TOF-MS) have several advantages. It does not require prior derivatization of sample to enhance the volatility of the lipids. The extent of fragmentation of MALDI-TOF-MS is low; thus, detection of molecular ion is possible in most cases. Schiller et al. [167] have successfully used a matrix of 2,5-dihydroxybenzoic acid to identify phosphatidylcholine and different PLs as their molecular ions (M + 1). Diacylglycerols were mainly detected as their corresponding sodium or potassium adducts but not as their protonated form. MALDI-TOF-MS can be used for direct investigation of lipid mixtures occurring, for example, in cell membranes because of its high sensitivity (up to picomolar concentrations).

MS in combination with GC and HPLC is also useful in structural determination of the individual lipid molecules. Le Quere et al. [168] have developed an online hydrogenation method that allows selective hydrogenation of all the unsaturated species after chromatographic separation for deducing structural information such as carbon skeleton and double-bond equivalents. Le Quere [169] has reviewed this methodology and the use of GC–MS and tandem MS for analysis of structural features of fatty acids. Introduction of delayed ion extraction (DE) to MALDI-TOF-MS has dramatically improved the resolution and accuracy of the mass spectroscopy for molecular identification. Fujiwaki et al. [170] have described DE coupled MALDI-TOF-MS for precise identification of lysosphingolipids and gangliosides.

D. Enzymatic Methods

Higgins [171] has described an enzymatic method for determining TAG content of samples. This involves reaction of the TAG with lipase in order to obtain glycerol and FFAs. The glycerol so produced is then converted to α-glycerophosphate using glycerol kinase. The α-glycerophosphate dehydrogenase is then used to reduce nicotinamide adenine diphosphate (NAD) to NADH. The resultant NADH is then measured by a colorimetric reaction.

Cholesterol oxidase is used for determining cholesterol concentration in blood plasma. Polyunsaturated fatty acids (PUFA) with *cis*-methylene groups between their double bonds (e.g., linoleic, linolenic, and arachidonic acids) can be quantitatively measured by reading the UV absorbance of conjugated diene hydroperoxides produced via lipoxygenase (lipoxidase)-catalyzed oxidation. The extinction coefficient of diene hydroperoxides at 234 nm is the same for all PUFAs. Fatty acid esters have to be saponified before such analysis. The phosphatidylcholine or lecithin content of foods (e.g., as a measure of the egg content of foods) can be made by catalyzing conversion of lecithin to phosphatidic acid and choline by lecithinase (phospholipase D) [5].

The method of stereospecific analysis of TAG described by Brockerhoff and Yurkowski [172] uses pancreatic lipase that eventually removes fatty acids from the *sn*-1 and *sn*-3 positions of the TAG. This procedure has recently been employed to determine the existing structural differences of fish oils and seal oil [173]. Phospholipase A_2 is used to release fatty acids at the *sn*-2 position of the synthesized phosphatide during the analytical procedure. Lands [174] used a different approach to determine steric positions of the fatty acids. TAG is hydrolyzed with lipase, and the products are separated by TLC. The *sn*-3 hydroxyl of glycerol is phosphorylated with diacylglycerol kinase to produce 3-phosphoryl monoacylglycerol. In the following step of analysis, phospholipase A_2 is used to remove the fatty acids only from the *sn*-2 position. The fatty acids in the *sn*-1 position can be released by saponifying the resultant 1-acyl lysophosphatidic acid. The fatty acids in positions 1 and 2 can be identified by GC analysis. The composition of fatty acids in position 3 can be calculated by comparing these results with those from total fatty acid composition determination of TAG.

E. IMMUNOCHEMICAL METHODS

Lipids are not generally very immunogenic. However, most glycolipids (except in the pure form) possess antibodies of high activity and specificity. Therefore, glycolipids to be administered to the animal are conjugated by covalent linking to a foreign protein (as a hapten) or using them as a part of the bilayer of a liposome to stimulate the production of specific antibodies [62,64]. Immunochemical methods have also been developed for the assay of PLs and TAGs (62). The steroid hormones when conjugated with serum albumin are sufficiently immunogenic to stimulate generation of antibodies with high activity, and this allows their detection.

Immunostaining of TLC plates for detection and assay of glycoproteins is widely done. The TLC chromatogram containing separated glycolipids is treated with a radiolabeled specific antibody (usually with ^{125}I) to stain only the glycolipid antigen even in the presence of overlapping glycolipids. Detection of ^{125}I may be achieved using autoradiography, and the chromatographic mobility and antibody staining serve to identify the glycolipid [92,100,175]. To overcome low sensitivity of the immunoradiolabeled detection of glycolipids, enzyme-linked immunosorbent assay (ELISA) was developed. For ELISA, lipid is usually bound to a solid phase and the antibody is measured either by virtue of itself carrying enzyme or by using a second antibody that carries an enzyme [92,175].

V. SUMMARY

Lipids are integral components and building blocks of biological materials. To understand their constituents, chemistry, and biological functions, lipids have to be isolated and studied. Therefore, an extensive knowledge of the extraction and analysis of lipids is essential to carry out studies on lipids. This chapter provided comprehensive information on methods available for extraction and analysis of lipids from biological materials with examples when necessary. More details of a particular topic could be obtained from the references listed.

REFERENCES

1. H.D. Belitz and W. Grosch. *Food Chemistry*. Springer-Verlag, New York, NY, 1987.
2. D.E. Carpenter, J.N. Ngvainti, and S. Lee. Lipid analysis. In: *Methods of Analysis for Nutrition Labeling* (D.M. Sulivan and D.E. Carpenter, eds.). AOAC Press, Arlington, VA, 1993, pp. 85–104.
3. W.R. Bloor. Outline of a classification of the lipids. *Proc. Soc. Exp. Biol. Med.* 17:138–140 (1920).
4. W.W. Christie. *Lipid Analysis*. Pergamon Press, Oxford, 1982.
5. Y. Pomeranz and C.L. Meloan. *Food Analysis; Theory and Practice,* 4th edn. AVI, Westport, CT, 1994.
6. F. Pernet and R. Tremblay. Effect of ultrasonication and grinding on the determination of lipid class content of microalgae harvested on filters. *Lipids* 38:1191–1195 (2003).
7. B.N. Ametaj, G. Bobe, Y. Lu, J.W. Young, and D.C. Beitz. Effect of sample preparation, length of time, and sample size on quantification of total lipids from bovine liver. *J. Agric. Food Chem.* 51:2105–2110 (2003).
8. R.S. Kirk and R. Sawyer. *Pearson's Composition and Analysis of Foods,* 9th edn. Longman, London, 1992, pp. 22–26.
9. D.B. Min and D.F. Steenson. Crude fat analysis. In: *Food Analysis* (S.S. Neilson, ed.). Aspen, Gaithersburg, MD, pp. 201–215.
10. J. Folch, M. Lees, and G.H.S. Stanley. A simple method for the isolation and purification of total lipids from animal tissues. *J. Biol. Chem.* 226:497–509 (1957).
11. E.G. Bligh and W.J. Dyer. A rapid method of total lipid extraction and purification. *Can. J. Biochem. Physiol.* 37:911–917 (1959).

12. G.J. Nelson. Isolation and purification of lipids from biological matrices. In: *Analysis of Fats, Oils and Lipoproteins* (E.G. Perkins, ed.). AOCS Press, Champaign, IL, 1991, pp. 20–59.
13. C.N. Lee, B. Trevino, and M. Chaiyawat. A simple and rapid solvent extraction method for determining total lipids in fish tissues. *J. AOAC Int.* 79:487–492 (1996).
14. F. Smedes. Determination of total lipid using non-chlorinated solvents. *Analyst* 124:1711–1718 (1999).
15. A. Hara and N.S. Radin. Lipid extraction of tissues with low toxicity solvent. *Anal. Chem.* 90:420–426 (1978).
16. N.S. Radin. Extraction of tissue lipids with solvent of low toxicity. In: *Methods of Enzymology, Vol. 72* (J. Lowenstein, ed.). Academic Press, New York, NY, 1981, pp. 5–7.
17. H. Gunnlaugsdottir and R.G. Ackman. Three extraction methods for determination of lipids in fish meal; alternative to chloroform-based methods. *J. Sci. Food Agric.* 61:235–240 (1993).
18. I. Undeland, N. Harrod, and H. Lingnert. Comparison between methods using low-toxicity solvents for the extraction of lipids from heming (*Clupea harengus*). *Food Chemistry* 61:355–365 (1998).
19. G.W. Burton, A. Webb, and K.U. Ingold. A mild rapid and efficient method of lipid extraction for use in determining vitamin E/lipid ratios. *Lipids* 20:29–39 (1985).
20. H. Swaczyna and A. Montag. Estimation of cholesterol fatty acid esters in biological materials. *Fette Seifen Anstrimittel.* 86:436–446.
21. M.G.C.B. Soares, K.M.O. da Silva, and L.S. Guedes. Lipid extraction; a proposal of substitution of chloroform by dichloromethane in the method of Folch Lees and Solane. *Aquat. Biol. Technol.* 35:655–658 (1992).
22. M.E. Honeycutt, V.A. Mcfarland, and D.D. Mcsant. Comparison of three lipid extraction methods for fish. *Bull. Environ. Contam. Toxicol.* 55:469–472 (1995).
23. L.A. Johnson and E.W. Lusas. Comparison of alternative solvents for oil extraction. *J. Am. Oil Chem. Soc.* 60:181A–194A (1983).
24. E.W. Lusas, L.R. Watkins, and K.C. Rhee. Separation of fats and oils by solvent extraction; nontraditional methods. In: *Edible Fats and Oil Processing; Basic Principles and Modern Practices* (D.R. Erikson, ed.). AOCS Press, Champaign, IL, 1990, pp. 56–58.
25. E.W. Lusas, L.R. Watkins, and S.S. Koseoglu. Isopropyl alcohol to be tested as solvent. *Inform* 2:970–976 (1991).
26. W.R. Morrison and A.M. Coventry. Solvent extraction of fatty acids from amylase inclusion complexes. *Starch* 41:24–27 (1989).
27. G. Isaac, M. Waldeback, U. Eriksson, G. Odhm, and K. Markides. Total lipid extraction of homogenized and intact lean fish muscles using pressurized fluid extraction and batch extraction techniques. *J. Agric. Food Chem.* 53:5506–5512 (2005).
28. B.E. Ritcher, B.A. Jones, J.L. Ezzell, N.L. Porter, N. Avdalorie, and C. Pohl. Accelerated solvent extraction: a technique for sample preparation. *Anal. Chem.* 68:1033–1039 (1996).
29. K.C. Shafer. Accelerated solvent extraction of lipids for determining the fatty acid composition of biological material. *Anal. Chim. Acta* 358:69–77 (1998).
30. J. Poerschmann and R. Carlson. New fractionation scheme for lipid classes based on "in cell fractionation" using sequential pressurized liquid extraction. *J. Chromatogr. A* 1127:18–25 (2006).
31. C. Leary, T. Grcie, G. Guthier, and M. Bnouham. Microwave oven extraction procedure for lipid analysis in biological samples. *Analysis* 23:65–67 (1995).
32. A. Batista, W. Vetter, and B. Lukas. Use of focused open vessel microwave-assisted extraction as prelude for the determination of the fatty acid profile of fish—a comparison with results obtained after liquid–liquid extraction according to Bligh and Dyer. *Eur. Food Res. Technol.* 212:377–384 (2001).
33. J.R.J. Pare, G. Matni, J.M.R. Belanger, K. Li, C. Rule, and B. Thibert. Use of the microwave assisted process in extraction of fat from meat, dairy and egg products under atmospheric pressure condition. *J. AOAC Int.* 80:928–933 (1997).
34. F. Temelli. Extraction of triglycerides and phospholipids from canola with supercritical carbon dioxide and ethanol. *J. Food Sci.* 57:440–442 (1992).
35. M.J. Cocero and L. Calvo. Supercritical fluid extraction of sunflower seed oil with CO_2–ethanol mixtures. *J. Am. Oil Chem. Soc.* 73:1573–1578 (1996).
36. L. Montarini, J.W. King, G.R. List, and K.A. Rennick. Selective extraction of phospholipid mixture by supercritical CO_2 solvent. *J. Food Sci.* 61:1230–1233 (1996).

37. F. Diorisi, B. Aeschlimann Hug, J.M. Aeschimann, and A. Houlemar. Supercritical CO_2 extraction for total fat analysis of food products. *J. Food Sci.* 64:612–615 (1999).
38. P. Lembke and H. Engelhardt. Development of a new supercritical fluid extraction method for rapid determination of total fat content of food. *Chromatographia* 35:509–516 (1993).
39. M. Fattori, N.R. Bulley, and A. Meisen. Fatty acid and phosphorus contents of canola seed extracts obtained with supercritical carbon dioxide. *J. Agric. Food Chem.* 35:739–743 (1987).
40. J.W. King, J.H. Johnson, and J.P. Friedrich. Extraction of fat tissue from meat products with supercritical carbon dioxide. *J. Agric. Food Chem.* 37:951–954 (1989).
41. J.W. King, F.J. Eller, J.N. Snyder, J.H. Johnson, F.K. Mckeith, and C.R. Stites. Extraction of fat from ground beef for nutrient analysis using analytical supercritical fluid extraction. *J. Agric. Food Chem.* 44:2700–2704 (1996).
42. J.W. Hampson, K.C. Jones, T.A. Foglia, and K.M. Kohout. Supercritical fluid extraction of meat lipids: an alternative approach to the identification of irradiated meats. *J. Am. Oil Chem. Soc.* 73:717–721 (1996).
43. V.J. Barthet and J.K. Daun. An evaluation of supercritical fluid extraction as an analytical tool to determine fat in canola, flax and solin and mustard. *J. Am. Oil Chem. Soc.* 79:245–251 (2002).
44. N. Devineni, P. Mallikarjunan, M.S. Chinnan, and R.D. Phillips. Supercritical fluid extraction of lipids from deep fat fried food products. *J. Am. Oil Chem. Soc.* 74:1517–1523 (1997).
45. N.T. Durnford and F. Temelli. Extraction condition and moisture content of canola flakes related to lipid composition of supercritical CO_2 extracts. *J. Food Sci.* 62:155–159 (1997).
46. N.T. Durnford, F. Temelli, and E. Leblanc. Supercritical CO_2 extraction of oil and residual protein from Atlantic mackerel (*Scomber scombrus*) as affected by moisture content. *J. Food Sci.* 62:289–294 (1997).
47. M. Snyder, J.P. Friedrich, and D.D. Christianson. Effect of moisture and particle size on the extractability of oils from seeds with supercritical CO_2. *J. Am. Oil Chem. Soc.* 61:1851–1856 (1984).
48. D.D. Christianson, J.P. Friedrich, G.R. List, K. Warner, E.B. Bagley, A.C. Stringfellow, and G.E. Inglett. Supercritical fluid extraction of dry milled corn germ with CO_2. *J. Food Sci.* 49:229–232, 272 (1984).
49. Y. Ikushima, N. Saito, K. Hatakeda, S. Ito, T. Asano, and T. Goto. A supercritical CO_2 extraction from mackerel (*Scomber japonicas*) powder: experiment and modelling. *Bull Chem. Soc. Jpn.* 59:3709–3713 (1986).
50. S.L. Taylor, J.W. King, and G.R. List. Development of oil content in oilseeds by analytical supercritical fluid extraction. *J. Am. Oil Chem. Soc.* 70:437–439 (1993).
51. A.I. Hopia and V.-M. Ollilainen. Comparison of the evaporative light scattering detector, ELSD and refractive index detector (RID) in lipid analysis. *J. Liq. Chromatogr.* 16:2469–2482 (1993).
52. AOAC, *Association of Official Analytical Chemists' Official Methods of Analysis.* AOAC Press, Arlington, VA, 2000, pp. 66–68.
53. A.I. Carrapiso and C. Garcia. Development in lipid analysis; some new extraction techniques and *in situ* transesterification. *Lipids* 35:1167–1177 (2000).
54. L.D. Bergelson. *Lipid Biochemical Preparations.* Elsevier/North-Holland Biomedical Press, Amsterdam, 1990, pp. 1–36.
55. M.A. Wells and J.C. Dittmer. The use of Sephadex for the removal of non-lipid contaminants from lipid extracts. *Biochemistry* 2:1259–1263 (1963).
56. R.E. Wuthier. Purification of lipids from non-lipid contaminants on Sephadex bead columns. *J. Lipid Res.* 7:558–565 (1966).
57. D.C. Zimmerman. The relationship between seed density and oil content in flax. *J. Am. Oil Chem. Soc.* 39:77–78 (1962).
58. W.H. Hunt, M.H. Neustadt, J.R. Hardt, and L. Zeleny. A rapid dielectric method for determining the oil content of soybean. *J. Am. Oil Chem. Soc.* 29:258–261 (1952).
59. J.L. Rodrigneuz-Otero, M. Hermida, and A. Cepeda. Determination of fat protein and total solids in cheese by near IR reflectance spectroscopy. *J. AOAC Int.* 78:802–807 (1995).
60. M.H. Lee, A.G. Carinato, D.M. Mayod, and B.A. Rasco. Noninvasive short wavelength neon IR spectroscopic method to exhibit the crude lipid content in the muscle of intact rainbow trout. *J. Agric. Food Chem.* 40:2176–2179 (1992).
61. J.P. Wold, T. Jokebsen, and L. Krane. Atlantic salmon average fat content estimated by neon IR transmittance spectroscopy. *J. Food Sci.* 61:74–78 (1996).
62. D.A. Biggs. Milk analysis with the infrared milk analyzer. *J. Dairy Sci.* 50:799–803 (1967).

63. T.M. Eads and W.R. Croasmun. NMR application to fats and oils. *J. Am. Oil Chem. Soc.* 65:78–83 (1988).
64. T.F. Cornway and F.R. Earle. Nuclear magnetic resonance for determining oil content of seeds. *J. Am. Oil Chem. Soc.* 40:265–268 (1963).
65. D.E. Alexander, L. Silvela, I. Collins, and R.C. Rodgers. Analysis of oil content of maize by wide-line NMR. *J. Am. Oil Chem. Soc.* 44:555–558 (1967).
66. I. Collins, D.E. Alexander, R.C. Rodgers, and L. Silvela. Analysis of oil content of soybean by wide-line NMR. *J. Am. Oil Chem. Soc.* 44:708–710 (1967).
67. J.A. Robertson and W.R. Windham. Comparative study of methods of determining oil content of sunflower seed. *J. Am. Oil Chem. Soc.* 58:993–996 (1981).
68. D.B. Min. Crude fat analysis. In: *Introduction to the Chemical Analysis of Foods* (S.S. Neilson, ed.). Jones and Bartlett, London, 1994, pp. 81–192.
69. C. Beauvallet and J.-P. Renou. Application of NMR spectroscopy in meat research. *Trends Food Sci. Technol.* 3:197–199 (1992).
70. M.J. Gidley. High-resolution solid state NMR of food materials. *Trends Food Sci. Technol.* 3:231–236 (1992).
71. G. Haugaard and J.D. Pettinati. Photometric milk fat determination. *J. Dairy Sci.* 42:1255–1275 (1959).
72. I. Katz, M. Keeney, and R. Bassette. Caloric determination of fat in milk and saponification number of a fat by the hydroxamic acid reaction. *J. Dairy Sci.* 42:903–906 (1959).
73. I. Stern and B. Shapiro. A rapid and simple determination of esterified fatty acids and for total fatty acids in blood. *J. Clin. Pathol.* 6:158–160 (1953).
74. M.R. Hojjati and X.C Jiang. Rapid, specific and sensitive measurements of plasma sphingomyelin and phosphatidylcholine. *J. Lipid Res.* 47:673–676 (2006).
75. J.W. Fitzgerald, G.R. Rings, and W.C. Winder. Ultrasonic method for measurement of fluid non-fat and milk fat in fluid milk. *J. Dairy Sci.* 44:1165 (1961).
76. D.H. Kropf. New rapid methods for moisture and fat analysis: a review. *J. Food Qual.* 6:199–210 (1984).
77. F.Y. Iskander. Determination of iodine value by bromine: instrumental neutron activation analysis. *J. AOAC Int.* 72:498–500 (1989).
78. R.R. Allen. Determination of unsaturation. *J. Am. Oil Chem. Soc.* 32:671–674 (1955).
79. F.R. van de Voort, J. Sedman, G. Emo, and A.A. Ismail. Rapid and direct iodine value and saponification number determination of fats and oils by attenuated total reflectance/Fourier transform infrared spectroscopy. *J. Am. Oil Chem. Soc.* 69:1118–1123 (1992).
80. Y.B. Che Man, G. Setiowatry, and F.R. van de Voort. Determination of iodine value of palm oil by Fourier transform infrared spectroscopy. *J. Am. Oil Chem. Soc.* 76:693–699 (1999).
81. H. Li, F.R. van de Voort, A.A. Ismail, J. Sedman, R. Cox, C. Simard, and H. Burjs. Discrimination of edible oil products and quantitative determination of their iodine value by Fourier transform–neon infrared spectroscopy. *J. Am. Oil Chem. Soc.* 77:29–36 (2000).
82. R.R. Lowery and L.J. Tinsley. Rapid colorimetric determination of free fatty acids. *J. Am. Oil Chem. Soc.* 53:470–472 (1975).
83. Y.B. Che Man, N.H. Moh, and F.R. van de Voort. Determination of free fatty acids in crude palm oil and refined bleached deodorized palm olein using Fourier transform infrared spectroscopy. *J. Am. Oil Chem. Soc.* 76:485–490 (1999).
84. A.A. Ismail, F.R. van de Voort, and J. Sedman. Rapid quantitation determination of free fatty acids in fats and oils by FTIR spectroscopy. *J. Am. Oil Chem. Soc.* 70:335–341 (1993).
85. K. van Putte and J.C. van den Enden. Pulse NMR as quick method for the determination of the solid fat content in partially crystallized fats. *J. Phys. Eng. Sci. Instrum.* 6:910–912 (1973).
86. J.C. van den Enden, A.J. Haigton, K. van Putte, L.F. Vermaas, and D. Waddington. A method for the determination of the solid phase content of fats using pulse nuclear magnetic resonance. *Fette Seifen Anstrichmittel* 80:180–186 (1978).
87. J.C. van den Enden, J.B. Rossel, L.F. Vermaas, and D. Waddington. Determination of the solid fat content of hard confectionary butters. *J. Am. Oil Chem. Soc.* 59:433–439 (1982).
88. V.K.S. Shukla. Studies on the crystallization behaviour of the cocoa butter equivalents by pulsed nuclear magnetic resonance. Part I. *Fette Seifen Anstrichmittel* 85:467–471 (1983).
89. F.R. van de Voort, P. Memon, J. Sedman, and A.A. Ismail. Determination of solid fat index by FTIR spectroscopy. *J. Am. Oil Chem. Soc.* 73:411–416 (1996).

90. M.P. Purdon. Application of HPLC to lipid separation and analysis: sample preparation. In: *Analysis of Fats, Oils, and Lipoproteins* (E.G. Perkins, ed.). AOCS Press, Champaign, IL, 1991, pp. 166–192.

91. M. Kates. *Techniques of Lipidology,* 2nd edn. Elsevier, New York, NY, 1986.

92. F.W. Hemming and J.N. Hawthrone. *Lipid Analysis*. BIOS Scientific, Oxford, 1996.

93. J.L. Sebedio, C. Septier, and A. Grandgirard. Fractionation of commercial frying oil samples using Sep-Pak cartridges. *J. Am. Oil Chem. Soc.* 63:1541–1543 (1986).

94. G.D. Wachob. Solid phase-extraction of lipids. In: *Analysis of Fats and Lipoproteins* (E.G. Perkins, ed.). Champaign, IL, 1991, pp. 122–137.

95. S.E. Ebeler and J.D. Ebeler. Solid phase extraction methodologies for separation of lipids. *Inform* 7:1094 (1996).

96. M.A. Kaluzny, L.A. Duncan, M.V. Merritt, and D.E. Epps. Rapid separation of lipid classes in high field and purity using bonded phase column. *J. Lipid Res.* 26:135 (1985).

97. J.A. Prieto, A. Ebri, and C. Collar. Optimized separation of non-polar and polar lipid classes from wheat flour by solid phase extraction. *J. Am. Oil Chem. Soc.* 69:387–391 (1992).

98. M.N. Vaghela and A. Kilara. A rapid method for extraction of total lipids from whey protein concentrates and separation of lipid classes with solid phase extraction. *J. Am. Oil Chem. Soc.* 72:1117–1120 (1995).

99. H.G. Bateman and T.C. Perkins. Method for extraction and separation by solid phase extraction of neutral lipids free fatty acids and polar lipid from minced microbial cultures. *J. Agric. Food Chem.* 45:132–134 (1997).

100. P. Alaupovic and E. Koren. Immunoaffinity chromatography of plasma lipoprotein particles. In: *Analysis of Fats, Oils, and Lipoproteins* (E.G. Perkins, ed.). AOCS Press, Champaign, IL, 1991, pp. 599–622.

101. A. Kuksis. GLC and HPLC of neutral glycerolipids. In: *Lipid Chromatographic Analysis* (T. Shibamoto, ed.). Dekker, New York, NY, 1994, pp. 177–222.

102. F.D. Gunstone. *Fatty Acid and Lipid Chemistry*. Blackie, London, UK, 1996.

103. A. Kuksis, L. Marai, J.J. Myher, Y. Habashi, and S. Pind. Application of GC/MS, LC/MS, and FAB/MS to determination of molecular species of glycerolipids. In: *Analysis of Fats, Oils, and Lipoproteins* (E.G. Perkins, ed.). AOCS Press, Champaign, IL, 1991, pp. 464–495.

104. M.V. Calvo, L. Ramos, and J. Fontecha. Determination of cholesterol oxides content in milk products by solid phase extraction and gas chromatography–mass spectrometry. *J. Sep. Sci.* 26:927–931 (2003).

105. M.J. Petron, J.A. Garcia-Regueiro, L. Martin, E. Muriel, and T. Antequera. Identification and quantification of cholesterol and cholesterol oxidation products in different types of Iberian hams. *J. Agric. Food Chem.* 51:5786–5791 (2003).

106. M. Hayakawa, S. Sugiyama, and T. Osawa. HPLC analysis of lipids; analysis of fatty acids and their derivatives by a microcolumn HPLC system. In: *Lipid Chromatographic Analysis* (T. Shibamoto, ed.). Dekker, New York, NY, 1994, pp. 270–273.

107. E.W. Hammond. *Chromatography for the Analysis of Lipids*. CRC Press, Boca Raton, FL, 1993.

108. R.A. Moreau. Quantitative analysis of lipids by HPLC with a FID or an evaporative light-scattering detector. In: *Lipid Chromatographic Analysis* (T. Shibamoto, ed.). Dekker, New York, NY, 1994, pp. 251–273.

109. P. Laakso and W.W. Christie. Chromatographic resolution of chiral diacylglycerol derivatives: potential in the stereospecific analysis of triacyl-*sn*-glycerols. *Lipids* 25:349–353 (1990).

110. N. Sehat, M.P. Yurawecz, J.A.G. Roach, M.M. Mossoba, J.K.G. Kramer, and Y. Ku. Silver-ion high performance liquid chromatographic separation and identification of conjugated linoleic acid isomers. *Lipids* 33:217–221 (1998).

111. W.W. Christie, B. Nikolva-Damyanova, P. Laakso, and B. Herslof. Stereospecific analysis of triacyl-*sn*-glycerols *via* resolution of diasteromeric diacylglycerol derivatives by high performance liquid chromatography on silica gel. *J. Am. Oil Chem. Soc.* 68:695–701 (1991).

112. T. Takagi. Chromatographic resolution of chiral lipid derivatives. *Prog. Lipid Res.* 29:277–298 (1990).

113. T. Takagi and Y. Ando. Stereospecific analysis of triacyl-*sn*-glycerols by chiral high performance liquid chromatography. *Lipids* 26:542–547 (1991).

114. Y. Ando, K. Nishimura, N. Aoyansgi, and T. Takagi. Stereospecifc analysis of fish oil acyl-*sn*-glycerols. *J. Am. Oil Chem. Soc.* 69:417–424 (1992).

115. M.D. Grieser and J.N. Gesker. High performance liquid chromatography of phospholipids with flame ionization detection. *J. Am. Oil Chem. Soc.* 66:1484–1487 (1989).

116. J. Becart, C. Chevalier, and J.P. Biesse. Quantitative analysis of phospholipids by HPLC with a light scattering evaporating detector: application to raw materials for cosmetic use. *J. High Resolut. Chromatogr.* 13:126–129 (1990).

117. W.S. Letter. A rapid method for phospholipid class separation by HPLC using an evaporative light scattering detector. *J. Liq. Chromatogr.* 15:253–266 (1992).

118. A. Avalli and G. Contarini. Determination of phospholipids in dairy products by SPE/HPLC/ELSD. *J. Chromatogr.* A. 1071:185–190 (2005).

119. W.W. Christie and W.R. Morrison. Separation of complex lipids of cereals by high-performance liquid chromatography with mass detection. *J. Chromatogr.* 436:510–513 (1988).

120. C.N. Christopoulou and E.G. Perkins. High performance size exclusion chromatography of monomer, dimer, and trimer mixtures. *J. Am. Oil Chem. Soc.* 66:1338–1343 (1989).

121. I.C. Burkov and R.J. Henderson. Analysis of polymers from autoxidized marine oils by gel permeation HPLC using light-scattering detector. *Lipids* 26:227–231 (1991).

122. L.G. Blomberg, M. Demirbuker, and M. Anderson. Characterization of lipids by supercritical fluid chromatography and supercritical fluid extraction. In: *Lipid Analysis in Oils and Fats* (R.J. Hamilton, ed.). Blackie Academics & Professional, London, 1998, pp. 34–58.

123. K. Matsumoto and M. Taguchi. Supercritical fluid chromatographic analysis of lipids. In: *Lipid Chromatographic Analysis* (T. Shibamoto, ed.). Dekker, New York, NY, 1994, pp. 365–396.

124. R. Huopalathi, P. Laakso, J. Saaristo, R. Linko, and H. Kallio. Preliminary studies on triacylglycerols of fats and oils by capillary SFC. *J. High Resolut. Chromatogr. Chromatogr. Commun.* 11:899–902 (1988).

125. F.O. Geiser, S.G. Yocklovich, S.M. Lurcott, J.W. Guthrie, and E.J. Levy. Water as a stationary phase modified in packed-column supercritical fluid chromatography I. Separation of free fatty acids. *J. Chromatogr.* 459:173–181 (1988).

126. A. Nomura, J. Yamada, K. Tsunoda, K. Sakaki, and T. Yokochi. Supercritical fluid chromatographic determination of fatty acids and their esters on an ODS-silica gel column. *Anal Chem.* 61:2076–2078 (1989).

127. M. Demirbuker and G. Blomberg. Group separation of triacylglycerols on micropacked argentation columns using supercritical media as mobile phases. *J. Chromatogr. Sci.* 28:67–72 (1990).

128. M. Demirbüker and G. Blomberg. Separation of triacylglycerols: supercritical fluid argentation chromatography. *J. Chromatogr.* 550:765–774 (1991).

129. T. Greibrokk. Application of supercritical fluid extraction in multidimensional system. *J. Chromatogr.* A. 703:523–536 (1995).

130. N.C. Shantha and G.E. Napolitano. Lipid analysis using thin layer chromatography and Iatroscan. In: *Lipid Analysis in Oils and Fats* (R.J. Hamiton, ed.). Blackie Academics & Professional, London, 1998, pp. 1–33.

131. V.P. Skipski, M. Barclary, E.S. Reichman, and J. Good. Separation of acidic phospholipids by one dimensional thin-layer chromatography. *Biochem. Biophys. Acta* 137:80–89 (1967).

132. H.W. Gardner. Preparative isolation of monogalactosyl and digalactosyl diglycerides by thin layer chromatography. *J. Lipid Res.* 9:139–141 (1968).

133. J.E. Stony and B. Tuckley. Thin layer chromatography of plasma lipids by single development. *Lipids* 2:501–502 (1967).

134. B. Fried. Lipids. In: *Handbook of Thin Layer Chromatography* (C.J. Sharma and B. Fried, eds.). Dekker, New York, NY, 1996, pp. 704–705.

135. D. Kozutsumi, A. Kawashima, M. Adachi, M. Takami, N. Taketomo, and A. Yonekubo. Determination of docosahexaenoic acid in milk using affinity solid-purification with argentous ions and modified thin layer chromatography. *Int. Dairy J.* 13:937–943 (2003).

136. N.C. Shantha. Thin-layer chromatography–flame ionization detection Iatroscan system. *J. Chromatogr.* 624:21–35 (1992).

137. C.C. Parrish and R.G. Ackman. Chromarod separations for the analysis of marine lipid classes by Iatroscan thin-layer chromatography–flame ionization detection. *J. Chromatogr.* 262:103–112 (1983).

138. R.G. Ackman. Problems in introducing new chromatographic techniques for lipid analyses. *Chem. Ind.* 17:715–722 (1981).

139. M. Ranny. *Thin-Layer Chromatography with Flame Ionization Detection*. Reidel, Dordrecht, 1987.

140. R.G. Ackman, C.A. McLeod, and A.K Banerjee. An overview of analysis by Chromarod-Iatroscan TLC-FID. *J. Planar Chromatogr.* 3:450–490 (1990).

141. J.K. Kaitaranta and P.J. Ke. TLC-FID assessment of lipid oxidation as applied to fish lipids rich in triglycerides. *J. Am. Oil Chem. Soc.* 58:710–713 (1981).

142. J.L. Sebedio and R.G. Ackman. Chromarods modified with silver nitrate for the quantitation of isomeric unsaturated fatty acids. *J. Chromatogr. Sci.* 19:552–557 (1981).

143. M. Tanaka, K. Takase, J. Ishi, T. Itoh, and H. Kaneko. Application of a thin layer chromatography–flame ionization detection system for the determination of complex lipid constituents. *J. Chromatogr.* 284:433–440 (1984).

144. C.G. Walton, W.M.N. Ratnayake, and R.G. Ackman. Total sterols in sea foods: Iatroscan-TLC/FID versus the Kovacs GLC/FID method. *J. Food Sci.* 54:793–795 (1989).

145. W.M. Indrasena, A.T. Paulson, C.C. Parrish, and R.G. Ackman. A comparison of alumina and silica gel chromarods for the separation and characterization of lipid classes by Iatroscan TLC-FID. *J. Planar Chromatogr.* 4:182–188 (1991).

146. J.K.G. Kramer, R.C. Fouchard, F.D. Sauer, E.R. Farnworth, and M.S. Wolynetz. Quantitating total and specific lipids in a small amount of biological sample by TLC-FID. *J. Planar Chromatogr.* 4:42–45 (1991).

147. C.C. Parrish, G. Bodennec, and P. Gentien. Separation of polyunsaturated and saturated lipids from marine phytoplankton on silica gel coated chromarods. *J. Chromatogr.* 607:92–104 (1992).

148. R. Pryzbylski and N.A.M. Eskin. Two simplified approaches to the analysis of cereal lipids. *Food Chem.* 51:231–235 (1994).

149. F.R. van de Voort and J. Sedman. FTIR spectroscopy: the next generation of oil analysis methodologies? *Inform* 11:614–620 (2000).

150. F.R. van de Voort. Fourier transform infrared spectroscopy applied to food analysis. *Food Res. Int.* 25:397–403 (1992).

151. J. Sedman, A.A. Ismail, A. Nicodemo, S. Kubow, and F. van de Voort. Application of FTIR/ATR differential spectroscopy for monitoring oil oxidation and antioxidant efficiency. In: *Natural Antioxidants: Chemistry, Health, and Applications* (F. Shahidi, ed.). AOCS Press, Champaign, IL, 1996, pp. 358–378.

152. D. Chapman and F.M. Goni. Physical properties: optical and spectral characteristics. In: *The Lipid Handbook*, 2nd edn. (F.D. Gunstone, J.L. Harwood, and F.B. Padley, eds.). Chapman and Hall, London, 1994, pp. 487–504.

153. J.N. Schoolery. Some quantitative applications of carbon-13 NMR spectroscopy. *Prog. Nucl. Magn. Reson. Spectrosc.* 11:79–93 (1977).

154. T.Y. Shiao and M.S. Shiao. Determination of fatty acid composition of triacylglycerols by high resolution NMR spectroscopy. *Bot. Bull. Acad. Sinica* 30:191–199 (1989).

155. S. Ng and W.L. Ng. ^{13}C NMR spectroscopic analysis of fatty acid composition of palm oil. *J. Am. Chem. Soc.* 60:266–268 (1983).

156. K.F. Wallenberg. Quantitative high resolution ^{13}C NMR of the olefinic and carbonyl carbons of edible vegetable oils. *J. Am. Oil Chem. Soc.* 67:487–494 (1990).

157. F.D. Gunstone. The C-13 NMR spectra of oils containing γ-linolenic acid. *Chem. Phys. Lipids* 56:201–207 (1990).

158. F.D. Gunstone. High resolution NMR studies of fish oils. *Chem. Phys. Lipids* 59:83–89 (1991).

159. M. Aursand, L. Jørgensen, and H. Grasdalen. Positional distribution of ω3 fatty acids in marine lipid triacylglycerols by high resolution ^{13}C NMR spectroscopy. *J. Am. Oil Chem. Soc.* 72:293–297 (1995).

160. A. Haase. Introduction to NMR imaging. *Trends Food Sci. Technol.* 3:206–207 (1992).

161. C. Simoneau, M.J. McCarthy, D.S. Reid, and J.B. German. Measurement of fat crystallization using NMR imaging and spectroscopy. *Trends Food Sci. Technol.* 3:208–211 (1992).

162. J.M. Halloin, T.G. Cooper, E.J. Potchen, and T.E. Thompson. Proton magnetic resonance imaging of lipids of pecan embryos. *J. Am. Oil Chem. Soc.* 70:1259–1262 (1993).

163. J.R. Heil, W.E. Perkins, and M.J. McCarthy. Use of magnetic resonance procedures for measurement of oils in French-style dressings. *J. Food Sci.* 55:763–766 (1990).

164. G.M. Pilhofer, H.C. Lee, M.J. McCarthy, P.S. Tong, and J.B. German. Functionality of milk fat in foam formation and stability. *J. Dairy Sci.* 77:55–60 (1994).

165. J.M. Tingely, J.M. Pope, P.A. Baumgartner, and V. Sarafis. Magnetic resonance imaging of fat and muscle distribution in meat. *Int. J. Food Sci. Technol.* 30:437–441 (1995).

166. R.P. Singh. Heat and mass transfer in foods during deep fat frying. *Food Technol.* 49: 134–137 (1995).

167. J. Schiller, J. Arnhold, S. Benard, M. Muller, S. Reichl, and K. Arnold. Lipid analysis by matrix-assisted laser desorption and ionization mass-spectrometry. A methodological approach. *Anal. Biochem.* 267:46–56 (1999).

168. J.-L. Le Quere, E. Semon, B. Lanher, and J.L. Sebedio. On-line hydrogenation in GC–MS analysis of cyclic fatty acid monomers isolated from heated linseed oil. *Lipids* 24:347–350 (1989).

169. J.-L. Le Quere. Gas chromatography mass spectrometry and tandem mass spectrometry. In: *New Trends in Lipids and Lipoprotein Analysis* (J.-L. Sebedio and E.G. Parkins, eds.). AOCS Press, Champaign, IL, 1995.

170. T. Fujiwaki, S. Yamaguchi, M. Tasaka, N. Sakura, and T. Taketomi. Application of delayed extraction-matrix assisted laser desorption ionization time-of-flight mass spectrometry for analysis of sphingolipids in pericardial fluid, peritoneal fluid and serum from Gaucher disease patients. *J. Chromatogr.* B. 776:115–123 (2002).

171. T. Higgins. Evaluation of a colorimetric triglyceride method on the KDA analyzer. *J. Clin. Lab. Autom.* 4:162–165 (1984).

172. H. Brockerhoff and M. Yurkowski. Stereospecific analysis of several vegetable fats. *J. Lipid Res.* 7:62–64 (1966).

173. U.N. Wanasundara and F. Shahidi. Positional distribution of fatty acids in triacylglycerols of seal blubber oil. *J. Food Lipids* 4:51–64 (1997).

174. W.E.M. Lands. Lipid metabolism. *Annu. Rev. Biochem.* 34:313–346 (1965).

175. M. Saito and R.K. Yu. TLC immunostaining of glycolipids. *Chem. Anal.* 108:59–68 (1990).

6 Methods for *trans* Fatty Acid Analysis

Magdi M. Mossoba and Richard E. McDonald

CONTENTS

I. INTRODUCTION

Trans fatty acids are present in a variety of food products and dietary supplements; some are derived from natural sources, such as dairy products, but most come from products that contain commercially hydrogenated fats. Margarines used to be the major source of *trans* fats; however, in recent years processed foods, such as snacks and fast foods, are more likely to be the major sources of dietary *trans* fats [1,2]. The nutritional properties of *trans* fatty acids have been debated for many years, particularly with respect to the amounts of low-density and high-density lipoprotein (LDL, HDL) contained in serum. Some studies have shown that *trans* fatty acids elevate levels of serum LDL cholesterol and lower HDL cholesterol [3–6]. Such results drew a great deal of attention, which eventually led to the mandatory labeling of *trans* fatty acids on food products in the United States, Canada, and many other countries [7–10]. For nutrition labeling purposes, *trans* fats are defined as the sum of all unsaturated fatty acids that contain one or more isolated, nonconjugated, double bonds with a *trans* double-bond geometric configuration. However, conjugated fatty acids with a *trans* double bond, such as conjugated linoleic acid (CLA) isomers, are excluded from this definition of *trans* fats. In the United States, label declaration of *trans* fats are not required for products that contain less than 0.5 g *trans* fat per reference amount and per labeled serving, as long as no claims are made for fat, fatty acids, or cholesterol. In Canada, to claim that a product is *trans* fat free, it must contain less than 0.2 g of *trans* fat per serving and per reference amount. Unlike other countries, Denmark imposed an upper limit on the amount of *trans* fats in foods (Danish Food Act, Executive order No. 160 of 11 March 2003), namely oils and fats must contain less than 2 g of *trans* fatty acids per 100 g of total fat, and fat-free labels would be allowed if a food product contains less than 1.0 g of *trans* fat per 100 g of the fat or oil.

Although partially hydrogenated vegetable oils have been reported to present a possible risk factor for coronary heart disease [11], some *trans* conjugated fatty acids in ruminant fat have been reported to have several beneficial physiological effects in experimental animals [12]. *trans*-Vaccenic acid (*trans*-11-18:1), which is the major *trans* fatty acid isomer present in meat and dairy products from ruminants [13,14], has been shown to be converted to *cis*-9,*trans*-11-18:2, a CLA isomer [15–17], by the action of $\Delta 9$ desaturase present in mammalian tissue [18,19]. The relationship between *trans*-18:1 fatty acid isomers and CLA isomers has been a promising and an increasingly active area of research [20–22].

The mandatory requirement in many countries [7,8], to declare the amount of *trans* fat present in food products and dietary supplements, has led to a need for validated official methods that are both sensitive and accurate for the rapid quantitation of total *trans* fatty acids. There have been active research programs to improve the various methodologies to determine *trans* fatty acids in various products. Analytical procedures used to quantify and identify fatty acids have been reviewed [9,10,23–25]. This chapter discusses several techniques and the latest developments in analytical methodologies, including validated gas chromatographic (GC) and infrared (IR) spectroscopic official methods, as well as new powerful techniques in mass spectrometry (MS), near-infrared (NIR) spectroscopy, and silver ion chromatography (thin-layer chromatography (TLC) and high-performance liquid chromatography (HPLC)), hyphenated techniques (GC–IR and GC–MS), and procedures based on supercritical fluid chromatography (SFC). The analysis of *trans* fatty acid isomers is extremely challenging and complex. Various combinations of techniques have effectively been used to determine the quantity and confirm the identity of individual *trans* and *cis* isomers. However, improved methods are still needed to accurately and conveniently determine the total *trans* fat content as well as specific *trans* fatty acid isomers in foods and dietary supplements for regulatory compliance.

II. ANALYSIS USING SPECTROSCOPY

Traditional, informative IR transmission methods as well as those based on novel internal reflection approaches for the determination of total *trans* fat will be presented next.

A. INFRARED SPECTROSCOPY

IR spectroscopy is a widely used technique for determining nonconjugated *trans* unsaturation in both natural and processed fats and oils [9,24]. It is not applicable to materials that have functional groups with absorption bands close to 966 cm^{-1}, which is the strong absorption band arising from the C–H deformation about a *trans* double bond. This absorption band is absent in natural vegetable oils that are composed of saturated fatty acids and fatty acids with only *cis*-unsaturated double bonds. For increased accuracy, oil samples have traditionally been converted to methyl esters before analysis. This eliminated interfering absorptions associated with the carboxyl groups of free fatty acids and the glycerol backbone of triacylglycerols.

1. Conventional IR Methods

The early Official Method of the American Oil Chemists' Society (AOCS), Cd 14–61 [26], for the determination of *trans* fatty acid concentrations in fats and oils was based on a comparison of the absorption at 966 cm^{-1} for standards and unknowns. Test samples and standards were diluted in carbon disulfide and placed in an absorption cell so that the transmittance or absorbance could be measured in an IR spectrophotometer. The quantitation of the *trans* concentration was based on Beer's law:

$$A = abc, \tag{6.1}$$

where
$A = \text{absorbance} = \log \dfrac{1}{\text{transmittance}}$
$a = \text{absorptivity}$
$b = \text{path length}$
$c = \textit{trans} \text{ concentration}$

Disadvantages of using this method include (1) the need to make methyl ester derivatives at *trans* levels less than 15%, (2) the use of the toxic, volatile solvent carbon disulfide, (3) the high bias found for triacylglycerols, and (4) low accuracy obtained for *trans* levels less than 5%. Therefore, there has been a great deal of interest in improving methods to determine *trans* fatty acid concentrations.

A classical study [27] was conducted that showed that the AOCS method produced *trans* values that were 2%–3% too high, whereas derivatized methyl esters produced values that were 1.5%–3% too low. In 1965, correction factors were suggested that were incorporated into AOAC method 965.35 [28]. The percentage of *trans* fat was calculated with a procedure similar to AOCS method Cd 14–61. Several formulas using correction factors were proposed to calculate *trans* concentration as methyl esters or triacylglycerols. Different formulas were also used for oils containing long-chain or short-/medium-chain fatty acids.

In another study [29], the concentration of *trans* fatty acids was determined based on the ratio of the IR absorptions at 965 and 1170 cm^{-1}. Huang and Firestone [30] proposed a dual-beam differential spectrophotometry procedure with a zero *trans*-containing vegetable oil in the reference cell. This procedure, which resulted in a 1%–2% high bias for both triacylglycerols and methyl esters, was the first attempt to eliminate the sloping background (Figure 6.1) of the *trans* IR band.

A two-component calibration curve was also proposed [31] in an attempt to overcome some of the drawbacks of the earlier procedures. A calibration curve was developed by means of different levels of the *trans* monoene, methyl elaidate (ME), and methyl linoleate dissolved in carbon disulfide. The calibration standards and the test samples in carbon disulfide were scanned against a carbon disulfide background and recorded from 900 to 1500 cm^{-1}. After a baseline had been drawn as a tangent from about 935 and 1020 cm^{-1} at the peak minima, the corrected absorbance of the calibration standards *trans* peaks, at 966 cm^{-1}, was obtained. The baselines for the test samples spectra were obtained by overlaying the spectra of the calibration standards spectra at corresponding concentrations. This method compensates for the low bias of other methods and eliminates the need for correction factors.

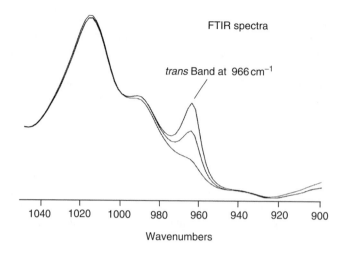

FIGURE 6.1 Observed conventional infrared absorption spectra for *trans* fatty acid methyl esters in carbon disulfide. At low *trans* levels, the *trans* band is reduced to a shoulder.

The introduction of Fourier transform infrared spectroscopy (FTIR) instruments also facilitated more accurate and rapid determination of *trans* fatty acids. A laser light source was used for wavelength accuracy. Interferometers were used that allowed all wavelengths of light to be measured simultaneously. Since FTIR spectrometers are computerized, multiple spectral scans could be averaged in a few minutes. An early attempt in which an absorption band–height ratio procedure [32] was used with an attenuated total reflection (ATR) cell allowed the use of neat samples and eliminated the need to rely on volatile toxic solvents. An FTIR spectrometer equipped with a thin (0.1 mm) transmission flow cell was used to develop an automated procedure for calculating the percentage of *trans* in fats and oils [33].

Postmeasurement spectral subtraction manipulations were also used to try to correct for the highly sloping background in the FTIR absorption spectrum of hydrogenated vegetable oil fatty acid methyl esters (FAMEs) [34]. Determination of *trans* content, however, required additional IR measurement of an appropriate reference material and the digital subtraction of this reference absorption spectrum from that of the test portion. FTIR spectroscopy, in conjunction with a transmission flow cell, was used to rapidly determine the *cis* and *trans* content of hydrogenated oil simultaneously [35].

The 1995 official method of the AOCS for determination of the *trans* fatty acid Cd 14–95 [36] utilized two standard curves [31]. The choice of curve depended on the *trans* concentration. This method is reported to be accurate to determine the *trans* content of fats with *trans* levels of 0.5% or greater. Test samples and standards are converted to methyl esters, diluted in 10 mL of carbon disulfide, and then placed in a transmission IR cell before measuring the transmittance or absorbance over the range 1050–900 cm^{-1} in an IR spectrophotometer. Typical IR absorption spectra showing the contrast in band shape between oil containing 2% and 70% *trans* were published [36]. The baseline-corrected absorbance (A_c) is determined by subtracting the absorbance of the baseline at the peak maximum (A_B) from the maximum absorbance at the peak (A_p). Two plots are then constructed by using standards to cover samples with low levels of *trans* (1%–10%) and moderate to high *trans* levels (10%–70%), and two regression equations are generated. The percentage of *trans* as ME of the unknown is then obtained by referring to the appropriate calibration data ($\leq$10% or >10%) and solving the following equations:

$$A_c = (A_p - A_B),\qquad\qquad (6.2)$$

$$\text{Methyl elaidate weight equivalents(g)} = \frac{A_c - \text{intercept}}{\text{slope}}, \tag{6.3}$$

$$\%trans = \frac{\text{Methyl elaidate weight equivalents(g)}}{\text{Sample weight(g/10 mL of CS}_2)} \times 100. \tag{6.4}$$

The major advantage of this method is its accuracy at low *trans* levels. However, methyl ester derivatization and the use of carbon disulfide were still required.

More accurate results than those produced by the current official methods were claimed for an IR procedure, which uses a partially hydrogenated vegetable oil methyl ester mixture as the calibration standard [37]. The reportedly improved results were attributed to the assortment of *trans* monoene and polyene isomers in the calibration standard with different absorbtivities relative to that of ME.

2. Ratioing of Single-Beam FTIR Spectra

Figure 6.1 indicates that the 966 cm^{-1} *trans* band is only a shoulder at low levels (near 2% of total fat). This is due to the overlap of the *trans* band with other broad bands in the spectrum, which produces a highly sloped background that diminishes the accuracy of the *trans* analysis. Many reports in the literature have proposed changes to the procedures above, including minor refinements to major modifications aimed at overcoming some of the limitations already discussed. These studies have resulted in the development of procedures that use spectral subtraction to increase accuracy, as well as means of analyzing neat samples to eliminate the use of solvents.

Mossoba et al. [38] described a rapid IR method that uses a Fourier transform IR spectrometer equipped with an ATR cell for quantitating *trans* levels in neat fats and oils. This procedure measured the 966 cm^{-1} *trans* band as a symmetric feature on a horizontal background (Figure 6.2). The ATR cell was incorporated into the design to eliminate one potential source of error: the weighing of test portions and their quantitative dilution with the volatile CS$_2$ solvent. The high bias previously found for triacylglycerols has been attributed to the overlap of the *trans* IR band at

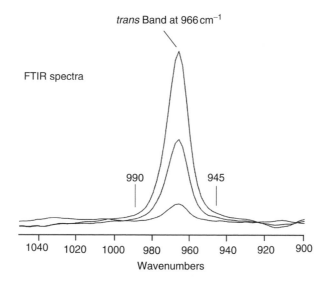

FIGURE 6.2 Observed ATR-FTIR infrared absorption spectra for neat (without solvent) fatty acid methyl esters. Symmetric bands were obtained because the spectrum of a *trans*-free fat was subtracted from those of a *trans* fats.

966 cm^{-1} with ester group absorption bands. Errors for the determination of *trans* concentrations below 5% that resulted from this overlap could result in relative standard deviation values greater than 50% [39]. The interfering absorption bands were eliminated, and baseline-resolved *trans* absorption bands at 966 cm^{-1} were obtained by ratioing [38] the FTIR single-beam spectrum of the oil or fat being analyzed against the single-beam spectrum of a reference material (triolein, a mixture of saturated and *cis*-unsaturated triacylglycerols or the corresponding unhydrogenated oil). This approach was also applied to methyl esters. Ideally, the reference material should be a *trans*-free oil that has an otherwise similar composition to the test sample being analyzed.

The simplified method just outlined allowed the analysis to be carried out on neat (without solvent) fats or oils that are applied directly to the ATR crystal with little or no sample preparation. With this method, the interference of the ester absorptions with the 966 cm^{-1} *trans* band and the uncertainty associated with the location of the baseline were eliminated. Figure 6.2 shows the symmetric spectral bands that were obtained when different concentrations of ME in methyl oleate (MO) were ratioed against MO. A horizontal baseline was observed, and the 966 cm^{-1} band height and area could be readily measured. The minimum identifiable *trans* level and the lower limit of quantitation were reported [38] to be 0.2% and 1%, respectively, in hydrogenated vegetable oils.

Further refinement of this procedure by means of single-bounce, horizontal attenuated total reflection (SB-HATR) IR spectroscopy was subsequently reported [40]. Using this procedure, only 50 µL (about 2–3 Pasteur pipette drops) of neat oil (either triacylglycerols or methyl esters) is placed on the horizontal surface of the zinc selenide element of the SB-HATR IR cell. The absorbance values obtainable were within the linearity of the instrument. The test portion of the neat oil can easily be cleaned from the IR crystal by wiping with a lint-free tissue before the next neat sample is applied. The method is accurate for *trans* concentrations greater than 1%. The SB-HATR FTIR procedure was used to determine the *trans* content of 18 food products [40].

This internal reflection FTIR procedure was voted official method AOCS Cd 14d-99 by the AOCS in 1999 [41] and official method 2000.10 by AOAC International in 2000 [42] after testing and validation in a 12 laboratory international collaborative study. Analytical ATR-FTIR results exhibited high accuracy relative to the gravimetrically determined values. Comparison of test materials with similar levels of *trans* fatty acids indicated that the precision of the current ATR-FTIR method was superior to that of the two most recently approved transmission IR official methods: AOAC 965.34 [43] and AOAC 994.14 [44].

The ATR-FTIR method was also evaluated for use with matrices of low *trans* fat and low total fat contents such as milk [45] and human adipose tissue [46]. Preliminary results indicated that the presence of low levels (<1%) of conjugated *cis*/*trans* dienes in these matrices, with absorbance bands near 985 and 947 cm^{-1}, interfered with the accurate determination of total isolated (nonconjugated) *trans* fatty acids [45,46]. Attempts to eliminate interfering absorbance peaks by use of spectral subtraction techniques [45,46] were not satisfactory. In order to overcome the effect of interferences, the ATR-FTIR method was modified by inclusion of the standard addition technique [47]. This modification was applied to several food products, namely, dairy products, infant formula, and salad dressing. This standard addition modification minimized the interfering conjugation absorbance bands. The presence of <1% CLA in two butter and two cheese products containing 6.8%, 7.5%, 8.5%, and 10.4% *trans* fatty acids (as a percentage of total fat) would have resulted in errors of −11.6%, 10.4%, 17.6%, and 34.6%, respectively, in *trans* fat measurements using the unmodified method.

These ATR-FTIR official methods [41,42,48] eliminated the baseline offset and slope, but were only partly successful in improving accuracy. This is because finding a reference fat that is absolutely *trans*-free and whose composition would closely match every unknown test sample is impossible. This factor also had a negative impact on sensitivity.

To improve sensitivity and accuracy, another ATR-FTIR procedure that measures the height of the negative second derivative of the *trans* absorption band relative to air (Figure 6.3) was recently published [49,50]. Reference standards consisting of the *trans* monoene trielaidin (TE, *trans*-18:1)

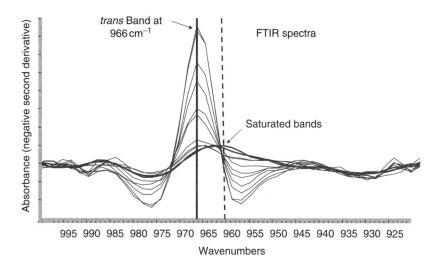

FIGURE 6.3 Negative second derivative for observed ATR-FTIR infrared absorption spectra of neat (without solvent) triacylglycerols. The presence of newly discovered [50] weak infrared bands attributed to saturated fats and oils could be identified by this procedure.

diluted in triolein (TO, *cis*-18:1) were initially used to generate calibration data. The negative second derivative procedure totally eliminated both the baseline offset and slope of the *trans* IR band as well as the requirement to use a *trans*-free reference fat. The second derivative of an absorbance spectrum enhanced the resolution of IR bands, and made it possible to notice small shifts in IR band position and the presence of interferences.

These advantages offered by the negative second derivative ATR-FTIR procedure made it possible to improve sensitivity and resolve a discrepancy in accuracy between GC and IR at low *trans* levels [50]; a hydrogenated soybean oil (HSBO) sample was found to have no *trans* fat by GC, but was reported by IR to have a low and significant *trans* level of 1.2% of total fat [49]. The negative second derivative ATR-FTIR spectrum observed for the HSBO test sample [49] indicated that the band position shifted to 960 cm^{-1} from the expected position of the *trans* band absorption at 966 cm^{-1}. It was subsequently recognized that for this fully HSBO, the weak band at 960 cm^{-1} should have been attributed to tristearin (TS, 18:0) [50], a saturated rather than a *trans* fat. This result was confirmed by showing that a reference sample of TS exhibited an identical weak band at 960 cm^{-1}.

Another finding reported in the negative second derivative study [49] was the presence of 0.5%, as determined by GC, mainly TE contaminant in the commercially available TO. Therefore, it was recommended that commercially available TO should no longer be used to prepare calibration standard mixtures. Instead, several saturated fats, in addition to TS, were screened for possible interferences with the *trans* band at 966 cm^{-1} [50]. The saturated fats trilaurin (TL, 12:0), trimyristin (TM, 14:0), tripalmitin (TP, 16:0), and triarachidin (TA, 20:0) were measured by IR and were all found to exhibit similar weak absorption bands with varying degrees of interferences. TP exhibited a band at 956 cm^{-1}, which was farthest from the *trans* absorption at 966 cm^{-1}. This least interfering saturated fat also had the lowest absorptivity, and TP was therefore used as a substitute for TO to prepare calibration standard mixtures of TE in TP as low as approximately 0.5% of total fat [50]. This negative second derivative procedure is currently being validated in an international collaborative study.

Once validated, the *trans* fat lower limit of quantitation of this negative second derivative procedure will be determined and is expected to be found near 0.5%–1% of total fat. Therefore, for fats and oils with a high content of saturated fats and only a trace amount (≤0.1% of total fat as determined by GC) of *trans* fat, such as coconut oil and cocoa butter, the weak bands observed at energies slightly lower than 966 cm^{-1} (Figure 6.3) must not be mistaken for, and erroneously reported as, *trans* bands [50]. Such recognition of potential interferences from saturated fats would result in the correct interpretation of IR spectra for unknown *trans* fats and oils, improve the accuracy of the IR determination at low *trans* levels (particularly ≤1% of total fat), and make this relatively sensitive negative second derivative IR procedure suitable for the rapid determination of total *trans* fats and the labeling of food products and dietary supplements.

B. FTIR–Partial Least Squares Regression

The quantitative analysis of *trans* unsaturation was also investigated by using a transmission FTIR–partial least squares (PLS) [51] procedure. This approach offered significantly reduced analysis time compared with conventional GC methods (see later) and was effective for a wide range of *trans* values. Improved results were reported for raw materials and food products [51].

A heated SB-HATR sampling accessory was developed and used to develop a method for the simultaneous determination of iodine value (IV) and *trans* content for neat fats and oils [52]. PLS regression was employed for the development of the calibration models, and a set of nine pure triacylglycerol test samples served as the calibration standards. Satisfactory agreement (SD < 0.35) was obtained between the predictions from the PLS calibration model and *trans* determinations by the SB-HATR FTIR method. In another study [53], isolated and conjugated *trans* fatty acids were simultaneously determined by chemometric analysis of their FTIR spectra [53]. Similarly, PLS analysis was used in establishing calibration models. The authors developed two calibration models using the IR spectral profiles of the mixtures of isolated *trans* and conjugated *trans* fatty acids (CLA) in virgin olive oil and their IR spectral profiles in the region 1000–650 cm^{-1} where the CH bending bands of the *trans* fatty acids and CH deformation bands of CLA isomers were observed. Two calibration models were generated for *trans* and CLA isomers in the ranges 1%–2.5% and 0%–30% in olive oil. The cross-validation procedure was used for the validation of calibration models. These models were then used to determine the *trans* and CLA contents of unknown test samples.

C. Near-Infrared Spectroscopy

A generalized PLS calibration was developed for determination of the *trans* content of edible fats and oils by Fourier transform near-infrared (FT-NIR) spectroscopy [54]. The *trans* reference data, determined by using the mid-IR SB-HATR FTIR official method, were used in the development of the generalized FT-NIR calibration. The FT-NIR *trans* predictions obtained using this generalized calibration were in good agreement with the SB-HATR results. The authors concluded that FT-NIR provided a viable alternative to the SB-HATR [54].

Recent advances in FT-NIR have made it possible to determine not only the total *trans* fatty acid content of a fat or oil, but also its fatty acid composition [55,56], and without the need for prior derivatization to volatile derivatives as required for GC analysis. Quantitative FT-NIR models were developed by comparing accurate GC results (for FAMEs) to FT-NIR measurements (for neat fats and oils) and using chemometric analysis. The FT-NIR spectra showed unique fingerprints for saturated FA, *cis*- and *trans*-monounsaturated FA, and all *n*-6 and *n*-3 polyunsaturated fatty acids within triacylglycerols to permit qualitative and quantitative comparisons of fats and oils. The quantitative models were based on incorporating accurate GC fatty acid composition data obtained for different fats and oils and FT-NIR spectral data into the calibration models. FT-NIR classification models were developed based on chemometric analyses of 55 fats, oils, and fat/oil mixtures that were used in the identification of similar test samples. Three calibration models were developed: one was used for the analysis of common fats and oils with low levels of *trans* FA and the other two were used for fats and oils with intermediate to high levels of *trans* FA. This FT-NIR procedure exhibited the potential to rapidly determine FA composition of unknown fats and oils in their neat form and without derivatization [56]. It should be noted that the FT-NIR methodology is matrix dependent [55,56]. Thus, it is highly dependent on factors that could influence the NIR absorption spectra. It is therefore expected that external contaminations such as residual solvent or the presence of conjugated *cis/trans* 18:2 and *trans/trans* isomers or other FAs in the fats and oils could alter the absorption spectrum and ultimately the FA analysis, unless they are accounted for and robust calibration models are developed.

III. ANALYSIS USING CHROMATOGRAPHY

The chromatography of lipids involves the separation of individual components of a lipid mixture as they pass through a medium with a stationary matrix. This matrix may be packed or bound to a column, as in GC and HPLC, or bound to a glass plate as in TLC. The mobile phase in GC is usually an inert gas such as helium or nitrogen. For HPLC or TLC, the mobile phase may be an aqueous or organic solvent. Intact triacylglycerols or fatty acids can be separated, but many chromatography applications consist of separating the methyl ester or other derivatives of individual fatty acids.

A. Gas Chromatography

The most significant occurrence in the chromatography of lipids has been the development of GC. It is today the industry standard. FAMEs, including saturates and unsaturates, were first successfully separated in 1956 with a 4 ft long column packed with Apiezon M vacuum grease [57]. Separations were demonstrated based on chain length, degree of unsaturation, and *cis* or *trans* geometric configuration.

1. Separations Using Packed Columns

In general, retention times of fatty acid derivatives on nonpolar columns are based on volatility and, therefore, separation occurs primarily by carbon chain length. Retention times of fatty acid derivatives on polar columns are mainly determined on the basis of polarity and chain length. These columns, therefore, are more effective at resolving unsaturated fatty acids with different degrees of unsaturation. The chromatographic characteristics of several packed columns (Silar 10C, Silar 9CP, SP 2340, and OV-275), made available in the mid-1970s as a result of the then recently developed silicone stationary phases, were compared [58]. The development of these highly polar, temperature-resistant columns made it possible to separate the geometric isomers of fatty acids. A column packed with 12% Silar 10C as the stationary phase was able to baseline separate the all-*cis* from the all-*trans* components of octadecadiethylenic FAMEs. All eight of the geometric isomers of linolenic acid were also partially resolved. However, it was not possible to effectively separate individual positional *cis* and *trans* isomers using a 20 ft long packed column [58]. This drawback of packed columns was due largely to practical considerations that limited separation efficiencies, even with highly selective stationary phases [59]. Although larger numbers of theoretical plates could be produced, the need to compensate for back pressure prevented the operation of these columns at more than 6000 theoretical plates. However, a packed column was successfully used to determine *trans* concentration in a collaborative study [60]. This early study concluded that GC analysis using an OV-275 packed column was as effective as IR spectroscopy in quantitating total *trans* unsaturation.

2. Separations Using Capillary Columns

The development of flexible fused silica columns in the early 1980s [61] led to the popularity of capillary columns, which dramatically increase the number of effective plates, thus improving

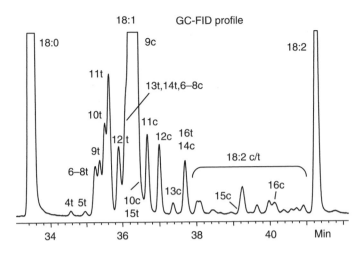

FIGURE 6.4 Partial GC chromatogram observed for *trans* and *cis* fatty acid methyl ester isomers from a milk fat test sample using a 100 m fused silica capillary column. (From Kramer, J.K.G., Blackadar, C.B. and Zhou, J., *Lipids*, 37, 823, 2002; Kramer, J.K.G., Cruz-Hernandez, C. and Zhou, J., *Eur. J. Lipid Sci. Technol.*, 103, 600, 2001.)

column separation efficiencies. The number of effective plates could be raised from 40,000 to 250,000 by increasing the column length from 15 to 100 m.

The increased availability of long capillary columns [61–81] with a variety of diameters and stationary phases has decreased the use of packed columns in the last 30 years. As Figure 6.4 indicates, monoene and diene [79,80] FAMEs and their isomers from milk fat were only partially separated in a single run on a 100 m capillary column. The *trans* triene isomers present in partially HSBO were first separated into four peaks on a 100 m capillary GC column [64]. Fractional crystallization and reversed-phase HPLC, followed by GC analysis of hydrazine reduction products, served to identify and quantify these triene isomers.

Today, most laboratories use long capillary columns to quantitate individual fatty acid isomers found in hydrogenated oils. The application of capillary GC analysis in the separation and identification of positional and geometric isomers of unsaturated fatty acids has been reviewed [65]. This GC analysis becomes even more complex for geometrical and positional isomers found in hydrogenated fish oils or milk fat, which contain many more fatty acid isomers than are present in hydrogenated vegetable oils.

An extensive analytical study of hydrogenated menhaden fish oil using capillary GC was conducted [66,67]. At an IV of 84.5, the most unsaturated isomers were eliminated, although 13.1% diene, 8.3% triene, and 0.4% tetraene isomers were still present. The 20 carbon isomers were analyzed further to determine the *cis* and *trans* bond positions along the fatty acid chain. A wide range of monoene, diene, and triene *cis*- and *trans*-positional isomers were identified at virtually every position of the carbon chain.

The content of *trans*-C16:1 in human milk, as determined by GC, is usually high due to the overlap with peaks attributed to C17 fatty acids [68]. Isolation by silver ion (see Section III.B) TLC ($AgNO_3$-TLC) followed by GC on a highly polar 100 m capillary column allowed the determination of the average content of total *trans*-C16:1 to be 0.15% ± 0.04% from 39 test samples of human milk fat [63]. The C16:1 positional isomers *trans* $\Delta 4$, $\Delta 5$, $\Delta 6/7$, $\Delta 8$, $\Delta 9$, $\Delta 10$, $\Delta 11$, $\Delta 12$, $\Delta 13$, and $\Delta 14$ were reportedly quantified in 15 test samples. The mean relative contents were 2.6%, 3.5%, 7.6%, 7.2%, 24.7%, 10.4%, 10.1%, 14.3%, 8.4%, and 11.3% (as a percentage of total *trans*-C16:1), respectively [68]. As Figure 6.5 indicates, the *trans* and *cis*-18:1 fractions from milk fat were isolated by silver ion TLC and almost completely separated by GC on a highly polar 100 m

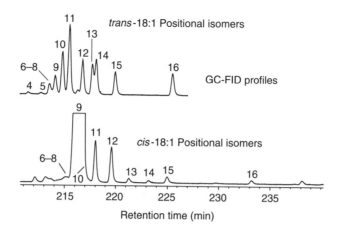

FIGURE 6.5 GC chromatograms demonstrating the overlap between *trans*-18:1 and *cis*-18:1 fatty acid methyl ester isomeric fractions from a milk fat test sample using a 100 m fused silica capillary column, after fractionation of the geometric isomers by silver ion TLC. (From Kramer, J.K.G., Blackadar, C.B. and Zhou, J., *Lipids*, 37, 823, 2002; Kramer, J.K.G., Cruz-Hernandez, C. and Zhou, J., *Eur. J. Lipid Sci. Technol.*, 103, 600, 2001.)

capillary column [79,80]. Note the presence of the Δ6–9 *trans*-18:1 positional isomers (unlabeled peaks) in the *cis* fraction.

Wilson et al. [69] reported an AgNO₃-TLC combined with capillary GC technique for the quantitative analysis of *trans* monoenes derived from partially hydrogenated fish oil containing *trans* isomers with C-20 and C-22 carbons. FAMEs were separated into saturates (R-f 0.79), *trans* monoenes (R-f 0.4), *cis* monoenes (R-f 0.27), dienes (R-f 0.10), and polyunsaturated fatty acids with three or more double bonds remaining at the origin. These authors reported that direct GC analysis underestimated the *trans* content of margarines by at least 30%. In this study, C-20 and C-22 *trans* monoenes were found in relatively large quantities averaging 13.9% and 7.5%, respectively, in margarines [69].

Using a combination of Ag-TLC and GC, the frequency distributions of the CLA isomer c9t11-C18:2, and the isolated or nonconjugated fatty acids *trans*-C18:1, *trans*-C18:2, t11-C18:1 (vaccenic acid), t11c15-C18:2 (major mono-*trans*-C18:2), as well as of total *trans* fatty acids were reported [70] for more than 2000 test samples of European bovine milk fats. These were obtained under conditions different from normal feeding conditions: barn feeding, pasture feeding, and feeding in the transition periods in spring and late autumn. The average contents were (c9t11-18:2) 0.76%, (*trans*-C18:1) 3.67%, (*trans*-C18:2) 1.12%, and (total *trans* fatty acids) 4.92%. High correlation coefficients (*r*) were reported between the content of the CLA isomer c9t11-18:2 and the contents of *trans*-C18:1, *trans*-C18:2, total *trans* fatty acids, and C18:3 of 0.97, 0.91, 0.97, and 0.89, respectively. A probable metabolic pathway in the biohydrogenation of linolenic acid was also reported: c9c12c15 → c9t11c15 → t11c15 → t11 [70].

3. Equivalent Chain Length

It is difficult to identify peaks in gas chromatograms, mainly because of the complexity of the fatty acid composition of the various matrices. The major identification problems are a result of the large number of fatty acid isomers present, and the fact that some polyunsaturated fatty acids can have longer retention times than the next longer-chain fatty acid. This happens most frequently on highly polar columns used to separate fatty acid isomers.

One classical way to predict where a certain fatty acid will elute on a GC column is based on the so-called equivalent chain length (ECL). The use of ECL to identify fatty acid isomers on GC columns has been reviewed [71]. It was reported that ECLs are constants for a specific carrier gas and column and are independent of experimental conditions [72]. A specific fatty acid can be characterized by obtaining its ECL on both polar and nonpolar columns. The ECL consists of one or two integers indicating the positional chain length, and two numbers after the decimal that indicate the fractional chain length (FCL). The ECL values for fatty acids on a specific column are determined by first using semilog paper to plot the log of the retention times (y axis) against those of the saturated fatty acids (e.g., $16:0 = 16.00$, $18:0 = 18.00$, $20:0 = 20.00$) on the x axis observed under isothermal conditions. Using the same column, the retention time (y-axis value) of an unsaturated fatty acid is then plotted to find its ECL on the x axis. For example, the ECL of linoleic acid ($18:2n$-6) was found to be 18.65. The FCL value was 0.65 (18.65–18.00). Using these data, the ECL for $20:2n$-6 was predicted to be 20.65 ($20.00 + 0.65$). The actual ECL of $20:2n$-6 was experimentally determined to be 20.64 [71]. ECL data from a GC analysis of rapeseed oil showed that the *cis*-11 20:1 fatty acid coelutes with *trans* triene isomers at oven temperatures above 160°C [73].

4. Relative Response Factors

Area normalization is often used to report the relative concentration of fatty acid isomers in an oil mixture. This method is accurate only if all isomers exhibit the same response in the GC detector. However, all the fatty acid isomers do not give the same relative response in the GC flame ionization detector (FID). Relative response factors are in use today; they were first proposed in 1964 [74] and were found to depend on the weight percent of all carbons except the carbonyl carbon in the fatty acid chain. A later study confirmed these findings and concluded that the required corrections were not insignificant, even though they were often not used [75]. Failure to use a correction factor could result in an error of about 6% (relative to 18:0) in the case of 22:6.

5. Official Capillary GC Methods

Capillary GC has been the most widely used analytical method to analyze FAME [76–80]. A successful GC determination of total *trans* FAME composition depends on the experimental conditions dictated by the method used, as well as sound judgment by the analyst to correctly identify peaks attributed to *trans* FAME and their positional isomers. The most recent GC methods to determine *trans* FAME describe separations that require long capillary columns with highly polar stationary phases. Under these conditions, a separation is based on the chain length of the fatty acid, degree of unsaturation, and the geometry and position of double bonds. The expected elution sequence for specific fatty acids with the same chain length on highly polar columns is as follows: saturated, monounsaturated, diunsaturated, etc. *trans*-Positional isomers are followed by *cis*-positional isomers, but today there is still extensive overlap of the geometric isomers.

The improved resolution of the large number of peaks attributed to *trans*- and *cis*-positional isomers obtained under today's improved experimental conditions leads to the partial or complete overlap of many peaks belonging to the two different groups of geometric isomers [76–81]. This is because the retention time range for late eluting *trans*-18:1 positional isomers, starting at Δ12 (or Δ13 depending on experimental conditions used), is the same as that for the *cis*-18:1 positional isomers, Δ6–Δ14. In addition, *cis/trans*-18:2 methylene- and nonmethylene-interrupted fatty acid retention time range also overlaps with that of the *cis*-18:1 positional isomers.

As stated earlier, elimination of GC peak overlap usually requires prior separation of the *cis* and *trans*-18:1 geometric isomers by silver ion-TLC [76–81] or other methods. GC analysis of the isolated *trans* and *cis* fractions [79,80] clearly demonstrates the extent of overlap (Figure 6.5). Most individual 18:1 isomers can be completely resolved by GC under isothermal conditions at significantly lower temperatures [76–81].

The effect of this GC peak overlap on the accuracy of *trans* FAME determinations for various matrices depends on (1) the nature of the *trans* fat matrix that is analyzed (vegetable oil, fish oil, milk fat) as they would exhibit vastly different isomeric distributions, (2) the GC experimental conditions (for instance, the length and age of a column, nature of the carrier gas, and temperature program), and (3) the analyst's experience, ability, and skill in optimizing the performance of the gas chromatograph, and subsequently (4) correctly identifying all the observed GC peaks in the widely different and complex fatty acid profiles.

Currently available GC official methods include AOAC 996.06 [82], which is appropriate for the determination of fat in food products, and AOCS Ce 1f-96 [83], which is applicable to the determination of *cis* and *trans* fatty acids in hydrogenated and refined vegetable oils and fats. AOCS is in the process of planning a new collaborative study to validate AOCS method Ce 1k-07 titled "Direct methylation of lipids for the determination of total fat, saturated, *cis*-monounsaturated, *cis*-polyunsaturated and *trans* fatty acids by GC" [84]. Although GC experimental results performed in a laboratory by a single analyst usually yield good repeatability, the same level of reproducibility was not obtained by analysts in different laboratories as demonstrated by the most recent GC validation study that was approved in 2005 as AOCS Official Method Ce 1h-05 [85]. For example, for a *trans* fatty acid level of 1%, the interlaboratory reproducibility percent relative standard deviation was high, approximately 20% or twice as high as the intralaboratory repeatability percent relative standard deviation. Moreover, no GC lower limit of quantitation was reported in this multilaboratory collaborative study [85].

B. THIN-LAYER CHROMATOGRAPHY

TLC has been widely used to separate classes of lipids on layers of silica gel applied to glass plates. TLC is simple to use and does not necessarily require sophisticated instrumentation. In most cases, TLC using silica gel does not separate fractions on the basis of number or configuration of double bonds in fatty acid mixtures [86]. The most effective TLC separation of *cis* and *trans* isomers has been achieved by means of argentation: layers of silica impregnated with silver nitrate.

Argentation is the general term used to describe methods that use the long-known principle that silver ion forms complexes with *cis* more strongly than with *trans* double bonds. In argentation chromatography the separation of *cis* and *trans* isomers depends on the relative interaction strength of the π electrons of double bonds in each isomer with silver ions. The actual interaction of each fatty acid isomer depends on the geometry, number, and position of double bonds. Ag-TLC has been used for several years to separate unsaturated lipids on thin-layer plates [87].

The use of argentation methods, including TLC, countercurrent distribution (CCD), and liquid column chromatography, has been reviewed [87,88]. Several classes of fatty acid isomers can be isolated using preparative argentation TLC. These isolated fractions can then be analyzed further by using ozonolysis [66,67,89–91] and GC. Good resolution can be achieved for a variety of FAMEs using two-dimensional argentation TLC separation and reversed-phase chromatography on a single plate [92].

In recent studies by Kramer and his coworkers [80,93], silver ion TLC was applied to the separation of saturated, *trans*-monounsaturated, *cis*-monounsaturated, as well as *cis*-/*trans*-conjugated diene fatty acids. Silica G plates (20 × 20 cm, 0.25 mm thick; Fisher Scientific) were washed with 50:50 methanol/chloroform (v/v), activated at 110°C for 1 h, impregnated with 5% silver nitrate solution in acetonitrile (w/v), and activated again before use. The plates were developed with 90:10 hexane/diethyl ether solution (v/v), and the bands were identified under UV light at 243 nm after spraying with 2′,7′-dichlorofluorescein (2% in methanol, v/v). The bands containing the isolated fatty acids were scraped and collected in test tubes, and the fatty acids were extracted with organic solvent and separated by GC.

Lately, a new silver ion solid phase extraction (SPE) cartridge (Supelco, Bellefonte, PA; www.sigmaaldrich.com/supelco/bulletin/t406062.pdf) reportedly allows the rapid fractionation of FAMEs including *cis*/*trans* geometric isomers. However, its benefits and limitations have yet to be evaluated and compared with established procedures.

C. High-Performance Liquid Chromatography

For most applications, HPLC does not exhibit the separation efficiencies offered by long capillary GC columns. It does, however, offer several advantages. For example, the mild conditions used in HPLC work enable heat-sensitive components to be separated, a feature that reduces the possibility of isomerization taking place during the analysis of unsaturated fatty acids. Another advantage is that fatty acid fractions can be collected and analyzed further using hyphenated techniques, such as GC–FTIR and GC–MS.

A variety of detectors are used to identify lipid solutes as they elute from HPLC columns. The most common HPLC detector has been UV-based equipment that allows one to monitor the column effluent at about 200 nm. The lack of strong absorbing lipid chromophores limits the sensitivity of this detection method unless fatty acid derivatives that absorb strongly in the UV range are prepared. Refractive index (RI) detectors are also commonly used but cannot be paired with solvent gradients and are sensitive to temperature fluctuations. Evaporative light-scattering detectors (ELSD) can be used with solvent gradients, are at least as sensitive as the best RI detectors, and can be used for quantitation. The advantages of these and other detectors useful for HPLC analysis have been reviewed [94].

1. Reversed-Phase HPLC

Recently, a reverse-phase HPLC procedure for the fractionation of *cis*-18:1 and *trans*-18:1 fatty acids was proposed by using two semipreparative reverse-phase C18 columns in series before GC quantitation [95]; however, the observed resolution of the *cis* and *trans* geometric isomers was poor and further optimization was needed.

2. Silver Ion HPLC

The many advantages of using argentation HPLC columns over TLC plates include reproducible separation of analytes, column reusability, short run times, and high recoveries. Until recently, a lack of stable columns with controlled silver levels limited the use of silver ion HPLC separations. The silver ions would bleed from the silica adsorbent, cause unpredictable results, shorten the column life, and hinder the reproducibility of the results. A more successful argentation HPLC procedure was developed by linking the silver ions via ionic bonds to a silica–phenylsulfonic acid matrix [96]. This column gave excellent reproducible separations for triacylglycerols, fatty acids, and their positional and geometric isomers [97]. This column was also used for the separation, collection, and quantification of all eight geometric isomers of linolenic acid phenacyl esters, with a mobile phase ranging from 5% methanol in dichloromethane to a 50:50 solvent mixture [98].

Commercial silver ion HPLC columns were introduced approximately a decade ago and have dramatically increased the use of this technique. A commercially available (Chrompack Ltd.) argentation HPLC column with an acetonitrile–hexane mobile phase was used to separate the *cis/trans* fatty acid isomers of MO, methyl linoleate, methyl linolenate, and 15 of the 16 *cis/trans* methyl arachidonate positional isomers [99]. In other research, this column was used with a 0.15% acetonitrile in hexane isocratic mobile phase to obtain four fractions from hydrogenated vegetable oil that were subsequently analyzed by capillary GC [100].

Silver ion HPLC has since been repeatedly shown to be a powerful technique for the separation of geometric and positional isomers of fatty acids [101,102] and FAMEs [103,104]. Lately, Delmonte et al. [105] investigated the silver ion HPLC separation and quantitation of *trans*-18:1 fatty acid positional isomers (from Δ6 to 15) (Figure 6.6) as well as their dependence on parameters such as elution temperature; they also discussed the many limitations of this procedure and the factors that affected the reproducibility of this separation.

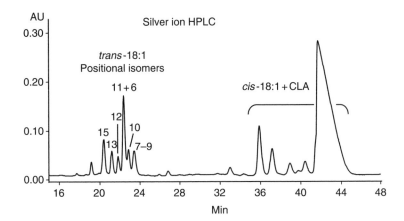

FIGURE 6.6 Silver ion HPLC chromatogram showing the separation between *trans*-18:1 and *cis*-18:1 and CLA fatty acid methyl ester positional isomers from a milk fat test sample using three chromsphere five lipids columns in series. (From Delmonte, P., Kramer, J.K.G. and Yurawecz, M.P., *Analysis of Trans 18:1 Fatty Acids by Silver Ion HPLC in Lipid Analysis and Lipidomics*, M.M. Mossoba, J.K.G. Kramer, J.T. Brenna, and R.E. McDonald, eds, AOCS Press, Champaign, IL, 2006. With permission.)

IV. ANALYSIS USING OFF-LINE COMBINED TECHNIQUES

As stated earlier, the direct quantitation of all fatty acid isomers in hydrogenated oils by GC is nearly impossible; there is overlap of *cis/trans* monoene and diene positional isomers even with the high efficiency of long capillary columns. Several analytical techniques have been combined with GC to determine accurate fatty acid profiles of hydrogenated oils including ozonolysis and IR spectroscopy.

A. OZONOLYSIS/GAS CHROMATOGRAPHY

Ozonolysis followed by GC can be used to determine the double-bond position of *cis* and *trans* isomers. Ozonolysis cleaves the hydrocarbon chain of fatty acids at positions of unsaturation. This includes reductive cleavage to form aldehydes, aldehyde esters, and other fragments, which are identified according to their retention times on GC columns compared with known standards. Ozonolysis has been used to determine the double-bond position of both monoene [106] and diene isomers [107,108]. Even though the double-bond position can be determined by ozonolysis and GC, the geometric configuration cannot. Computer programs involving solutions of several simultaneous equations were used to calculate fatty acid bond position from ozonolysis results [108]. A published review discusses the use of ozonolysis and other chemical methods for lipid analysis [109].

B. SILVER ION CHROMATOGRAPHY/NMR

^{13}C NMR spectroscopy can be an effective tool for the identification of fatty acid isomers. This tool is most useful if purified fatty acid fractions are first obtained. Silver ion chromatography on both TLC plates and HPLC columns have been used to obtain purified fractions for ^{13}C NMR analysis. HSBO was separated into six bands with preparative silver ion TLC [110]. The isomers in each band were identified by capillary GC and ^{13}C NMR analysis. The presence of specific *trans* diene isomers was confirmed by observing their unique chemical shifts. Silver ion HPLC was used to obtain a purified *trans* monoene fraction from HSBO [111].^{13}C NMR spectroscopic analysis of this fraction confirmed the presence of the minor $\Delta6$ and $\Delta7$ *trans* monoene positional isomers. The presence of

the Δ5 *trans* monoene positional isomer was inferred. This HPLC fraction had been analyzed by capillary GC, but several minor isomers could not be identified [111].

C. GAS CHROMATOGRAPHY/IR SPECTROSCOPY

GC provides useful information concerning the total fatty acid composition of hydrogenated oils. However, the overlap of some of the *trans* with *cis* monoene peaks makes it difficult to get accurate GC determinations of total *cis* and total *trans* monoene content of food products. A combined GC and IR method to determine *cis* and *trans* fatty acid monoenes that coelute on GC columns was developed and was adopted as an AOAC official method [112] after completion of a collaborative study [113]. In this method, the total *trans* content of the oil was first determined by modifying a published IR procedure [114] that used a two-component calibration plot. Then the weight percentages of all the *trans* diene (18:2*t* and 18:2*tt*) and triene (18:30*t*) isomers were determined by means of a highly polar 50 or 100 m capillary GC column. The weight percentage of *trans* monoenes (18:1*t*) was then determined by the following formula with the appropriate correction factors (0.84 and 1.74):

$$\text{IR } trans = \%18{:}1t + (0.84 \times \%18{:}2t) + (1.74 \times \%18{:}2tt) + (0.84 \times \%18.3t). \quad (6.5)$$

After the *trans* monoenes had been calculated, the *cis* monoenes were determined by finding the difference between the total monoenes determined by GC and the *trans* monoenes that were calculated:

$$\%18{:}1c = (\text{total}\%18{:}1t \text{ by GC} + 18{:}1c \text{ by GC}) - (\%18{:}1t \text{ by calculation}). \quad (6.6)$$

V. ANALYSIS USING ONLINE HYPHENATED TECHNIQUES

The lack of standards for many fatty acids and their isomers and the problems of coelution of peaks may lead to the misidentification of fatty acids in gas chromatograms. Some publications had misidentified *trans* monoene positional isomers [100,115], as it was later determined [111]. Another study determined the level of *trans,trans*-18:2 isomers in margarine to be an order of magnitude too high (3% instead of 0.3%) also due to GC peak misidentification [116]. Similarly, the published literature contains an incorrect report that liquid canola shortening was contaminated with fatty acids found in animal fat [117].

To help identify peaks, a GC column can be interfaced to another instrument such as an IR spectrometer or a mass spectrometer. Hyphenated techniques use online detection to confirm the identity of peaks in a chromatogram. Griffiths et al. [118,119] have documented the performance of interfaces between gas, supercritical fluid, and high-performance liquid chromatography and Fourier transform spectrometry (GC–FTIR, SFC–FTIR, and HPLC–FTIR, respectively). Some of these hyphenated techniques were first used to elucidate the structure of unusual fatty acid isomers, including cyclic fatty acid monomers (CFAM) that often contain *trans* double bonds in the hydrocarbon chain and *cis* double bonds in five- or six-membered rings. The formation and the biological effects of these cyclic compounds have been reviewed [120].

A. GC–FTIR

For many GC–FTIR instruments, the effluent from the GC column flows continuously through a light pipe (LP) gas cell [118]. LP instruments generally have detection limits of 10–50 ng for unknown complex mixtures of FAMEs. Online IR spectra of CFAM peaks eluting from a gas chromatogram were obtained with an LP GC–FTIR [121]. The results were used to show which of these CFAM contained *cis* and *trans* double bonds. Several minor peaks in this mixture could not be identified because of the limited sensitivity of the method.

Bourne et al. [122] developed an improved GC–FTIR technique using a matrix isolation (MI) interface that increases the sensitivity of the GC–FTIR determination by an order of magnitude. GC–MI–FTIR is extremely useful for quantitating peak area, determining peak homogeneity, and obtaining structural information on a compound. Although the LP GC–FTIR technique can confirm the identity of intact molecules, its detection limits are unfortunately higher than those of other GC detectors (e.g., flame ionization). MI is a technique in which analytes and an inert gas (argon) are rapidly frozen at cryogenic temperatures (12 K) and are trapped as a solid matrix on the outer rim of a moving gold-plated disk. The IR spectra of these molecules are free from bands due to intermolecular hydrogen bonding and other band-broadening effects. These combined benefits yielded greater sensitivity that equals, for many applications, that of gas chromatography–mass spectrometry (GC–MS) [123].

GC–MI–FTIR was used to quantitate low levels of saturated, *trans* monoene [124], diene [125,126], triene, and conjugated diene fatty acids and their isomers [126] in hydrogenated vegetable and fish oils. Figures 6.7 through 6.10 show IR spectra that exhibit the different stretching and out-of-plane deformation absorption bands for *cis* and *trans* double bonds in straight-chain fatty acids (Figure 6.7), CFAM (Figure 6.8), as well as for conjugated diene geometric isomers (Figures 6.9 and 6.10). The characteristic absorption bands shown were used for identification. The conjugated *trans,trans* diene isomer had an out-of-plane deformation absorption band at 990 cm^{-1}, whereas the *cis,trans* isomer had absorption bands at 950 and 986 cm^{-1}. The absorption band for the corresponding methylene-interrupted *trans* diene and the *trans* monoene was at 971 cm^{-1}. There were also unique absorption bands for *trans* and *cis* diene isomers in the area of the spectrum typical of the carbon–hydrogen stretch vibrations (3035/3005 and 3018 cm^{-1}, respectively) and in the area typical of the carbon–hydrogen out-of-plane deformation absorption bands (972 and 730 cm^{-1}, respectively).

GC–MI–FTIR is also effective in determining the concentration of individual FAMEs without having to consider the relative response of the gas chromatograph's FID. It is possible to quantitate *trans* isomers even with partial GC peak overlap from *cis* isomers. Quantitation of *trans* diene isomers was based on measurement of the height of the observed C–H out-of-plane deformation band at 971 cm^{-1} for *trans* groups and that of the CH$_2$ asymmetric stretching band at 2935 cm^{-1} for the 17:0 internal standard [125]. Calibration plots of absorbance versus nanograms injected were generated for the range of 2–33 ng. Recovery (on cryogenic disk) was based on the determination of the internal standard. The amount of analyte present in injected aliquots was calculated from the observed absorbance values and the corresponding calibration plot.

When GC–MI–FTIR was used to quantitate the *trans* monoene isomers in margarine, the results had a high bias relative to the GC FID's response [125,126]. The higher GC–MI–FTIR values presumably resulted from its higher specificity. This is because the MI-FTIR determination is based on a unique discriminatory feature (971 cm^{-1} absorbance band) that is observed only for *trans* species. The intensity of this band is not affected by *cis* isomers even when they chromatographically overlap. These results confirm that quantitation by means of GC peak areas can result in an underestimation of *trans* monoenes.

GC–FTIR analysis was used to determine the geometric configuration of FAMEs separated by silver ion HPLC [127]. This was achieved with the more recent direct deposition (DD) interface that condenses GC eluates on a moving zinc selenide window cooled to near liquid nitrogen temperature. This DD instrumentation is even more sensitive than the MI interface [118] because the analytes are condensed on a track that is about 100 μm wide, and microscope objectives are used to collect and focus the IR beam. Unlike GC–MI–FTIR, during GC–DD–FTIR operations, the analytes are not diluted in argon or any other matrix.

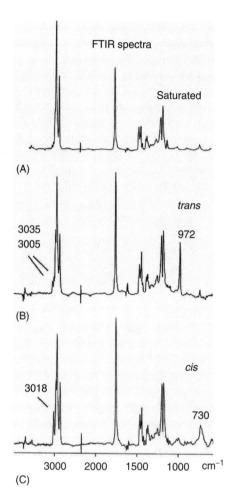

FIGURE 6.7 Sharp matrix isolation-FTIR spectral bands observed at 4 cm^{-1} resolution for (A) 18:0, (B) methylene-interrupted *trans,trans*-18:2, and (C) methylene-interrupted *cis,cis*-18:2 straight-chain fatty acid methyl esters after separation by capillary GC and cryogenic trapping under vacuum. (From Mossoba, M.M., McDonald, R.E., Chen, J.Y.T., Armstrong, D.J. and Page, S.W., *J. Agric. Food Chem.*, 38, 86, 1990. With permission.)

B. GC–EIMS

MS can be a very effective tool when used in combination with GC to determine the location of double bonds in fatty acids and their positional isomers. The major problem of analyzing mass spectra of FAMEs is the tendency of the double bonds to migrate during electron ionization. The mass spectra exhibit low-mass ions that do not provide structural information. Some of the methods used to overcome this problem include soft ionization and derivatization [128]. Chemical ionization (CI) methods for the determination of double- and triple-bond positions were reviewed [129]. A CI–MS procedure was used to determine the double-bond positions in fatty acids from marine organisms [130].

One successful approach has been to derivatize the carboxyl group to a nitrogen-containing compound. Common derivatizing agents include pyrrolidide, picolinyl ester, and 4,4-dimethylox-azoline (DMOX). A recent review of these derivatives indicates that the most useful ones by far are the picolinyl ester and DMOX derivatives [131]. With these derivatives, double-bond ionization and

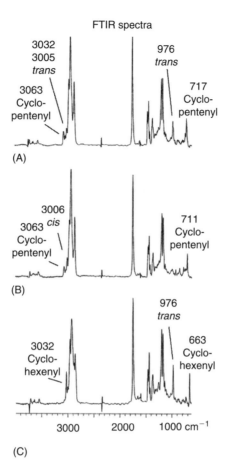

FIGURE 6.8 Sharp matrix isolation-FTIR spectral bands observed at 4 cm^{-1} resolution for methyl ester derivatives of C18 cyclic fatty acid monomers having structures consistent with (A) a cyclopentenyl ring and a *trans* double bond, (B) a cyclopentenyl ring and a *cis* double bond, and (C) a cyclohexenyl ring and a *trans* double bond along the hydrocarbon chain. Data collected after separation by capillary GC and cryogenic trapping under vacuum. (From Mossoba, M.M., Yurawecz, M.P., Roach, J.A.G., Lin, H.P., McDonald, R.E., Flickinger, B.D. and Perkins, E.G., *J. Am. Oil Chem. Soc.*, 72, 721, 1995. With permission.)

migration are minimized. Simple radical-induced cleavage occurs at each C–C bond along the chain. Therefore, for unsaturated fatty acids containing up to several double bonds, there is decreased abundance of low-mass ions and an increase in a series of ions resulting from carbon–carbon bond scission. Diagnostic ions occur wherever there is a functional group in the chain that interrupts the pattern of cleavage from C–C bonds. A C=C bond or a five- or six-membered ring might be responsible for such disruptions.

GC–EIMS analysis of picolinyl ester derivatives can identify polyunsaturated fatty acids [132,133]. However, the GC resolution of the picolinyl fatty acid esters was not as good as that of other derivatives (such as DMOX). Reversed-phase HPLC fractionation of the picolinyl fatty acid esters before identification by GC–MS was necessary to obtain acceptable results for hydrogenated samples [134]. A total of 39 fatty acid components in cod liver oil were identified using this method. In a more recent study, silver ion HPLC was used to fractionate CFAMs before converting them to the picolinyl ester derivatives [135]. Some of the GC problems associated with picolinyl esters could be overcome by using high-temperature, low-bleed, cross-linked polar columns [136].

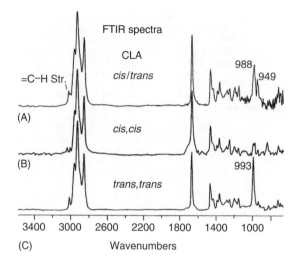

FIGURE 6.9 Direct deposition-FTIR spectral bands observed at 4 cm^{-1} resolution for conjugated (A) *cis/trans* or *trans/cis*-18:2, (B) *cis,cis*-18:2, and (C) *trans,trans*-18:2 fatty acid methyl ester CLA isomers after separation by capillary GC and cryogenic trapping under vacuum. (From Mossoba, M.M., Yurawecz, M.P., Roach, J.A.G., McDonald, R.E., Flickinger, B.D. and Perkins, E.G., *Lipids*, 29, 893, 1994. With permission.)

Mossoba et al. [137] reported that 2-alkenyl-4,4-dimethyloxazoline derivatives of diunsaturated CFAMs exhibited distinctive mass spectral fragmentation patterns that could be used to pinpoint the positions of double bonds and of 1,2-disubstituted, unsaturated, five- and six-membered rings along the hydrocarbon chain. One of the advantages of using the DMOX derivatives was the good

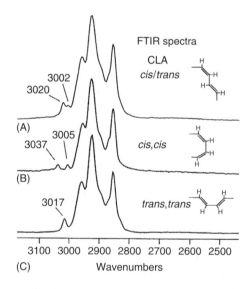

FIGURE 6.10 Direct deposition-FTIR spectral bands observed at 4 cm^{-1} resolution for conjugated (A) *cis/trans* or *trans/cis*-18:2, (B) *cis,cis*-18:2, and (C) *trans,trans*-18:2 fatty acid methyl ester CLA isomers after separation by capillary GC and cryogenic trapping under vacuum. Expanded IR spectral range showing C–H stretching bands for conjugated *cis–trans* (*top* spectrum) and *trans–trans* (*bottom* spectrum) 18:2 dienes. (From Mossoba, M.M., Yurawecz, M.P., Roach, J.A.G., McDonald, R.E., Flickinger, B.D. and Perkins, E.G., *Lipids*, 29, 893, 1994. With permission.)

TABLE 6.1

Infrared Bands (cm^{-1}) Attributed to Unsaturation Sites in Cyclic Fatty Acid Monomers

GC Peak	Ring, Five Membered	Chain, trans	Ring, Six Membered	Chain			Ring		
				cis	trans	trans	Six Membered	Five Membered	Six Membered
1	3061	3035			3003	970		719	
2	3061	3035			3003	970		719	
3 + 3'	3061			3005					716	
4	3063	3032			3005	979		716	
6	3063			3006					711	
7	3063			3006					711	
8			3032		3000	976				663
9			3032		3005	972				664
10			3032		3005	972				664
11			3032		3004	975				663
12			3031					723		664
13			3031					725		662
14			3025							
15			3025							

Source: From Mossoba, M.M., Yurawecz, M.P., Roach, J.A.G., Lin, H.P., McDonald, R.E., Flickinger, B.D. and Perkins, E.G., *J. Am. Oil Chem. Soc.*, 72, 721, 1995.

Note: Microscale supercritical fluid extraction (SFE) can be directly coupled to a capillary SFC column. This procedure was used to analyze the fatty acid composition of a 1 mg sample of cottonseed kernel [149].

chromatographic resolution: sometimes higher than that observed for derivatives of FAMEs [138,139]. Most CFAMs in heated flaxseed oil were identified, and the double-bond configurations (*cis* or *trans*) were unequivocally established by using GC–MI–FTIR (Table 6.1).

Mossoba et al. [127] confirmed the identity of individual *trans* monoene fatty acid positional isomers in partially HSBO. FAMEs were fractionated by silver ion HPLC and then analyzed by GC–DD–FTIR to determine geometric configuration and by GC–EIMS on DMOX derivatives to determine double-bond position [127]. GC peak resolution obtained with a 100 m capillary column was higher for the DMOX derivatives than for that of the FAMEs. Figure 6.11 demonstrates the excellent resolution that was obtained for DMOX derivatives of *trans* monoene positional isomers. The bottom GC profile, with about twice the amount of injected sample, shows evidence of overload in the early part of the trace but enhanced response for the Δ13 and Δ14 positional isomers. The double-bond positions for nine individual *trans* monoene positional isomers were confirmed by their unique DMOX mass spectra. Most significantly, this was the first report of the capillary GC separation of the Δ13 and Δ14 *trans* monoene positional isomers in hydrogenated vegetable oil. Figure 6.12 presents the mass spectral data that confirmed their bond position. The identity of the *trans* Δ13 isomer was further confirmed by comparison with a standard. Hence, the double-bond position and configuration could be readily established for a complex distribution of monounsaturated fatty acid isomers found in dietary fat by using the two hyphenated techniques, GC–DD–FTIR and GC–EIMS.

Mass spectral methods that permit direct analysis of FAMEs are highly desirable because GC analysis of FAMEs remains the method of choice for the determination of fatty acid composition. Since 1999, Brenna and coworkers [140–144] have published a series of papers demonstrating a purely mass spectrometric technique for double bonds localization in FAME isomers that relies on gas phase derivatization. He introduced the acronym CACI, Covalent-Adduct Chemical Ionization,

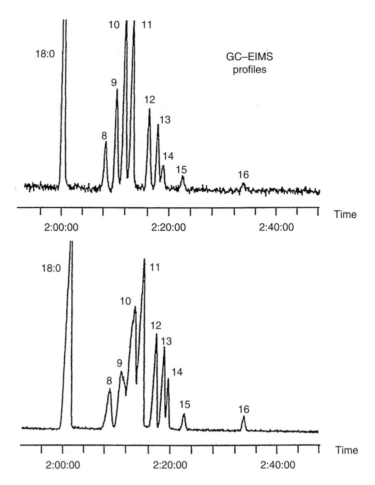

FIGURE 6.11 GC–EIMS chromatographic data for fatty acid DMOX derivatives of *trans*-18:1 positional isomers using an SP-2560 100 m fused silica capillary column at 140°C. Injection volume: *top* trace, 0.4 μL; *bottom* trace, 1.0 μL. (From Mossoba, M.M., McDonald, R.E., Roach, J.A.G., Fingerhut, D.D., Yurawecz, M.P., and Sehat, N., *J. Am. Oil Chem. Soc.*, 74, 125, 1997. With permission.)

to describe analytical methods that rely on an ion–molecule reaction followed by tandem mass spectral analysis. He applied acetonitrile CACI for the analysis of nonconjugated and conjugated FAMEs. Observed highly reproducible mass spectra were readily interpreted and used to identify double-bond position and *cis/trans* geometry and for quantitative analysis. Brenna's work demonstrated the advantage of applying the acetonitrile CACI procedure for double-bond localization.

C. SFC–FTIR

Capillary SFC is an effective tool for separating nonpolar to moderately polar complex mixtures of natural products having molecular weights of 100–1000 Da. SFC uses a gas compressed above its critical temperature and pressure to carry analytes through a chromatographic column. The many applications of SFC to the analysis of lipids have been reviewed [145–149]. SFC can be used to separate some compounds at lower temperatures than those required for GC, and a simultaneous interface with FID and FTIR detectors can be readily achieved. Different SFC columns can be used to separate lipids according to carbon number or degree of unsaturation. However, capillary GC and silver ion HPLC columns give better peak resolution when separating fatty acid derivatives from hydrogenated oils.

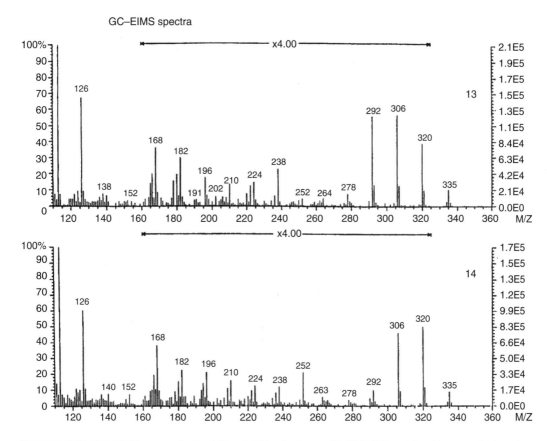

GC–EIMS spectra

FIGURE 6.12 GC–EIMS spectra for the Δ13 and Δ14 pair of *trans*-18:1 DMOX positional isomers; the ions due to allyllic cleavage are at the following positions: Δ13: 238, 293, and 306 m/z; Δ14: 252, 306, and 320 m/z. (From Mossoba, M.M., McDonald, R.E., Roach, J.A.G., Fingerhut, D.D., Yurawecz, M.P. and Sehat, N., *J. Am. Oil Chem. Soc.*, 74, 125, 1997. With permission.)

SFC has been used to obtain structural information, backed up by MS for detection [150,151]. The main limitation of early applications has been the lack of suitable commercial interfaces [152].

SFC with online FTIR detection (SFC–FTIR) has been used to elucidate structural information from fatty acid isomer mixtures. Standards as well as fatty acids were separated from coconut oil, Ivory soap, soybean oil, and butterfat by means of a packed SFC column [153]. The presence of unsaturated fatty acids was verified by an IR absorption maximum at 3016 cm⁻¹. SFC-IR spectroscopy was used to analyze triacylglycerols and free fatty acids under similar conditions [154]. One study featured an SFC coupled to an FTIR spectrophotometer equipped with an LP flow cell to determine the level of *trans* unsaturation in partially hydrogenated soybean oil [155]. The IR absorption band at 3016 cm⁻¹ indicated the presence of linoleic acid, whereas the 972 cm⁻¹ absorption band indicated the presence of *trans* unsaturation.

A packed microcolumn argentation SFC system was used to separate and quantitate triacylglycerols in vegetable, fish, and hydrogenated oils [156]. This system gave excellent separations for these complex samples and was as effective as HPLC.

VI. *TRANS* ISOMERS IN COMMERCIAL PRODUCTS

Although the relationship of dietary lipids to human health is a complicated issue that has sparked some controversy, the importance of controlling fat intake to help maintain an active and healthy lifestyle has been recognized for many years. Consumer health concerns about the types and content of dietary fat has resulted in a great deal of research. *Trans* fatty acids have received negative publicity in the media with respect to their effects on serum cholesterol. In recent years, some food products have been reformulated to reduce their total fat and *trans* fatty acid content.

Processed foods (for instance, bakery products, fast foods, fried foods, and savory snacks) are the major (over 60%) source of dietary *trans* fats [1,157,158] in the North American diet because they are prepared with partially hydrogenated vegetable oils. The total fat content and *trans* levels can vary widely in various food products and even within a food category.

In recent years, the *trans* fat content of margarines has been relatively reduced by blending partially hydrogenated vegetable oils with unhydrogenated liquid oils. However, some margarines are still a significant (11%) source of *trans* fats in the diet; their *trans* fatty acid composition consists mostly of *trans*-18:1 isomers (85%), as well as mono-*trans* isomers of linoleic (9 *cis*, 12 *trans*-18:2 and 9 *trans*, 12 *cis*-18:2) and α-linolenic (9 *cis*, 12 *cis*, 15 *trans*-18:3, 9 *cis*, 12 *trans*, 15 *cis*-18:3, and 9 *trans*, 12 *cis*, 15 *cis*-18:3) acids. The total *trans* levels vary widely between 0% and 40% (as percent of total fat) depending on the country and the type of margarine. Moreover, zero-*trans* margarine products [159] have been produced by the complete replacement of partially hydrogenated oils with interesterified liquid oils as well as small quantities of oils high in saturated fat (palm-kernel oil, palm oil, or coconut oil) in order to achieve the solid texture and higher melting point characteristics of margarines; their total *trans* fat content is low (0%–2%, as percent of total fat) and their *trans* fatty acid composition is similar to those observed for refined unhydrogenated vegetable oils, including the absence of unnatural positional isomers of *cis*-9 18:1.

The predominant *trans* isomers in refined (unhydrogenated) edible oils used in salads and cooking are present at low levels (near or below 2%, as percent of total fat) and consist of the mono-*trans* geometrical isomers of linoleic and α-linolenic acids [160–162].

Dairy products are another important source of *trans* fats in some countries because milk fat may contain up to 8% *trans* fat (as percent of total fat) [163,164]. The predominant *trans* fatty acid among the t-18:1 isomers found in dairy fat is vaccenic acid (11t-18:1), and as stated earlier, a precursor of CLA.

VII. CONCLUSIONS

As long as the nutritional effects of *trans* fatty acids are debated, there will be an important need to rapidly and accurately determine the total *trans* content by simple and rapid methods. As reviewed in this chapter, the validated gas chromatographic and IR methodologies have been, and are currently being, further optimized to increase their accuracy and reliability in the measurement of low levels of *trans* in food products and dietary supplements for regulatory compliance purposes.

In addition, the nutritional effects of many positional and geometric isomers of fatty acids have not been extensively studied. The identification and quantitation of individual fatty acid isomers will be a critical step in any future study designed to determine the nutritional properties of various individual *cis* and *trans* geometric and positional isomers with one or more double bonds (non-conjugated or conjugated). The complex isomer mixtures in various food matrices inevitably contain overlapping *cis* and *trans* GC peaks that make identification difficult. Capillary GC on long, highly polar columns will continue to be the industry standard for separating fatty acid isomers. However, as discussed in this chapter, GC alone cannot separate all fatty acid isomers in hydrogenated oils and other complex food matrices. Therefore, a combination of off-line or online techniques have been used to separate and identify individual fatty acid isomers. The availability of commercial silver ion HPLC columns should increase their use, and silver ion chromatography is expected to continue to be an important chromatographic tool that can be combined with other techniques. Complementary mass spectral analyses based on the GC–EIMS DMOX derivatives approach or the more recent Covalent-Adduct Chemical Ionization procedures for methyl esters are also expected to continue to play a critical role in the analysis of both isolated and conjugated *trans* fatty acids.

The accurate determination of the total *trans* content and of individual *trans* isomers in complex mixtures of both natural and hydrogenated oils and fats is still a challenging task today.

REFERENCES

1. S. Satchithanandam, C.J. Oles, C.J. Spease, M.M. Brandt, M.P. Yurawecz, and J.I. Rader. Trans, saturated, and unsaturated fat in foods in the United States prior to mandatory trans-fat labeling. *Lipids* 39:11–18 (2004).
2. U.S. Department of Agriculture/Agricultural Research Service, USDA Nutrient Database for Standard Reference, Release 14. Nutrient Data Laboratory Home Page, http://www.nal.usda.gov/fnic/foodcomp 2001.
3. R.P. Mensink and M.B. Katan. Effect of dietary *trans* fatty acids on high-density and low-density lipoprotein cholesterol levels in healthy subjects. *N. Engl. J. Med.* 323:439 (1990).
4. J.T. Judd, B.A. Clevidence, R.A. Muesing, J. Wittes, M.E. Sunkin, and J.J. Podczasy. Dietary *trans* fatty acids: Effects on plasma lipids and lipoproteins of healthy men and women. *Am. J. Clin. Nutr.* 59:861 (1994).
5. K. Almendingen, O. Jordal, P. Kierulf, B. Sandstad, and J.I. Pedersen. Effects of partially hydrogenated fish oil, partially hydrogenated soybean oil, and butter on serum lipoproteins and Lp(a) in men. *J. Lipid Res.* 36:1370 (1995).
6. P.M. Kris-Etherton (ed.). *Trans* fatty acids and coronary heart disease risk: Report of the expert panel on *trans* fatty acids and coronary heart disease. *Am. J. Clin. Nutr.* 62:655S (1995).
7. Department of Health and Human Services, FDA Food Labeling; Trans Fatty Acids in Nutrition Labeling; Nutrient Content Claims, and Health Claims; Final Rule, Federal Register 68, No. 133, July 11, 2003, pp. 41434–41506 (2003).
8. W.M.N. Ratnayake and C. Zehaluk. Trans fatty acids in foods and their Labeling regulations. In: *Healthful Lipids* (C.C. Akoh and O.-M. Lai, eds.). AOCS Press, Champaign, IL, 2005, pp. 1–32.
9. R.E. McDonald and M.M. Mossoba. *Trans* fatty acids: Labeling, nutrition, and analysis. In: *Food Lipids and Health* (R.E. McDonald and D.B. Min, eds.). Dekker, New York, NY, 1996, p. 161.
10. M.M. Mossoba, V. Milosevic, M. Milosevic, J.K.G. Kramer, and H. Azizian. Determination of total trans fats and oils by infrared spectroscopy for regulatory compliance. *Anal. Bioanal. Chem.* 389:87–92 (2007).
11. A. Aro. Epidemiological studies of *Trans* fatty acids and coronary heart disease. In: *Trans Fatty Acids in Human Nutrition* (J.L. Sébédio and W.W. Christie, eds.). The Oily Press, Dundee, Scotland, 1998, pp. 235–260.
12. P.W. Parodi. Cows' milk fat components as potential anticarcinogenic agents. *J. Nutr.* 127:1055–1060 (1997).
13. M. Henninger and F. Ulberth. *Trans* fatty acid content of bovine milk fat. *Milchwissenschaft* 49:555–558 (1994).
14. R.L. Wolff. Content and distribution of *trans*-18:1 acids in ruminant milk and meat fats, their importance in European diets and their effect on human milk. *J. Am. Oil Chem. Soc.* 72:259–272 (1995).
15. P.W. Parodi. Conjugated octadecadienoic acids of milk fat. *J. Dairy Sci.* 60:1550–1553 (1977).
16. S.F. Chin, W. Liu, J.M. Storkson, Y.L. Ha, and M.W. Pariza. Dietary sources of conjugated dienoic isomers of linoleic acid, a newly recognized class of anticarcinogens. *J. Food Compost. Anal.* 5:185–197 (1992).
17. M.P. Yurawecz, J.A.G. Roach, N. Sehat, N.M.M. Mossoba, J.K.G. Kramer, J. Fritsche, H. Steinhart, and Y. Ku. A new conjugated linoleic acid isomer, 7 *trans*,9 *cis*-octadecadienoic acid, in cow milk, cheese, beef and human milk and adipose tissue. *Lipids* 33:803–809 (1998).
18. M.R. Pollard, F.D. Gunstone, A.T. James, and L.J. Morris. Desaturation of positional and geometric isomers of monoenoic fatty acids by microsomal preparations from rat liver. *Lipids* 15:306–314 (1980).
19. R.O. Adlof, S. Duval, and E.A. Emken. Biosynthesis of conjugated linoleic acid in humans. *Lipids* 35:131–135 (2000).
20. M.P. Yurawecz, M.M. Mossoba, J.K.G. Kramer, M.W. Pariza, and G.J. Nelson (eds.). *Advances in CLA Research, Vol. 1.* American Oil Chemists' Society, Champaign, IL, 1999.

21. J.-L. Sebedio, W.W. Christiee, and R. Adlof (eds.). *Advances in CLA Research, Vol. 2.* American Oil Chemists' Society, Champaign, IL, 2003.

22. M.P. Yurawecz, J.K.G. Kramer, O. Gudmundsen, M.W. Pariza, and S. Banni (eds.). *Advances in CLA Research, Vol. 3.* American Oil Chemists' Society, Champaign, IL, 2006.

23. M.M. Mossoba, J.K.G. Kramer, P. Delmonte, M.P. Yurawecz, and J.I. Rader. Official Methods for the determination of *trans* fats. In: *Trans Fats Alternatives* (D.R. Kodali and G.R. List, eds.). AOCS Press, Champaign, IL, 2005, pp. 47–70.

24. M.M. Mossoba and D. Firestone. New methods for fat analysis in foods. *Food Test Anal.* 2(2):24 (1996).

25. N. Hinrichsen and H. Steinhart. Techniques and applications in lipid analysis. In: *Trans Fats Alternatives* (D.R. Kodali and G.R. List, eds.). AOCS Press, Champaign, IL, 2005, pp. 1–20.

26. American Oil Chemists' Society. *Official Methods and Recommended Practices, Official Method Cd 14–61*, Reapproved, 1989 (D. Firestone, ed.). AOCS, Champaign, IL, 1989.

27. D. Firestone and P. LaBouliere. Determination of isolated *trans* isomers by infrared spectrophotometry. *J. Assoc. Off. Anal. Chem.* 48:437 (1965).

28. K. Helrich (ed.). In: *Official Methods of Analysis, 15th edn.* Method 965.35. Association of Official Analytical Chemists, Champaign, IL, 1990.

29. R.R. Allen. The determination of *trans* isomers in GLC fractions of unsaturated esters. *Lipids* 4:627 (1969).

30. A. Huang and D. Firestone. Determination of low level isolated *trans* isomers in vegetable oils and derived methyl esters by differential infrared spectrophotometry. *J. Assoc. Off. Anal. Chem.* 54:47 (1971).

31. B.L. Madison, R.A. DePalma, and R.P. d'Alonzo. Accurate determination of *trans* isomers in shortenings and edible oils by infrared spectrophotometry. *J. Am. Oil Chem. Soc.* 59:178 (1982).

32. P.S. Belton, R.H. Wilson, H. Sadeghi-Jorabchi, and K.E. Peers. A rapid method for the estimation of isolated *trans* double bonds in oils and fats using FTIR combined with ATR. *Lebensm. Wiss. Technol.* 21:153 (1988).

33. R.T. Sleeter and M.G. Matlock. Automated quantitative analysis of isolated (nonconjugated) *trans* isomers using FTIR spectroscopy incorporating improvements in the procedure. *J. Am. Oil Chem. Soc.* 66:933 (1989).

34. T.G. Toschi, P. Capella, C. Holt, and W.W. Christie. A comparison of silver ion HPLC plus GC with FTIR spectroscopy for the determination of *trans* double bonds in unsaturated fatty acids. *J. Sci. Food Agric.* 61:261 (1993).

35. F.R. van de Voort, A.A. Ismail, and J. Sedman. Automated method for the determination of *cis* and *trans* content of fats and oils by Fourier transform infrared spectroscopy. *J. Am. Oil Chem. Soc.* 72:873 (1995).

36. American Oil Chemists' Society. *Official Methods and Recommended Practices, Official Method Cd-14–95* (D. Firestone, ed.). AOCS, Champaign, IL, 1995.

37. W.M.N. Ratnayake and G. Pelletier. Methyl esters from a partially hydrogenated vegetable oil is a better infrared external standard than methyl elaidate for the measurement of total *trans* content. *J. Am. Oil Chem. Soc.* 73:1165 (1996).

38. M.M. Mossoba, M.P. Yurawecz, and R.E. McDonald. Rapid determination of the total *trans* content of neat hydrogenated oils by attenuated total reflection spectroscopy. *J. Am. Oil Chem. Soc.* 73:1003 (1996).

39. *Smalley Series Report.* American Oil Chemists' Society, Champaign, IL, 1995.

40. L.H. Ali, G. Angyal, C.M. Weaver, J.I. Rader, and M.M. Mossoba. Determination of total *trans* fatty acids in foods: Comparison of capillary column gas chromatography and single bounce horizontal attenuated total reflection infrared spectroscopy. *J. Am. Oil Chem. Soc.* 73:1699 (1996).

41. American Oil Chemists' Society. *Official Methods and Recommended Practices, Official Method Cd 14–99* (D. Firestone, ed.). AOCS, Champaign, IL, 1999.

42. P. Cunniff (ed.). *Official Methods of Analysis, 17th edn.* Method 2000.10. AOAC International, Gaithersburg, MD, 2000.

43. K. Helrich (ed.). *Official Methods of Analysis, 16th edn.* Method 965.34. AOAC International, Gaithersburg, MD, 1997.

44. K. Helrich (ed.). *Official Method of Analysis, 15th edn.* Method 994.14. Association of Official Analytical Chemists, Champaign, IL, 1994.

45. J.K.G. Kramer, V. Fellner, M.E.R. Dugan, F.D. Sauer, M.M. Mossoba, and M.P. Yurawecz. Evaluating acid and base catalysts in the methylation of milk and rumen fatty acids with special emphasis on conjugated dienes and total trans fatty acids. *Lipids* 32:1219–1228 (1997).

46. J. Fritsche, M.M. Mossoba, M.P. Yurawecz, J.A.G. Roach, and H. Steinhart. *Trans* fatty acids in human adipose tissue. *J. Chromatogr. B* 705:177–182 (1998).

47. M.M. Mossoba, J.K.G. Kramer, J. Fritsche, M.P. Yurawecz, K.D. Eulitz, Y. Ku, and J.I. Rader. Application of standard addition to eliminate conjugated linoleic acid and other interferences in the determination of total trans fatty acids in selected food products by infrared spectroscopy. *J. Am. Oil Chem. Soc.* 78:631–634 (2001).

48. M.M. Mossoba, M. Adam, and T. Lee. *Rapid Determination of Total Trans Fat Content. An Attenuated Total Reflection Infrared Spectroscopy International Collaborative Study, AOAC International,* 84:1144–1150 (2001).

49. M. Milosevic, V. Milosevic, J.K.G. Kramer, H. Azizian, and M.M. Mossoba. Determination of low levels of *trans* fatty acids in foods using an improved ATR-FTIR procedure. *Lipid Technol.* 16:252–255 (2004).

50. M.M. Mossoba, J.K.G. Kramer, V. Milosevic, M. Milosevic, and H. Azizian. Interference of saturated fats in the determination of low levels of *trans* fats (below 0.5%) by infrared spectroscopy. *J. Am. Oil Chem. Soc.* 84:339–342 (2007).

51. N. Dupuy, C. Wojciechowski, and J.P. Huvenne. Quantitative analysis of edible *trans* fats by Fourier transform infrared, spectroscopy from raw materials to finished products. *Sci. Aliments* 19(6):677–686 (1999).

52. J. Sedman, F.R. van de Voort, and A.A. Ismail. Simultaneous determination of iodine value and *trans* content of fats and oils by single-bounce horizontal attenuated total reflectance Fourier transform infrared spectroscopy. *J. Am. Oil Chem. Soc.* 77(4):399–403 (2000).

53. A.A. Christy, P.K. Egeberg, and E.T. Østensen. Simultaneous quantitative determination of isolated *trans* fatty acids and conjugated linoleic acids in oils and fats by chemometric analysis of the infrared profiles. *Vib. Spectrosc.* 33:37–48 (2003).

54. H. Li, F.R. van de Voort, A.A. Ismail, J. Sedman, and R. Cox. *Trans* determination of edible oils by Fourier transform near-infrared spectroscopy. *J. Am. Oil Chem. Soc.* 77(10):1061–1067 (2000).

55. H. Azizian, J.K.G. Kramer, A.R. Kamalian, M. Hernandez, M.M. Mossoba, and S.L. Winsborough. Quantification of trans fatty acids in food products by GC, ATR-FTIR and FT-NIR methods. *Lipid Technol.* 16:229–231 (2004).

56. H. Azizian and J.K.G. Kramer. A rapid method for the quantification of fatty acids in fats and oils with emphasis on trans fatty acids using Fourier transform near infrared spectroscopy (FT-NIR). *Lipids* 40:855–867 (2005).

57. A.T. James and A.J.P. Martin. Gas–liquid chromatography: The separation and identification of the methyl esters of saturated and unsaturated acids from formic acid to *n*-octadecanoic acid. *Biochem. J.* 63:144 (1956).

58. H. Heckers, K. Ditmar, F.W. Melcher, and H.O. Kalinowski. Silar 10C, Silar 9 CP, Sp2340 and OV-275 in the gas–liquid chromatography of fatty acid methyl esters on packed columns: Chromatographic characteristics and molecular structures. *J. Chromatogr.* 135:93 (1977).

59. D.M. Ottenstein, L.A. Witting, P.H. Silvis, D.J. Rometchko, and N. Pelick. Column types for the chromatographic analysis of oleochemicals. *J. Am. Oil Chem. Soc.* 61:390 (1984).

60. L. Gildenberg and D. Firestone. Gas chromatographic determination of *trans* unsaturation in margarine: Collaborative study. *J. Assoc. Off. Anal. Chem.* 68:46 (1985).

61. R.G. Ackman. Application of gas–liquid chromatography to lipid separation and analysis: Qualitative and quantitative analysis. In: *Analysis of Fats, Oils and Lipoproteins* (E.G. Perkins, ed.). American Oil Chemists' Society, Champaign, IL, 1991, p. 270.

62. W.M.N. Ratnayake. Determination of *trans* unsaturation by IR spectrometry and determination of fatty acid composition of partially hydrogenated vegetable oils and animal fats by gas chromatography/infrared spectrophotometry: Collaborative study. *J. Assoc. Off. Anal. Chem.* 78:783 (1995).

63. W.M.N. Ratnayake and G. Pelletier. Positional and geometrical isomers of linoleic isomers in partially hydrogenated oils. *J. Am. Oil Chem. Soc.* 69:69 (1992).

64. E.G. Perkins and C. Smick. Octadecatrienoic fatty acid isomers of partially hydrogenated soybean oil. *J. Am. Oil Chem. Soc.* 64:1150 (1987).

65. R.L. Wolff. Recent applications of capillary gas–liquid chromatography to some difficult separations of positional or geometric isomers of unsaturated fatty acids. In: *New Trends in Lipid and Lipoprotein Analysis* (J.L. Sebedio and E.G. Perkins, eds.). American Oil Chemists' Society, Champaign, IL, 1995, p. 147.

66. J.L. Sebedio and R.G. Ackman. Hydrogenation of a menhaden oil: I. Fatty acid and C20 monoethylenic isomer compositions as a function of the degree of hydrogenation. *J. Am. Oil Chem. Soc.* 60:1986 (1983).

67. J.L. Sebedio and R.G. Ackman. Hydrogenation of a menhaden oil: II. Formation and evolution of the C20 dienoic and trienoic fatty acids as a function of the degree of hydrogenation. *J. Am. Oil Chem. Soc.* 60:1992 (1983).

68. D. Precht and J. Molkentin. Identification and quantitation of *cis/trans* C16:1 and C17:1 fatty acid positional isomers in German human milk lipids by thin-layer chromatography and gas chromatography/ mass spectrometry. *Eur. J. Lipid Sci. Technol.* 102(2):102–113 (2000).

69. R. Wilson, K. Lyall, J.A. Payne, and R.A. Riemersma. Quantitative analysis of long-chain *trans*-monoenes originating from hydrogenated marine oil. *Lipids* 35(6):681–687 (2000).

70. D. Precht and J. Molkentin. Frequency distributions of conjugated linoleic acid and *trans* fatty acid contents in European bovine milk fats. *Milchwissenschaft* 55(12):687–691 (2000).

71. W.W. Christie. Equivalent chain-lengths of methyl ester derivatives of fatty acids on gas chromatography—A reappraisal. *J. Chromatogr.* 447:305 (1988).

72. T.K. Miwa, K.L. Mikolajczak, F.R. Earle, and I.A. Wolff. Gas chromatographic characterization of fatty acids: Identification constants for mono- and dicarboxylic methyl esters. *Anal. Chem.* 32:1739 (1960).

73. R.L. Wolff. Analysis of alpha-linolenic acid geometrical isomers in deodorized oils by capillary gas–liquid chromatography on cyanoalkyl polysiloxane stationary phases: A note of caution. *J. Am. Oil Chem. Soc.* 71:907 (1994).

74. R.G. Ackman and J.C. Sipos. Application of specific response factors in the gas chromatographic analysis of methyl esters of fatty acids with flame ionization detectors. *J. Am. Oil Chem. Soc.* 41:377 (1964).

75. C.D. Bannon, J.D. Craske, and A.E. Hilliker. Analysis of fatty acid methyl esters with high accuracy and reliability. V. Validation of theoretical relative response factors of unsaturated esters in the flame ionization detector. *J. Am. Oil Chem. Soc.* 63:105 (1986).

76. R.L. Wolff and D. Precht. A critique of 50-m CP-Sil 88 capillary columns used alone to assess trans–unsaturated FA in foods: The case of TRANSFAIR study. *Lipids* 37:627–629 (2002).

77. D. Precht and J. Molkentin. C18:1, C18:2 and C18:3 trans and cis fatty acid isomers including conjugated cis-9,trans-11 linoleic acid (CLA) as well as total fat composition of German human milk lipids. *Nahrung* 43:233–244 (1999).

78. W.M.N. Ratnayake. Analysis of dietary trans fatty acids. *J. Oleo Sci.* 50:73–86 (2001).

79. J.K.G. Kramer, C.B. Blackadar, and J. Zhou. Evaluation of two GC columns (60-m Supelcowax 10 and 100-m CP Sil 88) for analysis of milkfat with emphasis on CLA, 18:1, 18:2, and 18:3 isomers, and short- and long-chain FA. *Lipids* 37:823–835 (2002).

80. J.K.G. Kramer, C. Cruz-Hernandez, and J. Zhou. Conjugated linoleic acids and octadecenoic acids: Analysis by GC. *Eur. J. Lipid Sci. Technol.* 103:600–609 (2001).

81. M. Buchgraber and F. Ulberth. Determination of trans octadecenoic acids by silver-ion chromatography–gas liquid chromatography: An intercomparison of methods. *J. AOAC Int.* 84:1490–1498 (2001).

82. *Official Methods of Analysis, 17th edn.* Method 996.06, revised 2001. AOAC International, Gaithersburg, MD, 1997.

83. AOCS. Official Method Ce 1f-96, revised 2002. *American Oil Chemists' Society, Official Methods and Recommended Practices, 5th edn.* (Firestone, D., ed.) Champaign, IL, 2004.

84. AOCS, Official Method Ce 1h-07, 2007. *American Oil Chemists' Society, Official Methods and Recommended Practices, 5th edn.* (Firestone, D., ed.) Champaign, IL, 2004.

85. AOCS, Official Method Ce 1h-05, 2005, *American Oil Chemists' Society, Official Methods and Recommended Practices, 5th edn.* (Firestone, D., ed.) Champaign, IL, 2004.

86. B. Nikolova-Damyanova. Silver ion chromatography and lipids. In: *Advances in Lipid Methodology, Vol. 1* (W.W. Christie, ed.). The Oily Press, Ayr, Scotland, 1992, p. 181.

87. C.R. Scholfield. Analysis and physical properties of the isomeric fatty acids. In: *Geometrical and Positional Fatty Acid Isomers* (E.A. Emken and H.J. Dutton, eds.). American Oil Chemists' Society, Champaign, IL, 1979, p. 17.

88. G. Dobson, W.W. Christie, and B. Nikolova-Damyanova. Silver ion chromatography of lipids and fatty acids. *J. Chromatogr. B* 671:197 (1995).

89. C.M. Marchand and J.L. Beare-Rogers. Complementary techniques for the identification of *trans,trans*-18:2 isomers in margarines. *Can. Inst. Food Sci. Technol. J.* 15:54 (1982).

90. D.L. Carpenter, B.S. Lehmann, B.S. Mason, and H.T. Slover. Lipid composition of selected vegetable oils. *J. Am. Oil Chem. Soc.* 53:713 (1976).

91. C.R. Caughman, L.C. Boyd, M. Keeney, and J. Sampugna. Analysis of positional isomers of monounsaturated fatty acids by high performance liquid chromatography of 2,4-dinitro-phenylhydrazones of reduced ozonides. *J. Lipid Res.* 28:338 (1987).

92. D.A. Kennerly. Two dimensional thin-layer chromatographic separation of phospholipid molecular species using plates with both reversed-phase and argentation zones. *J. Chromatogr.* 454:425 (1988).

93. C. Cruz-Hernandez, Z. Deng, H. Zhou, A.R. Hill, M.P. Yurawecz, P. Delmonte, M.M. Mossoba, M.E. Dugan, and J.K.G. Kramer. Methods for analysis of conjugated linoleic acids and trans-18:1 isomers in dairy fats by using a combination of GC, silver ion TLC/GC, and silver ion LC. *J. AOCS Int.* 87(2):545–562 (2004).

94. W.W. Christie. Separation of molecular species of triacylglycerols by high-performance liquid chromatography with a silver ion column. *J. Chromatogr.* 454:272 (1988).

95. P. Juaneda. Utilization of reverse-phase HPLC as alternative to silver ion chromatography for the separation of cis- and trans-C18:1 fatty acid isomers. *J. Chromatogr. A* 954(1–2):285–289 (2002).

96. W.W. Christie. A stable silver-loaded column for the separation of lipids by high performance liquid chromatography. *J. High Resolut. Chromatogr. Chromatogr. Commun.* 10:148 (1987).

97. W.W. Christie and G.H.M. Breckenridge. Separation of *cis* and *trans* isomers of unsaturated fatty acids by HPLC in the silver ion mode. *J. Chromatogr.* 469:261 (1989).

98. P. Juaneda, J.L. Sebedio, and W.W. Christie. Complete separation of the geometric isomers of linolenic acid by high performance liquid chromatography with a silver ion column. *J. High Resolut. Chromatogr. Chromatogr. Commun.* 17:321 (1994).

99. R.O. Adlof. Separation of *cis* and *trans* unsaturated fatty acid methyl esters by silver ion high-performance liquid chromatography. *J. Chromatogr.* 659:95 (1994).

100. R.O. Adlof, L.C. Copes, and E.A. Emken. Analysis of the monoenoic fatty acid distribution in hydrogenated vegetable oils by silver-ion high-performance liquid chromatography. *J. Am. Oil Chem. Soc.* 72:571 (1995).

101. R. Cross and H. Zackari. Ag + HPLC of conjugated linoleic acids on a silica-based stationary phase. III; model compounds. *J. Sep. Sci.* 25:897–903 (2002).

102. R. Cross and H. Zackari. Ag + HPLC of conjugated linoleic acids on a silica based stationary phase. Part IV: A reference stationary phase and retention mechanisms. *J. Sep. Sci.* 26:480–488 (2003).

103. C. Cruz-Hernandez, Z. Dena, J. Zhou, et al. Methods for analysis of conjugated linoleic acids and trans-18:1 isomers in dairy fats by using a combination of gas chromatography, silver-ion thin-layer chromatography/gas chromatography, and silver-ion liquid chromatography. *J. AOAC Int.* 87:545–560 (2004).

104. P. Delmonte, M. Yurawecz, M. Mossoba, et al. Improved identification of conjugated linoleic acid isomers using silver-ion HPLC separations. *J. AOAC Int.* 87:563–568 (2004).

105. P. Delmonte, J.K.G. Kramer, and M.P. Yurawecz. *Analysis of Trans 18:1 Fatty Acids by Silver Ion HPLC in Lipid Analysis and Lipidomics* (M.M. Mossoba, J.K.G. Kramer, J.T. Brenna, and R.E. McDonald, eds.). AOCS Press, Champaign, IL, 2006.

106. B.W. Smallbone and M.R. Sahasrabudhe. Positional isomers of *cis*- and *trans*-octa-decenoic acids in hydrogenated vegetable oils. *Can. Inst. Food Sci. Technol. J.* 18:174 (1985).

107. A.E. Johnston, H.J. Dutton, C.R. Scholfield, and R.O. Butterfield. Double bond analysis of dienoic fatty acids in mixtures. *J. Am. Oil Chem. Soc.* 55:486 (1978).

108. H.J. Dutton, S.B. Johnson, R.J. Purch, M.S.F. Lieken Jie, F.D. Gunstone, and R.T. Homan. Composition of mixed octadecadienoates via ozonolysis, chromatography and computer solution of linear equations. *Lipids* 68:481 (1988).

109. J.L. Sebedio. Classical chemical techniques for fatty acid analysis. In: *New Trends in Lipid and Lipoprotein Analysis* (J.L. Sebedio and E.G. Perkins, eds.). American Oil Chemists' Society, Champaign, IL, 1995, p. 277.

110. R.E. McDonald, D.J. Armstrong, and G.P. Kreishman. Identification of trans-diene isomers in hydrogenated soybean oil by gas chromatography, silver nitrate–thin layer chromatography, and 13C-NMR spectroscopy. *J. Agric. Food Chem.* 37:637 (1989).

111. E.P. Mazzola, J.B. McMahon, R.E. McDonald, N. Sehat, and M.M. Mossoba. ^{13}C-NMR Spectral confirmation of $\Delta 6$ and $\Delta 7$ *trans*-18:1 positional isomers. *J. Am. Oil Chem. Soc.* 74:1335–1337 (1997).

112. W.M.N. Ratnayake. Determination of trans unsaturation by IR spectrometry and determination of fatty acid composition of partially hydrogenated vegetable oils and animal fats by gas chromatography/infrared spectrophotometry: Collaborative study. *J. Assoc. Off. Anal. Chem.* 78:783 (1995).

113. Association of Official Analytical Chemists. *Official Methods of Analysis of AOAC International, 16th edn.* (K. Helrich, ed.). AOAC, Arlington, VA, 1995, Section 994.14.

114. B.L. Madison, R.A. DePalma, and R.P. d'Alonzo. Accurate determination of trans isomers in shortenings and edible oils by infrared spectrophotometry. *J. Am. Oil Chem. Soc.* 59:178 (1982).

115. D. Precht. Variation of *trans* fatty acids in milk fats. *Z. Ernahrungswiss.* 34:27 (1995).

116. M.R. Sahasrabudhe and C.J. Kurian. Fatty acid composition of margarines in Canada. *Can. Inst. Food Sci. Technol. J.* 12:140 (1979).

117. R.G. Ackman. Misidentification of fatty acid methyl ester peaks in liquid canola shortening. *J. Am. Oil Chem. Soc.* 67:1028 (1978).

118. P.R. Griffiths, A.M. Haefner, K.L. Norton, D.J.J. Fraser, D. Pyo, and H. Makishima. FT-IR interface for gas, liquid, and supercritical fluid chromatography. *J. High Resolut. Chromatogr.* 12:119 (1989).

119. D.A. Heaps and P.R. Griffiths. Reduction of detection limits of the direct deposition GC/FT-IR interface by surface-enhanced infrared absorption. *Anal. Chem.* 77:5965–5972 (2005).

120. J.L. Sebedio and A. Grandgirard. Cyclic fatty acids: Natural sources, formation during heat treatment, synthesis and biological properties. *Prog. Lipid Res.* 28:303 (1989).

121. J.L. Sebedio, J.L. LeQuere, E. Semon, O. Morin, J. Prevost, and A. Grandgirard. A heat treatment of vegetable oils: II. GC–MS and GC–FTIR spectra of some isolated cyclic fatty acid monomers. *J. Am. Oil Chem. Soc.* 64:1324 (1987).

122. S. Bourne, G. Reedy, P. Coffey, and D. Mattson. Matrix isolation GC/FTIR. *Am. Lab.* 16(6):90 (1984).

123. G.T. Reedy, D.C. Ettinger, J.F. Schneider, and S. Bourne. High-resolution gas chromatography/matrix isolation infrared spectrometry. *Anal. Chem.* 57:1602 (1985).

124. M.M. Mossoba, R.E. McDonald, and A.R. Prosser. Gas chromatography/matrix isolation/Fourier transform infrared spectroscopic determination of *trans*-monounsaturated and saturated fatty acid methyl esters in partially hydrogenated menhaden oil. *J. Agric. Food Chem.* 41:1988 (1993).

125. M.M. Mossoba, R.E. McDonald, J.Y.T. Chen, D.J. Armstrong, and S.W. Page. Identification and quantitation of *trans*-9,*trans*-l2 octadecadienoic acid methyl ester and other geometric isomers in hydrogenated soybean oil and margarines by capillary gas chromatography/matrix isolation/Fourier transform infrared spectroscopy. *J. Agric. Food Chem.* 38:86 (1990).

126. M.M. Mossoba, R.E. McDonald, D.J. Armstrong, and S.W. Page. Identification of minor C18 triene and conjugated diene isomers in hydrogenated soybean oil and margarine by gas chromatography/matrix isolation/Fourier transform infrared spectroscopy. *J. Chromatogr. Sci.* 29:324 (1991).

127. M.M. Mossoba, R.E. McDonald, J.A.G. Roach, D.D. Fingerhut, M.P. Yurawecz, and N. Sehat. Spectral confirmation of *trans* monounsaturated C18 fatty acid positional isomers. *J. Am. Oil Chem. Soc.* 74:125–130 (1997).

128. B. Schmitz and R.A. Klein. Mass spectrometric localization of carbon–carbon double bonds: A critical review of recent methods. *Chem. Phys. Lipids.* 39:285 (1986).

129. H. Budzikiewicz. Structure elucidation by ion–molecule reactions in the gas phase: The location of C, C-double and triple bonds (1). *Fresenius J. Anal. Chem.* 321:150 (1985).

130. A. Brauner, H. Budzikienwicz, and W. Boland. Studies in chemical ionization mass spectrometry. *Org. Mass Spectrom.* 17:161 (1982).

131. G. Dobson and W.W. Christie. Structural analysis of fatty acids by mass spectrometry of picolinyl esters and dimethyloxazoline derivatives. *Trends Anal. Chem.* 15:130 (1996).

132. W.W. Christie, E.Y. Brechany, S.B. Johnson, and R.T. Holman. A comparison of pyrrolidide and picolinyl ester derivatives for the identification of fatty acids in natural samples by gas chromatography–mass spectrometry. *Lipids* 21:657 (1986).

133. W.W. Christie, E.Y. Brechany, and R.T. Holman. Mass spectra of the picolinyl esters of isomeric mono- and dienoic acids. *Lipids* 22:224 (1987).

134. W.W. Christie and K. Stefanov. Separation of picolinyl ester derivatives of fatty acids by high-performance liquid chromatography for identification by mass spectrometry. *J. Chromatogr.* 392:259 (1987).

135. W.W. Christie, E.Y. Brechany, J.L. Sebedio, and J.L. LeQuere. Silver ion chromatography and gas chromatography–mass spectrometry in the structural analysis of cyclic monoenoic acids formed in frying oils. *Chem. Phys. Lipids.* 66:143 (1993).

136. D.J. Harvey. Mass spectrometry of picolinyl and other nitrogen-containing derivatives of lipids. In: *Advances in Lipid Methodology, Vol. 1* (W.W. Christie, ed.). The Oily Press, Ayr, Scotland, 1992, p. 19.

137. M.M. Mossoba, M.P. Yurawecz, J.A.G. Roach, R.E. McDonald, B.D. Flickinger, and E.G. Perkins. Rapid determination of double bond configuration and position along the hydrocarbon chain in cyclic fatty acid monomers. *Lipids* 29:893 (1994).

138. G. Dobson, W.W. Christie, E.Y. Brechany, J.L. Sebedio, and J.L. LeQuere. Silver ion chromatography and gas chromatography–mass spectrometry in the structural analysis of cyclic dienoic acids formed in frying oils. *Chem. Phys. Lipids* 75:171 (1995).

139. M.M. Mossoba, M.P. Yurawecz, J.A.G. Roach, H.P. Lin, R.E. McDonald, B.D. Flickinger, and E.G. Perkins. Elucidation of cyclic fatty acid monomer structures. Cyclic and bicyclic ring sizes and double bond position and configuration. *J. Am. Oil Chem. Soc.* 72:721 (1995).

140. C.K. Van Pelt, B.K. Carpenter, and J.T. Brenna. Studies of structure and mechanism in acetonitrile chemical ionization tandem mass spectrometry of polyunsaturated fatty acid methyl esters. *J. Am. Soc. Mass Spectrom.* 10:1253–1262 (1999).

141. C.K. Van Pelt, M.C. Huang, C.L. Tschanz, and J.T. Brenna. An octane fatty acid, 4,7,10,13,16,19,22,25-octacosaoctaenoic acid (28:8n-3), found in marine oils. *J. Lipid Res.* 40:1501–1505 (1999).

142. A.L. Michaud, G.Y. Diau, R. Abril, and J.T. Brenna. Double bond localization in minor homoallylic fatty acid methyl esters using acetonitrile chemical ionization tandem mass spectrometry. *Anal. Biochem.* 307:348–360 (2002).

143. A.L. Michaud, M.P. Yurawecz, P. Delmonte, B.A. Corl, D.E. Bauman, and J.T. Brenna. Identification and characterization of conjugated fatty acid methyl esters of mixed double bond geometry by acetonitrile chemical ionization tandem mass spectrometry. *Anal. Chem.* 75:4925–4930 (2003).

144. A.L. Michaud, P. Lawrence, R.O. Adlof, and J.T. Brenna. On the formation of conjugated linoleic acid diagnostic ions with acetonitrile chemical ionization tandem mass spectrometry. *Rapid Commun. Mass Spectrom.* 19(3):363–368 (2005).

145. P. Laakso. Supercritical fluid chromatography of lipids. In: *Advances in Lipid Methodology, Vol. 1* (W.W. Christie, ed.). The Oily Press, Ayr, Scotland, 1992, p. 81.

146. J.W. King. Applications of capillary supercritical fluid chromatography–supercritical fluid extraction to natural products. *J. Chromatogr. Sci.* 28:9 (1990).

147. J.W. King. Analysis of fats and oils by SFE and SFC. *Inform* 4:1089 (1993).

148. W.E. Artz. Analysis of lipids by supercritical fluid chromatography. In: *Analysis of Fats, Oils and Derivatives* (E.G. Perkins, ed.). American Oil Chemists' Society Press, Champaign, IL, 1993, p. 270.

149. P. Sandra and F. David. Basic principles and the role of supercritical fluid chromatography in lipid analysis. In: *Supercritical Fluid Chromatography in Oil and Lipid Analysis* (J.W. King and G.R. List, eds.). American Oil Chemists' Society Press, Champaign, IL, 1996, p. 321.

150. A. Medvedovici and F. David. Comprehensive pSFC × pSFC-MS for the characterization of triglycerides in vegetable oils, *LC GC Eur.* 16:32–345 (2003).

151. P. Sandra, A. Medvedovici, Y. Zhao, and F. David. Characterization of triglycerides in vegetable oils by silver-ion packed-column supercritical fluid chromatography coupled to mass spectroscopy with atmospheric pressure chemical ionization and coordination ion spray. *J. Chromatogr. A* 974:231–241 (2002).

152. R.D. Smith, H.T. Kalinowski, and H.R. Udseth. Fundamentals and practice of supercritical fluid chromatography–mass spectrometry. *Mass Spectrom. Rev.* 6:445 (1987).

153. J.W. Heligeth, J.W. Jordan, L.T. Taylor, and M. Ashraf Khorassani. Supercritical fluid chromatography of free fatty acids with on-line FTIR detection. *J. Chromatogr. Sci.* 24:183 (1986).

154. E.M. Calvey and L.T. Taylor. Supercritical extraction of foods with FT-IR detection. In: *Abstracts of Papers, AGFD, 196th National Meeting of the American Chemical Society*, Los Angeles, LA, 1988, p. 194.

155. E.M. Calvey, R.E. McDonald, S.W. Page, M.M. Mossoba, and L.T. Taylor. Evaluation of SFC/FT-IR for examination of hydrogenated soybean oil. *J. Agric. Food Chem.* 39:542 (1991).

156. L.G. Blomberg, M. Demirbüker, and P.E. Andersson. Argentation supercritical fluid chromatography for quantitative analysis of triacylglycerols. *J. Am. Oil Chem. Soc.* 70:939 (1993).

157. U.S. Department of Agriculture/Agricultural Research Service, USDA Nutrient Database for Standard Reference, Release 14. Nutrient Data Laboratory Home Page, http://www.nal.usda.gov/fnic/foodcomp 2001.

158. E.A. Emken. Physiochemical properties, intake, and metabolism. *Am. J. Clin. Nutr.* 62:659S–669S (1995).

159. W.M.N. Ratnayake, G. Pelletier, R. Hollywood, S. Bacler, and D. Leyte. Trans fatty acids in Canadian margarines: Recent trends. *J. Am. Oil Chem. Soc.* 75:1587–1594 (1998).
160. P. Denecke. About the formation of trans fatty acids during deodorization of rapeseed oil. *Eur. J. Med. Res.* 1:109–114 (1995).
161. S. O'Keefe, S. Gaskins-Wright, V. Wiley, and I.-C. Chen. Levels of trans geometrical isomers of essential fatty acids in some unhydrogenated U.S. vegetable oils. *J. Food Lipids.* 1:165–176 (1994).
162. W. De Greyt, A. Kint, M. Kellens, and A. Huyghebaert. Determination of low trans levels in refined oils by Fourier transform infrared spectroscopy. *J. Am. Oil Chem. Soc.* 75:115–118 (1998).
163. R.L. Wolff, C.C. Bayard, and R.J. Fabien. Evaluation of sequential methods for the determination of butterfat fatty acid composition with emphasis on trans-18:1 acids. Application to the study of seasonal variations in French butters. *J. Am. Oil Chem. Soc.* 72:1471–1483 (1995).
164. D. Precht. Variation in trans fatty acids in milk fat. *Z. Ernahrung.* 34:27–29 (1995).

7 Chemistry of Frying Oils

Kathleen Warner

CONTENTS

I. INTRODUCTION

Deep-fat frying imparts desired sensory characteristics of fried food flavor, golden brown color, and crisp texture in foods. During frying, at ~190°C, as oils thermally and oxidatively decompose, volatile and nonvolatile products are formed that alter functional, sensory, and nutritional qualities of oils. During the past 30 years, scientists have reported extensively on the physical and chemical changes that occur during frying and on the wide variety of decomposition products formed in frying oils. A small amount of oxidation in frying oils is important to develop the delicious deep-fried flavor characteristic of fried foods. However, as oils breakdown further because of the processes of oxidation, hydrolysis, and polymerization, compounds are formed that can cause off-flavors and may even be toxic if formed in high concentrations. Hydrogenated oils have been commonly used for commercial deep-fat frying in the United States since the 1950s, but concerns about *trans* fatty acids in hydrogenated oils have encouraged the use of alternative oils. Sometimes, these alternatives are less oxidatively stable oils and subsequent problems developed because of the use of unstable oils for frying. The chemistry of frying is especially important to understand, as less stable alternative oils are investigated as potential substitutes for hydrogenated oil. This chapter will review the physical and chemical changes in oils during frying, including reactions that occur in the frying process. In addition to discussing the degradation products formed, their effects on oil stability and quality of fried food will be included. Methods to measure oil deterioration will be discussed in terms of their significance, advantages, and limitations.

II. PROCESS OF FRYING

A. CHANGES IN OILS DURING HEATING AND FRYING

1. Physical Changes

Deep-fat frying is a process of cooking and drying in hot oil with simultaneous heat and mass transfer. As heat is transferred from the oil to the food, water is evaporated from the food and oil is absorbed by the food [1] (Figure 7.1). Steam from the dehydration process is vaporized. Factors affecting this heat and mass transfer include thermal and physical properties of the food and oil, shape and size of the food, and oil temperature [2]. Blumenthal [3] described the physical mechanisms that take place as a french-fried potato is fried, and recommended procedures to optimize the frying process in order to increase frying oil life and decrease oil absorption into food. Blumenthal also proposed conducting basic frying studies using model foods that have surface-to-volume ratios characteristic of some fried food, including cotton balls (all interior volume, crispy exterior surface to represent battered and breaded chicken); french-fried potatoes (significant interior volume, significant external surface), and potato chips (large surface area, little interior volume. Alexander et al. [4] compared corn, peanut oil, and partially hydrogenated soybean oils heated in open containers with those heated in a pressurized deep-fat frying system and reported that pressurized deep frying with fats resulted in less deterioration than open-vat heating, as shown by chemical analyses. Pokorny [5] reported that oil is absorbed into the fried material and adsorbed on its surface and that oil losses depend on the type of fried food; for example, potatoes absorb more oil than meat. Sun and coworkers [2,6,7] have extensively studied the fundamental mechanisms of deep-fat frying using tortilla chips as model food and have suggested approaches to reducing chip fat content from a food engineering approach.

Physical and chemical changes in oils that occur during heating and frying are presented in Figure 7.1. Physical changes in the oil include increases in color, foaming, and viscosity. Methods exist to measure these changes; however, qualitative changes can also be determined subjectively by visual inspection. Although these practices are not recommended, many small-scale oil users, such as restaurants, discard frying oils when frying causes excessive foaming of oil or when the oil color darkens. Of course, instrumental or chemical analyses are preferable to visual inspection in

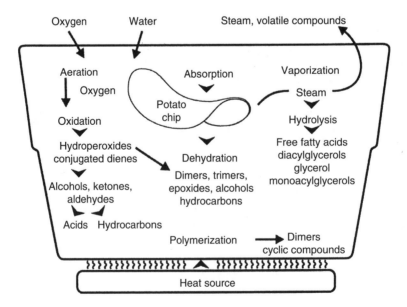

FIGURE 7.1 Physical and chemical reactions that occur during frying.

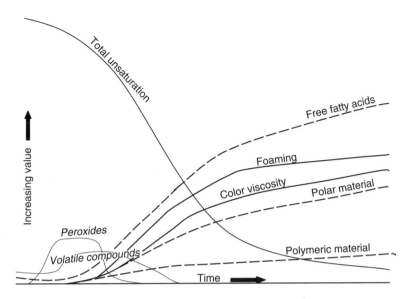

FIGURE 7.2 Formation and degradation of compounds during frying.

measuring frying oil deterioration, especially in research applications. Formation of nonvolatile decomposition products produces these physical changes in frying oil, such as increases in viscosity, color, and foaming [8]. Figure 7.2 shows the increases in these characteristics over frying time. Chemical changes during frying increase the concentration of free fatty acids as well as polar materials and polymeric compounds, and decrease fatty acid unsaturation. The effect of chemical changes on flavor quality of fried food and on the stability of the oil will be discussed in this chapter; however, potential toxic effects of degradation products on health will not be included because this topic is well covered in the literature [4,9].

2.   Chemical Changes

Frying oils not only transfer heat to cook foods but also produce characteristic deep-fried food flavor and, unfortunately, undesirable off-flavors if the oil is allowed to deteriorate. During deep-fat frying the deteriorative chemical processes of hydrolysis, oxidation, and polymerization occur. The oil breaks down to form volatile products and nonvolatile monomeric and polymeric compounds (Figure 7.1). A small amount of oxidation is necessary to form volatile compounds such as 2,4-decadienal that are responsible for the deep-fried flavor. This aldehyde forms during the decomposition of linoleic acid, so oils with low levels of linoleic acid do not produce much of this flavor. Although some oxidation is needed, continued heating and frying further decompose the fatty acids until breakdown products accumulate to levels that produce off-flavors and potentially toxic effects, making the oil unacceptable. The amounts of the compounds that are formed and their chemical structures depend on many factors, including oil and food types, frying conditions, and oxygen availability [10]. In addition, these chemical reactions—hydrolysis, oxidation, and polymerization—are interrelated, producing a complex mixture of products. The individual processes of hydrolysis, oxidation, and polymerization and their degradation products are described later.

a.   Hydrolysis
As food is placed in oil at frying temperatures, air and water initiate a series of interrelated reactions. Water and steam hydrolyze triglycerides, which produce mono- and diglycerides, and eventually free fatty acids and glycerol (Figure 7.3). Glycerol partially evaporates because it volatilizes above 150°C, and the reaction equilibrium is shifted in favor of other hydrolysis products [5]. The extent of hydrolysis depends on factors such as oil temperature, interface area between the oil and the

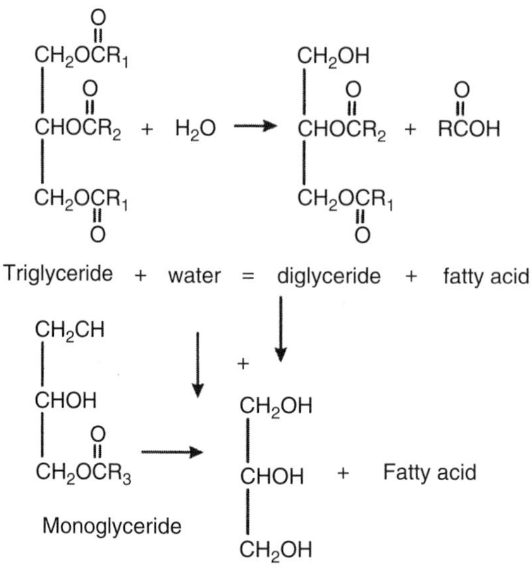

FIGURE 7.3 Hydrolysis reactions in frying oils.

aqueous phases, and amount of water and steam because water hydrolyzes oil more quickly than steam [5]. Free fatty acids and low molecular weight acidic products arising from fat oxidation enhance the hydrolysis in the presence of steam during frying [5]. Hydrolysis products, like all oil degradation products, decrease the stability of frying oils and can be used to measure oil fry life, for example, free fatty acids (Figure 7.2).

b. Oxidation

Oxygen is present in the fresh frying oil and more oxygen is added into the frying oil at the oil surface and by addition of food. Oxygen activates a series of reactions involving formation of many compounds including free radicals, hydroperoxides, aldehydes, ketones, and conjugated dienoic acids. The chemical reactions that occur during the oxidation process contribute to the formation of both volatile and nonvolatile decomposition products. For example, ethyl linoleate oxidation leads to the formation of conjugated hydroperoxides that can form noncycling long-chain products or they can cyclize and form peroxide polymers [8]. The oxidation mechanism in frying oils is similar to autoxidation at 25°C; however, the unstable primary oxidation products—hydroperoxides—decompose rapidly at 190°C into secondary oxidation products such as aldehydes and ketones (Figures 7.2 and 7.4). Secondary oxidation products, such as aldehydes, that are volatile, significantly contribute to the odor of the oil and flavor of the fried food [11,12]. If the secondary oxidation products are unsaturated aldehydes, such as 2,4-decadienal, 2,4-nonadienal, 2,4-octadienal, 2-heptenal, or 2-octenal, they contribute to the characteristic deep-fried flavor in oils and are considered desirable [12]. However, saturated and unsaturated aldehydes, such as hexanal, heptanal, octanal, nonanal, and 2-decenal, contribute undesirable off-odors and off-flavors. Oils that are unstable to oxidation, such as polyunstaturated ones, form the highest amounts of unacceptable degradation products. Therefore, oils that contain linoleic and linolenic acids, such as soy, sunflower, and canola, have significant off-flavor formation under frying conditions. On the other hand, oxidatively more stable oils, such as high oleic sunflower, that have little linoleic or linolenic acids, form only low amounts of oxidation products. The primary decomposition products from the oxidation of oleic acid produce off-flavors such as fruity and plastic. These off-odors/flavors of heated high oleic oils can be attributed primarily to heptanal, octanal, nonanal, and 2-decenal [11].

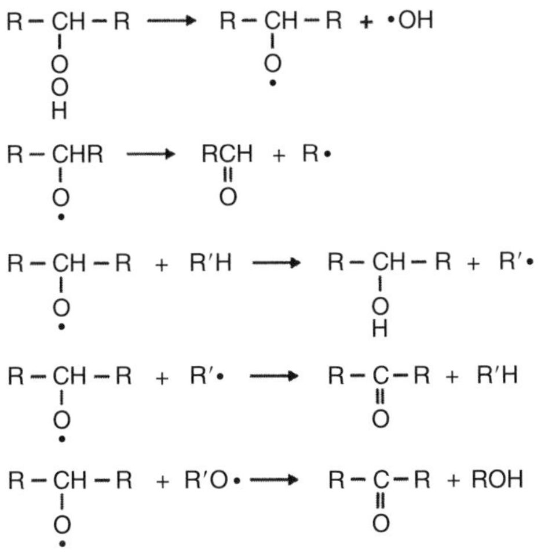

$$R-\underset{\underset{H}{\overset{|}{O}}}{\overset{|}{C}H}-R \longrightarrow R-\underset{\underset{\bullet}{\overset{|}{O}}}{\overset{|}{C}H}-R + {}^{\bullet}OH$$

$$R-\underset{\overset{|}{O}\bullet}{\overset{|}{C}HR} \longrightarrow \underset{\overset{\|}{O}}{RCH} + R\bullet$$

$$R-\underset{\overset{|}{O}\bullet}{\overset{|}{C}H}-R + R'H \longrightarrow R-\underset{\underset{H}{\overset{|}{O}}}{\overset{|}{C}H}-R + R'\bullet$$

$$R-\underset{\overset{|}{O}\bullet}{\overset{|}{C}H}-R + R'\bullet \longrightarrow R-\underset{\overset{\|}{O}}{\overset{|}{C}}-R + R'H$$

$$R-\underset{\overset{|}{O}\bullet}{\overset{|}{C}H}-R + R'O\bullet \longrightarrow R-\underset{\overset{\|}{O}}{\overset{|}{C}}-R + ROH$$

FIGURE 7.4 Oxidation reactions in frying oils.

Analysis of primary oxidation products, such as hydroperoxides, at any point in the frying process provides little information because their formation and decomposition fluctuate quickly and are not easily predicted (Figure 7.2). During frying, oils with polyunsaturated fatty acids, such as linoleic acid, have a distinct induction period of hydroperoxides followed by a rapid increase in peroxide values, then a rapid destruction of peroxides [8,13]. Measuring levels of polyunsaturated fatty acids, such as linoleic acid, can help determine extent of thermal oxidation. Wessels [10] reported that oxidative degradation produced oxidized triglycerides containing hydroperoxide, epoxy, hydroxy, and keto groups and dimeric fatty acids or dimeric triglycerides. Volatile degradation products are usually saturated and monounsaturated hydroxy, aldehydic, keto, and dicarboxylic acids; hydrocarbons; alcohols; aldehydes; ketones; and aromatic compounds [13].

c. Polymerization
Polymerization of frying oil results in the formation of compounds with high molecular weight and polarity (Figure 7.5). Polymers can form from free radicals or triglycerides by the Diels–Alder reaction. Cyclic fatty acids can form within one fatty acid; dimeric fatty acids can form between two fatty acids, either within or between triglycerides; and polymers with high molecular weight are obtained as these molecules continue to cross-link. As polymerized products increase in the frying oil, viscosity of the oil also increases and the color of the oil darkens.

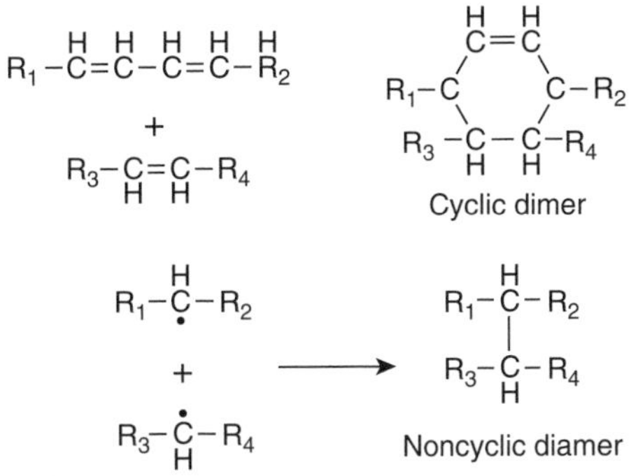

FIGURE 7.5 Polymerization reactions in frying oils.

B. FACTORS AFFECTING OIL DECOMPOSITION

Thermal decomposition of a frying oil is affected by many variables, such as unsaturation of fatty acids, oil temperature, oxygen absorption, metals in substrates and in the oil, type of oil, and nature of the fried food [14]. A list of factors that affect the processes of hydrolysis, oxidation, and polymerization and frying oil deterioration are presented in Table 7.1. Frying oil degradation can be limited by controlling these factors. For example, all frying should start with a good initial quality oil that is not oxidized and has low amounts of catalyzing metals. Oils with low amounts of polyunsaturated fatty acids will have the most oxidative stability. Also, additions of antioxidants and antifoam additives may also help maintain oil quality [15–19]. The type of food being fried affects the resulting composition of the frying oil as fatty acids are released from fat-containing foods, such as chicken, and their concentration in the frying oil increases with continued use. Fat from fish will also change the fatty acid composition of the oil and decrease the frying oil stability. Breaded and battered food can degrade frying oil more quickly than nonbreaded food and decrease stability as well. However, even foods such as potatoes degrade oil stability because of the increased aeration produced as the food is added to the frying oil. Food particles accumulating in the oil also deteriorate oil; therefore, filtering oils through adsorbents will help remove these particles along with other oxidation products to enhance oil fry life.

The extent of these degradation reactions can be limited by carefully managing frying conditions such as temperature and time, exposure of oil to oxygen, continuous or intermittent frying, oil filtration, and turnover of oil. Frying protocols of intermittent or continuous frying affect fry life. Perkins and Van Akkeren [20] found that cottonseed oil intermittently heated for 62 h had significantly more polar material than oil heated continuously for 166 h. They suggested that this may be caused by increased amounts of fatty acyl peroxides, which decompose upon repeated heating and cooling, causing further damage to the oil. Adding fresh makeup oil to the fryer is commonly done in most frying operations; however, in the snack food industry where more makeup oil is added than in restaurant-style frying, a complete turnover of the oil in the first 8–12 h of the frying cycle can be achieved so the oil reaches an equilibrium state [2]. Levels of the degradation products in frying oil can also be affected by absorption into the fried food and evaporation [21]. However, accumulation of degradation products in the frying medium and their eventual incorporation in fried foods is of concern mainly when commercial or industrial frying operations are carried out under abusive conditions [9]. Fritsch [1] stated that combinations of these factors (Table 7.1) determine the rate at which individual reactions take place. For example, in one operation, the rate of hydrolysis may be twice that of the rate of oxidation, whereas in another operation with different conditions, the reverse may occur.

TABLE 7.1
Factors Affecting Frying Oil Decomposition

Oil/Food/Additives	Process
Unsaturation of fatty acids	Oil temperature
Oil	Frying time
Fried food	Aeration/oxygen absorption
Metals in oil/food	Frying equipment
Initial oil quality	Continuous or intermittent heating or frying
Degradation products in oil	Frying rate
Antioxidants	Heat transfer
Antifoam additives	Turnover rate; addition of makeup oil
	Filtering of oil/fryer cleaning

III. DECOMPOSITION PRODUCTS

During frying, oils degrade to form volatile and nonvolatile decomposition products. Foods fried in deteriorated oils may contain a significant amount of decomposition products to cause potential adverse effects to safety, flavor, oxidative stability, color, and texture of the fried food. Although volatile compounds are primarily responsible for flavor—both positive and negative—thermal polymers do not affect flavor directly. Therefore, thermal polymers may exist in an edible product, but the conditions leading to their formation are not usually encountered in commercial practice, because precautions are taken to inhibit their formation. In a research laboratory situation, Chang and coworkers [22] isolated the nonvolatile fraction as a brownish, transparent viscous liquid, which is indicative of a considerable amount of decomposition products in their highly deteriorated oil.

A. VOLATILE DECOMPOSITION COMPOUNDS

In deep-fat frying, as oil is continuously or intermittently heated in the presence of air, thermal and oxidative decomposition of the oil occurs producing both volatile and nonvolatile decomposition products (Table 7.2). Selke et al. [23] identified volatile odor constituents and their precursors from heated soybean oil, using model triglycerides [pure triolein, mixture of triolein (25%)–tristearin, and a randomly esterified triglyceride of stearic and 25% oleic acids] heated at 192°C in air for 10 min. Each model system produced the same major compounds, identified as heptane, octane, heptanal, octanal, nonanal, 2-decenal, and 2-undecenal. These seven compounds were unique to the oxidation of the oleate fatty acid in each triglyceride sample. Later, Selke and coworkers [24] analyzed pure trilinolein and mixtures of trilinolein–tristearin, trilinolein–triolein, and trilinolein–triolein–tristearin heated to 192°C in air. Major volatiles included pentane, acrolein, pentanal, 1-pentanal, hexanal, 2- or 3-hexanal, 2-heptenal, 2-octenal, 2,4-decadienal, and 4,5-epoxide-2-enal.

B. NONVOLATILE DECOMPOSITION COMPOUNDS

Nonvolatile degradation products (Table 7.2) in deteriorated frying oils include polymeric triacylglycerols, oxidized triacylglycerol derivatives, cyclic substances, and breakdown products [13]. Polymeric triacylglycerols result from condensation of two or more triacylglycerol molecules to form polar and nonpolar high molecular weight compounds. The nonpolymerized part of the oil contains mainly unchanged triacylglycerols in combination with their oxidized derivatives. In addition, it contains mono- and diacylglycerols, partial glycerols containing chain scission products, triacylglycerol with cyclic or dimeric fatty acids, and any other nonvolatile products. Rojo and Perkins [21] classified the degradation products as polar and nonpolar polymeric fatty acid methyl esters and monomeric fatty acid methyl esters with unchanged, changed (oxidized, cyclized,

TABLE 7.2
Volatile and Nonvolatile Decomposition Products from Frying Oil

Nonvolatile	Volatile
Monoacylglycerols	Hydrocarbons
Diacylglycerols	Ketones
Oxidized triacylglycerols	Aldehydes
Triacylglycerol dimers	Alcohols
Triacylglycerol trimers	Esters
Triacylglycerol polymers	Lactones
Free fatty acids	

isomerized, etc.), and fragmented fatty acid esters. Clark and Serbia stated that large declines in iodine values are needed for a significant amount of these polymers to form [9]. This is not usually a problem in the snack food industry because iodine values do not change much as a result of a high oil turnover rate. However, oil used in small-scale batch frying operations such as restaurants, where oil turnover is low, may be more deteriorated. Chang et al. [22] studied the nonvolatile decomposition products from pure trilinolein, triolein, and tristearin produced under simulated deep-fat frying conditions at 185°C for 74 h. Chromatographic, chemical, and spectrometric analysis indicated the presence of dimers in all three triacylglycerol mixtures. Similarly, Christopoulou and Perkins [25] isolated and characterized dimers in heated soybean oil. Dimers and higher polymers isolated from heated cottonseed oil at 225°C in the presence of air contained moderate amounts of carbonyl and hydroxyl groups [26].

IV. MEASUREMENT OF DECOMPOSITION PRODUCTS: SIGNIFICANCE, ADVANTAGES, LIMITATIONS

The physical and chemical changes occurring in frying oils and the many compounds formed in deteriorated frying oil have been extensively reported. Although these compounds often are used to measure the amount of oil degradation, many of the existing methods are based on measuring nonspecific compounds that may or may not relate to oil degradation or fried food quality. Therefore, it is not surprising that frying is often described as more of an art than a science. In fact, the frying industry is still searching for the ultimate criteria to evaluate frying stability of oils and fried food flavor quality and stability. White [27] reviewed existing analyses to measure formation of volatile and nonvolatile components in order to detect deterioration in frying oils, including the standard methods of polar components, conjugated dienes, and fatty acids, as well as rapid analyses such as dielectric constant. Croon et al. [28] compared various methods to evaluate 100 frying oil samples using four quick test methods (Foodoil Sensor–dielectric constant, RAU Test, Fritest, and spot test) and two laboratory methods (free fatty acids and chromatographic analysis of triglyceride dimers) with a standard column chromatographic determination of polar compounds. The RAU Test is a colorimetric test kit that contains redox indicators, reacting with the total amount of oxidized compounds. Fritest (E Merck, Darmstadt) is a colorimetric test kit sensitive to carbonyl compounds and the spot test assays the free fatty acids to indicate hydrolytic degradation and free fatty acids. The Foodoil Sensor correlated with polar compounds more than did the RAU Test, Fritest, and spot test. The amount of free fatty acids was found to be an unreliable indication of deteriorated frying oil. Probes that measure the dielectric constant in oils are available and report the amount of total polar compounds in the oil. These probes must be calibrated accurately against the total polar compound levels measured by the column chromatography (CC) methods.

Fritsch [1] noted that commercial and industrial frying oil operators want to know the answer to one primary question: When should frying oil be discarded? Since there are many variables that affect oil degradation (Table 7.1), a specific method may be ideal for one operation but completely useless in another. The determination of the end point of a frying oil is dependent on good judgment and knowledge of the particular frying operation, as well as on the type of frying oil and the analytical measurements used [1]. Some of the methods used to measure degradation products in frying oil are listed in Table 7.3 and are discussed in the following section. Volatile compounds can be collected by several techniques, including direct injection, static head space, dynamic or purge-and-trap head space, and solid phase microextraction, and analyzed by capillary gas chromatography.

A. NONVOLATILE DECOMPOSITION PRODUCTS

Paradis and Nawar [34] reported that nonvolatile higher molecular weight compounds are reliable indicators of fat deterioration because their accumulation is steady and they are not volatile

TABLE 7.3

Methods to Measure Decomposition Products in Frying Oil

Nonvolatile Compounds and Related Processes	Method/Reference
Iodine value	AOCS Cd 1–25/93 [29]; AOAC 28.023 [30]
Fatty acid composition	AOCS Ce 1–6293 [29]
Total polar compounds	AOCS Cd 20–91 [29]
High-performance size exclusion chromatography	[31]
Free fatty acids	AOCS Ca 5a–40/93 [29]
Dielectric constant	[1]
Nonurea adduct–forming esters	[26]
Color	AOCS Td 3a–64/93 [29]
Viscosity	[32]
Smoke point	AOCS Cc 9a–48/93 [29]
Foam height	[32]
Volatile Compounds and Related Processes	
Peroxide value	AOCS Cd 8–53 [29]
Conjugated dienes	AOCS Ti 1a–64 [29]
Volatile compounds	AOCS Cg 4–94 [29]
Sensory analysis of odor and flavor	[33]

(Figure 7.2). As mentioned earlier, the formation and accumulation of nonvolatile compounds are responsible for physical changes in frying oil [8]. Most methods for assessing deterioration of frying oils are then based on these changes. Nonspecific methods for measuring nonvolatile compounds in deteriorated frying oil include free fatty acids [1], iodine value [34], nonurea adduct–forming esters [26], viscosity [32], and petroleum ether–insoluble oxidized fatty acids [35]. White stated that none of these methods has proved to be a good measure of oil deterioration [27]. Melton et al. [36] noted that nonvolatile decomposition products are a better measure of degradation of a frying oil than volatile products are and concluded that more research is needed to determine the total polar component levels at which different frying oils should be discarded and to relate those levels to fried-food quality for each oil type. A total polar compound level of 24%–26% is used in Europe to determine the end point of frying oil in restaurant frying. A much lower level is required when fried food needs to be stable in shelf-life storage.

Smith et al. [37] evaluated 65 samples of partially hydrogenated soybean oil used for frying battered chicken and french-fried potatoes in fast-service restaurants. Frying times correlated highly with increases in dielectric constant, polar materials, and free fatty acids. Oleic and linoleic acids increased in the oils with increasing hours of use, whereas stearic acid decreased because of contamination with chicken fat. Collected oil samples that had been discarded before 100 h of frying time had values of 4.0 for the Foodoil Sensor, 1% free fatty acids, and 27% polar materials, which have been suggested as end points for discarding frying oil. Perkins [8] measured nonvolatile decomposition products in cottonseed oil and tallow to show that polymers increased with increasing heating time, and that cottonseed oil was deteriorated more by intermittent heating and added water than by continuous heating and no water addition. Cuesta et al. [31] measured polar components by high-performance size exclusion chromatography (HPSEC) to investigate the thermooxidative and hydrolytic changes in frying oils. These researchers were able to quantitate triacylglycerol polymers and dimers, oxidized triacylglycerols, diacylglycerols, and free fatty acids.

Christopoulou and Perkins [25] recommended model systems such as pure fatty acids and triglycerides oxidized under simulated deep-fat frying conditions to control the various factors (Table 7.1) affecting the thermal–oxidative reactions and to facilitate the structure elucidation of the decomposition products such as thermal and oxidative dimers. Arroyo et al. [14] reported a linear correlation of $r = 0.99$ between number of fryings and amount of decomposition products, including total polar compounds, triacylglycerol polymers, and triacylglycerol dimers. Although diacylglycerol levels were significantly correlated ($r = 0.945$), free fatty acids were not significantly correlated ($r = 0.27$) with the number of fryings. Arroyo and coworkers also found that hydrolytic changes paralleled thermoxidative changes, as evidenced by high correlations between levels of triglyceride polymers and triglyceride dimers (thermoxidative process) and diglycerides (hydrolytic process) with the number of fryings [14]. Dobarganes et al. [38] measured triglyceride species and polar compound level and distribution and reported no significant differences in the frying oils and lipids extracted from fried food for either total polar compounds or polar compound distribution. Thus, the study results indicated no preferential adsorption of altered oil compounds on the fried potato surface. Billek et al. [35] compared four methods to assess frying oils and reported good correlations between results with gel permeation chromatography (GPC), liquid chromatography (LC) on a silica gel column, polar and nonpolar components CC on silica gel, and petroleum ether–insoluble oxidized fatty acids. However, they found that measuring petroleum ether–insoluble oxidized fatty acids was time consuming and inaccurate. The GPC method was able to determine dimeric and oligomeric triacylglycerols in frying oil irrespective of the presence of oxidized compounds, whereas the LC method indicated the total amount of polar and oxidized compounds. Separation of polar and nonpolar components by CC was simple and quick.

Wessels [10] reported that methods to analyze frying oils, including measurement of peroxide value, benzidine value, petroleum ether–insoluble oxidized fatty acids, acid value, smoke point, UV absorbance, refractive index, iodine value, viscosity, color, and fatty acid composition, were of limited significance. Abdel-Aal and Karara [39] measured changes occurring in corn oil during heating and during frying of potato chips and onion rings by refractive index, acid value, peroxide value, total carbonyls, benzidine value, and oil color (which all increased) and iodine value (which decreased). These changes were more pronounced in oil that was used intermittently rather than continuously. Furthermore, onion rings were more detrimental to the oil than potato chips, possibly because of the breading material that accumulated in the oil. These investigators observed significant differences in the physiochemical changes of the oil extracted from the fried foods and the frying oil.

B. Volatile Decomposition Products

Since many of the volatile decomposition products volatilize during frying, it is difficult to get an accurate representation of oil deterioration by instrumental and chemical analyses of these compounds. Methods that measure volatile compounds directly or indirectly include peroxide value, gas chromatographic volatile compound analysis, and sensory analysis (Table 7.3).

1. Peroxides

Fritsch [1] stated that peroxide value is not a good measure of heat abuse in frying oils because peroxides are unstable at frying temperature (Figure 7.2). Usuki et al. [40] confirmed this observation with pan frying (thin-film heating) of soybean oil, which produced high peroxide values at 230°C. No thermostable peroxides were detected after the oil was fractionated by silicic acid CC.

2. Volatile Compounds

The fatty acid composition of frying oils has a major effect on the volatile compounds detected in the oil and on the flavor of the fried food. Although frying oils are complex mixtures of triacylglycerols, a wide variety of fatty acids, and many minor constituents, degradation compounds are primarily from the fatty acids [5]. Chang et al. [22] found that 79 of 93 compounds identified in corn oil and 64 of 100 compounds identified in hydrogenated cottonseed oil were also detected in pure triolein and trilinolein after all were heated under simulated frying conditions. Even though the fatty acid composition of the two oils differed greatly, approximately half of the compounds were the same. As expected, unsaturated fatty acids contribute significantly more to the formation of volatile compounds than those from the more stable saturated fatty acids, such as palmitic and stearic.

Chang et al. [22] concluded that identifying volatile compounds in frying oil and fried food is important because these compounds help in understanding the chemical reactions that occur during frying and because flavor of deep-fried food is partly attributable to the volatile compounds. These researchers identified 220 volatile compounds from corn oil, hydrogenated cottonseed oil, trilinolein, and triolein after simulated frying conditions. Macku and Shibamoto [41] collected volatile compounds formed in the head space from heated corn oil and identified 18 aldehydes, 15 heterocyclic compounds, 13 hydrocarbons, 11 ketones, 4 alcohols, 3 esters, and 7 miscellaneous compounds. Takeoka et al. [42] isolated and identified volatile constituents of unidentified frying oils by simultaneous distillation–extraction and fractionation by silica gel CC as 1-pentanol, hexanal, furfural alcohol, (*E*)-2-heptanal, 5-methyl furfuranal, 1-octen-3-ol, octanal, 2-pentylfuran, (*E*)-2-octenal, nonanal, (*E*)-2-nonenal, and hexadecanoic acid. Chung et al. [43] used gas chromatography/mass spectrometry (GC–MS) to identify 99 volatile compounds in the head space of peanut oil heated to 50°C, 100°C, 150°C, or 200°C for 5 h, including 42 hydrocarbons, 22 aldehydes, 11 fatty acids, 8 alcohols, 4 furans, 2 esters, and 2 lactones. Total amounts of all identified volatiles increased as oil temperature increased. Chang and coworkers [44] identified 53 volatile flavor compounds from potato chips, including 8 nitrogen compounds, 2 sulfur compounds, 14 hydrocarbons, 13 aldehydes, 2 ketones, 1 alcohol, 1 phenol, 3 esters, 1 ether, and 8 acids.

Twenty-six volatile compounds were identified and quantified from aged potato chips fried in partially hydrogenated canola oil or cottonseed oil [45]. The chip samples fried in cottonseed oil had higher concentrations of aldehydes but heterocyclic compounds levels were not different. No differences in peroxide values were found between oil type. Neff et al. [11] reported that 32 volatile compounds in triolein heated for up to 6 h at 190°C had identifiable undesirable odors by olfactometry–GC–MS, whereas only 18 volatile compounds had identifiable undesirable odors in heated trilinolein. For the same heated oils, Warner et al. found that triolein had four volatile compounds with the desirable deep-fried odor but that trilinolein had seven compounds with the deep-fried food odor; however, >800 ppm of these compounds was found in trilinolein and only 30 ppm in triolein [12]. These findings help explain why high oleic oils that are also low in linoleic acid develop only low amounts of the deep-fried flavor. Care should be taken in interpreting data on volatile compounds in used frying oil because of the fluctuations in formation and degradation of the compounds at frying temperature (Figure 7.2).

3. Sensory

Frying oil affects the flavor of fried food because these oils undergo chemical reactions and the reaction products contribute to the distinctive deep-fried flavor [5,12] as well as to undesirable odors in deteriorated oils [11]. Flavor quality of fried food is affected by oil type, frying conditions, and degradation products. Gere [46] noted the positive relationship of initial freshness of frying oil and sensory properties of food fried in the oil. However, deep-fried flavor is not optimal at the start of frying [33,47–49]. When oils are tasted before heating, they usually have little flavor if properly processed. This low intensity of flavor continues during the early portions of the frying cycle because the typical deep-fried flavor develops as heating and frying time increase. Food processors often heat oils or fry preliminary batches of food to condition the oil to develop this flavor. Some oils develop this characteristic deep-fried flavor more quickly than others depending on the fatty acid composition of the oil. For example, Warner et al. [33] found in previous research that cottonseed oil with high (50%–55%) linoleic acid produces significantly higher intensity of deep-fried flavor in potato chips and french-fried potatoes than do oils with low (10%) linoleic acid, such as high (80%–90%) oleic oils. As the fatty acids decompose in high temperature conditions, the volatile degradation products produce characteristic flavors. Some oxidation products, such as 2,4-decadienal, that breakdown from linoleic acid are important in the formation of deep-fried flavor. Flavor improves after the first stage of frying and becomes less acceptable during the last stage. Frying conditions should be adjusted so that optimal flavor characteristics are maintained for as long as possible during the frying cycle.

Sensory evaluation is still the method most often used by different countries to determine when to discard a frying oil [36]. Billek et al. [35] reported that scientific groups in Germany used sensory assessment of a used frying oil; however, if this method did not give a clear indication that the oil was deteriorated, then instrumental or chemical analysis was used to support a final decision on oil quality. More recently, proceedings from the Third International Symposium on Deep Fat Frying recommended that the principle quality index for deep-fat frying be sensory parameters of the fried food [50]. To further confirm oil abuse, total polar materials should be <24% and polymeric triglycerides <12% [50] in restaurant-type frying. Sensory analysis of frying oil and fried food quality may be conducted by analytical descriptive/discriminative panels using trained, experienced panelists [33,47,48,51] or by consumer panels using untrained judges [36]. Melton et al. used consumer panels to find that the flavor likability of fried food, which is dependent on consumer perception and is affected by the type of oil used for frying [36]. In further studies, Melton and coworkers could find no differences in fresh chip flavor or likability scores between potato chips fried in partially hydrogenated canola oil or in cottonseed oil when evaluated by a consumer sensory panel [45]. On the other hand, Warner found that a trained, experienced, analytical descriptive panel could detect differences ($P < 0.05$) in the type and intensity of flavors in fried food prepared in various oil types [33]. More research is needed to understand the relationship between fried food flavor and the volatile and nonvolatile decomposition compounds produced in frying oils.

REFERENCES

1. Fritsch, C.W. Measurements of frying fat deterioration: A brief view. *J. Am. Oil Chem. Soc.* 58:272 (1981).
2. Moreira, R., Palau, J. and Sun, X. Simultaneous heat and mass transfer during the deep fat frying of tortilla chips. *J. Food Proc. Eng.* 18:307 (1995).
3. Blumenthal, M.M. A new look at the chemistry and physics of deep fat frying. *Food Tech.* 45:68 (1991).
4. Alexander, J.C., Chanin, B.E. and Moran, E.T. Nutritional effects of fresh, laboratory heated, and pressure deep fry fats. *J. Food Sci.* 48:1289 (1983).
5. Pokorny, J. Flavor chemistry of deep fat frying in oil, in *Flavor Chemistry of Lipid Foods*, Min, D.B. and Smouse, T.H., Eds., American Oil Chemists' Society, Champaign, IL, 1989, pp. 113–115.
6. Sun, X. and Moreira, R. Oil distribution in tortilla chips during the deep fat frying. In *Proceedings of American Society of Agricultural Engineering*, Atlanta, GA, 1994.
7. Moreira, R., Palau, J. and Sun, X. Deep fat frying of tortilla chips: An engineering approach. *Food Tech.* 49:307 (1995).
8. Perkins, E.G. Formation of non-volatile decomposition products in heated fats and oils. *Food Tech.* 21:125 (1967).
9. Clark, W.L. and Serbia, G.W. Safety aspects of frying fats and oils. *Food Tech.* 45:84 (1991).
10. Wessels, H. Determination of polar compounds in frying fats. *Pure Appl. Chem.* 55:1381 (1983).
11. Neff, W.E., Warner, K. and Byrdwell, W.C. Odor significance of undesirable degradation compounds in heated triolein and trilinolein. *J. Am. Oil Chem. Soc.* 77:1303–1313 (2000).
12. Warner, K., Neff, W.E., Byrdwell, W.C. and Gardner, H.W. Effect of oleic and linoleic acids in the production of deep fried odor in heated triolein and trilinolein. *J. Agric. Food Chem.* 49:899–905 (2001).
13. Perkins, E.G. Lipid oxidation of deep fat frying, in *Food Lipids and Health*, McDonald, R.E. and Min, D.B., Eds., Dekker, New York, 1996, p. 139.
14. Arroyo, R., Cuesta, C., Garrido-Polonio, C., Lopez-Varela, S. and Sanchez-Muniz, F.J. High-performance size-exclusion chromatographic studies on polar components formed in sunflower oil used for frying. *J. Am. Oil Chem. Soc.* 69:557 (1992).

15. Gordon, M.H. and Kourimska, L. The effects of antioxidants on changes in oils during heating and deep frying. *J. Sci. Food Agric.* 68:347 (1995).

16. Carlson, B.L. and Tabacchi, M.H. Frying oil deterioration and vitamin loss during food service operation. *J. Food Sci.* 51:218 (1986).

17. Warner, K., Mounts, T.L. and Kwolek, W.F. Effects of antioxidants, methyl silicone, and hydrogenation on room odor of soybean cooking oils. *J. Am. Oil Chem. Soc.* 62:1483 (1985).

18. Frankel, E.N., Warner, K. and Moulton, K.J. Effects of hydrogenation and additives on cooking oil performance of soybean oil. *J. Am. Oil Chem. Soc.* 62:1354 (1985).

19. Snyder, J.M., Frankel, E.N. and Warner, K. Headspace volatile analysis to evaluate oxidative and thermal stability of soybean oil. Effect of hydrogenation and additives. *J. Am. Oil Chem. Soc.* 63:1055 (1986).

20. Perkins, E.G. and Van Akkeren, L.A. Heated fats. IV. Chemical changes in fats subjected to deep fat frying process: Cottonseed oil. *J. Am. Oil Chem. Soc.* 42:782 (1965).

21. Rojo, J. and Perkins, E. Cyclic fatty acid monomer formation in frying fats. *J. Am. Oil Chem. Soc.* 64:414 (1987).

22. Chang, S.H., Peterson, R. and Ho, C.T. Chemical reactions involved in deep fat frying of foods. *J. Am. Oil Chem. Soc.* 55:718 (1978).

23. Selke, E., Rohwedder, W.K. and Dutton, H.J. Volatile components from triolein heated in air. *J. Am. Oil Chem. Soc.* 54:62 (1977).

24. Selke, E., Rohwedder, W.K. and Dutton, H.J. Volatile components from trilinolein in air. *J. Am. Oil Chem. Soc.* 57:25 (1980).

25. Christopoulou, C.N. and Perkins, E.G. Isolation and characterization of dimers formed in used soybean oil. *J. Am. Oil Chem. Soc.* 66:1360 (1989).

26. Firestone, D., Horwitz, W., Friedman, L. and Shue, G.M. Heated fats. I. Studies of the effects of heating on the chemical nature of cottonseed oil. *J. Am. Oil Chem. Soc.* 38:253 (1961).

27. White, P.J. Methods for measuring changes in deep-fat frying oils. *Food Tech.* 45:75 (1991).

28. Croon, L.B., Rogstad, A., Leth, T. and Kiutamo, T. A comparative study of analytical methods for quality evaluation of frying fat. *Fette Seifen Anstrichmittel* 88:87 (1986).

29. *Official Methods and Recommended Practices of the American Oil Chemists' Society*, 4th ed. Champaign, IL, 1989.

30. Association of Official Analytical Chemists. *Official Methods of Analysis*, 14th ed. Washington, D.C., 1984.

31. Cuesta, C., Sanchez-Muniz, F.J., Garrido-Polonio, C., Lopez-Varela, S. and Arroyo, A. Thermoxidative and hydrolytic changes in sunflower oil used in fryings with a fast turnover of fresh oil. *J. Am. Oil Chem. Soc.* 70:1069 (1993).

32. Stevenson, S.G., Vaisey-Genser, M. and Eskin, N.A.M. Quality control in the use of deep frying oils. *J. Am. Oil Chem. Soc.* 61:1102 (1984).

33. Warner, K., Orr, P. and Glynn, M. Effect of fatty acid composition of oils on flavor and stability of fried food. *J. Am. Oil Chem. Soc.* 74:347–356 (1997).

34. Paradis, A.J. and Nawar, W.W. Evaluation of new methods for assessment of used frying oils. *J. Food Sci.* 46:449 (1981).

35. Billek, G., Guhr, G. and Waibel, J. Quality assessment of used frying fats: A comparison of four methods. *J. Am. Oil Chem. Soc.* 55:728 (1978).

36. Melton, S.L., Jafar, S., Sykes, D. and Trigiano, M.K. Review of stability measurements for frying oils and fried food flavor. *J. Am. Oil Chem. Soc.* 71:1301 (1994).

37. Smith, L.M., Clifford, A.J., Hamblin, C.L. and Creveling, R.K. Changes in physical and chemical properties of shortenings used for commercial deep-fat frying. *J. Am. Oil Chem. Soc.* 63:1017 (1986).

38. Dobarganes, C.M., Marquez-Ruiz, G. and Perez-Camino, M. Thermal stability and frying performance of genetically modified sunflower seed (*Helianthus annuus* L.) oils. *J. Agric. Food Chem.* 41:678 (1993).

39. Abdel-Aal, M.H. and Karara, H.A. Changes in corn oil during deep fat frying of foods. *Leben-Wiss. Tech.* 19:323 (1986).

40. Usuki, R., Fukui, H., Kamata, M. and Kaneda, T. Accumulation of peroxides in pan-frying oil. *Fette Seifen Anstrichmittel* 82:494 (1980).

41. Macku, C. and Shibamoto, T. Headspace volatile compounds formed from heated corn oil and corn oil with glycine. *J. Agric. Food Chem.* 39:1265 (1991).

42. Takeoka, G., Perrino, Jr., C. and Buttery, R. Volatile constituents of used frying oils. *J. Agric. Food Chem.* 44:654 (1996).

43. Chung, T.Y., Eiserich, J.P. and Shibamoto, T. Volatile compounds identified in headspace samples of peanut oil heated under temperatures ranging from 50 to 200°C. *J. Agric. Food Chem.* 41:1467 (1993).

44. Lee, S., Reddy, B.R. and Chang, S. Formation of a potato chip-like flavor from methionine under deep-fat frying conditions. *J. Food Sci.* 38:788 (1973).

45. Melton, S.L., Trigiano, M.K., Penfield, M.P. and Yang, R. Potato chips fried in canola and/or cottonseed oil maintain high quality. *J. Food Sci.* 58:1079 (1993).

46. Gere, A. Decrease in essential fatty acid content of edible fats during the frying process. *Z. Ernahrungs-wiss.* 21:191 (1982).

47. Warner, K. and Mounts, T.L. Frying stability of soybean and canola oils with modified fatty acid compositions. *J. Am. Oil Chem. Soc.* 70:983 (1993).

48. Warner, K., Orr, P. and Glynn, M. Effect of frying oil composition on potato chip stability. *J. Am. Oil Chem. Soc.* 71:1117 (1994).

49. Warner, K. Impact of high-temperature food processing on fats and oils, in *Impact of Processing on Food Safety*, Jackson, L.S., Knize, M.G. and Morgan, J.N., Eds., Plenum Publishers, New York, NY, 1999, pp. 67–77.

50. Anonymous. Recommendations of the 3rd International Symposium on Deep Fat Frying. *Eur. J. Lipid Sci. Technol.* 102:594 (2000).

51. Warner, K. Sensory evaluation of oils and fat-containing foods, in *Methods to Assess Quality and Stability of Oils and Fat-Containing Foods*, Warner, K. and Eskin, N.A.M., Eds., American Oil Chemists' Society, Champaign, IL, 1995.

Part II

Processing

8 Recovery, Refining, Converting, and Stabilizing Edible Fats and Oils

Lawrence A. Johnson

CONTENTS

I. INTRODUCTION

Processing seeds or animal tissues into edible oils can be broken into four sets of operations: recovery, refining, conversion, and stabilization. Oil recovery is often referred to as *extraction* or *crushing* when processing plant sources and *rendering* in the case of processing animal tissues. Oil extraction involves pressing the oil-bearing material to separate crude oil from the solids high in protein or washing flaked or modestly pressed material with solvent, almost always hexane. The defatted solids after pressing are known as *cake* and after solvent extraction as *meal*. The oil, *crude oil*, because it contains undesirable components, such as pigments, phosphatides, free fatty acids, and off-flavors and off-odors, must be refined to remove these contaminants and produce high-quality edible oils. Refined oils consist primarily (>99%) of triglycerides and can be *converted*, usually by hydrogenation; but winterizing, fractional crystallization, and interesterification should also be considered conversion processes because they achieve different properties from the original oil such as converting liquid oil into semisolid or solid fats. Plasticizing, tempering, and stehling are operations designed to stabilize crystal–oil mixtures used for shortenings and margarines.

II. OIL RECOVERY

For several thousand years, fats and oils have been recovered from oil-bearing seeds, fruits, and fatty animal tissues, and used for food, cosmetics, lubricants, and lighting fluids. Of the more than several hundred plants and animals that produce fats and oils in sufficient quantities to warrant processing into edible products, only 11 sources are commercially significant in the United States (Table 8.1) [1,2]. In other countries, the importance of each source may be different, and other sources not in this list may also be important such as olive oil being consumed in large amounts in Mediterranean countries.

All oil recovery processes are designed to obtain triglycerides as free as possible from undesirable impurities; to obtain a yield as high as possible consistent with economics of the process; and to produce cake, meal, or flour (finely ground meal), usually high in protein content, of maximum value [3]. Three general types of processes are used to crush oilseeds: *hard pressing, prepress solvent extraction*, and *direct solvent extraction*. The extraction process of choice depends primarily on the oil content of the source material, the amount of residual oil in the meal allowed, the amount of protein denaturation allowed, the amount of investment capital available, and local environmental laws concerning emissions of volatile organic compounds (VOCs).

TABLE 8.1

Major Edible Fats and Oils in the United States and Methods of Processing

Source	U.S. Oil Consumption[a] (Million Pounds)	Oil Content (%)	Prevalent Method of Recovery
Soybean	15,655	19	Direct solvent extraction
Corn (germ)	1,397	40	Wet or dry milling and prepress solvent extraction
Tallow (edible tissue)	1,362	70–95	Wet or dry rendering
Canola	1,264	42	Prepress solvent extraction
Coconut (dried copra)	1,021	66	Hard pressing
Cottonseed	772	19	Hard pressing or prepressing or direct solvent extraction
Lard (edible tissue)	988	70–95	Wet or dry rendering
Palm	260	47	Hard pressing
Palm kernel	390	48	Hard pressing
Sunflower	320	40	Prepress solvent extraction
Peanut (shelled)	230	47	Hard pressing or prepress solvent extraction

Source: From Johnson, L.A. in *Technology and Solvents for Extracting Non-Petroleum Oils*, P.J. Wan and P. J. Wakelyn, eds., AOCS Press, Champaign, IL, 1997 except as otherwise noted.

[a] Data from U.S.D.A. *Oil Crops Yearbook*. USDA, Economic Research Service, Washington, D.C., 2000.

A. EXTRACTION OF OIL FROM OILSEEDS

The oldest oil recovery method is hard pressing, where the seed is pressed, usually after various pretreatments to enhance oil recovery, to squeeze the oil from the solids known as *cake* (Figure 8.1). In the early years, lever presses and screw-operated presses, often driven by oxen or other work animals, were used; then, during the Industrial Revolution, batch hydraulic presses were introduced, which evolved at the turn of the century into continuous screw presses [4] connected to line shafts driven by steam engines. Today, continuous screw pressing and direct solvent extraction have become the preferred processing methods because the oil is more completely recovered. For a long time, the rule of thumb has been that materials containing >30% oil require pressing, either hard pressing or prepressing, before solvent extraction. Hard pressing involves squeezing as much oil as possible, whereas prepress solvent extraction involves squeezing out only part of the oil before subjecting the partially deoiled material to more complete extraction with solvent. The recent adoption of the expander has largely done away with this rule, and even high oil content materials can be solvent-extracted today with no or little prior oil extraction [5,6]. Direct solvent extraction, without any prior pressing or expanding, has long been the most widely practiced method, the oil is more completely recovered, and the oil and meal are economically recovered undamaged by heat. Both prepress solvent extraction and direct solvent extraction are depicted in Figure 8.2.

1. Seed Storage

Oilseeds are often harvested at moisture contents higher than levels that allow for long-term storage and must be dried for safe storage. Increasingly, farmers are storing oilseeds, particularly soybeans, on the farm to take advantage of higher prices that are paid later in the crop year.

Storage for extended periods at moisture contents exceeding critical moisture levels will damage oilseeds, reducing the yields of oil and protein, and diminishing the quality of the oil (notably darker color and higher refining loss). Seeds at harvest are alive and respire, converting seed mass to CO_2 and other metabolites, albeit at low rates when the moisture content is below the critical moisture

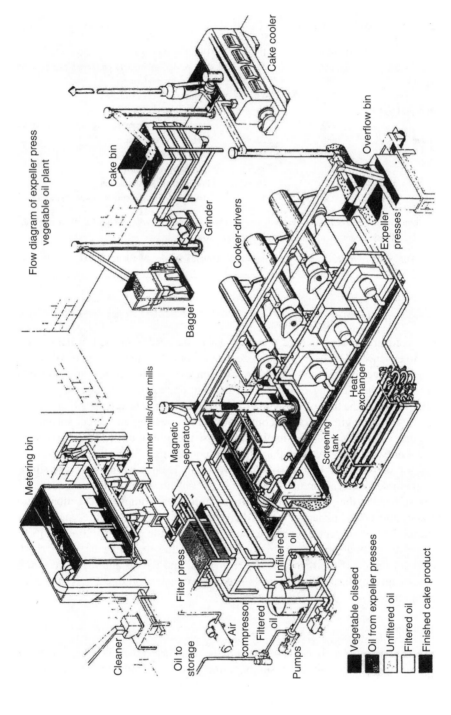

FIGURE 8.1 Depiction of hard screw pressing. (Diagram courtesy of Anderson International, Cleveland, OH.)

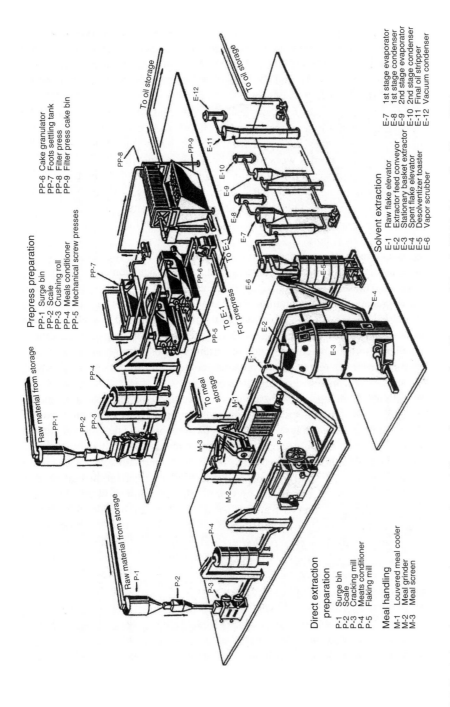

Direct extraction preparation

P-1 Surge bin
P-2 Scale
P-3 Cracking mill
P-4 Meats conditioner
P-5 Flaking mill

Meal handling

M-1 Louvered meal cooler
M-2 Meal grinder
M-3 Meal screen

Prepress preparation

PP-1 Surge bin
PP-2 Scale
PP-3 Crushing roll
PP-4 Meals conditioner
PP-5 Mechanical screw presses

PP-6 Cake granulator
PP-7 Foots settling tank
PP-8 Filter press
PP-9 Filter press cake bin

Solvent extraction

E-1 Raw flake elevator
E-2 Extractor feed conveyor
E-3 Stationary basket extractor
E-4 Spent flake elevator
E-5 Desolventizer toaster
E-6 Vapor scrubber

E-7   1st stage evaporator
E-8   1st stage condenser
E-9   2nd stage evaporator
E-10 2nd stage condenser
E-11 Final oil stripper
E-12 Vacuum condenser

FIGURE 8.2 Depiction of prepress solvent extraction and direct solvent extraction. (Courtesy of French Oil Machinery Co., Piqua, OH.)

level. The critical moisture level for safe storage varies with the seed species: usually, the higher the oil content, the lower is the critical moisture value (Table 8.1). At moisture content exceeding the critical moisture level, respiration rate increases, and the seed can even germinate and become subject to fungi attack. Respiration and germination liberate heat and, when there is insufficient aeration, the heat further accelerates these reactions. Under extreme conditions, the seed may become scorched or even catch on fire (especially cottonseed). Modern seed storage facilities employ temperature-monitoring systems to alert elevator and storage operators when seed temperatures exceed critical set points. Then the seed is moved to another bin to disperse hot spots and or aerated (blowing air through the seed). The percentage of seed that is heat-damaged is often a factor in the U.S. grades and standards (e.g., soybeans).

Overdrying can increase seed fragility, leading to excessive breakage during handling, storing, and processing. In the case of soybeans (a dicot), the cotyledon is prone to splitting into halves when the hull becomes separated from the cotyledon (meat) during conveying and transporting, a problem that becomes worse when the beans are overdried. *Splits* are undesirable because they are difficult to separate from foreign matter, and the oil deteriorates at a faster rate. Oil from soybean splits is higher in free fatty acids, phosphatides, iron, and peroxides due to activation of catabolic enzymes [7]. Oils from field- and storage-damaged seeds are usually poor in flavor [8]. For these reasons, damaged kernels and splits are also factors in the U.S. grades and standards for soybeans.

When oilseeds are received at the crushing plant, samples are often taken for analysis of moisture, foreign matter, damaged seed, oil, protein, and free fatty acid contents. Shipments arriving with similar values for these analyses may be segregated based on actions that are required to minimize further degradation. Seed containing more than the critical moisture content is immediately processed or dried. Seed with excessive foreign matter is transferred to scalping operations to reduce the level of contaminants. Foreign matter usually contains more moisture than the seed and the foreign matter tends to become concentrated at certain locations during placement into storage bins. Removing foreign matter reduces bulk moisture content and improves storability. Usually, foreign matter is also a factor in U.S. grading systems.

Both inside and outside storage systems are used. Metal and concrete bins are used for inside storage. Cottonseed is often stored in special buildings having roof slopes the same as the angle of repose for the seed (Muskogee buildings). Some oilseeds are stored outside on concrete pads, and the piles may be covered with tarpaulins and/or equipped with air distribution systems.

Any handling, including normal harvesting, conveying, and transporting, increases seed damage. Nature has given seeds a high degree of subcellular organization that is affected by seed damage. Seed endosperm is composed of many cells containing oil and storage protein, which supply energy, nitrogen reserves, and other metabolites to support the growth of the embryo during germination. Storage protein (the greatest proportion of seed nitrogen) is concentrated in discrete bodies known as *protein bodies*; and lipids (predominantly triglycerides) are stored in *spherosomes*. Phospholipids are largely associated with pseudomembranes around protein bodies and spherosomes. Enzymes and cellular metabolites are present in the cytoplasm. Some of these enzymes, notably lipase, hydrolyze triglycerides increasing free fatty acid content, which must be removed by refining. Phospholipase activity may render some phospholipids nonhydratable and difficult to remove from the oil by means of normal refining procedures. Lipoxygenases, unique to soybeans and other legumes, oxidize linoleic and linolenic acids causing painty, green, beany flavors in protein; these enzymes may also reduce the oxidative stability of the oil.

In intact seeds, these enzymes are kept away from the oil by natural compartmentalization within the cells of the seeds. However, damage by weather, harvesting, and/or handling bruises the seed, breaking cell walls and membranes, allowing the oil and enzymes to come into contact with one another. This accelerates adverse reactions whose rates are dependent on temperature, moisture, and extent of damage.

2. Cleaning

The first step in processing oilseeds usually involves cleaning the seed (Figure 8.3). Unless removed by cleaning operations, foreign matter reduces oil and protein contents and increases wear and damage to expensive processing equipment; in addition, foreign matter may adversely affect oil quality (especially color). Magnets are placed in chutes just ahead of processing equipment to prevent damage by tramp iron. Vibrating and/or shaker screens with or without aspiration are used to remove stems, pods, leaves, splits, broken grain, dirt, and extraneous seeds.

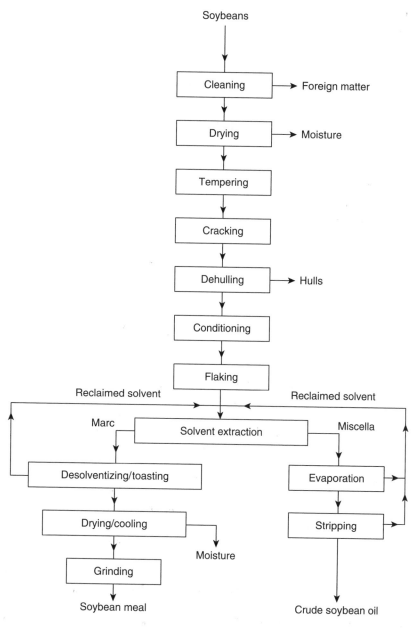

FIGURE 8.3 Steps in processing soybeans into meal and edible oil.

3.   Dehulling

Usually, it is desirable to remove the hull or seed coat that surrounds the oilseed meat. Hulls always contain much less oil and protein than do meats. Removing the hull reduces the amount of material that must be handled, extracted, and desolventized, thus increasing downstream plant capacity. The protein content of the meal is raised by removing the hull. Oftentimes, as with corn and sunflower seed, the hull contains waxes on the outer surface as a natural protective mechanism of the seed. These waxes must be removed during latter oil-refining steps, often by winterization or cold centrifugation; otherwise, the wax may become insoluble at cold temperatures and make the oil unattractively cloudy and cause emulsions such as mayonnaise to separate. Removing the hulls can alleviate these wax problems. On the other hand, hulls can be helpful at times, such as by providing fiber to allow easier hard pressing or to enhance solvent drainage, and thus they are not always removed.

Dehulling must be done carefully to ensure that the meat is not broken into too many small pieces, which would be difficult to separate from the hull. In addition, crushing meats during dehulling causes oil cells to be ruptured, freeing the oil. This damage should be minimized to prevent absorption of liberated oil by hulls that are removed. Either one of these problems increases the oil content of the hulls and reduces the yield of oil. Some seeds, such as soybeans, are often dried, dried and conditioned, or heated before dehulling to help free the meat from the hull.

Corrugated roller mills or bar mills, which shear the seed, and impact mills, which shatter the brittle seed coat, are used. The equipment cuts or breaks the hull to free the meat, a step often referred to as *decortication*. Sometimes, the combination of sizing the seed with shaker screens before decortication and using mills with different settings optimized for each seed size increases the efficiency of decortication. Hull/meat separation systems comprising one or more shaker screens, aspirators, and gravity tables are effective because the hull is often larger and almost always lower in density and more buoyant in an airstream than the oil-rich meat. Hulls may be blended back with meal to control protein level, sold as a separate coproduct for cattle roughage, or burned in boilers to generate steam and electricity in the process known as *cogeneration*.

When making edible flours, more complete removal of hulls is required. More than 90% of the hulls from soybeans must be removed to assure that the minimum specification of 50% protein is met. Even in the livestock feed industry, the trend is toward meals containing higher protein and lower fiber contents. It is becoming increasingly difficult to produce soybean meal with high protein content because increasing farm yields are depressing protein content.

4.   Hard Screw Pressing

Since flakes transfer heat more rapidly than do cracked meats, the seed is usually flaked (0.38–0.50 mm [0.015–0.020 in.] thickness), before cooking at 115°C over a 60 min before hard pressing (Figure 8.1). Initial stages of cooking should be done with moist heat (seed moisture maintained at 10%), injecting steam or spraying water into the top deck. Cooked flakes are dried in lower trays and should exit the cooker at <2.5% moisture [9]. The cooked and dried flakes are then conveyed to screw presses. If proper cooking methods are employed, in conjunction with a well-maintained, modern, screw press, as low as 3%–4% residual oil content can be achieved. Many screw press plants are quite small and do not optimally, if at all, flake, cook, and dry before screw pressing; so as much as 6%–10% oil often remains with the cake. Ultimately, hard pressing with more and more pressure is self-defeating because application of pressure causes capillaries to be reduced in volume, sheared, and eventually sealed by coagulation of protein. This places the lowest practical limit at about 3% residual oil.

A screw press (Figure 8.4) is basically a continuous screw auger designed to accept feed material and subject it to gradually increasing pressure, as it is conveyed through the barrel cage. The barrel is composed of bars surrounding the screw and oriented parallel to the screw axis. The bars are separated by spacers decreasing in size toward the solids discharge end, which allow the oil to drain. A plug of compressed oil-lean solids, the *cake*, forms at the discharge end. Increasing

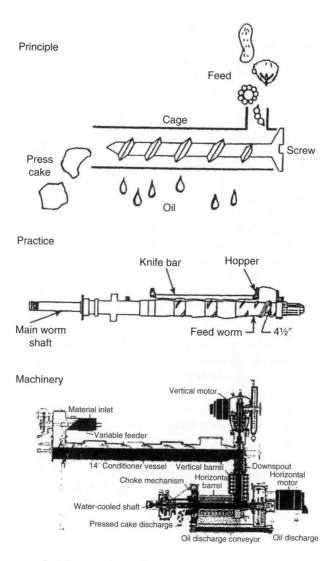

FIGURE 8.4 Screw press principles, practice, and machinery.

pressure down the length of the barrel is achieved by increasing the root diameter of the screw, decreasing the pitch of the screw flights, and controlling the opening for the discharging cake by means of a choke. This design causes fresh material to be rammed against the plug [10,11].

Screw presses are composed of three sections: feeding, ramming, and plugging. The meats or flakes fall into a rapidly rotating feed screw, which feeds them into the pressing cage to expel entrapped air and squeeze out the easily removed oil. In some screw presses, the feeding section may be a separate vertical screw, or it may be mounted directly to the ramming screw. In either case, the feeding screw turns faster than the ramming screw. Knife bars with projecting nibs are clamped between the half cages to prevent cake slippage and rotation. Maximum pressure is developed in the ramming section as partially deoiled cake is rammed against the deoiled plug. The plug section provides the resistance against which the ramming occurs. The friction in the barrel generates heat that must be removed to achieve low residual oil. Some screw presses recycle cooled pressed oil over the cage to remove excess heat, whereas others use water-cooled shafts and bar cages.

In hard pressing, screw presses are choked to put maximum pressure on flaked, cooked seed, while maintaining cake discharge. The oil drains from the screw press cage and is pumped or flows by gravity to a basin to allow settling of cellular debris (foots) that was removed with the oil. The cake is then ground into meal.

In the United States, hard pressing is largely limited to minor oilseeds (e.g., peanuts and rapeseed), or in areas where supplies are not sufficient for large-scale solvent plants (in cottonseed). There is interest in hard screw pressing oilseeds to produce organically grown or certifiable nongenetically modified vegetable oils or to comply with local laws that prevent construction of new solvent plants such as in California. A small amount of soybeans is hard-screw-pressed, where the meal has particularly high value (e.g., high-rumen bypass meal for dairy cattle feed). Screw pressing copra, peanuts, sesame, and cocoa butter also remains popular in developing countries where investment capital is limited.

5. Prepress Solvent Extraction

In prepress solvent extraction (Figure 8.2), part of the oil that is easily removed is pressed out as described earlier. The press is choked such that less pressure is developed. Consequently, less oil is extracted and throughput is increased. Usually, the oil content of prepress cake is 15%–18%, and the partially deoiled cake is then extracted with solvent. The cake may be broken into pieces and even flaked to increase bulk density and extractor capacity, and to speed extraction. The remaining steps are the same as for direct solvent extraction, which is described in the next section.

6. Direct Solvent Extraction

Cell walls are impermeable to oil and nearly so to extraction solvents. Consequently, cell walls and membranes must be *distorted* or ruptured to get the oil out, regardless of whether solvent extracting or screw pressing is done. This requirement often calls for reducing the size of seed particles. Small pieces also transfer heat and moisture more readily during conditioning or cooking. But excessive size reduction reduces mechanical distortion during flaking. Flaking mills distort larger particles more than smaller particles [12]. Of course, particles must be small enough to pass into the nip between rolls of the mill. Heating before flaking reduces oil viscosity, inactivates enzymes, coagulates protein, ruptures some cell walls and membranes, and makes the seed particle plastic for subsequent flaking or pressing. Proper plastic texture is necessary to produce thin, nonfragile flakes with minimal fines and maximal cell distortion. The flaked material must have tenacious, thin structure, with porosity that allows transport of the oil or *miscella* (solvent–oil mixture). The flake thickness influences the rate of oil extraction. Oilseeds are conditioned before flaking by using vertical stack cookers or rotary steam tube driers to heat the seed 70°C–80°C for 20–30 min, while maintaining 10.5%–11.5% moisture [13,14]. This treatment makes the meats soft and pliable enough to be flaked with smooth-surfaced roller mills to low thickness 0.25 mm (0.010–0.012 in.) without producing excessive fines, which adversely affect other operations.

Extraction has been likened to cleaning paint from a brush [15]. As such, key to getting the brush clean is getting good contact and penetration of the solvent; then there must be enough clean solvent to dissolve the solute (e.g., the paint), and enough time and heat to quickly dissolve more solute. However, oilseeds extraction does not follow the single mechanism of leaching as the brush example implies and others have often assumed. Instead, oilseeds extraction involves a combination of leaching, diffusion, and dialysis [12,16–19], which results in an ever-decreasing rate of extraction as the relative importance of each mechanism changes during the course of extraction [20]. For flakes, the larger proportion of readily extractable oil is derived from ruptured cells, especially near the surface. The transfer of oil from distorted interior cells probably is governed by capillary flow, and the rate of oil transfer is partly dependent on viscosity of the miscella. A portion of the slowly extracted oil is contained within intact undistorted cells and must be transferred by osmosis. This transfer is very slow [21]. Presumably, the process of extruding flaked meats, known as *expanding*, shifts the relative importance toward leaching because nearly all the cells are ruptured and the collet structure is quite porous.

Another portion of the slowly extracted oil relates to slowly soluble extractable materials, such as phosphatides, free fatty acids, nonsaponifiables, and pigments, which contribute to refining loss. The best quality oil, high in triglyceride content, is extracted first, while with more exhaustive extraction, poorer quality oil is extracted. Thus, at low residual oil levels, the proportions of free fatty acids and phosphatides extracted are greater, and thus the refining loss is greater. However, current industry practice is to strive for the most complete extraction possible. Typically, residual oil contents range from 0.5% to 1.0%.

Flake thickness [22–25] and solvent temperature [25,26] have profound effects on extraction rate, and empirical relationships to extraction time have been observed [27]. While these factors are easy to control by adept operators, general lack of understanding and appreciation often exist in practice. The moisture content of the flakes is another factor affecting the rate of solvent extraction [28]. In most cases, 9%–11% moisture is ideal. Hexane and water are immiscible, and higher moisture contents interfere with the penetration of hexane. Lower moisture levels reduce the structural strength of the flakes, leading to additional fines.

To reduce the amount of solvent used in extractors, countercurrent flow of the solvent to the flakes is used (Figure 8.5) [29]. That is, the freshest flakes contact the oldest solvent and progress through the process until nearly oil-free flakes contact fresh solvent. Flakes enter the extractor through a plug vapor seal that allows the material to enter while keeping hexane vapors from escaping. The extractor is an enclosed vessel designed to wash, extract, and drain flakes.

Two principal types of extractors have been employed over the years: *immersion extractors* and *percolation extractors*. An immersion extractor immerses and soaks the material in solvent (an industrial example is the Hildebrandt U-tube extractor, and a laboratory example is the Soxhlet extractor). Generally, more solvent usage is required by immersion extractors. Few immersion extractors processing oilseeds remain; percolation extractors now dominate. In a percolation extractor, the solvent percolates by gravity through a bed of material (a laboratory example of a percolation extractor is the Goldfisch). The solvent flows over the surface of the particles and diffuses through the material during its downward circuitous travel. Miscella flows in successive passes through the bed, while the solvent spray and the bed move in opposite directions to each

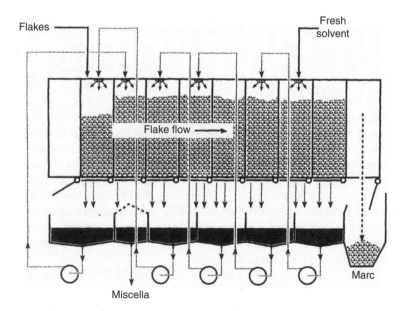

FIGURE 8.5 Flake flow relative to solvent flow. (Redrawn from Milligan, E.D., *J. Am. Oil Chem. Soc.*, 53, 286, 1976.)

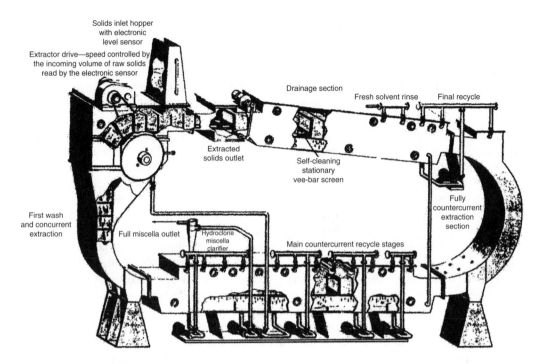

FIGURE 8.6 Schematic drawing of one type of commonly used extractor. (Courtesy of Crown Iron Works Co., Minneapolis, MN.)

other. Percolation extractors also have larger extraction capacities in less space, and fewer operating problems are associated with these devices than with immersion extractors.

Most modern extractors are of the percolation type (Figure 8.6). The extraction principles employed by most extractors are the same, but there are different methods of achieving countercurrent flow of solvent to flakes. The shallow-bed chain extractor (Figure 8.6) is one of the today's popular extractors and resembles a full-loop conveyor [30]. Flakes fed into an inlet hopper are conveyed down the first leg of the loop, where they are washed with moderately dilute miscella to extract surface oil and penetrate the cells. As the flake bed moves into the bottom horizontal section, full miscella is recycled through the bed for filtering, and then to a liquid cyclone for removing fines and, finally, to the evaporation system. Flakes are conveyed counterclockwise, through progressively more dilute miscella washes, until a final wash with fresh solvent is used in the top horizontal section of the loop. The latter half of the top loop is used for drainage, after which solvent-laden spent flakes pass to the desolventizer. The following advantages are claimed: the shallow bed (usually about 1 m) promotes drainage and, thus, low solvent carryover to the meal desolventizer and uniform contact of the flakes with solvent, since the bed is turned over while moving up the right vertical leg.

Today, hexanes (a blend of about 60% *n*-hexane with other hexane isomers) are the solvent of choice, although many other solvents were used during development of solvent extraction technologies. Alternative extraction solvents were extensively reviewed [31,32], but no suitable alternative has been developed. The primary disadvantage of hexane is its flammability and a price structure and supply that are tied to petroleum prices.

Flakes are conveyed to the extractor, where they are extracted for 30–60 min. Generally, <1% residual oil in the extracted material is achieved, and the amount is lower for soybeans (about 0.5%). The miscella contains 22%–30% oil, and the solvent is separated from the crude oil by distillation and stripping columns. The miscella is heated under vacuum to evaporate the solvent, which is usually done in two stages. The first-stage evaporator concentrates the oil to about 90% and uses reclaimed heat from heated solvent vapors from meal desolventization. Steam is used to heat the second-stage evaporator, where the oil is concentrated to >99%. Most of the remaining solvent is removed in a disc-and-doughnut stripping column, where evaporation is promoted by means of heat, vacuum (450–500 mmHg absolute pressure), and steam sparging. Crude oil leaves the stripping column with <0.15% moisture and hexane [33]. Trading specifications require the oil to have a flash point greater than 250°C, which is equivalent to no more than 800 ppm of hexane.

The marc generally contains 30%–32% solvent holdup, which must be recovered and recycled. Heat must be used to evaporate the solvent holdup from the meal. Live steam is also injected to aid heat transfer and to provide moisture vapor to strip the solvent. Regardless of the type of extractor used, the extracted flakes (*spent flakes*) must be drained of the solvent held by the material. The solvent that will not drain is referred to as *solvent holdup*, and the solvent-laden flakes are called *marc*. Solvent holdup should be minimized because this solvent must be removed by evaporation using heat. Greater solvent holdup increases the energy required for desolventizing the meal.

Toasting is often needed for feed meals to efficiently denature trypsin inhibitors (protease inhibitors in soybeans affecting protein digestibility) and the enzyme urease (soybeans), bind gossypol to protein (cottonseed), and improve protein digestibility. Of course, none of these objectives can be achieved without considerable protein denaturation and the accompanying loss of water solubility by the protein. However, depending on the method used, meals with great differences in protein solubilities or dispersibilities can be produced.

The preponderance of meal is used for feed, where extensive heat treatment is necessary to maximize feed conversion efficiency by livestock. A conventional desolventizer/toaster (DT) (Figure 8.7) is usually composed of about six stacked trays, all with indirect heating. The first two employ live steam injection through nozzles within the sweep arms to evaporate the majority of the solvent. Meal advances down through the trays, and a series of gates and floats control the levels in each tray. The lower four trays are essentially toasting/drying sections, where the meal is held at a minimum temperature of 100°C, and the meal is dried to a value suitable for dryers that follow the DT. Drying at normal DT conditions to <17% moisture is detrimental to available lysine. However, meal should not leave the DT at >22% moisture, for this would result in prohibitive drying energy requirements.

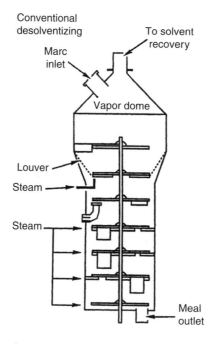

FIGURE 8.7 Meal desolventizing/toasting equipment.

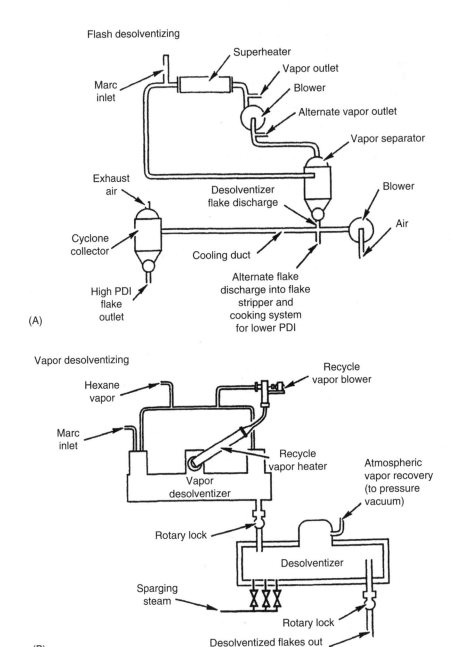

FIGURE 8.8 Setups for (A) flash desolventizing and (B) vapor desolventizing.

In recent years, the Schumacher-type desolventizer/toaster/dryer/cooler has become widely accepted. This device consists of four trays: the top tray is for predesolventizing; the second for desolventizing–toasting with injection of steam through its perforated bottom (achieving countercurrent use of steam relative to solvent evaporation); the third for drying, with hot air blown through its perforated bottom; and the fourth tray is for cooling by blowing cold air through its perforated bottom.

The *flash desolventizer* and the *vapor desolventizer* (Figure 8.8) were developed to reduce protein denaturation and produce highly soluble protein food ingredients (e.g., protein isolates) from soybeans [34]. Integrating these systems with cooking systems produces edible protein flours with a broad spectrum of protein dispersibility characteristics. The system includes a desolventizing tube, a flake separator, a circulating blower, and a vapor heater. These units are arranged in a closed loop in which hexane vapor is superheated under pressure and continuously circulated. Solvent-laden flakes from dehulled soybeans are fed into the system and conveyed by the high-velocity circulating vapor stream. The turbulent superheated vapor flow (157°C–166°C) elevates the temperature of the flakes to 77°C–88°C, well above the boiling point of hexane (65°C), in <3 s. Because the flakes enter the flash desolventizer at low moisture for a very short period and no steam is injected into the vapor stream, little denaturation of protein occurs. As the flakes travel through the tube to the cyclone separator, the greatest portion of the entrained hexane is evaporated. At this point, if care is taken during conditioning, the protein dispersibility index (PDI) of the meal protein will be 2%–5% of the untreated seed (native protein). The substantially desolventized flakes are removed from the system through a cyclone with a vapor-tight, rotary air lock and move to deodorizers.

Vapor desolventizing is similar to flash desolventizing in that superheated hexane vapor furnishes the required heat energy. Flakes are contacted with hot hexane vapor in a horizontal drum equipped with an agitator/conveyor. Flakes from either system usually enter a deodorizer to be stripped of hexane traces using only indirect heat. A slow moving agitator gently tumbles the spent flakes. The PDI may be further reduced by up to 10% units. The final PDI is controlled at the flake stripper. Sparge steam may be used to minimize solvent loss and produce low PDI products (50%–65% PDI). If only indirect steam is used, medium-range PDI products are produced (60%–75% PDI). If the stripper is bypassed or operated without any heat or steam, highly dispersible products can be produced (75%–90% PDI). However, as PDI increases, more hexane remains with the flakes as they exit the system, 0.5%–1.2% hexane for high PDI products.

7. Meal Grinding

Desolventized meal is generally ground, so that 95% passes through a U.S. 10-mesh screen and a maximum of 3%–6% passes through a U.S. 80-mesh screen. Meal for edible purposes is ground, sized, and sold as grits in a wide variety of sizes and as flour (<U.S. 100 mesh).

8. Oil and Meal Storage

Both oil and meal must be cooled before placing into storage because degradation reactions are accelerated at higher temperatures. Neither product should be stored any longer than necessary. Oil degrades through oxidation, although crude oil is more stable than refined oil because crude oil contains natural antioxidants that are removed in refining steps. Preventing water from contacting the oil is also important to prevent hydrolysis, which increases refining losses. Thus, to extend storability of oil, protection against water, heat, and air is important.

B. EXTRACTION OF OIL-BEARING FRUITS

Oil palms and olives are two examples of commercially important fruits providing important edible oils. Palm fruit can provide two distinctly different fats, one from the fleshy mesocarp and the other from the seed kernels. While olives and palms are processed slightly differently, space allows discussion of the palm oil production system only (Figure 8.9).

Palm trees are perennials grown on plantations, particularly in Malaysia. It takes 3 years for the plantings to mature sufficiently to bear fruit, and they produce for about 25 years. Palm fruits are hand-harvested and immediately transported to the mill, where they are quickly *sterilized*. The fruits are processed within hours after harvesting because the oil immediately begins to degrade after harvesting. The fruits are sterilized by heating under steam pressure (145°C) for 1 h to inactivate the enzyme lipase, which otherwise would quickly hydrolyze the oil and increase refining loss [35]. Sterilization also aids in stripping the fruit from bunch stalks and preconditions the material for subsequent steps.

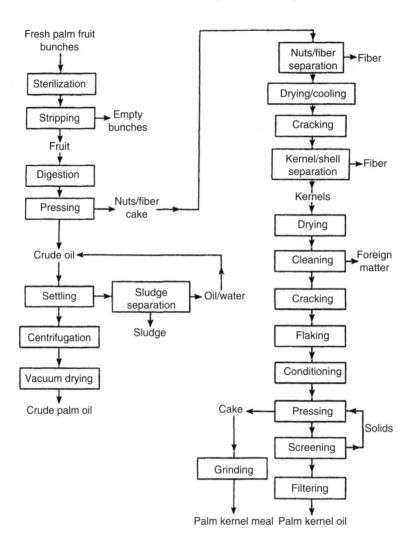

FIGURE 8.9 Steps in processing palm fruits.

The sterilized fruits are stripped from bunch stalks with drum-type strippers. The material is then sent to a digester where it is reheated (95°C–100°C for 2 min) to loosen the pericarp from the nuts and to break the oil cells. The material is conveyed to continuous screw presses similar to, but not quite the same as, those used for oilseeds, to extract oil from the fruit flesh, but not the kernel. The liquid extract, *press liquor*, from the screw press contains about two-thirds oil, one-quarter water, and one-tenth solids. The press cake contains fruit flesh fiber and nuts. Water must be added to the press liquor to facilitate satisfactory settling of solids after screening, an operation referred to as *clarification*. Oil is skimmed off the top and passed to a centrifuge (clarifier) and then to a vacuum-dryer. The crude oil is cooled and placed in storage.

The press cake is conveyed by means of a breaking conveyor to an aspirator (a vertical air column), where the nuts fall into a rotating polishing drum at the bottom, and the fruit fiber is blown to a cyclone, where it is separated from discharge air. The fiber is used to fuel the steam boiler. The nuts are conditioned by drying to loosen the kernels from the shell and cool the nuts to harden the shells. The nuts are cracked in an impact mill into two or more pieces. The shells are separated from the kernel with winnowing columns and by hydrocloning or clay bathing. All three of these operations separate shells from kernels based on density differences. The kernels are dried and screw-pressed or solvent-extracted to produce palm kernel oil and meal [36]. For every 10 t of palm oil produced, 1 t of palm kernel oil is produced.

C. RECOVERY OF ANIMAL FATS AND MARINE OILS

Animal fats and marine oils are recovered from fatty tissues by the cooking process known as *rendering*. Both edible and inedible fats are produced; the inedible tallow and grease being the majorities in the United States are used as an energy source in livestock feeds. Raw materials include animal offal, bones and trimmings from meat processors, fish species unsuitable for marketing as fillets and other fish products (menhaden, pilchard, herring, etc.), and fish cannery wastes. Until recent years, edible tallow and lard were used for deep-fat frying in fast-food restaurants, but recent consumer concerns over cholesterol and saturated fats have reduced sales in these markets. Increasingly, larger proportions are used in margarine and bakery shortenings. The defatted solid material is high in excellent quality protein that can be sold for use in livestock feeds as *meat and bone meal* (45%–54% protein), *meat meal* (52%–60% protein), *poultry by-product meal* (58%–62% protein), and *fish meal* (60%–65% protein). Fish meal commands high prices because it is especially valued in poultry diets and aquaculture diets.

Both *wet rendering* and *dry rendering* methods are used. Regardless of the process used, the material is conveyed on receipt to a crusher or prebreaker to break the material into small pieces (2–5 cm). The broken material is conveyed to either batch or continuous cookers, where heating and grinding evaporate the moisture, break down the fat cells, and release the fat, action not too dissimilar from frying bacon.

1. Wet Rendering

Wet rendering is the older method and involves cooking the material (in the presence of water) by steam under pressure (172–516 kPa [25–75 psi]) for 90–150 min [37]. When the added water comes only from steam, the process is known as *steam rendering* and this process is used to produce *prime steam lard*. The water, denatured protein, and other solids settle to the bottom, while the fat, being less dense, floats on top of the liquid. Water, known as *stick water*, is drained off, and the remaining *tankage* goes to a press for fat removal.

The presses may be either hydraulic batch type or continuous screw presses similar to those used for processing oilseeds. The high-protein solids portion is known as *cracklings* and typically contains 6%–10% residual fat. The cracklings are hammer-milled and screened, with oversized particles being recycled to the mill, thus producing *meal*. The fat discharged from the press must be centrifuged and/or filtered. Most fish, such as anchovy and menhaden, are processed by wet rendering.

2. Dry Rendering

In the newer and more efficient dry rendering process, the material is cooked in its own fat (115°C–120°C) in agitated, steam-jacketed vessels for 1.5–4 h, until the moisture has evaporated [37]. No steam or water is added. The cooked material is then passed across a screen to allow the free fat to drain. The remaining tankage is sent to a press, and the remaining steps are the same as for wet rendering.

III. REFINING

A. BACKGROUND

Consumers usually want bland-flavored or flavor-neutral, light-colored, and physically and oxidatively stable oils. Crude oils are not usually considered to be edible until numerous nonglyceride compounds have been removed through operations collectively known as *refining*. However, some oils, such as olive, tallow, and lard, have been consumed without refining. Undesirable components

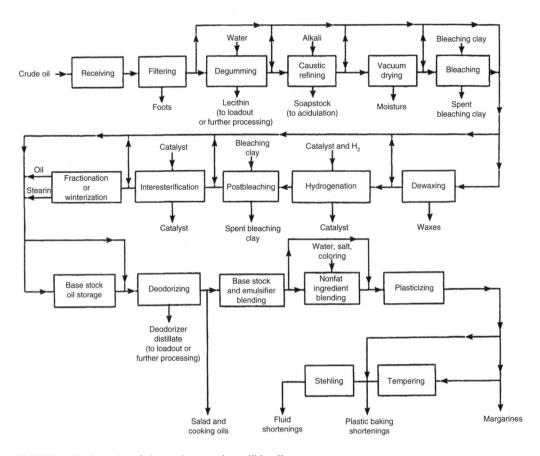

FIGURE 8.10 Steps in refining and converting edible oils.

of crude oils include small amounts of protein or other solids, phosphatides, undesirable natural flavors and odors, free fatty acids, pigments, waxes, sulfur-containing compounds (canola and rapeseed), trace solvent residue, and water. However, not all nonglyceride compounds are deleterious. Tocopherols protect the oil against autoxidation and provide vitamin E activity, and β-carotene provides vitamin A activity. Other phenolic compounds, such as sesamol in sesame oil, act as natural antioxidants. Unfortunately, some of the refining operations are not perfectly selective and also remove some beneficial compounds along with the targeted undesirable ones.

There are two major types of refining: chemical and physical. The major steps involved in chemical refining include degumming, neutralizing, bleaching, and deodorizing (Figure 8.10). Physical refining removes free fatty acids and flavors by distillation, to combine the steps of neutralization and deodorization into one operation. *RBD oil* refers to oil that has been alkali-refined, bleached, and deodorized or oil that has been physically refined.

B. DEGUMMING

Degumming (Figure 8.11) is a water-washing process to remove phosphatides. Unless removed, phosphatides can spontaneously hydrate from moisture in the air during storage or in the headspace. Degumming may be conducted either as a separate operation or simultaneously with neutralization. In the cases of oils rich in phosphatides, such as soybean and canola oils, degumming is usually a separate operation. Hydration makes phosphatides insoluble in the oil, and they precipitate, yielding an oil that is unattractive because of unsightly sludge or *gums*. Phosphatides can degrade and cause

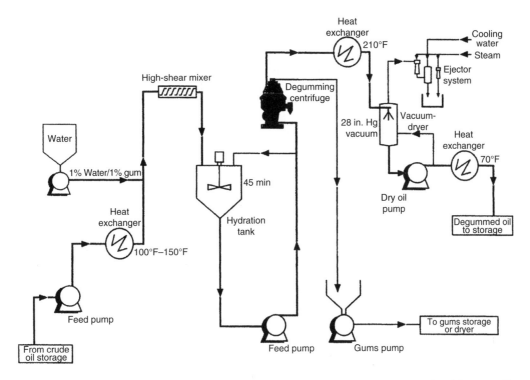

FIGURE 8.11 Process flow sheet for degumming. (Redrawn from diagram provided by Delaval Separator Co., Sullivan Systems, Inc., Larkspur, CA. With permission.)

dark colors when the oil is heated as in the later deodorization step. All soybean oil in the export trade is degummed [38]. Phosphatides are also surfactants and, if present in frying oils, can cause dangerous foaming. When hot oil foams up and spills over the rim of a cooking vessel, it may burn the user; if it contacts a flame, it will catch fire. Phosphatides, also known as *lecithin*, are important food emulsifiers and, in the case of soybean oil, oftentimes become economical to recover.

The gums are rendered insoluble in oil by hydrating them with 1%–3% water. As a general rule, the amount of water should be equivalent to the hydratable phosphatide content of the oil. If *single-bleached lecithin* is to be recovered, then hydrogen peroxide may be added to the water; if *double-bleached lecithin* is to be recovered, then benzoyl peroxide may also be added [39]. Alternatively, the degummed lecithin alone may be treated with hydrogen peroxide or both. The mixture is intensively mixed and then agitated for 30–60 min at 60°C–80°C in a slow mixing vessel (hydration tank) to allow the phosphatides to become fully hydrated and to coalesce. Hydration of the phosphatides is not instantaneous, and adequate time must be allowed. Higher temperatures solubilize more phosphatides, and lower temperatures increase oil viscosity; either one reduces the efficacy of degumming [40]. The gums, being denser than oil, can be removed by settling or filtering; more often, however, they are centrifuged out. The wet-degummed oil is either dried (as described later) or immediately neutralized. Usually, about 90% of the phosphatides are removed by this process. The gums typically contain 25% water and 75% oil-soluble substances (of which one-third is neutral oil).

There are both hydratable and nonhydratable phosphatides. Of the 1%–3% phosphatides in soybean oil, 0.2%–0.8% is generally regarded nonhydratable. The phosphatides are composed of phosphatidylcholine, phosphatidylinositol, phosphatidylethanolamine, and phosphatidic acid. The first two are always hydratable, but the latter two can complex with divalent metal ions, rendering them nonhydratable. *Acid degumming* and *superdegumming* make more of the phosphatides hydratable. Nonhydratable phosphatides remain oil soluble. The nonhydratable phosphatides are believed to be calcium and magnesium salts of phosphatidylethanolamine and, especially, phosphatidic acid, that arise from the enzymatic action of phospholipases when the cellular structure of the seed is damaged [41]. Nonhydratable phosphatides are particularly problematic in soybean oil.

Acid degumming is an improvement over conventional degumming described earlier and has become the usual practice in the U.S. soybean industry. A small amount (0.05%–0.2%) of concentrated phosphoric acid (75%) is added to warm oil (70°C) followed by stirring for 5–30 min and degumming as described in connection with conventional degumming. Longer mixing times are often substituted for lower reaction temperatures. Phosphoric acid is added to make the phosphatides more hydratable by binding calcium and magnesium ions before adding water. Phosphoric acid pretreatment also partially removes chlorophyll from the oil.

Phosphatide content varies widely in vegetable oils but is highest in crude soybean oil [42–46] (Table 8.2). Soybean oil is the only oil that is regularly degummed. The use of expanders in preparing soybeans for extraction almost doubles the usual phosphatide content of soybean oil. Only about half of the phosphatide content of soybeans is extracted with hexane when preparing soybeans by flaking alone. Sometimes, the degumming operation is conducted at the mill so that the gums may be added back to the meal. Gums contribute digestible energy to livestock. The available U.S. supply of soy lecithin is about twice the volume that can be economically sold. The gums for lecithin production are dried and may be further purified and/or bleached. Soybean lecithin is a mixture of about 40% phosphatides (16% phosphatidylcholine, 14% phosphatidylethanolamine, and 10% phosphatidylinositol), 35% oil, 17% phytoglycolipids, 7% carbohydrate, and 1% moisture [49].

Recently, *superdegumming* processes have been developed in which more of the phosphatides are rendered hydratable. A strong solution of citric acid is added to warm oil (70°C), and the mixture is stirred and cooled to 25°C to precondition the gums. Then water is added with stirring for an additional 3 h to hydrate the gums. This process causes the phosphatides to form liquid phospholipid crystals, which are easily removed during centrifugation.

Another variation of degumming, *dry degumming*, is occasionally applied to oils relatively low in phosphatide content, such as palm, coconut, and peanut oils. The oil is treated with concentrated acid to agglomerate the gums. The gums are then separated from the oil by being adsorbed to bleaching earth during subsequent steps of bleaching and filtering.

C. NEUTRALIZATION (ALKALI REFINING)

The term *neutralization* comes from neutralizing the natural acidity of the oil emanating from the presence of free fatty acids. Some use the term *refining* to refer to neutralization. Neutralization is the most important operation in refining edible oils (Figure 8.12). An improperly neutralized oil will present problems in subsequent refining steps of bleaching and deodorizing, and in conversion operations of hydrogenation and interesterification.

Neutralization is achieved by the reaction of the free fatty acid with caustic soda (sodium hydroxide) to form soap referred to as *soapstock. Saponification* refers to reactions between glycerides and sodium hydroxide also to form soaps. Neutralization must be done correctly or some of the glycerides will be saponified, resulting in increased refining loss. The oil, low in acid value, is termed *neutral oil*. Removing the soapstock must also be done carefully to prevent high losses of entrained neutral oil, a second means of increasing refining loss.

Soapstock is a coproduct of refineries in that it can be acidulated with sulfuric acid to produce a salable product. Once reacidified, the fatty acids (95% fatty acids) will separate in settling basins as 35%–40% free liquid, the so-called *acid oil*, from an emulsified layer (high in phosphatides) and a water layer. Most acid oil is used as a high-energy ingredient in livestock feed, but when market prices are attractive, it is sold to fatty acid producers, who distill it to produce feedstocks for various oleochemicals (e.g., surfactants and detergents).

TABLE 8.2
Properties of Some Crude and Refined, Bleached, Deodorized (RBD) Oils[a]

	Oils									
	Soybean[b]		Cottonseed[c]		Canola[d]		Palm[e]		Sunflowerseed[f]	
Properties	Crude	RBD	Crude	RBD	Crude	RBD	Crude	RBD	Crude	RBD
Triglycerides (%)	95–97	>99	NA	>99	NA	>99	NA	>99	NA	>99
Phosphatides (%)	1.5–2.5	0.003–0.045	0.7–0.9	NA	2.7–3.5	NA	0.006–0.013	0.012	0.5–1.0	NA
Unsaponifiable matter (%)	1.6	0.3	NA	NA	0.5–1.2	NA	NA	NA	<1.3	NA
Plant sterols	0.33	0.13	0.37	NA	NA	NA	0.036–0.062	0.011–0.016	NA	NA
Tocopherols	0.15–0.21	0.11–0.18	0.11	0.06	0.06	NA	0.06–0.10	0.04–0.06	0.05	NA
Hydrocarbons (squalene)	0.014	0.01	NA	NA	NA	NA	0.02–0.05	NA	NA	NA
Free fatty acids (%)	0.3–0.7	<0.05	0.9–37	<0.05	0.4–1.0	<0.05	2.0–5.0	<0.10	0.8–2.4	<0.05
Trace metals										
Iron (ppm)	1–3	0.1–0.3	NA	NA	1.5	<0.1	5–10	0.12	NA	NA
Copper (ppm)	0.03–0.05	0.02–0.06	NA	NA	0.10	<0.01	0.05	0.05	NA	NA

[a] NA, data not available.
[b] From Ref. [42].
[c] From Ref. [43].
[d] From Refs. [44,45].
[e] From Refs. [35,46,47].
[f] From Ref. [48].

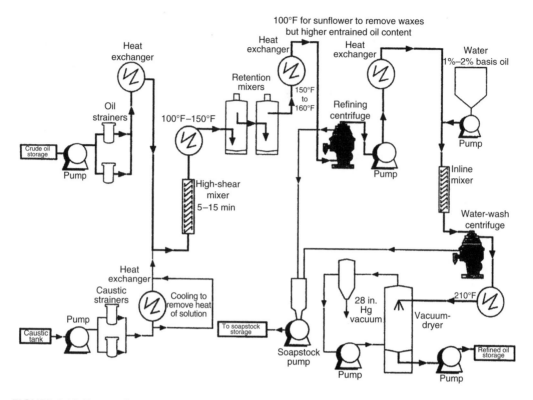

FIGURE 8.12 Process flow sheet for alkali refining. (Redrawn from diagram provided by Delaval Separator Co., Sullivan Systems, Inc., Larkspur, CA. With permission.)

In the case of cottonseed oil, proper and timely neutralization is important to achieve adequate removal of gossypol and oil that is low in red color. For reasons that are not clear, gossypol is adsorbed onto soapstock particles even though gossypol is unsaponifiable.

The amount and strength of sodium hydroxide used depend on the amount of free fatty acids present in the oil. Nearly all oils, other than soybean and rapeseed oils, are simultaneously degummed and neutralized. Free fatty acids form water-soluble sodium soaps, and any phosphatides become hydrated and water insoluble. The amount of sodium hydroxide used is termed *treat*. The proper treat produces adequately refined oil with the lowest refining loss. Excessive treat can saponify triglycerides and reduce the yield of refined oil. The proper treat is determined by titrating the oil to determine the free fatty acid content and using industry tables, such as those published in the *Official Methods and Recommended Practices of the American Oil Chemists' Society* [50].

Proper neutralization is dependent on using the proper amount of sodium hydroxide, proper mixing, proper temperature, adequate contact time, and efficient separation. As in acid degumming, some oils are preconditioned with phosphoric acid. That is, before neutralizing, the oil is treated with 0.02%–0.5% phosphoric acid at 60°C–90°C for 15–30 min, making the phosphatides less soluble in the oil, and is thus more easily removed. The proper amount of caustic is proportionately metered into the warm oil stream with good mixing and sent to retention, or dwell mixers (5–10 min mixing time). The emulsion is then thermally shocked by heating to about 75°C to break out the soapstock. Soapstock is removed from the oil by using continuous, disk-type centrifuges. Refined oil is then washed with soft water (10%–20%) at 90°C and recentrifuged to remove most of the soap. The remaining soap is removed during bleaching. The presence of excessive soap moving into the bleaching operation can reduce the effectiveness in removing colors.

D. Miscella Refining (Neutralization)

Alkali refining or neutralizing in the presence of hexane is known as *miscella refining*. In the case of cottonseed, it is desirable to carry out alkali refining as quickly as possible after extraction (about 6 h) at the extraction plant; otherwise, gossypol may become fixed in the oil, hence unremovable [51]. In addition, carrying out alkali refining in the presence of hexane reduces viscosity of the oil phase and increases the density difference between the oil phase and the water/soap phase. This improves the separation efficiency and reduces refining loss. Usually, the oil content of the miscella is concentrated to 40%–60% oil. The oil is mixed with sodium hydroxide with high-shear mixers, sometimes using high-pressure piston pumps with homogenizing valves. The mixture is heated to 65°C to melt the soapstock and then cooled to 45°C, and the aqueous and oil phases are separated by centrifuging. Water washing is not required in miscella refining. The neutralized oil miscella must then be evaporated and the oil stripped, dried, and cooled. Miscella refining produces oil with better color. The soapstock is usually added back to the meal by way of the desolventizer/toaster and hence contributes digestible energy to the meal.

The oil then is marketed as *once-refined oil*. Once-refined oil is re-refined as any other oil after arriving at the vegetable oil refinery, but much less caustic is required because most of the free fatty acid content has already been neutralized. Although any oil can be subjected to miscella refining, the additional capital investment in the safety features allowing centrifuges to work with hexane is justified only for cottonseed.

E. Drying

The water saturation level in edible fats and oils is about 0.8%, but oils should contain <0.3%. Vegetable oils must be dried before heating for prolonged periods to high temperatures as in hydrogenation and deodorization; otherwise, hydrolysis can occur, recreating free fatty acids. Drying is accomplished by spraying the hot oil (115°C) into a vacuum tower (15 mmHg absolute vacuum). The moisture content of the degummed and neutralized oil is reduced to <0.1%. The gums for lecithin production are also vacuum-dried in this manner to 0.5% moisture.

F. Bleaching

The primary purpose of bleaching (Figure 8.13) is to improve oil color by removing pigments with neutral clays, activated earths, synthetic silicates, silica gel, and carbon black. Other benefits of bleaching are the breakdown of peroxides and cleanup of residual traces of soaps and phosphatides. The primary pigments of concern are those that give red-brown (carotenoids, xanthophyll, gossypol, etc.) or green colors (chlorophyll). The process is generally done under vacuum because the usual bleaching clays can catalyze oxidation in the presence of air (or oxygen). Adsorbent is mixed with hot oil (80°C–110°C) for 15–30 min to form a slurry. Mixing enhances oil contact with the adsorbent. The pigments are adsorbed onto the surfaces of various clays or earths (some may be activated by treatment with acid), even sometimes activated carbon, and the solids are removed by filtration. Activated earths are made from certain bentonites, specifically montmorillonite. Acid activation is believed to be achieved by replacing aluminum ions in the clay structure [52] with hydrogen ions by treating with sulfuric acid; excess acid is removed with water, and the activated earth is dried and milled [38]. The hard-to-bleach oils are normally done so with acid-activated clays.

Bleaching power seems to be a function of the clay's bound acidity, and clays with high total acidity and a reasonable level of acidity are preferred. If a bleaching clay is washed completely free of residual acid, bleaching power is greatly reduced. Clay from which water has been removed gives better results than clays containing adsorbed water; but if the earth is dried to <10% moisture, its internal structure will collapse, reducing surface area and thus adsorptive power. This partly explains why heating the oil/clay slurry is important to remove water that is adsorbed in the clay lattices.

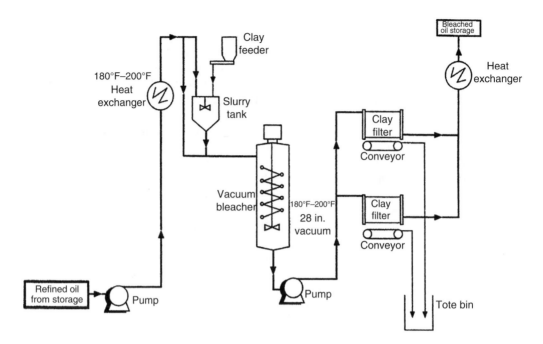

FIGURE 8.13 Process flow sheet for vacuum bleaching. (Redrawn from diagram provided by Delaval Separator Co., Sullivan Systems, Inc., Larkspur, CA. With permission.)

About 0.2%–2% bleaching clay is usually used; the precise amount depends upon the amount of pigments present. At the low end, 0.2%–0.4% is used for soybean oil, whereas rice bran oil requires considerably more (3%–5%). In addition to removing pigments and residual soap, bleaching takes out trace metals and some oxidation products. It is important to remove any residual soap as completely as possible (typically <10 ppm) because soaps can *poison* hydrogenation and interesterification catalysts, reducing their activities. Bleaching may be done batchwise or continuously.

Usually, the earth is mixed with a small amount of the oil at cool temperatures (80°C), while the bulk of the oil is deaerated and heated to bleaching temperature (100°C–110°C). Time is not as critical as temperature, and usually only 15–20 min is required. Once the proper temperature and vacuum have been achieved, the oil/earth slurry is allowed to enter and become mixed with the bulk oil. Good mixing is important to allow for contact with the oil.

Filters are usually precoated with diatomaceous earth to enhance removal of the bleaching earth by leaf filters. Exhaustive removal of the earth is very important to oil stability because the earth acts as a *prooxidant*. Used bleaching medium is called *spent earth*. Spent bleaching earth contains some entrained oil, as much as 30%–50%. The spent bleaching earth is blown with steam to reduce the oil content of the cake to about 20%. Some processors wash the cake with solvent to reduce the oil content to about 3%. Because of the large surface area that may be exposed to air and the catalytic effect of bleaching earth, spent bleaching earth is prone to spontaneous combustion and is regarded as a hazardous material by landfill operators. Disposal of spent bleaching earth is becoming a problem, and there is a considerable interest in regenerating bleaching earth by extracting contaminants.

Edible oils should be pale yellow. Color is measured by the Lovibond tintometer, usually in red and yellow terms. Most finished edible oils are less than 10 yellow and 2.5 red, with high-grade shortenings being less than 1.0 red.

G. Dewaxing

Waxes can harm the appearance of bottled oil by causing unsightly cloudiness or sediments. Corn, rice bran, safflower, sesame, and sunflower seed oils are notorious for problematic high wax contents (0.2%–3.0%) and must undergo dewaxing; occasionally canola oil also has wax problems.

Waxes can be removed by cooling the oil to 6°C–8°C and filtering or centrifuging at cold temperatures, a process similar to winterization (described in more detail in Section IV.B) [52]. To get wax crystals large enough to ease separation, cooling must be done slowly over 4 h, and the crystals should be allowed to mature for another 6 h [40]. The oil is then carefully heated to 18°C and filtered.

Sometimes, dewaxing is accomplished simultaneously with removing the gums and/or soapstock by carrying out the centrifugation at cool temperatures. Sunflower seed oil is often predewaxed (from 1500 to 400 ppm) by cooling to 25°C for 24 h and then degumming with a centrifuge at this temperature. Alternatively, sunflower seed oil may be simultaneously dewaxed and alkalirefined. After neutralization, as already described, the oil–soapstock mixture is cooled to 5°C–8°C and held there for 4–5 h under gentle mixing, and then the oil is mixed with 4%–6% of water heated to 18°C. The soapy water phase wets and causes the small wax crystals to form a heavy suspension in soapy water. The soapy suspension is centrifuged to produce a wax/soapstock fraction and a refined, dewaxed oil [40]. Sometimes, sodium lauryl sulfate is added to help wet the crystals [53].

H. Deodorization

The final step in refining fats and oils is deodorization (Figure 8.14). Oils that are converted by a variety of processes are done so before deodorizing. The primary objective of deodorization is to

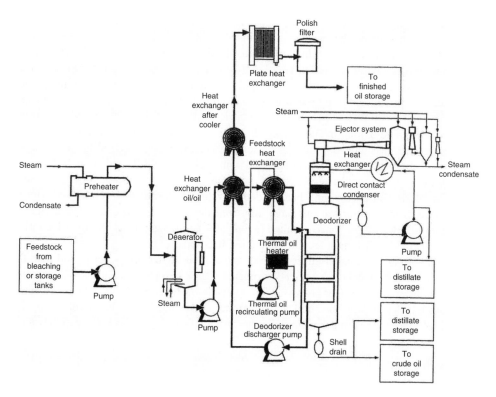

FIGURE 8.14 Process flow sheet for deodorization. (Redrawn from diagram provided by Delaval Separator Co., Sullivan Systems, Inc., Larkspur, CA. With permission.)

remove compounds responsible for undesirable odors and flavors, such as residual free fatty acids (especially low molecular weight fatty acids), aldehydes, ketones, and alcohols. Deodorization also removes peroxide decomposition products; freshly deodorized oil should have a peroxide value of zero and free fatty acid content of <0.03%. These compounds are more volatile than are triglycerides and are preferentially removed. Lard is one fat that is often not deodorized, since the better grades have mild, unobjectionable flavors. Some losses of monoglycerides, sterols, sterol esters, tocopherols, and other natural antioxidants also result. The tocopherols are potent natural antioxidants and contribute significantly to the greater oxidative stability of crude oil compared with deodorized oil.

Deodorization is essentially steam distillation performed at high temperatures (180°C–270°C) and under high vacuum (3–8 mmHg absolute pressure). Steam is sparged to carry away the volatiles and to provide agitation. Vacuum is usually provided by three to five stages of steam ejectors connected in series. Because deodorization is a mass transfer operation, deodorizers are designed to provide large surface areas and shallow oil depths. Only minor amounts of triglycerides are lost. The concentration of materials to be removed is in the range of 0.1%–1% for most oils, and the original values should be reduced by over 99% [54]. The usual loss of oil weight is 0.2%–0.8%. The amount of stripping steam ranges from 10 to 50 kg of steam per 100 kg of oil.

Factors that affect the efficacy of deodorization are the vapor pressures of the materials to be removed, the product flow rate, the intimacy of steam mixing with the oil, the absolute pressure achieved during deodorization, the temperature of deodorization (which controls the vapor pressure of the materials being removed), the sparge steam rate, and the time of deodorization [54,55].

Usually, steam alone is not used to heat the oil because very high steam pressures would be required to heat the oil sufficiently. Rather, a eutectic mixture of diphenyl and diphenyl oxide, known by the trade name *Dowtherm A*, is used. This product has a boiling point of 258°C and at 304°C generates only 110 kPa (16 psi) pressure.

Deodorization may be conducted either in batch, in semibatch, or in continuous vessels. The type of process largely depends on the volume and number of different products being processed. Batch deodorizers have cycle times of 6–8 h. Continuous deodorizers are most suitable when a limited number of products are manufactured in very large volume. There are five stages in deodorization: deaeration, heating, deodorization/steam stripping, heat recovery/cooling, and final cooling. Stripping steam is provided through sparging rings and airlift pumps. After the deodorized oil has cooled, a small amount (0.005%–0.01%) of citric acid is added to chelate metal cations; so they would not promote oxidation and reduce shelf life. Crude oils usually have greater oxidative stability than refined oils. Indeed, many processors do not expose deodorized fats and oils to the atmosphere, but discharge oil into tanks blanketed with nitrogen and fill bottled oil under nitrogen.

Deodorization also removes any residual pesticide and hexane. Some pigments, such as β-carotene, are destroyed by the high heat in deodorization and, thus, the yellow and sometimes red colors are reduced. Deodorized oils have improved flavor, odor, and color. The high temperatures used in deodorization cause limited geometric isomerization. Although deodorization removes most peroxides, it cannot reclaim rancid oxidized oils.

Deodorizer distillate is condensed, recovered, and sold at higher prices per pound than the oil itself. Soybean deodorizer distillate typically contains 12.3% tocopherols and 21.9% sterols [49]. Deodorizer distillate may be further processed into valuable fractions rich in tocopherols (vitamin E), which are in high demand by the food and pharmaceutical industries. The sterols may also be purified and sold into the pharmaceutical industry for manufacturing various synthetic hormones.

I. PHYSICAL REFINING

Physical refining (Figure 8.15) is also known as *steam refining*. These terms are applied for the removal of the free fatty acids from the oil rather than their reaction with alkali, as well as for the removal of the compounds normally targeted by deodorization. Physical refining combines both

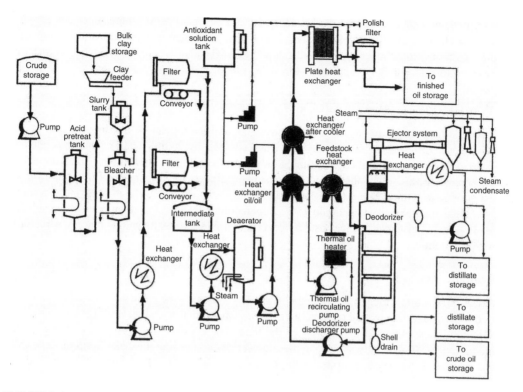

FIGURE 8.15 Depiction of physical refining. (Redrawn from diagram provided by Delaval Separator Co., Sullivan Systems, Inc., Larkspur, CA. With permission.)

neutralization and deodorization into one operation. Physical refining is always preceded by degumming and bleaching steps. The major advantage of physical refining is that the yield of oil is improved because there is none of the neutral oil loss that accompanies the production of soapstock. This process also affords the possibility of recovering fatty acids for the oleochemical industry without the need for acidulating soapstock and attendant wastewater production. The equipment used for physical refining is similar to deodorization, but with additional steam sparging trays.

Not all oils are suitable for physical refining. The oil must be low in phosphorus content (most comes from phosphatides). The exact upper limit for phosphorus in oil suitable for physical refining is a long-standing controversy, but it may be <5 ppm. Higher levels result in dark-colored oils. Physical refining is also more appropriate for those higher in free fatty acid contents because the benefit of higher yield justifies the cost of the process. Most regard soybean oil as unsuitable for physical refining, or at best difficult to physically refine, whereas palm, rice bran, and coconut oils and animal fats are well suited for physical refining.

IV. CONVERSION

A. BACKGROUND

Oftentimes, it is desirable or necessary to convert or transform highly unsaturated refined oils into more saturated forms; there may be the requirement for greater oxidative stability or altered physical forms of plastic and solid fats, or for less saturated forms to provide greater physical stability at cold temperatures (Figure 8.10). The term *conversion* is often restricted to mean only *hydrogenation*, a process by which hydrogen is added to highly reactive, unsaturated, double carbon–carbon bonds. However, winterization, fractional crystallization, and interesterification, particularly directed interesterification, should also be regarded as modes of conversion, since these processes can also significantly change oxidative stability, physical stability, and functional properties of fats and oils, and their fractions. Stehling, votating, and tempering might also be considered to be conversion processes inasmuch as they are applied to the stabilization of fluid and plastic fats.

B.   Winterization and Fractional Crystallization

The saturated fatty acids are not randomly distributed, and the oil can be fractionated into two or more fractions, differing in saturation level (as reflected by iodine value), and thus melting characteristics, oxidative stability, and functional properties. Both winterization and fractionation involve three stages: cooling the liquid oil to supersaturation to form nuclei for crystallization, gradual cooling to remove latent heat of crystallization as the crystals grow in size, and separating the crystalline fraction from the liquid [56]. Supercooling to very low temperatures establishes excessive amounts of nuclei, which results in promotion of very small crystals that are difficult to recover by filtration.

1.   Winterization

Winterization was first widely practiced on cottonseed oil. During the early years, the oil was stored in outdoor tanks, exposed to cold temperatures in the winter months. The oil would cloud, and as the crystals grew, they settled to the bottom of the tank. The crystals are composed of triglycerides containing more saturated fatty acids than the triglycerides composing the clear liquid oil. The clear oil was pumped off the lower, crystal-rich fraction, called *stearin*. Typically, 20%–25% of the cottonseed oil comprised the crystallized stearin [43]. Cottonseed stearin was blended with other high-melting fats to be used as shortening or margarine, where the high palmitic acid content contributes to improved crystal formation with improved functionality. Unwinterized cottonseed oil would cloud when stored under refrigerated temperatures and was unsuitable for salad oils and mayonnaise (the emulsion would break when crystals formed).

Today, winterization consists of cooling bleached oil under refrigeration by passing the oil through continuous, chilling heat exchangers to cool the oil to 4°C–7°C, and then to tanks with slow agitation. The amount of agitation is critical: agitation is required to remove latent heat of crystallization (fusion), but too much agitation breaks up the crystals, making them more difficult to remove. Sometimes, compounds that act as crystal inhibitors are added to increase crystal size and aid filtration [57]. After allowing the crystals to grow for several days, they are removed, usually with vacuum filters, but sometimes with plate-and-frame filters or centrifuges. The winterized oil (called *winter salad oil*) is then deodorized and packaged or bulk-stored.

Some processors have carried out winterization in solvents, such as hexane. The solvent reduces viscosity, improves the efficiency of filtration, and increases yield of winterized oil. After filtration, the solvent must be removed from the two fractions. This process is called *miscella winterization*.

In addition to cottonseed oil, canola oil is often winterized. Sunflower seed oil may also be winterized, not to remove saturated triglycerides, but to remove waxes that cause similar problems. Partially or lightly hydrogenated soybean oil that is winterized to remove saturated triglycerides (GS_3) may be used for salad oils, which will have acceptable cold test values.

In a variation of winterization, some fats are pressed to achieve fractionation. Here, a hydraulic press squeezes liquid oil from solid fat crystals. Hard butters for cocoa butter substitution in confectionery products and some specialty fats are produced by pressing of palm kernel and coconut oils [58].

2.   Fractional Crystallization

The *dry fractionation* is a form of fractional crystallization very similar to winterization, but the term is usually reserved for more saturated fats (palm oil, palm kernel oil, animal fats), as opposed to liquid oils, where one or more cuts of crystallized fats, usually termed *oleins* (high in oleic acid), are removed from higher melting triglycerides, usually termed *stearines* (high in stearic acid). Temperatures higher than those normally reserved for winterization are used. For instance, crystallization of palm oil is carried out at 20°C to produce *palm olein* in about 70% yield and *palm stearin* (for shortening and margarine). A second fractionation at a lower temperature may be carried out to produce *super-palm olein* (for frying and cooking oils) and *palm mid-fraction* (for cocoa butter substitutes). Fractional crystallization is used to make a number of cocoa butter substitutes from animal fats and from palm, palm kernel, and coconut oils.

Wet fractionation in various solvents (hexane, acetone, isopropanol, and 2-nitropropane) is also carried out on palm oil and hydrogenated soybean and cottonseed oils. Some confectionary fats and oils high in oxidative stability are produced by means of this form of fractional crystallization.

A third fractionation process is occasionally used; an aqueous detergent phase is mixed into preferentially wet, partially crystallized fat. The aqueous detergent phase contains 0.5% sodium lauryl sulfate, plus magnesium sulfate, as an electrolyte. The crystals become suspended in the aqueous phase and are removed from the liquid oil by centrifugation. The water is removed from the crystals by heating the mixture and centrifuging. Both phases are washed with water to remove detergent and vacuum-dried to remove traces of water.

C. HYDROGENATION

1. Process

Hydrogenation was first used industrially to hydrogenate chemical feedstocks; it was first applied to whale oils in 1903 [59]. In 1909, hydrogenation was patented for use in producing shortening from cottonseed oil to replace lard. Hydrogenation is used for two purposes: to improve oxidative stability (by hydrogenating some of the double bonds to saturated ones) and to convert liquid oils or soft fats into plastic or hard fats (facilitating uses for which less saturated forms are unsatisfactory). Liquid oils are converted into shortenings and margarine fats; oils, prone to rapid oxidation, such as soybean oil, are partially hydrogenated for use as salad and frying oils.

Unsaturated double bonds are converted to saturated bonds by the addition of hydrogen (H_2). The reaction between the liquid oil and H_2 gas is accelerated by using a suitable solid catalyst; thus the reaction is *heterogeneous* involving three phases. Hydrogenation is exothermic, and heats about 1.7°C per unit drop in iodine value (IV) [60].

For successful hydrogenation, many of the crude oil impurities must be removed, and refining operations must be correctly carried out beforehand. Many of the contaminants (soaps, gums, sulfur, magnesium, potassium, chromium, zinc, and mercury) can poison the catalyst, reducing its activity. Canola and rapeseed oils are particularly notorious for high levels of natural sulfur content, which can cause problems in hydrogenation.

Both batch and continuous processes are used. For hydrogenation to occur, gaseous H_2, liquid oil, and solid catalyst must be brought together at a suitable temperature. Thus, hydrogenation is a mass transfer issue, and mass transfer of reactants is the rate-limiting factor (Figure 8.16). H_2 is first dispersed as bubbles; then it must dissolve in the bulk oil, diffuse to the catalyst particle, and from there diffuse to the catalyst surface. Triglycerides must also diffuse to the catalyst surface, receive the H_2, and then diffuse out into the bulk oil. Therefore, a higher rate of reaction is achieved by raising the temperature and increasing agitation to disperse the gas in bubbles as small as possible. Small bubbles have more surface area for transfer per unit of mass for diffusion into the bulk oil. Agitation also keeps the film thicknesses small. Increasing gas flow rate helps to keep dissolved H_2 at high levels. The reaction is carried out in pressurized reactors because increased pressure increases the saturation concentration of H_2, hence the driving force for dissolution. Other factors that influence the reaction are the quantity and activity of the catalyst.

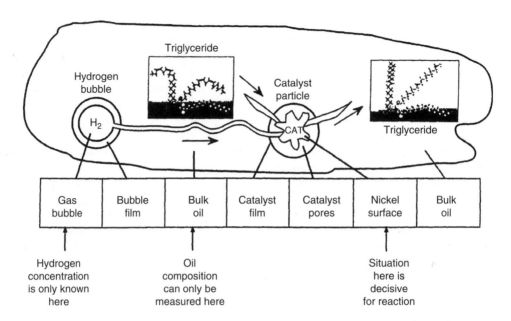

FIGURE 8.16 Hydrogenation reaction mechanism. (Modified from Coenen, J.W.E., *J. Am. Oil Chem. Soc.*, 53, 382, 1976; photographs from Beckmann, H.J., *J. Am. Oil Chem. Soc.*, 60, 234A, 1982. With permission.)

Pressurized reaction vessels are used to contain H_2, which is highly explosive should H_2, air (oxygen), and ignition come together. The reaction is normally conducted at 250°C–300°C over 40–60 min. The catalyst is mixed with a small part of the oil at room temperature. The bulk of the oil is pumped into the reactor; a vacuum is established, deaerating the oil; and the oil is heated. Finally, the oil/catalyst slurry is added. The extent of reaction is followed by monitoring the refractive index (RI) of the oil, which is directly related to IV.

A catalyst increases the rate of reaction without being consumed in the reaction. Small amounts of catalyst are effective; in the case of hydrogenation, the usual amount is in the range of 0.01%–0.02% of the weight of oil. Reduced nickel is the most widely used catalyst by the vegetable oils industry, but copper, platinum, palladium, and ruthenium are also effective. Nickel is the catalyst of choice because of good activity, selectivity, filterability, reusability, and economical use. Copper has been occasionally used, but it must be completely removed to prevent accelerating oxidation. The catalytic activity is a surface phenomenon, inducing pore formation increases surface area and thus activity [61]. However, the size and shape of pores in the catalyst are important: pores must be at least 10 nm in diameter to accommodate the transfer of triglycerides into them [60].

Catalyst promoters are substances that enhance the activities of catalysts without having catalytic activity for themselves. It is believed that the promoter function is structural, that is, these substances somehow permit a larger number of active sites on the catalyst particle. For this reason, hydrogenation catalysts are commonly supported on siliceous materials, aluminum oxide, chromium oxide, cobalt oxide, or copper oxide.

Catalysts are expensive, but they can be reused numerous times. Hydrogenation catalysts can also catalyze oxidation. For these reasons, it is important to efficiently remove the catalyst from the hydrogenated oil and then recover it, which is accomplished by filtration. While surface area per unit weight increases as the size of the catalyst particle decreases, so does the relative difficulty of removing the catalyst by filtration. For that reason, the catalyst is often incorporated onto a support, such as 17%–25% nickel fixed onto kieselguhr [60]. The catalyst is removed with leaf filters, and

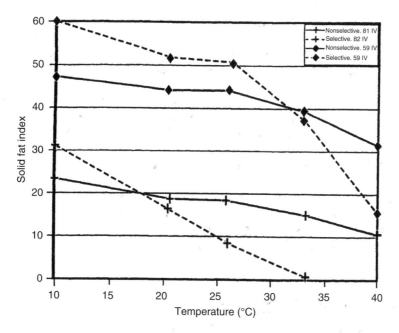

FIGURE 8.17 SFI properties of soybean oil hydrogenated under selective and nonselective conditions. (Redrawn from Stauffer, C.E., *Fats and Oils*, Eagan Press, St. Paul, MN, 1996. With permission.)

the filtered, hydrogenated oil is bleached to assure removal of all catalyst before deodorizing. This process, though termed *postbleaching*, is carried out in the same manner as normal bleaching.

2. Selectivity

Selectivity refers to the relative rates of hydrogenation of specific fatty acids; selective hydrogenation is the opposite of random hydrogenation. Ideally, the most unsaturated fatty acids are hydrogenated before the hydrogenation of the less unsaturated fatty acids (Figure 8.17). High selectivity gives high oxidative stability for a given IV. For many applications, one wants as much oxidative stability as is consistent with producing oil having no solid fat crystals at normal usage temperatures. Selectivity is always relative because perfect selectivity (i.e., all linolenic acid (18:3) converted to linoleic acid (18:2) before any linoleic acid is converted to oleic acid (18:1)) has not been achieved. However, producing catalysts with ever-increasing selectivity is a goal of catalyst suppliers.

3. Isomerization

Two types of isomerization spontaneously occur during hydrogenation: *geometrical and positional*. Only the extent to which isomerization occurs can be affected by processing conditions and catalyst selection. Geometrical isomerization refers to conversion of only *cis* double bonds to *trans* double bonds. Most unhydrogenated plant oils have only *cis* double bonds in their constituent fatty acids; however, milk fat and animal depot fats may have modest amounts of natural *trans* fatty acids, usually attributed biological hydrogenation by rumen bacteria.

Positional isomerization refers to the shift of double-bond position within the chain. If hydrogen atoms exist on the catalyst surface, the double bond opens, and hydrogen atoms add to the carbon atoms at either end [62]. When the catalyst has sufficient hydrogen atoms, another hydrogen atom is added such that the original double bond is converted to a saturated, single bond between carbons. When there are not sufficient hydrogen atoms to cover the catalyst, a hydrogen atom may be removed from either side of the partially saturated bond. This produces a new double bond, which may form in its original position or moved one carbon away, either up or down the chain. When the double bond re-forms, it may take either the *cis* or the *trans* configuration. These new double bonds may be half-hydrogenated and then dehydrogenated to re-form double bonds even further from the original position.

If hydrogenation is carried out at high pressure, low temperature, high agitation, high gassing rate, and low catalyst levels, then the catalyst is more saturated or covered with free hydrogen atoms, and there is little geometrical and positional isomerization. In addition, under these conditions the selectivity is low. However, under *starved conditions* of limited hydrogen on the catalyst surface, when the reaction could proceed much faster if more hydrogen were present, both isomerization and selectivity are high.

Hydrogenation cannot be conducted without producing some geometrical and positional isomerization, but a wide range in the extent of isomerization is possible. The more times the catalyst is reused, the less selective it becomes, and the greater is its tendency to produce isomerization. This has been attributed to natural poisoning of the catalyst by contaminants even in well-refined oil.

4. Effects on Physical and Functional Properties

Hydrogenation raises the melting point and reduces the IV of the triglycerides, usually to convert liquid oils at room temperature to semisolid plastic fats. Completely hydrogenated fats (IV < 1) are solid and brittle at room temperature and are used to add solids in baking shortenings. An advantage of hydrogenation is that a wide range of physical properties can be achieved, depending on how much H_2 is reacted with the oil. Thus, the processor and the food scientist have great flexibility in developing new products. Solid fat index (SFI) as determined by dilatometry and solid fat content (SFC) as determined by pulsed nuclear magnetic resonance (NMR) allow the analyst to describe the relative amounts of solid fat versus liquid oil over a temperature range usually reported at 10°C, 21.2°C, 26.7°C, 33.3°C, and 37.8°C, thus affording a measure of plasticity (especially important to margarines, baking shortenings, and confectionery fats).

Since *trans* fatty acids melt at considerably higher temperatures than do *cis* fatty acids, the former contribute considerably to melting and plastic properties of the fats. In some products, a high level of *trans* fatty acids (as high as 55%) is desired for functional reasons, often to achieve proper melting characteristics (e.g., in cocoa butter substitutes). Margarines and shortenings usually contain substantial *trans* fatty acids (12%–33%) [44]. However, there continues to be considerable debate about the effect of *trans* fatty acids on health. However, it is now believed that *trans* fatty acids are not as atherogenic as saturated fatty acids but not as healthy as unsaturated fatty acids [63].

Some catalysts are treated (e.g., H_2S, SO_2, CS_2, and CO) to reduce sites available to hydrogen to increase *trans* fatty acid production. There is also interest in developing catalysts that produce lower amounts of *trans* isomers. Some of the semiprecious metals have this tendency (Figure 8.18), but their costs are greater and recovery becomes even more important. As some semiprecious metals have higher catalytic activity than nickel, equivalent hydrogenation rates can be achieved at lower temperatures, reducing formation of *trans* fatty acids, as well as energy usage [64]. Precious metals are normally supported on carbon. Once the catalytic activity is spent, the catalysts are returned to the manufacturer for reclamation of the precious metal and remanufacturing of catalysts.

D. Interesterification

Interesterification, especially directed interesterification, can also be used to convert oils into more and/or less saturated fractions and ones with different functional properties from the original fat or oil or blend of two oils. Interesterification is described in much greater detail in Chapters 10 and 30.

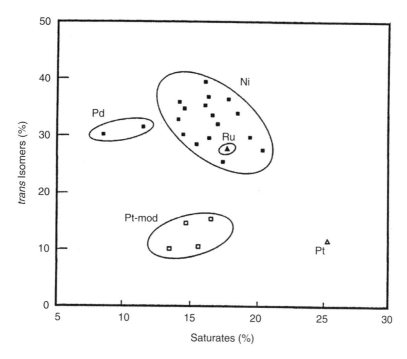

FIGURE 8.18 Effect of catalyst type on *trans* fatty acid content of hydrogenated soybean oil. (Redrawn from Okonek, D.V., Berben, P.H., and Martelli, G., *Proceedings of the Latin American Congress and Exhibition on Fats and Oils Processing*, D. Barrera-Arellano, M. Regitano d'Arce, and L. Goncalves, eds., Campinas, Brazil, September 25–28, 1995. With permission.)

V. STABILIZING PHYSICAL FORMS

Shortenings and margarines are *plastic*, that is, they have the appearance of solids in that they resist small stresses but yield to a deforming stress above a certain minimal value (the yield stress) to flow like liquid [65]. Shortenings and margarines are plastic at room temperature because they consist of two phases—solid fat crystals and liquid oil—and the solid crystals are sufficiently finely dispersed to be held together by internal cohesive forces (Figure 8.19). Votating is conducted to produce the fine dispersion of solid crystals and entrained liquid oil. The dominant controlling factor for hardness is the relative proportion of solid crystals to liquid oil, followed by crystal size and *polymorphic* form

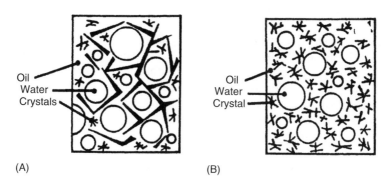

FIGURE 8.19 Plasticization of margarine. (A) Firm, with yield value of 1000 g/mL and (B) soft, with yield value of 200 g/mL. (Redrawn from Haighton, A.J., *J. Am. Oil Chem. Soc.*, 53, 397, 1976. With permission.)

(different crystal forms). The greater the proportion of crystals and the smaller the crystal size, the firmer is the product, because there is greater opportunity for crystals to interassociate. The purpose of tempering is to facilitate transformation of the fat into the correct crystalline form.

A. CRYSTALLIZING

1. Plasticizing

Plasticizing is the most commonly used process for chilling and plasticizing margarines and shortenings. *Votator* is one manufacturer's name for a swept-surface heat exchanger used to plasticize fat and thus the term *votation*. Molten fat is pumped into the first swept-surface heat exchanger known as an A unit, where the fat is supercooled to 15°C–25°C in 10–20 s, depending on the product. Ammonia or *Freon* products are used as the heat-exchange media in the exchanger jackets by means of a direct expansion refrigerant system. Oftentimes, 5–25 vol% of nitrogen, or in some cases air, is injected along with the melted fat and whipped or dispersed in the crystallized fat. The entrained gas makes shortenings white and may provide some physical leavening in bakery products.

The supercooled fat is then passed through one or more worker tubes, known as B units. Here the fat crystals are sheared, while the heat of crystallization is dissipated. The temperature of the fat can rise to 5°C–10°C, as a result of the heat of crystallization. The fat at this point may be sent to a filling unit, or it may be subjected to a second but special heat exchanger known as a C unit. The final product will have a temperature of 13°C–24°C.

Firm table margarine is made with quiescent A units (lacking an internal agitator) replacing worker B units, whereas worker B units are used to produce softer margarines.

2. Stehling

Stehling is a process that is often used to produce fluid shortenings, which are suspensions of fine crystals of solid fat in liquid oil. These products are pourable and pumpable, characteristics that are important to liquid frying shortenings, bread and cake shortenings, and nondairy products. Stehling is nothing but stirring the oil/fat blend (and emulsifiers) after votating for a prolonged period at a precise temperature, to induce the formation of very small crystals that will remain dispersed in the liquid oil and not separate during reasonable transporting and storage periods.

B. POSTPROCESSING TEMPERING

After plasticizing and packaging, shortenings are often moved to tempering rooms for 2–4 days, where the temperature is controlled at 25°C–35°C, depending on the product type and tempering time [66]. During tempering, the fat transforms from one crystal form to another more stable form, a phenomenon known as *polymorphism*. Polymorphism refers to the property of fats to exist in several different crystalline forms, depending on the orientation of the molecules in crystallized fat. The crystals spontaneously transform from one crystal type to another, successively to the most stable form, higher in melting temperature. Both the rate and the extent of transformation depend on the fatty acid composition, the votating conditions, and the temperature and duration of storage. Tempering expedites transformation to the most stable form. The least stable form is α, and no fats are stable in this form. The next form is β′, and oils high in palmitic acid are stable in this form. This crystal form, characterized by smooth, small, fine crystals, is highly functional in cake shortenings and margarines. The highest melting form is β, which is characterized by large, coarse, grainy crystals and is preferred in bread shortenings and confectionery fats.

Relative to untempered fats, tempering causes the fat to become slightly firmer above the tempering temperature and slightly softer at temperatures below the tempering temperature. The mechanism by which tempering occurs is not well understood, but it is believed to involve melting and recrystallization. Tempering serves to extend the plastic range of a shortening, rendering the product more functional. Untempered shortening may be grainy, brittle, and lacking in proper spreading qualities. Thus, tempering improves plasticity, creaming properties, and performance of baking margarines and shortenings.

VI. PROSPECTS FOR IMPROVED PROCESSES

Even though most regard the fats and oils industry to be mature, exciting new developments are in the process of adoption or testing. These developments involve advances in both processing techniques and equipment. In addition, new feedstocks modified through genetic engineering and traditional breeding could reduce or eliminate the need for some processing.

A. EXPANDER PREPARATION

In recent years, expanders, an innovation pioneered in Brazil, have found acceptance for preparing flaked soybeans for extraction. Whereas with cottonseed, all the flakes are expanded, most soybean processors expand only part of the extractor feed, generally about 30%. As a result, part of the meal experiences more heat treatment than the unexpanded portion, but these differences are believed to have no practical significance. Expanding produces a porous collet for more rapid extraction, more complete drainage of solvent from the marc, higher miscella concentrations for lower evaporation costs, and higher bulk density for greater extractor capacity. Approximately 70% of the soybean mills in the United States now employ expanders.

In extrusion/pressing or expanding/pressing, whole soybeans are extruded or expanded, usually with an autogenous dry extruder, followed by screw pressing. This nontraditional process is used commercially in some unusual situations. In developing countries, for example, local processing of soybeans into crude cooking oils is attractive. Other applications of these extrusion/expansion techniques arise when specialty oilseeds are crushed for high-value oils, such as some genetically modified soybeans, and when high-value feed ingredients are produced (e.g., high-rumen bypass protein for dairy animals). The advantage is low capital investment, but a significant disadvantage is relatively low yield: about 6%–7% residual oil is achieved compared with 0.5% residual oil for solvent extraction. Today, about 1%–2% of the U.S. soybean crop is screw-pressed, but the percentage could increase if regulations on hexane emissions become more restrictive and mandated levels become more difficult to achieve.

B. ALTERNATIVE SOLVENTS FOR EXTRACTION

The 1990 Clean Air Act is causing much concern over hexane emissions. n-Hexane, the main component of commercial hexane, is one of 189 hazardous air pollutants listed in the Clean Air Act, and hexane will be regulated as both a *criteria pollutant* and a *hazardous air pollutant*. The emission limit as a criteria pollutant is 100 t/year and as a hazardous air pollutant is 10 t/year. To exceed either limit, a processor must obtain a federal operating permit, and there is an annual fee based on hexane consumption. Great strides have been made recently in reducing hexane loss, and today, the average loss in a soybean plant would be about 0.2 gal/t, down from 1.0 gal/t 30 years ago. For a 2000 t/day plant; however, this loss still translates into 460 t of hexane per year.

While the industry continues to reduce hexane loss through engineering advances, a number of laboratories are researching alternative solvents (e.g., acetone, ethanol, isopropanol, isohexane, heptane) with less serious health and environmental risks. Enthusiasm for supercritical carbon dioxide has waned because of high capital costs and engineering problems in moving large quantities of solids through a reactor operating at very high pressure.

In a recent cottonseed plant trial, isohexane, a major component of commercial hexane, reportedly performed well and resulted in 38% steam savings [67]. But isohexane is even more expensive than hexane.

Work on ethanol and isopropanol has been going on for several years. The solubility of oil in these two solvents is temperature-dependent, and this property can be used for advantage. High solvent-to-meal ratios are required because of low solubility compared with hexane. Acceptable energy usage is achieved by reducing evaporation costs by first chill separating. The full miscella is chilled to separate a heavier oil-rich phase containing >90% oil, which is stripped, and a lighter oil-lean phase that is recycled to the extractor. The extracted flakes are partially desolventized with mechanical presses followed by heat. Both solvent streams are used to wash the flakes. Although polar solvents usually extract poorer quality oil, the use of alcohols with chill separation produces good quality oil.

C. MEMBRANE FILTRATION

Membrane technology to separate materials on the basis of molecular size has greatly improved in recent years [68]. *Microfiltration, ultrafiltration, nanofiltration*, and *reversed osmosis* are terms to designate different molecular weight separations. Membranes have been developed that are now stable to solvents and have greater selectivity. Membranes are explored to concentrate oil in the miscella before evaporating.

Membrane filtration has considerable potential in degumming. When hydrated, the phosphatide molecule becomes oriented with the hydrophilic portion sequestered in the water droplet. In the nonaqueous environment of degumming, reverse miscelles are formed. Micelles are large compared with the triglyceride molecules in which they are dispersed, and they are relatively easy to separate. The gums and pigments are concentrated in the 5% retentate, and high-quality oil is recovered in the 95% permeate.

Membranes are also explored to remove free fatty acids, and this approach appears to work. Perhaps more importantly, membrane-degummed oil is suitable for physical refining. Other applications for using membrane technologies in a vegetable oil refinery showing promise include miscella bleaching and hydrogenation catalyst removal.

D. ENZYMATIC DEGUMMING

Both soybean and canola oils have high levels of nonhydratable phosphatides. Recently, enzymatic degumming for the conversion of nonhydratable phosphatides to hydratable forms has been perfected [69]. Water-degummed oil is treated with the enzyme phospholipase A_2 after adjusting the pH to 5 with citric acid. Phospholipase A_2 hydrolyzes the *sn*-2 fatty acid to form lysolecithin, which is easily hydrated and removed. The oil is sufficiently low in phosphatides to be suitable for physical refining.

E. SUPERCRITICAL FLUID REFINING

Supercritical fluids have been used to extract oil from oilseeds. The extracted oil is usually lower in phosphatides and free fatty acids. Although some research report insufficient selectivity, others claim success in using supercritical fluids to refine vegetable oils. Liquids are more suitable for supercritical fluid technologies than are solids in terms of material handling properties. Pumping against the high pressure is relatively easy.

F. BLEACHING WITH SILICA GEL

Bleaching clays are easily poisoned by residual soaps and phosphatides. Silica gel can be used to preserve adsorption capacity for pigments, especially problematic chlorophyll [70]. In the preferred method, the oil is contacted with silica gel before contact with clay occurs. When silica gel is incorporated, clay levels can be lowered by as much as 50%, reducing solid waste and neutral oil loss.

REFERENCES

1. L.A. Johnson. Theoretical, comparative and historical analyses of alternative technologies for oilseeds extraction. In: *Technology and Solvents for Extracting Non-Petroleum Oils* (P.J. Wan and P.J. Wakelyn, eds.). AOCS Press, Champaign, IL, 1997.
2. U.S. Department of Agriculture. *Oil Crops Yearbook*. USDA, Economic Research Service, Washington, D.C., 2000.
3. F.A. Norris. Extraction of fats and oils. In: *Bailey's Industrial Oil and Fat Products*, Vol. 2, 4th edn. (D. Swern, ed.). Wiley-Interscience, New York, 1982.
4. V.D. Anderson. U.S. Patent 647,354 (1990).
5. M.A. Williams. Extrusion preparation for solvent extraction. *Oil Mill Gaz.* 91(5):24–29 (1986).
6. M.A. Williams. Expanders for high-oil seeds. *Oil Mill Gaz.* 94(3):10–12 (1989).
7. T.L. Mounts, G.R. List, and A.J. Heakin. Postharvest handling of soybeans: Effects on oil quality. *J. Am. Oil Chem. Soc.* 56:883–885 (1979).
8. G.R. List, C.D. Evans, K. Warner, R.E. Beal, W.F. Kwolek, L.T. Black, and K.J. Moulton. Quality of oil from damaged soybeans. *J. Am. Oil Chem. Soc.* 54:8–14 (1977).
9. D.K. Bredeson. Mechanical extraction. *J. Am. Oil Chem. Soc.* 55:762–764 (1978).
10. J.A. Ward. Processing high oil content seeds in continuous screw presses. *J. Am. Oil Chem. Soc.* 53:261–264 (1976).
11. D.K. Bredeson. Mechanical pressing. *J. Am. Oil Chem. Soc.* 54:489A–490A (1977).
12. R.D. Good. Theory of soybean extraction. *Oil Mill Gaz.* 75(3):14–17 (1970).
13. R.P. Hutchins. Continuous solvent extraction of soybeans and cottonseed. *J. Am. Oil Chem. Soc.* 53:279–282 (1976).
14. F. McDonald. Soybeans: Conditioning, cracking, flaking and solvent extraction. *Oil Mill Gaz.* 84(12):25–27 (1980).
15. G. Anderson. Solvent extraction of soybeans. In: *Soybean Extraction and Oil Processing*, Practical Short Course Series (L.R. Watkins, E.W. Lusas, and S.S. Koseoglu, eds.). Texas A&M University, College Station, 1987.
16. K.W. Becker. Solvent extraction of soybeans. *J. Am. Oil Chem. Soc.* 55:754–761 (1978).
17. J.R. Karnofsky. The theory of solvent extraction. *J. Am. Oil Chem. Soc.* 26:564–569 (1949).
18. J.R. Harrison. Review of extraction process: Emphasis cottonseed. *Oil Mill Gaz.* 82(4):16–24 (1977).
19. H.B. Coats and G. Karnofsky. Solvent extraction: II. The soaking theory of extraction. *J. Am. Oil Chem. Soc.* 27:51–53 (1950).
20. E. Bernardini. Batch and continuous solvent extraction. *J. Am. Oil Chem. Soc.* 53:275–278 (1976).
21. D.F. Othmer and J.C. Agarwal. Extraction of soybeans: Theory and mechanism. *Chem. Eng. Prog.* 51:372–378 (1959).
22. H.B. Coats and M.R. Wingard. Solvent extraction: III. The effect of particle size on extraction rate. *J. Am. Oil Chem. Soc.* 27:93–96 (1950).
23. N.W. Myers. Solvent extraction in the soybean industry. *J. Am. Oil Chem. Soc.* 54:491A–493A (1977).
24. J. Fawbush. How to condition and flake to produce good quality. *Oil Mill Gaz.* 85(11):39–42 (1981).
25. G. Anderson. 5-Point processing efficiency for best results with known handicaps. *Oil Mill Gaz.* 81(8):10–13 (1977).
26. M.R. Wingard and R.C. Phillips. Solvent extraction: IV. The effect of temperature on extraction rate. *J. Am. Oil Chem. Soc.* 28:149–152 (1951).
27. K.W. Becker. Critical operating problems of solvent extraction plants. *Oil Mill Gaz.* 84(9):20–24 (1980).
28. L.K. Arnold and P.J. Patel. Effect of moisture on the rate of solvent extraction of soybeans and cottonseed meats. *J. Am. Oil Chem. Soc.* 30:216–218 (1953).
29. E.D. Milligan. Survey of current solvent extraction equipment. *J. Am. Oil Chem. Soc.* 53:286–290 (1976).
30. G.D. Brueske. Solvent extraction processes for product improvement. *Oil Mill Gaz.* 81(4):8–13 (1976).
31. R.J. Horn, S.P. Koltun, and A.V. Graci. Biorenewable solvents for vegetable oil extraction. *J. Am. Oil Chem. Soc.* 59:674A–684A (1982).
32. L.A. Johnson and E.W. Lusas. Comparison of alternative solvents for oils extraction. *J. Am. Oil Chem. Soc.* 60:181A–194A (1983).
33. K.W. Becker. Processing of oilseeds to meal and protein flakes. *J. Am. Oil Chem. Soc.* 48:299–304 (1971).

34. K.W. Becker. Distillation and solvent recovery for soybeans and other oilseed plants. *J. Am. Oil Chem. Soc.* 48:110A–112A, 124A–126A (1971).

35. Y. Basiron. Palm oil. In: *Bailey's Industrial Oil and Fat Products*, Vol. 5, 5th edn. (Y.H. Hui, ed.). Wiley, New York, 1995.

36. T.S. Tang and P.K. Teoh. Palm kernel oil extraction—The Malaysian experience. *J. Am. Oil Chem. Soc.* 62:254–258 (1985).

37. L.S. Tufft. Rendering. In: *Bailey's Industrial Oil and Fat Products*, Vol. 5, 5th edn. (Y.H. Hui, ed.). Wiley, New York, 1995.

38. L.H. Wiedermann. Degumming, refining and bleaching soybean oil. *J. Am. Oil Chem. Soc.* 58:159–165 (1981).

39. R.A. Carr. Degumming and refining particles in the U.S. *J. Am. Oil Chem. Soc.* 53:347–352 (1976).

40. G. Haroldsson. Deguming, dewaxing and refining. *J. Am. Oil Chem. Soc.* 60:203A–207A (1983).

41. D.R. Erickson. *Practical Handbook of Soybean Processing and Utilization*. AOCS Press, Champaign, IL, 1995.

42. E.H. Pryde. Composition of soybean oil. In: *Handbook of Soy Oil Processing and Utilization* (D. Erickson, E. Pryde, O.L. Brekke, T. Mounts, and R.A. Falb, eds.). American Oil Chemists' Society, Champaign, IL, 1980, pp. 13–31.

43. L.A. Jones and C.C. King. *Cottonseed Oil*. National Cottonseed Products Association and the Cotton Foundation, Memphis, TN, 1990.

44. M. Vaisey-Genser and N.A.M. Eskin. *Canola Oil: Properties and Performance*. Publication No. 60. Canola Council of Canada, Winnipeg, Manitoba, 1982.

45. T.K. Mag. Canola oil processing in Canada. *J. Am. Oil Chem. Soc.* 60:332A–336A (1983).

46. P.A.T. Swoboda. Chemistry of refining. *J. Am. Oil Chem. Soc.* 62:287–292 (1976).

47. S.H. Goh, Y.M. Choo, and S.H. Ong. Minor constituents of palm oil. *J. Am. Oil Chem. Soc.* 62:237–240 (1985).

48. E.J. Campbell. Sunflower oil. *J. Am. Oil Chem. Soc.* 60:339A–392A (1976).

49. T.L. Mounts. Chemical and physical effects of processing fats and oils. *J. Am. Oil Chem. Soc.* 58:51A–54A (1981).

50. American Oil Chemists' Society. *Official Methods and Recommended Practices of the American Oil Chemists's Society*, 4th edn. AOCS, Champaign, IL, 1993.

51. G.C. Cavanagh. Miscella refining. *J. Am. Oil Chem. Soc.* 53:361–363 (1976).

52. J.C. Cowan. Degumming, refining, bleaching and deodorization theory. *J. Am. Oil Chem. Soc.* 53:344–346 (1976).

53. C.E. Stauffer. *Fats and Oils*. Eagan Press, St. Paul, MN, 1996.

54. A.M. Gavin. Edible oil deodorization. *J. Am. Oil Chem. Soc.* 55:783–791 (1978).

55. C.T. Zehnder and C.E. McMichael. Deodorization: Principles and practices. *J. Am. Oil Chem. Soc.* 44:478A, 480A, 508A, 510A–512A (1967).

56. H.P. Kreulen. Fractionation and winterization of edible fats and oils. *J. Am. Oil Chem. Soc.* 53:393–396 (1976).

57. R.C. Hastert. Practical aspects of hydrogenation and soybean salad oil manufacture. *J. Am. Oil Chem. Soc.* 58:169–174 (1981).

58. Institute of Shortening and Edible Oils. *Food Fats and Oils*. Washington, D.C., 1994.

59. W. Normann. British Patent 1515 (1903).

60. L. Faur. Transformation of fat for use in food products. In: *Oils and Fats Manual*, Vol. 2 (A. Karleskind, ed.). Lavoisier Publishing, Secaucus, NJ, 1996, pp. 897–1024.

61. J.W.E. Coenen. Hydrogenation of edible oils. *J. Am. Oil Chem. Soc.* 53:382–389 (1976).

62. R.R. Allen. Hydrogenation. *J. Am. Oil Chem. Soc.* 58:166–169 (1981).

63. Council for Agricultural Science and Technology. *Food Fats and Health*. Task Force Report No. 118. CAST, Ames, IA, 1991.

64. D.V. Okonek, P.H. Berben, and G. Martelli. Precious metal catalysis for fats and oils applications. In: *Proceedings of the Latin American Congress and Exhibition on Fats and Oils Processing* (D. Barrera-Arellano, M. Regitano d'Arce, and L. Goncalves, eds.). Campinas, Brazil, September 25–28, 1995.

65. B.A. Greenwell. Chilling and crystallization of shortenings and margarines. *J. Am. Oil Chem. Soc.* 58:206–207 (1981).

66. A.J. Haighton. Blending, chilling and tempering of margarines and shortenings. *J. Am. Oil Chem. Soc.* 53:397–399 (1976).

67. P.J. Wan, R.J. Hron, M. Dowd, S. Kuk, and E.J. Conkerton. Isohexane as an alternative hydrocarbon solvent. *Oil Mill Gaz.* 100(12):28–32 (1995).

68. S. Koseoglu. Use of membrane technology in edible oil processing: Current status and future prospects. In: *Proceedings of the Latin American Congress and Exhibition on Fats and Oils Processing* (D. Barrera-Arellano, M. Regitano d'Arce, and L. Gonçalves, eds.). Campinas, Brazil, September 25–28, 1995.

69. H. Buchold. EnzyMax—A state of the art degumming process and its applications to the oil industry. In: *Proceedings of the Latin American Congress and Exhibition on Fats and Oils Processing* (D. Barrera-Arellano, M. Regitano d'Arce, and L. Gonçalves, eds.). Campinas, Brazil, September 25–28, 1995.

70. G.A. Bravo. Silica refining: Environmental, economic and quality benefits. In: *Proceedings of the Latin American Congress and Exhibition on Fats and Oils Processing* (D. Barrera-Arellano, M. Regitano d'Arce, and L. Gonçalves, eds.). Campinas, Brazil, September 25–28, 1995.

9 Crystallization and Polymorphism of Fats

Patrick J. Lawler and Paul S. Dimick

CONTENTS

I. CRYSTALLIZATION: GENERAL PRINCIPLES

Crystallization can be considered a subset of overall solidification. Solids are crystalline, semicrystalline, or amorphous. The crystallization process from solution first requires supersaturation; supercooling is a prerequisite for crystallizing from a melt. These phenomena lead to nucleation and crystal growth. This chapter emphasizes fat crystallization from the melt.

Once formed, crystals can have different shapes, called habits or morphologies. Stable crystals modify their habit, whereas metastable ones undergo phase transitions [1]. Both these processes result in polymorphic behavior, a behavior common to fats and other lipids. Further, most crystals ripen and disappear, as a result of changes in the degree of supersaturation [1]. Supersaturation evolves as crystal growth proceeds, and the liquid phase becomes less supersaturated. This reduced supersaturation results in a stability requirement for larger crystals, since crystals below a critical size will return to solution or the melt.

A. SUPERCOOLING

Crystallization requires a solute concentration greater than the concentration of the saturated solution. Observation of the schematic saturation–supersaturation curve in Figure 9.1 aids in understanding this phenomenon. The solid line represents a saturation or solubility curve. Below this, crystallization is impossible. Above this continuous line, the system is supersaturated: for example, at point 2, crystallization is possible. Although crystallization is possible at point 2, it will not occur without agitation or seeding. This contrasts with point 3, where crystallization is spontaneous. The saturation curve results from thermodynamic factors, whereas the position of the unstable boundary (dashed line in Figure 9.1) depends on kinetic factors, particularly the rate of cooling [2].

Review of the following equations, relative to melt systems, provides a conceptual understanding of how solubility (i.e., saturation) depends on the size of the solute. From the following equation:

$$\ln X = \frac{2\tau V}{r \text{RT}},$$ (9.1)

where
  X is the solubility increase versus crystals of infinite size
  τ represents the surface tension
  V is the molar volume
  r represents the spherical crystal radius
  RT is the room temperature,

it is clear that larger crystals would be less soluble. The following equation,

$$\Delta T = \frac{2\tau V T}{r \Delta H},$$ (9.2)

where
  ΔT is supercooling
  r is the smallest size of nuclei possible at a specific crystallization temperature $T\text{–}\Delta T$,

indicates that the more a melt is supercooled, the smaller the critical radius necessary for stability.

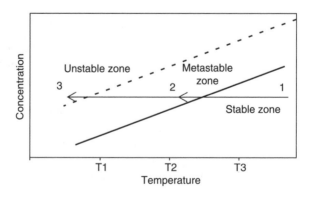

FIGURE 9.1 Saturation–supersaturation diagram.

B. NUCLEATION

Crystallization centers or nuclei are formed only when sufficient supercooling has been achieved. A crystal nucleus is the smallest crystal that can exist in a solution with defined temperature and concentration [1]. Crystal nuclei are further distinguished from crystal embryos as molecular aggregates that continue to grow rather than redissolve [3].

Generally, it is accepted that both homogeneous and heterogeneous forms of nucleation occur. The former occurs in the absence of any foreign particle surfaces. The latter type occurs in practical systems when nuclei develop on the surface of solid impurities that are present. Chemical nucleation has also been referred to as a third, distinct type of nucleation [4]. Unlike heterogeneous nucleation, chemical nucleation results when organic agents added to a system dissolve and chemically react, activating the polymer. The ionic chain ends, which develop from polymer chain scission, aggregate and form nuclei. Therefore, it appears that chemical nucleation simply refers to indirect chemical induction of heterogeneous nuclei rather than direct physical addition of solid heterogeneous nuclei.

Both primary and secondary nucleation have been described by many authors [1–3]. Secondary nucleation is relatively simple to understand. It results from fracture of growing crystals into smaller stable crystal nuclei. Nuclei stability is a function of nuclei solubility. This solubility, in turn, depends on the nuclei size at a given temperature. Nucleation theory addresses the concept of critical size regarding the development of stable primary nuclei.

An elegant discussion of these concepts was presented by Timms [2]. Two opposing actions exist when molecules attempt to aggregate. The first, energy evolution due to the heat of crystallization favors crystallization, as energy is released from the aggregating embryo. The second action, that of molecular surface enlargement requires energy input to overcome surface tension or pressure. Stable nuclei will form only when the heat of crystallization is greater than the energy required to overcome surface energy. This relationship is represented in the following equation:

$$\Delta G_{\text{embryo}} = 4\pi r^2 \lambda - \frac{4\pi r^3 \Delta G_{\upsilon}}{3V_{\text{m}}}, \tag{9.3}$$

where
   the first term on the right represents the surface energy contributions
   the second depicts the volume contributions of the heat of fusion
   ΔG_{embryo} is the Gibbs free energy of the embryo
   r is the radius of the nucleus
   λ is the surface free energy per unit surface area
   ΔG_{υ} is the molar free energy change resulting from the melt–solid phase change
   V_{m} is the molar volume [3]

The overall free energy will reach a maximum at some critical embryo size. Free energy will then tend toward a minimum in all embryos. This occurs through melting of smaller embryos or continued growth of embryos greater than the critical size.

C. CRYSTAL GROWTH

Crystal growth continues as the properly configured crystallizing molecule diffuses to the proper place on the growing crystal surface. The rate of growth is directly proportional to supercooling and varies inversely with viscosity, since molecular diffusion is reduced as melt viscosity increases. Equation 9.2 predicts that small amounts of supercooling will lead to larger crystals, whereas greater supercooling leads to smaller crystals. These larger crystals will be relatively perfect, since the crystal attachment surface can become more precisely configured with the slower crystallization rate. Greater supercooling, by contrast, affords faster crystallization (provided the viscosity remains adequate for molecular diffusion) but also results in more crystal faults. Faults can arise as molecules from the melt attach to the crystal surface and have insufficient time to become optimally arranged before new attachments are made.

Boistelle [1] described one effect of the variable consequences resulting from different amounts of supercooling—the formation of metastable phases. These are simply thermodynamically unstable phases formed as a consequence of kinetic preference. In crystallizing materials, the stable phase rarely forms first; rather, the free energy of the first crystal is closest to the free energy of the original melt. The kinetic preference for the unstable phase is a consequence of a lower difference in surface free energy [λ in Equation 9.3] between the melt and the crystal surface when compared with the difference in the stable phase versus the melt. As with nucleation, the rate of growth is proportional to the degree of supersaturation. Since the solubility of a stable phase is less than that of a metastable phase, the nucleation and growth rates of the stable phase are expected to be faster. However, as discussed earlier, solubility curves are developed from thermodynamic data, whereas the initial nucleation and growth are kinetically governed.

As crystallization continues, the degree of supersaturation in the system decreases. This then causes the critical crystal size to become greater. Therefore, smaller crystals will dissolve and only larger crystals will grow. Eventually, only one crystal size will be stable. This process, called Ostwald ripening, occurs over a very long time possibly for years.

Sintering has been described as the formation of crystal bridges between preformed crystals in a semisolid dispersion of crystals and liquid [5]. In fats, the bridges are fat crystals having different thermodynamic and kinetic crystallization parameters relative to the precrystallized material. The bridge material can result from fractional crystallization and can consist of triacylglycerols or minor lipid compounds that fail to crystallize with the predominant lipids. Sintering has been measured in flocculation studies that assumed the adhesion between preformed crystals (sintering) resulted in greater flocculate volumes [5]. Sintering is important especially relative to the final texture and consistency of fat-based foods.

The process of sintering, like Ostwald ripening, occurs over a relatively long time. As such, both sintering and Ostwald ripening are often called postcrystallization processes. Although each describes crystallization phenomena, each is more precisely a crystal maturation process.

D. Crystal Geometry

Crystals are solids with atoms arranged in a periodic three-dimensional pattern [6]. Representation of the arrangement as a point lattice (Figure 9.2) depicts each point having identical surroundings. All cells of a particular lattice are identical; therefore, any one cell's dimensions and angles describe the lattice constants or lattice parameters of the unit cell (Figure 9.3). The unit cell is the repeating unit of the whole structure. This is compared to the concept of the subcell. As the name implies, a subcell is a smaller periodic structure within the real unit cell which defines the repeating unit of the whole structure [7]. This has particular relevance to long-chain hydrocarbons, where the real unit cell is large and the subcell geometry is measured with x-ray diffraction. Only seven different cells are necessary to include all the possible point lattices. These correspond to the seven crystal systems into which all crystals can be classified (Table 9.1).

E. Crystal Polymorphism and Habit

Solids of the same composition that can exist in more than one form are referred to as polymorphic. The discussion of metastable versus stable in Section I.C alluded to this phenomenon. Crystal habit has simply been defined as the overall shape of the crystal [2]. From a crystallographic perspective, habit reflects the directional growth within a crystal, whereas morphology describes the set of faces determined by means of the symmetry elements of the crystal [1]. This subtle distinction allows crystals of the same morphology to nevertheless occupy different habits.

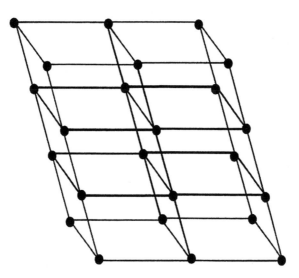

FIGURE 9.2 A point lattice. (Adapted from Cullity, B.D., ed., *Elements of X-Ray Diffraction*, Addison-Wesley, Reading, MA, 1978.)

Polymorphism is defined in terms of an ability to reveal different unit cell structures resulting from varied molecular packing. Polytypism refers to altered stacking direction of the crystal lamellae [8]. Each lamella is configured as identical polymorphs, but the direction of tilt of the hydrocarbon axis from the methyl end-group plane varies (Figure 9.4).

II. FAT CRYSTALLIZATION

A. Lipid Classification

All fats are lipids, but the converse is not true. Therefore, a distinction must be drawn to provide an accurate perspective when considering fat crystallization as discussed in this chapter. Fats are defined [9] as follows: "A glyceryl ester of higher fatty acids. . . . Such esters are solids at room temperature and exhibit crystalline structure. . . . The term 'fat' usually refers to triacylglycerols specifically, whereas 'lipid' is all-inclusive." Included in the more general term lipid are hydrocarbons, steroids, soaps, detergents, all acylglycerols, phospholipids, gangliosides, and lipopolysaccharides [9]. Since all these lipids crystallize with many degrees of complexity, this chapter primarily addresses the crystallization of triacylglycerols, making only limited reference to other lipids.

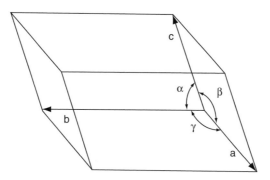

FIGURE 9.3 A unit cell. (Adapted from Cullity, B.D., ed., *Elements of X-Ray Diffraction*, Addison-Wesley, Reading, MA, 1978.)

TABLE 9.1
The Seven Crystal Systems

System	Angles and Axial Lengths
Triclinic	All axes unequal and none at right angles $a \neq b \neq c$ and $\alpha \neq \beta \neq \gamma$ and $\neq 90°$
Orthorhombic	All axes unequal and all at right angles $a \neq b \neq c$ and $\alpha = \beta = \gamma$ and $= 90°$
Hexagonal	Two axes $= 120°$ and the third at $90°$ relative to them $a = b \neq c$ and $\alpha = \beta = 90°$ and $\gamma = 120°$
Cubic	All axes equal and all at right angles $a = b = c$ and $\alpha = \beta = \gamma$ and $= 90°$
Tetragonal	Two of three axes equal and all at right angles $a = b \neq c$ and $\alpha = \beta = \gamma$ and $= 90°$
Rhombohedral	All axes equal and none at right angles $a = b = c$ and $\alpha = \beta = \gamma$ and $\neq 90°$
Monoclinic	Three unequal axes having one pair not $= 90°$ $a \neq b \neq c$ and $\alpha = \gamma = 90° \neq \beta$

Source: From Cullity, B.D., ed., *Elements of X-Ray Diffraction*, Addison-Wesley, Reading, MA, 1978, 32–80.

Lipids have been classified based on their interaction with water [10]. Triacylglycerols belong to class I polar lipids. These are insoluble, nonswelling amphiphiles. Triacylglycerols will spread at the aqueous interface and form a stable monolayer. They have a low affinity for water compared with other class I polar lipids such as diacylglycerol and cholesterol. Class II polar lipids include many of the phospholipids, as well as glycolipids and monoacylglycerol. These, too, are insoluble; however, they swell because water is soluble in their polar moieties. A result of this interaction with water is the ability of class II polar lipids to undergo lyotropic mesomorphism and develop into liquid crystals. It has been suggested that mesomorphism of phospholipids influences the nucleation and solidification behavior of cocoa butter [11].

B. TRIACYLGLYCEROL CRYSTAL PACKING STRUCTURE

Both the technical and biological functions of lipids are better understood with knowledge of their structural composition. x-ray analysis provides much of the structural information known regarding all lipids. Shipley [7] presented a brief history of the x-ray analysis of triacylglycerol single crystals. One of the first direct studies performed [12], the examination of the triclinic, β form of trilaurin, set the groundwork for determining the conformational nature of other, β-form triacylglycerols.

Of the seven crystal systems referred to in Section I.D, three predominate in the crystalline triacylglycerols [13]. Usually, the most stable form of triacylglycerols has a triclinic subcell with parallel hydrocarbon-chain planes ($T_{||}$). A second common subcell is orthorhombic with perpendicular chain phases ($O_\perp$). The third common subcell type is hexagonal (H) with no specific chain plane conformation [14]. Therefore, the hexagonal form exhibits the lowest stability and has the highest Gibbs free energy, closest to the original melt. Figure 9.5 contains subcell representations of these three common triacylglycerol conformations.

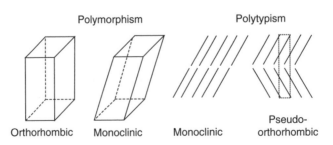

FIGURE 9.4 Polymorphism versus polytypism. (Reproduced from Sato, K. and Garti, N., *Crystallization and Polymorphism of Fats and Fatty Acids*, N. Garti and K. Sato, eds., Dekker, New York, 1988. With permission.)

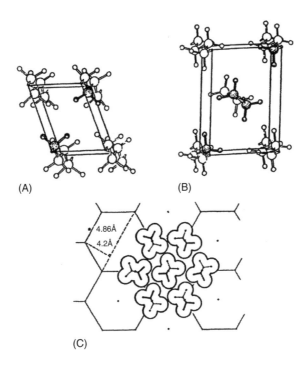

(A)   (B)

4.86Å
4.2Å

(C)

FIGURE 9.5 Schematic representations of (A) triclinic parallel, (B) orthorhombic perpendicular, and (C) hexagonal subcells. (Reproduced from Small, D.M., ed., *The Physical Chemistry of Lipids*, Plenum Press, New York, 1986. With permission.)

Interpretation of x-ray crystallography data from trilaurin and tricaprin [15–17] resulted in representation of triacylglycerols in a tuning fork conformation when crystalline. The fatty acid esterified at the *sn*-1 and *sn*-2 positions of glycerol are extended and almost straight. The *sn*-3 ester projects 90° from *sn*-1 and *sn*-2, folds over at the carboxyl carbon, and aligns parallel to the *sn*-1 acyl ester. Molecules are packed in pairs, in a single layer arrangement, with the methyl groups and glycerol backbones in separate regions (Figure 9.6).

The schematic view presented in Figure 9.6 represents simple, monoacid triacylglycerols. The polymorphic structures described for these simple triacylglycerols are valid for natural fats that contain complex triacylglycerols, given their common x-ray short spacings (axes normal to the chain direction) [13]. The structure depicted in Figure 9.6 illustrates a bilayer arrangement of the fatty acyl chains, which is the common packing structure for natural fats. However, this bilayer structure does not exist in all triacylglycerols [18–20]. Indeed, a trilayer structure has been demonstrated [21]. This trilayer structure occurs when the *sn*-2 position of the triacylglycerol contains a fatty acid ester that is either cis-unsaturated or of a chain length different by four or more carbons from those on the *sn*-1 and *sn*-3 positions. Also, it was predicted that the trilayer structure would arise if the *sn*-2 position contained a saturated acyl ester with unsaturated moieties occupying the *sn*-1 and *sn*-3 positions [22]. Figure 9.7 depicts these varied layered structures.

When unsaturation results in a *trans* configuration around the carbon–carbon double bond, the crystal structure exhibits the normal bilayer appearance. The *trans* carbon–carbon double bond results in a linear chain configuration, unlike the bent chain configuration observed in the *cis*-unsaturated molecule. In fact, trielaidin (*trans*-C18:1) has the same polymorphic configuration as tristearin (C18:0) [13], although the phase transition temperatures of trielaidin are 30°C below that of tristearin. Synthesized, stereospecific 1,2-dioleoyl-3-acyl-*sn*-glycerides with even carbon-saturated fatty acyl chains of 14–24 carbons have a trilayer packing structure [23].

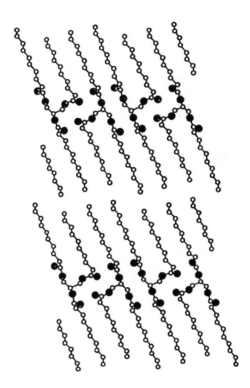

FIGURE 9.6 Schematic representation of the crystal structure of the triclinic form of tricaprin projected on the *bc* plane. (Reproduced from Sato, K. and Garti, N. eds., *Crystallization and Polymorphism of Fats and Fatty Acids*, N. Garti and K. Sato, eds., Dekker, New York, 1988. With permission.)

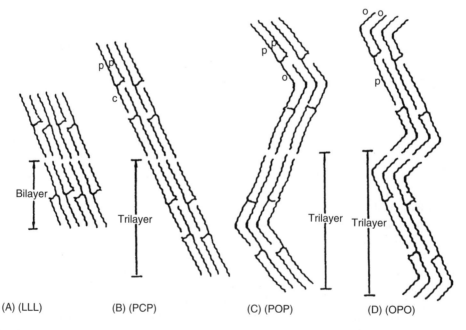

FIGURE 9.7 Schematic comparison of triacylglycerol bilayer and trilayler structures: (A) trilaurin (LLL), (B) 2-caproyldipalmitin (PCP), (C) 2-oleyldipalmitin (POP), and (D) 2-palmitoyl-diolein (OPO). (Reproduced from Small, D.M. ed., *The Physical Chemistry of Lipids*, Plenum Press, New York, 1986. With permission.)

III. POLYMORPHISM AND PHASE BEHAVIOR OF NATURAL FATS

As described earlier, solids with the same composition that exhibit different structural geometry are said to be polymorphic. The geometry of a particular polymorph confers unique physical properties beyond that of x-ray diffraction, although the x-ray diffraction pattern provides for unequivocal polymorph assignment [24]. Intramolecular and intermolecular conformation can also be inferred from infrared, Raman, and other forms of vibrational spectroscopy. Dilatometry and melting behavior are also used to evaluate the polymorphic nature of fats. Most multicomponent fats exhibit monotropic not enantiotropic polymorphism, and transformations proceed only from less stable to more stable forms [13].

A. NOMENCLATURE

The complicated melting behavior and polymorphism of fats resulted in confusion concerning terminology. Chapman [25] provided a good review of this controversy. It is now accepted that fats and triacylglycerols primarily occur in any of three basic polymorphic forms. The reference in Section II.B to the triclinic parallel ($T_\parallel$), orthorhombic perpendicular ($O_\perp$), and hexagonal (H) subcells addressed polymorphic stability. The most stable and highest melting, $T_\parallel$ is the β polymorph. Another polymorph, with variable stability and a melting point lower than β, is β'. Phases with the hexagonal subcell have the lowest melting point and represent the α polymorph.

X-ray diffraction provides not only the characteristic short spacing measurements, which define the lateral chain packing and subsequent polymorph assignment, but also measurements of the lamella layer thickness (the "d-spacing"). The d-spacing depends on the length of the molecule and the angle of tilt between the chain axis and the basal lamellar plane. Although three basic polymorphs exist, subtle variations in d-spacing lead to more than three polymorphic designations. Several of these were illustrated [26] and are shown in Figure 9.8. Figure 9.9 illustrates the orientation of the three primary polymorphs. The rarely isolated sub-α form is very unstable and can be formed only at very low temperatures. The existence of more than one β' form results from chain tilt relative to the basal plane. This can be imagined by observation of the right-most projection of Figure 9.9 and the β' polymorph. A second β' polymorph would have a different interlamellar angle of tilt. The β' designation is maintained, since the subcell is still orthorhombic perpendicular ($O_\perp$); however a longer d-spacing results in a lower melting point and therefore a polymorphic designation β'-2. As discussed earlier, this difference may also be referred to as polytypism.

Within groups having the same subcell, lower melting polymorphs are designated with a progressively higher subscript. The objective classification of acylglycerol polymorphs originally proposed [27] is also used for other lipids. This scheme (Table 9.2) refers largely to subcell packing and gives rise to subscripted forms having similar subcell dimensions. Finally, the bilayer or trilayer structure of triacylglycerol is designated with 2 or 3 following the polymorph description. For example, β'_2-2 designates a bilayer structure of a β' polymorph with the second highest melting point.

Application of the nomenclature to the most stable forms of various triacylglycerol systems, with sn positions designated A, B, and C for sn-1, 2, and 3, respectively, leads to the following:

where A = B = C and all are saturated or *trans*-unsaturated, the form is β-2;

where A = C and are saturated and B is *cis*-unsaturated or has 4 or more carbons less than A and B, the form is β-3;

where, in mixed saturated/unsaturated systems, A = B or B = C (asymmetrical), the form is β'-3;

where, in mixed saturated/unsaturated systems, A = C (symmetrical), the form is β-3.

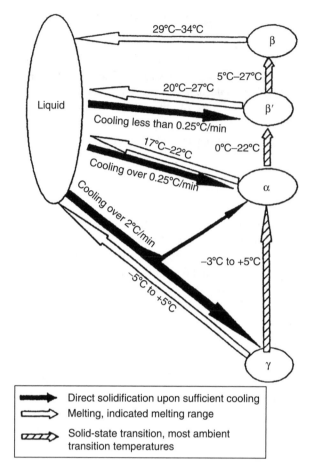

FIGURE 9.8 Polymorphic transitions of cocoa butter. (Reproduced from van Malssen, K., Peschar, R., Brito, C., and Schenk, H., *J. Am. Oil Chem. Soc.* 73, 1225, 1996. With permission.)

The packing requirements for β′ are less stringent than for β; therefore, mixed fatty acid triacylglycerols, such as those in lauric fats, tend to be β′-stable [28].

B. PHASE BEHAVIOR

Timms [24,28] provided a complete review of the phase behavior of fats. A phase is a physical state of matter that is homogeneous and separated from other phases by a definite boundary. Therefore, if only the solid and liquid states of matter are considered, solid–solid, liquid–solid, and liquid–liquid phases can exist. Triacylglycerol mixtures are ideally miscible when liquid and therefore show no heat or volume changes when mixed. As a consequence, distinct liquid–liquid phases are not apparent. Many natural fats clearly exhibit several distinct solid phases with partial miscibility, which leads to compound crystals and solid solutions. Identification of these solid solutions with x-ray powder diffraction is effective, particularly with fats in equilibrium or relatively stable metastable forms. Thermal analysis using differential scanning calorimetry (DSC) provides for observation of the effect of changing temperature [28].

An idealized set of DSC thermograms is depicted in Figure 9.10. Melting profiles can be correlated to definitive x-ray determination, which allows subsequent estimation of polymorphic form using DSC alone. However, polymorph analysis using DSC alone depends on thermal history, and other techniques may be necessary to sort out the complexity [22].

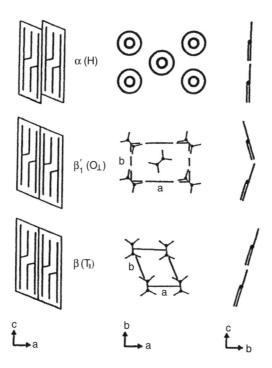

FIGURE 9.9 Schematic projections of triacylglycerol polymorphs viewed in the planes depicted. (Reproduced from Hagemann, J.W. *Crystallization and Polymorphism of Fats and Fatty Acids*, N. Garti and K. Sato, eds., Dekker, New York, 1988. With permission.)

C. MILK FAT

At least 168 different triacylglycerol species have been identified in milk fat [29]. This complexity results in three distinct endotherms on heating of samples held at 26°C. As with many natural fats, the temperature at which crystallization occurs influences milk fat firmness, crystalline conformation, and percentage of solid fat [30]. Shear effects on milk fat crystallization have also been investigated [31]. Shear rate was directly related to crystal growth rate at crystallization temperature of 15°C and 20°C. At 30°C, shear was negatively related to growth rate, indicating the possibility of different growth mechanisms. However, the lower amount of supercooling at 30°C necessitates a larger crystal diameter for stability. Therefore, shear may simply interrupt the crystal growth, preventing development of the critical crystal size.

TABLE 9.2

Spectroscopic Parameter Used to Define Acylglycerol Polymorphs

Polymorph	IR Bands (cm^{-1})	X-Ray Short Spacing (nm)	Subcell
α	720	0.415 (st)[a]	Hexagonal
B′	719, 727	0.420, 0.380	O$_\perp$
B	717	0.460 (st), 0.385 (st), 0.370 (st)	T$_\parallel$

Source: From Hagemann, J.W., *Crystallization and Polymorphism of Fats and Fatty Acids*, N. Garti and K. Sato, eds., Dekker, New York, 1988, 97–137.

[a] Strong line.

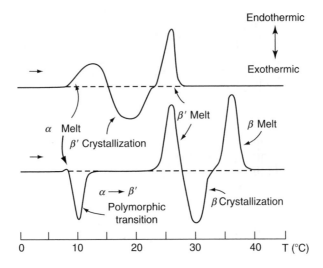

FIGURE 9.10 Schematic representation of polymorph identification by means of differential scanning calorimetry. (Reproduced from Small, D.M. ed., *The Physical Chemistry of Lipids*, Plenum Press, New York, 1986. With permission.)

Fats may be thermally treated with a process referred to as tempering to produce specific attributes in a finished product. Tempering normally results in the formation of stable crystals having the proper size and in the proper amount [32,33]; Hardness variability in milk fat results from different thermal treatments. Less solid fat results when dairy butter is cooled slowly [34]. This is better understood considering the presence of three milk fat fractions with largely independent solid solutions [35]. A DSC thermogram (Figure 9.11) depicts these independent fractions. The fractions are neatly defined as high-, middle-, and low-melting fractions (HMF, MMF, LMF, respectively). LMF is liquid at ambient temperature. Stable polymorphs of MMF and HMF were found to be a mixture of β'-2 + β'-3 and β'-2, respectively. The role of LMF in the polymorphic transformations in milk

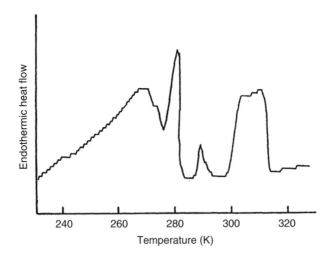

FIGURE 9.11 Differential scanning calorimetry melting curve of milk fat showing independent solid solutions.

fat fractions was investigated [36]. LMF was found to facilitate the transformation of HMF to the β-2 form but has no effect on MMF, which stayed in the β′ form.

The various milk fat fractions and their different stable forms provide opportunities regarding applications to food and nonfood systems. Anhydrous milk fat fractions were evaluated regarding their effect on chocolate tempering [37]. Chocolate with HMF addition required normal chocolate tempering temperatures (<30°C) to produce adequate temper for molding, even though a high-melting endotherm existed after tempering at 31.1°C. This high-melting endotherm was attributed to the high-melting acylglycerols of HMF, which crystallized separately from cocoa butter. The incompatibility of the HMF with cocoa butter reported by the authors is not surprising, given the β-2 stable conformation for HMF and a β-3 stable form for cocoa butter.

D. PALM OIL

Palm oil is expressed from the pulp of the oil palm (*Elaeis guineensis*) fruit. It contains about 12 major triacylglycerols and is unique among vegetable oils because of the large percentage (10%–15%) of saturated acyl esters at the *sn*-2 position. The free fatty acid composition is almost 5%, and there is a positive relationship between percentage of free fatty acid and hardness. At room temperature the oil appears as a slurry of crystals in oil. These phases persist even subsequent to tempering as measured by DSC [27]. The lower melting solid solution consists predominantly of 1-palmitoyl-2,3-dioleoyl-*sn*-glycerol (POO) and the higher melting solution is dominated by l,3-palmitoyl-2-oleoyl-*sn*-glycerol (POP). Three polymorphs were determined to be in palm oil [38]. These were β'_2 (a sub-α form found only on rapid cooling to subzero temperatures), α-2, and the stable β'_1 form. The stable β′ form is not surprising, given the heterogeneous triacylglycerol composition. The β′ stability has resulted in the addition of palm oil to oils destined for shortening or margarine, since β-tending fats can result in gritty textures. Small amounts of a palm oil β form have been produced using thermocycling [39]. As expected, thermocycling of the stearin (high-melting) fraction led to almost 40% β crystals after 36 days at 5°C.

E. LAURIC FATS

Coconut oil and palm kernel oil (PKO) are the two predominant lauric fats. The term "lauric fat" refers to the high percentage of lauric acyl esters in the triacylglycerol. The triacylglycerol composition of coconut oil is over 84% trisaturated, although at least 79 different triacylglycerol species have been identified. Less than 10% of the fatty acid composition is unsaturated [29,40]. The chain length of the acyl esters varies from C8 through C16. The polymorphic nature of coconut oil and PKO as well as hydrogenated PKO and PKO stearin is similar and relatively simple. Two polymorphs have been identified: α and β′-2. The α form is fleeting and can be recognized only after rapid cooling, as it quickly transforms into the β′-2 polymorph [41–43]. This too is not surprising, given the mixed chain length and asymmetry of the triacylglycerol molecules. The melting point of these fats is sharp at 22°C for coconut oil and 25°C for PKO [29].

F. LIQUID OILS

Evaluations of polymorphism in oils (fats that are liquid at room temperature) are limited. Cotton-seed and peanut oils exhibit no α form and no β-2 form. These oils simply crystallized in the sub-α form, sometimes referred to as β'_2 because of its orthorhombic perpendicular subcell arrangement [43]. This form results only at very low temperatures. The sub-α form of cottonseed and peanut oils transformed into a stable β'_1-2 form. Four other oils (corn, safflower, sunflower, and soybean) showed polymorphism similar to that of peanut and cottonseed, but these four fats developed a stable β-2 form.

G. Hydrogenated Fats

Complete hydrogenation eliminates the asymmetry, often leading to β′ stable polymorphs. Therefore, soybean, peanut, sunflower, corn, and sesame oils, having a large composition of C18
unsaturated fatty acids, are converted to hydrogenated fats having stearoyl esters and consequently
show the stable β-2 form. More highly saturated fats, such as cottonseed, olive, palm, and cocoa,
are converted to fats containing 1,3-dipalmitoyl-2-stearoyl-sn-glycerol (PStP), 1,3-distearoyl-2-
palmitoyl-sn-glycerol (StPSt), or 1,-palmitoyl-2,3-distearoyl-sn-glycerol (PStSt) [42,44]. The
rearrangement of PStP into a stable β form is hindered by misalignment of the methyl end plane
of the β′ unit cell [13], and a fat rich in this triacylglycerol will stay in the β′ form. Hydrogenated
fats rich in StPSt can transform into a stable β form, since the interlamellar methyl end plane can
rearrange more easily. The high PStSt fats have equally stable β′ and β forms, and any transformation to the β form occurs over a long period of time.

H. Cocoa Butter

Two reviews of the composition of cocoa butter provide a good understanding of the heterogeneity
of this natural fat [45,46]. More than 98% of cocoa butter is simple lipid. More than 95% of this is
triacylglycerol. Three triacylglycerols predominate: 1-palmitoyl-2-oleoyl-3-stearoyl-sn-glycerol
(POS) composes about 40%, 1,3-distearoyl-2-oleoyl-sn-glycerol (SOS) makes up 27.5%, and
1,3-dipalmitoyl-2-oleoyl-sn-glycerol (POP) makes up about 15%. At least 15 minor triacylglycerols
have also been identified. Free fatty acid values are reported to be 1.5%, and the concentration of
mono- and diacylglycerol is about 2% of the simple lipid fraction. The concentration of phospholipid and glycolipids in cocoa butter varies by climate and method of analysis. These complex
lipids constitute about 1% of original cocoa butter.

1. Cocoa Butter Polymorph Nomenclature

More is known about the phase behavior of cocoa butter than that of any other fat [28]. X-ray
diffraction techniques have been used to define six polymorphs, each having a distinctive melting
point [47]. Interestingly, refined cocoa butter was used in this work, and the possible effects of the
compounds removed by refining appear to have been discounted. The existence of 6 polymorphs in
a study with 12 different cocoa butters was confirmed [48]. Thermal analysis alone was used to
evaluate the polymorphism of cocoa butter and mixtures of cocoa butter with other fats [49]. That
study led to some of the confusion with terminology regarding cocoa butter polymorphs in that the
previously defined [47] designations (from lowest, I, to highest, VI, melting point) were reversed
regarding melting temperature. Also, six cocoa butter polymorphs were found in work combining
microscopy and thermal analyses [50]. Table 9.3 lists the nomenclature and melting points of the
cocoa butter polymorphs as determined by several groups. The definitions of cocoa butter polymorphs follow several conventions [27,47,52]. Consistency throughout this chapter requires the
Wille and Lutton "I through VI" system [47] if thermal data are referenced, and the Hernqvist [13]
system if x-ray data are referenced. When appropriate, only the Wille and Lutton system is used.

2. Polymorphic Formation and Transformation in Pure Cocoa Butter

A good description of the procedures necessary to form the six cocoa butter polymorphs is available
[50]. The microscopic analyses used in this work also provide insight into the crystal habit of the
forms. Each crystal formation procedure required a different method of tempering. Other than rapid
cooling of the melt to 0°C or lower, no reference was made by any of the authors regarding the rate
of cooling. Neglect of this crystallization parameter or the assumption that the crystallization
temperature was most important is one explanation for the lack of reference to cooling rate. In
fact, van Malssen et al. [53] found that crystallization temperature, not rate, determined which form

TABLE 9.3

Nomenclature and Melting Point (°C) of Cocoa Butter Polymorphs

Willie and Lutton [47]		Lovegren et al. [49]		Hicklin et al. [50]		Davis and Dimick [51]		Hernqvist [52][a]	Larsson [27]
Form	mp	Form	mp	Form	mp	Form	mp		
I	17.3	VI	13.0	I	17.9	I	13.1	sub-α	β_2' or γ
II	23.3	V	20.0	II	24.4	II	17.1	α	A
III	25.5	IV	23.0	III	27.7	III	22.4	B_2'	$\alpha + \beta$
IV	27.3	III	25.0	IV	28.4	IV	26.4	B_1'	β'
V	33.8	II	30.0	V	33.0	V	30.7	β	β_2
VI	36.3	I	33.5	VI	34.6	VI	33.8	β	B

[a] From StOSt x-ray diffraction data and PStP x-ray diffraction data.

would crystallize from the melt. However, their data indicate a possible discrepancy. They stated at 0.25°C/min cooling, β' crystals formed between 22°C and 26°C. At a cooling rate of 1°C/min, the α form crystallized at less than 23°C. Further, at a cooling rate of 6°C/s, both α and γ (sub-α) forms crystallized at less than 3°C. These authors' observation of no β crystallization after 10 days at 28°C is supported by recent real-time x-ray data, which showed no direct crystallization of β (form VI) from cocoa butter melts [54]. As discussed earlier, cocoa butter transforms monotropically, from the least stable to the more stable polymorphs. Figure 9.8 depicts the polymorphic development in cocoa butter.

One unique consequence of the polymorphism of cocoa butter is called fat "bloom." Investigations of bloom formation were published as early as 1937 [55]. Bloom occurs in chocolate, where it appears as a thin coating or scattered white patches on the surface. Neville et al. [56] attempted to describe the mechanism of bloom formation in terms of physical expansion of melting fat pushing higher melting solid fat to the surface. Full [57] presented a good review of the current understanding of bloom. Three possible mechanisms of bloom formation have been described [58] and are listed in Table 9.4. Schlichter-Aronhime and Garti [59] argued that because bloom results from a form IV to V conversion and from a V to VI conversion, the condition is caused by liquefaction of fat and does not depend strictly on the appearance of a specific polymorphic form. Regardless of the actual mechanism or mechanisms of bloom formation, the polymorphism and phase behavior of cocoa butter underlie the phenomenon.

TABLE 9.4

Possible Mechanisms of Chocolate Fat Bloom Formation

Procedure	Bloom Mechanism
Poor tempering with good cooling and storage at room temperature	Type IV crystals rapidly transform to type V, resulting in excessive bloom
Good tempering and good cooling and storage at temperature >23°C and/or with thermocycling	Type V crystals form and transform to type VI crystals, resulting in variable levels of bloom, usually over a longer time
Mixed triacylglycerol systems leading to solvation and recrystallization after tempering	Liquefied fat recrystallizing on chocolate surface during and after cooling, resulting in a variable rate of bloom formation.

Source: From Full, N.A., Physical and sensory properties of milk chocolate made with dry-fractionated milk fat. M.S. thesis, University Park, PA, 1995.

3. Natural Fat Mixtures and Polymorphism

Although fat mixtures are not used in the confectionery industry alone, this chapter emphasizes the confectionery fat blends specifically used in chocolate and chocolate-like products. Cocoa butter is the primary fat used in chocolate. Its expense has led to the development of other fats, used alone or in combination, to replace some or all cocoa butter in cocoa containing confections. The general term applied to these fats is "confectionery fat." Two general subclassifications of cocoa butter replacers (CBRs) exist: cocoa butter equivalents (CBE) and cocoa butter substitutes (CBS) [60]. Essentially, a CBE is a mixed fat that provides a fatty acid and triacylglycerol composition similar to those of cocoa butter. A CBS is a fat that provides some of the desired physical characteristics to a confection independent of its dissimilar chemical composition to that of cocoa butter. The development and use of these fats successfully resulted only after the phase behavior of cocoa butter alone and in fat mixtures had been assessed.

CBE fat blends must be tempered, since they will exhibit polymorphism similar to that of cocoa butter. The most stable form is β. Production of these fats is achieved by blending fractionated and natural fats to achieve an *sn*-2-oleoyl-disaturated triacylglycerol composition equal to cocoa butter. Ideally, blends of CBE and cocoa butter will be compatible in all proportions (compatibility refers to the phase behavior of the blend; complete compatibility implies no eutectic effect at any composition). A thermal eutectic is apparent when the melting point of a blend is lower than that of any of the pure fats in the system.

Unlike CBE, CBS blends have been developed to mimic the hardness and melting properties of cocoa butter only; they are chemically dissimilar to cocoa butter. Palm kernel oil contains a fraction with a triacylglycerol composition that forms a stable β'-2 polymorph but exhibits physical properties like that of cocoa butter. However, the β-3 stable form of cocoa butter precludes substantial mixing of cocoa butter and palm kernel oil, since the two fats develop incompatible polymorphic arrangements that lead to softening [61].

Other mixed-fat systems have been studied in detail. Cocoa butter was found to exhibit only minimal changes in polymorphism when mixed with up to 30% milk fat [41]. Although shifts in the x-ray spectra of these blends were minimal, physical softening of cocoa butter by milk fat and milk fat fractions has been shown [57]. Reddy et al. [62] used DSC to evaluate the polymorphic development and transitions in cocoa butter–milk fat systems. The HMF delayed polymorph transitions, while the LMF facilitated transformation (since the LMF is liquid at normal ambient temperatures). Similar effects in systems of olive oil and cocoa butter have also been observed [63]. It is probable that the liquid fat increases the mobility of the crystal matrix, allowing more rapid conformational changes. The same mechanism may act in LMF cocoa butter blends. The transformation inhibition by HMF supports earlier data showing that hydrogenated milk fat inhibited chocolate fat bloom [64].

4. Effects of Emulsifiers and Other Additives

Schlichter-Aronhime and Garti [59] compared the addition of low percentages of surfactants to cocoa butter to samples of cocoa butter blended with other fats. They noted that fats can change the melting ranges and the number and type of polymorphs, dependent on system compatibility, whereas surfactants may affect the rate of transformation without sensibly altering the crystal lattice. These authors' generalization of work regarding surfactants and polymorphism stated that surfactants stabilize metastable polymorphs, thereby delaying transformation to the most stable form.

Chocolate bloom was found to be inhibited by sorbitan monostearate [65]. Also, sorbitan tristearate was shown to inhibit the transformation of form V cocoa butter to form VI [59]. However, in the same series of experiments, these authors found that sorbitan tristearate hastened the transformation of the metastable polymorphs to form V. They reported that this occurs as a result of the liquefaction effect of the emulsifier and the liquid-mediated transformation process to form V. They reasoned that form V is stabilized because it transforms to form VI via the solid state, simply by expulsion of trapped liquid. Therefore, the emulsifier promotes only liquid-mediated transformations.

The effectiveness of the emulsifier relative to its state appears conflicting. Schlichter-Aronhime and Garti [59] stated that solid emulsifiers efficiently delay transformations, whereas liquid ones have no effect. However, in another publication the same authors [66] stated that "liquid emulsifiers ... will enhance the α to β transformation probably because of their weak structure compatibility with tristearin which causes a higher mobility of triacylglycerol molecules."

Hydrogenated canola oil, which is stable in the β form, can be preserved in the β′ form by adding 3%–5% of 1,2-diacylglycerol. However, 1,3-diacylglycerols were not effective [67–69]. The action of emulsifiers on the polymorphism of fats appears to be related to the physical structure, chemical composition, and thermal properties of both the fat and the additive.

Like the addition of emulsifiers to fats, the addition to fat of seeding materials, including triacylglycerol seeds, has also been investigated with regard to crystallization rates and polymorphic transitions. The possibility of forming a stable β form in cocoa butter with the addition of preformed seed crystals of cocoa butter and triacylglycerols was investigated [70], These data showed that a suitable seed must have a chemical composition close to that of the predominant cocoa butter fats. This as well as subsequent efforts [71] showed that the final cocoa butter polymorph depends on the crystallization temperature, not the polymorphic form of the added seed.

The effects of several triacylglycerol seeds on the polymorphism and solidification of cocoa butter in dark chocolate have also been investigated [72,73]. Regarding crystallization rate, it was concluded that thermodynamic stability and crystal structure similarity to cocoa butter are most effective. Specifically, 1,3-distearoyl-2-oleoylglyceride (StOSt) produced greater enhancement than 1,3-dibehenoyl-2-oleoyl-glyceride (BOB), which produced a much greater enhancement than 1,2,3-tristearoylglyceride (StStSt). It seems that a chain length similarity between the cocoa butter and the crystal seed better facilitates crystal growth. Interestingly, the β form of the StOSt produced a greater crystallization rate than a mixture of β′ and β. Therefore, although crystallization temperature alone determines the final cocoa butter form, the rate of crystallization may be a function of the polymorphic form of the seed.

5. Intrinsic Seeding and Cocoa Butter Crystallization

The crystal morphology of cocoa butter during isothermal crystallization has been detailed [74]. A higher concentration of StOSt was found in crystals that developed early versus the original StOSt concentration in the melt. It was hypothesized that subsequent crystallization occurred on these crystal seeds. Such fractional crystallization may produce crystal seeds, rich in StOSt, which have been shown to accelerate cocoa butter crystallization [72,73].

Compounds other than triacylglycerols, found in cocoa butter have also been investigated for their potential to act as crystal seeds. High-melting crystals isolated from crystallizing cocoa butter melts during the very early stages of crystallization were found to contain very high phospholipid and glycolipid concentrations relative to the original cocoa butter [75,76]. Dimick [77,78] proposed that these amphiphilic compounds may associate with the small amount of water in cocoa butter and serve as the nuclei for crystallization (Figure 9.12). The phospholipid species found in both the crystal seed and the original butter were identified [11,79], The data indicated that faster crystallizing cocoa butters contained a relatively high percentage of phosphatidylcholine and phosphatidylglycerol, whereas the slower crystallizing samples had more phosphatidylinositol and significantly less phosphatidylcholine.

The effects on cocoa butter crystallization from simple degumming and those due to added phospholipid seed material were subsequently investigated in the same laboratory [80,81]. Degummed Bahian and Côte d'Ivoire cocoa butters both had significantly slower crystallization

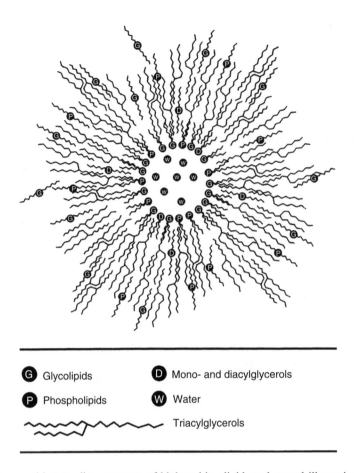

G Glycolipids   D Mono- and diacylglycerols

P Phospholipids   W Water

~~~~~~~~~~~~~~~~ Triacylglycerols

FIGURE 9.12 Proposed intermediate structure of high-melting lipid seed crystal illustrating polar core with double chain length saturated triacylglycerol surface. (From Dimick, P.S., *Proceedings of the International Symposium L'ALLIANCE 7–CEDUS*, The Crystallization of Food Products, Paris, Nov. 24, 1994; Dimick, P.S., *Physical Properties of Fats, Oils, and Emulsifiers*, N. Widlak, ed., AOCS Press, Champaign, IL, 2000.)

rates relative to the untreated butters. Addition of pure *sn*-1,2-distearoylphosphatidylcholine significantly inhibited the crystallization of the original butters. However, addition of 0.1% of this phospholipid to the degummed butters increased their crystallization rates to that equal to the untreated samples. After tempering, the original Côte d'Ivoire butter containing pure *sn*-1,2-distearoylphosphatidylcholine had greater solids at 30°C (via NMR) than the untreated butter. Pure *sn*-1,2-dioleoylphosphatidylcholine increased the crystallization rate of the Côte d'Ivoire butter but inhibited the crystallization of the Bahian butter. Also, as with *sn*-1,2-distearoylphosphatidylcholine, added *sn*-1,2-dioleoylphosphatidylcholine in the Côte d'Ivoire butter significantly increased the solids at 30°C. These data show that phospholipids can increase the crystallization rate and enhance the development of the more stable polymorphic form of cocoa butter.

Clearly, the compositional complexity of commercial cocoa butter complicates our understanding of its crystallization and polymorphic transitions. However, the phenomena observed with pure phospholipids in cocoa butter may be of general use in the fats and oils industry with regard to the control and manipulation of solidification during processing.

REFERENCES

1. R. Boistelle. Fundamentals of nucleation and crystal growth. In: *Crystallization and Polymorphism of Fats and Fatty Acids* (N. Garti and K. Sato, eds.), Dekker, New York, NY, 1988, pp. 189–226.
2. R.E. Timms. Crystallisation of fats. In: *Development of Oils and Fats* (R.J. Hamilton, ed.), Blackie Academic and Professional, Glasgow, 1995, pp. 204–223.
3. J. Garside. General principles of crystallization. In: *Food Structure and Behaviour* (J.V.M. Blanshard and P. Lillford, eds.), Academic Press, Orlando, FL, 1987, pp. 35–49.
4. J.P. Mercier. Nucleation in polymer crystallization: A physical or chemical mechanism? *Poly. Eng. Sci.* 30:270 (1990).
5. D. Johansson and B. Bergenstahl. Sintering of fat crystal networks in oil during post-crystallization processes. *J. Am. Oil Chem. Soc.* 72:911 (1995).
6. B.D. Cullity, ed. *Elements of X-Ray Diffraction.* Addison-Wesley, Reading, MA, 1978, pp. 32–80.
7. G.G. Shipley. X-ray crystallographic studies of aliphatic lipids. In: *The Physical Chemistry of Lipids* (D.M. Small, ed.), Plenum Press, New York, NY. 1986, pp. 97–147.
8. K. Sato and N. Garti. Crystallization and polymorphic transformation: An introduction. In: *Crystallization and Polymorphism of Fats and Fatty Acids* (N. Garti and K. Sato, eds.), Dekker, New York, NY, 1988, pp. 3–7.
9. R.J. Lewis, Sr., ed. *Hawley's Condensed Chemical Dictionary.* Van Nostrand Reinhold, New York, NY, 1993.
10. D.M. Small. Lipid classification based on interactions with water. In: *The Physical Chemistry of Lipids* (D.M. Small, ed.), Plenum Press, New York, NY, 1986, pp. 89–95.
11. C.M. Savage and P.S. Dimick. Influence of phospholipids during crystallization of hard and soft cocoa butters. *Manuf. Confect.* 11:127 (1995).
12. V. Vand and I.P. Bell. A direct determination of the crystal structure of the B form of trilaurin. *Acta Crystallogr.* 4:465 (1951). Quoted in G.G. Shipley, x-ray crystallographic studies of aliphatic lipids. In: *The Physical Chemistry of Lipids* (D.M. Small, ed.), Plenum Press, New York, NY, 1986, pp. 97–147.
13. J.W. Hagemann. Thermal behavior and polymorphism of acylglycerides. In: *Crystallization and Polymorphism of Fats and Fatty Acids* (N. Garti and K. Sato, eds.), Dekker, New York, NY, 1988, pp. 97–137.
14. C.W. Hoerr and F.R. Paulika. The role of x-ray diffraction in studies of the crystallography of monoacid saturated triglycerides. *J. Am. Oil Chem. Soc.* 45:793 (1968).
15. L.H. Jensen and A.J. Mabis. Crystal structure of β-tricaprin. *Acta Crystallogr.* 137:681 (1963). Quoted in G.G. Shipley, x-ray crystallographic studies of aliphatic lipids. In: *The Physical Chemistry of Lipids* (D.M. Small, ed.), Plenum Press, New York, NY, 1986, pp. 97–147.
16. L.H. Jensen and A.J. Mabi. Refinement of the structure of β-tricaprin. *Acta Crystallogr.* 21:770 (1966). Quoted in G.G. Shipley, x-ray crystallographic studies of aliphatic lipids. In: *The Physical Chemistry of Lipids* (D.M. Small, ed.), Plenum Press, New York, NY, 1986, pp. 97–147.
17. K. Larsson. The crystal structure of the β-form of triglycerides. *Proc. Chem. Soc.* 87 (1963). Quoted in G.G. Shipley, x-ray crystallographic studies of aliphatic lipids. In: *The Physical Chemistry of Lipids* (D.M. Small, ed.), Plenum Press, New York, NY, 1986, pp. 97–147.
18. E.S. Lutton. Triple chain-length structures of saturated triglycerides. *J. Am. Oil Chem. Soc.* 70:248 (1948).
19. K. Larsson. Molecular arrangement of triglycerides. *Fette Seifen Anstrichm.* 74:136 (1972). Quoted in D.J. Hanahan, ed., *The Physical Chemistry of Lipids.* Plenum Press, New York, NY, 1986.
20. T. Malkin. The polymorphism of glycerides. *Prog. Chem. Fats Other Lipids* 2:1 (1954). Quoted in D.J. Hanahan, ed., *The Physical Chemistry of Lipids.* Plenum Press, New York, NY, 1986.
21. S. de Jong, T.C. van Soest, and M.A. van Schaick. Crystal structures and melting points of unsaturated triacylglycerols in the β-phase. *J. Am. Oil Chem. Soc.* 68:371 (1991).
22. D.M. Small, ed. *The Physical Chemistry of Lipids.* Plenum Press, New York, NY, 1986.
23. D.A. Fahey, D.M. Small, D.R. Kodali, D. Atkinson, and T.G. Redgrave. Structure and polymorphism of 1,2-dioleoyl-3-acyl-*sn*-glycerols: 3-and 6-layered structures. *Biochemistry* 24:3757 (1985).
24. R.E. Timms. Physical chemistry of fats. In: *Fats in Food Products* (D.P.J. Moran and K.K. Rajah, eds.), Blackie Academic & Professional, Glasgow, 1994, pp. 1–27.
25. D. Chapman. X-ray diffraction studies. In: *The Structure of Lipids by Spectroscopic and X-Ray Techniques*, Wiley, New York, NY, 1965. Quoted in D.J. Hanahan, ed., *The Physical Chemistry of Lipids*, Plenum Press, New York, NY, 1986.

26. K. van Malssen, R. Peschar, C. Brito, and H. Schenk. Real-time x-ray powder diffraction investigations on cocoa butter: III. Direct β-crystallization of cocoa butter: Occurrence of a memory effect. *J. Am. Oil Chem. Soc.* 73:1225 (1996).

27. K. Larsson. Classification of glyceride crystal forms. *Acta Chem. Scand.* 20:2255–2260 (1966). Quoted in L. Hernqvist, Crystal structures of fats and fatty acids. In: *Crystallization and Polymorphism of Fats and Fatty Acids* (N. Garti and K. Sato, eds.), Dekker, New York, NY, 1988, pp. 97–137.

28. R.E. Timms. Phase behaviour of fats and their mixtures. *Prog. Lipid Res.* 23:1 (1984).

29. N.O.V. Sonntag. Composition and characteristics of individual fats and oils. In: *Bailey's Industrial Oil and Fat Products*, 4th edn. (D. Swern, ed.), Wiley, New York, NY, 1979, pp. 289–477.

30. J. Foley and J.P. Brady. Temperature-induced effects on crystallization behaviour, solid fat content and the firmness values of milk fat. *J. Dairy Res.* 51:579 (1984).

31. D.S. Grail and R.W. Hartel. Kinetics of butterfat crystallization. *J. Am. Oil Chem. Soc.* 69:741 (1992).

32. R. Easton, D. Kelly, and L. Barton. The use of cooling curves as a method of determining the temper of chocolate. *Food Technol.* 5:521 (1952).

33. R. Cook and E. Meursing. *Chocolate Production and Use.* Harcourt Brace Jovanovich, New York, NY, 1982.

34. H. Mulder. Melting and solidification of milk fat. *Neth. Milk Dairy J.* 7:149–176 (1953). Quoted in R.E. Timms, Phase behaviour of fats and their mixtures. *Prog. Lipid Res.* 23:1 (1984).

35. R.E. Timms. The phase behaviour of mixtures of cocoa butter and milk fat. *Lebensm. Wiss. Technol.* 13:61 (1980).

36. R.E. Timms. The phase behaviour and polymorphism of milk fat, milk fat fractions and fully hardened milk fat. *Austr. J. Dairy Technol.* 35:47 (1980).

37. S.Y. Reddy, N. Full, P.S. Dimick, and G.R. Ziegler. Tempering method for chocolate containing milk-fat fractions. *J. Am. Oil Chem. Soc.* 73:723 (1996).

38. U. Persmark, K.A. Melin, and P.O. Stahl. *Riv. Ital. Sost. Grasse* 53:306–310 (1976). Quoted in R.E. Timms, Phase behaviour of fats and their mixtures. *Prog. Lipid Res.* 23:1 (1984).

39. P.H. Yap, J.M. deMan, and L. deMan. Polymorphism of palm oil and palm oil products. *J. Am. Oil Chem. Soc.* 66:693 (1989).

40. J. Bezard, M. Bugaut, and G. Clement. *J. Am. Oil Chem. Soc.* 48:134–139 (1971). Quoted in N.O.V. Sonntag, Composition and characteristics of individual fats and oils. In: *Bailey's Industrial Oil and Fat Products*, 4th edn. (D. Swern, ed.), Wiley, New York, NY, 1979, pp. 289–477.

41. G.M. Chapman, E.E. Akehurst, and W.B. Wright. Cocoa butter and confectionery fats. Studies using programmed temperature x-ray diffraction and differential scanning calorimetry. *J. Am. Oil Chem. Soc.* 48:824 (1971).

42. A. Hvolby. *J. Am. Oil Chem. Soc.* 51:50–54 (1974). Quoted in R.E. Timms, Phase behaviour of fats and their mixtures. *Prog. Lipid Res.* 23:1 (1984).

43. U. Riiner. Investigation of the polymorphism of fats and oils by temperature programmed x-ray diffraction. *Lebensm. Wiss. Technol.* 3:101 (1970).

44. C.W. Hoerr. Morphology of fats, oils and shortenings. *J. Am. Oil Chem. Soc.* 37:539 (1960).

45. S. Chaiseri and P.S. Dimick. Cocoa butter—Its composition and properties. *Manuf. Confect.* 67:115 (1987).

46. V.K.S. Shukla. Cocoa butter properties and quality. *Lipid Technol.* 7:54 (1995).

47. R.L. Wille and E.S. Lutton. Polymorphism of cocoa butter. *J. Am. Oil Chem. Soc.* 43:491 (1966).

48. A. Huyghebaert and H. Hendrickx. *Lebens. Wiss. Technol.* 4:59 (1971). Quoted in R.E. Timms, Phase behaviour of fats and their mixtures. *Prog. Lipid Res.* 23:1 (1984).

49. N.V. Lovegren, M.S. Gray, and R.O. Feuge. Polymorphic changes in mixtures of confectionery fats. *J. Am. Oil Chem. Soc.* 53:83 (1976).

50. J.D. Hicklin, G.G. Jewel, and J.F. Heathcock. Combining microscopy and physical techniques in the study of cocoa butter polymorphs and vegetable fat blends. *Food Microstructure* 4:241 (1985).

51. T.R. Davis and P.S. Dimick. Solidification of cocoa butter. In: *Proceedings of the 40th Anniversary Production Conference, Pennsylvania Manufacturing Confectioners' Conference*, Drexel Hill, PA, 1986.

52. L. Hernqvist. Chocolate temper. In: *Industrial Chocolate Manufacture and Use* (S.T. Beckett, ed.), Blackie & Sons, London, 1988, p. 159.

53. K. van Malssen, R. Peschar, and H. Schenk. Memory effect in solidification of cocoa butter. *Voedingsmiddelentechnologie* 28:67 (1995).

54. K. van Malssen, R. Peschar, and H. Schenk. Real-time x-ray powder diffraction investigations on cocoa butter: II. The relationship between melting behavior and composition of β-cocoa butter. *J. Am. Oil Chem. Soc.* 73:1217 (1996).

55. W. Layton, S. Back, R.J. Johnson, and J.F. Morse. Physico-chemical investigation incidental to the study of chocolate fat bloom: I. *J. Soc. Chem. Ind.* 56:196 (1937). Quoted in I.J. Kleinert, Studies on the formation of fat bloom and methods of delaying it. Paper presented at 15th Production Conference, Pennsylvania Manufacturing Confectioners' Conference, Lancaster, PA, 1961.

56. H.A. Neville, N.R. Easton, and L.R. Barton. The problem of chocolate bloom. *Food Technol.* 4:439 (1950).

57. N.A. Full. Physical and sensory properties of milk chocolate made with dry-fractionated milk fat. *M.S. thesis*, University Park, PA, 1995.

58. D.J. Cebula and G. Ziegleder. x-ray diffraction studies of bloom formation in chocolates after long-term storage. *Fett. Wiss. Technol.* 95:340 (1993).

59. J. Schlichter-Aronhime and N. Garti. Solidification and polymorphism in cocoa butter and the blooming problems. In: *Crystallization and Polymorphism of Fats and Fatty Acids* (N. Garti and K. Sato, eds.), Dekker, New york, NY, 1988, pp. 363–393.

60. G.G. Jewell. Vegetable fats. In: *Industrial Chocolate Manufacture and Use* (S.T. Beckett, ed.), Blackie, Glasgow, 1988, pp. 227–235.

61. G. Talbot. Fat eutectics and crystallization. In: *Physico-chemical Aspects of Food Processing* (S.T. Beckett, ed.), Chapman-Hall, New York, NY, 1995, pp. 142–166.

62. S.Y. Reddy, P.S. Dimick, and G.R. Ziegler. Compatibility of milk fat fractions with cocoa butter determined by differential scanning calorimetry. *Inform* 5:522 (1994).

63. N.V. Lovegren, M.S. Gray, and R.O. Feuge. Effect of liquid fat on melting point and polymorphic behavior of cocoa butter and cocoa butter fraction. *J. Am. Oil Chem. Soc.* 53:108 (1976).

64. L. Campbell, D. Anderson, and P. Keeney. Hydrogenated milk fat as an inhibitor of the fat bloom defect in chocolate. *J. Dairy Sci.* 52:976 (1969).

65. J.W. DuRoss and W.H. Kightly. Relationship of sorbitan monostearate and polysorbate 60 to bloom resistance in properly tempered chocolate. *Manuf. Confect.* 45:51 (1965).

66. J. Schlichter-Aronhime, S. Sarig, and N. Garti. Mechanistic considerations of polymorphic transformations of tristearin in the presence of emulsifiers. *J. Am. Oil Chem. Soc.* 64:529 (1987).

67. N. Krog. Functions of emulsifiers in food systems. *J. Am. Oil Chem. Soc.* 54:124 (1977). Quoted in J. Aronhime, S. Sarig, and N. Garti, Emulsifiers as additives in fats: Effect on polymorphic transformations and crystal properties of fatty acids and triglycerides. *Food Struct.* 9:337 (1990).

68. A.N. Mostafa and J.M. deMan. Application of infrared spectroscopy in the study of polymorphism of hydrogenated canola oil. *J. Am. Oil Chem. Soc.* 62:1481 (1985). Quoted in J. Aronhime, S. Sarig, and N. Garti, Emulsifiers as additives in fats: Effect on polymorphic transformations and crystal properties of fatty acids and triglycerides. *Food Struct.* 9:337 (1990).

69. A.N. Mostafa, A.K. Smith, and J.M. deMan. Crystal structure of hydrogenated canola oil. *J. Am. Oil Chem. Soc.* 62:760 (1985): Quoted in J. Aronhime, S. Sarig, and N. Garti, Emulsifiers as additives in fats: Effect on polymorphic transformations and crystal properties of fatty acids and triglycerides. *Food Struct.* 9:337 (1990).

70. C. Giddey and E. Clerc. Polymorphism of cocoa butter and its importance in the chocolate industry. *Int. Choc. Rev.* 16:548 (1961).

71. I. Hachiya, T. Koyano, and K. Sato. Seeding effects on crystallization behavior of cocoa butter. *Agric. Biol. Chem.* 53:327 (1989).

72. I. Hachiya, T. Koyano, and K. Sato. Seeding effects on solidification behavior of cocoa butter and dark chocolate: I. Kinetics of solidification. *J. Am. Oil Chem. Soc.* 66:1757 (1989).

73. T. Koyano, I. Hachiya, and K. Sato. Fat polymorphism and crystal seeding effects on fat bloom stability of dark chocolate. *Food Struct.* 9:231 (1990).

74. P.S. Dimick and D.M. Manning. Thermal and compositional properties of cocoa butter during static crystallization. *J. Am. Oil Chem. Soc.* 64:1663 (1987).

75. T.R. Davis and P.S. Dimick. Lipid composition of high-melting seed crystals formed during cocoa butter solidification. *J. Am. Oil Chem. Soc.* 66:1494 (1989).

76. T.R. Davis and P.S. Dimick. Isolation and thermal characterization of high-melting seed crystals formed during cocoa butter solidification. *J. Am. Oil Chem. Soc.* 66:1488 (1989).

77. P.S. Dimick. Influence de la composition sur la cristallisation du beurre de cacao. In: *Proceedings of the International Symposium L'ALLIANCE 7–CEDUS*, The Crystallization of Food Products, Paris, Nov. 24 (1994).

78. P.S. Dimick. Compositional effects on crystallization of cocoa butter. In: *Physical Properties of Fats, Oils, and Emulsifiers* (N. Widlak, ed.), AOCS Press, Champaign, IL, 2000, pp. 140–163.

79. D.H. Arruda and P.S. Dimick. Phospholipid composition of lipid seed crystal isolates from Ivory Coast cocoa butter. *J. Am. Oil Chem. Soc.* 68:385 (1991).

80. P.J. Lawler and P.S. Dimick. Crystallization kinetics of cocoa butter as influenced by phosphatidylcholine and fractionated lecithin. *88th Annual Meeting of the American Oil Chemists' Society*, Champaign, IL, 1997, *Abstracts*.

81. P.J. Lawler and P.S. Dimick. Solidification of cocoa butter as influenced by phosphatidylcholine and fractionated lecithin. *88th Annual Meeting of the American Oil Chemists' Society*, Champaign, IL, 1997, *Abstracts*.

10 Chemical Interesterification of Food Lipids: Theory and Practice

Dérick Rousseau and Alejandro G. Marangoni

CONTENTS

I. INTRODUCTION

Interesterification, hydrogenation, and fractionation are three processes available to food manufacturers to tailor the physical and chemical properties of food lipids [1,2]. At present, roughly one-third of all edible fats and oils in the world are hydrogenated, whereas ~10% are either fractionated or interesterified [3]. Each operation is based on different principles to attain its goal. Fractionation is a physical separation process based on the crystallization behavior of triacylglycerols [4,5]. Hydrogenation, on the other hand, is a chemical process leading to the saturation of double bonds present in fatty acids to harden fats for use as margarine and shortening basestocks. Interesterification, also a chemical process, causes a fatty acid redistribution within and among triacylglycerol molecules, which can lead to substantial changes in lipid functionality. This chapter discusses the application of the theory of chemical interesterification to the production of edible fats and oils.

II. LIPID COMPOSITION

The chemical composition of a fat partly dictates its physical and functional properties [6]. The chemical nature of lipids is dependent on fatty acid structure and distribution on the glycerol backbone. Fatty acids vary in chain length and in the number, position, and configuration of double bonds [7]. Triacylglycerols composed of saturated fatty acids (e.g., myristic, palmitic, stearic) have high-melting points and are generally solid at ambient temperature, whereas triacylglycerols consisting of unsaturated (monoene, polyene) fatty acids (e.g., oleic, linoleic, linolenic) are usually liquid at room temperature. Butterfat, for example, contains ~70% saturated fatty acids, whereas many vegetable oils contain almost exclusively unsaturated fatty acids [4,8].

The fatty acid distribution within naturally occurring triacylglycerols is not random [9,10]. The taxonomic patterns of vegetable oils consist of triacylglycerols obeying the 1,3 random-2-random distribution, with saturated fatty acids being located almost exclusively at the 1,3-positions of triacylglycerols [8,11,12]. Conversely, fats from the animal kingdom (tallow, lard, etc.) are quite saturated at the *sn*-2 position [13].

The industrial applicability of a given fat is limited by its nonrandom distribution, which imparts a given set of physical and chemical properties. The objective of modification strategies, such as chemical interesterification, is the creation from natural fats of triacylglycerol species with new and desirable physical, chemical, and functional properties [14].

III. A BRIEF HISTORY

Interesterification reactions have been knowingly performed since the mid-1800s. The first published mention was by Pelouze and Gélis [15]. Duffy [16] performed an alcoholysis reaction between tristearin and ethanol. Later, Friedel and Crafts [17] generated an equilibrium interchange between ethyl benzoate and amyl acetate. Glyceride rearrangement was also reported by Grün [18], Van Loon [19,20], and Barsky [21]. The first publication demonstrating the chemical interesterification of edible lipids was presented by Normann [22]. Chemical interesterification has been industrially viable in the food industry since the 1940s to improve the spreadability and baking properties of lard [23,24]. In the 1970s, there was renewed interest in this process, particularly as a hydrogenation replacement for the manufacture of zero-*trans* margarines. Today it plays a key role in the production of low-calorie fat replacers, such as Proctor and Gamble's Olestra and Nabisco's Salatrim or Benefat [25,26].

IV. THE FOUR FACES OF INTERESTERIFICATION

Excellent reviews in the area of chemical interesterification include Sreenivasan [1], Rozenaal [11], Kaufmann et al. [27], Going [28], Hustedt [29], and Marangoni and Rousseau [30].

Interesterification can be divided into four classes of reactions: acidolysis, alcoholysis, glycerolysis, and transesterification [28,31,32].

Acidolysis involves the reaction of a fatty acid and a triacylglycerol. Reactions can produce an equilibrium mixture of reactants and products or can be driven to completion by physically removing one of the reaction products. For example, coconut oil and stearic acid can be reacted to partially replace the short chain fatty acids of coconut oil with higher melting stearic acid [28].

Alcoholysis involves the reaction of a triacylglycerol and an alcohol and has several commercial applications, primarily the production of monoacylglycerols and diacylglycerols. Alcoholysis must be avoided in the interesterification of food lipids; however, since monoacylglycerols and diacyl-glycerols are undesirable by-products [28]. Glycerolysis is an alcoholysis reaction in which glycerol acts as the alcohol [31].

Transesterification is the most widely used type of interesterification in the food industry. Hence, we concentrate on this reaction. Figure 10.1 shows the effects of interesterification on the fatty acid distribution of a putative triacylglycerol (1-stearoyl-2-oleoyl-3-linoleoyl glycerol) (SOL). In sequence, the ester bonds linking fatty acids to the glycerol backbone are split; then the newly liberated fatty acids are randomly shuffled within a fatty acid pool and reesterified onto a new position, either on the same glycerol (intraesterification) or onto another glycerol (interesterification) [1]. For reasons involving thermodynamic considerations, intraesterification occurs at a faster rate

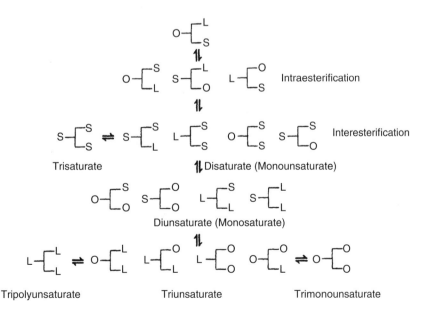

FIGURE 10.1 Triacylglycerol formation during interesterification: S, stearoyl; O, oleoyl; L, linoleoyl. (Adapted from Sreenivasan, B., *J. Am. Oil Chem. Soc.*, 55, 796, 1978.)

than interesterification [33]. Once the reaction has reached equilibrium, a complex, random mixture of triacylglycerol species is obtained (Figure 10.1).

The extent of the effects of interesterification on the properties of a fat will depend on the fatty acid and triacylglycerol variety of the starting material. If a single starting material (e.g., palm stearin) is randomized, the effects will not be as great as if a hardstock is randomized with a vegetable oil [34]. Furthermore, if a material has a quasi-random distribution prior to randomization (e.g., tallow), randomization will not lead to notable modifications.

The interesterification reactions consists of three main steps: catalyst activation, ester bond cleavage, and fatty acid interchange. We now examine each subject in detail.

V. INTERESTERIFICATION CATALYSTS

A. Is a Catalyst Necessary for Interesterification?

Interesterification can proceed without catalyst at high temperatures (~300°C); the desired results are not obtained, however, because equilibrium is slowly attained at such temperatures, and isomerization, polymerization, and decomposition reactions can occur [11,28,35]. In fact, polymerization has been shown to occur at 150°C [36]. Addition of catalyst significantly lowers reaction temperature and duration [37]. Other important considerations include the type and concentration of catalyst [38].

B. Available Catalysts for Interesterification

There are three groups of catalysts (acids, bases, and their corresponding salts and metals), which can be subdivided into high- and low-temperature groups [27]. High temperature catalysts include metals salts such chlorides, carbonates, oxides, nitrates, and acetates of zinc, lead, iron, tin, and cobalt [39]. Others include alkali metal hydroxides of sodium and lithium [40]. Most commonly used are low-temperature catalysts such as alkylates (methylate and ethylate) of sodium and sodium/potassium alloys; however, other bases, acids, and metals are also available [41]. Alkylates of sodium are simple to use and inexpensive, and only small quantities are required. Furthermore, they are active at low temperatures (<50°C). This last characteristic allows their use for directed interesterification [1,42].

C. Precautions

Performing a chemical interesterification reaction is a relatively straightforward process. However, a 100% reaction yield is never attainable [43]. Volatile fatty acid alkyl esters, formed in stoichiometric yields with the catalyst during the reaction, must be washed out, and a small amount of partial acylglycerols is always produced [44]. Trace amounts of moisture will inactivate alkylate catalysts by producing the corresponding alcohols. Hence, the fat or oil should contain less than 0.01% (w/w) water [29]. Free fatty acids and peroxides also impair catalyst performance, and levels should be maintained as low as possible, preferably below 0.1% and 1.0% (w/w), respectively [37]. The fat should be well refined, dried, and heated (120°C–150°C) under a nitrogen blanket before addition of catalyst [14]. Finally, sodium alkylates are toxic, highly reactive materials that should be handled with care. Their shelf life is a few months [11].

With a dry oil devoid of impurities, only trace amounts of catalyst [<0.4% (w/w)] are required [14]. Catalyst concentration should be minimized to prevent excessive losses due to saponification [42]. Experience has shown that above 0.4% catalyst, the addition of each additional 0.1% of catalyst results in the loss of ~1% neutral fat [29]. Konishi et al. [45], however, observed that ester interchange between soybean oil and methyl stearate in hexane was improved by using 10% (w/w) sodium methoxide in hexane. Proportions between 0.1% and 4.0% led to similar amounts of ester interchange.

It is also necessary to use the catalyst in a form that is easily and completely dispersed [46]. For example, if Na/K catalysts are not finely dispersed in a suitable solvent, a violent reaction with residual moisture may occur at the catalyst surface, followed by splitting of surrounding fat molecules to form a coating of soap. The heat generated by such a reaction is enough to decompose triacylglycerols and cause local charring [28].

D. The "Real" Catalyst

The real catalyst is believed to be a metal derivative of a diacylglycerol, and the aforementioned catalysts are most likely its precursor [11,42]. Upon catalyst addition to the lipid, a reddish brown color slowly develops (within a few minutes, depending on the application and reaction conditions) in the mixture, indicating the activation of the presumed true catalyst.

Some workers time the interesterification reaction from the appearance of the reddish brown color; others simply time the reaction from the moment of catalyst addition. Because it is impossible to predict reaction onset, it is difficult to obtain a partial interesterification. Most reactions are conducted until an equilibrium has been reached. Reaction times are longer in industrial settings, because the catalyst must be totally homogenized within the fat [43]. Preactivation is unnecessary if the catalyst is predissolved prior to addition to the substrate [35]. Placek and Holman [47] incorrectly attributed the induction period to the interaction between the catalyst and impurities. Although impurities are sometimes present, the induction period is not strictly due to their presence; rather, it is due to catalyst activation. As stated by Coenen [41] and many others, the activation energy for the catalyst is higher than for the reaction. A preactivation of 15 min has been found to accelerate the reaction itself [45]. Interestingly, Hustedt [29] stated that once the brown intermediate had appeared in the reaction mixture, interesterification was complete. No basis was given for this statement.

E. Reaction Termination

The interesterification reaction is allowed to continue for a predetermined time period and is stopped with addition of water and/or dilute acid. Going [28] described three patents dealing with catalyst removal techniques for minimizing fat loss. Generally, most of the catalyst can be washed out with water to a separate salt, or a soap-rich aqueous phase. Alternatively, reaction with phosphoric acid results in a solid phosphate salt, which can be filtered out. Both these methods result in substantial fat loss. A technique has been developed that minimizes loss by addition of CO_2 along with water. The system becomes buffered with sodium carbonate at a pH low enough to not split the fat [48].

VI. REACTION MECHANISMS

The exchange of fatty acids between triacylglycerol hydroxyl sites does not occur directly but via a series of alcoholysis reactions involving partial acylglycerols [49]. The proposed mechanisms of chemical interesterification depend on the inherent properties of the triacylglycerol ester carbonyl group (C=O). The carbonyl carbon is particularly susceptible to nucleophilic attack because of electronic and steric considerations. The electronegative oxygen pulls electrons away from the carbonyl carbon, leading to a partial positive charge on the carbon, and also increases the acidity of hydrogens attached to the carbon at a position α to the carbonyl group (Figure 10.2).

Steric considerations also come into play. The carbonyl carbon is joined to three other groups by σ bonds (sp^2 orbitals); hence they lie in a flat plane, 120° apart. The remaining p orbital from the carbon overlaps with a p orbital from the oxygen, forming a π bond. This flat plane and the absence of neighboring bulky groups permit easy access for nucleophiles to approach and react with the carbonyl carbon.

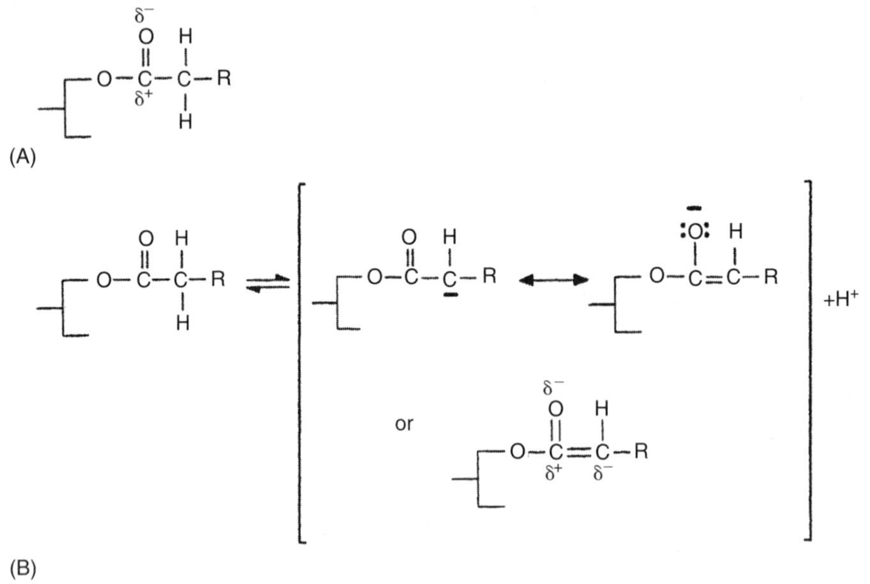

(A)

(B)

FIGURE 10.2 Carbonyl group properties in triacylglycerols. (A) Increased acidity of the α carbon to the carbonyl group due to resonance stabilization of the carbanion. (B) The carbonyl carbon is prone to nucleophilic attack because of the electronegativity of oxygen. (Adapted from Marangoni, A.G. and Rousseau, D., *Trends Food Sci. Technol.*, 6, 329, 1995.)

The transition state of the reaction is a relatively stable tetrahedral intermediate with a partial negative charge on the oxygen. As the reaction progresses, a group leaves and the structure reverts to the planar carbonyl structure. Strong evidence supports the cleavage of the carbonyl carbon–oxygen bond as the mechanism for the release of the leaving group.

For acid-catalyzed nucleophilic acyl substitution, a hydrogen easily associates with the carbonyl oxygen owing to the polarized nature of the carbonyl function and the presence of free electron pairs on the oxygen, imparting a positive charge to this atom [50]. The carbonyl carbon is then even more susceptible to nucleophilic attack, since oxygen can accept π electrons without gaining a negative charge. Acid-catalyzed interesterification is not discussed further because it is not used for the chemical interesterification of food lipids.

A. CARBONYL ADDITION MECHANISM

In alkaline conditions encountered during interesterification, the catalyst (which is nucleophilic) attacks the slightly positive carbonyl carbon at one of three fatty acid-glycerol ester bonds and forms a tetrahedral intermediate. The fatty acid methyl ester is then released, leaving behind a glycerylate anion (Figure 10.3A). Kinetics of base-catalyzed hydrolysis of esters shows that the reaction is dependent on both ester and base concentration (second-order kinetics). This newly formed glycerylate anion is the nucleophile for subsequent intra- and intermolecular carbonyl carbon attacks, which continue until a thermodynamic equilibrium has been reached (Figure 10.4).

During an attack, a new triacylglycerol is not necessarily formed. The transition complex (glycerylate + fatty acid) will decompose, either to regenerate the original species and active catalyst or to form a new triacylglycerol and a new active catalyst ion. This process continues until all available fatty acids have exchanged positions and an equilibrium composition of acylglycerol mixture has been achieved [1]. Support for this mechanism was provided by Coenen [41] who presented the kinetics between a simple mixture of S_3 and U_3. The kinetics were described with six

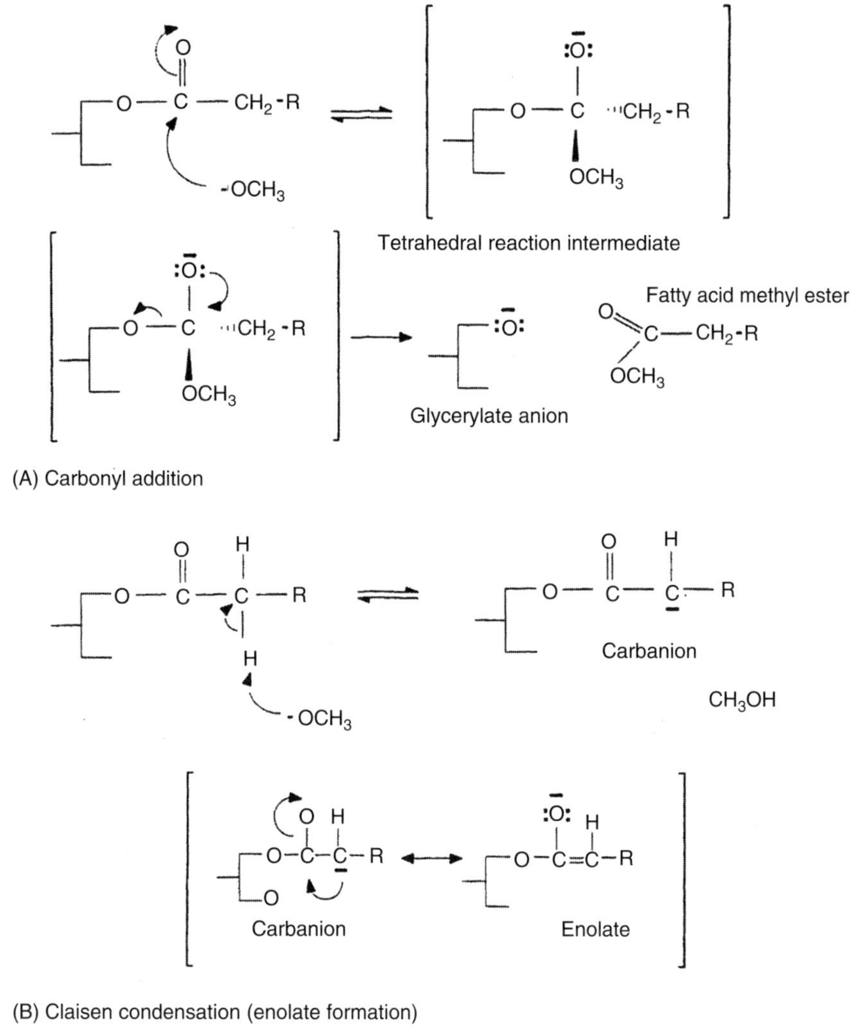

(A) Carbonyl addition

Tetrahedral reaction intermediate

Fatty acid methyl ester

Glycerylate anion

(B) Claisen condensation (enolate formation)

Carbanion

Enolate

Carbanion

CH_3OH

FIGURE 10.3 Proposed reaction mechanisms for chemical interesterification: carbonyl addition and Claisen condensation. (Adapted from Marangoni, A.G. and Rousseau, D., *Trends Food Sci. Technol.*, 6, 329, 1995.)

possible reactions between various triacylglycerol and diacylglycerol anions, with a rate constant $3k$ (Figure 10.5). Not all possible exchanges produced a net change in triacylglycerol composition, leading to $2k$ and k rate constants.

B. CLAISEN CONDENSATION

In Claisen condensations, the sodium methoxide removes an acidic hydrogen from the carbon α to the carbonyl carbon, yielding an enolate ester [23]. This reaction produces a carbanion, a powerful nucleophile (Figure 10.3B). This nucleophile will attack carbonyl groups, forming a β-keto ester intermediate and a glycerylate. The glycerylate is now free to attack other carbonyl carbons and exchange esters intra- and intermolecularly (Figure 10.6). Once this carbanion has been created, the same considerations as for the usual carbonyl carbon chemistry apply.

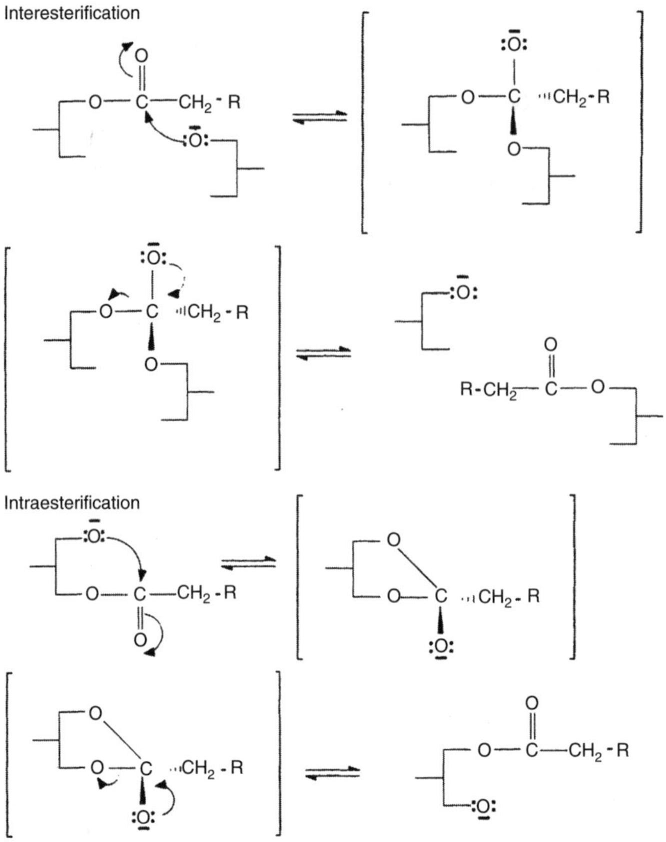

FIGURE 10.4 Reaction mechanisms for the chemical inter- and intraesterification of two triglycerides via the carbonyl addition mechanism. (Adapted from Marangoni, A.G. and Rousseau, D., *Trends Food Sci. Technol.*, 6, 329, 1995.)

$$S_3 + U_2ONa \underset{k}{\overset{3k}{\rightleftharpoons}} SU_2 + S_2ONa$$

$$U_3 + S_2ONa \underset{k}{\overset{3k}{\rightleftharpoons}} S_2U + U_2ONa$$

$$SU_2 + U_2ONa \underset{3k}{\overset{2k}{\rightleftharpoons}} U_3 + SUONa$$

$$S_2U + S_2ONa \underset{3k}{\overset{2k}{\rightleftharpoons}} S_3 + SUONa$$

$$S_2U + U_2ONa \underset{k}{\overset{2k}{\rightleftharpoons}} SU_2 + SUONa$$

$$SU_2 + S_2ONa \underset{k}{\overset{2k}{\rightleftharpoons}} S_2U + SUONa$$

FIGURE 10.5 Kinetics of interesterification via the carbonyl reaction mechanism (S, SS, SSS and U, UU, UUU: mono-, di-, and trisaturated and unsaturated, respectively). (Adapted from Coenen, J.W.E., *Rev. Fr. Corps Gras*, 21, 403, 1974.)

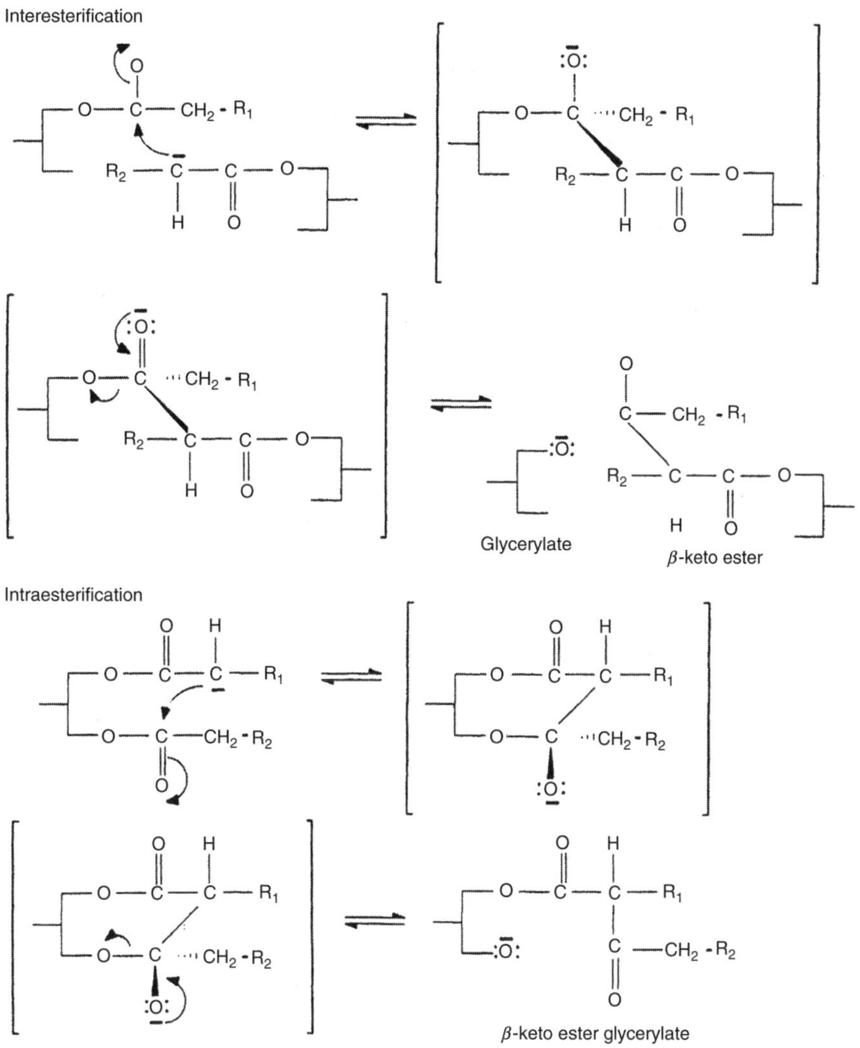

FIGURE 10.6 Reaction mechanism for the chemical inter- and intraesterification of two triacylglycerols via the Claisen condensation mechanism. (Adapted from Marangoni, A.G. and Rousseau, D., *Trends Food Sci. Technol.*, 6, 329, 1995.)

VII. RANDOM AND DIRECTED INTERESTERIFICATION

A. RANDOM INTERESTERIFICATION

Interesterification reactions performed at temperatures above the melting point of the highest melting component in a mixture result in complete randomization of fatty acids among all triacylglycerols according to the laws of probability [28,51].

The energy differences between the various combinations of triacylglycerols are insignificant and do not appear to lead to fatty acid selectivity [52]. Hence, random interesterification is entropically driven (randomization of fatty acids among all possible triacylglycerol positions) until an equilibrium is reached [41].

In ester–ester interchange, the fatty acid distribution is theoretically fully randomized, meaning that the resulting triacylglycerol structure can be predicted from the overall fatty composition of the mixture (Table 10.1) [11].

TABLE 10.1

Theoretical Triacylglycerol Compositions after Complete Interesterification of n Fatty Acids (A, B, C, D, ...) with Molar Fractions $a, b, c, d, ...$

Type	Quantity	Proportion
Monoacid (AAA, BBB, ...)	N	$a^3, b^3, ...$
Diacid (AAB, AAC, ...)	$n(n^- - 1)$	$3a^2b, 3a^2c, ...$
Triacid (ABC, DEF, ...)	$\dfrac{n(n-1)(n-2)}{6}$	$6abc, 6def, ...$
Total	$\dfrac{n(n+1)(n+2)}{6}$	

Source: Adapted from Rozenaal, A. *Inform*, 3, 1232, 1992.

In Table 10.1, a, b, and c are the molar concentrations of fatty acids A, B, and C. AAA, AAB, and ABC are triacylglycerols composed of one, two, or three different fatty acids, respectively. For AAB, there are three possible isomers, whereas for ABC there are six. For example, 1-stearoyl-2-oleoyl-3-linoleoyl glycerol results in the following fully randomized equilibrium mixture:

SSS	3.7%
OOO	3.7%
LLL	3.7%
SSO	11.1%
SSL	11.1%
SOO	11.1%
SLL	11.1%
OOL	11.1%
OLL	11.1%
SOL	22.2%

Gavriilidou and Boskou [53] found that a random distribution was obtained after chemical interesterification of olive oil–tristearin blends. They observed that trisaturate and triunsaturate proportions decreased markedly, whereas proportions of SSU and UUS increased (Table 10.2).

Not all workers agree that chemical interesterification is a purely random process. Kuksis et al. [54] found that the triacylglycerol composition of rearranged butter and coconut oils approached random distribution but deviated from true random distribution, even when experimental error was accounted for. This result was attributed to differences in the reactivity of the fatty acids and to possibly different esterification rates of the inner and outer hydroxyl sites on the glycerol backbone.

B. BATCH INTERESTERIFICATION

Random interesterification can be accomplished in either batch or continuous mode. A typical batch reactor (Figure 10.7) consists of a reaction vessel fitted with an agitator, heating/cooling coils, nitrogen sparger, and vacuum pump [3,14,28]. In a batch process, the raw lipid is heated to 120°C–150°C under vacuum in the reaction vessel to remove any trace of moisture [1,29]. As mentioned, moisture and peroxides deactivate the catalyst. Following the drying step, the mixture is cooled to 70°C–100°C. Catalyst is sucked into the reaction vessel and disperses to form a white slurry. The reaction is allowed to proceed for 30–60 min. When completion has been confirmed by analysis, the catalyst is neutralized in the reaction vessel. Processing losses can be minimized by

TABLE 10.2

Triacylglycerol Makeup for Olive Oil–Tristearin Blends before and after Interesterification

| | Olive Oil–Tristearin in Blend | | | |
| | 75%–25% (w/w) | | 80%–20% (w/w) | |
Species[a]	Initial	Randomized	Initial	Randomized
SSS	25.1	4.4	20.1	2.8
SSU	3.4	24.1	3.6	19.2
UUS	23.4	44.3	25	44.2
UUU	46.9	27.2	50	33.9

Source: From Gavriilidou, V. and Boskou, D. *Int. J. Food Sci. Technol.*, 26, 451, 1991.

[a] S, saturated; U, unsaturated.

using as little catalyst as possible and neutralizing with phosphoric acid or CO_2 prior to addition of water.

C. CONTINUOUS INTERESTERIFICATION

During continuous random interesterification, the fat is flash-dried and catalyst is continuously added. The fat then passes through elongated reactor coils with residence time determined by the coil length and the flow rate of the oil. The catalyst is then neutralized with water, separated from the oil by centrifugation, and dried [14].

Rozenaal [11] mentioned a continuous interesterification process in which a solution of sodium hydroxide and glycerol in water was used as precatalyst. Heated oil was mixed with the catalyst solution and subsequently spray-dried in a vacuum drier to obtain a fine dispersion and to remove the water. The reaction could be carried out in a coil reactor at 130°C. With this setup, the reaction took only a few minutes.

D. REGIOSELECTIVITY IN INTERESTERIFICATION

Elegant work by Konishi et al. [45] demonstrated that chemical interesterification can be regioselective. Sodium methoxide-catalyzed ester interchange between soybean oil and methyl stearate in

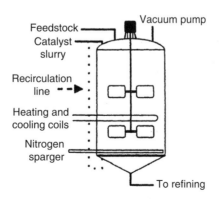

FIGURE 10.7 Batch random interesterification reaction vessel. (Adapted from Haumann, B.F. *Inform*, 5, 668, 1994; Laning, S.J., *J. Am. Oil Chem. Soc.* 62, 400, 1985; Going, L.H. *J. Am. Oil Chem. Soc.*, 44, 414A, 1967.)

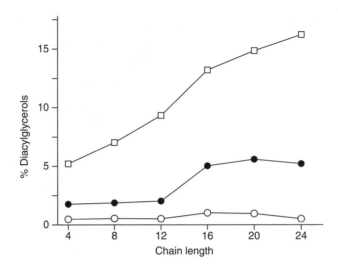

FIGURE 10.8 Composition of diacylglycerol content versus phase transfer catalyst chain length for sodium methoxide–catalyzed interesterification of trilaurin and methyl palmitate in the presence of seven quaternary ammonium salts. PPOH (○); LaPOH (●); LaLaOH (□). (Adapted from Cast, J., Hamilton, R.J., Podmore, J., and Porecha, L., *Chem. Ind.*, 763, 1991.)

hexane, at 30°C, revealed that fatty acid interchange at *sn*-1,3 positions progressed 1.7 times faster than at the *sn*-2 position after 24 h of reaction.

Cast et al. [55] demonstrated the regioselectivity of chemical interesterification in the presence of phase transfer catalysts (tetraalkyl ammonium bromides), with trilaurin (LaLaLa) and methyl palmitate in the presence of sodium methoxide. Under normal conditions, interesterification resulted in 17.7% LaLaP and 82.3% unreacted LaLaLa. Using tetrahexyl ammonium bromide, 11.9% LaLaP, 6.2% LaPP, and 81.2% LaLaLa were obtained. Seven quaternary ammonium salts were tested for their effect on the reaction: tetraethyl, tetrapropyl, tetrabutyl, tetrapentyl, tetrahexyl, and tetrahepryl ammonium bromides. The amount of LaLaOH diacylglycerol increased as the chain length increased up to tetrahexyl ammonium bromide (Figure 10.8). The amounts of LaPOH and PPOH diacylglycerol species increased up to tetraheptyl ammonium bromide. Reactions were performed for 48 h, with a fivefold increase in the amount of LaPOH diacylglycerols compared with the amount after 1 h of reaction.

Most importantly, following lipase hydrolysis and subsequent 2-monoacylgrycerol isolation, it was discovered that the reaction between trilaurin and methyl palmitate in the presence of tetrapentyl ammonium bromide contained 49% lauric acid and 51% palmitic acid, which represents an enrichment factor for the 2-position of 1.51 times. Hence, under these conditions, chemical interesterification was not a random process.

E. DIRECTED INTERESTERIFICATION

If the interesterification reaction is carried out at temperatures below the melting point of the highest melting component (most likely a trisaturated triacylglycerol species), the end result will be a mixture enriched in this component. This was first reported by Eckey [35] who, for the interester-ification of lard, discovered that certain catalysts were active below the melting point of the fat and that the reaction reached equilibrium within 30 min.

During directed interesterification, two reactions take place simultaneously. As the trisaturate is produced by interesterification, it crystallizes and falls from solution. Then, to regain equilibrium, the reaction equilibrium in the remaining liquid phase is pushed toward increased production of the crystallizing trisaturate [42,47,56]. Crystallization continues until all triacylglycerols capable of crystallizing have been eliminated from the reaction phase [43].

Early developments in the area of directed interesterification showed that the following factors determine the effectiveness of the reaction [47,56]:

Interesterification rate in the liquid phase
Rate of heat removal
Fat crystal nucleation rate
Trisaturate crystallization rate out of liquid phase

The rate of interesterification is an important factor, as the trisaturates will precipitate out of solution as quickly as they are formed.

Fat crystallization generates heat. Removal of this heat is hindered by the poor conductivity of fat and the low convection in viscous or plastic media. Heat removal directly affects the nucleation rate. Rapid cooling to temperatures much below the melting point of trisaturate increases the nucleation rate, hence crystallization. Trisaturate crystallization is also hindered by the viscosity of the lipid phase. Gently yet thorough agitation is helpful in speeding up crystallization.

For directed interesterification, Na/K alloy is the catalyst of choice, given its low-temperature activity compared with that of the metal alone or that of the alkylates [28,47]. Typically, the alloy is continuously metered in by a pump and well dispersed by means of a high shear agitator to provide the proper catalyst particle size, ensuring optimal activity. Initially, the fat is at least partially randomized at temperatures above the melting point of the highest melting triacylglycerol. When the fractional crystallization approach is used, the fat/catalyst slurry is chilled in conventional scraped-wall heat exchangers to specific temperatures in a series of steps designed to maintain the directed fractional crystallization process. Once chilled, the mixture is held under gentle agitation for a period of time so as to achieve the desired degree of crystal formation. Enhancements of the procedure include stepwise reduction of temperature and the use of temperature cycling [57,58].

Directed interesterification can be used to increase the solid fat content without affecting unsaturated fatty acids. Periodic drops in temperature accelerate the reaction. Kattenberg [58] applied this knowledge to interesterification of sunflower oil and lard blends and accelerated the reaction by a factor of 3. In another study, various oils, after directed interesterification, were chilled at 15°C for various durations (30–180 min), then subjected to further reaction at 23°C for 12–168 h. These treatments influenced solid fat content of the final product [59].

The effects of directed interesterification on cottonseed oil were reported by Eckey [35]. Cottonseed oil contains 25% saturated fatty acids. With random interesterification, only 1.5% trisaturates was obtained, whereas directed interesterification led to the production of 19% trisaturates.

A review by Huyghebaert et al. [60] showed that the directed interesterification of an SOL mixture resulted in the following proportions:

Solid	SSS	33.3%
Liquid	OOO	8.3%
	OOL	24.99%
	OLL	24.99%
	LLL	8.3%

The segregation of saturated fatty acids into trisaturated species is necessarily accompanied by a corresponding tendency for unsaturated fatty acids to form triunsaturated species [47].

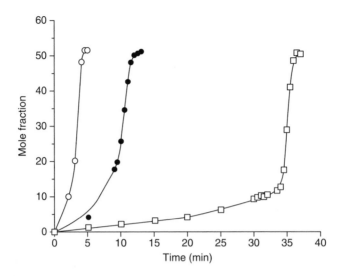

FIGURE 10.9 Theoretical interesterification kinetics of glycol esters of C8 and C10 as a function of time and temperature. 42°C (○); 37°C (●); 32°C (□). (Adapted from Coenen, J.W.E., *Rev. Fr. Corps Gras*, 21, 403, 1974.)

VIII.   KINETICS OF CHEMICAL INTERESTERIFICATION

Random interesterification is usually conducted until equilibrium has been reached. There are many conflicting reports in the literature concerning interesterification reaction rates. Coenen [41] stated that once a sufficient concentration of catalyst in solution had been reached in the reaction mixture, the actual interesterification reaction was extremely fast, requiring only a few minutes, unless operations had to proceed at very low temperatures. The kinetics were modeled in several ways to support this theory. The first example was a model system consisting of short chain fatty esters (C8, C10) of ethylene glycol (Figure 10.9). The induction period was long, yet the reaction itself was rapid, even at 32°C. In the second example, interesterification of palm oil was evaluated using solid fat content determinations (Figure 10.10). The reaction rate was faster at higher temperatures. These data confirm that an activation period is indeed required and agree with Weiss et al. [24] and Rozenaal [11], who reported that the catalyst formation phase was longer than the interesterification reaction, since the activation energy was higher for catalyst formation than for the interesterification reaction itself.

Lo and Handel [61] observed that interesterification of soybean oil and beef tallow was complete after 30 min (Figure 10.11). Reaction completion was determined by lipase hydrolysis analysis. Results by Konishi et al. [45] showed that in certain cases the interesterification reaction can progress for as long as 24 h, even with catalyst preactivation. Thus, depending on conditions, randomization can proceed for many hours.

Other factors that may influence interesterification onset include agitation intensity, catalyst particle size, and temperature. Studies by many, including Konishi et al. [45], Laning [14], and Wiedermann et al. [23], have shown that interesterification kinetics are temperature dependent (Figure 10.12).

IX.   ASSESSING THE EFFECTS OF INTERESTERIFICATION ON LIPID PROPERTIES

Fats and oils are usually modified to attain a certain functionality, such as improved spreadability, a specific melting point, or a particular solid fat content–temperature profile. However, changes in triacylglycerol structure may constitute the purpose of the reaction, as in the synthesis of a particular

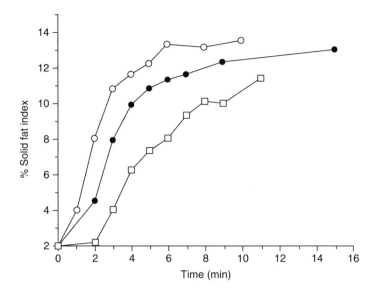

FIGURE 10.10 Practical example of interesterification kinetics with palm oil. The SFI (at 40°C) served as a measure of interesterification. Reaction temperatures 60°C (○), 52°C (●), and 45°C (□). (Adapted from Coenen, J.W.E., *Rev. Fr. Corps Gras*, 21, 403, 1974.)

structure. For that purpose, the fatty acid distribution constitutes the reaction goal. Methods described to assess physical properties include cloud point, Mettler dropping point, pulsed nuclear magnetic resonance, differential scanning calorimetry, cone penetrometry, x-ray diffraction, and polarized light microscopy. Chromatographic methods include thin-layer chromatography (TLC), high pressure liquid chromatography (HPLC), and gas–liquid chromatography (GLC). Other methods not discussed include mass spectroscopy [62] and stereospecific lipase hydrolysis [63,64].

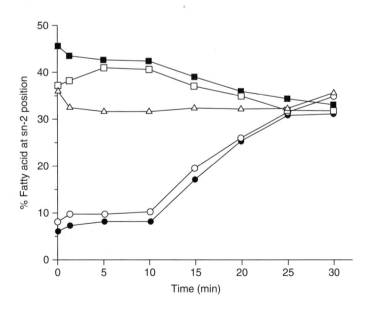

FIGURE 10.11 Changes in the fatty acid distribution at the *sn*-2 position during random interesterification of a 60:40 (% w/w) soybean oil–beef tallow mixture: ○, 16:0; ●, 18:0; □, 18:1; ■, 18:2, △, 18:3. (Adapted from Lo, Y.C. and Handel, A.D., *J. Am. Oil Chem. Soc.*, 60, 815, 1983.)

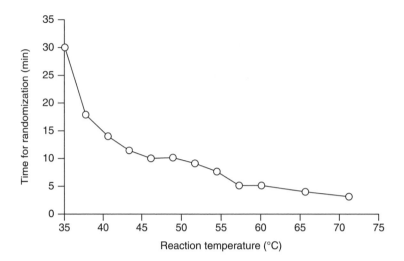

FIGURE 10.12 Influence of temperature on the interesterification reaction rate with glycerol/NaOH catalyst. (Adapted from Laning, S.J., *J. Am. Oil Chem. Soc.*, 62, 400, 1985.)

A. Physical Properties

Physical properties can be determined by examining thermal characteristics, rheological characteristics, or crystal habit.

1. Cloud Point

The cloud point, or temperature at which crystallization is induced, producing a crystal cloud, is one of the older indices used to study physical properties of fats. Eckey [35] monitored the change in cloud point during cottonseed oil interesterification. Generally, randomization increased the cloud point quickly at first and then more slowly until an increase of 13°C–15°C was reached, after which no change was observed, regardless of reaction duration. For Placek and Holman [47], a study of cloud point indicated the extent of interesterification of lard.

2. Dropping Point

The Mettler dropping point is a simple yet effective method of measuring the effect of interesterification on fats. In this procedure, liquefied fats are crystalized in sample cups and subsequently heated until they begin flowing under their own weight. Kaufmann and Grothues [65] performed a thorough study of the dropping points of hardstock and vegetable oil mixtures as an indicator of catalyst activity. Laning [14] demonstrated the effect of chemical interesterification on palm, palm kernel, and coconut oils (Table 10.3). A reduction in dropping point for saturated palm kernel oil (PKO) and saturated coconut oil was due to the lower average molecular weight of the triacylglycerol in the randomized fat. The reduction in dropping point reported for randomized PKO was due to an increase of triacylglycerol species with intermediate degrees of unsaturation. Cho et al. [66] used the dropping point as an indicator of the measure of reaction equilibrium. A blend of 70% hydrogenated canola oil, 10% palm stearin, and 20% canola oil had an initial dropping point of 37°C, which dropped to 35°C following 5 min of reaction and to 32°C after 20 min, remaining constant thereafter. List et al. [67] used dropping point as a verification of interesterification completion in the preparation of "*zero-trans*" soybean oil margarine base stock. Rousseau et al. [34] examined the effect of chemical interesterification and blending on butterfat–canola oil and found that a linear increase in the proportion of canola oil did not lead to a linear reduction in dropping point.

TABLE 10.3

Effect of Interesterification on Dropping Point in Palm, Palm Kernel, and Coconut Oil (°C)

Oil	Before Treatment	Random Treatment	Directed Treatment
Palm oil	39.4	42.7	51.1
Palm kernel (PKO)	28.3	26.9	30.0
Coconut oil (CO)	25.5	28.2	—
Saturated PKO	45.0	34.4	—
Saturated CO	37.8	31.6	—

Source: From Laning, S.J., *J. Am. Oil Chem. Soc.*, 62, 400, 1985.

3. Nuclear Magnetic Resonance

The amount of solid triacylglycerols in a lipid sample can be determined by means of NMR techniques. Laning [14] demonstrated the effect of chemical interesterification on palm, palm kernel, and coconut oils. Random interesterification resulted in modest solid fat content (SFC) changes, while directed interesterification produced more significant increases, attributable to the increase in trisaturated triacylglycerols. Generally speaking, interesterification results in more linear profiles, owing to the greater variety of triacylglycerol species [68].

Blending of butterfat with canola oil produced slight changes in the solid fat content of butterfat–canola oil blends, as exemplified by a contour profile (Figure 10.13) [34]. No changes greater than ±6% were evident. The biggest increase in solid fat content produced by interesterification was of

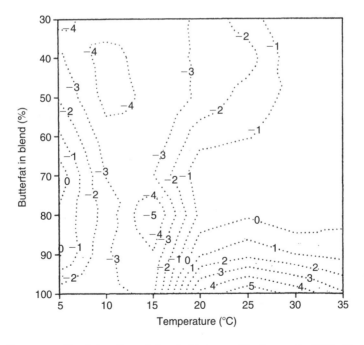

FIGURE 10.13 Contour profile of the effect of chemical interesterification on the SFC of butterfat–canola oil blends. Each line represents a 1% change in SFC. (Adapted from Rousseau, D., Forestière, K., Hill, A.R., and Marangoni, A.G. *J. Am. Oil Chem. Soc.*, 72, 973, 1996.)

butterfat at 25°C; SFC "valleys" were present for the 80% butterfat–20% canola oil blend at 15°C, while the largest decreases were present for the 40% butterfat–60% canola oil blend at 10°C.

4. Differential Scanning Calorimetry

DSC is used to measure the melting or crystallization profile and accompanying changes in enthalpy of fats. Rost [69] described the directed interesterification of palm oil. Calorimetry results indicated that the melting thermogram for noninteresterified palm oil consisted of two main peaks centered around 10°C and 19°C. Directed interesterification of palm oil led to a broader melting profile with no distinct peaks. Rossell [70] studied the effects of chemical interesterification on palm kernel oil crystallization. Randomization did not alter the shape of the crystallization curve; only peak temperatures were slightly lower. Because of the wide range of triacylglycerols that must be packed into fat crystals, interesterified fats generally show simpler melting curves with less polymorphism upon chemical interesterification [68].

Zeitoun et al. [71], who examined interesterified blends of hydrogenated soybean oil and various vegetable oils (1:1 w/w ratio), found that each oil influenced the melting and crystallization behavior of the interesterified blends differently as a result of initial variations in oil composition. Rousseau et al. [34] found that chemical interesterification of butterfat–canola oil blends also led to simpler, more continuous melting profiles. However, overall changes were minimal.

5. Cone Penetrometry

This is a rapid yet empirical method used in the evaluation of fat texture and rheology [72]. Jakubowski [43] found that interesterification doubled penetration depth of a blend of 35%–65% tallow–sunflower oil blends, at 15°C. Rousseau et al. [73] reported that interesterification substantially decreased the hardness index of blends of butterfat and canola oil. Other rheological measurements include viscoelasticity measurements [73].

6. X-Ray Diffraction

The polymorphic behavior of fats is important in many food systems (fat spreads, chocolate, etc.) [52]. Fat spread crystals exist as one of three primary forms: α, β', and β. The β modification is to be avoided in fat spreads because it results in a sandy texture [72]. The β' crystals are the most desirable form. Chemical interesterification alters the crystal morphology and structure of fats. Larsson [74] stated that a greater variety of fatty acids hinders β-crystal formation. Hence, upon interesterification of butterfat, which normally consists of a predominance of β' crystals and a slight proportion of β-crystals, the latter disappeared upon triacylglycerol randomization [68,75,76].

List et al. [77], while working with margarine oils, found that chemical interesterification and blending of vegetable oils and hydrogenated hardstock of soybean oil or cottonseed oil resulted in β'-crystal polymorphs.

Hernqvist et al. [52] interesterified mixtures of tristearin, triolein, and trielaidin. These mixtures were chosen to produce model systems for vegetable oil blends used in margarine. Polymorphic transitions of interesterified blends were studied, and depending on the blend, two to four polymorphs (sub-α, α, β', or β) were observed.

7. Polarized Light Microscopy

The morphology of the crystals comprising the three-dimensional fat crystal network is largely responsible for the appearance and texture of a fat and exerts a profound influence on its functional properties. Interesterification leads to noticeable modifications in crystal morphology, which can be examined in great detail with polarized light microscopy [78]. Prior to interesterification, lard consists of large crystals promoting graininess. Following interesterification, tiny delicate crystals, typical of the β' polymorph, are present [79]. Becker [80] performed an in-depth study on the influence of interesterification on crystal morphology of binary and ternary mixtures of trilaurin, triolein, and tristearin. He also found that fat crystals following interesterification were smaller than before randomization and had different morphologies. A study of butterfat–canola oil blends revealed that gradual addition of canola oil led to gradual spherulitic aggregation of the crystal structure [75].

B. CHEMICAL PROPERTIES

Changes in physical properties provide an arbitrary measure of interesterification structural modifications but give no real information on the compositional changes. Following these changes can be difficult unless simple substances are used [33]. Studies on molecular rearrangement of triacylglylcerol species provide a true indication of the chemistry of interesterification. The chemistry of interesterification can be followed with different chromatographic techniques: TLC, HPLC, and GLC.

Freeman [33], who examined the changes in monounsaturated triacylglycerols during the course of interesterification with TLC, found that intraesterification occurred at a faster rate than the general randomization that results from interesterification.

Chobanov and Chobanova [57] made extensive use of TLC to study the alteration in composition of 10 triacylglycerol groups during the monophasic interesterification of mixtures of sunflower oil with lard and tallow.

Parviainen et al. [81] studied the effects of randomization on milk fat triacylglycerol; they found an S_2U decrease in C36 and C38 species (45% and 52%, respectively) and an increase in trisaturated C44–C50 species. This combination led to a broader crystallization range and higher SFCs at temperatures above 25°C.

Herslöf et al. [82] used reversed phase HPLC and GC to analyze the interesterification reaction between fatty acid methyl esters and trilaurin and found that the theoretical and experimental compositions for the interesterified systems matched.

Rossell [70] measured the evolution in triacylglycerol species following chemical interesterification of palm kernel oil by means of GLC.

Huyghebaert et al. [60] and Rousseau et al. [34] used GLC to follow the evolution of butterfat triacylglycerol species as a result of interesterification. Typical results are shown in Figure 10.14.

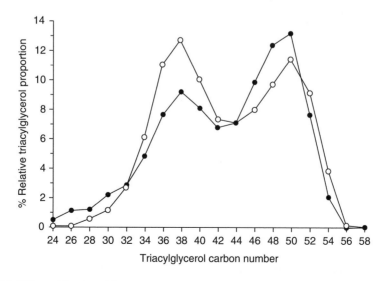

FIGURE 10.14 Effect of interesterification on butterfat triacylglycerols: ○, native butterfat; ●, randomized butterfat. (Adapted from Rousseau, D., Forestière, K., Hill, A.R., and Marangoni, A.G. *J. Am. Oil Chem. Soc.*, 72, 973, 1996.)

X. APPLYING INTERESTERIFICATION TO FOOD LIPIDS

Chemical interesterification is used industrially to produce fats and oils used in margarines, shortenings, and confectionery fats [60]. Because of legislation and for economic reasons, interesterification is a more common process in Europe than in North America. It is popular for many reasons. For example, little in the way of chemical properties is affected, and the fatty acid distribution is changed but the fatty acids' inherent properties are not. Moreover, unsaturation levels stay constant and there is no cis-trans isomerization [36,43]. Interesterification can improve the physical properties of fats and oils. Similar changes in physical properties may be obtained by means of blending, fractionation, or hydrogenation. Production costs, market prices, or raw material and nutritional concerns will determine the process to be used. Applications described include lard, margarines, palm oil and palm kernel oil, milk fat, and fat substitutes.

A. Shortening

Chemical interesterification has been successfully used for decades to improve the physical properties of lard. Ordinary lard has a grainy appearance, a poor creaming capacity, and a limited plastic range, which is not improved by plasticizing in a scraped-surface heat exchanger [43,47]. Addition of a hardstock plus plasticizing helps in these respects, but the product develops an undesirable graininess during storage [83]. Chemical interesterification halves the solid fat content of the lard at 20°C, improves the plastic range of the fat considerably, and prevents the development of graininess, which is due to the large proportion (64%) of palmitic acid at the sn-2 position [83,84]. This improvement in plasticity and stability is due to alterations in the polymorphic behavior, with interesterified lard crystallizing in a β'-2 form, characteristic of hydrogenated vegetable oil shortenings [83].

Chemically, the β' 3-tending disaturated OPS (1-oleoyl-2-palmitoyl-3-stearoyl-glycerol) (large crystals responsible for lard graininess) are exchanged for a mixture of disaturated triacylglycerols, with a lower melting point and greater intersolubility; the sn-2 palmitic acid concentration drops from 64% to 24% promoting β' behavior [83]. A detectable morphological change that accompanies these chemical changes is an increase in the relative proportion of small fat crystals [85].

Duterte [86] mentioned that the crystalline modifications were observed prior to the theoretical completion of randomization. Production of fine crystals extends lard's plastic range and gives it a smooth appearance [47]. Random interesterification helps to resolve the graininess problem, yet the limited plastic range problem is not fully resolved. The S_2U triacylglycerols in randomized lard give little plasticity at higher temperature. Directed interesterification resolves the plastic range problem [56].

B. Margarines

In the manufacture of margarine, the objective is to produce a fat mixture with a steep solid fat content curve to obtain a stiff product in the refrigerator that nevertheless spreads easily upon removal and melts quickly in the mouth. It should crystallize as a β' polymorph [87]. Depending on oil costs and availability, different treatments can be used.

As an alternative to hydrogenation for the production of margarine, Lo and Handel [61] chemically interesterified blends of 60% soybean oil with 40% beef tallow. Final results indicated properties similar to those of commercial tub margarine oil. Yet the interesterified blend contained less polyunsaturated fatty acids and more saturated fatty acids than commercial margarine oil.

According to Sonntag [31], short and medium chain fatty acids (C6–C14) have good melting properties whereas long chain fatty acids (C20–C22) can provide stiffening power in margarine. Acids of these two types can be combined with interesterification to produce triacylglycerols that provide blends with good spreadability, high temperature stability, and a pleasant taste.

Margarine oil with high proportions of lauric acid has a low melting point and narrow plastic range, which leads to a margarine that is hard in the fridge but partly melts at room temperature [87]. Decreasing the lauric acid concentration can rectify this problem of extremes. For example, coconut oil can be interesterified with an oil such as palm, and 60% of the interesterified mixture then blended with 40% oil, such as sunflower oil.

In the manufacture of zero-*trans* margarines, chemical interesterification of soybean oil–soy trisaturate using 0.2% (w/w) sodium methoxide at 75°C–80°C for 30 min resulted in a β′-crystallizing fat with good organoleptic properties [64].

List et al. [77] described the preparation of potential margarine and shortening bases by interesterification of vegetable oil and hardstocks (hydrogenated oil or stearin). They found that the interesterified fats possessed plasticity curves similar to those of commercial soft-tub margarine oils prepared by blending hydrogenated hardstocks or commercial all-purpose shortening oils. However, the commercial blends and interesterified blends differed with respect to crystallization behavior.

C. Palm Oil and Palm Kernel Oil

Palm oil has many applications in the food industry. Most often, interesterification of palm oil is combined with hydrogenation and/or fractionation to achieve the most desirable physical and functional properties [14]. Laning [14] described the applications of palm oil in cooking, frying, and salad oils. Corandomization of palm oil with other fats and oils, in combination with fractionation, produced a fluid salad oil.

Cocoa butter, used in the production of chocolate, is expensive and not always available, so substitutes are created, such as those that result from the blending of interesterified lauric acid with other fats. According to Sreenivasan [1], palm kernel oil is a hard butter that melts at 46°C and produces a waxy feel. With interesterification, the melting point is reduced to 35°C. Furthermore, by blending hydrogenated PKO and the randomized product, a whole series of hard butters with highly desirable melting properties is obtained. The effect of randomization on the melting properties of cocoa butter is shown in Figure 10.15.

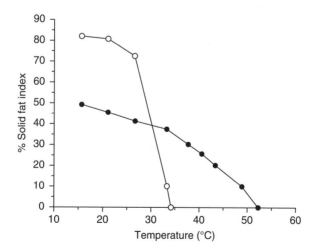

FIGURE 10.15 Solid content index of cocoa butter and randomized cocoa butter measured by dilatometry: ○, native cocoa butter; ●, randomized cocoa butter. (Adapted from Going, L.H., *J. Am. Oil Chem. Soc.*, 44, 414A, 1967.)

D. Milk Fat

Much research has been done on the chemical interesterification of milk fat. Milk fat, like most fats, does not have a random distribution, which conveys a predetermined set of physical properties. Butyric and caproic acids, for example, are predominantly located at *sn*-3, while palmitic acid is mostly at *sn*-1 and *sn*-2 [88]. Other fatty acids are not as specific. Interesterification of milk fat can be a powerful means of modifying its functional properties.

Weihe and Greenbank [89] presented the first paper dealing with the chemical interesterification of milk fat, with details appearing in Weihe [90]. These investigators performed randomization of milk fat at 40°C–45°C for 20 min to 6 h with 0.1%–0.3% Na/K alloy. For directed interesterification, xylene or hexane was added before the reaction, which was begun at 25°C–38°C and dropped in three to five steps to 10°C–25°C. Directed interesterification led to more substantial changes than random interesterification (e.g., on solid fat; Figure 10.16). Increases in melting point were greater in the presence of a solvent than without, and direct interesterification generated larger increases in melting point than random interesterification.

Interesterification increased the softening point of milk fat by 3.7°C–4.2°C, which was explained by higher proportion (5%–7%) of high-melting triacylglycerols, which translated into a higher hardness. Mickle [91], on the other hand, found that interesterification reduced the hardness of butter and also led to a rancid, metallic flavor. Refining (free fatty acid removal and steam injection under vacuum) removed the undesirable flavor, yet the final product was tasteless. Finally, an in-depth study by Mickle et al. [92] revealed the effects of three interesterification reaction parameters on the hardness of a semisolid resembling butter. All three parameters—duration (5–55 min), temperature (40°C–90°C), and catalyst concentration (0.5%–5%)—had statistically significant effects ($p < .05$), with catalyst concentration (at 1%–2%) having the greatest influence on hardness, which diminished 45%–55%. de Man [93] observed by means of polarized light microscopy that the crystal habit of interesterified milk fat was markedly changed from that of native milk fat. The effects of cooling procedures on consistency, crystal structure, and solid fat content of milk fat were also examined [94]. Parodi [95] examined the relationship between trisaturates and the

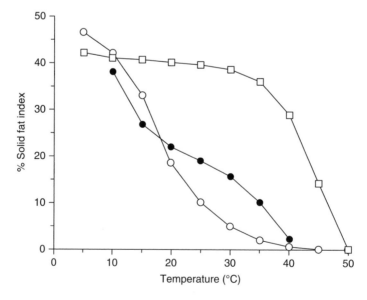

FIGURE 10.16 Proportion of soild fat of native butterfat (○), randomized butterfat (●), and butterfat subjected to direct interesterification (□) measured by dilatometry. (Adapted from Weihe, H.D., *J. Dairy Sci.*, 44, 944, 1961.)

softening point of milk fat. Interesterification increased the softening point from ~32.5°C to ~36.5°C. Timms [68] found that milk fat and beef tallow-interesterified blends lacked milk fat flavor. Timms and Parekh [96] explored the possibility of incorporating milk fat into chocolate. Interesterified milk fat appeared to be better suited to chocolate than noninteresterified milk fat, but the improvement gained did not compensate for the investment and loss of flavor from interesterification.

E. FAT SUBSTITUTES

Newer applications of chemical interesterification include the production of low-calorie fat substitutes such as Salatrim and Olestra. Salatrim/Benefat consists of chemically interesterified mixtures of short chain and long chain fatty acid triacylglycerols. The short chain fraction consists of triacetin, tripropionin, or tributyrin, while the long chain fractions consist of hydrogenated soybean oil [26].

Olestra is an acylated sucrose polyester with six to eight fatty acids obtained from vegetable oil (e.g., soybean, corn, sunflower). It is prepared by interesterifying sucrose and edible oil methyl esters in the presence of an alkali catalyst, at 100°C–140°C [97]. Olestra is nondigestible, hence noncaloric; it is also nontoxic, yet nutritional concerns potentially exist. Its functionality is dependent on the chain length and unsaturation of the esterified fats, as with normal lipids [98]. It can be exchanged for fats in products such as ice cream, margarine, cheese, and baked goods, and it can be blended with vegetable oil.

XI. OXIDATIVE STABILITY

The many advantages of chemical interesterification have been discussed in detail. Many authors have shown, however, that chemical interesterification can negatively influence the oxidative stability of fats and oils. Lau et al. [99] demonstrated that randomized corn oil oxidized three to four times faster than native corn oil. They concluded that the triacylglycerol structure probably was implicated, but the mechanisms remained unclear. Lo and Handel [61] showed that interesterified blends of soybean oil and beef tallow were more unstable following interesterification.

Gavriilidou and Boskou [100] examined the effects of chemical interesterification on the autoxidative stability of an 80% olive oil–20% tristearin blend. The randomized fats were less stable than the native mixtures (Figure 10.17). Addition of BHT stabilized the fats, resulting in a peroxide value similar to that for commercially processed hydrogenated vegetable oil used in margarine.

An important contribution to the literature was made by Zalewski and Gaddis [101], who investigated the effect of transesterification of lard on stability, antioxidant efficiency, and rancidity development. Interesterification of lard did not affect its resistance to oxidation, but changes in oxidative stability due to tocopherol decomposition and the formation of reducing substances were noted. In the absence of antioxidants, both interesterified and native lard had similar peroxide values. Furthermore, because of the position of unsaturated fatty acids at 1,3-positions or randomization toward the 2-position in pork fat triacylglycerols, there was no appreciable effect on initiation of oxidation and autoxidation rates.

Tautorus and McCurdy [102] demonstrated the effects of chemical and enzymatic randomization on the oxidative stability of vegetable oils stored at different temperatures. Noninteresterified and interesterified oils (canola, linseed, soybean, and sunflower) stored at 55°C demonstrated little difference to lipid oxidation, whereas noninteresterified samples were more stable at 28°C. Samples at 55°C underwent much greater oxidation than the samples at 28°C.

Park et al. [103] found that loss of tocopherols accelerated the autoxidation of randomized oils. α-Tocopherol was not detectable following interesterification, while γ-tocopherol and δ-tocoperol diminished 12% and 39%, respectively.

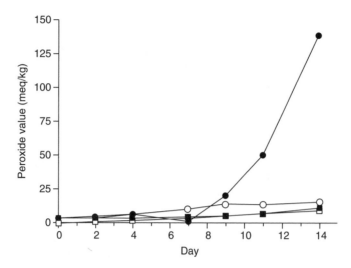

FIGURE 10.17 Change in peroxide value of an 80%:20% olive oil–tristearin blend before and after inter-esterification: ○, native blend; ●, randomized blend; □, hydrogenated blend; ■, randomized blend + BHT. (Adapted from Gavriilidou, V. and Boskou, D., *Food Flavours, Ingredients and Composition*, G. Charalambous, ed, Elsevier Science Publishers, Amsterdam, 1993, 313.)

Konishi et al. [104] found that regioselectively interesterified blends of methyl stearate and soybean oil had increased oxidative stability over both native and randomized blends, as monitored by peroxide value and volatiles analysis. The improved oxidative stability was presumably due to the regioselective incorporation of stearic acid at the *sn*-1(3) carbon sites of the triacylglycerol moiety, which stabilized the linoleic acid, predominantly located at the *sn*-2 position.

XII. NUTRITIONAL CONSEQUENCES OF INTERESTERIFICATION

Perhaps chemical interesterification's greatest advantage over hydrogenation lies in nutrition. At present, there are still unsettled nutritional concerns regarding *trans* fatty acids and their possible links to coronary heart disease [105–107]. *Trans* fatty acids are present in many edible fats and oils produced worldwide, yet these substances occur in great proportions in partially hydrogenated margarines. Barring nonhydrogenated margarines, literature data indicate that the typical *trans* fatty acid content of margarines is 10%–27% in the United States and 10%–50% in Canada [108,109].

It has been shown that randomization does not influence the nutritional value of unsaturated fatty acids [110]. However, not much is known about the potential importance of stereospecificity in the biological activity of dietary fatty acids [111]. In clinical trials, substitution of randomized butter for natural butter tended to reduce serum triacylglycerol and cholesterol concentrations [112]. Human infants absorbed 88% stearic acid when fed lard but only 40% when fed randomized lard. Hence, absorbability and pharmacological properties of fatty acids can be influenced by the molecular form in which they are absorbed [63].

De Schrijver et al. [111] examined lipid metabolism response in rats fed tallow, native, or randomized fish oil, and native or randomized peanut oil and found that randomized lard had no significant effects on any of the lipid measurements. Absorption of oleic acid and polyunsaturated fatty acids did not depend on the fatty acid profile of dietary fat. Kritchevsky [113] found that peanut oil's tendency to produce atherogenicity in rabbits disappeared following chemical inter-esterification.

It is known that human milk is well absorbed in part because of its proportion of long chain saturated fatty acids located at the *sn*-2 position. Lien et al. [114] found that mixtures of coconut oil and palm olein were better absorbed by rats if the proportion of long chain saturated fatty acids at the *sn*-2 position was increased by random chemical interesterification.

Mukherjee and Sengupta [115] found that interesterified soya–butterfat feeding significantly decreased serum cholesterol in humans and rats. The decrease was greater than when noninteresterified blends were fed. The lowering of serum cholesterol paralleled the decrease in concentration of trisaturates and the scattering of myristic acid away from *sn*-2 to *sn*-1 and *sn*-3 positions.

Finally, Koga et al. [116] examined the effects of randomization of partially hydrogenated corn oil on fatty acid and cholesterol absorption and on tissue lipid levels in rats. They found that interesterification did not lead to beneficial effects but rather enhanced the hypercholesterolemic tendency of *trans* fatty acids.

There appears to be some dispute as to the health effects of interesterification. The dietary concerns for avoiding *trans* fatty acids seem well documented, whereas the nutritional effects of fatty acid positional distribution are presently less clear-cut.

XIII. DISTINGUISHING CHEMICAL FROM ENZYMATIC INTERESTERIFICATION

Although great strides have been made with extracellular microbial lipases as catalysts for interesterification, most of the industry still relies on chemical interesterification. Each type of interesterification possesses advantages and disadvantages. Advantages of chemical interesterification over enzymatic transformations primarily involve cost recovery and initial investment. Chemical catalysts are much cheaper than lipases. Even with immobilization procedures, capital investment remains high. Second, chemical interesterification is a tried-and-true approach; it has been around for a long time, and industrial procedures and equipment are available [45].

Costs aside, does treatment by means of chemical or enzymatic interesterification in identical applications result in the same final product? Kalo et al. [117] compared the changes in triacylglycerol composition and physical properties of butterfat interesterified using either sodium methoxide or a nonspecific lipase from *Candida cylindracae* and found only small differences in both interesterified butterfats. The compositional changes induced by both chemical and enzymatic means were similar, with the trisaturate triacylglycerol content being slightly higher in the enzymatically modified product. In terms of physical properties, the chemically interesterified butterfat was slightly harder than its enzymatically modified counterpart, Hence, for randomization purposes, the methods appeared to yield similar results for the modification of butterfat. However, the product's butter flavor must be taken into account. The harsh process conditions of chemical interesterification result in loss of butter's fine flavor. For purposes where flavor is not a problem, the simpler, tried-and-true chemical process is preferable.

Enzymatic interesterification has many advantages, such as milder processing conditions and the possibility of regiospecificity and fatty acid specificity. This specificity permits structuring not possible by chemical means. For the production of nutritionally superior fats, enzymatic interesterification is ideally suited.

XIV. PERSPECTIVES

Chemical interesterification is likely to remain a force in the food industry for the foreseeable future. With the progressive demise of hydrogenation likely to continue, interesterification (both chemical and enzymatic) will gain greater prominence as a food lipid modification strategy.

ACKNOWLEDGMENTS

The authors acknowledge the financial assistance of the Ontario Ministry of Food and Rural Affairs (Ontario Food Processing Research Fund, and Food Program to the University of Guelph) and the Natural Sciences and Engineering Research Council (NSERC) of Canada.

REFERENCES

1. B. Sreenivasan. Interesterification of fats. *J. Am. Oil Chem. Soc.* 55:796 (1978).
2. F.V.K. Young. Interchangeability of fats and oils. *J. Am. Oil Chem. Soc.* 62:372 (1985).
3. B.F. Haumann. Tools: hydrogenation, interesterification. *Inform* 5:668 (1994).
4. E. Deffense. Milk fat fractionation today—A review. *J. Am. Oil Chem. Soc.* 70:1193 (1993).
5. J. Makhlouf, J. Arul, A. Boudreau, P. Verret, and M.R. Sahasrabudhe. Fractionnement de la matière grasse laitière par cristallisation simple et son utilisation dans la fabrication de beurres mous. *Can. Inst. Food Sci. Technol. J.* 20:236 (1987).
6. J.M. de Man. Physical properties of milk fat. *J. Dairy Sci.* 47:119 (1964).
7. W.W. Nawar. Chemistry. In *Bailey's Industrial Oil and Fat Products*, 5th ed. (Y.H. Hue, ed.), Wiley, Toronto, 1996, p. 397.
8. F.D. Gunstone. The distribution of fatty acids in natural glycerides of vegetable origin. *Chem. Ind.* 1214 (1962).
9. F.A. Norris and K.F. Mattil. A new approach to the glyceride structure of natural fats. *J. Am. Oil Chem. Soc.* 24:274 (1947).
10. T.P. Hilditch. *The Chemical Constitution of Foods*. Chapman and Hall, London, 1956, pp. 1–24.
11. A. Rozenaal. Interesterification of fats. *Inform* 3:1232 (1992).
12. P. Desnuelle and P. Savary. Sur la structure glycéridique de quelques corps gras naturels. *Fette Seifen Anstrichtm.* 61:871 (1959).
13. W.W. Nawar. Lipids. In *Food Chemistry*, 2nd ed. (O.E. Fennema, ed.), Dekker, New York, NY, 1985, p. 139.
14. S.J. Laning. Chemical interesterification of palm, palm kernel and coconut oil. *J. Am. Oil Chem. Soc.* 62:400 (1985).
15. J. Pelouze and A. Gélis. *Ann. Chim. Phys.* 10:434 (1844).
16. P.J.J. Duffy. *J. Chem. Soc.* 5:303 (1852).
17. C. Friedel and J.R. Crafts. *Annalen.* 133:207 (1865).
18. A. Grün. U.S. Patent 1,505,560 (1924).
19. C. Van Loon. U.S. Patent 1,744,596 (1929).
20. C. Van Loon. U.S. Patent 1,873,513 (1932).
21. G. Barsky. U.S. Patent 2,182,332 (1939).
22. Firma Oelwerke Germania GmbH and W. Normann. German Patent 417,215 (1920).
23. L.H. Wiedermann, T.J. Weiss, G.A. Jacobson, and K.F. Mattil. A comparison of sodium-methoxide treated lards. *J. Am. Oil Chem. Soc.* 38:389 (1961).
24. T.J. Weiss, G.A. Jacobson, and L.H. Wiedermann. Reaction mechanics of sodium methoxide treatment of lard. *J. Am. Oil Chem. Soc.* 38:396 (1961).
25. R.J. Jandacek and M.R. Webb. Physical properties of pure sucrose octaester. *Chem. Phys. Lipids* 22:163 (1978).
26. R.E. Smith, J.W. Finley, and G.A. Leveille. Overview of Salatrim, a family of low-calorie fats. *J. Agric. Food Chem.* 42:432 (1994).
27. H.P. Kaufmann, F. Grandel, and B. Grothues. Umesterungen auf dem Fettgebiet: 1. Theoretische Grundlagen und Schriften; die Hydrier-Umesterung. *Fette Seifen Anstrichtm.* 60:99 (1958).
28. L.H. Going. Interesterification products and processes. *J. Am. Oil Chem. Soc.* 44:414A (1967).
29. H.H. Hustedt. Interesterification of edible oils. *J. Am. Oil Chem. Soc.* 53:390 (1976).
30. A.G. Marangoni and D. Rousseau. Engineering triacylglycerols: The role of interesterification. *Trends Food Sci. Technol.* 6:329 (1995).
31. N.O.V. Sonntag. Glycerolysis of fats and methyl esters—Status, review and critique. *J. Am. Oil Chem. Soc.* 59:795A (1982).
32. R.O. Feuge. Derivatives of fats for use as foods. *J. Am. Oil Chem. Soc.* 39:521 (1962).
33. I.P. Freeman. Interesterification: I. Change of glyceride composition during the course of interesterification. *J. Am. Oil Chem. Soc.* 45:456 (1968).
34. D. Rousseau, K. Forestière, A.R. Hill, and A.G. Marangoni. Restructuring butterfat through blending and chemical interesterification: 1. Melting behavior and triacylglycerol modifications. *J. Am. Oil Chem. Soc.* 72:973 (1996).
35. E.W. Eckey. Directed interesterification of glycerides. *Ind. Eng. Chem. Res.* 40:1183 (1948).

36. H.P. Kaufmann and B. Grothues. Umesterungen auf dem Fettgebiet: III. Über den Einfluss verschiedener Umesterungs-Katalysoren auf ungesättigte Fettsäuren. *Fette Seifen Anstrichtm.* 61:425 (1959).

37. F. Joly and J.-P. Lang. Interestérification des corps gras. Mise au point de travaux sur les corps gras animaux. *Rev. Fr. Corps Gras.* 25:423 (1978).

38. K. Täufel, C. Franzke, and M. Achtzehn. Über die Umestrung unter Acyl-Austausch bei natürlichen Fetten und gehärteten Fetten sowie bei Fett-Mischungen. *Fette Seifen Anstrichtm.* 60:456 (1958).

39. K.W. Mattil and F.A. Norris. U.S. Patents 2,625,478 (1953).

40. W.E. Dominick and D. Nelson. U.S. Patent 2,625,485 (1953).

41. J.W.E. Coenen. Fractionnement et interestérification des corps gras dans la perspective du marché mondial des matières premières et des produits finis: II. Interestérification. *Rev. Fr. Corps Gras.* 21:403 (1974).

42. W.Q. Braun. Interesterification of edible fats. *J. Am. Oil Chem. Soc.* 37:598 (1960).

43. A. Jakubowski. L'interestérification entre corps gras animaux et huiles végétales. *Rev. Fr. Corps Gras.* 18:429 (1971).

44. M. Naudet. Modifications de structure des corps gras: B. Modifications affectant leur structure glycéridique. *Rev. Fr. Corps Gras.* 21:35 (1974).

45. H. Konishi, W.E. Neff, and T.L. Mounts. Chemical interesterification with regioselectivity for edible oils. *J. Am. Oil Chem. Soc.* 70:411 (1993).

46. E.W. Eckey. Esterification and interesterificaion. *J. Am. Oil Chem. Soc.* 33:575 (1956).

47. C. Placek and G.W. Holman. Directed interesterification of lard. *Ind. Eng. Chem.* 49:162 (1957).

48. G.W. Holman and V. Mills. U.S. Patent 2,886,578 (1959).

49. M. Naudet. Interestérification et estérification. Mécanismes réactionnels et conséquences sur la structure. *Rev. Fr. Corps Gras.* 23:387 (1976).

50. K.P.C. Vollhardt. *Organic Chemistry*, 4th edn. Freeman, New York. NY (2002).

51. L.P. Klemann, K. Aji, M.M. Chrysam, R.P. D'Amelia, J.M. Henderson, A.S. Huang, M.S. Otterburn, and R.G. Yarger. Random nature of triacylglycerols produced by the catalyzed interesterification of short- and long-chain fatty triglycerides. *J. Agric. Food Chem.* 42:442 (1994).

52. L. Hernqvist, B. Herslöf, and M. Herslöf. Polymorphism of interesterified triglycerides and triglyceride mixtures. *Fette Seifen Anstrichtm.* 86:393 (1984).

53. V. Gavriilidou and D. Boskou. Chemical interesterification of olive oil–tristearin blends for margarine. *Int. J. Food Sci. Technol.* 26:451 (1991).

54. A. Kuksis, M.J. McCarthy, and J.M.R. Beveridge. Triglyceride composition of native and rearranged butter and coconut oils. *J. Am. Oil Chem. Soc.* 41:201 (1963).

55. J. Cast, R.J. Hamilton, J. Podmore, and L. Porecha. Specificity in the inter-esterification process in the presence of phase transfer catalysts. *Chem. Ind.* 763 (1991).

56. H.K. Hawley and G.W. Holman. Directed interesterification as a new processing tool for lard. *J. Am. Oil Chem. Soc.* 33:29 (156).

57. D. Chobanov and R. Chobanova. Alterations in glyceride composition during interesterification of mixtures of sunflower oil with lard and tallow. *J. Am. Oil Chem. Soc.* 54:47 (1977).

58. H.R. Kattenberg. Beschleunigung der gelenkten Umestrung mittels periodischer variabler Temperaturführung. *Fette Seifen Anstrichtm.* 76:79 (1974).

59. M.M. Chakrabarty and K. Talapatra. Studies in interesterification: I. Production of edible fats from vegetable oils by directed interesterification. *Fette Seifen Anstrichtm.* 68:310 (1966).

60. A. Huyghebaert, D. Verhaeghe, and H. De Moor. In *Fats in Food Products* (D.P.J. Moran and K.K. Rajah, eds.), Blackie Academic and Professional, London, 1994, p. 319.

61. Y.C. Lo and A.D. Handel. Physical and chemical properties of randomly interesterified blends of soybean oil and tallow for use as margarine oils. *J. Am. Oil Chem. Soc.* 60:815 (1983).

62. R.A. Hites. Quantitative analysis of triglyceride mixtures by mass spectroscopy. *Anal. Chem.* 42:1736 (1970).

63. F.J. Filer, F.H. Mattson, and S.J. Fomon. Triglyceride configuration and fat absorption by the human infant. *J. Nutr.* 99:293 (1969).

64. G.R. List, E.A. Emken, W.F. Kwolek, T.D. Simpson, and H.J. Dutton. "*Zero trans*" margarines: Preparation, structure, and properties of interesterified soybean oil–soy trisaturate blends. *J. Am. Oil Chem. Soc.* 54:408 (1977).

65. H.P. Kaufmann and B. Grothues. Umesterungen auf dem Fettgebiet: IV. Tropfpunktsänderungen bei der Ein- und Mehrfett-Umesterung. *Fette Seifen Anstrichtm.* 62:489 (1960).

66. F. Cho, J.M. de Man, and O.B. Allen. Application of simplex-centroid design for the formulation of partially interesterified canola/palm blends. *J. Food Lipids* 1:25 (1993).

67. G.R. List, T. Pelloso, F. Orthoefer, M. Chrysam, and T.L. Mounts. Preparation and properties of *zero trans* soybean oil margarines. *J. Am. Oil Chem. Soc.* 72:383 (1995).

68. R.E. Timms. The physical properties of blends of milk fat with beef tallow and butter fat fraction. *Aust. J. Dairy Technol.* 34:60 (1979).

69. H.E. Rost. Untersuchungen über die gerichtete Umesterung von Palmöl. *Fette Seifen Anstrichtm.* 62:1078 (1960).

70. J.B. Rossell. Differential scanning calorimetry of palm kernel oil products. *J. Am. Oil Chem. Soc.* 52:505 (1975).

71. M.A.M. Zeitoun, W.E. Neff, G.R. List, and T.L. Mounts. Physical properties of interesterified fat blends. *J. Am. Oil Chem. Soc.* 70:467 (1993).

72. J.M. de Man. Consistency of fats: A review. *J. Am. Oil Chem. Soc.* 60:6 (1983).

73. D. Rousseau, K. Forestière, A.R. Hill, and A.G. Marangoni. Restructuring butterfat through blending and chemical interesterification: 3. Rheology. *J. Am. Oil Chem. Soc.* 72:983 (1996).

74. K. Larsson. *Lipids—Molecular Organization, Physical Functions and Technical Applications*, Oily Press, Dundee, 1994. p. 1.

75. D. Rousseau, K. Forestière, A.R. Hill, and A.G. Marangoni. Restructuring butterfat through blending and chemical interesterification: 2. Microstructure and polymorphism. *J. Am. Oil Chem. Soc.* 72:973 (1996).

76. I.L. Woodrow and J.M. de Man. Polymorphism in milk fat shown by x-ray diffraction and infrared spectroscopy. *J. Dairy Sci.* 51:996 (1961).

77. G.R. List, T.L. Mounts, F. Orthoefer, and W.E. Neff. Margarine and shortening oils by interesterification of liquid and trisaturated triglycerides. *J. Am. Oil Chem. Soc.* 72:379 (1995).

78. S.F. Herb, M.C. Audsley, and W. Riemenschneider. Some observations on the microscopy of lard and rearranged lard. *J. Am. Oil Chem. Soc.* 33:189 (1956).

79. C.W. Hoerr. Morphology of fats, oils and shortenings. *J. Am. Oil Chem. Soc.* 37:539 (1960).

80. E. Becker. Der Einfluss der Umesterung auf das Kristallizationsverhaltern von Felten bzw. Fettmischungen. *Fette Seifen Austrichtm.* 61:1040 (1959).

81. P. Parviainen, K. Vaara, S. Ali-Yrrkö, and M. Antila. Changes in the triglyceride composition of butterfat induced by lipase and sodium methoxide catalysed interesterification reactions. *Milchwissenschaft* 41:82 (1986).

82. B. Herslöf, M. Herslöf, and O. Podlaha. Simple small scale preparation of specific triglycerides and triglyceride mixtures. *Fette Seifen Anstrichtm.* 82:460 (1980).

83. E.S. Lutton, M.F. Mallery, and J. Burgers. Interesterification of lard. *J. Am. Oil Chem. Soc.* 39:233 (1962).

84. F.E. Luddy, S.G. Morris, P. Magidman, and R.W. Riemenschneider. Effect of catalytic treatment with sodium methylate on glycerin composition and properties of lard and tallow. *J. Am. Oil Chem. Soc.* 32:523 (1955).

85. G.W. Hoerr and D.F. Waugh. Some physical characteristics of rearranged lard. *J. Am. Oil Chem. Soc.* 32:37 (1950).

86. R. Duterte. Interestérification simple ou mixte. *Rev. Fr. Corps Gras.* 19:587 (1972).

87. F.D. Gunstone and F.A. Norris. In *Lipids in Foods, Chemistry, Biochemistry and Technology* (R. Maxwell, ed.), Pergamon Press, Oxford, 1983, p. 144.

88. A. Kuksis, L. Marai, and J.J. Myher. Triglyceride structure of milk fat. *J. Am. Oil Chem. Soc.* 50:193 (1973).

89. H.D. Weihe and G.R. Greenbank. Properties of interesterified butter oil. *J. Dairy Sci.* 41:703 (1958).

90. H.D. Weihe. Interesterified butter oil. *J. Dairy Sci.* 44:944 (1961).

91. J.B. Mickle. Flavor problems in rearranged milk fat. *J. Dairy Sci.* 43:436 (1960).

92. J.B. Mickle, R.L. Von Guten, and R.D. Morrison. Rearrangement of milk fat as a means for adjusting hardness of butterlike products. *J. Dairy Sci.* 46:1357 (1963).

93. J.M. de Man. Physical properties of milk fat: I. Influence of chemical modification. *J. Dairy Res.* 28:81 (1961).

94. J.M. de Man. Physical properties of milk fat: II. Some factors influencing crystallization. *J. Dairy Res.* 28:117 (1961).

95. P.W. Parodi. Relationship between trisaturated glyceride composition and the softening point. *J. Dairy Res.* 46:633 (1979).

96. R.E. Timms and J.V. Parekh. The possibilities for using hydrogenated, fractionated or interesterified milk fat in chocolate. *Aust. J. Dairy Technol.* 13:177 (1980).

97. F.D. Gunstone and J.L. Harwood. Synthesis. In *The Lipid Handbook*, 2nd ed. (F.D. Gunstone, J.L. Harwood, and F.B. Padley, eds.), Chapman and Hall, London, 1994, p. 359.

98. J. Stanton. Fat substitutes. In *Bailey's Industrial Oils and Fat Products, Vol. 1, Edible Oil and Fat Products: General Applications*, 5th ed. (Y.H. Hui, ed.), Wiley, Toronto, 1996, p. 281.

99. F.Y. Lau, G. Hammond, and P.F. Ross. Effects of randomization on the oxidation of corn oil. *J. Am. Oil Chem. Soc.* 59:407 (1982).

100. V. Gavriilidou and D. Boskou. Effect of chemical interesterification on the autoxidative stability of olive oil-tristearin blends. In *Food Flavours, Ingredients and Composition* (George Charalambous, ed.), Elsevier Science Publishers, Amsterdam, 1993, p. 313.

101. S. Zalewski and A.M. Gaddis. Effect of interesterification of lard on stability, antioxidant, synergist efficiency, and rancidity development. *J. Am. Oil Chem. Soc.* 44:576 (1967).

102. C.L. Tautorus and A.R. McCurdy. Effect of randomization on oxidative stability of vegetable oils at two different temperatures. *J. Am. Oil Chem. Soc.* 67:525 (1990).

103. D.K. Park, J. Terao, and S. Matsushita. Influence on interesterification on the autoxidative stability of vegetable oils. *Agric. Biol. Chem.* 47:121 (1983).

104. H. Konishi, W.E. Neff, and T.L. Mounts. Oxidative stability of soybean oil products obtained by regioselective chemical interesterification. *J. Am. Oil Chem. Soc.* 72:139 (1995).

105. P.L. Zock and M.B. Katan. Hydrogenation alternatives: Effects of *trans* fatty acids and stearic acid versus linoleic acid on serum lipid and lipoproteins in humans. *J. Lipid Res.* 33:399 (1992).

106. P.J. Nestel, M. Noakes, B. Belling, R. McArthur, P.M. Clifton, and M. Abbey. Plasma lipoprotein lipid and Lp[a] changes with substitution of elaidic acid for oleic acid in the diet. *J. Lipid Res.* 33:1029 (1992).

107. R.P. Mensink and M.B. Katan. Effect of dietary *trans* fatty acids on high-density and low-density lipoprotein cholesterol levels in healthy subjects. *New Engl. J. Med.* 323:43 (1990).

108. J.L. Weihrauch, C.A. Brignoli, J.B. Reeves III, and J.L. Iverson. Fatty acid composition of margarines, processed fats and oils. *Food Technol.* 31(2):80 (1977).

109. W.M.N. Ratnayake, R. Hollywood, and E. O'Grady. Fatty acids in Canadian margarines. *Can. Inst. Food Sci. Technol. J.* 24:81 (1991).

110. R.B. Alfin-Slater, L. Aftergood, H. Hansen, R.S. Morris, D. Melnick, and O. Gooding. Nutritional evaluation of interesterified fats. *J. Am. Oil Chem. Soc.* 43:110 (1966).

111. R. de Schrijver, D. Vermeulen, and E. Viane. Lipid metabolism responses in rates fed beef tallow, native or randomized fish oil and native or randomized peanut oil. *J. Nutr.* 121:94 (1991).

112. A. Christophe, G. Verdonk, A. Decatelle, A. Huyghebaert, L. Iliano, and A. Lauwers. Nutritional studies with randomized butter fat. In *Fats for the Future* (S.G. Brooker, A. Renwick, S.F. Hainan, and L. Eyres, eds.), Duromark Publishing, Auckland, New Zealand, 1983, p. 238.

113. D. Kritchevsky. Effects of triglyceride structure on lipid metabolism. *Nutr. Rev.* 46:17 (1981).

114. E.L. Lien, R.J. Yuhas, F.G. Boyle, and R.M. Tomarelli. Corandomization of fat improves absorption in rats. *J. Nutr.* 123:1859 (1993).

115. S. Mukherjee and S. Sengupta. Studies on lipid responses to interesterified soya oil–butterfat mixture in hypercholesterolemic rats and human subjects. *J. Am. Oil Chem. Soc.* 58:28 (1981).

116. T. Koga, T. Yamato, I. Ikeda, and M. Sugano. Effects of randomization of partially hydrogenated corn oil on fatty acid and cholesterol absorption, and tissue levels in rats. *Lipids* 30:935 (1995).

117. P. Kalo, P. Parviainen, K. Vaara, S. Ali-Yrkkö, and M. Antila. Changes in the triglyceride composition of butter fat induced by lipase and sodium methoxide catalysed interesterification reactions. *Milchwissenschaft* 41:82 (1986).

Part III

Oxidation and Antioxidants

11 Chemistry of Lipid Oxidation

Hyun Jung Kim and David B. Min

CONTENTS

I. INTRODUCTION

Lipid oxidation causes nutritional losses and produces undesirable flavor, color, and toxic compounds, which make foods less acceptable or unacceptable to consumers. Lipid oxidation can occur by either diradical triplet oxygen or nonradical singlet oxygen. Triplet oxygen oxidation has been extensively studied during the last 70 years to improve the oxidative stability of foods. However, triplet oxygen oxidation could not fully explain the initiation step of lipid oxidation [1]. Rawls and Van Santen [2] suggested that singlet oxygen is involved in the initiation of lipid oxidation because nonradical and electrophilic singlet oxygen can directly react with double bonds of food components without the formation of free radicals. Singlet oxygen oxidation of lipids is very significant because the rate of singlet oxygen oxidation is much greater than that of triplet oxygen oxidation. Singlet oxygen rapidly increases the oxidation rate of foods even at very low temperatures [2]. Singlet oxygen oxidation also produces new compounds, which are not found in ordinary triplet oxygen oxidation in foods [1,3].

This chapter reviews the chemistry of triplet and singlet oxygen, the important chemical mechanisms involved in lipid oxidation by triplet oxygen and singlet oxygen for the formation of volatile compounds, the effect of singlet oxygen on the development of reversion flavor in soybean oil, and quenching mechanisms of carotenoids and tocopherols in singlet oxygen oxidation. Singlet oxygen oxidation will be emphasized in that its importance to lipid oxidation has increased.

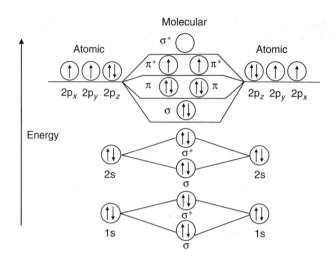

FIGURE 11.1 Molecular orbital of triplet oxygen. (From Min, D.B. and Boff, J.M., *Comp. Rev. Food Sci. Food Saf.*, 1, 58, 2002. With permission.)

II. CHEMISTRY OF TRIPLET AND SINGLET OXYGEN

Lipid oxidation occurs by the combination of triplet and singlet oxygen. The most abundant and stable oxygen is triplet oxygen which we breathe now. The differences in the chemical properties of triplet and singlet oxygen are best explained by their molecular orbitals (Figures 11.1 and 11.2). The spin multiplicity used to define spin states of molecules is defined as $2S + 1$, where S is the total spin quantum number. One spin is designated as $+1/2$. The total spin quantum number (S) of triplet oxygen is $1/2 + 1/2 = 1$. Triplet state oxygen has a spin multiplicity of 3, which can show three distinctive energy levels under magnetic field. Triplet oxygen is paramagnetic with diradical properties (Figure 11.1). Triplet oxygen reacts readily with other radical compounds in foods. However, most food compounds are nonradical and in the singlet state.

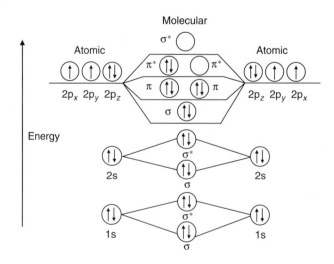

FIGURE 11.2 Molecular orbital of singlet oxygen. (From Min, D.B. and Boff, J.M., *Comp. Rev. Food Sci. Food Saf.*, 1, 58, 2002. With permission.)

TABLE 11.1

Comparison of Triplet and Singlet Oxygen

		Triplet	Singlet
Energy level		0	22.4 kcal/mol
Nature		Diradical	Nonradical, electrophilic
Reaction		Radical compound	Electron-rich compounds

Source: From Min, D.B. and Boff, J.M., *Comp. Rev. Food Sci. Food Saf.*, 1, 58, 2002.

The molecular orbital of singlet oxygen differs from that of triplet oxygen where electrons in the π-antibonding orbital are paired (Figure 11.2). The total spin quantum number (S) is $+1/2 - 1/2 = 0$ and the multiplicity of the state is 1 under magnetic field. Singlet oxygen violates Hund's rule, which states that one electron is placed into each orbital of equal energy one at a time before the addition of the second electron. Singlet oxygen is a highly energetic molecule. The resulting electronic repulsion can produce five excited state conformations. The $^1\Delta$ state of singlet oxygen among them is responsible for most singlet oxygen oxidation in foods and is generally referred to singlet oxygen. The most energetic π-antibonding electrons of the activated $^1\Delta$ state have opposite spins and lie in a single orbital (Figure 11.2). The energy of singlet oxygen is 22.4 kcal/mol above the ground state and exists long enough to react with other singlet state molecules [4,5]. Singlet oxygen is not a radical compound and reacts with nonradical, singlet state, and electron-rich compounds containing double bonds. The lifetime of singlet oxygen is from 50 to 700 µs, depending on the solvent system of foods. The temperature has little effect on the oxidation rate of singlet oxygen with foods due to the low activation energy of 0–6 kcal/mol [6]. Table 11.1 shows a summary of chemical properties of triplet and singlet oxygen.

The oxidation rate of lipid is dependent on temperature, the presence of inhibitors or catalysts, the nature of the reaction environment, and the nature of the compounds [7]. These factors are important in varying degrees to both triplet and singlet oxygen oxidation of lipids. Temperature has little effect on singlet oxygen oxidation but has a significant effect on triplet oxygen oxidation, which requires high activation energy. Polyunsaturated fatty acids are more susceptible to radical-initiated triplet oxygen oxidation than monounsaturated fatty acids, since the activation energy to initiate free-radical formation in polyunsaturated fatty acids is lower than that in monounsaturated fatty acids.

III. MECHANISMS OF TRIPLET OXYGEN OXIDATION

A. REACTION OF TRIPLET OXYGEN WITH FATTY ACIDS

Triplet oxygen is a diradical compound and can react with radical compounds. However, food compounds are not radical compounds and they should be in a radical state to react with diradical triplet oxygen for oxidation. The mechanism of triplet oxygen oxidation with linoleic acid is shown in Figure 11.3. The triplet oxygen oxidation has three steps: initiation, propagation, and termination. Initiation is a step for the formation of free alkyl radicals. Heat, light, metals, and reactive oxygen species facilitate the radical formation of food components. The initiation of radical formation in lipids occurs at the carbon that requires least energy for a hydrogen atom removal [7]. The carbon–hydrogen bond on C8 or C14 of linoleic acid is about 75 kcal/mol and the carbon–hydrogen bond on the saturated carbon without any double bond next to it is ~100 kcal/mol [8]. The energy required to break carbon–hydrogen bond on C11 of linoleic acid is about 50 kcal/mol [8]. The double bonds at C9 and C12 decrease the carbon–hydrogen bond at C11 by withdrawing electrons.

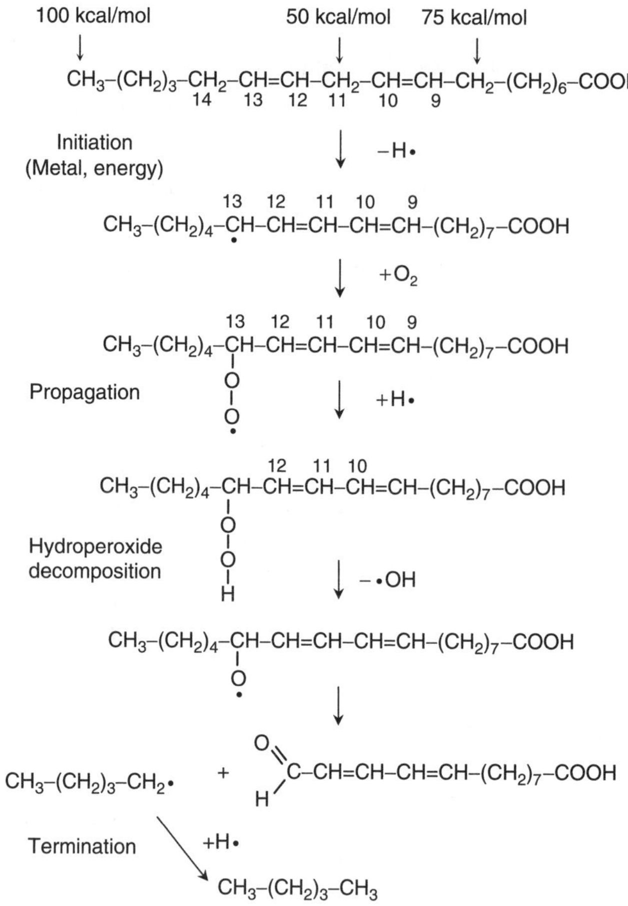

100 kcal/mol   50 kcal/mol   75 kcal/mol

↓   ↓   ↓

CH₃–(CH₂)₃–CH₂–CH=CH–CH₂–CH=CH–CH₂–(CH₂)₆–COOH
            14   13 12 11   10   9

Initiation   ↓   –H•
(Metal, energy)

              13   12   11 10   9
CH₃–(CH₂)₄–CH–CH=CH–CH=CH–(CH₂)₇–COOH
          •

                          ↓   +O₂

              13   12   11 10 9
CH₃–(CH₂)₄–CH–CH=CH–CH=CH–(CH₂)₇–COOH
          |
Propagation   O   ↓   +H•
          |
          O
          •

              12   11 10
CH₃–(CH₂)₄–CH–CH=CH–CH=CH–(CH₂)₇–COOH
          |
          O
Hydroperoxide |
decomposition O   ↓   –•OH
          |
          H

CH₃–(CH₂)₄–CH–CH=CH–CH=CH–(CH₂)₇–COOH
          |
          O
          •   ↓

                        O
                        ‖
CH₃–(CH₂)₃–CH₂•   +   C–CH=CH–CH=CH–(CH₂)₇–COOH
                       /
                      H

Termination   +H•

              CH₃–(CH₂)₃–CH₃

FIGURE 11.3 Mechanisms of triplet oxygen oxidation with linoleic acid.

The various strengths of carbon–hydrogen bond of fatty acids explain the differences of oxidation rates of oleic, linoleic, and linolenic acids (1:12:25) during triplet oxygen oxidation [8]. The weakest carbon–hydrogen bond of linoleic acid is the one at C11 and the hydrogen at C11 will be removed first to form radical at C11. The radical at C11 will be rearranged to form conjugated pentadienyl radical at C9 or C13 with *trans* double bond (Figure 11.4). Triplet oxygen can react with conjugated double bond radicals of linoleic acid and produce peroxyl radical at C9 or C13 (Figure 11.4).

The peroxyl radical with the standard one-electron reduction potential of 1000 mV easily abstracts hydrogen atom from other fatty acids and produces hydroperoxide and another alkyl radical [9]. The alkyl radical can also abstract hydrogen atom from other fatty acids. This chain reaction is called free-radical chain reaction and propagation step. The chain reactions of free alkyl radicals and peroxyl radicals accelerate the oxidation. The alkoxyl radicals react with other alkoxyl radicals or are decomposed to nonradical products. The formation of nonradical volatile and nonvolatile compounds at the end of oxidation is called the termination step (Figure 11.3).

B. Decomposition of Hydroperoxides

The primary oxidation products are lipid hydroperoxides, which are relatively stable at room temperature and in the absence of metals. At high temperature or in the presence of metals, hydroperoxides are readily decomposed to alkoxy radicals and then form aldehydes, ketones,

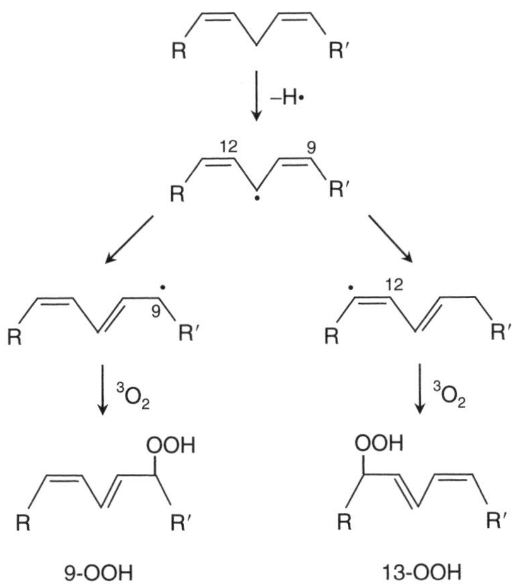

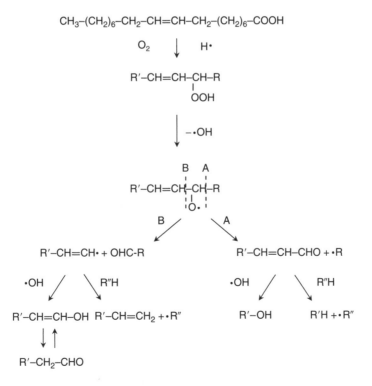

FIGURE 11.4 Conjugated hydroperoxide formation from linoleic acid by free-radical reaction of triplet oxygen. (From Choe, E. and Min, D.B., *Comp. Rev. Food Sci. Food Saf.*, 5, 169, 2006. With permission.)

acids esters, alcohols, and short-chain hydrocarbons. The decomposition of hydroperoxides to produce volatile compounds is shown in Figure 11.5. The most likely decomposition pathway of hydroperoxide is the cleavage between the oxygen and oxygen of R–O–O–H to produce R–O + O–H instead of R–O–O + H. Hiatt et al. [10] reported that the activation energy for cleaving between

$$CH_3-(CH_2)_6-CH_2-CH=CH-CH_2-(CH_2)_6-COOH$$

$$O_2 \quad \downarrow \quad H\bullet$$

$$\begin{array}{c} R'-CH=CH-CH-R \\ | \\ OOH \end{array}$$

$$\downarrow -\bullet OH$$

$$\begin{array}{c} B \quad A \\ R'-CH=CH-CH-R \\ O\bullet \end{array}$$

B ↙ ↘ A

R'–CH=CH• + OHC-R   R'–CH=CH–CHO + •R

•OH ↙ ↘ R"H   •OH ↙ ↘ R"H

R'–CH=CH–OH R'–CH=CH₂ + •R"   R'–OH   R'H + •R"

↓↑

R'–CH₂–CHO

FIGURE 11.5 Decomposition of hydroperoxides to produce volatile compounds.

TABLE 11.2
Secondary Oxidation Products of Fatty Acid Methyl Esters by Lipid Oxidation

	Oleic Acid	Linoleic Acid	Linolenic Acid
Aldehyde	Octanal	Pentanal	Propanal
	Nonanal	Hexanal	Butanal
	2-Decenal	2-Octenal	2-Butenal
	Decanal	2-Nonenal	2-Pentenal
		2,4-Decadienal	2-Hexenal
			3,6-Nonadienal
			Decatrienal
Carboxylic acid	Methyl heptanoate	Methyl heptanoate	Methyl heptanoate
	Methyl octanoate	Methyl octanoate	Methyl octanoate
	Methyl 8-oxooctanoate	Methyl 8-oxooctanoate	Methyl nonanoate
	Methyl 9-oxononanoate	Methyl 9-oxononanoate	Methyl 9-oxononanoate
	Methyl 10-oxodecanoate	Methyl 10-oxodecanoate	Methyl 10-oxodecanoate
	Methyl 10-oxo-8-decenoate		
	Methyl 11-oxo-9-undecenoate		
Alcohol	1-Heptanol	1-Pentanol	
		1-Octene-3-ol	
Hydrocarbon	Heptane	Pentane	Ethane
	Octane		Pentane

Source: From Choe, E. and Min, D.B., *Comp. Rev. Food Sci. Food Saf.*, 5, 169, 2006.

oxygen and oxygen of R–O–O–H was 44 kcal/mol, whereas the cleavage between the oxygen and hydrogen of R–O–O–H was 90 kcal/mol. Therefore, the hydroperoxide groups are cleaved by homolysis to yield an alkoxy and a hydroxy radical as shown in Figure 11.5.

The alkoxy radical formed from hydroperoxide is cleaved by the homolytic β-scission of a carbon–carbon bond to produce oxo compound and an alkyl or alkenyl radical. The homolytic β-scission is an important free-radical reaction that produces volatile compounds in edible oils during oxidation. The unsaturated alkoxy radical can be cleaved by β-scission in two mechanisms of cleavage A and cleavage B (Figure 11.5). Scission of the carbon–carbon bond on the side of the carbon atom with the oxygen will result in the formation of unsaturated oxo compounds and alkyl radical, whereas scission of the carbon–carbon bond between the double bond and the carbon atom with the oxygen will produce a 1-olefin radical and an alkyl oxo compounds. The alkyl radical can combine with hydroxy radical to produce an alcohol and the 1-olefin radical can cleaved to a 1-enol. The 1-enol will produce the corresponding oxo compound by tautomerization. An alternative reaction of the radicals eliminated by β-scission of the alkoxy radical is hydrogen abstraction from a compound RH. The secondary lipid oxidation products are mostly low-molecular-weight aldehydes, ketones, alcohols, and short-chain hydrocarbons as shown in Table 11.2.

IV.   MECHANISMS OF SINGLET OXYGEN OXIDATION

A.   Formation of Singlet Oxygen

Singlet oxygen can be formed chemically, enzymatically, and photochemically as shown in Figure 11.6 [9]. The important mechanism for the formation of singlet oxygen is by photosensitization. Photosensitizers such as chlorophyll, pheophytins, porphyrins, riboflavin, myoglobin, and synthetic colorants in foods can absorb energy from light and transfer it to triplet oxygen to form singlet oxygen [11–13]. The chemical mechanism for the formation of singlet oxygen in the

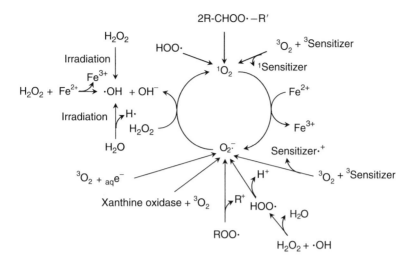

FIGURE 11.6 Singlet oxygen formation by chemical, photochemical, and biological methods. (From Choe, E. and Min, D.B., *J. Food Sci.*, 70, R142, 2005. With permission.)

presence of sensitizer, light, and triplet oxygen in foods is shown in Figure 11.7. The photosensitizer absorbs light energy rapidly and becomes an unstable, excited, and singlet state molecule (^{1}Sen*). The excited singlet sensitizer loses its energy by internal conversion, light emission, or intersystem crossing (Figure 11.7). Internal conversion involves the transformation from high-energy to low-energy state by releasing energy as heat. Emission of fluorescence converts the excited singlet sensitizer (^{1}Sen*) to ground state singlet sensitizer (^{1}Sen). The excited singlet sensitizer may also undergo an intersystem crossing to become an excited triplet state molecule (^{3}Sen*). The emission of phosphorescence converts ^{3}Sen* to ^{1}Sen. The lifetime of the ^{3}Sen* is greater than ^{1}Sen*. The ^{3}Sen* reacts with $^3O_2$ to form $^1O_2$ and ^{1}Sen by triplet–triplet annihilation mechanism. The sensitizer returns to ground state (^{1}Sen) and may begin the cycle again to generate singlet oxygen. Sensitizers may generate 10^3–10^5 molecules of singlet oxygen before becoming inactive [14].

The excited triplet sensitizer (^{3}Sen*) reacts with triplet oxygen and can produce superoxide anion by electron transfer [9]. Superoxide anion produces hydrogen peroxide by spontaneous dismutation. Hydrogen peroxide reacts with superoxide anion to form singlet oxygen by Haber–Weiss reaction [15]. Haber–Weiss reaction occurs in the presence of transition metals such as iron or copper.

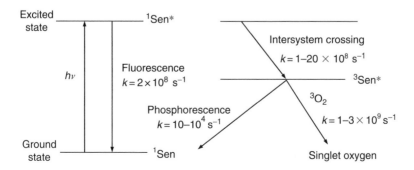

FIGURE 11.7 Chemical mechanism for the formation of singlet oxygen in the presence of sensitizer, light, and triplet oxygen. (From Min, D.B. and Lee, H.O., *Food Lipids and Health*, R.E. McDonald and D.B. Min, eds., Marcel Dekker, New York, 1996. With permission.)

$$O_2^- + O_2^- + 2H^+ \xrightarrow{\text{Dismutation}} H_2O_2 + O_2,$$

$$H_2O_2 + O_2^- \xrightarrow{\text{Haber-Weiss}} \cdot OH + OH^- + {}^1O_2.$$

Singlet oxygen is also produced by the Russell mechanism from peroxy radicals [15].

$$
\begin{array}{ccccccc}
\text{R-CH-R'} & & \text{R-CH-R'} & \xrightarrow{\text{Russell}} & \text{R-CH-R'} & & \text{R-C-R'} \\
| & & | & & | & & || \\
\text{O} & + & \text{O} & & \text{O} & + & \text{O} & + {}^1O_2 \\
| & & | & & | & & \\
\text{O} & & \text{O} & & \text{H} & & \\
\cdot & & \cdot & & & &
\end{array}
$$

B. Type I and Type II Pathways

Once ^{3}Sen* is formed, there are two major pathways: Type I and Type II. The excited triplet sensitizer (^{3}Sen*) may react directly with a compound (RH) such as linoleic acid or phenol compounds by donating and accepting hydrogen or electron and producing free radicals or free-radical ions as shown in Figure 11.8. This mechanism is known as Type I pathway [16,17]. The ^{3}Sen* acts as a photochemically activated free-radical initiator to form R. The Rzcan abstract hydrogen from other compounds to initiate the free-radical chain reaction or react with triplet oxygen to form peroxy radical. The ^{3}Sen* in Type I pathway can also react with ${}^3O_2$ to form superoxide anion by electron transfer to triplet oxygen. Less than 1% of the reaction of triplet sensitizer and triplet oxygen produces superoxide anion [18]. The rate of the Type I pathway is mostly dependent on the type and concentration of sensitizers and substrate compound. Readily oxidizable compounds such as phenols or reducible compounds such as quinines favor Type I pathway [4].

The excited triplet sensitizer (^{3}Sen*) may react with triplet oxygen to form singlet oxygen and singlet sensitizer by triplet sensitizer–triplet oxygen annihilation in Type II pathway (Figure 11.8). Energy is transferred from the high-energy excited triplet sensitizer to low-energy triplet oxygen (${}^3O_2$) to form high-energy singlet oxygen (${}^1O_2$) and low-energy ground state singlet sensitizer [17]. More than 99% of the reaction between triplet sensitizer and triplet oxygen produces singlet oxygen [18]. The rate of Type II pathway is mostly dependent on the solubility and concentration of oxygen in the food system. As the oxygen in a system becomes depleted, the shift from Type II to Type I mechanism is favored [19,20]. Oxygen is more soluble in nonpolar lipids than in water [21]. The chlorophyll-induced singlet oxygen formation in soybean oil favors the Type II pathway. In contrast, water-based food systems such as milk favor Type I pathway due to the reduced

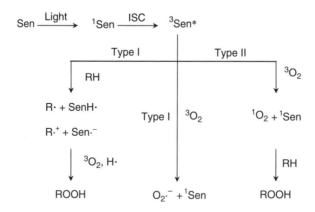

FIGURE 11.8 Formation of excited triplet sensitizer (^{3}Sen*) and its reaction with substrate via Type I and Type II reaction. (From Sharman, W.M., Allen, C.M., and van Lier, J.E., *Methods in Enzymology*, L. Packer and H. Sies, eds., Academic Press, New York, 2000. With permission.)

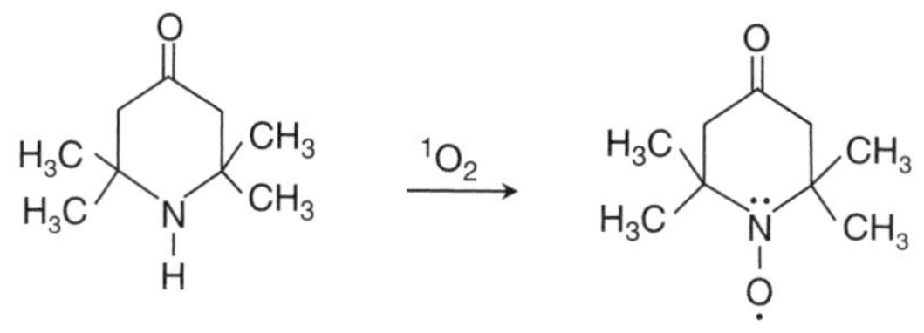

FIGURE 11.9 Reaction of 2,2,6,6-tetramethyl-4-piperidone with singlet oxygen to form 2,2,6,6-tetramethyl-4-piperidone-*N*-oxyl. (From Bradley, D.G., Lee, H.O., and Min, D.B., *J. Food Sci.*, 68, 491, 2003. With permission.)

availability of oxygen. The shift from Type I to Type II or vice versa is dependent on the concentration of oxygen and the types and concentration of compounds.

Type I and Type II reactions will enhance oxidation by either the formation of reactive radical compound species or the production of singlet oxygen. The competition between compound and triplet oxygen for the excited triplet sensitizer determines whether the reaction pathway is Type I or Type II. Photosensitized oxidation may change the types of pathway during the course of the reaction as the concentration of compounds and oxygen changes. In aqueous lipid biphasic systems, the longer half-life of singlet oxygen in the lipid phase favors Type II pathway.

C. DETECTION OF SINGLET OXYGEN BY ELECTRON SPIN RESONANCE SPECTROSCOPY

The singlet oxygen detection during photosensitized oxidation of foods is difficult due to the short lifetime of singlet oxygen. However, analytical techniques have been developed for the detection of singlet oxygen and the measurement of activity. One of the most common detection methods is electron spin resonance spectroscopy (ESR), which is highly sensitive for the detection of free radicals. A spin-trapping agent such as 2,2,6,6-tetramethyl-4-piperidone (TMPD) can react with singlet oxygen to form a stable nitroxide radical adduct, 2,2,6,6-tetramethyl-4-piperidone-*N*-oxyl (TAN), which is measured by ESR [17]. The reaction of TMPD with singlet oxygen to form TAN is shown in Figure 11.9. Although other reactive oxygen species such as superoxide and hydroxyl radicals can react with TMPD, they do not convert TMPD to TAN. This method is highly specific to singlet oxygen [22].

ESR detected the formation of singlet oxygen in meat [23] and milk [24] using a spin-trapping technique. Addition of TMPD to milk during illumination produced TAN after 5 min, confirming the formation of singlet oxygen in milk under light [24]. The limitation of this method is that concentration of TAN should remain over 10^{-8} M for detection and over 10^{-6} M for good spectral resolution. Since the lifetime of singlet oxygen is <1 μs, steady-state concentrations $>10^{-7}$ M are rarely maintained. Coupling spin trapping with ESR spectroscopy could improve this technique to measure singlet oxygen concentration. Effects of 0, 5, and 15 min illumination on ESR spectrum of TAN in a water solution of riboflavin and TMPD is shown in Figure 11.10 [25].

D. REACTION OF SINGLET OXYGEN WITH FATTY ACIDS

Singlet oxygen, an electrophilic molecule can directly react with the electron-rich double bonds of unsaturated molecule without the formation of free-radical intermediates [26]. Singlet oxygen oxidation reaction is very rapid in foods due to the low activation energy required for the chemical reaction. The reaction rates of singlet oxygen with oleic, linoleic, linolenic, and arachidonic acids are 0.74, 1.3, 1.9, and 2.4×10^5 M^{-1} s^{-1}, respectively, which is relatively proportional to the number of double bonds in the molecules [8]. The type of polyunsaturated fatty acids is not important in singlet oxygen oxidation. The total number of double bonds is more important in

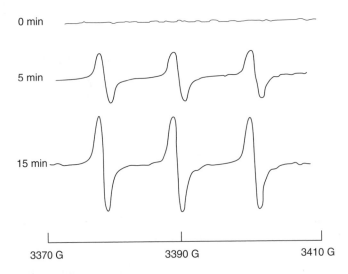

FIGURE 11.10 Effects of 0, 5, and 15 min illumination on ESR spectrum of 2,2,6,6-tetramethyl-4-piperidone-*N*-oxyl in water solution of riboflavin and 2,2,6,6-tetramethyl-4-piperidone. (From Min, D.B. and Boff, J.M., *Comp. Rev. Food Sci. Food Saf.*, 1, 58, 2002. With permission.)

singlet oxygen oxidation than the types of double bonds such as nonconjugated or conjugated dienes or trienes, which are important in free-radical triplet oxygen oxidation [27].

Singlet oxygen participates in reactions such as 1,4-cycloaddition to diene and heterocyclic compounds, "ene" reaction, and 1,2-cycloaddition to olefins, all of which involve direct reaction with double bonds as shown in Figures 11.11 and 11.12. Singlet oxygen reaction with linoleic or linolenic acid forms both conjugated and nonconjugated diene hydroperoxides (Figure 11.11). The linoleic and linolenic acid reactions with triplet oxygen produces only conjugated diene hydroperoxides. Direct reaction of singlet oxygen with double bonds forms hydroperoxides at positions 10 and 12 in linoleic acid and 10 and 15 in linolenic acid, which do not form in triplet

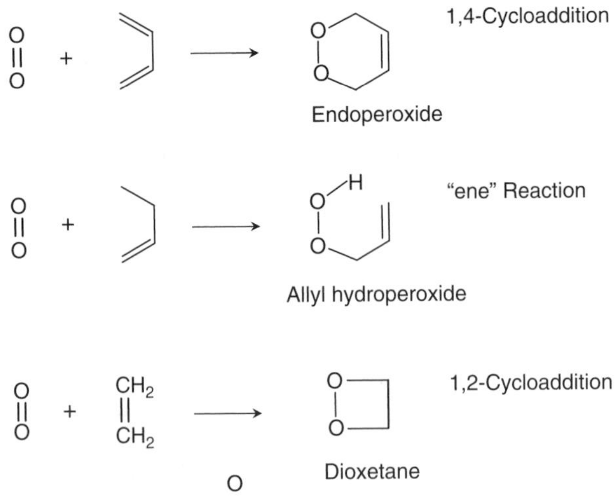

FIGURE 11.11 Reactions of singlet oxygen with olefins by 1,4-cycloaddition, "ene" reaction, and 1,2-cycloaddition. (From Bradley, D.G. and Min, D.B., *Crit. Rev. Food Sci. Nutr.*, 31, 211, 1992. With permission.)

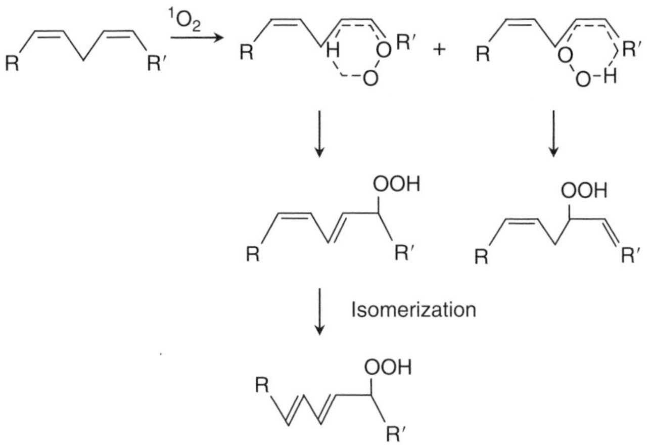

FIGURE 11.12 Conjugated and nonconjugated hydroperoxide formation from a diene fatty acid by the "ene" reaction of singlet oxygen. (From Min, D.B. and Boff, J.M., *Comp. Rev. Food Sci. Food Saf.*, 1, 58, 2002. With permission.)

oxygen oxidation [28]. These properties may be used to determine singlet oxygen activity in lipids [29] and produce compounds that are absent in triplet oxygen oxidation.

V. COMPARISON OF TRIPLET AND SINGLET OXYGEN OXIDATION

Neff and Frankel [30] compared triplet oxygen and photosensitized singlet oxygen oxidation of oleic, linoleic, and linolenic acids. The hydroperoxides formed from triplet oxygen or singlet oxygen and oleic, linoleic, or linolenic acids are shown in Table 11.3. Triplet oxygen oxidation produces only the

TABLE 11.3
Hydroperoxides of Fatty Acids Formed by Triplet and Singlet Oxygen Oxidation

		Oleate	Linoleate	Linolenate
$^3O_2$		8-OOH		
		9-OOH		
		10-OOH		
		11-OOH		
	Conjugated hydroperoxides		9-OOH	9-OOH
				12-OOH
			13-OOH	13-OOH
				16-OOH
$^1O_2$		9-OOH		
		10-OOH		
	Conjugated hydroperoxides		9-OOH	9-OOH
				12-OOH
			13-OOH	13-OOH
				16-OOH
	Nonconjugated hydroperoxides		10-OOH	10-OOH
			12-OOH	15-OOH

Source: From Frankel, E.N., Neff, W.E., and Bessler, T.R., *Lipids*, 14, 961, 1979.

TABLE 11.4

**Relative Oxidation Rates of Triplet and Singlet Oxygen
with Oleate, Linoleate, and Linolenate**

	Oleate	Linoleate	Linolenate
Triplet oxygen	1	27	77
Singlet oxygen	30,000	40,000	70,000

Source: From Min, D.B. and Lee, H.O. in *Food Lipids and Health*, R.E. McDonald
and D.B. Min, eds., Marcel Dekker, New York, 1996, 241.

conjugated diene hydroperoxides in linoleic and linolenic acids. The relative reaction ratios of triplet oxygen with oleic, linoleic, and linolenic acid for hydroperoxide formation are 1:12:25, which is dependent on the relative difficulty for the radical formation in the molecule [8,25]. The reaction rate of triplet oxygen with linolenic acid is about twice as fast as that of linoleic acid because linolenic acid has two pentadienyl groups in the molecule, compared with linoleic acid with one pentadienyl group. The hydroperoxides formed by singlet oxygen oxidation are at positions that formerly contained double bonds. Singlet oxygen produced conjugated and nonconjugated hydroperoxides from linoleic and linolenic acids. The relative reaction rates of triplet oxygen and singlet oxygen with oleic, linoleic, and linolenic acids are shown in Table 11.4. Singlet oxygen reacts with oleic, linoleic, and linolenic acids about 1,000–30,000 times faster than the triplet oxygen.

VI. FLAVOR PROPERTIES OF VOLATILE COMPOUNDS FROM OIL OXIDATION

The types of volatile compounds produced from the oxidation of edible oils are influenced by the composition of the hydroperoxides and the types of oxidative cleavage of double bonds in the fatty acids (Figure 11.5). Most decomposed products of hydroperoxides such as alcohols, aldehydes, furans, hydrocarbons, ketones, and acid compounds are responsible for the off-flavor in the oxidized oil. Aliphatic carbonyl compounds, such as alkanals, *trans, trans*-2,4-alkadienals, isolated alkadienals, isolated *cis*-alkenals, *trans, cis*-2,4-alkadienals, and vinyl ketones, have more influence on the oxidized oil flavor due to their low threshold values.

Frankel [7] reported that the significant compounds responsible for off-flavor are *trans, cis*-2,4-decadienal, *trans, trans*-2,4-decadienal, *trans, cis*-heptadienal, 1-octen-3-ol, *n*-butanal, and *n*-hexanal in decreasing order of importance calculated from the concentration and threshold value. Hexanal and 2-decenal, and 2-heptenal and *trans*-2-octenal were produced from soybean and corn oils at peroxide value of 5, respectively [31]. Pentane, hexanal, propenal, and 2,4-decadienal were formed in canola oil at 60°C [32].

To study the importance of volatile compounds in the flavor perception of oxidized oil, the concentration and threshold values of volatile compounds should be considered. It has been difficult for scientists to agree on flavor perceptions of compounds. Different people have used different terms for the same flavor compound of edible oil. The difficulties of sensory description of a compound are partially due to the changes in flavor perception according to concentration and evaluating conditions. However, the flavor perceptions of some compounds have been generally accepted by scientist in the field as shown in Table 11.5 [33].

VII. SINGLET OXYGEN OXIDATION OF SOYBEAN OIL

Soybean oil represents about 70% of all edible fats and oils consumed in the United States [34]. Soybean oil is inexpensive and widely available compared with other edible oils.

TABLE 11.5
Flavor Perceptions of Volatile Compounds Formed by Lipid Oxidation

Flavor Perception	Responsible Compounds
Cardboard	*trans, trans*-2,6-Nonadienal
Oily	Aldehydes
Painty	Pent-2-enal, aldehydes
Fishy	*trans, cis, trans*-2,4,7-Decatrienol, oct-1-en-3-one
Grassy	*trans*-2-Hexanal, nona-2,6-dienal
Deep-fried	*trans, trans*-2,4-Decandienal

Source: From Min, D.B. and Bradley, D.G. in *Encyclopedia of Food Science and Technology*, Y.H. Hui, ed, Wiley, New York, 1992, 828–832.

The development of reversion flavor, described as beany or grassy, is a unique defect to soybean oil and can be formed in soybean oils, which have low peroxide values [35]. To improve the flavor stability and quality of soybean oil, reversion flavor has been extensively studied in soybean oil since 1936 [35].

Smouse and Chang [36] identified 2-pentyl furan in reverted soybean oil and reported that it significantly contributed to the reversion flavor of soybean oil. Chang et al. [37] isolated and identified all four 2-pentenyl furan isomers in reverted soybean oil. Sensory evaluation showed that the addition of 2 ppm 2-pentyl furan to freshly deodorized and bland soybean oil produced the reversion flavor. The addition of 2 ppm 2-pentyl furan to deodorized cottonseed and corn oils also produced reversion flavor found in reverted soybean oil [38]. Ho et al. [35] reported that 2-(1-pentenyl) furan contributed to reversion flavor. Smagula et al. [39] reported that 2-(2-pentenyl) furan is also a contributor according to sensory evaluation. Flavor thresholds of the 2-pentenyl furan isomers were between 0.25 and 6 ppm.

Smouse and Chang [36] and Ho et al. [35] proposed the mechanisms for the formation of 2-pentyl furan from linoleic acid and 2-pentenyl furan isomers from linolenic acid using triplet oxygen, respectively. The proposed mechanisms for the formation of 2-pentyl furan from linoleic acid and isomers of 2-pentenyl furan from linolenic acid by triplet oxygen have been questioned. The formations of both the 2-pentyl furan and 2-pentenyl furan isomers require a hydroperoxide at carbon 10 of linoleic acid. The formation of 10-hydroperoxide in linoleic or linolenic acids by free-radical triplet oxygen oxidation is highly improbable but is very common in the singlet oxygen oxidation of linoleic or linolenic acids as shown in Table 11.3.

Soybean oil contains 1.0–1.5 ppm of chlorophyll [40], which is an excellent photosensitizer. Choe and Min [40] reported that headspace volatile compounds of soybean oil stored under light increased as added chlorophyll increased from 0 to 2, 4, or 8 ppm. Soybean oil containing no chlorophyll, which was removed by silicic acid liquid column chromatography, did not produce headspace volatile compounds under light at 10°C. The 0, 2, 4, 6, and 8 ppm added chlorophyll did not produce the volatile compounds in soybean oil in the dark storage. The formation of headspace volatile compounds in the soybean oil decreased inversely with the amount of added β-carotene, which quenches singlet oxygen [41]. The very rapid formation of volatile compounds in the soybean oil in the presence of chlorophyll, light, and oxygen was due to the singlet oxygen oxidation.

Lee and Min [41] identified 2-pentyl furan and 2-pentenyl furan in soybean oil containing 5 ppm chlorophyll *b* during 4 days of storage under light (Figure 11.13) and showed the involvement of singlet oxygen to form 2-pentyl furan and 2-pentenyl furan which are mainly responsible

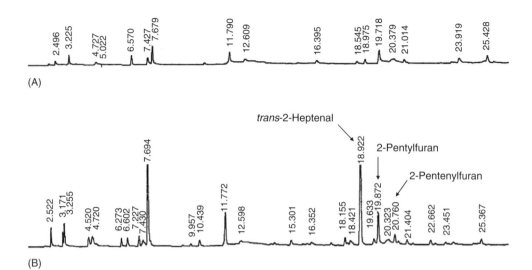

FIGURE 11.13 Gas chromatogram of soybean oil with 5 ppm chlorophyll (A) initially and (B) after storage under light for 4 days. (From Min, D.B., Callison, A.L., and Lee, H.O., *J. Food Sci.*, 68, 1175, 2003. With permission.)

for the reversion flavor. The chemical mechanisms for the formation of 2-(2-pentenyl) furan from linoleic acid and 2-pentyl furan from linolenic acid by singlet oxygen oxidation are shown in Figures 11.14 and 11.15, respectively. The chlorophyll in soybean oil should be carefully removed during the oil processing to minimize the formation of reversion during storage.

VIII. SINGLET OXYGEN QUENCHING MECHANISMS

Other than exclusion of light and reduction of oxygen present, the use of quenching agents is the best way to reduce singlet oxygen oxidation. Natural food components such as tocopherols, carotenoids, and ascorbic acid can act as effective singlet oxygen quenchers [42]. Quenching agents may be involved to minimize the development or activity of singlet oxygen at several stages in the oxidation of foods. Figure 11.16 shows the development of singlet oxygen and its subsequent reaction with substrate (A) to form the oxidized product (AO_2).

At every stage in this reaction, there is at least one alternate route, which, if taken, would minimize the oxidation of the compound (A). The first step represents the return of the excited singlet sensitizer (^{1}Sen*) to ground state (^{1}Sen) without intersystem crossing, which can form the excited triplet sensitizer (^{3}Sen*). The second represents reaction with a quencher (Q) at a rate K_Q, returning the excited triplet sensitizer (^{3}Sen*) to ground state (^{1}Sen) before reaction with triplet oxygen. The excited triplet sensitizer (^{3}Sen*) may react with triplet oxygen ($^3O_2$) to form singlet oxygen ($^1O_2$). Following its creation, there are three fates for singlet oxygen in foods: (1) it may naturally decay to the ground state at a rate k_d; (2) it may react with a singlet state compound (A) at a rate k_r forming the oxidized product (AO_2); and (3) it may be destroyed by a quencher either chemically at a rate k_{ox-Q} to form the product QO_2 or physically at a rate k_q to return to free triplet oxygen.

There are three points at which a quencher may act (Figure 11.16): one is quenching of the excited triplet sensitizer, and the other two are quenching of singlet oxygen either chemically or physically. Chemical quenching involves the reaction of singlet oxygen with the quencher to produce an oxidized product (QO_2). Physical quenching results in the return of singlet oxygen to triplet oxygen without the consumption of oxygen or the formation of oxidized products by either energy transfer or charge transfer. Therefore, triplet oxygen quenchers must be able either to donate

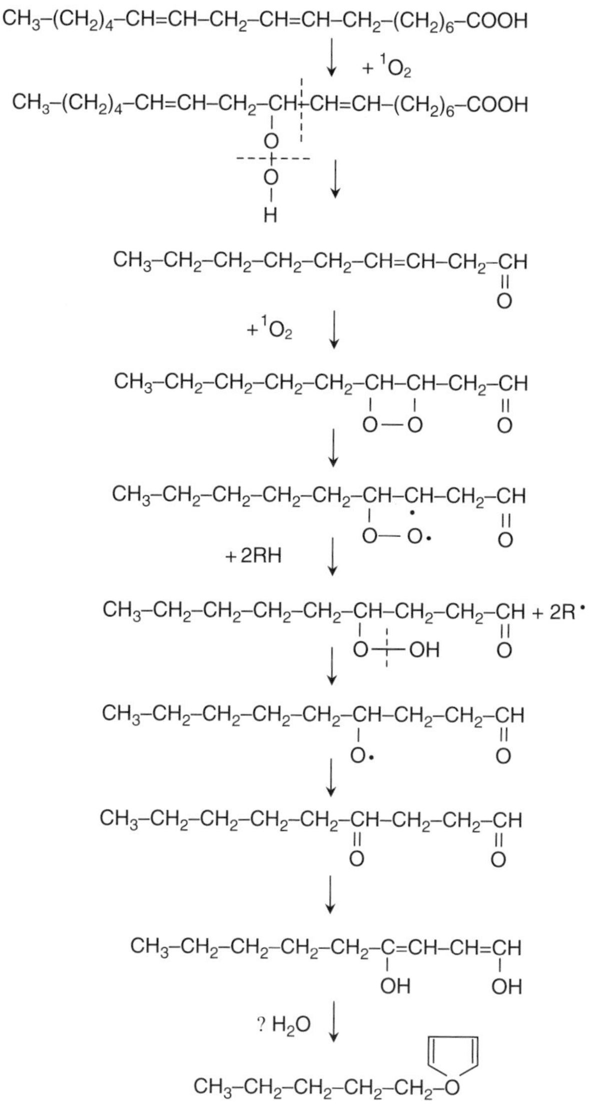

FIGURE 11.14 Mechanism for the formation of 2-pentyl furan from linoleic acid by singlet oxygen. (From Min, D.B., Callison, A.L., and Lee, H.O., *J. Food Sci.*, 68, 1175, 2003. With permission.)

electrons or to accept energy 22.4 kcal above the ground state. An example of the latter is β-carotene, which has a low singlet energy state, and it can therefore accept the energy from singlet oxygen [41]. Ascorbic acid is an example of a chemical that can quench the excited sensitizer. Table 11.6 lists quenching rates of several antioxidants.

A. QUENCHING MECHANISM OF CAROTENOIDS

Carotenoids, which are responsible for the yellow and red colors of many plants and animal products, have been known to minimize singlet oxygen oxidation [41,43]. Carotenoids include a class of hydrocarbons called carotenes and their oxygenated derivatives called xanthophylls. Foote [16] found that one molecule of β-carotene can quench 250–1000 molecules of singlet oxygen at a rate of 1.3×10^{10} M^{-1} s^{-1}.

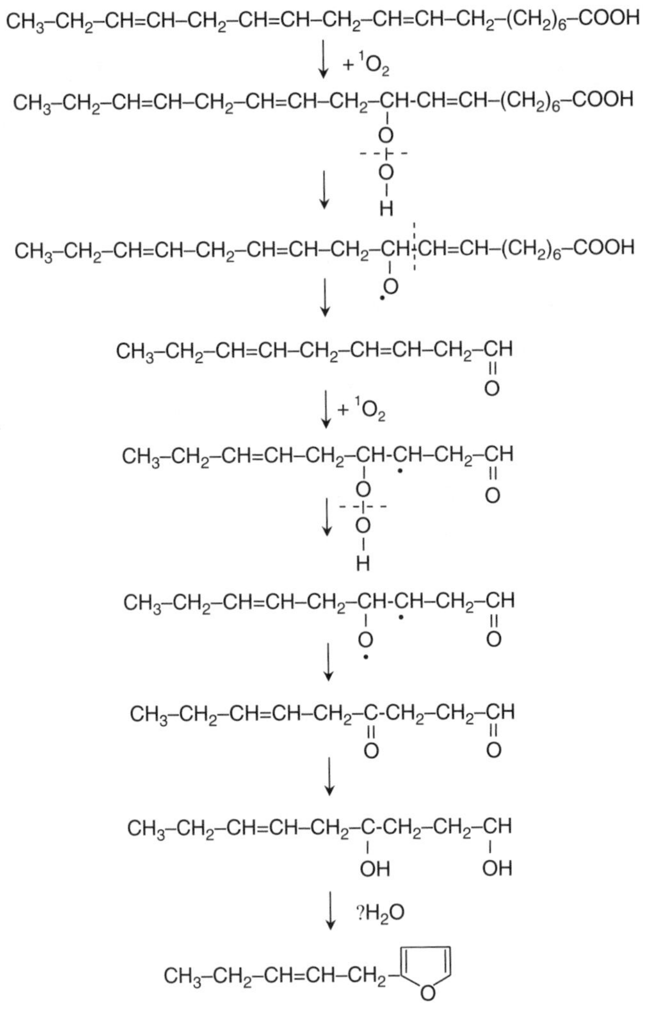

$$CH_3-CH_2-CH=CH-CH_2-CH=CH-CH_2-CH=CH-CH_2-(CH_2)_6-COOH$$

$$\downarrow \ +{}^1O_2$$

$$CH_3-CH_2-CH=CH-CH_2-CH=CH-CH_2-\underset{\overset{\displaystyle O}{\underset{\displaystyle H}{\overset{|}{\underset{|}{O}}}}}{CH}-CH=CH-(CH_2)_6-COOH$$

$$CH_3-CH_2-CH=CH-CH_2-CH=CH-CH_2-\underset{\overset{|}{\cdot O}}{CH}|CH=CH-(CH_2)_6-COOH$$

$$CH_3-CH_2-CH=CH-CH_2-CH=CH-CH_2-\underset{\overset{\parallel}{O}}{CH}$$

$$\downarrow \ +{}^1O_2$$

$$CH_3-CH_2-CH=CH-CH_2-\underset{\overset{|}{O}}{CH}-\underset{\cdot}{CH}-CH_2-\underset{\overset{\parallel}{O}}{CH}$$

$$\downarrow$$

$$CH_3-CH_2-CH=CH-CH_2-\underset{\overset{|}{\underset{\cdot}{O}}}{CH}-\underset{\cdot}{CH}-CH_2-\underset{\overset{\parallel}{O}}{CH}$$

$$CH_3-CH_2-CH=CH-CH_2-\underset{\overset{\parallel}{O}}{C}-CH_2-CH_2-\underset{\overset{\parallel}{O}}{CH}$$

$$\downarrow$$

$$CH_3-CH_2-CH=CH-CH_2-\underset{\overset{|}{OH}}{C}-CH_2-CH_2-\underset{\overset{|}{OH}}{CH}$$

$$\downarrow \ ?H_2O$$

$$CH_3-CH_2-CH=CH-CH_2-\text{[furan ring]}$$

FIGURE 11.15 Mechanism for the formation of 2-pentenyl furan from linolenic acid by singlet oxygen. (From Min, D.B., Callison, A.L., and Lee, H.O., *J. Food Sci.*, 68, 1175, 2003. With permission.)

Energy transfer mechanism is responsible for the minimization of singlet oxygen oxidation of lipids by β-carotene [43]. Electron excitation energy is transferred from singlet oxygen to singlet state carotenoid (CAR), producing triplet state carotenoid (^{3}CAR) and triplet oxygen, which is called

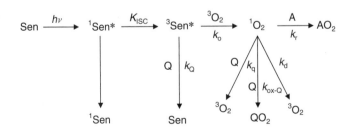

FIGURE 11.16 Formation of singlet oxygen and its reaction with substrate A to produce the oxidized product AO$_2$. (From Jung, Y.J., Lee, E., and Min, D.B., *J. Food Sci.*, 45, 183, 1991. With permission.)

TABLE 11.6
Singlet Oxygen Quenchers and Their Quenching Rates

Quenching Compound	Quenching Rate ($M^{-1} s^{-1}$)
β-Apo-8′-carotenal	3.1×10^9
β-Carotene	4.6×10^9
Lutein	5.7×10^9
Zeaxanthin	6.8×10^9
Lycopene	6.9×10^9
Isozeaxanthin	7.4×10^9
Astaxanthin	9.9×10^9
Canthaxanthin	11.2×10^9
α-Tocopherol	2.7×10^7
1,4-Diazabicyclo-(2,2,2)-octane	1.5×10^7
Dimethylfuran	2.6×10^7
Bis(di-n-butyldithiocarbamato)nickel chelate	1.2×10^9
{2,2′-Thiobis(4–1,1,3,3,-tetramethylbutyl) phenalto)}-n-butylamine)nickel chelate	3.7×10^7

Sources: From Min, D.B. and Lee, H.O. in *Food Lipids and Health*, R.E. McDonald and D.B. Min, eds., Marcel Dekker, New York, 1996, 241; Lee, S.H. and Min, D.B., *J. Agric. Food Chem.*, 38, 1630, 1990; Li, T.L., King, J.M., and Min, D.B., *J. Food Biochem.*, 24, 477, 2000; Foote, C.S., Chang, Y.C., and Denny, R.W., *J. Am. Chem. Soc.*, 92, 5216, 1970.

singlet oxygen quenching. Energy is also transferred from excited triplet state sensitizer (^{3}Sen*) to the singlet state carotenoid (^{1}CAR), which is called triplet sensitizer quenching. The triplet state carotenoid can easily return to the singlet state carotenoid dissipating as a heat.

$$^1O_2 + {}^1CAR \rightarrow {}^3O_2 + {}^3CAR,$$

$$^1CAR + {}^3Sen^* \rightarrow {}^3CAR + {}^1Sen,$$

$$^3CAR \rightarrow {}^1CAR.$$

The energy transfer from singlet oxygen (22.4 kcal/mol) to carotenoids with nine or more conjugated double bonds (<22.4 kcal/mol) is exothermic [44]. Foote [44] reported that the carotenoid with seven conjugated double bonds was effective to quench triplet chlorophyll. Carotenoids with fewer than nine conjugated double bonds have energies above that of singlet oxygen and are less efficient singlet oxygen quenchers. Carotenoids with 11 or more conjugated double bonds quench at a diffusion-controlled rate of singlet oxygen.

The rate of singlet oxygen quenching by carotenoids is dependent on the number of conjugated double bonds and the type and number of functional groups on the ring portion of the molecule. This is important in the solubility of carotenoid. Kobayashi and Sakamoto [45] compared the quenching activity of β-carotene with astaxanthin and found that the quenching activity of astaxanthin decreased with increasing hydrophobicity, whereas the quenching activity of β-carotene increased. Lee and Min [41] evaluated the effectiveness of five carotenoids, including lutein, zeaxanthin, lycopene, isozeaxanthin, and astaxanthin, in quenching chlorophyll-photosensitized oxidation of soybean oil and reported that the effectiveness increased with the number of double bonds in the carotenoid and the concentration of carotenoid added.

B. Quenching Mechanisms of Tocopherols

Tocopherols are the most abundant and prevalent antioxidants in nature and studied as free-radical scavengers. When present in systems that are vulnerable to singlet oxygen oxidation, tocopherols inhibit lipid peroxidation. Tocopherols were identified in soybean oil about 1100 ppm and exist in α-, β-, γ-, and δ-tocopherol at approximately 4%, 1%, 67%, and 29%, respectively [46]. Jung et al. [46] studied the effectiveness of α-, γ-, and δ-tocopherol in the chlorophyll-photosensitized oxidation of soybean oil and reported that α-tocopherol quenched singlet oxygen at the rate of 2.7×10^7 M^{-1} s^{-1}. Tocopherols as singlet oxygen quenchers involve charge transfer [44]. Tocopherols can form a charge transfer complex with singlet oxygen by electron donation from tocopherol to singlet oxygen. The transfer complex undergoes an intersystem crossing ultimately to form triplet oxygen and tocopherol. Since this does not involve chemical reactions between tocopherol and singlet oxygen, it is called physical quenching:

$$T + {}^1O_2 \rightarrow [T^+ - {}^1O_2]_1 \rightarrow [T^+ - {}^1O_2]_3 \rightarrow T + {}^3O_2.$$

The destruction of vitamin D_2 in a model system by singlet oxygen oxidation was reduced by the addition of α-tocopherol [47]. The rate of singlet oxygen quenching by α-tocopherol is similar to that of β-carotene.

IX. DETERMINING QUENCHING MECHANISMS

The quenching mechanism of photosensitized singlet oxygen oxidation can be determined by measuring the rate constant of total quenching, physical quenching, and chemical quenching. Quenchers work in numerous ways to inhibit the formation of oxidized products as has been previously described (Figure 11.16).

The quantum yield of a photochemical reaction is defined as the ratio of the number of molecules of a product formed to the number of photons of light absorbed. This value is used to measure the relative efficiency of a photochemical reaction. The quantum yield of oxidized product formation ($\emptyset AO_2$) can be defined by the equation:

$$\emptyset AO_2 = A \times B \times C, \tag{11.1}$$

where
A represents the partitioning of singlet sensitizer for singlet oxygen oxidation
B is the partitioning of triplet sensitizer for singlet oxygen formation
C is the formation of the oxidized product

The concentration of quencher necessary to inhibit a substantial amount of the singlet oxygen sensitizer is particularly high since the lifetime of singlet oxygen sensitizer is very short. For these reasons, the singlet sensitizer quenching is not considered in the steady-state equation. Therefore, term A (Equation 11.1) is a constant (K) that is equal to the quantum yield of intersystem crossing.

Term B (Equation 11.1) represents the rate of singlet oxygen formation, which is dependent on the triplet sensitizer quenching rate and the rate of triplet–triplet annihilation. Therefore, B is:

$$B = \frac{k_o[\text{oxygen}]}{k_o[\text{oxygen}] + k_Q[\text{oxygen}]}, \tag{11.2}$$

where
k_o is the reaction rate constant of triplet–triplet annihilation
k_Q is the reaction rate constant of triplet sensitizer quenching (Figure 11.16)

Term C (Equation 11.1) represents the formation of oxidized product, which is dependent on the concentration and nature of the compound, the physical and chemical quenching of singlet oxygen, as well as the natural decay rate of singlet oxygen. The assemblage of these factors generates the following equation:

$$C = \frac{k_r[\text{substrate}]}{k_r[\text{substrate}] + (k_{\text{ox-Q}} + k_q)[\text{chemical} + \text{physical quencher}] + k_d},$$ (11.3)

where

k_r is the reaction rate constant of the reaction between singlet oxygen and the substrate
$k_{\text{ox-Q}}$ is the reaction rate constant of chemical quenching
k_q is the reaction rate constant of physical quenching
k_d is the decay rate constant of singlet oxygen

If a quencher inhibits photosensitized oxidation by quenching singlet oxygen, the steady-state equation can be written as:

$$\text{ØAO}_2 = K \frac{k_o[^3O_2]}{k_o[^3O_2] + k_Q[Q]} \times \frac{k_r[A]}{k_r[A] + (k_{\text{ox-Q}} + k_q)[Q] + k_d},$$ (11.4)

where K is the quantum yield of intersystem crossing of the excited singlet sensitizer (A from Equation 11.1), and both B and C have been appropriately substituted with Equations 11.2 and 11.3, respectively.

In a given system, if there is only singlet oxygen quenching such that $k_Q[Q] \ll k_o[^3O_2]$, then B term is equal to 1. Therefore, the steady-state equation becomes:

$$\text{ØAO}_2 = K \frac{k_r[A]}{k_r[A] + (k_{\text{ox-Q}} + k_q)[Q] + k_d}.$$ (11.5)

This equation can be inverted to

$$[\text{ØAO}_2]^{-1} = K^{-1}\left[1 + \frac{(k_{\text{ox-Q}} + k_q)[Q] + k_d}{k_r}\right][A]^{-1},$$ (11.6)

so that it is in slope–intercept form.

Alternatively, if there is only triplet sensitizer quenching such that $(k_{\text{ox-Q}} + k_q)[Q] \ll k_r[A] + k_d$, then the slope–intercept form of the equation is

$$[\text{ØAO}_2]^{-1} = K^{-1}\left[1 + \frac{k_Q[Q]}{k_o[^3O_2]}\right]\left[1 + \frac{k_d}{k_r[A]}\right].$$ (11.7)

The significance of these two equations is the fact that one describes a system in which singlet oxygen quenching is dominant, and the other describes a system in which triplet sensitizer is dominant. A plot of

$$\left[\text{AO}_2[\text{ØAO}_2]^{-1}\right] = K^{-1}\left[1 + \frac{(k_{\text{ox-Q}} + k_q)[Q] + k_d}{k_r}\right][A]^{-1}\right]^{-1}$$

against $[A]^{-1}$ at different $[Q]$ will appear in one of two manners, dependent on which mechanism dominates a system. If singlet oxygen quenching is dominant (Equation 11.6), then the plots at

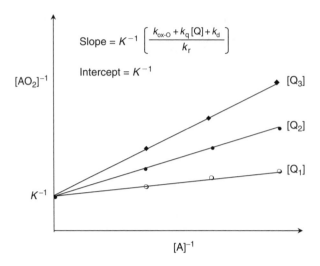

FIGURE 11.17 Characteristics plot of a singlet oxygen quenching mechanism. (From Li, T.L., King, J.M., and Min, D.B., *J. Food Biochem.*, 24, 477, 2000. With permission.)

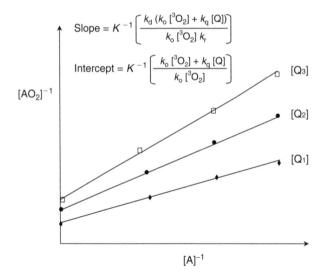

FIGURE 11.18 Characteristic plot of a triplet sensitizer quenching mechanism. (From Li, T.L., King, J.M., and Min, D.B., *J. Food Biochem.*, 24, 477, 2000. With permission.)

various [Q] will all have the same y-intercept but different slopes (Figure 11.17). If triplet sensitizer quenching is dominant (Equation 11.7), then both the intercept and the slope will vary (Figure 11.18).

REFERENCES

1. Frankel, E.N., Neff, W.E., and Selke, E., Analysis of autoxidized fats by gas chromatography-mass spectrometry: VII. Volatile thermal decomposition products of pure hydroperoxides from autoxidized and photosensitized oxidized methyl oleate, linoleate and linolenate, *Lipids*, 16, 279, 1981.
2. Rawls, H.R. and Van Santen, P.J., A possible role for singlet oxidation in the initiation of fatty acid autoxidation, *J. Am. Oil Chem. Soc.*, 47, 121, 1970.

3. Bradley, D.G. and Min, D.B., Singlet oxygen oxidation in foods, *Crit. Rev. Food Sci. Nutr.*, 31, 211, 1992.
4. Korycka-Dahl, M.B. and Richardson, T., Activated oxygen species and oxidation of food constituents, *Crit. Rev. Food Sci. Nutr.*, 10, 209, 1978.
5. Girotti, A.W., Lipid hydroperoxide generation, turnover, and effector action in biological systems, *J. Lipid Res.*, 39, 1529, 1998.
6. Yang, W.T. and Min, D.B., Chemistry of singlet oxygen oxidation of foods, in *Lipids in Food Flavors*, Ho, C.T. and Hartmand, T.G., Eds., American Chemical Society, Washington D.C., 1994, p. 15.
7. Frankel, E.N., Chemistry of autoxidation: mechanism, products and flavor significance, in *Flavor Chemistry of Fats and Oils*, Min, D.B. and Smouse, T.H., Eds., AOCS Press, Champaign, IL, 1985, p. 1.
8. Min, D.B. and Boff, J.M., Chemistry and reaction of singlet oxygen in foods, *Comp. Rev. Food Sci. Food Saf.*, 1, 58, 2002.
9. Choe, E. and Min, D.B., Chemistry and reactions of reactive oxygen species in foods, *J. Food Sci.*, 70, R142, 2005.
10. Hiatt, R., Mill, T., Irwin, K.C., Mayo, T.R., Gould, C.W., and Castleman, J.K., Homolytic decomposition of hydroperoxides, *J. Org. Chem.*, 33, 1416, 1968.
11. Foote, C.S. and Denny, R.W., Chemistry of singlet oxygen. VII: Quenching by b-carotene, *J. Am. Chem. Soc.*, 90, 6233, 1968.
12. Afonso, S.G., Fnriquez de Salamanca, R., and del C. Baflle, A.M., The photodynamic and non-photodynamic actions of porphyrins, *Braz. J. Med. Biol. Res.*, 32, 255, 1999.
13. Lledias, F. and Hansberg, W., Catalase modification as a marker for singlet oxygen, in *Methods in Enzymology*, Parks, O.W. and Sies, H., Eds., Academic Press, New York, 2000, p. 110.
14. Kochevar, I.E. and Redmond, R.W., Photosensitized production of singlet oxygen, in *Methods in Enzymology*, Packer, L. and Sies, H., Eds., Academic Press, New York, 2000, p. 20.
15. Halliwell, B. and Gutteridge, J.M.C., *Free Radicals in Biology and Medicine*, 3rd ed., Oxford University Press, New York, 2001.
16. Foote, C.S., Photosensitized oxidation and singlet oxygen: consequences in biological systems, in *Free Radicals in Biology*, Pryor, W.A., Ed., Academic Press, New York, 1976, p. 85.
17. Sharman, W.M., Allen, C.M., and van Lier, J.E., Role of activated oxygen species in photodynamic therapy, in *Methods in Enzymology*, Packer, L. and Sies, H., Eds., Academic Press, New York, 2000, p. 376.
18. Kepka, A. and Grossweiner, L.I., Photodynamic oxidation of iodide ion and aromatic amino acids by eosine, *Photochem. Photobiol.*, 14, 621, 1972.
19. He, Y.Y., An, J.Y., and Jiang, L.J., EPR and spectrophotometric studies on free radicals (O_2^*, Cysa-HB*) and singlet oxygen ($^1O_2$) generated by irradiation of cysteamine substituted hypocrellin B, *Int. J. Radiat. Biol.*, 74, 647, 1998.
20. Song, Y.Z., An, J.Y., and Jiang, L.J., ESR evidence of the photogeneration of free radicals (GDHB*, O_2^*) and singlet oxygen ($^1O_2$) by 15-deacetyl-13-glycine-substituted hypocrellin B, *Biochem. Biophys. Acta*, 1472, 307, 1999.
21. Ke, P.J. and Ackman, R.G., Bunsen coefficient for oxygen in marine oils at various temperatures determined by exponential dilution method with a polarographic oxygen electrode, *J. Am. Oil Chem. Soc.*, 50, 429, 1973.
22. Ando, T., Yoshikawa, F., Tanigawa, T., Kohno, M., Yoshida, N., and Motoharu, K., Quantification of singlet oxygen from hematoporphyrin derivative by electron spin resonance, *Life Sci.*, 61, 1953, 1997.
23. Whang, K. and Peng, I.C., Electron paramagnetic resonance studies of the effectiveness of myoglobin and its derivatives as photosensitizers in singlet oxygen generation, *J. Food Sci.*, 53, 1863, 1988.
24. Bradley, D.G., Lee, H.O., and Min, D.B., Singlet oxygen detection in skim milk by electron spin resonance spectroscopy, *J. Food Sci.*, 68, 491, 2003.
25. Min, D.B. and Lee, H.O., Chemistry of lipid oxidation, in *Food Lipids and Health*, McDonald, R.E. and Min, D.B., Eds., Marcel Dekker, New York, 1996, p. 241.
26. Adam, W., Singlet molecular oxygen and its role in organic peroxide chemistry, *Chem. Ztg.*, 99, 142, 1975.
27. Doleiden, F.H., Farenholtz, S.R., Lamola, A.A., and Rwozzolo, A.M., Reactivity of cholesterol and some fatty acids toward singlet oxygen, *Photochem. Photobiol.*, 20, 519, 1974.
28. Frankel, E.N., Neff, W.E., and Bessler, T.R., Analysis of autoxidized fats by gas chromatography-mass spectrometry: V. Photosensitized oxidation, *Lipids*, 14, 961, 1979.

29. Stratton, S.P. and Liebler, D.C., Determination of singlet oxygen-specific versus radical-initiated lipid peroxidation in photosensitized oxidation of lipid bilayers: effect of β-carotene and α-tocopherol, *Biochemistry*, 36, 12911, 1997.

30. Neff, W.E. and Frankel, E.N., Quantitative analysis of hydroxystearate isomers from hydroperoxides by high pressure liquid chromatography of autoxidized and photosensitized oxidized fatty esters, *Lipids*, 15, 587, 1980.

31. Steenson, D.F.M., Lee, J.H., and Min, D.B., Solid-phase microextraction of volatile soybean oil and corn oil compounds, *J. Food Sci.*, 67, 71, 2002.

32. Vaisey-Genser, M., Malcomson, L.J., Przybylski, R., and Eskin, N.A.M., Consumer acceptance of stored canola oils in Canada, The 12th project report research on canola, seed, and oil meal, Canada Canola Council, Canada, 1999, p. 189.

33. Min, D.B. and Bradley, D.G., Fats and oils: flavors, in *Encyclopedia of Food Science and Technology*, Hui, Y.H., Ed., Wiley, New York, 1992, pp. 828–832.

34. Golbitz, P., *Soya & Oilseed Bluebook*, Soyatech, Inc., Bay Harbor, Maine, 2000, p. 2001.

35. Ho, C.T., Smagula, M.S., and Chang, S.S., The synthesis of 2-(1-pentenyl) furan and its relationship to the reversion flavor of soybean oil, *J. Am. Oil Chem. Soc.*, 55, 233, 1978.

36. Smouse, T.H. and Chang, S.S., A systematic characterization of the reversion flavor of soybean oil, *J. Am. Oil Chem. Soc.*, 44, 509, 1967.

37. Chang, S.S., Shen, G.H., Tang, H., Jin, Q.Z., Shi, H., Carlin, J.T., and Ho, C.T., Isolation and identification of 2-pentenylfurans in the reversion flavor of soybean oil, *J. Am. Oil Chem. Soc.*, 60, 553, 1983.

38. Chang, S.S., Smouse, T.H., Krishnamurthy, R.G., Mookherjee, B.D., and Reddy, R.B., Isolation and identification of 2-pentyl-furan as contributing to the reversion flavour of soybean oil, *Chem. Ind. (London)*, 1926, 1966.

39. Smagula, M.S., Ho, C.T., and Chang, S.S., The synthesis of 2-(2-pentenyl) furans and their relationship to the reversion flavor of soybean oil, *J. Am. Oil Chem. Soc.*, 56, 516, 1978.

40. Choe, E. and Min, D.B., Mechanisms and factors for edible oil oxidation, *Comp. Rev. Food Sci. Food Saf.*, 5, 169, 2006.

41. Lee, S.H. and Min, D.B., Effects, quenching mechanisms, and kinetics of carotenoids in chlorophyll-sensitized photooxidation of soybean oil, *J. Agric. Food Chem.*, 38, 1630, 1990.

42. Li, T.L., King, J.M., and Min, D.B., Quenching mechanisms and kinetics of carotenoids in riboflavin photosensitized singlet oxygen oxidation of vitamin D_2, *J. Food Biochem.*, 24, 477, 2000.

43. Forss, D.A., Mechanism of formation of aroma compounds in milk and milk products, *J. Dairy Res.*, 46, 691, 1979.

44. Foote, C.S., Chang, Y.C., and Denny, R.W., Chemistry of singlet oxygen. X. Carotenoid quenching parallels biological protection, *J. Am. Chem. Soc.*, 92, 5216, 1970.

45. Kobayashi, M. and Sakamoto, Y., Singlet oxygen quenching ability of astaxanthan esters from the green alga *Haematococcus pluvialis*, *Biotechnol. Lett.*, 21, 265, 1999.

46. Jung, Y.J., Lee, E., and Min, D.B., α-, γ-, and δ-tocopherol effects on chlorophyll photosensitized oxidation of soybean oil, *J. Food Sci.*, 45, 183, 1991.

47. King, J.M. and Min, D.B., Riboflavin photosensitized singlet oxygen oxidation of Vitamin D, *J. Food Sci.*, 63, 31, 1998.

12 Lipid Oxidation of Muscle Foods

Marilyn C. Erickson

CONTENTS

I.   INTRODUCTION

Following storage at refrigerated or frozen temperatures lipid oxidation is one of the major causes of quality deterioration in muscle foods. Often seen in later stages of storage, quality losses are manifested through a variety of mechanisms, which are summarized in Table 12.1 [1–17]. Although lipid oxidation usually causes a decrease in consumer acceptance, in some cases lipid oxidation leads to enhancement of product quality. An example is the enzymatic production of fresh-fish aromas. This chapter reviews the fundamental mechanisms of lipid oxidation as they apply to muscle foods. Included in this chapter is a discussion of the impact of tissue structure and compositional factors on pathways, kinetics, and extent of oxidation. Also included is a section describing the effect of various food processing applications on lipid oxidation reactions. Throughout this chapter, the reader will be made aware of the multiple interactions among muscle constituents during the process of lipid oxidation. Therefore, this chapter (Section III.G) details how mathematical models may be used to account for these interactions and indicates how shelf-life predictions and conditions for optimal stability may be derived.

II.   BASIC CHEMISTRY OF LIPID OXIDATION

The two major components involved in lipid oxidation are unsaturated fatty acids and oxygen. In this process, oxygen from the atmosphere is added to certain fatty acids, creating unstable intermediates that eventually breakdown to form unpleasant flavor and aroma compounds. Although enzymatic and photogenic oxidation may play a role, the most common and important process by

TABLE 12.1
Consequences of Lipid Oxidation Activity

Consequence	References
Fresh flavors	[1,2]
Off-flavors (warmed over/rancid)	[3,4]
Cholesterol oxidation products with potentially detrimental health implications	[5–9]
Protein denaturation and functionality changes	[10–12]
Pigment changes	
Myoglobin (red) $\rightarrow$ metmyoglobin (brown)	[13–15]
Loss of red (carotenoid) pigmentation	[16,17]

which unsaturated fatty acids and oxygen interact is a free radical mechanism characterized by three main phases:

Initiation: $$In^{\bullet} + RH \rightarrow InH + R^{\bullet}$$

Propagation: $$R^{\bullet} + O_2 \rightarrow ROO^{\bullet}$$

$$ROO^{\bullet} + RH \rightarrow R^{\bullet} + ROOH$$

Termination: $$2ROO^{\bullet} \rightarrow O_2 + RO_2R$$

$$ROO^{\bullet} + R^{\bullet} \rightarrow RO_2R$$

Initiation occurs as hydrogen is abstracted from an unsaturated fatty acid, resulting in a lipid-free radical, which, in turn, reacts with molecular oxygen to form a lipid peroxyl radical. Although irradiation can directly abstract this hydrogen from lipids, initiation is frequently attributed in most foods, including muscle foods, to reaction of the fatty acids with active oxygen species. The propagation phase of oxidation is fostered by lipid–lipid interactions, whereby the lipid peroxyl radical abstracts hydrogen from an adjacent molecule, resulting in a lipid hydroperoxide and a new lipid-free radical. Interactions of this type continue 10 [18] to 100 times [19] before two free radicals combine to terminate the process. Additional magnification of lipid oxidation, however, occurs through branching reactions (also known as secondary initiation): $Fe^{2+} + LOOH \rightarrow LO^{\bullet} + OH^{\bullet}$. The radicals produced then proceed to abstract hydrogens from unsaturated fatty acids. Additional information describing these free radical processes is presented later in Sections I.A through I.C, on initiation, propagation, and termination, respectively.

By themselves, lipid hydroperoxides are not considered harmful to food quality; however, they are further degraded into compounds that are responsible for off-flavors. The main mechanism for the formation of aldehydes from lipid hydroperoxides is homolytic scission (β cleavage) of the two C–C bonds on either side of the hydroperoxy group [20]. This reaction proceeds via the lipid alkoxyl radical, with the two odd electrons produced on neighboring atoms forming a carbonyl double bond. Two types of aldehydes are formed from the cleavage of the carbon bond: aliphatic aldehydes derived from the methyl terminus of the fatty acid chain and aldehydes still bound to the parent lipid molecule. Since unsaturated aldehydes can be oxidized further, additional volatile products may be formed [20].

A. Initiation

The direct reaction of a lipid molecule with a molecule of oxygen is highly improbable because the lipid molecule is in a singlet electronic state and the oxygen molecule has a triplet ground state. To circumvent this spin restriction, oxygen can be activated by any of the following three initiation mechanisms: (1) formation of singlet oxygen; (2) formation of partially reduced or activated oxygen species such as hydrogen peroxide, superoxide anion, or hydroxyl radical; and (3) formation of active oxygen–iron complexes (ferryl iron or ferric–oxygen–ferrous complex). In addition, the oxidation of fatty acids may occur either directly or indirectly through the action of enzyme systems, of which three major groups are involved: microsomal enzymes, peroxidases, and dioxygenases, such as lipoxygenase or cyclooxygenase. That such chemical and enzymatic reactions exist in living tissue is evidenced by the occurrence in aerobic organisms of enzymes that can eliminate or detoxify these compounds (e.g., superoxide dismutase, catalase, glutathione peroxidase). Therefore, activated oxygen species are likely to be present in the food item even before it is harvested, not just produced during processing and storage. As for which mechanism of initiation is primarily responsible, a large volume of research has been published exploring this issue but no consensus has arisen. The reader is therefore encouraged to look in the reviews of Kanner et al. [21], Hsieh and Kinsella [22],

Kappus [23], and Bradley and Min [24] for a more in-depth look at mechanisms of initiation. However, additional information on specific sources of initiation will be presented in Section III.E.

B. PROPAGATION

Propagation reactions form the basis of the chain reaction process and in general include the following:

Radical coupling with oxygen: $\qquad R^{\bullet} + O_2 \rightarrow ROO^{\bullet}$

Atom or group transfer: $\qquad ROO^{\bullet} + RH \rightarrow ROOH + R^{\bullet}$

Fragmentation: $\qquad ROO^{\bullet} \rightarrow R^{\bullet} + O_2$

Rearrangement:

Cyclization:

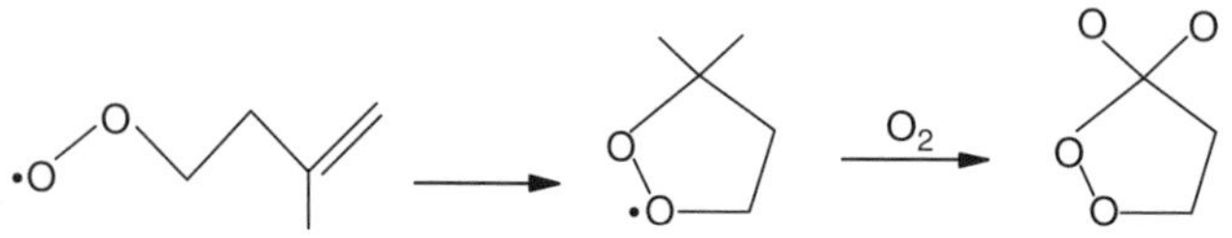

In oxygen radical coupling, molecular oxygen reacts with the carbon-centered free radical at or near the diffusion-controlled rate of approximately 10^9 M^{-1} s^{-1}. A major consequence of this reactivity is that the concentration of $R^{\bullet}$ is much smaller than that of $ROO^{\bullet}$. In atom transfer, a peroxyl radical, $ROO^{\bullet}$, will not readily abstract hydrogen from a saturated hydrocarbon, it will do so very readily from allylic and bisallylic C–H bonds of unsaturated fatty acids.

Lower bond energies for bisallylic and allylic hydrogens versus methylene hydrogens (75 and 88 versus 100 kcal/mol, respectively), as well as resonance stabilization of the radical intermediate, contribute to ease of abstraction from unsaturated fatty acids [25,26]. The newly formed hydroperoxy radical can, in turn, abstract hydrogen from an adjacent unsaturated fatty acid such that the reaction sequence goes through 8–14 propagation cycles before termination [27]. Conditions that determine the chain propagation length include initiation rate, structures of aggregates (increasing with increasing structure of the aggregates), temperature, presence of antioxidants, and chain branching. Chain branching involves the breakdown of fatty acid hydroperoxides to the lipid peroxyl or alkoxyl radical. Given the bond dissociation energies of LOO–H (about 90 kcal/mol) and LO–OH (about 44 kcal/mol), spontaneous decomposition is unlikely at refrigerated or freezing temperatures [28]. Instead, breakdown of hydroperoxides would be dominated by one-electron transfers from metal ions during low temperature storage.

$$Fe^{2+} + LOOH \rightarrow Fe^{3+} + LO^{\bullet} + OH^{-}$$

The major contributors to decomposition of lipid hydroperoxides in food and biological systems would be heme and nonheme iron, with reactions involving the ferrous ion occurring much more quickly than those involving ferric ion.

C. Termination

To break the repeating sequence of propagating steps, two types of termination reactions are encountered: radical–radical coupling and radical–radical disproportionation, a process in which two stable products are formed from A$^{\bullet}$ and B$^{\bullet}$ by an atom or group transfer process. In both cases, nonradical products are formed. However, the termination reactions are not always efficient. When coupling gives rise to tertiary tetroxides, they decompose to peroxyl radicals at temperatures above $-80°C$ and to alkoxyl radicals at temperatures above $-30°C$ [29]. Secondary and primary peroxyl radicals, on the other hand, terminate efficiently by a mechanism in which the tetroxide decomposes to give molecular oxygen, an alcohol, and a carbonyl compound.

III. MUSCLE COMPOSITION AND LIPID OXIDATION

A. Muscle Structure and Function

Before discussing individual constituents of muscle tissue, a review of the structural and chemical features that contribute to muscle tissue's oxidative stability will be taken. Within the animal body, there are more than 600 muscles varying widely in shape, size, and activity. However, at the cellular level there is close resemblance among muscles from a wide variety of organisms. The typical arrangement of a skeletal muscle in cross section consists of epimysium (connective tissue surrounding the entire muscle), perimysium (connective tissue separating the groups of fibers into bundles), and endomysia (sheaths of connective tissue surrounding each muscle fiber). Also surrounding each muscle fiber is the sarcolemma membrane, which periodically, along the length of the fiber, forms invaginations usually referred to as T tubules.

Within the muscle fiber, the sarcoplasm serves to suspend organelles such as mitochondria and lysosomes. Water constitutes 75%–80% of the sarcoplasm but in addition may contain lipid droplets, variable quantities of glycogen granules, ribosomes, proteins, nonprotein nitrogenous compounds, and a number of inorganic constituents.

Myofibrils, which are organelles found only in muscle cells, are long, thin, cylindrical rods that extend the entire length of the muscle fiber and constitute the contractile apparatus, which is composed of primarily myosin and actin proteins. Also intracellular in nature is the sarcoplasmic reticulum (SR). Forming a closely meshed network around each myofibril, SR membranes serve as the storage site for Ca^{2+} in resting muscle fibers.

The contractile process is a complex mechanism that starts with an action potential spreading through the T tubules and sarcolemma, finally reaching the SR. At the SR membrane, depolarization occurs, stimulating the release of calcium. The released calcium binds to troponin causing a conformational change in the protein, which, in turn, triggers myosin ATPase to hydrolyze ATP to ADP. The chemical energy released from the hydrolysis is utilized by the myosin and actin to initiate the sliding mechanism resulting in the contraction of the muscle. When the impulse that started the potential subsides, the sarcolemma polarizes, stimulating the energy-dependent sequestration of calcium by the SR membranes.

B. Biochemical Changes in Muscle Postmortem

After death of an animal, all circulation ceases—an event that rapidly brings about important changes in the muscle tissue (Table 12.2). The principal changes are attributable to a lack of oxygen (anaerobic conditions) and the accumulation of certain waste products, especially lactate and H^+. In a short time, the mitochondrial system ceases to function in all but surface cells because internal oxygen is rapidly depleted. Anaerobic glycolysis continues to regenerate ATP and lactate, but eventually the decrease in pH caused by the presence of lactate disrupts glycolytic activity. When ATP is depleted, the immediate response seen is the onset of rigor whereby actin and myosin remain in a contracted state as a result of the absence of a plasticizing agent (ADP or ATP). A more

TABLE 12.2

Biochemical Changes and Negative Consequences in Postmortem Muscle Food

Biochemical Change	Consequence
Decrease in ATP	Loss of energy source needed for reduction of many compounds
Increase in hypoxanthine	Off-flavor produced by ATP degradation
Conversion of hypoxanthine dehydrogenase to xanthine oxidase	Xanthine oxidase can initiate oxidation when molecular oxygen is present in system
Decrease in ascorbate, glutathione	Loss of secondary antioxidants and cofactors
Increase in low molecular weight iron	Initiator of oxidation
Oxidation of myoglobin to porphyrin radical	Can react with hydrogen peroxide to produce ferryl oxene (Fe^{4+}), which can initiate lipid oxidation
Loss of tocopherol	Loss of primary antioxidant
Disintegration of membranes	Could cause hydrolysis of phospholipids, uneven maintenance of ions
Loss of Ca^{2+} sequestration	Increased calcium ion content in aqueous phase causes many inactive processes to become active

important response with regard to lipid oxidation is the cellular membranes' inability to maintain their integrity. Consequently, lysosomal enzymes, such as phospholipase and lipase, may be released, affecting, in turn, the susceptibility of lipids to oxidize. Calcium leakage is a noted response to the increased membrane permeability of SR and mitochondria [30,31]. Increased calcium concentrations, in turn, could activate enzymic systems, such as phospholipase, which would not normally be turned on.

C. Variability between and within Muscles

Although the muscle function of locomotion is similar throughout the animal kingdom, compositional differences exist between species and even within different muscles of the same species. Red and white muscles present a classic example, with white muscles (i.e., muscles dominated by white fibers) presumably best suited for vigorous activity for a short period and red muscles for sustained activity. Red fibers tend to be smaller than white fibers and contain more mitochondria, possess greater concentrations of myoglobin and lipid, have a thicker sarcolemma and a much less extensive and more poorly developed SR, and have less sarcoplasm; however, red fibers are more generously supplied with blood than are white fibers. All of these differences lead to differences in oxidative stability.

Fish muscle is noticeably different from avian or mammalian muscles. Containing a larger percentage of myofibrillar protein than mammalian skeletal muscle, fish are characterized by a large percentage of unsaturated fatty acids. Changes in composition in response to environment are common. For example, fish that live in a low temperature environment have a larger fraction of dark muscle than fish that live in a warmer environment [32]. Caution is necessary when sampling fish for oxidative stability as local differences exist [32–35]. The most susceptible portion of herring fillets to oxidation during ice and frozen storage was found under the skin and was attributed to the large proportion of dark muscle in the sample compared with the middle and inner parts [33,34]. However, isolation and subsequent analysis of white and dark muscles will not always ensure a uniform response to oxidation. Although white muscle is considered to be very uniform in composition no matter where it is located on the fish, dark muscle varies in composition as a function of its location, containing more lipid in the anterior part of the fish and more water and protein in the posterior part [32].

To circumvent the inherent variability in composition that occurs between meat cuts, experimental studies often isolate subcellular membranes from muscles and measure their response to

TABLE 12.3

Phospholipid Composition (wt %) of Subcellular Membranes in Muscle

	Sarcolemma	Sarcoplasmic Reticulum	Mitochondria
Phosphatidylcholine	45.5	58.2	48.9
Phosphatidylethanolamine	22.4	29.4	39.3
Phosphatidylserine	17.6	9.4	9.0
Sphingomyelin	14.4	3.2	2.9

Source: From Fiehn, W., Peter, J.B., Mead, J.F., and Gan-Elepano, M., *J. Biol. Chem.*, 246, 5617, 1971.

oxidative catalysts. Depending on the degree of purification, wide variability may exist among isolated membranes. For instance, a microsomal fraction may contain membranes from SR, mitochondria, Golgi bodies, lysosomes, etc. A comparison of the lipid content of several of these subcellular membranes (Table 12.3) shows clearly that as the percentage of one membrane type changes in the isolate, the batch lipid composition also changes [36]. Even within one membrane type, such as SR, fractions isolated from cisternal and longitudinal SR [37] have revealed different ratios of phospholipid to protein [38]. Here also the lipid composition of the isolated membranes changes if more of one fraction is isolated. Ultimately, these changes in lipid content lead to differences in the oxidative susceptibility of the membrane isolate. The next section reviews the effects of lipid composition on oxidative stability.

D. LIPID SUBSTRATE

1. Fatty Acid Unsaturation

Carbon–hydrogen bond dissociation energies of a fatty acid are lowest at bisallylic methylene positions. These are the positions between adjacent double bonds [26,39]. Consequently, these positions are the thermodynamically favored sites for attack by lipid peroxyl radicals in polyunsaturated fatty acids (PUFAs). In studies involving the use of homogeneous solutions of purified lipids, a linear correlation has been found between the number of bisallylic methylene positions and the oxidizability of the lipids [40]. More recently, Wagner et al. [41] subjected cultured cells to oxidative stress following systematic alteration of the lipid unsaturation through supplementation of the growth medium with various PUFAs. In that study, the apparent oxidizability of the cellular lipids correlated exponentially with the number of bisallylic methylene positions in the cellular fatty acids. Different responses by the homogeneous and cellular systems to changes in PUFA content may be explained by a clustering of lipids within cell membranes that increases the apparent substrate concentration. Alternatively, Ursini et al. [42] suggested that unsaturated fatty acids are drawn into clusters of peroxidized lipids as part of a phase-compensating behavior. Such a process would feed the peroxyl radical-propagating reactions within the clusters of peroxidized lipid. In any event, Wagner et al. [41] found no apparent effect on the rate or extent of radical formation with fatty acid chain length, whereas Yin and Faustman [43] found with their liposomal model that both increased unsaturation and increased chain length resulted in greater phospholipid and oxymyoglobin oxidation. Location of methylene-interrupted double bonds also appears to affect the rate of oxidation as n-3 fatty acids autoxidized faster than n-6 fatty acids [44].

Despite the appearance of oxidative products in stored muscle foods, in some studies investigators were not able to detect losses of unsaturated fatty acids [45–47]. Failure to observe measurable losses in PUFAs during frozen storage of carp may have been due to considerable fish-to-fish variation in lipid composition [48]. Even a paired-fillet technique designed to minimize the variability in triacylglycerol composition of mackerel or catfish was insufficient to detect losses in total lipids or the triacylglycerol fraction [49]. However, focusing on specific lipid fractions rather than total lipid did lead to greater sensitivity in detection of losses of unsaturated fatty acids, with losses occurring primarily in the phospholipid fraction [46,50,51].

2. Lipid Composition of Muscle

A wide degree of variation in lipid content exists among muscles from different species (Table 12.4) [50,52,53]. Typically, in these muscles, the quantity of phospholipids is 500 mg/100 g muscle, directly paralleling the actual amount of membrane [54]. The remainder of the lipid may therefore be considered to be primarily nonpolar triacylglycerols. However, the level of triacylglycerol does not determine the oxidative susceptibility of that sample. Rather, the relative reactivity and accessibility to catalysts and inhibitors constitute the major determinants for identification of the critical site of oxidation. Membrane lipids are distributed throughout the tissue, but the triacylglycerols or storage lipids often are not. In chicken and trout muscle, adipose cells containing the triacylglycerols were primarily found in peripheral subcutaneous fat; in red meat muscle, adipose cells were found both between and within muscle fibers; and in salmon muscle, adipose cells were mainly distributed in the myosepta and, to a lesser extent, in the connective tissue surrounding bundles of white muscle fibers [55,56]. One must keep in mind, though, that adipose cells are bounded by membrane phospholipids. Oxidation of either fraction may then spread to the other lipid site. When such spreading occurred for a peroxidizing fish microsomal fraction system, emulsified lipids added to the system were oxidized [57]. Carbon-centered lipid radicals, which have been found in the extracellular medium of oxidizing cells, may be the vehicle by which oxidation of other adjacent lipid structures occurs [58].

As shown in Table 12.4, the degree of polyunsaturation in the muscle varies depending on the species. To understand the importance of this component to oxidative stability, comparisons between species or strains have been conducted and related to its fatty acid composition. In some cases, such as for two strains of channel catfish, the stabilities reflected the level of polyunsaturation [50], whereas in the case of frozen raw beef, pork, or chicken, they did not [53]. Heme pigments and levels of catalase instead were suggested to be the important determinants differentiating the stabilities of those meats.

Studies involving dietary modification, on the other hand, have consistently shown that oxidative stability reflects unsaturation in the muscle tissue. When the level of unsaturation in the diet was

TABLE 12.4
Lipid Composition of Raw Muscle Tissues

	Muscle/Strain/Variety	% Fat	% PUFA	References
Channel	Aqua	3.7	24.4	[50]
	LSU	5.4	22.1	[50]
Tilapia	Red	1.7	26.1	[52]
	Blue	1.9	24.5	[52]
Beef	Longissimus dorsi	4.4	4.9	[53]
	Semimembranosus	3.5	6.6	[53]
Pork	Longissimus dorsi	4.5	5.3	[53]
	Semimembranosus	3.3	9.2	[53]
Chicken	Breast	1.4	18.9	[53]
	Thigh	6.0	15.5	[53]

TABLE 12.5
Studies Exploring the Influence of Diet on Muscle Lipid Composition and Oxidative Stability

Source of Muscle	Primary Lipid Source in Diets	References
Lamb	Sunflower seed	[59]
Pork	Canola oil	[60]
Pork	Safflower oil or tallow	[61]
Pork	Oxidized corn oil	[62]
Rainbow trout	Fish oil or swine fat	[63]
Chicken	Olive oil, coconut oil, linseed oil, or partially hydrogenated soybean oil	[64]
Chicken	Oxidized corn oil	[65]
Chicken	Full-fat flax seed	[66]
Chicken	Fish meal	[67]
Chicken	Conjugated linoleic acid	[68]
Chicken	α-Linolenic acid	[69]

increased, lipid oxidation occurred to a greater extent (Table 12.5) [59–69]. Decreased lipid oxidation has been observed with feeding diets containing increased levels of saturated, monounsaturated, or conjugated linoleic fatty acids. Feeding diets containing oxidized dietary oil may or may not affect stability. Although Monahan et al. [62] did not see a significant influence on oxidation in pork chops when oxidized oil was incorporated into the diet of the animal, other investigators have observed an increased oxidative lability by lipids in broilers [65,70]. In addition to altering oxidative susceptibilities, the fatty acid composition also affects the types of oxidative volatiles produced [71,72]. Using principal component analysis on data of volatile compounds in samples, Meynier et al. [72] distinguished oxidized breast muscles from turkeys fed a diet containing 6% tallow, rapeseed oil, or soya oil.

3. Susceptibility of Lipid Classes to Oxidize

In vitro studies have been valuable tools in defining the contribution of individual lipid classes (triacylglycerols, phospholipids, free fatty acids) to the oxidative stability of a food system. Through in vitro studies, free fatty acids have been shown to oxidize faster than triacylglycerols [73], whereas the reactivity of membrane lipids is greater than that of emulsified triacylglycerols [74], apparently because arrangement of phospholipids in the membrane facilitates propagation. Proximity of the phospholipids to catalytic sites of oxidation (enzymatic lipid peroxidation, heme-containing compounds) also may contribute to the importance of membrane lipids in tissue oxidation along with the high degree of polyunsaturation in phospholipids [75]. Evidence that phospholipids are the major contributors to the development of warmed-over flavor (WOF) in meat from several different species of animals has been provided [76–81]. However, levels of total lipid seemed to be the major contributor to WOF in pork [76].

The method used to measure degree of oxidation may influence a study's conclusions. For example, a greater concentration of volatiles produced by the triacylglycerol fraction often does not have the impact on flavor that a smaller concentration of volatiles produced by the phospholipid fraction would have. This response arises because solubility in the lipid and flavor threshold of many of the volatiles increases as the level of fat increases. Hence, Roozen et al. [82,83] demonstrated that lowering the fat content in model systems increases the chance of flavor defects by reducing the concentration of volatiles retained in the fat.

The time frame under which an investigator examines the relative contributions of lipid classes to oxidation may also be a determining factor in the reported results. At earlier stages of oxidation, the peroxide value of raw sardine fillets was attributed to preferential oxidation of phospholipids and in later stages to oxidation of triacylglycerols [84]. Igene et al. [85] also showed that triacylglycerols in model meat systems were slow to oxidize and as such did not serve as a source of oxidative products until late in storage. In contrast, Erickson [50] suggested that free fatty acids released from the triacylglycerols served as the major site of oxidation in early stages of frozen storage of channel catfish and that phospholipids were only major contributors in later stages of storage. Thus, accessibility of lipid to hydrolytic enzymes could be an important factor for determining the oxidative susceptibility of a lipid class.

4. Susceptibility to Oxidation of Membrane Lipids

Variations in oxidative susceptibility within the individual phospholipid classes can be ascribed to the nature of the polar head group (choline, ethanolamine, serine, or inositol) and the degree of fatty acid unsaturation of the individual phospholipid [43,86]. With chicken meat, Pikul and Kummerow [87] identified phosphatidylinositol as containing the highest malonaldehyde levels and largest percentage of PUFAs, followed by phosphatidylethanolamine (PE), phosphatidylserine (PS), phosphatidylcholine (PC), cardiolipin, lysophosphatidylcholine, and sphingomyelin. In model meat systems, PE consistently changed more than PC in polyunsaturation during frozen storage, reflecting the higher initial levels of unsaturated fatty acids in PE [85]. Even when fatty acid composition was held constant in a liposome model system, PE-based liposomes exhibited a greater increase in both lipid and oxymyoglobin oxidation than PC-based liposomes [43]. In contrast, saturated PS added to PC-based liposomes inhibited oxidation through modification of surface charge and subsequent trapping of iron [88,89]. Similarly, inhibition by plasmalogen phospholipids (glycerophospholipids that contain a vinyl ether moiety at the *sn*-1 position) has been ascribed to their binding of iron [90] and to decreased propagation via oxidation of the vinyl ether bond [91]. On the other hand, possible causes of the inhibition of membrane phospholipid oxidation on incorporation of cholesterol include an alteration in packing of lipids in the membrane and chemical trapping of oxygen on conversion of cholesterol into nonradical oxide derivatives [92–94]. Cholesterol oxidation, in turn, is reduced by incorporation of sphingomyelin into the membrane bilayer [95].

5. Hydrolysis of Lipids and Associated Effects on Lipid Oxidation

Disintegration of lysosomal membranes in muscle tissue may occur on mincing or storage of nonheated muscle foods. As a result, muscle lipids may be exposed to lipolytic enzymes that are released from these organelles. In support of this statement is a report that both lipase and phospholipase activities were found in frozen fish [96]. Although phospholipases have been shown to be heat-inactivated more quickly than lipases [97], responses by these enzymes to frozen storage temperatures have been variable. When oyster was stored at −35°C, the activity of lipase was suppressed much more than that of phospholipase [46]. In contrast, in frozen cape hake mince, phospholipids were hydrolyzed faster than the neutral lipids above −12°C, whereas neutral lipids were hydrolyzed faster than phospholipids below −12°C [98].

As opposed to short-chain free fatty acids in dairy products, long-chain free fatty acids released in muscle foods do not contribute directly to rancid aromas. In general, further oxidation of these fatty acids is necessary to generate volatile products that are associated with sensory deterioration of the product. This mode of deterioration is distinct from the sensory deterioration described by Refsgaard et al. [99] for salmon. In these samples, hydrolysis of neutral lipids generated free fatty acids during frozen storage that contributed directly to an increased intensity of trained oil taste, bitterness, and metal taste by panelists.

The source of the free fatty acids determines whether lipid hydrolysis has an accelerating or inhibiting effect on subsequent rates of lipid oxidation. Free fatty acids originating from triacylglycerols accelerate oxidation [100,101], whereas free fatty acids hydrolyzed from phospho-lipids have been shown to inhibit oxidation [57,101,102]. In the latter case, it was suggested that free fatty acids disrupted the fatty acid alignment that facilitates free radical chain propagation in membranes [103]. Alternatively, Borowitz and Montgomery [104] concluded that the response to phospholipase may be dependent on the time of application of phospholipase. They found that when peroxidation preceded phospholipase A_2 (PLA$_2$) activity, the hydrolysis facilitated propagation of the peroxidative process. In contrast, if PLA$_2$ was activated before initiation of oxidation, oxidation was inhibited. This sequence would account for the acceleration of oxidation during frozen storage when fish had been held before mincing [105].

Although membrane lipid hydrolysis modifies the degree of lipid oxidation, the extent of hydrolytic activity may depend on the extent of membrane lipid oxidation. Oxidized fatty acids attached to phospholipids have been found to be more susceptible to hydrolysis by PLA$_2$ than those that were not oxidized [106]. Supporting this linkage between oxidation and hydrolysis, Han and Liston [107] also found increases in both activities when ferric iron was added to fish muscle.

E. Catalysts

Catalysts of lipid oxidation in muscle foods include both enzymic and nonenzymic sources (Table 12.6), but by and large, the bulk of the research to date has focused on the contribution of heme and nonheme iron to promotion of lipid oxidation.

1. Transition Metal Ions

Iron heme proteins, including myoglobin and hemoglobin, are abundant in muscle tissue [108]. Relative concentrations depend on species and muscle type. Beef, lamb, and pork generally contain more myoglobin than hemoglobin, whereas chicken contains a greater amount of hemoglobin [109]. The ability of heme proteins to promote lipid peroxidation has been demonstrated by many researchers [110–112]. One mechanism of this activation involves decomposition of preformed fatty acid hydroperoxides to peroxyl radicals, which in the presence of oxygen propagate lipid peroxidation [113]. Results of experiments with inhibitors suggest that the major pathway of peroxyl radical production involves high valence state iron complexes in a reaction analogous to the classical peroxidase pathway [114]:

$$Fe^{3+}(porphyrin) + ROOH \rightarrow Fe^{4+} = O(porphyrin)^{+\bullet} + ROH$$
$$Fe^{4+} = O(porphyrin)^{+\bullet} + ROOH \rightarrow Fe^{4+} = O(porphyrin) + ROO^{\bullet} + H^+$$

Alternatively, heme may donate its reducing equivalents to low molecular weight iron and copper complexes [115,116] and thereby contribute to catalysis of lipid oxidation. Heme compounds,

---

TABLE 12.6

Potential Catalysts of Lipid Oxidation in Muscle Foods

Nonenzymic
    Transition low molecular weight metal ions
    Metmyoglobin–H_2O_2
    Porphyrin compounds (sensitizers for the generation of singlet oxygen)
Enzymic
    Lipoxygenase
    Myeloperoxidase
    Membrane enzymic systems that reduce iron

---

particularly metmyoglobin and oxy-/deoxymyoglobin, also activate lipid oxidation through an intermediate species following interaction of the heme moiety with hydrogen peroxide [117–120]. An ongoing debate attempts to ascertain whether this intermediate species activates oxidation of PUFAs through ferryl oxygen [121,122], through a tyrosine peroxyl radical on the heme compound [123,124], or through radical transfer to other proteins that generate long-lived radicals [125,126]. Enhanced formation of this activated heme species has been observed when the heme compound is preincubated in the presence of the secondary lipid oxidation product 4-hydroxynonenal [127]. Decreased formation of the activated species, on the other hand, occurred in the presence of the free fatty acid linoleate [128] and was attributed to the fatty acid anion forming a hemichrome species that could not be activated by hydrogen peroxide. Based on studies using myoglobins with different affinities for heme, released hemin has been suggested as the critical entity that drives heme protein-mediated lipid oxidation in washed fish muscle during degradation of the heme ring and release of nonheme iron diminishes the contribution of myoglobin to promote lipid oxidation [129]. Given that exposure to hydrogen peroxide has also led to release of nonheme iron from myoglobin and hemoglobin [130–132], the contribution of heme compounds to initiation of oxidation could diminish with time.

In the nonheme form, iron participates in the production of the reactive oxygen species, the hydroxyl radical, via the chemical Fenton reaction: $Fe^{2+} + H_2O_2 \rightarrow Fe^{3+} + OH^{\bullet} + OH^-$. This reaction is effective when Fe^{3+} can be recycled to Fe^{2+} by various reducing agents.

Levels of low molecular weight nonheme iron are initially low, being only 2.4%–3.9% of total muscle iron in beef, lamb, pork, and chicken [109] and 6.7%–13.9% of total iron in flounder and mackerel muscle [133]. However, in muscles that have been processed and stored, increases in the catalytic low molecular weight iron fraction have been found [133–135]. Potential sources of nonheme iron are dislodgment of iron from the heme pocket by cooking [132,136] and release of iron from the iron storage protein, ferritin, by the reducing agents cysteine and ascorbate [137]. These increases are significant because concentrations of iron (2.2 μM) and copper (1.4 μM) that were found in the low molecular weight fractions of fish muscle were shown to catalyze lipid oxidation in fish muscle SR model systems [138]. Comparative evaluation of these ions in cuttlefish muscle during frozen storage, however, revealed that Fe(II) was a more effective catalyst of lipid oxidation than Cu(II) or Cd(II), and the prooxidative effect of Fe(II) was concentration dependent [139]. Hence, it was not surprising that modifications in dietary iron level altered the development of lipid oxidation in turkey dark muscle and in pork muscles [140–142].

Three general approaches have been taken to decipher the contribution of heme and nonheme iron to lipid oxidation of muscle foods:

1. Evaluation of the levels of heme and nonheme in the muscle food and relation of these values to the muscle's oxidative stability during storage [53,143].
2. Evaluation of the improvement in oxidative stability on addition of inhibitor/chelator that cancel out the contribution of one of the components [144–147].
3. Evaluation of the level of oxidation induced on addition of one of the iron sources to muscle or muscle model systems [144–150].

From these studies, the following conclusions may be drawn:

1. In raw red meat and dark muscle fish, heme iron is the major catalyst.
2. In both raw and cooked white flesh fish, nonheme iron is the major catalyst.

However, conflicting results make it difficult to draw a conclusion on the role of heme and nonheme iron in cooked red meat samples. Some factors that may modify the response are as follows:

1. Concentration of catalysts used in the study.
2. Distribution of catalysts in muscle system. Johns et al. [148] suggested that conflicting results from model system studies were due in part to the difficulty of evenly dispersing the catalysts in the system.
3. Concentration of reducing substances in system. Nonheme iron is more active in the reduced state, whereas heme iron is more active in the oxidized state [151].
4. pH of system. Heme catalysis is influenced less by increasing pH than nonheme iron [152]. Therefore, contribution of nonheme iron would increase with decreasing pH.
5. Amount of heat applied to the cooked system.
6. Presence of H_2O_2 in system.
7. Presence of chelators. Endogenous chelators, such as citrate, phosphate, and nucleotides, modify the reactivity of low molecular weight iron by modifying its redox potential to different extents [153].
8. Presence of one or more lipid classes. Oxidative response of chicken muscle model systems differed depending on the class of lipids present [154].

2. Singlet Oxygen Generation Systems

Although many different mechanisms have been proposed for the production of singlet oxygen, it is believed that the two most common mechanisms involve nonenzymatic photosensitization of natural pigments [155] or direct enzymatic production of singlet oxygen [22]. Nonenzymatic production of singlet oxygen, in turn, involves two different pathways. The type 1 pathway is characterized by hydrogen atom transfer or electron transfer between an excited triplet sensitizer and a substrate, resulting in the production of free radicals or free radical ions. These free radicals may then react with triplet oxygen to produce oxidized compounds, which readily breakdown to form free radicals that can initiate free radical chain reactions. In the second pathway for production of nonenzymatic production of singlet oxygen (type II), the excited triplet sensitizer reacts with triplet oxygen via a triplet–triplet annihilation mechanism. Enzymatic production of singlet oxygen, on the other hand, has been shown to be a direct or indirect consequence of the action of certain microsomal oxidases, lipoxygenase, and prostaglandin synthetase.

Evidence that singlet oxygen can initiate lipid oxidation can be obtained from analysis of the oxidative products. In the reaction of singlet oxygen with unsaturated fatty acids, one end of the singlet oxygen molecule reacts with the α-olefinic carbon, while the other end abstracts the γ-allylic hydrogen. As a result of this six-membered ring transition state, both conjugated and nonconjugated hydroperoxides are formed [156], whereas free radical autoxidation of lipid produces only nonconjugated hydroperoxides.

Pigments present in muscle foods that may act as photosensitizers (because their conjugated double-bond system easily absorbs visible light energy) include hematoporphyrins and riboflavin. In model systems containing myoglobin and its derivatives, Whang and Peng [157] demonstrated through electron paramagnetic resonance spectroscopy coupled with a spin trapping technique that dissociated hematin and especially the protoporphyrin IX ring exerted a photosensitizing function. Moreover, the location of porphyrin within membranes affected the extent of photodamage with increased residence times of singlet oxygen generated at deeper locations [158]. The participation of photosensitization in meat systems is supported by several storage studies: (1) ground turkey, ground pork, and shrimp samples exposed to light had higher levels of oxidative products than samples stored in the dark [159–161]; (2) incorporation of a UV light absorber in the packaging of pork patties prevented light-induced lipid oxidation [161]; and (3) incorporation of a singlet oxygen quencher (2,2,6,6-tetramethyl-4-piperidone) reduced the prooxidant effect of light in turkey meat [159].

3. Enzymic Initiation Systems

Several enzyme systems capable of initiating lipid oxidation have been identified in muscle foods. Among these, lipoxygenase stereoselectively absorbs a hydrogen atom from an active methylene group in 1,4-pentadiene structures of PUFA and releases a stereospecific conjugated diene hydroperoxy fatty acid product. Although it has greater recognition for its off-flavor development in vegetables and legumes, lipoxygenase has also been found in fish gill tissue [162,163], chicken muscle [164], and sardine skin [165]. However, according to Kanner et al. [166], these enzymes are not true initiators, since preformed hydroperoxides are necessary for their activation. Despite this requirement, products of lipoxygenase have been associated with fresh flavors of fish [163]. Their contribution to off-flavor generation in muscle foods during storage, on the other hand, remains debatable. While Grossman et al. [164] detected little loss of lipoxygenase activity in chicken muscle stored at −20°C for 12 months, German et al. [163] noted that lipoxygenases were unstable, being inactivated by 50% within 3 h at 0°C or completely inactivated with a single freeze–thaw cycle.

Membrane systems that reduce iron constitute another enzymic system of importance in the process of lipid oxidation because they not only generate active catalysts but do so in an environment consisting of highly unsaturated membrane lipids. Membrane systems in muscle capable of generating active oxygen species in the presence of NAD(P)H and ferric iron include SR [167] and mitochondria [168]. Although the enzymic systems in beef and chicken utilize NADPH preferentially [169,170], the enzymic system in fish utilizes NADH preferentially [167]. In the latter case, both NADH-cytochrome b_5 reductase and cytochrome b_5 have been associated with reduction of low and high concentrations of ferric-histidine and low concentrations of ferric-ATP [171]. During storage, as NADH concentrations drop to levels that are maximal to the enzymic system [167], stimulation by this enzymic system would be expected to increase with time postmortem.

Myeloperoxidase is another enzyme that may be present in muscle systems postmortem and capable of initiating lipid oxidation [172]. Normally found in neutrophils of blood, myeloperoxidase may contaminate muscle tissues following slaughter and spreading of blood over the surface of the product. Even when the blood has been washed off, residual concentrations of myeloperoxidase may be sufficient to accelerate oxidation in stored foods through the following reactions:

$$H_2O_2 + Cl^- \rightarrow HOCl + OH^-$$
$$HOCl + O_2^{-\bullet} \rightarrow O_2 + Cl^- + {}^{\bullet}OH$$

In lipoproteins and phospholipid liposomes, hypochlorous acid (HOCl) has been found to initiate lipid oxidation [173] and may do so by interacting with organic peroxides in vivo to form reactive radicals that subsequently initiate lipid oxidation [174].

F. Antioxidants

By far the most important defense mechanism for lipid oxidation is the presence of antioxidants, which can delay or slow the rate of oxidation of autoxidizable materials. Inhibition may take two forms: a reduction in the rate at which the maximal level of oxidation is approached and a reduction in the maximal level of oxidation. This section focuses primarily on antioxidants endogenously present in muscle tissues, with only limited discussion of antioxidants applied exogenously during processing.

1. Tocopherol

The main lipid-soluble antioxidant present in muscle tissue is tocopherol. Tocopherol is actually used as a generic description for mono-, di-, and trimethyl tocols that contain a 6-chromanol ring structure with different numbers of methyl groups at the 5, 7, and 8 positions and a saturated or

TABLE 12.7

Structure of Tocopherol and Tocotrienol

Trivial Name	Structure
α-Tocopherol	5,7,8-Trimethyltocol
β-Tocopherol	5,8-Dimethyltocol
γ-Tocopherol	7,8-Dimethyltocol
δ-Tocopherol	8-Methyltocol
α-Tocotrienol	5,7,8-Trimethyltocotrienol
β-Tocotrienol	5,8-Dimethyltocotrienol
γ-Tocotrienol	7,8-Dimethyltocotrienol
δ-Tocotrienol	8-Methyltocotrienol

unsaturated 16-carbon isoprenoid side chain (Table 12.7). α-Tocopherol is the predominant form in muscle tissue of beef, pork, chicken, and fish, although depending on the diet composition, γ-tocopherol and α-tocotrienol may also be present to varying degrees [52,175–178]. Studies indicate that when γ-tocopherol is supplied continuously in the diet, it accumulates in the muscle but to a much lesser extent than that when rats are fed similar levels of α-tocopherol [179,180]. Under these conditions, γ-tocopherol may instead tend to be accumulated in fat deposits.

In general, the mechanism of antioxidant action is believed to involve competition by the antioxidant for the peroxyl radical [181]. When tocopherol acts as the antioxidant, a series of chain-breaking oxidative reactions can occur:

Tocopherol

↓↑

Tocopheroxyl radical

↓↑

Tocopherone

↓

Tocopherylquinone

In this mechanism, α-tocopherol donates a phenolic hydrogen atom to a lipid peroxyl radical, which forms relatively unreactive intermediates. Although tocopherylquinone is a stable oxidized product, the tocopheroxyl radical and tocopherone may be reduced back to tocopherol by ascorbic acid [182] and ubiquinone [183]. Other mechanisms postulated to be involved in tocopherol's antioxidant action including scavenging of carbon-centered and hydrogen radicals [184], scavenging of singlet oxygen [185], and complexation of iron in the presence of ascorbate [186]. In another mechanism, Dmitriev et al. [187] suggested that reduction of the peroxyl radical back to the lipid molecule occurred along with the formation of the oxidized tocopherol and either molecular oxygen or the superoxide anion radical.

There are two approaches to identifying the contribution of tocopherol to muscle stability: monitoring of tocopherol concentrations in stored samples and comparison of muscle stability of animals that have had dietary tocopherol supplements and animals that have not. Pfalzgraf et al. [188] incorporated both approaches into their study but found that although supplemented samples had improved stability compared with unsupplemented (basal) samples, α-tocopherol levels in muscle tissue did not change during refrigerated storage for either supplemented or basal samples. In contrast, losses of tocopherol have been observed in frozen pork samples [189] while losses observed in frozen fish samples were preceded by a lag [50,190–192]. In these latter studies, accelerated degradation of tocopherol corresponded to an increased production of oxidative products. Similarly, the ratio of tocopherol to its oxidized product, tocopherolquinone, has been significantly related to the extent of oxidation produced in postmortem fish [193]. That a shift in rate of tocopherol degradation and oxidative product generation occurred within similar time periods suggests that when a critical concentration of tocopherol is reached, effective competition by tocopherol for peroxyl radicals is no longer possible and the oxidative propagation of lipids proceeds unchecked. The critical concentration varies depending on the level of phospholipid PUFAs present in the sample. Likewise, fatty acid and tocopherol compositions were considered to be the determinant factors for oxidative stability in an in vitro system [194]. Consequently, since fatty acid composition in most studies has not changed in response to varying tocopherol levels [195–198], dietary supplementation with tocopherol has been found overwhelmingly to improve oxidative stability of both lipids and proteins in muscle foods as a result of deposition of the dietary tocopherol within cellular membranes [196–209]. Due to the nonlinear relationship between dietary α-tocopherol levels and oxidative activity in muscle tissue [210], however, the most cost-effective dietary level depends on the treatment of the muscle and the type of muscle [211]. Postmortem supplementation of tocopherol, in contrast, has not proven as effective as dietary supplementation when levels in tissue were comparable [212], since exogenous tocopherol does not reside in the tissue membranes. However, deposition via dietary supplementation is not uniform throughout the muscles. Levels of α-tocopherol were found in beef to be highest in oxidative muscles (m. psoas major and m. gluteus medius) and lowest in glycolytic muscles (m. longissimus thoracis and m. longissimus lumborum) [213]. In turkey, α-tocopherol levels were higher in leg muscle than in breast muscle [214], whereas in salmon they were higher in the ventral area than in the midline [215], and in herring they were lower under the skin than in other parts of the fillet [34]. The degree to which tocopherol exerts an antioxidant effect consequently depends on other compositional parameters of the muscle. For example, oxidative stabilities were improved in breast meat but not in thigh meat when chickens were supplemented with α-tocopheryl acetate [216]. The variable deposition of tocopherol homologs into muscle tissue also factors into the changes in oxidative stability that are measured in response to dietary supplementation (α-tocopherol is preferentially deposited compared with γ-tocopherol [217]). For example, a decreased improvement in oxidative stability was found when diets included a natural source of tocopherol (RRR-α, γ-, and δ-tocopherol) compared with a synthetic source (all-rac-α-tocopherol) [218]. Such responses may account for the reduced effectiveness on cholesterol oxidation of free-range feeding of pigs compared with supplementation with dietary α-tocopheryl acetate [219].

2. Ascorbic Acid

Exogenous addition of ascorbic acid and its derivatives controls rancidity in a number of muscle tissues [220–225]. Within these tissues, ascorbic acid may function as an antioxidant through a variety of mechanisms: it may act as an oxygen scavenger [226], it may scavenge free radicals generated in the aqueous phase [227], it may maintain heme compounds in a reduced noncatalytic state [228], and it may regenerate tocopherol [229]. At the same time, ascorbic acid can act as a prooxidant by maintaining Fe(II) in its reduced state [228]. In an investigation on ground mullet tissue, Deng et al. [230] found that ascorbic acid tended to function as a prooxidant with small quantities and an antioxidant at high concentrations, with dark muscle requiring lower concentrations of ascorbic acid for the shift to occur than light muscle. Hence, dietary supplementation with ascorbic acid would only be advantageous if the levels increase in the tissue beyond these critical concentrations. In fact, several investigators have found the absence of any enhancement in oxidative stability for broilers fed ascorbic acid–supplemented diets [207,231]. Enhancement by dietary ascorbic acid, on the other hand, has been found when the diet also contained an elevated level of α-tocopherol [231,232]. Under those conditions, the ratio of ascorbic acid to dehydroascorbic acid could serve as a marker of oxidative susceptibility as it did for Iglesias et al. [233] in postmortem fish muscle. In another study, Erickson [234] demonstrated how distribution site can impact effectiveness of ascorbic acid. In that study, vacuum tumbling was used for the exogenous application of the antioxidant for intercellular distribution in channel catfish fillets, whereas an ascorbic acid bath was used to deliver the antioxidant to live channel catfish for absorption and intracellular distribution in the muscle tissue [235]. Final muscle concentrations of ascorbic acid with both treatments were twice what they were before treatment, but responses differed. In the case of intercellularly distributed ascorbic acid, the reducing agent protected membrane phospholipid but accelerated oxidation of triacylglycerols. In contrast, intracellularly distributed ascorbic acid did not accelerate oxidation of triacylglycerols but again protected membrane phospholipids. Considerations of compartmentation as well as the antioxidant polarity and target membrane charge must therefore be taken into account to develop more effective treatments for enhancement of tissue stability.

3. Carotenoids

The carotenoids are a group of fat-soluble pigments characterized by a linear, long-chain polyene structure. In addition to their roles in pigmentation and vitamin A activity, carotenoids may have a function similar to that of α-tocopherol, that is, to protect tissues from oxidative damage through scavenging of singlet oxygen and scavenging of peroxyl radicals [181]. In a manner exclusive of hydrogen abstraction, carotenoids are postulated to scavenge peroxyl radical through addition of the radical to the conjugated system such that the resulting carbon-centered radical is stabilized by resonance. When oxygen concentrations are low, a second peroxyl radical is added to the carbon-centered radical to produce a nonradical polar product. At high oxygen pressures, however, carotenoids act as prooxidants because the carbon-centered radical may add oxygen in a reversible reaction, resulting in an unstable chain-carrying peroxyl radical, which can further degrade to radicals and nonradical polar products with no net inhibition of oxidation [236]. As to performance of carotenoids in muscle tissue, the response to supplementation has varied. In postmortem supplementation, Lee and Lillard [237] observed similar levels of oxidation in cooked control and β-carotene-mixed hamburger patties following refrigerated storage. On the other hand, Clark et al. [238] reported that dietary canthaxanthin delayed formation of oxidative products in minced trout flesh during refrigerated storage. Similarly, Bjerkeng and Johnsen [239] observed a positive antioxidant effect in rainbow trout fillets. However, dietary supplementation has not always had a positive effect. Sigurgisladottir et al. [240] and Maraschiello et al. [241] did not observe any antioxidant effects by dietary carotenoids in salmon and broiler muscle tissue, respectively. Ruiz et al. [242] contended that sufficient levels of α-tocopherol must be present to demonstrate an antioxidant effect by carotenoids. In cases where the proportion of carotenoid/α-tocopherol is too high, the carotenoid competes with tocopherol for absorption, and the levels of α-tocopherol in the muscle decline. Variability in distribution of the carotenoid in the muscle tissue [215] could also account for different responses in dietary supplement studies if sample size is inadequate. As a final note, deposited carotenoids in muscle have different stabilities during storage (i.e., astaxanthin is more stable than canthaxanthin [243]), and these stabilities should be considered in supplementation of muscles to increase oxidative stability of the tissue.

4. Glutathione

Although the traditional approach to understanding the role of glutathione in relation to lipid oxidation has focused on its action as a substrate for detoxification enzymes, evidence is accumulating that glutathione in and of itself acts to control lipid oxidation in several ways. In the first case, glutathione may reduce the initiator, ferryl myoglobin, back to metmyoglobin [244]. In the second case, glutathione may serve to reduce oxidized sulfhydryl groups nonenzymatically [245]. Based on standard one-electron reduction potentials, Buettner [246] contends that it is thermodynamically feasible for glutathione to be oxidized to glutathione disulfide nonenzymatically, with the simultaneous reduction of a hydroxy, peroxy, or lipid radical to a hydroperoxide. Buettner's pecking order of oxidative activity [246] is supported by the finding that in frozen minced fish, glutathione declined at a faster rate than did ascorbic acid, which, in turn, declined faster than α-tocopherol [247].

Similarly, glutathione and ascorbate declined faster than α-tocopherol and ubiquinone in both light and dark muscle of mackerel [248]. However, caution must be taken in viewing glutathione degradation uniquely as a response to inhibition because in the oxidation of glutathione to its disulfide, superoxide anions can be produced. If the superoxide anion is not removed quickly through the action of superoxide dismutase, the net effect of the reduction of harmful radicals may be minimal owing to the formation of an active oxygen species that could lead to initiation or promotion of propagation in lipid oxidation. Alternatively, in the presence of low concentrations of oxygen, thiol radicals could have a greater tendency to react with each other such that oxidized glutathione formation without the intermediate production of superoxide anion and hydrogen peroxide could occur.

The extent to which glutathione is capable of inhibiting cellular lipid oxidation in muscle foods is questionable. Although Murai et al. [249] indicated that reduced glutathione was effective in limiting the adverse effects of oxidized fish oil in the diet of yellowtail fish, dietary supplementation of fingerling channel catfish did not affect fish performance, body composition, or stability of fillet samples [250].

5. Carnosine

Carnosine (β-alanyl-L-histidine) is an endogenously synthesized dipeptide present in beef, pork, chicken, and fish skeletal muscle at concentrations ranging from 0 to 70 mM [251–253]. Its exogenous addition to salted ground pork and beef muscle inhibited lipid oxidation [254,255], supporting model system studies demonstrating the inhibition of oxidation promoted by iron [256], hydrogen peroxide-activated hemoglobin [256], lipoxygenase [256], singlet oxygen [257], peroxyl radicals [258,259], and hydroxyl radicals [260]. Inhibitory action by carnosine may be related to its ability to chelate copper ions [261], scavenge free radicals [262], or trap volatile aldehydes [263,264]. Although carnosine at a 0.09% level in the diet has increased the oxidative stability of muscle [265], its high price makes its use as a feed additive impractical. Supplementation of diets with the carnosine precursors histidine and β-alanine has been attempted; however, it did not prove to be an efficient method for improving the oxidative stability of pork [266].

6. Flavonoids and Phenolic Acids

Flavonoids are secondary products of plant metabolism and include more than 4000 individual compounds divided into six subclasses: flavones, flavonones, isoflavones, flavonols, flavanols, and anthocyanins. Phenolic acids, structurally related to flavonoids, serve as precursors of flavonoid biosynthesis. Phenolic acids include hydroxycinnamic (caffeic, p-coumaric, ferulic, and sinapic acids), hydroxycoumarin (scopoletin), and hydroxybenzoic acids (4-hydroxybenzoic, ellagic, gallic, gentisic, protcatechuic, salicylic, and vanillic acids). The effectiveness of flavonoids and phenolic acids in retarding lipid oxidation in foods appears to be related not only to their chelating capacity but also to their ability to act as free radical acceptors. However, economic and regulatory hurdles prevent the use of the purified forms of these phenolics; consequently, plant extracts have been examined for their antioxidant potential in numerous studies (Table 12.8). In general, these plant extracts are inhibitory to lipid oxidation in the muscle food systems but are not as effective as synthetic antioxidant treatments [268,271]. Variables that have affected the response of muscle systems to plant extracts include: treatment concentration [274,275], stage at which treatment is applied [274,276], type of muscle system [277,278], and sample time [279]. Fractionation of grape and cranberry extracts has also revealed different potencies with flavonol oligomers and flavonol aglycones being the most potent inhibitors of lipid oxidation in frozen fish and cooked pork systems, respectively [280,281]. In membrane model system studies, antioxidant efficiency of flavonoids has been found to be dependent not only on their redox properties but also on their ability to interact with biomembranes [282,283]. Consequently, dietary supplementation of flavonoids could be a means to deposit the flavonoid in or near membranes. Unfortunately, dietary application of tea

TABLE 12.8
Selected Studies Evaluating Antioxidant Effectiveness of Exogenous Application of Plant Extracts to Muscle Food Systems

Plant Extract	Type of System	References
Green tea	Fish meat model system	[267]
	Raw and cooked chicken breast meat	[268]
Black pepper	Ground pork	[269]
Dried spices	Cooked minced meat patties	[270]
Grape seed	Raw and cooked chicken breast meat	[268]
Cocoa leaves	Mechanically deboned chicken meat	[271]
Cloudberry, beetroot, or willow herb	Cooked pork patties	[272]
Potato peel	Irradiated lamb	[273]

catechins or rosemary extract did not significantly improve lipid stability of beef, whereas exogenous application had inhibited lipid oxidation during refrigerated storage of patties [284]. Similarly, in pigs, there was only a tendency toward reduction of lipid oxidation noted in oregano-fed pork [285]. In contrast, dietary supplementation with tea catechins, rosemary, or oregano has proven effective in inhibiting lipid oxidation of turkey or chicken meat but to varying degrees [286–288]. When only one flavonoid was used in the supplement, dietary tea catechins were as effective as tocopherol, however, dietary rosemary and oregano were less effective. Dietary combinations of oregano and rosemary [289], or oregano and tocopherol [287,288], on the other hand, have proven to be more effective than dietary tocopherol alone in chicken muscle tissues.

7. Antioxidant Enzymes

Superoxide dismutase, catalase, and glutathione peroxidase are enzymes present in muscle tissues that may be classified as preventive antioxidants. Superoxide dismutase converts superoxide anion to hydrogen peroxide; catalase converts the hydrogen peroxide to water and oxygen; and glutathione peroxidase converts hydroperoxides to alcohols, thereby eliminating their potential decomposition by Fe^{2+}. Comparison of these enzyme activities with the metabolic activities in different fish species suggests that levels of glutathione peroxidase and superoxide dismutase reflect the degree of oxidative activity in the tissue [290]. Similarly, antioxidant enzyme activity was higher in the oxidative sartorius muscle of turkey than in the glycolytic pectoralis major muscle [291]. Nutritional status of the animal before slaughter may also play an important role in dictating the enzyme levels in the flesh [292,293]. For example, Maraschiello et al. [294] found that glutathione peroxidase activity in chicken thighs decreased as the level of α-tocopherol in the diet increased. In contrast, homeostatic compensation did not occur for glutathione peroxidase activity in turkey muscles on vitamin E supplementation [291]. Addition of dietary selenium was considered to be very effective in maintaining glutathione peroxidase activities in chicken muscle tissue during storage [295]; however, dietary selenium addition translated into only minor improvements in the oxidative stability [296]. Further evidence that the role of antioxidative enzymes is minimal is that their exogenous addition to raw and cooked meats produced only low or moderate inhibition of oxidative activity [297,298]. Such limited usefulness may be due to the susceptibility of these enzymes to inactivation by active oxygen species [299,300,380].

G. Mathematical Modeling

The pathways and relationships involved in lipid oxidation are complex. Consequently, there is a need to have methods to quantitatively link product composition to oxidative stability. Mathematical modeling is a tool that allows the synthesis of data from one or many experiments into an integrated system from which quantitative changes in many components may be calculated. Critical to any mathematical model is the endpoint selected for shelf life. Oftentimes, this endpoint is arbitrarily selected, but it should be based on consumer acceptability.

Aside from studies exploring variations in moisture and oxygen concentrations [301], few studies have attempted to model lipid oxidation, and even fewer focus on lipid oxidation in muscle foods. Kurade and Baranowski [302] reported that the shelf life of frozen and minced fish meat might be predicted by measuring total iron and myoglobin levels and the time for extracted lipids to gain 1% weight. Using these three variables, deviation from the actual shelf life measured 7.38%. On the other hand, the estimate for frozen shelf life of fish samples obtained by Ke et al. [303] gave an average deviation of 17% when only the last variable had been used. In either case, these models are flawed as predictors of the contribution of membrane lipids to oxidative stability. From isolated model system studies, it is known that the membrane environment is a major factor in the high susceptibility of phospholipids, and extraction of the phospholipids eliminates that factor. Experimental support was provided by Ke et al. [303] who observed that the polar lipids oxidized more slowly than the neutral lipids.

Using a slightly different approach to model lipid oxidation, Tappel et al. [304] incorporated into their model some of the major chemical features associated with lipid oxidation, including peroxidizability of polyunsaturated lipids, activation of inducers and their initiation of lipid peroxidation, concurrent autoxidation, inhibition of lipid peroxidation by vitamin E, reduction of some of the hydroperoxides by glutathione peroxidase, and formation of thiobarbituric acid-reactive substances (TBARS). The equations used to model the reactions were first brought into agreement with published information on these reactions by the determination of kinetic factors: activation degradation factor, inducer loss factor, antioxidant use factor, autoxidation factor, and hydroperoxide reduction factor. Subsequently, when the simulation program was applied to tissue slice and microsomal peroxidizing systems, the results of the simulation were in agreement with experimental data.

Babbs and Steiner [305] also used a computation model based on reactions involved in initiation, propagation, and termination. Their model incorporated 109 simultaneous enzymatic and free radical reactions, and rate constants were adjusted to account for the effects of phase separation of the aqueous and membrane lipid compartments. Computations from this model suggested that substantial lipid peroxidation occurred only when cellular defense mechanisms were weakened or overcome by prolonged oxidative stress. Consequently, although this model was developed to understand the contribution of free radical reactions to disease states of living organisms, useful insights may also be gleaned from it or similar models in understanding the oxidative stability of muscle food systems. In this manner, the variability that is inherent in the composition of foods may be factored into shelf-life predictions, and conditions for optimal stability may be derived.

IV. EFFECT OF PROCESSING TREATMENTS ON OXIDATION

As foods are subjected to various processing treatments before storage, the opportunity arises to modify their pattern of oxidation. Table 12.9 summarizes typical responses exhibited by muscle tissue during storage following various treatments [281–349]. Each of these treatments is covered in more detail in Sections IV.A through IV.P.

A. Bleeding

Slaughter of animals or fish is a necessary first step in converting the living organism to food. Slaughter methods and the accompanying bleeding step, however, may affect lipid oxidation through alteration in the removal of hemoglobin catalysts. For example, hemoglobin content and

TABLE 12.9

Typical Oxidative Response by Muscle Foods to Processing Treatments

Processing Treatment	Typical Oxidative Response	References
Bleeding	Inhibits	[306,307]
Rinses incorporating oxidizing agents	Promotes	[308–311]
Washing	Variable	[312–315]
Skinning	Variable	[34,316]
Mincing	Promotes	[317–325]
Salting	Promotes	[325–328]
Curing	Inhibits	[329–333]
Smoking	Inhibits	[334,335]
Cooking	Promotes	[336–348]
Deep-fat frying	Promotes	
High pressure	Promotes	[349–353]
Vacuum drying	Promotes	[354,355]
Irradiation	Promotes	[356–363]
Glazing/edible coatings	Inhibits	[332–336]
Freezing	Inhibits	[161,364–367]
Packaging	Inhibits	[358,360,361,368–379]

oxidative content of stored muscle tissue were significantly lower in skipjack killed by an instantaneous mechanical bleeding step than killed by holding in iced sea water [306]. In contrast, although no significant differences were found in residual hemoglobin contents of breast muscle subjected to an electrical stunning and bleeding treatment and a carbon dioxide stunning and bleeding treatment, TBARS content was higher in samples from the latter treatment [307].

B. RINSES

Chlorine rinses on muscle food constitute one processing step that could dramatically alter the site to which lipid oxidation is directed. Chlorine rinses, based primarily on sodium hypochlorite, are commonly used to reduce microbial loads in muscle foods. Such rinses have led to incorporation of chlorine into beef, pork, chicken, and shrimp [381–384] with phospholipids incorporating more chlorine per mole of lipid than neutral lipids. Lipid chlorohydrins, formed by the reaction of HOCl with unsaturated fatty acids [385], would likely disrupt the membrane's physical organization and reduce free radical chain oxidation of lipids [386]. This activity could be the basis for the reduced development of WOF in cooked and stored breast patties prepared from chickens rinsed in a chlorine bath versus patties prepared from nonrinsed chickens [387]. However, the capacity to form lipid chlorohydrins in tissue samples would be diminished in samples having high concentrations of thiol and amino groups because these groups display a much greater reactivity with HOCl than unsaturated fatty acids [388,389]. Preferential scavenging of chlorine by thiols and amino groups in dark chicken meat could account for the lack of significant sensory differences in cooked and stored thigh patties prepared from nonrinsed and chlorine-rinsed chickens [387].

Owing to health concerns about trihalomethanes and other chlorination reaction products generated during interaction of organics and aqueous chlorine, efforts have been made to explore alternatives. In one example, immersion chilling of chicken carcasses in tap water or 50 ppm monochloramine resulted in no subsequent differences in chlorine content or oxidative stability of breast, thigh, or skin [308]. In the case of chlorine dioxide, however, treated salmon and red grouper fillets had elevated levels of oxidative products [309]. Similarly, channel catfish fillets rinsed in ozonated water or water containing hydrogen peroxide had higher levels of oxidative products than untreated fillets [310]. Despite the elevated levels of oxidative products in both of these fish studies, the products were still considered acceptable. However, it is unclear at this time whether the increased concentration of oxidative products initially would impact the progression of lipid oxidation during storage. In the event sanitation rinses accelerate lipid oxidation of stored muscle tissues, the treatments could also include antioxidants. For example, Mahmoud et al. [311] rinsed skinless carp fillets with both electrolyzed water (main active agent is chlorine) and essential oils (0.5% carvacrol + 0.5% thymol) and found that this treatment significantly suppressed lipid oxidation.

C. WASHING

Washing (as opposed to rinsing) is designed to remove chemical constituents of the tissue rather than microbial contaminants. Used in the process of making surimi, washing removes not only soluble proteins but fat, prooxidants, antioxidants, and oxidative products as well. Muscle oxidative stability ultimately depends on the relative levels of these components removed. For example, hydroperoxides and TBARS are found in the wash water during surimi processing, and the last dewatering stage is critical to removing them and ensuring surimi quality [312]. Undeland et al. [313] also demonstrated that washing removed prooxidative enzymes and low molecular weight iron and copper catalysts from minced herring. Despite the removal of these components, however, washing decreased the lipid stability of the product pointing to the simultaneous removal of antioxidants. Consequently, to offset this antioxidant loss, investigators have recommended washing in antioxidant solutions [314]. For example, Richards et al. [315] demonstrated that an antioxidant wash improved the stability of mackerel fillets from stage 1 rigor but not stage 3 rigor. The improved response by the fresher fish was attributed to the greater removal of uncoagulated blood in those samples. As storage progresses and the pH drops, hemoglobin coagulation and binding to SR increase [390].

D. SKINNING

Accessibility of tissues to oxygen is considered one of the most important factors contributing to oxidative instability. Although filleting of fish is a common practice, it has been shown that the skin protects underlying areas from oxidation [34]. If the skin has to be removed early in the processing chain, deep skinning is an alternative to normal skinning and has been shown to improve the cold storage stability of saithe fillets [316]. Improvement is warranted since the highest rate of oxidation is observed in the under-skin layer lipids [34].

E. MINCING

Mincing muscle tissue disrupts cellular integrity and exposes more of the lipids to the oxidative catalysts; it also dilutes the antioxidants and increases the exposure of the tissue to oxygen [391]. In particular, the intracellular location of the triacylglycerols could provide protection against hydrolysis in the intact muscle; but on mincing, this protection is minimized if not eliminated. Mechanical disruption of the tissue also induces membrane lipids to form much smaller vesicles, and the increased surface area accelerates their degradation [317]. Further promotion of oxidation occurs in mechanically deboned flesh in response to release of hemoglobin and lipids from the bone marrow [318–320] and release of nonheme iron from the iron parts of the deboner [318,321]. Although meat grinder wear and degree of stress applied during deboning have not led to significant variations in oxidative stability of stored sample [321,322], the head pressure used in the mechanical deboning of roaster breasts affected the chemical composition of the product and, in turn, its susceptibility to lipid oxidation [320].

Hot boning (prerigor excision of muscle or muscle systems from animals) and further processing of prerigor meat have economic advantages represented by reductions in refrigeration costs, space requirements, processing delays, and product turnover time. Studies generally have found that product was less susceptible to oxidation when ground prerigor than postrigor, a result that was attributed to the higher ultimate pH in prerigor ground meat [323,324]. In a case of prerigor meat that was more unstable than postrigor meat, the increased initiation reactions in the prerigor meat may have been due to higher product temperature during grinding [325].

F. Salting

Salt (sodium chloride) is added to muscle foods for a variety of reasons, such as adding flavor and inhibiting microbial growth. Nevertheless, an accelerating effect on lipid oxidation has been found with salt in a variety of meats, including beef, pork, chicken, and fish [325–328]. Although Ellis et al. [392] did not attribute the prooxidative effect of salt to the chloride ion, Osinchak et al. [393] identified chloride as the active component of salt in a liposomal model system and suggested that it may operate through release of iron from ferritin or modification in bilayer organization. Concentration of the chloride anion in the system also appears to affect the response, with low concentrations elevating and high levels inhibiting lipid oxidation. Another factor that influences the response to the chloride anion is the associated cation in the salt. Divalent cations were more stimulatory than monovalent cations at equivalent concentrations of chloride up to 0.22 M chloride [393]. Wettasinghe and Shahidi [394] concluded that mediation in response to anions by cations is through their ability to participate in ion pairing interactions with anion counterparts.

G. Application of Curing Mixtures

1. Nitrite-Based Curing Agents

Meat preservation by means of curing is typically obtained by application of mixtures containing nitrite as the key ingredient. Other ingredients in the curing mixture include sodium chloride, sugars, ascorbate, polyphosphates, and spices. Nitrite imparts multiple functional roles to cured products, inhibiting spore germination of *Clostridium botulinum* when added in combination with sodium chloride, producing the characteristic cured meat color, contributing to the characteristic cured meat flavor, and inhibiting the development of WOF in cooked cured meats. Mechanisms proposed for the antioxidative activity of nitrite include formation of a strong complex with heme pigments (thereby preventing the release of nonheme iron and its subsequent catalysis of lipid oxidation), complexation of nonheme iron (which is catalytically less active than noncomplexed iron), and reaction with membrane-unsaturated lipids (which stabilizes the lipids) [329–331].

2. Nitrite-Free Curing Agents

A particular concern with the use of nitrite for the curing of meat has been the formation of *N*-nitrosamines, which are known carcinogens. Given the unlikelihood of finding a single compound that could perform all the functions of sodium nitrite, research directed to the elimination of the use of nitrites focused on formulation of multicomponent alternatives [332,333]. During this search, dinitrosyl ferrohemochrome was synthesized from hemin and nitric oxide and found to be capable of imparting a characteristic cured color to meat [395]. Subsequently, this natural cooked, cured meat pigment (CCMP) was demonstrated to accentuate the antioxidant activity of the ingredients used for flavor preservation in nitrite-free curing compositions [333]. In a β-carotene–linoleate model system, the antioxidant properties of CCMP were found to be concentration dependent and were hypothesized to involve quenching of free radicals [396].

H. Smoking

Smoking is a process that combines the effects of brining, heating, drying, and finally application of smoke to the product. The effectiveness of smoke as an antioxidant in processed meats is attributed to the phenols generated during thermal decomposition of phenolic acids and lignin [334]. In addition, smoke flavor may mask rancid flavor, thus requiring greater degrees of oxidation to render the product unacceptable [335].

I. Heating/Cooking

Another processing treatment that modifies lipid oxidation is the application of heat. Dislodgement of iron from heme compounds, disruption of cellular integrity, breakdown of preexisting hydroperoxides, and inactivation of lipases, phospholipases, lipoxygenase, and other enzymes associated with lipid peroxidation are consequences of heating and as a general rule lead to an acceleration in oxidation of stored precooked product. For example, heat-processed dark ground mackerel muscle oxidized faster during refrigerated storage than its raw counterpart; however, the opposite trend observed for the light ground muscle exemplified the exception to the rule [336].

The response of a product to heat is dependent on the endpoint temperature and the overall amount of heat applied [313,337]. A mathematical model derived to predict the development of WOF in minced beef during chill storage under various heating conditions estimated increasing levels of TBARS with increasing endpoint temperature (60°C–80°C) [338]. Spanier et al. [339] similarly found that higher core temperatures (68.3°C versus 51.7°C) in beef miniroasts caused higher levels of TBARS. On the other hand, Smith et al. [340] determined that acceleration in oxidation with increasing temperatures occurred only above a threshold temperature, which in the case of chicken breast was 74°C. The existence of a threshold temperature would explain results of Mast and MacNeil [341] who pasteurized mechanically deboned poultry at 59°C–60°C for up to 6 min and found that the treatment did not lead to acceleration in lipid oxidation during subsequent frozen storage at −18°C. Wang et al. [342] also found in the heating of lake herring that there existed a breakpoint in the amount of heat applied below which inactivation of lipoxygenase-like enzymes occurred and above which factors contributing to nonenzymatic oxidation increased. As temperature is further increased, another breakpoint develops in response to the generation of antioxidative Maillard reaction products. According to Hamm [343], the Maillard reaction in meats begins at about 90°C and increases with increased temperature and heating time. Later results support this statement: Huang and Greene [344] reported that beef subjected to high temperatures and long periods of heating developed lower TBA numbers than did samples subjected to lower temperatures for shorter periods of time. Differences previously found in method of cooking [345,346] therefore reflected differences in quantity of heat applied. In other cases, differences found to be characteristic of various cooking methods reflected variability in water loss rather than differences in the extent of oxidation. When TBA numbers were expressed per gram tissue, Pikul et al. [347] calculated that microwave cooking led to lower numbers than convection oven cooking. When expressed per gram of fat, however, the differences in results between microwave and convection cooking were not significant.

Length of refrigeration before cooking also affects the response of muscle tissue to cooking. Erickson [348] found that cooking of minced fish stored for 5 days generated greater amounts of either TBARS or fluorescent pigments than cooking of product stored for 7 days. A decreased loss of α-tocopherol in cooked 7 day product compared with cooked 2 or 5 day product gave further support to the conclusion that lipid oxidation had been inhibited during cooking in the 7 day sample. Change in pH, polyamines, Maillard reaction products, microbial removal of hydroperoxides and secondary oxidative products [397], and phospholipase activity were postulated as potential factors involved in the decreased response of refrigerated product to cooking.

J. DEEP-FAT FRYING

A specialized form of cooking is deep-fat frying, in which the product is immersed in hot cooking oil for some period of time to totally or partially cook it. Not only does the oil serve as the medium for heat transfer, it reacts with the protein and carbohydrate components of the food, developing unique flavors and odors that have definite appeal to the consumer. For breaded products, deep-fat frying also sets the batter, which binds the breading to the product surface.

A number of variables in addition to time and temperature of frying may impact the susceptibility to oxidation of a fried muscle product. Oil quantity is one of these variables. During frying, the product loses moisture and absorbs oil. When this absorbed oil is oxidized to any extent, it accelerates degradation of the product. Breading is another variable affecting oxidative susceptibility through attenuation of the response to oil quality. Breading inhibits the loss of moisture and absorption of frying oil [398], with finely ground breading material decreasing oil absorption to a greater extent than coarsely ground material [399]. Since batter and breading are in intimate contact with surface lipids, the catalysts, activators, and inhibitors present in the batter and breading also have the potential to affect lipid oxidation of the fried product.

K. HIGH PRESSURE

Increasing attention has been directed in recent years to the application of high hydrostatic pressures (up to 800 Mpa) to inactivate microorganisms [400]. Another distinct advantage claimed for pressurization is that heat-labile compounds undergo limited degradation compared with heat processing [401]. Although high hydrostatic pressure treatment prevented hydrolysis of phospholipids, a slight acceleration in lipid oxidation in cod and mackerel muscle occurred immediately following treatment [349] but became more pronounced with storage of pressurized dry-cured Iberian hams [350]. Similarly, oxidation of both sardine and chicken muscle has been accelerated by a high-pressure treatment, and the extent of oxidation has been related to intensity and duration of the treatment [351,352]. Inhibition of lipid oxidation by the addition of ethylenediamine-tetraacetic acid (EDTA) to pressure-treated minced pork indicated that transition metal ions were probably released from complexes and became available to catalyze oxidation in the treated samples [353].

L. DRYING

Drying has been a process applied to muscle foods for hundreds of years to extend the product's shelf life through reduction in water and subsequent inhibition of microbial growth. Freeze drying is an extension of this process; however, the porosity and surface areas of its products are higher than those of traditional dried products and therefore oxidation of freeze-dried products occurs more readily [354]. However, reduced surface areas of products subjected to controlled low temperature vacuum dehydration minimized the increased lipid oxidation associated with freeze-dried products [355].

M. IRRADIATION

Irradiation is the process of subjecting materials to electromagnetic radiation or electron beams of sufficient energy levels to sever chemical bonds. Although application of irradiation to muscle foods is intended to control pathogenic microorganisms, undesirable sensory changes in foods, especially development of off-flavors, may also arise, especially as the dose of radiation increases. For example, both Heath et al. [356] and Hashim et al. [357] reported that irradiation of uncooked chicken breast and thigh produced a characteristic bloody and sweet aroma that remained after the thighs were cooked but was not detectable after the breasts were cooked. These off-flavors are distinct from those associated with lipid oxidation and are believed to arise from protein oxidation.

To support this statement, aldehydes, ketones, and alcohols, typical volatile classes associated with lipid oxidation, did not increase following irradiation of pork loin and pork sausage [358,359], whereas irradiation increased the production of sulfur-containing volatiles (carbon disulfide, mercaptomethane, dimethyl sulfide, methyl thioacetate, and dimethyl disulfide) in pork loin [359]. The majority of these sulfur-containing volatiles dissipate from aerobically packaged samples while oxidative volatiles increase [359]. Acceleration in lipid oxidation of stored irradiated samples may be circumvented by storage in vacuum packaging [358,360,361] or by antioxidant addition/supplementation to muscles [358,362,363].

N. GLAZING

Glazing is a popular technique applied to fish products that will be frozen. Many types of glazes have been used, but the main glaze of commercial importance is a layer of ice, usually applied by immersing the product in water or by spraying it with water. The ice layer that is formed retards dehydration by preventing moisture from leaving the product and delays oxidation by preventing air contact with the product. However, as storage time progresses, sublimation occurs, thus decreasing effectiveness of such glazes as inhibitors. Various chemicals, including disodium acid phosphate, sodium carbonate, sodium lactate, corn syrup solids, cellulose gums, and pectinates, have at times been added to glazes so as to reduce the brittleness or the evaporative rate of the glaze; however, success has been limited [399,402].

Edible coatings (formed directly on foods) and films (preformed, then placed on foods) can also function to prevent quality losses associated with lipid oxidation by acting as oxygen barriers. Materials displaying these barrier properties include polysaccharides (alginates, pectins, agars, carrageenans, cellulose derivatives, amylose, starches, chitin, etc.) and proteins (casein, whey proteins, wheat gluten, soy proteins, corn zein, gelatin, collagen derivatives, etc.). Studies demonstrating the effectiveness of these coatings include those conducted on frozen turkey [403], frozen king salmon [404], frozen cooked ham and bacon pieces [405], and refrigerated beef patties [406]. In some cases, the coatings were no more effective than ice glazes [402]. Inclusion of green tea leaf extract in coatings applied to pork patties and irradiated, on the other hand, improved the stability of the product during subsequent refrigerated storage more than that if the product had been coated with only the pectin-based film material [407].

O. FREEZING

An excellent method of preserving the quality of meat and fish for long periods is by freezing the product. At temperatures below $-10°C$, both enzymatic and nonenzymatic reactions associated with lipid oxidation are decreased. In the range $0°C$ to $-10°C$, however, decreased oxidative stabilities have been noted. When water is removed as ice in this temperature range, an accelerating effect due to increased concentration of reactants is observed and it offsets the temperature-induced deceleration. Illustrating this accelerating effect of ice crystal formation on lipid oxidation, Apgar and Hultin [364] incubated microsomal membrane fractions in the presence and absence of miscible solvents (alcohols) to prevent freezing at temperatures below $0°C$. When the ratio of the reaction rate at $6°C$ was compared with that at $-12°C$, the rate of lipid oxidation was found to decrease less in the presence of ice than in the presence of alcohols.

In general, rate of freezing has been found to have little influence on the oxidative stability of frozen products unless the rate is very slow [365]. Pressure-shift freezing (freezing in conjunction with high pressure), however, has generated carp fillets that are more stable to oxidation during frozen storage than is the case with air blast freezing [366]. Frozen storage temperatures also play a dominant role in dictating the stability of muscle foods. This is particularly true when temperature fluctuations occur during storage, since the extracellular formation of ice crystals is accelerated and cellular disruption is enhanced, thus facilitating the interaction of catalysts with lipid substrates

[161]. Order of time/temperature holding treatments, on the other hand, markedly influenced development of rancidity. Lamb held at temperatures of $-5°C$ to $-10°C$ before storage at $-35°C$ developed more rancidity than lamb stored at $-35°C$ first, followed by storage at $-5°C$ or $-10°C$ [367].

P. PACKAGING

The stability of muscle foods during storage may be influenced by the packaging system. Vacuum packaging, which restricts the oxygen concentration, has been shown in numerous studies to extend the shelf life of muscle foods [358,360,361,368–373]. Similarly, oxidation was inhibited when an oxygen scavenger was incorporated into the packaging [374]. In contrast, application of modified atmospheres has produced variable results. Nolan et al. [375] found only minimal improvement in rancidity and WOF scores for precooked ground meat packaged in CO_2 or N_2 compared with packaging in air, whereas Hwang et al. [376] found that cooked beef loin slices packaged in an 80% N_2 and 20% CO_2 gas mixture responded similarly to vacuum packaging. Since residual oxygen concentrations in ground meat would be expected to be greater than intact samples, the responses from these two studies support those of model systems [408], which suggest that oxygen concentrations must be extremely low (<100 mmHg) before significant reduction in rates of lipid oxidation can be achieved.

Packaging materials present other opportunities to inhibit rates of lipid oxidation in stored samples. Antioxidants incorporated into the packaging inhibited lipid oxidation of fish muscle and beef steaks during refrigerated storage [377,378]. Similarly, packaging materials that incorporated a UV light absorber prevented light-induced lipid oxidation in pork patties [160]. Alternatively, holding packages in light with low emission in the blue band (warm-tone lamp) lowered lipid oxidation of turkey meat compared with holding under a daylight lamp [379].

V. SUMMARY

During refrigerated and frozen storage, many muscle foods are susceptible to degradation by lipid oxidative mechanisms. This chapter has reviewed the compositional factors in muscle that have been identified as contributing to or modifying lipid oxidation reactions. Interactions abound among these factors and are responsible for the inconsistencies in shelf-life evaluation of muscle foods. To fill the void, a greater emphasis should be placed on generation of mathematical models that can be used to link a muscle's oxidative stability to its composition. Ultimately, these models should be integrated with models simulating the effects of process variables. This type of approach would facilitate shelf-life prediction of muscle foods and derivation of conditions for optimal stability without the need to conduct accelerated storage tests on each batch of material being processed.

REFERENCES

1. D.B. Josephson, R.C. Lindsay, and D.A. Stuiber. Enzymic hydroperoxide initiated effects in fresh fish. *J. Food Sci.* 52:596–600 (1987).
2. C. Karahadian and R.C. Lindsay. Role of oxidative processes in the formation and stability of fish flavors. In: *Flavor Chemistry. Trends and Developments* (R. Teranishi, R.G. Buttery, and R. Shahidi, eds.). American Chemical Society, Washington, DC, 1989, pp. 60–75.
3. A.J. St. Angelo, J.R. Vercellotti, M.G. Legendre, C.H. Vinnett, J.W. Kuan, C. James, Jr., and H.P. Dupuy. Chemical and instrumental analyses of warmed-over flavor in beef. *J. Food Sci.* 52:1163–1168 (1987).
4. J. Kerler and W. Grosch. Odorants contributing to warmed-over flavor (WOF) of refrigerated cooked beef. *J. Food Sci.* 61:1271–1274, 1284 (1996).
5. S.W. Park and P.B. Addis. Cholesterol oxidation products in some muscle foods. *J. Food Sci.* 52:1500–1503 (1987).

6. M.P. Zubillaga and G. Maerker. Quantification of three cholesterol oxidation products in raw meat and chicken. *J. Food Sci.* 56:1194–1196, 1202 (1991).

7. N.J. Engeseth and J.I. Gray. Cholesterol oxidation in muscle tissue. *Meat Sci.* 36:309–320 (1994).

8. T. Ohshima, N. Li, and C. Koizumi. Oxidative decomposition of cholesterol in fish products. *J. Am. Oil Chem. Soc.* 70:595–600 (1993).

9. K. Osada, T. Kodama, L. Cui, K. Yamada, and M. Sugano. Levels and formation of oxidized cholesterols in processed marine foods. *J. Agric Food Chem.* 41:1893–1898 (1993).

10. D.M. Smith. Functional and biochemical changes in deboned turkey due to frozen storage and lipid oxidation. *J. Food Sci.* 52:22–27 (1987).

11. E.A. Decker, Y.L. Xiong, J.T. Calvert, A.D. Crum, and S.P. Blanchard. Chemical, physical, and functional properties of oxidized turkey white muscle myofibrillar proteins. *J. Agric. Food Chem.* 41:186–189 (1993).

12. L. Wan, Y.L. Xiong, and E.A. Decker. Inhibition of oxidation during washing improves the functionality of bovine cardiac myofibrillar protein. *J. Agric. Food Chem.* 41:2267–2271 (1993).

13. J.G. Akamittath, C.J. Brekke, and E.G. Schanus. Lipid oxidation and color stability in restructured meat systems during frozen storage. *J. Food Sci.* 55:1513–1517 (1990).

14. C. Faustman and R.G. Cassens. The biochemical basis for discoloration in fresh meat: a review. *J. Muscle Foods* 1:217–243 (1990).

15. C. Faustman, S.M. Specht, L.A. Malkus, and D.M. Kinsman. Pigment oxidation in ground veal: influence of lipid oxidation, iron and zinc. *Meat Sci.* 31:351–362 (1992).

16. H.J. Andersen, G. Bertelsen, A.G. Chrisophersen, A. Ohlen, and L.H. Skibsted. Development of rancidity in salmonoid steaks during retail display. A comparison of practical storage life of wild salmon and farmed rainbow trout. *Z. Lebensm. Unters. Forsch.* 191:119–122 (1990).

17. H.-M. Chen, S.P. Meyers, R.W. Hardy, and S.L. Biede. Color stability of astaxanthin pigmented rainbow trout under various packaging conditions. *J. Food Sci.* 49:1337–1340 (1984).

18. D.C. Borg and K.M. Schaich. Iron and hydroxyl radicals in lipid oxidation: Fenton reactions in lipid and nucleic acids co-oxidized with lipid. In: *Oxy-Radicals in Molecular Biology and Pathology: Proceedings* (P.A. Cerutti, I. Fridovich, and J.M. McCord, eds.). Alan R. Liss, New York, NY, 1988, pp. 427–441.

19. J.M.C. Gutteridge and B. Halliwell. The measurement and mechanism of lipid peroxidation in biological systems. *Trends Biochem. Sci.* 15:129–135 (1990).

20. E.N. Frankel. Volatile lipid oxidation products. *Prog. Lipid Res.* 22:1–33 (1982).

21. J. Kanner, J.B. German, and J.E. Kinsella. Initiation of lipid peroxidation in biological systems. *Crit. Rev. Food Sci. Nutr.* 25:317–364 (1987).

22. R.J. Hsieh and J.E. Kinsella. Oxidation of polyunsaturated fatty acids: mechanisms, products, and inhibition with emphasis on fish. *Adv. Food Nutr. Res.* 33:233–339 (1989).

23. H. Kappus. Lipid peroxidation: mechanisms and biological relevance. In: *Free Radicals and Food Additives* (O.I. Aruoma and B. Halliwell, eds.). Taylor and Francis, New York, NY, 1991, pp. 59–75.

24. D.G. Bradley and D.B. Min. Singlet oxygen oxidation of foods. *Crit. Rev. Food Sci. Nutr.* 31:211–236 (1992).

25. H.W. Gardner. Oxygen radical chemistry of polyunsaturated fatty acids. *Free Radic. Biol. Med.* 7:65–86 (1989).

26. W.H. Koppenol. Oxy radical reactions: from bond-dissociation energies to reduction potentials. *FEBS Lett.* 264:165–167 (1990).

27. G.S. Wu, R.A. Stein, and J.F. Mead. Autoxidation of fatty acid monolayers absorbed on silica gel: II. Rates and products. *Lipids* 12:971–978 (1977).

28. J. Terao. Reactions of lipid hydroperoxides. In: *Membrane Lipid Oxidation* (V.-P. Carmen, ed.). CRC Press, Boca Raton, FL, 1990, pp. 219–238.

29. N.A. Porter. Autoxidation of polyunsaturated fatty acids: initiation, propagation, and product distribution (basic chemistry). In: *Membrane Lipid Oxidation* (V.-P. Carmen, ed.), CRC Press, Boca Raton, FL, 1990, pp. 33–62.

30. D.R. Buege and B.B. Marsh. Mitochondrial calcium and postmortem muscle shortening. *Biochem. Biophys. Res. Commun.* 65:478–482 (1975).

31. M.L. Greaser, R.G. Cassens, W.G. Hoekstra, and E.J. Briskey. The effect of pH-temperature treatments on the calcium-accumulating ability of purified sarcoplasmic reticulum. *J. Food Sci.* 34:633–637 (1969).

32. R.M. Love. *The Chemical Biology of Fishes*, Vol. 2, *Advances 1968–1977*. Academic Press, London, 1980.

33. I. Undeland, G. Hall, and H. Lingnert. Lipid oxidation in fillets of herring (*Clupea harengus*) during ice storage. *J. Agric. Food Chem.* 47:524–532 (1999).

34. I. Undeland, M. Stading, and H. Lingnert. Influence of skinning on lipid oxidation in different horizontal layers of herring (*Clupea harengus*) during frozen storage. *J. Food Agric.* 78:441–450 (1998).

35. I. Undeland, B. Ekstrand, and H. Lingnert. Lipid oxidation in herring (*Clupea harengus*) light muscle, dark muscle, and skin, stored separately or as intact fillets. *J. Am. Oil Chem. Sci.* 75:581–590 (1998).

36. W. Fiehn, J.B. Peter, J.F. Mead, and M. Gan-Elepano. Lipids and fatty acids of sarcolemma, sarcoplasmic reticulum, and mitochondria from rat skeletal muscle. *J. Biol. Chem.* 246:5617–5620 (1971).

37. G. Salviati, P. Volpe, S. Salvatori, R. Betto, E. Damiani, A. Margreth, and I. Pasquali-Roncheti. Biochemical heterogeneity of skeletal-muscle microsomal membranes. Membrane origin, membrane specificity and fibre types. *Biochem. J.* 202:289–301 (1982).

38. M.G. Sarzala and M. Michalak. Studies on the heterogeneity of sarcoplasmic reticulum vesicles. *Biochim. Biophys. Acta* 513:221–235 (1978).

39. H.W. Gardner. Oxygen radical chemistry of polyunsaturated fatty acids. *Free Radic. Biol. Med.* 7:65–86 (1989).

40. J.D. Cosgrove, D.F. Church, and W.A. Pryor. The kinetics of the autoxidation of polyunsaturated fatty acids. *Lipids* 22:299–304 (1987).

41. B.A. Wagner, G.R. Buettner, and C.P. Burns. Free radical-mediated lipid peroxidation in cells: oxidizability is a function of cell lipid bis-allylic hydrogen content. *Biochemistry* 33:4449–4453 (1994).

42. F. Ursini, M. Maiorino, and A. Sevanian. Membrane hydroperoxides. In: *Oxidative Stress: Oxidants and Antioxidant* (H. Sies, ed.). Academic Press, New York, NY, 1991, pp. 319–336.

43. M.-C. Yin and C. Faustman. Influence of temperature, pH and phospholipid composition upon the stability of myoglobin and phospholipid: a liposome model. *J. Agric. Food Chem.* 41:853–857 (1993).

44. S. Adachi, T. Ishiguro, and R. Matsuno. Autoxidation kinetics for fatty acids and their esters. *J. Am. Oil Chem. Soc.* 72:547–551 (1995).

45. A. Beltrán and A. Moral. Gas chromatographic estimation of oxidative deterioration in sardine during frozen storage. *Lebensm. Wiss. Technol.* 23:499–504 (1990).

46. Y. Jeong, T. Ohshima, C. Koizumi, and Y. Kanou. Lipid deterioration and its inhibition of Japanese oyster *Crassostrea gigas* during frozen storage. *Nippon Suisan Gakkai Shi.* 56:2083–2091 (1990).

47. S.M. Polvi, R.G. Ackman, S.P. Lall, and R.L. Saunders. Stability of lipids and omega-3 fatty acids during frozen storage of Atlantic salmon. *J. Food Process. Preserv.* 15:167–181 (1991).

48. J. Mai and J.E. Kinsella. Changes in lipid composition of cooked minced carp (*Cyprinus carpio*) during frozen storage. *J. Food Sci.* 44:1619–1624 (1979).

49. Y. Xing, Y. Yoo, S.D. Kelleher, W.W. Nawar, and H.O. Hultin. Lack of changes in fatty acid composition of mackerel and cod during iced and frozen storage. *J. Food Lipids* 1:1–14 (1993).

50. M.C. Erickson. Compositional parameters and their relationship to oxidative stability of channel catfish. *J. Agric. Food Chem.* 41:1213–1218 (1993).

51. K. Whang and I.C. Peng. Lipid oxidation in ground turkey skin and muscle during storage. *Poult. Sci.* 66:458–466 (1987).

52. M.C. Erickson. Lipid and tocopherol composition of two varieties of tilapia. *J. Aquat. Prod. Technol.* 1:91–109 (1992).

53. K.S. Rhee, L.M. Anderson, and A.R. Sams. Lipid oxidation potential of beef, chicken, and pork. *J. Food Sci.* 61:8–12 (1996).

54. W.W. Christie. The composition, structure and function of lipids in the tissues of ruminant animals. *Prog. Lipid Res.* 17:111–137 (1978).

55. J.B. German. Muscle lipids. *J. Muscle Foods* 1:339–361 (1990).

56. S. Zhou, R.G. Ackman, and C. Morrison. Storage of lipids in the myosepta of Atlantic salmon (*Salmo salar*). *Fish Physiol. Biochem.* 14:171–178 (1995).

57. B.M. Slabyj and H.O. Hultin. Oxidation of a lipid emulsion by a peroxidizing microsomal fraction from herring muscle. *J. Food Sci.* 49:1392–1393 (1984).

58. J.A. North, A.A. Spector, and G.R. Buettner. Cell fatty acid composition affects free radical formation during lipid peroxidation. *Am. J. Physiol.* 267:C177–C188 (1994).

59. H.A. Bremner, A.L. Ford, J.J. MacFarlane, D. Ratcliffe, and N.T. Russell. Meat with high linoleic acid content: oxidative changes during frozen storage. *J. Food Sci.* 41:757–761 (1976).

60. K.S. Rhee, Y.A. Ziprin, G. Ordonez, and C.E. Bohac. Fatty acid profiles of the total lipids and lipid oxidation in pork muscles as affected by canola oil in the animal diet and muscle location. *Meat Sci.* 23:201–210 (1988).

61. D.K. Larick, B.E. Turner, W.D. Schoenherr, M.T. Coffey, and D.H. Pilkington. Volatile compound content and fatty acid composition of pork as influenced by linoleic acid content of the diet. *J. Anim. Sci.* 70:1397–1403 (1992).

62. F.J. Monahan, A. Asghar, J.I. Gray, D.J. Buckley, and P.A. Morrissey. Effect of oxidized dietary lipid and vitamin E on the colour stability of pork chops. *Meat Sci.* 37:205–215 (1994).

63. S.M. Boggio, R.W. Hardy, J.K. Babbitt, and E.L. Brannon. The influence of dietary lipid source and alpha-tocopheryl acetate level on product quality of rainbow trout (*Salmo gairdneri*). *Aquaculture* 51:13–24 (1985).

64. C.F. Lin, J.I. Gray, A. Asghar, D.J. Buckley, A.M. Booren, and C.J. Flegal. Effects of dietary oils and α-tocopherol supplementation on lipid composition and stability of broiler meat. *J. Food Sci.* 54:1457–1460, 1484 (1989).

65. A. Asghar, C.F. Lin, J.I. Gray, D.J. Buckley, A.M. Booren, R.L. Crackel, and C.J. Flegal. Influence of oxidised dietary oil and antioxidant supplementation on membrane-bound lipid stability in broiler meat. *Br. Poult. Sci.* 30:815–823 (1989).

66. A.O. Ajuyah, D.U. Ahn, R.T. Hardin, and J.S. Sim. Dietary antioxidants and storage affect chemical characteristics of ω-3 fatty acid enriched broiler chicken meats. *J. Food Sci.* 58:43–46, 61 (1993).

67. S.F. O'Keefe, F.G. Proudfoot, and R.G. Ackman. Lipid oxidation in meats of omega-3 fatty acid–enriched broiler chickens. *Food Res. Intern.* 28:417–424 (1995).

68. M. Du, D.U. Ahn, K.C. Nam, and J.L. Sell. Influence of dietary conjugated linoleic acid on volatile profiles, color, and lipid oxidation of irradiated raw chicken meat. *Meat Sci.* 56:387–395 (2000).

69. D.U. Ahn, S. Lutz, and J.S. Sim. Effects of dietary α-linolenic acid on the fatty acid composition, storage stability and sensory characteristics of pork loin. *Meat Sci.* 43:291–299 (1996).

70. C. Jensen, R. Engberg, K. Jakobsen, L.H. Skibsted, and G. Bertelsen. Influence of the oxidative quality of dietary oil on broiler meat storage stability. *Meat Sci.* 47:211–222 (1997).

71. J.S. Elmore, D.S. Mottram, M. Enser, and J.D. Wood. Effect of the polyunsaturated fatty acid composition of beef muscle on the profile of aroma volatiles. *J. Agric. Food Chem.* 47:1619–1625 (1999).

72. A Meynier, C. Genot, and G. Gandemer. Oxidation of muscle phospholipids in relation to their fatty acid composition with emphasis on volatile compounds. *J. Sci. Food Agric.* 79:797–804 (1999).

73. T.P. Labuza, H. Tsuyuki, and M. Karel. Kinetics of linoleate oxidation in model systems. *J. Am. Oil Chem. Soc.* 46:409–416 (1969).

74. B.M. Slabyj and H.O. Hultin. Oxidation of a lipid emulsion by a peroxidizing microsomal fraction from herring muscle. *J. Food Sci.* 49:1392–1393 (1984).

75. H.O. Hultin, E.A. Decker, S.D. Kelleher, and J.E. Osinchak. Control of lipid oxidation processes in minced fatty fish. In: *Seafood Science and Technology* (E.G. Bligh, ed.). Fishing News Books, Oxford, 1992, pp. 93–100.

76. B.R. Wilson, A.M. Pearson, and F.B. Shorland. Effect of total lipids and phospholipids on warmed-over flavor in red and white muscle from several species as measured by thiobarbituric acid analysis. *J. Agric. Food Chem.* 24:7–11 (1976).

77. J.O. Igene and A.M. Pearson. Role of phospholipids and triglycerides in warmed-over flavor development in meat model systems. *J. Food Sci.* 44:1285–1290 (1979).

78. J.O. Igene, A.M. Pearson, and J.I. Gray. Effects of length of frozen storage, cooking and holding temperatures upon component phospholipids and the fatty acid composition of meat triglycerides and phospholipids. *Food Chem.* 7:289–303 (1981).

79. C. Willemot, L.M. Poste, J. Salvador, and D.F. Wood. Lipid degradation in pork during warmed-over flavour development. *Can. Inst. Food Sci. Technol. J.* 18:316–322 (1985).

80. C.Y.W. Ang. Comparison of broiler tissues for oxidative changes after cooking and refrigerated storage. *J. Food Sci.* 53:1072–1075 (1988).

81. T.C. Wu and B.W. Sheldon. Influence of phospholipids on the development of oxidized off flavors in cooked turkey rolls. *J. Food Sci.* 53:55–61 (1988).

82. J.P. Roozen, E.N. Frankel, and J.E. Kinsella. Enzymic and autoxidation of lipids in low fat foods: model of linoleic acid in emulsified hexadecane. *Food Chem.* 50:33–38 (1994).

83. J.P. Roozen, E.N. Frankel, and J.E. Kinsella. Enzymic and autoxidation of lipids in low fat foods: model of linoleic acid in emulsified triolein and vegetable oils. *Food Chem.* 50:39–43 (1994).

84. S.-Y. Cho, Y. Endo, K. Fujimoto, and T. Kaneda. Oxidative deterioration of lipids in salted and dried sardine during storage at 5°C. *Nippon Suisan Gakkai Shi* 55:541–544 (1989).

85. J.O. Igene, A.M. Pearson, L.R. Dugan, Jr., and J.F. Price. Role of triglycerides and phospholipids on development of rancidity in model meat systems during frozen storage. *Food Chem.* 5:263–176 (1980).

86. K. Fukuzawa, T. Tadokoro, K. Kishikawa, K. Mukai, and J.M. Gebicki. Site-specific induction of lipid peroxidation by iron in charged micelles. *Arch. Biochem. Biophys.* 260:146–152 (1988).

87. J. Pikul and F.A. Kummerow. Relative role of individual phospholipids on thiobarbituric acid reactive substances formation in chicken meat, skin and swine aorta. *J. Food Sci.* 55:1243–1248, 1254 (1990).

88. K. Yoshida, J. Terao, T. Suzuki, and K. Takama. Inhibitory effect of phosphatidylserine on iron-dependent lipid peroxidation. *Biochem. Biophys. Res. Commun.* 179:1077–1081 (1991).

89. Y. Tampo and M. Yonaha. Effects of membrane charges and hydroperoxides on Fe(II)-supported lipid peroxidation in liposomes. *Lipids* 31:1029–1038 (1996).

90. M. Zommara, N. Tachibana, K. Mitsui, N. Nakatani, M. Sakono, I. Ikeda, and K. Imaizumi. Inhibitory effect of ethanolamine plasmalogen on iron- and copper-dependent lipid peroxidation. *Free Radic. Biol. Med.* 18:599–602 (1995).

91. P.J. Sindelar, Z. Guan, G. Dallner, and L. Ernster. The protective role of plasmalogens in iron-induced lipid peroxidation. *Free Radic. Biol. Med.* 26:318–324 (1999).

92. J. Sunamoto, Y. Baba, K. Iwamoto, and H. Kondo. Liposomal membranes. XX. Autoxidation of unsaturated fatty acids in liposomal membranes. *Biochim. Biophys. Acta* 833:144–150 (1985).

93. J. Szebeni and K. Toth. Lipid peroxidation in hemoglobin-containing liposomes. Effects of membrane phospholipid composition and cholesterol content. *Biochim Biophys. Acta* 857:139–145 (1986).

94. T. Parasassi, A.M. Giusti, M. Raimondi, G. Ravagnan, O. Sapora, and E. Gratton. Cholesterol protects the phospholipid bilayer from oxidative damage. *Free Radic. Biol. Med.* 19:511–516 (1995).

95. R.M. Sargis and P.V. Subbaiah. Protection of membrane cholesterol by sphingomyelin against free radical-mediated oxidation. *Free Radic. Biol. Med.* 40:2092–2102 (2006).

96. A.J. de Koning, S. Milkovitch, and T.H. Mol. The origin of free fatty acids formed in frozen cape hake mince (*Merluccius capensis,* Castelnau) during cold storage at −18°C. *J. Sci. Food Agric.* 39:79–84 (1987).

97. I. Bosund and B. Ganrot. Effect of pre-cooking of Baltic herring on lipid hydrolysis during subsequent cold storage. *Lebensm. Wiss. Technol.* 2:59–61 (1969).

98. A.J. de Koning and T.H. Mol. Rates of free fatty acid formation from phospholipids and neutral lipids in frozen cape hake (*Merluccius* spp) mince at various temperatures. *J. Sci. Food Agric.* 50:391–398 (1990).

99. H.H.F. Refsgaard, P.M.B. Brockhoff, and B. Jensen. Free polyunsaturated fatty acids cause taste deterioration of salmon during frozen storage. *J. Agric. Food Chem.* 48:3280–3285 (2000).

100. T.P. Labuza. Kinetics of lipid oxidation in foods. *Crit. Rev. Food Sci. Technol.* 2:355–405 (1971).

101. R.L. Shewfelt. Fish muscle lipolysis. A review. *J. Food Biochem.* 5:79–100 (1981).

102. F. Mazeaud and E. Bilinski. Free fatty acids and the onset of rancidity in rainbow trout (*Salmo gairdneri*) flesh. Effect of phospholipase A. *J. Fish Res. Board Can.* 33:1297–1302 (1976).

103. R.L. Shewfelt and H.O. Hultin. Inhibition of enzymic and non-enzymic lipid peroxidation of flounder muscle sarcoplasmic reticulum by pretreatment with phospholipase A_2. *Biochim. Biophys. Acta* 751:432–438 (1983).

104. S.M. Borowitz and C. Montgomery. The role of phopholipase A_2 in microsomal lipid peroxidation induced with *t*-butyl hydroperoxide. *Biochem. Biophys. Res. Commun.* 158:1021–1028 (1989).

105. A. Kolakowska. The rancidity of frozen Baltic herring prepared from raw material with different initial freshness. *Refrig. Sci. Technol.* 4:341–348 (1981).

106. A. Sevanian and E. Kim. Phospholipase A_2 dependent release of fatty acids from peroxidized membranes. *Free Radic. Biol. Med.* 1:263–271 (1985).

107. T.J. Han and J. Liston. Correlation between lipid peroxidation and phospholipid hydrolysis in frozen fish muscle. *J. Food Sci.* 53:1917–1919 (1988).

108. D.J. Livingston and W.D. Brown. The chemistry of myoglobin and its reactions. *Food Technol.* 35 (5):244–252 (1981).

109. T.J. Hazell. Iron and zinc compounds in the muscle meats of beef, lamb, pork and chicken. *J. Sci. Food Agric.* 33:1049–1056 (1982).

110. J. Kendrick and B.M. Watts. Acceleration and inhibition of lipid oxidation by heme compounds. *Lipids* 4:454–458 (1969).

111. R.M. Kaschinitz and Y. Hatefi. Lipid oxidation in biological membranes. Electron transfer proteins as initiator of lipid autoxidation. *Arch. Biochem. Biophys.* 171:292–304 (1975).

112. J. Kanner, H. Mendel, and P. Budowski. Carotene oxidizing factors in red pepper fruits (*Capsicum annum* L.): peroxidase activity. *J. Food Sci.* 42:1549–1551 (1977).

113. A.L. Tappel. The mechanism of the oxidation of unsaturated fatty acids catalyzed by hematin compounds. *Arch. Biochem. Biophys.* 44:378–395 (1953).

114. M.J. Davies. Detection of peroxyl and alkoxyl radicals produced by reaction of hydroperoxides with heme-proteins by electron spin resonance spectroscopy. *Biochim. Biophys. Acta* 964:28–35 (1988).

115. K.S. Rhee, Y.A. Zibru, and G. Ordonez. Catalysts of lipid oxidation in raw and cooked beef by metmyoglobin-hydrogen peroxide, nonheme iron and enzyme systems. *J. Agric. Food Chem.* 35:1013–1017 (1987).

116. B.R. Schricker and D.D. Miller. Effects of cooking and chemical treatments on heme and nonheme iron in meat. *J. Food Sci.* 48:1340–1342, 1349 (1983).

117. S. Harel and J. Kanner. Muscle membranal lipid peroxidation initiated by H_2O_2-activated metmyoglobin. *J. Agric. Food Chem.* 33:1188–1192 (1985).

118. J. Kanner and S. Harel. Initiation of membranal lipid peroxidation by activated metmyoglobin and methemoglobin. *Arch. Biochem. Biophys.* 237:314–321 (1985).

119. W.K.M. Chan, C. Faustman, M. Yin, and E.A. Decker. Lipid oxidation induced by oxymyoglobin and metmyoglobin with involvement of H_2O_2 and superoxide anion. *Meat Sci.* 46:181–190 (1997).

120. M.P. Richards and H.O. Hultin. Effect of pH on lipid oxidation using trout hemolysate as a catalyst: a possible role for deoxyhemoglobin. *J. Agric. Food Chem.* 48:3141–3147 (2000).

121. D.J. Kelman, J.A. DeGray, and R.P. Mason. Reaction of myoglobin with hydrogen peroxide forms a peroxyl radical which oxidizes substrates. *J. Biol. Chem.* 269:7458–7463 (1994).

122. S.I. Rao, A. Wilks, M. Hamberg, and P.R. Ortiz de Montellano. The lipoxygenase activity of myoglobin. Oxidation of linoleic acid by the ferryl oxygen rather than protein radical. *J. Biol. Chem.* 269:7210–7216 (1994).

123. M.J. Davies. 1990. Detection of myoglobin-derived radicals on reaction of metmyoglobin with hydrogen peroxide and other peroxidic compounds. *Free Radic. Res. Commun.* 10:361–370 (1990).

124. P. Gatellier, M. Anton, and M. Renerre. Lipid peroxidation induced by H_2O_2-activated metmyoglobin and detection of a myoglobin-derived radical. *J. Agric. Food Chem.* 43:651–656 (1995).

125. J.A. Irwin, H. Østdal, and M.J. Davies. Myoglobin-induced oxidative damage: evidence for radical transfer from oxidized myoglobin to other proteins and antioxidants. *Arch. Biochem. Biophys.* 362:94–104 (1999).

126. H. Østdal, H.J. Andersen, and M.J. Davies. Formation of long-lived radicals on proteins by radical transfer from heme enzymes—a common process? *Arch. Biochem. Biophys.* 362:105–112 (1999).

127. M.P. Lynch and C. Faustman. Effect of aldehyde lipid oxidation products on myoglobin. *J. Agric. Food Chem.* 48:600–604 (2000).

128. C.P. Baron, L.H. Skibsted, and H.J. Andersen. Peroxidation of linoleate at physiological pH: hemichrome formation by substrate binding protects against metmyoglobin activation by hydrogen peroxide. *Free Radic. Biol. Med.* 28:549–558 (2000).

129. E.W. Grunwald and M.P. Richards. Studies with myoglobin variants indicate that released hemin is the primary promoter of lipid oxidation in washed fish muscle. *J. Agric. Food Chem.* 54:4452–4460 (2006).

130. A. Puppo and B. Halliwell. Formation of hydroxyl radicals from H_2O_2 in the presence of Fe. (Is Hb a biological Fenton reagent?) *Biochem. J.* 249:185–190 (1988).

131. L.A. Eguchi and P. Saltman. The aerobic reduction of Fe(III) complexes by hemoglobin and myoglobin. *J. Biol. Chem.* 259:14337–14338 (1984).

132. K. Hegetschweiler, P. Saltman, C. Dalvit, and P.E. Wright. Kinetics and mechanisms of the oxidation of myoglobin by Fe(III) and Cu(II) complexes. *Biochim. Biophys. Acta* 912:384–397 (1987).

133. E.A. Decker and H.O. Hultin. Factors influencing catalysis of lipid oxidation by the soluble fraction of mackerel muscle. *J. Food Sci.* 55:947–950, 953 (1990).

134. J. Kanner, B. Hazan, and L. Doll. Catalytic "free" iron ions in muscle foods. *J. Agric. Food Chem.* 36:412–415 (1988).

135. J.V. Gomez-Basuri and J.M. Regenstein. Processing and frozen storage effects on the iron content of cod and mackerel. *J. Food Sci.* 57:1332–1336 (1992).

136. J.O. Igene, J.A. King, A.M. Pearson, and J.I. Gray. Influence of heme pigments, nitrite, and non-heme iron on development of warmed-over flavor (WOF) in cooked meat. *J. Agric. Food Chem.* 27:838–842 (1979).

137. E.A. Decker and B. Welch. Role of ferritin as a lipid oxidation catalyst in muscle food. *J. Agric. Food Chem.* 38:674–677 (1990).

138. E.A. Decker, C.-H. Huang, J.E. Osinchak, and H.O. Hultin. Iron and copper: role in enzymic lipid oxidation of fish sarcoplasmic reticulum at in situ concentrations. *J. Food Biochem.* 13:179–186 (1989).

139. A. Thanonkaew, S. Benjakul, W. Vissessanguan, and E.A. Decker. The effect of metal ions on lipid oxidation, colour, and physicochemical properties of cuttlefish (*Sepia pharaonis*) subjected to multiple freeze-thaw cycles. *Food Chem.* 95:591–599 (2006).

140. J. Kanner, I. Bartov, M.-O. Salan, and L. Doll. Effect of dietary iron level on in situ turkey muscle lipid peroxidation. *J. Agric Food Chem.* 38:601–604 (1990).

141. D.K. Miller, J.V. Gomez-Basuri, V.L. Smith, J. Kanner, and D.D. Miller. Dietary iron in swine rations affects nonheme iron and TBARS in pork skeletal muscles. *J. Food Sci.* 59:747–750 (1994).

142. D.K. Miller, V.L. Smith, J. Kanner, D.D. Miller, and H.T. Lawless. Lipid oxidation and warmed-over aroma in cooked ground pork from wine fed increasing levels of iron. *J. Food Sci.* 59:751–756 (1994).

143. K.S. Rhee and Y.A. Ziprin. Lipid oxidation in retail beef, pork and chicken muscles as affected by concentrations of heme pigments and nonheme iron and microsomal enzymic lipid peroxidation activity. *J. Food Sci.* 11:1–15 (1987).

144. J.O. Igene, J.A. King, A.M. Pearson, and J.I. Gray. Influence of heme pigments, nitrite, and non-heme iron on development of warmed-over flavor (WOF) in cooked meat. *J. Agric. Food Chem.* 27:838–842 (1979).

145. F. Shahidi and C. Hong. Role of metal ions and heme pigments in autoxidation of heat-processed meat products. *Food Chem.* 42:339–346 (1991).

146. J. Fischer and J.C. Deng. Catalysis of lipid oxidation: a study of mullet (*Mugil cephalus*) dark flesh and emulsion model system. *J. Food Sci.* 42:610–614 (1977).

147. J.Z. Tichivangana and P.A. Morrissey. Factors influencing lipid oxidation in heated fish muscle systems. *Ir. J. Food Sci. Technol.* 8:47–57 (1984).

148. A.M. Johns, L.H. Birkinshaw, and D.A. Ledward. Catalysts of lipid oxidation in meat products. *Meat Sci.* 25:209–220 (1989).

149. S. Apte and P.A. Morrissey. Effect of water-soluble haem and non-haem iron complexes on lipid oxidation of heated muscle systems. *Food Chem.* 26:213–222 (1987).

150. F.J. Monahan, R.L. Crackel, J.I. Gray, D.J. Buckley, and P.A. Morrissey. Catalysis of lipid oxidation in muscle model systems by haem and inorganic iron. *Meat Sci.* 34:95–106 (1993).

151. B.E. Greene and L.G. Price. Oxidation-induced color and flavor changes in meat. *J. Agric. Food Chem.* 23:164–167 (1975).

152. J.Z. Tichivangana and P.A. Morrissey. The influence of pH on lipid oxidation in cooked meats from several species. *Ir. J. Food Sci. Technol.* 9:99–106 (1985).

153. Y. Yoshida, S. Furuta, and E. Niki. Effects of metal chelating agents on the oxidation of lipids induced by copper and iron. *Biochim. Biophys. Acta* 1210:81–88 (1993).

154. R.V. Sista, M.C. Erickson, and R.L. Shewfelt. Lipid oxidation in a chicken muscle model system: oxidative response of lipid classes to iron ascorbate or methemoglobin catalysis. *J. Agric. Food Chem.* 48:1421–1426 (2000).

155. D.B. Min and H.-O. Lee. Chemistry of lipid oxidation. In: *Food Lipids and Health* (R.E. McDonald and D.B. Min, eds.). Marcel Dekker, New York, NY, 1996, pp. 241–268.

156. M.B. Korycka-Dahl and T. Richardson. Activated oxygen species and oxidation of food constituents. *Crit. Rev. Food Sci. Nutr.* 10:209–241 (1978).

157. K. Whang and I.C. Peng. Electron paramagnetic resonance studies of the effectiveness of myoglobin and its derivatives as photosensitizers in singlet oxygen generation. *J. Food Sci.* 53:1863–1865, 1893 (1988).

158. I. Bronshtein, K.M. Smith, and B. Ehrenberg. The effect of pH on the topography of porphyrins in lipid membranes. *Photochem. Photobiol.* 81:446–451 (2005).

159. K. Whang and I.C. Peng. Photosensitized lipid peroxidation in ground pork and turkey. *J. Food Sci.* 53:1596–1598, 1614 (1988).

160. H.J. Andersen and L.H. Skibsted. Oxidative stability of frozen pork patties. Effect of light and added salt. *J. Food Sci.* 56:1182–1184 (1991).

161. L.S. Bak, A.B. Andersen, E.M. Andersen, and G. Bertelsen. Effect of modified atmosphere packaging on oxidative changes in frozen stored cold water shrimp (*Pandalus borealis*). *Food Chem.* 64:169–175 (1999).

162. J.B. German and J.E. Kinsella. Lipid oxidation in fish tissue. Enzymatic initiation via lipoxygenase. *J. Agric. Food Chem.* 33:680–683 (1985).

163. J.B. German, H. Zhang, and R. Berger. Role of lipoxygenases in lipid oxidation in foods. In: *Lipid Oxidation in Food* (A.J. St. Angelo, ed.). American Chemical Society, Washington, DC, 1992, pp. 74–92.

164. S. Grossman, M. Bergman, and D. Sklan. Lipoxygenase in chicken muscle. *J. Agric. Food Chem.* 36:1266–1270 (1988).

165. S. Mohri, S.-Y. Cho, Y. Endo, and K. Fujimoto. Linoleate 13(S)-lipoxygenase in sardine skin. *J. Agric. Food. Chem.* 40:573–576 (1992).

166. J. Kanner, J.B. German, and J.E. Kinsella. Initiation of lipid peroxidation in biological systems. *Crit. Rev. Food Sci. Nutr.* 25:317–364 (1987).

167. R.E. McDonald and H.O. Hultin. Some characteristics of the enzymic lipid peroxidation system in the microsomal fraction of flounder skeletal muscle. *J. Food Sci.* 52:15–21, 27 (1987).

168. S.-W. Luo. NAD(P)H-dependent lipid peroxidation in trout muscle mitochondria. Ph.D. thesis, University of Massachusetts, Amherst, MA, 1987.

169. T.-S. Lin and H.O. Hultin. Enzymatic lipid peroxidation in microsomes of chicken skeletal muscle. *J. Food Sci.* 41:1488–1489 (1976).

170. K.S. Rhee, T.R. Dutson, and G.C. Smith. Enzymic lipid peroxidation in microsomal fractions from beef skeletal muscle. *J. Food Sci.* 49:675–679 (1984).

171. M.-X. Yang and A.I. Cederbaum. Role of cytochrome b_5 in NADH-dependent microsomal reduction of ferric complexes, lipid peroxidation, and hydrogen peroxide generation. *Arch. Biochem. Biophys.* 324:282–292 (1995).

172. J. Kanner and J.E. Kinsella. Lipid deterioration initiated by phagocytic cells in muscle foods: β-carotene destruction by a myeloperoxidase-hydrogen peroxide-halide system. *J. Agric. Food Chem.* 31:370–376 (1983).

173. O.M. Panasenko, S.A. Evgina, E.S. Driomina, V.S. Sharov, V.I. Sergienko, and Y.A. Vladimirov. Hypochlorite induces lipid peroxidation in blood lipoproteins and phospholipid liposomes. *Free Radic. Biol. Med.* 19:133–140 (1995).

174. O.M. Panasenko, J. Arnhold, and J. Schiller. Hypochlorite reacts with an organic hydroperoxide forming free radicals, but not singlet oxygen, and thus initiates lipid peroxidation. *Biochem. Moscow* 62:951–959 (1997).

175. E.-L. Syväoja, K. Salminen, V. Piironen, P. Varo, O. Kerojoki, and P. Koivistoinen. Tocopherol and tocotrienols in Finnish foods: fish and fish products. *J. Am. Oil Chem. Soc.* 62:1245–1248 (1985).

176. V. Piironen, E.-L. Syväoja, P. Varo, K. Salminen, and P. Koivistoinen. Tocopherols and tocotrienols in Finnish foods: meat and meat products. *J. Agric. Food Chem.* 33:1215–1218 (1985).

177. M.C. Erickson. Lipid and tocopherol composition of farm-raised striped and hybrid striped bass. *Comp. Biochem. Physiol.* 101A:171–176 (1992).

178. M.C. Erickson. Variation of lipid and tocopherol composition in three strains of channel catfish (*Ictalurus punctatus*). *J. Sci. Food Agric.* 59:529–536 (1992).

179. W.A. Behrens and R. Madere. Mechanisms of absorption, transport and tissue uptake of RRR-α-tocopherol and d-γ-tocopherol in the white rat. *J. Nutr.* 117:1562–1569 (1987).

180. M. Clément, L. Dinh, and J.-M. Bourre. Uptake of dietary *RRR*-α- and *RRR*-γ-tocopherol by nervous tissues, liver and muscle in vitamin-E-deficient rats. *Biochim. Biophys. Acta* 1256:175–180 (1995).

181. G.W. Burton, L. Hughes, D.O. Foster, and E. Pietrzak. Antioxidant mechanisms of vitamin E and β-carotene. In: *Free Radicals: From Basic Science to Medicine* (G. Poli, E. Albano, and M.U. Dianzani, eds.). Birkhäuser Verlag, Basel, 1993, pp. 388–399.

182. K. Mukai, M. Nishimura, A. Nagano, K. Tanaka, and E. Niki. Kinetic study of the reaction of vitamin C derivatives with tocopheroxyl (vitamin E radical) and substituted phenoxyl radicals in solution. *Biochim. Biophys. Acta* 993:168–173 (1989).

183. V. Kagan, E. Serbinova, and L. Packer. Antioxidant effects of ubiquinones in microsomes and mitochondria are mediated by tocopherol recycling. *Biochem. Biophys. Res. Commun.* 169:851–857 (1990).

184. M. Hiramatsu, R.D. Velasco, and L. Packer. Decreased carbon centered and hydrogen radicals in skeletal muscle of vitamin E supplemented rats. *Biochem. Biophys. Res. Commun.* 179:859–864 (1991).

185. S. Kaiser, P. Di Mascio, M.E. Murphy, and H. Sies. Physical and chemical scavenging of singlet molecular oxygen by tocopherols. *Arch. Biochem. Biophys.* 277:101–108 (1990).

186. D.A. Stoyanovsky, V.E. Kagan, and L. Packer. Iron binding to alpha-tocopherol-containing phospholipid liposomes. *Biochem. Biophys. Res. Commun.* 160:834–838 (1989).

187. L.F. Dmitriev, M.V. Ivanova, and V.Z. Lankin. Interaction of tocopherol with peroxyl radicals does not lead to the formation of lipid hydroperoxides in liposomes. *Chem. Phys. Lipids* 69:35–39 (1994).

188. A. Pfalzgraf, M. Frigg, and H. Steinhart. α-Tocopherol contents and lipid oxidation in pork muscle and adipose tissue during storage. *J. Agric. Food Chem.* 43:1339–1342 (1995).

189. E.H. Kim, C.K. Han, K.S. Sung, C.S. Yoon, N.H. Lee, D.Y. Kim, and C.J. Kim. Effect of vitamin E on the lipid stability of pork meat. *Korean J. Anim. Sci.* 36:285–291 (1994).

190. M.C. Erickson. Ability of chemical measurements to differentiate oxidative stabilities of frozen minced muscle tissue from farm-raised striped bass and hybrid striped bass. *Food Chem.* 48:381–385 (1993).

191. M.C. Erickson and S.T. Thed. Comparison of chemical measurements to differentiate oxidative stability of frozen minced tilapia fish muscle. *Int. J. Food Sci. Technol.* 29:585–591 (1994).

192. M. Pazos, M.J. Gonzalez, J.L. Gallardo, J.L. Torres, and I. Medina. Preservation of the endogenous antioxidant system of fish muscle by grape polyphenols during frozen storage. *Eur. Food Res. Technol.* 220:514–519 (2005).

193. M. Pazos, L. Sanchez, and I. Medina. α-Tocopherol oxidation in fish muscle during chilling and frozen storage. *J. Agric. Food Chem.* 53:4000–4005 (2005).

194. M.-C. Yin and C. Faustman. The influence of microsomal and cytosolic components on the oxidation of myoglobin and lipid in vitro. *Food Chem.* 51:159–164 (1994).

195. S. Sigurgisladottir, C.C. Parrish, S.P. Lall, and R.G. Ackman. Effects of feeding natural tocopherols and astaxanthin on Atlantic salmon (*Salmo salar*) fillet quality. *Food Res. Intern.* 27:23–32 (1994).

196. F.J. Monahan, D.J. Buckley, P.A. Morrissey, P.B. Lynch, and J.I. Gray. Influence of dietary fat and α-tocopherol supplementation on lipid oxidation in pork. *Meat Sci.* 31:229–241 (1992).

197. M. Frigg, A.L. Prabuck, and E.U. Ruhdel. Effect of dietary vitamin E levels on oxidative stability of trout fillets. *Aquaculture* 84:145–158 (1990).

198. P. Akhtar, J.I. Gray, T.H. Cooper, D.L. Garling, and A.M. Booren. Dietary pigmentation and deposition of α-tocopherol and carotenoids in rainbow trout muscle and liver tissue. *J. Food Sci.* 64:234–239 (1999).

199. A. Asghar, J.I. Gray, A.M. Booren, E.A. Gomaa, M.M. Abouzied, and E.R. Miller. Effects of supranutritional dietary vitamin E levels on subcellular deposition of α-tocopherol in the muscle and on pork quality. *J. Sci. Food Agric.* 57:31–41 (1991).

200. J.E. Cannon, J.B. Morgan, G.R. Schmidt, R.J. Delmore, J.N. Sofos, G.C. Smith, and S.N. Williams. Vacuum-packaged precooked pork from hogs fed supplemental vitamin E: chemical, shelf-life and sensory properties. *J. Food Sci.* 60:1179–1182 (1995).

201. P. Dirinck, A. De Winne, M. Casteels, and M. Frigg. Studies on vitamin E and meat quality. 1. Effect of feeding high vitamin E levels on time-related pork quality. *J. Agric. Food Chem.* 44:65–68 (1996).

202. C.F. Lin, J.I. Gray, A. Asghar, D.J. Buckley, A.M. Booren, and C.J. Flegal. Effects of dietary oils and α-tocopherol supplementation on lipid composition and stability of broiler meat. *J. Food Sci.* 54:1457–1460, 1484 (1989).

203. D.M. Gatlin III, S.C. Bai, and M.C. Erickson. Effects of dietary vitamin E and synthetic antioxidants on composition and storage quality of channel catfish, *Ictalurus punctatus*. *Aquaculture* 106:323–332 (1992).

204. Q. Liu, K.K. Scheller, D.M. Schaefer, S.C. Arp, and S.N. Williams. Dietary α-tocopherol acetate contributes to lipid stability in cooked beef. *J. Food Sci.* 59:288–290 (1994).

205. N.J. Engeseth, J.I. Gray, A.M. Booren, and A. Asghar. Improved oxidative stability of veal lipids and cholesterol through dietary vitamin E supplementation. *Meat Sci.* 35:1–15 (1993).

206. C. Jensen, C. Lauridsen, and G. Bertelsen. Dietary vitamin E: quality and storage stability of pork and poultry. *Trends Food Sci. Technol.* 9:62–72 (1998).
207. P.A. Morrissey, P.J.A. Sheehy, K. Galvin, J.P. Kerry, and D.J. Buckley. Lipid stability in meat and meat products. *Meat Sci.* 49(Suppl):73–86 (1998).
208. P. Gatellier, Y. Mercier, E. Rock, and M. Renerre. Influence of dietary fat and vitamin E supplementation on free radical production and on lipid and protein oxidation in turkey muscle extracts. *J. Agric. Food Chem.* 48:1427–1433 (2000).
209. Y. Mercier, P. Gatellier, A. Vincent, and M. Renerre. Lipid and protein oxidation in microsomal fraction from turkeys: Influence of dietary fat and vitamin E supplementation. *Meat Sci.* 58:125–134 (2001).
210. L. Cortinas, A. Barroeta, C. Villaverde, J. Galobart, F. Guardiola, and M.D. Baucells. Influence of the dietary polyunsaturation level on chicken meat quality: lipid oxidation. *Poult. Sci.* 84:48–55 (2005).
211. K. Eder, G. Grunthal, H. Kluge, F. Hirche, J. Spilke, and C. Brandsch. Concentrations of cholesterol oxidation products in raw, heat-processed and frozen-stored meat of broiler chickens fed diets differing in the type of fat and vitamin E concentrations. *Br. J. Nutr.* 93:633–643 (2005).
212. M. Mitsumoto, R.N. Arnold, D.M. Schaefer, and R.G. Cassens. Dietary versus postmortem supplementation of vitamin E on pigment and lipid stability in ground beef. *J. Anim. Sci.* 71:1812–1816 (1993).
213. A. Lynch, J.P. Kerry, M.G. O'Sullivan, J.B.P. Lawlor, D.J. Buckley, and P.A. Morrissey. Distribution of α-tocopherol in beef muscles following dietary α-tocopheryl acetate supplementation. *Meat Sci.* 56:211–214 (2000).
214. F.M. Higgins, J.P. Kerry, D.J. Buckley, and P.A. Morrissey. Effect of dietary α-tocopheryl acetate supplementation on α-tocopherol distribution in raw turkey muscles and its effect on the storage stability of cooked turkey meat. *Meat Sci.* 50:373–383 (1998).
215. H.H.F. Refsgaard, P.B. Brockhoff, and B. Jensen. Biological variation of lipid constituents and distribution of tocopherols and astaxanthin in farmed Atlantic salmon (*Salmo salar*). *J. Agric. Food Chem.* 46:808–812 (1998).
216. K. Galvin, P.A. Morrissey, and D.J. Buckley. Influence of dietary vitamin E and oxidised sunflower oil on the storage stability of cooked chicken muscle. *Br. Poult. Sci.* 38:499–504 (1997).
217. M.P.M. Parazo, S.P. Lall, J.D. Castell, and R.G. Ackman. Distribution of α- and γ-tocopherols in Atlantic salmon (*Salmo salar*) tissues. *Lipids* 33:697–704 (1998).
218. C. Jensen, L.H. Skibsted, K. Jakobsen, and G. Bertelsen. Supplementation of broiler diets with *all-rac-α-* or a mixture of natural source *RRR-α-*, γ-, δ-tocopherol acetate. 2. Effect on the oxidative stability of raw and precooked broiler meat products. *Poult. Sci.* 74:2048–2056 (1995).
219. A.I. Rey, C.J. Lopez-Bote, and J.D. Buckley. Effect of feed on cholesterol concentration and oxidation products development of longissimus dorsi muscle from Iberian pigs. *Ir. J. Agric. Food Res.* 43:69–83 (2004).
220. K.H. Moledina, J.M. Regenstein, R.C. Baker, and K.H. Steinkraus. Effects of antioxidants and chelators on the stability of frozen stored mechanically deboned flounder meat from racks after filleting. *J. Food Sci.* 42:759–764 (1977).
221. E. Bilinski, R.E.E. Jonas, and M.D. Peters. Treatments affecting the degradation of lipids in frozen Pacific herring, *Clupea harengus pallasi*. *Can. Inst. Food Sci. Technol. J.* 14:123–127 (1981).
222. K.T. Hwang and J.M. Regenstein. Protection of menhaden mince lipids from rancidity during frozen storage. *J. Food Sci.* 54:1120–1124 (1988).
223. D.W. Freeman and C.W. Shannon. Rancidity in frozen catfish fillets as influenced by antioxidant injection and storage. *J. Aquat. Food Prod. Technol.* 3:65–76 (1994).
224. S.D. Shivas, D.H. Kropf, M.C. Hunt, C.L. Kastner, J.L.A. Kendall, and A.D. Dayton. Effects of ascorbic acid on display life of ground beef. *J. Food Prot.* 47:11–15 (1984).
225. R.C. Benedict, E.D. Strange, and C.E. Swift. Effect of lipid antioxidants on the stability of meat during storage. *J. Agric. Food Chem.* 23:167–173 (1975).
226. W.M. Cort. Ascorbic acid chemistry. In: *Ascorbic Acid: Chemistry, Metabolism and Uses* (P.A. Seib and B.M. Tolbert, eds.). American Chemical Society, Washington, DC, 1982, pp. 533–550.
227. D.E. Cabelli and B.H.J. Bielski. Kinetics and mechanism for the oxidation of ascorbic acid/ascorbate by HO_2/O_2^- radicals. A pulse radiolysis and stopped-flow photolysis study. *J. Phys. Chem.* 87:1809–1812 (1983).
228. E.A. Decker and H.O. Hultin. Nonenzymic catalysts of lipid oxidation in mackerel ordinary muscle. *J. Food Sci.* 55:951–953 (1990).

229. J.E. Packer, T.F. Slater, and R.L. Willson. Direct observation of a free radical interaction between vitamin E and vitamin C. *Nature (London)*. 278:737–738 (1979).

230. J.C. Deng, M. Watson, R.P. Bates, and E. Schroeder. Ascorbic acid as an antioxidant in fish flesh and its degradation. *J. Food Sci.* 43:457–460 (1978).

231. A. Grau, R. Codony, S. Grimpa, M.D. Baucells, and F. Guardiola. Cholesterol oxidation in frozen dark chicken meat: influence of dietary fat source, and α-tocopherol and ascorbic acid supplementation. *Meat Sci.* 57:197–208 (2001).

232. C. Castellini, A. Dal Bosco, and M. Bernardini. Improvement of lipid stability of rabbit meat by vitamin E and C administration. *J. Sci. Food Agric.* 81:46–53 (2000).

233. J. Iglesias, M.J. Gonzalez, and I. Medina. Determination of ascorbic and dehydroascorbic acid in lean and fatty fish species by high-performance liquid chromatography with fluorometric detection. *Eur. Food Res. Technol.* 223:781–786 (2006).

234. M.C. Erickson. Oxidative susceptibility of catfish lipids as affected by the microenvironment of ascorbic acid. *J. Food Lipids* 4:11–22 (1997).

235. S.T. Thed, M.C. Erickson, and R.L. Shewfelt. Ascorbate absorption by live channel catfish as a function of ascorbate concentration, pH, and duration of exposure. *J. Food Sci.* 58:75–78 (1993).

236. K. Jørgensen and L.H. Skibsted. Carotenoid scavenging of radicals. Effect of carotenoid structure and oxygen partial pressure on antioxidative activity. *Z. Lebensm. Unters. Forsch.* 196:423–429 (1993).

237. K.-T. Lee and D.A. Lillard. Effects of tocopherols and β-carotene on beef patties oxidation. *J. Food Lipids* 4:261–268 (1997).

238. T.H. Clark, C. Faustman, W.K.M. Chan, H.C. Furr, and J.W. Riesen. Canthaxanthin as an anti-oxidant in a liposome model system and in minced patties from rainbow trout. *J. Food Sci.* 64:982–986 (1999).

239. B. Bjerkeng and G. Johnsen. Frozen storage quality of rainbow trout (*Oncorhynchus mykiss*) as affected by oxygen, illumination, and fillet pigment. *J. Food Sci.* 60:284–288 (1995).

240. S. Sigurgisladottir, C.C. Parrish, S.P. Lall, and R.G. Ackman. Effects of feeding natural tocopherols and astaxanthin on Atlantic salmon (*Salmo salar*) fillet quality. *Food Res. Intern.* 27:23–32 (1994).

241. C. Maraschiello, E. Esteve, and J.A. García Regueiro. Cholesterol oxidation in meat from chickens fed α-tocopherol- and β-carotene-supplemented diets with different unsaturation grades. *Lipids* 33:705–713 (1998).

242. J.A. Ruiz, A.M. Pérez-Vendrell, and E. Esteve-García. Effect of β-carotene and vitamin E on oxidative stability in leg meat of broilers fed different supplemental fats. *J. Agric. Food Chem.* 47:448–454 (1999).

243. I. Gobantes, G. Choubert, and R. Gómez. Quality of pigmented (astaxanthin and canthaxanthin) rainbow trout (*Oncorhynchus mykiss*) fillets stored under vacuum packaging during chilled storage. *J. Agric. Food Chem.* 46:4358–4362 (1998).

244. D. Galaris, E. Cadenas, and P. Hochstein. Glutathione-dependent reduction of peroxides during ferryl- and met-myoglobin interconversion: a potential protective mechanism in muscle. *Free Radic. Biol. Med.* 6:473–478 (1989).

245. E. Beutler. Nutritional and metabolic aspects of glutathione. *Annu. Rev. Nutr.* 9:287–302 (1989).

246. G.R. Buettner. The pecking order of free radicals and antioxidants: lipid peroxidation, α-tocopherol, and ascorbate. *Arch. Biochem. Biophys.* 300:535–543 (1993).

247. R.G. Brannan and M.C. Erickson. Quantification of antioxidants in channel catfish during frozen storage. *J. Agric. Food Chem.* 44:1361–1366 (1996).

248. D. Petillo, H.O. Hultin, J. Krzynowek, and W.R. Autio. Kinetics of antioxidant loss in mackerel light and dark muscle. *J. Agric. Food Chem.* 46:4128–4137 (1998).

249. T. Murai, T. Akiyama, H. Ogata, and T. Suzuki. Interaction of dietary oxidized fish oil and glutathione on fingerling yellowtail *Seriola quinqueradiata*. *Nippon Suisan Gakkai Shi* 54:145–149 (1988).

250. D.M. Gatlin III and S.C. Bai. Effects of dietary lipid and reduced glutathione on composition and storage quality of channel catfish, *Ictalurus punctatus* (Rafinesque). *Aquacult. Fish. Manage.* 24:457–463 (1993).

251. K.G. Crush. Carnosine and related substances in animal tissues. *Comp. Biochem. Physiol.* 34:3–30 (1970).

252. J.E. Plowman and E.A. Close. An evaluation of a method to differentiate the species of origin of meats on the basis of the contents of anserine, balenine and carnosine in skeletal muscle. *J. Sci. Food Agric.* 45:69–78 (1988).

253. R.P. Wilson, W.E. Poe, and E.H. Robinson. Leucine, isoleucine, valine, and histidine requirements of fingerling channel catfish. *J. Nutr.* 110:627–633 (1980).

254. E.A. Decker and A.D. Crum. Inhibition of oxidative rancidity in salted ground pork by carnosine. *J. Food Sci.* 56:1179–1181 (1991).

255. B.J. Lee, D.G. Hendricks, and D.P. Cornforth. Antioxidant effects of carnosine and phytic acid in a model beef system. *J. Food Sci.* 63:394–398 (1998).

256. E.A. Decker and H. Faraji. Inhibition of lipid oxidation by carnosine. *J. Am. Oil Chem. Soc.* 67:650–652 (1990).

257. T.A. Dahl, W.R. Midden, and P.E. Hartman. Some prevalent biomolecules as defenses against singlet oxygen damage. *Photochem. Photobiol.* 47:357–362 (1988).

258. R. Kohen, Y. Yamamoto, K.C. Cundy, and B. Ames. Antioxidant activity of carnosine, homocarnosine, and anserine present in muscle and brain. *Proc. Natl. Acad. Sci. U.S.A.* 85:3175–3179 (1988).

259. M. Salim-Hanna, E. Lissi, and L. Videla. Free radical scavenging activity of carnosine. *Free Radic. Res. Commun.* 14:263–270 (1991).

260. O. Aruoma, M.J. Laughton, and B. Halliwell. Carnosine, homocarnosine and anserine: could they act as antioxidants in vivo? *Biochem. J.* 264:863–869 (1989).

261. E.A. Decker, A.D. Crum, and J.T. Calvert. Differences in the antioxidant mechanism of carnosine in the presence of copper and iron. *J. Agric. Food Chem.* 40:756–759 (1992).

262. W.K.M. Chan, E.A. Decker, J.B. Lee, and D.A. Butterfield. EPR spin-trapping studies of the hydroxyl radical scavenging activity of carnosine and related dipeptides. *J. Agric. Food Chem.* 42:1407–1410 (1994).

263. G. Kansci, C. Genot, A. Meynier, and G. Gandemer. The antioxidant activity of carnosine and its consequences on the volatile profiles of liposomes during iron/ascorbate induced phospholipid oxidation. *Food Chem.* 60:165–175 (1997).

264. S. Zhou and E.A. Decker. Ability of carnosine and other skeletal muscle components to quench unsaturated aldehydic lipid oxidation products. *J. Agric. Food Chem.* 47:51–55 (1999).

265. W.K.M. Chan, E.A. Decker, C.K. Chow, and G.A. Boissonneault. Effect of dietary carnosine on plasma and tissue antioxidant concentrations and on lipid oxidation in rat skeletal muscle. *Lipids* 29:461–466 (1994).

266. L. Mei, G.L. Cromwell, A.D. Crum, and E.A. Decker. Influences of dietary β-alanine and histidine on the oxidative stability of pork. *Meat Sci.* 49:55–64 (1998).

267. Y. He and F. Shahidi. Antioxidant activity of green tea and its catechins in a fish meat model system. *J. Agric Food Chem.* 45:4262–4266 (1997).

268. T.M. Rabbah, K.I. Ereifej, M.A. Al-Mahasneh, and M.A. Al-Rababah. Effect of plant extracts on physicochemical properties of chicken breast meat cooked using conventional electric oven or microwave. *Poult. Sci.* 85:148–154 (2006).

269. N. Tipsrisukond, L.N. Fernando, and A.D. Clarke. Antioxidant effects of essential oil and oleoresin of black pepper from supercritical carbon dioxide extractions in ground pork. *J. Agric. Food Chem.* 46:4329–4333 (1998).

270. S.S.L.A.A. El-Alim, A. Lugasi, J. Hóvári, and E. Dworschák. Culinary herbs inhibit lipid oxidation in raw and cooked minced meat patties during storage. *J. Sci. Food Agric.* 79:277–285 (1999).

271. O. Hassan and L.S. Fan. The anti-oxidation potential of polyphenol extract from cocoa leaves on mechanically deboned chicken meat (MDCM). *LWT-Food Sci. Technol.* 38:315–321 (2005).

272. A.I. Rey, A. Hopia, R. Kivikari, and M. Kahkonen. Use of natural food/plant extracts: cloudberry (*Rubus chamaemorus*), beetroot (*Beta Vulgaris* "Vulgaris") or willow herb (*Epilobium angjustifolium*) to reduce lipid oxidation of cooked pork patties. *LWT-Food Sci. Technol.* 38:363–370 (2005).

273. S.R. Kanatt, R. Chaner, P. Radhakrishna, and A. Sharma. Potato peel extract—a natural antioxidant for retarding lipid peroxidation in radiation processed lamb meat. *J. Agric. Food Chem.* 53:1499–1504 (2005).

274. M.B. Mielnik, E. Olsen, G. Vogt, D. Adeline, and G. Skrede. Grape seed extract as antioxidant in cooked, cold stored turkey meat. *LWT-Food Sci. Technol* 39:191–198 (2006).

275. M. Estevez and R. Cava. Effectiveness of rosemary essential oil as an inhibitor of lipid and protein oxidation: contradictory effects in different types of frankfurters. *Meat Sci.* 72:348–355 (2006).

276. C.W. Balentine, P.G. Crandall, C.A. O'Bryan, D.Q. Duong, and F.W. Pohlman. The pre- and post-grinding application of rosemary and its effects on lipid oxidation and color during storage of ground beef. *Meat Sci.* 73:413–421 (2006).

277. M. Estevez, S. Ventanas, R. Ramirez, and R. Cava. Influence of the addition of rosemary essential oil on the volatiles pattern of porcine frankfurters. *J. Agric. Food Chem.* 53:8317–8324 (2005).

278. B. Stodolak, A. Starzynska, M. Czyszczon, and K. Zyla. The effect of phytic acid on oxidative stability of raw and cooked meat. *Food Chem.* 101:1041–1045 (2007).

279. M. Perez-Mateos, T.C. Lanier, and L.C. Boyd. Effects of rosemary and green tea extracts on frozen surimi gels fortified with ω3 fatty acids. *J. Sci. Food Agric.* 86:558–567 (2006).

280. M. Pazos, J.M. Gallardo, J.L. Torres, and I. Medina. Activity of grape polyphenols as inhibitors of the oxidation of fish lipids and frozen fish muscle. *Food Chem.* 92:547–557 (2005).

281. C.H. Lee, J.D. Reed, and M.P. Richards. Ability of various polyphenolic classes from cranberry to inhibit lipid oxidation in mechanically separated turkey and cooked ground pork. *J. Muscle Foods* 17:248–266 (2006).

282. A. Saija, M. Scalese, M. Lanza, D. Marzullo, F. Bonina, and F. Castelli. Flavonoids as antioxidant agents: importance of their interaction with biomembranes. *Free Radic. Biol. Med.* 19:481–486 (1995).

283. M. Pazos, S. Lois, J.L. Torres, and I. Medina. Inhibition of hemoglobin- and iron-promoted oxidation in fish microsomes by natural phenolics. *J. Agric. Food Chem.* 54:4417–4423 (2006).

284. M.N. O'Grady, M. Maher, D.J. Troy, A.P. Moloney, and J.P. Kerry. An assessment of dietary supplementation with tea catechins and rosemary extract on the quality of fresh beef. *Meat Sci.* 73:132–143 (2006).

285. J.A.M. Janz, P.C.H. Morel, B.H.P. Wilkinson, and R.W. Purchas. Preliminary investigation of the effects of low-level dietary inclusion of fragrant essential oils and oleoresins on pig performance and pork quality. *Meat Sci.* 75:350–355 (2007).

286. S.Z. Tang, J.P. Kerry, D. Sheehan, D.J. Buckley, and P.A. Morrissey. Antioxidative effect of dietary tea catechins on lipid oxidation of long-term frozen stored chicken meat. *Meat Sci.* 57:331–336 (2001).

287. P. Florou-Paneri, I. Giannenas, E. Christaki, A. Govaris, and N. Botsoglou. 2006. Performance of chickens and oxidative stability of the produced meat as affected by feed supplementation with oregano, vitamin C, vitamin E and their combinations. *Archiv fur Geflugelkunde* 70:232–240 (2006).

288. I.A. Giannenas, P. Florou-Paneri, N.A. Botsoglou, E. Christaki, and A.B. Spais. Effect of supplementing feed with oregano and/or α-tocopheryl acetate on growth of broiler chickens and oxidative stability of meat. *J. Anim. Feed Sci.* 14:521–535 (2005).

289. H. Basmacioglu, O. Tokusoglu, and M. Ergul. The effect of oregano and rosemary essential oils or α-tocopheryl acetate on performance and lipid oxidation of meat enriched with n-3 PUFA's in broilers. *S. Afr. J. Anim. Sci.* 34:197–210 (2004).

290. A. Aksnes and L.R. Njaa. Catalase, glutathione peroxidase and superoxide dismutase in different fish species. *Comp. Biochem. Physiol.* 69B:893–896 (1981).

291. M. Renerre, K. Poncet, Y. Mercier, P. Gatellier, and B. Métro. Influence of dietary fat and vitamin E on antioxidant status of muscles of turkey. *J. Agric. Food Chem.* 47:237–244 (1999).

292. C.J. Lammi-Keefe, P.B. Swan, and P.V.J. Hegarty. Effect of level of dietary protein and total or partial starvation on catalase and superoxide dismutase activity in cardiac and skeletal muscles in young rats. *J. Nutr.* 114:2235–2240 (1984).

293. C.-J. Huang and M.-L. Fwu. Degree of protein deficiency affects the extent of the depression of the antioxidative enzyme activities and the enhancement of tissue lipid peroxidation in rats. *J. Nutr.* 123:803–810 (1993).

294. C. Maraschiello, C Sárraga, and J.A. García Regueiro. Glutathione peroxidase activity, TBARS, and α-tocopherol in meat from chickens fed different diets. *J. Agric. Food Chem.* 47:867–872 (1999).

295. T. Arai, M. Sugawara, T. Sako, S. Motoyoshi, T. Shimura, N. Tsutsui, and T. Konno. Glutathione peroxidase activity in tissues of chickens supplemented with dietary selenium. *Comp. Biochem. Physiol.* 107A:245–248 (1994).

296. Y.C. Ryu, M.S. Rhee, K.M. Lee, and B.C. Kim. Effects of different levels of dietary supplemental selenium on performance, lipid oxidation, and color stability of broiler chicks. *Poult. Sci.* 84:809–815 (2005).

297. S.K. Lee, L. Mei, and E.A. Decker. Lipid oxidation in cooked turkey as affected by added antioxidant enzymes. *J. Food Sci.* 61:726–728, 795 (1996).

298. T. Hoac, C. Daun, U. Trafikowska, J. Zackrisson, and B. Akesson. Influence of heat treatment on lipid oxidation and glutathione peroxidase activity in chicken and duck meat. *Innov. Food Sci. Emerg. Technol.* 7:88–93 (2006).

299. E. Pigeolet, P. Corbisier, A. Houbion, D. Lambert, C. Michiels, M. Raes, M.-D. Zachary, and J. Remacle. Glutathione peroxidase, superoxide dismutase, and catalase inactivation by peroxides and oxygen derived free radicals. *Mech. Ageing Dev.* 51:283–297 (1990).

300. J.A. Escobar, M.A. Rubio, and E.A. Lissi. SOD and catalase inactivation by singlet oxygen and peroxyl radicals. *Free Radic. Biol. Med.* 20:285–290 (1996).

301. I.B. Simon, T.P. Labuza, and M. Karel. Computer-aided predictions of food storage stability: oxidative deterioration of a shrimp product. *J. Food Sci.* 36:280–286 (1971).

302. S.A. Kurade and J.D. Baranowski. Prediction of shelf-life of frozen minced fish in terms of oxidative rancidity as measured by TBARS number. *J. Food Sci.* 52:300–302, 311 (1987).

303. P.J. Ke, B.A. Linke, and B. Smith-Lall. Quality preservation and shelf life estimation of frozen fish in terms of oxidative rancidity development. *Lebensm. Wiss. Technol.* 15:203–206 (1982).

304. A.L. Tappel, A.A. Tappel, and C.G. Fraga. Application of simulation modeling to lipid peroxidation processes. *Free Radic. Biol. Med.* 7:361–368 (1989).

305. C.F. Babbs and M.G. Steiner. Simulation of free radical reactions in biology and medicine: a new two-compartment kinetic model of intracellular lipid peroxidation. *Free Radic. Biol. Med.* 8:471–485 (1990).

306. T. Sakai, S. Ohtsubo, T. Minami, and M. Terayama. Effect of bleeding on hemoglobin contents and lipid oxidation in the skipjack muscle. *Biosci. Biotechnol. Biochem.* 70:1006–1008 (2006).

307. C.Z. Alvarado, M.P. Richards, S.F. O'Keefe, and H. Wang. The effect of blood removal on oxidation and shelf life of broiler breast meat. *Poult. Sci.* 86:156–161 (2007).

308. S.P. Axtell, S.M. Russell, and E. Berman. Effect of immersion chilling of broiler chicken carcasses in monochloramine on lipid oxidation and halogenated residual compound formation. *J. Food Prot.* 69:907–911 (2006).

309. J. Kim, Y. Lee, S.F. O'Keefe, and C.-I. Wei. Effect of chlorine dioxide treatment on lipid oxidation and fatty acid composition in salmon and red grouper fillets. *J. Am. Oil Chem. Soc.* 74:539–542 (1997).

310. T.J. Kim, J.L. Lilva, R.S. Chamul, and T.C. Chen. Influence of ozone, hydrogen peroxide, or salt on microbial profile, TBARS and color of channel catfish fillets. *J. Food Sci.* 65:1210–1213 (2000).

311. B.S.M. Mahmoud, K. Yamazaki, K. Miyashita, I.I. Shin, and T. Suzuki. A new technology for fish preservation by combined treatment with electrolyzed NaCl solutions and essential oil compounds. *Food Chem.* 99:656–662 (2006).

312. S. Eymard, E. Carcouet, M.J. Rochet, J. Dumay, C. Chopin, and C. Genot. Development of lipid oxidation during manufacturing of horse mackerel surimi. *J. Sci. Food Agric.* 85:1750–1756 (2005).

313. I. Undeland, B. Ekstrand, and H. Lingnert. Lipid oxidation in minced herring (*Clupea harengus*) during frozen storage. Effect of washing and precooking. *J. Agric. Food Chem.* 46:2319–2328 (1998).

314. S.D. Kelleher, H.O. Hultin, and K.A. Wilhelm. Stability of mackerel surimi prepared under lipid-stabilizing processing conditions. *J. Food Sci.* 59:269–271 (1994).

315. M.P. Richards, S.D. Kelleher, and H.O. Hultin. Effect of washing with or without antioxidants on quality retention of mackerel fillets during refrigerated and frozen storage. *J. Agric. Food Chem.* 46:4363–4371 (1998).

316. B. Dulavik, N.K. Sorensen, H. Barstad, O. Horvli, and R.L. Olsen. Oxidative stability of frozen light and dark muscles of saithe (*Pollachius virens* L.). *J. Food Lipids* 5:233–245 (1998).

317. D.A. Silberstein and D.A. Lillard. Factors affecting the autoxidation of lipids in mechanically deboned fish. *J. Food Sci.* 43:764–766 (1978).

318. J. McNeill, Y. Kakuda, and C. Findlay. Influence of carcass parts and food additives on the oxidative stability of frozen mechanically separated and hand-deboned chicken meat. *Poult. Sci.* 67:270–274 (1988).

319. S. Barbut, H.H. Draper, and P.D. Cole. Effect of mechanical deboner head pressure on lipid oxidation in poultry meat. *J. Food Prot.* 52:21–25 (1989).

320. C.M. Lee and R.T. Toledo. Degradation of fish muscle during mechanical deboning and storage with emphasis on lipid oxidation. *J. Food Sci.* 42:1646–1649 (1977).

321. M.P. Wanous, D.G. Olson, and A.A. Kraft. Oxidative effects of meat grinder wear on lipids and myoglobin in commercial fresh pork sausage. *J. Food Sci.* 54:545–548, 552 (1989).

322. A. Culbertson. Hot processing of meat: a review of the rationale and economic implications. In: *Development in Meat Science*, Vol. 1 (R. Lawrie, ed.). Applied Science Publishers, London, 1980, p. 61.

323. D.L. Drerup, M.D. Judge, and E.D. Aberle. Sensory properties and lipid oxidation in prerigor processed fresh pork sausage. *J. Food Sci.* 46:1659–1661 (1981).

324. K.S. Rhee, J.T. Keeton, Y.A. Ziprin, R. Leu, and J.J. Bohac. Oxidative stability of batter-breaded restructured nuggets processed from prerigor pork. *J. Food Sci.* 53:1047–1050 (1988).

325. A. Takiguchi. Effect of NaCl on the oxidation and hydrolysis of lipids in salted sardine fillets during storage. *Nippon Suisan Gakkai Shi* 55:1649–1656 (1989).

326. M.D. Judge and E.D. Aberle. Effect of prerigor processing on the oxidative rancidity of ground light and dark porcine muscles. *J. Food Sci.* 45:1736–1739 (1980).

327. C.-C. Chen, A.M. Pearson, J.I. Gray, and R.A. Merkel. Effects of salt and some antioxidants upon the TBA number of meat. *Food Chem.* 14:167–172 (1984).

328. E. Torres, A.M. Pearson, J.I. Gray, A.M. Booren, and M. Shimokomaki. Effect of salt on oxidative changes in pre- and post-rigor ground beef. *Meat Sci.* 23:151–163 (1988).

329. P.A. Morrissey and J.Z. Tichivangana. The antioxidant activities of nitrite and nitrosylmyoglobin in cooked meats. *Meat Sci.* 14:175–190 (1985).

330. J.O. Igene, K. Yamauchi, A.M. Pearson, and J.I. Gray. Mechanisms by which nitrite inhibits the development of warmed-over flavour (WOF) in cured meat. *Food Chem.* 18:1–18 (1985).

331. L.A. Freybler, J.I. Gray, A. Asghar, A.M. Booren, A.M. Pearson, and D.J. Buckley. Nitrite stabilization of lipids in cured pork. *Meat Sci.* 33:85–96 (1993).

332. J. Yun, F. Shahidi, L.J. Rubin, and L.L. Diosady. Oxidative stability and flavour acceptability of nitrite-free meat-curing systems. *Can. Inst. Food Sci. Technol. J.* 20:246–251 (1987).

333. F. Shahidi, L.J. Rubin, and D.F. Wood. Stabilization of meat lipids with nitrite-free curing mixtures. *Meat Sci.* 22:73–80 (1988).

334. Food and Agriculture Organization of the United Nations. *Prevention of Losses in Cured Fish,* Fisheries Technical Paper No. 219, FAO, Rome, 1981.

335. A. Zotos, M. Hole, and G. Smith. The effect of frozen storage of mackerel (*Scomber scombrus*) on its quality when hot-smoked. *J. Sci. Food Agric.* 67:43–48 (1995).

336. F. Shahidi and S.A. Spurvey. Oxidative stability of fresh and heat-processed dark and light muscles of mackerel (*Scomber scombrus*). *J. Food Lipids* 3:13–25 (1996).

337. E.R. Kingston, F.J. Monahan, D.J. Buckley, and P.B. Lynch. Lipid oxidation in cooked pork as affected by vitamin E, cooking and storage conditions. *J. Food Sci.* 63:386–389 (1998).

338. M.M. Mielche and G. Bertelsen. Effects of heat treatment on warmed-over flavour in ground beef during aerobic chill storage. *Z. Lebensm. Unters. Forsch.* 197:8–13 (1993).

339. A.M. Spanier, J.R. Vercellotti, and C. James, Jr. Correlation of sensory, instrumental and chemical attributes of beef as influenced by meat structure and oxygen exclusion. *J. Food Sci.* 57:10–15 (1992).

340. D.M. Smith, A.M. Salih, and R.G. Morgan. Heat treatment effects on warmed-over flavor in chicken breast meat. *J. Food Sci.* 52:842–845 (1987).

341. M.G. Mast and J.H. MacNeil. Physical and functional properties of heat pasteurized mechanically deboned poultry meat. *Poult. Sci.* 55:1207–1213 (1976).

342. Y.-J. Wang, L.A. Miller, and P.B. Addis. Effect of heat inactivation of lipoxygenase on lipid oxidation in lake herring (*Coregonus artedii*). *J. Am. Oil Chem. Soc.* 68:752–757 (1991).

343. R. Hamm. Heating of muscle systems. In: *The Physiology and Biochemistry of Muscle as a Food* (E.J. Briskey, R.G. Cassens, and J.C. Trautman, eds.). University of Wisconsin Press, Madison, WI, 1966, p. 363.

344. W.H. Huang and B.E. Greene. Effect of cooking method on TBA numbers of stored beef. *J. Food Sci.* 43:1201–1203, 1209 (1978).

345. J.N. Gros, P.M. Howat, M.T. Younathan, A.M. Saxton, and K.W. McMillin. Warmed-over flavor development in beef patties prepared by three dry heat methods. *J. Food Sci.* 51:1152–1155 (1986).

346. V.T. Satyanarayan and K.-O. Honikel. Effect of different cooking methods on warmed-over flavor development in pork. *Z. Lebensm. Unters. Forsch.* 194:422–425 (1992).

347. J. Pikul, D.E. Leszczynski, P.J. Bechtel, and F.A. Kummerow. Effects of frozen storage and cooking on lipid oxidation in chicken meat. *J. Food Sci.* 49:838–843 (1984).

348. M.C. Erickson. Changes in lipid oxidation during cooking of refrigerated minced channel catfish muscle. In: *Lipid Oxidation in Food* (A.J. St. Angelo, ed.). American Chemical Society, Washington, DC, 1992, pp. 344–350.

349. T. Ohshima, T. Nakagawa, and C. Koizumi. Effect of high hydrostatic pressure on the enzymatic degradation of phospholipids in fish muscle during storage. In: *Seafood Science and Technology* (E.G. Bligh, ed.). Fishing News Books, Oxford, 1992, pp. 64–75.

350. A.I. Andres, C.E. Adamsen, J.K.S. Moller, J. Ruiz, and L.H. Skibsted. High-pressure treatment of dry-cured Iberian ham. Effect on colour and oxidative stability during chill storage packed in modified atmosphere. *Eur. Food Res. Technol.* 222:486–491 (2006).

351. M. Tanaka, Z. Xueyi, Y. Nagashima, and T. Taguchi. Effect of high pressure on the lipid oxidation in sardine meat. *Nippon Suisan Gakkai Shi* 57:957–963 (1991).

352. N. Braganolo, B. Danielsen, and L.H. Skibsted. Combined effect of salt addition and high-pressure processing on formation of free radicals in chicken thigh and breast muscle. *Eur. Food Res. Technol.* 223:669–673 (2006).

353. P.B. Cheah and D.A. Ledward. Catalytic mechanism of lipid oxidation following high pressure treatment in pork fat and meat. *J. Food Sci.* 62:1135–1138 (1997).

354. F. Martinez and T.P. Labuza. Rate of deterioration of freeze-dried salmon as a function of relative humidity. *J. Food Sci.* 33:241–247 (1968).

355. V.A.-E. King and J.-F. Chen. Oxidation of controlled low-temperature vacuum dehydrated and freeze-dried beef and pork. *Meat Sci.* 48:11–19 (1998).

356. J.J. Heath, S.L. Owens, S. Tesch, and K.W. Hannah. Effect of high-energy electron irradiation of chicken meat on thiobarbituric acid values, shear values, odor, and cooked yield. *Poult. Sci.* 69:313–319 (1990).

357. I.B. Hashim, A.V.A. Resurreccion, and K.H. McWatters. Descriptive sensory analysis of irradiated frozen or refrigerated chicken. *J. Food Sci.* 60:664–666 (1995).

358. C. Jo and D.U. Ahn. Volatiles and oxidative changes in irradiated pork sausage with different fatty acid composition and tocopherol content. *J. Food Sci.* 65:270–275 (2000).

359. D.U. Ahn, K.C. Nam, M. Du, and C. Jo. Volatile production in irradiated normal, pale soft exudative (PSE) and dark firm dry (DFD) pork under different packaging and storage conditions. *Meat Sci.* 57:419–426 (2001).

360. D.U. Ahn, D.G. Olson, C. Jo, X. Chen, C. Wu, and J.I. Lee. Effect of muscle type, packaging, and irradiation on lipid oxidation, volatile production, and color in raw pork patties. *Meat Sci.* 49:27–39 (1998).

361. D.U. Ahn, K.C. Nam, M. Du, and C. Jo. Effect of irradiation and packaging conditions after cooking on the formation of cholesterol and lipid oxidation products in meats during storage. *Meat Sci.* 57:413–418 (2001).

362. X. Chen, C. Jo, J.I. Lee, and D.U. Ahn. Lipid oxidation, volatiles and color changes of irradiated pork patties as affected by antioxidants. *J. Food Sci.* 64:16–19 (1999).

363. K. Galvin, P.A. Morrissey, and D.J. Buckley. Effect of dietary α-tocopherol supplementation and gamma-irradiation on α-tocopherol retention and lipid oxidation in cooked minced chicken. *Food Chem.* 62:185–190 (1998).

364. M.E. Apgar and H.O. Hultin. Lipid peroxidation in fish muscle microsomes in the frozen state. *Cryobiology* 19:154–162 (1982).

365. M.C. Tomás and M.C. Añón. Study on the influence of freezing rate on lipid oxidation in fish (salmon) and chicken breast muscles. *Int. J. Food Sci. Technol.* 25:718–721 (1990).

366. A. Sequeira-Munoz, D. Chevalier, B.K. Simpson, A. Le Bail, and H.S. Ramaswamy. Effect of pressure-shift freezing versus air-blast freezing of carp (*Cyprinus carpio*) fillets: a storage study. *J. Food Biochem.* 29:504–516 (2005).

367. C.J. Hagyard, A.H. Keiller, T.L. Cummings, and B.B. Chrystall. Frozen storage conditions and rancid flavour development in lamb. *Meat Sci.* 35:305–312 (1993).

368. R.S. Miles, F.K. McKeith, P.J. Bechtel, and J. Novakofski. Effect of processing, packaging and various antioxidants on lipid oxidation of restructured pork. *J. Food Prot.* 49:222–225 (1986).

369. M. Bhattacharya, M.A. Hanna, and R.W. Mandigo. Lipid oxidation in ground beef patties as affected by time–temperature and product packaging parameters. *J. Food Sci.* 53:714–717 (1988).

370. K. Satomi, A. Sasaki, and M. Yokoyama. Effect of packaging materials on the lipid oxidation in fish sausage. *Nippon Suisan Gakkai Shi* 54:517–521 (1988).

371. E.E.M. Santos and J.M. Regenstein. Effects of vacuum packaging, glazing, and erythorbic acid on the shelf-life of frozen white hake and mackerel. *J. Food Sci.* 55:64–70 (1990).

372. M.S. Brewer and C.A.Z. Harris. Effect of packaging on color and physical characteristics of ground pork in long-term frozen storage. *J. Food Sci.* 56:363–366, 370 (1991).

373. M.D. Fernández-Esplá and E. O'Neill. Lipid oxidation in rabbit meat under different storage conditions. *J. Food Sci.* 58:1262–1264 (1993).

374. B.Y. Jeong, T. Ohshima, C. Koizumi, and Y. Kanou. Lipid deterioration and its inhibition of Japanese oyster *Crassostrea gigas* during frozen storage. *Nippon Suisan Gakkai Shi* 56:2083–2091 (1990).

375. N.L. Nolan, J.A. Bowers, and D.H. Kropf. Lipid oxidation and sensory analysis of cooked pork and turkey stored under modified atmospheres. *J. Food Sci.* 54:846–849 (1989).

376. S.-Y. Hwang, J.A. Bowers, and D.H. Kropf. Flavor, texture, color, and hexanal and TBA values of frozen cooked beef packaged in modified atmosphere. *J. Food Sci.* 55:26–29 (1990).

377. C.-H. Huang and Y.-M. Weng. Inhibition of lipid oxidation in fish muscle by antioxidant incorporated polyethylene film. *J. Food Process Preserv.* 22:199–209 (1998).

378. C. Nerin, L. Tovar, D. Djenane, J. Camo, J. Salafranca, J.A. Beltran, and P. Roncales. Stabilization of beef meat by a new active packaging containing natural antioxidants. *J. Agric. Food Chem.* 54:7840–7846 (2006).

379. E. Boselli, M.F. Caboni, M.T. Rodriguez-Estrada, T.G. Toschi, M. Daniel, and G. Lercker. Photoxidation of cholesterol and lipids of turkey meat during storage under commercial retail conditions. *Food Chem.* 91:705–713 (2005).

380. D.U. Aim, F.H. Wolfe, and J.S. Sim. The effect of metal chelators, hydroxyl radical scavengers, and enzyme systems on the lipid peroxidation of raw turkey meat. *Poult. Sci.* 72:1972–1980 (1993).

381. H.M. Cunningham and G.L. Lawrence. Effect of exposure of meat and poultry to chlorinated water on the retention of chlorinated compounds and water. *J. Food Sci.* 42:1504–1505, 1509 (1977).

382. H.A. Ghanbari, W.B. Wheeler, and J.R. Kirk. Reactions of aqueous chlorine and chlorine dioxide with lipids: chlorine incorporation. *J. Food Sci.* 47:482–485 (1982).

383. H.A. Ghanbari, W.B. Wheeler, and J.R. Kirk. The fate of hypochlorous acid during shrimp processing: a model system. *J. Food Sci.* 47:185–187, 197 (1982).

384. J.J. Johnston, H.A. Ghanbari, W.B. Wheeler, and J.R. Kirk. Chlorine incorporation in shrimp. *J. Food Sci.* 48:668–670 (1983).

385. C.C. Winterbourn, J.J.M. van den Berg, E. Roitman, and F.A. Kuypers. Chlorohydrin formation from unsaturated fatty acids reacted with hypochlorous acid. *Arch. Biochem. Biophys.* 296:547–555 (1992).

386. C.C. Winterbourn, H.P. Monteiro, and C.F. Galilee. Ferritin-dependent lipid peroxidation by stimulated neutrophils: inhibition by myeloperoxidase-derived hypochlorous acid but not by endogenous lactoferrin. *Biochim. Biophys. Acta* 1055:179–185 (1990).

387. M.C. Erickson. Flavor quality implications in chlorination of poultry chiller water. *Food Res. Intern.* 32:635–641 (1999).

388. J.M. Albrich, C.A. McCarthy, and J.K. Hurst. Biological reactivity of hypochlorous acid: implications for microbicidal mechanisms of leukocyte myeloperoxidase. *Proc. Natl. Acad. Sci. USA.* 78:210–214 (1981).

389. C.C. Winterbourn. Comparative reactivities of various biological compounds with myeloperoxidase–hydrogen peroxide-chloride, and similarity of the oxidant to hypochlorite. *Biochim. Biophys. Acta* 840:204–210 (1985).

390. C. Thongraung, S. Benjakul, and H.O. Hultin. Effect of pH, ADP and muscle soluble components on cod hemoglobin characteristics and extractability. *Food Chem.* 97:567–576 (2006).

391. H.O. Hultin. Potential lipid oxidation problems in fatty fish processing. In: *Proceedings of the Fatty Fish Utilization Symposium*. Raleigh, NC, 1988, pp. 185–223.

392. R. Ellis, G.T. Currie, F.E. Thornton, N.C. Bollinger, and A.M. Gaddis. Carbonyls in oxidizing fat. 11. The effect of the pro-oxidant activity of sodium chloride on pork tissue. *J. Food Sci.* 33:555–561 (1968).

393. J.E. Osinchak, H.O. Hultin, O.T. Zajicek, S.D. Kelleher, and C.-H. Huang. Effect of NaCl on catalysis of lipid oxidation by the soluble fraction of fish muscle. *Free Radic. Biol. Med.* 12:35–41 (1992).

394. M. Wettasinghe and F. Shahidi. Oxidative stability of cooked comminuted lean pork as affected by alkali and alkali-earth halides. *J. Food Sci.* 61:1160–1164 (1996).

395. F. Shahidi, L.J. Rubin, L.L. Diosady, and D.F. Wood. Preparation of the cooked cured-meat pigment, dinitrosyl ferrohemochrome, from hemin and nitric oxide. *J. Food Sci.* 50:272–273 (1985).

396. M. Wettasinghe and F. Shahidi. Antioxidant activity of preformed cooked cured-meat pigment in a β-carotene/linoleate model system. *Food Chem.* 58:203–207 (1997).

397. K.S. Rhee, L.M. Krahl, L.M. Lucia, and G.R. Acuff. Antioxidative/antimicrobial effects and TBARS in aerobically refrigerated beef as related to microbial growth. *J. Food Sci.* 62:1205–1210 (1997).

398. W. Nawar, H. Hultin, Y. Li, Y. Xing, S. Kelleher, and C. Wilhelm. Lipid oxidation in seafoods under conventional conditions. *Food Rev. Int.* 6:647–660 (1990).

399. F.W. Wheaton and T.B. Lawson. *Processing Aquatic Food Products*. Wiley, New York, NY, 1985.

400. D.G. Hoover, C. Metrick, A.M. Papineau, D.F. Farkas, and D. Knorr. Biological effects of high hydrostatic pressure in food microorganisms. *Food Technol.* 43(3):99–107 (1989).

401. R. Hayashi, Y. Kawamura, T. Nakasa, and O. Okinaka. Application of high pressure to food processing: pressurization of egg white and yolk, and properties of gels formed. *Agric. Biol. Chem.* 53:2935–2939 (1989).

402. K.E. Ijichi. Evaluation of an alginate coating during frozen storage of red snapper and silver salmon. MS thesis, University of California, Davis, 1978.

403. A.A. Klose, E.P. Mechi, and H.L. Hanson. Use of antioxidants in the frozen storage of turkeys. *Food Technol.* 6(7):308–311 (1952).

404. Y.M. Stuchell and J.M. Krochta. Edible coatings on frozen king salmon: effect of whey protein isolate and acetylated monoglycerides on moisture loss and lipid oxidation. *J. Food Sci.* 60:28–31 (1995).

405. R. Villegas, T.P. O'Connor, J.P. Kerry, and D.J. Buckley. Effect of gelatin dip on the oxidative and colour stability of cooked ham and bacon pieces during frozen storage. *Int. J. Food Sci. Technol.* 34:385–389 (1999).

406. Y. Wu, J.W. Rhim, C.L. Weller, F. Hamouz, S. Cuppett, and M. Schnepf. Moisture loss and lipid oxidation for precooked beef patties stored in edible coatings and films. *J. Food Sci.* 65:300–304 (2000).

407. H.J. Kang, C. Jo, J.H. Kwon, J.H. Kim, H.J. Chung, and M.W. Byun. Effect of a pectin-based edible coating containing green tea powder on the quality of irradiated pork patty. *Food Control* 18:430–435 (2007).

408. S.C. Salaris and C.F. Babbs. Effect of oxygen concentration on the formation of malondialdehyde-like material in a model of tissue ischemia and reoxygenation. *Free Radic. Biol. Med.* 7:603–609 (1989).

13 Polyunsaturated Lipid Oxidation in Aqueous System

Kazuo Miyashita

CONTENTS

I. INTRODUCTION

Oxidative deterioration of polyunsaturated lipids is one of the most important problems in food chemistry, because lipid oxidation products cause undesirable flavors and lower the nutritional quality and safety of lipid-containing foods. Docosahexaenoic acid (DHA; 22:6n-3) and arachidonic acid (AA; 20:4n-6) have beneficial health and physiological effects. The functions of these polyunsaturated fatty acids (PUFAs) have attracted consumer attention and both are used in functional foods and nutraceuticals. In the course of investigation on the dietary effects of these n-3 and n-6 PUFA, lipid peroxidation has also received considerable attention because of its possible contribution to the potential damage of biological systems [1–6].

Lipid oxidation proceeds through a free radical chain reaction consisting of chain initiation, propagation, and termination processes (Figure 13.1) [7–12]. The rate-limiting step in the reaction is abstraction of hydrogen radical (H•) from substrate lipids (LH) to form lipid free radicals (L•). Since this hydrogen abstraction occurs at the bis-allylic positions (CH=CH—CH2—CH=CH) present in PUFA and the susceptibility of PUFA to oxidation depends on the availability of bis-allylic hydrogens, oxidative stability of each PUFA is inversely proportional to the number of bis-allylic positions in the molecule or the degree of unsaturation of the PUFA. Thus, when the relative oxidative stabilities of typical PUFA are compared in air (bulk phase) or in organic solvents, DHA (22:6n-3) is most rapidly oxidized, followed by eicosapentaenoic acid (20:5n-3; EPA), AA (20:4n-6), α-linonelic acid (18:3n-3; α-LN), and linoleic acid (18:2n-6; LA), respectively [13–19]. From these results, it is generally accepted that lipid oxidation of DHA and EPA is a major problem in the utilization to make food materials and that a high intake of these PUFA may increase the oxidative stress of biological systems.

On the other hand, it has been reported that DHA ingestion did not increase lipid peroxides to the level we would expect from the "peroxidizability" index of the tissue total lipids [20]. In particular, lipid peroxide levels in the brain and testis decreased when DHA was given to the

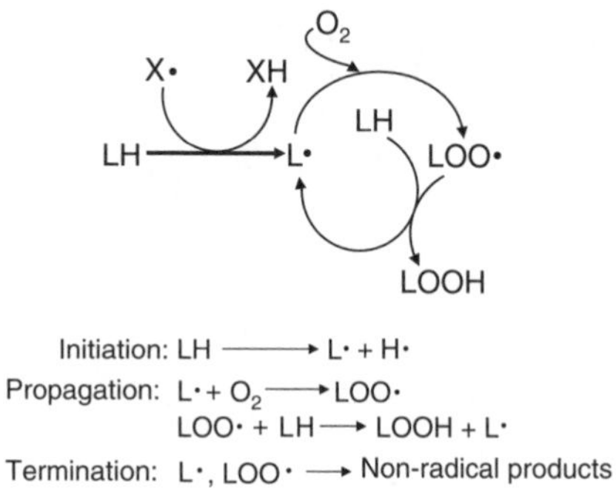

Initiation: LH $\longrightarrow$ L• + H•
Propagation: L• + O_2 $\longrightarrow$ LOO•
LOO• + LH $\longrightarrow$ LOOH + L•
Termination: L•, LOO• $\longrightarrow$ Non-radical products

FIGURE 13.1 Oxidation mechanism of polyunsaturated lipids. H•: hydrogen radical; L•: lipid free radicals; LOO•: peroxy radical; LH: substrate. (From Miyashita, K., *Lipid Tech. Newslett.*, 8, 35, 2002. With permission.)

animals. Ando et al. [21] also examined the effects of fish oil on lipid peroxidation of rat organs and found that levels of phospholipid hydroperoxides and thiobarbituric acid reactive substances (TBARS) in rat organs fed on a fish oil diet were similar to those of the safflower-oil diet group. Wander and Du [22] measured plasma lipid peroxidation after supplementation with EPA and DHA from fish oil and tocopherol in postmenopausal women. They found that neither the concentration of plasma TBARS nor protein oxidation changed after fish oil supplementation. These results indicate a difference in the oxidative stability of highly unsaturated fatty acids such as DHA and AA between biological systems and bulk phases.

Many foods and biological systems are complex, multicomponent, and heterogeneous systems, in which lipids are present with various types of other components in aqueous medium. It follows that the lipid oxidation in aqueous solutions is very important if we are to fully understand the factors that affect lipid oxidation in foods and in biological systems. In emulsions, the membranes surrounding the emulsion droplets consist of surface-active substances such as emulsifiers and/or proteins. They provide a protective barrier to the penetration and diffusion of metals or radicals that initiate lipid oxidation. The chemical and physical nature of emulsifier and proteins, therefore, is an important factor for protection of PUFA against oxidation in aqueous phase [12,23–35]. The different behavior of antioxidants has also been reported between in bulk phase and emulsion systems [36–39]. In bulk phase, the relative oxidative stabilities of PUFA increase with decreasing the degree of unsaturation. On the other hand, the reverse result has been obtained in micelles [40]. There are significant differences in the relative oxidative stability of PUFA between in bulk phase and in aqueous phases such as food and biological systems.

The aim of this article is to highlight the relative order of oxidative stability of lipids having different number of unsaturation in aqueous phases.

II. OXIDATIVE STABILITY OF PUFA IN MICELLES

The oxidative stability of PUFA in aqueous micelles is markedly different from that in bulk phase [40–42]. When the oxidative stabilities of six types of typical PUFA were compared in micelles (Figure 13.2), LA was the most susceptible to oxidation—as much as 50% of the substrate was lost after only 13 h of oxidation. In contrast, DHA and EPA were very stable—even after 2000 h of oxidation, more than 80% and 90% of the substrates remained unchanged [40]. Also, a marked difference in oxidative stability was observed between α-LN and γ-linolenic acid (γ-LN) in micelles, while no difference in the oxidation rate was found between them during oxidation in bulk phase. Moreover, when a mixture of DHA and LA was oxidized in aqueous micelles, the stability increased with increasing molar ratio of DHA to LA in the mixture [40].

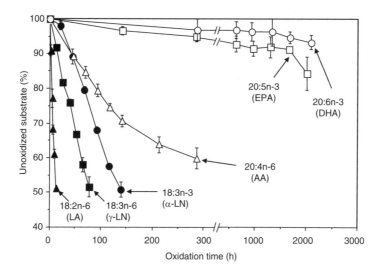

FIGURE 13.2 Oxidative stability of six types of typical PUFA in aqueous micelles. The stability was determined by following the decrease in unoxidized substrate content. Each PUFA (1 mM) was incubated in the dark at 37°C with Fe(II) (1.0 μM) and ascorbic acid (20.0 μM) in 10.05 mL of a phosphate buffer (pH 7.4) containing 1.0% (w/v) of Tween 20. LA: linoleic acid; γ-LN: γ-linolenic acid; α-LN: α-linolenic acid; AA: arachidonic acid; EPA: eicosapentaenoic acid; DHA: docosahexaenoic acid. (From Miyashita, K., Nara, E., and Ota, T., *Biosci. Biotechnol. Biochem.*, 57, 1638, 1993. With permission.)

Another striking difference between behavior in aqueous micelles and bulk phase is shown by the order of oxidation rate of PUFA in Figure 13.2—it indicates that the oxidative stability of PUFA in aqueous micelles rises with increasing number of bis-allylic positions in each molecule, and this relationship is the reverse of that in bulk phase or in organic solvents. This unusual order of oxidative stability is closely related to the conformation of PUFA in aqueous micelles. Kato et al. [43] have reported that n-3 double bonds of DHA and EPA reacted with *N*-bromosuccinimide to convert to the corresponding bromohydrin with 87% and 89% selectivity, respectively. They demonstrated that this high selectivity is due to the coiled configurations of DHA and EPA in an aqueous medium.

The conformation and kinetics of a molecule or part of a molecule are reflected in NMR relaxation times, i.e., spin–lattice relaxation (T_1) and spin–spin relaxation (T_2) time. Proton NMR relaxation times of each signals of PUFA (Figure 13.3) are shown in Table 13.1 [44]. When comparing the protons in PUFA forming micelle with those of PUFA in chloroform solution, two characteristic trends were observed. First, protons in PUFA forming micelles had shorter relaxation times than those in chloroform solution, except for methylene adjacent to methyl terminal (c) of DHA. Second, T_2 of protons in micelles was much shorter than their T_1, whereas T_1 and T_2 were almost the same in chloroform solution. The decrease of T_2 by micelle formation indicates that lipid molecules in micelles are mutually rigidly associated so molecular motion is more restricted than in non-associated lipid molecules in the chloroform solution. The shorter T_2 in micelle compared to T_1 means the aqueous system is closer to a solid state.

Proton T_2 gives more information for the difference in the molecular conformation of each PUFA between chloroform solution and aqueous micelles. There is little difference between the T_2 of corresponding protons on PUFA in the chloroform solution, but those of methyl protons (a) of DHA in micelles had a much larger value than those of LA, and that of AA were intermediate. Olefin protons (d) and bisallylic protons (e) showed a similar tendency as for methyl protons (a). For the proton on the carboxyl terminal (h), LA had a slightly longer T_2 than DHA. The mobility of the hydrophobic part of the DHA molecule is thus considered higher than that of LA when forming

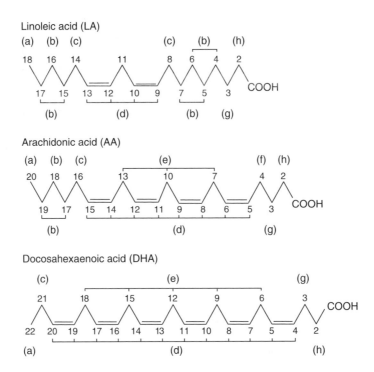

FIGURE 13.3 Molecular structure of three polyunsaturated fatty acids. (From Kobayashi, H., Yoshida, M., Maeda, I., and Miyashita, K., *J. Oleo Sci.*, 56, 105, 2004. With permission.)

micelles. DHA molecules in micelles may be packed more loosely than in LA, and the hydrophobic moiety of DHA may move more freely in micelles. Micelles of AA appear to have flexibility between those of DHA and LA; this flexibility may allow water molecules to permeate micelles. The penetration of water molecule to acyl moieties in micelles inhibits the hydrogen abstraction from bis-allylic positions of unoxidized fatty acid by peroxy radical of adjacent oxidized fatty acid in the propagation stage of free radical oxidation.

When free fatty acids are dispersed in water, they become arranged in the form of micelles, in which the polar head-groups lie in the aqueous phase, and non-polar tails form the oil phase. Emulsifier molecules also arrange themselves with free fatty acids so that the polar head-groups are located at the surface and non-polar tails are located in the interior (Figure 13.4), therefore, the interaction of emulsifier and lipid at the micelles affects the oxidative stability of lipids. When sodium salts of LA and DHA were used as substrates, the oxidative stability of DHA was markedly increased by addition of Tween 20 (polyoxyethylenesorbitan monolaurate) as an emulsifier (Figure 13.5). In contrast, the stability of LA with Tween 20 was slightly less than that of LA without Tween 20 [45]. This specific effect of Tween 20 to protect DHA against oxidation in aqueous micelles is associated with its different interaction with two acids. DHA may be protected from attack by free radicals and/or oxygen by forming a packed micellar conformation with Tween 20.

Yazu et al. [46] have also demonstrated the higher oxidative stability of EPA compared with LA in aqueous micelles. They explained the characteristic oxidative stability of EPA in aqueous solution on the basis of the formation of polar peroxy radicals—epidioxy-peroxy radicals—which are easily derived from peroxy radicals of EPA but not from those of LA. The epidioxy-peroxy radicals easily diffuse from the core to the surface in micelles because of their higher polarity. This localization of peroxy radicals may render the EPA less easily oxidized in aqueous micelles by enhancing the termination reaction rate for peroxy radicals and by reducing the rate of propagation.

TABLE 13.1
T_1 and T_2 for PUFA

		(a)	(b)	(c)	(d)	(e)	(f)	(g)	(h)
T_1									
Chloroform	LA	3.6 ± 0.2	2.0 ± 0.2	2.0 ± 0.3	3.5 ± 0.3	1.9 ± 0.1	2.0 ± 0.3	2.0 ± 0.3	1.8 ± 0.2
Solution	AA	3.3 ± 0.2	2.1 ± 0.3	2.2 ± 0.3	3.2 ± 0.5	1.6 ± 0.1	2.2 ± 0.3	1.8 ± 0.2	2.0 ± 0.4
	DHA	3.3 ± 0.4	—	1.5 ± 0.1	2.1 ± 0.3	3.5 ± 0.4	—	2.1 ± 0.2	2.1 ± 0.2
Aqueous	LA	1.6 ± 0.4	1.0 ± 0.3	0.91 ± 0.42	1.2 ± 0.3	0.92 ± 0.10	0.91 ± 0.42	0.82 ± 0.21	0.76 ± 0.11
Micelles	AA	1.6 ± 0.2	1.3 ± 0.2	1.0 ± 0.6	1.7 ± 0.3	0.95 ± 0.15	1.0 ± 0.2	0.95 ± 0.22	0.97 ± 0.10
	DHA	2.3 ± 0.3	—	2.1 ± 0.3	1.0 ± 0.2	1.9 ± 0.3	—	0.89 ± 0.10	1.1 ± 0.2
T_2									
Chloroform	LA	3.4 ± 0.3	1.8 ± 0.2	1.7 ± 0.3	3.3 ± 0.4	1.8 ± 0.2	1.7 ± 0.3	1.7 ± 0.3	1.6 ± 0.3
Solution	AA	3.0 ± 0.3	1.8 ± 0.2	2.1 ± 0.2	3.0 ± 0.4	1.4 ± 0.1	2.0 ± 0.3	1.5 ± 0.2	1.8 ± 0.3
	DHA	3.2 ± 0.5	—	1.1 ± 0.1	2.1 ± 0.2	3.4 ± 0.4	—	2.0 ± 0.3	2.0 ± 0.1
Aqueous	LA	0.25 ± 0.10	0.21 ± 0.10	0.15 ± 0.08	0.39 ± 0.11	0.10 ± 0.05	0.15 ± 0.08	0.19 ± 0.04	0.48 ± 0.12
Micelles	AA	0.65 ± 0.18	0.64 ± 0.15	0.68 ± 0.24	0.55 ± 0.09	0.49 ± 0.07	0.70 ± 0.18	0.25 ± 0.10	0.19 ± 0.06
	DHA	1.4 ± 0.3	—	1.3 ± 0.2	0.86 ± 0.12	1.1 ± 0.2	—	0.28 ± 0.07	0.28 ± 0.09

Source: From Kobayashi, H., Yoshida, M., Maeda, I., and Miyashita, K., *J. Oleo Sci.*, 56, 105, 2004. With permission.

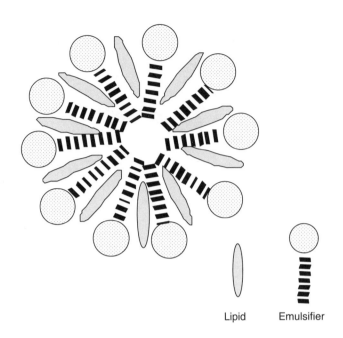

FIGURE 13.4 Aqueous structure of micelle. (From Miyashita, K., *Lipid Tech. Newslett.*, 8, 35, 2002. With permission.)

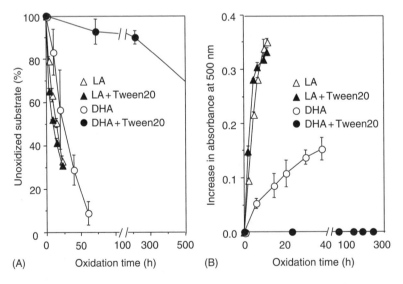

FIGURE 13.5 Effect of Tween 20 on the oxidative stability of LA and DHA in aqueous micelles. Changes in the amounts of unoxidized substrate (A) and development of total peroxides (B) during the oxidation of LA and DHA in a buffer with (solid circle and solid triangle) or without (open circle and open triangle) Tween 20. Each sodium salt (1.0 mM) was incubated at 37°C with $FeSO_4$ (1.0 μM) and ascorbic acid (20.0 μM) in a phosphate buffer (pH 7.4 at 37°C) with or without Tween 20. The unoxidized substrate content was determined by GC as previously described. Total peroxide formation is expressed as the absorbance at 500 nm. (From Miyashita, K., Inukai, N., and Ota, T., *Biosci. Biotechnol. Biochem.*, 61, 716, 1997. With permission.)

However, this concept based on the difference in the polarity of peroxy radicals cannot explain the higher oxidative stability of α-LN compared with γ-LN in aqueous micelles and the specific effect of Tween 20 on the oxidation of DHA in aqueous micelles.

III. AQUEOUS OXIDATION OF ETHYL ESTERS

The oxidative stability of ethyl esters of PUFA in aqueous micelles is also quite different from that in bulk phase or in organic solvent. When a small amount of ethyl linoleate (ethyl LA), ethyl linolenate (ethyl LN), and ethyl docosahexaenoate (ethyl DHA) were oxidized in aqueous micelles with Fe(II)-ascorbic acid, the highest stability was shown by ethyl DHA, followed by ethyl LN and ethyl LA. The reverse order of oxidative stability was obtained when these esters were oxidized in chloroform or ethanol with 2,2′-azobis(2,4-dimethyl-valeronitrile) (AMVN) as oxidation initiator [47]. The rate-limiting step of free radical lipid oxidation is the abstraction of hydrogen radical (H•) from substrate lipids (LH) (Figure 13.1). This abstraction can be initiated by, for example, alkoxy (LO•) or hydroperoxy (LOO•) radicals formed by the decomposition of hydroperoxides [7–12]. The decomposition rate in air is directly proportional to the number of bis-allylic positions in present in PUFA hydroperoxides [18,48,49]. The higher oxidative stability of ethyl DHA is partly due to the stability of DHA monohydroperoxides in aqueous micelles.

The oxidation products formed in aqueous micelles differ from those in bulk phase. PUFA containing more than three double bonds can yield hydroperoxy epidioxides as primary oxidation products; they are formed by a rapid 1,3-cyclization and further oxidation from inner monohydroperoxides [8–12]. Reversed-phase high performance liquid chromatography (HPLC) of oxidized ethyl LN in aqueous micelles showed the formation of hydroperoxy epidioxides, but the ratio of epidioxides to monohydroperoxides was lower than for ethyl LN oxidized in bulk phase [47]. Furthermore, no peaks corresponding to epidioxides could be detected by HPLC of oxidized ethyl DHA in aqueous micelles. The lower accumulation of epidioxides in aqueous micelles may be a result of inhibitory effect of penetration of water to acyl moieties and/or the steric hindrance in the packed conformation of ethyl DHA and emulsifier.

The oxidative stability of PUFA monoacylglycerols (MG) and triacylglycerols (TGs) in aqueous phase also increases with increasing degree of unsaturation, when their concentrations were very low. Monodocosahexaenoin was the most oxidatively stable, followed by monoarachidonin, mono-linolenin, and monolinolein when each MG (1 mM) was oxidized with Fe(II)-ascorbic acid as an initiator [50]. The same relationship was found in the aqueous oxidation of tridocosahexaenoin and trilinolein [50]. The characteristic oxidative stability of acylglycerols in aqueous solution was also found in TG from natural oils [51]. Figure 13.6 compares the oxidative stability of soybean oil TG (SoyTG) and tuna oil TG (TunaTG), in which major PUFA in these TG was LA (54.9%) and DHA (17.0%), respectively [52]. The average number of bis-allylic positions in the TunaTG molecule was calculated from its fatty acid composition to be 3.72, whereas the number in SoyTG molecule was 1.94. Since the oxidative stability of polyunsaturated lipids in bulk phase decrease with the increasing number of bis-allylic positions present in their molecules, TunaTG was oxidized more rapidly than SoyTG in air. However, in aqueous solution, TunaTG was much more stable to oxidation than SoyTG. This result was confirmed by the determination of oxidation products.

IV. OXIDATIVE STABILITY OF TG IN EMULSION

Knowledge of the characteristic oxidative stability of PUFA or their esters in aqueous phase is important for better understanding of the oxidation of PUFA in aqueous food systems. But, in such systems lipids are mainly dispersed as emulsions and it is important to compare the oxidative stability of different polyunsaturated lipids in emulsion. Food emulsions are complex multiphase systems, in which different molecular species interact with each other. The various molecules in an emulsion system become distributed according to their polarity and surface activity between

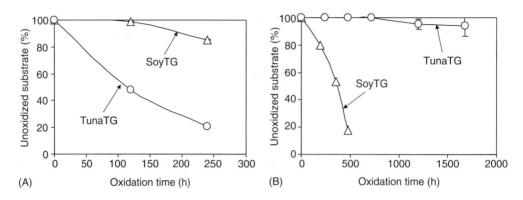

FIGURE 13.6 Oxidative stability of soybean oil TG (SoyTG) and tuna oil TG (TunaTG). Each TG (2g) was oxidized in air in the dark at 37°C (A). For aqueous oxidation (B), each TG (0.025% [w/v]) was incubated in the dark at 37°C in a phosphate buffer (pH 7.4) containing 0.5% (w/v) of Triton X-100 (*t*-octylphenoxy-polyethoxyethanol). The stability was determined by following the decrease in unoxidized substrate content. (From Miyashita, K., *Lipid Tech. Newslett.*, 8, 35, 2002. With permission.)

different phases, which include the oil phase, the water phase, and the interfacial region. Lipid oxidation in such systems is an interfacial phenomenon that is greatly influenced by the nature of interface. The aqueous structure of interface in emulsion (Figure 13.7) is almost the same as that of micelles (Figure 13.4). Since the lipid oxidation generally proceeds from the interface to the interior of the oil droplet in oil-in-water emulsions, the oxidation of lipids at the interface is an important factor to predicting the oxidative stability of lipids in emulsion. In aqueous micelles, DHA or DHA ester is oxidatively more stable than LA or LA ester, although DHA is chemically more susceptible to oxidation than LA. This is because DHA itself—or DHA and emulsifier—would take on a more protective confirmation against oxidative attack of free radicals and/or oxygen.

The interface area increases with decreasing droplet size. The larger the area of the interface and the smaller the interior oil phase, the more markedly the effects of the interface on the lipid oxidation will be recognized. Therefore, when the lipid concentration is very low and droplet size is relatively small, there is a possibility that DHA containing lipids is more oxidatively stable than LA containing lipids in emulsion. For example, in emulsions that were each dispersed with one of three different sucrose esters [52] (Figure 13.8), TGs containing DHA (DHATG) were more stable to oxidation than soybean oil TG (SoyTG). This is despite the fact that the average number of bis-allylic positions in DHATG (DHA: 39.3 mol%) and SoyTG (LA: 54.2 mol%) was calculated to

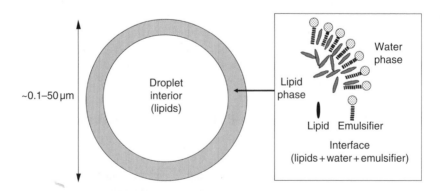

FIGURE 13.7 Aqueous structure of emulsion. (From Miyashita, K., *Lipid Tech. Newslett.*, 8, 35, 2002. With permission.)

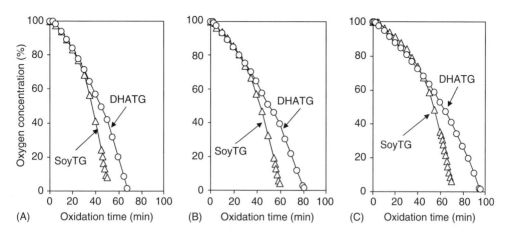

FIGURE 13.8 Oxidative stability of SoyTG and DHATG in emulsions dispersed with S570 (HLB:5) (A), S1170 (HLB:11) (B), and S1670 (HLB: 16) (C). The stability was followed by measuring the decrease in oxygen concentration in the aqueous solution. The fatty acid composition of each emulsifier was the same and consists of palmitic (30%) and stearic (70%) acids. Each TG (1.0% [w/v]) was incubated in the dark at 37°C in a phosphate buffer (pH 7.4) containing 0.1% (w/v) of emulsifier. Oxidation was induced by the addition of 2,2′-azobis(2-amidino-propane) dihydrochloride (AAPH) (1.0 mM). (From Miyashita, K., *Lipid Tech. Newslett.*, 8, 35, 2002. With permission.)

be 6.97 and 1.86, respectively, and DHATG was oxidized more rapidly than SoyTG in bulk phase. DHATG was also oxidatively more stable than SoyTG in the emulsion dispersed with polyglycerol ester, DS10 (Figure 13.9). On the other hand, DHATG was less stable to oxidation than SoyTG in the emulsion dispersed with HS10 and there was little difference in the stability between both TGs in the emulsion dispersed with SWA20D (Figure 13.9).

The difference in effect of these emulsifiers on the oxidative stability of both TG can be explained by the different interaction of TG and emulsifier at the interface. The interface is a narrow

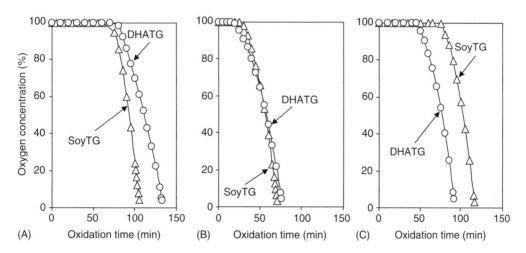

FIGURE 13.9 Oxidative stability of SoyTG and DHATG in emulsions dispersed with DS10 (A), SWA20D (B), and HS10 (C). The stability was followed by measuring the decrease in oxygen concentration in the aqueous solution. The reaction conditions are the same as those in Figure 13.8. (From Miyashita, K., *Lipid Tech. Newslett.*, 8, 35, 2002. With permission.)

region between the lipid and water phases that consists of a mixture of lipid, water, and emulsifier (Figure 13.7). The lipid oxidation at the interface is affected by many factors such as the size and concentration of the emulsion droplets, the thickness, electrical charge, the packing degree of emulsifier and lipid at the interface, and composition of the interface, and the extent of droplet–droplet interactions. In particular, the oxidative stability of lipids in emulsion is strongly affected by altering the droplet size and the packing of the emulsifier and lipid molecules at the interface. When the concentrations of lipid and emulsifier are the same, the area of interface increases with decreasing droplet size. The opportunity for the attack by oxidation inducer such as free radicals and metal ions on lipids at the interface increases with increasing area of interface. However, there was little difference in the droplet size between emulsions of SoyTG and DHATG dispersed with three kinds of sucrose esters or polyglycerol esters (Figures 13.8 and 13.9). Therefore, the different effect of emulsifier on the oxidative stability of SoyTG and DHATG is expected to be derived from the difference in conformation of emulsifier and each TG molecule at the interface.

Oxidative stability was compared among TGs dispersed with three kinds of sucrose esters having the same fatty acyl composition, the stability increased with increasing their hydrophile–lipophile balance (HLB) in both TGs, although these emulsifiers had no effect on the oxidative stability of SoyTG and DHATG in bulk phase [27]. When the HLB of sucrose ester was the same, the oxidative stability increased with increasing acyl chain length of fatty acid esterified [27]. These effects of HLB or fatty acyl composition may also be explained by the different packing of emulsifier at the interface. Polyglycerol esters also affected the oxidative stability of both types of TG in emulsion (Figure 13.9), but the magnitude of effect of polyglycerol esters was less than that of sucrose ester. In addition, the difference in the oxidative stability of TG dispersed with polyglycerol ester cannot be explained by the relationship with HLB or the acyl composition.

The characteristic oxidation behavior of different polyunsaturated lipids at the interface, as described in this chapter, provides us with the basic and important information needed to fully understand the lipid oxidation in food emulsion and to give an effective method for protection of lipids against oxidation in emulsion. That is the importance of chemical and physical properties of interface for prevention of PUFA against oxidation. This information will also help us to investigate the oxidative stability of PUFA in biological systems, because PUFAs are usually present in such systems in an aqueous medium with other components. In phosphatidylcholine (PC) liposomes, the model systems of biological membranes, the degree of unsaturation of PUFA had little effect on the oxidative stability, whereas the stability was affected by the positional distribution of PUFA in the PC molecule [53–56]. This is a fundamental and important result for a better understanding of the oxidation of PUFA in biological systems. Lipid oxidation in PC liposomes and in cellular model systems is described later.

V. POSITIONAL DISTRIBUTION OF MONOHYDROPEROXIDES ISOMERS

The key event in lipid peroxidation is the formation of a lipid radical by the abstraction of a hydrogen radical from the bis-allylic positions. The resulting pentadienyl radical reacts with oxygen at both ends to form two kinds of conjugated diene monohydroperoxides (MHP). The specific conformation of PUFAs in aqueous micelles may affect the rate of hydrogen abstraction from the particular bis-allylic position and/or the selective attack of oxygen at pentadienyl radicals.

In the free radical oxidation of LA, hydrogen abstraction occurs at the C11 position, which resulted in production of a pentadienyl radical between C9 and C13. Then the radical reacts at either end with oxygen to produce a mixture of 9-MHP and 13-MHP. Since AA and DHA have two or more bis-allylic methylene groups, there are several possible positions for hydrogen abstraction; carbons 7, 10, or 13 for AA and carbons 6, 9, 12, 15, or 18 for DHA, respectively. Since oxygen can attack carbons at either end of the pentadienyl radical, the resulting MHP isomers were those with hydroperoxide substitution on carbons 5, 9, 8, 12, 11, and 15 for AA, and 4, 8, 7, 11, 10, 14, 13, 17, 16, 20 for DHA.

MHP isomer distribution from the oxidation of PUFA ethyl esters in chloroform and aqueous solution is shown in Table 13.2. MHP was determined by GC–MS after extraction of ethyl esters from the solution, hydrogenation, and trimethylsilylation [57]. When the distribution of LA ethyl esters was compared in aqueous solution and chloroform solution, an even distribution between the two MHP isomers (9- and 13-MHP) was found in the oxidation of both in chloroform and aqueous solution. On the other hand, relatively higher percentage was found for 11- and 15-MHP in the oxidation of AA in aqueous solution than in chloroform solution (Table 13.2). The difference in the MHP distributions between the oxidation in chloroform and aqueous emulsion was also observed for DHA. As shown in Table 13.2, the proportion of 10- and 16-MHP in aqueous oxidation was much higher than that in chloroform solution. The proportion of 14-MHP in aqueous oxidation was also higher than that in chloroform solution. On the other hand, the formation of 7- and 20-MHP in aqueous emulsion was less than that in chloroform solution.

Oxidation of PUFA containing more than two double bonds such as AA or DHA produces a significant amount of hydroperoxy epidioxides as the main oxidation products, other than MHP, at an early stage of oxidation [8–12]. The external MHP (5-MHP and 15-MHP for AA, and 4-MHP

TABLE 13.2
Isomeric Distribution of MHP Isomers Formed in the Oxidations of PUFA Esters

	MHP Formation		Positional Distribution			
PUFA	Hydrogen Abstraction	Resulted MHP	In Chloroform[a] Solution	In Aqueous[a] Micelles	In PC[b] Liposome	In Cellular[c] PL
LA	C-13	9-MHP	49.3 ± 1.5	48.9 ± 2.3	49.2 ± 3.6	49.7 ± 1.5
		13-MHP	50.7 ± 1.5	51.1 ± 2.3	50.8 ± 3.6	50.3 ± 1.5
AA	C-7	5-MHP	19.4 ± 1.1	17.1 ± 3.3	9.6 ± 0.7	23.7 ± 3.7
		9-MHP	15.3 ± 0.4	15.2 ± 1.5	23.8 ± 2.0	6.6 ± 1.3
	C-10	8-MHP	15.2 ± 0.5	13.9 ± 1.6	5.1 ± 1.1	12.9 ± 2.3
		12-MHP	15.0 ± 1.1	11.7 ± 1.3	26.3 ± 2.4	23.2 ± 3.9
	C-13	11-MHP	15.1 ± 0.5	20.0 ± 0.5	7.4 ± 2.4	3.4 ± 2.3
		15-MHP	19.9 ± 1.3	22.1 ± 1.3	27.9 ± 2.1	30.2 ± 3.4
DHA	C-6	4-MHP	17.7 ± 1.3	21.7 ± 1.1	17.9 ± 2.4	25.7 ± 2.2
		8-MHP	12.9 ± 1.7	9.8 ± 0.6	11.2 ± 1.1	12.8 ± 1.6
	C-9	7-MHP	10.5 ± 0.7	5.2 ± 0.8	6.2 ± 1.1	ND[d]
		11-MHP	7.2 ± 1.0	1.9 ± 1.4	2.1 ± 0.5	ND[d]
	C-12	10-MHP	4.2 ± 0.5	10.1 ± 1.5	13.0 ± 1.1	10.2 ± 1.3
		14-MHP	4.3 ± 0.6	9.5 ± 1.0	4.6 ± 0.8	4.9 ± 4.5
	C-15	13-MHP	3.5 ± 1.0	2.0 ± 0.6	5.5 ± 1.0	ND[d]
		17-MHP	4.5 ± 0.7	3.9 ± 0.3	3.1 ± 0.6	ND[d]
	C-18	16-MHP	8.7 ± 1.4	21.9 ± 2.4	21.3 ± 4.0	24.6 ± 1.5
		20-MHP	26.5 ± 1.1	14.1 ± 1.5	16.4 ± 1.3	21.9 ± 4.3

Source: From Kobayashi, H., Yoshida, M., and Miyashita, K., *Chem. Phys. Lipids*, 126, 111, 2003. With permission.

[a] The ethyl ester of each PUFA was oxidized in chloroform and in aqueous solution. The oxidation was induced by 2,2′-azobis(2,4-dimethyl-valeronitrile) (AMVN) or 2,2′-azobis(2-amidino-propane) dihydrochloride (AAPH) for chloroform and aqueous oxidation, respectively.

[b] 1-Palmitoyl-2-linoleoyl-phosphatidylcholine (LA-PC), 1-palmitoyl-2-arachidonoyl-phosphatidylcholine (AA-PC), and 1-palmitoyl-2-docosahexaenoyl-phosphatidylcholine (DHA-PC) were oxidized in liposomes. AAPH was used as oxidation inducer.

[c] Each PUFA was supplemented to the cell (HepG2). Supplementation with PUFA resulted in their incorporation into cellular lipids. Cellular oxidation was accelerated by the addition of H_2O_2.

[d] Not detected.

and 20-MHP for DHA), formed by hydroperoxidation on the sp^2 carbon of PUFA closest to the methyl terminal or the carbonyl terminal, does not result in hydroperoxy epidioxide. Only the internal MHP, which are formed by hydroperoxidation on an sp^2 carbon of PUFA adjacent to a bis-allylic position, can undergo 1,3-cyclization to form hydroperoxy epidioxides. The formation of hydroperoxy epidioxides affects the positional distribution of MHP as the amount of internal MHP decreases with an increase in the formation of epidioxides. As for the oxidation of AA in chloroform solution, the proportion of internal MHP, 8-, 9-, 11-, and 12-MHP, was lower than that of external MHP isomers, 5- and 15-MHP (Table 13.2). This result could be due to the tendency of the inner-peroxy radicals to undergo rapid 1,3-cyclization. The lower proportion of internal MHP than external MHP was also found in the oxidation of DHA in chloroform solution (Table 13.2). The same tendency was observed in the oxidation of these PUFAs in aqueous solution, except in the case of the methyl-terminal bis-allylic position, which gave amounts of 11-MHP for AA and 16-MHP for DHA comparable to those of the corresponding external MHPs (Table 13.2). In the oxidation of DHA in aqueous solution, 10-MHP was also formed at a relatively high proportion. These high proportions of selective internal MHPs would be due to the low rate of 1,3-cyclization of the MHP and/or the high rate of oxygen attack at the positions, which may be caused by the specific conformation of these PUFAs, its flexibility and permeability in aqueous solution.

VI. OXIDATION OF PC IN LIPOSOMES

When three kinds of 1-palmitoyl (PA; 16:0)-2-PUFA-PC, that is 1-PA-2-linoleoyl (LA;18:2n-6)-PC (PA-LA), 1-PA-2-arachidonyl (AA; 20:4n-6)-PC (PA-AA), and 1-PA-2-docosahexaenoyl (DHA; 22:6n-3)-PC (PA-DHA), were oxidized in bulk phase (Figure 13.10), PA-LA was oxidatively the most stable, followed by PA-AA and PA-DHA, in that order [58]. The oxidative stability also increased with decreasing the number of bis-allylic positions in each PC molecule in t-BuOH solution [57] (Figure 13.11). The stability of a 1:1 mixture of 1,2-dipalmitoyl-PC (diPA)+1,2-dilinoleoyl-PC (diLA) was also higher than that of diPA+1,2-diarachidonyl-PC (diAA) in bulk phase (Figure 13.10). When the oxidative stability of PA-LA or PA-AA was compared with that of diPA+diLA or diPA+diAA, respectively, 1-PA-2-PUFA-PC showed much higher stability than the 1:1 mixture of diPA-PC+diPUFA-PC in both cases (Figure 13.10), although the degree of

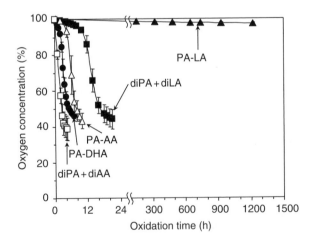

FIGURE 13.10 Oxidative stability of PC in bulk phase. Oxidation was done by incubation of each PC at 37°C in the dark. The oxidative stability was evaluated by the analysis of the decrease in the oxygen concentration in the headspace gas of the vial by GC. (From Araseki, M., Yamamoto, K., and Miyashita, K., *Biosci. Biotechnol. Biochem.*, 66, 2573, 2002. With permission.)

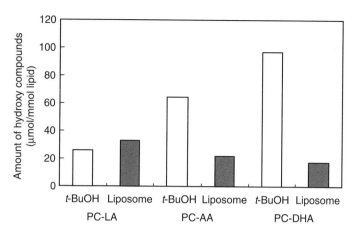

FIGURE 13.11 Amounts of total hydroxy compounds derived from oxidation products of PUFA in PC, formed in *t*-BuOH solution and liposomes. AAPH was used as oxidation inducer. The oxidation products were extracted with chloroform/methanol, then hydrogenated, transmethylated, trimethylsilylated (TMS), and then subjected to GC–MS analysis for the quantitative comparison of TMS-derivatives from mono-, di-, and tri-hydroxy compounds. These hydroxyl compounds are derived from monohydroperoxides, hydroperoxy epidioxides, and other oxidation products. (From Kobayashi, H., Yoshida, M., and Miyashita, K., *Chem. Phys. Lipids*, 126, 111, 2003. With permission.)

unsaturation of each 1-PA-2-PUFA-PC was the same as that of the corresponding mixture of diPA-PC + diPUFA-PC. The same tendency was observed in the PC oxidation in chloroform (Figure 13.12). Figure 13.12 clearly shows that the oxidative stability of 1-PA-2-PUFA-PC was higher than that of the corresponding diPA-PC + diPUFA-PC mixture. The comparison of the stabilities of three kinds of 1-PA-2-PUFA-PC or two kinds of diPA-PC + diPUFA-PC also showed that the oxidative stability of these PCs decreased with increasing degree of unsaturation.

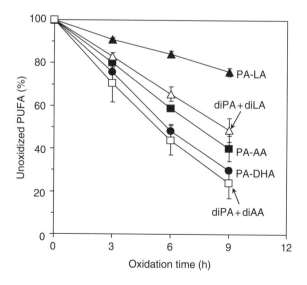

FIGURE 13.12 Oxidative stability of PC in chloroform solution. Each PC (0.5 mM) in chloroform was incubated at 37°C in the dark under AMVN (1.0 mM) as an oxidation inducer. The oxidative stability was evaluated by the analysis of the decrease in the unoxidized PUFA in each PC molecule. (From Araseki, M., Yamamoto, K., and Miyashita, K., *Biosci. Biotechnol. Biochem.*, 66, 2573, 2002. With permission.)

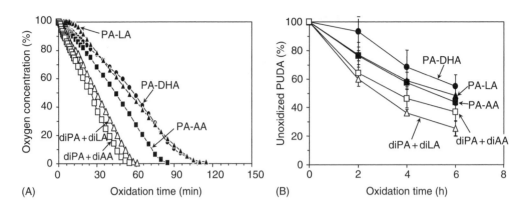

FIGURE 13.13 Oxidative stability of PC in liposomes. PC (0.5 mM) was oxidized in liposomes in the presence of AAPH (3.0 mM). The oxidative stability was evaluated by the analysis of the decrease in oxygen concentration in the solution (A) or the decrease in unoxidized PUFAs in the PC molecule (B). (From Araseki, M., Yamamoto, K., and Miyashita, K., *Biosci. Biotechnol. Biochem.*, 66, 2573, 2002. With permission.)

The higher oxidative stability of 1-PA-2-PUFA-PC than that of corresponding mixture of diPA-PC + diPUFA-PC was also found in the oxidation of PC in liposomes (Figure 13.13). However, the effect of the degree of unsaturation on the oxidative stability of PC in liposomes (Figures 13.11 and 13.13) was different from that in bulk (Figure 13.10), in chloroform solution (Figure 13.12), or in *t*-BuOH solution (Figure 13.11). As shown in Figure 13.11, analysis of oxidation products showed that PA-DHA was most oxidatively stable, followed by PA-AA and PA-LA. On the other hand, oxygen consumption analysis (Figure 13.13A) showed that PA-DHA and PA-LA were more oxidatively stable than PA-AA in liposomes. The difference in the oxidative stability was comparatively small in both cases. Furthermore, the stability of diPA + diLA in liposomes was almost the same as that of diPA + diAA (Figure 13.13A). When the oxidative stability of PC in liposomes was evaluated by using the GC method (Figure 13.13B), PA-DHA was also oxidatively more stable than PA-AA or PA-LA, and little difference in the stability was found between PA-AA and PA-LA. In Figure 13.13A, there was little difference in the oxidative stability of diPA + diAA and diPA + diLA, but, GC analysis of the decrease in unoxidized PUFA showed that diPA + diAA was more stable than diPA + diLA (Figure 13.13B).

It is generally accepted that the oxidative stability of PUFA decreases with increasing degrees of unsaturation, however, as shown in Figures 13.11 and 13.13, such a relationship was not found in PC containing DHA, AA, or LA in liposomes. On the other hand, the stability of 1-PA-2-PUFA-PC was higher than that of the corresponding PC mixture of 1,2-diPA-PC + 1,2-diPUFA-PC in liposomes (Figure 13.13). These results suggest that the degree of unsaturation was not the main factor affecting the oxidative stability of PC in liposomes. When oil droplets are dispersed in an aqueous solution, the surface area of the droplets increased with decreasing droplet size at the same oil concentration. The chance of an attack of an oxidation inducer such as AAPH on the interface of the oil droplet increases with increase in the surface area; therefore, the smaller droplet size reduces the oxidative stability. However, as shown in Table 13.3, the sizes of 1-PA-2-PUFA-PC were smaller than those of 1,2-diPA-PC + 1,2-diPUFA-PC, although 1-PA-2-PUFA-PC was oxidatively more stable than the corresponding mixture of 1,2-diPA-PC + 1,2-diPUFA-PC in liposomes (Figure 13.13). Liposomes are hollow vesicles composed of single or multiple bilayers and the surface area may not be correlated with the size of liposomes but with the packing degree of the PC bilayers.

The higher oxidative stability of 1-PA-2-PUFA-PC than that of 1,2-diPA-PC + 1,2-diPUFA-PC will also be affected by the conformation of each PUFA and the intermolecular conformation of the fatty acyl chain at sn-1 and sn-2 positions in liposomes. Furthermore, the higher stability of 1-PA-2-PUFA-PC may also be explained by the idea that an intramolecular free radical chain

TABLE 13.3
Mean Vesicle Diameter in PC Liposomes

PC	Mean Diameter (mm)
PA-DHA	2.043 ± 0.012
PA-LA	2.651 ± 0.013
PA-AA	2.057 ± 0.021
diPA + diLA	4.433 ± 0.106
diPA + diAA	4.221 ± 0.313

Source: From Araseki, M., Yamamoto, K., and Miyashita, K., *Biosci. Biotechnol. Biochem.*, 66, 2573, 2002. With permission.

The values are expressed as the mean $\pm$ SD of the results from three separate analysis.

reaction between PUFA of esters occurs more rapidly than the intermolecular chain reaction. A 1:1 mixture of 1,2-diPUFA-PC + 1,2-diPA-PC was also oxidatively less stable than corresponding 1-PA-2-PUFA-PC in bulk (Figure 13.10) and in an organic solvent (Figure 13.12). The difference in the oxidative stability of both types of PC would be due to the different rate of hydrogen abstraction by free radicals from intermolecular and intramolecular acyl groups not to the PC conformation, because PC takes no packed conformation in bulk or in organic solvent systems.

In the propagation stage of autoxidation, fatty alkyl radicals react with molecular oxygen to form peroxy radicals. The peroxy radical abstracts a hydrogen atom from another unsaturated fatty compound to form a hydroperoxide and an alkyl radical. The latter reacts with molecular oxygen in a repetition of the first propagation reaction. The initially formed hydroperoxide may decompose subsequently to yield free radicals such as alkoxy and hydroxy radicals. These radicals serve as initiators for these reactions. The intramolecular hydrogen abstraction from an unsaturated acyl group by a free radical of another acyl group in the same ester molecule will occur more rapidly than the intermolecular hydrogen abstraction, because acyl groups bonded to the same ester molecule are nearer to one another. A similar effect of the esterified position of PUFA was observed in the oxidation of triacylglycerols [59]. When the oxidative stability of a 1:2 (mol/mol) mixture of trieicosapentanoylglycerol (EEE) and tripalmitoylglycerol (PPP) was compared with that of synthetic 1,2(or 2,3)-dipalmitoyl-3(or 1) eicosapentaenoylglycerol (PPE), PPE was much more oxidatively stable than PPP + EEE.

Table 13.2 shows proportion of MHP isomers formed in the oxidation of PC in liposomes. The uneven distribution of MHP in PC-AA liposomes indicated the preferential oxygen attack on the methyl-terminal side of the radical. However, in case of PC-DHA, the oxygen attack favored the carboxy terminal side of the pentadienyl radical. In PC-AA oxidation, amounts of 5-MHP + 9-MHP, 8-MHP + 12-MHP, and 11-MHP + 15-MHP were almost the same. This result indicates the same rate of hydrogen abstraction at the three bis-allylic positions (C7, 10, and 13) and difficulty in the 1,3-cyclization of internal MHP. On the other hand, MHP distribution in PC-DHA revealed the favorable hydrogen abstraction at C18, C6, and C12. PC in liposomes with highly unsaturated fatty acids, such as AA and DHA, is also more permeable and shows more flexibility in fatty acid chains than those formed from PC containing less unsaturated fatty acids such as linoleic. NMR analysis and molecular dynamics simulation of PC containing DHA in liposomes indicates the wide variety of DHA conformation—including back-bended, helical, and angle-iron conformations—occurring in liposome systems [60–63]. This variety in the DHA chain conformation gives looser packing of the lipid chains [64,65]. The looser packing of the membrane at the lipid–water interface brings about the high water permeability [62]. Molecular dynamics simulation also indicates the remarkable overlapping of water molecules with double bond regions of the DHA chain.

The presence of water molecules near a DHA molecule will lower the density of the bis-allylic hydrogen and reduce the chain-carrying reaction of lipid peroxidation. The higher water permeability of DHA and its specific conformation may be a reason for the uneven hydrogen abstraction from the DHA molecule to produce the characteristic distribution of DHA-MHP isomers.

VII. OXIDATION OF PUFA IN HEPG2 CELLS

When LA, AA, or DHA were added to the culture medium of HepG2 cells and incubated, remarkable incorporation of the respective PUFA into cellular total lipids (TL) was observed (Table 13.4) [66]. The increase in LA, AA, or DHA was compensated by a decrease in monoenoic fatty acids such as 18:1n-9 and 18:1n-7. The same change in fatty acid composition was also found in cellular neutral lipids (NL) (Table 13.5) [66]. The content of LA, AA, or DHA in cellular NL after the addition of each PUFA was significantly higher than that in the NL of the control cells. The content of each PUFA in the cellular phospholipids (PL) after the addition of each PUFA was also significantly higher than that in the PL of the control cells (Table 13.6) [66]. When the peroxidation level in the cells was evaluated by fluorescence analysis, the incorporation of LA, AA, or DHA enhanced the cellular lipid peroxidation (Figure 13.14) [66]. However, the degree of unsaturation of the added PUFA had little effect on the cellular lipid peroxidation level. Igarashi and Miyazawa [67] reported an increase in membrane phospholipid hydroperoxide after incorporation of LA in HepG2 cells. The fluorescent analysis of lipid peroxidation was based on the oxidation of DPPP to DPPP oxide by organic peroxides present in the cell membranes [68,69]. Therefore, the result in Figure 13.14 represents the peroxidation level of cell membrane PL. The average number of bis-allylic positions per 1 mol of fatty acid in cellular PL can be calculated from the fatty acid composition (Table 13.6) to be 0.57 for control cells, 0.64 for LA-supplemented cells, 0.77 for AA-supplemented cells, and 0.81 for DHA-supplemented cells. It is known that the oxidative stability of PUFA decreases with increasing number of bis-allylic positions. However, as shown in Figure 13.14, there was little difference in the lipid peroxidation level in the LA-, AA-, and DHA-supplemented cells.

Figure 13.14 also shows that the addition of H_2O_2 (0.5 mM) enhanced cellular lipid peroxidation levels in control, LA-, and AA-supplemented cells when compared with those without H_2O_2. The enhancement of the lipid peroxidation level by the addition of H_2O_2 has been reported in mouse

TABLE 13.4
Incorporation of Each PUFA into the TL of Cells Incubated with LA, AA, and DHA for 24 h

	Cell			
Fatty Acid (wt%)	Control	+LA	+AA	+DHA
16:0	25.1 ± 0.4^a	23.7 ± 0.2^a	23.3 ± 1.0^a	25.3 ± 0.6^a
18:0	5.8 ± 0.4^a	5.7 ± 0.2^a	6.9 ± 0.6^a	7.1 ± 0.1^a
16:1n-7	7.2 ± 0.1^a	4.4 ± 0.0^b	3.5 ± 0.2^c	4.6 ± 0.1^b
18:1n-7	12.9 ± 0.2^a	9.4 ± 0.2^b	7.9 ± 0.7^c	$8.1 \pm 0.1^{b,c}$
18:1n-9	22.9 ± 0.4^a	17.0 ± 0.2^b	15.8 ± 1.6^b	17.2 ± 0.4^b
18:2n-6	2.6 ± 0.1^a	15.2 ± 0.5^b	2.2 ± 0.6^a	1.9 ± 0.0^a
20:4n-6	4.9 ± 0.2^a	5.0 ± 0.1^a	20.3 ± 3.6^b	4.1 ± 0.1^a
22:6n-3	3.3 ± 0.1^a	2.6 ± 0.1^a	2.9 ± 0.0^a	17.3 ± 1.2^b

Source: From Araseki, M., Kobayashi, H., Hosokawa, M., and Miyashita, K., *Biosci. Biotechnol. Biochem.*, 69, 483, 2005. With permission.

Data are expressed as mean $\pm$ SD ($n = 3$). Values not sharing a common superscript are significantly different at $P < 0.01$.

TABLE 13.5

Incorporation of Each PUFA into the NL of Cells Incubated with LA, AA, and DHA for 24 h

Fatty Acid (wt%)	Cell			
	Control	+LA	+AA	+DHA
16:0	19.3 ± 2.1^a	18.3 ± 1.0^a	19.7 ± 1.0^a	21.8 ± 0.8^a
18:0	4.4 ± 1.3^a	3.3 ± 0.1^a	3.4 ± 0.6^a	3.2 ± 0.2^a
16:1n-7	$3.5 \pm 0.4^{a,b}$	3.0 ± 0.2^a	$3.0 \pm 0.1^{a,b}$	4.0 ± 0.2^b
18:1n-7	8.5 ± 0.8^a	6.6 ± 0.2^b	6.4 ± 0.7^b	$7.8 \pm 0.3^{a,b}$
18:1n-9	21.2 ± 2.0^a	17.2 ± 0.6^a	17.5 ± 1.6^a	19.3 ± 0.6^a
18:2n-6	1.6 ± 0.3^a	22.5 ± 1.4^b	2.3 ± 0.6^a	1.1 ± 0.0^a
20:4n-6	0.8 ± 0.1^a	1.7 ± 0.1^a	14.1 ± 3.6^b	0.6 ± 0.1^a
22:6n-3	2.1 ± 0.1^a	2.4 ± 0.2^a	2.7 ± 0.0^a	23.1 ± 2.6^b

Source: From Araseki, M., Kobayashi, H., Hosokawa, M., and Miyashita, K., *Biosci. Biotechnol. Biochem.*, 69, 483, 2005. With permission.

Data are expressed as mean $\pm$ SD ($n = 3$). Values not sharing a common superscript are significantly different at $P < 0.01$.

polymorphonuclear leukocytes [68] and in human monocytic leukemia cells (U-937) [69] at a range of 0.1–1.0 mM of H_2O_2. On the other hand, induction of lipid peroxidation by H_2O_2 was not observed in DHA-supplemented cells. The lower cellular oxidation level in DHA-supplemented cells after H_2O_2 addition was also observed by the analysis of total MHP content in the cellular PL (Table 13.7) [66]. Table 13.7 also shows that the main source for MHP was LA in all cases, and a significant amount of AA-MHP was observed only in AA-supplemented cells. A small amount of DHA-MHP was also observed in DHA-supplemented cells, but not in other cells. PUFA containing more than two double bonds, such as AA and DHA, produces further oxidation products from MHP

TABLE 13.6

Incorporation of Each PUFA into the PL of Cells Incubated with LA, AA, and DHA for 24 h

Fatty Acid (wt%)	Cell			
	Control	+LA	+AA	+DHA
16:0	22.2 ± 1.0^a	22.4 ± 0.7^a	22.1 ± 0.8^a	23.2 ± 0.7^a
18:0	9.0 ± 0.2^a	11.2 ± 0.1^b	12.4 ± 0.1^c	$11.8 \pm 0.2^{b,c}$
16:1n-7	5.4 ± 0.2^a	3.6 ± 0.2^b	2.9 ± 0.3^c	4.1 ± 0.4^b
18:1n-7	9.4 ± 0.1^a	7.5 ± 0.1^b	7.0 ± 0.6^b	7.3 ± 0.1^b
18:1n-9	16.7 ± 0.1^a	13.1 ± 0.1^b	11.4 ± 0.7^c	13.8 ± 0.1^b
18:2n-6	2.8 ± 0.1^a	12.2 ± 0.1^b	1.5 ± 0.1^a	3.2 ± 1.0^a
20:4n-6	9.8 ± 0.1^a	10.9 ± 0.1^a	19.7 ± 0.3^b	8.3 ± 0.5^c
22:6n-3	4.9 ± 0.0^a	$3.9 \pm 0.1^{a,b}$	3.2 ± 0.2^b	10.5 ± 0.8^c

Source: From Araseki, M., Kobayashi, H., Hosokawa, M., and Miyashita, K., *Biosci. Biotechnol. Biochem.*, 69, 483, 2005. With permission.

Data are expressed as mean $\pm$ SD ($n = 3$). Values not sharing a common superscript are significantly different at $P < 0.01$.

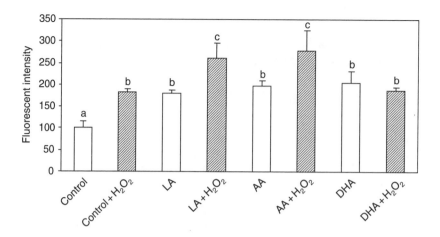

FIGURE 13.14 Formation of MHP in HepG2 cells treated with or without H_2O_2. Values not sharing a common superscript are significantly different at $P < 0.01$. Values are mean $\pm$ SD ($n = 3$). (From Araseki, M., Kobayashi, H., Hosokawa, M., and Miyashita, K., *Biosci. Biotechnol. Biochem.*, 69, 483, 2005. With permission.)

such as hydroperoxy epidioxides. The relatively lower contribution of AA and DHA to MHP formation found in Table 13.7 might be due to the further oxidation of AA-MHP and DHA-MHP.

It is generally accepted that the oxidative stability of PUFA in the bulk phase and in organic solution is inversely proportional to the degree of unsaturation. But, the oxidative stability of PUFA and its ester in an aqueous environment increased with increasing degree of unsaturation [40,47,57]. PC containing DHA (PC-DHA) was a little more oxidatively stable than PC-AA and PC-LA in liposomes as a model of biological membrane [58]. The difference in the oxidative stability of PC in liposomes and in the bulk phase or an organic solvent is due to the specific conformation of PC bilayers in the liposomes. The order of lipid peroxidation levels of the cellular PL found in Figure 13.14 and Table 13.7 was almost the same as that in the oxidative stability of PC in liposomes, suggesting that the characteristic cellular lipid peroxidation is also correlated with the PUFA conformation in the membrane PL.

TABLE 13.7
Amounts of MHP Formed in the Oxidation of Cellular PL with or without H_2O_2

	Cell	Amount of MHP (μmol/mmol Lipid)			
		LA-MHP	AA-MHP	DHA-MHP	Total MHP
Without H_2O_2	Control	1.80 ± 0.43	0.15 ± 0.05	ND	1.95 ± 0.48
	+LA	3.06 ± 0.40	0.17 ± 0.08	ND	3.23 ± 0.35
	+AA	2.17 ± 0.35	1.23 ± 0.11	ND	3.40 ± 0.29
	+DHA	2.37 ± 0.31	0.21 ± 0.06	0.36 ± 0.05	2.93 ± 0.20
With H_2O_2	Control	3.10 ± 0.62	0.32 ± 0.09	ND	3.42 ± 0.71
	+LA	3.50 ± 0.46	0.17 ± 0.06	ND	3.67 ± 0.41
	+AA	2.47 ± 0.47	1.55 ± 0.57	ND	4.02 ± 1.03
	+DHA	2.27 ± 0.47	0.20 ± 0.09	0.33 ± 0.10	2.79 ± 0.65

Source: From Araseki, M., Kobayashi, H., Hosokawa, M., and Miyashita, K., *Biosci. Biotechnol. Biochem.*, 69, 483, 2005. With permission.

Data are expressed as mean $\pm$ SD ($n = 3$). ND: not detected.

With the oxidation of PC-AA and PC-DHA in *t*-BuOH solution, there was little difference in MHP isomeric distributions [57]. However, in PC-AA liposome, the distribution of 9-, 12-, and 15-MHP was much larger than that of 5-, 8-, or 11-MHP (Table 13.2). In PC-AA oxidation in liposome, the distribution of MHP, derived from hydrogen abstraction, at C7 (5-MHP + 9-MHP), C10 (8-MHP + 12-MHP), and C13 (11-MHP + 15-MHP) was almost the same. These results in PC-AA liposome were very similar to those of AA-MHP isomeric distributions of cellular PL oxidation, except for the higher distribution of 5-MHP in the cells (Table 13.2). In the case of PC-DHA liposome, the MHP derived from hydrogen abstraction at C18 (16-MHP + 20-MHP), at C6 (4-MHP + 8-MHP), at C12 (10-MHP + 14-MHP), at C15 (13-MHP + 17-MHP), and at C9 (7-MHP + 11-MHP) were 37.7%, 29.1%, 17.6%, 8.6%, and 8.3%, respectively. This tendency was also consistent with the result of DHA-MHP composition of cellular PL in cells supplemented with DHA (Table 13.2). In the oxidation of DHA in the cellular PL, DHA-MHP was formed only by abstraction at C6, C12, and C18, but not at the C9 or C15 position. This characteristic distribution of DHA-MHP isomers found in the cellular PL oxidation indicates uneven hydrogen abstraction from DHA molecules in the cell membrane lipid, which might also be derived from the specific conformation of the DHA molecule and higher water permeability in the membrane lipids. Hence, further studies are necessary to reveal the relationship between the conformation and physicochemical properties of AA and DHA and their oxidative stability in biological membrane lipids. In addition, it is important to determine the activity of endogenous membrane antioxidants or free radicals on AA and DHA, which can take specific conformation in the membrane lipids.

VIII.   CONCLUSION

This chapter shows the difference in the relative oxidative stability of PUFA, such as AA and DHA, between bulk phase and aqueous or biological systems. The oxidative stability of AA and DHA in aqueous systems is similar to those in models for biological systems such as PC in liposomes or cellular lipids. A similar or greater relative stability was found for AA and DHA in these systems when compared with the stability of LA. This can be explained by the high flexibility and permeability of molecular conformation of AA and DHA in these aqueous circumstances. On the other hand, MHP distribution is different from each aqueous system. The distribution is strongly affected by the specific conformation of AA and DHA in aqueous and biological systems. The conformation and its dynamics are different in each system and may result in different MHP distribution.

We know that there is little influence on the lipid peroxidation in animals fed DHA or DHA containing fish oil, showing that the effect of highly unsaturated fatty acids such as DHA in biological systems would be different from the bulk systems or organic solutions. Previous studies [20–22] in animals and humans indicate the difference in the oxidative stability of highly unsaturated fatty acids such as AA and DHA between biological systems and bulk phases. Protection of cellular EPA against oxidation has also been reported using *E. coli* [70]. The present information will be important for better understanding of lipid peroxidation in biological systems, which are complex, multicomponent, and heterogeneous, with lipids present alongside other components in aqueous medium.

REFERENCES

1. Piche, L.A., Draper, H.H., and Cole, P.D., Malondialdehyde excretion by subjects consuming cod liver oil vs a concentrate of n-3 fatty acids, *Lipids*, 23, 370, 1988.
2. Garrido, A., Garrido, F., Guerra, R., and Valenzyela, A., Ingestion of high doses of fish oil increases the susceptibility of cellular membranes to the induction of oxidation stress, *Lipids*, 24, 833, 1989.
3. Hu, M.-L., Frankel, E.N., Leibovitz, B.E., and Tappel, A.L., Effect of dietary lipids and vitamin E on in vitro lipid peroxidation in rat liver and kidney homogenates, *J. Nutr.*, 119, 1574, 1989.

4. Burns, C.P. and Wagner, B.A., Heightened susceptibility of fish oil polyunsaturated-enriched neoplastic cells to ethane generation during lipid oxidation, *J. Lipid Res.*, 32, 79, 1991.
5. Frankel, E.N., Recent advances in lipid oxidation, *J. Sci. Food Agric.*, 54, 495, 1991.
6. Fritsche, K.L. and Johnston, P.V., Rapid autoxidation of fish oil in diets without added antioxidants, *J. Nutr.*, 118, 425, 1988.
7. Frankel, E.N., Lipid oxidation, *Prog. Lipid Res.*, 19, 1, 1980.
8. Frankel, E.N., Volatile lipid oxidation products, *Prog. Lipid Res.*, 22, 1, 1982.
9. Frankel, E.N., Lipid oxidation: Mechanisms, products and biological significance, *J. Am. Oil Chem. Soc.*, 61, 1908, 1984.
10. Frankel, E.N., Chemistry of free radical and singlet oxidation of lipids, *Prog. Lipid Res.*, 23, 197, 1985.
11. Porter, N.A., Galdwell, S.E., and Mills, K.A., Mechanisms of free radical oxidation of unsaturated lipids, *Lipids*, 30, 277, 1995.
12. Frankel, E.N., *Lipid Oxidation*, The Oily Press, Dundee, Scotland, 1998.
13. Gunstone, F.D. and Hilditch, T.P., The union gaseous oxygen with methyl oleate, linoleate, and linolenate, *J. Chem. Soc.*, 24, 127, 1945.
14. Holman, R.T. and Elmer, O.C., The rates of oxidation of unsaturated fatty acids and esters, *J. Am. Oil Chem. Soc.*, 24, 127, 1947.
15. Miyashita, K. and Takagi, T., Study on the oxidative rate and prooxidant activity of free fatty acids, *J. Am. Oil Chem. Soc.*, 63, 1380, 1986.
16. Cosgrove, J.P., Church, D.F., and Pryor, W.A., The kinetics of the autoxidation of polyunsaturated fatty acids, *Lipids*, 22, 299, 1987.
17. Cho, S.-Y., Miyashita, K., Miyazawa, T., Fujimoto, K., and Kaneda, T., Autoxidation of ethyl eicosapentaenoate and docosahexaenoate, *J. Am. Oil Chem. Soc.*, 64, 876, 1987.
18. Cho, S.-Y., Miyashita, K., Miyazawa, T., Fujimoto, K., and Kaneda, T., Autoxidation of ethyl eicosapentaenoate and docosahexaenoate under light irradiation, *Nippon Suisan Gakkaishi*, 53, 813, 1987.
19. Miyashita, K., Frankel, E.N., Neff, W.E., and Awl, R.A., Autoxidation of polyunsaturated triacylglycerols. III. Syntheytic triacylglycerols containing linoleate and linolenate, *Lipids*, 25, 48, 1990.
20. Kubo, K., Saito, M., Tadokoro, T., and Maekawa, A., Dietary docosahexaenoic acid does not promote lipid peroxidation in rat tissue to the extent expected from peroxidizability index of the lipids, *Biosci. Biotechnol. Biochem.*, 62, 1698, 1998.
21. Ando, K., Nagata, K., Yoshida, R., Kikugawa, K., and Suzuki, M., Effect of n-3 polyunsaturated fatty acid supplementation on lipid peroxidation of rat organs, *Lipids*, 35, 401, 2000.
22. Wander, R.C. and Du, S.-H., Oxidation of plasma proteins is not increased after supplementation with eicosapentaenoic and docosahexaenoic acids, *Am. J. Clin. Nutr.*, 72, 731, 2000.
23. Rhee, K.S., Ziprin, Y.A., and Rhee, K.C., Water-soluble antioxidant activity of oilseed protein derivatives in model lipid peroxidation systems of meat, *J. Food Sci.*, 44, 1132, 1979.
24. Allen, J.C. and Wrieden, W.L., Influence of milk proteins on lipid oxidation in aqueous emulsion. I. Casein, whey protein and α-lactalbumin, *J. Dairy Res.*, 49, 239, 1982.
25. Mei, L., McClements, D.J., and Decker, E.A., Lipid oxidation in emulsions as affected by charge status of antioxidants and emulsion, *J. Agric. Food Chem.*, 47, 2267, 1999.
26. Mancuso, J.R., McClements, D.J., and Decker, E.A., The effects of surfactant type, pH, and chelators on the oxidation of salmon oil-in-water emulsions, *J. Agric. Food Chem.*, 47, 4112, 1999.
27. Kubouchi, H., Kai, H., Miyashita, K., and Matsuda, K., Effects of emulsifiers on the oxidative stability of soybean oil TAG in emulsions, *J. Am. Oil Chem. Soc.*, 79, 567, 2002.
28. Hirose, A., Sato, N., and Miyashita, K., Effects of alkylamines and PC on the oxidative stability of soybean oil TAG in milk casein emulsions, *J. Am. Oil Chem. Soc.*, 80, 431, 2003.
29. Min, H., McClements, D.J., and Decker, E.A., Impact of whey protein emulsifiers on the oxidative stability of salmon oil-in-water emulsions, *J. Agric. Food Chem.*, 51, 1435, 2003.
30. Min, H., McClements, D.J., and Decker, E.A., Lipid oxidation in corn oil-in-water emulsions stabilized by casein, whey protein isolate, and soy protein isolate, *J. Agric. Food Chem.*, 51, 1696, 2003.
31. Ogawa, S., Decker, E.A., and McClements, D.J., Influence of environmental conditions on the stability of oil in water emulsions containing droplets stabilized by lecithin-chitosan membranes, *J. Agric. Food Chem.*, 51, 5522, 2003.
32. Osborn-Barnes, H.T. and Akoh, C.C., Copper-catalyzed oxidation of a structured lipid-based emulsion containing alpha-tocopherol and citric acid: Influence of pH and NaCl, *J. Agric. Food Chem.*, 51, 6851, 2003.

33. Faraji, H., McClements, D.J., and Decker, E.A., Role of continuous phase protein on the oxidative stability of fish oil-in-water emulsions, *J. Agric. Food Chem.*, 52, 4558, 2004.
34. Faraji, H. and Lindsay, R.C., Characterization of the antioxidant activity of sugars and polyhydric alcohols in fish oil emulsions, *J. Agric. Food Chem.*, 52, 7164, 2004.
35. Kellerby, S., McClements, D.J., and Decker, E.A., Role of proteins in oil-in-water emulsions on the stability of lipid hydroperoxides, *J. Agric. Food Chem.*, 54, 7879, 2006.
36. Porter, W.L., Black, E.D., and Drolet, A.M., Use of polyamide oxidative fluorescence test on lipid emulsions: Contrast in relative effectiveness of antioxidants in bulk versus dispersed systems, *J. Agric. Food Chem.*, 37, 615, 1989.
37. Frankel, E.N., Huang, S.-W., Kanner, J., and German, J.B., Interfacial phenomena in the evaluation of antioxidants: Bulk oils vs emulsions, *J. Agric. Food Chem.*, 42, 1054, 1994.
38. Huang, S.-W., Frankel, E.N., and German, J.B., Antioxidant activity of α- and γ-tocopherols in bulk oils and in oil-in-water emulsions, *J. Agric. Food Chem.*, 42, 2108, 1994.
39. Frankel, E.N., Huang, S.-W., Aeschbach, R., and Prior, E., Antioxidant activity of a rosemary extract and its constituents, carnosic acid, carnosol, and rosmarinic acid, in bulk oil and oil-in-water emulsion, *J. Agric. Food Chem.*, 44, 131, 1996.
40. Miyashita, K., Nara, E., and Ota, T., Oxidative stability of polyunsaturated fatty acids in an aqueous solution, *Biosci. Biotechnol. Biochem.*, 57, 1638, 1993.
41. Miyashita, K., Tateda, N., and Ota, T., Oxidative stability of free fatty acid mixtures from soybean, linseed, and sardine oils in an aqueous solution, *Fisheries Sci.*, 60, 315, 1994.
42. Miyashita, K., Azuma, G., and Ota, T., Oxidative stability of geometric and positional isomers of unsaturated fatty acids in aqueous solution, *J. Jpn., Oil Chem. Soc.*, 44, 425, 1995.
43. Kato, T., Hirukawa, T., and Namiki, K., Selective terminal olefin of n-3 polyunsaturated fatty acids, *Tetrahedron Lett.*, 33, 1475, 1992.
44. Kobayashi, H., Yoshida, M., Maeda, I., and Miyashita, K., Proton NMR relaxation times of polyunsaturated fatty acids in chloroform solutions and aqueous micelles, *J. Oleo Sci.*, 56, 105, 2004.
45. Miyashita, K., Inukai, N., and Ota, T., Effect of Tween 20 on the oxidative stability of sodium linoleate and sodium docosahexaenoate, *Biosci. Biotechnol. Biochem.*, 61, 716, 1997.
46. Yazu, K., Yamamoto, Y., Ukegawa, K., and Niki, E., Mechanism of lower oxidative stability of eicosapentaenoate than linoleate in aqueous micells, *Lipids*, 31, 337, 1996.
47. Hirano, S., Miyashita, K., Ota, T., Nishikawa, M., Maruyama, K., and Nakayama, S., Aqueous oxidation of ethyl linoleate, ethyl linolenate, and ethyl docosahexaenoate, *Biosci. Biotechnol. Biochem.*, 161, 281, 1997.
48. Frankel, E.N., Hydroperoxides in *Symposium on Foods: Lipids and Their Oxidation*, Shultz, H.W., Day, E.A., and Sinnhuber, R.O. (Eds.), AVI Pub., Westport, CT, 1962, pp. 51.
49. Yamamoto, Y., Niki, E., and Kamiya, Y., Oxidation of lipids. I. Quantitative determination of the oxidation of methyl linoleate and methyl linolenate, *Bull. Chem. Soc. Jpn.*, 55, 1548, 1982.
50. Miyashita, K., Hirano, S., Itabashi, Y., Ota, T., Nishikawa, M., and Nakayama, S., Oxidative stability of polyunsaturated monoacylglycerol and triacylglycerol in aqueous micelles, *J. Jpn. Oil Chem. Soc.*, 46, 205, 1997.
51. Miyashita, K., Hirao, M., Nara, E., and Ota, T., Oxidative stability of triglycerides from orbital fat of tuna and soybean oil in an emulsion, *Fisheries Sci.*, 61, 273, 1995.
52. Miyashita, K., Polyunsaturated lipids in aqueous systems do not follow our preconceptions of oxidative stability, *Lipid Tech. Newslett.*, 8, 35, 2002.
53. Miyashita, K., Nara, E., and Ota, T., Comparative study on the oxidative stability of phosphatidylcholines from salmon egg and soybean in aqueous solution, *Biosci. Biotechnol. Biochem.*, 58, 1772, 1994.
54. Nara, E., Miyashita, K., and Ota, T., Oxidative stability of PC containing linoleate and docosahexaenoate in an aqueous solution with or without chicken egg albumin, *Biosci. Biotechnol. Biochem.*, 59, 2319, 1995.
55. Nara, E., Miyashita, K., and Ota, T., Oxidative stability of liposomes prepared from soybean PC, chicken egg PC, and salmon egg PC, *Biosci. Biotechnol. Biochem.*, 61, 1736, 1997.
56. Nara, E., Miyashita, K., Ota, T., and Nadachi, Y., The oxidative stabilities of polyunsaturated fatty acids in salmon egg phosphatidylcholine liposomes, *Fisheries Sci.*, 64, 282, 1998.
57. Kobayashi, H., Yoshida, M., and Miyashita, K., Comparative study of the product components of lipid oxidation in aqueous and organic systems, *Chem. Phys. Lipids*, 126, 111, 2003.
58. Araseki, M., Yamamoto, K., and Miyashita, K., Oxidative stability of polyunsaturated fatty acid in phosphatidylcholine liposomes, *Biosci. Biotechnol. Biochem.*, 66, 2573, 2002.

59. Endo, Y., Hoshizaki, S., and Fujimoto, K., Oxidation of synthetic triacylglycerols containing eicosapentaenoic and docosahexaenoic acids: Effect of oxidation system and triacylglycerol structure, *J. Am. Oil Chem Soc.*, 74, 1041, 1997.

60. Everts, S. and Davis, J.H., ^{1}H and ^{13}C NMR of multilamellar dispersions of polyunsaturated (22:6) phospholipids, *Biophys. J.*, 79, 885, 2000.

61. Feller, S.E., Garwrish, K., and MacKerell, A.D. Jr., Polyunsaturated fatty acids in lipid bilayers intrinsic and environmental contributions to their unique physical properties, *J. Am. Chem. Soc.*, 124, 318, 2001.

62. Saiz, L. and Klein, M.L., Structural properties of a highly polyunsaturated lipid bilayer from molecular dynamics simulations, *Biophys. J.*, 81, 204, 2001.

63. Huber, T., Rajamoorthi, K., Kurze, V.F., Beyer, K., and Brown, M.F., Structure of docosahexaenoic acid-containing phospholipid bilayers as studied by ^{2}H NMR and molecular dynamics simulations, *J. Am. Chem. Soc.*, 124, 298, 2002.

64. Huster, D., Jin, A.J., Arnold, K., and Gawrisch, K., Water permeability of polyunsaturated lipid membranes measured by ^{17}O NMR, *Biophys. J.*, 73, 855, 1997.

65. Olbrich, K., Rawicz, W., Needham, D., and Evans, E., Water permeability and mechanical strength of polyunsaturated lipid bilayers, *Biophys. J.*, 79, 321, 2000.

66. Araseki, M., Kobayashi, H., Hosokawa, M., and Miyashita, K., Lipid peroxidation of a human cell line (HepG2) after incorporation of linoleic acid, arachidonic acid, and docosahexaenoic acid, *Biosci. Biotechnol. Biochem.*, 69, 483, 2005.

67. Igarashi, M. and Miyazawa, T., The growth inhibitory effect of conjugated linoleic acid on a human hepatoma cell line, HepG2, is induced by a change in fatty acid metabolism, but not the facilitation of lipid peroxidation in the cells, *Biochim. Biophys. Acta*, 1530, 162, 2001.

68. Okimoto, Y., Watanabe, A., Niki, E., Yamashita, T., and Noguchi, N., A novel fluorescent probe diphenyl-1-pyrenylphosphine to follow lipid peroxidation in cell membranes, *FEBS Lett.*, 474, 137, 2000.

69. Takahashi, M., Shibata, M., and Niki, E., Estimation of lipid peroxidation of live cells using a fluorescent probe, diphenyl-1-pyrenylphosphine, *Free Radic. Biol. Med.*, 31, 164, 2000.

70. Nishida, T., Orikasa, Y., Ito, Y., Yu, R., Yamada, A., Watanabe, K., and Okuyama, H., *Eschericha coli* engineered to produce eicosapentaenoic acid becomes resistant against oxidative damages, *FEBS Lett.*, 580, 2731, 2006.

14 Methods for Measuring Oxidative Rancidity in Fats and Oils

Fereidoon Shahidi and Udaya N. Wanasundara

CONTENTS

I. INTRODUCTION

Autoxidation is a natural process that takes place between molecular oxygen and unsaturated fatty acids. Autoxidation of unsaturated fatty acids occurs via a free radical chain mechanism consisting of basic steps of initiation (Equation 14.1), propagation (Equations 14.2 and 14.3), and termination (Equations 14.4 through 14.6). Initiation starts with the abstraction of a hydrogen atom adjacent to a double bond in a fatty acid (RH) molecule/moiety, and this may be catalyzed by light, heat, or metal ions to form a free radical. The resultant alkyl free radical ($R^\bullet$) reacts with atmospheric oxygen to form an unstable peroxy free radical, which may in turn abstract a hydrogen atom from another unsaturated fatty acid to form a hydroperoxide (ROOH) and a new alkyl free radical. The new alkyl free radical initiates further oxidation and contributes to the chain reaction. The chain reaction (or propagation) may be terminated by formation of nonradical products resulting from combination of two radical species.

Initiation:

$$RH \xrightarrow{\text{initiator}} R^\bullet + H^\bullet \tag{14.1}$$

Propagation:

$$R^\bullet + O_2 \rightarrow ROO^\bullet \tag{14.2}$$

$$ROO^\bullet + RH \rightarrow ROOH + R^\bullet \tag{14.3}$$

Termination:

$$R^\bullet + R^\bullet \rightarrow RR \tag{14.4}$$

$$R^\bullet + ROO^\bullet \rightarrow ROOR \tag{14.5}$$

$$ROO^\bullet + ROO^\bullet \rightarrow ROOR + O_2 \tag{14.6}$$

The mechanism of lipid autoxidation has been postulated by Farmer et al. [1], Boland and Gee [2], and Bateman et al. [3]. The propagation step in the autoxidation process includes an induction period when hydroperoxide formation is minimal [4,5]. The rate of oxidation of fatty acids increases in relation to their degree of unsaturation. The relative rate of autoxidation of oleate, linoleate, and linolenate is in the order of 1:40–50:100 on the basis of oxygen uptake and 1:12:25 on the basis of peroxide formation [6]. Therefore, oils that contain relatively high proportions of polyunsaturated fatty acid (PUFA) may experience stability problems. The breakdown products of hydroperoxides, such as alcohols, aldehydes, ketones, and hydrocarbons, generally possess offensive off-flavors. These compounds may also interact with other food components and change their functional and nutritional properties [7].

II. MEASUREMENT OF OXIDATIVE RANCIDITY

There are various methods available for measurement of lipid oxidation in foods. Changes in chemical, physical, or organoleptic properties of fats and oils during oxidation may be monitored to assess the extent of lipid oxidation. However, there is no uniform and standard method for detecting all oxidative changes in all food systems. The available methods to monitor lipid oxidation in foods and biological systems may be divided into two groups. The first group measures primary oxidative changes and the second determines secondary changes that occur in each system.

A. PRIMARY CHANGES

1. Changes in Reactants

Methods that measure primary changes of lipids may be classified as those that quantify loss of reactants (unsaturated fatty acids). Measurement of changes in fatty acid composition is not widely used in assessing lipid oxidation because it may require total lipid extraction from food and subsequent conversion to derivatives suitable for gas chromatographic analysis. Separation of lipids into neutral, glycolipid, phospholipid, and other classes may also be necessary. However, it has been proven that this method serves as a useful technique to identify class of lipids and fatty acids that are involved in the oxidative changes [8,9] and also to assess lipid oxidation induced by different metal complexes that afford a variety of products [10]. On the other hand, changes of fatty acid composition cannot be used in more saturated oils because this indicator reflects only the changes that occur in unsaturated fatty acids during oxidation [11]. Therefore, oxidative changes in marine oils and highly unsaturated vegetable oils may be monitored using this indicator. Similarly, changes in iodine value due to loss of unsaturation during accelerated oxidation studies may be used as an index of lipid oxidation [12].

2. Weight Gain

It is generally accepted that addition of oxygen to lipids and formation of hydroperoxides is reasonably quantitative during initial stages of autoxidation. Therefore, the measurement of induction period from weight gain data is theoretically sound. In this method, oil samples (about 2.0 g) are weighed into Petri dishes; then traces of water are removed by placing the samples overnight in a vacuum oven at 35°C and over a desiccant. Samples are then reweighed and stored in an oven at a set temperature. The weight gain of the samples may be recorded at different time intervals.

Olcott and Einset [13] reported that marine oils exhibit a fairly sharp increase in their weight at the end of the induction period and are rancid by the time they gain 0.3%–0.5% in weight (at 30°C–60°C). Ke and Ackman [14] reported that this method is simple, has a satisfactory reproducibility, and may be used to compare oxidation of lipids from different parts of fish. Recently, Wanasundara and Shahidi [15,16] used this method to compare storage stability of vegetable (Figure 14.1) and marine oils as affected by added antioxidants and were able to compare relative activity of antioxidants employed. However, surface exposure of the sample to air is an important variable in determining the rate of oxidation. Therefore, use of equal size containers to store samples is essential when carrying out such experiments.

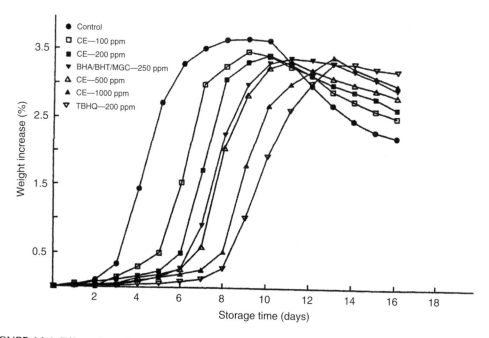

FIGURE 14.1 Effect of canola extracts (CEs) and commercial antioxidants on the weight gain of canola oil stored at 65°C. BHA, butylated hydroxyanisole; BHT, butylated hydroxytoluene; MGC, monoacylglycerol citrate; TBHQ, *tert*-butylhydroquinone.

The weight gain method also suffers from certain disadvantages: (1) the weighing frequency hinders monitoring of fast kinetics (a higher frequency would involve nocturnal weighing), and low or moderate temperatures require long analysis times for stable samples; (2) discontinuous heating of the sample (which must be cooled before weighing) may give rise to nonreproducible results, so the heating and cooling intervals must be accurately controlled; (3) the method involves intensive human participation; and (4) the working conditions (sample size, shape of container, and temperature) may influence the results. Nevertheless, this method offers advantages, such as low instrumentation cost as well as unlimited capacity and speed for sample processing.

3. Hydroperoxides

In the oxidation of fats and oils, the initial rate of formation of hydroperoxides exceeds their rate of decomposition, but this is reversed at later stages. Therefore, monitoring the amount of hydroperoxides as a function of time indicates whether a lipid is in the growth or decay portion of the hydroperoxide concentration curve. This information can be used as a guide for considering the acceptability of a food product with respect to the extent of product deterioration. By monitoring the induction period before the appearance of hydroperoxides, one can assess the effectiveness of added antioxidants on the stability of a food lipid.

4. Peroxide Value

The classical method for quantitation of hydroperoxides is the determination of peroxide value (PV). The hydroperoxide content, generally referred to as PV, is determined by an iodometric method. This is based on the reduction of the hydroperoxide group (ROOH) with iodide ion (I^-). The amount of iodine (I_2) liberated is proportional to the concentration of peroxide present. Released I_2 is assessed by titration against a standardized solution of sodium thiosulfate ($Na_2S_2O_3$) using a starch indicator. Chemical reactions involved in PV determination are given below:

$$ROOH + 2H^+ + 2KI \rightarrow I_2 + ROH + H_2O + 2K^+$$
$$I_2 + 2Na_2S_2O_3 \rightarrow Na_2S_4O_6 + 2NaI$$

Potential drawbacks of this method are absorption of iodine at unsaturation sites of fatty acids and liberation of iodine from potassium iodide by oxygen present in the solution to be titrated [17]. Results may also be affected by the structure and reactivity of peroxides as well as reaction temperature and time. The iodometric method for determination of PV is applicable to all normal fats and oils, but it is highly empirical and any variation in procedure may affect the results. This method also fails to adequately measure low PV because of difficulties encountered in determination of the titration end point. Therefore, the iodometric titration procedure for measuring PV has been modified in an attempt to increase the sensitivity for determination of low PV. The modification involves the replacement of the titration step with an electrochemical technique in which the liberated iodine is reduced at a platinum electrode maintained at a constant potential. PV ranging from 0.06 to 20 meq/kg have been determined in this manner, but it is essential to deaerate all solutions to prevent further formation of peroxides.

Several other chemical methods have also been suggested for monitoring PV. Colorimetric methods based on the oxidation of Fe^{2+} to Fe^{3+} and determination of Fe^{3+} as ferric thiocyanate, and a 2,6-dichlorophenol-indophenol procedure are reported in the literature [18]. In studies on the oxidation of biological tissues and fluids, measurement of fatty acid hydroperoxides is more common than measurement of their decomposition products. Fatty acid hydroperoxides can be analyzed by high-performance liquid chromatography (HPLC) or their corresponding hydroperoxy acid reduction products may be determined by gas chromatography–mass spectrometry (GC–MS) [19]. Fluorescence methods have also been developed to determine hydro-peroxides by allowing them to react with substances such as luminol and dichlorofluorescein, which form fluorescent products [17]. Although determination of PV is common, its usefulness is generally limited to the initial stages of lipid oxidation.

5. Active Oxygen and Oil Stability Index/Rancimat Methods

The Active Oxygen Method (AOM), also referred to as the Swift test of the American Oil Chemists' Society, is a common accelerated method used for assessing oxidative stability of fats and oils. This method is based on the principle that aging and rancidification of a fat is greatly accelerated by aeration in a tube held at a constant elevated temperature. In this method, air is bubbled through a heated oil at 98°C–100°C for different time intervals and the PVs are determined. The PVs are then plotted against time and the induction period determined from the graph. Even though this method has been used extensively over the years, its inherent deficiencies and difficulties have also been identified. These include the following: (1) the end point is determined by the amount of peroxides in the oxidized oil; peroxides are unstable and decompose readily to more stable secondary products. (2) During the rapid oxidation phase, the reaction is extremely susceptible to variations in the oxygen supply. Automated versions of the AOM apparatus, known as the Oil Stability Instrument (OSI) and Rancimat, are now available. The Rancimat method uses a commercial apparatus marketed by Metrohm Ltd. (Herisau, Switzerland). The OSI, a computer-assisted instrument developed by Archer Daniels Midland (ADM), is now produced commercially by Omnion Inc. (Rockland, MA). These methods may be considered as automated AOM since both employ the principle of accelerated oxidation. However, the OSI and Rancimat tests measure the changes in conductivity caused by ionic volatile organic acids, mainly formic acid, automatically and continuously, whereas in the AOM, peroxide values are determined. Organic acids are stable oxidation products that are produced when an oil is oxidized by a stream of air bubbled through it. In the OSI and Rancimat methods, oxidation proceeds slowly at first because during the induction period formic acid is released slowly. The end point is selected where the rapid rise in conductance begins. The Rancimat became available in the early 1980s and is capable of running only eight samples simultaneously; however, OSI is capable of running up to 24 samples at the same time. In addition, instruments that monitor the drop in the overhead pressure of an oil during heating might be used. An example of this sort of equipment is the Oxidograph, which is commercially produced by Mikrolab (Aarhus, Denmark). In Oxidograph, a sample of oil or fat is exposed to oxygen or air at elevated temperatures. Heating is done in an aluminum block. As the sample absorbs oxygen, the pressure change in the reaction vessel is measured electronically by means of pressure transducers. The rate of oxygen consumption during the early stages of storage of lipids also provides an ideal parameter for shelf life prediction. Wewala [20] used this method for prediction of shelf life of dried whole milk and found very good correlation between the head space oxygen content and storage time (Figure 14.2).

6. Conjugated Dienes

Oxidation of polyunsaturated fatty acids is accompanied by an increase in the ultraviolet absorption of the product. Lipids containing methylene-interrupted dienes or polyenes show a shift in their double-bond position during oxidation due to isomerization and conjugate formation [21]. The resulting conjugated dienes exhibit an intense absorption at 234 nm; similarly conjugated trienes absorb at 268 nm.

Farmer and Sutton [22] indicated that the absorption increase due to the formation of conjugated dienes and trienes is proportional to the uptake of oxygen and formation of peroxides during the early stages of oxidation. St. Angelo et al. [23] studied the autoxidation of peanut butter by measuring the PV and absorption increase at 234 nm due to the formation of conjugated dienes. Shahidi et al. [24] and Wanasundara et al. [25] found that conjugated dienes and PV of marine and vegetable oils correlate well during their oxidation (Figure 14.3). These authors concluded

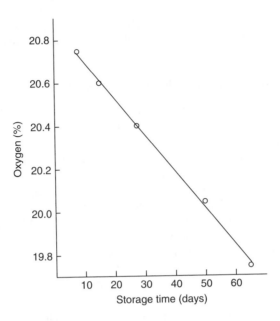

FIGURE 14.2 Change of the headspace oxygen content of stored dried whole-milk samples.

that the conjugated diene method may be used as an index of stability of lipids in place of, or in addition to, PV. However, carotenoid-containing oils may give high absorbance values at 234–236 nm because of the presence of double bonds in the conjugated structure of carotenoids. The conjugated diene method is faster than PV determination, is much simpler, does not depend on chemical reactions or color development, and requires a smaller sample size. However, presence of compounds absorbing in the same region may interfere with such determinations.

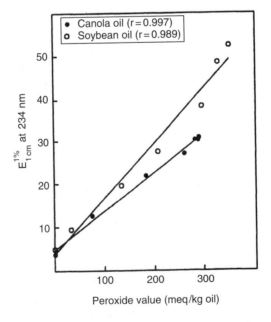

FIGURE 14.3 Relationship between peroxide values and conjugated diene values of oxidized vegetable oils.

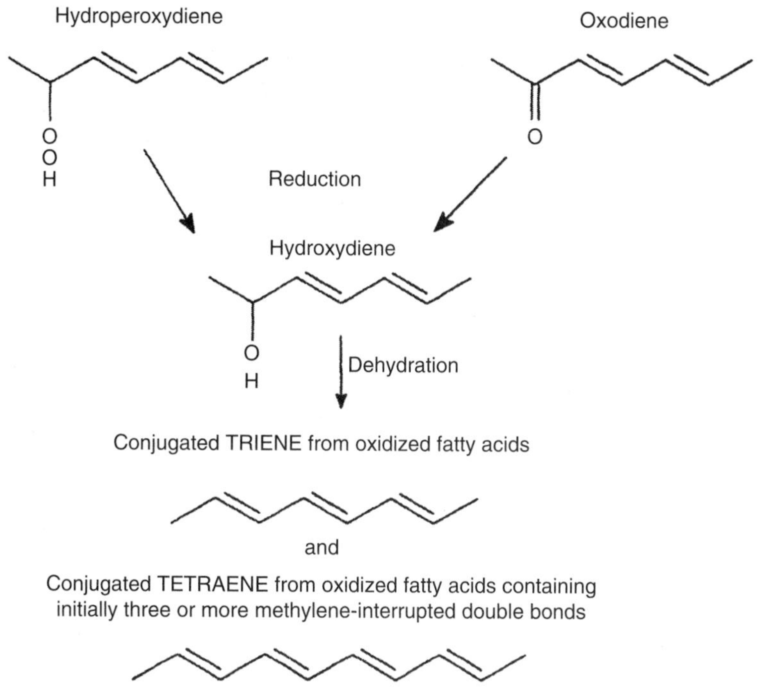

FIGURE 14.4 Chemical reaction steps in the assay of conjugable oxidation products.

Parr and Swoboda [26] have described an alternate spectroscopic method to determine lipid oxidation of stored oils. In this assay, hydroperoxides of polyenoic fatty acids as well as hydroxy and carbonyl compounds derived from them are converted to more conjugated chromophores by two chemical reaction steps, namely reduction and then dehydration (Figure 14.4). These yield *conjugable oxidation products* (COPs), which are measured and expressed as COP values. The first step of the analytical procedure involves reduction of the carbonyl group by sodium borohydride, which results in the disappearance of the characteristic ultraviolet absorption of carbonyl compounds of oxidized polyenoic fatty acids (oxodienes). The decrease in the absorption at 275 nm is known as *oxodiene value*. The next step of the COP assay involves changes in the spectrum of the reduced compound to its dehydrated counterpart, which exhibits absorption maxima at 268 and 301 nm. The sum of these absorbance changes at 268 and 301 nm yields the COP value, whereas their relative proportions define the COP ratio. For the calculation of oxodiene and COP results the concentration of the final lipid solution also has to be taken into account.

B. Secondary Changes

The primary oxidation products (hydroperoxides) of fats and oils are transitionary intermediates that decompose into various secondary products. Measurement of secondary oxidation products as indices of lipid oxidation is more appropriate since secondary products of oxidation are generally odor-active, whereas primary oxidation products are colorless and flavorless. Secondary oxidation products include aldehydes, ketones, hydrocarbons, and alcohols, among others. The following sections describe common methods used for measuring secondary oxidation products of lipids.

1. 2-Thiobarbituric Acid Value

One of the oldest and most frequently used tests for assessing lipid oxidation in foods and other biological systems is the 2-thiobarbituric acid (TBA) test. The extent of lipid oxidation is reported as

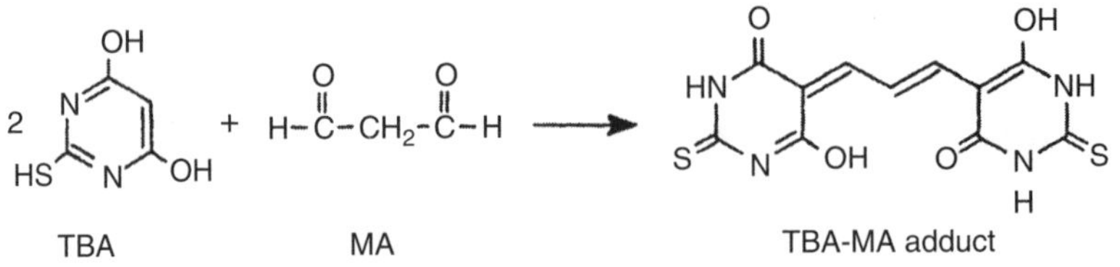

FIGURE 14.5 Reaction of 2-thiobarbituric acid (TBA) and malonaldehyde (MA).

the TBA value and is expressed as milligrams of malonaldehyde (MA) equivalents per kilogram sample or as micromoles MA equivalents per gram sample. MA is a relatively minor product of oxidation of polyunsaturated fatty acids that reacts with the TBA reagent to produce a pink complex with an absorption maximum at 530–532 nm [27]. The adduct is formed by condensation of two molecules of TBA with one molecule of MA (Figure 14.5). Other products of lipid oxidation, such as 2-alkenals and 2,4-alkadienals, also react with the TBA reagent. However, the exact mechanism of their reaction with the TBA reagent is not well understood. There are several procedures for the determination of TBA values. The TBA test may be performed directly on the sample, its extracts, or distillate. In case of the distillation method, volatile substances are distilled off with steam. Then the distillate is allowed to react with the TBA reagent in an aqueous medium. The advantage of the distillation method is the absence of interfering substances. In the extraction method, TBA-reactive substances (TBARSs) are extracted from food material into an aqueous medium (i.e., aqueous trichloroacetic acid) prior to color development with the TBA reagent. The main disadvantages of both of these methods are long assay time and possibility of artifact formation. In the direct assay method, lipid sample (oil) reacts with the TBA reagent and the absorbance of the colored complex so prepared is recorded. The direct assay method is simple and requires less time for sample preparation.

There are certain limitations when using the TBA test for evaluation of the oxidative state of foods and biological systems because of their chemical complexity. Dugan [28] reported that sucrose and some compounds in wood smoke react with the TBA reagent to give a red color that interferes with the TBA test. Baumgartner et al. [29] also found that a mixture of acetaldehyde and sucrose when subjected to the TBA test produced a 532 nm absorbing pigment identical to that produced by MA and TBA. Modifications of the original TBA test have been reported by Marcuse and Johansson [30], Ke and Woyewoda [31], Robbles-Martinez et al. [32], Pokorny et al. [33], Shahidi et al. [34,35], Thomas and Fumes [36], and Schmedes and Holmer [37]. However, it has been suggested that TBARS values produce an excellent means for evaluating the relative oxidative state of a system as affected by storage condition or process variables [38]. Nonetheless, it is preferable to quantitate the extent of lipid oxidation by a complementary analytical procedure in order to verify the results.

Several attempts have been made to establish a relationship between TBA values and the development of undesirable flavors in fats and oils. It has been shown that flavor threshold values correlate well with the TBA results of vegetable oils, such as those of soybean, cottonseed, corn, safflower [17], and canola [5].

2. Oxirane Value

The oxirane oxygen or epoxide groups are formed during autoxidation of fats and oils. The epoxide content is determined by titrating the oil sample with hydrobromic acid (HBr) in acetic acid and in the presence of crystal violet, to a bluish green end point. This method has been standardized by the American Oil Chemists' Society in their tentative method (Cd 9–57) [39], but it is not sensitive and lacks specificity. The HBr may also attack α,β-unsaturated carbonyls and conjugated dienals, and the reaction is not quantitative with some *trans*-epoxides. Fioriti et al. [40] found that picric acid was the best of several acidic chromophores in its reaction with epoxides. Despite a nonquantitative reaction, the product concentration followed Beer's law. This method has been found to be particularly well suited for the determination of epoxides in heated fats and oils, where the oxirane content is often less than 0.1%.

3. *p*-Anisidine Value

p-Anisidine value (*p*-AnV) is defined as 100 times the optical density measured at 350 nm in a 1.0-cm cell of a solution containing 1.0 g of oil in 100 mL of a mixture of solvent and reagent, according to the IUPAC method [41]. This method determines the amount of aldehyde (principally 2-alkenals and 2,4-alkadienals) in animal fats and vegetable oils. Aldehydes in an oil react with the *p*-anisidine reagent under acidic conditions. The reaction of *p*-anisidine with aldehydes affords yellowish products, as shown in Figure 14.6. List et al. [42] reported a highly significant correlation between *p*-AnV and flavor acceptability scores of salad oils processed from undamaged soybeans.

4. TOTOX Value

The *p*-AnV is often used in the industry in conjunction with PV to calculate the so-called total oxidation or TOTOX value:

$$\text{TOTOX value} = 2PV + p - AnV$$

The TOTOX value is often considered to have the advantage of combining evidence about the past history of an oil (as reflected in the *p*-AnV) with its present state (as evidenced in the PV). Therefore, determination of TOTOX value has been carried out extensively to estimate oxidative deterioration of food lipids [43]. However, despite its practical advantages, TOTOX value does not have any sound scientific basis because it combines variables with different dimensions. Recently, Wanasundara and Shahidi [44] defined $TOTOX_{TBA}$ as $2PV + TBA$ since determination of *p*-AnV may not be always feasible.

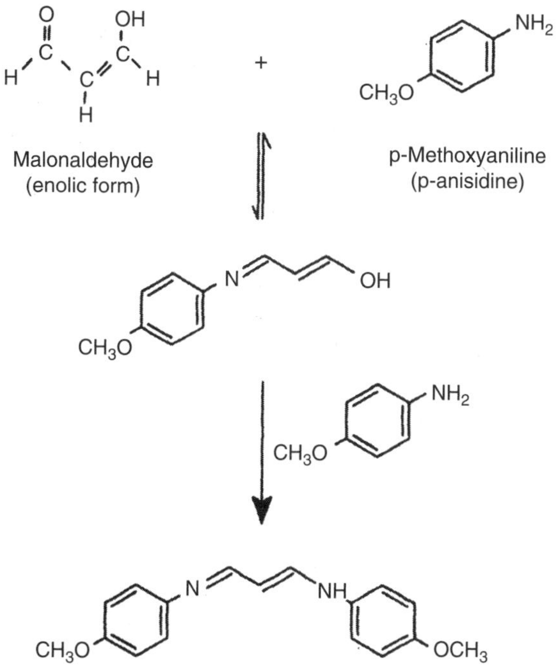

FIGURE 14.6 Possible reactions between *p*-anisidine reagent and malonaldehyde.

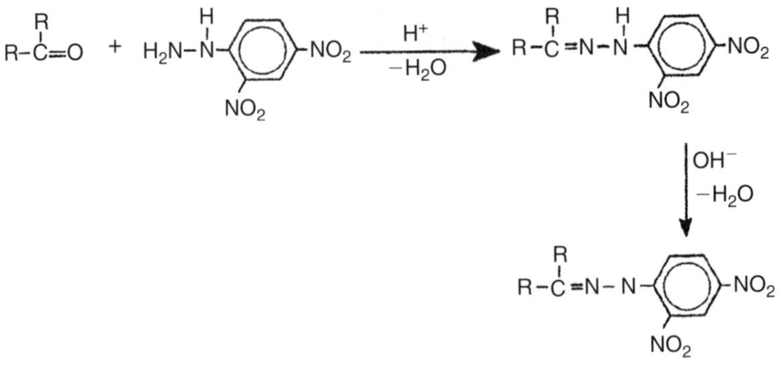

FIGURE 14.7 Reaction steps in the production of hydrozones from carbonyls and 2,4-dinitrophenylhydrazine.

5. Carbonyls

An alternative approach for monitoring the extent of lipid oxidation in fats and oils is to measure the total or individual volatile carbonyl compounds formed from degradation of hydroperoxides. One of the more reliable methods for total carbonyl analysis is based on the absorbance of the quinoidal ion, a derivative of aldehydes and ketones. This ion is formed from the reaction of 2,4-dinitrophenylhydrazine (2,4-DNDH) with an aldehyde or ketone, followed by the reaction of the resulting hydrozones with alkali (Figure 14.7), which is then analyzed spectroscopically at a given wavelength. Many variations of this spectroscopic method have been reported [45,46]. Each method offers an alternative solvent, wavelength, or workup to analyze the quinoidal ion.

The analysis of individual carbonyl compounds is another method which has recently gained popularity. Hexanal, one of the major secondary products formed during the oxidation of linoleic or other ω-6 fatty acids in lipid-containing foods [47,48], has been used to follow lipid oxidation. Shahidi and Pegg [47] reported that a linear relationship existed between hexanal content, sensory scores, and TBA numbers of cooked ground pork, whereas St. Angelo et al. [49] established a similar correlation for cooked beef. O'Keefe et al. [50] have used hexanal as an indicator to assess oxidative stability of meat from broiler chickens fed fish meal. Supplementation of high amounts of fish meal to the diet increased the hexanal content of the thigh meat during storage. However, recent studies have shown that during oxidation of marine oils, which are rich in polyunsaturated fatty acids of the ω-3 type, large amounts of propanal are formed and that a good correlation exists between the content of propanal and the amount of TBARS in such samples [51,52]. Therefore, it is essential to use appropriate indicators when assessing stability of food lipids. We recommend that hexanal be used when oils under investigation are rich in ω-6 fatty acids while propanal would serve as a reliable indicator when oils high in ω-3 fatty acids are being considered.

6. Hydrocarbons and Fluorescent Products

Studies of oxidized methyl linoleate and soybean oil [53] have revealed that saturated hydrocarbons could be detected when aldehydes are either absent or undetectable. Snyder et al. [54] have reported that ethane, propane, and pentane are predominant short chain hydrocarbons formed through thermal decomposition of soybean oil. Correlations of flavor acceptability scores and pentane content determined by GC techniques have been used to assess rancidity of fats and oils [55]. Significant correlations existed between the amount of pentane produced and the number of rancid descriptions of stored vegetable oils [56]. Correlation of headspace pentane concentrations and sensory scores of stored freeze-dried pork samples was reported by Coxon [8].

Another secondary change that occurs during autoxidation of biological systems is the formation of fluorescent products from the reaction of MA with amino compounds such as proteins and nucleotides [57]. This method has been used to determine the extent of lipid oxidation in biological

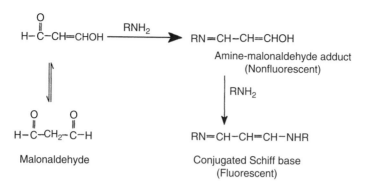

FIGURE 14.8 Production of fluorescent chromophores from the reaction of lipid oxidation products and amines.

tissues. It has been established that fluorescent compounds with a general structure of 1-amino-3-iminopropane may develop through the reaction of an amino group with carbonyl compounds, mainly MA [58–60] (Figure 14.8).

Kikugawa and Beppu [61] reported that the development of fluorescence depends not only on the formation of condensation products between MA and free amino groups, but also on the nature of the substituents of the latter compounds. Different excitation and emission maxima were observed for different condensation products. Advantages of the fluorescence method as a means of measuring lipid oxidation have been reported by Dillard and Tappel [58]. This method can detect fluorescent compounds at concentrations as low as parts per billion levels and is found to be 10–100 times more sensitive than the TBA assay.

III. MEASUREMENT OF FRYING FAT DETERIORATION

Deep frying is a popular method for food preparation, especially in fast food restaurants. Although vegetable oils are used primarily as a heat exchange medium for cooking, when used for deep frying, they contribute to the quality of fired products. In the process of deep fat frying, a complex series of chemical reactions take place. These reactions are characterized by a decrease in the total unsaturation content of the fat with a concurrent increase in the amount of free fatty acids, cyclic fatty acids (Figure 14.9), foaming, color, viscosity, and formation of polar matter and polymeric compounds. As these reactions proceed, the functional, sensory, and nutritional quality of frying fats change and may reach a point where high-quality foods can no longer be prepared. Therefore, it is essential to determine when the frying fat is no longer usable.

Quality evaluation of frying fats may be carried out in many ways. The first attempt to define a deteriorated frying fat was made by the German Society for Fat Research in 1973. It recommended that "a used frying fat is deteriorated if, without doubt, its odor and taste were unacceptable; or if in case of doubtful sensory assessment, the concentration of petroleum ether–insoluble oxidized fatty acids in it was 0.7% or higher and its smoke point was lower than 170°C; or if the concentration of petroleum ether–insoluble oxidized fatty acids was 1.0% or higher." Although sensory evaluation of foods is the most important quality assessment, taste evaluations are not practical for routine quality control. It is always preferred to have a quantitative method for which rejection point could be established by sensory means. Peroxide values provide an indication of frying fat quality if they are used in a very specific way. However, peroxides generally decompose at about 150°C and hence at frying temperatures (usually 180°C–190°C) no accumulation of peroxides occurs. Free fatty acids from frying fats can be determined by direct titration with a standardized base in ethanol. Fritsch [62] has shown that in most deep fat frying operations the amount of free fatty acids produced by hydrolysis is too small to affect the quality of foods. However, industrially, frying oil quality is usually checked by the measurement of

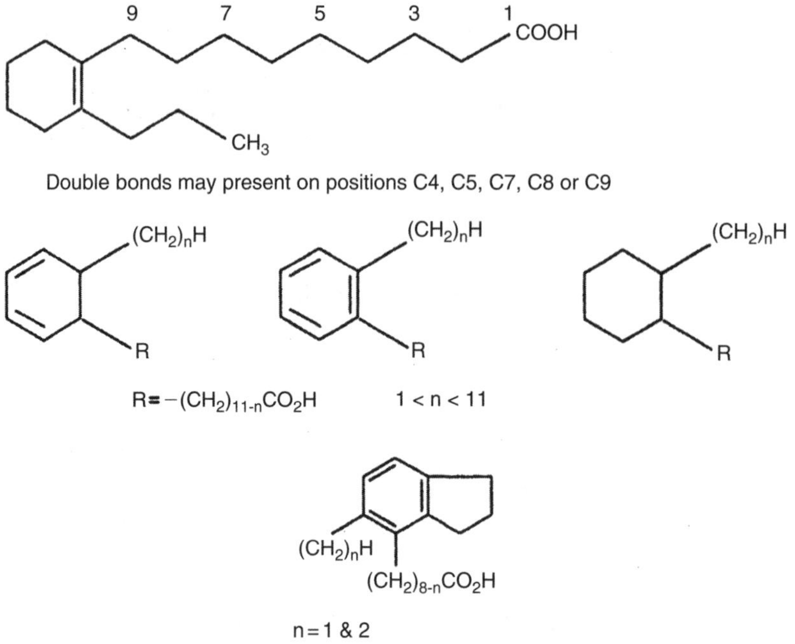

FIGURE 14.9 Structures of cyclic fatty acids formed during frying.

color or free fatty acid content in order to tell an operator when a fat is ending its useful life. The foaming characteristics of used fats would also lead one to the same conclusion.

A quick colorimetric test kit is now available for measuring oil quality [63]. Blumenthal et al. [64] developed a spot test to measure free fatty acids in which drops of used fat are placed on a glass covered with silica gel containing a pH indicator in order to give a three-color test scale of blue, green, and yellow. This may indicate the amount of free fatty acids in a sample. Northern States Instrument Corp. (Lino Lakes, MN) has developed an instrument that measures the dielectric constant of insulating liquids. The instrument is a compact unit, relatively inexpensive, simple to operate, and requires only a few drops of oil for each measurement. For evaluation of frying fats, the instrument must be calibrated first with a fresh oil sample prior to its use in frying operations.

Determination of total polar matter in frying fats appears to be emerging as a reliable method for assessing the useful life of fats and oils subjected to frying and is an official method in Europe. Total polar matter is determined by dissolving the fat in a relatively nonpolar solvent, such as toluene or benzene, and running through a silica gel column that adsorbs the polar compounds. After evaporation of the solvent, the nonpolar fat can be weighed and the total polar matter calculated from the weight difference data or determined directly by their elution from the column with diethyl ether or a mixture of chloroform and methanol. Sebedio et al. [65] illustrated that polar and nonpolar fractions of fried oils can be quantitatively estimated using Iatroscan thin-layer chromatography–flame ionization detection (TLC-FID) system with Chromarod SII. This method requires a very small sample and is much faster than silicic acid column separation.

IV. RECENT DEVELOPMENTS FOR QUANTITATION OF LIPID OXIDATION

A. ESR Spectroscopy

Lipid oxidation in foods and biological systems has conventionally been tested by monitoring either primary or secondary oxidation products. Over the last 20 years or so, advances in pulse radiolysis [66] and electron spin resonance (ESR) [67] techniques have facilitated the detection and study of short-lived free radical intermediates. ESR spectroscopy allows selective detection of free radicals. The technique depends on the absorption of microwave energy (which arises from the promotion of an electron to a higher energy level) when a sample is placed in a variable magnetic field. A major limitation in the detection of free radicals by ESR is the requirement that radical concentrations remain higher than 10^{-8} M. Radical lifetimes in solution are very short (<1 ms), and steady-state concentrations generally remain well below 10^{-7} M. Several approaches have been developed to overcome this problem, either by enhancing the rate of radical production or by diminishing the rate of its disappearance. These techniques include rapid freezing, lyophilization, or spin trapping [68]. Although application of ESR spectroscopy as a precise method to study lipid oxidation in animal tissues and other biological systems is commonplace, its application to foods is relatively new.

Yen and Duh [69] and Chen and Ho [70] have reported that inhibition of free radical formation by different antioxidants can be measured using very stable free radicals such as 1,1-diphenyl-2-picrylhydrazyl (DPPH). The mechanism of the reaction of antioxidant with DPPH radical is as follows:

$$(DPPH)^{\bullet} + HO\text{---}R\text{---}OH \rightarrow (DPPH)\text{:}H + HO\text{---}R\text{---}O^{\bullet}$$

$$HO\text{---}R\text{---}O^{\bullet} + (DPPH) \rightarrow (DPPH)\text{:}H + O = R = O$$

DPPH radical, with a deep violet color, receives a hydrogen atom from the antioxidant and is converted to a colorless molecule. Using this reagent, the free radical scavenging ability of the antioxidant can be determined by spectrophotometric methods.

B. INFRARED SPECTROSCOPY

Infrared (IR) spectroscopy has also been used for measurement of rancidity, and it is of particular value in recognition of unusual functional groups and in studies of fatty acids with trans double bonds. Production of hydroperoxides during oxidation of lipids gives rise to an absorption band at about 2.93 μm, whereas the disappearance of a band at 3.20 μm indicates the replacement of a hydrogen atom on a double bond, or polymerization. It has also been suggested that the appearance of an additional band at 5.72 μm, due to $C = O$ stretching, indicates the formation of aldehydes, ketones, or acids. Furthermore, changes in the absorption bands in the 10- to 11-μm region indicates cis–trans isomerization and probably formation of conjugated bonds. Determination of oxidative deterioration of lipids using IR method is simple, rapid, and requires small amounts of sample (20 mg).

van de Voort et al. [71] and Sedman et al. [72] have investigated the feasibility of employing Fourier transform infrared (FTIR) spectroscopy to assess the oxidative status or forecast the oxidative stability of oils. These authors constructed a spectral library by recording the FTIR spectra of oils spiked with various compounds representative of common oil oxidation products. Table 14.1 shows that each of the various types of oxidation products gives rise to discernible and characteristic absorptions in the FTIR spectrum. Similar absorption bands were detected in the spectra of oils oxidized under accelerated conditions and monitored in real time by FTIR spectroscopy. On the basis of the results of this study, the authors proposed a quantitative approach whereby the oxidative status of an oil could be determined through calibrations developed with oils spiked with appropriate compounds representative of the functional groups associated with typical oxidative end products. These concepts were subsequently put into practice with the development of a calibration for the determination of peroxide value. A similar approach may be used to develop a parallel method for evaluating *p*-anisidine values.

C. CHEMILUMINESCENCE SPECTROSCOPY

Burkow et al. [73] reported that hypochlorite-activated chemiluminescence could provide a useful means for evaluation of antioxidants in edible oils. Because of high sensitivity and ability to detect

TABLE 14.1

Peak Positions of the Functional Group Absorptions of Reference Compounds Representative of Products Formed in Oxidized Oils

Compounds	Vibration	Frequency (cm^{-1}) at Peak Maximum
Water	νOH	3650 and 3550
	δHOH	1625
Hexanol	νROH	3569
tert-Butyl hydroperoxide	$\nu ROOH$	3447
Hexanal	$\nu RHC = O$	2810 and 2712
	$\nu RHC = O$	1727
2-Hexenal[a]	$\nu RHC = O$	2805 and 2725
	$\nu RHC = O$	1697
	$\nu RC = CH—HC = O$	1640
	$\delta RC = CH—HC = O$	974
2,4-Decadienal[a]	$\nu RHC = O$	2805 and 2734
	$\nu RHC = O$	1689
	$\nu RC = CH—HC = O$	1642
	$\delta RC = CH—HC = O$	987
4-Hexen-3-one[a]	$\nu RC(= O)HC = CHR$	1703 and 1679
	$\nu RC(= O)HC = CHR$	1635
	$\delta RC(= O)HC = CHR$	972
Oleic acid	$\nu RCOOH$	3310
	$\nu RC(= O)OH$	1711

[a] All double bonds in the *trans* form.

small changes in the degree of oxidation of lipids, this method may be employed to evaluate the effects of antioxidants on oils during low temperature storage (about 35°C) within a 24 h period. Chemiluminescence generally originates from electronically excited stages, such as singlet molecular oxygen in lipid peroxidation [74]. The chemiluminescence method has been tested for estimating the degree of deterioration of edible oils containing antioxidants [75] as well as for shelf life dating of fish samples [76].

D. NMR Spectroscopy

High-resolution nuclear magnetic resonance (NMR) spectroscopy makes it possible to determine various types of hydrogen atoms (protons, ^{1}H) in triacylglycerol (TAG) molecules. This is due to the fact that hydrogen atoms in a strong magnetic field absorb energy, in the radiofrequency range, depending on their molecular environment. During oxidation of food lipids, changes occur in the environment in which protons in an oxidizing TAG molecule are located. These changes may be monitored by employing ^{1}H NMR spectroscopy [77–80]. For this purpose, the oil is dissolved in $CDCl_3$ and its NMR spectrum recorded (Figure 14.10). The sharp signal at the extreme right side of the spectrum (high applied field) is due to tetramethylsilane (TMS) added to the solution to serve as an internal standard. The spectrum shows eight groups of signals labeled *a–h*. These signals are assigned: *a*, hydrogens directly attached to double-bonded carbons (olefinic protons) and the methine proton in the glyceryl moiety (δ 5.1–5.4); *b*, hydrogens in the two methylene groups in the glyceryl moiety (δ 4.0–4.4); *c*, hydrogens in the CH_2 groups attached to two double-bonded carbon atoms (diallylmethylene protons) ($=HC—CH_2—CH=$; δ 2.6–2.9); *d*, hydrogens in the three CH_2 groups alpha to the carboxyl groups (α-CH_2; δ 2.2–2.4); *e*, hydrogens in the CH_2 groups attached to saturated carbons and double-bonded carbon atoms ($—CH_2—C=$; δ 1.8–2.2); *f*;

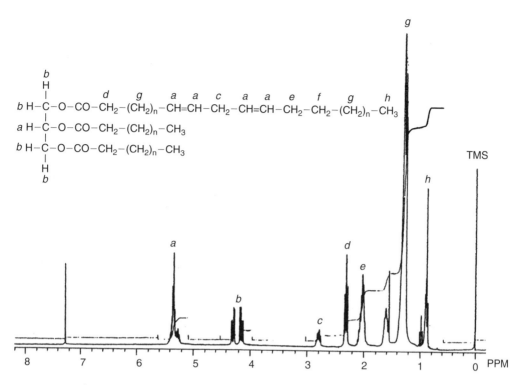

FIGURE 14.10 ^{1}H NMR spectrum of oxidized canola oil (peak at δ 0.00 and 7.26 are for TMS and CHCl$_3$ protons, as impurities in CDCl$_3$, respectively).

hydrogens in the CH$_2$ groups attached to the saturated carbon atoms (=C—CH$_2$=CH$_2$; δ 1.45–1.8); g, hydrogens in the CH$_2$ groups bonded to two saturated carbon atoms ([CH$_2$]$_n$; δ 1.1–1.45); and h, hydrogens in the three terminal CH$_3$ groups (δ 0.7–1.0). The relative number of protons in each group is calculated based on the integration of methylene protons of the glyceryl moiety (δ 4.0–4.4) of the TAG (four protons in the two methylene groups of the TAG moiety) molecules. The area per proton is obtained as

$$\text{Area per each proton} = \text{area of } b\text{–type protons}/4$$

Since area per proton is known, one may calculate the number of protons belonging to each and every individual signal by dividing the integration number of individual signals by the area per proton. As an example, the total number of diallylmethylene protons equals the area of c-type protons/(the area of b-type protons/4).

The total number of aliphatic, olefinic, and diallylmethylene protons are calculated, from which ratios of aliphatic to olefinic protons (R_{ad}) and aliphatic to diallylmethylene (R_{ad}) protons may be obtained. These ratios increase steadily during the storage and oxidation of oils. Shahidi [79] and Wanasundara and Shahidi [80] have shown that the ratio of olefinic to aliphatic protons, measured by NMR, decreases continuously as long as the oxidation reaction proceeds. They suggested that the NMR technique could be useful for measuring oxidative deterioration of oils containing PUFAs, even at stages beyond the point at which PV profile reaches a maximum. Saito and Udagawa [78] have used this method to evaluate oxidative deterioration of brown fish mean and suggested that NMR methodology is suitable for comparing the storage conditions of the fish meal as well as estimating the effect of antioxidants in both fish meal and fish oil. These authors reported good correlations between peroxide values and NMR data. However, Wanasundara and Shahidi [80] found that linear relationships between peroxide values and NMR data were not as suitable as those of TOTOX values and NMR data. It is obvious that TOTOX values correlate better with R_{ao} and R_{ad} than peroxide values since both TOTOX and NMR data estimate overall changes that occur in fatty acid profiles as reflected in both primary and secondary oxidation products of lipids. Thus, NMR methodology offers a rapid, nondestructive, and reliable technique for estimating the oxidative state of edible oils during processing and storage.

E. CHROMATOGRAPHIC TECHNIQUES

Different chromatographic techniques have been developed and applied to quantitate oxidation products in variety of substances, including model compounds, oils, and food lipids, subjected to oxidation under very different conditions, from room to frying temperatures. Separation based on reversed phase or size exclusion chromatography (SEC) and detection systems based on UV absorption, infrared, refractive index, flame ionization, or evaporative light scattering (ESLD) are used for assessment of lipid oxidation products. For quantification of free MA, reversed phase HPLC using ion-pairing reagent or size exclusion separation followed by monitoring the absorbance at 267 nm has been described [81–83]. MA and 4-hydroxynonenal (4-HNE) can also be derivatized with 2,4-DNDH at room temperature to form dinitrophenylhydrozone (DNP) derivatives. The DNP derivatives could be solubilized in organic solvents and separated on a reversed phase HPLC and detected at 300–330 nm, depending on the type of hydrozone formed [84–88]. An HPLC–fluorescence method that employs postcolumn detection can be used for the analysis of hydroperoxide mixtures containing conjugated and nonconjugated diene structures. Lipid hydroperoxides can react with diphenyl-1-1pyrenylphosphine (DPPP) to form DPPP oxides that have excitation at 352 nm and emission at 380 nm [89].

Oils heated at high temperatures (oil used for frying) and TAGs oxidize and form polymeric TAG and hydrolytic products (e.g., diacylglycerols and fatty acids). Solid phase extraction with silica could be used to separate polar and nonpolar fractions of oxidized oils. The polar fraction can be analyzed by high-performance size exclusion column (highly cross-linked styrene–divinylbenzene copolymer) using a refractive index detector. Polar compounds are separated as an inverse order of their molecular weight; TAG polymers, TAG dimers, oxidized TAG monomers, diacylglycerols, monostearine, and fatty acids [90]. It has been observed that oxidized monomers show a progressive increase during early stages of oxidation. According to Marquez-Ruiz et al. [91], who used trilinolein (LLL) as the model compound, during early stages of oxidation LLL-oxidized monomers increase paralleled that of peroxide value, as primarily hydroperoxides were formed. The peroxides that are labile products readily degrade to a multitude of secondary products, such as oxygenated side products of the same chain length as the parent hydroperoxides. The oxidized TAGs comprise those monomeric TAGs containing at least one oxidized fatty acyl group (e.g., a peroxide group or any other oxygenated function, such as epoxy, keto, hydroxy). Therefore, determination of oxidized TAG monomers may provide a measure of both primary and secondary products of lipid oxidation. High-performance size exclusion (HPSEC) has also been employed to analyze dimers, trimers, oligomers, partial acylglycerols and cyclic fatty acids of heated, thermally oxidized lipids such as frying fats [92]. Mass spectrometry (MS) coupled with normal- or reversed-phase HPLC permits direct characterization of hydroperoxides and nonvolatile high molecular weight secondary oxidation products of triacylglycerols, cholesterol, and phospholipids without precolumn derivatization. The necessity of removing solvents from labile compounds separated from HPLC is not a requirement when interfaced with chemical ionization (CI) or atmospheric pressure chemical ionization (APCI). These soft MS techniques are mainly qualitative and allow identification of the molecules. For quantitative analysis standardization of spectra with authentic reference compounds to interpret fragmentation patterns may be necessary. Lipid oxidation products such as polyunsaturated hydroperoxides are difficult to synthesize and are not readily available.

GC methods to quantify MA have also been reported. The advantages of the GC methods are increased sensitivity, particularly when used with MS detection, and the possibility for simultaneous analysis of several aldehydes. Reduction of MA to 1,3-propanediol with borane trimethylamine [93] forms a butyldimethylsilyl ether, which can be analyzed by GC–MS with an HP-5 capillary column (25 m long), temperature programmed from 115°C to 165°C and [^{2}H$_8$] propanediol as the internal standard. MA can also be converted to 1-methylpyrozole by reaction with *N*-methylhydrazine at room temperature for a 1 h period; this derivative could be recovered by extraction with dichloromethane and analyzed on a DB-Wax capillary column (30 m long), temperature programmed from 30°C to 200°C, and a nitrogen–phosphorus detector [94,95].

Oxidized fatty acid methyl esters (FAMEs) could also be analyzed with a combination of silica column chromatography and high-performance size exclusion separation. The combined chromatographic analysis permits quantitation of groups of compounds (nonpolar fatty acid monomers, dimers, oxidized fatty acid monomers, and fatty acid polymers) differing in polarity of molecular weight [96]. The HPSEC–separated fractions of these oxidized fatty acids could be further analyzed on GC–MS for detection of their structural identities. The fraction of oxidized fatty acid monomers includes epoxides, ketones, and hydroperoxides as well as polyoxygenated monomeric compounds. Marquez-Ruiz and Dobarganes [96] also described that GC–MS coupled with DB-wax column and AEI–MS was useful in identifying short chain aldehydes resulting from the break-down of lipid hydroperoxides.

REFERENCES

1. E.H. Farmer, G.F. Bloomfield, A. Sundaralingam, and D.A. Sutton. The course and mechanism of autoxidation reactions in olefinic and polyolefinic substances including rubber. *Trans. Faraday Soc.* 38:348–356 (1942).
2. J.L. Boland and G. Gee. Kinetics in the chemistry of rubber and related materials. *Trans. Faraday Soc.* 42:236–243 (1946).
3. L. Bateman, H. Hughes, and A.L. Morris. Hydroperoxide decomposition in relation to the initiation of radical chain reactions. *Disc. Faraday Soc.* 14:190–194 (1953).
4. T.P. Labuza. Kinetics of lipid oxidation in foods. *CRC Crit. Rev. Food Technol.* 2:355–405 (1971).
5. Z.J. Hawrysh. Stability of canola oil. In: *Canola and Rapeseed: Production, Chemistry, Nutrition and Processing Technology* (F. Shahidi, ed.). Van Nostrand Reinhold, New York, 1990, pp. 99–122.
6. R.J. Hsieh and J.E. Kinsella. Oxidation of polyunsaturated fatty acids: Mechanisms, products and inhibition with emphasis on fish. *Adv. Food Nutr. Res.* 33:233–241 (1989).
7. E.R. Sherwin. Oxidation and antioxidants in fat and oil processing. *J. Am. Oil Chem. Soc.* 55:809–814 (1978).
8. D. Coxon. Measurement of lipid oxidation. *Food Sci. Technol. Today* 1:164–166 (1987).
9. J.I. Gray and F.J. Monahan. Measurement of lipid oxidation in meat and meat products. *Trends Food Sci. Technol.* 3:320–324 (1992).
10. J.M.C. Gutteride and B. Halliwell. The measurement and mechanism of lipid peroxidation in biological systems. *Trends Biochem. Sci.* 15:129–135 (1990).
11. F. Shahidi, U.N. Wanasundara, Y. He, and V.K.S. Shukla. Marine lipids and their stabilization with green tea and catechins. In: *Flavor and Lipid Chemistry of Seafoods* (F. Shahidi and K. Cadwallader, eds.). American Chemical Society, Washington, DC, 1997.
12. B.J.F. Hudson. Evaluation of oxidative rancidity technique. In: *Rancidity of Foods* (J.C. Allen and J. Hamilton, eds.). Applied Science Publishers, London, 1983, pp. 47–58.
13. H.S. Olcott and E. Einset. A weighing method for measuring the induction period marine and other oils. *J. Am. Oil Chem. Soc.* 35:161–162 (1958).
14. P.J. Ke and R.G. Ackman. Metal-catalyzed oxidation in mackerel skin and meat lipids. *J. Am. Oil Chem. Soc.* 53:636–640 (1976).
15. U.N. Wanasundara and F. Shahidi. Canola extracts as an alternative antioxidant for canola oil. *J. Am. Oil Chem. Soc.* 71:817–822 (1994).

16. U.N. Wanasundara and F. Shahidi. Stabilization of seal blubber and menhaden oils with green tea catechins. *J. Am. Oil Chem. Soc.* 73:1183–1190 (1996).

17. J.I. Gray. Measurement of lipid oxidation. *J. Am. Oil Chem. Soc.* 55:539–546 (1978).

18. A. Lips, R.A. Chapman, and W.D. McFarlane. The application of ferric thiocyanate method to the determination of incipient rancidity in fats and oils. *Oil Soap* 20:240–243 (1943).

19. S.E. Ebeler and T. Shibamoto. Gas and high-performance liquid chromatographic analysis of lipid peroxidation products. In: *Lipid Chromatographic Analysis* (T. Shibamoto, ed.). Dekker, New York, 1994, pp. 223–249.

20. A.R. Wewala. Prediction of oxidative stability of lipids based on the early stage oxygen consumption rate. In: *Natural Antioxidants: Chemistry, Health Effects and Applications* (F. Shahidi, ed.). AOCS Press, Champaign, IL, 1997, pp. 331–345.

21. M.K. Logani and R.E. Davies. Lipid oxidation: Biological effects and antioxidants. *Lipids* 15:485–495 (1980).

22. E.H. Farmer and D.A. Sutton. Peroxidation in relation to olefinic structure. *Trans. Faraday Soc.* 42:228–232 (1946).

23. A.J. St. Angelo, R.L. Ory, and L.E. Brown. A comparison of minor constituents in peanut butter as possible source of fatty acid peroxidation. *Am. Peanut Res. Educ. Assoc.* 4:186–196 (1972).

24. F. Shahidi, U.N. Wanasundara, and N. Brunet. Oxidative stability of oil from blubber of harp seal (*Phoca groenlandica*) as assessed by NMR and standard procedures. *Food Res. Int.* 27:555–562 (1994).

25. U.N. Wanasundara, F. Shahidi, and C.R. Jablonski. Comparison of standard and NMR methodologies for assessment of oxidative stability of canola and soybean oils. *Food Chem.* 52:249–253 (1995).

26. L.J. Parr and P.A.T. Swoboda. The assay of conjugable oxidation products applied to lipid deterioration in stored foods. *J. Food Technol.* 11:1–12 (1976).

27. B.G. Tarladgis, A.M. Pearson, and L.R. Dugan. Chemistry of the 2-thiobarbituric acid test for determination of oxidative rancidity in foods. *J. Sci. Food Agric.* 15:602–607 (1964).

28. L.R. Dugan. Stability and rancidity. *J. Am. Oil Chem. Soc.* 32:605–609 (1955).

29. W.A. Baumgartner, N. Baker, V.A. Hill, and E.T. Wright. Novel interference in thiobarbituric acid assay for lipid peroxidation. *Lipids* 10:309–311 (1975).

30. R. Marcuse and L. Johansson. Studies on the TBA test for rancidity grading: II. TBA reactivity of different aldehyde classes. *J. Am. Oil Chem. Soc.* 50:387–391 (1973).

31. P.J. Ke and A.D. Woyewoda. Microdetermination of thiobarbituric acid values in marine lipids by a direct spectrophotometric method with mono-phasic reaction systems. *Anal. Chem. Acta* 106:279–284 (1979).

32. C. Robbles-Martinez, E. Cervantes, and P.J. Ke. Recommended method for testing the objective rancidity development in fish based on TBARS formation. *Can. Tech. Rep. Fish Aqu. Sci.* No. 1089.

33. J. Pokorny, H. Valentova, and J. Davidek. Modified determination of 2-TBA value in fats and oils. *Die Nahrung* 29:31–38 (1985).

34. F. Shahidi, L.J. Rubin, L.L. Diosady, and D.F. Wood. Effect of sulfanimide on the TBA values of cured meats. *J. Food Sci.* 50:274–275 (1985).

35. F. Shahidi, J. Yun, L.J. Rubin, and D.F. Wood. Control of lipid oxidation in cooked ground pork with antioxidants and dinitrosyl ferrohemochrome. *J. Food Sci.* 52:564–567 (1987).

36. M.C. Thomas and J. Fumes. Application of 2-thiobarbituric acid reaction to exudates of frozen and refrigerated meats. *J. Food Sci.* 52:575–579 (1987).

37. A. Schmedes and G. Holmer. A new thiobarbituric acid (TBA) method for determining free malonaldehyde and hydroperoxides selectively as a measure of lipid peroxidation. *J. Am. Oil Chem. Soc.* 66:813–817 (1989).

38. J.I. Gray and A.M. Pearson. Rancidity and warmed-over flavor. In: *Advances in Meat Research: Restructured Meat and Poultry Products* (A.M. Pearson and T.R. Dutson, eds.). Van Nostrand Reinhold, New York, 1987, pp. 219–229.

39. *Official and Tentative Methods of the American Oil Chemists' Society*, 3rd edn. American Oil Chemists' Society, Champaign, IL, 1981.

40. J.A. Fioriti, A.P. Bentz, and R.J. Sims. The reaction of picric acid with epoxides: A colorimetric method. *J. Am. Oil Chem. Soc.* 43:37–41 (1966).

41. *Standard Methods for the Analysis of Oils and Fats and Derivatives*, 7th edn. (C. Paquot and A. Hautfenne, eds.). Blackwell Scientific Publishers Ltd., Oxford, 1987.

42. G.R. List, C.D. Evans, W.K. Kwolek, K. Warner, and B.K. Bound. Oxidation and quality of soybean oil: A preliminary study of the anisidine test. *J. Am. Oil Chem. Soc.* 51:17–21 (1974).

43. J.B. Rossell. Measurement of ransidity. In: *Rancidity in Foods* (J.C. Allen and J. Hamilton, eds.). Applied Science Publishers, London, 1983, pp. 21–45.

44. U.N. Wanasundara and F. Shahidi. Storage stability of microencapsulated seal blubber oil. *J. Food Lipids* 2:73–86 (1995).

45. N. Yukawa, H. Takamura, and T. Matoba. Determination of total carbonyl compounds in aqueous media. *J. Am. Oil Chem. Soc.* 70:881–884 (1993).

46. S.R. Meyer and L. Robrovic. The spectroscopic quinodal ion method for the analysis of carbonyl compounds. *J. Am. Oil Chem. Soc.* 72:385–387 (1995).

47. F. Shahidi and R.B. Pegg. Hexanal as an indicator of meat flavor deterioration. *J. Food Lipids* 1:177–186 (1994).

48. F. Shahidi and R.B. Pegg. Hexanal as an indicator of the flavor deterioration of meat and meat products. In: *Lipids in Food Flavors* (C.T. Ho and T.G. Hartman, eds.). ACS Symposium Series 558, American Chemical Society, Washington DC, 1994, pp. 256–279.

49. A.J. St. Angelo, J.R. Vercellotti, M.G. Legengre, C.H. Vinnelt, J.W. Kuan, C. Janies, and H.P. Duppy. Chemical and instrumental analysis of warmed-over flavor in beef. *J. Food Sci.* 52:1163–1168 (1987).

50. S.F. O'Keefe, F.G. Proudfoot, and R.G. Ackman. Lipid oxidation in meats of omega-3 fatty acid-enriched broiler chickens. *Food Res. Int.* 28:417–424 (1995).

51. U.N. Wanasundara and F. Shahidi. Stability of edible oils as reflected in their propanal and hexanal contents. *Paper presented at the Canadian Section of American Oil Chemists' Society's Annual Meeting*, Guelph, ON, November 15–16 (1995).

52. F. Shahidi and S.A. Spurvey. Oxidative stability of fresh and heat-processed dark and white muscles of mackerel (*Scomber scombrus*). *J. Food Lipids* 3:13–25 (1996).

53. E. Selke, H.A. Moser, and W.K. Rohwedder. Tandem gas chromatography–mass spectrometry analysis of volatiles from soybean oil. *J. Am. Oil Chem. Soc.* 47:393–397 (1970).

54. J.M. Snyder, E.N. Frankel, and E. Selke. Capillary gas chromatographic analysis of headspace volatiles from vegetable oils. *J. Am. Oil Chem. Soc.* 62:1675–1679 (1985).

55. P.K. Jarvi, G.D. Lee, D.K. Erickson, and E.A. Butkus. Determination of the extent of rancidity of soybean oil by gas chromatography compared with peroxide values. *J. Am. Oil Chem. Soc.* 48:121–124 (1971).

56. K. Warner, C.D. Evans, G.R. List, B.K. Boundly, and W.F. Kwolek. Pentane formation and rancidity in vegetable oils and potato chips. *J. Food Sci.* 39:761–765 (1974).

57. W.R. Bidlack and A.L. Tappel. Fluorescent products of phospholipids during lipid peroxidation. *Lipids* 8:203–207 (1973).

58. C.J. Dillard and A.L. Tappel. Fluorescent products of lipid peroxidation of mitochondria and microsomes. *Lipids* 6:715–721 (1971).

59. C.J. Dillard and A.L. Tappel. Fluorescent products from reaction of peroxidizing polyunsaturated fatty acids with phosphatidyl ethanolamine and phenyamine. *Lipids* 8:183–189 (1973).

60. F. Shahidi, R.B. Pegg, and R. Harris. Effect of nitrite and sulfanilamide on the 2-thiobarbituric acid (TBA) values in aqueous model and cured meat systems. *J. Muscle Foods* 2:1–9 (1991).

61. K. Kikugawa and M. Beppu. Involvement of lipid oxidation products in the formation of fluorescent and cross-linked proteins. *Chem. Phys. Lipids* 44:277–296 (1987).

62. C.W. Fritsch. Measurement of frying fat deterioration. *J. Am. Oil Chem. Soc.* 58:272–274 (1981).

63. M.M. Blumenthal and J.R. Stockier. Isolation and detection of alkaline contaminant materials in frying oils. *J. Am. Oil Chem. Soc.* 63:687–688 (1986).

64. M.M. Blumenthal, J.R. Stockler, and P.J. Summers. Alkaline contaminant materials in frying oils, a new quick test. *J. Am. Oil Chem. Soc.* 62:1373–1374 (1985).

65. J.L. Sebedio, P.O. Astorg, A. Septier, and A. Grandgirard. Quantitative analysis of polar components in frying oils by the iatroscan thin-layer chromatography–flame ionization detection technique. *J. Chromatogr.* 405:371–378 (1987).

66. M.G. Simic. Kinetic and mechanistic studies of peroxy, vitamin E and antioxidant free radicals by pulse radiolysis. In: *Autoxidation in Food and Biological Systems* (M.G. Simic and M. Karel, eds.). Plenum Press, New York, 1980, pp. 17–26.

67. K.M. Schaich and D.C. Borgi. EPR studies in autooxidation. In: *Autoxidation in Food and Biological Systems* (M.G. Simic and M. Karel, eds.). Plenum Press, New York, 1980, pp. 45–70.

68. M.J. Davies. Application of electron spin resonance spectroscopy to the identification of radicals produced during lipid peroxidation. *Chem. Phys. Lipids* 44:149–173 (1987).

69. G.C. Yen and P.D. Duh. Scavenging effect of methanolic extracts of peanut hulls on free radical and active-oxygen species. *J. Agric. Food Chem.* 42:629–632 (1994).

70. C.W. Chen and C.T. Ho. Antioxidant properties for polyphenols extracted from green and black teas. *J. Food Lipids* 2:35–46 (1995).

71. F.R. van de Voort, A.A. Ismail, J. Sedman, and G. Emo. Monitoring the oxidation of edible oils by Fourier transform infrared spectroscopy. *J. Am. Oil Chem. Soc.* 71:243–253 (1994).

72. J. Sedman, A.A. Ismail, A. Nicodema, S. Kubow, and F.R. van der Voort. Application of FTIR.ATR differential spectroscopy for monitoring oil oxidation and antioxidant efficacy. In: *Natural Antioxidants: Chemistry, Health Effects and Applications* (F. Shahidi, ed.). AOCS Press, Champaign, IL, 1997, pp. 358–378.

73. I.C. Burkow, P. Moen, and K. Overbo. Chemiluminescence as a method for oxidative rancidity assessment of marine oils. *J. Am. Oil Chem. Soc.* 69:1108–1111 (1992).

74. I. Neeman, D. Joseph, W.H. Biggeley, and H.H. Seliger. Induced chemiluminescence of oxidized fatty acids and oils. *Lipids* 20:729–734 (1985).

75. I.C. Burkow, L. Vikersveen, and K. Saarem. Evaluation of antioxidants for cod liver oil by chemiluminescence and rancimat methods. *J. Am. Oil Chem. Soc.* 72:553–557 (1995).

76. T. Miyazawa, M. Kikuch, K. Fujimoto, Y. Endo, S-Y. Cho, R. Usuki, and T. Keneda. Shelf-life dating of fish meats in terms of oxidative rancidity as measured by chemiluminescence. *J. Am. Oil Chem. Soc.* 68:39–43 (1991).

77. H. Saito and K. Nakamura. Application of NMR method to evaluate the oxidative deterioration of crude and stored fish oils. *Agric. Biol. Chem.* 54:533–534 (1990).

78. H. Saito and M. Udagawa. Use of NMR to evaluation the oxidative deterioration of *Niboshi*, boiled and dried fish. *Biosci. Biotech. Biochem.* 56:831–832 (1992).

79. F. Shahidi. Current and novel methods for stability testing of canola oil. *Inform* 3:543 (1992).

80. U.N. Wanasundara and F. Shahidi. Application of NMR spectroscopy to assess oxidative stability of canola and soybean oils. *J. Food Lipids* 1:15–24 (1993).

81. H.S. Lee, D.W. Shoeman, and A.S. Csallany. Urinary response to in vivo lipid peroxidation induced by vitamin E deficiency. *Lipids* 27:124–128 (1992).

82. P. Cogrel, I. Morel, G. Lescoat, M. Chevanne, P. Brissot, P. Cillard and J. Cillard. The relationship between fatty acid peroxidation and α-tocopherol consumption in isolated normal and transformed hepatocytes. *Lipids* 28:115–119 (1993).

83. M. Kinter. Analytical technologies for lipid oxidation products analysis. *J. Chomatogr. B* 671:223–236 (1995).

84. H. Kosugi, T. Kojima, and K. Kikugawa. Characteristics of the thiobarbituric acid reactivity of human urine as a possible consequence of lipid oxidation. *Lipids* 28:337–343 (1993).

85. G.A. Cordis and N. Maulik. Estimation of the extent of lipid peroxidation in the ischemic and reperfused heart by monitoring lipid metabolic products with the aid of high-performance liquid chromatography. *J. Chromatogr.* 632:97–103 (1993).

86. D. Bagchi, M. Bagchi, E.A. Hassoun, and S.J. Stohs. Detection of paraquat-induced in vivo lipid peroxidation by gas chromatography/mass spectrometry and high pressure liquid chromatography. *J. Anal. Toxicol.* 17:411–414 (1993).

87. D. Bagchi, M. Bagchi, E.A. Hassoun, and S.J. Stohs. Effect of carbon-tetrachloride, menadoin and paraquat on the urinary extraction of malonaldehyde, formaldehyde, acetaldehyde and acetone in rats. *J. Biochem. Toxicol.* 8:101–106 (1993).

88. D. Bagchi, M. Bagchi, E.A. Hassoun, and S.J. Stohs. Carbon-tetrachloride-induced urinary extraction of formaldehyde, malondialdehyde, acetaldehyde and acetone in rats. *Pharmacology* 47:209–216 (1993).

89. E. Frankel. *Lipid Oxidation*. The Oily Press, Bridgewater, England, 2005, pp. 99–164.

90. G. Marquez-Ruiz, N. Jorge, M. Martin-Polvillo, and M.C. Dobarganes. Rapid quantitative determination of polar compounds in fats and oils by solid-phase extraction and exclusion chromatography using monostearine as internal standard. *J. Chromatogr.* 749:55–60 (1996).

91. G. Marquez-Ruiz, M. Martin-Polvillo, and M.C. Dobarganes. Quantitation of oxidized triglyceride monomers and dimers as an useful measurement for early and advanced stages of oxidation. *Grasas y Aceites* 47:48–53 (1996).

92. G. Marquiz-Ruiz, N. Jorge, M. Martin-Polvillo, and M.C. Dobarganes. Rapid quantitative determination of polar compounds in fats and oils by solid-phase extraction and exclusion chromatography using monostearin as internal standard. *J. Chromatogr.* 749:55–60 (1996).
93. F.P. Corongiu and S. Banni. Detection of conjugated dienes by second derivative ultraviolet spectrophotometry. *Meth. Enzymol.* 233:303–306 (1994).
94. H. Tamura and T. Shibamoto. Gas chromatographic analysis of malonaldehyde and 4-hydroxy-2-(*E*)-nonenal produced from arachidonic acid and linoleic acid in a lipid peroxidation model system. *Lipids* 26:170–173 (1991).
95. D. Bagchi, M. Bagchi, E.A. Hassoun, and S.J. Stohs. Endrin-induced urinary excretion of formaldehyde. Acetaldehyde, malondialdehyde and acetone in rats. *Toxicology* 75:81–89 (1992).
96. G. Marquez-Ruiz and M.C. Dobarganes. Analysis of lipid oxidation products by combination of chromatographic techniques. In: *New Techniques and Applications in Lipid Analysis* (R.E. McDonald and M.M. Mossoba, ed.). AOCS Press, Champaign, IL, 1997, pp. 216–233.

15 Antioxidants

David W. Reische, Dorris A. Lillard,
and Ronald R. Eitenmiller

CONTENTS

I. INTRODUCTION

Lipid oxidation in foods is a serious problem, difficult to overcome often, and leads to loss of shelf life, palatability, functionality, and nutritional quality. Loss of palatability is due to the generation of off-flavors that arise primarily from the breakdown of unsaturated fatty acids during autoxidation. The high reactivity of the carbon double bonds in unsaturated fatty acids makes these substances primary targets for free radical reactions. Autoxidation is the oxidative deterioration of unsaturated fatty acids via an autocatalytic process consisting of a free radical chain mechanism [1]. The chain of reaction includes initiation, propagation, and termination. Propagation reactions are primarily responsible for the autocatalytic nature of autoxidation.

Autoxidation must be induced by preformed or primary hydroperoxides. The source of preformed hydroperoxides is either photosynthesized oxidation or lipoxygenase catalysis. Photosynthesized oxidation (photooxidation) involves direct reaction of light-activated, singlet oxygen with unsaturated fatty acid and the subsequent formation of hydroperoxides. Lipoxygenase catalysis involves enzymatic oxidation of unsaturated fatty acids to their corresponding hydroperoxides. Formation of primary hydroperoxides is catalyzed by either light, metals, singlet oxygen, and sensitizers, or preformed hydroperoxide decomposition products.

II. ANTIOXIDANTS

In foods containing lipids, antioxidants delay the onset of oxidation or slow the rate at which it proceeds. These substances can occur as natural constituents of foods, but they can also be intentionally added to products or formed during processing. Their role is not to enhance or improve the quality of foods, but they do maintain food quality and extend shelf life. Antioxidants for use in food processing must be inexpensive, nontoxic, effective at low concentrations, stable, and capable of surviving processing (carry-through effect); color, flavor, and odor must be minimal. The choice of which antioxidant to use depends on product compatibility and regulatory guidelines [2].

Antioxidants not only extend shelf life of the products but also reduce raw material waste, reduce nutritional losses, and widen the range of fats that can be used in specific products [3]. By extending maintaining quality and increasing the number of oils that can be used in food products, antioxidants allow processors to use more available and/or less costly oils for product formulation.

A. CLASSIFICATION

Antioxidants can be broadly classified by mechanism of action as primary antioxidants and secondary antioxidants. Some antioxidants exhibit more than one mechanism of activity and are often referred to as multiple-function antioxidants. Chemical modes of action vary greatly because these substances are able to function at all stages of the free radical reaction.

1. Primary Antioxidants

Primary, type 1, or chain-breaking antioxidants are free radical acceptors that delay or inhibit the initiation step or interrupt the propagation step of autoxidation. Initiation of autoxidation occurs when an α-methylenic hydrogen molecule is abstracted from an unsaturated lipid to form a lipid (alkyl) radical (R) (Equation 15.1).

$$RH \rightarrow R\bullet + H\bullet. \tag{15.1}$$

This highly reactive lipid radical can then react with oxygen to form a peroxy radical (ROO$\bullet$) in a propagation reaction (Equation 15.2).

$$R\bullet + O_2 \rightarrow ROO\bullet + H\bullet. \tag{15.2}$$

During propagation, peroxy radicals react with lipid to form a hydroperoxide and a new unstable lipid radical (Equation 15.3). This lipid radical will then react with oxygen to produce another peroxy radical, resulting in a cyclical, self-catalyzing oxidative mechanism (Equation 15.4).

$$ROO\bullet + RH \rightarrow ROOH + R\bullet, \tag{15.3}$$

$$R\bullet + O_2 \rightarrow ROO\bullet + H\bullet. \tag{15.4}$$

Hydroperoxides are unstable and can degrade to produce radicals that further accelerate propagation reactions. These reactions are typically referred to as branching steps (Equations 15.5 and 15.6).

$$ROOH \rightarrow RO\bullet + OH\bullet, \tag{15.5}$$

$$RO\bullet + RH \rightarrow ROH + R\bullet. \tag{15.6}$$

Hydroperoxide degradation leads to the undesirable odors and flavors associated with rancidity in later stages of oxidation.

Primary antioxidants react with lipid and peroxy radicals and convert them to more stable, nonradical products. Primary antioxidants donate hydrogen atoms to the lipid radicals and produce lipid derivatives and antioxidant radicals (A•) that are more stable and less readily available to further promote autoxidation. Like hydrogen donors, primary antioxidants have higher affinities for peroxy radicals than lipids [4]. Therefore, peroxy and oxy free radicals formed during the propagation (Equations 15.2 and 15.4) and branching (Equations 15.5 and 15.6) steps of autoxidation are scavenged by primary antioxidants (Equations 15.7 and 15.8). Antioxidants may also interact directly with lipid radicals (Equation 15.9).

$$ROO\bullet + AH \rightarrow ROOH + A\bullet, \tag{15.7}$$

$$RO\bullet + AH \rightarrow ROH + A\bullet, \tag{15.8}$$

$$R\bullet + AH \rightarrow RH + A\bullet. \tag{15.9}$$

The antioxidant radical produced by hydrogen donation has a very low reactivity with lipids. This low reactivity reduces the rate of propagation, since reaction of the antioxidant radical with oxygen or lipids is very slow. The antioxidant radical is stabilized by delocalization of the unpaired electron around a phenol ring to form stable resonance hybrids. Antioxidant radicals are capable of participating in termination reactions with peroxy (Equation 15.10), oxy (Equation 15.11), and other antioxidant radicals (Equation 15.12). The formation of antioxidant dimers (dimerization) is prominent in fats and oils and indicates that phenolic antioxidant radicals readily undergo termination reactions. This effectively stops the autocatalytic free radical chain mechanism as long as the antioxidant is present in its nonradical form.

$$ROO\bullet + A\bullet \rightarrow ROOA, \tag{15.10}$$

$$RO\bullet + A\bullet \rightarrow ROA, \tag{15.11}$$

$$A\bullet + A\bullet \rightarrow AA. \tag{15.12}$$

Before initiation of autoxidation, there must be an induction period in which antioxidants are consumed and free radicals are generated. Therefore, primary antioxidants are most effective if they are added during the induction and initiation stages of oxidation when the cyclical propagation steps have not occurred. Addition of antioxidants to fats that already contain substantial amounts of peroxides will quickly result in loss of antioxidant function [5]. In addition to radical scavenging, primary antioxidants can reduce hydroperoxides to hydroxy compounds. However, the main antioxidative mechanism of primary antioxidants is radical scavenging.

Primary antioxidants are mono- or polyhydroxy phenols with various ring substitutions. Substitution with electron-donating groups ortho and para to the hydroxyl group of phenol increases the antioxidant activity of the compound by an inductive effect. These hindered phenolic antioxidants decrease the reactivity of the hydroxyl group by increasing its electron density. Substitution with butyl or ethyl groups para to the hydroxyl enhances the antioxidant activity. Because of steric hindrance, however, the presence of longer chain or branched alkyl groups at the para positions can decrease antioxidant effectiveness [6]. Substitutions of branched alkyl groups at ortho positions enhance the phenolic antioxidant's ability to form stable resonance structures and further reduce the antioxidant radical's ability to participate in propagation reactions.

The most commonly used primary antioxidants in foods are synthetic compounds. Examples of important primary phenolic antioxidants include butylated hydroxyanisole (BHA), butylated hydroxytoluene (BHT), propyl gallate (PG), and tertiary butylhydroquinone (TBHQ). However, a few natural components of food also act as primary antioxidants and are commonly added to foods. Tocopherols are the most commonly used natural primary antioxidants. Carotenoids are another group of natural compounds that have primary antioxidant activity, although the mechanisms differs from the phenolics.

2. Secondary Antioxidants

Secondary, preventive, or type 2, antioxidants act through numerous possible mechanisms. These antioxidants slow the rate of oxidation by several different actions, but they do not convert free radicals to more stable products. Secondary antioxidants can chelate prooxidant metals and deactivate them, replenish hydrogen to primary antioxidants, decompose hydroperoxides to nonradical species, deactivate singlet oxygen, absorb ultraviolet radiation, or act as oxygen scavengers. These antioxidants are often referred to as synergists because they promote the antioxidant activity of type 1 antioxidants. Citric acid, ascorbic acid, ascorbyl palmitate, lecithin, and tartaric acid are good examples of synergists. Some of the more important types of secondary antioxidant mechanisms are discussed in the following sections.

a. Chelators

Several heavy metals with two or more valence states (Fe, Cu, Mn, Cr, Ni, V, Zn, A1) promote oxidation by acting as catalysts of free radical reactions. These redox-active transition metals transfer single electrons during changes in oxidation states.

Two mechanisms of oxidation promotion by metals have been proposed. Metals are believed to either interact with hydroperoxides or to react directly with lipid molecules. Metals are able to promote oxidation by interacting directly with unsaturated lipids (Equation 15.13) and lowering the activation energy of the initiation step of autocatalysis. However, because of thermodynamic constraints, spin barriers, and an extremely slow reaction rate, this direct interaction of metals with lipid moieties is not the main mechanism of metal catalysis [7–9].

$$M^{(n-1)+} + RH \rightarrow M^{n+} + H^+ + R\bullet. \tag{15.13}$$

Metals are known to interact with hydroperoxides and promote oxidation. Moreover, it is thought that a metal–hydroperoxide complex forms and subsequently decomposes to produce free radicals. Metals enhance the rate of decomposition of hydroperoxides and the generation of free radicals. Two metal–hydroperoxide reactions are possible.

$$M^{(n+1)+} + ROOH \rightarrow M^{n+} + H^+ + ROO\bullet, \tag{15.14}$$

$$M^{n+} + ROOH \rightarrow M^{(n+1)+} + OH^- + RO\bullet. \tag{15.15}$$

Equation 15.15 is less significant in aqueous solution, since metals in their lower oxidation states accelerate hydroperoxide degradation more than metals in their higher oxidation states [10]. Even trace amounts of these metals promote electron transfer from lipids or hydroperoxides because the Equations 15.14 and 15.15 can be cyclical with regeneration of the lower oxidation state of the metal. Nevertheless, Equation 15.15 occurs much more slowly than Equation 15.14 [7]. Although the metal–hydroperoxide mechanisms are generally accepted as the most important mechanisms for metal catalysis of autoxidation, it is unclear whether redox-active transition metals promote lipid peroxidation directly through the formation of metal–lipid complexes or by forming peroxy and oxy radicals.

Chelation of metals by certain compounds decreases their prooxidant effect by reducing their redox potentials and stabilizing the oxidized form of the metal. Chelating compounds may also sterically hinder formation of the metal hydroperoxide complex. Citric acid (and its lipophilic, monoglyceride ester), phosphoric acid (and its polyphosphate derivatives), and ethylenediaminetetraacetic acid (EDTA) can chelate metals. EDTA forms a thermodynamically stable complex with metal ions. The metal-chelating ability of oligophosphate increases with phosphate group number up to six residues. Carboxyl groups of citric acid are thought to be responsible for binding with metals and forming complexes. Malic, tartaric, oxalic, and succinic acids bind metals in the same manner.

In addition to their antioxidant activity, many of these compounds have other unique functions as food additives. Citric acid, malic acid, and tartaric acid are important food acidulants. Phosphates are added as buffers, emulsifiers, and acidulants and water-binders. Chelating antioxidants are also referred to as synergists because they enhance the activity of phenolic antioxidants. This synergism is sometimes referred to as acid synergism when the chelator is citric or other acids.

b. Oxygen Scavengers and Reducing Agents

Ascorbic acid, ascorbyl palmitate, erythorbic acid, sodium erythorbate, and sulfites prevent oxidation by scavenging oxygen and acting as reductants. Oxygen scavenging is useful in products with headspace or dissolved oxygen. Reducing agents function by donating hydrogen atoms. Ascorbic acid and sulfites react directly with oxygen and eliminate it from the food product [11]. L-Ascorbic acid has strong reducing properties, and its most significant chemical property is its ability to oxidize through one- or two-electron transfers [12,13]. One-electron reactions involve an L-ascorbic acid radical (semidehydroascorbic acid). A proton is lost, whereon a bicyclic radical, which is the intermediate leading to dehydroascorbic acid, is formed. A two-electron transfer occurs when transition metals catalyze ascorbate autoxidation. In this process, L-ascorbate and oxygen form a ternary complex with the metal catalyst. Two π electrons from L-ascorbate shift to oxygen through the transition metal. Oxidation of ascorbic acid has been reviewed by Liao and Seib [14].

Ascorbic acid represents a truly multifunctional antioxidant. Schuler [15] identified the following classes of reactions as significant to the antioxidant action of ascorbic acid in food systems: quenching of singlet oxygen, reductions of free radicals and primary antioxidant radicals, and removal of molecular oxygen in the presence of metal ions.

c. Singlet Oxygen Quenchers

Singlet oxygen is a high-energy molecule that is responsible for photooxidation of unsaturated fats and the subsequent generation of hydroperoxides. Singlet oxygen quenchers deplete singlet oxygen of its excess energy and dissipate the energy in the form of heat. Carotenoids, including β-carotene, lycopene, and lutein, are active singlet oxygen quenchers at low oxygen partial pressure. Figure 15.1 gives an overview of lipid oxidation and the interaction of antioxidants.

III. SYNTHETIC AND NATURAL ANTIOXIDANTS

Consumers are concerned about the safety of their food and about potential effects of synthetic additives on their health. Despite the superior efficacy, low cost, and high stability of synthetic antioxidants in foods, the suspicion that these compounds may act to promote carcinogenicity has led to a decrease in their use [16]. A trend toward the use of "natural" food additives in the food industry has been apparent for quite some time—a result of consumer demand. Some natural preservatives exist inherently in foods; others can be added to the product or can arise as a result of processing or cooking. Natural food antioxidants such as citric acid and ascorbic acid are

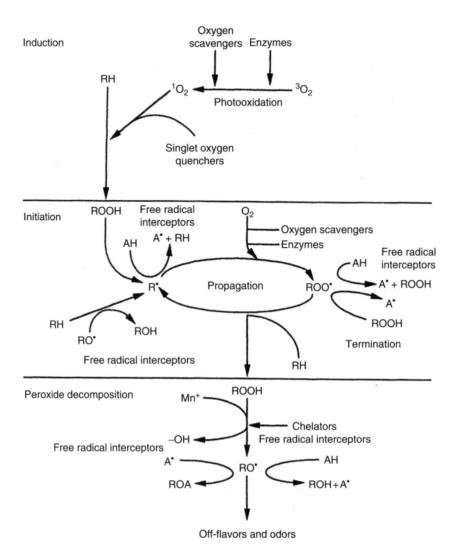

FIGURE 15.1 Overview of lipid oxidation and the interaction of antioxidants.

used widely in the food industry. Recent research has focused on isolation and identification of effective antioxidants of natural origin.

A. SYNTHETIC ANTIOXIDANTS

Synthetic antioxidants are intentionally added to foods to inhibit lipid oxidation. Synthetic antioxidants approved for use in food include BHA, BHT, PG (also octyl and dodecyl gallate), ethoxyquin, ascorbyl palmitate, and TBHQ. The synthesis of novel antioxidants for food use is limited by rising costs of research and development, costs associated with safety assessment, and the time required to obtain regulatory approval of additives [17]. These restrictions, as well as growing consumer preference for natural food additives, have led industry to emphasize natural materials as a source of novel antioxidants.

Phenolic compounds represent some of the oldest and most frequently used antioxidants in foods. The differences in antioxidant activity of the phenolic antioxidants are due to variations in structure that directly influence physical properties. Phenols in which the aromatic ring contains alkyl groups (hindered phenols) are extremely effective antioxidants. Hindered phenols are also effective antimicrobials in foods. The characteristics of a good product ultimately determine the selection of the phenolic antioxidant. BHA and BHT are fairly heat stable and are used in heat-processed foods. PG decomposes at 148°C and is inappropriate for high temperature processing. Therefore, heat-stable TBHQ is useful in frying applications. BHA and BHT are strongly lipophilic and are used extensively in oil-in-water emulsions. BHA and BHT are also typically used together in mixtures, acting synergistically. A summary of the physical properties and applications of phenolic antioxidants is provided in Table 15.1.

Titles 9 and 21 of the U.S. Code of Federal Regulations (CFR) govern the use of antioxidants in meat and poultry products and in foods, respectively. Phenolic antioxidants are effective at low concentration and are often used at levels <0.01% [6]. The level of phenolic antioxidants permitted in the United States varies according to the product. As specified in 21 CFR, 172.110 and 172.115, limitations for BHA and BHT alone or in combination are as follows for specific products: potato granules, 10 ppm; dehydrated potato shreds, dry breakfast cereals, potato flakes, sweet potato flakes, 50 ppm; and emulsion stabilizers for shortenings, 200 ppm. At high levels, phenolic compounds become prooxidants because of their high reactivity and participation in the initiation process. Allowable limits vary greatly depending on the food product and the antioxidant. Regulations concerning synthetic antioxidants vary greatly from country to country and complicate marketing of products internationally.

Commercial antioxidant preparations are available in solid and liquid blends. Liquid blends are convenient because the antioxidant is solubilized for each addition during processing. Solvents include vegetable oils, propylene glycol, glyceryl monooleate, ethanol, and acetylated monoglycerides [5]. Antioxidant preparations typically contain mixtures of phenolic antioxidants, a synergist, and a solvent system. Some of the most important synthetic antioxidants are discussed in the following sections. Structures are provided in Figure 15.2.

1. Butylated Hydroxyanisole

BHA is typically used as a 9:1 mixture of 3-BHA and 2-BHA isomers [6]. The 3-isomer shows higher antioxidant activity than the 2-isomer [5]. It is a waxy, monophenolic, white solid that is fat soluble. It is effective in preventing oxidation of animal fats, but ineffective for vegetable fats. BHA has good carry-through in baking, but volatilizes during frying. It is commonly added to packaging materials.

2. Butylated Hydroxytoluene

BHT is a widely used monophenolic antioxidant. This fat-soluble white, crystalline solid is appropriate for high temperature processing but is not as stable as BHA. It will volatilize and has less carry-through. BHA and BHT act synergistically to provide greater antioxidant activity than either antioxidant alone. Therefore, foods typically contain BHA/BHT mixtures at levels up to 0.02%. The postulated synergistic mechanism of BHA and BHT involves the interactions of BHA with peroxy radicals to produce a BHA phenoxy radical. The BHA phenoxy radical is then believed to abstract a hydrogen from the hydroxyl group of BHT. BHT effectively acts as a hydrogen replenisher of BHA, allowing BHA to regenerate its effectiveness. The BHT radical can then react with a peroxy radical and act as a chain terminator [7].

3. Tertiary Butylhydroxyquinone

TBHQ is a beige diphenolic powder. It is used in frying applications with highly unsaturated vegetable oils, where it has good carry-through. It is generally considered to be more effective in vegetable oils than BHA or BHT [5]. Citric acid and TBHQ show excellent synergism in vegetable oils. TBHQ is not permitted for use in foods in Canada or the European Economic Community. In the United States, TBHQ is not permitted to be combined with PG.

TABLE 15.1

Characteristics of Phenolic Antioxidants and Relevant Chapter of U.S. Code of Federal Regulations (CFR)

Compound	Molecular Weight	Appearance	Boiling Point (°C)	Melting Point (°C)	Solubility	Functionality	21 CFR (9 CFR)
Butylated hydroxyanisole (BHA)	180.25	Waxy solid	264–270	48–63	Insoluble in water Slightly soluble in glycerol and mineral oil Soluble in fats, alcohol, propylene glycol, petroleum, ether; glyceryl monooleate; paraffin	Good carry-through in baked and fried products Volatile, distillable Most effective in animal fats Used in packaging materials Synergistic with other antioxidants Slight phenol odor Less expensive than BHA Synergistic with BHA and other antioxidants Less expensive than BHA Volatile, distillable	172.110 182.3169 172.615 172.515 165.175 166.110 164.110 (381.147) (318.7) 171.115 182.3173
Butylated hydroxytoluene (BHT)	220.356	White crystals	265	70	Insoluble in water, glycerol, and propylene glycol Slightly soluble in mineral oil Soluble in fats, paraffin, glyceryl monooleate, alcohol, petroleum, ether, most organic solvents	Less carry-through in baked and fried products than BHA More sterically hindered than BHA Most effective in animal fats Slight phenol odor Poor carry-through properties Synergistic with other antioxidants Decomposes at frying temperatures antioxidants Decomposes at frying temperatures	172.615 137.350 165.175 166.110 164.110 (381.147) (318.7) 184.1660 172.615

Name	Molecular weight	Appearance	Melting point (°C)	Boiling point (°C)	Solubility	Characteristics	m/z
Propyl gallate (PG)	212.20	White crystals	150	Decomposes above 148	Slightly soluble in water, fats, mineral oil, glyceryl monooleate; Soluble in alcohol, glycerol, propylene glycol	Stability of octyl and dodecyl forms greater than propyl; Discolors in the presence of metals; always used in combination with a chelator; Less soluble in fats than BHA and BHT; More effective in vegetable oils than BHA and BHT; More effective in vegetable oils than BHA and BHT; Poor carry-through in baking, but good carry-through in frying; Synergistic with other antioxidants	165.175; 166.110; 164.110 (318.7); (381.47); 171.185; 165.175; 166.110; 164.110 (318.7)
Tertiary butylhydroquinone (TBHQ)	166.22	White to tan crystals	126.5–128.5	300	Slightly soluble in water; Moderately soluble in fats, propylene glycol, glyceryl monooleate; Soluble in alcohol	Excellent antioxidant in vegetable oils; Does not discolor in the presence of metals; Little odor; Poor carry-through in baking and frying	
Trihydroxybutyrophenone (THBP)	196	Tan powder	149–153	—	Slightly soluble in water; Moderately soluble in fats; Soluble in alcohol, propylene glycol, paraffin; Insoluble in water	Synergistic with other antioxidants; Used in packaging materials; Turns brown in the presence of metals; Used extensively in animal rations	172.190
Ethoxyquin	217.31	Yellow liquid		123–125	Soluble in most organic solvents	Effective in pigment retention	172.140

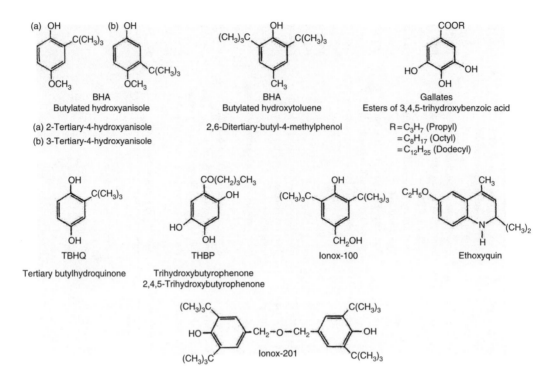

FIGURE 15.2 Structures of some synthetic antioxidants.

4. 6-Ethoxy-1,2-Dihydro-2,2,4-Trimethylquinoline (Ethoxyquin)

Ethoxyquin may be used as an antioxidant for the preservation of chili powder, paprika, and ground chili at levels not to exceed 100 ppm (CFR 21, 172.140). It can be used in animal products at much lower levels. A primary role of ethoxyquin in the feed industry is the protection of carotenoids. Similarly, it stabilizes the color in paprika and chili powder. In oils, it exists primarily in the form of a radical that acts as a free radical terminator. Dimerization of this radical occurs in oil and will inactivate the antioxidant.

5. Gallates

Propyl gallate, octyl gallate, and dodecyl gallate are approved for use as antioxidants in foods. PG, octyl gallate, and dodecyl gallate are, respectively, the n-propyl, n-octyl, and n-dodecyl esters of 3,4,5-trihydroxybenzoic acid. Commercially, PG is the only gallate used in substantial quantity. A slightly water-soluble, white crystalline powder, PG is used widely in foods for which lipid-soluble BHA, BHT, and TBHQ are not suitable. PG is not stable at high temperatures, degrading at 148°C, and it is not suitable for frying applications. Octyl and dodecyl gallate are more lipid soluble, more heat stable, and have better carry-through. Gallates are sold as mixtures with metal chelators because they will form undesirable, dark-colored complexes with iron and copper [3]. Gallates act synergistically with primary antioxidants and some secondary antioxidants, and are often included in mixed antioxidant preparations.

6. Tocopherols

α-, β-, γ-, δ-Tocopherols and the corresponding tocotrienols (vitamin E homologs) are natural, monophenolic antioxidant constituents of vegetable oils. Tocopherols are described in detail in Section III.B.

7. Erythorbic Acid and Ascorbyl Palmitate

Erythorbic acid (D-ascorbic acid) is often used as an antioxidant in fruits and as a curing accelerator in cured meats [2]. It is highly soluble in water and insoluble in oil. It has Generally Recognized as Safe (GRAS) status with the U.S. Food and Drug Administration, but unlike L-ascorbic acid, it is not a natural constituent of foods. It has minimal vitamin C activity.

Ascorbyl palmitate and ascorbyl stearate are synthetic derivatives of ascorbic acid. Ascorbyl palmitate is used in fat-containing foods because its solubility in hydrophobic media is superior to that of ascorbic acid and its salts; it is still fairly insoluble, however, and requires the aid of solubilizing magnets and/or high temperatures for solubilization [3]. Ascorbyl palmitate is usually used in combination with tocopherols. Ascorbyl palmitate has GRAS status, and the United States imposes no restrictions on usage levels. Ascorbyl palmitate is hydrolyzed by the digestive system to provide nutritionally available ascorbic acid and palmitic acid, but health claims cannot be made for its vitamin C contribution [13].

B. NATURAL ANTIOXIDANTS

Extensive research has been dedicated to identification of antioxidants from various natural sources. Ascorbic acid and tocopherols are the most important commercial natural antioxidants. Other sources of natural antioxidants include carotenoids, flavonoids, amino acids, proteins, protein hydrolysates, Maillard reaction products (MRPs), phospholipids, and sterols. Numerous naturally occurring phenolic antioxidants have been identified in plant sources and vegetable extracts. Enzymes also play important roles as antioxidants. Processing of foods can induce the formation of antioxidants. MRPs, protein hydrolysates, fermentation products, and nitrosyl compounds from curing have been reported to possess antioxidant activity. Natural antioxidants allow food processors to produce stable products with "clean" labels that tout all-natural ingredients. However, these products can have several drawbacks, including high usage levels, undesirable flavor and/or color contributions, and lack of stability due to low antioxidant efficiency.

The safety of natural antioxidants should not be taken for granted. Cautions must be heeded, since numerous natural products are potential carcinogens, mutagens, or teratogens, and the safety of many natural compounds with antioxidant activity has not been established. A case in point is nordihydroguaiaretic acid (NDGA), which was used extensively as an antioxidant earlier in this century. NDGA is a natural constituent of the creosote bush, which was removed from GRAS status when unfavorable toxicological results were reported. Regardless of the politics surrounding the issue of safety and "natural" additives, natural antioxidant products are commercially important and desired by the consumer. The main advantage of substances naturally present in foods is that the burden of proof of safety may be less rigorous than that required for synthetic products. No safety testing is required if the antioxidant is a natural constituent of GRAS ingredients. In addition, some natural antioxidants derived from spices, herbs, and Maillard reactions can be listed as flavorants rather than antioxidants, a technical distinction that serves to exempt the substances from safety testing requirements.

1. Tocopherols and Tocotrienols

Tocopherols and tocotrienols (Figure 15.3) comprise the group of chromanol homologs that possess vitamin E activity in the diet. They are natural monophenolic compounds with varying antioxidant activities. Eight naturally occurring homologs are included in the vitamin E family [12]. They are fat-soluble 6-hydroxychroman compounds. The α-, β-, γ-, and δ-tocopherols are characterized by a saturated side chain consisting of three isoprenoid units. The corresponding tocotrienols (α, β, γ, and δ) have double bonds at the $3'$, $7'$, and $11'$ positions of the isoprenoid side chain. Only *RRR* isomers are found naturally. Synthetic α-tocopherol (all-*rac*-α-tocopherol) consists of eight stereoisomers found in equal amounts in the synthetic mixture. Biologically, *RRR*-α-tocopherol is the most active vitamin E homolog. Tocopherols and tocotrienols are widely distributed in the plant

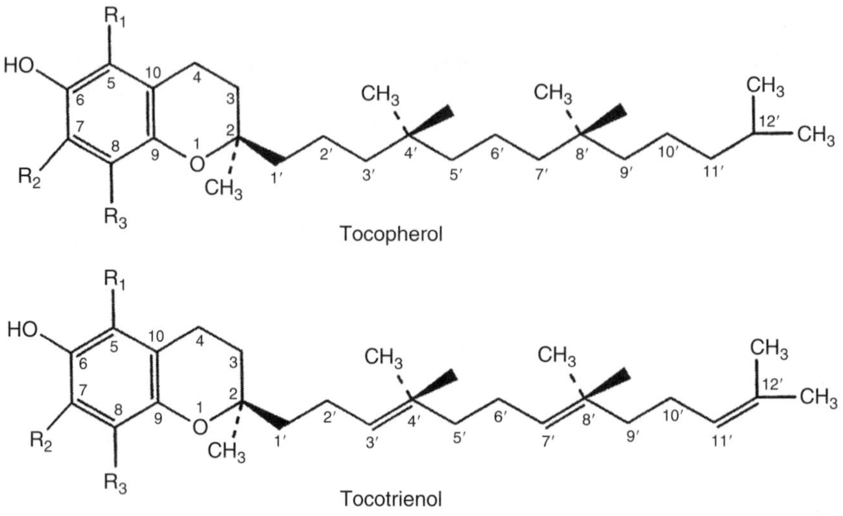

FIGURE 15.3 Structures of tocopherols and tocotrienols.

Tocopherol or Tocotrienol	R1	R2	R3
α-5,7,8-Trimethyl	CH3	CH3	CH3
β-5,8-Dimethyl	CH3	H	CH3
γ-7,8-Dimethyl	H	CH3	CH3
δ-8-Methyl	H	H	CH3

kingdom, with vegetable oils providing the most concentrated source of vitamin E. Tocotrienols are less common but are present in palm oil, rice bran oil, cereals, and legumes. Palm oil has a unique vitamin E profile, providing tocotrienols in higher concentrations than other food sources. Tocopherols and tocotrienols are retained throughout the edible oil refining process, although there is some loss during the deodorization step. Worldwide, the main commercial source of natural tocopherols is in the soybean oil refining industry. Commercial natural antioxidant preparations prepared from soybean oil typically consist of >80% γ- and δ-tocopherol [2]. Synthetic tocopherols are commercially available and vary in isomeric form. Tocopherols provide a useful natural antioxidant source for foods marketed under an "all-natural" label. They are permitted in foods according to GMP regulations (21 CFR 182.3890). Natural tocopherols are limited to 0.03% (300 ppm in animal fats) (9 CFR 318.7). Because of the high vitamin E content of most vegetable oils, addition of tocopherols can lead to prooxidant activity [18].

The antioxidative mechanism of α-tocopherol is well understood. α-Tocopherol donates a hydrogen to a peroxy radical resulting in a α-tocopheryl semiquinone radical (Equation 15.16). This radical may further donate another hydrogen to produce methyltocopherylquinone (Equation 15.17) or react with another α-tocopheryl semiquinone radical to produce an α-tocopherol dimer (Equation 15.18). Higher polymeric forms can then form.

$$\text{ROO} \bullet + \alpha\text{-tocopherol} \rightarrow \text{ROOH} + \alpha\text{-tocopheryl semiquinone} \bullet, \qquad (15.16)$$

$$\alpha\text{-Tocopherylsemiquinone} \bullet + \text{ROO} \bullet \rightarrow \text{ROOH} + \text{methyltocopherylquinone}, \qquad (15.17)$$

$$\alpha\text{-Tocopherylsemiquinone} \bullet + \alpha\text{-tocopherylsemiquinone} \bullet \rightarrow \alpha\text{-tocopheryl dimer}. \qquad (15.18)$$

The methyltocopherylquinone is unstable and will yield α-tocopherylquinone. The α-tocopheryl dimer continues to possess antioxidant activity. Numerous other decomposition products with various degrees of antioxidant activity can arise from oxidation of tocopherols [15]. The antioxidant activity of the tocols and tocotrienols increases from α through δ. α-Tocopherol and α-tocotrienol are fully substituted benzoquinone derivatives and are the most effective antioxidants. α-Tocopherol is highly reactive with peroxy radicals and prevents them from participating in propagation reactions. The α-tocopheroxyl radical is stable because a resonance structure forms on the benzoquinone ring. A comprehensive review of the fundamental antioxidant chemistry of tocopherols and tocotrienols has been compiled by Kamal-Eldin and Appelqvist [19].

2. Ascorbic Acid and Ascorbate Salts

L-Ascorbic acid, or vitamin C, is ubiquitous (Figure 15.4) in nature as a component of plant tissues and is produced synthetically in large quantities. Ascorbic acid is attractive as an antioxidant

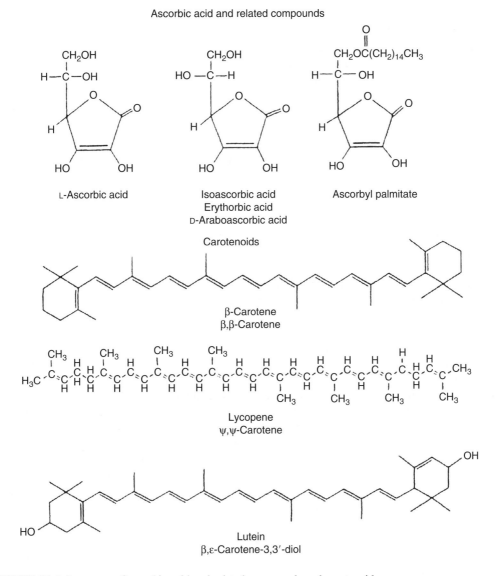

FIGURE 15.4 Structures of ascorbic acid and related compounds and carotenoids.

because it has GRAS status with no usage limits, is a natural or nature-identical product, and is highly recognized as an antioxidant nutrient by the consumer. In some food products, it is also a flavorant and acidulant. However, in foods that are heat-treated, ascorbic acid can participate in nonenzymatic browning and may be degraded through reductone reactions.

Ascorbic acid acts as a primary or a secondary antioxidant. In vivo, ascorbic acid donates hydrogen atoms as a primary antioxidant. Ascorbic acid is also capable of scavenging radicals directly by converting hydroperoxides into stable products. Ascorbic acid is an important antioxidant in plant tissue and is essential for the prevention of oxidative cellular damage by hydrogen peroxide [20]. In foods, ascorbic acid is a secondary antioxidant with multiple functions. Ascorbic acid can scavenge oxygen, shift the redox potential of food systems to the reducing range, act synergistically with chelators, and regenerate primary antioxidants [21]. The vitamin is commonly used as a synergist to donate hydrogen to primary antioxidants such as tocopherol. Tocopheroxyl radicals are reduced back to tocopherol by ascorbic acid (Equation 15.19).

$$\text{Tocopheroxyl radical} + \text{ascorbic acid} \rightarrow \text{tocopherol} + \text{dehydroascorbic acid.} \tag{15.19}$$

Ascorbic acid oxidizes through one- or two-electron transfers that are due to its enediol structure [12,13]. It is a reductone and has a high affinity for oxygen. The 2 and 3 positions of ascorbic acid are unsubstituted. Oxidation of ascorbic acid occurs in two steps, with monohydroperoxide as an intermediate to the formation of dehydroascorbic acid.

Ascorbic acid and its salts (sodium ascorbate and calcium ascorbate) are water soluble and are not applicable as antioxidants for oils and fats. They are used extensively to stabilize beverages. In the United States, ascorbyl palmitate is used in fat-containing foods because its lipid solubility is superior to that of ascorbic acid. Ascorbyl palmitate has GRAS status, and there are no restrictions on usage levels.

3. Carotenoids

Carotenoids (Figure 15.4) are yellow, orange, and red lipid-soluble pigments that are ubiquitous in green plants and in fruits and vegetables. They are 40-carbon isoprenoids or tetraterpenes with varying structural characteristics. The two classes of carotenoids are carotenes and xanthophylls. Carotenes (e.g., β-carotene and lycopene) are polyene hydrocarbons that vary in degree of unsaturation. Xanthophylls (e.g., astaxanthin and canthaxanthin) are synthesized from carotenes by hydroxylation and epoxidation reactions and therefore contain oxygen groups [22]. About 10% of the 600 or so-identified carotenoids have the biological activity of vitamin A and are referred to as provitamin A compounds. β-Carotene is the most abundant of the provitamin A carotenoids found in food and also has the highest vitamin A activity [23]. β-Carotene is a polyene, synthesized from eight isoprene units. It has an intense orange-red color and, as a food additive, is used primarily as a colorant for oils and fats. Carotenoids can act as primary antioxidants by trapping free radicals or as secondary antioxidants by quenching singlet oxygen. Carotenoids are typically secondary antioxidants in foods. However, in the absence of singlet oxygen (low oxygen partial pressure), carotenoids may also prevent oxidation by trapping free radicals and acting as chain-breaking antioxidants.

Singlet oxygen is unstable and can react with lipids to produce free radicals. In the presence of β-carotene, singlet oxygen will preferentially transfer energy to β-carotene to produce triplet state β-carotene (Equation 15.20). The transfer of energy from singlet oxygen to carotenoid takes place through an exchange electron transfer mechanism [22].

Triplet state β-carotene releases energy in the form of heat, and the carotenoid is returned to its normal energy state (Equation 15.21). Carotenoids are very effective quenchers: one carotenoid molecule is able to interact with numerous singlet oxygen. β-Carotene, for example, can quench up to 1000 molecules of singlet oxygen [23].

$$^{1}O_2 + \beta\text{-carotene} \rightarrow {}^{3}\beta\text{-carotene*} + {}^{3}O_2, \tag{15.20}$$

$$^{3}\beta\text{-Carotene*} \rightarrow \beta\text{-carotene} + \text{heat}. \tag{15.21}$$

The ability of carotenoids to quench singlet oxygen is directly related to the number of carbon double bonds in the compound. Carotenoids with nine or more conjugated double bonds are very effective antioxidants. β-Carotene, isozeaxanthin, and lutein are all effective singlet oxygen quenchers. Xanthophylls are not efficient in scavenging singlet oxygen because of the addition of functional groups to the hydrocarbon structure.

Carotenoids are known to scavenge free radicals at low oxygen pressures (<150 mmHg) and to act as primary antioxidants in vitro [10,21]. The conjugated double bonds of carotenoids are very susceptible to attack by peroxy radicals. β-Carotene is capable of reacting with peroxy radicals to produce a resonance-stabilized carotene product (Equation 15.22). The unsaturated structure of carotene allows delocalization of electrons in the radical. This carotene radical can then participate in termination reactions (Equation 15.23) and divert damaging peroxy radicals to less deleterious side reactions [23].

$$\text{Carotene} + \text{ROO•} \rightarrow \text{carotene•}, \tag{15.22}$$

$$\text{Carotene•} + \text{ROO•} \rightarrow \text{termination product}. \tag{15.23}$$

Antioxidant activity of carotenoids has been the subject of much research. Lutein, lycopene, and β-carotene have been reported to inhibit photooxidation of purified oils [24,25]. Annatto color, containing the carotenoid bixin, has been shown to have antioxidant activity [26]. Combinations of carotenoids and tocopherols act synergistically [26,27]. Carotenoids are very unstable, and care must be taken when they are added to processed foods. β-Carotene is difficult to apply as an antioxidant because it is not easily soluble in most common solvents and is very highly reactive [13]. Carotenoid stability is affected by oxygen, heat, pH, light, and the presence of metals. With improved stability, carotenoids could see widespread use as antioxidants.

4. Enzymatic Antioxidants

Glucose oxidase, superoxide dismutase, catalase, and glutathione peroxidase act as antioxidants by removing from the lipid environment either oxygen or highly oxidative species. The enzymes just named act biologically to eliminate cellular free radicals, to keep reactive oxygen species at low concentrations, and to catalyze the destruction of hydrogen peroxide. Thus, they constitute an important biological defense mechanism against free radical damage. Glucose oxidase is an enzyme that removes oxygen by using it to produce gluconic acid and hydrogen peroxide from glucose. Commercial glucose oxidase systems include catalase to hydrolyze the hydrogen peroxide [6].

Superoxide dismutase removes superoxide radicals ($O_2^{•-}$) by converting them to triplet oxygen ($^{3}O_2$) (Equation 15.24):

$$O_2^{•-} + 2H^+ \rightarrow H_2O_2 + {}^{3}O_2. \tag{15.24}$$

Catalase then converts the hydrogen peroxide to water (Equation 15.25):

$$H_2O_2 \rightarrow 2H_2O + {}^{3}O_2. \tag{15.25}$$

Glutathione dehydrogenase catalyzes the oxidation of glutathione (γ-Glu–Cys–Gly, GSH) in the presence of dehydroascorbic acid. Dehydroascorbic acid acts as a hydrogen acceptor and is returned to its active, free radical scavenging form, ascorbic acid. Glutathione peroxidase oxidizes glutathione in the presence of hydroperoxide. The sulfhydryl group on one GSH will react with the sulfhydryl group of another GSH to produce oxidized glutathione (GSSG) with a disulfide bond (Equation 15.26) [7].

$$ROOM + 2GSH \rightarrow ROH + H_2O + GSSG. \tag{15.26}$$

Glucose oxidase, catalase, and superoxide dismutase are used commercially as antioxidants in various foods.

5. Proteins and Related Substances

Numerous amines, amino acids, peptides, and protein hydrolysates have antioxidant activity. A comprehensive list of all the proteinaceous antioxidant sources is not presented here. Amines have been shown to possess antioxidant activity. Recently, spermine and spermidine isolated from fish sources were used to inhibit fish oil oxidation [28]. Numerous amines such as hypoxanthine and xanthine can be readily isolated from marine sources. Amino acids have chelating abilities, but also exhibit antioxidant activity when used alone. Glycine, methionine, histidine, tryptophan, proline, and lysine are effective antioxidants in oil [6].

Proteins and protein hydrolysates possess antioxidative factors. Iron-binding proteins such as ferritin and transferritin have antioxidant function. Histidine-containing peptides such as carnosine [29], as well as synthetic peptides [30] and peptides obtained from protein hydrolysis [31,32], possess antioxidant activity. Glutathione tripeptide (γ-Gly–Cys–Gly, GSH) has antioxidant activity, which is mediated by the sulfhydryl group of cysteine and glutathione peroxidase. The sulfhydryl group on one GSH will react with the sulfhydryl group of another GSH to produce oxidized glutathione (GSSG) with a disulfide bond. Glutathione/glutathione peroxidase was effective in preventing lipid oxidation in a minced mackerel system [33]. In addition, GSH can form disulfides with proteins or other thiols [34]. It is capable of acting as a free radical scavenger by forming a thiyl radical. GSH also donates electrons for the reduction of dehydroascorbate to ascorbic acid in the presence of glutathione dehydrogenase. Ascorbic acid is then returned to its active, free radical scavenging form. The enzyme has been purified from wheat flour, which has a relatively high activity [7].

6. Maillard Reaction Products

MRPs are an excellent example of natural, process-induced oxidation inhibitors that arise as a result of cooking [35]. MRPs are formed during the cooking of low-moisture foods at temperatures above 80°C. They are produced from the reaction of amines and reducing sugars. Lipids, vitamins, and other food constituents also participate in Maillard reactions. MRPs are presumed to be safe because they occur naturally as products in cooked foods. They are often used as bases for flavorants and gravies. The brown pigment associated with MRPs can be an advantage in cooked products where brown color is desirable. Use of MRPs as a source of antioxidants in foods has been intensively studied, but our understanding of the compounds responsible for the antioxidant activity is incomplete and the mechanisms of action are unknown. Identification of the compounds responsible for antioxidant activity has proved difficult because of the complexity of the Maillard reaction, the vast number and variety of MRPs, and the diversity of the model systems that can be studied. MRPs have been shown to have antioxidant activity in model systems as well as in some fat-containing foods [36–39]. Lingnert and Eriksson [39] used processing parameters and the Maillard reaction to prevent oxidation in cookies, and Sato et al. [40] inhibited warmed-over flavor in cooked beef with MRPs.

Because of the conflicting views present in the literature, it is difficult to state conclusively which of the numerous MRPs are actually responsible for antioxidant activity. It is even more difficult to attempt to describe the mechanism of action of these suspected antioxidants. Investigators have shown a correlation between colorless, low-molecular-weight, intermediate MRPs and antioxidant activity [41–44]. Evans et al. [45] showed a correlation between antioxidant activity and reductone levels in MRP mixtures. In contrast, Kirigaya et al. [36] showed that antioxidant activity was proportional to the color intensity of MRP and proposed that nondialyzable, high-molecular-weight melanoidins, which inhibit hydroperoxide and carbonyl compound formation, were responsible for antioxidant activity. By completely oxidizing the reductones in the MRP solution with DPI butanol, these investigators also demonstrated that reductones contributed little antioxidant activity in their model system [36]. Other researchers found a direct relationship between color intensity of Maillard reaction solutions and antioxidant activity [38]. Yamaguchi et al. [38] also showed increases in antioxidant activity with increases in melanoidin formation.

Theories on the mechanism of antioxidant activity of MRPs conflict as well. Kawakishi et al. [46] hypothesized that the protective effects of melanoidins against autoxidation were likely to depend on their ability to chelate metals. Amadori compounds may behave like reductones, which inhibit autoxidation. Eichner [44] believes that MRP intermediates may scavenge oxygen. These conflicting reports likely reflect the different reaction conditions used in the experiments and the multiple antioxidative functions exhibited by MRPs with different mechanisms of action. Because of the overwhelming complexity of even model systems of the Maillard reaction, it would be imprudent to discount any of these theories as to the nature and activity of antioxidant MRPs.

7. Phospholipids

The antioxidative action of phospholipids is not well understood. It is likely that antioxidant activity differs among the various phospholipids as a result of the wide variance in functional groups and structures. Possible actions include regeneration of primary antioxidants, metal chelation, and decomposition of hydroperoxides. Phospholipids have been shown to be synergists. Moreover, phosphatidylcholine (PC), phosphatidylethanolamine (PE), and phosphatidylserine (PS) display antioxidant activity that is possibly linked to chelating ability [47]. Lecithin, once an important commercial antioxidant, now sees limited use because of inefficiency as an antioxidant and poor heat stability. Burkow et al. [48] found lecithin to have antioxidant activity in cod liver oil.

8. Sterols

Sterols have been documented to have antioxidant activity. It is thought that sterols interact with oil surfaces and inhibit oxidation. Sterols may be oxidized at oil surfaces and inhibit propagation by acting as hydrogen donors. Maestroduran and Borjapadilla [49] have reviewed sterol antioxidants and provided information about recent patents.

9. Sulfur Dioxide and Other Sulfites

Sulfites are reducing agents that are weak antioxidants in foods. Sulfites such as sulfur dioxide, sodium sulfite, and sodium, potassium, and metabisulfites are used to prevent flavor and color degradation in beverages and fruits. Sulfites react with molecular oxygen to form sulfates. They also act as reducing agents that promote the formation of phenols from quinones, thereby preventing browning reactions.

10. Gums

Polysaccharides have been studied for their antioxidant effects. Gums are primarily used for their texture-enhancing effects, but they also possess antioxidant activity, which may be due to metal chelation and oxygen consumption, and their viscosity-increasing effects [50,51]. Guaiaconic acids present in the resin of the *Guajacum officinale* L. tree are responsible for the antioxidant activity of this food additive [15]. Guaiac gum was commonly used to preserve refined animal fats but has limited use now. It is not as potent as synthetic antioxidants, has poor heat stability, and is fairly expensive. Xanthan gum, pectin, guar gum, and tragacanth gum are recognized as antioxidants.

11. Antioxidants in Plants

Antioxidant components of plants include vitamin E homologs, carotenoids, proteins, and many other compounds. Plants produce a diverse assortment of phenolic metabolites that readily undergo oxidation and have the potential to minimize effects of autoxidation. Several phenolics in addition to vitamin E (Section III.B.1) have shown potential for use as food antioxidants or are already serving as such. Common plant phenolic antioxidants include gallic acid (as a constituent of polymeric gallotannins and ellagitannins) and protocatechuic acids, phenylpropanoids, and mixed-pathway metabolites such as alkyl ferulates, flavonoids, and suberins [52]. Gallic acid is typically found as a constituent of polymers or glycosidically linked as a galloyl ester. Gallate esters are often used as food antioxidants. Gallo- and ellagitannins are found in the leaves, fruits, pods, and galls of dicotyledonous plants [52]. These are often called hydrolyzable tannins and range in structure from simple esters to polyesters. Phenylpropanoids are derived from cinnamic acid and *p*-coumaric acid or their derivatives. Phenylpropanoids include lignins, lignans, neolignans, monolignols, coumarins, and hydroxycinnamic acids and their derivatives [52]. These compounds are ubiquitous in plants. Lignans are the most abundant phenylpropanoids.

a. Flavonoids and Phenolic Acids

Flavonoids (Figure 15.5) are secondary products of plant metabolism and consist of anthocyanins, catechins, flavones, flavonols, isoflavones, and proanthocyanidins. Several of the flavonoids have antioxidant activity related to their ability to chelate metals. Polyvalent phenol structures in flavonoids can form complexes with metal ions. Flavonoids also act as primary antioxidants and

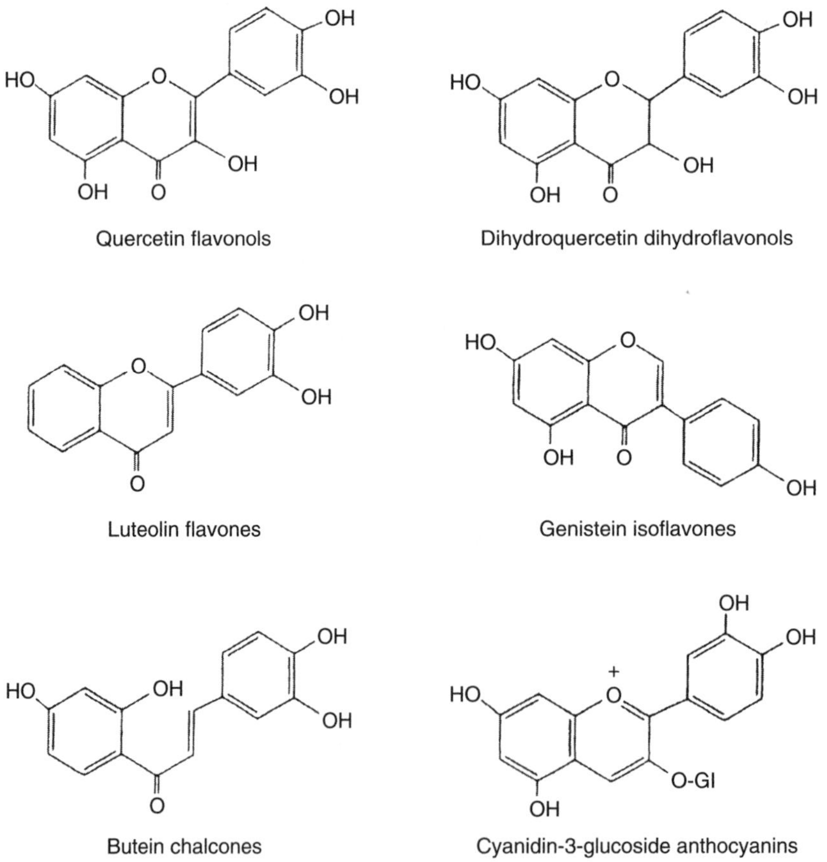

Quercetin flavonols

Dihydroquercetin dihydroflavonols

Luteolin flavones

Genistein isoflavones

Butein chalcones

Cyanidin-3-glucoside anthocyanins

FIGURE 15.5 Structures of some flavonoids.

superoxide anion scavengers [6]. These compounds are responsible for the antioxidant activity reported in many plant and spice extracts. Phenolic acids are structurally related to flavonoids and serve as precursors of their biosynthesis. Phenolic acids such as hydroxycinnamic (caffeic, *p*-coumaric, ferulic, and sinapic acids), hydroxycoumarin (scopoletin) and hydroxybenzoic acids (4-hydroxybenzoic, ellagic, gallic, gentisic, protocatechuic, salicylic, and vanillic acids) are phenolic compounds that can form metal complexes. Antioxidant activity of these compounds varies greatly and is also dependent on the food system. Hydroxycinnamic acid esters were found to be more active than the free acids in model systems involving linoleic acid [53].

Flavones (apigenin, chrysoeriol, diosmetin, isovitexin, luteolin, and nobiletin) and flavonols (gossypetin, isorhamnetin, kaempferol, myricetin, robinetin, and quercetin) occur in fruits as glycosides. These compounds are also prevalent in vegetables, tea, and wine [54]. Quercitin has gained attention as a very potent antioxidant [55]. It has been shown to be very effective in linoleic model systems [56]. Phenolic acids, flavonoids, and other phenolics have potential as food antioxidants. Contents of specific phenolic compounds in plants can be very low, requiring large amounts of raw material to obtain sufficient amounts of these antioxidants. Usage can be limited because the compounds are often present in the form of glycosides and are not soluble in oil. In addition, some flavonoids are toxic.

b.   Other Natural Sources of Antioxidants

Numerous plants have been identified as sources of phenolic compounds with antioxidant activity. The list of natural antioxidants is growing as a result of the amount of research that is conducted to isolate and identify these compounds in plants. A comprehensive listing of all the sources of antioxidant compounds identified in plant materials is beyond the scope of this chapter. Recent reviews by Pratt [56] and Pratt and Hudson [57] contain detailed information on numerous phenolic compounds. Some recent research findings are given later. The diversity of plant sources and compounds considered to be antioxidants is evident even from this brief description.

For example, flavonoid derivatives have been recently identified as potent antioxidants found in apples [58,59] and chrysanthemum [60]. Antioxidant activity has been correlated to flavonoids found in rice, buckwheat, barley, and malt [61–64]. Other examples of antioxidants include tannins from bark, lignans from papua mace, and capsaicin from peppers. Compounds with antioxidant activity isolated from high alpine plant species [65], marine sources [66], and wood smoke [2] have also been characterized.

Tea. Tea extracts are a source of natural antioxidants. Tea catechins (Figure 15.6) have potent antioxidant activity [67–71]. Extracts of green and black tea contain epicatechin, epicatechin gallate, epigallocatechin, epigallocatechin gallate, and gallocatechins [72]. During the fermentation process required to produce black tea, catechins are oxidized to produce flavins and flavin gallates that have antioxidant activity [73]. Tea antioxidants have been patented for use in several food products [74–76].

Sesame seed compounds. Several antioxidative compounds have been isolated and identified from sesame seed oil (Figure 15.7). The compounds are lignan phenols such as pinoresinol, sesaminol, sesamol (and its dimer), and sesamolinol [77–79]. Sesame seed oil has a much longer shelf life than many edible oils because of the presence of these phenolic compounds [80].

Soybean. Soybeans have several antioxidative constituents that include soy proteins [81]. Soy and other vegetable hydrolysates contain phenolic compounds with antioxidant activity. Fermented soy products contain isoflavones and genistein, which inhibit oxidation. Diets containing soy provide genestin, which is a phytoestrogen. Dietary estrogens are structurally similar to endogenous estrogens and mimic their action by binding to the estrogen receptor [82]. Dietary estrogen can produce biological effects similar to those of the endogenous estrogen. The role of dietary estrogens such as genistein in preventing or enhancing the progression of chronic diseases, such as breast cancer, heart disease, or bone loss, has not been clarified at this time. However, the interest in the biological effects of dietary estrogens is great both in the research community and among consumers.

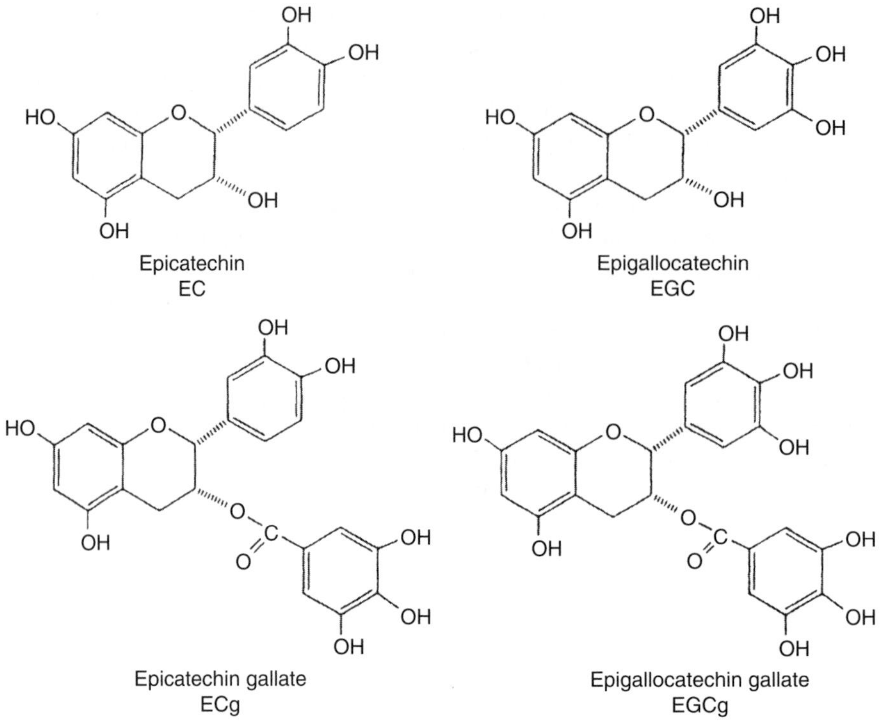

FIGURE 15.6 Structures of green tea antioxidants.

Herbs and spices/spice extracts. Spices and herbs have been used as flavorants for thousands of years. The antioxidant activity of spices and herbs is thought to be primarily due to the presence of phenolic compounds and especially phenolic acids and flavonoids. The strong flavor of spices and herbs precludes their use in many food products. Researchers have tried to identify and isolate specific antioxidant components of spices and herbs that do not contribute undesirable flavor or color to foods. To date, only rosemary and sage (*Perilla* plants) are commercially available as flavorless, odorless, and colorless antioxidant extracts. Thyme, fenugreek [83], and turmeric [84] possess antioxidative components and may provide sources for commercial products.

Rosemary antioxidants (Figure 15.8) have been used in processed foods for decades. Currently, concentrated rosemary extracts are available that do not impart flavor or color to foods. The use of these products is increasing significantly with the rising consumer demand for natural food additives [2]. These extracts are substantially more expensive than synthetic alternatives and require higher usage levels because of lower efficiencies. The diterpene, phenolics, carnosol, and carnosic acid have been shown to be major antioxidant constituents of rosemary extracts [85,86]. These compounds are as active as α-tocopherol in bulk oils [87]. Carnosic acid was even more potent than BHA and BHT in soybean oil [88].

12. Nutritional Aspects

Antioxidant nutrients including vitamin C (ascorbic acid), vitamin E, carotenoids (such as β-carotene), and natural phenolic constituents of foods are touted as free radical scavengers that may act to prevent cancer, heart disease, and cataracts [89]. Some researchers suggest that synthetic antioxidants in the diet may also prevent cancer formation. Free radicals, produced in the body as a natural product of oxidative reactions, can cause oxidation of cell lipids and DNA damage

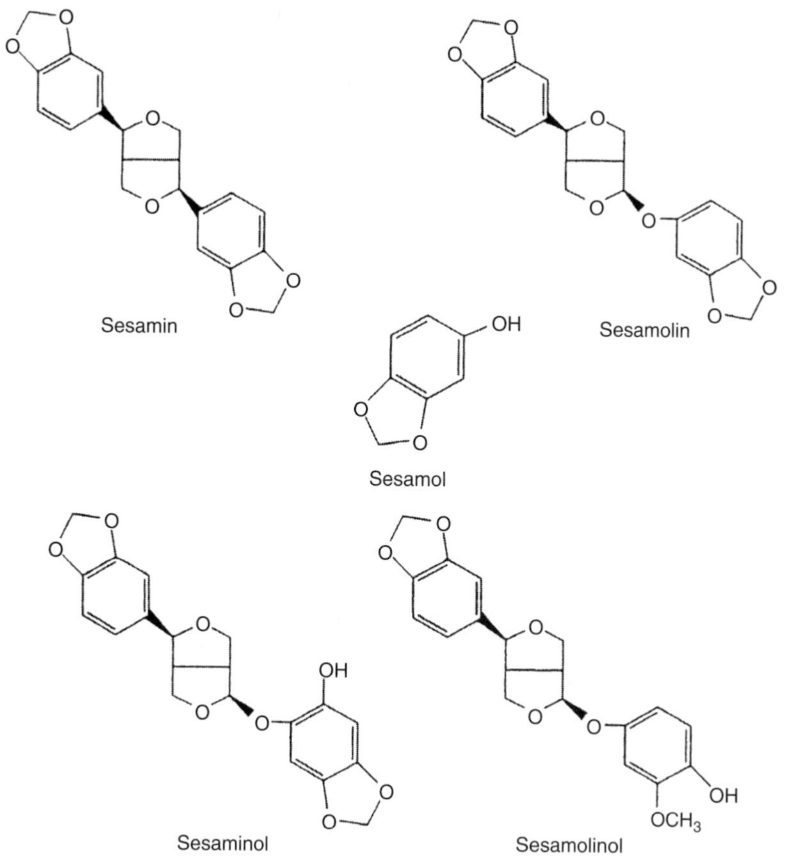

FIGURE 15.7 Structures of sesame antioxidants.

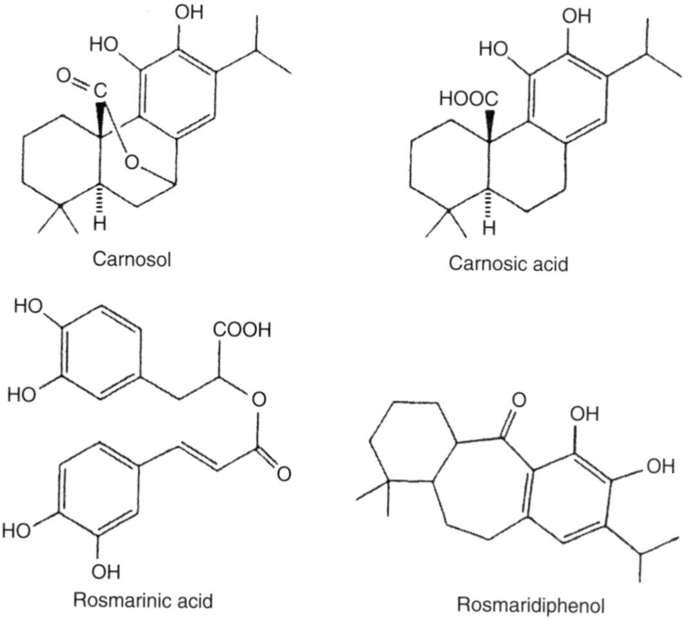

FIGURE 15.8 Structures of rosemary antioxidants.

that may lead to serious diseases. Dietary antioxidants are thought to scavenge these free radicals, prevent them from damaging cells and DNA, and possibly reduce oxidized fatty acids or mutagens that lead to heart disease or cancer. Epidemiological studies have shown trends suggesting that antioxidants may be beneficial in disease prevention. However, a cause-and-effect relationship cannot be determined in humans because lifestyle factors interfere with interpretation of research results.

REFERENCES

1. W.W. Nawar. Lipids. In: *Food Chemistry*, 2nd edn. (O.R. Fennema, ed.). Marcel Dekker, New York, 1985, p. 139.
2. J. Giese. Antioxidants: Tools for preventing lipid oxidation. *Food Technol.* 50:72 (1996).
3. P.P. Coppen. The use of antioxidants. In: *Rancidity in Foods* (J.C. Alenn and R.J. Hamilton, eds.). Applied Science Publishers, New York, 1983, p. 67.
4. W.L. Porter. Recent trends in food applications of antioxidants. In: *Autoxidation in Food and Biological Systems* (M.G. Simic and M. Karel, eds.). Plenum Press, New York, 1980, p. 143.
5. D.F. Buck. Antioxidants. In: *Food Additives User's Handbook* (J. Smith, ed.). Blackie & Son, London, 1991, p. 1.
6. D. Rajalakshmi and S. Narasimhan. Food antioxidants: Sources and methods of evaluation. In: *Food Antioxidants: Technological, Toxicological, and Health Perspectives* (D.L. Madhavi, S.S. Deshpande, and D.K. Salunkhe, eds.). Marcel Dekker, New York, 1996, p. 65.
7. H.D. Belitz and W. Grosch. *Food Chemistry*. Springer-Verlag, New York, 1987, p. 128.
8. G. Minotti. Sources and role of iron in lipid peroxidation. *Chem. Res. Toxicol.* 6:134 (1993).
9. S.D. Aust, L.A. Morehouse, and C.E. Thomas. Role of metals in oxygen radical reactions. *J. Free Radic. Biol. Med.* 1:3 (1985).
10. M.H. Gordon. The mechanism of antioxidant action in vitro. In: *Food Antioxidants* (B.J.F. Hudson, ed.). Elsevier Applied Science, New York, 1990, p. 1.
11. S.J. Jadhave, S.S. Nimbalkar, A.D. Kulkarni, and D.L. Madhavi. Lipid oxidation in biological and food systems. In: *Food Antioxidants: Technological, Toxicological, and Health Perspectives* (D.L. Madhavi, S.S. Deshpande, and D.K. Salunkhe, eds.). Marcel Dekker, New York, 1996, p. 5.
12. R.R. Eitenmiller and W.O. Landen Jr. Vitamins. In: *Analyzing Food for Nutrition Labeling and Hazardous Contaminants* (I.J. Jeon and W.G. Ikins, eds.). Marcel Dekker, New York, 1995, p. 195.
13. L.E. Johnson. Food technology of the antioxidant nutrients. *Crit. Rev. Food Sci. Nutr.* 35:149 (1995).
14. M.T. Liao and P.A. Seib. Selected reactions of L-ascorbic acid related to foods. *Food Technol.* 41:104 (1987).
15. P. Schuler. Natural antioxidants exploited commercially. In: *Food Antioxidants* (B.J.F. Hudson, ed.). Elsevier Applied Science, New York, 1990, p. 99.
16. M. Namiki. Antioxidants/antimutagens in food. *Crit. Rev. Food Sci. Nutr.* 29:273 (1990).
17. B. Haumann. Antioxidants: Firms seeking products they can label as "natural." *INFORM* 1:1002 (1990).
18. J. Loliger. Natural antioxidants. In: *Rancidity in Foods* (J.C. Alenn and R.J. Hamilton, eds.). Applied Science Publishers, New York, 1983, p. 89.
19. A. Kamal-Eldin and L.A. Appelqvist. The chemistry and antioxidant properties of tocopherols and tocotrienols. *Lipids* 31:671 (1996).
20. C.H. Foyer. Ascorbic acid. In: *Antioxidants in Higher Plants* (R.G. Alscher and J.L. Hess, eds.). CRC Press, Boca Raton, FL, 1993, p. 31.
21. D.L. Madhavi, R.S. Singhal, and P.R. Kulkarni. Technological aspects of food antioxidants. In: *Food Antioxidants: Technological, Toxicological, and Health Perspectives* (D.L. Madhavi, S.S. Deshpande, and D.K. Salunkhe, eds.). Marcel Dekker, New York, 1996, p. 159.
22. K.E. Pallett and A.J. Young. Carotenoids. In: *Antioxidants in Higher Plants* (R.G. Alscher and J.L. Hess, eds.). CRC Press, Boca Raton, FL, 1993, p. 59.
23. S.S. Deshpande, U.S. Deshpande, and D.K. Salunkhe. Nutritional and health aspects of food antioxidants. In: *Food Antioxidants: Technological, Toxicological, and Health Perspectives* (D.L. Madhavi, S.S. Deshpande, and D.K. Salunkhe, eds.). Marcel Dekker, New York, 1996, p. 361.
24. E.C. Lee and D.B. Min. Quenching mechanism of β-carotene on the chlorophyll sensitized photooxidation of soybean oil. *J. Food Sci.* 53:1894 (1988).

25. S.H. Lee and D.B. Min. Effects, quenching mechanisms, and kinetics of carotenoids on the chlorophyll sensitized photooxidation of soybean oil. *J. Agric. Food Chem.* 38:1630 (1990).

26. K.M. Haila, S.M. Lievenon, and M.I. Heinonen. Effects of lutein, lycopene, annatto, and γ-tocopherol on autoxidation of triglycerides. *J. Agric. Food Chem.* 44:2096 (1996).

27. P. Palozza and N.I. Krinsky. β-Carotene and α-tocopherol are synergistic antioxidants. *Arch. Biochem. Biophys.* 297:184 (1992).

28. S. Sasaki, T. Ohta, and E.A. Decker. Antioxidant activity of water-soluble fractions of salmon spermary tissue. *J. Agric. Food Chem.* 44:1682 (1996).

29. G. Kansci, C. Genot, and G. Gandemer. Evaluation of antioxidant effect of carnosine on phospholipids by oxygen-uptake and TBA test. *Sci. Aliments* 14:663 (1994).

30. H.M. Chen, K. Muramoto, F. Yamaguchi, and K. Nokihara. Antioxidant activity of designed peptides based on the antioxidative peptide isolated from digests of a soybean protein. *J. Agric. Food Chem.* 44:2619 (1996).

31. H.M. Chen, K. Muramoto, and F. Yamauchi. Structural analysis of antioxidative peptides from soybean β-conglycinin. *J. Agric. Food Chem.* 43:574 (1995).

32. F. Shahidi and R. Amarowilz. Antioxidant activity of protein hydrolysates from aquatic species. *J. Am. Oil Chem. Soc.* 73:1197 (1996).

33. T.D. Jia, S.D. Kelleher, H.O. Hultin, D. Petillo, R. Maney, and J. Krzynowek. Comparison of quality loss and changes in the glutathione antioxidant system in stored mackerel and bluefish muscle. *J. Agric. Food Chem.* 44:1196 (1996).

34. A. Hausladen and R.G. Alscher. Glutathione. In: *Antioxidants in Higher Plants* (R.G. Alscher and J.L. Hess, eds.). CRC Press, Boca Raton, FL, 1993, p. 1.

35. C.E. Eriksson. Lipid oxidation catalysts and inhibitors in raw materials and processed foods. *Food Chem.* 9:3 (1982).

36. N. Kirigaya, H. Kato, and M. Fujimaki. Studies on antioxidant activity of nonenzymic browning reaction products: 1. Relations of color intensity and reductones with antioxidant activity of browning reaction products. *Agric. Biol. Chem.* 32:287 (1968).

37. H. Lingnert, C.E. Eriksson, and G.R. Waller. Characterization of antioxidative Maillard reaction products from histidine and glucose. In: *The Maillard Reaction in Foods and Nutrition* (G.R. Waller and M.S. Feather, eds.). ACS Symposium Series 215, American Chemical Society, Washington, D.C., 1983, p. 335.

38. Y. Yamaguchi, Y. Koyama, and M. Fujimaki. Fractionation and antioxidative activity of browning reaction products between D-xylose and glycine. *Prog. Food Nutr. Sci.* 5:429 (1981).

39. H. Lingnert and C.E. Eriksson. Antioxidative Maillard reaction products: I. Products from sugars and free amino acids. *J. Food Process. Preserv.* 4:161 (1980).

40. K. Sato, G.R. Hegarty, and H.K. Herring. The inhibition of warmed-over flavor in cooked meats. *J. Food Sci.* 38:398 (1973).

41. C.K. Park and D.H. Kim. Relationship between fluorescence and antioxidant activity of ethanol extracts of a Maillard browning mixture. *J. Am. Oil Chem. Soc.* 60:98 (1983).

42. R.S. Farag, Y. Ghali, and M.M. Rashed. Linoleic acid oxidation catalyzed by Amadori compounds in aqueous media. *Can. Inst. Food Sci. Technol. J.* 15:179 (1982).

43. A. Huyghebaert, L. Vandewalle, and G.V. Landschoot. Comparison of the antioxidant activity of Maillard and caramelisation reaction products. In: *Recent Developments in Food Analysis* (W. Baltes, P.B. Czedik-Eysenberg, and W. Pfannhauser, eds.). Verlag Chemie, Weinheim, 1982, p. 409.

44. K. Eichner. Antioxidative effect of Maillard reaction intermediates. *Prog. Food Nutr. Sci.* 5:441 (1981).

45. C.D. Evans, H.A. Moser, P.M. Cooney, and J.E. Hodge. Amino-hexose-reductones as antioxidants: 1. Vegetable oils. *J. Am. Oil Chem. Soc.* 35:84 (1958).

46. S. Kawakishi, Y. Okawa, and T. Hayashi. Interaction between melanoidin and active oxygen producing system. In: *Trends in Food Science: Proceedings of the 7th World Congress of Food Science and Technology* (A.H. Ghee, L.W. Sze, and F.C. Woo, eds.). Singapore Institute of Food Science and Technology, Singapore, 1987, p. 15.

47. M.F. King, L.C. Boyd, and B.W. Sheldon. Antioxidant properties of individual phospholipids in a salmon oil model system. *J. Am. Oil Chem. Soc.* 69:545 (1992).

48. I.C. Burkow, L. Vikersveen, and K. Saarem. Evaluation of antioxidants for cod-liver oil by chemiluminescence and the Rancimat method. *J. Am. Oil Chem. Soc.* 72:553 (1995).

49. R. Maestroduran and R. Borjapadilla. Antioxidant activity of natural sterols and organic acids. *Grasas Aceites* 44:208 (1993).

50. K. Shimada, H. Muta, Y. Nakamura, H. Okada, K. Matsuo, S. Yoshioka, T. Matsudaira, and T. Nakamura. Iron-binding property and antioxidative activity of xanthan on the autoxidation of soybean oil in emulsion. *J. Agric. Food Chem.* 42:1607 (1994).

51. K. Shimada, H. Okada, K. Matsuo, and S. Yoshioka. Involvement of chelating action and viscosity in the antioxidative effect of xanthan in an oil/water emulsion. *Biosci. Biotech. Biochem.* 60:125 (1996).

52. N.G. Lewis. Plant phenolics. In: *Antioxidants in Higher Plants* (R.G. Alscher and J.L. Hess, eds.). CRC Press, Boca Raton, FL, 1993, p. 135.

53. M. Foti, M. Piattelli, M.T. Baratta, and G. Ruberto. Flavonoids, coumarins, and cinnamic acids as antioxidants in a micellar system. Structure–activity relationship. *J. Agric. Food Chem.* 44:497 (1996).

54. C.A. Rice-Evans and N.J. Miller. The relative antioxidant activities of flavonoids as bioactive components of food. *Biochem. Soc. Trans.* 24:790 (1996).

55. C.A. Rice-Evans, N.J. Miller, P.G. Bolwell, P.M. Bramley, and J.B. Pridham. The relative antioxidant activities of plant-derived polyphenolic flavonoids. *Free Radical Res.* 22:375 (1995).

56. D.E. Pratt. Natural antioxidants from plant material. In: *Phenolic Compounds in Foods and Their Effects on Health* (M.-T. Huang, C.-T. Ho, and C.Y. Lee, eds.). ACS Symposium Series, American Chemical Society, Washington, D.C., 1992, p. 54.

57. D.E. Pratt and B.J.F. Hudson. Natural antioxidants not exploited commercially. In: *Food Antioxidants* (B.J.F. Hudson, ed.). Elsevier Applied Science, New York, 1990, p. 171.

58. T. Ridgway, J. O'Reilly, G. West, G. Tucker, and H. Wiseman. Potent antioxidant properties of novel apple-derived flavonoids with commercial potential as food additives. *Biochem. Soc. Trans.* 24:S391 (1996).

59. N.J. Miller, A.T. Diplock, and C.A. Rice-Evans. Evaluation of the total antioxidant activity as a marker of the deterioration of apple juice oil storage. *J. Agric. Food Chem.* 43:1794 (1995).

60. Y. Chuda, H. Ono, M. Ohnishi-Kameyama, T. Nagata, and T. Tsushida. Structural identification of two antioxidant quinic acid derivatives from Garland. *J. Agric. Food Chem.* 44:2037 (1996).

61. N. Ramarathnam, T. Osawa, M. Namiki, and S. Kawakishi. Chemical studies on novel rice hull antioxidants: I. Isolation, fractionation, and partial characterization. *J. Agric. Food Chem.* 36:732 (1988).

62. N. Ramarathnam, T. Osawa, M. Namiki, and S. Kawakishi. Chemical studies on novel rice hull antioxidants: II. Identificaiton of isovitexin, a C-glycosyl flavonoid. *J. Agric. Food Chem.* 37:316 (1989).

63. B.D. Oomah and G. Mazza. Flavonoids and antioxidative activity in buckwheat. *J. Agric. Food Chem.* 44:1746 (1996).

64. M.N. Maillard, M.H. Soum, P. Boivin, and C. Berset. Antioxidant activity of barley and malt—Relationship with phenolic content. *Lebensm. Wiss. Technol.* 29:238 (1996).

65. B. Wildi and C. Lutz. Antioxidant composition of selected high alpine plant species from different altitudes. *Plant Cell Environ.* 19:138 (1990).

66. T.A. Seymour, S.J. Li, and M.T. Morrissey. Characterization of a natural antioxidant from shrimp shell waste. *J. Agric. Food Chem.* 44:682 (1996).

67. G.C. Yen and H.Y. Chen. Antioxidant activity of various tea extracts in relation to their antimutagenicity. *J. Agric. Food Chem.* 43:27 (1995).

68. Y.B. Tripathi, S. Chaurasia, M. Sharma, S. Shukla, V.P. Singh, and P. Tripathi. Tea—A strong antioxidant. *Curr. Sci.* 68:871 (1995).

69. H. Wiseman, P. Plitzanopoulou, and J. Orielly. Antioxidant properties of ethanolic and aqueous extracts of green tea compared to black tea. *Biochem. Soc. Trans.* 24:S390 (1996).

70. B. Zhao, X. Li, R.G. He, S.J. Cheng, and X. Wenjuan. Scavenging effect of extracts of green tea and natural antioxidants on active oxygen radicals. *Cell Biophys.* 14:175 (1989).

71. G. Cao, E. Sofic, and R.L. Prior. Antioxidant capacity of tea and common vegetables. *J. Agric. Food Chem.* 44:3426 (1996).

72. T.L. Lunder. Catechins of green tea—Antioxidant activity. *ACS Symp. Ser.* 507:114 (1992).

73. N.J. Miller, C. Castelluccio, L. Tijburg, and C. Rice-Evans. The antioxidant properties of theaflavins and their gallate esters—Radical scavengers or metal chelators. *FEBS Lett.* 392:40 (1996).

74. J. Mai, L.J. Chambers, and R.E. McDonald. U.S. Patent 4,891,231 (1989).

75. J. Mai, L.J. Chambers, and R.E. McDonald. U.S. Patent 4,925,681 (1990).

76. P.H. Todd. U.S. Patent 5,527,552 (1996).

77. A. Mimura, K. Takebayashi, M. Niwana, Y. Takahara, T. Osawa, and H. Tokuda. Antioxidative and anticancer components produced by cell-culture of sesame. *ACS Symp. Ser.* 547:281 (1994).

78. Y. Fukuda and M. Namiki. Recent studies on sesame seed and oil. *Nippon Shokuhin Kogyo Gakkaishi* 35:552 (1988).

79. Y. Fukuda, T. Osawa, S. Kawakishi, and M. Namiki. Chemistry of lignan antioxidants in sesame seed and oil. *ACS Symp. Ser.* 547:264 (1994).

80. M. Namiki. The chemistry and physiological functions of sesame. *Food Rev. Int.* 11:281 (1995).

81. S.Y. Wu and M.S. Brewer. Soy protein isolate antioxidant effect on lipid-peroxidation of ground-beef and microsomal lipids. *J. Food Sci.* 59:702 (1994).

82. B. Helferich. Dietary estrogens: A balance of risks and benefits. *Food Technol.* 50:158 (1996).

83. N.S. Hettiarachchy, K.C. Glenn, R. Gnanasambandam, and M.G. Johnson. Natural antioxidant extract from fenugreek (*Trigonella foenumgraecum*) for ground-beef patties. *J. Food Sci.* 61:516 (1996).

84. R. Selvam, L. Subramanian, R. Gayathri, and N. Angayarkanni. The antioxidant activity of turmeric (*Curcuma longa*). *J. Ethnopharmacol.* 47:59 (1995).

85. Q. Chen, H. Shi, and C.T. Ho. Effects of rosemary extracts and major constituents on lipid oxidation and soybean lipoxygenase activity. *J. Am. Oil Chem. Soc.* 69:999 (1992).

86. M.E. Cuvelier, H. Richard, and C. Berset. Antioxidant activity and phenolic composition of pilot-plant and commercial extracts of sage and rosemary. *J. Am. Oil Chem. Soc.* 73:645 (1996).

87. A.I. Hopia, S.-W. Huang, K. Schwartz, J.B. German, and E.N. Frankel. Effect of different lipid systems on antioxidant activity of rosemary constituents carnosol and carnosic acid with and without α-tocopherol. *J. Agric. Food Chem.* 44:2030 (1996).

88. S.L. Richheimer, M.W. Bernart, G.A. King, M.C. Kent, and D.T. Bailey. Antioxidant activity of lipid-soluble phenolic diterpenes from rosemary. *J. Am. Oil Chem. Soc.* 73:507 (1996).

89. D.L. Madhavi, S.S. Deshpande, and D.K. Salunkhe. Summary, conclusions, and future research need. In: *Food Antioxidants: Technological, Toxicological, and Health Perspectives* (D.L. Madhavi, S.S. Deshpande, and D.K. Salunkhe, eds.). Marcel Dekker, New York, 1996, p. 471.

16 Tocopherol Stability and Prooxidant Mechanisms of Oxidized Tocopherols in Lipids

Hyun Jung Kim and David B. Min

CONTENTS

I. INTRODUCTION

Lipid oxidation in foods is a free radical chain reaction of unsaturated fatty acids. The high reactivity of the carbon double bonds in unsaturated fatty acids makes these substances primary targets for free radical reactions. The chain reaction includes initiation, propagation, and termination step. The propagation step is a slow step and responsible for the autocatalytic nature of oxidation [1,2]. The reaction between free radicals and lipids in the food system contributes to the deterioration of flavor, discoloration, and the formation of potential toxic products [3–5].

Lipid oxidation in foods can be prevented by adding antioxidants. Antioxidants delay the onset of oxidation or slow the rate of lipid oxidation in foods [6]. The antioxidant activity depends on temperature, pH, the degree and number of unsaturated fatty acids, the availability of oxygen and transition metal ions [6–9].

Tocopherols are the most common and major lipid-soluble antioxidants in nature [10–13]. However, tocopherols themselves are rapidly degraded in the presence of molecular oxygen and free radicals in foods [7] and lose the antioxidant activity or become prooxidants [14–18]. The concentration of tocopherol influences the antioxidant activity of tocopherols in foods. The antioxidant activity is generally greatest at low concentrations and decreases or becomes prooxidant at high concentrations [14,17]. The ability of tocopherol to be an antioxidant, neutral, or prooxidant is related to their complex function, chemical behavior, and environment. This chapter covers the important chemical mechanisms involved in the prooxidant effect of tocopherol.

II. TOCOPHEROL AS ANTIOXIDANT

Tocopherols compete with unsaturated fatty acids for lipid peroxy radicals. Tocopherols donate a hydrogen atom at the 6-hydroxy group on its chroman ring to lipid peroxy radical. Tocopherol (TH) with a reduction potential of 300–400 mV easily donates hydrogen to lipid peroxy radical (ROO·) with a reduction potential of 1000 mV and produces lipid hydroperoxide (ROOH) and tocopheroxy radical (T·).

$$TH + ROO· \rightarrow T· + ROOH.$$

Tocopheroxy radicals (T·) are resonance structures, which are more stable than lipid peroxy radicals (ROO·). The reaction rate of α-tocopherol with lipid peroxy radical is 10^7 $M^{-1}s^{-1}$ [19,20] and 10^5–10^6 times faster than that of unsaturated lipid with lipid peroxy radical [19,21]. Tocopherols take away the radicals from the oxidizing fatty acids to prevent from further radical chain reactions. One tocopherol molecule can protect about 10^3–10^8 polyunsaturated fatty acid molecules at low peroxide value [9]. Tocopheroxy radical (T·) can interact with other compounds or each other depending on the lipid oxidation rates. Tocopheroxy radicals may react with lipid peroxy radicals and form nonradical products [9].

$$T· + ROO· \rightarrow T - OOR.$$

The effectiveness of tocopherols as antioxidants depends on the chemical and physical characteristics.

A. STRUCTURAL CHARACTERISTICS OF TOCOPHEROL

Tocopherols comprise four types of 6-hydroxy chromanol homologs that possess vitamin E activity in diet [22]. α-, β-, γ-, and δ-Tocopherols consist of one hydroxyl group at 6 position and one or more methyl groups at the 5, 7, or 8 position of the chromanol ring with a 16-carbon saturated phytyl group as shown in Figure 16.1. The phytyl group has three chiral centers at carbons 2, 4', and 8'. All naturally occurring tocopherol isomers have R configuration at all three positions in their phytyl group, RRR-α-tocopherol (2D, 4'D, 8'D) [7]. The positions of methyl groups on the chromanol ring determine the forms of α-, β-, γ-, and δ-tocopherols (Figure 16.1). α-Tocopherol has three methyl groups on the chromanol ring at the 5, 7, and 8 positions, whereas β- and

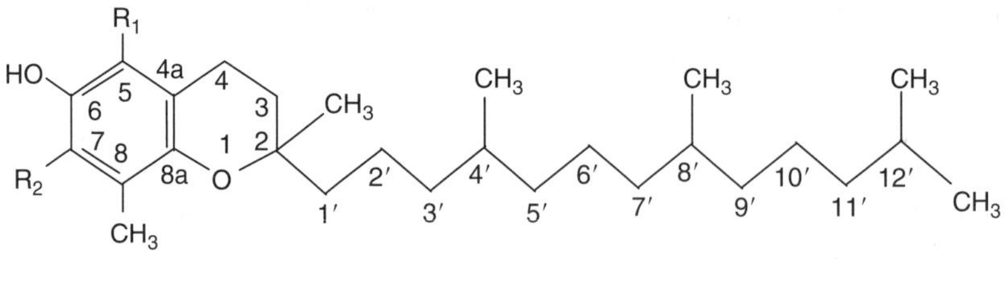

FIGURE 16.1 Structure of tocopherol.

Trivial Name	Chemical Name	R_1	R_2
α-Tocopherol	5,7,8-Trimethyltocopherol/tocotrienol	CH_3	CH_3
β-Tocopherol	5,8-Dimethyltocopherol/tocotrienol	CH_3	H
γ-Tocopherol	7,8-Dimethyltocopherol/tocotrienol	H	CH_3
δ-Tocopherol	8-Methyltocopherol/tocotrienol	H	H

γ-tocopherols have two at the 5 and 8 positions and 7 and 8 positions, respectively. δ-Tocopherol have only one methyl group at the 8 position. The chromanol ring only is responsible for the antioxidant activity of tocopherols. The phytyl group which is very lipophilic has no effect on the chemical reactivity of antioxidant but it is important for proper positioning in foods [23].

The antioxidant activity of α-, β-, γ-, and δ-tocopherols in vegetable oil is in the order of α-tocopherol > β- or γ-tocopherol > δ-tocopherol [17,24]. Isnardy et al. [25] reported that α-tocopherol degraded faster than γ- or δ-tocopherol in purified rapeseed oil. α-Tocopherol was completely destroyed during the oxidation of soybean oil, whereas most of γ- and δ-tocopherols were remained [26]. Tocopherols are the chain-breaking antioxidants by donating their phenolic hydrogen to lipid free radicals [5]. As the antioxidant activity or hydrogen-donating ability of tocopherol is high, the stability of tocopherol in vegetable oil is low [14]. The reaction rates of α-, γ-, and δ-tocopherols with lipid peroxy radical at 30°C were 2.4×10^6, 1.6×10^6, and $0.7 \times 10^6 \, M^{-1}s^{-1}$, respectively [11]. α-Tocopherol had the highest antioxidant activity in vegetable oil with the least stability during storage [9,27].

B. BOND DISSOCIATION ENERGY

The antioxidant activities of α-, β-, γ-, and δ-tocopherols can be explained by bond dissociation energy (BDE). The BDE measures the bond strength in a chemical bond. The BDE in antioxidant reactivity indicates the driving force for the hydrogen transfer from the phenolic antioxidant to lipid radical. The BDE depends on the strength of O–H bond in the phenolic antioxidant. The lower BDE means the easier donation of hydrogen to lipid peroxy radical by cleaving the O–H bond. The BDE of water, phenol, and α-, β-, γ-, and δ-tocopherols are shown in Table 16.1. The O–H bond energy for phenol (87 kcal/mol) is considerably lower than water (119.3 kcal/mol). The BDE of hydroxyl group on the chromanol ring of α-, β-, γ-, and δ-tocopherols are 75.8, 77.7, 78.2, and 79.8 kcal/mol, respectively (Table 16.1) [30]. The O–H bond of hydroxyl group in α-tocopherol can be more easily cleaved to donate a hydrogen atom to lipid peroxy radical than that of β-, γ-, or δ-tocopherol. α-Tocopherol is a fully methylated chromanol (Figure 16.1), which is more sterically hindered than β-, γ-, or δ-tocopherol. Sterically hindered phenols are the most active antioxidants for scavenging lipid peroxy radicals by easily donating hydrogen atoms [30]. The fully methylated α-tocopherol having lower BDE could be more potent as a hydrogen donor than β-, γ-, or δ-tocopherol [9,30]. The BDE also indicates the stabilization of the resulting radical. The most stable radicals can be derived from their parent compounds with the lowest BDE, which will be the

---

TABLE 16.1
Bond Dissociation Energy (BDE) of Water, Phenol, and Tocopherols

Compound	BDE (kcal/mol)
Water[a]	119.3
Phenol[b]	87.0
α-Tocopherol[c]	75.8
β-Tocopherol[c]	77.7
γ-Tocopherol[c]	78.2
δ-Tocopherol[c]	79.8

[a] Ref. [28].
[b] Ref. [29].
[c] Ref. [30].

TABLE 16.2

Standard One-Electron Reduction Potentials for Common Free Radicals

Compounds	Half-Cell	Standard Reduction Potential (mV)
·OH	H^+/H_2O	2310
RO·[a,b]	H^+/ROH	1600
ROO·[a,b]	$H^+/ROOH$	1000
R·[a]	H^+/RH	600
α-Tocopheryl·	H^+/α-Tocopherol	270
β-Tocopheryl·	H^+/β-Tocopherol	345
γ-Tocopheryl·	H^+/γ-Tocopherol	350
δ-Tocopheryl·	H^+/δ-Tocopherol	405

Sources: Adapted from Kamal-Eldin, A. and Appelqvist, L.A., *Lipids*, 31, 671, 1996; Choe, E. and Min, D.B., *J. Food Sci.*, 70, R142, 2005.

[a] RO·, ROO·, and R· are lipid alkoxyl, peroxy, and alkyl radical, respectively.

[b] RO· and ROO· forms in tocopherol are tocopheryl oxy radical and tocopheryl peroxy radical, respectively, and the structures are shown in Figures 16.2 through 16.5.

most efficient hydrogen atom donors. α-Tocopheroxy radical will be the most stable radicals among tocopherol homologs.

C.   Reduction Potential

One-electron reduction potential is a thermodynamic property of the oxidized antioxidant. The reduction potential can provide a rank of antioxidants as reducing agents. As the reduction potential of a compound is low, the compound can donate an electron or hydrogen atom to the other which has a high reduction potential [1]. The standard one-electron reduction potentials of common free radicals and α-, β-, γ-, and δ-tocopherols are listed in Table 16.2. The reduction potentials of lipid alkyl, peroxy, and alkoxyl radicals are 600, 1000, and 1600 mV, respectively. To work as an antioxidant and prevent lipid oxidation, the reduction potential of a free radical scavenger should be lower than 600 mV which is a reduction potential of lipid alkyl radical. Tocopherols have lower standard one-electron reduction potentials than lipid alkyl, peroxy, and alkoxyl radicals (Table 16.2). Therefore, tocopherols can be easily oxidized by donating a hydrogen atom to lipid peroxy radical which will become a reduced form [1]. The reduction potentials of α-, β-, γ-, and δ-tocopherols are 270, 345, 350, and 405 mV, respectively. α-Tocopherol is the best reducing agent to donate a hydrogen atom.

III.   TOCOPHEROL AS PROOXIDANT

Tocopherols can act as antioxidants or prooxidants depending on temperature, pH, concentration, the presence of other compounds near to tocopherols, and their chemical characteristics [6–9].

The prooxidant effect of α-tocopherol has been proposed to be induced by hydrogen abstraction between the tocopheroxyl radical (T·) and lipid molecules or lipid hydroperoxides [31].

$$RH + T· \rightarrow TH + R·,$$

$$ROOH + T· \rightarrow TH + ROO·.$$

The reaction rate constants of α-tocopheroxyl radicals with polyunsaturated fatty acids or with hydroperoxides of polyunsaturated fatty acids were reported in the range of 10^{-5}–10^{-1} $M^{-1}s^{-1}$ [9].

The reactions are very slow compared with the antioxidative reactions of α-tocopherol and the termination reaction of lipid autoxidation. The resulting antioxidant radical must not propagate the chain reaction and will not undergo hydrogen abstraction reactions or react with oxygen to form another peroxy radical. Thus, these reactions cannot totally explain the prooxidant effect of α-tocopherol.

The alternative prooxidant mechanism of tocopherol has been suggested to be more significant in the presence of high levels of hydroperoxides [32]. This reaction mechanism involves hydrogen bonding between tocopherol (TO–H) and lipid hydroperoxide (RO–O–H). The peroxide abstracts a hydrogen from the tocopherol and cleaves the O–O bond in the peroxide. As a result of this reaction, the alkoxyl radical (RO·) is formed and then propagates lipid oxidation due to its high reduction potential.

$$\text{TOH} \;+\; \text{ROOH} \longrightarrow \text{ROO}\!\!\begin{smallmatrix}\text{H}\\ \cdots\\ \text{H}\end{smallmatrix} \longrightarrow \text{RO}\cdot \;+\; \text{H}_2\text{O} \;+\; \text{TO}\cdot$$

Tocopherols are degraded in the presence of molecular oxygen and produce oxidized products resulting in the loss of antioxidant activity. The oxidized tocopherol products act as prooxidants in lipids. The addition of oxidized α-, γ-, and δ-tocopherols to soybean oil lowered the oxidative stability of soybean oil [33]. The oxidation of tocopherols by strong oxidizing agents such as chromic acid, nitric acid, and ferric chloride generally produce lactones, quinines, and many degradation products [9]. α-Tocopherolquinone, α-tocopherolhydroquinone, 4a,5-epoxy-α-tocopherolquinone, and 7,8-epoxy-α-tocopherolquinone have been reported as the oxidation products of α-tocopherol in beef and bovine muscle microsomes [34,35], in triolein [8] and triolein and tripalmitin mixture [36], and in fish muscle [37]. During the formation of α-tocopherolquinone, α-tocopherolhydroquinone, 4a,5-epoxy-α-tocopherolquinone, and 7,8-epoxy-α-tocopherolquinone from α-tocopherol oxidation, many intermediate compounds could be produced. Rietjens et al. [18] suggested that increased levels of oxidized α-tocopherol could result in increased levels of inter-mediate radicals, which can initiate lipid oxidation.

Tocopherol concentration is important whether it can be an antioxidant or prooxidant. Gener-ally, the antioxidant activity of tocopherol is greatest at lower concentration and decreases at higher concentration. The optimum concentrations of α-, γ-, and δ-tocopherols to increase the oxidative stability of oil were 100, 250–500, 500–1000 ppm, respectively [14,17]. α-Tocopherol at high concentrations acts as a prooxidant during the oxidation of lipids resulting in the increase of hydroperoxide levels and conjugated dienes [14,15,17,38]. It could be assumed that the higher the concentration of tocopherols in lipids, the higher the amounts of intermediate radicals formed from tocopherol oxidation during the storage of lipids. These intermediate compounds such as alkyl, alkoxyl, and peroxy radicals can initiate the processes of lipid oxidation. It is important to prevent the oxidation of tocopherol and remove the oxidized tocopherols. The detailed prooxidant mechanisms of oxidized tocopherols will be discussed.

IV. PROOXIDANT MECHANISM OF OXIDIZED TOCOPHEROL

Oxidized tocopherols act as prooxidants [33]. α-Tocopherolquinone, α-tocopherolhydroquinone, epoxy-α-tocopherolquinone, and α-tocopherol hydroperoxide formed during the oxidation of α-tocopherol were identified during the oxidation of α-tocopherol [8,34,36,37]. The mech-anisms for the formation of α-tocopherolquinone, 4a,5-epoxy-α-tocopherolquinone, 7,8-epoxy-α-tocopherolquinone, and α-tocopherolquinone hydroperoxide to study the prooxidant mechanism of oxidized tocopherols are postulated in Figures 16.2 through 16.5. The standard reduction potentials for common free radicals are shown in Table 16.2. The reaction rates of lipid oxidation in the presence of tocopherol are shown in Table 16.3. When the published data on the reaction rates and

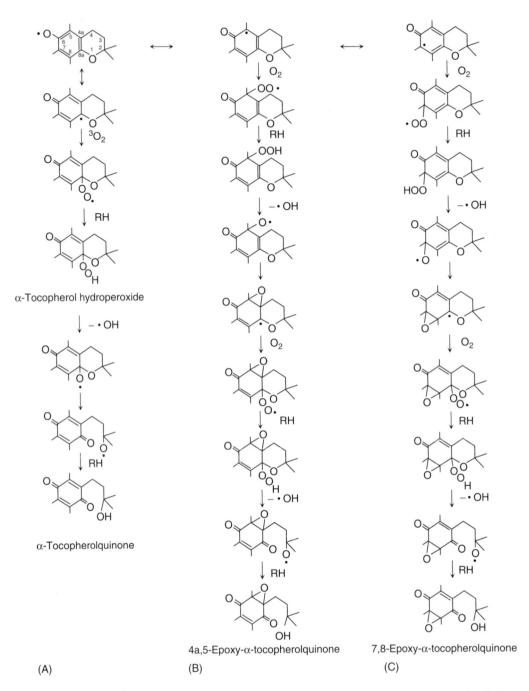

FIGURE 16.2 Possible mechanism for the formation of α-tocopherolquinone (A), 4a,5-epoxy-α-tocopherolquinone (B), and 7,8-epoxy-α-tocopherolquinone (C) from α-tocopheryl radical with triplet oxygen.

reduction potentials of α-tocopherol derivatives were not available, the estimated values based on the data shown in Tables 16.2 and 16.3 were used to discuss the reaction mechanisms and kinetics of α-tocopherol derivatives as shown in Figures 16.2 through 16.5.

The formations of α-tocopherol peroxide, α-tocopherolquinone, 4a,5-epoxy-α-tocopherolqui-none, and 7,8-epoxy-α-tocopherolquinone from α-tocopheroxy radical and triplet ordinary oxygen

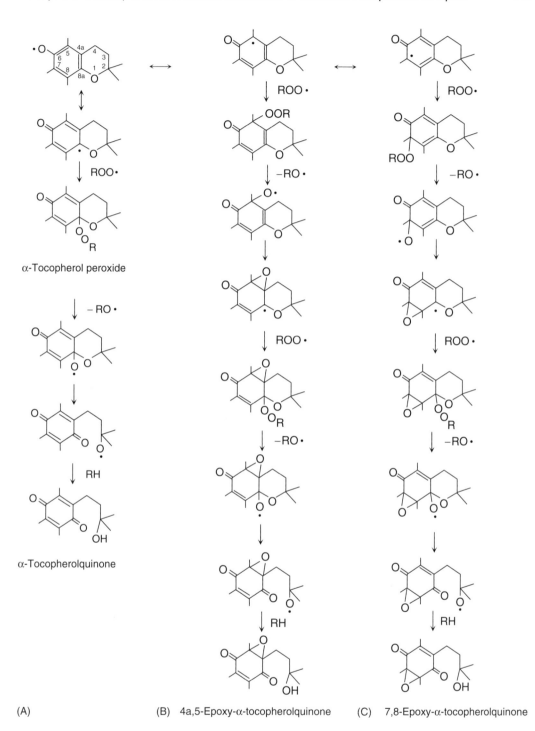

(A)   (B) 4a,5-Epoxy-α-tocopherolquinone   (C) 7,8-Epoxy-α-tocopherolquinone

FIGURE 16.3 Possible mechanism for the formation of α-tocopherolquinone (A), 4a,5-epoxy-α-tocopherolquinone (B), and 7,8-epoxy-α-tocopherolquinone (C) from α-tocopheryl radical with acyl peroxy radical.

are shown in Figure 16.2. Figure 16.2A shows the formation of α-tocopherolquinone from α-tocopheroxy radical. The α-tocopheroxy radical formed from α-tocopherol by donating hydrogen to unsaturated lipid peroxy radical forms 8a-carbon-centered α-tocopheryl radical by resonance.

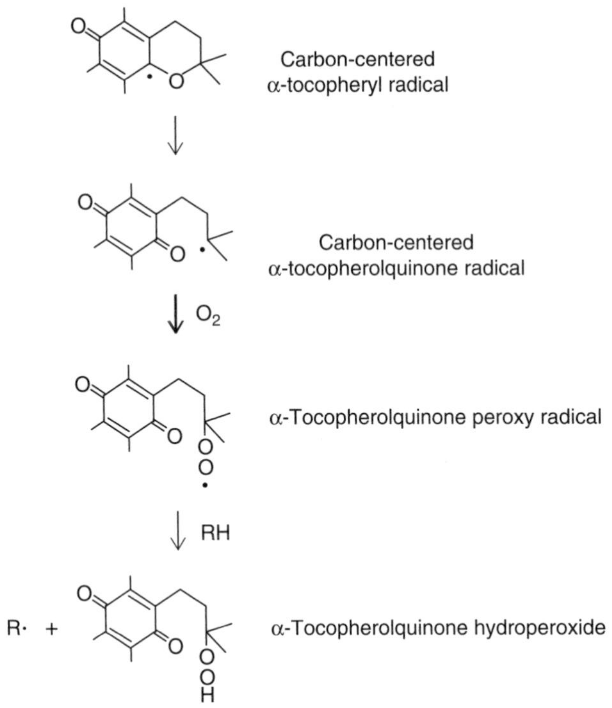

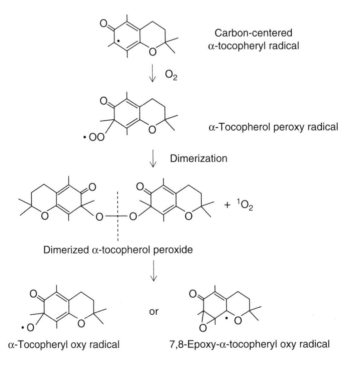

FIGURE 16.4 Possible mechanism for the formation of α-tocopherolquinone hydroperoxide from α-tocopheryl radical with acyl peroxy radical.

FIGURE 16.5 Possible mechanism for the formation of dimerized α-tocopherol peroxide, α-tocopheryl oxy radical, and 7,8-epoxy-α-tocopheryl oxy radical from α-tocopheryl radical.

TABLE 16.3

Reaction Rates of the Lipid (RH) and Tocopherol (TH)

Reaction	k ($M^{-1}s^{-1}$)	References
$R\cdot + {}^3O_2 \rightarrow ROO\cdot$	3×10^8	[39]
	5×10^6	[40]
$RO\cdot + RH \rightarrow R\cdot + ROH$	1×10^7	[41]
$ROO\cdot + RH \rightarrow R\cdot + ROOH$	10–100	[19,21]
$R\cdot + ROOH \rightarrow ROO\cdot + RH$	1×10^5	[21]
$T\cdot + RH \rightarrow TH + R\cdot$	0.07	[42]
	0.5	[21]
$T\cdot + ROOH \rightarrow TH + ROO\cdot$	0.1–0.5	[43]
$TH + ROO\cdot \rightarrow T\cdot + ROOH$	1×10^6	[19]
$2TOO\cdot \rightarrow TOOT + {}^1O_2$	—	—
$(2ROO\cdot \rightarrow ROOR + {}^1O_2)$	1×10^5	[44]
$TOOT \rightarrow 2TO\cdot$	—	—
$TO\cdot + RH \rightarrow TOH + R\cdot$	—	—
$ROO\cdot + T\cdot \rightarrow ROOT$	2.5×10^6	[10,45]
$R\cdot + TOO\cdot \rightarrow ROOT$	—	—
$ROOT \rightarrow RO\cdot + TO\cdot$	—	—
$TOO\cdot + T \rightarrow TOOH + T\cdot$	—	[15]
$TOOH \rightarrow TO\cdot + \cdot OH$	—	—

The 8a-carbon-centered α-tocopheryl radical reacts with oxygen to form α-tocopherol peroxy radical at the rate of 10^7 $M^{-1}s^{-1}$ (Table 16.3). The α-tocopherol peroxy radical has the reduction potential of ~1000 mV (Table 16.2) and can abstract a hydrogen from unsaturated fatty acids to form α-tocopherol hydroperoxide at the rate of 10^1–10^2 $M^{-1}s^{-1}$ (Table 16.3). The peroxide bond strength of α-tocopherol hydroperoxide is about 44 kcal/mol [46]. The α-tocopherol hydroperoxide is cleaved to form α-tocopherol oxy radical and hydroxy radical (·OH). The hydroxy radical is very reactive with the high reduction potential of 2300 mV (Table 16.2). As soon as the hydroxy radical is formed, it can abstract hydrogen from other compounds very easily. The α-tocopherol oxy radical forms α-tocopherolquinone oxy radical at the 2 position by cleaving the bond between 1 and 8a. The α-tocopherolquinone oxy radical has the reduction potential of about 1600 mV (Table 16.2) and forms α-tocopherolquinone by abstracting a hydrogen from unsaturated lipids at the rate of 10^7 $M^{-1}s^{-1}$ (Table 16.3).

Figure 16.2B shows the formation of 4a,5-epoxy-α-tocopherolquinone from the carbon-centered α-tocopheryl radical at the 5 position. The 5-carbon-centered α-tocopheryl radical reacts with oxygen to form α-tocopherol peroxy radical and then forms α-tocopherol hydroperoxide by abstracting hydrogen from unsaturated lipids. The α-tocopherol hydroperoxide is cleaved to form α-tocopherol oxy radical and hydroxy radical (·OH). The α-tocopherol oxy radical at the 5 position is cyclized to form 4a,5-epoxy-α-tocopheryl radical at the 8a position. The 4a,5-epoxy-α-tocopheryl radical reacts with oxygen to form 4a,5-epoxy-α-tocopherol peroxy radical and then forms 4a,5-epoxy-α-tocopherol hydroperoxide. The cleavage of 4a,5-epoxy-α-tocopherol hydroperoxide forms 4a,5-epoxy-α-tocopherolquinone oxy radical at the carbon 2 position and hydroxy radical (·OH). The 4a,5-epoxy-α-tocopherolquinone oxy radical forms 4a,5-epoxy-α-tocopherolquinone as shown in Figure 16.2B.

Figure 16.2C shows the formation of 7,8-epoxy-α-tocopherolquinone from the carbon-centered α-tocopheryl radical at the 7 position. The 7-carbon-centered α-tocopheryl radical reacts with oxygen to form α-tocopherol peroxy radical and then forms α-tocopherol hydroperoxide by abstracting hydrogen. The α-tocopherol hydroperoxide produces α-tocopherol oxy radical and hydroxy radical (·OH). The α-tocopherol oxy radical at the 7 position forms 7,8-epoxy-α-tocopheryl radical at the 8a position and then produces 7,8-epoxy-α-tocopherol peroxy radical by reacting with oxygen. The 7,8-epoxy-α-tocopherol peroxy radical forms 7,8-epoxy-α-tocopherol hydroperoxide. The 7,8-epoxy-α-tocopherol hydroperoxide forms 7,8-epoxy-α-tocopherolquinone oxy radical at the carbon 2 position and hydroxy radical (·OH). The 7,8-epoxy-α-tocopherolquinone oxy radical forms 7,8-epoxy-α-tocopherolquinone as shown in Figure 16.2C. α-Tocopherol peroxy radical, α-tocopherol oxy radical, α-tocopherolquinone oxy radical, and hydroxy radical formed from the oxidation of α-tocopherol are prooxidants due to high reduction potentials (Table 16.2).

The formations of α-tocopherol peroxide, α-tocopherolquinone, 4a,5-epoxy-α-tocopherolquinone, and 7,8-epoxy-α-tocopherolquinone from α-tocopheroxy radical and lipid peroxy radical (ROO·) are shown in Figure 16.3. Figure 16.3A shows the formation of α-tocopherolquinone from α-tocopheroxy radical. The α-tocopheroxy radical forms 8a-carbon-centered α-tocopheryl radical by resonance. The 8a-carbon-centered α-tocopheryl radical reacts with lipid peroxy radical (ROO·) to form α-tocopherol peroxide at the rate of 10^6 $M^{-1}s^{-1}$ (Table 16.3). The α-tocopherol peroxide can produce α-tocopherol oxy radical at the 8a carbon and alkoxy radical (RO·) by cleaving peroxide bond. The α-tocopherol oxy radical at the 8a position forms α-tocopherolquinone oxy radical by cleaving the bond between 1 and 8a positions. The α-tocopherolquinone oxy radical forms α-tocopherolquinone (Figure 16.3A).

Figure 16.3B shows the formation of 4a,5-epoxy-α-tocopherolquinone from the carbon-centered α-tocopheryl radical at the 5 position. The 5-carbon-centered α-tocopheryl radical reacts with lipid peroxy radical (ROO·) to form α-tocopherol peroxide. The breakdown of α-tocopherol peroxide forms α-tocopherol oxy radical at the carbon 5 and alkoxy radical (RO·). The α-tocopherol oxy radical is cyclized to form 4a,5-epoxy-α-tocopheryl radical at the 8a position. The 4a,5-epoxy-α-tocopheryl radical can react with lipid peroxy radical (ROO·) to form 4a,5-epoxy-α-tocopherol peroxide. The 4a,5-epoxy-α-tocopherol peroxide produces 4a,5-epoxy-α-tocopherol oxy radical at the carbon 8a and alkoxy radical (RO·). The 4a,5-epoxy-α-tocopherol oxy radical at the carbon 8a forms 4a,5-epoxy-α-tocopherolquinone oxy radical at the carbon 2 and then produces 4a,5-epoxy-α-tocopherolquinone by abstracting a hydrogen from unsaturated lipids (Figure 16.3B).

Figure 16.3C shows the chemical mechanism for the formation of 7,8-epoxy-α-tocopherolquinone from the carbon-centered α-tocopheryl radical at the carbon 7. The 7-carbon-centered α-tocopheryl radical reacts with lipid peroxy radical (ROO·) to form α-tocopherol peroxide which is cleaved to α-tocopherol oxy radical and alkoxy radical (RO·). The α-tocopherol oxy radical is cyclized to form 7,8-epoxy-α-tocopheryl radical at the 8a position and then forms 7,8-epoxy-α-tocopherol peroxide by reacting with lipid peroxy radical (ROO·). The 7,8-epoxy-α-tocopherol peroxide forms 7,8-epoxy-α-tocopherol oxy radical and alkoxy radical (RO·) by cleaving the peroxide bond. The 7,8-epoxy-α-tocopherol oxy radical forms 7,8-epoxy-α-tocopherolquinone oxy radical at the 2 position which generates 7,8-epoxy-α-tocopherolquinone by abstracting a hydrogen from unsaturated lipids (Figure 16.3C). α-Tocopherol peroxy radical, α-tocopherol oxy radical, and α-tocopherolquinone oxy radical are prooxidants (Table 16.2).

The formation of α-tocopherolquinone peroxy radical from carbon-centered α-tocopheryl radical is shown in Figure 16.4. The carbon-centered α-tocopherolquinone radical at the carbon 2 reacts with triplet oxygen to form α-tocopherolquinone peroxy radical at the rate of about 10^7 $M^{-1}s^{-1}$ (Table 16.3). The α-tocopherolquinone peroxy radical with the reduction potential of about 1000 mV (Table 16.2) can abstract hydrogen atom from unsaturated lipids to form α-tocopherolquinone hydroperoxide at the rate of 10^1–10^2 $M^{-1}s^{-1}$ (Table 16.3). The α-tocopherolquinone peroxy radical with 1000 mV of reduction potential is prooxidant.

Figure 16.5 shows the singlet oxygen formation from the carbon-centered α-tocopheryl radical. The carbon-centered α-tocopheryl radical reacts with triplet oxygen and forms α-tocopherol peroxy radical. Two moles of α-tocopherol peroxy radical form dimerized α-tocopherol peroxide and singlet oxygen at the rate of 10^5 $M^{-1}s^{-1}$ [44,47]. Singlet oxygen formed from α-tocopheryl radical is a strong prooxidant and reacts directly with electron-rich compounds containing double bonds even at very low temperature due to its low activation energy for reaction [47,48]. The reaction rates between singlet oxygen and unsaturated fatty acids are $10^5 M^{-1}s^{-1}$ [47].

The tocopherol peroxy radical (TOO·), tocopherol oxy radical (TO·), α-tocopherolquinone oxy radical, α-tocopherolquinone peroxy radical, alkoxy radical (RO·), hydroxy radical (·OH), and singlet oxygen ($^1O_2$) are formed during the oxidation of α-tocopherol. The reduction potentials of these intermediate compounds are high to easily oxidize food compounds (Table 16.2). They all act as strong prooxidants.

The oxidized α-tocopherol compounds have polar hydroxyl and nonpolar hydrocarbon groups. Yoon et al. [49] and Mistry and Min [50,51] reported that thermally oxidized lipid compounds with polar hydroxyl and nonpolar hydrocarbons in the same molecule were prooxidants in soybean oil during storage. They reported that the oxidized lipids with hydroxyl and/or carbonyl groups were less soluble in the soybean oil and moved to the surface of oil. The oxidized oils having polar and nonpolar groups decreased the surface tension between air and oil and increased the transportation of oxygen from air to oil [49–51]. The oxidized oils accelerated the oxidation of oil. The oxidized α-tocopherol compounds with the polar groups and nonpolar hydrocarbons in the same molecule would reduce the surface tension between headspace air and oil and accelerate the oxidation of oil.

The prooxidant mechanisms of oxidized α-tocopherol may be mainly due to α-tocopherol peroxy radical, α-tocopherol oxy radical, α-tocopherolquinone oxy radical, hydroxy radical, and singlet oxygen which are formed during the oxidation of α-tocopherol in foods during storage. The α-tocopherol at higher concentration may produce more α-tocopherol peroxy radical, α-tocopherol oxy radical, α-tocopherolquinone oxy radical, hydroxy radical, and singlet oxygen and promote the oxidation of oils. The oxidized α-tocopherol compounds with polar hydroxyl and nonpolar hydrocarbons in the same molecule may also contribute to the oxidation of oil by reducing the surface tension of oil and increasing the diffusion of oxygen from air to oil.

REFERENCES

1. Buettner, G., The pecking order of free radicals and antioxidants: lipid peroxidation, α-tocopherol, and ascorbate, *Arch. Biochem. Biophys.*, 300, 535, 1993.
2. Lee, J.H., Koo, N., and Min, D.B., Reactive oxygen species, aging, and antioxidative nutraceuticals, *Comp. Rev. Food Sci. Food Saf.*, 3, 21, 2004.
3. Frankel, E.N., Chemistry of free radical and singlet oxidation of lipids, *Prog. Lipid Res.*, 23, 197, 1984.
4. Min, D.B. and Smouse, T.H., *Flavor Chemistry of Fats and Oils*, AOCS Press, Champaign, IL, 1985.
5. Nawar, W., Lipids, in *Food Chemistry*, 3rd edition, Fennema, O.R., Ed., Marcel Dekker, New York, 1996, p. 225.
6. Reische, D.W., Lillard, D.A., and Eitenmiller, R.R., Antioxidants, in *Food Lipids*, 2nd edition, Akoh, C.C. and Min, D.B., Eds., Marcel Dekker, New York, 2002, p. 489.
7. Gregory, J.F., Vitamins, in *Food Chemistry*, 3rd edition, Fennema, O.R., Ed., Marcel Dekker, New York, 1996, p. 553.
8. Verleyen, T., Verhe, R., Huyghebaert, A., Dewettinck, K., and De Grey, W., Identification of α-tocopherol oxidation products in triolein at elevated temperature, *J. Agric. Food Chem.*, 49, 1508, 2001.
9. Kamal-Eldin, A. and Appelqvist, L.A., The chemistry and antioxidant properties of tocopherols and tocotrienols, *Lipids*, 31, 671, 1996.
10. Kitts, D.D., An evaluation of the multiple effects of the antioxidant vitamins, *Trends Food Sci. Technol.*, 8, 198, 1997.
11. Belitz, H.D. and Grosch, W., *Food Chemistry*, 2nd edition, Springer, Berlin, 1999.
12. Warner, K., Effects on the flavor and oxidative stability of stripped soybean and sunflower oils with added pure tocopherols, *J. Agric. Food Chem.*, 53, 9906, 2005.
13. Choe, E. and Min, D.B., Mechanisms and factors for edible oil oxidation, *Comp. Rev. Food Sci. Food Saf.*, 5, 169, 2006.

14. Jung, M.Y. and Min, D.B., Effects of α-, γ-, and δ-tocopherols on the oxidative stability of soybean oil, *J. Food Sci.*, 55, 1464, 1990.

15. Bowry, V.W. and Stocker, R., Tocopherol-mediated peroxidation: the prooxidant effect of vitamin E on the radical-initiated oxidation of human low-density lipoprotein, *J. Am. Chem. Soc.*, 115, 6029, 1993.

16. Huang, S.-W., Frankel, E.N., and German, J.B., Effects of individual tocopherols and tocopherol mixtures on the oxidative stability of corn oil triglycerides, *J. Agric. Food Chem.*, 43, 2345, 1995.

17. Evans, J.C., Kodali, D.R., and Addis, P.B., Optimal tocopherol concentrations to inhibit soybean oil oxidation, *J. Am. Chem. Soc.*, 79, 47, 2002.

18. Rietjens, I.M.C.M., Boersma, M.G., de Haan, L., Spenkelink, B., Awad, H.M., Cnubben, N.H.P., van Zanden, J.J., van der Woude, H., Alink, G.M., and Koeman, J.H., The prooxidant chemistry of the natural antioxidants vitamin C, vitamin E, carotenoids and flavonoids, *Environ. Toxicol. Pharmcol.*, 11, 321, 2002.

19. Niki, E., Satio, T., Kawakami, A., and Kamiya, Y., Inhibition of oxidation of methyl linoleate in solution by vitamin E and vitamin C, *J. Biol. Chem.*, 259, 4177, 1984.

20. Choe, E. and Min, D.B., Chemistry and reactions of reactive oxygen species in foods, *J. Food Sci.*, 70, R142, 2005.

21. Naumov, V.V. and Vasil'ev, R.F., Antioxidant and prooxidant effects of tocopherol, *Kinet. Catal.*, 44, 101, 2003.

22. Azzi, A. and Stocker, A., Vitamin E: non-antioxidant roles, *Prog. Lipid Res.*, 39, 231, 2000.

23. Suzuki, Y.J. and Packer, L., Inhibition of NF-kappa B activation by vitamin E derivatives, *Biochem. Biophys. Res. Comm.*, 193, 277, 1993.

24. Burton, G.H. and Ingold, K.U., Autoxidation of biological molecules. 1. The antioxidant activity of vitamin E and related chain-breaking phenolic antioxidants in vitro, *J. Am. Chem. Soc.*, 103, 6472, 1981.

25. Isnardy, B., Wagner, K.-H., and Elmadfa, I., Effects of α-, γ-, and δ-tocopherols on the autoxidation of purified rapeseed oil triacylglycerols in a system containing low oxygen, *J. Agric. Food Chem.*, 51, 7775, 2003.

26. Player, M.E., Kim, H.J., Lee, H.O., and Min, D.B., Stability of α-, γ-, or δ-tocopherol during soybean oil oxidation, *J. Food Sci.*, 71, C456, 2006.

27. Huang, S.-W., Frankel, E.N., and German, J.B., Antioxidant activity of α- and γ-tocopherols in bulk oils and in oil-in-water emulsions, *J. Agric. Food Chem.*, 42, 2108, 1994.

28. Berkowitz, J., Ellison, G.B., and Gutman, D., Three methods to measure Rh bond-energies, *J. Phys. Chem.*, 98, 2744, 1994.

29. Bordwell, F.G. and Liu, W.Z., Solvent effects on hemolytic bond dissociation energies of hydroxylic acids, *J. Am. Chem. Soc.*, 118, 10819, 1996.

30. Wright, J.S., Johnson, E.R., and DiLabio, G.A., Predicting the activity of phenolic antioxidants: theoretical method, analysis of substituent effects, and application to major families of antioxidants, *J. Am. Chem. Soc.*, 123, 1173, 2001.

31. Terao, J. and Matsushita, S., The peroxidizing effect of a-tocopherol on autoxidation of methyl linoleate in bulk phase, *Lipids*, 21, 255, 1986.

32. Hicks, M. and Gebicki, J.M., Inhibition of peroxidation in linoleic acid membranes by nitroxide radicals, butylated hydroxy toluene and α-tocopherol, *Arch. Biochem. Biophys.*, 210, 56, 1981.

33. Jung, M.Y. and Min, D.B., Effects of oxidized α-, γ-, and δ-tocopherols on the oxidative stability of purified soybean oil, *Food Chem.*, 43, 183, 1992.

34. Faustman, C., Liebler, D.C., and Burr, J.A., α-Tocopherol oxidation in beef and in bovine muscle microsomes, *J. Agric. Food Chem.*, 47, 1396, 1999.

35. Liebler, D.C., Burr, J.A., Philips, L., and Ham, A.J.L., Gas chromatography-mass spectrometry analysis of vitamin E and its oxidation products, *Anal. Biochem.*, 236, 27, 1996.

36. Verleyen, T., Kamal-Eldin, A., Dobarganes, C., Verhe, R., Dewettinck, R., and Huyghebaert, A., Modeling of α-tocopherol loss and oxidation products formed during thermoxidation in triolein and tripalmitin mixture, *Lipids*, 36, 719, 2001.

37. Pazos, M., Sanchez, L., and Medina, I., α-Tocopherol oxidation in fish muscle during chilling and frozen storage, *J. Agric. Food Chem.*, 53, 4000, 2005.

38. Yoshida, Y., Niki, E., and Noguchi, N., Comparative study on the action of tocopherols and tocotrienols as antioxidant: chemical and physical effects, *Chem. Phys. Lipids*, 123, 63, 2003.

39. Hasegawa, K. and Patterson, L.K., Pulse radiolysis in model lipid systems: formation and behavior of peroxyl radicals in fatty acids. *Photochem. Photobiol.*, 28, 817, 1987.

40. Kasaikina, O.T., Kortenska, V.D., and Yanishlieva, N.V., Effect of chain transfer and recombination/disproportionation of inhibitor radicals on inhibited oxidation of lipids, *Russ. Chem. Bull.*, 48, 1891, 1999.

41. Small, R.D.J., Scaiano, J.C., and Patterson, L.K., Radical processes in lipids. A laser photolysis study of *t*-butoxy radical reactivity toward fatty acids, *Photochem. Photobiol.*, 29, 49, 1979.

42. Remorova, A.A. and Roginsky, V.A., Rate constants for the reaction of a-tocopherol phenoxy radicals with unsaturated fatty acid esters and the contribution of this reaction to the kinetics of inhibition of lipid peroxidation, *Kinet. Catal.*, 32, 726, 1991.

43. Mukai, K., Morimoto, H., Okauchi, Y., and Nagaoka, S., Kinetic study of reactions between tocopheroxyl radicals and fatty acids, *Lipids*, 28, 753, 1993.

44. Barclay, L.R.C., Baskin, K.A., Locke, S.J., and Vinquist, M.R., Absolute rate constants for lipid peroxidation and inhibition in model biomembranes, *Can. J. Chem.*, 68, 2258, 1989.

45. Kaouadji, M.N., Jore, D., Ferradini, C., and Patterson, L.K., Radiolytic scanning of vitamin E/vitamin C oxidation reduction mechanisms, *Bioelectrochem. Bioenerg.*, 18, 59, 1987.

46. Hiatt, R., Mill, T., Irwin, K.C., Mayo, T.R., Gould, C.W., and Castleman, J.K., Homolytic decomposition of hydroperoxides, *J. Org. Chem.*, 33, 1416, 1968.

47. Min, D.B. and Boff, J.M., Chemistry and reaction of singlet oxygen in foods, *Comp. Rev. Food Sci. Food Saf.*, 1, 58, 2002.

48. Bradley, D.G. and Min, D.B., Singlet oxygen oxidation in foods, *Crit. Rev. Food Sci. Nutr.*, 31, 211, 1992.

49. Yoon, S.H., Jung, M.Y., and Min, D.B., Effects of thermally oxidized triglycerides on the oxidative stability of soybean oil, *J. Am. Oil Chem. Soc.*, 65, 1652, 1988.

50. Mistry, B.S. and Min, D.B., Prooxidant effects of monoglycerides and diglycerides in soybean oil, *J. Food Sci.*, 53, 1896, 1988.

51. Mistry, B.S. and Min, D.B., Isolation and identification of minor components and their effects on flavor stability of soybean oil, in *Frontiers of Flavor*, Elsevier Science, Holland, 1988, p. 499.

17 Effects and Mechanisms of Minor Compounds in Oil on Lipid Oxidation

Eunok Choe

CONTENTS

I. INTRODUCTION

Lipid oxidation produces low molecular weight off-flavor compounds and affects the quality and shelf life of foods [1]. The off-flavor compounds decrease the consumer acceptability or industrial use of lipid as a food ingredient. Lipid oxidation destroys essential fatty acids and produces toxic compounds and oxidized polymers [2]. Lipid oxidation is influenced by many factors: energy input such as light or heat, food composition, types of oxygen, and minor compounds such as free fatty acids, mono- and diacylglycerols, metals, peroxides, thermally oxidized compounds, chlorophylls, carotenoids, tocopherols (TOH), and other phenolic compounds. Some of them accelerate the oil oxidation and others act as antioxidants. The major mechanisms of lipid oxidation during food processing and storage are autoxidation and photosensitized oxidation. The improvement of oxidative stability of lipid foods can be achieved by thorough understanding of chemical mechanisms of lipid oxidation and functions of some compounds present in oil, naturally or added on purpose, other than triacylglycerols.

II. MECHANISMS OF LIPID AUTOXIDATION

Lipid autoxidation is a free radical chain reaction, which requires lipids in radical forms. Lipids are normally in nonradical singlet state, and their reaction with atmospheric triplet oxygen ($^3O_2$) is

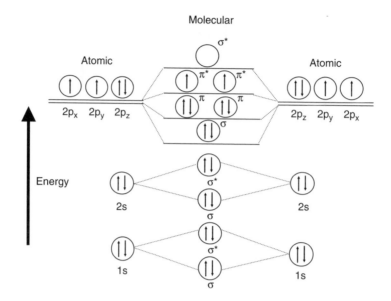

FIGURE 17.1 Molecular orbital of triplet oxygen, $^3O_2$.

thermodynamically unfavorable. The atmospheric oxygen has two unpaired electrons in $2p\pi*$ orbitals (Figure 17.1) and hardly reacts with nonradical lipids [3]. The autoxidation requires removal of hydrogen atom in lipids with production of lipid radicals. Heat, metal catalysts, and light accelerate formation of lipid radicals. The energy required to remove hydrogen from lipid molecules is dependent on the hydrogen position in the molecules. Allylic hydrogen, especially hydrogen attached to the carbon between two double bonds, is easily removed. Hydrogen at C8, C11, and C17 of linoleic acid is removed at 314, 209, and 418 kJ/mol, respectively [4]. On formation of lipid radicals by hydrogen removal, the double bond adjacent to the carbon radical in linoleic and linolenic acids shifts to the more stable next carbon, resulting in conjugated dienes and trienes. The shifted double bond mostly takes *trans* form instead of natural *cis* form because of lower energy.

The lipid radical reacts with triplet oxygen very quickly at normal oxygen pressure with a rate constant of ca. $2-8 \times 10^9$ $M^{-1}s^{-1}$,[5] and forms lipid peroxy radical. This process is controlled by oxygen migration which has activation energy of 24 kJ/mol in unsaturated lipids [5]. Lipid peroxy radical can also react with triplet oxygen; however, this reaction is slower than its production from lipid radical and oxygen, and thus the concentration of lipid peroxy radical is higher than that of lipid radical [6]. The lipid peroxy radical abstracts hydrogen from other lipid molecules and forms lipid hydroperoxide and another lipid radical. The radicals catalyze the reaction and the autoxidation is called free radical chain reaction. Figure 17.2 shows a formation of C9- (48% to 53%) and C13-hydroperoxide (48% to 53%) in the autoxidation of linoleic acid. Oleic acid produces C8- (26%–28%), C9- (22%–25%), C10- (22%–24%), and C11-hydroperoxides (26%–28%), and linolenic acid produces C9- (28%–35%), C12- (8%–13%), C13- (10%–13%), and C16-hydroperoxides (28%–35%) by the autoxidation [7].

Lipid hydroperoxides, the primary oxidation products, are relatively stable at room temperature. However, in the presence of metals or at high temperature they are readily decomposed to alkoxy radicals via homolytic cleavage in the oxygen–oxygen bond which requires 192 kJ/mol [8]. The alkoxy radical is unstable and undergoes homolytic β-scission of carbon–carbon bond to produce ultimately low molecular weight carbonyl compounds, alcohols, and short-chain hydrocarbons (Figure 17.3). Decomposition products of oleic, linoleic, and linolenic acids are listed in Table 17.1. Decomposition of lipid hydroperoxides is affected by temperature. Crude herring oil

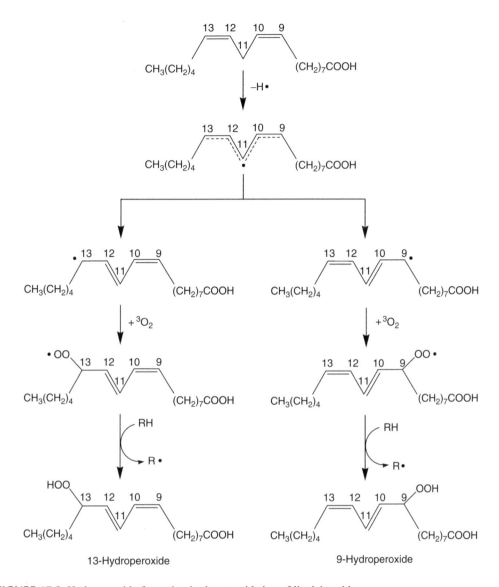

FIGURE 17.2 Hydroperoxide formation in the autoxidation of linoleic acid.

showed higher rate of hydroperoxide decomposition than that of hydroperoxide formation at 50°C in the dark, but the reverse phenomenon was observed in the same oil at 0°C or 20°C [6].

Most decomposition products of lipid hydroperoxides are responsible for the off-flavor in the oxidized oil. No single compound is responsible for the oxidized flavor of oils. Aliphatic carbonyl compounds have lower threshold values and more influence on the oil flavor than hydrocarbons; threshold values for hydrocarbons, alkanals, 2-alkenals, and 2,4-alkadienals are 90–2150, 0.04–1, 0.04–2.5, and 0.04–0.3 ppm, respectively [7]. *trans*-2-Hexenal, and *trans, cis, trans*-2,4,7-deca-trienal and 1-octen-3-one give grasslike and fishlike flavor in oxidized soybean oil, respectively [3]. The volatile compounds frequently detected in autoxidized oil are propenal, butanal, pentane, hexanal, 2,4-heptadienal, 1-octen-3-ol, 2-decenal, and 2,4-decadienal [7,9,10]. Pentane, hexanal, and 2,4-decadienal are mostly used as indicators to determine the extent of oil oxidation [11–14].

The autoxidation rate depends greatly on the formation of lipid radicals, which depends mainly on the types of lipids. Autoxidation occurs more quickly in more unsaturated lipids than in less

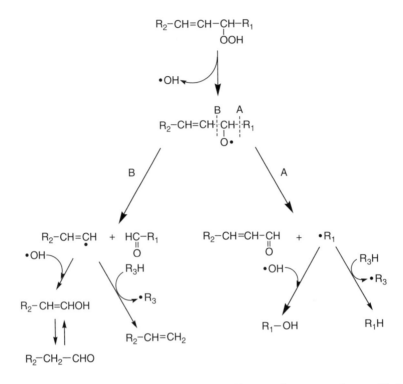

FIGURE 17.3 Mechanisms of hydroperoxide decomposition to form secondary oxidation products (R, alkyl group).

TABLE 17.1
Decomposition Products of Hydroperoxides of Fatty Acid Methyl Ester by Autoxidation

Fatty Acid	Aldehydes	Carboxylic Acid	Alcohol	Hydrocarbons
Oleic acid	Octanal, nonanal, 2-decenal, decanal	Methyl heptanoate, methyl octanoate, methyl 8-oxooctanoate, methyl 9-oxononanoate, methyl 10-oxodecanoate, methyl 10-oxo-8-decenoate, methyl 11-oxo-9-undecenoate	1-Heptanol	Heptane, octane
Linoleic acid	Pentanal, hexanal, 2-octenal, 2-nonenal, 2,4-decadienal	Methyl heptanoate, methyl octanoate, methyl 8-oxooctanoate, methyl 9-oxononanoate, methyl 10-oxodecanoate	1-Pentanol, 1-octene-3-ol	Pentane
Linolenic acid	Propanal, butanal, 2-butenal, 2-pentenal, 2-hexenal, 3,6-nonadienal, decatrienal	Methyl heptanoate, methyl octanoate, methyl nonanoate, methyl 9-oxononanoate, methyl 10-oxodecanoate		Ethane, pentane

Source: From Frankel, E.N. in *Flavor Chemistry of Fats and Oils*, D.B. Min and T.H. Smouse, eds, American Oil Chemists' Society, Champaign, 1985.

unsaturated oils [15,16]. Soybean, safflower, or sunflower oil (iodine values > 130) stored in the dark showed a significantly ($p < 0.05$) shorter induction period than coconut or palm kernel oil whose iodine value is less than 20 [17]. The activation energies in the autoxidation of methyl linoleate, linolenic acid, trilinolein, and trilinolenin are 84 ± 8.4 kJ/mol [18], 65 ± 4 kJ/mol [19], 34 ± 8 kJ/mol [5], and 9 ± 2 kJ/mol [5], respectively. The relative autoxidation rate of oleic, linoleic, and linolenic acids was reported as 1:40 to 50:100 on the basis of oxygen uptake [3].

Oxygen concentration affects the rate of lipid oxidation. When oxygen content is low, for example less than 4% in the headspace, the oxidation rate is dependent on oxygen concentration and is independent of lipid concentration [20,21]. When the oxygen is present at sufficiently high concentration, the rate of oil autoxidation is independent of oxygen concentration and directly dependent on the lipid concentration [22,23]. The rate constants for oxygen disappearance in soybean oil containing 2.5, 4.5, 6.5, and 8.5 ppm dissolved oxygen were 0.049, 0.058, 0.126, and 0.162 ppm/h, respectively, in the autoxidation at 55°C [24]. Przybylski and Eskin [12] reported that usual content of dissolved oxygen in edible oil is sufficient enough to oxidize the oil to a peroxide value of ca. 10 meq/kg in the dark. Concentration of dissolved oxygen in soybean oil was reported as 55 ppm at room temperature [21]. Autoxidation of oils increases as temperature increases [25–27]. The effect of oxygen concentration on the oxidation of oil also increases at high temperature [21]. The oxygen is transported into the oil by diffusion at low temperature because there is little stirring in the oil. However, at high temperature the oxygen can penetrate into the oil from the surface by both diffusion and by convection caused by increased molecular movement in the oil, which increases migration of oxygen into the oil.

III. MECHANISMS OF PHOTOSENSITIZED OXIDATION OF LIPIDS

Light accelerates the lipid oxidation, especially in the presence of photosensitizers. Photosensitizers are normally in singlet state (^{1}Sen) and become excited on absorption of light energy in picoseconds [28]. Excited singlet state sensitizers (^{1}Sen*) can return to their ground state via emission of light ($k = 2 \times 10^8$ s^{-1}) or intersystem crossing ($k = 1$–20×10^8 s^{-1}). Intersystem crossing results in excited triplet state of sensitizers (^{3}Sen*). Excited triplet sensitizers may accept hydrogen or an electron from the substrate (RH) and produce radicals (type I), R$^•$ and RH$^{•+}$, as shown in Figure 17.4. Readily oxidizable phenols and amines or readily reducible quinones are substrates that favor the type I process [29]. Excited triplet sensitizers can also react with triplet oxygen, and produce superoxide anion radicals ($O_2^{-•}$) by electron transfer or form singlet oxygen by energy transfer (type II). The excited triplet sensitizers return to their ground singlet state. Kochevar and Redmond [30] reported that one molecule of sensitizer may generate 10^3–10^5 molecules of singlet oxygen before becoming inactive. Unsaturated hydrocarbons and aromatic compounds are not so readily oxidized or reduced and more often favor the type II process [29]. Chlorophylls are photosensitizers that take type II pathway. Photosensitized oxidation of lipid follows the singlet oxygen oxidation pathway (type II) [31], and Lee and Min [32] suggested that singlet oxygen is involved in the initiation of the oil oxidation.

Singlet oxygen is electrophilic due to a completely vacant orbital (Figure 17.5); it directly reacts with high electron density double bonds via six-membered ring without lipid radical formation [33], and the resulting hydroperoxides are both conjugated and nonconjugated (Figure 17.6). Production of nonconjugated hydroperoxides is not observed in the autoxidation. Singlet oxygen oxidation of oleic, linoleic, and linolenic acids produces C9- and C10-hydroperoxides, C9-, C10-, C12-, and C13-hydroperoxides, and C9-, C10-, C12-, C13-, C15-, and C16-hydroperoxides, respectively (Table 17.2).

Lipid hydroperoxides formed by singlet oxygen oxidation are decomposed by the same mechanisms as the autoxidation; however, kinds and amounts of decomposition products are slightly different. Singlet oxygen oxidation of oleate produces more 2-decenal and octane than the autoxidation does, whereas contents of octanal and 10-oxodecanoate are higher in autoxidized oleate than in

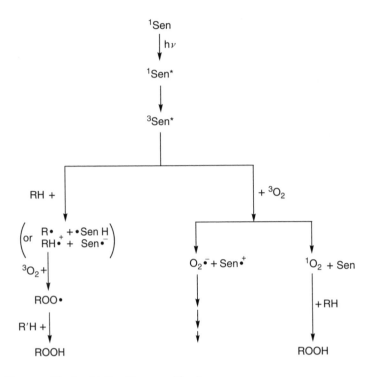

FIGURE 17.4 Photosensitized oxidation (Sen, sensitizer).

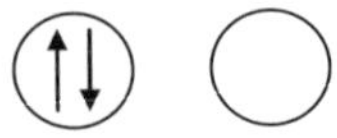

FIGURE 17.5 Electronic configuration of $2p\pi$ antibonding orbital of singlet oxygen, $^1O_2$.

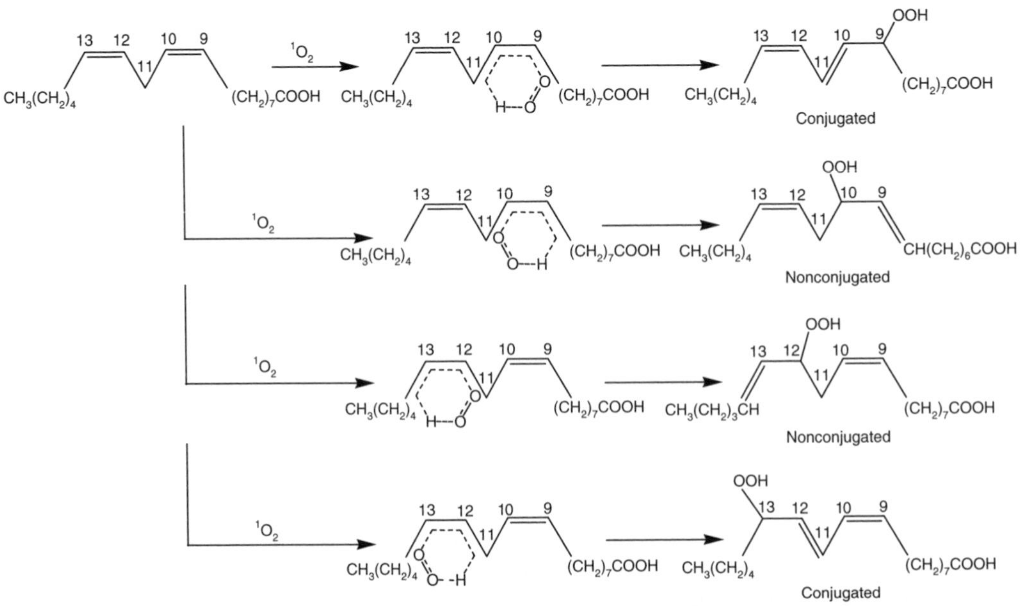

FIGURE 17.6 Hydroperoxide formation in linoleic acid oxidation by singlet oxygen.

TABLE 17.2

Hydroperoxides of Fatty Acids by Singlet Oxygen Oxidation

	Oleic Acid	Linoleic Acid	Linolenic Acid
Conjugated hydroperoxides		C9-OOH (32%)	C9-OOH (23%)
		C13-OOH (34%)	C12-OOH (12%)
	C9-OOH (48%)[a]		C13-OOH (14%)
	C10-OOH (52%)		C16-OOH (25%)
Nonconjugated hydroperoxides		C10-OOH (17%)	C10-OOH (13%)
		C12-OOH (17%)	C15-OOH (13%)

Source: From Frankel, E.N. in *Flavor Chemistry of Fats and Oils*, D.B. Min and T.H. Smouse, eds, American Oil Chemists' Society, Champaign, IL, 1985.

[a] Relative content.

singlet oxygen-oxidized oleate [7]. Singlet oxygen-oxidized linoleic and linolenic acids produce noticeable amount of 2-heptenal and 2-butenal which are negligible in autoxidized lipids [7]. Figure 17.7 shows a formation of 2-heptenal by the singlet oxygen oxidation of linoleic acid. Min and others [34] showed that 2-pentylfuran and 2-pentenylfuran (Figure 17.8) are formed from linoleic and linolenic acids, respectively, in the presence of chlorophyll under light. 2-Pentylfuran and 2-pentenylfuran cause beany flavor [35–38].

The reaction rate of singlet oxygen with lipid is much higher than that of triplet oxygen; the reaction rates of singlet oxygen and triplet oxygen with linoleic acid are 1.3×10^5 and 8.9×10^1 $M^{-1}s^{-1}$, respectively [31]. The rate of singlet oxygen oxidation depends on the kinds of lipids, and the rate differences among different lipids are lower in singlet oxygen oxidation than in autoxidation. The reaction rates of singlet oxygen with fatty acids are in the order of 10^4 $M^{-1}s^{-1}$, and the ratio of the rates among stearic, oleic, linoleic, and linolenic acids is 1.2: 5.3: 7.3: 10.0 [39]. Soybean oil reacts with singlet oxygen at the rate of 1.4×10^5 $M^{-1}s^{-1}$ in methylene chloride at 20°C [40]. Conjugation of double bonds in dienes or trienes does not affect the rate of singlet oxygen oxidation [41].

Since singlet oxygen oxidation does not require high activation energy (0–25 kJ/mol), temperature has little effect on singlet oxygen oxidation [41,42] and light is more affective. Singlet oxygen oxidation is higher under shorter wavelength light than under longer wavelengths [43]. Light effect on the lipid oxidation becomes less as temperature increases [27].

IV. EFFECTS OF MINOR COMPOUNDS IN OIL ON THE LIPID OXIDATION

Edible oil consists of mostly triacylglycerols, but it also contains minor compounds such as free fatty acids, mono- and diacylglycerols, metals, phospholipids, thermally oxidized compounds and peroxides, chlorophylls, carotenoids, TOH, and other phenolic compounds. Table 17.3 shows minor compounds present in crude and RBD (refined, bleached, and deodorized) soybean oil. These minor compounds accelerate or decelerate the lipid oxidation individually or interactively.

A. FREE FATTY ACIDS

Crude oil contains free fatty acids and most of them are removed by oil refining. Crude and refined soybean oil contains free fatty acids at about 0.7% and 0.02%, respectively [44]. Roasted sesame oil is consumed without refining and contains higher amount of free fatty acids (0.72%) [45] than other RBD oils. Free fatty acids are more susceptible to the autoxidation than esterified fatty acids [46]. Rate constants for the autoxidation of linolenic acid and soy lecithins at 100°C are 1.6×10^{-2} s^{-1}

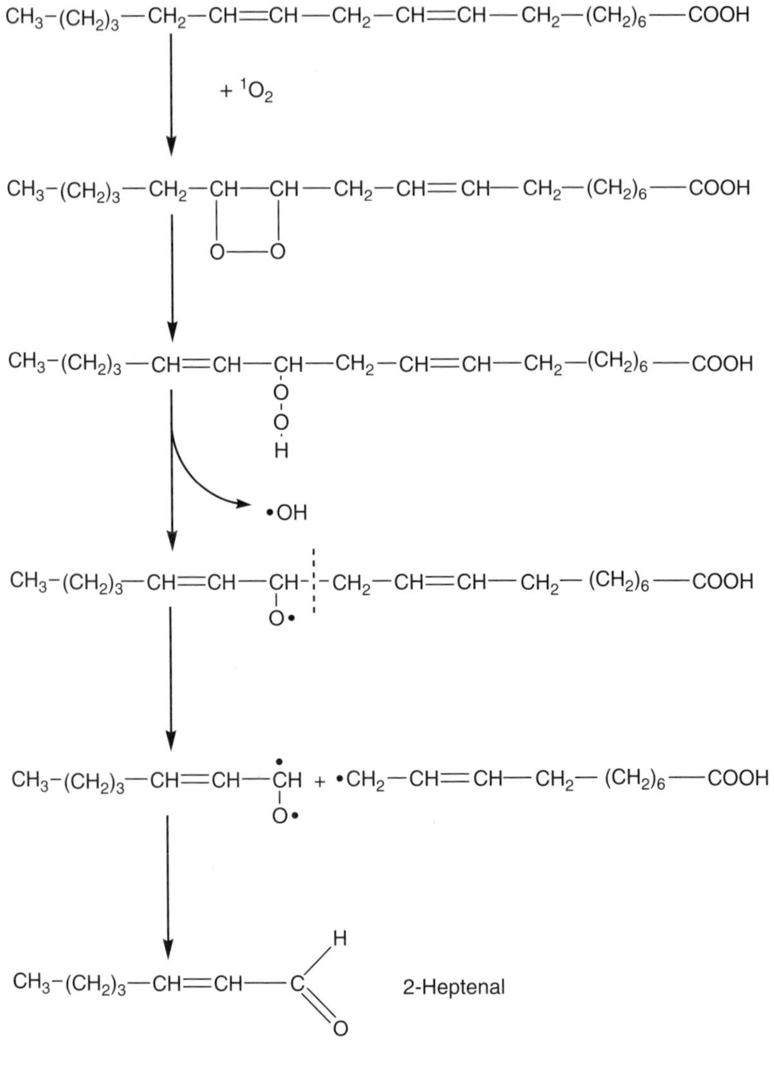

$CH_3-(CH_2)_3-CH_2-CH=CH-CH_2-CH=CH-CH_2-(CH_2)_6-COOH$

$+\ ^1O_2$

$CH_3-(CH_2)_3-CH_2-CH-CH-CH_2-CH=CH-CH_2-(CH_2)_6-COOH$

$O-O$

$CH_3-(CH_2)_3-CH=CH-CH-CH_2-CH=CH-CH_2-(CH_2)_6-COOH$

O
O
H

$\cdot OH$

$CH_3-(CH_2)_3-CH=CH-CH-CH_2-CH=CH-CH_2-(CH_2)_6-COOH$

$O\cdot$

$CH_3-(CH_2)_3-CH=CH-CH\ +\ \cdot CH_2-CH=CH-CH_2-(CH_2)_6-COOH$

$O\cdot$

$CH_3-(CH_2)_3-CH=CH-C$ 2-Heptenal

FIGURE 17.7 Formation of 2-heptenal from linoleic acid by singlet oxygen.

and $1.8 \times 10^{-3}\ s^{-1}$, respectively [19]. Free fatty acids act as prooxidants in the oxidation of edible oil [47,48]. The autoxidation rate of soybean oil at 55°C was $2.55 \times 10^{-2}\ h^{-1}$ and the addition of stearic acid at 1% increased the rate to $2.91 \times 10^{-2}\ h^{-1}$ [49]. Free fatty acids have hydrophilic carboxy group and hydrophobic hydrocarbons in the same molecule. The carboxy group will not easily dissolve in the hydrophobic oil, which makes free fatty acid be concentrated on the surface. Mistry and Min [48] reported that free fatty acids decreased the surface tension of soybean oil, and increased the diffusion rate of oxygen from the headspace into the oil accelerate the oil oxidation.

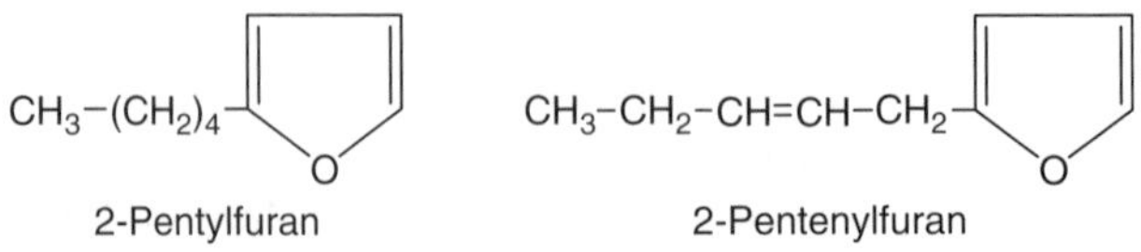

$CH_3-(CH_2)_4$ $CH_3-CH_2-CH=CH-CH_2$

2-Pentylfuran   2-Pentenylfuran

FIGURE 17.8 Structures of 2-pentylfuran and 2-pentenylfuran.

TABLE 17.3

Minor Compounds in Soybean Oil

Oil	Free Fatty Acids (%)	Hydroperoxides (POV, meq/kg)	Phospholipids (ppm)	Iron (ppm)	Chlorophylls (ppm)	Tocopherols (ppm)
Crude	0.74	2.4	510.0	2.90	0.30	1670
RBD	0.02	0	1.0	0.27	0	1138

Source: From Jung, M.Y., Yoon, S.H., and Min, D.B., *J. Am. Oil Chem. Soc.*, 66, 118, 1989.

B. MONO- AND DIACYLGLYCEROLS

Mono- and diacylglycerols are found in edible oils at 1% to 10%; rapeseed, soybean, safflower, olive, and palm oil contain 0.8%, 1.0%, 2.1%, 5.5%, and 5%–11% of diacylglycerols [50–52], respectively. Monoacylglycerol content of soybean oil ranges from 0.07% to 0.11%. Mono- and diacylglycerols increased the soybean oil oxidation at 55°C in the dark [53,54]. They have hydrophilic hydroxy groups and hydrophobic hydrocarbons, which makes them be positioned on the surface of oil. Mono- and diacylglycerols accelerate the oil oxidation by decreasing surface tension of oil and increasing the oxygen diffusion into the oil. The oxidative stability of oil can be improved by removal of mono- and diacylglycerols from the oil during refining process [54].

C. METALS

Crude oil contains transition metals such as iron and copper. Table 17.4 shows contents of copper and iron in some vegetable oils. Crude soybean oil contains 13.2 ppb of copper and 2.80 ppm of iron; however, the quantity decreases by oil refining [55,56], and refined soybean oil contains 2.5 ppb of copper and 0.2 ppm of iron, respectively [57]. Edible oils consumed without refining contain relatively high amounts of copper and iron; sesame and virgin olive oils contain 16 and 9.8 ppb of copper and 1.2 and 0.7 ppm of iron, respectively [58]. Metal increases the rate of oil oxidation by reducing the activation energy of initiation step in the lipid autoxidation [59]. Micciche and others [60] reported that the initiation of the oxidation of methyl linoleate requires the coexistence of ferrous (Fe^{2+}) and ferric ions (Fe^{3+}), and the oxidation is pH dependent. Metals accelerate the lipid oxidation by participating in producing lipid radicals and some of reactive oxygen species such as singlet oxygen and hydroxy radicals. Metals directly react with lipids and produce lipid radicals. They also produce superoxide anion radicals by the reaction with triplet oxygen [28]. Superoxide anion can produce hydrogen peroxide by dismutation, and the reaction between hydrogen peroxide

TABLE 17.4

Copper and Iron Contents in Edible Oils

Oils	Copper (ppb)	Iron (ppm)
Crude soybean oil[a]	13.2	2.80
Refined soybean oil[a]	2.5	0.20
Cold-pressed sesame oil[b]	16 (3.0 to 38)	1.2 (0.18–1.52)
Virgin olive oil[b]	9.8 (1.0 to 79)	0.7 (nd[c]–9.79)

Source: [a] Adapted from Sleeter, R.T., *J. Am. Oil Chem. Soc.*, 58, 239, 1981.
[b] adapted from MAFF, http://archive.food.gov.uk/maff
[c] Not detected.

and superoxide anion produces hydroxy radical, a strong oxidant, and singlet oxygen (Haber–Weiss reaction) [61]. The rate of Haber–Weiss reaction is increased in the presence of transition metals [62].

$$O_2^{-\bullet} + O_2^{-\bullet} + 2H^+ \rightarrow O_2 + H_2O_2 \text{ (dismutation)}$$

$$H_2O_2 + O_2^{-\bullet} \rightarrow HO^\bullet + OH^- +^1 O_2 \text{ (Haber–Weiss reaction)}$$

Hydroxy radical is also produced from decomposition of hydrogen peroxide in the presence of metals, so-called Fenton reaction [63]. Acceleration of hydrogen peroxide decomposition by copper is higher than that by iron, and ferrous ion acts 100 times faster than ferric ion [28].

Metals increase the decomposition of lipid hydroperoxide (ROOH) to alkoxy (RO$^\bullet$) or lipid peroxy (ROO$^\bullet$) radicals, which can accelerate the lipid oxidation [64].

$$ROOH + Fe^{2+} \rightarrow RO^\bullet + Fe^{3+} + OH^-$$

$$ROOH + Fe^{3+} \cdot \rightarrow ROO^\bullet + Fe^{2+} + H^+$$

Decomposition rate of lipid hydroperoxides by ferrous ion, 1.5×10^3 M^{-1}s^{-1} [65], is higher than that by ferric ion [66,67]. Ferric ion can degrade phenolic compounds such as caffeic acid in olive oil and decreases the oxidative stability of oil [68]. Lactoferrin binds the iron in fish oil or soybean oil and decreases the prooxidant activity of iron in the oil autoxidation at 50°C–120°C [69].

D. Phospholipids

Crude oil contains phospholipids such as phosphatidylethanolamine, phosphatidylcholine, phosphatidylinositol, and phosphatidylserine, but most of them are removed by oil processing such as degumming. The oils that are consumed without refining contain higher amount of phospholipids. Crude soybean oil contains phosphatidylcholine and phosphatidylethanolamine at 501 and 214 ppm, respectively; however, RBD soybean oil contains 0.86 and 0.12 ppm of phosphatidylcholine and phosphatidylethanolamine, respectively [70]. Unroasted sesame oil contains 690 ppm of phospholipids [71]. Extra virgin olive oil contains 34–156 ppm phospholipids and filtration of the oil lowers the contents to 21–124 ppm [72].

Phospholipids act as antioxidants and prooxidants depending on the presence of metals and concentration. Phosphatidylcholine decreased the oxidation of docosahexaenoic acid (DHA) at 25°C–30°C in the dark [73]. Peroxide formation in the autoxidation of soybean oil at 50°C for 8 weeks was decreased by addition of phosphatidylcholine, phosphatidylethanolamine, and phosphatidylserine at 0.03%–0.05% [74]. The egg yolk phospholipids at 0.031%–0.097% decreased the autoxidation of DHA-rich oil and squalene, and the antioxidant activity of egg yolk phosphatidylethanolamine was higher than that of phosphatidylcholine [75]. The mechanism of antioxidative effects of phospholipids has not been yet elucidated in detail. The polar group in phospholipids plays an important role; choline and ethanolamine which are degradation products of phosphatidylcholine and phosphatidylethanolamine, respectively, acted as antioxidants in the autoxidation of herring oil at 40°C [76]. King and others [77] reported that nitrogen-containing phospholipids such as phosphatidylcholine and phosphatidylethanolamine were efficient antioxidants under most conditions. The nitrogen moieties can donate hydrogen or electron to antioxidant radicals such as tocopheroxy radical and regenerate the antioxidants. Phospholipids also act as metal scavengers [72]; they chelate prooxidative metals and decrease the lipid oxidation.

Soybean oil oxidation was decreased with addition of 5–10 ppm phospholipids, but higher amount of phospholipids acted as prooxidants [78]. Phosphatidylcholine (300 ppm) added to purified soybean oil increased the volatile compound formation and oxygen consumption in the oil [78]. The prooxidant activity of phospholipids was suggested to be due to the presence of hydrophilic and hydrophobic groups in the molecule. The hydrophilic groups such as choline or ethanolamine are on the surface of oil and hydrophobic hydrocarbons are in the oil. Phospholipids decrease the surface tension of oil and increase the diffusion rate of oxygen from the headspace to the oil, which accelerates the oil oxidation. Prooxidant activity of phospholipids at 300 ppm was not observed when 1 ppm iron was copresent in the oil, rather phospholipids decreased the soybean oil oxidation [78]. Phosphatidic acid and phosphatidylethanolamine showed the highest antioxidant effect on soybean oil oxidation in the dark at 60°C in the presence of 1 ppm iron, followed by phosphatidylcholine, phosphatidylglycerol, and phosphatidylinositol [78].

Phospholipids having polyunsaturated fatty acids as well as phosphate group in the structure can be oxidized in the presence of iron, especially ferric ion. Ferric ion binds to the lipid phosphate, and this bound iron catalyzes the breakdown of hydroperoxides. Phospholipids having monounsaturated fatty acid are not oxidized. The Fe^{+3}-catalyzed oxidation of polyunsaturated phospholipids occurs selectively adjacent to specific double bonds such as C9 or C11 [79].

E. THERMALLY OXIDIZED COMPOUNDS AND HYDROPEROXIDES

Since crude oil is usually processed at high temperature, thermally oxidized compounds are produced and edible oils contain them. Dimers and trimers joined through carbon-to-oxygen linkage and hydroxy dimers are examples of oxidized compounds produced during oil processing. The RBD soybean oil contains 1.2% thermally oxidized compounds [80]. Thermally oxidized compounds accelerate the oil autoxidation, and the acceleration increases with their concentration [80]. Lipid hydroperoxides increase the headspace oxygen consumption and volatile formation in soybean oil at 55°C in the dark [81]. Prooxidant activity of oxidized compounds and peroxides is related to their structure. Oxidized compounds have both hydrophilic carbonyl group and hydrophobic hydrocarbons, and can act as an emulsifier. The emulsifiers lower the surface tension of oil and increase oxygen diffusion into the oil to accelerate the oil oxidation [82].

F. CHLOROPHYLLS

Some crude oils contain chlorophylls; virgin olive and rapeseed oils contain chlorophylls at 10 ppm and 5–35 ppm, respectively [83]. Chlorophylls are generally removed during the oil processing, especially the bleaching process. Crude canola oil contains 26 ppm chlorophylls; however, only 5% of chlorophylls (1.3 ppm) remained in the bleached oil [84], as shown in Table 17.5.

Chlorophylls and their degradation products, pheophytins and pheophorbides, act as antioxidants in the autoxidation of lipid [85–87]. Chlorophylls decrease the contents of free radicals in oil possibly by donating hydrogen to free radicals [86], which can break the chain reaction of lipid oxidation. Porphyrin was proposed to be an essential chemical structure for the antioxidant activity of chlorophylls [88]. The antioxidative activity of chlorophylls depends on the derivatives present,

TABLE 17.5
Chlorophyll Contents of Canola Oil during Processing (ppm)

Oil	Chlorophylls	Pheophytins	Pyropheophytins	Total (Relativity Based on Crude Oil)
Crude	1.88	4.65	19.70	26.23 (100)
Degummed	0.27	8.23	11.24	19.74 (75)
Refined	0.22	7.39	10.92	18.51 (71)
Bleached	—	0.88	0.46	1.34 (5)

Source: From Suzuki, K. and Nishioka, A., *J. Am. Oil Chem. Soc.*, 70, 837, 1993.

the lipid as a substrate, and temperature. Chlorophyll *a* showed a higher antioxidant activity in the autoxidation of rapeseed and soybean oils at 30°C than pheophytin [85]. Pheophytin *a* increased the induction period in the autoxidation of virgin olive oil at 60°C and 80°C, and the antioxidation was concentration dependent [89]. Copresence of pyropheophytin *a* improves the antioxidant activity of pheophytin *a*, and the formation of pyropheophytin from pheophytin is favored at higher temperature [89].

Chlorophylls increase the lipid oxidation under light. Purified soybean oil that contained no chlorophylls did not produce volatiles in the headspace under light at 10°C. However, the oil added with chlorophyll *a* produced headspace volatiles under the same experimental conditions, and the volatile formation increased as the concentration of chlorophyll increased [32]. The oxidation of virgin olive oil containing pheophytin increased by the illumination of fluorescent light [41]. Chlorophylls and their degradation products act as sensitizers in the presence of light and produce singlet oxygen by transferring the energy to atmospheric triplet oxygen, which accelerates the lipid oxidation [90–92]. Pheophytin has a higher sensitizing activity than chlorophyll, but lower than that of pheophorbide [41,93].

G. CAROTENOIDS

Carotenoids are tetraterpenoid compounds consisting of isoprenoid units. Double bonds in carotenoids are conjugated and in all *trans* forms. Carotenes, lutein, zeaxanthin, canthaxanthin, and astaxanthin are common carotenoids found in unrefined oil. Carotene is frequently found carotenoid in oils, and β-carotene is one of the most studied carotenoids. Crude palm oil and red palm olein contain high amount of carotenoids of 500–700 ppm [94]. Virgin olive oil contains 1.0–2.7 ppm β-carotene as well as 0.9–2.3 ppm lutein [89].

β-Carotene decreases lipid oxidation by scavenging free radicals, filtering out the light, or quenching singlet oxygen. A reduction potential of β-carotene (1.06 V) is not high enough to donate hydrogen to lipid radical ($E^{0'} = 0.6$ V) or lipid peroxy radicals ($E^{0'} = 0.77$–1.44 V). β-Carotene can donate hydrogen to hydroxy radical (HO$^{\bullet}$) whose reduction potential is 2.31 V, and become a carotene radical (Car$^{\bullet}$). Carotene radical is more stable than hydroxy radical due to its delocalization of unpaired electron through the conjugated polyene system. Carotene radical may react with lipid peroxy radicals (ROO$^{\bullet}$) at low oxygen concentration and form nonradical products, carotene peroxide [95,96], which can terminate the radical chain reaction.

$$Car + HO^{\bullet} \rightarrow Car^{\bullet} + H_2O$$

$$Car^{\bullet} + ROO^{\bullet} \rightarrow Car - OOR$$

β-Carotene added to purified olive oil which did not contain chlorophylls decreased hydroperoxide formation and headspace oxygen consumption in the oil under light at 25°C by light-filtering effect [90]. β-Carotene absorbs light energy, mainly between 400 and 500 nm. The higher the concentration of β-carotene in purified olive oil was, the lower the oil oxidation [90].

β-Carotene decreases photooxidation of soybean oil by quenching singlet oxygen [32,97]. Singlet oxygen quenching by β-carotene does not involve the oxidation of β-carotene, and is achieved physically by energy transfer [98]. High energy of singlet oxygen (93.6 kJ/mol) is transferred to β-carotene whose energy level (88 kJ/mol) is just below that of singlet oxygen [96], and results in less reactive triplet oxygen. One mole of β-carotene can quench 250–1000 molecules of singlet oxygen at a rate of 1.3×10^{10} M^{-1}s^{-1} [99].

$$^{1}O_2 + {}^{1}Car \rightarrow {}^{3}O_2 + {}^{3}Car^{*}$$

$$^{3}Car^{*} \rightarrow Car + Heat$$

The singlet oxygen quenching activity of carotenoids increases with the number of conjugated double bonds in the structure [4,96,100]. The substituents in the β-ionone ring also affect the singlet oxygen quenching activity [101]. β-Carotene and lycopene which have 11 conjugated double bonds are more effective singlet oxygen quenchers than lutein having 10 conjugated double bonds [102]. Presence of oxo and conjugated keto groups, or cyclopentane ring in the structure increases the singlet oxygen quenching ability, and β-ionone ring substituted with hydroxy, epoxy, or methoxy groups is less effective [102]. Antioxidant activity of carotenoids in the photooxidation of oil is concentration dependent; there was no significant difference in the antioxidant activity between lycopene and β-carotene at 10 ppm, but β-carotene was a better antioxidant than lycopene at 20 ppm, and the reverse phenomenon was observed at 40 ppm [102].

Carotenoids having less than nine conjugated double bonds are not good singlet oxygen quenchers; they act as sensitizer quenchers. Carotenoids inactivate excited photosensitizers (^{3}Sen*) by absorbing energy from them. The excited triplet carotenoids return to their singlet state by transferring the energy to the surrounding [103].

$$^1Carotenoid + {}^3Sen^* \rightarrow {}^3Carotenoid + Sen$$

$$^3Carotenoid \rightarrow {}^1Carotenoid$$

Although carotenoids can decrease the lipid oxidation, their prooxidant activity is also shown. β-Carotene acts as a prooxidant in soybean oil oxidation in the dark by the electron transfer mechanism [104]. β-Carotene may donate electrons to free radicals and become a β-carotene cation radical [105,106]. β-Carotene may undergo the addition reaction to lipid peroxy radical at high oxygen concentration, for example, higher than 150 mmHg of oxygen, and produce carotene peroxy radical (ROO – Car$^•$) [95]. β-Carotene peroxy radical reacts with triplet oxygen and then with lipid molecules (R'H), and the resulting lipid radicals accelerate the chain reaction of lipid oxidation [107].

$$Car + ROO^• \rightarrow ROO - Car^•$$

$$ROO - Car^• + {}^3O_2 \rightarrow ROO - Car - OO^•$$

$$ROO - Car - OO^• + R'H \rightarrow ROO - Car - OOH + R'^•$$

H. TOCOPHEROLS AND TOCOTRIENOLS

TOH are the most important natural antioxidants present in oil, especially soybean, canola, sunflower, and corn oils, and the contents are affected by the cultivar [108,109]. Animal lipids contain lower amount of TOH than vegetable oils; most vegetable oils contain more than 500 ppm TOH [110–112], but beef tallow and lard contain only 34 and 18 ppm (Table 17.6) [113,114], respectively. Palm oil contains low amount of TOH (107 ppm), but it has high concentration of tocotrienols; concentrations of α-, γ-, and δ-tocotrienols in palm oil are 211, 353–372, and 56–67 ppm, respectively [115]. Safflower oil contains tocotrienols at 12–15 ppm in addition to 397–540 ppm TOH [116]. The oil refining, especially deodorization, reduces TOH contents in oils [44,117,118]. Crude, bleached, and deodorized soybean oil contains TOH at 1670, 1467, and 1138 ppm, respectively [44]. Crude sunflower oil contains 755 ppm TOH, and bleaching and deodorization reduced the contents of TOH to 97% and 75%, respectively [119].

TOH decrease the lipid oxidation by scavenging free radicals. TOH have a standard reduction potential of 0.5 V, and can donate hydrogens to the alkyl, alkoxy, and peroxy radicals of lipids whose standard reduction potential is 0.6, 1.6, and 1.0 V, respectively [120]. TOH react with lipid peroxy radicals (ROO$^•$) to produce tocopheroxy radicals (TO$^•$) and lipid hydroperoxide (ROOH) at a rate of 10^4–10^9 M^{-1}s^{-1}, which is higher than the rate between lipid peroxy radical

TABLE 17.6
Tocopherol Contents of Edible Oil

Oil	Tocopherol (ppm)				
	α	β	γ	δ	Total
Soybean[a]	116.0	34.0	737.0	275.0	1162
Canola[a]	272.1	0.1	423.2	—	695.4
Sunflower[a]	613.0	17.0	18.9	—	648.9
Corn[a]	134.0	18.0	412.0	39.0	603
Roasted sesame[b]	4	—	584	9	597
Rapeseed[c]	252	—	314	—	566
Safflower[d]	386 to 520	8.6 to 12.4	2.4 to 7.7	—	397 to 540
Olive[e]	168 to 226	—	—	—	168 to 226
Palm	89	—	18	—	107
Beef tallow[f]	30.4	—	3.8	—	34.2
Lard[g]	18.0	—	—	—	18.0

Source: [a]Adapted from Przybylski, R., *Canola Council of Canada*, 1, 2001; [b]adapted from Kamal-Eldin, A. and Andersson, R., *J. Am. Oil Chem. Soc.*, 74, 375, 1997; [c]adapted from Velasco, J., Andersen, M.L., and Skibsted, L.H., *Food Chem.*, 77, 623, 2003; [d]adapted from Lee, Y.C. et al., *Food Chem.*, 84, 1, 2004; [e]adapted from Salvador, M.D. et al., *Food Chem.*, 74, 267, 2001; [f]adapted from Choe, E. and Lee, J., *Korean J. Food Sci. Technol.*, 30, 288, 1998; [g]adapted from Drinda, H. and Baltes, W., *Z. Lebensm. Unters. Forch.*, 4, 270, 1999.

and lipids (10–$60 \text{ M}^{-1} \text{ s}^{-1}$). This results in decrease in lipid radical formation and lipid oxidation. The tocopheroxy radicals have a resonance structure as shown in Figure 17.9, and are more stable than alkyl, alkoxy, and peroxy radicals of lipids. Tocopheroxy radicals may undergo reactions such as scavenging another radical to give a stable nonradical product, reacting with reducing agent such as ascorbic acid to regenerate TOH, or attacking lipids to produce new reactive lipid radicals that may initiate a new oxidation chain reaction. The relative importance of these secondary reactions of tocopheroxy radical influences the total antioxidant capacity of TOH [121]. When

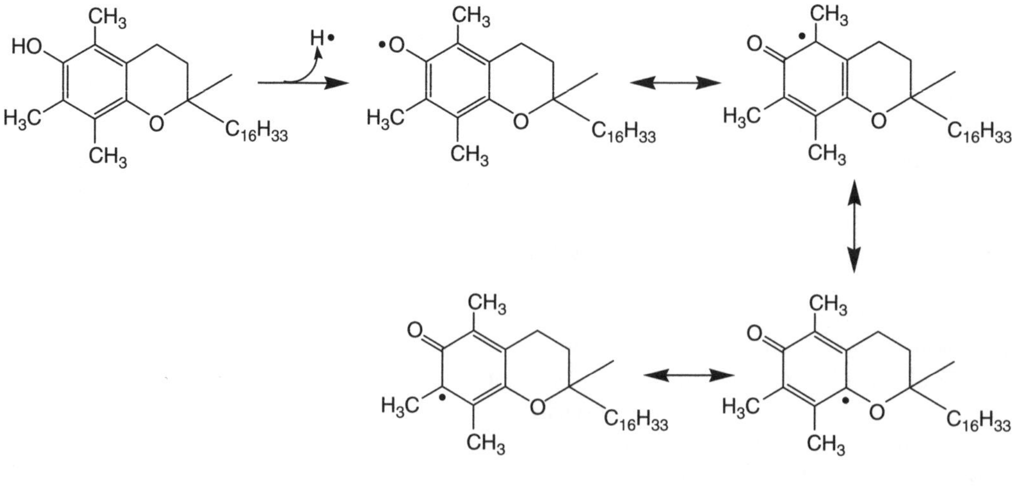

FIGURE 17.9 Resonance stabilization of α-tocopheroxy radical.

there are not enough lipid peroxy radicals at low lipid oxidation rate, tocopheroxy radicals react with each other and produce tocopheryl quinone and TOH. At higher lipid oxidation rate, tocopheroxy radicals can react with lipid peroxy radicals and produce TOH–lipid peroxy complexes, [TO-OOR], which are then hydrolyzed to tocopheryl quinone and lipid hydroperoxide (ROOH) [122]. Tocopheryl quinone was reported to decrease the malonaldehyde formation in the oxidation of 5, 8, 11-octadecatrienoic acid at room temperature [123].

$$TOH + ROO^{\bullet} \rightarrow TO^{\bullet} + ROOH$$

$$TO^{\bullet} + TO^{\bullet} \rightarrow tocopheryl\ quinone + TOH$$

$$TO^{\bullet} + ROO^{\bullet} \rightarrow [TO - OOR] \rightarrow tocopheryl\ quinone + ROOH$$

Antioxidant activity of TOH differs among the isomers. δ-Tocopherol generally has the highest free radical-scavenging activity followed by γ-, β-, and α-tocopherols [117]. However, Yanishlieva and others [124] reported that the antioxidant activity of α-tocopherol was higher than that of γ-tocopherol at low concentration (<400–700 ppm) in purified soybean and sunflower oil at 100°C. The optimal concentration of TOH as antioxidants is dependent on their oxidative stability; the isomer having lower oxidative stability generally shows lower optimal concentration for the maximal antioxidant activity. α-Tocopherol, the least stable isomer, shows the maximal antioxidant activity at 100 ppm in the autoxidation of soybean oil at 55°C, whereas the optimal concentrations of more stable γ- and δ-tocopherols are 250 and 500 ppm, respectively [125].

TOH, particularly α-tocopherol, act as prooxidants when present at high concentrations in oil [125–127]. α-Tocopherol at 60–70 ppm acted as a prooxidant in virgin olive oil autoxidation [109]. Prooxidant activity of TOH is more obvious when the concentration of lipid peroxy radicals is very low. Addition of 100 ppm α-tocopherol increased the oxidation of purified olive oil at the early stage of autoxidation; however, the same concentration of α-tocopherol added to moderately oxidized purified olive oil or lard (peroxide value = 15 meq/kg) significantly decreased the oil oxidation [128]. Tocopheroxy radical abstracts hydrogen from the lipid and produces TOH and lipid radical although the reaction rate is very low. Formation of lipid radicals by tocopheroxy radicals accelerates lipid oxidation, and this is called TOH-mediated peroxidation [129,130]. α-Tocopherol shows the highest prooxidant activity followed by γ- and δ-tocopherols in soybean oil autoxidation at 500–1000 ppm [125]. Prooxidant activity of α-tocopherol decreases as the oxidation temperature increases [131]. Ascorbic acid can provide hydrogen to tocopheroxy radical and prevent TOH-mediated peroxidation [130].

In addition to free radical-scavenging activity, TOH decrease lipid oxidation under light by singlet oxygen quenching [132], but their singlet oxygen quenching activity is 50-fold less than carotenoids [101]. Singlet oxygen quenching by TOH is physically achieved by charge transfer. Tocopherols (T) donate electron to singlet oxygen and then form a charge transfer complex ($[T^{+}-{}^{1}O_2]_1$) with singlet oxygen. The singlet state of tocopherol–$^{1}O_2$ complex undergoes inter-system crossing into the triplet state ($[T^{+}-{}^{1}O_2]_3$) and then dissociates to TOH and triplet oxygen. TOH can deactivate 40–120 molecules of singlet oxygen before they are destroyed [133].

Singlet oxygen quenching rate depends on the TOH isomers. The rates of physical quenching of singlet oxygen by α-, β-, γ-, and δ-TOH were 4.2×10^7, 2.3×10^7, 1.1×10^7, and 0.5×10^7 M^{-1} s^{-1}, respectively [134]. The difference in singlet oxygen quenching activity among TOH isomers is affected by concentration of TOH; when TOH are present at 1×10^{-3} M, singlet oxygen quenching activity is in the decreasing order of α-, γ-, and δ-tocopherols, however, there was no significant difference among TOH isomers at 4×10^{-3} M [135].

In addition to physical quenching of singlet oxygen, TOH can quench singlet oxygen by reacting irreversibly with singlet oxygen, and this is called chemical quenching [4]. The reaction between singlet oxygen and TOH produces oxidized compounds such as TOH hydroperoxydienone, tocopheryl quinone, and tocopheryl quinone epoxide (Figure 17.10). The oxidation rate of

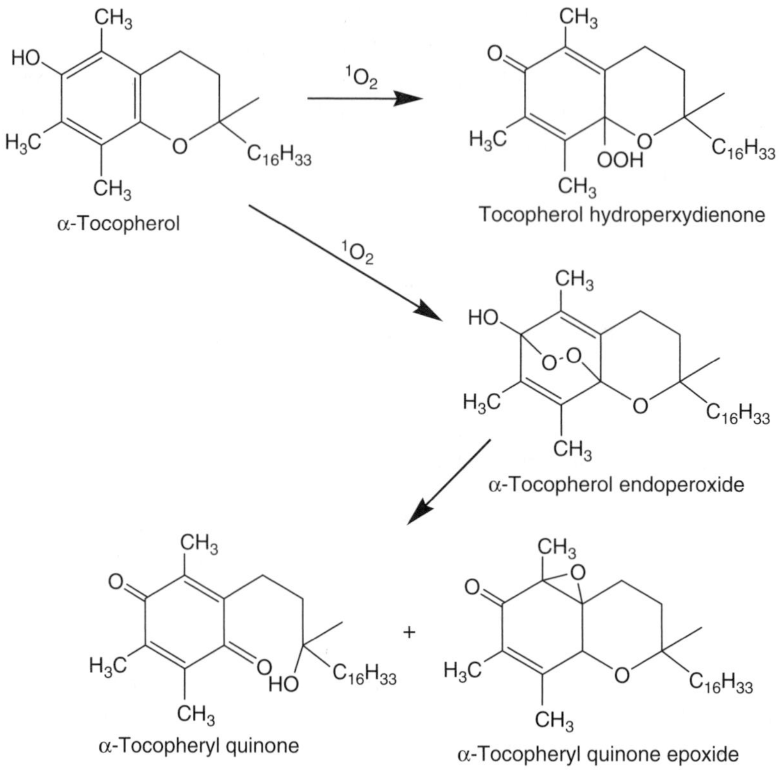

FIGURE 17.10 Singlet oxygen oxidation of α-tocopherol.

TOH by singlet oxygen differs among the isomers. α-Tocopherol reacts with singlet oxygen at the highest rate of 2.1×10^8 M^{-1} s^{-1}, followed by β- (1.5×10^8 M^{-1} s^{-1}), γ- (1.4×10^8 M^{-1} s^{-1}), and δ-tocopherols (5.3×10^7 M^{-1} s^{-1}) [136]. The chemical reaction proceeds through an intermediate hydroperoxide that decomposes to tocopheryl quinone and tocopheryl quinone epoxide [133]. Singlet oxygen-oxidized TOH at 250–1000 ppm increased the oxygen consumption and peroxide formation in soybean oil at 55°C in the dark, and the prooxidant activity was the highest in oxidized α-tocopherol followed by oxidized γ- and δ-tocopherols [137]. Prevention of TOH oxidation and removal of oxidized TOH during oil processing are strongly recommended to improve the oxidative stability of oil.

I. OTHER PHENOLIC COMPOUNDS

Oils contain antioxidative phenolic compounds other than TOH. Lignans in flaxseed and sesame oil, polyphenols such as tyrosol, hydroxytyrosol, catechol, and luteolin in olive oil, chlorogenic and caffeic acids in sunflower oil, and sinapic acid in rapeseed oil are good examples [138,139]. Figure 17.11 shows chemical structures of some phenolic compounds present in edible oils. Flaxseed lignans are mainly secoisolariciresinol and secoisolariciresinol diglucoside, and sesamin, sesamol, sesamolin, sesaminol, and sesamolinol are lignan compounds found in sesame oil. Roasting of sesame seeds causes the hydrolysis of sesamolin to sesamol [140,141] and increases the content of sesamol. Roasted sesame oil contains 36 ppm of sesamol [45], whereas unroasted sesame oil contains less than 7 ppm [142,143]. Contents of tyrosol, hydroxytyrosol, and phenolic acids in olive oil are 34.9, 37.8, and 36.3 ppm, respectively [68]. Phenolic compounds in oils are removed during the refining process; the alkaline refining eliminates hydroxytyrosol, catechol, and luteolin in olive oil, and tyrosol is removed during deodorization at 240°C [144], as shown in Table 17.7.

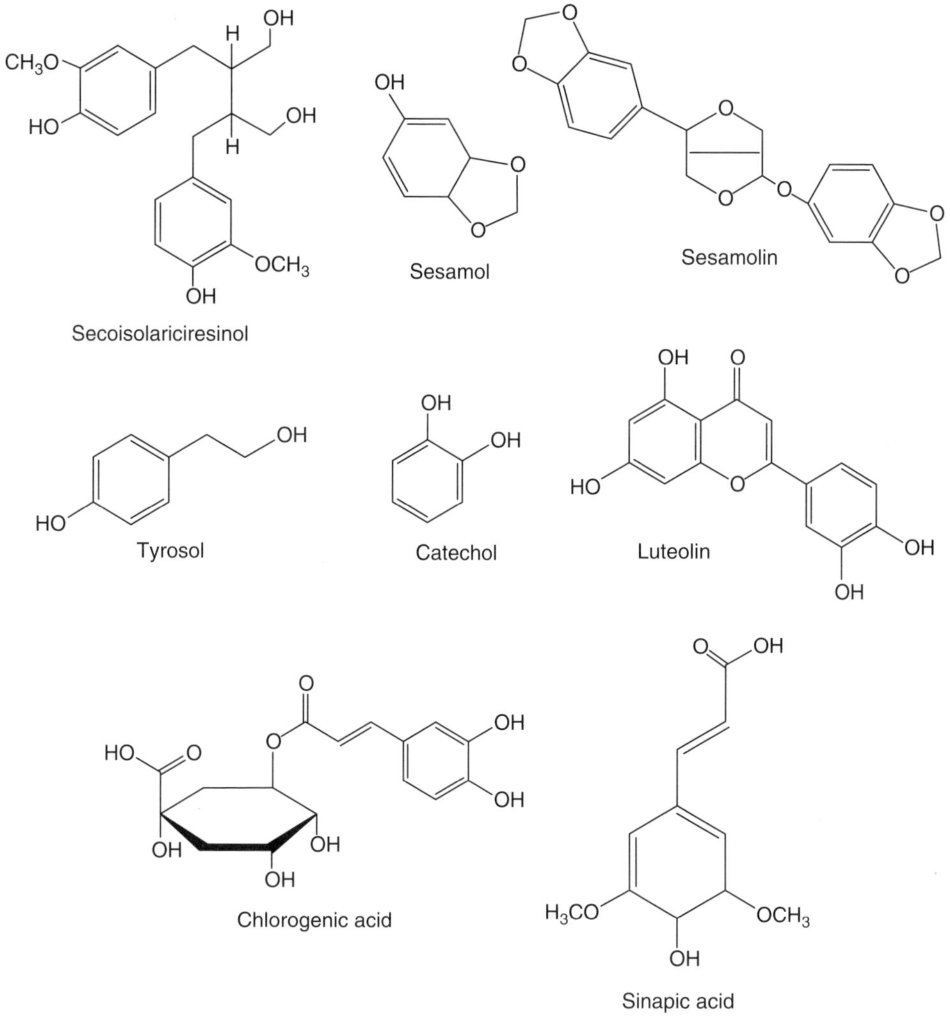

FIGURE 17.11 Phenolic compounds present in oils.

Lignans and polyphenols decrease the lipid oxidation by scavenging free radicals [121] or chelating metals [68]. Polyphenols donate phenolic hydrogen to radicals and produce semiquinone radicals. The semiquinone radical may scavenge another radical to give a quinone, disproportionate

TABLE 17.7
Effects of the Refining on the Phenolic Concentration (ppm) of Olive Oil

Polyphenol	Crude	Alkali Refined	Deodorized
Hydroxytyrosol	2.1	0.1	—
Catechol	4.1	0.6	—
Tyrosol	3.0	2.6	—
Luteolin	3.1	—	—

Source: From Garcia, A. et al., *J. Am. Oil Chem. Soc.*, 83, 159, 2006.

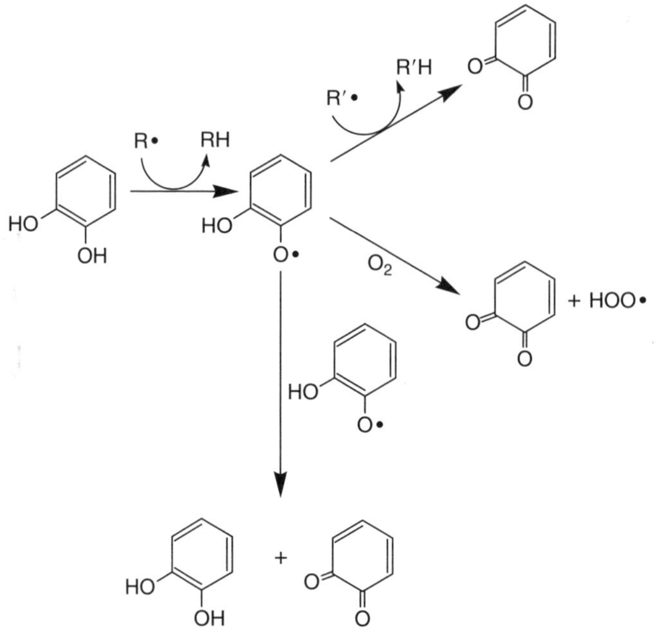

FIGURE 17.12 Reaction of catechol with lipid radicals to produce quinones ($R^\bullet$ and $R'^\bullet$, lipid radicals).

with another semiquinone radical to give the parent compound and quinone, or react with oxygen to produce quinone and a hydroperoxy radical as shown in Figure 17.12 [121]. Sesame oil shows a good oxidative stability because of sesame lignans [17,145–148] although it contains high amount of unsaturated fatty acids (iodine value = 109). Roasted sesame oil is more stable to the autoxidation than unroasted sesame oil [149,150]. The antioxidant activity of sesamol and sesaminol by scavenging radicals is higher than that of sesamin in sunflower oil autoxidation [143]. Flaxseed lignans also decrease the lipid autoxidation; secoisolariciresinol (300 ppm) increased the induction time from 10.4 to 11.1 h in the autoxidation of canola oil at 23°C for 30 days and the oxidation decreased as the lignan concentration and oxidation time increased; after 120 days oxidation, the induction time increased from 7.1 h in the oil without flaxseed lignan to 10.1 h in the oil having lignan [151].

Olive oil is very stable to the autoxidation due to free radical-scavenging and metal-chelating activity of polyphenols [152–155]. Polyphenols in olive oil play an antioxidant role mainly at the initial stage of autoxidation [109,156]. Hydroxytyrosol is the most effective in decreasing the autoxidation of olive oil [152,153,157]. Caffeic acid reacts with ferric ions and is oxidized to quinones, which is not effective in inhibiting iron-dependent free radical chain reactions any more [68]. However, hydroxytyrosol, tyrosol, vanillic acids, and p-coumaric acid are not oxidized by the ferric ions, and can exert antioxidant activity [68]. The antioxidant capacity of polyphenols decreases as the lipid autoxidation increases [121].

The singlet oxygen quenching activity of sesamol in chlorophyll-sensitized photooxidation of soybean oil was lower than that of α-tocopherol, similar to that of δ-tocopherol, and higher than that of DABCO (diazabicyclo[2,2,2]octane) at the same molar concentration [158]. Sesamol acts as a singlet oxygen quencher with a rate of 1.9×10^7 M^{-1} s^{-1} at 20°C [158].

Osborne and Akoh [159] reported that polar phenolic acids such as quercetin and gallic acid increased the reduction of iron and accelerated the oxidation of canola oil and caprylic acid structured lipid in the presence of iron at pH 3.0. Sitosterol, one of phenolic phytosterols present in sesame and corn oil [143]. may compete with lipid radicals at the oil surface and slightly decrease the oil oxidation [160,161]. Yanishlieva and Schiller [162] reported prooxidant activity of sitosterol by increasing the oxygen solubility in oil.

J. ANTIOXIDANT INTERACTIONS

Most oils contain antioxidants more than one kind and show interactions in decreasing lipid oxidation. Synergism is a phenomenon in which a net interactive antioxidant effect is higher than the sum of the individual effects. Synergistic antioxidant activity has been often observed in the copresence of metal chelator and free radical scavengers, because metal chelators mainly act at the initiation step of lipid oxidation and radical scavengers at the propagation step [28]. TOH are the most frequently encountered antioxidants in foods, and the synergistic antioxidation has been studied mostly with TOH.

α-Tocopherol shows synergistic effects with β-carotene to decrease the autoxidation [163] and photosensitized oxidation of soybean oil [164] possibly because TOH protect β-carotene from degradation. β-Carotene (0.75 M) in oleic acid sharply disappeared from the beginning of the reaction and mostly consumed within 100 h in the absence of TOH; however, copresence of α-tocopherol at 3.8×10^{-3} M increased the time to 1500 h [165].

Phosphatidylethanolamine shows a significant synergism with TOH to decrease the oxidation of trilinolein at 37°C by scavenging lipid peroxy radicals [166] and thus sparing TOH [133]. Phosphatidylcholine, which contains a tertiary amine, does not delay oxidation of TOH [167]. Phosphatidylethanolamine donates hydrogen to tocopheroxy radicals, which slows down the oxidation of TOH to tocopheryl quinone [78]. Phosphatidylinositol acts as a synergist with TOH in decreasing lipid oxidation mainly by forming inactive complexes with prooxidative metals [133]. Kago and Terao [168] proposed that phospholipids form microemulsion in oils and active phenolic group of TOH is positioned near the polar region where lipid peroxy radicals are concentrated, resulting in synergistic effects of phospholipids with TOH.

Ascorbic acid is another synergist with TOH in decreasing lipid oxidation by regenerating TOH from tocopheroxy radicals or oxidation products [130]. Phenolic compounds synergistically decrease the olive oil autoxidation with TOH [27]. Sinapic acid, main phenolic compound of rapeseed, showed synergism with TOH in decreasing the hydroperoxide and propenal formation in the oxidation of rapeseed oil at 40°C [169]. The synergistic antioxidant activity of sinapic acid with TOH was higher at low concentration of TOH (50 μM) than at high concentration of TOH (1000 μM). Sesamol and sesaminol show synergistic antioxidant activities with γ-tocopherol in the autoxidation of sunflower oil [143].

Synergistic effects of antioxidants is affected by hydroperoxide concentration in lipid; at low concentration of hydroperoxides, for example, less than 20 meq/kg α-tocopherol (100 ppm) decreased the antioxidant activity of 3,4-dihydroxyphenylacetic acid (40 ppm) in purified olive oil at 40°C [128]. Polar phenolic compounds such as dihydroxyphenylacetic acid are more important for the antioxidant activity in the initial stage of autoxidation, and TOH becomes effective when lipid hydroperoxides reach a critical concentration.

To minimize the lipid oxidation to improve the quality of lipid foods, it is recommended to monitor the minor compounds carefully in foods during processing and storage; metals and oxidized compounds are to be removed and appropriate amounts of antioxidants such as TOH and phenolic compounds should be selected. Heat, light, and oxygen should be also excluded as much as possible during handling, processing, and storage of foods containing high amount of lipids.

REFERENCES

1. Hamilton, R.J., The chemistry of rancidity in foods, in *Rancidity in Foods*, 3rd edn., Allen, J.C. and Hamilton, R.J., Eds., Blackie Academic & Professional, London, 1, 1994.
2. Aruoma, O.I., Free radicals, oxidative stress, and antioxidants in human health and disease, *J. Am. Oil Chem. Soc.*, 75, 199, 1998.

3. Min, D.B. and Bradley, G.D., Fats and oils: flavors, in *Wiley Encyclopedia of Food Science and Technology*, Hui, Y.H., Ed., John Wiley & Sons, New York, NY, 828, 1992.
4. Min, D.B. and Boff, J.M., Lipid oxidation of edible oil, in *Food Lipids*, Akoh, C.C. and Min, D.B., Eds., Marcel Dekker, New York, NY, 335, 2002.
5. Zhu, J. and Sevilla, M.D., Kinetic analysis of free-radical reactions in the low-temperature autoxidation of triglycerides, *J. Phys. Chem.*, 94, 1447, 1990.
6. Aidos, I. et al., Stability of crude herring oil produced from fresh byproducts: influence of temperature during storage, *J. Food Sci.*, 67, 3314, 2002.
7. Frankel, E.N., Chemistry of autoxidation: mechanism, products and flavor significance, in *Flavor Chemistry of Fats and Oils*, Min, D.B. and Smouse, T.H., Eds., American Oil Chemists' Society, Champaign IL, 1, 1985.
8. Hiatt, R. et al., Homolytic decomposition of hydroperoxides, *J. Org. Chem.*, 33, 1416, 1968.
9. Vaisey-Genser, M. et al., Consumer acceptance of stored canola oils in canola, 12th project report 'Research on Canola, Seed, Oil Meal', *Canada Canola Council*, Canada, 189, 1999.
10. Steenson, D.F.M., Lee, J.H., and Min, D.B., Solid-phase microextraction of volatile soybean oil and corn oil compounds, *J. Food Sci.*, 67, 71, 2002.
11. Warner, K. et al. Flavor score correlation with pentanal and hexanal contents of vegetable oil, *J. Am. Oil Chem. Soc.*, 55, 252, 1978.
12. Przybylski, R. and Eskin, N.A.M., Methods to measure volatile compounds and the flavor significance of volatile compounds, in *Methods to Assess Quality and Stability of Oils and Fat-Containing Foods*, Warner, K. and Eskin, N.A.M., Eds., American Oil Chemists' Society, Champaign, IL, 107, 1995.
13. Choe, E., Effects of heating time and storage temperature on the oxidative stability of heated palm oil, *Korean J. Food Sci. Technol.*, 29, 407, 1997.
14. Heinonen, M. et al., Inhibition of oxidation in 10% oil-in-water emulsions by β-carotene with α- and γ-tocopherols, *J. Am. Oil Chem. Soc.*, 74, 1047, 1997.
15. Parker, T.D. et al., Fatty acid composition and oxidative stability of cold-pressed edible seed oils, *J. Food Sci.*, 68, 1240, 2003.
16. Martin-Polvillo, M., Marquez-Ruiz, G., and Dobarganes, M.C., Oxidative stability of sunflower oils differing in unsaturation degree during long-term storage at room temperature, *J. Am. Oil Chem. Soc.*, 81, 577, 2004.
17. Tan, C.P. et al., Comparative studies of oxidative stability of edible oils by differential scanning calorimetry and oxidative stability index methods, *Food Chem.*, 76, 385, 2002.
18. Labuza, T.P. and Ragnarsson, J.O., Kinetic history effect on lipid oxidation of methyl linoleate in a model system, *J. Food Sci.*, 50, 145, 1985.
19. Ulkowski, M., Musialik, M., and Litwinienko, G., Use of differential scanning calorimetry to study lipid oxidation. 1. Oxidative stability of lecithin and linolenic acid, *J. Agric. Food Chem.*, 53, 9073, 2005.
20. Karel, M., Kinetics of lipid oxidation, in *Physical Chemistry of Foods*, Schwarzberg, H.G. and Hartel, R.W., Eds., Marcel Dekker, New York, NY, 651, 1992.
21. Andersson, K., *Influence of Reduced Oxygen Concentrations on Lipid Oxidation in Food During Storage* [Ph.D. thesis]. Chalmers Reproservice, Sweden, Chalmers University of Technology and the Swedish Institute for Food and Biotechnology, 1998.
22. Labuza, T.P., Kinetics of lipid oxidation in foods, *CRC Crit. Rev. Food Sci. Technol.*, 2, 355, 1971.
23. Kacyn, L.J., Saguy, I., and Karel, M., Kinetics of oxidation of dehydrated food at low oxygen pressures, *J. Food Proc. Preserv.*, 7, 161, 1983.
24. Min, D.B. and Wen, J., Effects of dissolved free oxygen on the volatile compounds of oil, *J. Food Sci.*, 48, 1429, 1983.
25. Shahidi, F. and Spurvey, S.A., Oxidative stability of fresh and heated-processed dark and light muscles of mackerel (*Scomber scombrus*), *J. Food Lipids*, 3, 13, 1996.
26. St. Angelo, A.J., Lipid oxidation in foods, *Crit. Rev. Food Sci. Nutr.*, 36, 175, 1996.
27. Velasco, J. and Dobarganes, C., Oxidative stability of virgin olive oil, *Eur. J. Lipid Sci. Technol.*, 104, 661, 2002.
28. Choe, E. and Min, D.B., Mechanisms and factors for edible oil oxidation, *Comp. Rev. Food Sci. Food Safety*, 5, 169, 2006.
29. Chen, Y. et al., Active oxygen generation and photo-oxygenation involving temporin (*m*-THPC), *Dyes Pigments*, 51, 63, 2001.

30. Kochevar, I.E. and Redmond, R.W., Photosensitized production of singlet oxygen, *Meth. Enzymol.*, 319, 20, 2000.

31. Rawls, H.R. and Van Santen, P.J., A possible role for singlet oxidation in the initiation of fatty acid autoxidation, *J. Am. Oil Chem. Soc.*, 47, 121, 1970.

32. Lee, E.C. and Min, D.B., Quenching mechanism of beta-carotene on the chlorophyll- sensitized photo-oxidation of soybean oil, *J. Food Sci.*, 53, 1894, 1988.

33. Gollnick, K., Mechanism and kinetics of chemical reactions of singlet oxygen with organic compounds, in *Singlet Oxygen*, Ranby, B. and Rabek, J.F., Eds., John Wiley & Sons, New York, NY, 111, 1978.

34. Min, D.B., Callison, A.L., and Lee, H.O., Singlet oxygen oxidation for 2-pentylfuran and 2-pentenylfuran formation in soybean oil, *J. Food Sci.*, 68, 1175, 2003.

35. Smouse, T.H. and Chang, S.S., A systematic characterization of the reversion flavor of soybean oil, *J. Am. Oil. Chem. Soc.*, 44, 509, 1967.

36. Ho, C.T., Smagula, M.S., and Chang, S.S., The synthesis of 2-(1-pentenyl)furan and its relationship to the reversion flavor of soybean oil, *J. Am. Oil Chem. Soc.*, 55, 233, 1978.

37. Smagula, M.S., Ho, C.T., and Chang, S.S., The synthesis of 2-(2-pentenyl) furans and their relationship to the reversion flavor of soybean oil, *J. Am. Oil Chem. Soc.*, 56, 516, 1979.

38. Chang, S.S. et al., Isolation and identification of 2-pentenylfurans in the reversion flavor of soybean oil, *J. Am. Oil Chem. Soc.*, 60, 553, 1983.

39. Vever-Bizet, C. et al., Singlet molecular oxygen quenching by saturated and unsaturated fatty-acids and by cholesterol, *Photochem. Photobiol.*, 50, 321, 1989.

40. Lee, S.H. and Min, D.B., Effects, quenching mechanisms, and kinetics of nickel chelates in singlet oxygen oxidation of soybean oil, *J. Agric. Food Chem.*, 39, 642, 1991.

41. Rahmani, M. and Saari Csallany, A., Role of minor constituents in the photooxidation of virgin olive oil, *J. Am. Oil Chem. Soc.*, 75, 837, 1998.

42. Yang, W.T.S. and Min, D.B., Chemistry of singlet oxygen oxidation of foods, in *Lipids in Food Flavors*, Ho, C.T. and Hartmand, T.G. Eds., American Chemical Society, Washington DC, 15, 1994.

43. Sattar, A., DeMan, J.M., and Alexander, J.C., Effect of wavelength on light-induced quality deterioration of edible oils and fats, *Can. Inst. Food Sci. Technol. J.*, 9, 108, 1976.

44. Jung, M.Y., Yoon, S.H., and Min, D.B., Effects of processing steps on the contents of minor compounds and oxidation stability of soybean oil, *J. Am. Oil Chem. Soc.*, 66, 118, 1989.

45. Kim, I. and Choe, E., Effects of bleaching on the properties of roasted sesame oil. *J. Food Sci.*, 70, C48, 2005.

46. Kinsella, J.E., Shimp, J.L., and Mai, J., The proximate composition of several species of freshwater fishes, *Food Life Sci. Bull.*, 69, 1, 1978.

47. Miyashita, K. and Takagi, T., Study on the oxidative rate and prooxidant activity of free fatty acids, *J. Am. Oil Chem. Soc.*, 63, 1380, 1986.

48. Mistry, B.S. and Min, D.B., Effects of fatty acids on the oxidative stability of soybean oil, *J. Food Sci.*, 52, 831, 1987.

49. Colakoglu, A.S., Oxidation kinetics of soybean oil in the presence of monolinolein, stearic acid and iron, *Food Chem.*, 101, 724, 2007.

50. Dialonnzo, R.P., Kozarek, W.J., and Wade, R.L., Glyceride composition of processed fats and oils as determined by gas chromatography, *J. Am. Oil Chem. Soc.*, 59, 392, 1982.

51. Perez-Camino, M.C., Moreda, W., and Cert, A., Determination of diacylglycerol isomers in vegetable oils by solid-phase extraction followed by gas chromatography on a polar phase, *J. Chromatogr. A.*, 721, 305, 1996.

52. Gupta, M.K., Frying oils, in *Bailey's Industrial Oil and Fat Products*, 6th edn., Shahidi, F., Ed., John Wiley and Sons, New York, NY, Vol. 4, chap. 1, 2005.

53. Mistry, B.S. and Min, D.B., Isolation of sn-a-monolinolein from soybean oil and its effects on oil oxidative stability, *J. Food Sci.*, 52, 786, 1987.

54. Mistry, B.S. and Min, D.B., Prooxidant effects of monoglycerides and diglycerides in soybean oil, *J. Food Sci.*, 53, 1896, 1988.

55. Leonardis, A.D. and Macciola, V., Catalytic effect of the Cu(II)- and Fe(III)-cyclohexanebutyrates on olive oil oxidation measured by rancimat, *Eur. J. Lipid Sci. Technol.*, 104, 156, 2002.

56. Debruyne, I., Soybean oil processing: quality criteria and flavor reversion, *Oil Mill. Gazetteer.*, 110, 10, 2004.

57. Sleeter, R.T., Effect of processing on quality of soybean oil, *J. Am. Oil Chem. Soc.*, 58, 239, 1981.
58. MAFF, *Metals in Cold-Pressed Oils*, Ministry of Agriculture, Fisheries and Food of United Kingdom, Food surveillance information sheet 138, 1997, http://archive.food.gov.uk/maff
59. Jadhav, S.J. et al., Lipid oxidation in biological and food systems, in *Food Antioxidants*, Madhavi, D.L., Deshpande, S.S., and Salunkhe, D.K., Eds., Marcel Dekker, New York, NY, 5, 1996.
60. Micciche, F. et al., Oxidation of methyl linoleate in micellar solutions induced by the combination of iron (II)/ascorbic acid and iron(II)/H_2O_2, *Arch. Biochem. Biophys.*, 443, 45, 2005.
61. Kehrer, J.P., The Haber–Weiss reaction and mechanisms of toxicity, *Toxicology*, 149, 43, 2000.
62. Hu, Y.Z. and Jiang, L.J., Generation of semiquinone radical anion and reactive oxygen ($^1O_2$, $O_2^{-\bullet}$, and $^\bullet OH$) during the photosensitization of a water-soluble perylenequinone derivative, *J. Photochem. Photobiol.*, 33, 51, 1996.
63. Salem, I.A., El-Maazawi, M., and Zaki, A.B., Kinetics and mechanisms of decomposition reaction of hydrogen peroxide in presence of metal complexes, *Int. J. Chem. Kinet.*, 32, 643, 2000.
64. Benjelloun, B. et al., Oxidation of rapeseed oil: effect of metal traces, *J. Am. Oil Chem. Soc.*, 68, 210, 1991.
65. Halliwell, B. and Gutteridge, J.M.C., *Free Radicals in Biology and Medicine*, 3rd edn., Oxford University Press, New York, NY, 297, 2001.
66. Mei, L. et al., Iron-catalyzed lipid oxidation in emulsions as affected by surfactant, pH, and NaCl, *Food Chem.*, 61, 307, 1998.
67. Kilic, B. and Richards, M.P., Lipid oxidation in poultry donor kebap: prooxidative and antioxidative factors, *J. Food Sci.*, 68, 686, 2003.
68. Keceli, T. and Gordon, M.H., Ferric ions reduce the antioxidant activity of the phenolic fraction of virgin olive oil, *J. Food Sci.*, 67, 943, 2002.
69. Shiota, M. et al., Utilization of lactoferrin as an iron-stabilizer for soybean and fish oil. *J. Food Sci.*, 71, C120, 2006.
70. Yoon, S.H. et al., Analyses of phospholipids in soybean oils by HPLC, *Korean J. Food Sci. Technol.*, 19, 66, 1987.
71. Yen, G.C., Influence of seed roasting process on the changes in composition and quality of sesame (*Sesame indicum*) oil, *J. Sci. Food Agric.*, 50, 563, 1990.
72. Koidis, A. and Boskou, D., The contents of proteins and phospholipids in cloudy (veiled) virgin olive oils, *Eur. J. Lipid Sci. Technol.*, 108, 323, 2006.
73. Lyberg, A.M., Fasoli, E., and Adlercreutz, P., Monitoring the oxidation of docosahexaenoic acid in lipids, *Lipids*, 40, 969, 2005.
74. Koo, B.S. and Kim, J.S., Effect of individual phospholipid components treating on storaging and frying stability in soybean oil, *Korean J. Food Cult.*, 20, 451, 2005.
75. Sugino, H. et al., Antioxidative activity of egg yolk phospholipids, *J. Agric. Food Chem.*, 45, 551, 1997.
76. Saito, H. and Ishihara, K., Antioxidant activity and active sites of phospholipids as antioxidants, *J. Am. Oil Chem. Soc.*, 74, 1531, 1997.
77. King, M.F., Boyd, L.C., and Sheldon, B.W., Effects of phospholipids on lipid oxidation of a salmon oil model system, *J. Am. Oil Chem. Soc.*, 69, 237, 1992.
78. Yoon, S.H. and Min, D.B., Roles of phospholipids in the flavor stability of soybean oil, *Korean J. Food Sci. Technol.*, 19, 23, 1987.
79. Morrill, G.A. et al., Interaction between ferric ions, phospholipid hydroperoxides, and the lipid phosphate moiety at physiological pH, *Lipids*, 39, 881, 2004.
80. Yoon, S.H., Jung, M.Y., and Min, D.B., Effects of thermally oxidized triglycerides on the oxidative stability of soybean oil, *J. Am. Oil Chem. Soc.*, 65, 1652, 1988.
81. Hahm, T.S., Effects of initial peroxide contents on the oxidative stability of soybean oil to prevent environmental pollution, *J. Environ. Res.*, 1, 112, 1988.
82. Min, D.B. and Jung, M.Y., Effects of minor components on the flavor stability of vegetable oils, in *Flavor Chemistry of Lipid Foods*, Min, D.B. and Smouse, T.H., Eds., AOCS Press, Champaign, IL, 242, 1989.
83. Salvador, M.D. et al., Cornicabra virgin olive oil: a study of five crop seasons. Composition, quality and oxidative stability, *Food Chem.*, 74, 267, 2001.
84. Suzuki, K. and Nishioka, A., Behavior of chlorophyll derivatives in canola oil processing, *J. Am. Oil Chem. Soc.*, 70, 837, 1993.

85. Endo, Y., Usuki, R., and Kaneda, T., Antioxidant effects of chlorophyll and pheophytin on the autoxidation of oils in the dark. I. Comparison of the inhibitory effects, *J. Am. Oil Chem. Soc.*, 62, 1375, 1985.

86. Endo, Y., Usuki, R., and Kaneda, T., Antioxidant effects of chlorophyll and pheophytin on the autoxidation of oils in the dark. II. The mechanism of antioxidative action of chlorophyll, *J. Am. Oil Chem. Soc.*, 62, 1387, 1985.

87. Francisca, G. and Isabel, M., Action of chlorophylls on the stability of virgin olive oil, *J. Am. Oil Chem. Soc.*, 69, 866, 1992.

88. Hoshina, C., Tomita, K., and Shioi, Y., Antioxidant activity of chlorophylls: its structure-activity relationship, in *Photosynthesis, Mechanisms and Effects*, Garab, G. Ed., Kluwer Academic Publishers, Dordrecht, The Netherlands, Vol. IV, 3281, 1998.

89. Psomiadou, E. and Tsimidou, M., Stability of virgin olive oil. 1. Autoxidation studies, *J. Agric. Food Chem.*, 50, 716, 2002.

90. Fakourelis, N., Lee, E.C., and Min, D.B., Effects of chlorophyll and β-carotene on the oxidation stability of olive oil, *J. Food Sci.*, 52, 234, 1987.

91. Whang, K. and Peng, I.C., Electron paramagnetic resonance studies of the effectiveness of myoglobin and its derivatives as photosensitizers in singlet oxygen generation, *J. Food Sci.*, 53, 1863, 1988.

92. Gutierrez-Rosales, F. et al., Action of chlorophylls and the stability of virgin olive oil, *J. Am. Oil Chem. Soc.*, 69, 866, 1992.

93. Endo, Y., Usuki, R., and Kaneda, T., Prooxidant activities of chlorophylls and their decomposition products on the photooxidation of methyl linoleate, *J. Am. Oil Chem. Soc.*, 61, 781, 1984.

94. Bonnie, T.Y.P. and Choo, Y.M., Valuable minor constituents of commercial red palm olein: carotenoids, vitamin E, ubiquinone and sterols, *J. Oil Palm Res.*, 12, 14, 2000.

95. Burton, G.W. and Ingold, K.U., β-carotene: an unusual type of lipid antioxidant, *Science*, 224, 569, 1984.

96. Beutner, S. et al., Quantitative assessment of antioxidant properties of natural colorants and phytochemicals: carotenoids, flavonoids, phenols and indigoids. The role of β-carotene in antioxidant functions, *J. Sci. Food Agric.*, 81, 559, 2001.

97. Azeredo, H.M.C., Faria, J.A.F., and Silva, M.A.A.P., Minimization of peroxide formation rate in soybean oil by antioxidant combinations, *Food Res. Int.*, 37, 689, 2004.

98. Psomiadou, E. and Tsimidou, M., Stability of virgin olive oil. 2. Photo-oxidation studies. *J. Agric. Food Chem.*, 50, 722, 2002.

99. Foote, C., Photosensitized oxidation and singlet oxygen: consequences in biological systems, in *Free Radicals in Biology*, Pryor, W.A., Ed., Academic Press, New York, NY, 85–133, 1976.

100. Foss, B.J. et al., Direct superoxide anion scavenging by a highly water-dispersible carotenoid phospholipids evaluated by electron paramagnetic resonance (EPR) spectroscopy, *Bioorg. Med. Chem. Lett.*, 14, 2807, 2004.

101. Di Mascio, P., Kaiser, S., and Sies, H., Lycopene as the most efficient biological carotenoid singlet oxygen quencher, *Arch. Biochem. Biophys.*, 274, 532, 1989.

102. Viljanen, K. et al., Carotenoids as antioxidants to prevent photooxidation, *Eur. J. Lipid Sci. Technol.*, 104, 353, 2002.

103. Stahl, W. and Sies, H., Physical quenching of singlet oxygen and cis-trans isomerization of carotenoids, *Ann. N.Y. Acad. Sci.*, 691, 10, 1992.

104. Lee, J.H., Ozcelik, B., and Min, D.B., Electron donation mechanisms of β-carotene as a free radical scavenger, *J. Food Sci.*, 68, 861, 2003.

105. Liebler, D.C., Baker, P.F., and Kaysen, K.L., Oxidation of vitamin E: evidence for competing autoxidation and peroxyl radical trapping reaction of the tocopheroxyl radical, *J. Am. Chem. Soc.*, 112, 6995, 1990.

106. Mortensen, A., Skibsted, L.H., and Truscott, T.G., The interaction of dietary carotenoids with radical species, *Arch. Biochem. Biophys.*, 385, 13, 2001.

107. Iannone, A. et al., Antioxidant activity of carotenoids: an electron-spin resonance study on β-carotene and lutein interaction with free radicals generated in a chemical system, *J. Biochem. Mol. Toxicol.*, 12, 299, 1998.

108. Mohamed, H.M.A. and Awatif, I.I., The use of sesame oil unsaponifiable matter as a natural antioxidants, *Food Chem.*, 62, 269, 1998.

109. Deiana, M. et al., Novel approach to study oxidative stability of extra virgin olive oils: importance of α-tocopherol concentration, *J. Agric. Food Chem.*, 50, 4342, 2002.

110. Przybylski, R., Canola oil: physical and chemical properties, *Canola Council of Canada*, 1, 2001.

111. Kamal-Eldin, A. and Andersson, R., A multivariate study of the correlation between tocopherol content and fatty acid composition in vegetable oils, *J. Am. Oil Chem. Soc.*, 74, 375, 1997.

112. Velasco, J., Andersen, M.L., and Skibsted, L.H., Evaluation of oxidative stability of vegetable oils by monitoring the tendency to radical formation. A comparison of electron spin resonance spectroscopy with the Rancimat method and differential scanning calorimetry, *Food Chem.*, 77, 623, 2003.

113. Choe, E. and Lee, J., Thermooxidative stability of soybean oil, beef tallow and palm oil during frying of steamed noodles, *Korean J. Food Sci. Technol.*, 30, 288, 1998.

114. Drinda, H. and Baltes, W., Antioxidant properties of lipoic and dihydrolipoic acid in vegetable oils and lard, *Z. Lebensm. Unters. Forch.*, 4, 270, 1999.

115. Al-Saqer, J.M. et al., Developing functional foods using red palm olein. IV. Tocopherols and tocotrienols, *Food Chem.*, 85, 579, 2004.

116. Lee, Y.C. et al., Chemical composition and oxidative stability of safflower oil prepared from safflower seed roasted with different temperatures, *Food Chem.*, 84, 1, 2004.

117. Reische, D.W., Lillard, D.A., and Eitenmiller, R.R., Antioxidants, in *Food Lipids*, Akoh, C.C. and Min, D.B., Eds., Marcel Dekker, New York, NY, 489, 2002.

118. Eidhin, D.N., Burke, J., and O'Beirne, D., Oxidative stability of ω3-rich camelina oil and camelina oil-based spread compared with plant and fish oils and sunflower spread, *J. Food Sci.*, 68, 345, 2003.

119. Alpaslan, M., Tepe, S., and Simsek, O., Effect of refining processes on the total and individual tocopherol content in sunflower oil, *Int. J. Food Sci. Technol.*, 36, 737, 2001.

120. Buettner, G.R., The pecking order of free radicals and antioxidants: lipid peroxidation, α-tocopherol and ascorbate, *Arch. Biochem. Biophys.*, 300, 535, 1993.

121. Niki, E. and Noguchi, N., Evaluation of antioxidant capacity. What capacity is being measured by which method?, *Life*, 50, 323, 2000.

122. Liebler, D.C., Antioxidant reactions of carotenoids, *Ann. N.Y. Acad. Sci.*, 691, 20, 1993.

123. Gavino, V.C. et al., Effect of polyunsaturated fatty acids and antioxidants on lipid peroxidation in tissue cultures, *J. Lipid Res.*, 22, 763, 1981.

124. Yanishlieva, N.V. et al., Kinetics of antioxidant action of α- and γ-tocopherols in sunflower and soybean triacylglycerols, *Eur. J. Lipid Sci. Technol.*, 104, 262, 2002.

125. Jung, M.Y. and Min, D.B., Effects of α-, γ-, and δ-tocopherols on the oxidative stability of soybean oil, *J. Food Sci.*, 55, 1464, 1990.

126. Cillard, J., Cillard, P., and Cormier, M., Effect of experimental factors on the prooxidant behavior of α-tocopherol, *J. Am. Oil Chem. Soc.*, 57, 255, 1980.

127. Terao, J. and Matsushita, S., The peroxidizing effect of α-tocopherol on autoxidation of methyl linoleate in bulk phase, *Lipids*, 21, 255, 1986.

128. Blekas, G., Tsimidou, M., and Boskou, D., Contribution of α-tocopherol to olive oil stability, *Food Chem.*, 52, 289, 1995.

129. Bowry, V.W. and Stocker, R., Tocopherol-mediated peroxidation. The prooxidant effect of vitamin E on the radical-initiated oxidation on human low-density lipoprotein, *J. Am. Chem. Soc.*, 115, 6029, 1993.

130. Yamamoto, Y., Role of active oxygen species and antioxidants in photoaging, *J. Dermatol. Sci.*, 27, Suppl 1, 1, 2001.

131. Marinova, E.M. and Yanishlieva, N.V., Effect of temperature on the antioxidative action of inhibitors in lipid autoxidation, *J. Sci. Food Agric.*, 60, 313, 1992.

132. Min, D.B. and Lee, E.C., Factors affecting singlet oxygen oxidation of soybean oil, in *Frontiers of Flavor*, Charalambous, G., Ed., Elsevier, New York, NY, 473, 1988.

133. Kamal-Eldin, A. and Appelqvist, L.A., The chemistry and antioxidant properties of tocopherols and tocotrienols, *Lipids*, 31, 671, 1996.

134. Neely, W.C., Martin, J.M., and Barker, S.A., Products and relative reaction rates of the oxidation of tocopherols with singlet molecular oxygen, *Photochem. Photobiol.*, 48, 423, 1988.

135. Jung, M.Y., Choe, E., and Min, D.B., Effects of α-, β-, γ-, and δ-tocopherols on the chlorophyll photosensitized oxidation of soybean oil, *J. Food Sci.*, 56, 807, 1991.

136. Mukai, K. et al., Structure-activity relationship in the quenching reaction of singlet oxygen by tocopherol (vitamin E) derivatives and related phenols. Finding of linear correlation between the rates of quenching of singlet oxygen and scavenging of peroxyl and phenoxyl radicals in solution, *J. Org. Chem.*, 56, 4188, 1991.

137. Jung, M.Y. and Min, D.B., Effects of oxidized α-, γ- and δ-tocopherols on the oxidative stability of purified soybean oil, *Food Chem.*, 45, 183, 1992.
138. Servili, M. and Montedoro, G.F., Contribution of phenolic compounds to virgin olive oil quality, *Eur. J. Lipid Sci. Technol.*, 104, 602, 2002.
139. Leonardis, A.D., Macciola, V., and Di Rocco, A., Oxidative stabilization of cold-pressed sunflower oil using phenolic compounds of the same seeds, *J. Sci. Food Agric.*, 83, 523, 2003.
140. Fukuda, Y. et al., Studies on antioxidative substances in sesame seed, *Agric. Biol. Chem.*, 49, 301, 1985.
141. Osawa, T. et al., Sesamolinol, a novel antioxidant isolated from sesame seeds, *Agric. Biol. Chem.*, 49, 3351, 1985.
142. Fukuda, Y. et al., Contribution of lignans analogues to antioxidative activity of refined unroasted sesame seed oil, *J. Am. Oil Chem. Soc.*, 63, 1027, 1986.
143. Dachtler, M. et al., On-line LC-NMR-MS characterization of sesame oil extracts and assessment of their antioxidant activity, *Eur. J. Lipid Sci.Technol.*, 105, 488, 2003.
144. Garcia, A. et al., Effect of refining on the phenolic composition of crude olive oils, *J. Am. Oil Chem. Soc.*, 83, 159, 2006.
145. Kikugawa, K., Arai, M., and Kurechi, T., Participation of sesamol in stability of sesame oil, *J. Am. Oil Chem. Soc.*, 60, 1528, 1983.
146. Fukuda, Y. and Namiki, M., Recent studies on sesame seed and oil, *Nippon Shokuhin Kogyo Gakkaishi*, 35, 552, 1988.
147. Yoshida, H., Composition and quality characteristics of sesame seed (*Sesamum indicum*) oil roasted at different temperatures in electric oven, *J. Sci. Food Agric.*, 65, 331, 1994.
148. Namiki, N., The chemistry and physiological functions of sesame, *Food Rev. Int.*, 11, 281, 1995.
149. Yoshida, H. and Takagi, S., Effects of seed roasting temperature and time on the quality characteristics of sesame (*Sesamum indicum*) oil, *J. Sci. Food Agric.*, 75, 19, 1997.
150. Yoshida, H., Kirakawa, Y.T., and Takagi, S., Roasting influences on molecular species of triacylglycerols in sesame seeds (*Sesamun indicum*), *J. Sci. Food Agric.*, 80, 1495, 2000.
151. Hosseinian, F.S. et al., Antioxidant capacity of flaxseed lignans in two model systems, *J. Am. Oil Chem. Soc.*, 83, 835, 2006.
152. Papadopoulos, G. and Boskou, D., Antioxidant effect of natural phenols on olive oil, *J. Am. Oil Chem. Soc.*, 68, 669, 1991.
153. Tsimidou, M., Papadopoulos, G., and Boskou, D., Phenolic compounds and stability of virgin olive oil—Part I, *Food Chem.*, 45, 141, 1992.
154. Gutierrez, F., Arnaud, T., and Garrido, A., Contribution of polyphenols to the oxidative stability of virgin olive oil, *J. Sci. Food Agric.*, 81, 1463, 2001.
155. Guillen, M.D. and Cabo, N., Fourier transform infrared spectra data versus peroxide and anisidine values to determine oxidative stability of edible oils, *Food Chem.*, 77, 503, 2002.
156. Chimi, H. et al., Peroxyl and hydroxyl radical scavenging activity of some natural phenolic antioxidants, *J. Am. Oil Chem. Soc.*, 68, 307, 1991.
157. Baldioli, M. et al., Antioxidant activity of tocopherols and phenolic compounds of virgin olive oil, *J. Am. Oil Chem. Soc.*, 73, 1589, 1996.
158. Kim, J.Y., Choi, D.S., and Jung, M.Y., Antiphoto-oxidative activity of sesamol in methylene blue- and chlorophyll-sensitized photo-oxidation of oil, *J. Agric. Food Chem.*, 51, 3460, 2003.
159. Osborne, H.T. and Akoh, C.C., Effects of natural antioxidants on iron-catalyzed lipid oxidation of structured lipid-based emulsions, *J. Am. Oil Chem. Soc.*, 80, 847, 2003.
160. Maestroduran, R. and Borjapadilla, R., Antioxidant activity of natural sterols and organic acids, *Grasas Y. Aceites*, 44, 208, 1993.
161. Brimberg, U.I. and Kamal-Eldin, A., On the kinetics of the autoxidation of fats: influence of pro-oxidants, antioxidants and synergists, *Eur. J. Lipid Sci. Technol.*, 105, 83, 2003.
162. Yanishlieva, N. and Schiller, H., Effect of sitosterol on autoxidation rate and product composition in a model lipid system, *J. Sci. Food Agric.*, 35, 219, 1983.
163. Palozza, P. and Krinsky, N.I., β-Carotene and α-tocopherol are synergistic antioxidants, *Arch. Biochem. Biophys.*, 297, 184, 1992.
164. Choe, E. and Min, D.B., Interaction effects of chlorophyll, β-carotene and tocopherol on the photooxidative stabilities of soybean oil, *Foods Sci. Biotechnol.*, 1, 104, 1992.
165. Shibasaki-Kitakawa, N. et al., Oxidation kinetics of β-carotene in oleic acid solvent with addition of an antioxidant, α-tocopherol, *J. Am. Oil Chem. Soc.*, 81, 389, 2004.

166. Ohshima, T., Fujita, Y., and Koizumi, C.J., Oxidative stability of sardine and mackerel lipids with reference to synergism between phospholipids and α-tocopherol, *J. Am. Oil Chem. Soc.*, 70, 269, 1993.
167. Lambelet, P., Saucy, F., and Loliger, J., Radical exchange reactions between vitamin E, vitamin C and phospholipids in autoxidizing polyunsaturated lipids, *Free Radic. Res.*, 20, 1, 1994.
168. Kago, T. and Terao, J., Phospholipids increase radical scavenging activity of vitamin E in a bulk oil model system, *J. Agric. Food Chem.*, 43, 1450, 1995.
169. Thiyam, U., Stockmann, H., and Schwarz, K., Antioxidant activity of rapeseed phenolics and their interactions with tocopherols during lipid oxidation, *J. Am. Oil Chem. Soc.*, 83, 523, 2006.

18 Antioxidant Mechanisms

Eric A. Decker

CONTENTS

I. INTRODUCTION

Krinsky [1] has defined biological antioxidants as "compounds that protect biological systems against the potentially harmful effects of processes or reactions that cause extensive oxidations." While food lipids are derived from biological systems, the ultimate purpose of food antioxidants is different, since they are used to inhibit oxidative reactions that cause deterioration of quality (e.g., of flavor, color, nutrient composition, texture). With this goal in mind, food antioxidants can be defined as any compounds serving to inhibit oxidative processes that deteriorate the quality of food lipids. Antioxidant mechanisms that fit this definition include free radical scavenging, inactivation of peroxides and other reactive oxygen species, chelation of metals, and quenching of secondary lipid oxidation products that produce rancid odors.

Reactive oxygen species and free radicals are produced by both enzymic and nonenzymic reactions. Therefore, foods usually contain endogenous antioxidants to protect against oxidative damage. These antioxidant systems often contain several distinctively different antioxidants for protection against different prooxidative compounds, including transition metals, heme-containing proteins, photosensitizers, and numerous sources of free radicals. Since prooxidants are both water and lipid soluble, endogenous antioxidant systems in foods are usually biphasic. Such multicomponent and biphasic antioxidants represent nature's own hurdle technology antioxidant system.

This chapter covers the basic mechanisms by which antioxidants influence oxidative reactions: inactivation of free radicals, control of oxidation catalysts, inactivation of oxidation intermediates, and interactions between antioxidants and secondary lipid oxidation products.

II. INACTIVATION OF FREE RADICALS

Antioxidants can slow lipid oxidation by inactivating or scavenging free radicals, thus inhibiting initiation and propagation reactions. Free radical scavengers (FRS) or chain-breaking antioxidants are capable of accepting a radical from oxidizing lipid species such as peroxyl (LOO·) and alkoxyl (LO·) radicals by the following reactions [2]:

$$\text{LOO· or LO· + FRS} \rightarrow \text{LOOH or LOPH + FRS·.}$$

FRS primarily react with peroxyl radicals for several reasons because (1) propagation is a slow step in lipid oxidation, meaning that peroxyl radicals are often found in the greatest concentration of all radicals in the systems; (2) peroxyl radicals have lower energies than radicals such as alkoxyl radicals [3] and thus react more readily with the low-energy hydrogens of FRS than with polyunsaturated fatty acids; and (3) FRS, being generally found at low concentrations, do not compete effectively with initiating radicals (e.g., OH) [4]. An FRS thus inhibits lipid oxidation by more effectively competing with other compounds (especially unsaturated fatty acids) for peroxyl radicals.

Chemical properties, including hydrogen bond energies, resonance delocalization, and susceptibility to autoxidation, will influence the antioxidant effectiveness of an FRS. Initially, antioxidant efficiency is dependent on the ability of the FRS to donate hydrogen to the free radical. As the hydrogen bond energy of the FRS decreases, the transfer of the hydrogen to the free radical is more energetically favorable and thus more rapid. The ability of an FRS to donate hydrogen to a free radical can be predicted from standard one-electron reduction potentials [3].

Any compound that has a reduction potential lower than the reduction potential of a free radical (or oxidized species) is capable of donating hydrogen to that free radical unless the reaction is kinetically unfeasible (Table 18.1). For example, FRS including α-tocopherol ($E' = 500$ mV), urate ($E' = 590$ mV), catechol ($E' = 530$ mV), and ascorbate ($E' = 282$ mV) all have reduction potentials below peroxyl radicals ($E' = 1000$ mV) and are therefore capable of donating a hydrogen atom to the peroxyl radical to form a peroxide. Standard reduction potentials can also be used to predict the ease with which a compound can donate its hydrogen to a radical. For instance, the hydrogen of

---

TABLE 18.1

Standard One-Electron Reduction Potentials for Common Free Radical Processes

Couple	E' (mV)
HO·, H$^+$/H$_2$O	2310
RO·, H$^-$/ROH	1600
ROO·, H$^+$/ROOH	1000
PUFA·, H$^+$/PUFA-H	600
Urate$^-$, H$^+$/urate$^-$-H	590
Catechol-O·, H$^+$/catechol-OH	530
α-Tocopheroxyl·, H$^+$ α-tocopherol	500
Ascorbate$^-$·, H$^+$/ascorbate$^-$	282

Source: Adapted from Buettner, G.R., *Arch. Biochem. Biophys.*, 300, 535, 1993.

the hydroxyl group on α-tocopherol has a lower reduction potential than the methylene-interrupted hydrogen of a polyunsaturated fatty acid ($E' = 600$ mV), thus allowing the α-tocopherol to react with peroxyl radicals more rapidly than is possible for unsaturated fatty acids.

The efficiency of the FRS is also dependent on the energy of the resulting free radical scavenger radical (FRS·). If the FRS is a low-energy radical, then the likelihood that the FRS will catalyze the oxidation of other molecules decreases. The most efficient FRS has low-energy radicals as a result of resonance delocalization (Figure 18.1) [5,6]. This can again be seen in standard reduction potentials, where FRS such as α-tocopherol and catechol have lower reduction potentials than polyunsaturated fatty acids and therefore do not efficiently abstract hydrogens from unsaturated fatty acids (Table 18.1) [3]. Efficient FRS also produces radicals that do not react rapidly with oxygen to form peroxides. When a radical scavenger forms peroxides during oxidation, it is likely that it will autoxidize, thus depleting the system of the free radical scavenger. In addition, FRS peroxides can decompose into additional radical species, which could further promote oxidation. Thus, formation of FRS peroxides can result in consumption of the antioxidant with no net decrease in free radical numbers [4].

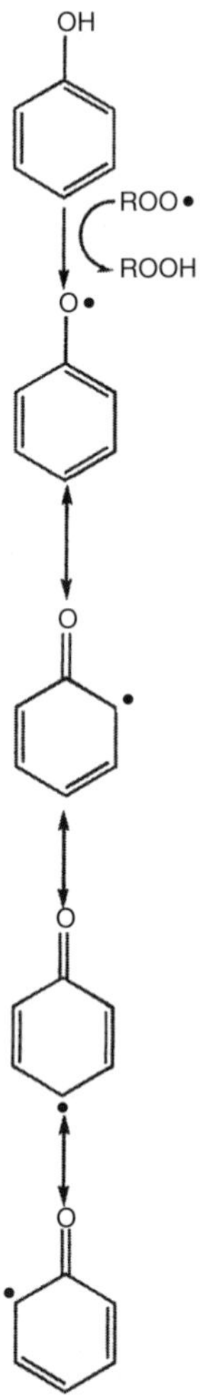

FIGURE 18.1 Resonance stabilization of a free radical by a phenolic. (Adapted from Shahidi, F. and Wanasundara, J.P.K., *Crit. Rev. Food Sci. Nutr.*, 32, 67, 1992.)

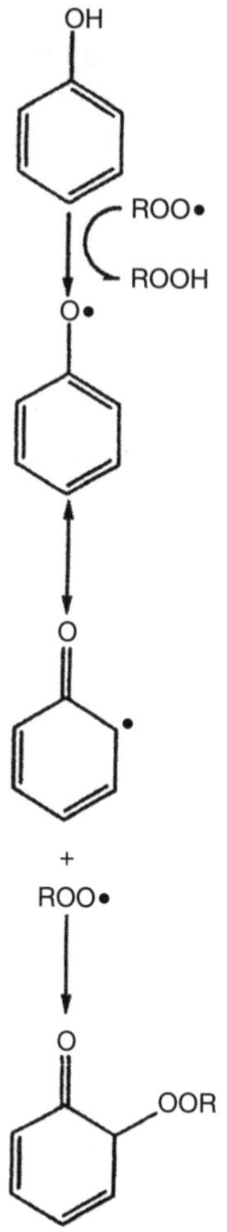

FIGURE 18.2 Mechanism by which one phenolic free radical scavenger can inactivate two peroxyl radicals.

FRS radicals may undergo additional reactions that remove radicals from the system; examples include termination reactions with other FRS or lipid radicals to form nonradical species (Figure 18.2). This means that each FRS is capable of inactivating at least two free radicals, the first being inactivated when the FRS interacts with the peroxyl radicals and the second when the FRS enters a termination reaction with another peroxyl radical.

Phenolics possess many of the properties of an efficient FRS. Hydrogen donation generally occurs through the hydroxyl group, and the radical subsequently formed is stabilized by resonance delocalization throughout the phenolic ring structure. The effectiveness of phenolic FRS can be increased by substitution groups. Alkyl groups in the ortho and para positions enhance the reactivity of the hydroxyl hydrogen toward lipid radicals; bulky groups at the ortho position increase the stability of phenoxy radicals; and a second hydroxy group at the ortho or para position stabilizes the phenoxy radical through an intermolecular hydrogen bond [5]. In foods, the efficiency of phenolic FRS depends on additional factors. Besides chemical reactivity, factors such as volatility, pH sensitivity, and polarity can influence the retention and activity of the FRS in stored and processed foods [6].

A. Tocopherols

Tocopherols are a group of phenolic FRS isomers originating in plants and eventually ending up in animal foods via the diet [7]. Interactions between tocopherols and lipid peroxyl radicals lead to the formation of a hydroperoxide and several resonance structures of tocopheroxyl radicals (Figure 18.3) [6]. Tocopheroxyl radicals can interact with other compounds or with each other to form a variety of

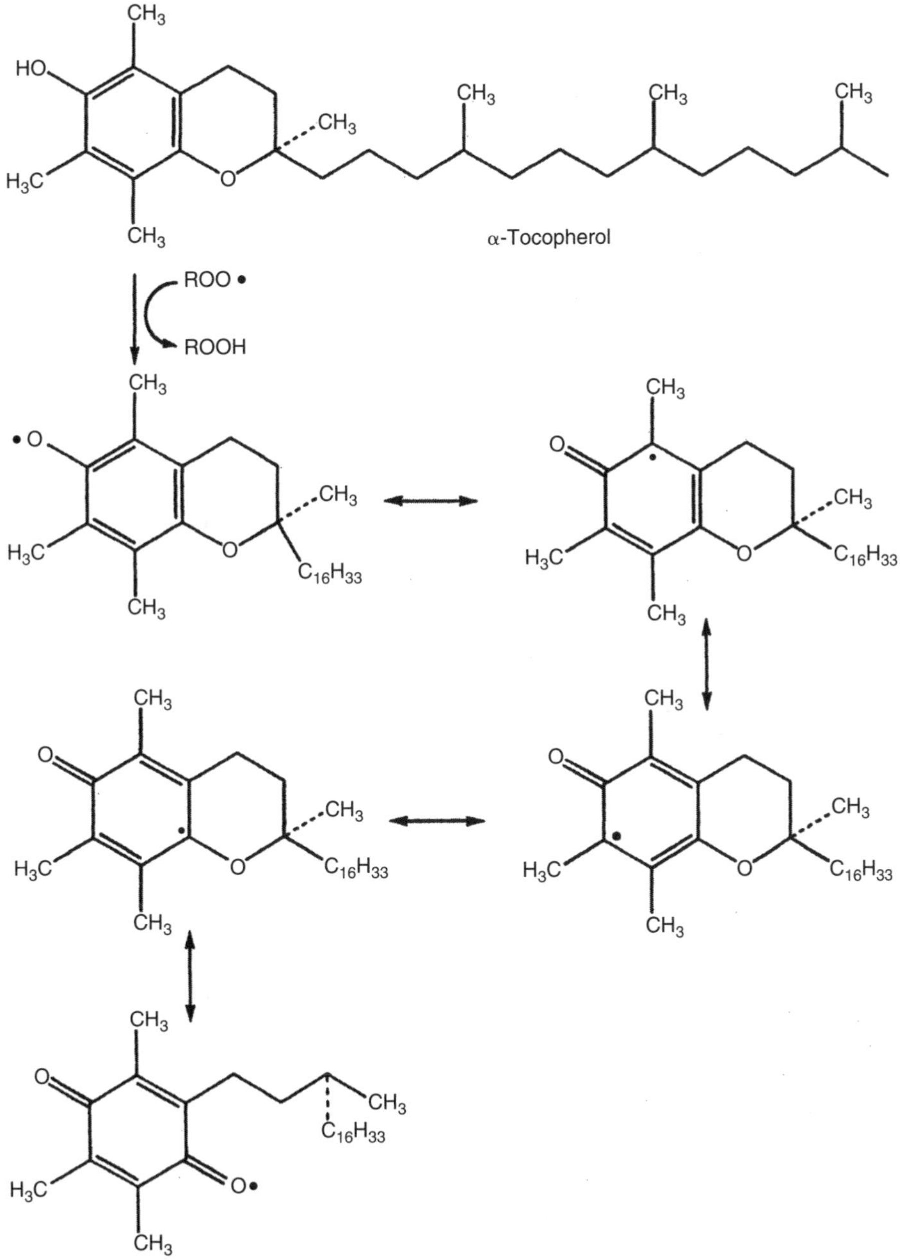

FIGURE 18.3 The different resonance structures of the α-tocopherol radical. (Adapted from Nawar, W.W., *Food Chemistry*, 3rd edn, O. Fennema, ed, Dekker, New York, 1996.)

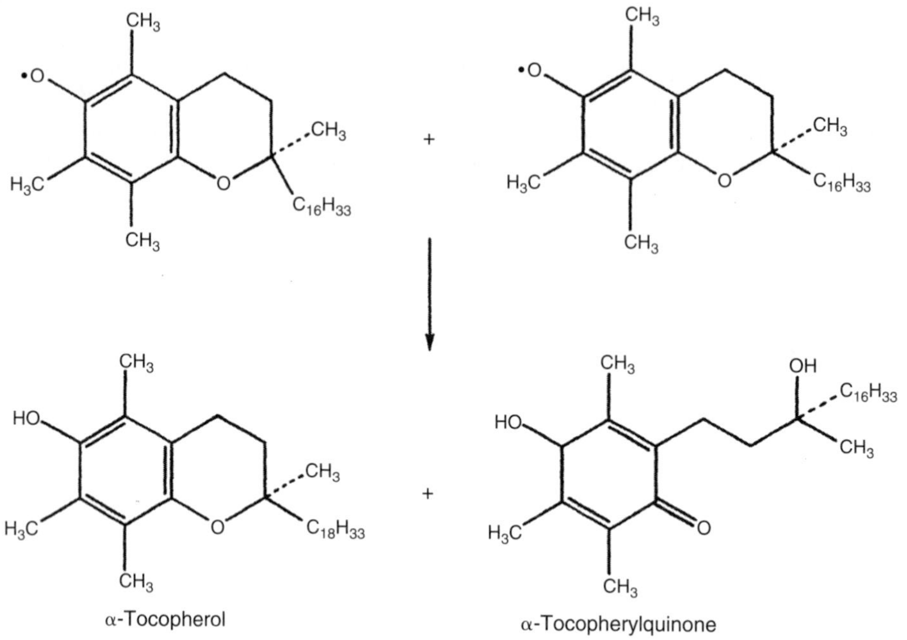

FIGURE 18.4 Formation of α-tocopherol and tocopherylquinone from two α-tocopherol radicals.

products. The types and amounts of these products depend on oxidation rates, radical species, lipid state (e.g., bulk vs. membrane lipids), and tocopherol concentration.

Under conditions of low oxidation rates in lipid membrane systems, tocopheroxyl radicals primarily convert to tocopherylquinone. Tocopherylquinone can form when the interaction of two tocopheroxyl radicals leads to the formation of tocopherylquinone and the regeneration of tocopherol (Figure 18.4) [6]. Formation of tocopherylquinone is also thought to occur by the transfer of an electron from a tocopheroxyl radical to a phospholipid peroxyl radical to form a phospholipid peroxyl anion and a tocopherol caution. The tocopherol caution hydrolyzes to 8α-hydroxytocopherone, which rearranges to tocopherylquinone (Figure 18.5) [8].

Under condition of more extensive oxidation, high concentrations of peroxyl radicals can favor the formation of tocopherol–peroxyl complexes. These complexes can hydrolyze to tocopherylquinone. Of less importance are interactions between tocopheroxyl and peroxyl radicals, which form an addition product ortho to the phenoxyl oxygen followed by elimination of an alkoxyl radical, addition of oxygen, and abstraction of hydrogen to form two isomers of epoxy-8α-hydroperoxy tocopherone.

Subsequent hydrolysis leads to the formation of epoxyquinones (see Figure 18.6 for an example of this reaction) [9,10]. Formation of epoxide derivatives of tocopherol represents no net reduction of radicals (because an alkoxyl radical forms) and a loss of tocopherol from the system, whereas any tocopherylquinone that is formed can be regenerated back to tocopherol in the presence of reducing agents (e.g., ascorbic acid and glutathione; see Section VII). An additional reaction that can occur is the interaction of two tocopheroxyl radicals to form tocopherol dimers [11].

B. SYNTHETIC PHENOLICS

Phenolic antioxidants for use in foods include synthetic compounds (Figure 18.7). Synthetic phenolic antioxidants exhibit varying polarity, with butylated hydroxytoluene (most nonpolar) > butylated hydroxyanisole > tertiary butylhydroquinone > propyl gallate. The antioxidant mechanism of the synthetic phenolics involves the formation of a resonance-stabilized phenolic radical that

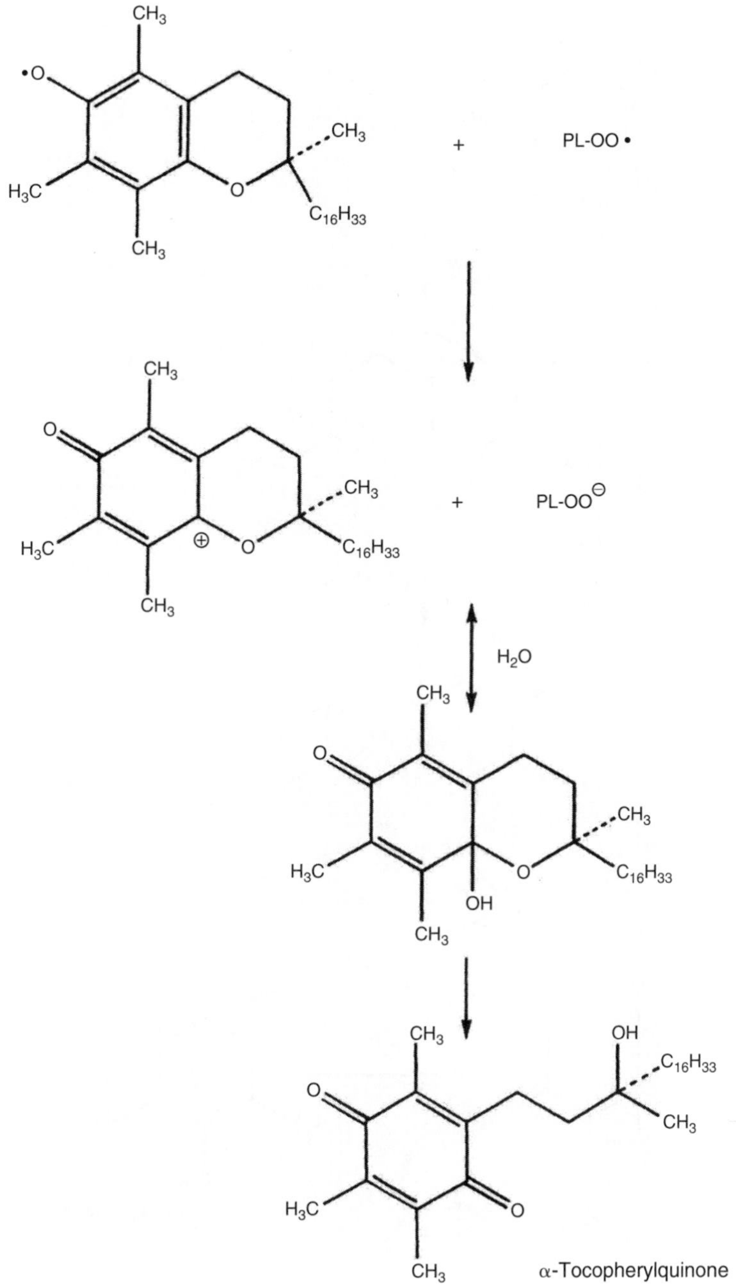

FIGURE 18.5 Proposed mechanism for the formation of α-tocopherylquinone from the interaction of a α-tocopherol radical and a phospholipid peroxyl radical. (Adapted from Liebler, D.C., *Crit. Rev. Toxicol.*, 23, 147, 1993.)

neither rapidly catalyzes the oxidation of other molecules nor reacts with oxygen to form antioxidant peroxides that autoxidize [5].

Synthetic phenolic radicals can potentially react with each other by means of mechanisms similar to that of α-tocopherol. These include reactions of two phenolic radicals to form a hydroquinone and a regenerated phenolic, as well as the formation of phenolic dimers. The phenolic

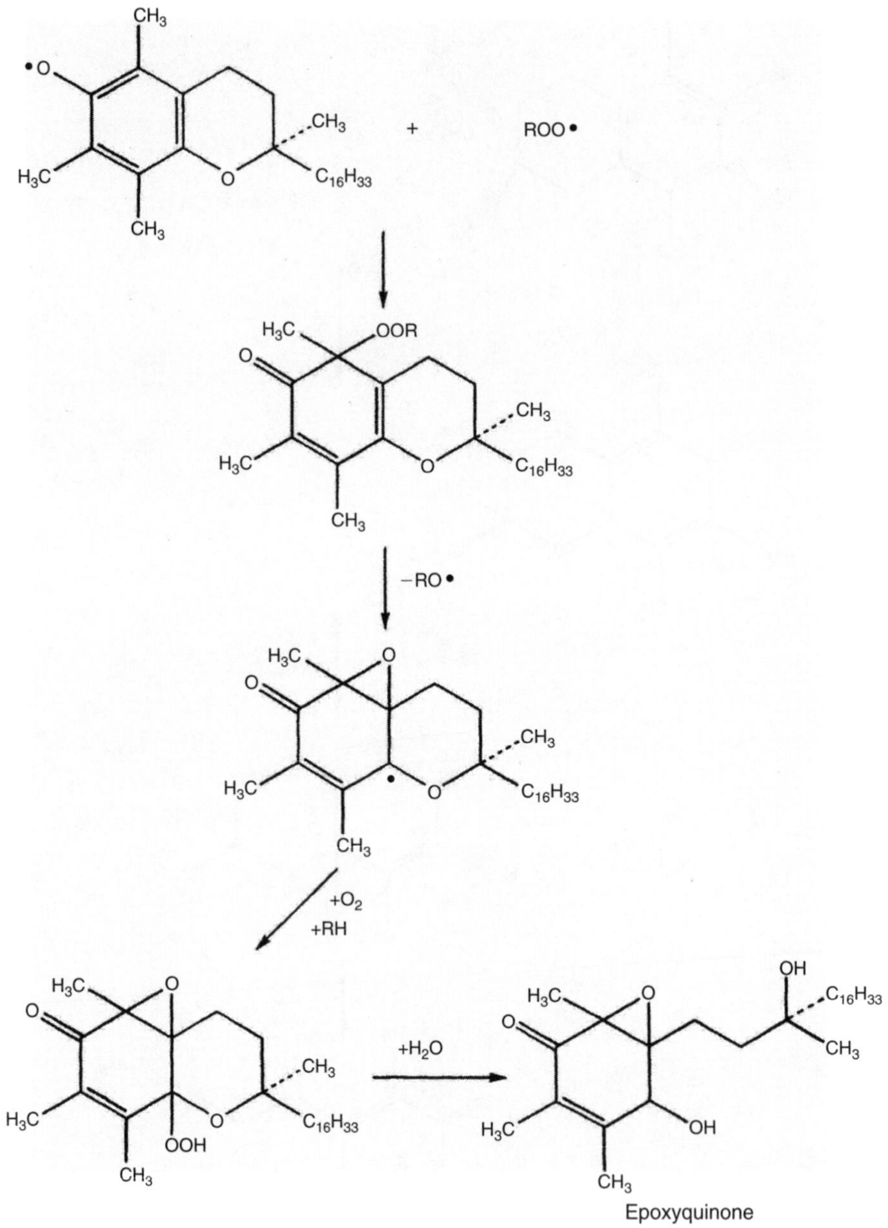

FIGURE 18.6 Formation of an epoxyquinone and an alkoxyl radical from the interaction of an α-tocopherol radical with a peroxyl radical. (Adapted from Liebler, D.C., *Crit. Rev. Toxicol.*, 23, 147, 1993.)

radicals can also react with other peroxyl radicals in termination reactions resulting in the formation of phenolic–peroxyl species adducts. In addition, oxidized synthetic phenolics undergo numerous degradation reactions (for review, see Ref. [5]). Since many of these degradation products still contain active hydroxyl groups, the products may retain antioxidant activity. Therefore, the net antioxidant activity of synthetic phenolics in food actually represents the activity of the original phenolic plus some of its degradation products. Synthetic phenolics are effective in numerous food systems; however, their use in the food industry has recently declined, reflecting safety concerns and consumer demand for all-natural products.

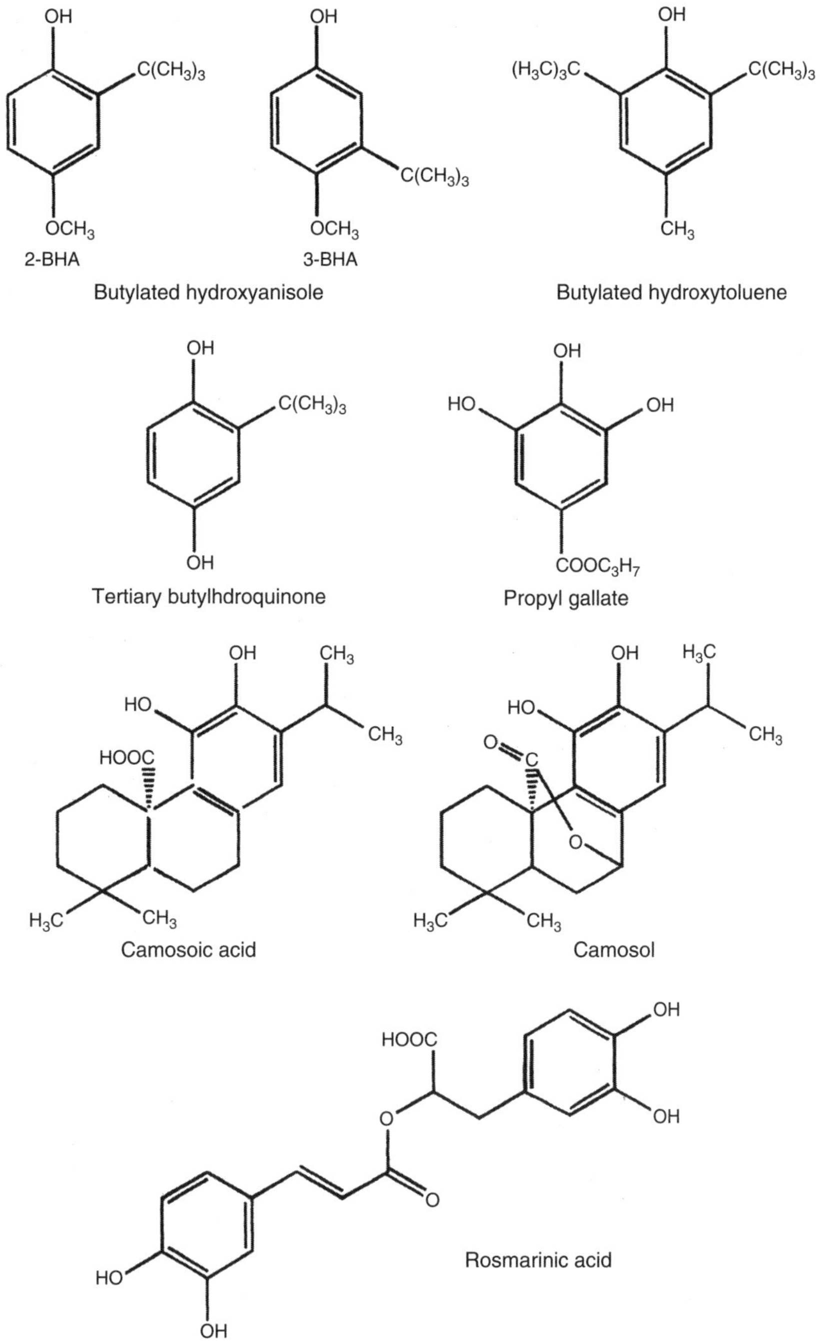

FIGURE 18.7 Structures of several important phenolic free radical scavengers used in foods.

C. UBIQUINONE

Ubiquinone, or coenzyme Q, is a phenolic conjugated to an isoprenoid chain. Ubiquinone is primarily associated with the mitochondrial membrane [12]. Reduced ubiquinone is capable of inactivating peroxyl radicals, but its radical scavenging activity is less than that of α-tocopherol

[13]. The lower free radical scavenging activity of reduced ubiquinone has been attributed to internal hydrogen bonding, which makes hydrogen abstraction more difficult [13]. Despite its lower radical scavenging activity, reduced ubiquinone has been found to inhibit lipid oxidation in liposomes [14] and low-density lipoprotein [15]. Presumably, it could be an important endogenous antioxidant in many foods.

D. PLANT PHENOLICS

Plants contain a diverse group of phenolic compounds including simple phenolics, phenolic acids, anthocyanins, hydrocinnamic acid derivatives, and flavonoids. Widely distributed in plant foods such as fruits, spices, tea, coffee, seeds, and grains, these phenolics have been estimated to be consumed in amounts >1 g/day.

All the phenolic classes have the structural requirements of FRS (see Figure 18.7 for several examples). However, the antioxidant activity of these compounds varies greatly, and some even exhibit prooxidant activity. Factors influencing the antioxidant activity of plant phenolics include position and degree of hydroxylation, polarity, solubility, reducing potential, stability of the phenolic to food processing operations, and stability of the phenolic radical. In addition, many phenolics contain acid or ring groups that may participate in metal chelation. These metal chelation properties, in addition to high reducing potentials, can accelerate metal-catalyzed oxidative reactions, leading to the prooxidative activity of plant phenolics under certain conditions [16,17].

Herbs and spices are sources of phenolic antioxidants used in foods. Rosemary extracts are the most commercially important source of an antioxidant ingredient containing plant phenolics. Carnosic acid, carnosol, and rosmarinic acid are the major antioxidant phenolics in rosemary extracts (Figure 18.7) [18]. Crude rosemary extracts have been found to inhibit lipid oxidation in a wide variety of food products including meats, bulk oils, and lipid emulsions [18–20]. Utilization of phenolic antioxidants from crude herb extracts such as rosemary is often limited by the presence of highly flavorable monoterpenes. Use of more purified forms of herbal phenolics is restricted by both economic and regulatory hurdles.

E. CAROTENOIDS

Carotenoids are a diverse group (>600 compounds) of yellow to red polyenes consisting of 3–13 double bonds and in some cases 6-carbon-hydroxylated ring structures at one or both ends of the molecule [21]. Carotenoids may be important biological antioxidants and are thought to play a role in controlling oxidatively induced diseases such as cancer and atherosclerosis [22]. The antioxidant properties of carotenoids depend on environmental conditions and the nature of oxidation catalyst. Carotenoids can be effective antioxidants in the presence of singlet oxygen (see Section III.A). However, when peroxyl radicals are the initiating species, the antioxidant efficiency of carotenoids depends on oxygen concentrations.

β-Carotene, the most extensively studied carotenoid antioxidant, reacts with lipid peroxyl radicals, resulting in the formation of a carotenoid radical. Burton and Ingold [23] found that under conditions of high oxygen tension, the antioxidant activity of β-carotene is diminished. They proposed that increasing oxygen results in increased formation of carotenoid peroxyl radicals, thus favoring autoxidation of β-carotene over inactivation of lipid peroxyl radicals. Under conditions of low oxygen tension, the lifetime of the carotenoid radical is long enough to permit reaction with another peroxyl radical, thus forming a nonradical species and effectively inhibiting oxidation by removing radicals from the system.

Incubation of β-carotene with peroxyl radical generators in organic solvents at high (atmospheric) oxygen tensions leads to addition reactions to form carotenoid peroxyl adducts (Figure 18.8). Addition of a peroxyl radical to the cyclic end group or the polyene chain followed by loss of an alkoxyl radical leads to the formation of 5,6- and 15,15'-epoxides. Elimination of the alkoxyl radical from the 15,15' positions can also cause cleavage of the polyene chain, resulting in the formation of

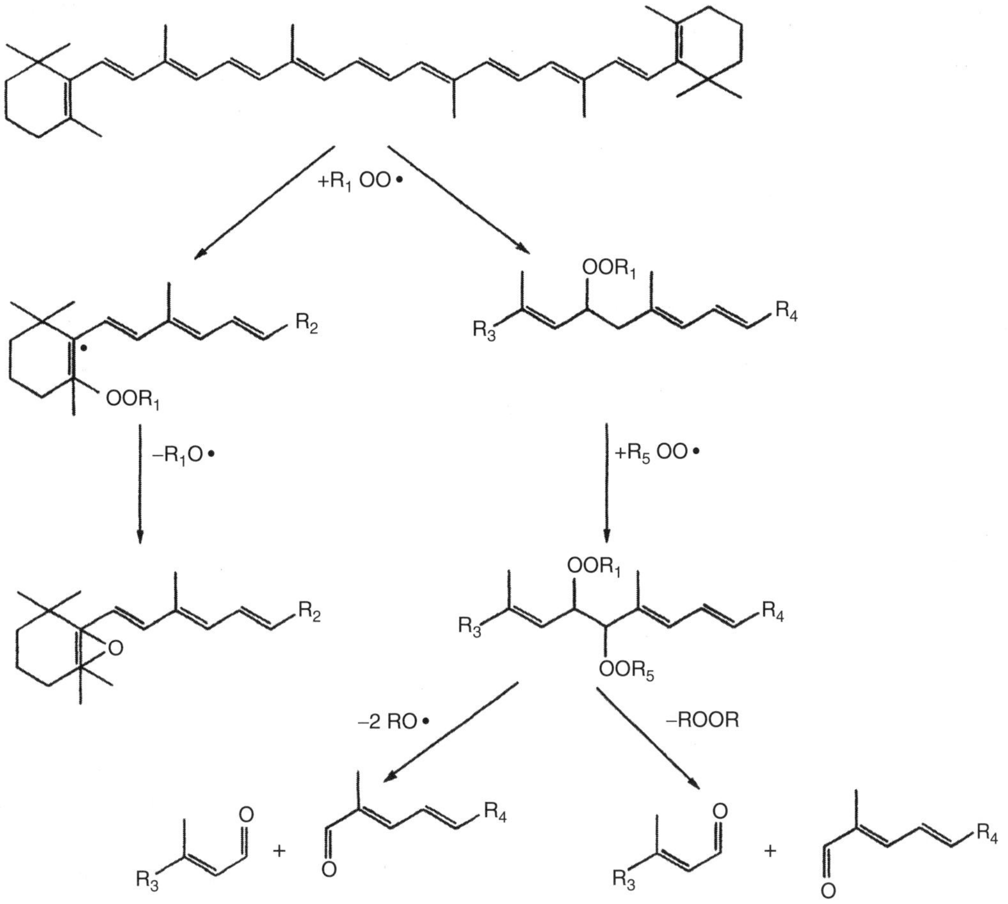

FIGURE 18.8 Products formed from the oxidation of β-carotene by a peroxyl radical. (Adapted from Liebler, D.C., *Ann. N.Y. Acad. Sci.*, 691, 20, 1992.)

aldehydes. Since the formation of β-carotene epoxides from the addition of peroxyl radicals results in the formation of an alkoxyl radical, the net change in radical number is zero; thus an antioxidant effect is not expected [24].

β-Carotene is capable of donating an electron to peroxyl radicals to produce a β-carotene cation radical and a peroxyl anion. The β-carotene cation radical is resonance stabilized and does not readily react with oxygen to form peroxides. However, the β-carotene cation radical appears to be strong enough to oxidize other lipophilic hydrogen donors, including tocopherols and ubiquinone [24]. Additional research is needed to identify the oxidation products that form from carotenoids under low oxygen partial pressures. Identification of these products may help determine the exact mechanism by which carotenoids act as FRS when oxygen concentrations are low. Such knowledge would make it easier to predict when carotenoids will exhibit antioxidant activity.

F. WATER-SOLUBLE FREE RADICAL INACTIVATORS

Free radicals are generated in the water phase of foods by processes such as the Fenton reaction, which produces hydroxyl radicals from hydrogen peroxide [25,26]. Since free radicals are found in the aqueous phase, biological systems contain water-soluble compounds capable of free radical inactivation. Ascorbic acid and glutathione scavenge free radicals, resulting in the formation of low-energy ascorbate and radicals [3]. While ascorbate and glutathione form low-energy radicals, other factors influence whether these compounds will act as antioxidants. Both ascorbate and glutathione are strong reducing compounds. Ascorbate, and in some cases glutathione, will catalyze the reduction of transition metals, which in turn can react with hydrogen and lipid peroxides to form radicals [27,28]. Ascorbate also causes the release of iron, which is sequestered to proteins such as ferritin [29]. Therefore, ascorbate and glutathione can potentially exhibit prooxidative activity in the presence of free transition metals or iron-binding proteins. In addition, in the presence of oxygen, glutathione radicals are capable of forming high-energy peroxides that can potentially catalyze the oxidation of lipids [3].

Thiols besides glutathione can inactivate free radicals. Cysteine is capable of scavenging free radicals. The energy of the resulting thio radical is high, however, suggesting that it may promote oxidation [3]. Thioctic acid is another thiol that can inactivate peroxyl radicals [30]. However, the reduced state of thioctic acid, dihydrolipoic acid [31], and cysteine [32] can be prooxidative because their reducing potential, and thus their ability to stimulate metal-catalyzed oxidation, is strong.

Several nitrogenous compounds can inactivate free radicals. Uric acid inactivates both hydroxyl and lipid radicals and inhibits lipid oxidation at physiological concentrations [33]. Uric acid is an important antioxidant in blood plasma [33–35]. Since uric acid is produced in postmortem skeletal muscle via ATP metabolism [36], it might possibly serve as an active endogenous antioxidant in muscle foods.

Amino acids, peptides, and proteins can interact with free radicals. Amino acids, including histidine, tyrosine, phenylalanine, tryptophan, cysteine, proline, and lysine, are capable of inactivating free radicals [37–40]. Blood proteins have been estimated to provide 10%–50% of the peroxyl radical trapping activity of plasma [34,41]. Serum albumin scavenges carbon-based free radicals partially through the involvement of its free sulfhydryl groups [42]. Amino acids, peptides, and proteins have been reported to inhibit lipid oxidation in bulk and emulsified lipid systems as well as in food products [43–47].

While protein, peptides, and amino acids often possess the structural characteristics needed to both scavenge radicals and inhibit lipid oxidation, the concentrations required for activity are often higher than other FRS. This is likely due to the fact that only a small percentage of the amino acids in a protein are surface exposed where they can interact with free radicals. Hydrolysis of proteins into peptides increases free radical scavenging activity presumably due to increased solvent exposure of amino acids [48]. Peptides have excellent potential as food antioxidant; however, issues exist with their potential allergenicity and bitter off-flavors.

III. CONTROL OF LIPID OXIDATION CATALYSTS

Lipid oxidation rates in foods often depend on catalyst concentrations and activity. Control of lipid oxidation catalysts can therefore be a very important factor in controlling oxidative rancidity. Both endogenous and added antioxidants help control the activity of transition metals, singlet oxygen, and enzymes.

A. CONTROL OF PROOXIDANT METALS

Transition metals accelerate lipid oxidation reactions by hydrogen abstraction and peroxide decomposition, resulting in the formation of free radicals [26]. The activity of prooxidative metals is influenced by chelators or sequestering agents. Transition metals such as iron exhibit low solubility at pH values near neutrality [25]. Therefore, in food systems, transition metals often exist chelated to other compounds. Many compounds will form complexes with metals, resulting in changes in catalytic activity. Some metal chelators increase oxidative reactions by increasing metal solubility and/or altering the redox potential [49]. Chelators also increase the prooxidant activity of transition metal activity by making them more nonpolar, thereby increasing their solubility in lipids [50]. Chelators that exhibit antioxidative properties inhibit metal-catalyzed reactions by one or more of the following properties: prevention of metal redox cycling, occupation of all metal coordination sites, formation of insoluble metal complexes, and stearic hindrance of interactions between metals and lipids or oxidation intermediates (e.g., peroxides) [51]. The prooxidative–antioxidative properties can depend on both metal and chelator concentrations. For instance, EDTA is prooxidative when ratios of EDTA to iron are 1 or less and antioxidative when EDTA–iron $\geq$ 1 [49].

The most common metal chelators used in foods contain multiple carboxylic acids (e.g., EDTA and citric acid) or phosphate (e.g., polyphosphates and phytate) groups. Chelators are typically water soluble but some will exhibit solubility in lipids (e.g., citric acid), thus allowing the chelator to inactivate metals in the lipid phase [52]. Chelator activity depends on pH, since the chelator must be ionized to be active. Therefore, as pH approaches the pK_a of the ionizable groups, chelator activity decreases. Chelator activity is also decreased by the presence of other chelatable ions (e.g., calcium), which will compete with the prooxidative metals for binding sites.

Although most food-grade chelators are unaffected by food processing operations and subsequent storage, polyphosphates are an exception. Polyphosphates are stronger chelators and antioxidants than mono- and diphosphates [53]. However, some foods contain phosphatases, which hydrolyze polyphosphates, thus decreasing their antioxidant effectiveness. This can be observed in muscle foods, where polyphosphates are relatively ineffective in raw meats that contain high levels of phosphatase activity [54] but are highly effective in cooked meats, where the phosphatases have been inactivated [55]. Nutritional implications should also be considered when chelators are used as food antioxidants, since chelators influence mineral bioavailability. For instance, EDTA enhances iron bioavailability, whereas phytate decreases iron, calcium, and zinc absorption [56].

Prooxidant metal activity is also controlled in biological systems by proteins. Proteins with strong binding sites include transferrin, ovotransferrin (conalbumin), lactoferrin, and ferritin. Transferrin, ovotransferrin, and lactoferrin are structurally similar proteins consisting of a single polypeptide chain with a molecular weight ranging from 76,000 to 80,000. Transferrin and lactoferrin bind two ferric ions apiece, whereas ovotransferrin has been reported to bind three [29,57,58]. Ferritin is a multisubunit protein (molecular weight 450,000) with the capability of storing up to 4500 ferric ions [59]. Transferrin, ovotransferrin, lactoferrin, and ferritin inhibit iron-catalyzed lipid oxidation by binding iron in its inactive ferric state and possibly by sterically hindering metal–peroxide interactions [29,60]. Reducing agents (ascorbate, cysteine, superoxide anion) and low pH can cause the release of iron from the proteins, resulting in an acceleration of lipid oxidation reactions [29,61]. The activity of copper can also be controlled by binding to proteins. Serum albumin binds one cupric ion [62] and ceruloplasmin binds up to six cupric ions [63].

Amino acids and peptides can chelate metals in a manner that decreases their reactivity. Both the chelating and the antioxidant activities of the skeletal muscle dipeptide carnosine depend on metal ion type [64–66]. Carnosine more effectively inhibits the oxidation of phosphatidylcholine liposomes catalyzed by copper than by iron. Decker et al. [66] found that the carnosine can chelate and inhibit the prooxidant activity of copper but more effectively than its constituent amino acid histidine. Phosphorylated peptides arising from casein have also been found to be strong iron chelators that can inhibit lipid oxidation [67].

Ceruloplasmin is a copper-containing enzyme that catalyzes the oxidation of ferrous ions:

$$4Fe^{2+} + 4H^+ + O_2 \rightarrow 4Fe^{3+} + 2H_2O.$$

This ferroxidase activity inhibits lipid oxidation by maintaining iron in its oxidized, inactivity state [63]. Since ceruloplasmin is primarily a constituent of blood, one would not expect to find it in most foods, other than muscle foods. Addition of ceruloplasmin to muscle foods in a pure form or as part of blood plasma has been found to effectively inhibit lipid oxidation [68].

B. Control of Singlet Oxygen

Singlet oxygen is an excited state of oxygen in which two electrons in the outer orbitals have opposite spin directions. Initiation of lipid oxidation by singlet oxygen is due to its electrophilic nature, which leads to the formation of lipid peroxides from unsaturated fatty acids [69].

Singlet oxygen can be inactivated by both chemical and physical quenching. Chemical quenching of singlet oxygen by β-carotene will lead to the formation of carotenoid breakdown products containing aldehyde and ketone groups as well as β-carotene-5,8-endoperoxide. β-Carotene-5,8-endoperoxide, which occurs mainly during the oxidation of β-carotene by singlet oxygen, therefore may provide a unique marker that could be used to monitor singlet oxygen–carotenoid interactions in foods and biological systems [70]. Tocopherols can chemically quench singlet oxygen in reactions that lead to the formation of tocopherol peroxides and epoxides [69]. Other compounds, including amino acids, peptides, proteins, phenolics, urate, and ascorbate, can chemically quench singlet oxygen, but little is known about the resulting oxidation products [69,71,72].

While carotenoids are capable of chemically inactivating singlet oxygen, these reactions cause carotenoid autoxidation, leading to loss of antioxidant activity. Therefore, the major mechanism of singlet oxygen inactivation by carotenoids is physical quenching. The most common energy states of singlet oxygen are 22.4 and 37.5 kcal above the ground state [69]. Carotenoids physically quench singlet oxygen by a transfer of energy to the carotenoid to produce an excited state of the carotenoid and ground state, triplet oxygen. Energy is dissipated from the excited carotenoid by vibrational and rotational interactions with the surrounding solvent to return the carotenoid to the ground state [73]. Nine or more conjugated double bonds are necessary for physical quenching [74]. The presence of six-carbon-oxygenated ring structures at the end of the polyenes increases the ability of carotenoids to physically quench singlet oxygen [74]. While it is generally believed that the physical quenching of singlet oxygen by carotenoids does not cause destruction of the carotenoid, these reactions may result in trans or cis isomer conversions [73].

Tocopherols and amines can physically quench singlet oxygen by a charge transfer mechanism. In this reaction, the electron donor (tocopherol or amine) forms a charge transfer complex with the electron-deficient singlet oxygen molecule [70]. An intersystem energy transfer occurs in the complex, resulting in a dissipation of energy and the eventual release of triplet oxygen.

C. Control of Lipoxygenasses

Lipoxygenases are active lipid oxidation catalysts found in plants and some animal tissues. Lipoxygenase activity can be controlled by heat inactivation and plant breeding programs that decrease the concentrations of these enzymes. Phenolics are capable of indirectly inhibiting lipoxygenase activity by acting as free radical inactivators, but also by reducing the iron in the active site of the enzyme to the catalytically inactive ferrous state [16].

IV. INACTIVATION OF OXIDATION INTERMEDIATES

Several compounds that can exist in foods indirectly influence lipid oxidation rates. While these substances do not always directly interact with lipids, they may interact with metals or oxygen to form reactive species. Examples of such compounds include superoxide anion, peroxides, and photosensitizers.

A. Superoxide Anion

Superoxide anion is produced by the addition of an electron to molecular oxygen. Superoxide participates in oxidative reactions by maintaining transition metals in their reduced, active states, by promoting the release of metals bound to proteins such as ferritin, and by the pH-dependent formation of its conjugated acid, the perhydroxyl radical, which can directly catalyze lipid oxidation

[26]. As superoxide anion participates in oxidative reactions, biological systems contain superoxide dismutase (SOD).

Two forms of SOD are found in eukaryotic cells, one in the cytosol and the other in the mitochondria [75]. Cytosolic SOD contains copper and zinc in the active site, whereas mitochondrial SOD contains manganese. Both forms of SOD catalyze the conversion of superoxide anion to hydrogen peroxide by the following reaction:

$$2O_2^- + 2H^+ \rightarrow O_2 + H_2O_2.$$

Other compounds can also possess SOD-like activity. The most notable of these are complexes of amino acids and peptides with transition metals. Cupric ions complexed to lysine, tyrosine, and histidine are capable of catalyzing the dismutation of superoxide [76]. Histidine-containing peptides complexed to nickel [77], copper [78], and zinc [78,79] also contain SOD-like activity. It should be noted that the SOD-like activity of metal–amino acid or peptide complexes generally has orders of magnitude lower than those of proteinaceous SOD.

B. Peroxides

Peroxides are important intermediates of oxidative reactions because they decompose via transition metals, irradiation, and elevated temperatures to form free radicals. Hydrogen peroxide exists in foods as a result of direct addition (e.g., aseptic processing operations) and formation in biological tissues by mechanisms including the dismutation of superoxide by SOD and the activity of peroxisomes [80]. Hydrogen peroxide is rapidly decomposed by the reduced state of transition metals (e.g., Fe and Cu) to the hydroxyl radical. The hydroxyl radical is an extremely reactive free radical that can oxidize most biological molecules at diffusion-limited reaction rates. Therefore, removal of hydrogen peroxide from biological materials is critical to the prevention of oxidative damage.

Catalase (CAT) is a heme-containing enzyme that catalyzes the following reaction [81]:

$$2H_2O_2 \rightarrow 2H_2O + O_2.$$

Hydrogen peroxide in higher plants and algae may be scavenged by ascorbate peroxidase. Ascorbate peroxidase inactivates hydrogen peroxide in the cytosol and chloroplasts by the following mechanism [82]:

$$2 \text{ Ascorbate} + H_2O_2 \rightarrow 2 \text{ Monodehydroascorbate} + 2H_2O.$$

Two ascorbate peroxidase isozymes which differ in molecular weight (57,000 vs. 34,000), substrate specificity, pH optimum, and stability have been described in tea leaves [83].

In addition to catalase, many biological tissues contain glutathione peroxidase (GSH-Px) to help control peroxides. GSH-Px differs from CAT in that it is capable of reacting with both lipid and hydrogen peroxides. GSH-Px is a selenium-containing enzyme that uses reduced glutathione (GSH) to catalyze hydrogen or lipid (LOOH) peroxide reduction [80]:

$$H_2O_2 + 2GSH \rightarrow 2H_2O + GSSG$$

or

$$LOOH + 2GSH \rightarrow LOH + H_2O + GSSG,$$

where GSSG is oxidized glutathione and LOH is a fatty acid alcohol. Two types of GSH-Px exist in biological tissues, and one shows high specificity for phospholipid hydroperoxides [80,84].

Thiodipropionic acid and dilauryl thiodipropionate are approved food additives capable of decomposing peroxides and peracids. The concentration allowed in foods is ≤ 200 ppm; however, they are relatively ineffective antioxidants and are therefore rarely used [52]. Methionine, which has been found to be antioxidative in some lipid systems, is thought to decompose peroxides by mechanisms similar to those of thiodipropionic acid and dilauryl thiodipropionate [52].

C. Photoactivated Sensitizers

In foods, light is capable of activating sensitizers such as chlorophyll, riboflavin, and heme-containing proteins to an excited state. These photoactivated sensitizers can promote oxidation by directly interacting with an oxidizable substrate to produce free radicals, by transferring energy to triplet oxygen to form singlet oxygen, or by transferring of an electron to triplet oxygen to form superoxide anion [69]. Carotenoids inactivate photoactivated sensitizers by physically absorbing their energy to form the excited state of the carotenoid, which then returns to the ground state by transfer of energy into the surrounding solvent [22,73].

V. ALTERATIONS IN LIPID OXIDATION BREAKDOWN PRODUCTS

Oxidation of fatty acids eventually leads to the formation of breakdown products via β-scission reactions. These reactions lead to a multitude of different oxidation products, known as secondary lipid oxidation products, which affect both the sensory characteristics and the functional properties of foods. Rancid odors arise from the production of secondary products such as aldehydes, ketones, and alcohols [85]. Secondary lipid oxidation products, and particular aldehydes, also impact food quality and nutritional composition through interaction with the amino groups of proteins and vitamins. Secondary products arising from lipid oxidation have been found to alter the function of proteins, enzymes, biological membranes, lipoproteins, and DNA [86–89].

Since aldehydes and other secondary products arising from lipid oxidation are potentially damaging, biological systems seem to have developed mechanisms to control their activity. Sulfur- and amine-containing compounds have the ability to interact with aldehydes. This may help explain why many proteins, peptides, amino acids, phospholipids, and nucleotides display antioxidant activity when secondary products are used to measure lipids oxidation. Carnosine and anserine, which can make up over 1% of the wet weight of muscle tissue, are capable of forming complexes with aldehydes produced from oxidizing lipids [90]. Carnosine is more effective in forming adducts with aldehydes than its constituent amino acids, histidine and β-alanine [91,92]. Glutathione is also very effective in binding aldehydes; but at the concentrations found in muscle foods, carnosine seems more likely to be the major aldehyde-binding component [92].

VI. SURFACE-ACTIVE ANTIOXIDANTS AND PHYSICAL EFFECTS

Lipids in food systems often have interfacial surfaces at which oxidative reactions are prevalent. Examples include oil-in-water emulsions, water-in-oil emulsions, the air–lipid interface of bulk oils and solid fats, and the water–lipid interface of biological membranes. Oxidation is prevalent at these interfaces as a result of increased contact with oxygen, the presence of aqueous phase free radicals, the presence of reactive oxygen generating systems and prooxidative metals, and possibly the migration of the more polar lipid peroxides out of the hydrophobic lipid core toward the more polar interface.

The effectiveness of phenolic antioxidants is often dependent on their polarity. Porter et al. [93] used the term "antioxidant paradox" to describe how polar antioxidants are most effective in bulk lipids, whereas nonpolar antioxidants are most effective in dispersed lipids. In bulk tocopherol-stripped corn oil, Trolox (a water-soluble analog of α-tocopherol) more effectively inhibited lipid peroxide formation than α-tocopherol. However, when tocopherol-stripped corn oil was emulsified with Tween 20, α-tocopherol inhibited peroxide formation more effectively than Trolox. The observed increase in activity of α-tocopherol compared with Trolox in emulsified oil was attributed to its retention in the oil and possibly to its ability (due to its surface activity) to concentrate at the oil–water interface. The lower activity of Trolox in emulsions was due to its partitioning into the water phase, where it was not able to inhibit autoxidation of the corn oil [94]. Similar effects have been observed for the phenolic antioxidants in rosemary extracts, with the more polar compounds (carnosic and rosmarinic acids) being most effective in bulk oils and the less polar compounds (carnosol) are more effective in emulsified lipids [18]. Similarly, the antioxidant activity of carnosic acid was improved in emulsified corn oil when it was made nonpolar by methylation [95].

The charge of dispersed lipids also influences oxidation rates, especially in the presence of transition metals. Since iron and other transition metals are common contaminants in most water systems, their ability to catalyze oxidation at the oil–water interface of dispersed lipids could be important. When the surface charge of dispersed lipids as either micelles [96,97] or phospholipid vesicles [98] is negative, iron-catalyzed lipid oxidation rates are much higher than they are at positively charged interfaces. This effect presumably exists because iron can bind to the interface of the dispersed lipid. The inhibitory effect of positively charged lipid micelles can be partially overcome by nitrilotriacetic acid, which forms negatively charged iron chelates, and by the addition of lipid-soluble peroxides [96,97]. Positively charged emulsifiers are uncommon in foods. However, proteins at pHs below their pIs produce positively charged lipid emulsion droplets. Oil-in-water emulsions stabilized by soy, whey protein isolate, or casein are more oxidatively stable at low pH when they produce cationic emulsion droplets than at high pH when the emulsion droplet is anionic [99]. Cationic oil-in-water emulsion droplets produced by absorption of chitosan onto phosphatidylcholine also increase the oxidative stability of oils [100].

Lipid oxidation can be inhibited by encapsulation. Potential mechanisms of inhibition include physical inhibition of oxygen diffusion into the lipid, chemical (e.g., free radical scavenging) and physical (e.g., chelation) antioxidant properties of the encapsulating agents, and possibly interaction of lipid oxidation products with the encapsulating material. Both protein- and carbohydrate-encapsulating agents have been found to retard oxidation rates. The effectiveness of these encapsulating agents depends on the factors such as concentration of the encapsulating agent [101], method of encapsulation (which affects the porosity of the encapsulating layer) [102], and the environmental relative humidity under which the encapsulated lipid is stored [103].

Some research indicates that encapsulation does inhibit oxygen diffusion into the lipid, but the same research also indicates that oxygen diffusion is not the only mechanism by which encapsulation inhibits oxidation [104]. More research is needed to determine the antioxidant mechanisms of encapsulation, since this technique could be an effective way to increase the stability of oxidatively labile lipids, thereby increasing their incorporation into foods.

Another factor that may influence oxidation rates is the physical state of the lipid. Lipids in foods often exist as a combination of both liquid and crystalline states, a condition that depends on both fatty acid composition and temperature. The influence of liquid fat concentration on oxidation rates was investigated in liposomes, where arachidonic acid oxidation rates were found to be greater at temperatures below the solid–liquid phase transition temperature of the host lipid [105]. The increase in oxidation rates was attributed to phase separation of the most unsaturated fatty acids, which increases the concentrations of oxidizable substrate into localized domains [106]. Little is known about how transition temperatures influence oxidation rates in food lipids.

VII. ANTIOXIDANT INTERACTIONS

Biological food systems usually contain multicomponent antioxidant systems. The numerous existing antioxidants have different potential functions, including inhibition of prooxidants of different types (e.g., metals, reactive oxygen species, enzymes); inactivation of free radicals and prooxidants in aqueous, interfacial, and lipid phases; and inactivation of compounds at different stages of oxidation (e.g., initiating species (·OH), propagating species (peroxides), lipid oxidation decomposition products [aldehydes]). In addition, multicomponent antioxidant systems are beneficial because direct interactions occur between antioxidants.

Combinations of chelators and FRS often result in synergistic inhibition of lipid oxidation [6]. Synergistic interaction most likely occurs by a "sparing" effect provided by the chelator. Since the chelator will decrease oxidation rates by inhibiting metal-catalyzed oxidation, fewer free radicals will be generated in the system. This means that the eventual inactivation of the FRS through reactions such as termination or autoxidation will be slower, thus making its concentration greater at any given time. The combination of chelator and FRS thus decreases free radical generation and increases radical scavenging potential.

Synergistic antioxidant activity can also be observed by the combination of two or more different FRS. This occurs when one FRS reacts more rapidly with free radicals than the other as a result of differences in bond dissociation energies and/or stearic hindrance of FRS–ROO·interactions [6]. These differences will result in one antioxidant being consumed faster than the other. However, it may be possible for this FRS to be regenerated by transfer of its radical to a different scavenger. In the system consisting of α-tocopherol and ascorbic acid [106], for example, α-tocopherol is the primary FRS because it is present in the lipid phase. Ascorbic acid then regenerates the tocopheroxyl radical or possibly tocopherylquinone back to α-tocopherol plus the semihydroascorbyl radical [4], which dismutases to dehydroascorbate [3]. In turn, dehydroascorbate may be regenerated by enzymes that use NADH or NADPH as reducing equivalents [107].

Since multicomponent antioxidant systems can inhibit oxidation at many different phases of oxidation, the resulting antioxidant activity can be synergistic. This suggests that the most effective antioxidant systems for foods would contain antioxidants with different mechanisms of action and/or physical properties. Determining which antioxidants would be most effective depends on the factors such as types of oxidation catalyst, physical state of lipid (bulk vs. emulsified), and factors that influence the activity of the antioxidants themselves (e.g., pH, temperature, the ability to interact with other compounds in foods).

REFERENCES

1. N.I. Krinsky. Mechanism of action of biological antioxidants. *Proc. Soc. Exp. Biol. Med.* 200:248 (1992).
2. J.L. Bolland and P. ten Have. Kinetic studies in the chemistry of rubber and related materials: IV. The inhibitory effect of hydroquinone on the thermal oxidation of methyl linoleate. *Trans. Faraday Soc.* 42:201 (1947).
3. G.R. Buettner. The pecking order of free radicals and antioxidants: Lipid peroxidation, α-tocopherol, and ascorbate. *Arch. Biochem. Biophys.* 300(2):535 (1993).
4. D.C. Liebler. The role of metabolism in the antioxidant functions of vitamin E. *Crit. Rev. Toxicol.* 23(2):147 (1993).
5. F. Shahidi and J.P.K. Wanasundara. Phenolic antioxidants. *Crit. Rev. Food Sci. Nutr.* 32:67 (1992).
6. W.W. Nawar. Lipids. In: *Food Chemistry*, 3rd edn. (O. Fennema, ed.). Dekker, New York, 1996, p. 225.
7. R.S. Parker. Dietary and biochemical aspects of vitamin E. *Adv. Food Nutr. Res.* 33:157 (1989).
8. D.C. Liebler and J.A. Burr. Oxidation of vitamin E during iron-catalyzed lipid per-oxidation: Evidence for electron-transfer reactions of the tocopheroxyl radical. *J. Biochem.* 31:8278 (1992).
9. D.C. Liebler, P.F. Baker, and K.L. Kaysen. Oxidation of vitamin E: Evidence for competing autoxidation and peroxyl radical trapping reaction of the tocopheroxyl radical. *J. Am. Chem. Soc.* 112:6995 (1990).
10. D.C. Liebler, K.L. Kaysen, and J.A. Burr. Peroxyl radical trapping and autoxidation reactions of α-tocopherol in lipid bilayers. *Chem. Res. Toxicol.* 4:89 (1991).
11. H.H. Draper, A.S. Csallany, and M. Chiu. Isolation of a trimer of alpha-tocopherol from mammalian liver. *Lipids* 2:47 (1966).
12. G. Zubay. Aerobic production of ATP: Electron transport. In: *Biochemistry* (B. Rogers, ed.). Addison-Wesley, Reading, MA, 1984.

13. K.U. Ingold, V.W. Bowry, R. Stocker, and C. Walling. Autoxidation of lipids and antioxidation by α-tocopherol and ubiquinol in homogeneous solution and in aqueous dispersions of lipids: Unrecognized consequences of lipid particle size as exemplified by oxidation of human low density lipoprotein. *Proc. Natl. Acad. Sci. U.S.A.* 90:45 (1993).

14. B. Frei, M.C. Kim, and B.N. Ames. Ubiquinol-10 is an effective lipid soluble antioxidant at physiological concentrations. *Proc. Natl. Acad. Sci. U.S.A.* 87:4879 (1990).

15. R. Stocker, V.W. Bowry, and B. Frei. Ubiquinol-10 protects human low density lipoprotein more efficiently against lipid peroxidation better than does α-tocopherol. *Proc. Natl. Acad. Sci. U.S.A.* 88:1646 (1991).

16. M.J. Laughton, P.J. Evans, M.A. Moroney, J.R.S. Hoult, and B. Halliwell. Inhibition of mammalian 5-lipoxygenase and cyclo-oxygenase by flavonoids and phenolic dietary additives. *Biochem. Pharm.* 42(9):1673 (1991).

17. W. Bors, C. Michel, and S. Schikora. Interaction of flavonoids with ascorbate and determination of their univalent redox potentials: A pulse radiolysis study. *Free Radic. Biol. Med.* 19(1):45 (1995).

18. E.N. Frankel, S.-W. Huang, R. Aeschbach, and E. Prior. Antioxidant activity of a rosemary extract and its constituents, carnosic acid, carnosol, and rosmarinic acid, in bulk oil and oil-in-water emulsion. *J. Agric. Food Chem.* 44:131 (1996).

19. O.I. Aruoma, B. Halliwell, R. Aeschbach, and J. Löligers. Antioxidant and pro-oxidant properties of active rosemary constituents: Carnosol and carnosic acid. *Xenobiotica* 22(2):257 (1992).

20. M.M. Mielche and G. Bertelsen. Approaches to the prevention of warmed-over flavour. *Trends Food Sci. Technol.* 5:322 (1994).

21. J.A. Olson. Vitamin A and carotenoids as antioxidants in a physiological context. *J. Nutr. Sci. Vitaminol.* 39:857 (1993).

22. P. Palozza and N.I. Krinsky. Antioxidant effects of carotenoids in vivo and in vitro: An overview. *Methods Enzymol.* 213:403 (1992).

23. G.W. Burton and K.U. Ingold. β-Carotene: An unusual type of lipid antioxidant. *Science* 224:569 (1984).

24. D.C. Liebler. Antioxidant reactions of carotenoids. *Ann. N.Y. Acad. Sci.* 691:20 (1992).

25. H.B. Dunford. Free radicals in iron-containing systems. *Free Radic. Biol. Med.* 3:405 (1987).

26. J. Kanner, J.B. German, and J.E. Kinsella. Initiation of lipid peroxidation in biological systems. *Crit. Rev. Food Sci. Nutr.* 25:317 (1987).

27. E.A. Decker and H.O. Hultin. Lipid oxidation in muscle foods via redox iron. In: *Lipid Oxidation in Foods* (A.J. St. Angelo, ed.). ACS Symposium Series 500, American Chemical Society, Washington, D.C., 1992, p. 33.

28. J. Kanner. Mechanism of nonenzymic lipid peroxidation in muscle foods. In: *Lipid Oxidation in Foods* (A.J. St. Angelo, ed.). ACS Symposium Series 500, American Chemical Society, Washington, D.C., 1992, p. 55.

29. B. Halliwell and J.M.C. Gutteridge. Oxygen free radicals and iron in relation to biology and medicine. Some problems and concepts. *Arch. Biochem. Biophys.* 246:501 (1986).

30. V.E. Kagan, A. Shvedova, E. Serbinova, S. Khan, C. Swanson, R. Powell, and L. Packer. Dihydrolipoic acid—A universal antioxidant both in the membrane and in the aqueous phase. *Biochem. Pharmacol.* 44(8):1637 (1992).

31. A. Bast and G.R.M.M. Haenen. Interplay between lipoic acid and glutathione in the protection against microsomal lipid peroxidation. *Biochim. Biophys. Acta* 963:558 (1988).

32. J. Kanner, S. Harel, and B. Hazan. Muscle membranal lipid oxidation by an iron redox cycle system: Initiation by oxy radicals and site-specific mechanism. *J. Agric. Food Chem.* 34:506 (1986).

33. B.N. Ames, R. Cathcart, E. Schwiers, and P. Hochstein. Uric acid provides an antioxidant defense in humans against oxidant- and radical-caused aging and cancer: A hypothesis. *Proc. Natl. Acad. Sci. U.S.A.* 78(11):6858 (1981).

34. D.D.M. Wayner, G.W. Burton, K.U. Ingold, L.R.C. Barclay, and S.J. Locke. The relative contributions of vitamin E, urate, ascorbate and proteins to the total peroxyl radical-trapping antioxidant activity of human blood plasma. *Biochim. Biophys. Acta* 924:408 (1987).

35. B. Frei, R. Stocker, and B.N. Ames. Antioxidant defenses and lipid peroxidation in human blood plasma. *Proc. Natl. Acad. Sci. U.S.A.* 85:9748 (1988).

36. E.A. Foegeding, T.C. Lanier, and H.O. Hutlin. Characteristics of muscle tissue. In: *Food Chemistry*, 3rd edn. (O. Fennema, ed.). Dekker, New York, 1996, p. 879.

37. S. Gebicki and J.M. Gebicki. Formation of peroxides in amino acids and proteins exposed to oxygen free radicals. *Biochem. J.* 289:743 (1993).

38. K. Kikugawa, T. Kato, and A. Hayasaka. Formation of dityrosine and other fluorescent amino acids by reaction of amino acids with lipid hydro peroxides. *Lipids* 26(11):922 (1991).

39. R. Kohen, Y. Yamamoto, K.C. Cundy, and B. Ames. Antioxidant activity of carnosine, homocarnosine, and anserine present in muscle and brain. *Proc. Natl. Acad. Sci. U.S.A.* 85:3175 (1988).

40. M. Karel, K. Schaich, and R.B. Roy. Interaction of peroxidizing methyl linoleate with some proteins and amino acids. *J. Agric. Food Chem.* 23(2):159 (1975).

41. E. Niki, Y. Yamamoto, M. Takahashi, K. Yamamoto, Y. Yamamoto, E. Komuro, M. Miki, H. Yasuda, and M. Mino. Free radical-mediated damage of blood and its inhibition by antioxidants. *J. Nutr. Sci. Vitaminol.* 34:507 (1988).

42. M. Soriani, D. Pietraforte, and M. Minetti. Antioxidant potential of anaerobic human plasma: Role of serum albumin and thiols as scavengers of carbons radicals. *Arch. Biochem. Biophys.* 312(1):180 (1994).

43. N. Yamaguchi, Y. Yokoo, and M. Fufimaki. Studies on antioxidative activities of amino compounds on fats and oils: II. Antioxidative activities of dipeptides and their synergistic effects on tocopherol. *Jpn. Food Ind. J.* 22:425 (1975).

44. K.S. Rhee and G.C. Smith. A further study of effects of glandless cottonseed flour on lipid oxidation and color changes in raw ground beef containing salt. *J. Food Prot.* 46:787 (1983).

45. N. Tsuge, U. Eikawa, Y. Nomura, M. Yamamoto, and K. Sugisawa. Antioxidative activity of peptides prepared by enzymatic hydrolysis of egg-white albumin. *Nippon Nogeikagaki Kaishi* 65(11):1635 (1991).

46. Z.Y. Chen and W.W. Nawar. The role of amino acids in the autoxidation of milk fat. *J. Am. Oil Chem. Soc.* 68(1):47 (1991).

47. N.C. Shantha, A.D. Crum, and E.A. Decker. Conjugated linoleic acid concentrations in cooked beef containing antioxidants and hydrogen donors. *J. Food Lipids* 2:57 (1995).

48. R.J. Elias, J.D. Bridgewater, R.W. Vachet, T. Waraho, D.J. McClements, and E.A. Decker. Antioxidant mechanisms of enzymatic hydrolysates of β-lactoglobulin in food lipid dispersions. *J. Agric. Food Chem.* 54:9565 (2006).

49. J.R. Mahoney and E. Graf. Role of α-tocopherol, ascorbic acid, citric acid and EDTA as oxidants in a model system. *J. Food Sci.* 51:1293 (1986).

50. C. Ruben and K. Larsson. Relations between antioxidant effect of α-tocopherol and emulsion structure. *J. Dispersion Sci. Technol.* 6:213 (1985).

51. E. Graf and J.W. Eaton. Antioxidant functions of phytic acid. *Free Radic. Biol. Med.* 8:61 (1990).

52. R.C. Lindsay, Food additives. In: *Food Chemistry*, 3rd edn. (O. Fennema, ed.). Dekker, New York, 1996, p. 767.

53. J.N. Sofos. Use of phosphates in low-sodium meat products. *Food Technol.* 40(9):52 (1986).

54. W. Li, J.A. Bowers, J.A. Craig, and S.K. Perng. Sodium tripolyphosphate stability and effect in ground turkey meat. *J. Food Sci.* 58:501 (1993).

55. G.R. Trout and G.R.S. Dale. Prevention of warmed-over flavor in cooked beef: Effect of phosphate type, phosphate concentration, a lemon juice/phosphate blend, and beef extract. *J. Agric. Food Chem.* 38:665 (1990).

56. D.D. Miller. Minerals. In: *Food Chemistry*, 3rd edn. (O. Fennema, ed.). Dekker, New York, 1996, p. 617.

57. J.E. Kinsella and D.M. Whitehead. Proteins in whey: Chemical, physical, and functional properties. *Adv. Food Nutr. Res.* 33:343 (1989).

58. W.D. Powrie and S. Nakai. Characteristics of edible fluids of animal origin: Eggs. In: *Food Chemistry*, 2nd edn. (O. Fennema, ed.). Dekker, New York, 1985, p. 829.

59. J.M. LaCross and M.C. Linder. Synthesis of rat muscle ferritins and function in iron metabolism of heart and diaphragm. *Biochim. Biophys. Acta* 633:45 (1980).

60. D.A. Baldwin, E.R. Jenny, and P. Aisen. The effect of human serum transferrin and milk lactoferrin on hydroxyl radical formation from superoxide and hydrogen peroxide. *J. Biol. Chem.* 259(21):13391 (1984).

61. E.A. Decker and B. Welch. The role of ferritin as a lipid oxidation catalyst in muscle foods. *J. Agric. Food Chem.* 38:674 (1990).

62. D.W. Appleton and B. Sarkar. The absence of specific copper (II) binding site in dog albumin. *Biol. Chem.* 246(16):5040 (1971).

63. J.M.C. Gutteridge. Antioxidant activity of ceruloplasmin. In: *CRC Handbook, Methods of Oxygen Radical Research*, Vol. 1 (R.A. Greenwald, ed.). CRC Press, Boca Raton, FL, 1985, p. 283.

64. C.E. Brown and W.E. Antholine. Multiple forms of the cobalt (II)–carnosine complex. *Biochem. Biophys. Res. Commun.* 88(2):529 (1979).

65. C.E. Brown. Interactions among carnosine, anserine, ophidine and copper in biochemical adaptation. *J. Theor. Biol.* 88:245 (1981).

66. E.A. Decker, V. Ivanov, B.Z. Zhu, and B. Frei. Inhibition of low density lipoprotein oxidation by carnosine and histidine. *J. Agric. Food Chem.* 49:511 (2001).

67. M. Daiz and E.A. Decker. Antioxidant mechanisms of caseinophosphopeptides and casein hydrolysates and their application in ground beef. *J. Agric. Food Chem.* 52:8208 (2005).

68. J. Kanner, F. Sofer, S. Harel, and L. Doll. Antioxidant activity of ceruloplasmin in muscle membrane and in situ lipid oxidation. *J. Agric. Food Chem.* 36:415 (1988).

69. D.G. Bradley and D.B. Min. Singlet oxygen oxidation of foods. *Crit. Rev. Food Sci. Nutr.* 31(3):211 (1992).

70. S.P. Stratton, W.H. Schaefer, and D.C. Liebler. Isolation and identification of singlet oxygen oxidation products of β-carotene. *Chem. Res. Toxicol.* 6(4):542 (1993).

71. J.R. Kanofsky. Quenching of singlet oxygen by human plasma. *Photochem. Photobiol.* 51(3):299 (1990).

72. T.A. Dahl, W.R. Midden, and P.E. Hartman. Some prevalent bioimolecules as defenses against singlet oxygen damage. *Photochem. Photobiol.* 47:357 (1988).

73. W. Stahl and H. Sies. Physical quenching of singlet oxygen and cis–trans isomerization of carotenoid. *Ann. N.Y. Acad. Sci.* 691:10 (1992).

74. P. DiMascio, P.S. Kaiser, and H. Sies. Lycopene as the most efficient biological carotenoid singlet oxygen quencher. *Arch. Biochem. Biophys.* 274:532 (1989).

75. I. Fridovich. Superoxide dismutases. *Adv. Enzymol.* 41:35 (1974).

76. K.E. Joester, G. Jung, U. Weber, and U. Weser. Superoxide dismutase activity of Cu^{2+} amino acid chelates. *FEBS Lett.* 25(1):25 (1972).

77. E. Nieboer, R.T. Tom, and F.E. Rossetto. Superoxide dismutase activity and novel reactions with hydrogen peroxide of histidine-containing nickel(II)-oligopeptide complexes and nickel(II)-induced structural changes in synthetic DNA. *Biol. Trace Elem. Res.* 21:23 (1989).

78. R. Kohen, R. Misgav, and I. Ginsburg. The SOD-like activity of copper:carnosine, copper:anserine and copper:homocarnosine complexes. *Free Radic. Res. Commun.* 12/13:179 (1991).

79. T. Yoshikawa, Y. Naito, T. Tanigawa, T. Yoneta, M. Yasuda, S. Ueda, H. Oyamada, and M. Kondo. Effect of zinc-carnosine chelate compound (Z-103), a novel antioxidant, on acute gastric mucosal injury induced by ischemia-reperfusion in rats. *Free Radic. Res. Commun.* 14(4):289 (1991).

80. W.A. Günzler and L. Flohé. Glutathione peroxidase. In: *CRC Handbook, Methods of Oxygen Radical Research*, Vol. 1 (R.A. Greenwald, ed.). CRC Press, Boca Raton, FL, 1985, p. 285.

81. A. Claiborne. Catalase activity. In: *CRC Handbook, Methods of Oxygen Radical Research*, Vol. 1 (R.A. Greenwald, ed.). CRC Press, Boca Raton, FL, 1985, p. 283.

82. K. Asada. Ascorbate peroxidase—A hydrogen peroxide-scavenging enzyme in plants. *Physiol. Plant* 85:235 (1992).

83. G.-X. Chen and K. Asada. Ascorbate peroxidase in tea leaves: Occurrence of two isozymes and the differences in their enzymatic and molecular properties. *Plant Cell Physiol.* 30(7):987 (1989).

84. M. Maiorino, C. Gregolin, and F. Ursini. Phospholipid hydroperoxide glutathione peroxidase. *Methods Enzymol.* 186:448 (1990).

85. J.R. Vercellotti, O.E. Mills, K.L. Belt, and D.L. Sullen. Gas chromatographic analysis of lipid oxidation volatiles in foods. In: *Lipid Oxidation in Foods* (A.J. St. Angelo, ed.). ACS Symposium Series 500, American Chemical Society, Washington, D.C., 1992, p. 232.

86. B. Halliwell and J.M. Gutteridge. Role of free radicals and catalytic metal ions in human disease: An overview. *Methods Enzymol.* 186:1 (1990).

87. Y.L. Xioing and E.A. Decker. Alterations in muscle protein functionality by oxidative and antioxidative processes. *J. Muscle Foods* 6:139 (1995).

88. H.P. Diegner, E. Friedrich, H. Sinn, and H.A. Dresel. Scavenging of lipid peroxidation products from oxidizing LDL by albumin alters the plasma half-life of a fraction of oxidized LDL particles. *Free Radic. Res. Commun.* 16(4):239 (1992).

89. P. Agerbo, B.M. Jøgensen, B. Jensen, T. Børresen, and G. Hølmer. Enzyme inhibition by secondary lipid autoxidation products from fish oil. *J. Nutr. Biochem.* 3:549 (1992).

90. A.A. Boldyrev, A.M. Dupin, E.V. Pindel, and S.E. Severin. Antioxidative properties of histidine-containing dipeptides from skeletal muscles of vertebrates. *Comp. Biochem. Physiol.* 89B(2):245 (1988).

91. S. Zhou and E.A. Decker. Ability of amino acids, dipeptides, polyamines and sulfhydryls to quench hexanal, a saturated aldehyde lipid oxidation product. *J. Agric. Food Chem.* 47:1932 (1999).

92. S. Zhou and E.A. Decker. Ability of carnosine and other skeletal muscle components to quench unsaturated aldehydic lipid oxidation products. *J. Agric. Food Chem.* 47:51 (1999).

93. W.L. Porter, E.D. Black, and A.M. Drolet. Use of polyamide oxidative fluorescence test on lipid emulsions: Contrasts in relative effectiveness of antioxidants in bulk versus dispersed systems. *J. Agric. Food Chem.* 37:615 (1989).

94. S.-W. Huang, A. Hopia, K. Schwarz, E.N. Frankel, and J.B. German. Antioxidant activity of α-tocopherol and Trolox in different lipid substrates: Bulk oils vs. oil-in-water emulsions. *J. Agric. Food Chem.* 44:444 (1996).

95. S.-W. Huang, E.N. Frankel, K. Schwarz, R. Aeschbach, and J.B. German. Antioxidant activity of carnosic acid and methyl carnosate in bulk oils and oil-in-water emulsions. *J. Agric. Food Chem.* 44:2951 (1996).

96. K. Fukuzawa and T. Fujii. Peroxide dependent and independent lipid peroxidation: Site-specific mechanisms of initiation by chelated iron and inhibition by α-tocopherol. *Lipids* 27(3):227 (1992).

97. Y. Yoshida and E. Niki. Oxidation of methyl linoleate in aqueous dispersions induced by copper and iron. *Arch. Biochem. Biophys.* 295(1):107 (1992).

98. K. Fukuzawa, K. Soumi, M. Lemura, S. Goto, and A. Tokumura. Dynamics of xanthine oxidase- and Fe^{3+}-APP-dependent lipid peroxidation in negatively charged phospholipid vesicles. *Arch. Biochem. Biophys.* 316(1):83 (1995).

99. M. Hu, D.J. McClements, and E.A. Decker. Lipid oxidation in corn oil-in-water emulsions stabilized by casein, whey protein isolate and soy protein isolate. *J. Agric. Food Chem.* 51:1696 (2003).

100. U. Klinkesorn, P. Sophanodora, P. Chinachoti, D.J. McClements, and E.A. Decker. Stability of spray-dried tuna oil emulsions encapsulated with two-layered interfacial membranes. *J. Agric. Food Chem.* 53:8365 (2005).

101. D.L. Moreau and M. Rosenberg. Oxidative stability of anhydrous milk fat microencapsulated in whey proteins. *Food Sci.* 61(1):39 (1996).

102. K. Taguchi, K. Iwami, F. Ibuki, and M. Kawabata. Oxidative stability of sardine oil embedded in spray-dried egg white powder and its use for *n*-3 unsaturated fatty acid fortification of cookies. *Biosci. Biotech. Biochem.* 56(4):560 (1992).

103. K. Iwami, M. Hattori, T. Yasumi, and F. Ibuki. Stability of gliadin-encapsulated unsaturated fatty acids against autoxidation. *J. Agric. Food Chem.* 36:160 (1988).

104. R. Matsuno and S. Adachi. Lipid encapsulation technology—Techniques and applications to food. *Trends Food Sci. Technol.* 4:256 (1993).

105. L.R. McLean and K.A. Hagaman. Effect of lipid physical state on the rate of peroxidation of liposomes. *Free Radic. Biol. Med.* 12:113 (1992).

106. P.B. McCay. Vitamin E: Interactions with free radicals and ascorbate. *Annu. Rev. Nutr.* 5:323 (1985).

107. S. Englard and S. Seifter. The biochemical functions of ascorbic acid. *Annu. Rev. Nutr.* 6:365 (1986).

Part IV

Nutrition

19 Fats and Oils in Human Health

David Kritchevsky

CONTENTS

I. FATS AND OILS IN HUMAN HEALTH

Most discussions of fats and health focus on the deleterious effects of the essential nutrients like fats and oils. What we are really discussing in that case is the possibly harmful effects of an excess of fats and oils. Fats (lipids) supply energy, support structural aspects of the body, and provide substances that regulate physiological processes.

Adipose tissue, which is the repository of most of our body fats, serves as an energy reservoir (fat supplies 9 cal/g compared with 4 cal/g for protein or carbohydrate), as a heat conserver, and as a shock absorber. Lipids contain essential fatty acids, such as linoleic and linolenic acids. These are metabolized eventually to provide eicosanoids, substances that possess hormone-like activity and thus may regulate many body functions. Fat is also the transport vehicle for vitamins A, D, E, and K.

Cholesterol, which has absorbed the brunt of the antifat attack, is a compound that is essential for life. It is not essential in the sense of essential fatty acids since the body can synthesize it, but it is a crucially important component of our biological economy. Cholesterol comprises about 0.2% of normal body weight. Most of it (about 33%) is in the brain and nervous system, where its function has not been probed beyond suggesting that its major function is as an insulator. Almost another one-third of the body's cholesterol is in muscle where it is a structural component. Every cell membrane contains cholesterol and phospholipid, another fatty substance. The esterified cholesterol found in muscle may represent a storage compartment. The percentage of cholesterol ester in muscle increases with age. Cholesterol is the parent substance for vitamin D_2, bile acids, adrenocortical hormones, and sex hormones. Thus, it is one of the most important biological substances. Fat also contributes to the palatability and flavor of food and hence contributes to the enjoyment of eating.

The two major causes of death in the developed world are heart disease and cancer. Both have been described as lifestyle diseases, and effects of diet fit under that rubric. Since diet is one of the easiest lifestyle factors to investigate and possibly change, its role has been pursued with vigor. However, dietary data are not as simple to obtain as one might expect. Population-based data, derived from availability statistics, do not account for individual variations. Recall may be flawed by habitual underreporting of intake [1,2]. These methods provide useful data but their shortcomings should be kept in mind.

II. LIPIDS AND CARDIOVASCULAR DISEASE

The major difficulty in assessment of the roles of diet and other factors is the absence of a clear, unequivocal, antemortem diagnosis. Failing that, the data have been analyzed to provide "risk" factors (called "odds" in Las Vegas) for indication of susceptibility. Among the major risk factors for coronary heart disease (CHD), also termed cardiovascular disease (CVD), are cigarette smoking, elevated cholesterol level, elevated blood pressure, obesity, and maleness. The first four risk factors are correctable and the last is an unalterable fact of life, but the others may, to some extent, be amenable to nutritional intervention. However, it is oversimplification to regard atherosclerosis as a consequence of diet or aging or both. The molecular mechanisms underlying the atherosclerotic process are elucidated, and increasing aspects of the disease exhibit a genetic component. We are discovering new molecular and immunological factors related to this disease. Fatal outcome is associated with plaque rupture and thrombosis, and the notion of the disease being due to simple accretion of cholesterol in the arteries is no longer tenable. Lusis [3] has published an elegant description of factors involved in atherogenesis.

Although an experimental relationship between dietary cholesterol and atherosclerosis had been adduced in 1913 when Anitschkow was able to produce aortic lesions in rabbits [4], interest in fat in the diet and its relation to this disease began to blossom in the 1950s. In 1950, Gofman et al. [5] developed a method for the ultracentrifugal separation of plasma lipoproteins, showed how these fractionated particles could be related to heart disease, and implicated diet as a factor. In 1953, Keys [6] described what was the beginning of his Seven Countries Study and indicated atherosclerosis as a new public health problem.

The lipoproteins are lipid–protein agglomerates rather than real chemical compounds. They are described by their hydrated densities (a physical property) but may differ in size and composition. Thus, although the chemical analyses of low-density lipoprotein (LDL), high-density lipoprotein (HDL), and so on are often published, they represent average values and are not as precise indicators of identity as melting point or spectrum. As research continues, we continue to discover subfractions of lipoproteins that affect CHD risk. Lipoprotein (a) [Lp(a)], which was described first by Berg [7], is an LDL particle in which apoprotein B is linked to an apoprotein unit [apoprotein (a)] via a disulfide bridge. Elevated Lp(a) levels have been associated with a high risk of CHD [8–10]. Lp(a) interferes with fibrinolysis [11], and its levels in the blood appear not to be affected by diet [12] or drugs [13].

Krauss and his coworkers [14,15] have identified subpopulations of LDL particles that differ in size and composition (Table 19.1).

The smaller, denser particles may be associated more strongly with the risk of coronary disease [16–18], and their levels may be determined genetically [19]. Animal studies had shown earlier that large lipoprotein molecules do not enter the arterial wall [20,21].

TABLE 19.1
Classification of LDL Particles

Class	Subfraction Density (g/mL)	Particle Diameter (nm)
LDL I	1.025–1.035	26–27
LDL II	1.032–1.038	25.5–26
LDL IIIA	1.038–1.050	24.7–25.6
LDL IIIB		24.2–24.6
LDL IVA	1.048–1.065	23.3–24.2
LDL IVB		21.8–23.2

Source: From Musliner, T.A. and Krauss, R.M., *Clin. Chem.*, 34, B78, 1988.

Note: LDL receptor activity and antioxidant content highest in LDL I and II. Triglyceride content increases with decreasing size.

Cholesterol has assumed a central role in experimental and human atherosclerosis, and the public is exhorted to know its "cholesterol number." However, cholesterol levels tend to vary diurnally and with season [22–24], and single measurements may not be an accurate indicator of risk. This is especially true if the single determined value is near one of the accepted cut points [25]. Low levels of cholesterol may lead to increased risk of noncardiovascular death [26–28]. Low cholesterol may become a problem at levels below 160 mg/dL [28] or 180 mg/dL [26].

Since ingested cholesterol has been shown to be atherogenic in some animal species, since elevated levels of cholesterol are a risk factor, and since it is relatively easy to measure, cholesterol has borne the brunt of the attack on CHD. The effect of dietary cholesterol on levels of blood cholesterol appears to be small. In 1950, Gertler et al. [29] made a large study on coronary disease with 4 groups of 10 men each. They were the men with lowest or highest serum cholesterol and those who ingested the most or the least dietary cholesterol. They were compared with similar groups selected from the control subjects. In every group, the men with coronary disease exhibited significantly higher cholesterol levels than the controls, but in no case was any relation to cholesterol intake seen. Early in the Framingham study, it was found that plasma cholesterol levels were unrelated to diet [30], a finding also reported from the Tecumseh study [31]. Several groups have reported that addition of eggs to the diet of free-living subjects did not affect their cholesterol level [32–34]. Gordon et al. [35] analyzed and compared the diets of men who did or did not have coronary disease in three large prospective coronary disease studies—Framingham, Puerto Rico, and Honolulu. Men who had coronary disease ingested fewer total calories, less carbohydrate, and less alcohol. Intakes of cholesterol and the P/S ratio of their dietary fat were similar for men who did or did not have coronary disease. McNamara [36] reviewed data from 68 clinical studies relating to effect of dietary cholesterol on plasma cholesterol. He concluded that there was a mean rise of 2.3 ± 0.2 mg/dL of plasma cholesterol for every 100 mg of ingested cholesterol. Hopkins [37] described the complexity of the association between cholesterol intake and plasma cholesterol. He found that the magnitude of the change in plasma cholesterol as a function of dietary cholesterol is influenced by baseline cholesterol intake. A recent epidemiological overview [38] concluded that after one considered dietary confounders there was no association between egg consumption at levels up to $1 +$ egg per day and the risk of CHD in nondiabetic men and women.

In contrast to dietary cholesterol, there is little question that the saturation of dietary fat exerts a profound influence on blood cholesterol levels. Ahrens et al. [39] fed a number of subjects a liquid formula diet containing 45% of energy as fat. In general, plasma cholesterol rose as the fat saturation rose. Keys et al. [40] and Hegsted et al. [41] studied the effects of changes in dietary fat on change in blood cholesterol levels in humans and offered formulas to predict changes in cholesterolemia based on changes in dietary fat. Both groups found fat saturation to have the greatest effect. Stearic acid did not appear to fit the formula, and direct experiments in human subjects have shown this to be true [42,43]. Hayes and Khosla [44] have hypothesized that the two most important fatty acids related to cholesterol levels are myristic acid (which raises cholesterol levels at every concentration) and linoleic acid (which exerts an increasing hypocholesterolemic effect until it reaches a dietary level of 6%–7% of energy). Hayes [45] has reviewed these data recently. McNamara et al. [46] fed normal subjects diets high or low in cholesterol and containing saturated or unsaturated fat. The major factor determining cholesterolemia was the saturation of the fat, the influence of which was about four times greater than that of dietary cholesterol (Table 19.2). McNamara [47] has reviewed exhaustively the connection between dietary cholesterol and atherosclerosis. He cites large epidemiological studies that indicate little connection between cholesterol intake and risk of CHD [48,49], emphasizes the role of saturated fat, and suggests that diets very high in cholesterol reflect an unbalance between intake of fats and of grains, vegetables, and fruits.

In addition to fatty acid saturation, the position of a specific fatty acid in the triglyceride molecule is important [50]. In an effort to test cholesterolemic effects of specific fatty acids, McGandy et al. [51] fed human subjects diets that contained fats into which high levels of specific saturated fatty acids (lauric, myristic, palmitic, or stearic) had been incorporated by

TABLE 19.2

Plasma Cholesterol Levels in Subjects Fed High or Low Levels of Cholesterol with Saturated or Unsaturated Fat

Fat in Diet	Fat (P/S)	Cholesterol	
		Dietary (mg)	Plasma (mg/dL)
Low cholesterol			
Saturated	0.31 ± 0.18	288 ± 64	243 ± 50
Unsaturated	1.90 ± 0.90	192 ± 60	218 ± 46
High cholesterol			
Saturated	0.27 ± 0.15	863 ± 161	248 ± 51
Unsaturated	1.45 ± 0.50	820 ± 102	224 ± 46

Source: From McNamara, D.J., Kolb, R., Parker, T.S., Samuel, P., Brown, C.D., and Ahrens, E.H. Jr., *J. Clin. Invest.*, 79, 1729, 1987.

interesterification. They found no differences. Earlier, a similar study had been carried out in rabbits with no effects on cholesterolemia or atherosclerosis [52].

During the randomization process, the structure of fats is changed so that each component fatty acid is present in each position of the triglyceride at one-third of its total concentration. This change appears to influence the atherogenicity of a fat. Tallow and lard each contain about 24% of total fatty acid as palmitic acid. In lard more than 90% of the component palmitic acid is at the 2-position of the triglyceride, whereas in tallow no more than 15% is. Lard is significantly more atherogenic for rabbits than tallow. When the fats are randomized, each contains 8% of palmitic acid at the 2-position and they are of equivalent atherogenicity [53] (Table 19.3). Studies with randomized cottonseed oil [54]

TABLE 19.3

Influence of Native and Randomized Tallow and Lard on Atherosclerosis in Rabbits

	Group			
	Tallow	Randomized Tallow	Lard	Randomized Lard
No.	7/8	7/8	8/8	8/8
% 16:0 at Sn2	3.7[a]	8.5[a]	20.8[b]	7.6[b]
Plasma				
Cholesterol (mg/dL)	1177 ± 156	1189 ± 166	926 ± 184	834 ± 153
Aorta				
Arch	1.29 ± 0.24	1.50 ± 0.53	2.69 ± 0.28[c]	1.50 ± 0.28[c]
Thoracic	0.79 ± 0.26	0.79 ± 0.28	1.75 ± 0.28[d]	0.69 ± 0.19[d]

Source: From Kritchevsky, D., Tepper, S.A., Kuksis, A., Eghtedary, K., and Klurfeld, D.M., *J. Nutr. Biochem.*, 9, 582, 1998.

Note: Rabbits were fed semipurified diet containing 0.4% cholesterol for 60 days ($p \leq 0.05$).

[a] Of 24.8% total.
[b] Of 21.4% total.
[c] Graded on 0–4 scale.
[d] Lard vs. randomized lard.

or palm oil [55] and with synthetic triglycerides [56] all indicate that increased presence of palmitic acid at the 2-position of a triglyceride enhances its atherogenic effect. There are no differences in serum lipids or in lipoprotein size in sera of rabbits that were fed native or randomized fat, nor are serum lipids affected. Since the presence of palmitic acid at the 2-position increases fat absorption [57,58], perhaps feeding fats with palmitic acid at the 2-position is the equivalent of a higher fat diet.

Every decade concern surfaces about the atherogenic effects of fats containing *trans* unsaturated double bonds (*trans* fat). The *trans* fats are generally metabolized in a manner similar to that of their *cis* counterparts. They accumulate in tissues on feeding and disappear relatively soon after cessation of feeding [59,60]. In rabbits, they lead to increased cholesterolemia but not more severe atherosclerosis; this is true in rabbits fed cholesterol [61] or a cholesterol-free atherogenic diet [62]. In men, fed diets high in elaidic acid, Lp(a) levels are increased [63,64], but this effect is not seen when the dietary *trans* fat is low in elaidic acid [65,66]. These findings suggest specific metabolic roles for specific *trans* fatty acids. In rats, *trans* fats exert no deleterious effects if the diet contains sufficient levels of linoleic acid [67]. In man, the cholesterolemic effects of *trans* fats seem to be related to the ratio of dietary *trans* fat to linoleic acid, the cholesterol levels rising as the ratio falls. Several studies have shown that the tissue levels of *trans* fatty acids are not higher in subjects with coronary disease than in controls [68,69]. Houtsmüller [70] in 1978 suggested that *trans* fat be regarded as a quasi-saturated fat.

In 1985, the Life Sciences Research Office of the Federation of American Societies of Experimental Biology published a report which concluded that there was little reason for health concerns at the reported intake level of *trans* fat (8 g/person/day) [71]. Two years later, the British Nutrition Foundation published a report with the same conclusion [72]. A decade later, the International Life Sciences Institute [73] and the British Nutrition Foundation [74] have found no reasons to alter their previous conclusions in their reports. All the reports contained the safe suggestion that more research was needed. Two reviews of *trans* fat effects have appeared recently [75,76]. The findings that *trans* fats may elevate plasma LDL cholesterol levels, thus increasing risk, suggest that it might be prudent to replace them when possible. It should also be noted that between 1960 and 1985, levels of *trans* fats in the American diet were fairly constant (7.63 ± 0.08 g/person/day) [77], whereas total age-adjusted mortality and deaths from heart disease and strokes fell by 28%, 37%, and 59%, respectively [78].

Several new players have appeared on the heart disease stage and they may ultimately affect our views of fat and cholesterol as major players in the CHD arena. Over 30 years, McCully [79] suggested that homocysteinemia could be a major risk for coronary disease, and this is recognized today [80–82]. Studies comparing lipid levels in European countries show little relation of these levels to CHD mortality. Ischemic heart disease mortality is four times higher in Belfast, Ireland than in Toulouse, France, despite general similarities in their diets (the French ingest significantly more cholesterol and alcohol) and in their risk factor profiles [83]. Similarly, mortality from heart disease in 50–54 year old men is four times higher in Vilnius, Lithuania than in Linköping, Sweden, despite the fact that differences in traditional risk factors are small [84]. In the years 1985–1987, age-specific mortality from ischemic heart disease for men aged 45–54 years was $237/100,000$ in Belfast and $56/100,000$ in Toulouse. In men aged 55–64, the rates per 100,000 were 761 in Belfast and 175 in Toulouse. In the younger age group, total cholesterol levels were significantly lower in the French (230 ± 41 mg/dL vs. 240 ± 41 mg/dL) and HDL cholesterol levels were higher (54 ± 15 mg/dL vs. 47 ± 12 mg/dL). There were no differences in the older men. Energy intake was virtually the same in the two groups. The French ingested significantly more protein and cholesterol and significantly less carbohydrate while fat intake was similar. In 1977, the CHD mortality per 100,000 men aged 50–54 was 300 in Lithuania and 220 in Sweden. In 1994, mortality was 445 in Lithuania (102% increase) and 110 in Sweden (50% decrease). Data from the Russian Lipid Research Clinics show an appreciable number of deaths in men with low levels of LDL and high levels of HDL [85].

The isolation of *Chlamydia pneumoniae* from atherosclerotic, but not normal, arteries [86,87] and the finding of cytomegalovirus in diseased arteries [88] may shed light on mechanisms underlying the onset of the disease.

The role of conjugated linoleic acid (CLA) in atherogenesis is under study. The CLA present in the diet (in diary products and meat of ruminant animals) is primarily octadeca-c9,t11-dienoic acid, but the commercial product used in most studies contains equal amounts (40%–45%) of the c9,t11 and t10,c12 modifications. CLA has been shown to inhibit atherogenesis in cholesterol-fed rabbits [89] and hamsters [90]. Of greater interest is the observation that feeding 1% CLA to rabbits with preestablished atherosclerosis leads to significant regression [91] of lesions.

III. LIPIDS AND CANCER

In 1930, Watson and Mellanby [92] showed that the incidence of coal tar–induced skin tumors in mice rose from 34% to 57% when 12.5%–25.0% butter was added to the basal diet that normally contained 3% fat. Baumann et al. [93] found that high-fat diets increased the yield of ultraviolet radiation–induced or chemically induced skin tumors in mice. They also found saturated fats to be less cocarcinogenic than unsaturated fats [94]. Carroll and Khor [95] made a similar observation. The reason that unsaturated fat enhanced carcinogenesis was found by Ip et al. [96] to be due to the requirement of tumor for linoleic acid as a growth factor. This finding may explain why rats fed fish oils [97] or fats high in *trans* unsaturated [98,99] fatty acids also show a reduced incidence of tumors.

Armstrong and Doll [100] published a thorough review correlating cancer incidence in over 30 countries with diet, gross national product (GNP), physician density, population density, and use of solid or liquid fuel. They found positive associations between breast and colorectal cancers and total fat consumption. They also found a strong association between these cancers and GNP. In their conclusion, they state, "It is clear that these and other correlations should be taken only as suggestions for further research and not as evidence of causation or as bases for preventive action." Several studies carried out in the 1970s found correlations between fat intake and risk of breast cancer [101,102]. The association between dietary fat and risk of breast cancer appears to be weakening [103]. Goodwin and Boyd [104] reviewed a large number of studies and found that 7 of 13 international comparisons found a correlation between fat intake and breast cancer risk, but only 1 of 14 case-control studies did. Hirohata et al. found no association between dietary fat intake and breast cancer in Japan [105] or Hawaii [106]. The NHANES I reported on 99 cases of breast cancer, as opposed to 5386 noncases, and found no differences in fat or fatty acid intake [107]. Willett et al. [108] studied a cohort of >89,000 American women whose fat intake ranged from 32% to 44% of calories and found a slight decrease in relative risk with increasing fat intake.

As in the case of breast cancer, local (case-control) studies of fat intake and colon cancer risk show minimal correlations with fat intake [109–111], whereas international studies find strong correlation [100,112]. Rogers and Longnecker [113] reviewed diet and cancer and in a summary of 24 cases of colon cancer found a small, but inconsistent, association between fat intake and risk of colon cancer. Stocks and Karn [114] in an early (1933) study of colon cancer and diet in England found dairy foods to be negatively correlated with risk. Jensen et al. [115] in studying colon cancer and diet in Finland and Denmark found saturated fat to be inversely correlated with risk. Stemmermann et al. [116] made a similar observation in Hawaii. Others [117,118] reported that risk increased with increasing intake of saturated fat. Tuyns [119] in a Belgian study suggested that the dietary factor leading to increased risk was oligosaccharides and not fat. It should be evident from the foregoing that there is no consensus with regard to the effects of dietary fat on colon cancer risk.

Is it fat or the calories it provides that affects cancer risk? In 1927, Hoffman [120] suggested that the increase in cancer incidence seen then was due to overnutrition. About 50 years later, Berg [121]

TABLE 19.4

Effects of Calories and Fat on Incidence of Methylcholanthrene-Induced Skin Tumors in Mice

Level of		Tumor
Calories	Fat	Incidence (%)
Low	Low	0
High	Low	54
Low	High	28
High	High	66

Source: From Lavik, P.S. and Baumann, C.A., *Cancer Res.*, 3, 749, 1943.

made a similar suggestion, namely, that increasing risk of hormone-related tumors was due to increased caloric intake. Moreschi [122] showed in 1909 that transplanted tumors did not grow as well in underfed mice as in freely fed ones. A few years later, Rous [123] demonstrated that neither spontaneous nor transplanted tumors showed optimal growth in rodents whose food intake was restricted. Lavik and Baumann [124] studied dietary effects on chemically induced skin tumors in mice. When the diet was low in both calories and fat, tumor incidence was nil. Tumor incidence in mice fed diets low in fat but high in calories was 48% higher than in those whose diet was low in calories but high in fat and only 18% lower than that in mice fed a high-fat, high-calorie diet (Table 19.4). Our work [125–128] showed that caloric restriction by 40% led to significantly reduced incidence of chemically induced breast or colon tumors in rats even when the restricted diet contained twice as much fat. At 10% caloric restriction, incidence of induced breast tumors in rats was unchanged, but tumors per tumor-bearing rat were reduced by 36% and total weight of tumors by 47%. Rats fed a diet containing 26.7% corn oil but restricted by 25% exhibited lower tumor incidence, fewer tumors, and smaller tumors than rats freely fed a diet containing only 5% fat (Table 19.5). In humans, both colon [129,130] and gastric [131] cancer have been correlated positively with caloric intake. Overweight in humans is clearly correlated with increasing cancer risk [132–134].

The effects of fat in cancer need to be stratified by fat type, fat quantity, and total caloric intake. There is too frequently a rush to judgment concerning effects of specific nutrients. Much of the recent interest in diet and cancer can be traced to the major work by Doll and Peto [135] in which they suggested that as many as 35% of deaths due to cancer might be associated with diet. They then made the following comment: "It must be emphasized that the figure chosen (of 35% of cancers related to diet) is highly speculative and chiefly refers to dietary factors which are not yet reliably identified." They also stated that "there is no evidence of any generalized increase (in deaths due to cancer) other than that due to tobacco."

A fatty acid that has been known for decades has recently emerged as a potent inhibitor of carcinogenesis. CLA has been shown to inhibit chemically induced forestomach tumors in mice when given intragastrically [136] and mammary tumors when included in the diet of rats [137]. It inhibits the growth and metastasis of human tumors when injected into immune-deficient mice [138] and inhibits growth of tumor cells in vitro [139]. The effects of CLA on carcinogenesis have been reviewed recently [140]. Its mode of action is unknown at this writing.

We have accumulated reams of data relating to effects of specific macro- and micronutrients on cancer risk in humans and cancer development in experimental animals. We must now begin to

TABLE 19.5
Influence of 25% Caloric Restriction on Dimethylbenz(a)-anthracene-Induced Tumors in Rats Fed High-Fat Diets

Diet	Incidence (%)	Multiplicity[a]	Wt (g)[b]	Burden[c]
Ad libitum				
5% Fat	65	1.9 ± 0.3	2.0 ± 0.7	4.2 ± 1.9
15% Fat	85	3.0 ± 0.6	2.3 ± 6.7	6.6 ± 2.7
20% Fat	80	4.1 ± 0.6	2.9 ± 0.5	11.8 ± 3.2
Restricted				
20% Fat	60	1.9 ± 0.4	0.8 ± 0.2	1.5 ± 0.5
26.7% Fat	30	1.5 ± 0.3	1.4 ± 1.0	2.3 ± 1.6
p	<0.005	<0.001	<0.001	<0.001

Source: From Klurfeld, D.M., Welch, C.B., Lloyd, L.M., and Kritchevsky, D., *Int. J. Cancer*, 43, 922, 1989.

[a] Tumors per tumor-bearing rat.
[b] Average tumor weight per rat.
[c] Weight of all tumors per rat.

examine the interactions of nutrients as they affect the major degenerative diseases—heart disease and cancer. It is also becoming evident that dietary patterns may be more indicative of risk than any particular dietary components.

ACKNOWLEDGMENT

Supported in part by a Research Career Award (HL-00734) from the National Institutes of Health.

REFERENCES

1. W. Mertz, J.C. Trui, J.T. Judd, S. Reiser, J. Hallfrisch, E.R. Morris, P.D. Steele, and E. Lashley. What are people really eating? The relation between energy intake derived from diet records and intake to maintain body weight. *Am. J. Clin. Nutr.* 54:291 (1991).
2. W. Mertz. Food intake measurements: Is there a "gold standard?" *J. Am. Diet. Assoc.* 92:1463 (1992).
3. A.J. Lusis. Atherosclerosis. *Nature* 407:233 (2000).
4. N. Anitschkow. Uber die veränderungen der kaninchenaorta bei experimenteller cholesterinsteatose. *Beitr. Pathol. Anat. Allg. Pathol.* 56:379 (1913).
5. J.W. Gofman, F. Lindgren, H. Elliott, W. Mantz, J. Hewitt, B. Strisower, V. Herring, and T.P. Lyon. The role of lipids and lipoproteins in atherosclerosis. *Science* 111:166 (1950).
6. A. Keys. Atherosclerosis: A problem in new public health. *J. Mt. Sinai Hosp. N.Y.* 20:118 (1953).
7. K. Berg. A new serum type system in man: The Lp(a) system. *Acta Pathol. Microbiol. Scand.* 59:369 (1963).
8. G.M. Kostner, P. Avogaro, G. Cazzolato, E. Marth, G. Bittolo-Bon, and G.B. Quinci. Lipoprotein Lp(a) and the risk of myocardial infarction. *Atherosclerosis* 38:51 (1981).
9. G.G. Rhoades, G. Dahlen, K. Berg, N.E. Morton, and A.L. Dannenberg. Lp(a) lipoprotein as a risk factor for myocardial infarction. *J. Am. Med. Assoc.* 256:2540 (1986).
10. J. Loscalzo. Lipoprotein(a): A unique risk factor for atherothrombotic disease. *Arteriosclerosis* 10:672 (1990).
11. J.M. Edelberg and S.V. Pizzo. Lipoprotein(a): The link between impaired fibrinolysis and atherosclerosis. *Fibrinolysis* 5:135 (1991).

12. S.A. Brown, J. Morrisett, J.R. Patch, R. Reeves, A.M. Gotto Jr., and W. Patsch. Influence of short-term dietary cholesterol and fat on human plasma Lp(a) and LDL levels. *J. Lipid Res.* 32:1281 (1991).

13. H.G. Fieseler, V.W. Armstrong, E. Wieland, J. Thiery, E. Schütz, A.K. Walli, and D. Seidel. Lp(a) concentrations are unaffected by treatment with the HMG-CoA reductase inhibitor provastatin: Results of a 2-year investigation. *Clin. Chim. Acta* 204:291 (1991).

14. R.M. Krauss, F.T. Lindgren, and R.M. Ray. Interrelationships among subgroups of serum lipoproteins in normal human subjects. *Clin. Chim. Acta* 104:275 (1980).

15. T.A. Musliner and R.M. Krauss. Lipoprotein subspecies and risk of coronary disease. *Clin. Chem.* 34: B78 (1988).

16. M.A. Austin, J.L. Breslow, C.H. Hennekens, J.E. Buring, W.C. Willett, and R.M. Krauss. Low-density lipoprotein subclass patterns and risk of myocardial infarction. *J. Am. Med. Assoc.* 260:1917 (1988).

17. P.S. Roheim and B.F. Asztalos. Clinical significance of lipoprotein size and risk of coronary athero-sclerosis. *Clin. Chem.* 41:147 (1995).

18. I. Rajman, S. Maxwell, R. Cramb, and M. Kendall. Particle size: The key to atherogenic lipoprotein? *Q. J. Med.* 87:709 (1994).

19. M.A. Austin. Genetic epidemiology of dyslipidaemia and atherosclerosis. *Ann. Med.* 28:459 (1996).

20. S. Stender and D.B. Zilversmit. Transfer of plasma lipoprotein components and of plasma proteins into aortas of cholesterol-fed rabbits. Molecular size as a determinant of plasma lipoprotein influx. *Arterio-sclerosis* 1:38 (1981).

21. B.G. Nordestgaard and D.B. Zilversmit. Large lipoproteins are excluded from the arterial wall in diabetic cholesterol-fed rabbits. *J. Lipid Res.* 29:1491 (1988).

22. D. Kritchevsky. Variation in serum cholesterol levels. In: *Nutrition Update* (J. Wein-inger and G.M. Briggs, eds.). John Wiley & Sons, New York, 1985, p. 91.

23. D.J. Gordon, J. Hyde, D.C. Trost, F.S. Whaley, P.J. Hannan, D.R. Jacobs, and L.G. Ekelund. Cyclic seasonal variation in plasma lipid and lipoprotein levels: The lipid research clinics coronary primary prevention trial placebo group. *J. Clin. Epidemiol.* 41:679 (1988).

24. D. Robinson, E.A. Bevan, S. Hinohara, and T. Takahashi. Seasonal variation in serum cholesterol levels—Evidence from the UK and Japan. *Atherosclerosis* 95:15 (1992).

25. D.M. Hegsted and R.J. Nicolosi. Individual variation in serum cholesterol levels. *Proc. Natl. Acad. Sci. U.S.A.* 84:6259 (1987).

26. L.D. Cowan, D.L. O'Connell, M.H. Criqui, E. Barrett-Connor, T.L. Bush, and R.B. Wallace. Cancer mortality and lipid and lipoprotein levels. The lipid research clinics program mortality follow-up study. *Am. J. Epidemiol.* 131:468 (1990).

27. S.B. Kritchevsky and D. Kritchevsky. Serum cholesterol and cancer risk: An epidemiologic perspective. *Annu. Rev. Nutr.* 12:391 (1992).

28. D. Jacobs, H. Blackburn, M. Higgins, D. Reed, H. Iso, G. McMillan, J. Neaton, J. Nelson, J. Potter, B. Rifkind, J. Rossouw, R. Shekelle, and S. Yusuf. Report of the conference on low blood cholesterol: Mortality associations. *Circulation* 86:1046 (1992).

29. M.M. Gertler, S.M. Garn, and P.D. White. Serum cholesterol and coronary artery disease. *Circulation* 2:696 (1950).

30. W.B. Kannel and T. Gordon, eds. The Framingham study: Diet and regulation of serum cholesterol. Section 24. In: *The Framingham Study: An Epidemiological Investigation of Cardiovascular Disease.* U.S. Govt. Printing Office, Washington, D.C., 1970.

31. A.B. Nichols, C. Ravenscroft, D.E. Lamphear, and L.D. Ostrander. Daily nutritional intake and serum lipids: The Tecumseh study. *J. Am. Med. Assoc.* 236:1948 (1976).

32. G. Slater, J. Mead, G.M. Dhopeshwarkar, S. Robinson, and R. Alfin-Salter. Plasma cholesterol and triglycerides in men with added eggs in the diet. *Nutr. Rep. Int.* 14:249 (1976).

33. M.W. Porter, W. Yamanaka, S.D. Carlson, and M.A. Flynn. Effect of dietary egg on serum cholesterol and triglycerides in human males. *Am. J. Clin. Nutr.* 30:490 (1977).

34. T.R. Dawber, R.J. Nickerson, F.N. Brand, and J. Pool. Eggs vs. serum cholesterol and coronary disease. *Am. J. Clin. Nutr.* 36:617 (1982).

35. T. Gordon, A. Kagan, M. Garcia-Palmieri, W.B. Kannel, W.J. Zukel, J. Tillotson, P. Sorlie, and M. Hjortland. Diet and its relation to coronary heart disease and death in three populations. *Circulation* 63:500 (1981).

36. D.J. McNamara. Relationship between blood and dietary cholesterol. In: *Advances in Meat Research* (A.M. Pearson and T.R. Dutson, eds.). Elsevier Applied Science, London, 1990, p. 63.

37. P.N. Hopkins. Effects of dietary cholesterol on serum cholesterol: A metaanalysis and review. *Am. J. Clin. Nutr.* 55:1060 (1992).

38. S.B. Kritchevsky and D. Kritchevsky. Egg consumption and coronary heart disease: An epidemiologic overview. *J. Am. Coll. Nutr.* 19:549S (2000).

39. E.H. Ahrens Jr., W. Insull Jr., R. Blomstrand, J. Hirsch, T.T. Tsaltas, and M.L. Peterson. The influence of dietary fats on serum lipid levels in man. *Lancet* 1:943 (1951).

40. A. Keys, J.T. Anderson, and F. Grande. Serum cholesterol response of changes in the diet. IV. Particular saturated fatty acids in the diet. *Metabolism* 14:776 (1965).

41. D.M. Hegsted, R.B. McGrandy, M.L. Myers, and F.J. Stare. Quantitative effects of dietary fat on serum cholesterol in man. *Am. J. Clin. Nutr.* 17:281 (1965).

42. F. Grande, J.T. Anderson, and A. Keys. Comparison of effects of palmitic and stearic acids in the diet on serum cholesterol in men. *Am. J. Clin. Nutr.* 23:1184 (1970).

43. A. Bonanome and G.M. Grundy. Effect of dietary stearic acid on plasma cholesterol and lipoprotein levels. *N. Engl. J. Med.* 319:1244 (1988).

44. K.C. Hayes and P.R. Khosla. Dietary fatty acid thresholds and cholesterolemia. *FASEB J.* 6:2600 (1992).

45. K.C. Hayes. Dietary fatty acids, cholesterol and lipoprotein profile. *Br. J. Nutr.* 84:397 (2000).

46. D.J. McNamara, R. Kolb, T.S. Parker, P. Samuel, C.D. Brown, and E.H. Ahrens Jr., Heterogeneity of cholesterol homeostasis in man: Responses to changes in dietary fat quality and cholesterol quantity. *J. Clin. Invest.* 79:1729 (1987).

47. D.J. McNamara. Dietary cholesterol and atherosclerosis. *Biochim. Biophys. Acta* 1529:310 (2000).

48. A. Ascherio, E.B. Rimm, E.L. Giovannucci, D. Spiegelman, M. Stampfer, and W.C. Willett. Dietary fat and risk of coronary heart disease in men: Cohort follow-up study in the United States. *Br. Med. J.* 313:84 (1996).

49. F.B. Hu, M.J. Stampfer, J.E. Manson, E. Rimm, G.A. Colditz, B.A. Rosner, C.H. Hennekens, and W.C. Willett. Dietary fat intake and the risk of coronary heart disease in women. *New Engl. J. Med.* 337:1491 (1997).

50. D. Kritchevsky. Effect of triglyceride structure on lipid metabolism. *Nutr. Rev.* 46:177 (1988).

51. R.B. McGandy, D.M. Hegsted, and M.L. Meyers. Use of semisynthetic fats in determining effects of specific dietary fatty acids on serum lipids in man. *Am. J. Clin. Nutr.* 23:1288 (1970).

52. D. Kritchevsky and S.A. Tepper. Cholesterol vehicle in experimental atherosclerosis. X. Influence of specific fatty acids. *Exp. Mol. Pathol.* 6:394 (1967).

53. D. Kritchevsky, S.A. Tepper, A. Kuksis, K. Eghtedary, and D.M. Klurfeld. Cholesterol vehicle in experimental atherosclerosis. 21. Native and randomized lard and tallow. *J. Nutr. Biochem.* 9:582 (1998).

54. D. Kritchevsky, S.A. Tepper, S. Wright, A. Kuksis, and T.A. Hughes. Cholesterol vehicle in experimental atherosclerosis. 20. Cottonseed oil and randomized cottonseed oil. *Nutr. Res.* 18:259 (1998).

55. D. Kritchevsky, S.A. Tepper, A. Kuksis, S. Wright, and S.K. Czarnecki. Cholesterol vehicle in experimental atherosclerosis. 22. Refined, bleached, deodorized (RED) palm oil, randomized palm oil and red palm oil. *Nutr. Res.* 20:887 (2000).

56. D. Kritchevsky, S.A. Tepper, S.C. Chen, G.W. Meijer, and R.M. Krauss. Cholesterol vehicle in experimental atherosclerosis. 23. Effects of specific synthetic triglycerides. *Lipids* 35:621 (2000).

57. R.M. Tomarelli, B.J. Meyer, J.R. Weaber, and F.W. Bernhart. Effect of positional distribution on the absorption of fatty acids of human milk and infant formulas. *J. Nutr.* 95:583 (1968).

58. L.J. Filer Jr., F.H. Mattson, and S.J. Fomon. Triglyceride configuration and fat absorption by the human infant. *J. Nutr.* 99:293 (1969).

59. C.E. Moore, R.B. Alfin-Slater, and L. Aftergood. Incorporation and disappearance of trans fatty acids in rat tissues. *Am. J. Clin. Nutr.* 33:2318 (1980).

60. D. Kritchevsky, L.M. Davidson, M. Weight, N.J.P. Kriek, and J.P. duPlessis. Effect of trans-unsaturated fats on experimental atherosclerosis in Vervet monkeys. *Atherosclerosis* 51:123 (1984).

61. G.C. McMillan, M.D. Silver, and B.I. Weigensberg. Elaidinized olive and cholesterol atherosclerosis. *Arch. Pathol.* 76:118 (1963).

62. H. Ruttenberg, L.M. Davidson, N.A. Little, D.M. Klurfeld, and D. Kritchevsky. Influence of trans unsaturated fats on experimental atherosclerosis in rabbits. *J. Nutr.* 113:835 (1983).

63. R.P. Mensink and M.B. Katan. Effect of dietary trans fatty acids on high density and low density lipoprotein cholesterol levels in healthy subjects. *N. Engl. J. Med.* 323:439 (1990).

64. P.J. Nestel, M. Noakes, and B. Belling. Plasma lipoprotein lipid and Lp(a) changes with substitution of elaidic acid for oleic acid in the diet. *J. Lipid Res.* 33:1029 (1992).

65. A.H. Lichtenstein, L.M. Susman, W. Carrasco, J.L. Jenner, J.M. Ordovas, and E.J. Shaefer. Hydrogenation impairs the hypolipidemic effects of corn oil. *Arterioscler. Thromb.* 13:154 (1993).

66. R. Wood, K. Kubena, B. O'Brien, S. Tseng, and G. Martin. Effect of butter, mono- and polyunsaturated fatty acid-enriched butter, trans fatty acid margarine and zero trans fatty acid margarine on serum lipids and lipoproteins in healthy men. *J. Lipid Res.* 34:1 (1993).

67. J.L. Zevbergen, U.M.T. Houtsmüller, and J.J. Gottenbos. Linoleic acid requirement of rats fed trans fatty acids. *Lipids* 23:178 (1988).

68. H. Heckers, M. Korner, T.W.L. Tuschen, and F.W. Melcher. Occurrence of individual trans-isomeric fatty acids in human myocardium, jejunum and aorta in relation to degrees of atherosclerosis. *Artherosclerosis* 28:389 (1977).

69. A. Aro, A.F.M. Kardinaal, I. Salminen, J.D. Kark, R.A. Riemersma, M. Delgado-Rodriguez, J. Gomez-Arcena, J.K. Huttienen, L. Kohlmeier, B.C. Martin, J.M. Martin-Moreno, V.P. Niagaev, J. Ringstad, M. Thamm, P. van't Veer, and F.J. Kok. Adipose tissue isomeric trans fatty acids and risk of myocardial infarction in nine countries: The Euramic study. *Lancet* 345:273 (1995).

70. U.M.T. Houtsmüller. Biochemical aspects of fatty acids with trans double bonds. *Fette. Seifen. Anstrichm.* 80:162 (1978).

71. F.R. Senti (ed.). *Health Aspects of Dietary trans Fatty Acids.* LSRO, Bethesda, MD, 1985.

72. British Nutrition Foundation. *Trans Fatty Acids.* London, 1987.

73. P.M. Kris Etherton (ed.). Trans fatty acids and coronary heart disease risk. *Am. J. Clin. Nutr.* 62:655S (1995).

74. British Nutrition Foundation Task Force. *Trans Fatty Acids.* London, 1995.

75. D. Kritchevsky. Trans fatty acids and cardiovascular risk. *Prostaglandins Leukot. Essent. Fatty Acids* 57:399 (1997).

76. A.H. Lichtenstein. Trans fatty acids and cardiovascular disease risk. *Curr. Opin. Lipidol.* 11:37 (2000).

77. J.E. Hunter and T.H. Applewhite. Reassessment of trans fatty acid availability in the U.S. diet. *Am. J. Clin. Nutr.* 54:363 (1991).

78. Health United States, 1994. DHHS Publ. No. (PH5 95–1232). National Center for Health Statistics, Hyattsville, MD, 1995.

79. K.S. McCully. Vascular pathology of homocysteinemia: Implications for the pathogenesis of atherosclerosis. *Am. J. Pathol.* 56:111 (1969).

80. M.R. Malinow. Plasma homocyst(e)ine: A risk factor for arterial disease. *J. Nutr.* 126:1238S (1996).

81. M.R. Nehler, L.M. Taylor Jr., and J.M. Porter. Homocysteinemia as a risk factor for atherosclerosis: A review. *Cardiovasc. Pathol.* 6:1 (1997).

82. G. Alfthan, A. Aro, and K.F. Gey. Plasma homocysteine and cardiovascular disease mortality. *Lancet* 349:397 (1997).

83. A.E. Evans, J.B. Ruidavets, E.E. McCrum, J.P. Cambou, R. McClear, P. Douste-Blazy, D. McMaster, A. Bingham, C.C. Patterson, J.L. Richard, J.M. Mathewson, and F. Cambien. Autres pays, autres cours? Dietary patterns, risk factors and ischaemic heart disease in Belfast and Toulouse. *Q. J. Med.* 88:469 (1995).

84. M. Kristenson, B. Zieden, Z. Kucinskienë, L.S. Elinder, B. Bergdall, B. Elwig, A. Abaravicius, L. Razinkovienë, H. Calkauskas, and A.G. Olsson. Antioxidant state and mortality from heart disease in Lithuanian and Swedish men: Concomitant cross sectional study of men aged 50. *Br. Med. J.* 314:629 (1997).

85. D.B. Shestov, A.D. Deev, A.N. Klimov, C.E. Davis, and H.A. Tyroler. Increased risk of coronary heart disease: Death in men with low total and low-density lipoprotein in the Russian lipid research clinics prevalence follow-up study. *Circulation* 88:846 (1993).

86. P. Saikku, M. Leinonen, K. Matilla, M.R. Ekman, M.S. Nieminen, P.H. Mäkelä, J.K. Huttunen, and V. Valtonen. Serological evidence of an association of a novel chlamydia, TWAR, with chronic coronary heart disease and acute myocardial infarction. *Lancet* 2:983 (1988).

87. D.H. Thom, J.T. Grayston, D.S. Siscovick, S.P. Wang, N.S. Weiss, and J.R. Dalang. Association of prior infection with *Chlamydia pneumoniae* and angiographically demonstrated coronary artery disease. *J. Am. Med. Assoc.* 268:68 (1992).

88. J.L. Melnick, E. Adam, and M.E. Debakey. Cytomegalovirus and atherosclerosis. *Eur. Heart J.* 14 (Suppl. K):30 (1993).

89. K.N. Lee, D. Kritchevsky, and M.W. Pariza. Conjugated linoleic acid and atherosclerosis in rabbits. *Atherosclerosis* 108:19 (1994).

90. T.A. Wilson, R.J. Nicolosi, M. Chrysam, and D. Kritchevsky. Conjugated linoleic acid reduces early aortic atherosclerosis greater than linoleic acid in hypercholesterolemic hamsters. *Nutr. Res.* 20:1795 (2000).

91. D. Kritchevsky, S.A. Tepper, S. Wright, P. Tso, and S.K. Czarnecki. Influence of conjugated linoleic acid (CLA) on establishment and progression of atherosclerosis in rabbits. *J. Am. Coll. Nutr.* 19:472S (2000).

92. A.F. Watson and E. Mellanby. Tar cancer in mice II. The condition of the skin when modified by external treatment or diet, as a factor in influencing the cancerous reaction. *Br. J. Exp. Pathol.* 11:311 (1930).

93. C.A. Baumann, H.P. Jacobi, and H.P. Rusch. The effect of diet on experimental tumor production. *Am. J. Hyg.* 30:1 (1939).

94. J.A. Miller, B.E. Kline, H.P. Rusch, and C.A. Baumann. The effect of certain lipids on the carcinogenicity of *p*-dimethylamino-azobenzene. *Cancer Res.* 4:756 (1944).

95. K.K. Carroll and H.T. Khor. Effect of level and type of dietary fat on incidence of mammary tumors induced in female Sprague–Dawley rats by 7,12-dimethylbenz(a)-anthracene. *Lipids* 6:415 (1971).

96. C. Ip, C.A. Carter, and M.M. Ip. Requirement of essential fatty acid for mammary tumorigenesis in the rate. *Cancer Res.* 45:1997 (1985).

97. J.J. Jurkowski and W.T. Cave Jr., Dietary effects of menhaden oil on the growth and membrane lipid composition of rat's mammary tumors. *J. Natl. Cancer Inst.* 74:1145 (1985).

98. S.L. Selenskas, M.M. Ip, and C. Ip. Similarity between trans fat and saturated fat in the modification of rat mammary carcinogenesis. *Cancer Res.* 44:1321 (1984).

99. M. Sugano, M. Watanabe, K. Yoshida, K. Tomioka, M. Mayamoto, and D. Kritchevsky. Influence of dietary cis and trans fats on DMH-induced colon tumors, steroid excretion and eicosanoid production in rats prone to colon cancer. *Nutr. Cancer* 12:177 (1989).

100. B. Armstrong and R. Doll. Environment factors and cancer incidence and mortality in different countries, with special reference to dietary practices. *Int. J. Cancer* 15:617 (1975).

101. G.E. Gray, M.C. Pike, and B.E. Henderson. Breast cancer incidence and mortality rates in different countries in relation to known risk factors and dietary practices. *Br. J. Cancer* 39:1 (1979).

102. G. Hems. The contribution of diet and childbearing to breast cancer rates. *Br. J. Cancer* 37:974 (1978).

103. T. Byers. Diet and cancer. Any progress in the interim? *Cancer* 62:1713 (1988).

104. P.J. Goodwin and N.F. Boyd. Critical appraisal of the evidence that dietary fat intake is related to breast cancer risk in humans. *J. Natl. Cancer Inst.* 79:473 (1987).

105. T. Hirohata, T. Shigematsu, A.M.Y. Nomura, A. Horie, and I. Hirohata. The occurrence of breast cancer in relation to diet and reproductive history. A case-control study in Fukuoka, Japan. *NCI Monogr.* 69:187 (1985).

106. T. Hirohata, A.M.Y. Nomura, J.H. Hankin, L.N. Kolonel, and J. Lee. An epidemiological study on the association between diet and breast cancer. *J. Natl. Cancer Inst.* 78:595 (1987).

107. D.Y. Jones, A. Schatzkin, S.B. Green, G. Block, L.A. Brinton, R.G. Ziegler, R. Hoover, and P.R. Taylor. Dietary fat and breast cancer in the national health and nutrition examination survey. I. Epidemiologic follow-up study. *J. Natl. Cancer Inst.* 79:465 (1987).

108. W.C. Willett, M.J. Stampfer, G.A. Colditz, B.A. Rosner, C.H. Hennekens, and F.E. Speizer. Dietary fat and the risk of breast cancer. *N. Engl. J. Med.* 316:22 (1987).

109. J. Higginson. Etiological factors in gastrointestinal cancer in man. *J. Natl. Cancer Inst.* 37:527 (1966).

110. S. Graham, H. Dayal, M. Swanson, A. Mittelman, and G. Wilkinson. Diet in the epidemiology of cancer of the colon and rectum. *J. Natl. Cancer Inst.* 61:709 (1978).

111. S. Bingham, D.R.R. Williams, T.J. Cole, and W.P.T. James. Dietary fibre and regional large-bowel cancer mortality in Britain. *Br. J. Cancer* 40:456 (1979).

112. E.G. Knox. Foods and disease. *Br. J. Prev. Soc. Med.* 31:71 (1977).

113. A.E. Rogers and M.P. Longnecker. Biology of disease. Dietary and nutritional influence on cancer: A review of epidemiological and experimental data. *Lab. Invest.* 59:729 (1988).

114. P. Stocks and M.K. Karn. A cooperative study of the habits, homelife, dietary and family histories of 450 cancer patients and an equal number of control patients. *Ann. Eugen (London)* 5:237 (1933).

115. O.M. Jensen, R. Maclennan, and J. Wahrendorf. Diet, bowel function, fecal characteristics and large bowel cancer in Denmark and Finland. *Nutr. Cancer* 4:5 (1982).

116. G.N. Stemmermann, A.M.Y. Nomura, and K.L. Heilbrun. Dietary fat and the risk of colorectal cancer. *Cancer Res.* 44:4633 (1984).

117. M. Jain, G.M. Cook, F.G. Davis, M. Grace, G.R. Howe, and A.B. Miller. A case-control study of diet and colorectal cancer. *Int. J. Cancer* 26:757 (1980).

118. M.J. Stampfer, W.C. Willett, G.A. Colditz, B. Rosner, C. Hennekens, and F.E. Speizer. A prospective study of diet and colon cancer in a cohort of women. *Fed. Proc.* 46:883 (1987).

119. A.J. Tuyns. A case-control study on colorectal cancer in Belgium: Preliminary results. *Med. Sociale. Prevent.* 31:81 (1986).

120. F.L. Hoffman. *Cancer Increase and Overnutrition.* Prudential Insurance Co., Newark, NJ, 1927.

121. J.W. Berg. Can nutrition explain the pattern of international epidemiology of hormone-dependent cancer? *Cancer Res.* 35:3345 (1975).

122. C. Moreschi. Beziehungen zwischen Ernährung und tumorwachstum. *Z. Immunitäts-forsch* 2:651 (1909).

123. P. Rous. The influence of diet on transplanted and spontaneous tumors. *J. Exp. Med.* 20:433 (1914).

124. P.S. Lavik and C.A. Baumann. Further studies on the tumor-promoting action of fat. *Cancer Res.* 3:749 (1943).

125. D. Kritchevsky, M.M. Weber, and D.M. Klurfeld. Dietary fat versus caloric content in initiation and promotion of 7,12-dimethylbenz(a) anthracene induced mammary tumorigenesis in rats. *Cancer Res.* 44:3174 (1984).

126. D.M. Klurfeld, M.M. Weber, and D. Kritchevsky. Inhibition of chemically induced mammary and colon tumor promotion by caloric restriction in rats fed dietary fat. *Cancer Res.* 47:2759 (1987).

127. D.M. Klurfeld, C.B. Welch, M.J. Davis, and D. Kritchevsky. Determination of degree of energy restriction necessary to reduce DMBA-induced mammary tumorigenesis in rats during the promotion phase. *J. Nutr.* 119:286 (1989).

128. D.M. Klurfeld, C.B. Welch, L.M. Lloyd, and D. Kritchevsky. Inhibition of DMBA-induced mammary tumorigenesis by caloric restriction in rats fed high fat diets. *Int. J. Cancer* 43:922 (1989).

129. M. Jain, G.M. Cook, F.G. Davis, M.G. Grace, G.R. Howe, and A.B. Miller. A case-control study of diet and colorectal cancer. *Int. J. Cancer* 26:757 (1980).

130. J.L. Lyon, A.W. Mahoney, D.W. West, J.W. Gardner, K.R. Smith, A.W. Sorenson, and W. Stanish. Energy intake: Its relation to colon cancer. *J. Natl. Cancer Inst.* 78:853 (1987).

131. S. Graham, B. Haughey, and J. Marshall. Diet in the epidemiology of gastric cancer. *Nutr. Cancer* 13:19 (1990).

132. E.A. Lew and L. Garfinkel. Variations in mortality by weight among 750,000 men and women. *J. Chronic Dis.* 32:563 (1979).

133. L. Garfinkel. Overweight and cancer. *Ann. Intern. Med.* 103:1034 (1985).

134. D. Albanes. Caloric intake, body weight and cancer: A review. *Nutr. Cancer* 9:199 (1987).

135. R. Doll and R. Peto. The causes of cancer. Quantitative estimates of avoidable risks of cancer in the United States today. *J. Natl. Cancer Inst.* 66:1191 (1981).

136. Y.L. Ha, J. Storkson, and M.W. Pariza. Inhibition of benzo(a)pyrene-induced mouse forestomach neoplasia by conjugated dienoic derivatives of linoleic acid. *Cancer Res.* 50:1097 (1990).

137. C. Ip, S.F. Chin, J.A. Scimeca, and M.W. Pariza. Mammary cancer prevention by conjugated dienoic derivatives of linoleic acid. *Cancer Res.* 51:6118 (1991).

138. S. Visonneau, A. Cesano, S.A. Tepper, J. Scimeca, D. Kritchevsky, and D. Santoli. Conjugated linoleic acid (CLA) suppresses growth of human breast carcinoma cells in SCID mice. *Anticancer Res.* 17:969 (1997).

139. S. Visonneau, A. Cesano, S.A. Tepper, J. Scimeca, D. Santoli, and D. Kritchevsky. Effect of different concentrations of conjugated linoleic acid (CLA) on tumor cell growth in vitro. *FASEB J.* 10:A182 (1996).

140. D. Kritchevsky. Antimutagenic and some other effects of conjugated linoleic acid. *Br. J. Nutr.* 83:459 (2000).

20 Unsaturated Fatty Acids

Steven M. Watkins and J. Bruce German

CONTENTS

I. INTRODUCTION

Fatty acids serve a wide variety of metabolic functions critical to all forms of life. They are a rich source of energy and carbon and well designed as a convenient unit for energy storage. However, the importance of fatty acids in human nutrition and physiology goes well beyond their role as a source of calories. Fatty acids provide the structure and hydrophobicity crucial to the maintenance of a semipermeable membrane barrier. Their structures can be modified by desaturation and elongation to produce a substantial variety of species with individual chemical and physical properties. Ester linkages to glycerides allow fatty acids to be easily exchanged for one another and allow cells to manipulate the physical properties of their membranes. Fatty acids also serve as precursors to active signal molecules such as eicosanoids, which are capable of producing potent biological effects. Evolution has produced a distinction between plants and animals in their capabilities for the metabolism of fatty acids. Higher animals are unable to synthesize all of the fatty acids required for certain tissue functions and are obligated to ingest fatty acids that are synthesized by plants. Animals have evolved a separate and distinct series of metabolic modifications of fatty acids, but are still unable to alter the original modifications inserted by plants. As a result, the membrane, signal, and storage lipids of animals vary widely according to their dietary intakes. In addition, the ability of an animal to produce a specific fatty acid relies either on an inherent mechanism for desaturation of saturated fatty acids or on the ingestion of a convertible precursor. The ingestion or metabolism of particular unsaturated fatty acids is necessary for a great variety of physiological and cellular functions. Inadequate intake or defective metabolism leads to various dysfunctions due to deficiencies of these fatty acids in particular cellular locations. In addition, dietary fatty acids have been well correlated with metabolic and physiological alterations associated with heart disease and cancer [1–3]. Unsaturated fatty acids in particular play an important role in these non-energy-producing metabolic functions.

Dietary fatty acids have the singular ability among macromolecules to incorporate into tissue intact, thereby altering tissue acyl compositions. Proteins and nucleic acids, while providing energy and building blocks for metabolism, are incapable of remodeling the protein and nucleic acid compositions of tissues in their own image. Consequently, fatty acids occupy a unique and important role in human nutrition. With the recent advances in basic knowledge of plant fatty acid biosynthesis and genomics, it is possible to produce virtually any fatty acid in significant quantities. The availability of fatty acids for supplementation and the ability to engineer agricultural products provide opportunities to significantly modify the lipid content of the food supply. Unfortunately, knowledge concerning fatty acid function in physiology lags far behind the ability to modify dietary fatty acid compositions. Developing an understanding of specific fatty acids and interactions among fatty acids and how they affect individual metabolism and health will be critical for nutrition and agriculture in the next decade.

II. FATTY ACID BIOSYNTHESIS

The primary product of fatty acid synthase in both plants and animals is palmitic acid. However, many plant and animal fatty acids are longer and more unsaturated than palmitic acid; consequently, acyl modification systems are a critical component in the regulation of cellular acyl composition.

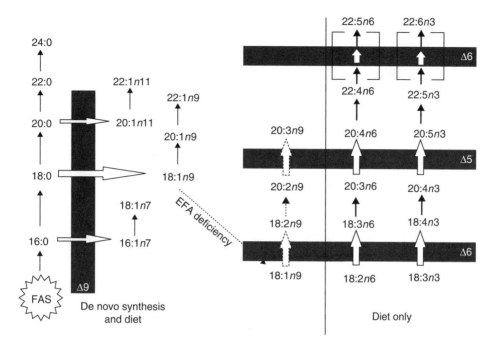

FIGURE 20.1 A generalized scheme for human fatty acid metabolism. Gray bars represent desaturase activities; black arrows represent elongase or chain-shortening activity; and the relative abundance of each fatty acid is represented by its size.

Products of fatty acid synthase or dietary fatty acids may be modified by a chain elongation or the insertion of double bonds. Both elongation and desaturation reactions are organism-, tissue-, and cell-specific, allowing individual cells to maintain their compositional identities largely independent of diet. The possibilities and limitations imposed on the fatty acid composition of a cell are intimately associated with these enzymatic reactions. Figure 20.1 provides an overview of the most common mammalian fatty acid modifications.

Before discussing the biosynthesis of fatty acids, it is useful to describe the standard nomenclature for fatty acids. The systematic method of naming fatty acids provides information on acyl-chain length and the degree and position of desaturation. Standard nomenclature describes a double bond occurring between the ninth and tenth carbons from the carboxyl end of a 16-carbon fatty acid as a $\Delta 9$ double bond (Figure 20.2). The same bond, when viewed from the methyl end of the fatty acid, is referred to as an n-7 double bond. The usefulness of two nomenclature systems has grown out of a need to view fatty acids from multiple perspectives. Customarily, the n system is used when fatty acids are discussed with respect to nutrition, whereas the Δ designation is more useful when observing the biochemical reactions of fatty acids. In many ways, the n system simplifies

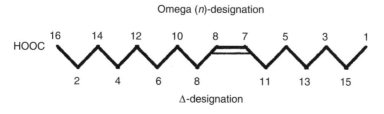

FIGURE 20.2 The omega (n) and delta (Δ) numbering systems for palmitoleic acid. Δ-Numbering starts at the carboxyl terminus carbon and n-numbering starts at the methyl terminus carbon.

investigation into the nutritional relevance of fatty acids. Animals are not capable of desaturating on the methyl side of a previously formed double bond; thus, in humans and other animals, fatty acids of a particular n designation will remain permanent. This greatly simplifies the analysis of fatty acid metabolism in animals. In contrast, because many organisms, including humans, are capable of acyl-chain elongation, the Δ designation for a particular double bond is subjected to a change in nomenclature with each elongation of the fatty acid. The Δ designation proves useful, however, when interpreting the chemical reactions associated with fatty acids. For instance, desaturases are referred to by the Δ designation because their catalytic action is consistent with the stereospecific insertion of a double bond from the carboxyl end of a fatty acid.

A. CARBON SOURCE AND DE NOVO SYNTHESIS

The biosynthesis of fatty acids is largely similar among plants and animals. Both are capable of producing fatty acids de novo from acetyl CoA via the concerted action of acetyl CoA carboxylase and fatty acid synthase. The first step in the de novo synthesis of fatty acids involves the production of malonyl CoA from acetyl CoA, a reaction catalyzed by acetyl CoA carboxylase. Acetyl CoA carboxylase carries out two partial reactions, each catalyzed at distinct sites, which first carboxylate the reaction cofactor biotin and then transfer the carboxyl group to acetyl CoA [4]. The net reaction is shown as follows:

$$\text{Acetyl CoA} + \text{HCO}_3^- + \text{ATP} \rightarrow \text{Malonyl CoA} + \text{ADP} + \text{Pi}.$$

In animals, this enzyme is soluble in the cytosol and appears to be regulated by a number of factors, including long-chain acyl CoA, providing sensitivity to both de novo production of acyl chains and diet. The second general step in the production of fatty acids is to activate both malonyl CoA and the primary unit of condensation, acetyl CoA, by transferring the acyl groups to an acyl carrier protein (ACP). These reactions are catalyzed by malonyl CoA–ACP transacylase and acetyl CoA–ACP transacylase, respectively. The malonyl ACP complex then enters a cycle of elongation catalyzed by the soluble enzyme complex fatty acid synthase. Fatty acid synthase lengthens the acyl chain by two carbons per cycle of activity, using acetyl CoA as the condensing unit. This series of reactions culminates in the production of palmitic and stearic acids. The cycle is terminated when acyl ACP thioesterase hydrolyzes the acyl ACP thioester and releases a fatty acyl CoA [5]. Because the products of fatty acid synthase are consistently palmitic and stearic acids, acyl ACP thioesterase is likely to be specific for the hydrolysis of 16- and 18-carbon acyl ACP complexes. Fatty acid synthesis is extensively reviewed by Goodridge [4] and Wakil et al. [6].

Although the general mechanisms of fatty acid synthesis are similar, there are several specific differences between plant and animal fatty acid synthesis. Plant fatty acid synthase products are complexed in acyl ACP, whereas animals produce acyl CoA. De novo synthesis of plant fatty acids occurs in the plastid where the products of fatty acid synthase are predominantly palmitoyl and stearoyl APC. These products are either used directly in the plastid as acyl ACP or translocated to the cytoplasm and converted to an acyl CoA complex [5]. The de novo synthesis of macromolecules in animals, including fatty acids, usually requires the transport of acetyl CoA into the cytoplasm, as acetyl CoA carboxylase and fatty acid synthetase are soluble cytoplasmic enzymes.

B. DESATURATION

Fatty acid desaturation of nutritional importance to humans is largely similar in plants and animals. The requirements for desaturation include molecular oxygen, a reduced pyridine nucleotide, an electron transfer system, a terminal desaturase enzyme, and a fatty acyl substrate [7]. A net reaction scheme for the $\Delta 9$ desaturation of stearoyl CoA is as follows:

$$\text{NADH} + \text{H}^+ + \text{Stearoyl CoA} + \text{O}_2 \rightarrow \text{NAD}^+ + \text{Oleoyl CoA} + \text{H}_2\text{O}.$$

Plant fatty acids provide a seminal source of polyunsaturated fatty acids (PUFAs) in human nutrition. This dependence on plant fatty acids is due to the fact that plants are capable of inserting double bonds into the $\Delta 12$ and $\Delta 15$ positions of fatty acids. The desaturation of fatty acids produced de novo by plants involves an initial $\Delta 9$ desaturation, followed by $\Delta 12$ and $\Delta 15$ desaturations. The insertion of double bonds on the methylene side of the previously unsaturated $\Delta 9$ bond is in opposition to animal metabolism, wherein additional double bonds are only inserted on the carboxyl side of the $\Delta 9$ double bond. This difference allows dietary fatty acids to be converted to forms not possible as a result of plant or animal metabolism alone. Plants express two pathways for the desaturation of fatty acids. The first, known as the prokaryotic pathway, is an array of desaturase activities that is present in the plastid. The second eukaryotic pathway requires the translocation of the fatty acids to the endoplasmic reticulum where desaturases act on the acyl groups. The products of the eukaryotic pathway are either used directly by the cell or translocated back to the plastid for further processing. Harwood [5] provides an extensive review on plant fatty acid desaturation. An interesting phenomenon of plant fatty acid biogenesis is that desaturation can use complex lipids as substrate. Both plastid- and endoplasmic reticulum-based desaturation introduce a double bond into fatty acids esterified in a phospholipid. In contrast, animals desaturate acyl CoA complexes.

Fatty acid desaturation in animals uses fatty acyl CoA thioesters as substrate. Despite multiple desaturase activities, cytochrome b_5 reductase acts as an electron donor common to all terminal desaturases. The desaturases are present on the cytoplasmic face of the endoplasmic reticulum. Animals possess $\Delta 9$, $\Delta 6$, and $\Delta 5$ desaturase activity. Like desaturases in plants, each of the desaturases has preferred substrates, which can display organism- and tissue-specific differences. Because double bonds are found inserted at the $\Delta 4$ position of 22:6n-3 and 22:5n-6, it was originally assumed that animals also possessed a $\Delta 4$ desaturase. Such a desaturase has not been identified, and work by Sprecher and colleagues [8,9] has demonstrated that 22:6n-3 is produced via a $\Delta 4$ desaturase-independent pathway. Fatty acids containing a $\Delta 4$ unsaturation are in fact the product of an additional elongation, a $\Delta 6$ desaturation, and a two-carbon chain shortening that takes place in the peroxisome [9–11]. An interesting phenomenon specific to this process is the coordination of the movement of fatty acid from the endoplasmic reticulum to the peroxisome and back again for acylation into lysophospholipids. How the cell recognizes the production of 22:6n-3 within the peroxisome and spares it from further chain shortening is not understood.

The first double bond inserted into the saturated acyl CoA products of fatty acid synthase is at the $\Delta 9$ position and is catalyzed by stearoyl CoA desaturase (SCD). In contrast to desaturation in plants, further desaturation occurs only on the carboxyl side of the initial unsaturation. Interestingly, most PUFAs are of plant origin, as monounsaturated n-9 fatty acids produced de novo are not further desaturated by animals except in times of essential fatty acid deficiency. However, fatty acids previously polyunsaturated by plants are readily desaturated to form familiar PUFAs, such as arachidonic acid and docosahexaenoic acid (DHA). These highly unsaturated fatty acids are products of both plant and animal unsaturation, and they have double bonds inserted on both sides of the original $\Delta 9$ double bond.

1. $\Delta 9$ Desaturase

The SCD introduces a double bond at the $\Delta 9$ position of stearoyl CoA, forming oleic acid and the n-9 family of fatty acids. This enzyme also adds a double bond at the n-7 position of palmitic acid, forming palmitoleic acid and the n-7 family of fatty acids. The enzyme appears to be a fundamental gene product in the regulation of a host of cellular processes. Its structure has been resolved to 2.6 Å and various aspects of the functional domains were identified in the plant enzyme [12]. The sequences of various SCD genes were reported, and more recent work has focused on the upstream regulatory regions of the genes from both microbial and animal sources [13–15]. From this work, a host of interesting regulatory sites have been identified, implying that the SCD genes are multiple-regulated in all cellular systems in which they are expressed. Primitive, single-cell organisms are known to alter SCD transcription in response to temperature, environmental shocks (pathogenic and osmotic), and substrate modification [16–19].

Not surprisingly, in addition to its many cellular actions, SCD appears to play multiple roles in higher plants and animals. Many reports have associated SCD with whole-tissue functions such as adipose accretion, lipid secretion, and tissue responses to stress. For example, plants induce SCD on thermal shock and senescence [20].

The most intriguing aspects of SCD in lipid regulation in animals are found in the apparent role of the SCD genes in lipogenesis. Much of this information was summarized by Ntambi [14]. Two highly homologous genes, described in mice, that code for SCD are termed, logically, *scd1* and *scd2*. *scd1* is one of the first genes induced during adipocyte differentiation and this induction is responsive to insulin, carbohydrate, and elevated cAMP [15]. Furthermore, inhibition of SCD prevents adipocyte differentiation and, in mature cells, reduces lipogenesis in adipocytes and in hepatocytes of avian [21] and mammalian cells. Genetically obese animals exhibit greater amounts of SCD, consistent with a pivotal role in adipocyte function [22]. The striking ability of PUFA to downregulate the enzyme activity is well described biochemically and recently was shown to be due to a substantial decrease in the stability of *scd1* mRNA [23]. It is not yet clear precisely what advantage such a regulatory control serves. However, various suggestions have been advanced that this could be a partial basis for reduced hepatic lipoprotein secretion during PUFA feeding and could possibly reduce adiposity in PUFA-fed animals. Thus, there is abundant evidence that unsaturated families of fatty acids interact with each other in highly complex ways.

The SCD gene is known to be sensitive to a variety of hormonal signals. For example, SCD is differentially regulated in females relative to males and its higher activity in females may be part of the spectrum of lipid metabolic changes that are a consequence of sex differences in the activity of growth hormone [24,25].

2. Δ12 Desaturase

The Δ12 desaturase of plants is responsible for the conversion of oleic acid to linoleic acid and is thus the molecular basis for the *n*-6 family of fatty acids. Although Δ12 desaturase is present solely in plants, its functional requirement in plants is not completely understood [26]. At the present time, the most apparent action of the enzyme activity is to improve thermal tolerance, especially of specific membrane compartments in plants [27]. However, the requirement by animals for the *n*-6 family of fatty acids is not solely related to thermal tolerance, and it is intriguing that a completely separate functionality has evolved in animals to take advantage of the *n*-6 family of PUFAs as precursors to more unsaturated species. The protein structure of the Δ12 desaturase at the molecular and sequence level is known, and the ability of the single gene product to affect Δ12 desaturation in transfection experiments with *Saccharomyces* is known [28]. Although structural studies have not reached the same level of understanding for the SCD, similar functional themes are apparent, such as analogous required histidine residues [29].

3. Δ15 Desaturase

The Δ15 or *n*-3 desaturase catalyzes the conversion of linoleic acid to α-linolenic acid (as the respective acyl CoA). This enzyme activity is widely distributed in plant tissues and is the basis for the abundance of its products, the *n*-3 family of fatty acids, in plants and animals. Although these fatty acids are conspicuously enriched in thylakoid photosynthetic membranes, the molecular advantage that the *n*-3 double bond provides has not yet been determined. At the molecular level, there are two isoforms of the gene, whose products are located in either the microsomal fraction or plastid [29]. The plastid form of the enzyme appears to be associated with photosynthesis because it is induced by light [30]. The gene product may also have a role in altering membrane composition as a stress response as one of the *n*-3 desaturase genes is induced by wounding [31].

4. Δ6 and Δ5 Desaturases

Although ubiquitous, the Δ6 and the Δ5 desaturase enzymes have only recently been cloned from animals [32,33]. Interestingly, although the Δ6 desaturase is exceedingly rare in the plant kingdom, it had previously been cloned and functionally characterized from borage [34]. Like animal SCD, these desaturases use a cytochrome b_5 domain as an electron transfer system [32,33]. The expression of the Δ6 desaturase was presumed to be limited to hepatic tissue; however, Northern blot analyses now demonstrate that many tissues—including heart, kidney, lung, skeletal muscle, and, notably, brain—express the desaturase [33]. There is a broad distribution of Δ5 desaturase mRNA expression among tissues as well, with Northern analyses identifying expression in lung, skeletal muscle, placenta, kidney, and pancreas [32]. Like the Δ6 desaturase, the Δ5 desaturase is expressed most abundantly in liver, brain, and heart [32]. The abundance of Δ6 and Δ5 desaturase mRNA in tissues not typically found to have desaturase activity may indicate a posttranslational regulation of desaturase activity in these tissues. However, it is still an open question as to whether tissues such as heart and brain are capable of fatty acid desaturation. Although surprisingly few functional characterizations of the Δ5 and Δ6 desaturase genes have been reported, there is evidence that the mRNA expression of both the Δ6 and Δ5 desaturases is regulated by nutrition. Hepatic Δ6 desaturase expression was suppressed in mice fed corn oil (containing high concentrations of linoleic acid) relative to those fed triolein [33]. Although oleic acid is not a typical substrate for the Δ6 desaturase and, by contrast, linoleic acid is an excellent substrate, these results may indicate that the Δ6 desaturase is regulated by the degree of membrane unsaturation. The Δ5 desaturase mRNA expression also appears to be regulated by dietary fatty acids. Mice fed triolein or a fat-free diet exhibited marked increases in hepatic Δ5 desaturase mRNA expression relative to mice fed safflower oil or fish oil [32].

Although the metabolism of n-3 and n-6, and even n-9, fatty acids in animals is thought to use the same Δ6 and Δ5 enzymes, there are specific differences in the desaturation efficiencies with regard to substrate. For instance, the most abundant n-6 fatty acids in animals are clearly linoleic and arachidonic acids. In contrast, the only n-3 fatty acid present at concentrations commensurate with these n-6 fatty acids is DHA, despite the fact that most dietary n-3 fatty acid exists as linolenic acid (18:3n-3). There is also the virtual nonexistence of the Δ6 and Δ5 desaturase products of endogenously produced or dietary oleic acid (18:1n-9) in animals. Clearly, the families of fatty acids are not simple competitors with each other for desaturation, and there must exist a complex regulation of desaturation. The availability of clones for the Δ5 and Δ6 desaturase genes heralds a truly exciting and productive future for understanding the role of PUFA metabolism in physiology and nutrition.

5. Regulation of Desaturase Activity

The regulation of desaturase activity is complex and appears to involve a number of signals. The activities of mammalian desaturating enzyme systems are sensitive to several metabolic signals [35–39]. These various effectors cause an inhibition in the net desaturation through the Δ6 and Δ5 enzymes. SCD activity is substantially suppressed in the fasted state [40] and is restored by refeeding or insulin administration. In culture, scd2 gene expression was suppressed by cholesterol [41]. However, in rats, scd expression was induced by dietary cholesterol [42]. The regulation of SCD by PUFAs and cholesterol was reviewed by Ntambi [13].

As direct effects, the desaturases are strongly inhibited by their products; hence, diets rich in PUFAs tend to suppress desaturase activity. Strikingly, fasting also decreases Δ6 desaturation, and a basic protocol of fasting and refeeding accelerates essential fatty acid deficiency [43].

Endocrine signaling has been variously reported to affect desaturation [35,43,44]. Glucagon, epinephrine, corticoids, and thyroxine all lower the activity of the Δ6 and Δ5 desaturases [43]. Diabetes in humans and in animal models is associated with lower concentrations of PUFAs, and this is paralleled by measurable decreases in the activity of the Δ6 and Δ5 desaturases. Consistently, insulin administration restores the desaturase activities in both humans and in animal models of diabetes and normalizes the content of PUFAs in membranes. As an illustration of the extent of control of these systems, insulin does not increase desaturation in normal individuals [45].

In studies designed to examine the mechanisms of modulatory effects, cAMP was implicated as causal to desaturase modification, and the effect can be mimicked by dibutyryl cAMP both in vivo [46] and in vitro [47]. Interestingly, corticosteroids and thyroxine depress both the $\Delta 6$ and $\Delta 5$ desaturase activities but increase the $\Delta 9$ desaturase activity [35,48].

To date, there is a paucity of metabolic systems in which the apparent activity of the $\Delta 5$ and $\Delta 6$ desaturases increases over controls. Growth hormone strongly induces both the $\Delta 6$ and $\Delta 5$ enzyme activities and their metabolic products [25,49]. Growth hormone also substantially downregulates the transcription of the SCD of liver and adipose.

The reciprocal response of the $\Delta 6$ and $\Delta 5$ compared with that of the $\Delta 9$ desaturase is intriguing and may reflect a truly interactive regulation. For example, it is not clear if the suppression of the $\Delta 9$ desaturase by growth hormone is an effect solely of growth hormone or also of the metabolic products of the enhanced desaturation by the $\Delta 6$ and $\Delta 5$ desaturases. Arachidonic acid, the product of the $\Delta 6$ and $\Delta 5$ desaturases, is known to downregulate directly the adipose SCD [23]. Such effects argue compellingly that the regulation of PUFA metabolism is sensitive to the products of metabolism whether formed de novo or ingested. This will likely be a focus of research in the future.

C. ELONGATION

The elongation of presynthesized fatty acids is critical to fatty acid metabolism in animals. C16 and C18 fatty acids are the primary products of both plant and animal fatty acid biosynthesis and, consequently, acyl chains longer than C18 must be elongated post-de novo synthesis. Many PUFAs critical to the structure and physiology of animals are longer than C18; hence, the need for an effective elongation system. In addition, the sequential desaturase activities involved in producing more unsaturated derivatives of *n*-fatty acids require intermediate elongation steps. This is not accomplished through further cycling of fatty acid synthase, but rather through independent activities located in the endoplasmic reticulum (ER) and the mitochondria. These systems are quite distinct and even use different substrates for condensation. Mitochondria and the ER add acetyl CoA and malonyl CoA as elongation substrate, respectively [50]. Although the mitochondria are very active in the production of acetyl CoA and β-oxidation, the ER appears to possess the majority of elongase activity. The ER system for acyl elongation appears to prefer unsaturated fatty acids for further elongation, although this preference varies among tissues [50]. There is evidence that a third elongase activity is present in the peroxisomes and that it is related to the peroxisome proliferation response [51]. Although the majority of elongation occurs in the liver, other tissues also express activity.

D. PEROXISOMAL PUFA SYNTHESIS

Although the majority of fatty acid modifications involve elongation and/or desaturation, some long-chain PUFAs are produced by the removal of two carbons by one cycle of oxidation. Long-chain unsaturated fatty acids are retroconverted in peroxisomes to produce acyl chains with two or four fewer carbon units. The most common example of this type of acyl modification involves the synthesis of the long-chain PUFA, DHA. DHA is produced by the elongation of 22:5*n*-3 to 24:5*n*-3, followed by a $\Delta 6$ desaturation and one cycle of β-oxidation to produce 22:6*n*-3 [8,9,52]. There is also evidence that the $\Delta 6$ desaturase involved in this pathway is distinct from the $\Delta 6$ desaturase associated with typical PUFA production [53]. Several other fatty acids are produced by this pathway, although it is not known if this is the primary source of their biosynthesis. Voss et al. [54] showed that arachidonic acid can be produced via the retroconversion pathway in rats injected with 22:4*n*-6. Hagve and Christophersen [55] demonstrated that isolated rat liver cells were capable of retroconverting 22:4*n*-6 and 22:6*n*-3 to 20:4*n*-6 and 20:5*n*-3, respectively, suggesting that retroconversion is actively used for the production of PUFAs in vivo. The semitoxic erucic acid was retroconverted to oleic acid in cultured human fibroblasts [56].

Although it appears clear that the retroconversion of fatty acids via peroxisomal oxidation plays a significant role in fatty acid biosynthesis, several important questions remain concerning the regulation of the process. It is known that retroconversion takes place in the peroxisomes and not in mitochondria where the majority of cellular β-oxidation occurs [9]. In addition, the same acyl CoA oxidase may be involved in all chain-shortening events [57]. A particularly intriguing aspect of the specific production of acyl chains by retroconversion is how the fatty acids are spared from further oxidation. Fatty acids with a Δ4 unsaturation are actively mobilized and returned to the ER where they are acylated into lysophospholipids [11]. Therefore, 22:6n-3 produced de novo in peroxisomes from an n-3 precursor is found acylated primarily in the phospholipids with remodeling pathways active at the ER. The distribution of 22:6n-3 in cellular phospholipids is dependent on whether 22:6n-3 is consumed intact or as a metabolic precursor [58]. A lack of peroxisomal retroconversion activity has been associated with several pathologies, including Zellweger's syndrome [9,59]. The accumulation of very long-chain fatty acids (C24–26) is a hallmark of Zellweger's syndrome, suggesting that peroxisomal oxidation is both a normal and critical component of fatty acid metabolism.

III. PUFA METABOLISM IN MEMBRANES

Knowledge of tissue or even cellular fatty acid composition is not sufficient to predict the effects of PUFAs in cell physiology. The realization that fatty acid location plays an important part in the use and function of specific fatty acids has been a major advancement in lipid metabolism. Unsaturated fatty acids, as well as their saturated counterparts, are esterified in phospholipids of cell membranes. There are a variety of phospholipid types, and each type has its own unique compositional fatty acid identity. In addition to phospholipid identities, individual cell membranes have unique phospholipid and fatty acid compositions. Although there are myriad data on the fatty acid composition of plant and animal tissues, these fatty acid profiles are subjected to important changes brought about by diet, disease state, or a variety of other factors. As a result of the variety of phospholipid pools in which fatty acids can be esterified, very slight changes in total cell fatty acid content can have significant effects on cell function. It is quite possible that lipids have been investigated and rejected as causal agents of a number of pathologies on the basis of largely unchanged fatty acid compositions. It is now clear that fatty acids exert their effects from specific locations and that their physiological effects are mediated in part by the movement of PUFAs into important phospholipid pools. The movement of fatty acids into phospholipid pools is catalyzed by fatty acid carrier proteins, and the acyl compositions of membranes and specific phospholipid pools are mediated by the specificities of these enzymes.

A. REMODELING

Dietary fatty acids can exert significant effects on the fatty acid compositions of phospholipid membranes. These effects must be attributed either to the synthesis of new phospholipids de novo or to the activities that change the composition of preexisting phospholipids. Quite commonly, it is the latter of the two mechanisms. Phospholipid fatty acid remodeling is both an important pathway for the incorporation of dietary fatty acids and a dynamic system for membrane property homeostasis. A generalized pathway for the remodeling of a phospholipid (Lands pathway) [60] involves (1) removal of the acyl group from the phospholipid via a lipase activity; (2) conversion of the fatty acid to an acyl CoA thioester; (3) a modification event, such as elongation or desaturation; and (4) reesterification of the acyl group or the insertion of a new acyl unit onto the lysophospholipid (Figure 20.3). The first step in the Lands pathway requires the hydrolysis of the acyl chain from the phospholipid and is catalyzed by the phospholipase A family of enzymes. Phospholipases have

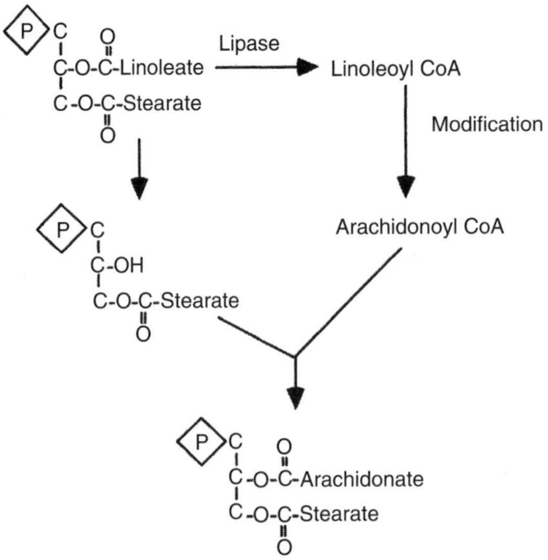

FIGURE 20.3 A generalized scheme for the fatty acid remodeling of phospholipids (the Lands pathway). (From Hill, E.E. and Lands, W.E.M., *Lipid Metabolism*, S.J. Wakil, ed, Academic Press, New York, 1970.)

specificities for both the phospholipid position and acyl chains, and these specificities contribute to the regulation of membrane remodeling. Phospholipases are well reviewed by Waite [61]. Once a phospholipase has acted on an acyl chain, the fatty acid must then be activated to an acyl CoA thioester. This is accomplished via the action of acyl CoA ligases. There are several distinct acyl CoA ligase activities, each with its own cellular locations and acyl preferences [62]. The activated acyl CoA complex can be converted by desaturation or elongation (see earlier) and reacylated into a lysophospholipid. The acyltransferase enzymes catalyze the esterification of acyl chains from acyl CoA thioesters into lysophospholipids. Two types of acyltransferase activities are relevant to membrane remodeling, and not surprisingly, they catalyze the insertion of a fatty acid into either the *sn*-1 or *sn*-2 position of phospholipids. The first type of enzyme, typified by glycerol-3-phosphate acyltransferase, is capable of inserting an acyl chain into the *sn*-1 position of glycerol-3-phosphate and has a general preference for saturated acyl CoA [63,64]. In fact, glycerol-3-phosphate acyltransferase is a number of distinct enzymes capable of carrying out the same reaction at different locations in the cell, and with varying acyl CoA preferences [62,63]. The primary acyltransferase responsible for esterifying acyl chains into the *sn*-2 position is lysophosphatidate acyltransferase [64]. Lysophosphatidate acyltransferase has a higher activity with unsaturated acyl CoA thioesters and is responsible for the virtual absence of saturated fatty acids at the *sn*-2 position in membrane phospholipids. As might be expected, the lysophosphatidate acyltransferase has a higher activity than the glycerol-3-phosphate acyltransferases [64], as the majority of the fatty acid remodeling in membranes occurs with unsaturated fatty acids. The combined preferences and activities of the acyltransferases are largely responsible for the positional and site-specific acyl compositions observed in phospholipid membranes.

1. Acyl-Specific Incorporation

Several phospholipid species are noteworthy for their high degree of incorporation of specific fatty acids. The ether-linked phospholipids, for example, are enriched at the *sn*-2 position by a CoA-independent transacylase relatively specific for arachidonic acid [65,66]. The most salient example of acyl-specific incorporation, however, can be found in the mitochondrial diphospholipid cardiolipin. Cardiolipin is characteristically enriched with linoleic acid to as much as 85% of its acyl species in vivo [67]. Cardiolipin is also selectively enriched in DHA to as much as 50% of its total fatty acid content [58,67,68]. The implications of the high degree of unsaturation of cardiolipin may involve its function as the membrane solvent for the electron transport enzymes. Other phospholipids incorporate specific fatty acids to varying degrees. Spector and Yorek [69] treated Y79 retinoblastoma cells with arachidonic acid, DHA, and oleic acid to determine the relative affinity of various phospholipid classes for unsaturated fatty acids. Phosphatidylcholine, phosphatidylethanolamine, and phosphatidylserine could all be enriched with the monounsaturated fatty acid oleic acid. Only phosphatidylethanolamine and phosphatidylinositol were substantially enriched in arachidonic acid, whereas phosphatidylcholine, phosphatidylethanolamine, and phosphatidylserine all incorporated DHA. The differences in acyl incorporation indicate that the synthesis and remodeling mechanisms described earlier have different activities, depending on both the fatty acid and the phospholipid substrates.

IV. DIETARY SOURCES OF UNSATURATED FATTY ACIDS

A. MICROBES

Most bacteria are fully capable of synthesizing all of the fatty acids required for their normal growth and reproduction. The fatty acid synthetase enzyme complex responsible for this activity adds acetate units to a final chain length of 16–18 carbons. These can be further desaturated (normally once to a monounsaturated fatty acid) after which the saturated and unsaturated fatty acids are esterified to yield membrane phospholipids. Bacteria in general do not store energy as fats; hence, triacylglycerols are not abundant forms of lipid in bacteria. They are thus not a quantitatively important fat source in foods. The unique metabolism of bacteria, especially the production of branched chain and odd-numbered fatty acids, can occasionally generate measurable quantities of unusual fatty acids in, for example, bovine milk, due to incorporation of the products of rumen fermentation. In the production of commercially viable oil sources, wax esters and eicosapentaenoic acid (EPA) have been produced at a commercially relevant scale in bacteria [70,71].

Microbes as a broad class of single-cell (or relatively undifferentiated collections of cells) organisms have the obligate biosynthetic capability of generating fatty acids necessary for membrane synthesis and other processes. Historically, these have not been considered an important source of fat in the diet, although it is recognized that these organisms add small quantities to certain foods, whose presence may be an important contribution, for example, for flavor. Nevertheless, even for microbes that produce large quantities of storage triacylglycerides, the economics of growing and obtaining the oils has been noncompetitive with the traditional sources of edible oils. However, recently this area has seen a considerable resurgence in interest both academically and in commercial application [72]. This change is largely due to the improved efficiency and capabilities of large-scale microbial fermentation, to the identification of therapeutically useful edible oils, and to the capability of microbes to produce unusual fatty acids or unusual concentrations of fatty acids and glycerides [73]. Fatty acids that have raised the ante, as it were, for edible oils include dihomo-γ-linoleic acid, EPA, and DHA, due to their ability to alter arachidonic acid metabolism and hence thrombosis, inflammation, cancer, and autoimmune diseases; DHA for inclusion in infant formulas; nervonic acid for its potential in treating neuropathies; long-chain monounsaturated fatty acids for adrenoleukodystrophy; and stearculic acid as a possible treatment for bowel cancer [74,75]. Several factors mitigate in favor of microbial production for high-value lipids. The greater potential for aggressive recombinant approaches to manipulate microbial lipid metabolism to obtain novel fatty acids is likely to increase the growth of this cottage industry for fatty acid production. Higher plants and animals are also somewhat limited in the glyceride forms that they will produce. For example, most plants do not place a saturated fatty acid in the sn-2 position of a triglyceride. Similarly, fish tend to place virtually all of the long-chain n-3 PUFAs in the sn-2 position. This limits both the total range of glycerides available using these plants and animals as sources of lipids and imposes structural effects on digestion and absorption of the fatty acids in nutritional applications. These limitations are both less well defined and more mutable in microbial fermentation applications. This area, though coming under intense regulatory scrutiny, may reach a significant segment of the food industry, at least in the short term [76,77].

Single-cell eukaryotes have, as a class, a remarkably wide variety of lipid metabolic capabilities. For example, some yeast produce only a single desaturase, the stearoyl or Δ9 enzyme, and neither produces nor requires PUFA for growth. At the other end of the spectrum, some fungi and algae can produce very high amounts of arachidonic, dihomo-γ-linolenic, eicosapentaenoic, and DHA [78]. An additional and synthetically useful attribute of these organisms is their ability to take up fatty acids from the medium and either to incorporate them into triacylglycerides and phospholipids (even with unusual stereospecificity) or to further metabolize them before esterification [78]. These various properties were known previously but were not thought to warrant commercialization. This is changing. Already, microbial lipid sources are proving to be a cost-effective feedstock for shrimp and fish aquaculture. The ability of the microbial feedstock to elaborate valuable pigments and antioxidants is also used to advantage, so this entire technology and its biotechnological elaborations are likely to increase in impact in the future.

B. AGRICULTURAL PRODUCTS

The extended metabolism of 18-carbon PUFAs to longer chain, more unsaturated fatty acids [frequently referred to as highly unsaturated fatty acids (HUFAs)] in animals means that even though animals and plants may contribute similar families of PUFAs to the diet, the precise form of these fatty acids will differ. As a result, an important consequence of consuming animal in contrast to vegetarian foods is that in the latter, linoleic and linolenic acids are the fatty acids ingested from the n-6 and n-3 families, whereas in animal foods, their metabolic products, preformed arachidonic, eicosapentaenoic, and DHA, are also ingested. It is now clear that these are significant nutritional, biochemical, and physiological differences.

Furthermore, whereas animal sources of fat are often grouped as similar, avian, aquatic, and ruminant or nonruminant mammalian storage lipids are very different in the quantity of depot triacylglycerols, their distribution and their fatty acids, as well as their composition and arrangement on the glycerol. The final content of fatty acids in storage triacylglycerols is the result of diet, metabolism, and de novo synthesis. In this respect, each of the major animal fat sources differs in important ways, which tends to distinguish each as a fat-rich commodity. These differences have important effects on the texture, flavor, and caloric density of the muscles as consumed directly [79–81] and also on processed foods prepared from them [82].

Although the differences among species in the quantity and distribution of fat are associated with the particular commodities, they are not necessarily all innate to them. These differences also reflect the historical development of the particular muscle food as a commodity. Even among ruminants, the fat content of modern beef muscle is higher and more saturated than that of comparable wild ruminant muscle [83]. Breeding and feeding practices allow for the production of meat at a specific fat concentration [84]. If different properties were perceived to be beneficial, the fat content could arguably be altered to various extents accordingly. Thus, when examining the content of storage fat in muscle tissue that is used as food, one is looking at a rather narrow window of a wide range of possibilities. As commodity needs become more defined and the fat functionality better understood, the means to arrive at these targets will need to be explored.

In addition to the differences in total quantity of fat and its tissue distribution, the composition of storage triacylglycerols in animal species differs as well. Red meats tend to be relatively higher in saturated fatty acids and lower in PUFAs than poultry or fish. Poultry and fish differ significantly in the chain length of monounsaturated fatty acids and in the content of n-3 PUFAs.

Once again, a consistent observation of the lipid content in different animal tissues is the variability within species. Even within ruminant animals in which the dietary PUFAs are largely hydrogenated by rumen flora, there is a significant range of composition. Among monogastric animals, the variability in fat composition within species due to muscle type, diet, environment, and age is typically greater than the differences noted among species [81].

An important question becomes, what unique properties of the metabolism of the three animal types lead to the observed or apparent differences in lipid composition and behavior? In all animals, the stored triacylglycerols both in adipose and individual muscle cells can be assembled from both dietary fatty acids and fatty acids synthesized de novo, primarily either in liver (in chickens and fish) or in adipose tissue (in pigs) [85,86]. In general, de novo synthesis of saturated fats is decreased by dietary fats [81]. Therefore, fats from the diet constitute the greatest source of variation in the composition of storage fats. Within this framework, metabolic control can be seen. For example, short- and medium-chain fatty acids are not incorporated into storage lipids of most animals. These pass into the liver where they are either elongated or oxidized for fuel [87]. Although monounsaturated and linoleic acid, 18:2, are readily incorporated, in most animals long-chain (greater than C18), highly unsaturated fatty acids are not esterified into triacylglycerols [85]. However, fish will accumulate PUFAs, notably the *n*-3 PUFAs 20:5 and 22:6, but only if they or their precursors are present in the diet and only at low water temperatures [88]. Fish actually require *n*-3 PUFAs in their diets but are unable to synthesize them [89,90]. Alternatively, very high concentrations (>50%) of saturated fats are not found in storage lipids due to the well-regulated activities of the Δ9 desaturase that produces oleic acid from stearic acid [85]. In ruminant animals, the rumen microorganisms hydrogenate unsaturated fatty acids in the diet, which has an overriding influence on the composition of the storage fats [91]. However, when PUFAs such as 18:2 are protected from ruminant microorganisms, they accumulate in storage lipids in beef comparably with accumulation in nonruminants [50]. Finally, mammals absorb fat into the lymph, whereas fish and poultry absorb fat directly into the portal vein. As a result, adipose tissue can access incoming fatty acids directly in mammals, but fat passes by liver first in avians and fish. Thus, there is considerably more hepatic metabolism of ingested fatty acids in avian and fish tissues.

C. Effect of Agriculture on the Composition of the Food Supply

Fatty acids occupy a unique position in nutrition in that they have the ability to survive digestion intact, enabling them to replace the fatty acid content of the consumer. It is reasonable then to expect that the lipid composition and, correspondingly, the physiology of individuals who consume particular fats and oils to be reflective of the fatty acid composition of their diet. Interestingly, it has been postulated that the dietary PUFA composition of an average human diet has changed markedly with modern advances in agriculture [92–94]. Wild foods are typically much higher in *n*-3 PUFAs than crops successfully developed by agriculture. There is a variety of evidence to suggest that the changing ratio of *n*-3 to *n*-6 fatty acids has affected human physiology adversely and that humans may have developed major classes of pathologies as a result of this change. The lower rates of coronary heart disease and cancer in populations consuming a higher *n*-3 to *n*-6 PUFA ratio are well documented [95–97]. Despite mounting evidence that human populations would benefit from an increased consumption of *n*-3 fatty acids, it is unlikely that this change will occur in the near future. The primary reason for this is the agricultural success of crops rich in *n*-6 fatty acids. The *n*-6 fatty acids are most typically found in seed crops, which are not only consumed directly but also used in animal feed. In addition, *n*-6-rich crops are generally more stable than *n*-3-rich crops, leading to their preferential cultivation and use as food ingredients. Thus, the increases in coronary artery disease, cancer, and autoimmunity may be a direct consequence of the advance of modern agriculture.

V. NUTRITIONAL EFFECTS OF UNSATURATED FATTY ACIDS

The field of PUFA biochemistry has only begun to develop convincing molecular models for the effects of fatty acids on physiology. Having lagged somewhat behind, lipid biochemistry is now poised to develop in the same way that protein and nucleic acid biochemistry has over the last 20 years. There is considerable information known about fatty acid synthesis as well as a massive collection of data on the fatty acid composition of foods. It has recently even become feasible to modify the fatty acyl content of foods through genetic manipulation. Yet there are very few data on how and why particular fatty acids modulate physiology. An understanding of the molecular basis of fatty acid nutrition will be important for the design of diets appropriate to the individual. PUFAs are thought to exert their physiologic effects through a variety of mechanisms, ranging from acting as precursors for signal molecule formation to modulating membrane structure. The remainder of this chapter will review what is known about the non-energy-producing functions of PUFA and how individual fatty acids modulate these functions.

A. ROLE OF PUFA IN CELL PHYSIOLOGY

Although all fatty acids contribute hydrophobicity to membranes, unsaturated fatty acids provide several unique functionalities. Unsaturated fatty acids and particularly PUFA can form a vast array of chemical structures, each of which has unique physiochemical properties. Cells can use fatty acids to modulate their membrane properties and the activities of membrane-associated enzymes, and for the production of potent signal molecules. Knowledge of the effect of fatty acids on membrane properties has been impeded by the fact that it is difficult to greatly modulate membrane fatty acid composition. Although the difficulty associated with modifying cell membrane composition suggests that membrane homeostasis is critical to cells, it makes investigation on the effects of individual fatty acids difficult. The true successes in this area have come from the discovery of small but highly active phospholipid pools and the enzymes that maintain them. The regulation of the ether-linked arachidonate-containing phospholipid pools is the best described example of how small changes in membrane compositions can have significant effects on cell physiology [98]. Future research in this area will likely have to focus on developing controllable models in which the contributions of specific fatty acids are determined.

1. Unsaturated Fatty Acids and Membrane Structure

A critical feature of the production of unsaturated fatty acids by eukaryotes is that the carbon–carbon double bonds exist in the *cis* configuration. Pi-bonded carbon–carbon double bonds may exist in two conformations, *cis* and *trans* (see Figure 20.4), yet nature has carefully preserved the production of the *cis* isomer to the virtual exclusion of *trans* isomers. The reason for this is best explained by viewing fatty acids as important structural components of cells. Brenner [99] and Cook [50] reviewed the key physiochemical features of double bonds in fatty acids. Most important among the changes imparted to a fatty acid by a *cis* double bond is the rigid "kink" or bend in the acyl chain.

Membranes are largely held together by London–van der Waals forces between adjacent fatty acyl chains [50]. Because these interaction forces are significantly diminished with even a slight increase in the distance between acyl chains, fatty acid packing plays a key role in membrane structure. The placement of unsaturated fatty acids into membranes is confined, with few exceptions, to the *sn*-2 position of phospholipids; as a result, the unsaturated fatty acid content of membranes rarely if ever exceeds 50 mol% unsaturated acyl chains. An advantage of including both an unsaturated and a saturated fatty acid on the same phospholipid molecule is that the two types of acyl chains cannot demix in the membrane. This ensures that slight depressions in the energy of association, such as those achieved between the spontaneously formed sphingomyelin–cholesterol in raft complexes, are not achieved with individual fatty acid types. The association between adjacent phospholipid molecules is entirely dependent on the van der Waals forces that attract their fatty acid components. These effects can be attributed to two primary structural properties of the fatty acid: (1) the length of the acyl chain and (2) the degree of unsaturation. These two structural elements can significantly affect both the relative volume the fatty acid

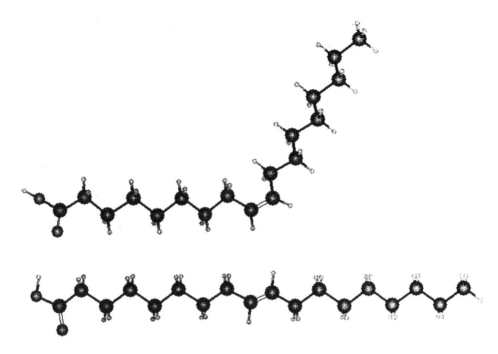

FIGURE 20.4 The *cis* and *trans* configuration of double bonds in fatty acids. Naturally occurring fatty acids are predominantly of the *cis* isomer.

occupies and the distance between a fatty acid and its neighboring acyl chain. The degree of unsaturation is particularly important for membrane properties because van der Waals forces are acutely sensitive to the distance between the interacting acyl chains [50].

Estimates of the spatial widths of fatty acids show significant differences between *cis* and *trans* double bonds. Whereas stearic acid is estimated to be 0.25 nm in diameter, Δ9-*cis*-octadecanoic acid (oleic) and Δ9-*cis*-octadecanoic acids have spatial widths of 0.72 and 0.31 nm, respectively [50]. Additionally, due to the fact that the acyl chain continues on the same side of the double bond, the *cis* configuration imparts a 30° bend in the fatty acid that is not relievable through any rotation of the single-bonded carbon atoms. In contrast, the structure of a *trans* fatty acid is relatively unaffected by the double bond [50]. The *cis* unsaturated bond thus interrupts a succession of London–van der Waals forces between membrane fatty acids by increasing the distance between adjacent fatty acids and lowers the membrane crystallization temperature. The more unsaturated the fatty acid, the less it is able to rotate around its carbon–carbon bonds and, consequently, the more it influences membrane acyl packing. Some calculations suggest that an increase in the number of double bonds provides diminishing returns in terms of the influence of fatty acid on van der Waals forces [99]. It is clear that acyl unsaturation can affect van der Waals interactions through either changes in the length of association or by increasing the distance of acyl separation. In addition, the position of a double bond in a fatty acid also plays a role in the modification of membrane structure. Double bonds at the methyl end of acyl chains are not particularly effective at modulating membrane bulk properties due to the fact that they induce a lesser degree of acyl separation. Interestingly, double bonds are typically first inserted at the Δ9 position, and fatty acids with double bonds present only near the methyl end of the chain are conspicuously absent from nature.

Cells take advantage of the large variety of fatty acid structures to maintain consistent bulk membrane properties in the face of changing temperature and pressure. Clearly, then, PUFAs are critical to cell function if viewed only from the perspective of membrane structure. In light of this fact, it is interesting that animals are largely incapable of producing PUFAs de novo.

2. Alterations in Membrane-Associated Enzyme Activity

Many membrane-associated enzymes are responsive to their fatty acid environment. There is considerable evidence that membrane unsaturated fatty acid content can modify the structure and therefore the functionality of membrane enzymes; and these effects are often cited when dietary fatty acids modulate physiologic functions. The direct action of membrane composition of membrane-associated enzyme activity is often difficult to establish. However, an interesting line of experimentation that involves measuring membrane-associated activities as a function of temperature has provided compelling evidence that the degree of membrane unsaturation is critical to cell function. Membrane-associated enzyme activity [100], molecular transport [101], and the insertion of proteins into membranes [102] all have maximal activity at temperatures just above the transition temperature of the membrane. Although clearly this does not indicate that cells use the changes in membrane unsaturation to modulate enzyme activities, it does provide strong evidence that diet-induced changes in membrane composition could play an important role in the modification of cell physiology.

For all of the difficulties in assigning a causal mechanism to changes in membrane-associated enzyme activities, it is worth noting that the effects themselves are largely incontrovertible. Dietary oils and fats have significant effects on many metabolic activities, including the activity of membrane-associated enzymes. An increasing body of literature has amassed concerning the modulation of membrane-bound enzyme activities in both cell and reconstituted systems. Brenner [99] provides an extensive review of the effects of unsaturated fatty acids enzyme kinetics.

VI. SYNTHESIS AND ABUNDANCE OF PUFA

Despite the considerable variation that is possible in PUFA structure, there are only about 20 unsaturated fatty acids of nutritional importance to humans. These consist primarily of monounsaturated and methylene-interrupted PUFAs in the *cis* configuration. Virtually all of the unsaturated fatty acids consumed in normal diets are members of the n-3, n-6, n-7, or n-9 families of fatty acids. Non-methylene-interrupted and *trans*-configured fatty acids were historically consumed in very small quantities, but the advent of modern food production may have enriched these fatty acids in the food supply. The primary unsaturated fatty acids in human nutrition are reviewed later.

A. N-7 Fatty Acids

1. Palmitoleic Acid (16:1n-7)

Palmitoleic acid is a minor component of both animal and vegetable lipids. Fish oil is particularly enriched in palmitoleic acid, and some seed oils also represent a significant source of the fatty acid. Palmitoleic acid is produced de novo by plants and animals by the $\Delta 9$ desaturation of palmitic acid.

2. Vaccenic Acid (18:1n-7)

Vaccenic acid is a major product of bacterial fatty acid synthesis and is also present at lower concentrations in plant and animal lipids. Vaccenic acid is produced by the elongation of palmitoleic acid. The true content of vaccenic acid in the diet may be underestimated due to difficulty in separating it from its n-9 isomer. In animals, vaccenic acid appears to be concentrated in the mitochondrial lipid cardiolipin [103].

B. N-9 Fatty Acids

1. Oleic Acid (18:1n-9)

Oleic acid is a $\Delta 9$ desaturase product of stearic acid and is produced de novo in plants, animals, and bacteria. Oleic acid is the most common unsaturated fatty acid and is the precursor for the production of most other PUFAs. Plants produce both *n*-3 and *n*-6 PUFAs from oleic acid, and animals can elongate and desaturate oleic acid into a variety of *n*-9 fatty acids. Olive oil is a particularly rich dietary source, and most foods, especially nuts and butter, are rich in oleic acid.

2. Erucic Acid (22:1*n*-9)

Erucic acid is a long-chain monounsaturated fatty acid found in plants, particularly in rapeseeds. It is an elongation product of oleic acid, and is an uncharacteristically long-chain unsaturated fatty acid for plants. Mildly toxic, erucic acid has been bred out of rapeseeds used for food oil production. In animals, dietary erucic acid can be retroconverted to form oleic acid via peroxisomal oxidation.

3. Mead Acid (20:3*n*-9)

Mead acid is a hallmark of essential fatty acid deficiency and has the distinction of being the only major PUFA produced de novo by animals. In the absence of dietary *n*-6 and *n*-3 fatty acids, the Δ6 desaturase converts oleic acid to 18:2*n*-9, which is further elongated and Δ5 desaturated to form mead acid [50]. It has been speculated that mead acid compensates for the loss of *n*-3 and *n*-6 PUFAs by increasing the unsaturation of animal cell membranes.

4. Other *n*-9 Fatty Acids

The family of *n*-9 fatty acids is derived exclusively from the production of oleic acid but can be converted by elongation, desaturation, β-oxidation, and so on. Other rare but naturally occurring *n*-9 fatty acids include 18:2, 20:1, and 22:3 [104].

C. N-6 FATTY ACIDS

1. Linoleic Acid (18:2*n*-6)

Linoleic acid, along with α-linolenic acid, is a primary product of plant PUFA synthesis. Linoleic acid is produced de novo by plants and in particular is enriched in seed oils. Although nature produces linoleic acid at concentrations fairly equitable with those of α-linolenic acid, modern agriculture has greatly enriched linoleic acid in the food supply. Although animals are incapable of producing linoleic acid, livestock are fed diets particularly rich in this fatty acid, and thus humans acquire a large portion of their linoleic acid from meats. Linoleic acid serves as a precursor for the production of the essential fatty acid arachidonic acid, as well as other *n*-6 acyl species.

2. γ-Linolenic Acid (18:3*n*-6)

γ-Linolenic acid (GLA) is produced in animals and lower plants by the Δ6 desaturation of linoleic acid. Natural sources include evening primrose oil, borage oil, and black currant oil, and minute amounts can be found in animal tissue [105]. In animals, dietary linoleic acid is desaturated by the Δ6 desaturase to produce GLA as an intermediate in the production of arachidonic acid. Interestingly, dietary GLA is accumulated in animal tissue largely as its direct elongation product 20:3*n*-6, and not substantially converted to arachidonic acid. There has been a great deal of recent interest in dietary GLA for its antagonistic action on arachidonic acid metabolism.

3. Dihomo-γ-Linolenic Acid (20:3*n*-6)

The elongation product of linoleic acid, dihomo-γ-linolenic acid (DGLA), is a minor component of animal phospholipids. DGLA serves as a precursor to the formation of the essential fatty acid arachidonic acid as well as for the prostaglandin G_1 series. Dietary DGLA does not appear to be rapidly converted to arachidonic acid, and because prostaglandins of the G_1 series have anti-inflammatory properties, DGLA has received attention as a potential therapeutic agent.

4. Arachidonic Acid (20:4*n*-6)

Arachidonic acid is the product of desaturation and elongation of linoleic acid in animals. Arachidonic acid is also produced in quantity in marine algae. Dietary linoleic acid is converted to arachidonic acid in animals by the concerted activity of the Δ6 desaturase, a microsomal elongase and the Δ5 desaturase. Arachidonic acid is referred to as an essential fatty acid for its action as the precursor for the production of eicosanoids. It is present in all tissues and is particularly enriched in phosphatidylcholine, and ether-linked phospholipid membrane pools.

5. Docosatetraenoic Acid (22:4*n*-6)

Docosatetraenoic acid is the direct elongation product of arachidonic acid and is present in minimal amounts in animal tissues. Docosatetraenoic acid is a substrate for peroxisomal retroconversion, resulting in the formation of arachidonic acid [55].

6. Other *n*-6 Fatty Acids

The family of *n*-6 fatty acids is derived exclusively from the production of linoleic acid but can be converted by elongation, desaturation, β-oxidation, and so on. Other rare but naturally occurring *n*-6 fatty acids include 16:2, 20:2, 22:2, 22:3, 24:2, 25:2, 26:2, and 30:4 [104].

D. N-3 Fatty Acids

1. α-Linolenic Acid (18:3*n*-3)

α-Linolenic acid is produced de novo by the Δ12 and Δ15 desaturations of oleic acid in plants. Along with linoleic acid, α-linolenic acid constitutes one of the two primary PUFA products of plant fatty acid biosynthesis. It is primarily present in the leaves of plants but is also a minor component of seed oils. α-Linolenic acid serves as the metabolic precursor for the production of *n*-3 fatty acids in animals. The success of agricultural seed oils has caused a significant shift in the natural balance of linoleic and linolenic acids, and over the last 100 years the average dietary content of α-linolenic acid has declined significantly [94].

2. Eicosapentaenoic Acid (20:5*n*-3)

EPA is produced de novo by marine algae and in animals by the desaturation/elongation of α-linolenic acid. EPA is the primary fatty acid of fish oil (~25%–20% by weight), although it is not produced de novo by fish. It has also been reported that significant EPA production can occur in animals by the β-oxidation chain shortening of DHA [55]. EPA has been investigated extensively for its action as a competitive inhibitor of arachidonic acid metabolism. Although eicosanoids can be produced from EPA, they appear to have either no activity or an activity that opposes arachidonic acid-derived eicosanoids.

3. Docosapentaenoic Acid (22:5*n*-3)

Docosapentaenoic acid is the elongation product of EPA and is present in most marine lipids. Docosapentaenoic acid can be converted to DHA via a three-step process involving a unique Δ6 desaturation in animals (see earlier) [8].

4. Docosahexaenoic Acid (22:6*n*-3)

DHA is produced de novo by marine algae and is a primary component of fish oil (~8%–20% by weight). The production of DHA in animals from linolenic acid occurs via the desaturation/elongation of α-linolenic acid to 24:5*n*-3. This very long-chain unsaturated fatty acid is desaturated by a

Δ6 desaturase (possibly a unique Δ6 desaturase enzyme) and the resulting fatty acid undergoes one cycle of β-oxidation to form DHA [8,9]. Animals appear to have a requirement for DHA for neural function and they rely on its production from *n*-3 precursors by elongation/desaturation cycles or through ingestion of the intact acid [106]. Although the exact role DHA plays in animal physiology is not understood, the great care with which the fatty acid is preserved in certain tissues implies that it may be an essential component of certain cells. Brain and retinal tissues are particularly enriched in DHA.

5. Other *n*-3 Fatty Acids

The family of *n*-3 fatty acids is derived from α-linolenic acid but can be modified by chain elongation, desaturation, β-oxidation, and so on. Naturally occurring but rare *n*-3 fatty acids include 16:3, 16:4, 18:4, 18:5, 20:2, 20:3, 20:4, 21:5, 22:3, 24:3, 24:4, 24:5, 24:6, 26:5, 26:6, 28:7, and 30:5 [104].

E. Unusual and Non-Methylene-Interrupted Fatty Acids

The vast majority of PUFAs contain multiple double bonds in a 1,4-pentadiene structure in which a single methylene carbon is positioned between the two double bonds. More double bonds are added as a direct result of the positional selectivity of the subsequent desaturase enzymes. Hence, from fatty acids containing two double bonds (linoleic acid 9,12 18:2) to those containing six double bonds (DHA; 4,7,10,13,16,19), all exhibit methylene interruption over the entire length of the molecule. However, there are certain naturally occurring fatty acids in which single double bonds exist at a distance. These are the non-methylene-interrupted fatty acids (NMIFAs). Various NMI-FAs have been described, including allenic, conjugated, allylic, enoic, acetylenic, cyclic, branched, hydroxylated, iso, and anti-iso fatty acids [104]. The most common NMIFAs are fatty acids in which one of the double bonds is ostensibly missing from the middle of a double bond system. Conifers were shown to contain up to 20% by weight in their seeds of the NMIFA 5,11,14-eicosatrienoic acid and 5,11,14,17-eicosatetraenoic acid [107]. In most reported studies, the basis of the synthesis of these fatty acids is through an active elongase enzyme that elongates 18–20-carbon fatty acids, bypassing the Δ6 desaturase. Indeed, it has been argued that the methylene-interrupted structure is not due to explicit enzyme specificity but rather the predisposition of substrate fatty acids [108]. Desaturation at the Δ5 position of the elongated fatty acid produces a 20-carbon fatty acid without the Δ8 double bond. This aggressive elongation has been argued to be the basis for the occurrence of small quantities of 5,11,14-eicosatrienoic acid in animal tissues. While most unusual fatty acids are not readily esterified into membrane phospholipids, for certain structures this is not true. 5,11,14-Eicosatrienoic acid, which is the structural analog to arachidonic acid with the absence of the Δ8 double bond, is esterified into several membranes, and even shows preference for incorporation into specific phospholipids including phosphatidylinositol [107,109]. Because the absence of the Δ8 double bond makes it impossible to synthesize prostaglandins or leukotrienes from NMIFA, replacement of arachidonic acid with NMIFA in phospholipids would be predicted to have substantial effects on eicosanoid cellular signaling proportional to the extent of displacement of arachidonic acid. This was shown to be true for isolated tissues [107]. Furthermore, in an animal model of genetic autoimmunity, animals fed diets containing these fatty acids exhibited a net reduced severity of disease consistent with a selective decrease in eicosanoid signaling [110]. Other examples of unusual NMIFA structures with potential biological actions include conjugated linoleic acid that has exhibited potent anticancer properties in animal models of carcinogenesis [111,112]. The true therapeutic value, as well as the potential toxicities, of these fatty acids will await the development of commercial sources in food grade quantities for larger studies. Nevertheless, the potential for modifying the biosynthetic capabilities of crop plants is already exploited for many nutritional and functional targets, and the possibility of producing fatty acids with unusual structures is likely to be limited only by the documented value of producing them.

VII. SUMMARY

The fatty acid nomenclature system based on n designation was developed to describe fatty acids in terms of their nutritional functions. Whereas this approach is of some use, it may lead to confusion over the true basis by which fatty acids modulate physiology. There is mounting evidence that each fatty acid has its own role in nutrition that is not dictated by its n designation. Chemically similar fatty acids often have widely ranging functionalities. This phenomenon is exemplified by the antagonistic relationship between arachidonic acid and EPA metabolism despite the fact that the fatty acids differ only by the additional double bond in EPA. The fundamental relation between members of the same n family is one of interconvertibility. Families are grouped only by their potential to be converted to longer chain members of the same n designation or, in a more practical sense, by their ability to act as precursors for the production of a particular fatty acid of interest. The enzymatic activities required for the conversion of fatty acids to longer, more unsaturated chain members of the same n family are redundant. Consequently, if cells were not capable of recognizing fatty acids based on their n designation and responding by modulating desaturase and elongase activities appropriately, animals would be completely at the mercy of their diets.

The regulation of acyl content appears to be critical to the function of a cell, as mammalian cells expend a substantial amount of energy in maintaining distinct and heterogeneous membrane fatty acid compositions. Cells are capable of preserving these compositional identities even when confronted with phospholipid diffusion and vesicular transport, suggesting that acyl composition is tightly regulated. In addition to overall membrane acyl content, phospholipid acyl composition is rigorously maintained in a positionally specific manner. The complex framework of enzymatic activities that upholds these compositions makes large-scale changes in membrane composition rare and, as a result, the effects specific fatty acids have on membrane structure and physiology are not well understood. However, the extraordinary selectivity of certain tissues for individual PUFAs strongly suggests a specific role for those fatty acids in cell function. For instance, brain and neural tissues are enriched in DHA, whereas adrenal glands have a high content of docosatetraenoic acid. Neither of these PUFAs can be produced de novo in humans and so their accumulation in cell membranes must be mediated by preferential absorption or specific desaturation reactions within the given tissue. PUFA specificity is also prevalent in subcellular organelles. Two examples include the inordinate enrichment of vaccenic acid in mitochondria and the conspicuous accumulation of arachidonic acid in the sn-2 position of nuclear membrane ether-linked phospholipids. Selective incorporation of PUFAs implies that fatty acids have some degree of functionality in the cell. In addition, these functions must be monitored by the cell in order for there to be a regulatory adaptation. How a cell senses its acyl composition is not understood. To date, the best data on adaptive acyl regulation in mammalian cells involve the essential fatty acid deficiency response. Cells deficient in arachidonic acid or its precursor n-6 fatty acids convert de novo-produced oleic acid to n-9 eicosatrienoic (mead) acid by the action of the $\Delta 5$ and $\Delta 6$ desaturases. The conversion mechanism is identical to that of the production of C20 n-6 and n-3 fatty acids. The unusual desaturation of oleic acid is commonly believed to be the result of the attempt of a cell to replace arachidonic acid in the membrane with a similarly unsaturated species. This is an example of a cross-family compensation and provides further evidence for the lack of an association between n designation and function. The regulation of mead acid production is intriguing in that it is not found in animals fed sufficient dietary n-6 fatty acids. This may be the consequence of a differential activity of the $\Delta 6$ desaturase on linoleic and oleic acids.

Given that each fatty acid is unique in terms of its chemistry and potentially its function, one interesting question concerning PUFA metabolism is, can animals recognize n designation? If not, how big of a role do dietary PUFAs play in physiology? How can animals regulate the production of important acyl compounds in the face of changing dietary n-3/n-6 ratios, given that animals have a redundant system of desaturation for both n families? Changing n-3/n-6 PUFA ratios appears to have significant effects on more pathologies in humans, including cardiovascular disease, immune function, and cancer. Because dietary acyl compositions vary tremendously throughout the world and over time, key questions must be answered concerning the regulatory factors and signals associated with the production of fatty acid and their metabolic products.

REFERENCES

1. A. Keys and R.W. Parlin. Serum cholesterol response to changes in dietary lipids. *Am. J. Clin. Nutr.* 19:175–181 (1966).
2. D.M. Hegsted, R.B. McGandy, M.L. Myers, and F.J. Stare. Quantitative effects of dietary fat on serum cholesterol in man. *Am. J. Clin. Nutr.* 17:281–295 (1965).
3. C.P. Caygill, A. Charlett, and M.J. Hill. Fat, fish, fish oil and cancer. *Br. J. Cancer* 74:159–164 (1996).
4. A.G. Goodridge. Fatty acid synthesis in eukaryotes. In: *Biochemistry of Lipids, Lipoproteins and Membranes* (D.E. Vance and J. Vance, eds.). Elsevier Science Publishers B.V., Amsterdam, 1991, pp. 111–139.
5. J.L. Harwood. Recent advances in the biosynthesis of plant fatty acids. *Biochim. Biophys. Acta* 1301:7–56 (1996).
6. S.J. Wakil, J.K. Stoops, and V.C. Joshi. Fatty acid synthesis and its regulation. *Annu. Rev. Biochem.* 52:537–579 (1983).
7. G.A. Thompson. *The Regulation of Membrane Lipid Metabolism.* CRC Press, Boca Raton, FL, 1992.
8. A. Voss, M. Reinhart, S. Sankarappa, and H. Sprecher. The metabolism of 7,10,13,16,19-docosapentae-noic acid to 4,7,10,13,16,19-docosahexaenoic acid in rat liver is independent of a 4-desaturase. *J. Biol. Chem.* 266:19995–20000 (1991).
9. S.A. Moore, E. Hurt, E. Yoder, H. Sprecher, and A.A. Spector. Docosahexaenoic acid synthesis in human skin fibroblasts involves peroxisomal retroconversion of tetracosahexaenoic acid. *J. Lipid Res.* 36:2433–2443 (1995).
10. S.P. Baykousheva, D.L. Luthria, and H. Sprecher. Peroxisomal–microsomal communication in unsaturated fatty acid metabolism. *FEBS Lett.* 367:198–200 (1995).
11. H. Sprecher and Q. Chen. Polyunsaturated fatty acid biosynthesis: A microsomal–per-oxisomal process. *Prostaglandins, Leukot. Essent. Fatty Acids* 60:317–321 (1999).
12. Y. Lindqvist, W. Huang, G. Schneider, and J. Shanklin. Crystal structure of delta9 stearoyl-acyl carrier protein desaturase from castor seed and its relationship to other di-iron proteins. *EMBO J.* 15:4081–4092 (1996).
13. J.M. Ntambi. Regulation of stearoyl-CoA desaturase by polyunsaturated fatty acids and cholesterol. *J. Lipid Res.* 40:1549–1558 (1999).
14. J.M. Ntambi. The regulation of stearoyl-CoA desaturase (SCD). *Prog. Lipid Res.* 34:139–150 (1995).
15. J.M. Ntambi, A.M. Sessler, and T. Takova. A model cell line to study regulation of stearoyl-CoA desaturase gene 1 expression by insulin and polyunsaturated fatty acids. *Biochem. Biophys. Res. Commun.* 220:990–995 (1996).
16. S. Nakashima, Y. Zhao, and Y. Nozawa. Molecular cloning of delta 9 fatty acid desaturase from the protozoan *Tetrahymena thermophila* and its mRNA expression during thermal membrane adaptation. *Biochem. J.* 317:29–34 (1996).
17. P.A. Meesters and G. Eggink. Isolation and characterization of a delta-9 fatty acid desaturase gene from the oleaginous yeast *Cryptococcus curvatus* CBS 570. *Yeast* 12:723–730 (1996).
18. S. Gargano, G. Di Lallo, G.S. Kobayashi, and B. Maresca. A temperature-sensitive strain of *Histoplasma capsulatum* has an altered delta 9-fatty acid desaturase gene. *Lipids* 30:899–906 (1995).
19. N. Murata and H. Wada. Acyl-lipid desaturases and their importance in the tolerance and acclimatization to cold of cyanobacteria. *Biochem. J.* 308:1–8 (1995).
20. M. Fukuchi-Mizutani, K. Savin, E. Cornish, et al. Senescence-induced expression of a homologue of delta 9 desaturase in rose petals. *Plant Mol. Biol.* 29:627–635 (1995).
21. P. Legrand, D. Catheline, M.C. Fichot, and P. Lemarchal. Inhibiting delta9-desaturase activity impairs triacylglycerol secretion in cultured chicken hepatocytes. *J. Nutr.* 127:249–256 (1997).
22. B.H. Jones, M.A. Maher, W.J. Banz, et al. Adipose tissue stearoyl-CoA desaturase mRNA is increased by obesity and decreased by polyunsaturated fatty acids. *Am. J. Physiol.* 271 (1 Pt 1):E44–E49 (1996).
23. A.M. Sessler, N. Kaur, J.P. Palta, and J.M. Ntambi. Regulation of stearoyl-CoA desaturase 1 mRNA stability by polyunsaturated fatty acids in 3T3-L1 adipocytes. *J. Biol. Chem.* 271:29854–29858 (1996).

24. K.N. Lee, M.W. Pariza, and J.M. Ntambi. Differential expression of hepatic stearoyl-CoA desaturase gene 1 in male and female mice. *Biochim. Biophys. Acta* 1304:85–88 (1996).

25. J.D. Murray, A.M. Oberbauer, K.R. Sharp, and J.B. German. Expression of an ovine growth hormone transgene in mice increases arachidonic acid in cellular membranes. *Transgenic Res.* 3:241–248 (1994).

26. Z. Gombos, H. Wada, Z. Varkonyi, D.A. Los, and N. Murata. Characterization of the Fad12 mutant of *Synechocystis* that is defective in delta 12 acyl-lipid desaturase activity. *Biochim. Biophys. Acta* 1299:117–123 (1996).

27. Y. Tasaka, Z. Gombos, Y. Nishiyama, et al. Targeted mutagenesis of acyl-lipid desaturases in *Synechocystis*: Evidence for the important roles of polyunsaturated membrane lipids in growth, respiration and photosynthesis. *EMBO J.* 15:6416–6425 (1996).

28. P.S. Covello and D.W. Reed. Functional expression of the extraplastidial *Arabidopsis thaliana* oleate desaturase gene (FAD2) in *Saccharomyces cerevisiae. Plant Physiol.* 111:223–226 (1996).

29. M.H. Avelange-Macherel, D. Macherel, H. Wada, and N. Murata. Site-directed mutagenesis of histidine residues in the delta 12 acyl-lipid desaturase of *Synechocystis. FEBS Lett.* 361:111–114 (1995).

30. T. Nichiuchi, T. Nakamura, T. Abe, H. Kodama, M. Nishimura, and K. Iba. Tissue-specific and light-responsive regulation of the promoter region of the *Arabidopsis thaliana* chloroplast omega-3 fatty acid desaturase gene (FAD7). *Plant Mol. Biol.* 29:599–609 (1995).

31. T. Hamada, T. Nishiuchi, H. Kodama, M. Nishimura, and K. Iba. cDNA cloning of a wounding-inducible gene encoding a plastid omega-3 fatty acid desaturase from tobacco. *Plant Cell Physiol.* 37:606–611 (1996).

32. H.P. Cho, M. Nakamura, and S.D. Clarke. Cloning, expression, and fatty acid regulation of the human delta-5 desaturase. *J. Biol. Chem.* 274:37335–37339 (1999).

33. H.P. Cho, M.T. Nakamura, and S.D. Clarke. Cloning, expression, and nutritional regulation of the mammalian delta-6 desaturase. *J. Biol. Chem.* 274:471–477 (1999).

34. O. Sayanova, M.A. Smith, P. Lapinskas, et al. Expression of a borage desaturase cDNA containing an N-terminal cytochrome b5 domain results in the accumulation of high levels of delta6-desaturated fatty acids in transgenic tobacco. *Proc. Natl. Acad. Sci. U.S.A.* 94:4211–4216 (1997).

35. R.R. Brenner. Endocrine control of fatty acid desaturation. *Biochem. Soc. Trans.* 18:773–775 (1990).

36. Y.S. Huang, D.E. Mills, R.P. Ward, V.A. Simmons, and D.F. Horrobin. Stress modulates cholesterol-induced changes in plasma and liver fatty acid composition in rats fed *n*-6 fatty acid-rich oils. *Proc. Soc. Exp. Biol. Med.* 195:136–141 (1990).

37. T. Nakada, I.L. Kwee, and W.G. Ellis. Membrane fatty acid composition shows delta-6-desaturase abnormalities in Alzheimer's disease. *Neuroreport* 1:153–155 (1990).

38. H. Sprecher. Enzyme activities affecting tissue lipid fatty acid composition. *World Rev. Nutr. Diet.* 66:166–176 (1991).

39. E. Wodtke and A.R. Cossins. Rapid cold-induced changes of membrane order and delta 9-desaturase activity in endoplasmic reticulum of carp liver: A time-course study of thermal acclimation. *Biochim. Biophys. Acta* 1064:343–350 (1991).

40. N. Oshino and R. Sato. The dietary control of the microsomal stearyl CoA desaturation enzyme system in rat liver. *Arch. Biochem. Biophys.* 149:369–377 (1972).

41. D.E. Tabor, J.B. Kim, B.M. Spiegelman, and P.A. Edwards. Transcriptional activation of the stearoyl-CoA desaturase 2 gene by sterol regulatory element-binding protein adipocyte determination and differentiation factor 1. *J. Biol. Chem.* 273:22052–22058 (1998).

42. M.L. Garg, A.A. Wierzbicki, A.B. Thomson, and M.T. Clandinin. Dietary cholesterol and/or *n*-3 fatty acid modulate delta 9-desaturase activity in rat liver microsomes. *Biochim. Biophys. Acta* 962:330–336 (1988).

43. R. Brenner. Factors influencing fatty acid chain elongation and desaturation. In: *The Role of Fats in Human Nutrition* (A.J. Vergroesen and M. Crawford, eds.). Academic Press, London, 1989.

44. A.B. Steffens, J.H. Strubbe, B. Balkan, and A.J. Scheurink. Neuroendocrine factors regulating blood glucose, plasma FFA and insulin in the development of obesity. *Brain Res. Bull.* 27:505–510 (1991).

45. I.N.T. de Gomez Dumm, M.J.T. de Alaniz, and R.R. Brenner. Effect of insulin on the oxidative desaturation of fatty acids in non-diabetic rats and in isolated liver cells. *Acta Physiol. Pharmacol. Latinoam.* 35:327–335 (1985).

46. I.N.T. de Gomez Dumm, M.J.T. de Alaniz, and R.R. Brenner. Effects of glucagon and dibutyryl 3'5' cyclic monophosphate on oxidative desaturation of fatty acids in the rat. *J. Lipid Res.* 16:264–268 (1975).

47. M.J.T. Alaniz, I.N.T. de Gomez Dumm, and R.R. Brenner. The action of insulin and dibutyryl cyclic AMP on the biosynthesis of polyunsaturated acids of alpha linolenic acid family in the cells. *Mol. Cell. Biochem.* 12:3–8 (1976).

48. I.N.T. de Gomez Dumm, M J.T. de Alaniz, and R.R. Brenner. Effect of thyroxine on delta 6 and delta 9 desaturation activity. *Adv. Exp. Med. Biol.* 83:609–616 (1977).

49. M.T. Nakamura, S.D. Phinney, A.B. Tang, A.M. Oberbauer, J.B. German, and J.D. Murray. Increased hepatic delta 6-desaturase activity with growth hormone expression in the MG101 transgenic mouse. *Lipids* 31:139–143 (1996).

50. H.W. Cook. Fatty acid desaturation and chain elongation in eukaryotes. In: *Biochemistry of Lipids, Lipoproteins and Membranes* (D.E. Vance and J. Vance, eds.). Elsevier Science Publishers B.V., Amsterdam, 1991, pp. 141–169.

51. M. Alegret, E. Cerqueda, R. Ferrando, et al. Selective modification of rat hepatic microsomal fatty acid chain elongation and desaturation by fibrates: Relationship with peroxisome proliferation. *Br. J. Pharm.* 114:1351–1358 (1995).

52. H. Sprecher and S. Baykousheva. The role played by beta-oxidation in unsaturated fatty acid biosynthesis. *World Rev. Nutr. Diet.* 75:26–29 (1994).

53. I. Marzo, M.A. Alava, A. Pineiro, and J. Naval. Biosynthesis of docosahexaenoic acid in human cells: Evidence that two different delta 6-desaturase activities may exist. *Biochim. Biophys. Acta* 1301:263–272 (1996).

54. A. Voss, M. Reinhart, and H. Sprecher. Differences in the interconversion between 20- and 22-carbon (n-3) and (n-6) polyunsaturated fatty acids in rat liver. *Biochim. Biophys. Acta* 1127:33–40 (1992).

55. T.A. Hagve and B.O. Christophersen. Evidence for peroxisomal retroconversion of adrenic acid (22:4(n-6)) and docosahexaenoic acids (22:6(n-3)) in isolated liver cells. *Biochim. Biophys. Acta* 875:165–173 (1986).

56. E. Christensen, T.A. Hagve, and B.O. Christophersen. The Zellweger syndrome: Deficient chain-shortening of erucic acid (22:1(n-9)) and adrenic acid (22:4(n-6)) in cultured skin fibroblasts. *Biochim. Biophys. Acta* 959:134–142 (1988).

57. E. Christensen, B. Woldseth, T.A. Hagve, et al. Peroxisomal beta-oxidation of polyunsaturated long chain fatty acids in human fibroblasts. The polyunsaturated and the saturated long chain fatty acids are retroconverted by the same acyl-CoA oxidase. *Scand. J. Clin. Lab. Invest. Suppl.* 215:61–74 (1993).

58. S.M. Watkins, T.Y. Lin, R.M. Davis, et al. Unique phospholipid metabolism in mouse heart in response to dietary docosahexaenoic or a-linolenic acid. *Lipids* 36:247–254 (2001).

59. M. Gronn, E. Christensen, T.A. Hagve, and B.O. Christophersen. The Zellweger syndrome: Deficient conversion of docosahexaenoic acid (22:6(n-3)) to eicosapentaenoic acid (20:5(n-3)) and normal delta 4-desaturase activity in cultured skin fibroblasts. *Biochim. Biophys. Acta* 1044:249–254 (1990).

60. E.E. Hill and W.E.M. Lands. *Lipid Metabolism* (S.J. Wakil, ed.). Academic Press, New York, 1970, p. 185.

61. M. Waite. Phospholipases. In: *Biochemistry of Lipids, Lipoproteins and Membranes* (D.E. Vance and J. Vance, eds.). Elsevier Science Publishers B.V., Amsterdam, 1991, pp. 269–295.

62. K. Waku. Origins and fates of fatty acyl-CoA esters. *Biochim. Biophys. Acta* 1124:101–111 (1992).

63. D.N. Brindley. Metabolism of triacylglycerols. In: *Biochemistry of Lipids, Lipoproteins and Membranes* (D.E. Vance and J. Vance, eds.). Elsevier Science Publishers B.V., Amsterdam, 1991, pp. 171–201.

64. J.L. Harwood. Lipid metabolism. In: *The Lipid Handbook* (F.D. Gunstone, J.L. Harwood, and F.B. Padley, eds.). Chapman & Hall, London, 1994, pp. 605–633.

65. F.H. Chilton, M. Cluzel, and M. Triggiani. Recent advances in our understanding of the biochemical interactions between platelet-activating factor and arachidonic acid. *Lipids* 26:1021–1027 (1991).

66. F.H. Chilton, A.N. Fonteh, C.M. Sung, et al. Inhibitors of CoA-independent transacylase block the movement of arachidonate into 1-ether-linked phospholipids of human neutrophils. *Biochemistry* 34:5403–5410 (1995).

67. S. Yamaoka, R. Urade, and M. Kito. Mitochondrial function in rats is affected by modification of membrane phospholipids with dietary sardine oil. *J. Nutr.* 118:290–296 (1988).

68. A. Berger, M.E. Gershwin, and J.B. German. Effects of various dietary fats on cardiolipin acyl composition during ontogeny of mice. *Lipids* 27:605–612 (1992).

69. A.A. Spector and M.A. Yorek. Membrane lipid composition and cellular function. *J. Lipid Res.* 26:1015–1035 (1985).

70. S.L. Neidelmen and J. Hunter-Cervera. Wax ester production by *Acinetobacter* sp. In: *Industrial Applications of Single Cell Oils* (D.J. Kyle and C. Ratledge, eds.). American Oil Chemists' Society Press, Champaign, IL, 1992, pp. 16–25.

71. K. Yazawa, K. Watanabe, C. Ishikawa, K. Kondo, and S. Kimura. Production of eicosapentaenoic acid from marine bacteria. In: *Industrial Applications of Single Cell Oils* (D.J. Kyle and C. Ratledge, eds.). American Oil Chemists' Society Press, Champaign, IL, 1992, pp. 29–50.

72. K.D.B. Boswell, R. Gladue, B. Prima, and D.J. Kyle. Production by fermentative microalgae. In: *Industrial Applications of Single Cell Oils* (D.J. Kyle and C. Ratledge, eds.). American Oil Chemists' Society Press, Champaign, IL, 1992, pp. 274–286.

73. C. Ratledge. Microbial routes to lipids. *Biochem. Soc. Trans.* 17:1139–1141 (1989).

74. C. Ratledge. Microbial lipids: Commercial realities or academic curiosities. In: *Industrial Applications of Single Cell Oils* (D.J. Kyle and C. Ratledge, eds.). American Oil Chemists' Society Press, Champaign, IL, 1992, pp. 287–300.

75. D.J. Kyle, V.J. Sicotte, J.J. Singer, and S.E. Reeb. Bioproduction of DHA by microalgae. In: *Industrial Applications of Single Cell Oils* (D.J. Kyle and C. Ratledge, eds.). American Oil Chemists' Society Press, Champaign, IL, 1992, pp. 287–300.

76. J.E. Vanderveen and W.H. Glinsmann. Fat substitutes. *Annu. Rev. Nutr.* 2:473–487 (1992).

77. C.S. Setser and W.L. Racette. Macromolecular replacers in food products. *Crit. Rev. Food Sci. Nutr.* 32:275–297 (1992).

78. H. Yamada, S. Shumizu, Y. Shinmen, K. Akimoto, H. Kawashima, and S. Jareonkitmongokol. Industrial applications of single cell oils. In: *Industrial Applications of Single Cell Oils* (D.J. Kyle and C. Ratledge, eds.). American Oil Chemists' Society Press, Champaign, IL, 1992, pp. 118–137.

79. G.C. Smith and Z.L. Carpenter. Eating quality of animal products and their fat content. In: *Proceedings of the Symposium on Changing the Fat Content and Composition of Animal Products*. National Academy of Science, Washington, D.C., 1974, pp. 147–182.

80. J.A. Dudek and E.R. Elkins. Effects of cooking on the fatty acid profiles of selected seafoods. In: *Health Effects of Polyunsaturated Fatty Acids in Seafoods* (A. Simopoulos, ed.). Academic Press, Orlando, FL, 1986, pp. 431–452.

81. E.T. Moran, E. Larmond, and J. Sommers. *Poult. Sci.* 52:1942–1946 (1973).

82. C.E. Allen and E.A. Foegeding. Some lipid characteristics and interactions in muscle foods—A review. *Food Tech.* 35:253–259 (1981).

83. G.J. Miller, R.A. Field, M.L. Riley, and J.C. Williams. Lipids in wild ruminant animals and steers. *J. Food Qual.* 9:331–335 (1986).

84. M. Enser. Pig adipose tissue consistency. In: *LIPIDFORM Symposium* (R. Marcuse, ed.). Oslo, 1987.

85. R. Jeffcoat and A.T. James. The regulation of desaturation of fatty acids in mammals. In: *Fatty Acid Metabolism and Its Regulation* (S. Numa, ed.). Elsevier Applied Science, Amsterdam, 1984.

86. R.J. Henderson and D.R. Tocher. The lipid composition and biochemistry of freshwater fish. *Prog. Lipid Res.* 26:281–347 (1987).

87. G.A. Thompson. *The Regulation of Membrane Lipid Metabolism.* CRC Press, Boca Raton, FL, 1980.

88. M.A. Sheridan. Lipid dynamics. *Fish Comp. Biochem. Physiol.* 90B:679–690 (1988).

89. T. Watanabe. Lipid nutrition. *Fish Comp. Biochem. Physiol.* 73B:3–15 (1982).

90. D.H. Greene and D.P. Selivonchek. Lipid metabolism. *Fish Prog. Lipid. Res.* 26:53–87 (1987).

91. J. Bitman. *Fat Content and Composition of Animal Products.* National Research Council, National Academy of Science, Washington, D.C., 1976, pp. 200–221.

92. O. Adam. Linoleic and linolenic acids intake. In: *Dietary n3 and n6 Fatty Acids: Biological Effects and Nutritional Essentiality* (C. Galli and A.P. Simopoulos, eds.). Plenum Press, New York, 1989, pp. 33–42.

93. K. Bloch. Early studies on the biosynthesis of polyunsaturated fatty acids. In: *Dietary n3 and n6 Fatty Acids: Biological Effects and Nutritional Essentiality* (C. Galli and A.P. Simopoulos, eds.). Plenum Press, New York, 1989, pp. 1–4.

94. M.A. Crawford, W. Doyle, P. Drury, et al. The food chain for *n*-6 and *n*-3 fatty acids with special reference to animal products. In: *Dietary n3 and n6 Fatty Acids: Biological Effects and Nutritional Essentiality* (C. Galli and A.P. Simopoulos, eds.). Plenum Press, New York, 1989, pp. 5–20.

95. H.O. Bang and J. Dyerberg. Fatty acid pattern and ischaemic heart disease [letter]. *Lancet* 1:633 (1987).

96. P. Bjerregaard and J. Dyerberg. Fish oil and ischaemic heart disease in Greenland [letter]. *Lancet* 2:514 (1988).

97. R.A. Karmali. *n*-3 Fatty acids and cancer. *J. Intern. Med. Suppl.* 225:197–200 (1989).
98. F.H. Chilton, A.N. Fonteh, M.E. Surette, M. Triggiani, and J.D. Winkler. Control of arachidonate levels within inflammatory cells. *Biochim. Biophys. Acta* 1299:1–15 (1996).
99. R.R. Brenner. Effect of unsaturated acids on membrane structure and enzyme kinetics. *Prog. Lipid Res.* 23:69–96 (1984).
100. J.C. Wilschut, J. Regts, H. Westenberg, and G. Scherphof. Action of phospholipases A2 on phosphatidylcholine bilayers. Effects of the phase transition, bilayer curvature and structural defects. *Biochim. Biophys. Acta* 508:185–196 (1978).
101. T.Y. Tsong, M. Greenberg, and M.I. Kanehisa. Anesthetic action of membrane lipids. *Biochemistry* 16:3115–3121 (1977).
102. W.T. Wickner. Role of hydrophobic forces in membrane protein asymmetry. *Biochemistry* 16:254–258 (1977).
103. R.L. Wolff, N.A. Combe, and B. Entressangles. Positional distribution of fatty acids in cardiolipin of mitochondria from 21-day-old rats. *Lipids* 20:908–914 (1985).
104. F.D. Gunstone. Fatty acid structure. In: *The Lipid Handbook*. Chapman & Hall, London, 1994, pp. 1–19.
105. F.B. Padley, F.D. Gunstone, and J.L. Harwood. Occurrence and characteristics of oils and fats. In: *The Lipid Handbook*. Chapman & Hall, London, 1994, pp. 47–223.
106. M.A. Crawford. The role of essential fatty acids in neural development: Implications for perinatal nutrition. *Am. J. Clin. Nutr.* 57:703S–709S; discussion 709S–710S (1993).
107. A. Berger, R. Fenz, and J.B. German. Incorporation of dietary 5,11,14-eicosatrienoate into various mouse phospholipid classes and tissues. *J. Nutr. Biochem.* 4:409–420 (1993).
108. Z. Cohen, D. Shiran, I. Khozin, and Y.M. Heimer. Fatty acid unsaturated in the red alga *Porphyridium cruentum*: Is the methylene interrupted nature of polyunsaturated fatty acids an intrinsic property of the desaturases? *Biochim. Biophys. Acta* 1344:59–64 (1997).
109. A. Berger, and J.B. German. Extensive incorporation of dietary delta-5,11,14 eicosatrienoate into the phosphatidylinositol pool. *Biochim. Biophys. Acta* 1085:371–376 (1991).
110. L.T. Lai, M. Naiki, S.H. Yoshida, J.B. German, and M.E. Gershwin. Dietary *Platycladus orientalis* seed oil suppresses anti-erythrocyte autoantibodies and prolongs survival of NZB mice. *Clin. Immunol. Immunopathol.* 71:293–302 (1994).
111. P.W. Parodi. Conjugated linoleic acid: An anticarcinogenic fatty acid present in milk fat. *Aust. J. Dairy Tech.* 49:93–97 (1994).
112. C. Ip, J.A. Scimeca, and H.J. Thompson. Conjugated linoleic acid. A powerful anticarcinogen from animal fat sources. *Cancer* 74:1050–1054 (1994).

21 Dietary Fats, Eicosanoids, and the Immune System

David M. Klurfeld

CONTENTS

I. EICOSANOIDS

There is a complex relationship among dietary fats, eicosanoids, and the immune system. Because long chain polyunsaturated fatty acids are the precursors for eicosanoids, these dietary components have the potential to modify levels of the products in the body. This is especially true if there is heavy reliance on a single fat in the diet. While this is the standard approach in nutrition studies in animals, it is also true that in some countries a single fat source provides as much as two-thirds of the population's total fat intake. In controlled feeding trials in humans, both the type and the amount of fat have been varied to permit an examination of the effects on eicosanoid production or changes in immune status.

Eicosanoids are only one of many possible mediators through which diet can influence the immune response. Eicosanoids are a large group of cyclized derivatives of the EFAs that have potent biological activities and always contain 20 carbon atoms. These compounds usually have very short half-lives (measured in seconds) and are derived from the precursor fatty acids via a series of enzymatic steps (Figure 21.1). In addition to the prostaglandins and leukotrienes, a variety of derivatives of 20 carbon fatty acids are produced. These include hydroxy and hydroperoxy fatty acids as well as hydroxylated or epoxydized derivatives. Cyclooxygenases add two oxygen molecules, lipoxygenases add a single oxygen molecule, and cytochrome P450s add one atom of oxygen to the fatty acid.

The principal characteristic of the essential fatty acids (EFAs) is the presence of two or more cis double bonds in the families of $\omega3$ or $\omega6$ fatty acids, which must be derived from the diet. Fatty acids containing trans double bonds do not have EFA activity and may increase the requirement for EFA. Endogenously synthesized unsaturated fatty acids are of the $\omega9$ series and cannot be converted to the EFA precursors. EFA-deficient animals have an increase in 20:3 $\omega9$ (derived from oleic acid)

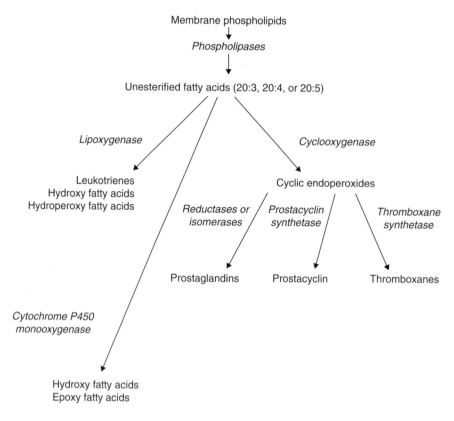

FIGURE 21.1 Pathways of metabolic fate of eicosanoic acids.

in tissue, and the triene/tetraene ratio in plasma lipids is often used clinically for diagnosis of this condition. This ratio essentially reflects the relative proportions of 20:3 ω9 to 20:4 ω6. Individuals with fat malabsorption (especially cystic fibrosis patients) or those on long-term parenteral nutrition are most likely to display signs of EFA deficiency. Diet-derived fatty acids are transported from the intestine to all tissues of the body, first via chylomicrons and then via other lipoprotein classes, usually becoming incorporated into the structurally important phospholipids of cell membranes.

Phospholipids of plasma membranes from mammals tend to be relatively enriched in long chain polyunsaturated fatty acids. Some of these serve as the precursors for synthesis of eicosanoids, principally arachidonic acid (20:4, ω6), eicosapentaenoic acid (EPA, 20:5 ω3), and dihomo-γ-linolenic acid (20:3, ω6). Although α-linolenic acid (18:3 ω3) is the first fatty acid of the ω3 series, the biological effects of the 20 carbon fatty acids are usually stronger, presumably because of their direct steric competition with arachidonic acid as substrates for the enzymes of further metabolism. Most, if not all, terrestrial animals have limited ability to synthesize EPA from linolenic acid except in tissues of the central nervous system. In contrast, fish are metabolically capable of these steps (they convert the 18:3 in plankton to 20:5), so they are a rich source. Another fatty acid concentrated in fish oil is docosahexaenoic acid (DHA, 22:6, ω3), which can potently inhibit cyclooxygenase. Thus, fish oil feeding has become an important paradigm in the study of the effects of dietary fat on eicosanoid metabolism and immune responses. It should be remembered that fish oils contain highly variable amounts of ω3 fatty acids, and many fish oils are high in saturated fatty acids [1].

Interest in eicosanoids, dietary fat, and the immune system is high because eicosanoid metabolism probably is one of the major mechanisms by which dietary fat modulates a variety of immune responses, although undoubtedly other mechanisms exist. Both the concentration of precursor fatty

TABLE 21.1

Relationship of Precursor Fatty Acids to Eicosanoid Products[a]

Fatty Acid	Prostaglandins		Thromboxanes	Leukotrienes
	E Series	F Series		
Dihomo-γ-linolenic acid, all-*cis*-8,11,14 C20:3 (ω6)	E_1	$F_{1\alpha}$	A_1	$A_3 C_3 D_3$
Arachidonic acid[b], all-*cis*-5,8,11,14 C20:4 (ω6)	E_2	$F_{2\alpha}$	A_2	$A_4 B_4 C_4 D_4 E_4$
Eicosapentaenoic aicd, all-*cis*-5,8,11,14,17 C20:5 (ω3)	E_3	$F_{3\alpha}$	A_3	$A_5 B_5 C_5$

[a] Subscripts indicate the number of double bonds.
[b] Can also be converted to prostacyclin (PGI_2).

acid in the phospholipids of cell membranes and their rate of metabolism are controlling factors in the amount of eicosanoids released by cells. Competition between ω3 and ω6 fatty acids for lipoxygenases and cyclooxygenases results in production of eicosanoids of different types and in varying amounts. In the prostaglandin family there are two major series, designated E and F. The E series has a ketone and hydroxyl group added at C-9 and C-11, while the F series has hydroxyl groups at both positions, eliminating the double bonds from these positions (Table 21.1). In addition to the letter designation, there are subseries indicated by numbers. Those prostaglandins derived from 20:3 are designated E_1 or $F_{1\alpha}$; those from 20:4 are E_2 or $F_{2\alpha}$; and those derived from 20:5 are E_3 or $F_{3\alpha}$. Therefore, the subscripts 1, 2, and 3 refer to the number of double bonds remaining in a prostaglandin molecule.

The most common method for quantitating eicosanoids is with an immunoassay. These methods depend on an antibody that recognizes the eicosanoids and does not cross-react with related compounds. Problems associated with the measurement of eicosanoids are the short half-life in tissues and body fluids as well as the cross-reactivity seen between some prostaglandins. For example, few methods can distinguish between prostaglandins E_1 and E_2, so they are often simply expressed as PGE. The short half-life of these compounds necessitates rapid isolation and chilling of samples, addition of a metabolic inhibitor, or all three. There is little doubt that many of the studies reporting levels of eicosanoids did not use conditions rigorous enough to ensure the accuracy of the analytical data. Moreover, since instability is a characteristic of some eicosanoids, stable analogs may be measured in place of the compound of interest.

The leukotrienes are not cyclized and, as the name implies, contain three double bonds in the acyl chain. Arachidonic acid is metabolized by 5-lipoxygenase to leukotriene A_4. This eicosanoid can be metabolized by an epoxide hydrolase to leukotriene B_4 or via glutathione *S*-transferase to leukotriene C_4. The latter reaction results in the addition to the fatty acid of cysteine, glycine, and glutamic acid. Leukotriene C_4 can then be metabolized to D_4 by γ-glutamyltransferase, which removes the glutamic acid. Finally, D_4 can be metabolized to leukotriene E_4 by cysteinyl-glycine dipeptidase, which removes the glycine and leaves only the cysteine residue on the acyl chain. These compounds are critical mediators of a variety of inflammatory responses (e.g., anaphylaxis, increased vascular permeability, attraction of leukocytes and their activation).

Other eicosanoid products include thromboxanes, which are synthesized by platelets and cause platelet aggregation and vascular constriction. Finally, prostacyclin (PGI_2) is produced by blood vessels and inhibits platelet aggregation. These categories of eicosanoids do not appear to participate in regulation of immune responses; rather, they are altered in response to the dietary manipulation that affects the prostaglandins and leukotrienes.

Complicating interpretation of the role of eicosanoids in modulating the immune response is the discovery that eicosanoids are produced by a wide range of eukaryotic organisms including fungi,

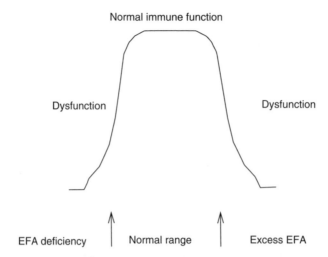

FIGURE 21.2 Immune responses as a function of dietary essential fatty acid. Note the broad and imprecise range of EFA adequacy for immune function; in particular, the upper limit of "normal" EFA varies for different immune functions.

protozoa, and helminthes [2]. Thus, some pathogens may secrete eicosanoids that directly modulate the immune response to infections characterized by chronicity and hypersensitivity.

Currently, there is a considerable debate about the quantitative requirements for ω3 fatty acids; it is fairly well established that the EFA requirement of both rodents and humans is on the order of 0.5% of energy. The EFA requirement is satisfied primarily by intake of linoleic acid (there is relatively little arachidonic acid in commonly consumed foods) plus consumption of a smaller amount of linolenic acid or its fatty acid metabolites, which are concentrated in fish oils. Alterations in some immune functions can be seen with both low and high intake of EFA. This can be conceptualized as a curve with a flattened top (Figure 21.2), in which the normal range of immune function is seen over a wide range of adequate EFA intake, whereas decreased immune responses are seen with either deficient or excessive intakes. The presumptive primary mechanisms are changes in the membrane microenvironment and/or precursors of the eicosanoids, although a number of other possible links have been suggested.

II. IMMUNE SYSTEM

A brief review of the components of the immune system will facilitate an understanding of the effects of dietary fat on the immune response. The immune system is the most dispersed network of cells in the body that are not in physical contact yet cooperate functionally. It is composed of both fixed and mobile cells that protect against invading organisms and the development of abnormal (malignant) cells. If, however, it overreacts to self, the result is autoimmunity; if it overreacts to nonself, there is an allergic response. Some of these immune cells have very short lives but others, which form the basis of immunological memory, exist for many years and perhaps for the lifetime of the host. Fixed cells of the reticuloendothelial system (RES) are found primarily in the liver, spleen, bone marrow, thymus, lymph nodes, lungs, and intestines. Mobile cells are found throughout the circulation in high numbers, which increase quickly in response to the presence of foreign antigens. It is estimated that 80% of mobile leukocytes are normally sequestered in organs of the RES, while the remainder float in the bloodstream.

Functionally, the components of the immune system are often divided into response elements. That is, humoral (or circulating) and cell-mediated immunity are the most commonly used concepts for describing immune functions, although there is communication between these facets of immune cells via a variety of soluble products called cytokines. In addition to the distinctions of humoral and cell-mediated immunity, a number of other components of the immune system play a role in protecting against pathogenic organisms. These include the polymorphonuclear leukocytes, which are divided into neutrophils, basophils, and eosinophils, based on their staining characteristics in a smear of blood treated with Giemsa or Wright's stain. Neutrophils participate in acute inflammatory reactions against bacteria and other foreign bodies. Basophils and eosinophils are components of allergic reactions. Eosinophils are elevated also in response to parasitic infections.

There is a wide range of normal immune functions and a wide range of normal numbers of circulating immune cells. For example, the normal range of leukocytes in the blood of humans is 5000 to 10,000/mm^3. Within this range there are no discernible differences in susceptibility to infection. In fact, there is little increased risk of infection unless total white blood cells drop below 1000/mm^3.

Functional terms used to describe the immune system include natural (innate) and specific (acquired). Natural immunity is a result of molecules such as cytokines or complement, or a result of the function of natural killer (NK) cells or phagocytes, or barriers such as mucosal surfaces. The natural immune system is able to attack a substance to which it has had no previous exposure; subsequent exposure to that stimulus does not increase the response. On the other hand, the specific immune response recognizes a specific feature (antigen) of the foreign substance or cell and generates an immunologic memory, which amplifies responses during subsequent exposures. The specific immune system can recruit portions of the innate system to function simultaneously. Lymphocytes are the primary effectors of the specific immune response. These cells "direct" much of the immunologic activity by the production of cytokines of many types.

Cytokines are soluble protein molecules that function as messengers to other leukocytes. They are produced and secreted primarily by monocytes, and their main targets are lymphocytes and other monocytes. Some of the important cytokines include interleukins, interferons, and tumor necrosis factors (TNFs). There are at least 15 distinct forms of interleukins, 3 interferons, and 2 tumor necrosis factors.

A. HUMORAL IMMUNITY

Humoral immunity is a result of the circulation in the plasma of antibodies and the activity of B lymphocytes. The B stands for bursa of Fabricius in birds, which is the immunological equivalent of the bone marrow in mammals. Humoral immunity characterizes the production of antibodies by the interaction of B lymphocytes with plasma cells. Circulating antibodies are the products of these cells, but they cannot be produced without interaction with T, or thymus-derived, cells. In addition to antibody production, the complement system is part of the immune response. Complement is a series of proteins in the plasma that participate in antigen-antibody reactions, displaying a wide range of actions ranging from aiding in phagocytosis by leukocytes to killing of tumor cells. Complement serves to stimulate various functions of the leukocytes.

B. CELL-MEDIATED IMMUNITY

Cell-mediated immunity is based on the interactions of T lymphocytes and monocytic cells. T cells have been divided into many subclasses based primarily on cell surface receptors, which vary in relation to the functions these cells perform. The generally accepted classification of T cells is the CD (cluster designation or cluster of differentiation) scheme, which depends diagnostically on antibodies to distinguish these cellular sets of lymphocytes. Therefore, T cells can be classified as CD4$^+$ and CD8$^+$, but these cells are often referred to by their effector functions as helper and suppressor cells, respectively. In addition, there are NK cells, LAK (lymphokine-activated killer) cells, and lymphocytes responsible for delayed-type hypersensitivity.

The monocytes and their derived macrophages are cells that protect against certain bacteria and tumor cells. Monocytes circulate in the blood, and once they have migrated into tissues they usually become fixed in place and are called macrophages. There are macrophages normally lining venules in the liver (Kupffer cells), spleen, and lymph nodes; these permanently fixed cells are termed histiocytes. The relative proportions of leukocytes in the blood differ across species. While the predominant circulating cells in rodents are lymphocytes, granulocytes are the most common in humans. This difference is relevant in comparing the experimental work done predominantly in rodent species with studies in humans. Macrophages are one of the most important sources of eicosanoids in the body, playing a major role in a variety of inflammatory and immunologic responses. Membranes of these cells are particularly enriched with arachidonic acid. Different inflammatory stimuli have the ability to induce secretion of eicosanoids of varied forms. Lymphocytes do not seem able to produce eicosanoids but can release arachidonic acid and certainly respond to the effects of these mediators. Many research studies describe use of mononuclear cells—these are mixed populations of lymphocytes and monocytes, usually obtained by differential centrifugation of blood or other biological fluids.

C. EICOSANOIDS AND IMMUNITY

Since eicosanoids are soluble mediators of the inflammatory response, their role is critical in many types of immune reaction. One of the most widely used drugs in the world, aspirin, is a cyclooxygenase inhibitor that reduces prostaglandin synthesis. The anti-inflammatory effect of aspirin and other nonsteroidal anti-inflammatory drugs (NSAIDs) is due to blocking of the metabolism of arachidonic acid by cyclooxygenase. Some animal studies suggest that high doses of other prostaglandin inhibitors, such as indomethacin, may enhance some mononuclear cell dependent responses. There are limited reports that humans with deficient immune responses respond to indomethacin treatment, whereas some normal individuals do not benefit from treatment with this drug and others show an increase in certain antibody responses [3]. The variation in response might be a function of which eicosanoid is involved in the altered immune responses. Burn injury leads to elevations of prostaglandin E and immune hyporesponsiveness. Administration of prostaglandin inhibitors lowers the PGE levels and partially corrects the immune response. NSAID administration appears to prevent colon cancer in both humans and experimental animals; in addition, prostaglandin synthesis is decreased by these drugs. It is unknown how these compounds inhibit the growth of colon cancer [4]. While there may be direct effects of these prostaglandin inhibitors on several aspects of colon carcinogenesis, including changes in cell metabolism, the cell cycle, and expression of tumor suppressor proteins, one cannot rule out alterations in immune response.

D. NONEICOSANOID MEDIATORS

Although it is clear that dietary fatty acids exert powerful effects on eicosanoids, which in turn profoundly modify some aspects of the immune response, alternate mechanisms by which dietary fat can alter immune responses have been proposed. These include changes in membrane microviscosity and dependent events, as well as the direct activation of protein kinase C (PKC) by arachidonic acid. The role of PKC is modulated by diacylglycerols (DAGs), whose affinity for PKC is altered by changes in the fatty acid moieties on the DAGs. Other potential mechanisms that have been implicated are activation of GTP-binding proteins by fatty acids and changes in phospholipase activity in the cell membrane. Finally, increased oxidation is suspected as one of the mediators of ω3 fatty acid effects on immune cells; an increase in dietary vitamin E that restored the in vitro responses of T cells from subjects fed fish oil was interpreted as supporting the possibility of such mediating activity [5]. Also, carotenoids exhibit both antioxidant and immunomodulatory roles that may be related. Feeding of carotenoids raises the number of circulating lymphocytes, increases proliferation of cytotoxic T cells, and enhances rejection of skin grafts in mice [6]. Similar data from humans were also seen:

a low-carotenoid diet led to decreased delayed-type skin hypersensitivity reactions (mediated by T cells), and supplementation with β-carotene restored the skin reactions to normal [7].

Although a variety of cell-based mechanisms that do not involve eicosanoids have been proposed for modulating the immune response, perhaps some of the strongest evidence comes from studies of general nutritional status. Both overnutrition and energy restriction have profound effects on the immune system's regulation. Overnutrition that results in obesity depresses the immune response, while energy restriction that avoids nutrient deficiencies enhances the immune response, particularly in older animals [8].

III. DIETARY LIPIDS

A. CHOLESTEROL

Dietary cholesterol has been studied extensively for its effects on immune functions, primarily in animals and in vitro. A fair number of studies in humans have been conducted, however. The rationale for this work is that cholesterol and fatty acids in the cell membrane are the primary determinants of membrane microviscosity, which in turn controls a number of events at the cell surface, including enzymatic activities such as those of phospholipase and cyclooxygenase. There is considerable variation in serum cholesterol concentrations among humans, but two factors have contributed to the relatively small number of studies showing a positive correlation between dietary cholesterol and changes in the immune response. First, dietary cholesterol is relatively weak in its ability to increase serum cholesterol. Second, the populations studied have not exhibited a broad enough range of serum cholesterol concentrations to permit observation of large effects on immunological functions in vivo (although some studies have noted differential responses in vitro). Therefore, experimental manipulations in animals or in vitro that entailed extreme differences in exposure to cholesterol have resulted in significant effects on immunologic responses.

It is impossible to singularly characterize the effect of cholesterol on immunologic responses. This is because cholesterol is delivered to cells in lipoproteins, complex aggregates of cholesterol, triglyceride, phospholipid, and one or more apoproteins to a variety of immunologic cells. It has been shown that isolated apolipoprotein E (apoE) inhibits lymphocyte proliferation, but its effectiveness depends on interaction with cholesterol and phospholipid. Recent interest in the role of oxidized lipoproteins in the development of atherosclerosis suggests that some of the immunologic effects attributed to cholesterol may have been due to oxidized cholesterol molecules. In fact, there is considerable evidence that oxidized low density lipoprotein (LDL) or cholesterol reduces immune responses by lymphocytes.

The change in immunologic response elicited by elevated cholesterol concentrations depends both on the concentration of cholesterol to which cells are exposed and on the type of immunologic response being measured. There is no single direction. Most studies in this area have looked at lymphocyte functions, and it can be concluded that results are primarily dependent on lipoprotein concentration [9].

Studies done in hypercholesterolemic rabbits and monkeys suggest that both B- and T-cell-dependent functions are elevated. However, hypercholesterolemic guinea pigs showed increased immune function with a doubling of baseline serum cholesterol but a significant reduction when concentrations of cholesterol were elevated fourfold. This suggests that there may be species specificity as well as specific ranges of serum cholesterol levels that modulate immune responses. Most human studies of lymphocyte function, in vivo or in vitro, have shown decreased functions in the presence of high total or LDL cholesterol [10]. However, the total number of circulating T-lymphocyte subsets CD3$^+$, CD4$^+$, and CD8$^+$ in hypercholesterolemic children correlates with LDL cholesterol concentrations over a threefold range [11]. Since T lymphocytes are found in significant numbers in atherosclerotic plaques, it is presumed that the elevated LDL levels are associated with both development of arterial lesions and changes seen with chronic inflammation.

Monocyte and macrophage functions are reduced in most studies that have examined the effect of excess cholesterol. Similarly, functions of the polymorphonuclear leukocyte have also been reduced in response to a surplus of cholesterol. The functions are, to a great extent, dependent on plasma membrane microviscosity; excess cholesterol incorporation will stiffen the membranes, thereby decreasing the ability of these cells to engulf microbes. A variety of other immune functions have been examined under the influence of different concentrations of cholesterol, and most studies show generally reduced immune cell responses; however, there is considerable disagreement in the research literature on the overall effect of cholesterol on immune reactions. Reports generally agree that there is reduced resistance to bacterial or viral infections, but some specific components of the immune system are suppressed while others are enhanced in the presence of hypercholesterolemia.

With the view that atherosclerosis shares many traits of a chronic inflammatory condition gaining widespread acceptance only in the late 1980s, the role of cholesterol in the human immune response as it relates to the development of arterial lesions has been of considerable interest. Since a large percentage of the fat-filled foam cells in atherosclerotic lesions are derived from macrophages, it is logical to assume that cytokines and classical inflammatory repair mechanisms are at work as part of the atherogenic process.

B. FATTY ACIDS

Just as cholesterol plays a major role in determining the physical state of the cell membrane, so do fatty acids. The more polyunsaturated a fatty acid molecule is, the more fluidity it imparts to the cell membrane. Also, the more polyunsaturated fat in the diet, the lower the immune response, within certain limits. That is, the immune response does not go down to zero in the presence of very high intake of polyunsaturated fat; it is, however, lower relative to that seen with ingestion of saturated fats. In both humans and animals, cell membrane phospholipids reflect dietary intake within limits of the membrane to accommodate a certain range of fatty acids. Dietary fat primarily alters storage of fatty acids in adipose tissue, but smaller changes in the membranes of many other cell types, including those of the RES, have been reported. A variety of experimental studies have demonstrated that both the amount and type of fatty acids consumed have roles to play in altering immune responses. Although it is easier to discuss specific fatty acids in the diet, it should be remembered that there are very few free fatty acids consumed and most are in triacylglycerol molecules, which are usually a mixture of fatty acids attached to a glycerol molecule. The majority of dietary fats usually supply four to six different fatty acids.

There is considerable evidence that both ω6 and ω3 polyunsaturated fatty acids (PUFAs) can modulate a variety of immunological activities. There was considerable interest in the clinical use of ω6-enriched vegetable oils during the 1970s for patients who had received kidney transplants, as well as those with multiple sclerosis and autoimmune disorders. Development of better immunosuppressive agents and other drugs for these conditions led investigators away from this area of research. However, there remains considerable interest in the effects of dietary fats on immune response, particularly in relation to development of cancer but also for cardiovascular disease and general health. It is fair to state that beyond the generally accepted immunosuppressive effects of polyunsaturated fats, there is considerable disagreement on whether the specific fatty acid, ratios between different fatty acids, or the quantity of dietary fat has the greatest impact on immunological responses. In general, when the long chain ω3 fatty acids increase in a cell membrane, there is a concomitant decrease in arachidonic acid concentration.

Sources of ω3 fatty acids used most commonly in studies of immune function include fish oils, evening primrose oil, and flaxseed oil. Plant-derived oils generally supply 18:3, while the fish oils have the longer chain fatty acids EPA and DHA. There is substantial debate concerning the optimal ratio of the ω6/ω3 ratio in the diet with many researchers favoring a range of 4.0–7.0 based on studies in rats. The U.S. Dietary Reference Intakes contain specific recommendations for EFAs; Adequate Intake levels of 12–17 g/d linoleic acid and 1.1–1.6 g/d α-linolenic acid were based on median intakes in the United States, which result in an $\omega6/\omega3$ ratio > 10. The committee acknowledged no signs of deficiency were observed with chronically lower intakes and that there were insufficient data to set Upper Level values [12].

One study of the immune response in rats fed blends of different proportions of sunflower oil (rich in linoleic acid) and flaxseed oil (rich in linolenic acid) found that the higher the $\omega3/\omega6$ ratio of fatty acids in the plasma, the greater the reduction of lymphocyte-dependent immune responses, including T-cell blastogenesis and NK cell activity [13]. These authors concluded that α-linolenic acid was as potent as fish oil for suppressing immune responses but this is not strongly supported by other studies.

While most animal studies simply use feeding of a single source of fat, or a blend, throughout an experiment, at least one study has tried to mimic the normal human intake pattern of $\omega3$ PUFA [14]. Mice were fed a diet containing safflower oil and switched to a sardine/olive oil mixture for periods of 1–7 days per week. Peritoneal macrophages had phospholipid compositions that reflected the diet. Synthesis of leukotrienes E_4 and C_4 decreased, while E_5 and C_5 levels increased with more frequent consumption of fish oil. Prostaglandin $F_{1\alpha}$ also decreased with increasing consumption of fish oil. The summary finding of this study—that fish oil must be consumed at least twice a week to produce significant changes from the control diet—touches on the current debate about how frequently a person has to consume fish to derive a health benefit. As with most questions on diet and risk of chronic diseases, there is no definitive answer, but some studies have suggested maximal benefit with consumption of at least two servings of fatty fish weekly.

A fundamental study on this topic looked at the effects of feeding a low-fat (26% of energy) diet that contained a high amount of fish in 22 subjects for 24 weeks or a low-fat diet without fish [15]. Fish intake ranged from 4 to 6 ounces daily. Therefore, this study used both type and amount of dietary fat as variables. Responses on these diets were compared with those of subjects who followed a diet with 35% of energy from fat. Feeding the low-fat, low-fish diet (which was enriched in plant PUFAs) increased the response of mononuclear leukocytes to the T-cell mitogen concanavalin A (con A), interleukin 1β levels, and tumor necrosis factor. No effects were seen on PGE_2 production, interleukin 6 levels, or delayed-type skin hypersensitivity. In contrast, the low-fat, fish-enriched diet resulted in a significant decrease in $CD4^+$ and a concurrent decrease in $CD8^+$ cells. There were significant reductions in the lymphocyte mitogenic response to con A, delayed-type hypersenitivity, interleukin 1β, interleukin 6, and tumor necrosis factor production by leukocytes. The practical implications of these immune alterations are not clear cut. It was conjectured that these decreases in immune responses would be favorable for atherosclerosis and inflammatory diseases but harmful for host defense against microorganisms. The issue of immune surveillance against cancer cells was not discussed but is a valid concern.

In another important study in this area, Purasiri et al. [16] found that a mixed $\omega3$ fat supplement of 4.8 g daily decreased the blood concentrations of a number of cytokines significantly (by 60%–80%) in individuals with colon cancer; these included several forms of interleukins, tumor necrosis factor α, and interferon-γ. Cytokines returned to baseline levels 3 months after the cessation of the supplements. This study clearly demonstrated a rapid and profound decline in a number of cytokines in response to an easily consumed amount of $\omega3$ fatty acids. The implications of this study are that regular consumption of $\omega3$ fatty acids might reduce the ability of the immune system to respond to infections or tumors.

Some studies have measured prostaglandin metabolites in urine as an index of whole-body metabolism. Supplementation of $\omega3$ fatty acids to a low-fat diet reduced thromboxane A_2 production, while a low-fat diet, with or without fish oil, resulted in reduced prostacyclin production in healthy men. Accompanying these dietary manipulations, however, were increases in thromboxane A_3 and prostacyclin A_3 [17].

There have been many studies in which fish oil supplements were administered to people with presumptive autoimmune inflammatory conditions such as rheumatoid arthritis. Most studies find that fish oil supplements are effective in reducing symptoms of these conditions.

However, there does not seem to be any benefit of ω3 fatty acid supplements over conventional NSAID therapy.

C. TOTAL DIETARY FAT

The early work on total dietary fat in the 1960s showed that animals fed high-fat diets were more susceptible to a wide range of spontaneous infections. These observations led to the experimental examination of specific parts of the immune response in animals fed different levels of dietary fat.

Differences in immune responses as a function of changes in total dietary fat have been studied in many animal models and in a few human trials. In general, the more fat in the diet, the lower the immune response. Many animal studies have shown that high-fat diets are immunosuppressive in comparison with low-fat diets. The basal level in most animal studies has been about 10% of energy from fat, and the high levels have ranged from 40% to 60% of energy from fat, usually of a single source. Most animal studies have fed a polyunsaturated vegetable oil, so it is difficult to distinguish between the effects of total fat and the effect of increased linoleic acid or other individual fatty acids.

A limited number of well-controlled human studies have examined the influence of amount of dietary fat on immune responses. In one experiment, seven healthy women lived in a metabolic suite and were fed a diet that provided 41% of energy from fat and 5% from PUFAs [18]. The subjects were divided into two groups that consumed either 26% of energy from fat with 3.2% from PUFAs or 31% of energy from fat with 9.1% PUFAs; the study was conducted in a crossover design. Both low-fat diets resulted in significant increases in serum complement fractions C3 and C4, as well as mitogenic responses of peripheral lymphocytes, phytohemagglutinin, con A, protein A, and pokeweed mitogen. The results are similar to those of Meydani et al. [15], cited above, namely, that lower fat diets increased blastogenic response of blood lymphocytes to con A. In another study, healthy women were fed either 3% or 8.3% of energy as linoleic acid [19]; increasing dietary 18:2 led to higher prostaglandin levels in urine but lower thrmoboxane B_2 and no change in prostaglandin $F_{1\alpha}$.

IV. CONCLUSIONS

Changes in fatty acids of the plasma or immune cells can affect eicosanoid levels and immune responses. However, other mechanisms also play a role. In recent years, changes in gene expression and transcription factors have been enumerated in response to dietary fat differences. The discovery of three cyclooxygenase (COX) pathways caused a boom in drug discovery affecting these routes, followed by widespread reports of adverse events for COX-2 inhibitors, leaving diet as perhaps the only current safe intervention of all COX isozymes. Systematic reviews of studies on ω3 fatty acids on immune-related endpoints such as asthma and rheumatoid arthritis have generally concluded that no effect was demonstrated, perhaps more because of a lack of consistent study design and quality than because of disparate results [20]. In contrast, many other reviews have concluded that long chain ω3 fatty acids from marine animals can modulate immune response and clinical endpoints, such as rheumatoid arthritis, if the dose is sufficient [21]. Because inflammation and related immune responses are now recognized in the pathogenesis of many chronic diseases not previously connected to these phenomena, it is possible that manipulation of dietary fatty acids will garner renewed interest for anti-inflammatory and immunomodulatory effects.

REFERENCES

1. E.H. Gruger. Fatty acid composition of fish oils. USDA Circular 276. U.S. Department of Agriculture, Washington, DC, 1967, p. 11.
2. M.C. Noverr, J.R. Erb-Downward, and G.B. Huffnagle. Production of eicosanoids and other oxylipins by pathogenic eukaryotic microbes. *Clin. Microbiol. Rev.* 16:517 (2003).
3. D.S. Hummell. Dietary lipids and immune function. *Prog. Food Nutr. Sci.* 17:287 (1993).

4. G.N. Levy. Prostaglandin H synthases, nonsteroidal anti-inflammatory drugs, and colon cancer. *FASEB J.* 11:234 (1997).

5. T.R. Kramer, N. Schoene, L.W. Douglass, J.T. Judd, R. Barbash, P.R. Taylor, H.N. Bhagavan, and P.P. Nair. Increased vitamin E intake restores fish oil-induced suppressed blastogenesis of mitogen-stimulated T lymphocytes. *Am. J. Clin. Nutr.* 54:896 (1991).

6. E. Seifter, G. Rettura, J. Padawer, and S.M. Levenson. Moloney murine sarcoma virus tumors in CBA/J mice: Chemopreventive and chemotherapeutic actions of supplemental beta carotene. *J. Natl. Cancer Inst.* 68:835 (1982).

7. C.J. Fuller, H. Faulker, A. Bendich, R.S. Parker, and D.A. Roe. Effect of β-carotene supplementation on photosuppression of delayed-type hypersensitivity in normal young men. *Am. J. Clin. Nutr.* 56:684 (1992).

8. G. Fernandes and J.T. Venkatraman. Dietary restriction: Effects on immunological function and aging. In: *Nutrition and Immunology* (D.M. Klurfeld, ed.). Plenum Press, New York, 1993, p. 91.

9. K.N. Traill, L.A. Huber, G. Wick, and G. Jurgens. Lipoprotein interactions with T cells: An update. *Immunol. Today* 11:411 (1990).

10. D.M. Klurfeld. Cholesterol as an immunomodulator. In: *Nutrition and Immunology* (D.M. Klurfeld, ed.). Plenum Press, New York, 1993, p. 79.

11. A. Sarria, L.A. Moreno, M. Mur, A. Lazarro, M.P. Lasierra, L. Roda, A. Giner, L. Larrad, and M. Bueno. Lymphocyte T subset counts in children with elevated low-density lipoprotein cholesterol levels. *Atherosclerosis* 117:119 (1995).

12. Panel on Macronutrients. Dietary fats: Total fat and fatty acids. In: *Dietary Reference Intakes for Energy, Carbohydrate, Fiber, Fat, Fatty Acids, Cholesterol, Protein, and Amino Acids*. National Academies Press, Washington, DC, 2005, p. 422.

13. N.M. Jeffery, P. Sanderson, E.J. Sherrington, E.A. Newsholme, and P.C. Calder. The ratio of *n*-6 to *n*-3 polyunsaturated fatty acids in the rat diet alters serum lipid levels and lymphocyte functions. *Lipids* 31:737 (1996).

14. H.S. Broughton and L.J. Morgan. Frequency of (*n*-3) polyunsaturated fatty acid consumption induces alterations in tissue lipid composition and eicosanoid synthesis in CD-1 mice. *J. Nutr.* 124:1107 (1994).

15. S.N. Meydani, A.H. Lichtenstein, S. Cornwall, M. Meydani, B.R. Goldin, H. Rasmussen, C.A. Dinarello, and E.J. Schaefer. Immunologic effects of national cholesterol education panel step-2 diets with and without fishderived *n*-3 fatty acid enrichment. *J. Clin. Invest.* 92:105 (1993).

16. P. Purasiri, A. Murray, S. Richardson, S.D. Heys, and O. Eremin. Modulation of cytokine production in vivo by dietary essential fatty acids in patients with colorectal cancer. *Clin. Sci.* 87:711 (1994).

17. A. Nordov, L. Hatcher, S. Goodnight, G.A. Fitzgerald, and W.E. Connor. Effects of dietary fat content, saturated fatty acids, and fish oil on eicosanoid production and hemostatic parameters in normal men. *J. Lab. Clin. Med.* 123:914 (1994).

18. D.S. Kelley, R.M. Dougherty, L.B. Branch, P.C. Taylor, and J.M. Iacono. Concentration of dietary *n*-6 polyunsaturated fatty acids and the human immune status. *Clin. Immunol. Immunopathol.* 62:240 (1992).

19. I.A. Blair, C. Prakash, M.A. Phillips, R.M. Dougherty, and J.M. Iacono. Dietary modification of omega 6 fatty acid intake and its effect on urinary eicosanoid excretion. *Am. J. Clin. Nutr.* 57:154 (1993).

20. K. Fritsche. Fatty acids as modulators of the immune response. *Annu. Rev. Nutr.* 20:45 (2006).

21. P.C. Calder. n-3 Polyunsaturated fatty acids, inflammation, and inflammatory diseases. *Am. J. Clin. Nutr.* 83:1515S (2006).

22 Dietary Fats and Coronary Heart Disease

Ronald P. Mensink and Jogchum Plat

CONTENTS

I. INTRODUCTION

Many factors are associated with increased risk for coronary heart disease (CHD), a major cause of morbidity and mortality in prosperous Western countries. Some of these factors, such as increasing age or a family history of premature CHD, are not amenable to preventive intervention, but other factors are. Three of these preventable factors—the distribution of plasma cholesterol over the low-density and high-density lipoproteins (LDLs and HDLs), the oxidizability of LDLs, and hemostasis—can be modified by changing the sources and amount of fats and oils in the diet. The purpose of this chapter is to review some of the most recent and important findings on the effects of dietary fatty acids on these three risk factors.

II. DIETARY FATS

Although fat and oils are complex mixture of fatty acids, each fat or oil has its characteristic fatty acid composition. Dairy fat, for example, is relatively rich in fatty acids with 14 or less carbon atoms, whereas olive oil has a high oleic acid content (Table 22.1). Sunflower oil, on the other hand, is rich in linoleic acid, although certain varieties exist that contain large amounts of oleic acid. In normal, regular diets, palmitic and stearic acids are the most prevailing saturated fatty acids, whereas oleic and linoleic acids are, respectively, the most widespread dietary monounsaturated and polyunsaturated fatty acids. About 30%–40% of total energy intake is provided by fat. For a person consuming 10 MJ (2400 kcal)/day, this corresponds to a fat intake of 80–107 g.

III. LIPOPROTEINS

The solubility of cholesterol in water is very low, approximately 5.2×10^{-3} mmol/L. The actual cholesterol concentration in the watery plasma of healthy subjects, however, is about 3.9–5.2 mmol/L,

---

TABLE 22.1
Major Fatty Acids in Some Edible Fats and Oils

Formula	Fatty Acid	Source
Saturated fatty acids		
	Medium-chain fatty acids	Dairy fat, coconut oil, palm kernel oil
C12:0	Lauric acid	Dairy fat, coconut oil, palm kernel oil
C14:0	Myristic acid	Dairy fat, coconut oil, palm kernel oil
C16:0	Palmitic acid	Palm oil, meat
C18:0	Stearic acid	Meat, cocoa butter
Monounsaturated fatty acids		
C18:1,*n*-9	Oleic acid	Olive oil, rapeseed oil, high oleic acid Sunflower oil
Polyunsaturated fatty acids		
C18:2,*n*-6	Linoleic acid	Sunflower oil, corn oil, soybean oil, corn oil
C18:3,*n*-3	α-Linolenic acid	Rapeseed oil, soybean oil
C20:5,*n*-3[a]	Timnodonic acid	Fatty fish, fish oil capsules
C22:5,*n*-3[b]	Cervonic acid	Fatty fish, fish oil capsules

[a] Trivial names, eicosapentaenoic acid (EPA).
[b] Trivial name, docosahexaenoic acid (DHA).

TABLE 22.2

Some Physical Characteristics and Mean Composition of Lipoprotein Fractions from Normotriglyceridemic Subjects

	Chylomicrons	VLDL	IDL	LDL	HDL
Density (g/mL)					
Lower limit	—	0.96	1.006	1.019	1.063
Upper limit	0.96	1.006	1.019	1.063	1.21
Size (nm)	75–1200	30–80	25–35	19–25	5–12
	% of Total Lipoprotein Mass				
Core components					
Triacylglycerols	87	52	29	6	6
Cholesterylester	3	9	28	40	21
Surface components					
Phospholipids	6	23	22	22	24
Free cholesterol	2	6	7	9	3
Apolipoprotein	2	10	14	23	46

and increases to more than 10 mmol/L in hypercholesterolemic people. This high degree of solubilization is achieved by the formation of lipoproteins.

Lipoproteins are globular, high molecular weight particles that are complex aggregates of lipid and protein molecules. A lipoprotein consists of a hydrophobic core, which mainly contains triacylglycerols and cholesterylesters, and a polar, hydrophilic coat composed of phospholipids, unesterified cholesterol, and specific apolipoproteins. In this way, the hydrophobic core is protected from the watery surrounding, and transport of large amounts of cholesterol and triacylglycerols through the blood vessels is possible.

Lipoproteins are a heterogeneous group, which can be divided into five major classes: chylomicrons, very low density lipoproteins (VLDLs), intermediate-density lipoproteins (IDLs), LDLs, and HDLs. Each class has its own characteristic lipid and apolipoprotein composition, size, and density, whereas each apolipoprotein has its own specific metabolic functions (Tables 22.2 and 22.3).

IV. METABOLISM OF DIETARY FATTY ACIDS AND LIPOPROTEINS

A. EXOGENOUS PATHWAY

The metabolism of dietary fatty acids, dietary cholesterol, and lipoproteins is depicted in Figure 22.1. In the duodenum, dietary triacylglycerols are dissolved with the help of bile salts, as well as small

TABLE 22.3

Some Major Apolipoprotein and Their Functions

Apolipoprotein	Lipoprotein	Function
ApoC-II	Chylomicrons, VLDL	Activator of LPL
ApoE	Chylomicron remnants, VLDL, IDL	Ligand for remnant and LDL receptors
ApoB-100	IDL, LDL	Ligand for LDL receptor
ApoA-I	Chylomicron, HDL	Cofactor for LCAT

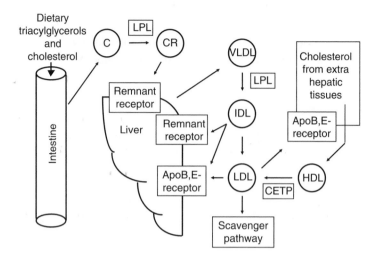

FIGURE 22.1 Overview of the metabolism of dietary fatty acids, dietary cholesterol, and plasma lipoproteins. Dietary triacylglycerols and cholesterol enter the blood circulation in chylomicrons (C). The triacylglycerols from the chylomicron core are hydrolyzed by lipoprotein lipase (LPL) and the so-formed chylomicron remnants (CR) are removed from the circulation by the hepatic remnant receptor. The liver excretes cholesterol and triacylglycerols into the circulation in VLDLs. These lipoproteins also interact with LPL and VLDL remnants—also called IDLs—are formed, which are taken up by the liver or converted into LDL. Most of the LDL is removed from the circulation by the (hepatic) LDL receptor pathway, whereas a smaller part is removed via the scavenger pathway. HDL binds cholesterol from tissues, which can be transferred with the assistance of cholesterol ester transfer protein (CETP) to LDL. HDL particles may also be taken up in the liver by a putative HDL receptor or lose a part of its content by the action of hepatic lipase and then reenter the circulation again (not drawn).

quantities of fatty acids and monoglycerides. The enzyme pancreatic lipase then hydrolyzes the dietary triacylglycerols into mono- and diglycerides, free fatty acids, and glycerol. The so-formed emulsion of lipids, which also contain the dietary cholesterol, passes the mucous membrane of the intestinal cells. Within the cell, further hydrolysis of lipids takes place and new triacylglycerols are formed by reesterification of the free fatty acids with glycerol. The newly synthesized triacylglycerols and cholesterylesters, derived from the dietary cholesterol, are incorporated into chylomicrons, which enter the lymph and subsequently the blood circulation in the subclavian vein. In the blood, the triacylglycerols from the chylomicron core are hydrolyzed by lipoprotein lipase (LPL), an enzyme adhered to the endothelial cells of the blood vessels, and activated by apolipoprotein C-II. The free fatty acids pass the endothelial cells and enter adipocytes or muscle cells. In these cells, the fatty acids are respectively stored as triacylglycerols or oxidized. The core of an emptied chylomicron mainly consists of cholesterylesters and is called a chylomicron remnant. These remnant particles are removed from the circulation by the hepatic remnant receptor, which has a high affinity for apolipoprotein E (apoE) from the chylomicron surface.

B. Endogenous Pathway

The liver excretes cholesterol and triacylglycerols into the circulation by the formation of VLDL particles. These lipoproteins also interact with LPL, triacylglycerols become hydrolyzed, and VLDL remnants (also called IDLs) are formed. A part of the IDL is taken up by the liver, whereas the remaining part is converted in the circulation into LDL. LDLs are nearly devoid of triacylglycerols and carry about 60%–70% of the total amount of cholesterol in the plasma. Most of the LDL is now removed from the circulation through the hepatic LDL receptor pathway, which recognizes apolipoprotein B-100 (apoB-100). LDL uptake from the blood by the LDL receptor-mediated pathway is highly controlled and this pathway will be downregulated if the amount of cholesterol in the cell becomes too high. A smaller part, however, is removed via the scavenger pathway. Uptake via this pathway is not saturable and is positively related with the LDL cholesterol concentration. Thus, the higher the blood LDL cholesterol concentration, the more LDL will be taken up via the scavenger pathway. When too much LDL is taken up via the scavenger pathway from macrophages, cells loaded with cholesterol are formed—so-called foam cells—which are frequently found in atherosclerotic lesions.

Cholesterol can also be transported out of tissues by reverse cholesterol transport. This system is mediated by HDL, lecithin cholesteryl acyltransferase (LCAT), and cholesterol ester transfer protein (CETP). HDL binds free cholesterol from tissues, which is esterified by LCAT, a protein associated with apolipoprotein A-I (apoA-I). The formed cholesterylesters move to the core of the HDL particle and the HDL is converted to a larger particle. The acquired cholesterylesters can now be transferred with the assistance of CETP to apoB-100-containing lipoproteins in exchange for triacylglycerols. The apoB-100-containing lipoproteins are further metabolized, as has already been described. The large HDL particles may also be taken up in the liver by a putative HDL receptor or lose a part of its content by the action of hepatic lipase and then reenter the circulation again.

V. PLASMA LIPOPROTEINS AND CORONARY HEART DISEASE

LDL and HDL have different effects on the risk for CHD. High concentrations of LDL are atherogenic, whereas high levels of HDL are negatively associated with the risk for CHD. As LDL carries most of the plasma cholesterol, the total plasma cholesterol is also a good index for the risk of CHD. It should be realized, however, that some people have high total cholesterol concentrations, due to high HDL levels. Therefore, the total cholesterol to HDL cholesterol might be the most efficient predictor for the risk for CHD [1]. In addition, high levels of triacylglycerols, which are in the fasting condition mainly found in the VLDLs, are positively related to the risk for CHD [2].

VI. DIETARY FATS AND PLASMA LIPOPROTEINS

A. Earlier Studies

In the 1950s Keys and coworkers started a series of well-controlled experiments to examine the effects of dietary fatty acids on plasma total cholesterol concentrations [3,4]. Groups of physically healthy men were fed diets that differed widely in the amount of fat and in dietary fatty acid composition. During the studies, individual allowances were adjusted weekly to keep body weight stable so that changes in plasma total cholesterol concentrations could be attributed solely to dietary changes. At the end of the studies, an empirical formula that could be used to predict for a group of subjects changes in plasma cholesterol concentrations from changes in dietary fatty acid composition was derived:

$$\Delta \text{ Plasma total cholesterol (mmol/L)} = 0.03 \times (2 \times \Delta\text{Sat}' - \Delta\text{Poly})$$

or

$$\Delta \text{ Plasma total cholesterol (mg/dL)} = 1.2 \times (2 \times \Delta\text{Sat}' - \Delta\text{Poly}),$$

where
 Sat$'$ is the percentage of energy provided by saturated fatty acids with 12, 14, or 16 carbon atoms (lauric, myristic, and palmitic acid, respectively)
 Poly refers to the amount of polyunsaturated fatty acids in the diet

How should this formula be interpreted? First, it should be realized that effects are expressed relative to those of carbohydrates. A hypercholesterolemic fatty acid is therefore defined as a fatty acid that causes an increase in the plasma cholesterol level when substituted in the diet for an isocaloric amount of carbohydrates. Thus, when 10% of energy from Sat' is replaced by carbohydrates, ΔSat' equals -10, and the expected decrease in plasma total cholesterol concentrations is $0.03 \times 2 \times -10 = -0.60$ mmol/L (-24 mg/dL). If this amount of carbohydrates is then replaced by linoleic acid, a further decrease of $-0.03 \times 10 = -0.30$ mmol/L (-12 mg/dL) is expected. Direct replacement of Sat' by Poly yields the sum of these two effects, a fall of 0.90 mmol/L (36 mg/dL). Further, this formula suggests that—because they are not part of the equation—the effects on plasma total cholesterol concentrations of saturated fatty acids with fewer than 12 carbon atoms, of stearic acid, and of monounsaturated fatty acids are similar to those of carbohydrates. Finally, it can be seen that the cholesterol-raising effect of Sat' is about twice the cholesterol-lowering effect of Poly.

Similar types of studies were carried out in the 1960s by Hegsted and colleagues [5]. Results were essentially similar, but it was also concluded that myristic acid was more cholesterolemic than palmitic and lauric acids.

These and other studies have led to recommendations that the most effective diet for lowering plasma total cholesterol concentration should contain a low proportion of the cholesterol-raising saturated fatty acids and a high proportion of linoleic acid. In addition, a reduction in cholesterol intake was advocated as dietary cholesterol increases plasma total cholesterol concentrations [5,6].

However, these earlier well-controlled studies were not specifically designed to examine the effects of specific dietary fatty acids on plasma cholesterol concentrations and over the various lipoproteins. Therefore, new studies were initiated that compared side-by-side effects of specific fatty acids on the plasma lipoprotein profile.

B. Recent Studies

1. Saturated Fatty Acids

To discuss the effects of saturated fatty acids on plasma lipid and lipoproteins, the saturated fatty acids are, in agreement with the results of Keys and colleagues [5], categorized into three classes: medium-chain fatty acids (MCFA), fatty acids with 12, 14, or 16 carbon atoms, and stearic acid.

a. Medium-Chain Fatty Acids

Saturated fatty acids with fewer than 12 carbon atoms are called short- and medium-chain saturated fatty acids and are found in relatively large amounts in coconut fat, palm kernel oil, and butterfat, but also in certain structured lipids, parenteral nutrition preparations, and sport drinks.

McGandy and coworkers [7] have carefully compared the effects of MCFAs, mainly capric acid, on plasma total and LDL cholesterol levels. Eighteen physically healthy men were fed several diets, each for 4 weeks. Diets contained several low-fat food items to which the experimental fats were added. It was shown that modest amounts of MCFAs in the diet have comparable effects on the plasma total and LDL cholesterol and on triacylglycerols concentrations as have carbohydrates. However, large amounts of MCFAs increased triacylglycerol concentrations. Results of two more recent studies [8,9], however, suggested that a mixture of MCFAs slightly increases LDL cholesterol concentrations relative to oleic acid (Figure 22.2). No effects on HDL cholesterol were found, whereas serum triacylglycerol concentrations were slightly increased.

b. Lauric, Myristic, and Palmitic Acids

Palm kernel oil, coconut oil, and dairy fat are rich in lauric acid, but contain also relatively high amounts of myristic acid. Therefore, it is hardly possible to study the specific effects of these two saturated fatty acids on plasma lipoproteins with natural fats. A diet enriched in palm kernel oil, for example, contains high amounts of both lauric and myristic acids, and it will subsequently be

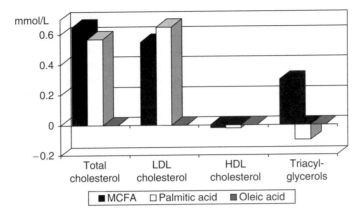

FIGURE 22.2 Effects of medium-chain fatty acids and palmitic acid on plasma total, LDL cholesterol, HDL cholesterol, and triacylglycerols concentrations relative to those of *cis*-monounsaturated fatty acids (oleic acid). Nine men received three mixed natural diets, each for 3 weeks, in random order. The composition of the diets was identical, except for 43% of daily energy intake, which was provided as either medium-chain fatty acids (C8:0 and C10:0), palmitic acid (C16:0), or oleic acid (*cis*-C18:1)). (From Cater, N.B., Heller, H.J., and Denke, M.A., *Am. J. Clin. Nutr.*, 65 41, 1997.)

impossible to ascribe the effects on the plasma lipoprotein profile to either lauric or myristic acid. To circumvent this problem, several studies have used synthetic fats to examine the cholesterolemic effects of these two saturated fatty acids. In this way, a fat with any desired fatty acid composition can be made. For example, when one interesterifies trilaureate with a high oleic acid sunflower oil, the resultant will be a high lauric acid fat without any myristic acid.

In a recent meta-analysis the effects of the individual saturated fatty acids on the serum lipoprotein profile have been estimated [10]. It was found that lauric, myristic, and palmitic acids all increased serum total and LDL cholesterol concentrations (Figure 22.3). These effects decreased with increasing chain length. Thus, lauric acid was more hypercholesterolemic than myristic acid,

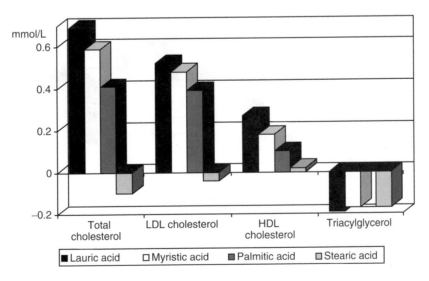

FIGURE 22.3 Effects of exchanging 10% of energy from carbohydrates by lauric (C12:0), myristic (C14:0), palmitic (C16:0), or stearic (C18:0) acid on plasma total, LDL cholesterol, HDL cholesterol, and triacylglycerols concentrations. (From Mensink, R.P. et al., *Am. J. Clin. Nutr.*, 77, 1146, 2003.)

which on its turn was more hypercholesterolemic acid than palmitic acid. For HDL cholesterol, effects did also depend on chain length. All three saturated fatty acids increased HDL cholesterol and effects of lauric acid were the strongest. Because the effects of lauric acid were proportionally higher on HDL than on LDL cholesterol, replacement of carbohydrates by lauric acid resulted in a significantly lower total to HDL cholesterol ratio. Myristic and palmitic acids did not affect the ratio of total to HDL cholesterol. Compared with carbohydrates, these three saturated fatty acids lowered triacylglycerol concentrations to the same extent.

c. Stearic Acid

Keys and coworkers already demonstrated that stearic acid did not increase plasma total cholesterol concentrations [4,11]. However, Bonanome and Grundy were the first to study the effects of stearic acid on the distribution of cholesterol over the various lipoproteins [12]. From that study it was concluded that stearic acid exerted similar effects on the plasma lipoprotein profile as oleic acid, a monounsaturated fatty acid. In fact, stearic acid significantly lowered total, LDL, and HDL cholesterol concentrations compared with the other saturated fatty acids (Figure 22.3). Furthermore, it has been reported that stearic acid did not change the total to HDL cholesterol ratio when compared with carbohydrates [10].

2. Monounsaturated Fatty Acids

The most abundant monounsaturated fatty acid in the human diet, oleic acid, has 18 carbon atoms and one double bond. Although olive oil probably is the most well-known source of oleic acid, animal fats are in many countries a major contributor to total oleic acid, but also to palmitic and stearic acids, intakes.

Effects of monounsaturated fatty acids on plasma total cholesterol levels are often described as neutral. This term is often misinterpreted. It does not mean that the plasma total cholesterol level does not change when monounsaturated fatty acids are added to the diet. Neutral indicates that monounsaturated fatty acids have the same effect on plasma total cholesterol as compared with an iso-caloric amount of carbohydrates. Although this may be correct for plasma total cholesterol levels, oleic acid and carbohydrates do not have similar effects on the distribution of cholesterol of the various lipoproteins [10,13] (Figure 22.4). This was also shown in a study with young healthy

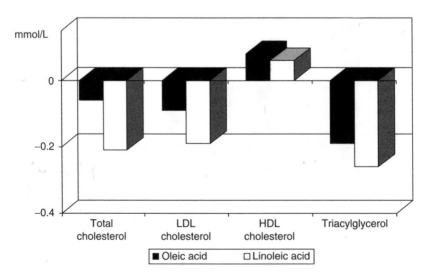

FIGURE 22.4 Effects of exchanging 10% of energy from carbohydrates by oleic (*cis*-C18:1) or linoleic (*cis*-C18:2) acid on plasma total, LDL cholesterol, HDL cholesterol, and triacylglycerols concentrations. (Mensink, R.P. et al., *Am. J. Clin. Nutr.*, 77, 1146, 2003.)

volunteers [14]. Increasing the intake of oleic acid at the expense of carbohydrates increased plasma HDL cholesterol concentrations and decreased those of triacylglycerols. The increase in HDL cholesterol was compensated for by a decrease in VLDL cholesterol. Effects of carbohydrates and oleic acid on plasma total and LDL cholesterol concentrations were comparable.

3. Polyunsaturated Fatty Acids

Polyunsaturated fatty acids in the diet belong to either the (n-6) or (n-3) family. About 90% of all polyunsaturated fatty acid in the diet is linoleic acid, which is found in vegetable oils like sunflower oil, corn oil, and soybean oil. The mean daily intake of fatty acids from the (n-3) family is only 1–3 g. These polyunsaturated fatty acids are either from vegetable or animal origin. α-Linolenic acid is found in rapeseed and soybean oils, whereas the very long chain fatty acids timnodonic or eicosapentaenoic acid (EPA) and cervonic or docosahexaenoic acid (DHA) are only present in fish oils.

a. n-6 Polyunsaturated Fatty Acids

Earlier studies found that linoleic acid was hypocholesterolemic as compared with carbohydrates and monounsaturated fatty acids [4,5]. The study of Mattson and Grundy [15], however, suggested that, as compared with monounsaturates, part of the cholesterol-lowering effect of linoleic acid was due to a decrease in HDL cholesterol. However, linoleic acid intake in that study was unrealistically high (28% of energy intake), which may have influenced the results. Studies at lower intakes found similar effects of linoleic and oleic acids on HDL cholesterol, but also on LDL cholesterol [16–18]. Thus, these more recent studies suggested that replacement of saturated fatty acids in the diet by monounsaturated fatty acids causes the same favorable change in plasma lipoprotein cholesterol levels as replacement by polyunsaturated fatty acids. However, it should be noted that in a recent meta-analysis effects of linoleic acid on plasma LDL cholesterol were slightly more favorable than that of oleic acid (Figure 22.4) [10].

b. n-3 Polyunsaturated Fatty Acids

The effects of α-linoleic acid on the plasma lipoprotein profile are similar to those of linoleic acid [19]. The highly unsaturated fatty acids from fish oils, however, have different effects. In normo-cholesterolemic subjects, these fatty acids do not change plasma LDL or HDL cholesterol concentrations, but do lower plasma triacylglycerols and the concentration of cholesterol in VLDL. In hyperlipidemic subjects, and in particular in patients with elevated triglyceride concentrations, fish oils also lower plasma triacylglycerols, but raise LDL and HDL cholesterol concentrations [20].

4. *Trans* Fatty Acids

Most unsaturated fatty acids found in nature have the *cis* configuration. This means that the two carbon side chains attached to the double bond point to the opposite (*cis*) direction. However, in some fatty acids the carbon side chain point to the same (*trans*) direction. In this way, two compounds are formed that have exactly the same number and type of atoms, but have different chemical, physical, and physiologic characteristics.

Trans fatty acids are formed when vegetable oils are hardened by hydrogenation. These hydrogenated fats are used for the production of certain types of margarines, frying fats, and foods prepared with these fats. Most *trans* fatty acids in the diet have 18 carbon atoms and 1 double bond (*trans*-C18:1). However, *trans* fatty acids are not only found in hydrogenated oils but also in milk fat and body fat from ruminants, formed from dietary polyunsaturated fatty acids by the action of bacteria in the rumen of these animals. The *trans* polyunsaturated fatty acids in the diet mainly originate from hydrogenated fish oils.

LDL cholesterol concentrations increase when *cis*-monounsaturated fatty acids in the diet are replaced by *trans*-monounsaturated fatty acids [21]. In most studies, a decrease in HDL cholesterol was also observed. Although the LDL cholesterol-raising effect of *trans*-monounsaturated fatty acids

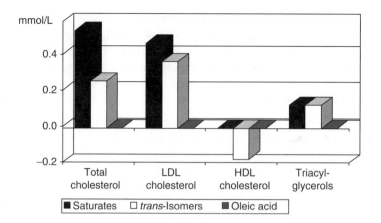

FIGURE 22.5 Effects of a mixture of saturated fatty acids and *trans*-isomers of oleic acid on plasma LDL cholesterol, HDL cholesterol, and triacylglycerols concentrations relative to those of *cis*-monounsaturated fatty acids (oleic acid). In random order, 25 men and 34 women received three mixed natural diets, each for 3 weeks. The composition of the diets was identical, except for 10% of daily energy intake, which was provided as either saturated fatty acids (mainly lauric and palmitic acids (C12:0 and C16:0, respectively)), *trans* isomers of oleic acid (*trans*-C18:1), or oleic acid (*cis*-C18:1). (Katan, M.B., Zock, P.L., and Mensink, R.P., *Annu. Rev. Nutr.*, 15, 473, 1995.)

is less than the effect of a mixture of saturated fatty acids, *trans*-monounsaturated fatty acids also lowered HDL cholesterol relative to a mixture of saturated fatty acids (Figure 22.5). Therefore, it was concluded that both types of fatty acids have an unfavorable effect on the plasma lipoprotein profile [10].

C. CONCLUSION

Dietary fatty acid composition affects the distribution of cholesterol over LDL and HDL. As compared with an isoenergetic amount of carbohydrates, lauric (C12:0), myristic (C14:0), and palmitic acids (C16:0) have a hypercholesterolemic effect, whereas stearic acid (C18:0) seems to be neutral. Linoleic (*cis,cis*-C18:2) and probably also oleic acid (*cis*-C18:1) have a small LDL cholesterol-lowering effect. *Trans* fatty acids have a strong plasma total and LDL cholesterol-increasing effect. Effects on LDL cholesterol levels are positively related to those on HDL cholesterol concentrations. An exception are *trans* fatty acids that do not have an effect on HDL as compared with carbohydrates. Thus, a reduction in the intake of the cholesterol-raising saturated fatty acids and *trans* fatty acids is more important for optimizing the plasma lipoprotein profile than a reduction in total fat intake per se.

VII. OXIDIZABILITY OF LOW-DENSITY LIPOPROTEINS

As has already been mentioned, elevated plasma LDL cholesterol concentrations are associated with increased risk for CHD. However, the atherogenicity of the LDL particle increases after oxidative modification of its polyunsaturated fatty acids.

A. LOW-DENSITY LIPOPROTEIN OXIDATION

Oxidation of LDL is a free radical-driven process that may initiate a cascade of reactions (Figure 22.6). For the in vivo situation, it is not clear where the initiating radical species is derived from, but several suggestions, based on in vitro experiments, have been made.

Some experiments have suggested that cellular production of superoxide anions (O_2^-) or hydroxyl radicals ($OH^\bullet$), which are intermediates in several metabolic processes from the mitochondrial

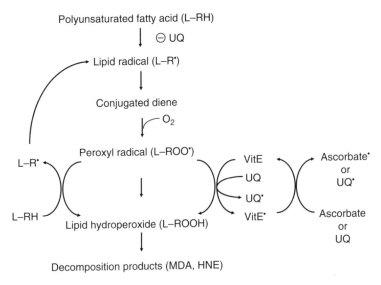

FIGURE 22.6 Schematic representation of lipid peroxidation. Polyunsaturated fatty acids are converted into lipid radicals, a process that can be inhibited by ubiquinol-10 (UQ). After molecular rearrangement, the lipid radical becomes a conjugated diene and then a peroxyl radical. This highly reactive species can attack other polyunsaturated fatty acids, thereby initiating a chain reaction. Vitamin E and ubiquinol, however, scavenge lipid peroxyl radicals, thereby breaking the chain reaction. Vitamin E can be regenerated by ascorbic acid (vitamin C) or ubiquinol-10.

respiratory chain or the cytochrome P450 system, initiate the lipid peroxidation reaction. Other experiments have proposed that lipid peroxidation is initiated by lipoxygenase activity, as this enzyme forms radicals as intermediate products in the formation of eicosanoids. These hypotheses are not necessarily contradictory because lipoxygenase activity might be particularly important in endothelial cells and peroxide initiation by superoxide anions and hydroxyl radicals in smooth muscle cells [22].

In vitro oxidation of LDL results in alterations in both the lipid and the protein components of LDL. The amount of unsaturated cholesterylester content decreases, especially cholesteryl arachidonate and cholesteryl linoleate. In addition, phosphatidylcholine—the main phospholipid in LDL—is converted to lysophosphatidylcholine after cleavage of a fatty acid from the *sn*-2 position by phospholipase A_2. It has been postulated that the released fatty acid is readily oxidized and might then become responsible for propagation of the lipid peroxidation chain reaction, as inhibitors of phospholipase A_2 block the generation not only of lysophospholipids but also of lipid peroxides [23].

After peroxidation, lipid peroxides decompose and breakdown products, such as malondialdehyde (MDA), and several aldehydes, such as 4-hydroxynonenal (4-HNE), are formed. These products can react with the ε-amino groups of apoB-100, which causes an irreversible modification of the apolipoprotein, as the number of free cysteine and charged lysine residues of apoB-100 decreases. This results in reduced recognition and uptake of LDL by the LDL receptor, since the affinity of the LDL receptor is based on binding of positively charged apoB-100 to the negatively charged binding domain of the LDL receptor. This reduced uptake is compensated for by an increased affinity of these modified LDLs to the acetyl or scavenger receptors on the cell surface of macrophages. This uptake is not downregulated and may lead to extensive lipid loading and the transformation of macrophages into foam cells (see also Section IV.B).

Peroxidation products are cytotoxic, and chronic irritation of endothelial cells results in lesions of the endothelial cell layer. In addition, lysophosphatidylcholine from oxidized LDLs and the expression of chemoattractant proteins like monocyte chemoattractant protein-1 (MCP-1), other chemoattractants, and inflammatory cytokines by damaged endothelial cells and leukocytes attract leukocytes from the circulation and initiate a local inflammation. Animal studies have shown that inflammatory cell recruitment and activation is critical for the development of atherosclerosis. For example, MCP-1 knockout mice and macrophage colony-stimulating factor (M-CSF) knockout mice are less susceptible for the development of atherosclerosis.

One of the earliest inflammatory steps in the atherogenesis is a slower rolling of leukocytes along the vascular endothelium, which proceeds by a subsequent attachment of rolling leukocytes to the vascular endothelium. In this process several adhesion molecules such as vascular cell adhesion molecule 1 (VCAM-1), intercellular adhesion molecule 1 (ICAM-1), and P-selectin play an important role. At least in vitro, these adhesion molecules are rapidly synthesized in response to several proinflammatory cytokines, such as tumor necrosis factor α (TNF-α) and interleukin-1 (IL-1). Moreover, HDL cholesterol particles can inhibit the cytokine-induced expression of adhesion molecules (VCAM-1 and E-selectin) on endothelial cells in vitro [24]. This finding may be a possible link with the antiatherogenic effect of high serum HDL cholesterol concentrations in vivo.

The presence of adhesion molecules on endothelial cells alone, however, is not enough for attachment of leukocytes to the endothelium. For this interaction leukocytes need to express ligands for these adhesion molecules. These ligands, expressed on lymphocytes and monocytes, are known as integrins. For example, very late antigen-4 (VLA-4), a β1-integrin is a ligand for VCAM-1 and lymphocyte function associated-1 (LFA-1) a β2-integrin for ICAM-1. After attachment, the proinflammatory leukocytes infiltrate into the activated endothelium, followed by further progressing of the inflammatory response. Interestingly, blocking VLA-4 by antibodies indeed decreased leucocyte entry and fatty streak formation in mice fed an atherogenic diet [25], which shows that this integrin plays a causal role during atherosclerosis. In conclusion, this process results in a continuous recruitment of new monocytes and T lymphocytes to the place of oxidation in the endothelium and in accumulation of macrophages filled with (oxidized) LDL in the arterial intima. In addition, several other processes are activated, which results in platelet aggregation, disturbance of eicosanoid homeostasis, and release of growth factors. These factors cause smooth muscle cell to proliferate and ultimately to migrate from the media to the intima. All these mechanisms together result in the formation of fatty streaks and atherosclerotic plaques (Figure 22.7).

1. Measurement of Low-Density Lipoprotein Oxidation

LDL oxidation is thought to be initiated and to proceed primarily in the endothelial layer, whereas oxidized LDL particles are rapidly removed from the circulation. Thus, it is very difficult to quantify the LDL oxidation process in vivo, and several in vitro assays have been developed to measure LDL oxidation tendency in vitro.

a. In Vitro Copper-Mediated Low-Density Lipoprotein Oxidation

Oxidative in vivo modification of LDL can be mimicked by exposure of LDL particles in vitro to redox-active metal ions, such as copper (Cu^{2+}), or to reactive oxygen species, such as superoxide anion.

Esterbauer et al. [26] have developed a method—frequently used in the earlier studies—to determine in vitro the susceptibility of LDL to oxidation by continuous monitoring of the formation of conjugated dienes, products formed after oxidation of the polyunsaturated fatty acids from the LDL particle. After LDL isolation, copper is added to the test tube to initiate the oxidation process, and the formation of conjugated dienes is then quantified by measuring the change in absorbance at 234 nm (Figure 22.8). This curve can be divided into three consecutive phases: the lag phase, the propagation phase, and the decomposition phase.

During the lag phase, LDL-bound lipophilic antioxidants protect the polyunsaturated fatty acids from oxidation. Tocopherols and β-carotene, for example, scavenge lipid peroxide radicals, thereby breaking the chain reaction. The antioxidant has now become a relatively stable radical, which does

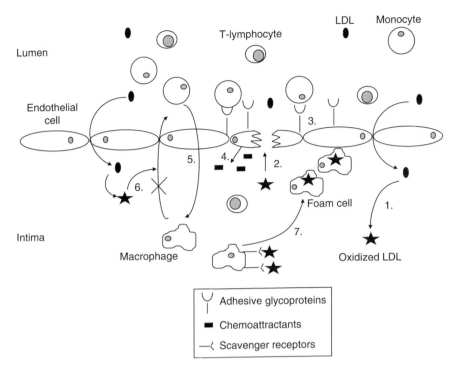

FIGURE 22.7 Schematic representation of the formation of an atherosclerotic plaque. (1) LDL enters the intimal layer and can be oxidized by several factors, such as lipoxygenase or reactive oxygen species. (2) Oxidized LDL is cytotoxic and causes endothelial damage, (3) which results in the expression of adhesive glycoproteins to which monocytes and T-lymphocytes attach. (4) The damaged endothelial cells excrete chemoattractants, which causes a continuous recruitment of monocytes and T-lymphocytes. (5) These cells pass the endothelial cell layer and monocytes may become macrophages. (6) Oxidized LDL prevents return of macrophages back to the lumen, and (7) the arrested macrophages absorb large amounts of oxidized LDL via the scavenger receptors and become foam cells, which may eventually lead to the formation atherosclerotic plaques.

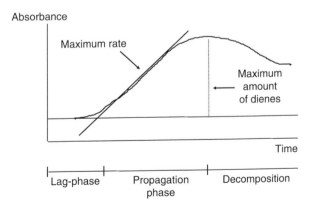

FIGURE 22.8 In vitro LDL oxidation. Formation of conjugated dienes during copper-catalyzed oxidation of LDL in vitro is monitored spectrophotometrically at 234 nm. The lag time before onset of rapid oxidation, the maximum rate of oxidation during the propagation phase, and the maximum amount of dienes formed are used to describe LDL oxidation characteristics.

not induce lipid peroxidation but is also not regenerated, as in vivo may happen. Addition of the water-soluble ascorbic acid (vitamin C) that is lost during LDL isolation leads to an increase in the lag phase because this antioxidant can regenerate tocopherols. Thus, at a certain stage the LDL particle becomes depleted of antioxidants; at this point, the peroxidation reaction shifts to an autocatalytic process. Now the propagation phase starts during which polyunsaturated fatty acids are oxidized and converted into conjugated dienes. The absorbance at 234 nm then reaches a maximum and may eventually decrease, since the produced conjugated dienes are labile and decompose to several products, such as MDA and 4-HNE.

The oxidation profile of LDL is described by the duration of the lag time, the maximal amount of dienes formed, and the maximal rate of oxidation. A short lag time and a large amount of dienes is considered to reflect a high oxidative susceptibility of the LDL particle. However, interpretation of the oxidation rate is not clear.

b. Thiobarbituric Acid-Reactive Substances

LDL oxidation is also estimated by measuring in plasma the amount of thiobarbituric acid-reactive substances (TBArs), such as MDA and MDA-like substances, which can be formed during the in vivo peroxidation process. The detection of TBArs in plasma and LDL might therefore be an indication of the possible occurrence of peroxidative injury. However, several authors regard this assay as nonspecific, as compounds that are not a product of lipid peroxidation are also measured [27].

c. Antibodies

Oxidation of LDL in vivo can be measured by the detection of autoantibodies against MDA-modified LDL in human plasma [28]. Another approach for the determination of in vivo LDL oxidation is the measurement of antibodies that cross-react with MDA-modified apoB-100, but not with native apoB-100 [29,30].

d. Isoprostanes

Isoprostanes are isomers of prostaglandins (PGs), which are produced in vivo primarily—if not exclusively—by free radical-mediated peroxidation of polyunsaturated fatty acids. Especially the F2-isoprostanes, which are isomers of the $PGF_{2\alpha}$, derived from peroxidation of arachidonic acid, are considered suitable markers for oxidative modifications in vivo [31]. F2-isoprostanes can be analyzed in the circulation as well as in the urine by gas chromatography/mass spectrometry (GC-MS) or an immunoassay.

2. Low-Density Lipoprotein Oxidation and Coronary Heart Disease

Although causality has not yet been proved, it would be unwise to ignore a possible role of LDL oxidation in the genesis of CHD.

Palinsky et al. [28] have demonstrated the existence of autoantibodies directed against MDA-LDL in the human circulation. Further, Ylä-Herttuala et al. [29] have shown that atherosclerotic lesions contain compounds that react with antibodies directed against MDA-modified LDL and 4-HNE-lysine, whereas those antibodies did not react with native LDL. In that same study it was found that LDL isolated from atherosclerotic lesions showed a substantial correspondence to the characteristics of in vitro oxidized LDL, such as an increased amount of lysophosphatidylcholine and a chemotactic activity for monocytes. Furthermore, patients with acute myocardial infarcts or with carotid atherosclerosis show significantly higher plasma concentrations of MDA-modified LDL than control subjects [30].

Urinary immunoreactive F2-isoprostanes were higher in hypercholesterolemic patients than in controls. Moreover, urinary F2-isoprostanes were inversely related to serum LDL cholesterol concentrations and LDL vitamin E levels [32]. In addition, in noninsulin-dependent diabetic patients urinary immunoreactive isoprostane levels were higher than those in controls, which could be counteracted by vitamin E supplementation [33].

TABLE 22.4

Fatty Acid Composition (g/100 g Fatty Acid) of an LDL Particle after Consumption of a Diet Rich in Oleic Acid or Linoleic Acids

| | Diet | |
Fatty Acid	Oleic Acid	Linoleic Acid
Saturated	24.7	24.8
Monounsaturated	25.0	20.4
Polyunsaturated	50.3	54.8
(*n*-6)	45.7	51.2
(*n*-3)	4.1	3.3

Source: From Abbey, M. et al., *Am. J. Clin. Nutr.*, 57, 391, 1993.

Not only oxidation of polyunsaturated fatty acids but also certain oxidized forms of cholesterol (oxysterols) are atherogenic and may play a role in plaque development [34]. Especially 7β-hydroxycholesterol may be a good marker for free radical-related lipid peroxidation. Several studies have now shown that in humans 7β-OH-cholesterol is associated with the risk for atherosclerosis [35,36]. In addition, feeding a mixture of various oxysterols to atherosclerosis-prone LDL receptor $(-/-)$ and apoE $(-/-)$ mice accelerated fatty streak lesion formation in comparison with cholesterol feeding alone [37]. For humans, however, so far no causal relation has been established between oxysteroids and lesion formation.

3. Dietary Effects

The fatty acid composition of the diet, and in particular the amount of polyunsaturated fatty acids, is reflected by the fatty acid composition of the LDL particle (Tables 22.4 and 22.5). As polyunsaturated fatty acids are more easily oxidized than monounsaturated or saturated fatty acids, it can be envisaged that LDL oxidizability is influenced by changing the dietary fatty acid composition. However, when interpreting the results it should be realized that diet always induces multiple

TABLE 22.5

Fatty Acid Composition (g/100 g Fatty Acid) of an LDL Particle after Consumption of a Diet Rich in (*n*-6) or (*n*-3) Polyunsaturated Fatty Acids from Respectively Corn or Fish Oils

| | Diet | |
Fatty Acid	Corn Oil	Fish Oil
Saturated	25.5	26.1
Monounsaturated	23.5	22.3
Polyunsaturated	48.9	49.3
(*n*-6)	46.4	40.7
(*n*-3)	2.5	8.6

Source: Adapted from Suzukawa, M. et al., *J. Lipid Res.*, 36, 473, 1995.

changes in the fatty acid composition of the LDL particle. For example, the proportion of linoleic acid in the LDL particle increases after enrichment of the diet with this fatty acid, whereas the proportion of oleic acid, arachidonic acid, palmitoleic acid, and palmitic acid decreases [38].

a. Effects of Linoleate-Rich versus Oleate-Rich Diets

Linoleic acid, an n-6 polyunsaturated fatty acid, contains more unsaturated bonds than oleic acid and is preferentially incorporated into tissue lipids. Accordingly, increasing the amount of linoleic acid in the diet at the expense of oleic acid leads to a higher proportion of linoleic acid in the LDL particle (Table 22.4). Therefore, it can be expected that, at least in vitro, LDL is more easily oxidized after consumption of linoleic acid-enriched diets. Indeed, several studies have demonstrated that replacement of oleic acid in the diet for linoleic acid may result in a decreased lag time, a higher production of dienes, and a reduced oxidation rate. In addition, LDL uptake by macrophages was increased after consumption of linoleic acid-enriched diets, suggesting that in vivo the LDL was modified to a greater extent [38–42].

b. Effects of n-6 versus n-3 Polyunsaturated Fatty Acids

Consumption of fish oils or fish oil capsules, which are rich in EPA and DHA, also affects the fatty acid composition of the LDL particle. The amount of these two n-3 polyunsaturated fatty acids increases mainly at the cost of n-6 polyunsaturated fatty acids (Table 22.5). Effects of fish oil relative to linoleic acid supplementation on LDL oxidation are contradictory.

Suzukawa et al. [43] found a reduction in lag time after in vitro copper-mediated LDL oxidation and an increased uptake of LDL by macrophages after supplementation with n-3 polyunsaturated fatty acids as compared with a corn oil-supplemented diet. Oostenbrug et al. [44] found dietary fish oils to increase the maximal amount of conjugated dienes formed during in vitro copper-mediated oxidation, whereas the lag time decreased. Several other studies demonstrated an increase in the amounts of TBArs in plasma and in LDL after fish oil supplementation [45–47]. However, other studies did not demonstrate an effect of dietary fish oil relative to oils rich in linoleic acid on in vitro LDL oxidation [48,49]. Currently, there is no explanation for these contradictory results.

c. Antioxidants

Fat-soluble antioxidants, especially α-tocopherol (vitamin E), are a main protecting factor against in vitro LDL oxidation. Supplementation of vitamin E increases the tocopherol content of LDL, resulting in a higher oxidation resistance of LDL, as evidenced from an increased lag time [50]. Other antioxidants, such as carotenoids, ascorbic acid, and ubiquinol-10, also have an important impact on the oxidative resistance of LDL.

Ascorbic acid, a hydrophilic antioxidant, and ubiquinol-10, a lipophilic antioxidant, are capable of regenerating tocopherols, and addition of these antioxidants to the test tube results in a longer lag time [51–53].

Oils rich in polyunsaturated fatty acids by nature contain relatively high concentrations of antioxidants. An exception are fish oils, which have relatively low levels of fat-soluble antioxidants, but their potentially harmful effects on LDL oxidation can be counteracted by addition of vitamin E to these oils.

B. CONCLUSION

Several studies now strongly suggest that oxidation of LDL in vivo takes place, which results in even more harmful LDL particles. Therefore, LDL oxidation may play an important role in the formation of atherosclerotic lesions.

Replacing saturated fatty acids in the diet with linoleic acid reduces plasma LDL cholesterol concentrations. However, this dietary intervention also increases the proportion of linoleic acid in the LDL particle thereby its in vitro oxidizability. As monosaturated fatty acids are less readily oxidized and have comparable beneficial effects on the plasma lipoprotein profile, oleic acid might be preferred over linoleic acid. Increased in vitro oxidizability of the LDL particle might also be observed when *n*-3 polyunsaturated fatty acids from fish oils are added to the diet, which can be overcome by simultaneously increasing the intake of vitamin E. Although these studies clearly demonstrate that diet affects in vitro LDL oxidizability, the importance of these findings for the in vivo situation are less evident. In fact, there is no good evidence that increased intakes of antioxidants protects against cardiovascular disease [54].

VIII. HEMOSTASIS

Hemostasis, derived from the Greek words for blood and standing, is a complex, delicately balanced system of interactions to keep the blood circulating as a fluid through the blood vessels. In case of imbalance, such as when a vessel is damaged, the blood stands, starts to clot, and a stable thrombus forms. This, of course, is necessary to stop a wound from bleeding. However, if a stable thrombus is formed in a small coronary artery, the artery becomes occluded, blood and oxygen supply are hampered, and heart attack results.

The hemostatic system involves interacting processes for the formation of a stable thrombus—platelet aggregation and blood clotting—but also a mechanism to dissolve the thrombus, that is, fibrinolysis (Figure 22.9). In vivo these processes are associated and the interplay defines the prethrombotic state of the blood.

Hemostatic factors are difficult to measure. Due to the venipuncture and subsequent blood sampling and plasma preparation, platelets might become activated, which makes it very difficult to obtain a true reflection of the in vivo situation. In addition, measurements are usually made in venous fasting blood, while one is interested in thrombotic tendency in the arteries. In addition, many different methods are used, which makes a comparison between studies difficult.

A. Platelet Aggregation

The activity of blood platelets is an important factor for thrombus formation. Aggregated platelets adhere to the injured blood vessel to form a hemostatic plug, excrete substances such as thrombin and calcium, and provide a phospholipid surface—all of which are important for blood coagulation.

Several mechanisms are being proposed to explain the effects of fatty acids on platelet aggregation. Differences in fatty acid composition can change the arachidonic acid content of platelet and endothelial phospholipids. Arachidonic acid acts as a substrate for thromboxane A_2 (TxA_2) in platelets and prostacycline (PGI_2) in endothelial cells, and the balance between these two eicosanoids affects platelet aggregation (Figure 22.10). Fatty acids have also been reported to

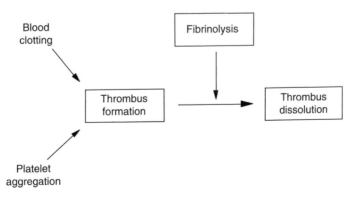

FIGURE 22.9 Processes involved in thrombus formation. The hemostatic system involves interacting processes for the formation of a stable thrombus—platelet aggregation and blood clotting—but also a mechanism to dissolve the thrombus, fibrinolysis.

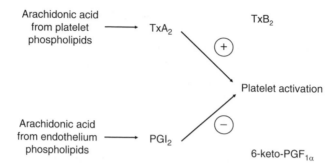

FIGURE 22.10 Schematic representation of platelet activation. Arachidonic acid from platelet acts as a substrate for thromboxane A_2 (TxA_2) and arachidonic acid from the endothelial cells for prostacycline (PGI_2). The balance between these two eicosanoids affects platelet activation. TxB_2 is the stable metabolite of TxA_2, and 6-keto-PGF$_{1\alpha}$ of PGI_2.

directly affect TxA_2 receptors on platelet membranes. A third mechanism is that differences in fatty acid composition can affect the cholesterol content of membranes, and consequently affect the fluidity of platelet membranes and platelet activation.

1. Measurement of Platelet Aggregation

A broad scale of methods is available to measure platelet aggregation in vitro. First, the blood sample needs to be anticoagulated to avoid clotting of the blood in the test tube or in the aggregometer. Different anticoagulants are used, such as citrate, which depletes the sample from calcium, and heparin or hirudin, which cause an inhibition of the conversion of prothrombin to thrombin. In vitro platelet aggregation can then be measured in whole blood, in platelet-rich plasma, or—to remove the influence of possible interfering constituents from the plasma—in a washed platelet sample. Finally, the aggregation reaction in the test tube can be triggered with many different compounds, such as collagen, adenosine diphosphate (ADP), archidonic acid, and thrombin.

2. Platelet Aggregation and Coronary Heart Disease

In vitro platelet aggregation has been reported to be an important marker for the prediction of reoccurrence of coronary events [55]. In addition, platelet aggregation measured in whole blood was strongly associated with the prevalence of ischemic heart disease [56].

3. Dietary Fats and Platelet Aggregation

a. Total Fat Content of Diets

Renaud and coworkers [57] studied nine groups of farmers from different areas in France and Britain, who differed with respect to dietary intakes of total and saturated fatty acids. In the groups with a high consumption of total and saturated fatty acids, an increased thrombin-induced aggregation in platelet-rich plasma was observed as compared with the groups of farmers with lower intakes. In a later intervention study [58], the diets of French farmers were reduced in saturated fat content by replacing the habitually consumed dairy fat by high linoleic acid margarine. A control group of farmers was advised not to change their diets. In the intervention group, total fat intake decreased along with the intake from saturated fatty acids, whereas the intake of dietary linoleic acid and α-linolenic acid was increased compared with initial values. In agreement with the previous study [57], a decreased thrombin-induced aggregation was observed in the intervention group, whereas platelet aggregation did not change in the control group. However, platelet aggregation induced by ADP was significantly increased in the intervention group. From this dietary intervention, however, it was not clear whether the changes of total fat content or the changed fatty acid composition of the diets were responsible for the changes observed in platelet aggregation.

b. Dietary Fatty Acid Composition

The effects of dietary fatty acid composition were further evaluated in well-controlled dietary intervention studies, in which total fat content of the diets was kept constant and only the dietary fatty acid composition changed. Results however are inconsistent.

When saturated fatty acids were replaced by oleic acid or linoleic acid, platelet aggregation was increased, decreased, or not changed. Comparable conflicting results have been found when oleic and linoleic acids were compared side by side. Other studies have examined the effects of different saturated fatty acids. Compared with oleic acid, MCFAs, lauric, myristic, or palmitic acids did not change collagen- or ADP-induced whole blood aggregation [59]. In another study, similar effects of stearic, oleic, and linoleic acids were found [60]. Thus, dietary fatty acids can modulate platelet aggregation, but the use of many different in vitro methods makes comparison between studies and extrapolation to the in vivo situation difficult. More consistent however are the effects of fish oils, which decrease platelet aggregation tendency [61].

Effects of specific saturated fatty acids on stable metabolites of TxA_2 and PGI_2—TxB_2 and 6-keto-$PGF_{1\alpha}$, respectively—in urine showed that lauric, myristic, palmitic, and stearic acids had similar effects on urinary thromboxane and PG excretion [62,63]. In addition, effects of *trans* fatty acids were comparable with those of stearic acid [64]. Furthermore, *n*-6 polyunsaturated fatty acids increased urinary 11-dehydro-TxB_2 excretion compared with saturated and monounsaturated fatty acids, which may be related to ADP-induced platelet aggregation [65].

B. COAGULATION

Several pathways for blood coagulation exist. The tissue factor pathway of blood coagulation, previously known as the extrinsic pathway of blood coagulation, appears to be the most important one. The factors involved in this pathway are depicted in a simplified scheme in Figure 22.11.

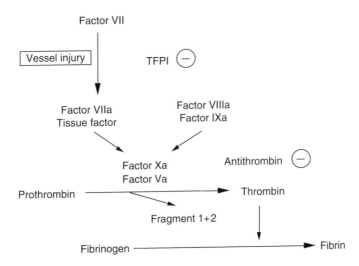

FIGURE 22.11 Schematic representation of the coagulation cascade. When factor VII contacts a tissue factor, expressed for example after vessel injury or inflammation, factor VII is rapidly activated into factor VIIa. The complex of factor VIIa with tissue factor initiates a cascade of reactions, which ultimately results—after cleavage of fragment 1 + 2—in the conversion prothrombin into thrombin. Thrombin cleaves fibrinogen into fibrin, which stabilizes a thrombus. Tissue factor pathway inhibitor (TFPI) and antithrombin-III inhibit the coagulation cascade.

Most coagulation factors are mainly present in inactivated form, except for factor VII, of which 1% circulates as activated factor VII (factor VIIa). However, Figure 22.11 only shows the activated coagulation factors, except for factor VII.

Thrombus formation in vivo is initiated when factor VII or factor VIIa contacts thromboplastin tissue factor, expressed, for example, after vessel injury or inflammation. A tissue factor, such as thromboplastin, is a procoagulant that is expressed only on activated endothelium. Once bound, factor VII is rapidly activated into factor VIIa. The complex of factor VIIa with tissue factor initiates a cascade of reactions, which ultimately results in the conversion of factor X into factor Xa and the generation of thrombin from prothrombin. Thrombin finally cleaves fibrinogen into fibrin, which stabilizes a thrombus. However, thrombin also inhibits some coagulation factors, and the coagulation cascade is thus inhibited by one of its end products so as to prevent uncontrolled formation of fibrin.

The coagulation cascade is further regulated by the action of coagulation inhibitors. An important inhibitor of coagulation is the tissue factor pathway inhibitor (TFPI), which inhibits the activity of the tissue factor–factor VIIa complex, and TFPI therefore prevents further activation of the coagulation cascade. The tissue factor–factor VIIa complex is also inhibited by antithrombin III, which also suppresses the activation of thrombin and other activated coagulation factors.

1. Measurement of Coagulation

Most assays measure the total amount (e.g., factor VII–antigen or fibrinogen concentrations) or the activity of circulating coagulation factors. However, these factors are normally present in large excess in the blood, and only a small percentage is converted to active enzymes under in vivo situations. Nowadays other assays are also available, which reflect actual in vivo coagulation. The plasma fragment $1 + 2$ concentration reflects the amount of prothrombin actually converted to thrombin, whereas fibrinopeptides A and B concentrations reflect the conversion of fibrinogen to fibrin. However, in healthy subjects, concentrations of most of these markers are very low and just above detection limits, which sometimes make it difficult to implement them in dietary studies.

Many methods are available to measure factor VII. In many studies, factor VII is measured with a coagulant assay (factor VII coagulant activity). Factor VII activity, however, is measured with a two-step chromogenic assay, which depends on the rate of generation of factor Xa from factor X by factor VIIa. Both methods may give different results and do not differentiate between factor VIIa and factor VII antigen concentrations.

Fibrinogen concentrations are usually measured with the method of Clauss [66]. A fixed surplus of thrombin is added to diluted platelet-poor plasma samples, the clotting time recorded, and the fibrinogen concentration read from a calibrator curve.

2. Coagulation and Coronary Heart Disease

Long-term prospective epidemiological studies have consistently reported that in healthy males factor VII coagulant activity and fibrinogen concentrations were higher in subjects who developed cardiovascular diseases at a later stage of the study. Factor VII coagulant activity was particularly associated with an increased risk of dying from cardiovascular disease [67,68]. In addition, from the Northwick Park Heart Study (NPHS) it has been reported that low and, unexpectedly, also high concentrations of antithrombin III were associated with increased deaths from CHD [69]. It must be noted that these prospective studies have only been carried out in males; whether associations for females are similar awaits confirmation.

3. Dietary Fats and Coagulation

a. Total Fat Content of Diets

Marckmann and colleagues [70] have investigated both short- and longer-term effects of low-fat/high-fiber diets on human blood coagulation. In an 8 month study it was found a low-fat/high-fiber diet significantly decreased plasma factor VII coagulant activity by 5%–10%, but only in the first 2 months and in the last months of dietary intervention. Absence of effects in the middle study period was explained by the fact that the subjects in this period did not follow the dietary guidelines strictly because of allowed study holidays. Plasma fibrinogen concentrations were not changed. The results of this study were confirmed in a study of shorter duration [71].

In another study, Marckmann et al. [72] investigated whether the low-fat or the high-fiber component of diets was responsible for decreased factor VII coagulant activity found in earlier studies. The experimental diets of this trial only differed in their fat content (39% versus 31% of energy) and carbohydrate content (47% versus 54% of energy). Factor VII coagulant activity was similar on the low-fat and high-fat diets. In addition, no changes in fibrinogen concentrations were found. This limited number of studies indicates reducing effects on factor VII coagulation activity of low-fat/high-fiber diets. However, more controlled studies are needed to definitely address whether the reduced fat content of low-fat diets, the increased fiber content, or a combination of these two dietary factors are responsible for the decreased factor VII coagulant activity of such diets.

b. Dietary Fatty Acid Composition

The effects of saturated compared with unsaturated fatty acids have been studied in several experiments. Irrespective of the marker used, effects were small, whereas results varied between studies. Diets enriched in *n*-3 fatty acids from fish did also not change factor VII coagulant activity [73,74] or fibrinogen concentrations [73–75].

A study of Almendingen et al. [76] investigated the effects of *trans* fatty acids in diets enriched with hydrogenated fish oil or hydrogenated soybean oil compared with a diet enriched in butterfat. The diet enriched in butterfat showed slightly increased fibrinogen concentrations as compared with the hydrogenated fish oil diet. No significant differences in the levels of factor VII or fibrin degradation products were observed.

Well-controlled studies investigating effects of specific saturated fatty acids have also been published. Tholstrup et al. [77] reported that diets rich in lauric plus myristic acids or palmitic acid increased factor VII coagulant activity as compared with a diet rich in stearic acid. It was also suggested that the turnover of prothrombin was increased on the lauric plus myristic acids diet as fragment 1 + 2 concentrations were higher on such a diet than on a stearic acid diet [78] (Figure 22.12). In a second

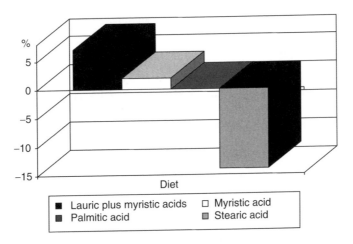

FIGURE 22.12 Effects of lauric plus myristic acid, myristic acid, and stearic acid on factor VII coagulant activity relative to those of palmitic acid. Healthy young men consumed diets enriched in lauric plus myristic acids (C12:0 and C14:0; 14% of energy), myristic acid (C14:0; 14% of energy), palmitic acid (C16; 15%–16% of energy), or stearic acid (C18:0; 14% of energy) for 3 weeks. (From Tholstrup, T. et al., *Am. J. Clin. Nutr.*, 59, 371, 1994; Tholstrup, T. et al., *Am. J. Clin. Nutr.*, 60, 919, 1994.)

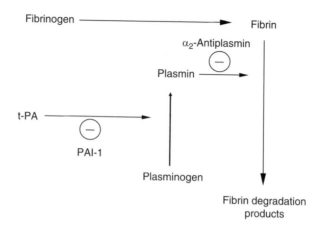

FIGURE 22.13 Schematic representation of the fibrinolytic pathway. Fibrin is degraded by plasmin. The conversion of plasmin from plasminogen is regulated by the action of tissue plasminogen activator (t-PA). t-PA activity is suppressed by plasminogen activator inhibitor type-1 (PAI-1), while plasmin is inhibited by α2-antiplasmin.

study, Tholstrup et al. [79] investigated diets rich in myristic or palmitic acid. On the myristic acid diet, subjects showed an increase in factor VII coagulant activity. Thus, these studies [77,79] suggest that saturated fatty acids, except for stearic acid and probably MCFA, increase factor VII coagulant activity.

C. Fibrinolysis

The process involved in thrombus dissolution, and thus the conversion of fibrin into fibrin degradation products, is called fibrinolysis. A simplified scheme of the fibrinolytic pathway is given in Figure 22.13. The central reaction in the fibrinolytic process is the conversion of plasminogen into plasmin, which is regulated by the action of tissue plasminogen activator (t-PA). The fibrinolytic capacity of blood is regulated by inhibiting t-PA activity by the action of plasminogen activator inhibitor type-1 (PAI-1), whereas plasmin is inhibited mainly by α_2-antiplasmin.

1. Measurement of Fibrinolytic Capacity of Plasma

The fibrinolytic capacity of plasma can be measured by global tests or more specific assays. The global tests include the dilute clot lysis time, the euglobulin clot lysis time, and the fibrin plate assay. The dilute and euglobulin clot lysis time measure total fibrinolytic capacity. The total blood sample or the insoluble protein sample (euglobulin fraction) is diluted with a buffer, clotted, and the lysis time of the clot recorded. In the fibrin plate assay, a standard volume of the euglobulin fraction of plasma is added to standardized plasminogen-rich fibrin plates and the amount of lysis is recorded. In more specific assays, total plasma concentrations of t-PA and PAI-1 can be determined with enzyme-linked immunosorbent assays. Plasma t-PA and PAI-1 activities can be estimated with a chromogenic assay.

2. Fibrinolysis and Coronary Heart Disease

From the NPHS has been reported that low fibrinolytic capacity of plasma, measured as clot lysis time, was significantly associated with increased CHD risk in men aged 40–54 [80]. In addition, Hamsten et al. [81] have reported that higher concentrations of PAI-1 were associated with increased risk of reoccurrence of coronary events.

3. Dietary Fats and Fibrinolysis

a. Total Fat Content of Diets

Marckmann et al. [70,71] found in both a short-term (2 week) and a longer-term trial (8 months) that plasma euglobulin fibrinolytic capacity and plasma t-PA activity were increased on low-fat/high-fiber diets as compared with high-fat/low-fiber diets (Figure 22.14). No changes were observed in t-PA and PAI-1 antigen concentrations. In agreement with these results, Mehrabian et al. [82] reported decreased plasminogen and PAI-1 activities, after consumption of a low-fat/high-fiber/low-cholesterol diet with less than 10% of energy from fat for 21 days. This was associated with decreased t-PA antigen concentrations. However, from these studies it is not possible to conclude whether the changes observed were due to the lower fat or cholesterol contents, or the higher fiber content of the diets.

In another study, Marckmann et al. [72] have also compared two diets, which differed in total fat content though fiber intake and the relative fatty acid composition were comparable. Their results indicated that plasma euglobulin fibrinolytic capacity on a high-fat diet (39% of energy from fat) did not change compared with a low-fat diet (31% of energy). It was concluded that an isolated reduction of total fat content of the diet does not affect the fibrinolytic capacity of the blood but that concomitant changes in the diet, as for example changes in fatty acid composition or fiber content, are necessary to provoke changes of fibrinolytic factors.

b. Dietary Fatty Acid Composition

Effects of different fatty acid composition on fibrinolytic capacity have been investigated extensively for fatty acids from fish, although recently additional data have become available for the effects of other fatty acids.

High intakes of fish may affect fibrinolytic capacity. Brown and Roberts [74] studied the effects of a daily consumption of 200 g of lean fish, with or without fish oil supplement. They observed an apparent enhancement of plasma euglobulin fibrinolytic capacity, which tended to be accentuated with fish oil supplementation. However, compared with meat diets, diets enriched with fatty fish have also been associated with an unfavorable increase in PAI-1 activity [73,75]. Increased PAI-1 activity has also been reported on diets enriched in partially hydrogenated soybean oil compared with a butterfat diet [76].

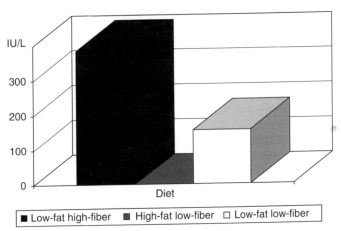

FIGURE 22.14 Effects of fat and fiber on the plasma euglobolin fibrinolytic activity relative to those of a high-fat/low-fiber diet. Healthy individuals consumed a low-fat/high-fiber (28% of energy from fat and 3.3 g/MJ fiber), a high-fat/low-fiber (39% of energy from fat and 2.1–2.3 g/MJ fiber), or a low-fat/low-fiber (31% of energy from fat and 2.2 g/MJ fiber) diet for 2 weeks. (From Marckmann, P., Sandström, B., and Jespersen, J., *Am. J. Clin. Nutr.*, 59, 935, 1994; Marckmann, P., Sandström, B., and Jespersen, J., *Arterioscler. Thromb.*, 12, 201, 1992.)

In other studies no unfavorable effects of lauric plus myristic acids, palmitic acid, or stearic acid diets relative to other unsaturated fatty acids were reported [60,77,79].

D. Conclusion

Results of platelet aggregation studies are difficult to interpret because of large differences in the methods used. Lowering dietary fat intake decreases in vitro platelet aggregation tendency induced by some antagonists but increases in vitro platelet aggregation by other antagonists. In addition, the effects of dietary fatty acid composition on in vitro platelet aggregation are contradictory.

To affect coagulation and fibrinolytic factors, it appears that lowering fat intake must be accompanied by changes in dietary fatty acid composition and increased fiber intakes. Though results between studies are less consistent, favorable effects on blood coagulation may be obtained by lowering the content of the major dietary saturated fatty acids and replacing them with unsaturated fatty acids.

REFERENCES

1. Kannel, W.B. and Wilson, P.W., Efficacy of lipid profiles in prediction of coronary disease, *Am. Heart J.*, 124, 768, 1992.
2. Hokanson, J.E. and Austin, M.A., Plasma triglyceride level is a risk factor for cardiovascular disease independent of high-density lipoprotein cholesterol level: a meta-analysis of population based prospective studies, *J. Cardiovasc. Risk*, 3, 213, 1996.
3. Keys, A., Anderson, J.T., and Grande, F., Prediction of serum-cholesterol responses of man to changes in fats in the diet, *Lancet*, 2, 959, 1957.
4. Keys, A., Anderson, J.T., and Grande, F., Serum cholesterol response to changes in the diet IV. Particular saturated fatty acids in the diet, *Metabolism*, 14, 776, 1965.
5. Hegsted, D.M. et al., Quantitative effects of dietary fat on serum cholesterol in man, *Am. J. Clin. Nutr.*, 17, 281, 1965.
6. Keys, A., Anderson, J.T., and Grande, F., Serum cholesterol response to changes in the diet II. The effect of cholesterol in the diet, *Metabolism*, 14, 759, 1965.
7. McGandy, R.B., Hegsted, D.M., and Myers, M.L., Use of semisynthetic fats in determining effects of specific dietary fatty acids on serum lipids in man, *Am. J. Clin. Nutr.*, 23, 1288, 1970.
8. Cater, N.B., Heller, H.J., and Denke, M.A., Comparison of the effects of medium-chain triacylglycerols, palm oil, and high oleic sunflower oil on plasma triacylglycerol fatty acids and lipid and lipoprotein concentrations in humans, *Am. J. Clin. Nutr.*, 65, 41, 1997.
9. Temme, E.H., Mensink, R.P., and Hornstra, G., Effects of medium chain fatty acids (MCFA), myristic acid, and oleic acid on serum lipoproteins in healthy subjects. *J. Lipid Res.*, 38, 1746, 1997.
10. Mensink, R.P. et al., Effects of dietary fatty acids and carbohydrates on the ratio of serum total to HDL cholesterol and on serum lipids and apolipoproteins: a meta-analysis of 60 controlled trials, *Am. J. Clin. Nutr.*, 77, 1146, 2003.
11. Grande, F., Anderson, J.T., and Keys, A., Comparison of effects of palmitic and stearic acids in the diet on serum cholesterol in man, *Am. J. Clin. Nutr.*, 23, 1184, 1970.
12. Bonanome, A. and Grundy, S.M., Effect of dietary stearic acid on plasma cholesterol and lipoprotein levels, *N. Engl. J. Med.*, 318, 1244, 1988.
13. Grundy, S.M., Comparison of monounsaturated fatty acids and carbohydrates for lowering plasma cholesterol, *N. Engl. J. Med.*, 314, 745, 1986.
14. Mensink, R.P. and Katan, M.B., Effect of monounsaturated fatty acids versus complex carbohydrates on high-density lipoproteins in healthy men and women, *Lancet*, 1, 122, 1987.
15. Mattson, F.H. and Grundy, S.M., Comparison of effects of dietary saturated, monounsaturated, and polyunsaturated fatty acids on plasma lipids and lipoproteins in man, *J. Lipid Res.*, 26, 194, 1985.
16. Mensink, R.P. and Katan, M.B., Effect of a diet enriched with monounsaturated or polyunsaturated fatty acids on levels of low-density and high-density lipoprotein cholesterol in healthy women and men, *N. Engl. J. Med.*, 321, 436, 1989.

17. Valsta, L.M. et al., Effects of a monounsaturated rapeseed oil and a polyunsaturated sunflower oil diet on lipoprotein levels in humans, *Arterioscler. Thromb.*, 12, 50, 1992.

18. Wahrburg, U. et al., Comparative effects of a recommended lipid-lowering diet vs a diet rich in monounsaturated fatty acids on serum lipid profiles in healthy young adults, *Am. J. Clin. Nutr.*, 56, 678, 1992.

19. Chan, J.K., Bruce, V.M., and McDonald, B.E., Dietary α-linolenic acid is as effective as oleic acid and linoleic acid in lowering blood cholesterol in normolipidemic men, *Am. J. Clin. Nutr.*, 53, 1230, 1991.

20. Harris, W.S., Fish oils and plasma lipid and lipoprotein metabolism in humans: a critical review, *J. Lipid Res.*, 30, 785, 1989.

21. Katan, M.B., Zock, P.L., and Mensink, R.P., Trans fatty acids and their effects on lipoproteins in humans, *Annu. Rev. Nutr.*, 15, 473, 1995.

22. Steinbrecher, U.P., Zhang, H., and Lougheed, M., Role of oxidatively modified LDL in atherosclerosis, *Free Radic. Biol. Med.*, 9, 155, 1990.

23. Steinberg, D. et al., Beyond cholesterol. Modifications of low-density lipoprotein that increase its atherogenicity, *N. Engl. J. Med.*, 320, 915, 1989.

24. Cockerill, G.W. et al., High density lipoproteins inhibit cytokine induced expression of endothelial cell adhesion molecules. *Arterioscler. Thromb. Vasc. Biol.*, 15, 1987–1991, 1995.

25. Shih, P.T. et al., Blocking very late antigen-4 integrin decreases leucocyte entry and fatty streak formation in mice fed and atherogenic diet, *Circ. Res.*, 84, 345–351, 1999.

26. Esterbauer, H. et al., Continuous monitoring of in vitro oxidation of human low density lipoprotein, *Free Radic. Res. Commun.*, 6, 67, 1989.

27. Janero, D.R., Malondialdehyde and thiobarbituric acid-reactivity as diagnostic indices of lipid peroxidation and peroxidative tissue injure, *Free Radic. Biol. Med.*, 9, 515, 1990.

28. Palinsky, W. et al., Low density lipoprotein undergoes oxidative modification in vivo, *Proc. Natl. Acad. Sci. U.S.A.*, 86, 1372, 1989.

29. Ylä-Herttuala, S. et al., Evidence for the presence of oxidatively modified low density lipoprotein in atherosclerotic lesions of rabbit and man, *J. Clin. Invest.*, 84, 1086, 1989.

30. Holvoet, P. et al., Malondialdehyde-modified low density lipoproteins in patients with atherosclerotic disease, *J. Clin. Invest.*, 95, 2611, 1995.

31. Pratico, D., F2-isoprostanes, sensitive and specific non-invasive indices of lipid peroxidation in vivo, *Atherosclerosis*, 147, 1–10, 1999.

32. Davi, G. et al., In vivo formation of 8-epi-prostaglandin F2a is increased in hypercholesterolemia, *Arterioscler. Thromb. Vasc. Biol.*, 17, 3230–3235, 1997.

33. Davi, G. et al., In vivo formation of 8-iso-prostaglandin f2alpha and platelet activation in diabetes mellitus: effects of improved metabolic control and vitamin E supplementation, *Circulation*, 99, 224–229, 1999.

34. Brown, A.J. and Jessup, W., Oxysterols and atherosclerosis, *Atherosclerosis*, 142, 1–28, 1999.

35. Salonen, J.T. et al., Lipoprotein oxidation and progression of carotid atherosclerosis, *Circulation*, 95, 840–845, 1997.

36. Zieden, B. and Diczfalusy, U., Higher 7b-hydroxycholesterol in high cardiovascular risk population, *Atherosclerosis*, 134, 172, 1997.

37. Staprans, I. et al., Oxidized cholesterol in the diet accelerates the development of atherosclerosis in LDL receptor- and apolipoprotein E-deficient mice, *Arterioscler. Thromb. Vasc. Biol.*, 20, 708–714, 2000.

38. Abbey, M. et al., Oxidation of low density lipoproteins, intraindividual variability and the effect of dietary linoleate supplementation, *Am. J. Clin. Nutr.*, 57, 391, 1993.

39. Reaven, P. et al., Feasibility of using an oleate-rich diet to reduce the susceptibility of low density lipoprotein to oxidative modification in humans, *Am. J. Clin. Nutr.*, 54, 701, 1991.

40. Parthasarathy, S. et al., Low density lipoprotein rich in oleic acid is protected against oxidative modification: implications for dietary prevention of atherosclerosis, *Proc. Natl. Acad. Sci. U.S.A.*, 87, 3894, 1990.

41. Reaven, P. et al., Effects of oleate rich and linoleate rich diets on the susceptibility of low density lipoprotein to oxidative modification in mildly hypercholesterolemic subjects, *J. Clin. Invest.*, 91, 668, 1993.

42. Croft, K.D. et al., Oxidation of low density lipoproteins: effect of antioxidant content, fatty acid composition and intrinsic phospholipase activity on susceptibility to metal ion-induced oxidation, *Biochim. Biophys. Acta*, 1254, 250, 1995.

43. Suzukawa, M. et al., Effects of fish oil fatty acids on low density lipoprotein size, oxidizability, and uptake by macrophages, *J. Lipid Res.*, 36, 473, 1995.

44. Oostenbrug, G.S., Mensink, R.P., and Hornstra, G., Effects of fish oil and vitamin E supplementation on copper-catalysed oxidation of human low density lipoprotein *in vitro*, *Eur. J. Clin. Nutr.*, 48, 895, 1994.

45. Whitman, S.C. et al., N-3 fatty acid incorporation into LDL particles renders them more susceptible to oxidation in vitro but not necessarily more atherogenic in vivo, *Arterioscler. Thromb.*, 14, 1170, 1994.

46. Lussier-Cacan, S. et al., Influence of probucol on enhanced LDL oxidation after fish oil treatment of hypertriglyceridemic patients, *Arterioscler. Thromb.*, 13, 1790, 1993.

47. Harats, D. et al., Fish oil ingestion in smokers and non smokers enhances peroxidation of plasma lipoproteins, *Atherosclerosis*, 90, 127, 1991.

48. Nenseter, M.S. et al., Effect of dietary supplementation with n-3 polyunsaturated fatty acids on physical properties and metabolism of low density lipoprotein in humans, *Arterioscler. Thromb.*, 12, 369, 1992.

49. Frankel, E.N. et al., Effect of n-3 fatty acid-rich fish oil supplementation on the oxidation of low density lipoproteins, *Lipids*, 29, 233, 1994.

50. Esterbauer, H. et al., Role of vitamin E in preventing the oxidation of low density lipoprotein, *Am. J. Clin. Nutr.*, 53, 314S, 1991.

51. Tribble, D.L. et al., Differing α-tocopherol oxidative lability and ascorbic acid sparing effects in buoyant and dense LDL, *Arterioscler. Thromb. Vasc. Biol.*, 15, 2025, 1995.

52. Jialal, I. and Grundy, S.M., Preservation of the endogenous antioxidants in low density lipoprotein by ascorbate but not probucol during oxidative modification, *J. Clin. Invest.*, 87, 597, 1991.

53. Stocker, R., Bowry, V.W., and Frei, B., Ubiquinol-10 protects human low density lipoprotein more efficiently against lipid peroxidation than does α-tocopherol, *Proc. Natl. Acad. Sci. U.S.A.*, 88, 1646, 1991.

54. Bjelakovic, G. et al., Mortality in randomized trials of antioxidant supplements for primary and secondary prevention: systematic review and meta-analysis, *JAMA*, 297, 842, 2007.

55. Trip, M.D. et al., Platelet hyperreactivity and prognosis in survivors of myocardial infarction, *N. Engl. J. Med.*, 322, 1549, 1990.

56. Elwood, P.C. et al., Whole blood platelet aggregometry and ischemic heart disease, *Atherosclerosis* 10, 1032, 1990.

57. Renaud, S. et al., Nutrients, platelet function and composition in nine groups of French and British farmers, *Atherosclerosis*, 60, 37, 1986.

58. Renaud, S. et al., Influence of long-term diet modification on platelet function and composition in Moselle farmers, *Am. J. Clin. Nutr.*, 43, 136, 1986.

59. Temme, E.H., Mensink, R.P., and Hornstra, G., Individual saturated fatty acids and effects on whole blood aggregation in vitro, *Eur. J. Clin. Nutr.*, 52, 697, 1998.

60. Thijssen, M.A., Hornstra, G., and Mensink, R.P., Stearic, oleic, and linoleic acids have comparable effects on markers of thrombotic tendency in healthy human subjects, *J. Nutr.*, 35, 2805, 2005.

61. Breslow, J.L., n-3 fatty acids and cardiovascular disease, *Am. J. Clin. Nutr.*, 83, 1477S, 2006.

62. Blair, I.A., Dougherty, R.M., and Iacono, J.M., Dietary stearic acid and thromboxane-prostacyclin biosynthesis in normal human subjects, *Am. J. Clin. Nutr.*, 60, 1054S, 1994.

63. Mustad, V.A. et al., Comparison of the effects of diets rich in stearic acid versus myristic acid and lauric acid on platelet fatty acids and excretion of thromboxane A2 and PGI2 metabolites in healthy young men, *Metabolism*, 42, 463, 1993.

64. Turpeinen, A.M. et al., Similar effects of diets rich in stearic acid or trans-fatty acids on platelet function and endothelial prostacyclin production in humans. *Arterioscler. Thromb. Vasc. Biol.*, 18, 316, 1998.

65. Lahoz, C. et al., Effects of dietary fat saturation on eicosanoid production, platelet aggregation and blood pressure, *Eur. J. Clin. Invest.*, 27, 780, 1997.

66. Clauss, A., Gerinnungsphysiologische schnellmethode zur bestimmung des fibrinogens, *Acta. Haematol.*, 17, 237, 1957.

67. Heinrich, J. et al., Fibrinogen and factor VII in the prediction of coronary risk, *Arterioscler. Thromb.*, 14, 55, 1993.

68. Meade, T.W. et al., Haemostatic function and ischaemic heart disease: principal results of the Northwick Park Heart study, *Lancet*, 2, 533, 1986.

69. Meade, T.W. et al., Antithrombin III and arterial disease, *Lancet*, 338, 850, 1991.

70. Marckmann, P., Sandström, B., and Jespersen, J., Favorable long-term effect of a low-fat/high-fiber diet on human blood coagulation and fibrinolysis, *Arterioscler. Thromb.*, 13, 505, 1993.

71. Marckmann, P., Sandström, B., and Jespersen, J., Low-fat, high-fiber diet favorably affects several independent risk markers of ischemic heart disease: observations on blood lipids, coagulation, and fibrinolysis from a trial of middle-aged Danes, *Am. J. Clin. Nutr.*, 59, 935, 1994.

72. Marckmann, P., Sandström, B., and Jespersen, J., Fasting blood coagulation and fibrinolysis of young adults unchanged by reduction in dietary fat content, *Arterioscler. Thromb.*, 12, 201, 1992.

73. Marckmann, P. et al., Effect of a fish diet versus a meat diet on blood lipids, coagulation and fibrinolysis in health young men, *J. Intern. Med.*, 229, 317, 1991.

74. Brown, A.J. and Roberts, D.C.K., Fish and fish oil intake: effect on haematological variables related to cardiovascular disease, *Thromb. Res.*, 64, 169, 1991.

75. Emeis, J.J. et al., A moderate fish intake increases plasminogen activator inhibitor type-1 in human volunteers, *Blood*, 74, 233, 1989.

76. Almendingen, K. et al., Effects of partially hydrogenated fish oil, partially hydrogenated soybean oil, and butter on haemostatic variables in men, *Arterioscler. Thromb. Vasc. Biol.*, 16, 375, 1996.

77. Tholstrup, T. et al., Fat high in stearic acid favorably affects blood lipids and factor VII coagulant activity in comparison with fats high in palmitic acid or high in myristic and lauric acids, *Am. J. Clin. Nutr.*, 59, 371, 1994.

78. Bladbjerg, E.M. et al., Dietary changes in fasting levels of factor VII coagulant activity, *Thromb. Haemost.*, 73, 239, 1995.

79. Tholstrup, T. et al., Effect on blood lipids, coagulation, and fibrinolysis of a fat high in myristic acid and a fat high in palmitic acid, *Am. J. Clin. Nutr.*, 60, 919, 1994.

80. Meade, T.W. et al., Fibrinolytic activity, clotting factors, and long-term incidence of ischaemic heart disease in the Northwick Park Heart Study, *Lancet*, 342, 1076, 1993.

81. Hamsten, A. et al., Plasminogen activator inhibitor in plasma: risk factor for recurrent myocardial infarction, *Lancet*, 2, 3, 1987.

82. Mehrabian, M. et al., Dietary regulation of fibrinolytic factors, *Atherosclerosis*, 84, 25, 1990.

23 Conjugated Linoleic Acids: Nutrition and Biology

Bruce A. Watkins and Yong Li

CONTENTS

I. INTRODUCTION

Conjugated linoleic acids (CLAs) are a family of positional and geometric isomers of octadecadienoic acid (18:2). Double bonds in CLA are conjugated and not separated by a methylene group ($-CH_2-$) as in linoleic acid (LA or 18:2n-6), an ω-6 essential fatty acid. The CLA isomers are found in many foods [1] but are predominant in products derived from ruminant sources (beef, lamb, and dairy) because of the process of bacterial biohydrogenation of polyunsaturated fatty acids (PUFAs) in the rumen [1–3].

The highest concentrations of CLA in food are present in dairy products [4,5] and fat in the meats of lamb, veal calves, and cattle [6]. In most cases, the *cis*-9, *trans*-11 isomer is the chief isomer of CLA found in food. Estimates of CLA intake range from 0.3 to 1.5 g/person/day and appear to be dependent on gender and the intake of food from animal and vegetable origins [4].

Since the finding that CLAs isolated from grilled beef inhibited chemically induced cancer [7–9], numerous studies have been initiated to investigate the physical, biochemical, and physiological properties of CLA isomers. The growing body of literature on CLA suggests that these isomeric conjugated fatty acids possess potent biological activities that may benefit human health.

II. CHEMISTRY

The acronym CLA is used to describe a family of octadecadienoic acid (18:2) isomers that possess a pair of conjugated double bonds along the alkyl chain from carbon 2 to 18. Octadecadienoic acids have been reported to contain conjugated double bonds at positions 7,9; 8,10; 9,11; 10,12; 11,13; and 12,14 (counting from the carboxyl end of the molecule) in chemically prepared CLA mixtures or natural products [10–13]. The positional conjugated diene isomers can occur in one or more of the following four geometric configurations: *cis,trans*; *trans,cis*; *cis,cis*; or *trans,trans*, which would give 24 possible isomers of CLA [10]. Many of these isomers were found in commercially available preparations of CLA produced under alkaline conditions from vegetable oils containing a high concentration of LA [14].

The most common CLA isomer found in meat from ruminant species and bovine dairy food products is octadeca-*c*9,*t*11-dienoic acid [15], even though minor components, such as the *t*7,*c*9; *t*8, *c*10; *t*10,*c*12; *t*11,*c*13; *c*11,*t*13; and *t*12,*t*14 isomers, and their *cis,cis*; *trans,trans* isomers were also reported in these products [11,12]. Two trivial names have been proposed for the *c*9,*t*11 isomer: bovinic acid [16] and rumenic acid [17]. The name bovinic acid is considered to be too restrictive for CLA because the *c*9,*t*11 isomer is not only produced in the rumen of the bovine but also produced by other ruminant animals by the same mechanism.

The CLA in ruminant meat and dairy products is believed to be formed by bacterial isomerization of LA and possibly α-linolenic acid (18:3*n*-3) from grains and forages to the *c*9,*t*11-18:2 in the rumen of these animals [1,2,18]. CLA may also be formed during cooking and processing of foods [18]. Presently CLA mixtures and pure isomers are available from various commercial sources. The composition of CLA products should be carefully checked before they are used in research work since the isomeric distribution varies largely between manufacturers and even between batches made by the same manufacturer [19,20].

III. ANALYSIS OF CLA IN FOOD AND BIOLOGICAL SAMPLES

As research continues on the actions of CLA, it is critical to have accurate compositional analyses of food and biological samples to correctly interpret data from investigations. The analysis of CLA in food and other biological samples follows the general guidelines for the determination of fatty acid composition with particular emphasis on the derivatization methods. The conjugated double bonds in CLA are less stable than the methylene-interrupted double bonds in LA. It is reported that the stability of CLA is similar to that of arachidonic acid (20:4*n*-6) and docosahexaenoic acid (DHA, 22:6*n*-3) [21,22]. Therefore, preventing isomerization and oxidation of these labile fatty acids is critical for a successful analysis.

The most common analyzed form of fatty acids is the fatty acid methyl ester (FAME). Both acid and alkaline catalysts can be used to prepare FAME for gas chromatography (GC) analysis. However, since acid catalysts, such as BF_3, will change the double-bond configuration in conjugated dienes and generate artifact CLA isomers [23–25], their use is not recommended in the analysis of CLA.

The alkaline catalysts perform best for lipids containing fatty acids with unique conjugated diene structures. Isomerization and artifacts are not produced when sodium methoxide or tetramethylguanidine (TMG) is used as transesterification agent; however, they do not methylate free fatty acids and N-linked (amide bond) fatty acids as those found in sphingolipids. Therefore, these methods are not suitable for samples with high acid values (high free fatty acid content). There is no single method that works optimally in all situations. The researcher needs to know the nature of the sample and select a suitable method accordingly.

Various instruments are adopted to separate and characterize the CLA isomers. A gas chromatograph equipped with a polar 30–100 m capillary column and a flame ionization detector is the most widely used instrument to characterize CLA isomers. Generally, GC columns have a limited capacity to fully separate all geometric and positional isomers of CLA; therefore, it is impossible to identify every individual CLA isomer by conventional capillary GC analysis. Mossoba et al. [26] reported that 10 CLA peaks were resolved in a commercial CLA preparation with a gas chromatograph equipped with a 100 m column (SP 2560, Supelco Inc., Bellefonte, PA or CP-Sil 88,

Chrompack, Bridgewater, NJ). In the procedure mentioned earlier, the 10 peaks resolved may actually represent more than 15 CLA isomers. A more powerful approach has been the application of augmentation (silver ion) high-performance liquid chromatography (Ag$^+$-HPLC) using a UV detector at a wavelength of 233 nm. Sehat et al. [10] described such a method that resolves CLA isomers according to geometric configuration and position of the conjugated diene structure. Moreover, Mossoba et al. [27] reported that 16 isomers of a CLA sample were resolved by an Ag$^+$-HPLC method. Although the Ag$^+$-HPLC method is more powerful than GC in resolving CLA isomers, it cannot be used to quantify other fatty acids because of the limitation of UV detectors and HPLC columns. Therefore, ideally the two methods should be combined for CLA analysis, using GC for the general fatty acid analysis and Ag$^+$-HPLC for the conjugated dienes. The results of these two procedures would provide a more complete profile of CLA isomers in food and biological samples.

IV. CLA CONTENT IN FOOD PRODUCTS AND BIOLOGICAL SAMPLES

Cheese is a chief source of dietary CLA in animal-derived products. The CLA concentrations in various dairy products (cheeses, milk, butter, buttermilk, sour cream, ice cream, and yogurt) range from 0.55 to 24 mg/g fat. The average CLA content in milk is about 10 mg/g milk fat [4,28]. The largest variation in the amount of CLA is found in various natural cheeses ranging from 0.55 to 24 mg/g of fat. Seven CLA peaks that could represent nine isomers were present in dairy products; among these c9,t11; t10,c12; t9,t11; and t10,t12 accounted for more than 89% [18]. The CLA content in cheeses is primarily dependent on the CLA content in the milk, which varies in CLA concentration because of seasonal variation, geography, nutrition of the cow, and management practices. In addition, CLA content of cheese, to a limited extent, is affected by the production process and maturation [29].

Reported values for CLA content in beef muscle vary considerably from 1.2 to 9.9 mg/g fat [1,18,30,31]. Fats and meats from ruminant species are a rich natural source of CLA, and the reported values ranged from 2.7 to 5.6 mg CLA/g fat in lamb, veal, and beef. Fritsche and Fritsche [32] reported that the amount of the c9,t11, 18:2 isomer in beef averaged 0.76% of total FAME for fat samples from bulls and 0.86% for fat from steers. Minor isomers, for example, t9,c11; c9,c11; and t9,t11, were also found in beef fat samples. Others have reported that the c9,t11 18:2 content in beef ranged from 1.7 to 6.5 mg/g fat [30] and 0.65% of total FAME in beef fillet [4].

CLA is also present in small amounts in other food products. Turkey meat has the highest CLA content of 2.5 mg/g fat for nonruminant species [18]. Chicken contained CLA (0.9 mg/g fat) as did pork (0.6 mg/g fat) with c9,t11 being the major isomer (84% and 82%, respectively) [1]. The amount of CLA in chicken egg yolk lipids ranges from 0 to 0.6 mg/g of fat [1,33–36]. CLA was found in plant oils (0.1–0.7 mg/g fat) and selected seafood (0.3–0.6 mg/g fat) in small quantities [1]. Unlike in ruminant derived food where c9,t11 is the chief isomer of CLA, this isomer accounts for only 38%–47% of the total CLA in plant oils and rather interestingly appears to be absent in seafood lipids. Banni et al. [37] carried out a series of analyses to characterize the fatty acids with conjugated dienes in partially hydrogenated oil (mixture of partially hydrogenated soybean oil and palm oil) and confirmed the presence of CLA isomers in these oils. Moreover, Mossoba and coinvestigators [26] reported that conjugated cis,$trans$ and $trans$,$trans$ 18:2 isomers were present in hydrogenated soybean oil and margarine.

CLA has been identified in various human tissues, such as adipose, serum, breast milk, and in bile and duodenal secretions [38]. Fogerty et al. [39] reported 5.8 mg/g fat of CLA in human milk for subjects consuming a normal Australian diet. Precht and Molkentin [40] reported a value of 3.8 mg/g fat (range 2.2–6.0 mg/g fat) in human milk obtained from 40 German women. McGuire et al. [16] analyzed 14 human milk samples from subjects in the Pacific Northwest and reported that CLA values ranged from 2.23 to 5.43 mg/g fat (mean 3.81 mg/g fat) or from 0.02 to 0.30 mg/g on a milk weight basis. In this last study, all milk samples contained 83%–100% of the c9,t11 isomer, and in 8 of the 14 samples the $c9,t11$ isomer was the only form observed. Jensen et al. [41] reported a lower level of CLA in human milk that ranged from 1.4 to 2.8 mg/g fat with an average of 1.8 mg/g fat. Based on these data, CLA in human milk for Western societies ranged from 1.4 to 5.8 mg/g fat.

The concentration of CLA found in human plasma and serum appears to be linked to dietary fat type and food consumption patterns. Herbel et al. [38] reported that plasma CLA concentration ranged from 6.4 to 7.3 μmol/L in human subjects that were given a high level of safflower oil for 6 weeks. The high LA intake from safflower oil did not increase the plasma level of CLA, indicating that dietary LA is not converted to CLA in these subjects. In another study, the effect of *trans* fatty acid intake on serum CLA was examined [42]. Eighty human subjects were put on a diet high in saturated fat mainly from dairy for 5 weeks. Then they were separated into two dietary groups: 40 subjects were assigned to a diet high in *trans* fatty acids from partially hydrogenated vegetable oil and the other 40 were given a similar diet high in stearic acid. At the termination of the study, serum samples from the *trans* diet group (CLA was 0.43% of total fatty acids) contained 30% more CLA than those samples obtained when on the dairy fat diet (0.32% CLA). Samples from those given the stearic acid diet had only half of the amount of CLA compared with those given the dairy fat diet. These data indicate that a possible relationship exists between *trans* fatty acid intake and serum CLA concentration. In another dietary intervention study of nine healthy men, Cheddar cheese was added to their diet at a level of 112 g/day for 4 weeks. Plasma CLA concentration increased from 7.1 to 9.6 μmol/L at the end of the cheese supplement period and was maintained at 7.8 μmol/L after another 4 weeks following intervention [43].

The CLA concentration and isomeric distribution have recently been characterized in human adipose tissue. Fritsche et al. [44] reported that human subcutaneous adipose tissue contained two major CLA isomers, both $c9,t11$ and $t9,t11$, and two minor isomers $t9,c11$ and $c9,c11$. The presence of CLA in human tissues is principally due to dietary intake since the amount of $c9,t11$–18:2 in human adipose tissue appears to be directly related to milk fat intake [45]. However, the possibility of endogenously produced CLA in human tissues could not be excluded. Brown and Moore [46] have reported the presence of CLA-producing bacterial strains of *Butyrivibrio fibrisolvens* isolated from human feces. Moreover, CLA is also produced in conventional, but not germ-free rats after consumption of LA [47]. On the other hand, consuming safflower oil that is high in triacylglycerol-esterified LA, the precursor for bioisomerization, did not increase the CLA concentrations in total lipids of human plasma [38]. Therefore, the contribution of colon-derived CLA to that found in human tissues and in nonruminant monogastric species is most likely negligible.

Another possible endogenous origin of tissue CLA is by the desaturation of *trans* vaccenic acid, which was demonstrated in rat liver microsomal preparations [48]. Studies performed in human subjects and pigs support this hypothesis. The results of Salminen et al. [42] support such a mechanism of endogenous CLA production by demonstrating that a high level of dietary *trans* fatty acids actually increased serum CLA content over stearic acid. Adlof et al. [49] showed that deuterated 11-*trans*-octadecenoate (fed as the triglyceride) was converted to 9-*cis*, 11-*trans* CLA via the Δ9 desaturase pathway at a CLA enrichment of about 30% in human subjects. The conversion of dietary C18:1 *trans* fatty acid to CLA was also demonstrated in a pig-feeding study using partially hydrogenated fat that was rich in C18:1 *trans* fatty acids [50]. This indicates that CLA could be produced by endogenous Δ9 desaturation of dietary *trans* vaccenic acid in pigs.

V. NUTRITION AND BIOLOGY OF CLA

CLA is the only known anticarcinogen tested in animals associated with foods originating from animal sources. Ha and coworkers [9] provided one of the earliest observations that CLA from beef was protective against chemically induced cancer. In that study, CLA isolated from extracts of grilled ground beef were found to reduce skin tumors in mice treated with 7,12-dimethylbenz[α] anthracene (DMBA), a known carcinogen [9]. Since then, numerous researchers have reported the effects of CLA isomers. The research on CLA has relied entirely on animal models and cell culture systems employing isomeric mixtures of CLA. Recent studies in human subjects have been negative [20,51,52]. The purported properties of CLA include anticarcinogenic [9,53–62] and antiatherosclerotic [63,64]. Other CLA effects include antioxidative [54,55,63,64] and immunomodulative [47,65–68].

More recently, preliminary data suggest that CLA may have a role in controlling obesity [69–71], reducing the risk of diabetes [72], and modulating bone metabolism [73,74]. In addition, some studies indicate that the biological effects of CLA are modulated by dietary sources of long-chain n-3 fatty acids [68,73,74].

Considerable research has been done to examine the anticarcinogenic properties of CLA. Isomers of CLA have been shown to reduce chemically induced tumorigenesis in rat mammary gland and colon [53,55–58,60,62,75]. Moreover, CLA modulated chemically induced carcinogenesis in mouse skin [9,76] and forestomach [54]. Sources of CLA also inhibited the growth of human tumor cell lines in culture [61,77,78] and in SCID (severe combined immunodeficiency) mice [59,79]. Recent investigations with CLA failed to show a benefit on rodent models of colon carcinogenesis [80] and prostate metastasis [81].

A. CANCER

In dietary preparations, CLA provided a potent anticarcinogenic effect in a rat mammary cancer model. Ip et al. [55] found that administering weanling rats CLA at 0.5%–1.5% of the diet for 2 weeks before DMBA induction (oral intubation of 10 mg in 1 mL corn oil) and continuing for 36 weeks resulted in a significant reduction in tumor incidence. The effect of CLA was maximized at a dietary level of 1% in this study. In another experiment, much lower doses of dietary CLA (0.05%–0.5%) were given to rats [53]. The dietary treatments induced a dose-dependent inhibition in mammary tumor yield with a lower dose of 5 mg of DMBA. In the 5 week short-term feeding experiment of 1% dietary CLA [53], rats were given CLA for 5 weeks and the carcinogens were introduced 1 week before the end of CLA treatment. Total tumor yield was reduced by 39% and 34% in the DMBA and methylnitrosourea (MNU) induction groups with CLA treatment, respectively, at the end of the experiment. The short-term feeding period corresponded to the maturation of the rat mammary gland to adult stage morphology, and the supplement of CLA was stopped shortly after the induction with the carcinogens. Therefore, these results suggest that CLA may have a direct modulating effect on susceptibility of the target organ to neoplastic transformation.

In a subsequent dietary study using the rat DMBA model, the inhibitory effect of CLA on induced mammary cancer was found to be independent of fat type (n-3 fatty acids were investigated in the study) provided in the diet [56]. When providing rats with a CLA supplement, the fatty acid isomers were incorporated into the mammary gland lipids, and the amount in the neutral lipids greatly exceeded that in phospholipids [57,82]. When CLA was removed from the diet after 4 and 8 weeks of feeding, neutral lipid- and phospholipid-CLA returned to basal values in about 4 and 8 weeks, respectively. The author observed that the rate of CLA disappearance in neutral lipids subsequent to CLA withdrawal closely paralleled the occurrence of new tumors in the rat target tissue. It appears that the tissue accumulation of CLA is important to its antitumor effect and that the concentration of CLA in neutral lipid may be a more sensitive biomarker of tumor protection than the amount in phospholipid for this model. The effect of CLA on cancer was maximized at a dietary level of 1% as shown in several studies, and increasing the dietary level of CLA did not afford additional protection [55,82].

The antitumor effect of naturally produced CLA was tested over a chemically prepared CLA mixture from LA, which was used in previous studies [83]. During the study, rats were given a diet with a high CLA content butter fat (equivalent to 0.8% CLA in the diet) and a low-level CLA diet that served as the control group (equivalent to 0.1% CLA in the diet) for 1 month before tumor induction using MNU. After tumor induction, rats were changed to a regular diet without CLA for 24 weeks before the termination of the experiment. CLA in butter inhibited the tumor yield by 53%.

CLA was also shown to reduce the population of mammary terminal end bud (TEB) cells, the cells that are the primary target of attack by carcinogens, and the proliferation of TEB cells. Similar effects reported by Banni et al. [62] revealed that providing rats with 0.5% and 1% CLA in the diet produced a graded and parallel reduction in TEB density and mammary tumor yield, and no further decrease was observed in either parameter in rats given 1.5% and 2% CLA. These studies support the hypothesis that exposure to CLA during the time of mammary gland maturation may modify the developmental potential of the target cells that are normally susceptible to carcinogen-induced transformation. In contrast to the previous experiments using a chemically prepared CLA mixture, the latter experiments, for the first time, showed that the natural form of CLA found in dairy foods is active in reducing the incidence of mammary tumors in rats.

The anticarcinogenic properties of CLA isomers were tested in various human cell culture models. Cunningham et al. [61] treated human normal mammary cells and MCF-7 breast cancer cells with LA, CLA, and eicosanoid synthesis inhibitors in cell culture. Fatty acids were complexed with bovine serum albumin (BSA) before introduction of treatments into the culture media at concentrations of $0–3.57 \times 10^5$ M. The results showed that CLA inhibited thymidine incorporation in normal and MCF-7 cancer cells. The treatment consisting of CLA plus nordihydroguaiaretic acid (NDGA), a lipoxygenase inhibitor, suppressed the growth of MCF-7 cancer cells synergistically. The authors suggested that the growth suppression was augmented by CLA through inhibition of leukotriene synthesis.

The effects of CLA on the growth of three different lung adenocarcinoma cell lines (A-427, SK-LU-1, A549) and one human glioblastoma cell line (A-172) were examined by Schonberg and Krokan [77]. There was a dose-dependent reduction in proliferation of the lung adenocarcinoma cell lines with A-427 being the most sensitive, but CLA had virtually no effect on the A-172 cell line. In addition, LA had no inhibitory effect on either of these cell lines. A significant increase in lipid peroxidation (measured by the formation of malondialdehyde, MDA) was observed after exposure of the lung adenocarcinoma cell lines to 40 μM CLA. The level of MDA was approximately twofold higher after exposure to 40 μM LA. Treatment with vitamin E (30 μM) totally abolished the formation of MDA; however, cell growth rates were only partially restored. These data might suggest that cytotoxic lipid peroxidation products are only in part responsible for the growth inhibitory effects of CLA. Further research is needed to elucidate a possible mechanism of CLA action in lipid peroxidation and disease.

The anticarcinogenic effect of CLA on a human prostatic cancer cell line was investigated in animal models. Cesano et al. [59] tested the effect of both CLA and LA against the human prostatic cancer cell line DU-145 in a mouse model. Mice were given a control diet and either LA or CLA at 1% of diet for 2 weeks before the subcutaneous inoculation of cancer cells. Mice were then maintained on the same dietary treatments for 10 more weeks before the termination of the study. Mice given the CLA-supplemented diet displayed not only smaller local tumors than those given the control diet, but also a drastic reduction in lung metastases. Mice given the LA diet had increased local tumor loads as compared with the CLA-treated and control groups. The studies mentioned earlier suggest that CLA and LA have distinct biological effects.

The anticarcinogenic properties of CLA isomers might be limited to certain types of cancers and may not be effective under some experimental conditions (including dietary sources of phytochemicals and nutraceutical fatty acids such as the n-3 PUFAs) [84]. Wong et al. [85] showed that providing mice diets containing from 0.1% to 0.9% CLA for 2 weeks before infusing with WAZ-2T metastatic mammary tumor cells into the right inguinal mammary gland did not affect mammary tumor latency, tumor incidence, or volume. Therefore, dietary CLA might be less effective on the growth of an established and aggressive mammary tumor.

The anticarcinogenic effects of CLA isomers have been exclusively tested in animal models and cell culture systems. There is no direct evidence that these fatty acids protect against carcinogenesis in humans. The most promising evidence to date that dietary sources of CLA are beneficial is from epidemiological studies that link milk consumption to reduced breast cancer. In one recent study,

Knekt et al. [86] found a significant inverse relationship between milk intake and breast cancer incidence among 4697 initially cancer-free Finish women over a 25 year follow-up period. It was found that the risk of breast cancer was halved in women who consumed more than 620 mL milk/day compared with those consuming less than 370 mL/day and suggested that CLA was the active component. The average amount of CLA in milk is approximately 10 mg/g of milk fat. Based on the average fat content of 35 g/L of milk [87], the daily intake of CLA for these women could be 217 mg for those consuming 620 mL milk/day and 130 mg for those consuming 370 mL milk/day. However, the calculated CLA intake level for these women is still far below the estimated CLA level of 1 g/day that would be significant for cancer prevention [53]. The results from the epidemiological studies and reduced incidence of breast cancer could suggest that dairy fats contain other factors that reduce cancer risk or other lifestyle and dietary factors associated with these populations, for example, fish intake (n-3 fatty acids) that would influence cancer incidence. Experiments designed for investigating the role of CLA in human cancer are needed to identify the relationships between these fatty acids and other lifestyle factors contributing to lower cancer risk.

B. CARDIOVASCULAR DISEASE

The potential health benefits of CLA may extend beyond cancer to the prevention of congestive cardiovascular diseases since these fatty acid isomers were shown to reduce atherogenesis in animal studies [63,64]. Lee et al. [63] assessed the effect of CLA on atherosclerosis in rabbits by supplementing 0.5 g/animal/day for 22 weeks. In comparison with control rabbits that were given a similar diet containing 14% fat and 0.1% cholesterol without CLA, total and LDL cholesterol and triacylglycerol (TAG) levels in blood were markedly lower in those given CLA. At the same time, the ratio of LDL cholesterol to HDL cholesterol and ratio of total cholesterol to HDL cholesterol were significantly reduced in the group given CLA. Rabbits given CLA also showed less atherosclerosis in the aorta. In another study by Nicolosi et al. [64], CLA effects on plasma lipoproteins and aortic atherosclerosis were examined. Fifty hamsters were divided into 5 groups of 10 and given 0 (control), 0.06 (low), 0.11 (medium), and 1.1 (high) percentage of total dietary energy of CLA or 1.1% of total energy as LA. Animals given the CLA-containing diets collectively had significantly reduced levels of plasma total cholesterol, non-HDL cholesterol (combined VLDL and LDL), and TAGs with no effect on HDL cholesterol, in comparison with the controls. For the CLA treatment, a tocopherol sparing effect was observed compared with the control group. Morphometric analysis of aortas revealed less early atherosclerosis in the CLA-treated hamsters compared with the control group. Interestingly, the LA treatment showed a similar effect compared with CLA in this study. Yeung et al. [88] reported that by feeding hamsters 20 g/kg dietary linoleic acid or CLA, serum total fasting cholesterol and TAG were significantly reduced compared with the control, which had no fatty acid supplementation. However, the cholesterol-lowering mechanism for LA and CLA appeared to be different. CLA lowered the activity of intestinal acyl CoA–cholesterol acyltransferase (ACAT), whereas LA had no effect on this enzyme, indicating that CLA could affect the absorption of dietary cholesterol. In rats fed (up to 1.5% dietary CLA for 60 days) the ratio of HDL cholesterol/TC and serum TAG concentrations were significantly elevated [89]. These studies showed that CLA was hypocholesterolemic and antiatherogenic; however, a recent experiment using C57BL/6 mice given atherogenic diets with added CLA (0.5% and 0.25% of the diet) demonstrated increased development of aortic fatty streaks despite a change in serum lipoprotein profiles that could be considered less atherogenic [90].

There is currently no conclusive evidence that CLA protects against early atherogenesis. It was initially proposed that CLA was able to protect LDL particles from oxidation [54,55,63,64]; however, a recent in vitro investigation by van den Berg et al. [21] showed no antioxidant properties attributable to CLA. In addition, CLA showed its positive effect on atherosclerosis in rabbits but not in mice, which could indicate a species difference for the response to dietary CLA supplementation.

Benito et al. [52] showed that although CLA isomers were incorporated into platelet lipids, antithrombotic properties such as blood clotting parameters and in vitro platelet aggregation were not affected by daily supplement of 3.9 g CLA to a typical Western diet for 63 days in healthy adult females.

In a human study involving 17 female subjects, CLA was given as a dietary supplement at 3.9 g/day for 93 days. Blood cholesterol or lipoprotein levels were not altered by CLA supplementation in healthy and normolipidemic subjects. In this investigation the short-term supplementation with CLA did not afford health benefits for the prevention of atherosclerosis [20].

C. BODY FAT AND LIPID METABOLISM

Numerous studies in growing animals demonstrated that CLA reduced fat deposition and increased lean body mass [69–71,91–94]. When CLA (0.5%–1% of diet) was given to AKR/J male mice (39 days old), it produced a rapid and marked decrease in fat accumulation and an increase in protein accumulation without any major effects on food intake [69]. Park et al. [93] provided diets with a lower dietary CLA level (5% corn oil + 0.5% CLA) to 6 week old ICR mice and observed a significantly reduced body fat content by 57% (male) and 60% (female), and increased lean body mass relative to control mice that were given a diet containing 5.5% corn oil. Similar effects of CLA on body fat accumulation were shown in rats. Yamasaki and colleagues [70] studied the effect of CLA on liver and different adipose tissues in 4 week old male rats given diets (AIN-93G with 7% safflower oil as a control) containing 1% and 2% dietary CLA (at the expense of the safflower oil) for 3 weeks. They observed reduced levels of TAGs and nonesterified fatty acids in the liver and white adipose tissue without significant changes in lipids of brown adipose tissue. In another study, Yamasaki et al. [95] reported that dietary CLA treatments (0%, 1%, and 2% of the diet) dose dependently accelerated the release of lipids in white adipose tissue and increased the clearance rate of serum nonesterified fatty acids. In addition, CLA treatment induced a nonsignificant increase in liver size, and liver TBAR (thiobarbituric acid reactive substances) levels were significantly higher in rats given the 2% CLA diet indicating a morphological change in liver caused by increased peroxidation.

The effect of CLA on body composition was also investigated in a randomized, double-blind, placebo-controlled study including 60 overweight or obese volunteers with body mass index (BMI) ranging from 27.5 to 39.0 kg/m^2 [96]. The subjects were divided into two groups receiving 3.4 g CLA or placebo (4.5 g olive oil) daily for 12 weeks. No difference in adverse events or other safety parameters was found between the treatment groups. Small changes in the laboratory safety data were not regarded as clinically significant. In the CLA group, mean weight was reduced by 1.1 kg (paired t-test, $p = 0.005$), whereas mean BMI was reduced by 0.4 kg/m^2 ($p = 0.007$). However, the overall treatment of CLA on body weight and BMI was not significant. The results indicate that CLA in the given dose did not adversely impact a healthy population based on the safety parameters investigated. In another study, Blankson et al. [97] reported that feeding overweight or obese human subjects (BMI 25–35 kg/m^2) up to 6.8 g of CLA/day for 12 weeks reduced the total body mass. The CLA effect on body fat mass reduction peaked at 3.4 g/day and higher dietary intake did not show a further benefit.

Repartitioning of fat to lean was reported in growing pigs fed CLA. Dugan et al. [91] demonstrated that in pigs (male and female) given a cereal-based basal diet containing either 2% CLA compared with sunflower oil led to reduced subcutaneous fat deposition and increased lean body mass. Pigs provided with CLA also had reduced feed intake (5.2%) and increased feed efficiency (5.9%) compared with pigs fed sunflower oil. In another study, finisher pigs were offered six treatments having from 0 to 10 g CLA/kg of diet for 8 weeks [92]. Dietary CLA treatments resulted in increased feed efficiency and lean tissue deposition and decreased fat deposition (decreased by 31% at the highest CLA level) in growing pigs.

The fat partitioning effect of CLA was further examined in the adipocyte 3T3-L1 cell culture. Park et al. [93] found that when added during fat accretion in the 3T3-L1 adipocyte culture, CLA

$(1 \times 10^4$ M complexed with albumin) reduced lipoprotein lipase activity and enhanced lipolysis leading to less fat deposition. In addition, skeletal muscle from mice fed CLA exhibited elevated carnitine palmitoyltransferase activity, which indicates elevated β oxidation of fatty acids. In a recent study using the same cell culture system [71], a specific CLA isomer, $t10,c12$-18:2, was found to reduce lipoprotein lipase activity, lower intracellular TAG and glycerol levels, and enhance the release of glycerol into the medium. CLA, especially the $t10,c12$ isomer, showed its antiobesity effect at 50–200 μM by inhibiting proliferation, suppressing triglyceride accumulation, and inducing apoptosis in 3T3-L1 preadipocyte cultures compared with albumin vehicle or linoleic treatments [98].

Lin et al. [99] found that the $t10,c12$ CLA isomer inhibited the activity of heparin-releasable lipoprotein lipase (HR-LPL) more strongly compared with the $c9,t11$ isomer, whereas both CLA isomers exhibited an inhibitory effect on HR-LPL compared with the LA treatment in 3T3-L1 adipocyte cultures. Yamasaki et al. [100] showed in a rat feeding study that 2% dietary CLA lowered serum leptin level after 1 week compared with the 8% safflower oil supplement control group. Leptin level in perirenal white adipose tissue was also low in animals fed CLA after 12 weeks of feeding. These findings suggest that the CLA mechanism for body fat reduction in mice, and possibly in other animals, is a result of inhibition of fat transportation and storage in adipocytes coupled with both elevated β oxidation in skeletal muscle and an increase in skeletal muscle mass [71].

D. BONE

Our laboratory recently investigated the effects of CLA isomers on bone modeling in growing male rats [73,74]. In our studies with rats, 1% dietary CLA combined with two ratios of n-6/n-3 fatty acids led to differences in CLA enrichment of various organs and tissues. Brain exhibited the lowest concentration of isomers, but bone tissue (periosteum and marrow) contained the highest amounts [73]. Both n-3 fatty acids and CLA lowered ex vivo prostaglandin E_2 (PGE_2) production in bone organ culture. The supplemental CLA isomers also reduced serum insulin-like growth factor type I (IGF-I) concentration and modulated IGF-binding protein (IGFBP) differentially depending on the ratio of n-6/n-3 fatty acids in the diet. Moreover, CLA increased IGFBP in rats given a high dietary level of n-6 fatty acids but decreased IGFBP in rats given a high level of n-3 fatty acids. In tibia, rats given CLA had markedly reduced mineral apposition rate (MAR) (3.69 versus 2.79 μm/day) and bone formation rate (BFR) (0.96 versus 0.65 μm/day) in comparison with those not given the CLA supplement [74]. Dietary lipid treatments did not affect serum intact osteocalcin or bone mineral content. These results showed that a mixture of CLA isomers at 1% of the diet modulated local factors that regulate bone metabolism and reduced BFRs. This response may be due to the total dietary level of CLA or varying effects of individual isomers of CLA on bone biochemistry and physiology.

In a subsequent study with rats, a 0.5% dietary level of CLA was provided with or without beef fat [101]. The dietary CLA treatments resulted in total CLA values ranging from 0.27% to 0.43% in the polar lipid fraction and from 2.02% to 3.37% in the neutral lipids in liver, bone marrow, and bone periosteum. We observed that CLA accumulated at a higher concentration in neutral lipids compared with polar lipids consistent with the findings of Ip et al. [56] for rat mammary gland. In rats, the $t10,c12$-18:2 isomer was incorporated into the phospholipid fraction of tissue lipid extracts at the same extent as was the $c9,t11$ isomer. The ratio of $c9,t11/t10,c12$ roughly reflected the isomeric distribution of these CLA isomers in the diet or supplement given to rats. Rat serum osteocalcin, a serum bone formation marker, was decreased in rats given CLA after 12 weeks of dietary treatment. Serum bone-specific alkaline phosphatase activity was also significantly decreased in rats given CLA. The fact that CLA lowered serum bone formation biomarkers, together with our previous finding that CLA lowered ex vivo PGE_2 production in bone organ culture and BFR, suggests that some CLA isomers may exert a downregulatory effect on bone metabolism in growing animals. Future research must be conducted to evaluate the effects of individual CLA isomers on osteoblast function and bone formation.

VI. MECHANISMS OF CLA ACTION

Continued research on CLA has focused on elucidating its mode of action. The CLA isomers have been demonstrated to possess antioxidant properties [54,55,63,64], inhibit carcinogen–DNA adduct formation [58,102,103], induce apoptosis [60], modulate tissue fatty acid composition and eicosanoid metabolism [61,62,67,68,73,74,104–106], and affect the expression and action of cytokines and growth factors [68]. Since most of the studies mentioned earlier were conducted with mixtures of several CLA isomers (mainly c9,t11-18:2 and t10,c12-18:2), these proposed mechanisms of action might be specific for individual CLA isomers.

A. Antioxidative Action

Antioxidants originating from both natural and synthetic sources with diverse structures have been known to demonstrate some anticarcinogenic activity [107]. Since CLA is proposed to possess antioxidative action, this could explain an important mechanism related to its anticarcinogenic activity. Several studies have been conducted using both in vivo and in vitro systems to clarify the role of CLA as an antioxidant; however, the results obtained from various test systems were conflicting. Ha et al. [54], for the first time, showed that CLA was a potent in vitro antioxidant. In an in vivo study, Ip et al. [55] reported that CLA was as effective as vitamin E in inhibiting the formation of TBARs in the mammary gland. Similar to the results of Ha et al. [54], Ip et al. observed no dose–response relationship in the dietary range of 0.25%–1.5% CLA. All doses tested produced a 30%–40% inhibition of peroxide formation.

A study with female rats given diets containing different lipids (20% corn oil or lard) or amounts of a fat blend (from 10% to 20% by weight) with or without 1% CLA [56] showed that CLA reduced MDA (a peroxidation product) production in rat mammary gland homogenate. In addition, CLA produced a greater reduction in rats fed a diet high in PUFA (corn oil, 35%) than in those given a high saturated fat lard diet (25%). CLA failed to show any inhibitory effect on 8-hydroxy-deoxyguanosine (8-OHdG, a marker of oxidative damage to DNA) level in rats. Leung and Liu [108] found that the t10,c12 CLA isomer exhibited a strong antioxidative property compared with the c9,t11 isomer and even α-tocopherol at a lower concentration of 2 and 20 μM in a total oxyradical scavenging capacity assay. The c9,t11 isomer yielded a weak antioxidant activity at lower concentrations of 2 and 20 μM, but at a higher concentration (200 μM) it performed as a strong prooxidant.

Although CLA has been shown to be antioxidative in both in vitro and in vivo studies, the anticancer effect of CLA in these studies cannot be satisfactorily explained based on the current findings. First, the maximal effective concentration of CLA in inhibiting peroxide formation did not agree with the most effective concentration in tumor inhibition [55]. Second, neither of the studies demonstrated a dose–response relationship between CLA concentration and its antioxidative efficacy, which is usually true for antioxygenic nutrients studied thus far.

van den Berg et al. [21] reinvestigated the antioxidative property of CLA using a lipid membrane system consisting of 1-palmitoyl-2-linoleoyl phosphatidylcholine. The results of this study indicated that CLA did not show any protective effect under the test conditions and was more susceptible to oxidative damage than LA and comparable with arachidonic acid. In agreement with the findings of van den Berg et al. [21], Banni et al. [109] in a recent study also showed that CLA was more prone to oxidation than LA, and no significant antioxidant effect of CLA was detected in the models tested. It would appear that CLA and its metabolites seem to behave like other PUFAs under conditions of oxidative stress.

In contrast to the antioxidative properties of CLA, results from experiments using human cancer cell lines suggested that CLA, because of its susceptibility to oxidative damage, could behave as a prooxidant in cell culture systems. Moreover, CLA may create an oxidatively stressed environment that is cytotoxic to cultured cells [77,110]. O'Shea et al. [110] used CLA dissolved in ethanol, which was added to a human cell culture (MCF-7 breast cancer line and SW-480 colon cancer cell line) media at concentrations of 0, 5, 10, 15, 20, and 30 ppm and incubated for 4, 8, and 12 days. The CLA treatment at 20 ppm increased lipid peroxidation and induced the expression and activity of antioxidant enzymes (superoxide dismutase, catalase, and glutathione peroxidase) in both cell lines. At 20 ppm, CLA also reduced ^{3}H-leucine incorporation into protein by 83%–91% and ^{3}H-uridine and ^{3}H-thymidine incorporation into RNA and DNA by 49%–91% and 86%–98%, respectively, compared with untreated control cells. O'Shea et al. [111] showed that milk fat enriched with CLA (primarily the triglyceride-bound $c9,t11$ isomer) induced the activities of superoxide dismutase (SOD), catalase, and glutathione peroxidase (GPx) in MCF-7 human breast cancer cell cultures and inhibited the growth and proliferation of these cells. Apparently, at least in part, the induced antioxidant enzyme system failed to protect these cells from peroxidative cytotoxicity. The controversial effect of CLA as an antioxidant or prooxidant agrees with previous knowledge that the balance between antioxidant and prooxidant activity is known to be a complex function dependent on the concentration of the testing material and oxygen partial pressure [112].

Basu et al. [113,114] reported the investigations of the urinary levels of 8-iso-PGF$_{2\alpha}$, a major isoprostane, and 15-keto-dihydro-PGF$_{2\alpha}$, a major metabolite of PGF$_{2\alpha}$, as indicators of nonenzymatic and enzymatic lipid peroxidation after dietary supplementation of CLA in healthy and obese human subjects. A significant increase of both 8-iso-PGF$_{2\alpha}$ and 15-ketodihydro-PGF$_{2\alpha}$ in urine was observed after 3 months and 1 month of daily CLA intake (4.2 g/day).

B. BIOCHEMICAL AND PHYSIOLOGICAL ACTIONS

Another potential action of CLA in preventing carcinogenesis is on DNA adduct formation. For example, dietary CLA (1%, 0.5%, 0.1% diet) inhibited PhIP (a mammary carcinogen)–DNA adduct formation in F344 rat liver and white blood cells in a dose-dependent manner [102]. Similar findings were also reported by Liew et al. [58] that when F344 rats were given CLA (0.5% diet equivalent) by gavage, the treatment significantly reduced IQ (2-amino-3-methylimidazo[4,5-f]quinoline)–DNA adduct formation in the colon. Schut et al. [103] used the same rat model to study CLA effects on both PhIP and IQ and found that CLA (0.1%–1%) inhibited PhIP–DNA adduct formation in the mammary gland and the colon. It was concluded that CLA could be a potential chemopreventive agent against PhIP- or IQ-induced tumors in rodents.

The isomers of CLA might also initiate apoptosis to protect mammary gland cells from chemically induced carcinogenesis. Ip et al. [60] showed in a primary culture of rat normal mammary epithelial organoids (MEOs) that CLA (0–128 μM), but not LA, inhibited growth of MEO, which was further shown to be mediated by a reduction in DNA synthesis and a stimulation of apoptosis. Ip et al. [115] showed that CLA was able to increase chromatin condensation and to induce DNA laddering—both evidence of apoptosis—in a rat mammary tumor cell line, indicating a potential mechanism of action of CLA. By inducing apoptosis in mammary gland epithelial cells, CLA could prevent breast cancer by reducing mammary epithelial cell density and inhibiting the outgrowth of initiated MEO.

Park et al. [116] showed that LA stimulated MCF-7 breast cancer cell growth, whereas CLA was inhibitory. The LA stimulated phospholipase C (PLC) activity and tended to increase membrane protein kinase C (PKC) activity. However, CLA supplementation did not modify membrane PLC or PKC activity. PGE$_2$ production was not influenced by LA or CLA addition in this experiment.

Isomers of CLA may exert their biochemical and physiological effects by modulating tissue or cellular eicosanoid metabolism [61,62,67,68,73,74,104–106]. In studies to determine the physiological action of CLA, investigators found reduced PGE$_2$ (a cyclooxygenase (COX)-catalyzed product of arachidonic acid) concentration in rat serum and spleen [67,104]. The amount of PGE$_2$

in cultured keratinocytes [106,117] and in ex vivo bone organ culture [73] were lowered with CLA treatment in cells and rats, respectively. Igarashi and Miyazawa [118] studied the inhibitory effect of CLA on the growth of a human hepatoma cell line, HepG2, and found that CLA's effect was interfered significantly by addition of LA or arachidonic acid. Addition of antioxidants, such as α-tocopherol and BHT, did not diminish CLA's inhibitory effect on cell proliferation, indicating that CLA's effect was not mediated by induction of lipid peroxidation but rather by changes in fatty acid metabolism. Moreover, CLA reduced the level of leukotriene B_4 (LTB_4) from the exudates of cells [67] and thromboxane A_2 (TXA_2) (another COX-catalyzed product of arachidonic acid) in platelet suspension [119]. Indirectly, CLA was found to suppress the growth of human MCF-7 breast cancer cells in culture synergistically with NDGA (a lipoxygenase inhibitor) [61], suggesting an inhibitory effect of CLA on the lipoxygenase pathway.

The mixed isomers of CLA were detected in numerous tissues examined in animals given a dietary supplement of CLA [73,104]. In one of our experiments, four groups of male rats were given a basal semipurified diet (AIN-93-G) containing 70 g/kg of added fat for 42 days [73,74]. The fat treatments were formulated to contain two levels of CLA (0% and 1% diet) and either n-6 (soybean oil having a ratio of n-6:n-3 fatty acids of 7.3) or n-3 fatty acids (menhaden oil + safflower oil having a ratio of n-6 to n-3 fatty acids of 1.8) following a 2×2 factorial design. The CLA isomers analyzed by GC were found in all rat tissues analyzed, although their concentrations varied with brain exhibiting the lowest concentration of CLA isomers and bone tissues (periosteum and marrow) containing the highest amounts. Dietary CLA decreased the concentrations of 16:1n-7, 18:1, total monounsaturates, and n-6 fatty acids, but increased the concentrations of n-3 fatty acids (22:5n-3 and 22:6n-3), total n-3, and saturates in the tissues analyzed. Ex vivo PGE_2 production in bone organ culture was decreased by the n-3 fatty acid and CLA treatments. In a subsequent study with rats, feeding 0.5% CLA resulted in total CLA concentrations ranging from 0.27% to 0.43% in the polar lipid fraction and from 2.02% to 3.37% in the neutral lipids in liver, bone marrow, and bone periosteum.

When CLA is incorporated into membrane phospholipids (induced by feeding a dietary supplement), it may compete with arachidonic acid and is likely to inhibit eicosanoid biosynthesis [74,120,121]. The reduction in the amount of n-6 fatty acids in peritoneal exudate cells and splenic lymphocyte total lipids by CLA seemed to be responsible, at least in part, for the reduced eicosanoid levels [67]. CLA could also affect the lipoxygenase pathway to reduce product formation. Cunningham et al. [61] showed that the addition of CLA and NDGA to MCF-7 cells resulted in synergistic growth suppression, suggesting that CLA effects were mediated through lipoxygenase inhibition.

The effect of dietary CLA on eicosanoid metabolism could be twofold. First, CLA isomers could directly compete with substrate concentration and activity or expression of COX or lipoxygenase. Second, CLA could be further desaturated and elongated to its 20-carbon equivalent and exert its effect by competing with other 20-carbon fatty acids, mainly arachidonic acid and eicosapentaenoic acid, to reduce the production of their corresponding eicosanoids. Experiments to evaluate the effects of CLA isomers on eicosanoid biosynthesis have not yet been conducted. We hypothesized that CLA depressed arachidonate-derived eicosanoid biosynthesis since dietary sources consistently reduced ex vivo PGE_2 production in rat bone organ culture and liver homogenate [74]. The reduction in PGE2 by CLA might be explained as a competitive inhibition of n-6 PUFA formation that results in lowered substrate availability of COX. Although there was a trend of reduced arachidonic acid concentration in bone tissues, the dramatic decrease in ex vivo PGE_2 production in bone organ culture could not be satisfactorily explained by a lack of substrate [74].

The biosynthesis of PGE_2 in bone (cells of the osteoblast lineage) is highly regulated by local and systemic factors [122–124]. Fatty acids have been shown to modulate the expression and activity of this key enzyme. For example, Nanji et al. [125] showed that saturated fat reduced peroxidation and decreased the levels of COX-2, the inducible form of COX, in rat liver. In a rat dietary study on colon tumorigenesis, a high-fat corn oil diet (rich in n-6 fatty acids) upregulated COX-2 expression, but a high-fat fish oil diet (rich in n-3 fatty acids) inhibited it; however, expression of COX-1, the constitutive enzyme, was not affected [126]. We speculate that CLA may influence PGE_2 production through the COX enzyme system, more likely on COX-2, to exert its physiological effects in bone and other tissues to influence bone metabolism as well as cancer.

Sebedio et al. [105] reported that CLA may be further desaturated and elongated to form conjugated 20:4 isomers that might block the access of arachidonic acid to COX. The unusual 20:4 isomers derived from CLA might also affect the activity of the COX enzymes. Further study with CLA is needed to confirm whether its isomeric analogs alter COX activity and expression as a primary mechanism of action and potential role in controlling cancer and inflammatory disease.

The isomers of CLA were shown to modulate the expression and activity of cytokines and growth factors. Buison et al. [127] reported that 1% CLA in a 40% (wt/wt) fat diet lowered circulating IGF-I level in obese female Wistar rats. We reported that dietary CLA lowered basal and lipopolysaccharide (LPS)-stimulated interleukin-6 production and basal tumor necrosis factor production by resident peritoneal macrophages in rats [68]. Furthermore, CLA reduced the release of LTB_4 [67], a strong bone resorption factor [128], from peritoneal exudate cells and splenic cells in response to the dietary CLA levels. Assuming that CLA would have similar effects on these cytokines in bone, together with the fact that CLA reduced the production of PGE_2 in bone tissue, one could hypothesize that a proper dietary level, the anti-inflammatory effects of CLA would be beneficial for the treatment of inflammatory bone diseases.

In our laboratory, the dietary CLA effects on serum concentrations of IGF-I and IGFBPs, and their subsequent impact on bone modeling, were examined in male rats [74]. The level of IGFBP in serum of rats was altered by n-6 and n-3 fatty acids, but CLA had variable effects. Interestingly, CLA increased IGFBP level in rats given a high dietary level of n-6 fatty acids but reduced it in those given a high level of n-3 fatty acids. Rats given the n-3 fatty acids had the highest serum level of IGFBP-3. This study also showed that CLA decreased the amount of IGF-I mRNA in liver of rats given n-3 fatty acids. In liver of rats, the expression of IGF-I mRNA appeared to be upregulated by n-3 fatty acids and downregulated by CLA. The lowering effect of CLA on growth factors was associated with reduced MAR and BFR in the tibia. These results showed that dietary PUFA type (and level) and CLA modulate growth factors that regulate bone metabolism and other aspects of health.

Different CLA isomers may also have their own unique mechanisms of action. Although limited information is available on the biological activity of individual CLA isomers, the data suggest that individual isomers exert different biological effects. Currently, the $c9,t11$ isomer has been shown to be effective in reducing mammary carcinogenesis and the $t10,c12$ isomer is more potent in inducing body compositional change. de Deckere et al. [129] treated hamsters with CLA preparations containing relatively pure $c9,t11$, $t10,c12$, or a mixture of both isomers and showed that the $t10,c12$ isomer was the most active form of CLA in inducing biological effects, such as increasing liver weight, decreasing fat deposition, and lowering LDL cholesterol. The authors concluded that $t10,c12$ appeared to be the physiologically active CLA isomer and the natural $c9,t11$ had little or no effect on lipid metabolism in hamsters.

Studies performed on mouse and cell culture also indicate differences between the two major CLA isomers in their clearance rate from the body. In skeletal muscle of mice treated with dietary CLA supplements, the $t10,c12$ isomer was cleared significantly faster than the $c9,t12$ isomer [94]. Yotsumoto et al. [130] recently showed that $t10,c12$, but not the $c9,c11$ isomer or LA inhibited cellular TAG synthesis and reduced apolipoprotein B secretion in HepG2 cell cultures. In the same study, the $c9,c11$ isomer inhibited cholesteryl ester synthesis but to a lesser extent than the $t10,c12$ isomer [130]. Information on the effects of individual CLA isomers is inadequate at present because of the limited supply of pure CLA isomers. With the advance in techniques for CLA preparation, individual CLA isomers will be available for future research.

C. Immune Function Modulation

Though CLA appears to have an anti-inflammatory effect, Turnock et al. [131] showed that feeding mice CLA for up to 4 weeks does not compromise their immune function against *Listeria monocytogenes*, an intracellular pathogen.

In a study evaluating the effect of CLA on immune competence, early-weaned pigs were given 0%–2% of a dietary CLA supplement [132]. On day 42, CLA induced a linear increase in percentages of CD8+ cytotoxic/suppressor T cells, indicating that CLA could be effective in inhibiting disease-associated growth suppression in pigs. Yamasaki et al. studied CLA and antibody production in vivo and in vitro [133]. In CLA-fed (0.05%–0.5%) Sprague–Dawley rats, the production of IgG, IgM, and IgA in spleen lymphocytes was dose dependently increased, whereas the serum concentration of these immunoglobulins were not affected. In an in vitro assay with spleen lymphocytes, CLA at 100 μM suppressed Ig production [133].

Dietary CLA supplement of 0.25 g/100 g of diet for a week significantly reduced histamine and PGE_2 release from female Hartley guinea pig trachea tissue superfusate when sensitized with antigens, indicating a reduced release of some inflammatory mediators during type 1 hypersensitivity reactions [134]. Feeding healthy young women subjects 3.9 g/day CLA did not change the indices of immune status, such as number of circulating white blood cells, granulocytes, monocytes, lymphocytes, and their subsets; lymphocyte proliferation in response to phytohemagglutinin; and influenza vaccine, serum influenza antibody titers, and DTH response [20,51].

VII. POTENTIAL ADVERSE EFFECTS OF CLA

Since CLA has become a widely advertised nutritional supplement for human use [135], it is important to study both its positive and potential negative effects on health using cell culture or animal models. Our recent dietary studies using an isomeric mixture of CLA revealed a negative effect on rat bone metabolism [74,101]. Rats given 1% CLA in the diet demonstrated decreased MAR and BFR in the tibia compared with rats not given CLA. In a follow-up experiment with rats, a lower level of CLA (0.5% diet) reduced serum osteocalcin and bone-specific alkaline phosphatase activity, both biomarkers of bone formation, after 12 weeks of dietary treatment [136]. The negative effect of CLA was likely due to a high dietary level of isomeric mixtures that does not reflect the usual food sources of CLA.

In a recent study by Belury et al. [137], CLA displayed the typical peroxisome proliferation response in rodent liver. Peroxisome proliferators may enhance tumorigenesis in liver, testes, and pancreas by acting as promoters, resulting in enhanced cell proliferation, altered cell differentiation, and inhibition of apoptosis in initiated cells [137]. This response might suggest that the chemoprotective effect of CLA in extrahepatic tissues [55,56,58,76] may be at the expense of enhanced hepatocarcinogenesis. Jones et al. [138] also reported that CLA lowered blood LDL cholesterol but increased VLDL cholesterol and resulted in liver hypertrophy in mice.

The results of Belury et al. [137] indicate that the peroxisome proliferation response to CLA may be greater in mice than in rats. These data suggest a species difference in the response to CLA, and such information might be relevant to a proper risk assessment for human consumption of isomeric mixtures of CLA. Thus far, the information on CLA action in humans is still very limited; additional research is necessary to clarify safety issues related not only to isomeric supplements of CLA but also to individual isomers that could behave differently from each other in various biological systems and physiological conditions.

Many CLA isomers present in commercial CLA supplements for human use do not exist in natural food products. Furthermore, even for the naturally occurring CLA isomers, human consumption without dietary supplementation is normally at a very low level. The possible negative effects of CLA at trace levels could not be easily detected. Therefore, before promoting the human use of dietary CLA supplements, thorough examination of the effects of individual CLA isomers is necessary to protect consumers from potential detrimental effects. Together with the fact that CLA lowered serum bone formation biomarkers and BFR, we believe that specific research must be implemented to evaluate the safety of these unique fatty acids.

REFERENCES

1. S.F. Chin, W. Liu, J.M. Storkson, Y.L. Ha, and M.W. Pariza. Dietary sources of conjugated dienoic isomers of linoleic acid, a newly recognized class of anticarcinogens. *J. Food Comp. Anal.* 5:185–197 (1992).
2. J.C. Bartlet and D.G. Chapman. Detection of hydrogenated fats in butter fat by measurement of cis-trans conjugated unsaturation. *J. Agric. Food Chem.* 9:50–53 (1961).
3. P.W. Parodi. Conjugated octadecadienoic acids of milk fat. *J. Dairy Sci.* 60:1551–1553 (1977).
4. J. Fritsche and H. Steinhart. Analysis, occurrence, and physiological properties of *trans* fatty acids (TFA) with particular emphasis on conjugated linoleic acid isomers (CLA)—a review. *Fett-Lipid* 100:190–210 (1998).
5. J. Molkentin. Bioactive lipids naturally occurring in bovine milk. *Nahrung* 43:185–189 (1999).
6. B.F. Haumann. Conjugated linoleic acid. *Inform* 7:152–159 (1996).
7. M.W. Pariza and W.A. Hargraves. A beef-derived mutagenesis modulator inhibits initiation of mouse epidermal tumors by 7,12-dimethylbenz[α]anthracene. *Carcinogenesis* 6:591–593 (1985).
8. M.W. Pariza, S.H. Ashoor, F.S. Chu, and D.B. Lund. Effects of temperature and time on mutagen formation in pan-fried hamburger. *Cancer Lett.* 7:63–69 (1979).
9. Y.L. Ha, N.K. Grimm, and M.W. Pariza. Anticarcinogens from fried ground beef: heat-altered derivatives of linoleic acid. *Carcinogenesis* 8:1881–1887 (1987).
10. N. Sehat, M.P. Yurawecz, J.A. Roach, M.M. Mossoba, J.K. Kramer, and Y. Ku. Silver-ion high-performance liquid chromatographic separation and identification of conjugated linoleic acid isomers. *Lipids* 33:217–221 (1998).
11. N. Sehat, J.K. Kramer, M.M. Mossoba, M.P. Yurawecz, J.A. Roach, K. Eulitz, K.M. Morehouse, and Y. Ku. Identification of conjugated linoleic acid isomers in cheese by gas chromatography, silver ion high performance liquid chromatography and mass spectral reconstructed ion profiles. Comparison of chromatographic elution sequences. *Lipids* 33:963–971 (1998).
12. N. Sehat, R. Rickert, M.M. Mossoba, J.K. Kramer, M.P. Yurawecz, J.A. Roach, R.O. Adlof, K.M. Morehouse, J. Fritsche, K.D. Eulitz, H. Steinhart, and Y. Ku. Improved separation of conjugated fatty acid methyl esters by silver ion-high-performance liquid chromatography. *Lipids* 34:407–413 (1999).
13. A.L. Davis, G.P. McNeill, and D.C. Caswell. Analysis of conjugated linoleic acid isomers by C-13 NMR spectroscopy. *Chem. Phys. Lipids* 97:155–165 (1999).
14. R.G. Ackman. Laboratory preparation of conjugated linoleic acids. *J. Am. Oil Chem. Soc.* 75:1227 (1998).
15. P.W. Parodi. Cows' milk fat components as potential anticarcinogenic agents. *J. Nutr.* 127:1055–1060 (1997).
16. M.K. McGuire, Y.S. Park, R.A. Behre, L.Y. Harrison, and T.D. Shultz. Conjugated linoleic acid concentrations of human milk and infant formula. *Nutr. Res.* 17:1277–1283 (1997).
17. J.K.G. Kramer, P.W. Parodi, R.G. Jensen, M.M. Mossoba, and M.P. Yurawecz. Rumenic acid: a proposed common name for the major conjugated linoleic acid isomer found in natural products. *Lipids* 33:835 (1998).
18. Y.L. Ha, N.K. Grimm, and M.W. Pariza. Newly recognized anticarcinogenic fatty acids: identification and quantification in natural and processed cheeses. *J. Agric. Food Chem.* 37:75–81 (1989).
19. R.O. Adlof, L.C. Copes, and E.L. Walter. Changes in conjugated linoleic acid composition within samples obtained from a single source. *Lipids* 36:315–317 (2001).
20. P. Benito, G.J. Nelson, D.S. Kelley, G. Bartolini, P.C. Schmidt, and V. Simon. The effect of conjugated linoleic acid on plasma lipoproteins and tissue fatty acid composition in humans. *Lipids* 36:229–236 (2001).
21. J.J. van den Berg, N.E. Cook, and D.L. Tribble. Reinvestigation of the antioxidant properties of conjugated linoleic acid. *Lipids* 30:599–605 (1995).
22. A. Zhang and Z.Y. Chen. Oxidative stability of conjugated linoleic acids relative to other polyunsaturated fatty acids. *J. Am. Oil Chem. Soc.* 74:1611–1613 (1997).

23. N.C. Shantha, E.A. Decker, and B. Hennig. Comparison of methylation methods for the quantification of conjugated linoleic acid isomers. *J. Am. Oil Chem. Soc. Int.* 76:644–649 (1993).

24. J.K. Kramer, V. Fellner, M.E. Dugan, F.D. Sauer, M.M. Mossoba, and M.P. Yurawecz. Evaluating acid and base catalysts in the methylation of milk and rumen fatty acids with special emphasis on conjugated dienes and total *trans* fatty acids. *Lipids* 32:1219–1228 (1997).

25. M.P. Yurawecz, J.A.G. Roach, M.M. Mossoba, D.H. Daniels, Y. Ku, M.W. Pariza, and S.F. Chin. Conversion of allylic hydroxy oleate to conjugated linoleic acid and methoxy oleate by acid-catalyzed methylation procedures. *J. Am. Oil Chem. Soc.* 71:1149–1155 (1994).

26. M.M. Mossoba, R.E. McDonald, D.J. Armstrong, and S.W. Page. Identification of minor C18 triene and conjugated diene isomers in hydrogenated soybean oil and margarine by GC-MI-FT-IR spectroscopy. *J. Chromatogr. Sci.* 29:324–330 (1991).

27. M.M. Mossoba, J.K.G. Kramer, M.P. Yurawecz, N. Sehat, J.A.G. Roach, K. Eulitz, J. Fritsche, M.E.R. Dugan, and Y. Ku. Impact of novel methodologies on the analysis of conjugated linoleic acid (CLA). Implications of CLA feeding studies. *Fett-Lipid* 101:235–243 (1999).

28. J. Jiang, L. Bjoerck, R. Fonden, and M. Emanuelson. Occurrence of conjugated cis-9,trans-11-octadecadienoic acid in bovine milk: effects of feed and dietary regimen. *J. Dairy Sci.* 79:438–445 (1996).

29. F. Lavillonniere, J.C. Martin, P. Bougnoux, and J.L. Sebedio. Analysis of conjugated linoleic acid isomers and content in French cheeses. *J. Am. Oil Chem. Soc.* 75:343–352 (1998).

30. N.C. Shantha, A.D. Crum, and E.A. Decker. Evaluation of conjugated linoleic acid concentrations in cooked beef. *J. Agric. Food Chem.* 42:1757–1760 (1994).

31. D.W. Ma, A.A. Wierzbicki, C.J. Field, and M.T. Clandinin. Conjugated linoleic acid in Canadian dairy and beef products. *J. Agric. Food Chem.* 47:1956–1960 (1999).

32. S. Fritsche and J. Fritsche. Occurrence of conjugated linoleic acid isomers in beef. *J. Am. Oil Chem. Soc.* 75:1449–1451 (1998).

33. D.U. Ahn, J.L. Sell, C. Jo, M. Chamruspollert, and M. Jeffrey. Effect of dietary conjugated linoleic acid on the quality characteristics of chicken eggs during refrigerated storage. *Poult. Sci.* 78:922–928 (1999).

34. M. Chamruspollert and J.L. Sell. Transfer of dietary conjugated linoleic acid to egg yolks of chickens. *Poult. Sci.* 78:1138–1150 (1999).

35. M. Du, D.U. Ahn, and J.L. Sell. Effect of dietary conjugated linoleic acid on the composition of egg yolk lipids. *Poult. Sci.* 78:1639–1645 (1999).

36. B.A. Watkins, A.A. Devitt, L. Yu, and M.A. Latour. Biological activities of conjugated linoleic acids and designer eggs. In: *Egg Nutrition and Biotechnology* (J.S. Sim, S. Nakai, and W. Guenter, eds.). CABI Publishing, Wallingford, UK, 1999, pp. 181–195.

37. S. Banni, B.W. Day, R.W. Evans, F.P. Corongiu, and B. Lombardi. Liquid chromatographic–mass spectrometric analysis of conjugated diene fatty acids in a partially hydrogenated fat. *J. Am. Oil Chem. Soc.* 71:1321–1325 (1994).

38. B.K. Herbel, M.K. McGuire, M.A. McGuire, and T.D. Shultz. Safflower oil consumption does not increase plasma conjugated linoleic acid concentrations in humans. *Am. J. Clin. Nutr.* 67:332–337 (1998).

39. A.C. Fogerty, G.L. Ford, and D. Svoronos. Octadeca-9,11-dienoic acid in foodstuffs and in the lipids of human blood and breast milk. *Nutr. Rep. Int.* 38:937–944 (1988).

40. D. Precht and J. Molkentin. C18:1, C18:2 and C18:3 *trans* and *cis* fatty acid isomers including conjugated *cis* delta 9, *trans* delta 11 linoleic acid (CLA) as well as total fat composition of German human milk lipids. *Nahrung* 43:233–244 (1999).

41. R.G. Jensen, C.J. Lammi-Keefe, D.W. Hill, A.J. Kind, and R. Henderson. The anticarcinogenic conjugated fatty acid, 9c,11t-18:2, in human milk: confirmation of its presence. *J. Hum. Lact.* 14:23–27 (1998).

42. I. Salminen, M. Mutanen, M. Jauhiainen, and A. Aro. Dietary *trans* fatty acids increase conjugated linoleic acid levels in human serum. *J. Nutr. Biochem.* 9:93–98 (1998).

43. Y.C. Huang, L.O. Luedecke, and T.D. Shultz. Effect of cheddar cheese consumption on plasma conjugated linoleic acid concentrations in men. *Nutr. Res.* 14:373–386 (1994).

44. J. Fritsche, M.M. Mossoba, M.P. Yurawecz, and J.A.G. Roach. Conjugated linoleic acid (CLA) isomers in human adipose tissue. *Z. Lebensm.-Unters. Forsch. A Food Res. Technol.* 205:415–418 (1997).

45. J. Jiang, A. Wolk, and B. Vessby. Relation between the intake of milk fat and the occurrence of conjugated linoleic acid in human adipose tissue. *Am. J. Clin. Nutr.* 70:21–27 (1999).

46. D.W. Brown and W.E.C. Moore. Distribution of *Butyrivibrio fibrisolvens* in nature. *J. Dairy Sci.* 43:1570–1574 (1960).

47. S.F. Chin, J.M. Storkson, W. Liu, K.J. Albright, and M.W. Pariza. Conjugated linoleic acid (9,11- and 10,12-octadecadienoic acid) is produced in conventional but not germ-free rats fed linoleic acid. *J. Nutr.* 124:694–701 (1994).

48. R. Pollard, F.D. Gunstone, A.T. James, and L.J. Morris. Desaturation of positional and geometric isomers of monoenoic fatty acids by microsomal preparations from rat liver. *Lipids* 15:306–314 (1980).

49. R.O. Adlof, S. Duval, and E.A. Emken. Biosynthesis of conjugated linoleic acid in humans. *Lipids* 35:131–135 (2000).

50. K.R. Glaser, M.R.L. Scheeder, and C. Wenk. Dietary C18:1 *trans* fatty acids increase conjugated linoleic acid in adipose tissue of pigs. *Eur. J. Lipid Sci. Technol.* 102:684–686 (2000).

51. D.S. Kelley, P.C. Taylor, I.L. Rudolph, P. Benito, G.J. Nelson, B.E. Mackey, and K.L. Erickson. Dietary conjugated linoleic acid did not alter immune status in young healthy women. *Lipids* 35:1065–1071 (2000).

52. P. Benito, G.J. Nelson, D.S. Kelley, G. Bartolini, P.C. Schmidt, and V. Simon. The effect of conjugated linoleic acid on platelet function, platelet fatty acid composition, and blood coagulation in humans. *Lipids* 36:221–227 (2001).

53. C. Ip, M. Singh, H.J. Thompson, and J.A. Scimeca. Conjugated linoleic acid suppresses mammary carcinogenesis and proliferative activity of the mammary gland in the rat. *Cancer Res.* 54:1212–1215 (1994).

54. Y.L. Ha, J. Storkson, and M.W. Pariza. Inhibition of benzo(α)pyrene-induced mouse forestomach neoplasia by conjugated dienoic derivatives of linoleic acid. *Cancer Res.* 50:1097–1101 (1990).

55. C. Ip, S.F. Chin, J.A. Scimeca, and M.W. Pariza. Mammary cancer prevention by conjugated dienoic derivative of linoleic acid. *Cancer Res.* 51:6118–6124 (1991).

56. C. Ip, S.P. Briggs, A.D. Haegele, H.J. Thompson, J. Storkson, and J.A. Scimeca. The efficacy of conjugated linoleic acid in mammary cancer prevention is independent of the level or type of fat in the diet. *Carcinogenesis* 17:1045–1050 (1996).

57. C. Ip, C. Jiang, H.J. Thompson, and J.A. Scimeca. Retention of conjugated linoleic acid in the mammary gland is associated with tumor inhibition during the post-initiation phase of carcinogenesis. *Carcinogenesis* 18:755–759 (1997).

58. C. Liew, H.A. Schut, S.F. Chin, M.W. Pariza, and R.H. Dashwood. Protection of conjugated linoleic acids against 2-amino-3-methylimidazo[4,5-f]quinoline-induced colon carcinogenesis in the F344 rat: a study of inhibitory mechanisms. *Carcinogenesis* 16:3037–3043 (1995).

59. A. Cesano, S. Visonneau, J.A. Scimeca, D. Kritchevsky, and D. Santoli. Opposite effects of linoleic acid and conjugated linoleic acid on human prostatic cancer in SCID mice. *Anticancer Res.* 18:1429–1434 (1998).

60. M.M. Ip, P.A. Masso-Welch, S.F. Shoemaker, W.K. Shea-Eaton, and C. Ip. Conjugated linoleic acid inhibits proliferation and induces apoptosis of normal rat mammary epithelial cells in primary culture. *Exp. Cell Res.* 250:22–34 (1999).

61. D.C. Cunningham, L.Y. Harrison, and T.D. Shultz. Proliferative responses of normal human mammary and MCF-7 breast cancer cells to linoleic acid, conjugated linoleic acid and eicosanoid synthesis inhibitors in culture. *Anticancer Res.* 17:197–203 (1997).

62. S. Banni, E. Angioni, V. Casu, M.P. Melis, G. Carta, F.P. Corongiu, H. Thompson, and C. Ip. Decrease in linoleic acid metabolites as a potential mechanism in cancer risk reduction by conjugated linoleic acid. *Carcinogenesis* 20:1019–1024 (1999).

63. K.N. Lee, D. Kritchevsky, and M.W. Pariza. Conjugated linoleic acid and atherosclerosis in rabbits. *Atherosclerosis* 108:19–25 (1994).

64. R.J. Nicolosi, E.J. Rogers, D. Kritchevsky, J.A. Scimeca, and P.J. Huth. Dietary conjugated linoleic acid reduces plasma lipoproteins and early aortic atherosclerosis in hypercholesterolemic hamsters. *Artery* 22:266–277 (1997).

65. M.E. Cook, C.C. Miller, Y. Park, and M. Pariza. Immune modulation by altered nutrient metabolism: nutritional control of immune-induced growth depression. *Poult. Sci.* 72:1301–1305 (1993).

66. C.C. Miller, Y. Park, M.W. Pariza, and M.E. Cook. Feeding conjugated linoleic acid to animals partially overcomes catabolic responses due to endotoxin injection. *Biochem. Biophys. Res. Commun.* 198:1107–1112 (1994).

.. Sugano, A. Tsujita, M. Yamasaki, M. Noguchi, and K. Yamada. Conjugated linoleic acid modulates tissue levels of chemical mediators and immunoglobulins in rats. *Lipids* 33:521–527 (1998).

6. J.J. Turek, Y. Li, I.A. Schoenlein, K.G.D. Allen, and B.A. Watkins. Modulation of macrophage cytokine production by conjugated linoleic acids is influenced by the dietary n-6:n-3 fatty acid ratio. *J. Nutr. Biochem.* 9:258–266 (1998).

69. J.P. DeLany, F. Blohm, A.A. Truett, J.A. Scimeca, and D.B. West. Conjugated linoleic acid rapidly reduces body fat content in mice without affecting energy intake. *Am. J. Physiol.* 276:R1172–R1179 (1999).

70. M. Yamasaki, K. Mansho, H. Mishima, M. Kasai, M. Sugano, H. Tachibana, and K. Yamada. Dietary effect of conjugated linoleic acid on lipid levels in white adipose tissue of Sprague–Dawley rats. *Biosci. Biotechnol. Biochem.* 63:1104–1106 (1999).

71. Y. Park, J.M. Storkson, K.J. Albright, W. Liu, and M.W. Pariza. Evidence that the *trans*-10,*cis*-12 isomer of conjugated linoleic acid induces body composition changes in mice. *Lipids* 34:235–241 (1999).

72. K.L. Houseknecht, J.P.V. Heuvel, S.Y. Moya-Camerena, C.P. Portocarrero, K.P. Nickel, and M.A. Belury. Dietary conjugated linoleic acid normalizes impaired glucose tolerance in the Zucker diabetic fatty fa/fa rat. *Biochem. Biophys. Res. Commun.* 244:678–682 (1998).

73. Y. Li and B.A. Watkins. Conjugated linoleic acids alter bone fatty acid composition and reduce ex vivo prostaglandin E_2 biosynthesis in rats fed n-6 or n-3 fatty acids. *Lipids* 33:417–425 (1998).

74. Y. Li, M.F. Siefert, D.M. Ney, M. Grahn, A.L. Grant, K.G. Allen, and B.A. Watkins. Dietary conjugated linoleic acids alter serum IGF-I and IGF binding protein concentrations and reduce bone formation in rats fed (n-6) or (n-3) fatty acids. *J. Bone Miner. Res.* 14:1153–1162 (1999).

75. H. Thompson, Z. Zhu, S. Banni, K. Darcy, T. Loftus, and C. Ip. Morphological and biochemical status of the mammary gland as influenced by conjugated linoleic acid: implication for a reduction in mammary cancer risk. *Cancer Res.* 57:5067–5072 (1997).

76. M.A. Belury, K.P. Nickel, C.E. Bird, and Y. Wu. Dietary conjugated linoleic acid modulation of phorbol ester skin tumor promotion. *Nutr. Cancer* 26:149–157 (1996).

77. S. Schonberg and H.E. Krokan. The inhibitory effect of conjugated dienoic derivatives (CLA) of linoleic acid on the growth of human tumor cell lines is in part due to increased lipid peroxidation. *Anticancer Res.* 15:1241–1246 (1995).

78. T.D. Shultz, B.P. Chew, and W.R. Seaman. Differential stimulatory and inhibitory responses of human MCF-7 breast cancer cells to linoleic acid and conjugated linoleic acid in culture. *Anticancer Res.* 12:2143–2145 (1992).

79. S. Visonneau, A. Cesano, S.A. Tepper, J.A. Scimeca, D. Santoli, and D. Kritchevsky. Conjugated linoleic acid suppresses the growth of human breast adenocarcinoma cells in SCID mice. *Anticancer Res.* 17:969–973 (1997).

80. J. Whelan, M. McEntee, P.M. Hansen, and M. Obuckowicz. Conjugated linoleic acid and intestinal tumorigenesis in Min/+ mice. *Inform* (Suppl. for the 92nd AOCS Annual Meeting & Expo):S56–S57 (2001).

81. K.K. Cornell, Y. Li, K. Boyd, D.J. Waters, and B.A. Watkins. Effects of dietary conjugated linoleic acid on prostate cancer growth in a subcutaneous microenvironment in the athymic mouse. *Inform* (Suppl. for the 92nd AOCS Annual Meeting & Expo):S57 (2001).

82. C. Ip and J.A. Scimeca. Conjugated linoleic acid and linoleic acid are distinctive modulators of mammary carcinogenesis. *Nutr. Cancer* 27:131–135 (1997).

83. C. Ip, S. Banni, E. Angioni, G. Carta, J. McGinley, H.J. Thompson, D. Barbano, and D. Bauman. Conjugated linoleic acid–enriched butter fat alters mammary gland morphogenesis and reduces cancer risk in rats. *J. Nutr.* 129:2135–2142 (1999).

84. D.P. Rose. Dietary fatty acids and prevention of hormone-responsive cancer. *Exp. Biol. Med.* 216:224–233 (1997).

85. M.W. Wong, B.P. Chew, T.S. Wong, H.L. Hosick, T.D. Boylston, and T.D. Shultz. Effects of dietary conjugated linoleic acid on lymphocyte function and growth of mammary tumors in mice. *Anticancer Res.* 17:987–993 (1997).

86. P. Knekt, R. Jarvinen, R. Seppanen, E. Pukkala, and A. Aromaa. Intake of dairy products and the risk of breast cancer. *Br. J. Cancer* 73:687–691 (1996).

87. Y.J. Kim and R.H. Liu. Selective increase in conjugated linoleic acid in milk fat by crystallization. *J. Food Sci.* 64:792–795 (1999).

88. C.H.T. Yeung, L. Yang, Y. Huang, J. Wang, and Z.Y. Chen. Dietary conjugated linoleic acid mixture affects the activity of intestinal acyl coenzyme A: cholesterol acyltransferase in hamsters. *Br. J. Nutr.* 84:935–941 (2000).

89. B. Szymczyk, P. Pisulewski, W. Szczurek, and P. Hanczakowski. The effects of feeding conjugated linoleic acid (CLA) on rat growth performance, serum lipoproteins and subsequent lipid composition of selected rat tissues. *J. Sci. Food Agric.* 80:1553–1558 (2000).

90. J.S. Munday, K.G. Thompson, and K.A. James. Dietary conjugated linoleic acids promote fatty streak formation in the C57BL/6 mouse atherosclerosis model. *Br. J. Nutr.* 81:251–255 (1999).

91. M.E.R. Dugan, J.L. Aalhus, A.L. Schaefer, and J.K.G. Kramer. The effect of conjugated linoleic acid on fat to lean repartitioning and feed conversion in pigs. *Can. J. Anim. Sci.* 77:723–725 (1997).

92. E. Ostrowska, M. Muralitharan, R.F. Cross, D.E. Bauman, and F.R. Dunshea. Dietary conjugated linoleic acids increase lean tissue and decrease fat deposition in growing pigs. *J. Nutr.* 129:2037–2042 (1999).

93. Y. Park, K.J. Albright, W. Liu, J.M. Storkson, M.E. Cook, and M.W. Pariza. Effect of conjugated linoleic acid on body composition in mice. *Lipids* 32:853–858 (1997).

94. Y. Park, K.J. Albright, J.M. Storkson, W. Liu, M.E. Cook, and M.W. Parzia. Changes in body composition in mice during feeding and withdrawal of conjugated linoleic acid. *Lipids* 34:243–248 (1999).

95. M. Yamasaki, K. Mansho, H. Mishima, G. Kimura, M. Sasaki, M. Kasai, H. Tachibana, and K. Yamada. Effect of dietary conjugated linoleic acid on lipid peroxidation and histological change in rat liver tissues. *J. Agric. Food Chem.* 48:6367–6371 (2000).

96. G. Berven, A. Bye, O. Hals, H. Blankson, H. Fagertun, E. Thom, J. Wadstein, and O. Gudmundsen. Safety of conjugated linoleic acid (CLA) in overweight or obese human volunteers. *Eur. J. Lipid Sci. Technol.* 102:455–462 (2000).

97. H. Blankson, J.A. Stakkestad, H. Fagertun, E. Thom, J. Wadstein, and O. Gudmundsen. Conjugated linoleic acid reduces body fat mass in overweight and obese humans. *J. Nutr.* 130:2943–2948 (2000).

98. M. Evans, C. Geigerman, J. Cook, L. Curtis, B. Kuebler, and M. McIntosh. Conjugated linoleic acid suppresses triglyceride accumulation and induces apoptosis in 3T3-L1 preadipocytes. *Lipids* 35:899–910 (2000).

99. Y.G. Lin, A. Kreeft, J.A.E. Schuurbiers, and R. Draijer. Different effects of conjugated linoleic acid isomers on lipoprotein lipase activity in 3T3-L1 adipocytes. *J. Nutr. Biochem.* 12:183–189 (2001).

100. M. Yamasaki, K. Mansho, Y. Ogino, M. Kasai, H. Tachibana, and K. Yamada. Acute reduction of serum leptin level by dietary conjugated linoleic acid in Sprague–Dawley rats. *J. Nutr. Biochem.* 11:467–471 (2000).

101. Y. Li, B.A. Watkins, and M.F. Siefert. Effects of dietary conjugated linoleic acid on tissue fatty acid composition, serum osteocalcin level and bone alkaline phosphatase activity in rats. *J. Bone Miner. Res.* 14:S315 (1999).

102. S. Josyula, Y.H. He, R.J. Ruch, and H.A. Schut. Inhibition of DNA adduct formation of PhIP in female F344 rats by dietary conjugated linoleic acid. *Nutr. Cancer* 32:132–138 (1998).

103. H.A. Schut, D.A. Cummings, M.H. Smale, S. Josyula, and M.D. Friesen. DNA adducts of heterocyclic amines: formation, removal and inhibition by dietary components. *Mutat. Res.* 376:185–194 (1997).

104. M. Sugano, A. Tsujita, M. Yamasaki, K. Yamada, I. Ikeda, and D. Kritchevsky. Lymphatic recovery, tissue distribution, and metabolic effects of conjugated linoleic acid in rats. *J. Nutr. Biochem.* 8:38–43 (1997).

105. J.L. Sebedio, P. Juaneda, G. Dobson, I. Ramilison, J.C. Martin, J.M. Chardigny, and W.W. Christie. Metabolites of conjugated isomers of linoleic acid (CLA) in the rat. *Biochim. Biophys. Acta* 1345:5–10 (1997).

106. C.J. Kavanaugh, K.L. Liu, and M.A. Belury. Effect of dietary conjugated linoleic acid on phorbol ester–induced PGE_2 production and hyperplasia in mouse epidermis. *Nutr. Cancer* 33:132–138 (1999).

107. N. Ito and M. Hirose. Antioxidants—carcinogenic and chemopreventive properties. *Adv. Cancer Res.* 53:247–302 (1989).

108. Y.H. Leung and R.H. Liu. *trans*-10,*cis*-12-Conjugated linoleic acid isomer exhibits stronger oxy-radical scavenging capacity than *cis*-9,*trans*-11-conjugated linoleic acid isomer. *J. Agric. Food Chem.* 48:5469–5475 (2000).

109. S. Banni, E. Angioni, M.S. Contini, G. Carta, V. Casu, G.A. Iengo, M.P. Melis, M. Deiana, M.A. Dessi, and F.P. Corongiu. Conjugated linoleic acid and oxidative stress. *J. Am. Oil Chem. Soc.* 75:261–267 (1998).

O'Shea, C. Stanton, and R. Devery. Antioxidant enzyme defence responses of human MCF-7 and ـW480 cancer cells to conjugated linoleic acid. *Anticancer Res.* 19:1953–1959 (1999).

. M. O'Shea, R. Devery, F. Lawless, J. Murphy, and C. Stanton. Milk fat conjugated linoleic acid (CLA) inhibits growth of human mammary MCF-7 cancer cells. *Anticancer Res.* 20:3591–3601 (2000).

112. G.W. Burton. Antioxidant action of carotenoids. *J. Nutr.* 119:109–111 (1989).

113. S. Basu, A. Smedman, and B. Vessby. Conjugated linoleic acid induces lipid peroxidation in humans. *FEBS Lett.* 468:33–36 (2000).

114. S. Basu, U. Riserus, A. Turpeinen, and B. Vessby. Conjugated linoleic acid induces lipid peroxidation in men with abdominal obesity. *Clin. Sci.* 99:511–516 (2000).

115. C. Ip, M.M. Ip, T. Loftus, S. Shoemaker, and W. Shea-Eaton. Induction of apoptosis by conjugated linoleic acid in cultured mammary tumor cells and premalignant lesions of the rat mammary gland. *Cancer Epidemiol. Biomarkers Prev.* 9:689–696 (2000).

116. Y. Park, K.G.D. Allen, and T.D. Shultz. Modulation of MCF-7 breast cancer cell signal transduction by linoleic acid and conjugated linoleic acid in culture. *Anticancer Res.* 20:669–676 (2000).

117. K.L. Liu and M.A. Belury. Conjugated linoleic acid reduces arachidonic acid content and PGE_2 synthesis in murine keratinocytes. *Cancer Lett.* 127:15–22 (1998).

118. M. Igarashi and T. Miyazawa. The growth inhibitory effect of conjugated linoleic acid on a human hepatoma cell line, HepG2, is induced by a change in fatty acid metabolism, but not the facilitation of lipid peroxidation in the cells. *Biochim. Biophys. Acta* 1530:162–171 (2001).

119. A. Truitt, G. McNeill, and J.Y. Vanderhoek. Antiplatelet effects of conjugated linoleic acid isomers. *Biochim. Biophys. Acta* 1438:239–246 (1999).

120. S.H. Abou-el-Ela, K.W. Prasse, R.L. Farrell, R.W. Carroll, A.E. Wade, and O.R. Bunce. Effects of D,L-2-difluoromethylomithine and indomethacin on mammary tumor promotion in rats fed high n-3 and/or n-6 fat diets. *Cancer Res.* 49:1434–1440 (1989).

121. J. Leyton, M.L. Lee, M. Locniskar, M.A. Belury, T.J. Slaga, D. Bechtel, and S.M. Fischer. Effects of type of dietary fat on phorbol ester-elicited tumor promotion and other events in mouse skin. *Cancer Res.* 51:907–915 (1991).

122. F.M. Maciel, P. Sarrazin, S. Morisset, M. Lora, C. Patry, R. Dumais, and A.J. de Brum-Fernandes. Induction of cyclooxygenase-2 by parathyroid hormone in human osteoblasts in culture. *J. Rheumatol.* 24:2429–2435 (1997).

123. W. Pruzanski, E. Stefanski, P. Vadas, B.P. Kennedy, and H. van den Bosch. Regulation of the cellular expression of secretory and cytosolic phospholipases A_2, and cyclooxygenase-2 by peptide growth factors. *Biochim. Biophys. Acta* 1403:47–56 (1998).

124. M. Suda, K. Tanaka, A. Yasoda, K. Natsui, Y. Sakuma, I. Tanaka, F. Ushikubi, S. Narumiya, and K. Nakao. Prostaglandin E_2 (PGE_2) autoamplifies its production through EP_1 subtype of PGE receptor in mouse osteoblastic MC3T3-E1 cells. *Calcif. Tissue Int.* 62:327–331 (1998).

125. A.A. Nanji, D. Zakim, A. Rahemtulla, T. Daly, L. Miao, S. Zhao, S. Khwaja, S.R. Tahan, and A.J. Dannenberg. Dietary saturated fatty acids down-regulate cyclooxygenase-2 and tumor necrosis factor α and reverse fibrosis in alcohol-induced liver disease in the rat. *Hepatology* 26:1538–1545 (1997).

126. J. Singh, R. Hamid, and B.S. Reddy. Dietary fat and colon cancer: modulation of cyclooxygenase-2 by types and amount of dietary fat during the postinitiation stage of colon carcinogenesis. *Cancer Res.* 57:3465–3470 (1997).

127. A. Buison, F. Ordiz, M. Pellizzon, and K.L.C. Jen. Conjugated linoleic acid does not impair fat regain but alters IGF-1 levels in weight-reduced rats. *Nutr. Res.* 20:1591–1601.

128. C. Garcia, B.F. Boyce, J. Gilles, M. Dallas, M. Qiao, G.R. Mundy, and L.F. Bonewald. Leukotriene B_4 stimulates osteoclastic bone resorption both in vitro and in vivo. *J. Bone Miner. Res.* 11:1619–1627 (1996).

129. E.A.M. de Deckere, J.M.M. van Amelsvoort, G.P. McNeill, and P. Jones. Effects of conjugated linoleic acid (CLA) isomers on lipid levels and peroxisome proliferation in the hamster. *Br. J. Nutr.* 82:309–317 (1999).

130. H. Yotsumoto, E. Hara, S. Naka, R.O. Adlof, E.A. Emken, and T. Yanagita. 10*trans*,12*cis*-Linoleic acid reduces apolipoprotein B secretion in HepG2 cells. *Food Res. Intern.* 31:403–409 (1998).

131. L. Turnock, M. Cook, H. Steinberg, and C. Czuprynski. Dietary supplementation with conjugated linoleic acid does not alter the resistance of mice to *Listeria monocytogenes* infection. *Lipids* 36:135–138 (2001).

132. J. Bassaganya-Riera, R. Hontecillas-Magarzo, K. Bregendahl, M.J. Wannemuehler, and D.R. Zimmerman. Effects of dietary conjugated linoleic acid in nursery pigs of dirty and clean environments on growth, empty body composition, and immune competence. *J. Anim. Sci.* 79:714–721 (2001).

133. M. Yamasaki, K. Kishihara, K. Mansho, Y. Ogino, M. Kasai, M. Sugano, H. Tachibana, and K. Yamada. Dietary conjugated linoleic acid increases immunoglobulin productivity of Sprague–Dawley rate spleen lymphocytes. *Biosci. Biotechnol. Biochem.* 64:2159–2164 (2000).

134. L.D. Whigham, E.B. Cook, J.L. Stahl, R. Saban, D.E. Bjorling, M.W. Pariza, and M.E. Cook. CLA reduces antigen-induced histamine and PGE_2 release from sensitized guinea pig tracheae. *Am. J. Physiol.* 280:R908–R912 (2001).

135. C. Steinhart. Conjugated linoleic acid—the good news about animal fat. *J. Chem. Educ.* 73:A302–A303 (1996).

136. Y. Li, M.F. Seifert, and B.A. Watkins. CLA rescued bone formation rate in the growing rat. *Inform* (Suppl. for the 92nd AOCS Annual Meeting & Expo):S58 (2001).

137. M.A. Belury, S.Y. Moya-Camarena, K.L. Liu, and J.P. Vanden Heuvel. Dietary conjugated linoleic acid induces peroxisome-specific enzyme accumulation and ornithine decarboxylase activity in mouse liver. *J. Nutr. Biochem.* 8:579–584 (1997).

138. P.A. Jones, L.J. Lea, and R.U. Pendlington. Investigation of the potential of conjugated linoleic acid (CLA) to cause peroxisome proliferation in rats. *Food Chem. Toxicol.* 37:1119–1125 (1999).

24 Dietary Fats and Obesity

Dorothy B. Hausman and Barbara Mullen Grossman

CONTENTS

I. INTRODUCTION

Obesity is variably defined as having a body weight of more than 20% over ideal body weight or a body mass index (weight (kg)/height (m)^2) of 30 or more [1]. Obesity is associated with many chronic diseases and alterations in physiological function, including cardiovascular disease, hypertension, diabetes mellitus, gallbladder disease, and certain types of cancer [2]. Obesity is a major public health problem in the United States and Europe and is becoming increasingly important in many other areas of the world [3]. The prevalence of obesity in adults in the United States has increased by 30% or more in the past decade, with increases in both genders and in all ethnic and racial populations and age groups [4]. It is now estimated that in the United States more than one-third of the total adult population and one-half of the adult African American and Hispanic population are obese [5].

ugy of human obesity is quite complex, involving genetic, metabolic, behavioral, and ital factors. Although obesity is believed to have a strong genetic component [6], the ⟍ incidence of obesity in specific population groups undergoing westernization indicates portance of dietary and lifestyle changes in the manifestation of this disease [7,8]. Among iry factors, both total energy intake and fat intake are significantly correlated with body mass dex in these population groups [7]. However, increased intake of fat energy is associated with a greater per unit increase in body mass than is increased intake of energy from nonfat sources. Therefore, much attention regarding dietary influences on obesity development or prevention has focused on high-fat diets.

II. DIETARY FAT AND BODY WEIGHT

The effect of dietary fat on increasing body weight has been well documented, especially in experimental animals. The literature in this area has been thoroughly reviewed [9–11], and for the most part, dietary fat at sufficient levels results in increased body weight. A genetic component to this effect is illustrated by studies with animals either resistant or susceptible to developing obesity when eating a high-fat diet [12]. The caloric density of the diet plays a major role in promoting weight gain. If fat density is about 25% or more, excess weight gain is more likely to occur [13].

Human studies, because they are more difficult to control, are less prevalent than studies with animals. However, human studies also demonstrate that dietary fat promotes weight gain. Lissner et al. [14] demonstrated that subjects consuming a high-fat diet gained weight, whereas those eating a low-fat diet lost weight. Others [15,16] have reported that switching from a high- to a low-fat diet promotes relative weight loss. Finally, indirect evidence provided from epidemiologic studies shows positive correlations between body weight and fat intake [17]. The Leeds Fat Study [18] reported that the proportion of obese individuals was 19 times higher among those who consumed higher-fat diets as compared with lower-fat diets. When Japanese men living in Honolulu were compared with those living in Hiroshima and Nagasaki, the prevalence of obesity of the men in Honolulu was greater than those in Japan, presumably due to the higher level of fat in the Honolulu diet, even though total energy intake was only slightly higher in the Honolulu men [19].

III. DIETARY FAT AND FOOD INTAKE

A. DOES DIETARY FAT AFFECT FOOD INTAKE?

Dietary fat can have a profound effect on energy balance and, ultimately, body weight. Studies designed to investigate the effect of dietary fat on food intake generally employ one of the two approaches: short-term studies examining the influence of dietary fat on meal size or frequency and long-term investigations of the effects of dietary fat on energy intake over days or weeks. Short-term effects are typically examined in one of the two ways: (1) fat is given as a preload, and subsequent feelings of hunger or food intake is reported or (2) fat is given as part of a mixed diet and concurrent food intake is measured. In the first instance, satiety, that is, the ability of a substance to suppress further eating, is measured by the time elapsed or amount of food eaten at the next meal. In the second case, satiation is assessed and is defined as the size of the current meal [20].

Many studies have been carried out examining the effect of a preload, often in the form of a liquid on subsequent intake, typically over a short time period. Others have investigated the effect of high- or low-fat meals or snacks, or meals supplemented with fat or carbohydrate, on subsequent feelings of fullness and food intake. The results of these studies have been quite variable (for review, see Blundell et al. [21]). Many reports indicate that with a preload, individuals do not fully compensate for the calories ingested, that is, they do not reduce their intake in accurate proportion to the calories previously consumed [22–25]. This seems to be truer for fat versus carbohydrate preloads [24–30], though some studies have indicated that fat has satiety value equal to carbohydrate [31,32] or even greater than carbohydrate [33,34]. In addition, compensation seems to occur initially but then decreases over time [31–33].

It is important that the volume, sensory characteristics, and protein content of the preloads be similar when investigating the satiety effect of fat. A number of studies have not controlled for all of these factors [23,30,35,36]. However, when preloads are similar, investigators have found that individuals vary substantially in their response to preloads and that body weight may play a role in these responses. For example, some have reported that males who are of normal weight and not concerned about their body weight or food intake (unrestrained eaters) appear to compensate adequately for the caloric content of a preload [30]. Porrini et al. [37] report that a high-protein food given as a snack 2 h before a meal exerts a higher effect on both intrameal satiation and postingestive satiety than a high-fat snack. When a first course is consumed as part of a meal, the sensory characteristics of the food play an important role in controlling subsequent food intake. Lawton et al. [26] have shown that in obese individuals, fat exerts only a weak action on satiety. Data such as these have led to the speculation that obesity may be the result of insensitivity to satiety signals generated by ingestion of fat.

In studies of satiation, dietary fat is an integral variable of the test diets. Studies show that caloric intake is greater with dietary fat than with carbohydrate. When subjects are fed diets in which fat, fiber, and simple sugars are manipulated to obtain low-energy versus high-energy diets, energy intake is greater on the high-energy diet than the low-energy diet [38]. Caputo and Mattes [39] reported that individuals consuming high-fat meals consume more calories than those consuming high- or low-carbohydrate or low-fat meals. Thomas et al. [40] have shown that individuals consuming high-fat or high-carbohydrate diets for 1 week consume more calories on the high-fat versus high-carbohydrate diets. This hyperphagic effect of dietary fat has been observed in many studies [41–46].

The excessive intake of dietary fat primarily occurs during a meal to increase meal size rather than between meals to increase meal frequency [43]. Overconsumption is particularly high when fat is combined with alcohol [47]. Studies indicate that palatability and energy density of high-fat diets play an important role in overfeeding, however, other factors may also influence this behavior [48]. Over the long term, many investigators have shown that this passive overconsumption of dietary fat can lead to obesity (for review, see [12,43,49,50]).

Relatively few long-term studies on humans have been conducted because of difficulties such as inability of the investigator to control for the subjects' current or past food intake, activity level, or genetic background. However, Kendall et al. [15] reported that women consuming either a low- or high-fat diet for two separate 11 week periods consume more calories on the high-fat diet. Lissner et al. [14] also showed that subjects eating either a low-, medium-, or a high-fat diet for 2 weeks consume the most calories on the high-fat diet.

In addition, animal studies have shown that the type of fat may influence the satiation effects of dietary fat. In both chickens [51] and rats [52], medium-chain triglycerides have been shown to have a greater satiating effect than long-chain triglycerides. However, when medium- or long-chain fatty acids were infused into the hepatic portal vein, the medium-chain fatty acids had no effect on feeding, whereas the long-chain fatty acids robustly inhibited feeding [53]. Nonetheless, many studies with rodent models have shown that a preference for dietary fat occurs when texture, olfaction, and postingestive cues are controlled (for review, see [54]). Pittman et al. [54] have suggested that orosensory cues play a major role in dietary fat preference in rats. In humans, Rolls et al. [55] and Stubbs and Harbron [56] report that substitution of long-chain triglycerides with medium-chain triglycerides depresses food intake. In addition, the physical form as well as the type of fat contributes to its satiation effects [33,57]. Overeating occurs with diets containing saturated fats [58–61] as well as mixed fats [62–65]. Lawton et al. [66] showed that in human subjects polyunsaturated fatty acids (PUFAs) may exert a stronger control over appetite than monounsaturated fatty acids (MUFA) or saturated fatty acids (SFA).

Some have suggested that the hyperphagic effect of high-fat diets is not due solely to the fat but is influenced by the presence of carbohydrate and overall caloric density [67]. Ramirez and Friedman [67] fed rats diets varying in carbohydrate, fat, cellulose, or caloric density and found that energy intake varies directly as a function of caloric density regardless of the fat or cellulose content of the diets. They concluded that high levels of fat, carbohydrate, and energy interact to produce overeating in animals fed high-fat diets. In support of this hypothesis, Emmet and Heaton [68] examined food records from 160 subjects who had weighed their food for 4 days. They reported that an increase in refined sugar intake is associated with a linear increase in the intake of fat combined with carbohydrate. This suggests that refined sugar may act as a vehicle for fat intake by increasing fat palatability.

Some have reported that there is an inverse relationship between consumption of sugar and fat, terming it the see-saw effect. This suggests that carbohydrates, per se, are protective against obesity, as sugars may displace fat energy from the diet [69–71]. When consumption of fat and sugar are expressed as a percentage of total intake, this inverse relationship is present. However, when expressed in absolute terms, there is a positive relationship between dietary fat and carbohydrate intake [72]. When high-fat and high-carbohydrate diets of equal caloric density are compared, both contribute to increased, uncompensated caloric intake [50].

B. Mechanisms for Fat-Induced Food Intake

1. Caloric Density

The higher caloric density of many high-fat diets may play a role in inducing this hyperphagic response [42,43,73]. In humans, Duncan et al. [38] reported that adult subjects eat almost twice as many calories on a high-density diet compared with a low-density diet. In studies using experimental animals, hyperphagia is typically observed only when the caloric density of the diets is high, greater than approximately 5.8 kcal/g of diet (for review, see Warwick and Schiffman [9] and Geary [74]). In addition, when caloric density is constant, rats fed a diet high in corn oil have a caloric intake similar to that of animals fed low-fat diets [67,75]. In addition, when the energy density of high-fat diets is reduced by adding nonnutritive fillers, energy intake and body weight typically do not increase (for review, see Geary [74]).

Several investigators have reported that a preference develops for a flavor that is paired with a high number of calories versus one paired with a low number of calories. A study by Johnson et al. [76] indicated that children report increased flavor pleasantness because of association with a high density of fat calories. In rats, a flavor associated with corn oil consumption is preferred over a flavor that is not paired with oil [77,78]. In studies in which oils or high-fat foods are given, animals often initially do not consume greater quantities of the fat. However, over time the rats do consume more of the fat as they learn about the associated postingestive consequences (i.e., greater caloric value) [79–83]. In contrast, others have suggested that dietary fat is overconsumed even when compared with an isoenergic carbohydrate diet of similar palatability [48,84,85]. Further, Lucas et al. [86] report that, relative to an isocaloric high-carbohydrate diet, the postingestive effect of high-fat diets stimulates overeating and conditions a stronger flavor preference in rats, suggesting that some quality in fat per se may be inducing intake.

2. Stomach Distention

Differences in stomach distention due to dietary fat versus carbohydrate or protein may also account for its hyperphagic effect [9]. Warwick and Schiffman [9] suggested that due to the greater caloric density of dietary fat, a high-fat meal has a smaller volume than an isocaloric high-carbohydrate meal, resulting in less stomach distention. This lesser distention would lead to an attenuation of satiety signals. In addition, Cunningham et al. [87] reported that the rate of stomach emptying increases when a high-fat diet is habitually consumed. When the energy density of high-fat diets is reduced by adding noncaloric fillers, the hyperphagia and weight gain does not occur [88–92]. Similar effects have been reported in humans [93,94].

Dietary fat has also been found to influence relative consumption of carbohydrate and protein. Crane and Greenwood [95] allowed rats to select from either high-carbohydrate or high-protein diets. Half of the diets contained 20% soybean oil as the fat component and the other half contained 20% lard as the fat component. Animals selecting from the soybean oil diets consumed more carbohydrate and less protein than animals choosing from the lard-based diets. Grossman et al. [96] further showed that rats gavaged with either beef tallow or corn oil 2 h before selecting from either high-carbohydrate or high-protein diets consume less carbohydrate and more protein if given tallow versus corn oil. They [96] also demonstrated that the hepatic vagus must be intact for this selection to occur and that mercaptoacetate (a fatty acid oxidation inhibitor) can blunt the effect.

3. Metabolic Signals

In an attempt to understand the mechanisms underlying the effects of dietary fat on appetite, investigators have examined absorptive or postabsorptive responses to dietary fat. Studies in rodents and humans have shown that a diet rich in diacylglycerols increases fatty acid oxidation and decreases food intake (for review, see Rudkowska et al. [97]). Scharrer and Langhans [98] demonstrated that in rats, consumption of a high-fat diet can be stimulated by inhibiting fatty acid oxidation with 2-mercaptoacetate. Friedman et al. [99] also reported a stimulation of food intake when fatty acid oxidation is inhibited with methyl palmoxirate. This feeding effect, labeled lipoprivic feeding, has been shown to be impaired by hepatic vagotomy [100] and subdiaphragmatic vagotomy [101]. In addition, Ritter and Taylor [102] reported that capsaicin can block this effect, implying that vagal sensory neurons appear to be involved in lipoprivic feeding. Studies utilizing brain lesions indicate that lipoprivic feeding involves the lateral parabrachial nucleus and possibly the area postrema/nucleus of the solitary tract [103]. Type of fat may influence this response as Wang et al. [104] reported that food intake is stimulated by mercaptoacetate in rats given corn oil but not tallow diets. Finally, Suzuki et al. [105] have shown that rats given mercaptoacetate exhibit reinforcing and palatability effects to sucrose, but not to corn oil, suggesting postingestive energy signals play a role in these behaviors.

Other investigators have examined aspects of fat metabolism as potential satiety signals by administering fat or fat metabolites centrally or peripherally. Central administration of oleic acid inhibits food intake, though shorter-chain fatty acids do not have this effect [106]. Some evidence suggests that malonyl CoA and long-chain fatty acid CoA are fuel sensors in the hypothalamus to help regulate feeding behavior [107]. However, in both animal and human studies, medium-chain triglycerides typically decrease food intake more than long-chain triglycerides (for review, see Jambor de Sousa et al. [108]). Arase et al. [109] reported that intracerebroventricular infusions of β-hydroxybutyrate (β-OHB) reduce food intake in Sprague–Dawley or Osborne–Mendel rats consuming either a high- or a low-fat diet. However, such infusions do not reduce food intake in S5B/PI rats, rats that are resistant to weight gain when consuming high-fat diets. Peripheral injections of β-OHB decrease food intake in S5B/PI rats but not Osborne–Mendel animals, and glycerol has no effect in either strain [110]. Peripheral injections of glycerol in Wistar rats [111] and β-OHB in Sprague–Dawley rats [112] also decrease food intake. In addition, central administration of inhibitors of fat synthesis results in decreased food intake in rodents [113,114].

Fat, when infused into the intestine, can suppress hunger, induce satiety, or delay gastric emptying [22,115–117]. However, as stated previously, dietary fat appears to be overconsumed and often can lead to obesity. The disparate effects of intraintestinal infusions of fat versus dietary fat have been termed the fat paradox [118]. It has been suggested that high-fat foods have a very high palatability and orosensory stimulation [119–121], leading to overconsumption before the nutrients can even enter the intestine to generate satiety signals. Pittman et al. [54] have shown that free fatty acids depolarize taste receptor cells and may increase concomitant tastants. In support of this hypothesis, satiety signals potentially arising from postabsorptive metabolic events appear to be blunted from fat in comparison with other nutrients. For example, it has been reported that carbohydrate and protein consumption is followed by an increase in their oxidation [91,122,123], whereas oxidation of fat is not generally stimulated until 3–7 days following consumption of a high-fat diet [124,125]. Furthermore, it has been reported that fat oxidation is especially limited in obese as compared with lean individuals [40,126].

4. Effects of Hormones and Pharmacological Agents

Use of hormones and pharmacological agents has further illuminated factors that may play a role in the effect of dietary fat on food intake. Pancreatic procolipase is a cofactor for lipase, an enzyme necessary for proper fat digestion. A pentapeptide produced by the cleavage of procolipase, Val–Pro–Asp–Pro–Arg, or enterostatin has been shown to reduce food intake in rats [127], especially when consuming a high-fat diet [128]. Peripheral injection or intracarotid injection [129] or injection into the lateral ventricle [130] suppresses fat intake in fat-adapted rats, suggesting both a gastrointestinal site and a central site of action. In addition, high-fat feeding and cholecystokinin-8 (CCK-8) increase intestinal enterostatin levels [131]. Lin et al. [132] report that β-casomorphin 1–7 stimulates intake of a high-fat diet in rats, and this effect is inhibited by enterostatin or naloxone.

Much research has been focused on CCK and its effects on satiety (for review, see Smith and Gibbs [133] and Geary [134]). It has been well documented that consumption of fat stimulates the release of CCK from endocrine cells located in the proximal small intestine, activating receptors in the stomach. This signal is transmitted along the vagus to the nucleus of the solitary tract, where it is forwarded to the hypothalamus. Vagotomy can block these effects of systematically administered CCK [135], as can CCK-A receptor antagonists [136]. CCK is also produced in the central nervous system (reviewed in Beinfeld [137]) and is released from the hypothalamus during feeding [138,139]. CCK administered into the cerebral ventricles inhibits food intake in primates [140].

Morphine has also been shown to have specific effects on intake of dietary fat. Rats given morphine injections subsequently increase fat intake while suppressing carbohydrate intake when given separate sources of macronutrients [141–143] or mixed diets [144]. Continuous infusion of morphine also stimulates fat intake [145]. An opioid agonist, butorphanol, also increases consumption of a high-fat diet [65]. Administration of opioid antagonists suppresses fat intake with little effect on protein or carbohydrate intake [146,147]. In humans, opioids also appear to play a role in regulating fat intake. Opiate antagonists cause a decrease in intake of fat calories with less effect on carbohydrate consumption [148–150]. A diet high in fat and sucrose increases gene expression of the opioid, dynorphin in the arcuate nucleus [151]. It appears that intake of highly palatable foods affects opioid activity.

Corticosterone has also been implicated in the regulation of fat intake. Castonguay et al. [152,153] reported that adrenalectomy reduces total caloric intake in rats, particularly fat intake, and that corticosterone can restore the fat consumption. Dallman et al. [154] have shown that insulin must be present for corticosterone to stimulate fat intake. Devenport et al. [155] have reported that the type 1 adrenocorticoid receptor mediates corticosterone's effect on fat appetite. Kumar et al. [156] have suggested that corticosterone acts to enhance carbohydrate rather than fat intake. It has been suggested [157] that the differing levels of micronutrients added to the diets in these studies may account for these disparate findings.

Investigators have also reported that the peptide galanin influences appetite for fat. Leibowitz [158] first demonstrated that galanin stimulates fat intake, especially at the end of the nocturnal cycle. Galanin is thought to work in concert with norepinephrine, which is colocalized with galanin in paraventricular neurons [159]. Smith et al. [160] report that centrally injected galanin induces fat intake only in fat-preferring rats, that is, baseline feeding preferences are important in determining the feeding response to galanin. A high-fat meal or increasing circulating lipids increases galanin in the parventricular nucleus [161].

Leptin, the gene product of *ob* gene, is shown to regulate body fat in mice and is produced in human adipose tissue as well (for review, see Harris [162]). Reports as to its relationship with fat intake in humans indicate that it is negatively associated with fat intake. Havel et al. [163] report that in women, high-fat/low-carbohydrate meals result in a lowering of 24 h circulating leptin concentration. Other research shows that there is a negative correlation between leptin levels and dietary fat (7 day records), when controlling for body weight [164]. Finally, Niskanen et al. [165] indicate that serum leptin concentrations in obese humans are inversely related to dietary fat intake. In mice, however, high-fat feeding or a high-fat diet increases serum leptin levels [166,167]. It has long been shown that high-fat feeding results in secretion of a protein, apo A-IV, by enterocytes, which in turn, can regulate secretion of triglycerides and inhibit food intake (for review, see Tso and Lui [168]). Doi et al. [169] have reported that increased levels of apo A-IV are attenuated by intravenous leptin infusions. Thus, it has been suggested that leptin may regulate fat-induced apo A-IV levels and food intake in animals.

Neuropeptide Y (NPY) increases intake of a preferred diet when given in the paraventricular nucleus or cerebral ventricles. When given in the amygdala to satiated rats, NPY causes a decrease in fat intake [170].

A variety of hypothalamic peptides and their dietary effects have been reviewed in the encyclopedic work of Leibowitz and Wortley [171].

IV. DIETARY FAT AND METABOLISM

Obesity is the final result of increased deposition of fat through increased de novo lipogenesis and increased fatty acid esterification relative to lipolysis and oxidation. In the following sections, the critical literature that characterize the role of dietary lipid in altering these metabolic events will be reviewed.

A. INFLUENCE OF DIETARY FAT ON LIPOGENESIS

Dietary lipid level influences the rate of lipogenesis. Early studies showed that de novo synthesis of fatty acids is decreased by high dietary lipid level [172,173]. Two key enzymes in the lipogenic pathway, fatty acid synthetase and acetyl CoA carboxylase, are reduced in animals receiving a high-fat diet. In addition, the pentose phosphate pathway and malic enzyme, both of which provide reducing equivalents for de novo lipogenesis, are also influenced by dietary lipid level. Malic enzyme, the pentose phosphate pathway, and the rate-limiting enzyme in this pathway, glucose-6-phosphate dehydrogenase, are decreased in rats fed a diet containing high levels of dietary fat and increased in diets high in carbohydrate [174–177].

In experimental animals, the lipid content and composition of the diet can cause a shift in the source of stored lipid. On a diet rich in lipids, rat adipose tissue fatty acids come mainly from dietary fat, whereas on a high-carbohydrate diet the fatty acids come from hepatic lipogenesis [178]. In genetically obese rats, both hepatic and adipose tissue lipogenic rates are decreased by a high-fat diet. Nonetheless, the animals still deposit more fat because of the increased uptake of fatty acids from the diet. High-fat diet-induced inhibition of lipogenesis is also influenced by dietary fatty acid type. For example, unsaturated fatty acids are better at inhibiting de novo lipogenesis than SFAs [179–181] with inhibitory effects further influenced by fatty acid chain length, degree of unsaturation, and double-bond location [182]. In addition, greater inhibition of lipogenesis is observed with *n*-3 as compared with *n*-6 PUFAs [177,183–185]. PUFA-mediated inhibition of lipogenesis may also be influenced by metabolic status as obese mice appear to be somewhat resistant to PUFA-induced feedback control of gene expression [186].

Mechanistic studies in experimental animals indicate that ingestion of *n*-6 or *n*-3 PUFA causes a rapid inhibition of the expression/activation of many enzymes involved in lipogenesis (see reviews by Jump and Clarke [176] and Clarke [177]) and a coordinate induction of genes encoding proteins involved in lipid oxidation and thermogenesis (discussed in Section IV.C). PUFA regulation of lipogenic genes is mediated at both the transcriptional level, as for pyruvate kinase, pyruvate dehydrogenase, acetyl CoA carboxylase, and fatty acid synthase, as well as the posttranscriptional level, as for glucose-6-phosphate dehydrogenase (see reviews by Jump and Clarke [176] and Clarke [177]). There is emerging evidence that the PUFA-induced suppression of lipogenic enzymes is mediated by changes in the expression and cellular localization of the transcription factor, sterol regulatory element-binding protein 1 (SREBP-1) [185,187,188]. Overexpression of this transcription factor in the liver leads to a marked elevation in the mRNAs encoding several lipogenic enzymes and to very high rates of de novo lipogenesis [189,190]. Consumption of diets containing n-6 (safflower oil) or n-3 (fish oil) fatty acids leads to a decrease in membrane content of SREBP-1 precursor and nuclear content of SREBP and a concomitant reduction in lipogenic gene expression in the liver [185], with all effects being greater for the n-3- than the n-6-containing diets [185]. Whether analogous regulatory processes govern n-3 fatty acid modulation of SREBPs and lipogenic genes in adipose tissue is unknown.

The effect of dietary fat on lipogenic processes in humans is not well studied, as until recently, de novo lipogenesis was believed to be an insignificant metabolic pathway particularly when consuming a Western, high-fat diet (for review, see Murphy [191]). Some studies in human subjects suggest a minor role of hepatic lipogenesis in energy balance but do not address the issue of extrahepatic lipogenesis (for review, see Hellerstein et al. [192]). Chascione et al. [193], comparing subjects fed a high-carbohydrate diet with those fed a high-fat diet after a period of energy restriction, suggested that adipose tissue may account for up to 40% of whole-body lipogenesis. Nonetheless, even with a high-carbohydrate intake, whole-body de novo lipogenesis was estimated to account for no more that 1 g/day of fatty acid synthesis (reviewed by Murphy [191]). This small contribution may have important metabolic consequences, however. Schwarz et al. [194] reported higher fasting triacylglycerol concentrations as well as ~4–5 fold higher rates of de novo lipogenesis in hyperinsulinemic obese subjects consuming a high-fat Western diet as compared with normoinsulinemic lean or obese subjects consuming the same diet. Change to a low-fat, high-carbohydrate diet resulted in an increase in triacylglycerol concentrations that was correlated with increased fractional rates of de novo lipogenesis in both normoinsulinemic lean and hyperinsulinemic obese subjects. These results confirm that, in humans, de novo lipogenesis is dependent on both dietary conditions and the health status of the subjects. They [194] further suggest that this increase in de novo lipogenesis may be a contributing factor to hypertriglyceridemia, which is in turn an exacerbating factor for insulin resistance.

B. Influence of Dietary Fat on Lipid Uptake

Lipoprotein lipase (LPL) has been called the gate keeper enzyme because it controls the rate of uptake of lipid by adipose cells [195]. This enzyme is elevated in association with genetic and diet-induced obesity in animals and humans. The ability of n-3 PUFA to lower serum triglycerides is thought by some to be because of an action of LPL. Increases in LPL gene expression [196] or activity are observed with n-3 fatty acid supplementation in both healthy and hypertriglyceridemic patients [196–198] and experimental animals [199,200]. In contrast, others observed no effect of n-3 fatty acids on LPL activity in humans [201,202] or rats [203]. Raclot et al. [204] reported a reduction in LPL expression in retroperitoneal fat depots, but not in subcutaneous fat depots, in rats fed the n-3 fatty acid docosahexaenoic acid (DHA) alone or in combination with the n-3 fatty acid eicosapentaenoic acid (EPA), suggesting site-specific effects. In addition, LPL activity may not necessarily correspond to mRNA levels, as posttranscriptional events, such as glycosylation and binding of LPL to cell surface heparan sulfate proteoglycans, modulate expression and activity of the enzyme [205]. Furthermore, although alterations in LPL activity or expression may function in the lowering of serum lipid levels associated with fish oil (n-3 fatty acid) consumption, it is at odds with the finding of reduced adipose tissue mass in animals fed the same diet. Conversely, the reduced obesity associated with diets high in *n*-3 PUFA may be due to the influence of these fatty acids on reducing hepatic fatty acid synthetase activity [199,200,206] and stimulating fatty acid oxidation (see Section IV.C).

Consumption of fat in excess of energy requirements leads to the uptake and storage of fat not only in adipose tissue but also in nonfat tissue such as skeletal muscle. Many investigators have reported increases in intramyocellular lipid (IMCL) content in skeletal muscle with both short- and long-term high-fat diet feeding regimes (for review, see Schrauwen-Hinderling et al. [207]). In some subjects, increased IMCL is associated with an increase in fat oxidation [207]. In contrast, many obese subjects are characterized by increased adipose tissue and plasma lipid concentrations, high IMCL, and low muscle oxidative capacity [208] (reviewed by Corcoran et al [209]). The obesity-associated increase in ICML is believed to be due to an imbalance between fat delivery to and oxidative capacity of skeletal muscle [207,209]. Excess accumulation of fat in skeletal muscle can be pathophysiological as high ICML has been indicated as an early marker in development of insulin resistance and type II diabetes [207]. Indeed, overexpression of muscle LPL has been associated, in some studies, with increased intramyocellular triacylglycerol concentrations and insulin resistance (for review, see Corcoran et al. [209]). Muscle fat accumulation and the subsequent development of insulin resistance can be influenced not only by the level but also the type of dietary fat, with some fatty acid types having potential protecting effects (for review, see Corcoran et al. [209], discussed further in Section IV.E).

C. Influence of Dietary Fat on Fatty Acid Oxidation and Energy Expenditure

It has been known since the early days of calorimetry that diets high in fat lower the respiratory quotient, an indicator of increased fatty acid oxidation [210]. Fatty acid oxidative rates are dependent in part on chain length and degree of unsaturation. Stable isotope studies in normal-weight men indicate that oxidation rates of individual fatty acids are highest for lauric acid, followed by PUFA and MUFA, and least for longer-chain SFA for which oxidation decreases with increasing chain length [211]. More recently, higher postprandial fat oxidation rates in response to a high-MUFA test meal as compared with a high-SFA test meal were detected by indirect calorimetry in both healthy nonobese to moderately obese men [212] and postmenopausal women [213]. Over time, this could impact body composition, as a follow-up 4 week study in which dietary SFA content was largely replaced with MUFA demonstrated a small but significant reduction in fat mass and body weight [214].

Dietary lipid stimulation of fatty acid oxidation is thought to act through the sympathetic nervous system [215] and stimulation of carnitine palmitoyltransferase activity [216], both of which are influenced by the source of dietary fat. For example, safflower oil-fed rats have the highest sympathetic activity when compared with coconut oil- or medium-chain triglyceride-fed rats [215]. Similarly, feeding fish oil causes a marked increase in carnitine acyltransferase activity in hepatic mitochondria as compared with corn oil feeding [216]. Peroxisomal oxidation is also increased by diets containing fish oils when compared with vegetable oils [183]. Studies in rodents have demonstrated that the *n*-3 fatty acids EPA and DHA both stimulate peroxisomal β oxidation in the liver, whereas EPA increases mitochondrial β oxidation [217]. Substitution of *n*-3 PUFA for a portion (~35%) of saturated fat in a high-fat diet (28% w/w) stimulates fatty acid oxidation in the liver and to a lesser extent in skeletal muscle [218], presumably due to the upregulation of genes encoding proteins for mitochondrial and peroxisomal enzymes (see reviews by Clarke [177] and Baillie et al. [219]). The recent demonstration of adipose depot-specific effects of dietary EPA and DHA on upregulating the expression of genes associated with mitochondrial biogenesis and β-oxidation [220] has important implications with regard to potential dietary approaches to the treatment and prevention of abdominal obesity and associated comorbidities.

There is mounting evidence that a reduced capacity for fat oxidation may be a contributing factor in the development of diet-induced obesity [221–226]. Ji and Friedman [221] observed that the whole-body fat oxidation rates of lean rats with varying susceptibilities to obesity predicted those that would subsequently develop obesity in response to high-fat feeding. They [222] also observed that treatment of obesity-prone rats with fenofibrate (a peroxisome proliferator-activated receptor alpha [PPARα] agonist that promotes fatty acid oxidation) increased whole-body fatty acid oxidation and reduced food intake, weight gain, and adiposity to levels seen in control obesity-resistant rats. Studies by Jackman et al. [223] demonstrated that while obesity-resistant rats increase fat oxidation in response to high-fat feeding, obesity-prone rats tend to preferentially partition dietary fat for storage in adipose tissue. Furthermore, Iossa et al. [224] observed that adult rats have a compromised ability to resist high-fat diet-induced obesity due, in part, to a reduced capacity to increase fat oxidation, an adaptive mechanism that counteracts obesity development in younger animals. A reduced capacity to oxidize fat after a high-fat meal has also been reported for obese insulin-resistant human subjects [225] and in individuals who are not, and have never been obese, but who have a strong familial predisposition to overweight [226]. Taken together, this suggests that a failure to adapt to a high-fat diet with an increase in fat oxidation may lead to an increased flux of fat to adipose tissue for storage and subsequent increase in fat mass.

In addition to fat oxidation per se, energy expenditure can also be influenced by dietary lipids. In general, high-fat diets have a lower heat increment than diets high in carbohydrate or protein [210]. This may lead to a decrease in dietary energy utilization and an increase in body weight gain when fed a high-fat diet. Although evidence exists for high-fat diets causing obesity, not all high-fat diets affect energy metabolism and body weight in the same manner. For example, diets high in essential fatty acids result in a lowering of body weight and increase in thermogenin content in rat brown adipose tissue [227]. Diets containing safflower oil cause an increase in thermogenesis [228] and uncoupling protein (UCP) content [229] of brown adipose tissue in comparison with diets containing beef tallow. As their name implies, uncoupling proteins dissociate mitochondrial oxidative phosphorylation from energy production, leading to energy loss as heat. n-3 PUFAs have been demonstrated to increase expression of UCPs beyond that observed with other types of dietary fats in several studies [203,219,230,231]. For instance, Takahashi and Ide [203] reported an increase in UCP-1 mRNA in brown adipose tissue of rats fed high-fat diets containing fish oil as compared with safflower oil. Likewise, Baillie et al. [219] reported an increase in skeletal muscle UCP-3 mRNA in fish oil-fed rats, which was inversely correlated with a 25% decrease in body fat mass. It has been postulated that one mechanism by which n-3 fatty acids decrease adipose tissue mass in rats and mice is by increasing thermogenesis. However, dissipation of consumed energy as heat does not explain the lack of variance in body weights in some studies.

D. INFLUENCE OF DIETARY FAT ON ADIPOSE TISSUE LIPOLYSIS

The quantity of fat stored in the adipose tissue is determined by the relative rates of the simultaneously occurring metabolic processes of triglyceride synthesis, lipid uptake, fatty acid oxidation, and lipolysis or triglyceride breakdown. The hydrolysis of triglycerides is catalyzed by hormone-sensitive lipase, an enzyme regulated by a complex cyclic AMP-dependent signal transduction cascade. A variety of hormones and neurotransmitters (e.g., adrenaline, glucagon, noradrenaline) stimulate various components of the signal transduction cascade and thereby increase lipolysis (see review by Vernon [232]). Insulin, in contrast, modulates the activity of cyclic AMP, thereby reducing lipolytic activity [232].

1. Level of Dietary Fat

There is considerable evidence that both hormone-stimulated lipolysis and the antipolytic effects of insulin are influenced by the quantity and type of dietary fat. However, the effect of dietary fat on lipolysis varies somewhat according to the species and specific adipose tissue depot being studied. For example, feeding rats high levels of dietary fat leads to a decrease in both catecholamine- [233–241] and glucagon- [233] stimulated lipolysis. The effect of high-fat feeding on adipose tissue lipolysis in the rat is believed to be due to changes in β-adrenoceptor number or to an uncoupling between the hormone receptor and adenylate cyclase, rather than to differences in hormone binding [234,235,241]. In contrast, several studies have indicated an increase in cyclic AMP-dependent signal transduction and lipolytic response with an increase in dietary fat [238,242–246]. More specifically, adenylate cyclase activity [242] and isoproterenol-stimulated lipolysis [243] are increased and phosphodiesterase activity is decreased in a depot-dependent manner [244] in pigs fed added fat diets. An increase in basal and hormone-stimulated lipolytic activity, positively correlated with fat cell size but not associated with sympathetic nervous system activity, is also observed in high-fat-fed rats [246]. There are few studies regarding the effect of high-fat feeding on adipose tissue lipolysis in humans. However, in one short-term study (7 days), Kather et al. [247] observed no differences in either sensitivity or response to catecholamines in adipose tissue of subjects eating fat-rich diets as compared with carbohydrate-rich diets. More recently, Suljkovicova et al. [248] reported no effect of dietary macronutrient composition on β-adrenergic responsiveness of adipose tissue to catecholamine action at rest, but a higher rate of lipolysis in adipose tissue of high-fat-fed subjects as compared with high-carbohydrate-fed subjects during exercise.

Fat feeding has been shown to influence the antilipolytic effects of insulin in adipose tissue and skeletal muscle from several species; however, the results are somewhat inconsistent. Smith et al. [236] observed a decreased response to the antilipolytic effects of insulin in adipose tissue from high-fat as compared with high-carbohydrate-fed rats. However, subsequent studies by Susini et al. [237] and Tepperman et al. [235] failed to observe an effect of dietary fat on the antilipolytic effects of insulin. In pigs, a decrease in the antilipolytic action of insulin in response to added fat feeding is observed in the subcutaneous fat depot but not for the perirenal site [249]. In contrast, a greater sensitivity to the antilipolytic action of insulin is observed in adipocytes of human subjects consuming fat-rich diets as opposed to energy-restricted diets [247]. Recent studies report contradictory effects of high-fat feeding on basal lipolysis in rat skeletal muscle, that is, an increase in one instance [250] and a decrease in the other [251]. Nonetheless, both observed a resistance to the antilipolytic action of insulin in skeletal muscle of the high-fat-fed animals [250,251].

2. Type of Dietary Fat

Lipolytic response may be influenced by the type as well as the level of fat included in the diet. The composition of dietary fat selectively influences fatty acid deposition in adipose tissue [252]. In turn, the composition of the fat tissue influences lipid mobilization and release of fatty acids into the circulation. Lipid mobilization from adipose tissue is not a random event, but instead is influenced by chain length, degree of saturation, and positional isomerization of the fatty acids [253–257]. The relative mobilization of fatty acids from adipose tissue is correlated positively with unsaturation and negatively with chain length [255]. The most easily mobilized fatty acids are those with 16–20 carbon atoms and 4 or 5 double bonds, whereas very long chain MUFAs and PUFAs are less readily mobilized [256]. Higher rates of lipolysis are also observed with *trans* as opposed to *cis* isomers of octadecenoic acid [254]. It has been suggested that the decreased fat accumulation in animals fed *trans* fatty acids may be associated with direct effects of the *trans* isomer on fat cell metabolism [254]. Likewise, a decreased visceral fat accumulation in fish oil-supplemented sucrose-fed rats is believed to be due, at least in part, to an effect of *n*-3 PUFA on increasing lipolytic responsiveness [200].

Adipose tissue from animals fed diets high in PUFA generally exhibit a greater lipolytic response to catecholamines and synthetic β-adrenoceptor agonists as compared with tissue from animals fed diets high in SFA [238,258–263]. The decreased responsiveness of fat cells from rats fed saturated fat diets is associated with reductions in adenylate cyclase, cyclic AMP phosphodiesterase, and hormone-sensitive lipase activity [258]. β-Adrenergic receptor binding is also lower in fat cell membranes from rats fed high-SFA diets than high-PUFA diets [259,264], due to decreased binding affinity rather than changes in receptor number [259,264]. The reduced binding affinity is in turn correlated with a reduction in membrane fluidity in cells from the rats fed high-SFA diets [259].

In contrast to the above studies, several investigators have failed to observe an effect of dietary fat type on lipolytic response [239,243,265,266]. Lipinski and Mathias [239] observed that norepinephrine-stimulated lipolysis in rat adipocytes is depressed by an increase in fat calories but is unaffected by the degree of saturation of the fat. Likewise, Mersmann et al. [243,265] reported an increase in the number of β-adrenergic receptors, with no change in receptor affinity or receptor-mediated function (i.e., lipolysis) in adipose tissue from pigs fed high levels of SFA. Portillo et al. [267] observed that under energy-controlled feeding conditions, various dietary fat regimes caused major changes in adipose tissue phospholipid composition, but no important changes in lipolysis, and thereby suggested that some changes in adipose tissue fatty acid composition may have little effect on overall physiological function.

3. Relationship to Obesity

Although the evidence is not totally consistent, it appears the rate of lipolysis may be influenced to some extent by both the level and the type of dietary fat. A reduction in lipolytic rate as is commonly observed in response to high-fat (particularly SFA) feeding could lead to an increased retention of stored triglycerides and thereby contribute to the development of obesity. However, this response may vary considerably according to species, age, adipose tissue depot site, and adipose tissue fatty acid composition. The relative contribution of lipolytic alterations to diet-induced obesity and the specific regulatory components of the lipolytic signal transduction cascade influenced by alterations in the level or type of dietary fat remain to be elucidated.

E. INFLUENCE OF DIETARY FAT ON INSULIN ACTION

1. Level of Dietary Fat

An association between high-fat diets and impaired insulin action has been observed in numerous in vivo and in vitro studies. Early studies in human subjects indicate that diets high in fat lead to a reduction in glucose tolerance [268,269]. More recent epidemiologic data suggest that individuals with higher fat intakes are more likely to develop disturbances of glucose metabolism, type 2 diabetes, and impaired glucose tolerance than individuals consuming lower amounts of fat, although obesity and physical inactivity may be confounding factors [270]. Studies in experimental animals indicate that high-fat feeding induces both a decline in insulin sensitivity [271–273] and the development of insulin resistance in a variety of tissues [274–280].

Cellular mechanisms responsible for the decline in insulin responsiveness in association with high-fat feeding have not been fully defined. Euglycemic, hyperinsulinemic clamp studies in human subjects and experimental animals indicate that high-fat feeding significantly impairs insulin action by a variety of mechanisms including skeletal muscle accumulation of lipids (reviewed by Corcoran et al. [209]), a reduction in skeletal muscle insulin signaling [281,282] and glucose metabolism [281], and a decreased ability of insulin to suppress hepatic glucose production [281,283]. Such studies also indicate that diet-associated development of peripheral insulin resistance may be modulated by age [284] and genetics (see reviews by Lopez-Miranda et al. [285] and Roche et al. [286]).

Reductions in insulin binding in tissues from rats fed high-fat diets as compared with high-carbohydrate diets have been observed by several groups of investigators [274,287–290]. However, other investigators failed to observe significant alterations in insulin binding in response to high-fat feeding [275,291,292]. Several postreceptor defects in insulin action [274,275,277,291–294] are observed in tissues from animals fed high-fat diets. Specifically, reductions in insulin receptor kinase activity [292], in the intracellular glucose transport system [277,291,293], and in the intracellular capacity to utilize glucose for lipogenesis [275] have all been reported in association with high-fat feeding.

Tissue expression and serum levels of resistin and tumor necrosis factor-alpha (TNFα), cytokines which oppose insulin action (see reviews by Steppan and Lazar [295] and Borst [296]), are frequently increased with high-fat feeding [297–299]; whereas expression or circulating concentrations of adiponectin, an insulin-sensitizing hormone, are decreased [299–301]. Recently, Pitombo et al. [299] observed that removal of visceral fat reversed the impairment in glucose homeostasis and insulin action in high-fat diet-induced obese rats and returned circulating levels of TNFα, adiponectin, and other cytokines to near-control levels. Taken together, this suggests that factors secreted from adipose tissue, particularly visceral fat, may contribute to peripheral insulin resistance observed with high-fat diet-induced and other forms of obesity.

2. Type of Dietary Fat

The effect of dietary fat on insulin action is greatly influenced by the type of fatty acid consumed (for review, see [209,302–305]). High-SFA intakes are consistently associated with insulin resistance, whereas MUFAs and PUFAs are less deleterious in this regard [209,270,302,304–307]. van Amelsvoort et al. [308] observed that insulin response is greater in epididymal fat cells from rats fed diets high in PUFA as compared with SFA. Likewise, diets with increasing ratios of PUFAs to SFAs induce alterations in the composition of adipocyte plasma membranes that are associated in a dose-dependent manner with increases in insulin binding, insulin receptor signaling, and glucose transporter activity [309–312]. Evidence from both animal and human studies suggest a link between a high-SFA intake and alterations in skeletal muscle plasma membrane and IMCL composition that interfere with cellular signaling and insulin-stimulated glucose uptake (see reviews by Corcoran et al. [209] and Haag and Dippenaar [303]). Accordingly, studies in both animal models [280,308,313,314] and humans [304,306,307,315] indicate that substitution of more saturated with less saturated dietary fat sources can improve or ameliorate high-fat diet-induced impairment in insulin function. In rats, substitution of safflower oil (high PUFA) for beef tallow (high SFA) in a moderate-fat diet leads to an increase in the glucose uptake response to insulin [280] and to alterations in gene expression of several insulin signal transduction pathway intermediates [316], but not to changes in insulin receptor mRNA or relative expression of insulin receptor mRNA isoforms [316]. In healthy men and women, isocaloric substitution of MUFAs for SFAs improves insulin sensitivity [306,307,315], with the effect dependent on both total dietary fat intake [306] and metabolic status of subjects [315]. Thus, Vessby et al. [306] recently observed that increasing dietary MUFA content improved insulin sensitivity in individuals with a moderate fat intake while having no beneficial effect in individuals with a high-fat intake (>37% energy). Lovejoy et al. [315] observed no impact of dietary fat type on insulin sensitivity in lean subjects, but a susceptibility to develop insulin resistance on the high-SFA diets in overweight individuals.

There is mounting evidence that n-3 PUFAs may afford a protective effect against diet-induced insulin resistance [209,270,305,312,317–320]. In animal studies, replacement of a small portion (6%–11%) of the fatty acids in a high-safflower oil diet with long-chain n-3 fatty acids from fish oil prevents the accumulation of intramuscular triglyceride and development of insulin resistance typically observed in association with consumption of a very high fat (59% calories) diet [314,315] (reviewed by Corcoran et al. [209]). Many studies have demonstrated that substitution of a portion of dietary fat with n-3 PUFA (fish oil) prevents or corrects the inhibitory effects of high-sucrose/high-fat feeding on insulin action [312,317,319,321], as evidenced by higher rates of insulin-stimulated glucose transport and other metabolic indices that were positively correlated with the fatty acid unsaturation index of adipocyte membrane phospholipids [321]. In contrast, Fickova et al. [322] observed a reduction in insulin-stimulated glucose transport and lower number of insulin binding sites on adipocytes from rats fed high levels of n-3 as compared with n-6 PUFAs. Likewise, Ezaki et al. [323] reported that although a substitution of a portion of the safflower oil in a high-fat diet with fish oil led to a transient (at week 1) increase in both insulin-stimulated glucose uptake and glucose transporter distribution in rat adipocytes, further fish oil feeding (4 weeks) resulted in a reappearance of insulin resistance and adipose cell enlargement comparable with that observed with the high-safflower diet.

Human studies have demonstrated promising effects of *n*-3 PUFA on improving lipid profiles, but less consistent effects on insulin resistance (reviewed by Riccardi et al. [304], Lombardo and Chicco [319], and Nettleton and Katz [320]). Epidemiologic evidence suggests that moderate fish oil consumption may have a protective effect against the development of impaired glucose tolerance and diabetes mellitus (see review by Lombardo and Chicco [319]). However, *n*-3 PUFA supplementation in patients with type 2 diabetes and obesity has produced inconsistent results [209,304,318,319,324,325]. Some studies [318] (reviewed in [319]) report an effect of *n*-3 PUFAs on improving insulin sensitivity. Thus in a recent hypocaloric low-fat dietary intervention that emphasized increased lean fish consumption in obese subjects, Haugaard et al. [318] observed that skeletal muscle membrane phospholipid *n*-3 PUFA content was an independent predictor of improved insulin sensitivity. Other studies in type 2 diabetics (reviewed by Corcoran et al. [209], Riccardi et al. [304], and Lombardo and Chicco [319]) report no effect of *n*-3 PUFA on improving insulin sensitivity. Still others [324,325] (reviewed by Lombardo and Chicco [319]) report a deterioration of glycemic control as evidenced by moderate increases in blood glucose concentrations and decreased insulin sensitivity in type 2 diabetics receiving large daily doses of fish oil. Although there is no definitive proof that *n*-3 fatty acids can reverse insulin resistance, it has been suggested [320] that regular consumption of modest amounts of *n*-3 PUFAs by persons with type 2 diabetes may result in beneficial lipid-lowering effects without adversely affecting glycemic control.

Recent epidemiological evidence reports an association between a high intake of *trans* fatty acids and an increased risk of developing type 2 diabetes [326]. Likewise, a 12 week diet study in rats [327] demonstrated a greater impairment in insulin sensitivity in adipocytes from *trans* fat-fed animals as compared with control-fed or SFA-fed animals. Several controlled intervention studies in humans (for review, see Riserus [328]) suggest that *trans* fatty acids may impair insulin sensitivity in insulin-resistant overweight subjects, while having had no effect on insulin sensitivity in lean healthy subjects. Clearly, additional studies are needed to confirm the effect of *trans* fatty acids on insulin sensitivity and to determine mechanisms of action.

3. Reversibility of Diet-Induced Alterations in Insulin Action

Prolonged consumption of a high-fat diet impairs insulin action and leads to the development of obesity. Conversely, a reduction in dietary fat content may improve insulin sensitivity and reduce obesity development. Harris and Kor [272] observed an impaired insulin response to a glucose challenge in rats fed high-fat (40% energy) diets for 8 weeks. This effect is reversed within 3 days following a modest reduction in dietary fat content (30% energy). As alterations in body weight or fat content are not observed until 14 days following the switch to the lower-fat diet, the improvement in insulin sensitivity is not believed to be secondary to a reduction in obesity [272]. Though these results are encouraging, an improvement in insulin response on reduction in dietary fat content has not been consistently observed. In rats previously fed high-fat (60% energy) diets for 6 months, Yakubu et al. [273] failed to detect either an improvement in insulin response or a reduction in body weight following subsequent consumption of a low-fat (20% energy) diet for 3 or 6 months. Thus, alterations in insulin response induced by a high-fat diet and the reversibility of these effects appear to be influenced by both level and type of dietary fat as well as the duration of the high-fat feeding.

V. DIETARY FAT AND ADIPOSE TISSUE CELLULARITY

The expansion of adipose tissue during growth or the development of obesity is achieved through an increase in adipose tissue size (cellular hypertrophy), an increase in adipose tissue number (cellular hyperplasia), or through a combination of both processes. Numerous studies provide evidence that variations in the level and type of fat included in the diet can lead to alterations in adipose cell size and number. As quantification of total adipose tissue cell number in human subjects is not readily obtainable, these studies have been primarily conducted in experimental animals.

A. Level of Dietary Fat

Many animal studies confirm that high-fat feeding leads to an expansion of adipose tissue mass through an increase in fat cell size and number and to the subsequent development of obesity. Studies of adipose tissue development in rodents indicate that increases in fat pad weight are typically associated with increases in both fat cell size and number until approximately 10–18 weeks of age [329–331]. Body and fat pad weights and fat cell size and number then plateau and remain fairly constant in the adult animal. High-fat feeding influences both the dynamic stage of adipose tissue development early in life [329,332–334] and also the more static phase associated with adulthood [329,335,336]. In mice fed high-fat diets from birth, increased fat pad weights are associated with a greater fat cell size through 18 weeks of age; followed by an increase in fat cell number through 52 weeks of age [329]. At that time fat pad weight is sixfold greater in the high-fat than the control-fed mice, whereas fat cell size and number are increased 2.3- and 2.5-fold, respectively. An even greater effect on adipose tissue cellularity is observed in young Osborne–Mendel rats, a strain susceptible to high-fat feeding [61], with a 4- to 16-fold increase in adipocyte number (dependent on the specific fat pad studied) observed between 24 and 105 days of age when the animals are fed a high-fat diet [329].

Dramatic alterations in adipose tissue cellularity are also observed in adult rats subjected to high-fat feeding [329,335,336]. In adult rats, high-fat diet-induced obesity is associated with increases in both fat cell size and fat cell number, with increases in cell size preceding changes in cell number. In 5 month old rats an increase in fat cell size is detected as early as 1 week after the introduction of a high-fat diet, followed by increases in cell number in the perirenal and epididymal fat pads after 2 and 8 weeks, respectively [335]. Likewise, Faust et al. [336] report increases in cell size in several fat pad depots after 3 weeks of high-fat feeding in adult rats but an increase in cell number only after 9 weeks of dietary treatment. Fat cell hyperplasia in response to high-fat feeding is also observed in adult genetically obese rodents, with the magnitude of response being depot dependent [337].

B. Type of Dietary Fat

It is generally accepted that a high level of fat in the diet may induce adipose cell hypertrophy and hyperplasia. However, the influence of dietary fat type on adipose tissue cellularity and the development of obesity is less definitive, particularly with respect to diet-induced alterations in fat cell hyperplasia. Several studies reported a greater effect of unsaturated as compared with saturated fat diets on increasing fat cell number [338–340], whereas other investigations report a greater degree of fat cell hyperplasia with saturated as compared with unsaturated fat diets [333,341,342]. Other studies [308,343–345] in various rodent models suggest that the alterations in fat cell size and number associated with high-fat feeding are not influenced by type of dietary fat. Thus, although long-term feeding of high-fat diets results in increases in fat pad weight and fat cell size in both guinea pigs [343] and mice [344], quite similar results are observed with inclusion of either beef tallow or corn oil in the high-fat diet formulation. It has been suggested [346] that although both the amount and type of dietary fat can influence body weight and fat deposition, the effects of fat type (i.e., saturated versus unsaturated) are generally less than those of dietary fat level.

In contrast to the studies reported above, more consistent effects on adipose cellularity are observed with dietary n-3 PUFA, which reportedly limit hypertrophy and hyperplasia in a depot-dependent manner [204,347]. Whether differences in fat cell size and number are observed appears to be dependent on the duration of the study. Cell size is decreased during short-term n-3 PUFA-feeding studies in rats [204,322,348,349], whereas fat cell number is also reduced during longer-term studies [347]. Recently, reductions in fat cell size in association with adipose tissue and dietary *n*-3 PUFA content [350] and with short-term *n*-3 PUFA (fish oil) supplementation [351] have been reported for human studies.

The mechanism by which *n*-3 PUFAs affect adipose tissue cellularity has not been totally defined. Alterations in both β-oxidation and lipolytic response (detailed in Sections IV.C and IV.D) are associated with the reduction in adipose tissue mass observed with *n*-3 PUFA diets. In addition, *n*-3 PUFA may influence adipose tissue cellularity indirectly by preventing adipose tissue matrix remodeling and adipocyte enlargement [352] or directly by inhibiting adipogenesis [334] (reviewed by Ailhaud et al. [353]). In vitro studies demonstrate potent effects of arachidonic acid (*n*-6 PUFA) on stimulating adipopogenesis, which are mediated through conversion to prostacyclin, increased cAMP production, and activation of the protein kinase A pathway [334] (reviewed by Ailhaud et al. [353]). Conversely, *n*-3 fatty acids, EPA, and to a lesser extent DHA inhibit the stimulatory effect of arachidonic acid on cAMP production and thereby attenuate arachidonic acid-stimulated adipogenesis [334]. Accordingly, mice born to dams consuming a high-fat linoleic (arachidonic acid precursor) acid-rich diet and reared to the same diet were 50% heavier at weaning and had an increased adipose tissue mass at 8 weeks, characterized by an increased adipose cell size as compared with pups from mothers fed a high-fat diet containing a mixture of linoleic and α-linolenic (*n*-3 fatty acid, DHA, and EPA precursor) acids [334]. Likewise, Ruzickova et al. [354] observed a reduction in adipose tissue cellularity and obesity development with the addition of *n*-3 PUFA of marine origin (EPA and DHA) to a high-fat linoleic acid-rich diet. Interestingly, studies in obese diabetic db/db mice [353] demonstrate an upregulation of adipose tissue expression of genes involved in matrix remodeling coupled with a parallel increase in adipocyte size, in animals fed high-fat diets rich in SFA and MUFA or *n*-6 PUFA. These effects were completely prevented by inclusion of *n*-3 PUFA, leading to the suggestion that prevention of high-fat diet-induced matrix remodeling and adipocyte enlargement could have important implications for preventing the development of type 2 diabetes in obese patients [352].

C. Interrelationship between Changes in Fat Cell Size and Number

When adipose tissue development in animals with genetic- or diet-induced obesity is monitored over time, it is generally observed that increases in fat cell size precede increases in fat cell number [329,331,335,336,347]. Lemmonier [329] suggested that there might be a maximum cell size that is constant regardless of animal sex or adipose depot site. Thus, the greater degree of hyperplasia observed in the perirenal compared with other fat depots in response to high-fat feeding may be due to the fact that cells of this depot are already near-maximal in size (for review, see [337]). Likewise, Faust et al. [336] suggested that if this critical fat cell size is reached during the development of high-fat diet obesity, a stimulus for new cell production or differentiation may be produced. Several studies have demonstrated an increased incorporation of [^{3}H]thymidine into the DNA of adipose tissue of high-fat-fed rats [355,356], providing evidence that the diet-induced increase in fat cell number is indeed a consequence of cellular proliferation. However, a direct link between an increase in cellular proliferation and the release of factors from adipose tissue triggering this effect has been difficult to demonstrate. For example, Bjorntorp et al. [357] reported that the addition of plasma from either chow-fed or high-fat-fed rats to adipocyte precursors in culture has similar effects on inducing preadipocyte proliferation and differentiation. Likewise, Shillabeer et al. [358] and Lau et al. [359] observed that although media conditioned by exposure to mature fat or cells obtained from adipose tissue influenced the proliferation and differentiation of cultured preadipocytes, similar effects on preadipocyte differentiation were observed with fat from rats fed differing levels and types of dietary fat [352]. In contrast, we observed that (1) rats fed high-fat diets have increased numbers of both very small and large cells within their adipose tissue depots [360], (2) media conditioned by exposure to the fat from these animals significantly increased the proliferation of preadipocytes in primary culture [360], (3) total tissue content of insulin-like growth factor type

I (IGF-I) is greater in the fat pads of high-fat diet rats as compared with their low-fat diet controls [361], and (4) adipogenic activity in the conditioned media from the high-fat-fed rats could be attenuated through neutralization with an IGF-I antibody [360]. Taken together, these studies support the concept that the enlargement of adipocytes during the development of diet-induced obesity may lead to the secretion of locally produced factors, such as IGF-I, capable of stimulating adipose cell proliferation and thus involved in the regulation of adipose tissue expansion. Interestingly, the recent demonstration by Crossno et al. [362] of increased trafficking of bone marrow-derived circulating progenitor cells to adipose tissue in response to high-fat feeding would implicate nonresident preadipocyte progenitors as an additional and novel mechanism for expanding adipose tissue mass during the development of diet-induced obesity.

D. CELLULARITY CHANGES: REVERSIBILITY OF OBESITY AND INSULIN RESISTANCE

The reversibility of obesity induced by high-fat feeding [336,363,364] may be dependent on the duration and severity of the dietary treatment. For example, diet-induced obesity is reversed when the animals are returned to a low-fat diet following 16 [363] but not 30 weeks [364] of high-fat feeding. Diet-induced alterations in adipose tissue cellularity may be a major factor dictating the reversibility of the obese state [336,363]. Thus, obesity associated with changes in fat cell size only is readily reversed on return to a low-fat regime [336,363]. Interestingly, diet-induced changes in fat cell number may have short-term beneficial effects, but long-term adverse consequences. For instance, an increased number of small insulin-sensitive fat cells may afford protection against insulin resistance and other obesity-associated comorbidities by expanding lipid storage and oxidation capacity and preventing further expansion of larger insulin-resistant adipocytes and ectopic accumulation of lipid [352,362]. Nonetheless, several animal studies indicate that diet-induced changes in fat cell number appear to be permanent, as switching from a high-fat to a low-fat diet leads to a reduction in body weight and fat cell size but not fat cell number [336,363,364]. Thus, an increased capacity for energy storage, although perhaps beneficial in the short-term, would increase the capability or capacity of adipose tissue mass to expand with subsequent episodes of increased caloric intake. If these observations are indeed applicable to humans, this would suggest that high-fat diet-induced obesity, particularly of extended duration, may be resistant to intervention.

REFERENCES

1. Whitney, E.N. and Rolfes, S.R. Energy balance and weight control. in *Understanding Nutrition*, Whitney, E.N. and Rolfes, S.R., Eds., West Publishing Co., New York, NY, 1993, p. 254.
2. National Research Council (NRC). *Diet and Health: Implications for Reducing Chronic Disease Risk*. National Academy Press, Washington, DC, 1989, p. 563.
3. Hubbard, V.S. Prevention of obesity: populations at risk, etiologic factors and intervention strategies. *Obes. Res.*, 3, 76S, 1995.
4. National Task Force on Obesity. Toward prevention of obesity: research directions. *Obes. Res.*, 2, 571, 1994.
5. Wickelgren, I. Obesity: how big a problem? *Science*, 280, 1364, 1998.
6. Andersson, L.B. Genes and obesity. *Ann. Med.*, 28, 5, 1996.
7. Popkin, B.M. et al. A review of dietary and environmental correlates of obesity with emphasis on developing countries. *Obes. Res.*, 3, 145S, 1995.
8. Hodge, A.M. et al. Prevalence and secular trends in obesity in Pacific and Indian Ocean island populations. *Obes. Res.*, 3, 77S, 1995.
9. Warwick, Z.S. and Schiffman, S.S. Role of dietary fat in calorie intake and weight gain. *Neurosci. Behav. Rev.*, 16, 585, 1992.
10. Bray, G.A., Fisler, J.S. and York, D.A. Neuroendocrine control of the development of obesity: understanding gained from studies of experimental animals models. *Front. Neuroendocrinol.*, 11, 128, 1990.

11. West, D.B. and York, B. Dietary fat, genetic predisposition and obesity: lessons from animal models. *Am. J. Clin. Nutr.*, 67(Suppl.), 505S, 1998.
12. Bray, G.A., Paeratakul, S. and Popkin, B.M. Dietary fat and obesity: a review of animal, clinical and epidemiological studies. *Physiol. Behav.*, 83, 549, 2004.
13. Cha, M.C. et al. High-fat hypocaloric diet modifies carbohydrate utilization of obese rats during weight loss. *Am. J. Physiol. Endocrinol. Metab. Gastrointest. Physiol.*, 280, E797, 2001.
14. Lissner, L. et al. Dietary fat and the regulation of energy intake in human subjects. *Am. J. Clin. Nutr.*, 46, 886, 1987.
15. Kendall, A. et al. Weight loss on a low-fat diet: consequence of the imprecision of the control of food intake in humans. *Am. J. Clin. Nutr.*, 53, 1124, 1991.
16. Prewitt, T.E. et al. Changes in body weight, body composition and energy intake in women fed high- and low-fat diets. *Am. J. Clin. Nutr.*, 54, 304, 1991.
17. James, W.P.T. Healthy nutrition: preventing nutrition-related diseases in Europe. *WHO Regional Publications, European Series* No. 24, 1988.
18. Blundell, J.E. and Macdiarmid, J.I. Passive overconsumption. Fat intake and short-term energy balance. *Ann. N. Y. Acad. Sci.*, 827, 392 1997.
19. Curb, J.D. and Marcus, E.B. Body fat and obesity in Japanese Americans. *Am. J. Clin. Nutr.*, 53(Suppl.), 1552S, 1991.
20. Blundell, J. Hunger, appetite and satiety-constructs in search of identities. in *Nutrition and Lifestyles,* Turner, M., Ed. Applied Sciences, London, 1979, p. 21.
21. Blundell, J.E. et al. Control of human appetite: implications for the intake of dietary fat. in *Annual Review of Nutrition,* McCormick, D.B., Bier, D.M., and Goodridge, A.G., Eds., Annual Reviews, Palo Alto, CA, 1996, p. 285.
22. Sepple, C.P. and Read, N.W. The effect of pre-feeding lipid on energy intake from a meal. *Gut*, 31, 158, 1990.
23. Birch, L.L. et al. Children's lunch intake: effects of mid-morning snacks varying in energy density and fat content. *Appetite*, 20, 83, 1993.
24. de Graaf, C. et al. Short term effects of different amounts of protein, fats and carbohydrates on satiety. *Am. J. Clin. Nutr.*, 55, 33, 1992.
25. Cotton, J.R. et al. Dietary fat and appetite: similarities and differences in the satiating effect of meals supplemented with either fat or carbohydrate. *J. Hum. Nutr. Diet.*, 7, 11, 1994.
26. Lawton, C.L. et al. Dietary fat and appetite control in obese subjects: weak effects on satiation and satiety. *Int. J. Obes. Relat. Metab. Disord.*, 17, 409, 1993.
27. van Amelsvoort, J.M.M. et al. Effects of varying the carbohydrate:fat ratio in a hot lunch on postprandial variables in male volunteers. *Br. J. Nutr.*, 61, 267, 1989.
28. Green, S.M., Burley, V.J. and Blundell, J.E. Effect of fat- and sucrose-containing foods on the size of eating episodes and energy intake in lean males: potential for causing overconsumption. *Eur. J. Clin. Nutr.*, 48, 547, 1994.
29. Blundell, J.E. et al. Dietary fat and the control of energy intake: evaluating the effects of fat on meal size and post-meal satiety. *Am. J. Clin. Nutr.*, 57, 772, 1993.
30. Rolls, B.J. et al. Satiety after preloads with different amounts of fat and carbohydrate: implications for obesity. *Am. J. Clin. Nutr.*, 60, 476, 1994.
31. Foltin, R.W. et al. Caloric compensation for lunches varying in fat and carbohydrate content by humans in a residential laboratory. *Am. J. Clin. Nutr.*, 52, 969, 1990.
32. Rolls, B.J. et al. Time course of effects of preloads high in fat or carbohydrate on food intake and hunger rating in humans. *Am. J. Physiol.*, 260, R756, 1991.
33. Hulshof, T., de Graaf, C. and Weststrate, J.A. The effects of preloads varying in physical state and fat content on satiety and energy intake. *Appetite*, 23, 273, 1993.
34. Foltin, R.W. et al. Caloric, but not macronutrient, compensation by humans for required-eating occasions with meals and snack varying in fat and carbohydrate. *Am. J. Clin. Nutr.*, 55, 331, 1992.
35. Geliebter, A.A. Effects of equicaloric loads of protein, fat and carbohydrate on food intake in the rat and man. *Physiol. Behav.*, 22, 267, 1979.
36. Rolls, B.J., Hetherington, M. and Burley, V.J. The specificity of satiety: the influence of foods of different macronutrient content on the development of satiety. *Physiol. Behav.*, 43, 145, 1988.
37. Porrini, M.A. et al. Weight, protein, fat, and timing of preloads affect food intake. *Physiol. Behav.*, 62, 563, 1997.

38. Duncan, K.H., Bacon, J.A. and Weinser, R.L. The effects of high and low energy density of diets on satiety, energy intake and eating time of obese and nonobese subjects. *Am. J. Clin. Nutr.*, 37, 763, 1983.

39. Caputo, F.A. and Mattes, R.D. Human dietary responses to covert manipulations of energy, fat and carbohydrate in a midday meal. *Am. J. Clin. Nutr.*, 56, 36, 1992.

40. Thomas, C.D. et al. Nutrient balance and energy expenditure during ad libitum feeding on high fat and high carbohydrate diets in humans. *Am. J. Clin. Nutr.*, 55, 934, 1992.

41. Tremblay, A. et al. Nutritional determinants of the increase in energy intake associated with a high-fat diet. *Am. J. Clin. Nutr.*, 53, 1134, 1991.

42. Stubbs, R.J. et al. Covert manipulation of the ratio of dietary fat to carbohydrate and energy density: effect on food intake and energy balance in free-living men eating ad libitum. *Am. J. Clin. Nutr.*, 62, 330, 1995.

43. Blundell, J.E. and MacDiarmid, J.I. Fat as a risk factor for overconsumption: satiation, satiety, and patterns of eating. *J. Am. Diet. Assoc.*, 97, S63, 1997.

44. Schrauwen, P. and Westerterp, K.R. The role of high-fat diets and physical activity in the regulation of body weight. *Br. J. Nutr.*, 84, 417, 2000.

45. Blundell, J.E. and Cooling, J. Routes to obesity: phenotypes, food choices and activity. *Br. J. Nutr.*, 83, S33, 2000.

46. Woods, S.C. et al. A controlled high-fat diet induces an obese syndrome in rats. *J. Nutr.*, 133, 1081, 2003.

47. Tremblay, A. and St-Pierre, S. The hyperphagic effect of a high-fat diet and alcohol intake persists after control for energy density. *Am. J. Clin. Nutr.*, 63, 479, 1996.

48. Warwick, Z.S. and Schiffman, S.S. Sensory evaluations of fat-sucrose and fat-salt mixtures: relationship to age and weight status. *Physiol. Behav.*, 48, 633, 1990.

49. Lissner, L. and Heitmann, B.L. Dietary fat and obesity: evidence from epidemiology. *Eur. J. Clin. Nutr.*, 49, 79, 1995.

50. Mazlan, N. et al. Effects of increasing increments of fat-and sugar-rich snacks in the diet on energy and macronutrient intake in lean and overweight men. *Br. J. Nutr.*, 96, 596, 2006.

51. Furuse, M. et al. Feeding behavior in chickens given diets containing medium chain triglyceride. *Br. Poult. Sci.*, 34, 211, 1993.

52. Furuse, M. et al. Feeding behavior in rats fed diets containing medium chain triglyceride. *Physiol. Behav.*, 52, 815, 1992.

53. de Sousa, U.L.J. et al. Hepatic-portal oleic acid inhibits feeding more potently than hepatic-portal carpylic acid in rats. *Physiol. Behav.*, 89, 329, 2006.

54. Pittman, D.W. et al. Linoleic and oleic acids alter the licking responses to sweet, salt, sour and bitter tastants in rats. *Chem. Senses*, 31, 835, 2006.

55. Rolls, B.J. et al. Food intake in dieters and nondieters after a liquid meal containing medium-chain triglycerides. *Am. J. Clin. Nutr.*, 48, 66, 1988.

56. Stubbs, R.J. and Harbron, C.G. Covert manipulation of the ratio of medium- to long-chain triglycerides in isoenergetically dense diets: effect on food intake in ad libitum feeding men. *Int. J. Obes. Relat. Metab. Disord.*, 20, 435, 1996.

57. Lucas, F., Ackroff, K. and Sclafani, A. Dietary fat-induced hyperphagia in rats as a function of fat type and physical form. *Physiol. Behav.*, 45, 937, 1989.

58. Corbett, S.W., Stern, J.S. and Keesey, R.E. Energy expenditure in rats with diet induced obesity. *Am. J. Clin. Nutr.*, 44, 173, 1986.

59. Louis-Sylvestre, J., Giachetti, I. and Le Magnen, J. Sensory versus dietary factors in cafeteria-induced overweight. *Physiol. Behav.*, 32, 901, 1984.

60. Pitts, G.C. and Bull, L.S. Exercise, dietary obesity, and growth in the rat. *Am. J. Physiol.*, 232, R38, 1977.

61. Schemmel, R., Mickelson, O. and Gill, J.L. Dietary obesity in rats: body weight and body fat accretion in seven strains of rats. *J. Nutr.*, 100, 1041, 1970.

62. Hill, J.O., Fried, S.K. and DiGirolamo, M. Effects of a high-fat diet on energy intake and expenditure in rats. *Life Sci.*, 33, 141, 1983.

63. Kaufman, L.N., Peterson, M.M. and Smith, S.M. Hypertension and sympathetic hyperactivity induced in rats by high-fat or glucose diets. *Am. J. Physiol.*, 260, E95, 1991.

64. Ramirez, I. Feeding a liquid diet increases energy intake, weight gain and body fat in rats. *J. Nutr.*, 117, 2127, 1987.

65. Romsos, D.R. et al. Effects of kappa opiate agonists, cholecystokinin and bombesin on intake of diets varying in carbohydrate-to-fat ratio in rats. *J. Nutr.*, 117, 976, 1987.

66. Lawton, G.L. et al. The degree of saturation of fatty acids influences post-ingestive satiety. *Br. J. Nutr.*, 83, 473, 2000.

67. Ramirez, I. and Friedman, M.I. Dietary hyperphagia in rats: role of fat, carbohydrate, and energy content. *Physiol. Behav.*, 47, 1157, 1990.

68. Emmett, P.M. and Heaton, K.W. Is extrinsic sugar a vehicle for dietary fat? *Lancet*, 345, 1537, 1995.

69. Bolton-Smith, C. and Woodward, M. Dietary composition and fat to sugar ratios in relation to obesity. *Int. J. Obes.*, 18, 820, 1994.

70. Gibney, M.J. et al. Consumption of sugars. *Am. J. Clin. Nutr.*, 62, 178S, 1995.

71. Hill, J.O. and Prentice, A.M. Sugar and body weight regulation. *Am. J. Clin. Nutr.*, 62, S264, 1995.

72. Macdiarmid, J.I., Cade, J.E. and Blundell, J.E. High and low fat consumers, their macronutrient intake and body mass index: further analysis of the National Diet and Nutrition Survey of British Adults. *Eur. J. Clin. Nutr.*, 50, 505, 1996.

73. Westerterp, K.R. et al. Dietary fat and body fat: an intervention study. *Int. J. Obes. Relat. Metab. Disord.*, 20, 1022, 1996.

74. Geary, N. Is the control of fat ingestion sexually differentiated? *Physiol. Behav.*, 83, 659, 2004.

75. Mullen, B.J. and Martin, R.J. Macronutrient selection in rats: effect of fat type and level. *J. Nutr.*, 120, 1418, 1990.

76. Johnson, S.L., McPhee, L. and Birch, L.L. Conditioned preferences: young children prefer flavors associated with high dietary fat. *Physiol. Behav.*, 50, 1245, 1991.

77. Elizalde, G. and Sclafani, A. Fat appetite in rats: flavor preferences conditioned by nutritive and non-nutritive oil emulsions. *Appetite*, 15, 189, 1990.

78. Tordoff, M.G., Tepper, B.J. and Friedman, M.I. Food flavor preferences produced by drinking glucose and oil in normal and diabetic rats: evidence for conditioning based on fuel oxidation. *Physiol. Behav.*, 41, 481, 1987.

79. Warwick, Z.S., Schiffman, S.S. and Anderson, J.J. Relationship of dietary fat content to food preferences in young rats. *Physiol. Behav.*, 48, 581, 1990.

80. Ackroff, K., Vigorito, M. and Sclafani, A. Fat appetite in rats: the response of infant and adult rats to nutritive and non-nutritive oil emulsions. *Appetite*, 15, 171, 1990.

81. Reed, D.R., Tordoff, M.G. and Friedman, M.I. Sham-feeding of corn oil by rats: sensory and post-ingestive factors. *Physiol. Behav.*, 47, 779, 1990.

82. Hamilton, C.L. Rat's preference for high fat diets. *J. Comp. Physiol. Psychol.*, 58, 459, 1964.

83. Deutsch, J.A., Molina, F. and Puerto, A. Conditioned taste aversion caused by palatable nontoxic nutrients. *Behav. Biol.*, 16, 161, 1976.

84. Romieu, I. et al. Energy intake and other determinants of relative weight. *Am. J. Clin. Nutr.*, 47, 406, 1988.

85. Miller, W.C. et al. Diet composition, energy intake, and exercise in relation to body fat in men and women. *Am. J. Clin. Nutr.*, 52, 426, 1990.

86. Lucas, F., Ackroff, K. and Sclafani, A. High-fat diet preference and overeating mediated by postingestive factors in rats. *Am. J. Physiol.*, 275, R1511, 1998.

87. Cunningham, K.M. et al. Gastrointestinal adaptation to diets of differing fat composition in human volunteers. *Gut*, 32, 483, 1991.

88. Geary, N. et al. Adaptation to high fat diet and carbohydrate-induced satiety in the rat. *Am. J. Physiol.*, 237, R139, 1979.

89. Peterson, A.D. and Baumgardt, B.R. Food energy intake of rats fed diets varying in energy concentration and density. *J. Nutr.*, 101, 1057, 1971.

90. Poppitt, S.D. Energy density of diets and obesity. *Int. J. Obes. Relat. Metab. Disord.*, 19(Suppl. 5), S20, 1995.

91. Prentice, A.M. Manipulation of dietary fat and energy density and subsequent effects on substrate flux and food intake. *Am. J. Clin. Nutr.*, 67, 535S, 1998.

92. Schemmel, R., Mickelsen, O. and Gill, J.L. Dietary obesity in rats: weight and body fat accretion in seven strains of rats. *J. Nutr.*, 100, 1041, 1970.

93. Rolls, B.J. and Bell, E.A. Intake of fat and carbohydrate: role of energy density. *Eur. J. Clin. Nutr.*, 53(Suppl. 1), S166, 1999.

94. Rolls, B.J. et al. Energy density but not fat content of foods affected energy intake in lean and obese women. *Am. J. Clin. Nutr.*, 69, 863, 1999.

95. Crane, S.B. and Greenwood, C.E. Dietary fat source influences neuronal mitochondrial monoamine oxidase activity and macronutrient selection in rats. *Pharmacol. Biochem. Behav.*, 27, 1, 1987.

96. Grossman, B.M. et al. Vagotomy and mercaptoacetate influence the effect of dietary fat on macronutrient selection by rats. *J. Nutr.*, 124, 804, 1994.

97. Rudkowska, I. et al. Diacylglyderol: efficacy and mechanism of action of an anti-obesity agent. *Obes. Res.*, 13, 1864, 2005.

98. Scharrer, E. and Langhans, W. Control of food intake by fatty acid oxidation. *Am. J. Physiol.*, 250, R1003, 1986.

99. Friedman, M.I., Tordoff, M.G. and Ramirez, I. Integrated metabolic control of food intake. *Brain Res. Bull.*, 17, 855, 1986.

100. Langhans, W. and Scharrer, E. Evidence for a vagally mediated satiety signal derived from hepatic fatty acid oxidation. *J. Auton. Nerv. Syst.*, 18, 13, 1987.

101. Ritter, S. and Taylor, J.S. Vagal sensory neurons are required for lipoprivic but not glucoprivic feeding in rats. *Am. J. Physiol.*, 258, R1395, 1990.

102. Ritter, S. and Taylor, J.S. Capsaicin abolishes lipoprivic but not glucoprivic feeding in rats. *Am. J. Physiol.*, 256, R1232, 1989.

103. Calingasan, N.Y. and Ritter, S. Lateral parabrachial subnucleus lesions abolish feeding induced by mercaptoacetate but not by 2-deoxy-d-glucose. *Am. J. Physiol.*, 265, R1168, 1993.

104. Wang, S.W. et al. Effects of dietary fat on food intake and brain uptake and oxidation of fatty acids. *Physiol. Behav.*, 56, 517, 1994.

105. Suzuki, A., Yamane, T. and Fishiki, T. Inhibition of fatty acid β-oxidation attenuates the reinforcing effects and palatability to fat. *J. Nutr.*, 22, 401, 2006.

106. Obici, S. et al. Central administration of oleic acid inhibits glucose production and food intake. *Diabetes*, 51, 271, 2002.

107. Obici, S. and Rossetti, L. Minireview: nutrient sensing and the regulation of insulin action and energy balance. *Endocrinology*, 144, 5172, 2003.

108. Jambor de Sousa, U.L. et al. Hepatic-portal oleic acid inhibits feeding more potently than heltic-portal caprylic acid in rats. *Physiol. Behav.*, 89, 329, 2006.

109. Arase, K. et al. Intracerebroventricular infusions of 3-OHB and insulin in a rat model of dietary obesity. *Am. J. Physiol.*, 255, R974, 1988.

110. Fisler, J.S., Egawa, M. and Bray, G.A. Peripheral 3-hydroxybutyrate and food intake in a model of dietary-fat induced obesity: effect of vagotomy. *Physiol. Behav.*, 58, 1, 1995.

111. Carpenter, R.G. and Grossman, S.R. Plasma fat metabolites and hunger. *Physiol. Behav.*, 30, 57, 1983.

112. Langhans, W., Egli, G. and Scharrer, E. Regulation of food intake by hepatic oxidative metabolism. *Brain Res. Bull.*, 15, 425, 1985.

113. Loftus, T.M. et al. Reduced food intake and body weight in mice treated with fatty acid synthetase inhibitors. *Science*, 288, 2379, 2000.

114. Makimura, H. et al. Cerulenin mimics effects of leptin on metabolic rate, food intake, and body weight independent of the melanocortin system, but unlike leptin, cerulenin fails to block neuroendocrine effects of fasting. *Diabetes*, 50, 733, 2001.

115. Houghton, L.A., Mangnall, Y.F. and Read, N.W. Effect of incorporating fat into a liquid test meal on the relation between intragastric distribution and gastric emptying in human volunteers. *Gut*, 31, 1226, 1990.

116. Read, N., French, S. and Cunningham, K. The role of the gut in regulating food intake in man. *Nutr. Rev.*, 52, 1, 1994.

117. Welch, I.M., Sepple, C.P. and Read, N.W. Comparisons of the effects on satiety and eating behaviour of infusion of lipid into the different regions of the small intestine. *Gut*, 29, 306, 1988.

118. Blundell, J.E. et al. The fat paradox: fat-induced satiety signals versus high fat overconsumption. *Int. J. Obes. Relat. Metab. Disord.*, 19, 832, 1995.

119. Mela, D. Sensory preference for fats: what, who, why? *Food Qual. Pref.*, 1, 71, 1991.

120. Drewnowski, A. and Greenwood, M.R. Cream and sugar: human preferences for high-fat foods. *Physiol. Behav.*, 30, 629, 1983.

121. Drewnowski, A. et al. Food preferences in human obesity: carbohydrates versus fats. *Appetite*, 18, 207, 1992.

122. Acheson, K.J., Flatt, J.P. and Jequier, E. Glycogen synthesis versus lipogenesis after a 500 gram carbohydrate meal in man. *Metabolism*, 31, 1234, 1982.

123. Schutz, Y., Acheson, K J. and Jequier, E. Twenty-four-hour energy expenditure and thermogenesis: response to progressive carbohydrate overfeeding in man. *Int. J. Obes.*, 9, 111, 1985.

124. Schrauwen, P. et al. Changes in fat oxidation in response to a high-fat diet. *Am. J. Clin. Nutr.*, 66, 276, 1997.

125. Jebb, S.A. et al. Changes in macronutrient balance during over- and underfeeding assessed by 12-d continuous whole-body calorimetry. *Am. J. Clin. Nutr.*, 64, 259, 1996.

126. Astrup, A. et al. Failure to increase lipid oxidation in response to increasing dietary fat content in formerly obese women. *Am. J. Physiol.*, 266, E592, 1994.

127. Erlanson-Albertsson, C. and Larsson, A. The activation peptide of pancreatic procolipase decreases food intake in rats. *Regul. Pept.*, 22, 325, 1988.

128. Erlanson-Albertsson, C. et al. Pancreatic procolipase propeptide, enterostatin, specifically inhibits fat intake. *Physiol. Behav.*, 49, 1191, 1991.

129. Lin, L., Bray, G. and York, D.A. Enterostatin suppresses food intake in rats after near-celiac and intracarotid arterial injection. *Am. J. Physiol. Regul. Integr. Comp. Physiol.*, 278, R1346, 2000.

130. Lin, L. and York, D.A. Chronic ingestion of dietary fat is a prerequisite for inhibition of feeding by enterostatin. *Am. J. Physiol.*, 275, R619, 1998.

131. Mei, J. et al. Identification of enterostatin, the pancreatic procolipase activation peptide in the intestine of rat: effect of CCK-8 and high-fat feeding. *Pancreas*, 8, 488, 1993.

132. Lin, L. et al. Beta-casomorphins stimulate and enterostatin inhibits the intake of dietary fat in rats. *Peptides*, 19, 325, 1998.

133. Smith, G.P. and Gibbs, J. Role of CCK in satiety and appetite control. *Clin. Neuropharmacol.*, 15(Suppl. 1) Pt A, 476A, 1992.

134. Geary, N. Endocrine controls of eating: CCK, leptin and ghrelin. *Physiol. Behav.*, 81, 719, 2004.

135. Smith, G.P., Jerome, C. and Norgren, R. Afferent axons in abdominal vagus mediate satiety effect of cholecystokinin in rats. *Am. J. Physiol.*, 249, R638, 1985.

136. Dourish, C.T. et al. Blockade of CCK-induced hypophagia and prevention of morphine tolerance by the CCK antagonist L364,718. in *CCK Antagonists*, Wang, R.Y. and Schoenfeld, T. Eds., Alan R. Liss, New York, NY, 1988, p. 307.

137. Beinfeld, M.C. Cholecystokinin in the central nervous system: a minireview. *Neuropeptides*, 3, 411, 1983.

138. DeFanti, B.A. et al. Lean (Fa/Fa) but not obese (fa/fa) Zucker rats release cholecystokinin at PVN after a gavaged meal. *Am. J. Physiol.*, 275, E1, 1998.

139. Schick, R.R. et al. Neuronal cholecystokinin-like immunoreactivity is postprandially released from primate hypothalamus. *Brain Res.*, 418, 20, 1987.

140. Figlewicz, D.P. et al. Intraventricular CCK inhibits food intake and gastric emptying in baboons. *Am. J. Physiol.*, 256, R1313, 1989.

141. Marks-Kaufman, R. Increased fat consumption induced by morphine administration in rats. *Pharmacol. Biochem. Behav.*, 16, 949, 1982.

142. Marks-Kaufman, R. and Kanarek, R.B. Morphine selectively influences macronutrient intake in the rat. *Pharmacol. Biochem. Behav.*, 12, 427, 1980.

143. Ottaviani, R. and Riley, A.L. Effect of chronic morphine administration on the self-selection of macronutrients in the rat. *Nutr. Behav.*, 2, 27, 1984.

144. Welch, C.C. et al. Preference and diet type affect macronutrient selection after morphine, NPY, norepinephrine, and deprivation. *Am. J. Physiol.*, 266, R426, 1994.

145. Gosnell, B.A. and Krahn, D.D. The effects of continuous morphine infusion on diet selection and body weight. *Physiol. Behav.*, 54, 853, 1993.

146. Marks-Kaufman, R. and Kanarek, R.B. Modifications of nutrient selection induced by naloxone in rats. *Psychopharmacology*, 74, 321, 1981.

147. Marks-Kaufman, R., Plager, A. and Kanarek, R.B. Central and peripheral contributions of endogenous opioid systems to nutrient selection in rats. *Psychopharmacology*, 85, 414, 1985.

148. Drewnowski, A. et al. Taste responses and preferences for sweet high-fat foods: evidence for opioid involvement. *Physiol. Behav.*, 51, 371, 1992.

149. Bertino, M., Beauchamp, G.K. and Engelman, K. Naltrexone, an opioid blocker, alters taste perception and nutrient intake in humans. *Am. J. Physiol.*, 261, R59, 1991.

150. Yeomans, M.R. and Wright, P. Lower pleasantness of palatable foods in nalmefene treated human volunteers. *Appetite*, 16, 249, 1991.

151. Welch, C.C. et al. Palatability-induced hyperphagia increases hypothalamic dynorphin peptide and mRNA levels. *Brain Res.*, 721, 126, 1996.

152. Castonguay, T.W., Dallman, M. and Stern, J. Corticosterone prevents body weight loss and diminished fat appetite following adrenalectomy. *Nutr. Behav.*, 2, 115, 1984.

153. Castonguay, T.W., Dallman, M.F. and Stern, J. Some metabolic and behavioral effects of adrenalectomy on obese Zucker rats. *Am. J. Physiol.*, 251, R923, 1986.

154. Dallman, M.F. et al. Minireview: glucocorticoids—food intake, abdominal obesity, and wealthy nations in 2004. *Endocrinology*, 145, 2633, 2004.

155. Devenport, L. et al. Macronutrient intake and utilization by rats: interactions with type I adrenocorticoid receptor stimulation. *Am. J. Physiol.*, 260, R73, 1991.

156. Kumar, B.A., Papamichael, M. and Leibowitz, S.F. Feeding and macronutrient selection patterns in rats: adrenalectomy and chronic corticosterone replacement. *Physiol. Behav.*, 42, 581, 1988.

157. Bligh, M.E. et al. Adrenal modulation of the enhanced fat intake subsequent to fasting. *Physiol. Behav.*, 48, 373, 1990.

158. Leibowitz, S.F. Hypothalamic neuropeptide Y, galanin, and amines. Concepts of coexistence in relation to feeding behavior. *Ann. N.Y. Acad. Sci.*, 575, 221, 1989.

159. Kyrkouli, S.E. et al. Stimulation of feeding by galanin: anatomical localization and behavioral specificity of this peptide's effects in the brain. *Peptides*, 11, 995, 1990.

160. Smith, B.K., York, D.A. and Bray, D.A. Effects of dietary preference and galanin administration in the paraventricular or amygdaloid nucleus on diet self-selection. *Brain Res. Bull.*, 39, 149, 1996.

161. Leibowitz, S.F. Regulation and effects of hypothalamic galanin: relation to dietary fat, alcohol ingestion, circulating lipids and energy homeostasis. *Neuropeptides*, 39, 327, 2005.

162. Harris, R.B.S. Leptin—Much more than a satiety signal. *Annu. Rev. Nutr.*, 20, 45, 2000.

163. Havel, P.J. et al. High-fat meals reduce 24-h circulating leptin concentrations in women. *Diabetes*, 48, 334, 1999.

164. Martin, L.J. et al. Serum leptin levels and energy expenditure in normal weight women. *Can. J. Physiol. Pharmacol.*, 76, 237, 1998.

165. Niskanen, L. et al. Serum leptin in relation to resting energy expenditure and fuel metabolism in obese subjects. *Int. J. Obes. Relat. Metab. Disord.*, 21, 309, 1997.

166. Surwit, R.S. et al. Low plasma leptin in response to dietary fat in diabetes- and obesity-prone mice. *Diabetes*, 46, 1516, 1997.

167. Frederich, R.C. et al. Leptin levels reflect body lipid content in mice: evidence for diet-induced resistance to leptin action. *Nat. Med.*, 1, 1311, 1995.

168. Tso, P. and Lui, M. Ingested fat and satiety. *Rev. Ingest. Sci.*, 81, 275, 2004.

169. Doi, T. et al. Effect of leptin on intestinal apolipoprotein AIV in response to lipid feeding. *Am. J. Physiol.*, 281, R753, 2001.

170. Primeaux, S.D., York, D.A. and Bray, G.A. Neuropeptide Y administration into the amygdala alters high fat food intake. *Peptides*, 27, 1644, 2006.

171. Leibowitz, S.F. and Wortley, K.E. Hypothalamic control of energy balance: different peptides, different functions. *Peptides*, 25, 473, 2004.

172. Bortz, W.M., Abraham, S. and Chaikoff, I.L. Localization of the block in lipogenesis resulting from feeding fat. *J. Biol. Chem.*, 238, 1266, 1963.

173. Romsos, D.R. and Leveille, G.A. Effect of diet on activity of enzymes involved in fatty acid and cholesterol synthesis. *Adv. Lipid Res.*, 12, 97, 1974.

174. Baltzell J.K. and Berdanier, C.D. Effect on the interaction of dietary carbohydrate and fat on the responses of rats to starvation-refeeding. *J. Nutr.*, 115, 104, 1985.

175. Goodridge, A.G. Dietary regulation of gene expression: enzymes involved in carbohydrate and lipid metabolism. *Annu. Rev. Nutr.*, 7, 157, 1987.

176. Jump, D.B. and Clarke, S.D. Regulation of gene expression by dietary fat. *Annu. Rev. Nutr.*, 19, 63, 1999.

177. Clarke, S.D. Polyunsaturated fatty acid regulation of gene transcription: a mechanism to improve energy balance and insulin resistance. *Br. J. Nutr.*, 83, S59, 2000.

178. Lemonnier, D. et al. Metabolism of genetically obese rats on normal or high-fat diet. *Diabetologia*, 10(Suppl.), 697, 1974.

179. Flick, P.K., Chen, J. and Vagelos, P.R. Effect of dietary linoleate on synthesis and degradation of fatty acid synthetase from rat liver. *J. Biol. Chem.*, 252, 4242, 1977.

180. Wilson, M.D. et al. Potency of polyunsaturated and saturated fats as short-term inhibitors of hepatic lipogenesis in rats. *J. Nutr.*, 120, 544, 1990.

181. Pan, J.S. and Berdanier, C.D. Thyroxine counteracts the effects of menhaden oil on fatty acid synthesis in BHE rats. *Nutr. Res.*, 10, 461, 1990.

182. Clarke, S.D. and Jump, D.B. Regulation of hepatic gene expression by dietary fats: a unique role for polyunsaturated fatty acids. in *Nutrition and Gene Expression*, Berdanier, C.D. and Hargrove, J.L. Eds., CRC Press, Boca Raton, FL, 1993, p. 228.

183. Mohan, P.F., Phillips, F.C. and Cleary, M.P. Metabolic effects of coconut, safflower, or menhaden oil feeding in lean and obese Zucker rats. *Br. J. Nutr.*, 66, 285, 1991.

184. Ikeda, I. et al. Effects of dietary alpha-linolenic, eicosapentaenoic and docosahexaenoic acids on hepatic lipogenesis and beta-oxidation in rats. *Biosci. Biotechnol. Biochem.*, 62, 675, 1998.

185. Xu, J. et al. Sterol regulatory element binding protein-1 expression is suppressed by dietary polyunsaturated fatty acids: a mechanism for the coordinate suppression of lipogenic genes by polyunsaturated fats. *J. Biol. Chem.*, 274, 23577, 1999.

186. Cheema, S.K. and Clandinin, M.T. Diet fat alters expression of genes for enzymes of lipogenesis in lean and obese mice. *Biochim. Biophys. Acta.*, 1299, 284, 1996.

187. Kim, H.J., Takahashi, M. and Ezaki, O. Fish oil feeding decreases mature sterol regulatory element-binding protein 1 (SREBP-1) by down-regulation of SREBP-lc mRNA in mouse liver: a possible mechanism for down-regulation of lipogenic enzyme m-RNAs. *J. Biol. Chem.*, 274, 25892, 1999.

188. Yahagi, N. et al. A crucial role of sterol regulatory element-binding protein-1 in the regulation of lipogenic gene expression by polyunsaturated fatty acids. *J. Biol. Chem.*, 274, 35840, 1999.

189. Shimano, H. et al. Overproduction of cholesterol and fatty acids causes massive liver enlargement in transgenic mice expressing truncated SREBP-1a. *J. Clin. Invest.*, 98, 1575, 1996.

190. Shimomura, I. et al. Nuclear sterol regulatory element-binding proteins activate genes responsible for the entire program of unsaturated fatty acid biosynthesis in transgenic mouse liver. *J. Biol. Chem.*, 273, 35299, 1998.

191. Murphy, E.J. Stable isotope methods for the in vivo measurement of lipogenesis and triglyceride metabolism. *J. Anim. Sci.*, 84, E94, 2006.

192. Hellerstein, M.K., Schwarz, J.M. and Neese, R.A. Regulation of hepatic de novo lipogenesis in humans. *Annu. Rev. Nutr.*, 16, 523, 1996.

193. Chascione, C. et al. Effect of carbohydrate intake on de novo lipogenesis in human adipose tissue. *Am. J. Physiol.*, 253, E664, 1987.

194. Schwarz, J.M. et al. Hepatic de novo lipogenesis in normoinsulinemic and hyperinsulinemic subjects consuming high-fat, low-carbohydrate and low-fat, high-carbohydrate isoenergetic diets. *Am. J. Clin. Nutr.*, 77, 43, 2003.

195. Greenwood, M.R. The relationship of enzyme activity to feeding behavior in rats: lipoprotein lipase as the metabolic gatekeeper. *Int. J. Obes.*, 9(Suppl. 1), 67, 1985.

196. Khan, S. et al. Dietary long-chain n-3 PUFAs increase LPL gene expression in adipose tissue of subjects with an atherogenic lipoprotein phenotype. *J. Lipid Res.*, 43, 979, 2002.

197. Harris, W.S. et al. Influence of n-3 fatty acid supplementation on the endogenous activities of plasma lipases. *Am. J. Clin. Nutr.*, 66, 254, 1997.

198. Zampelas, Z. et al. Postprandial lipoprotein lipase, insulin and gastric inhibitory polypeptide responses to test meals of different fatty acid composition: comparison of saturated, n-6 and n-3 polyunsaturated fatty acids. *Eur. J. Clin. Nutr.*, 48, 849, 1994.

199. Benhizia, F. et al. Effects of a fish oil-lard diet on rat plasma lipoproteins, liver FAS, and lipolytic enzymes. *Am. J. Physiol.*, 267, E975, 1994.

200. Peyron-Caso, E. et al. Dietary fish oil increases lipid mobilization but does not decrease lipid storage-related enzyme activities in adipose tissue of insulin-resistant, sucrose-fed rats. *J. Nutr.*, 133, 2239, 2003.

201. Nozaki, S. et al. Postheparin lipolytic activity and plasma lipoprotein response to omega-3 polyunsaturated fatty acids in patients with primary hypertriglyceridemia. *Am. J. Clin. Nutr.*, 53, 638, 1991.

202. Bagdade, J.D. et al. Effects of omega-3 fish oils on plasma lipids, lipoprotein composition, and postheparin lipoprotein lipase in women with IDDM. *Diabetes*, 39, 426, 1990.

203. Takahashi, Y. and Ide, T. Dietary n-3 fatty acids affect mRNA level of brown adipose tissue uncoupling protein 1, and white adipose tissue leptin and glucose transporter 4 in the rat. *Br. J. Nutr.*, 84, 175, 2000.

204. Raclot, T. et al. Site-specific regulation of gene expression by n-3 polyunsaturated fatty acids in rat white adipose tissues. *J. Lipid Res.*, 38, 1963, 1997.

205. Enerback, S. and Gimble, J.M. Lipoprotein lipase gene expression: physiological regulators at the transcriptional and post-transcriptional level. *Biochim. Biophys. Acta*, 1169, 107, 1993.

206. Herzberg, G.R. and Rogerson, M. Hepatic fatty acid synthesis and triglyceride secretion in rats fed fructose-or glucose-based diets containing corn oil, tallow or marine oil. *J Nutr.*, 118, 1061, 1988.

207. Schrauwen-Hinderling, V.B. et al. Intracellular lipid content in human skeletal muscle. *Obesity*, 14, 357, 2006.

208. Goodpaster, B.H. et al. Intramuscular lipid content is increased in obesity and decreased by weight loss. *Metabolism*, 49, 467, 2000.

209. Corcoran, M.P., Lamon-Fava, S. and Fielding, R.A. Skeletal muscle lipid deposition and insulin resistance: effect of dietary fatty acids and exercise. *Am. J. Clin. Nutr.*, 85, 662, 2007.

210. Kleiber, M. The respiratory quotient. in *The Fire of Life*, Kleiber, M., Ed., Wiley, New York, NY, 1961, p. 82.

211. DeLany, J.P. et al. Differential oxidation of individual dietary fatty acids in humans. *Am. J. Clin. Nutr.*, 72, 905, 2000.

212. Piers, L.S. et al. The influence of the type of dietary fat on postprandial fat oxidation rates: monounsaturated (olive oil) vs saturated fat (cream). *Int. J. Obes.*, 26, 814, 2002.

213. Soares, M.J. et al. The acute effects of olive oil v. cream on postprandial thermogenesis and substrate oxidation in postmenopausal women. *Br. J. Nutr.*, 91, 245, 2004.

214. Piers, L.S. et al. Substitution of saturated with monounsaturated fat in a 4-week diet affects body weight and composition of overweight and obese men. *Br. J. Nutr.*, 90, 717, 2003.

215. Young, J.B. and Walgren, M.C. Differential effects of dietary fatty acid on sympathetic nervous system activity in the rat. *Metabolism*, 43, 51, 1994.

216. Borgeson, C.E. et al. Effects of dietary fish oil on human mammary carcinoma and on lipid-metabolizing enzymes. *Lipids*, 24, 290, 1989.

217. Willumsen, N. et al. Docosahexaenoic acid shows no triglyceride-lowering effects but increases the peroxisomal fatty acid oxidation in liver of rats. *J. Lipid Res.*, 34, 13, 1993.

218. Ukropec, J. et al. The hypotriglyceridemic effect of dietary n-3 FA is associated with increased beta-oxidation and reduced leptin expression. *Lipids*, 38, 1023, 2003.

219. Baillie, R.A. et al. Coordinate induction of peroxisomal acyl-CoA oxidase and UCP-3 by dietary fish oil: a mechanism for decreased body fat deposition. *Prostaglandins Leukot. Essent. Fatty Acids*, 60, 351, 1999.

220. Flachs, P. et al. Polyunsaturated fatty acids of marine origin upregulate mitochondrial biogenesis and induce β-oxidation in white fat. *Diabetologia*, 48, 2365, 2005.

221. Ji, H. and Friedman, M.I. Fasting plasma triglyceride levels and fat oxidation predict dietary obesity in rats. *Physiol. Behav.*, 78, 767, 2003.

222. Ji, H., Outterbridge, L.V. and Friedman, M.I. Phenotype-based treatment of dietary obesity: differential effects of fenofibrate in obesity-prone and obesity-resistant rats. *Metabolism*, 54, 421, 2005.

223. Jackman, M.R. et al. Trafficking of dietary fat in obesity-prone and obesity-resistant rats. *Am. J. Physiol. Endocrinol. Metab.*, 291, E1083, 2006.

224. Iossa, S. et al. Effect of high-fat feeding on metabolic efficiency and mitochondrial oxidative capacity in adult rats. *Br. J. Nutr.*, 90, 953, 2003.

225. Blaak, E.E. et al. Fat oxidation before and after a high fat load in the obese insulin-resistant state. *J. Clin. Endocrinol. Metab.*, 91, 1462, 2006.

226. Giacco, R. et al. Insulin sensitivity is increased and fat oxidation after a high-fat meal is reduced in normal-weight healthy men with strong familial predisposition to overweight. *Int. J. Obes. Relat. Metab. Disord.*, 28, 342, 2004.

227. Nedergaard, J., Becker, W. and Cannon, B. Effects of dietary essential fatty acids on active thermogenin content in rat brown adipose tissue. *J. Nutr.*, 113, 1717, 1983.

228. Ide, T. and Sugano, M. Effects of dietary fat types on the thermogenesis of brown adipocytes isolated from rat. *Agric. Biol. Chem.*, 52, 511, 1988.

229. Matsuo, T., Komuro, M. and Suzuki, M. Beef tallow diet decreases uncoupling protein content in the brown adipose tissue of rats. *J. Nutr. Sci. Vitaminol.*, 42, 595, 1996.

230. Hun, C.S. et al. Increased uncoupling protein2 mRNA in white adipose tissue, and decrease in leptin, visceral fat, blood glucose, and cholesterol in KK-Ay mice fed with eicosapentaenoic and docosahexaenoic acids in addition to linolenic acid. *Biochem. Biophys. Res. Commun.*, 259, 85, 1999.

231. Oudart, H. et al. Brown fat thermogenesis in rats fed high-fat diets enriched with n-3 polyunsaturated fatty acids. *Int. J. Obes. Relat. Metab. Disord.*, 21, 955, 1997.

232. Vernon, R.G. Effects of diet on lipolysis and its regulation. *Proc. Nutr. Soc.*, 51, 397, 1992.

233. Gorman, R.R., Tepperman, H.M. and Tepperman, J. Effects of starvation, refeeding, and fat feeding on adipocyte ghost adenyl cyclase activity. *J. Lipid Res.*, 13, 276, 1972.

234. Gorman, R.R., Tepperman, H.M. and Tepperman, J. Epinephrine binding and the selective restoration of adenylate cyclase activity in fat-fed rats. *J. Lipid Res.*, 14, 279, 1973.

235. Tepperman, H.M., Dewitt, J. and Tepperman, J. Effect of a high fat diet on rat adipocyte lipolysis: responses to epinephrine, forskolin, methylisobutylxanthine, dibutyryl cyclic AMP, insulin and nicotinic acid. *J. Nutr.*, 116, 1984, 1986.

236. Smith, U., Kral, J. and Bjorntorp, P. Influence of dietary fat and carbohydrate on the metabolism of adipocytes of different size in the rat. *Biochim. Biophys. Acta*, 337, 278, 1974.

237. Susini, C., Lavau, M. and Herzog, J. Adrenaline responsiveness of glucose metabolism in insulin-resistant adipose tissue of rats fed a high-fat diet. *Biochem. J.*, 180, 431, 1979.

238. Chilliard, Y. Dietary fat and adipose tissue metabolism in ruminants, pigs, and rodents: a review. *J. Dairy Sci.*, 76, 3897, 1993.

239. Lipinski, B.A. and Mathias, M.M. Prostaglandin production and lipolysis in isolated rat adipocytes as affected by dietary fat. *Prostaglandins*, 16, 957, 1978.

240. Carr, R.E. et al. Lipolysis in rat adipose tissue in vitro is dependent on the quantity and not the quality of dietary fat. *Biochem. Soc. Trans.*, 23, 486S, 1995.

241. Portillo, M.P. et al. Effect of high-fat diet on lypolisis in isolated adipocytes from visceral and subcutaneous WAT. *Eur. J. Nutr.*, 38, 177, 1999.

242. Nicolas, C. et al. Dietary (n-6) polyunsaturated fatty acids affect beta-adrenergic receptor binding and adenylate cyclase activity in pig adipocyte plasma membrane. *J. Nutr.*, 121, 1179, 1991.

243. Mersmann, H.J. et al. Influence of dietary fat on beta-adrenergic receptors and receptor-controlled metabolic function in porcine adipocytes. *J. Nutr. Biochem.*, 6, 302, 1995.

244. Benmansour, N. et al. Effects of castration, dietary fat and adipose tissue sites on adipocyte plasma membranes cyclic AMP phosphodiesterase activity in the pig. *Int. J. Biochem.*, 23, 1205, 1991.

245. Jenkins, T.C., Thies, E.J. and Fotouhi, N. Dietary soybean oil changes lipolytic rate and composition of fatty acids in plasma membranes of ovine adipocytes. *J. Nutr.*, 124, 566, 1994.

246. Berger, J.J. and Barnard, R.J. Effect of diet on fat cell size and hormone-sensitive lipase activity. *J. Appl. Physiol.*, 87, 227, 1999.

247. Kather, H. et al. Influences of variation in total energy intake and dietary composition on regulation of fat cell lipolysis in ideal-weight subjects. *J. Clin. Invest.*, 80, 566, 1987.

248. Suljkovicova, H. et al. Effect of macronutrient composition of the diet on the regulation of lipolysis in adipose tissue at rest and during exercise: microdialysis study. *Metabolism*, 51, 1291, 2002.

249. Benmansour, N.M. et al. Effects of dietary fat and adipose tissue location on insulin action in young boar adipocytes. *Int. J. Biochem.*, 23, 499, 1991.

250. Guo, Z. and Zhou, L. Evidence for increased and insulin-resistant lipolysis in skeletal muscle of high-fat fed rats. *Metabolism*, 53, 794, 2004.

251. Kim, C.H. et al. Lipolysis in skeletal muscle is decreased in high-fat-fed rats. *Metabolism*, 52, 1586, 2003.

252. Perona, J.S. et al. Influence of different dietary fats on triacylglycerol deposition in rat adipose tissue. *Br. J. Nutr.*, 84, 765, 2000.

253. Raclot, T. and Oudart, H. Selectivity of fatty acids on lipid metabolism and gene expression. *Proc. Nutr. Soc.*, 58, 633, 1999.

254. Cromer, K.D., Jenkins, T.C. and Thies, E.J. Replacing cis octadecenoic acid with *trans* isomers in media containing rat adipocytes stimulates lipolysis and inhibits glucose utilization. *J. Nutr.*, 125, 2394, 1995.

255. Conner, W.E., Lin, D.S. and Colvis, C. Differential mobilization of fatty acids from adipose tissue. *J. Lipid Res.*, 37, 290, 1996.

256. Raclot, T. et al. Selectivity of fatty acid mobilization: a general metabolic feature of adipose tissue. *Am. J. Physiol.*, 269, R1060, 1995.

257. Raclot, T. Selective mobilization of fatty acids from white fat cells: evidence for a relationship to the polarity of triacylglycerols. *Biochem. J.*, 322, 483, 1997.

258. Awad, A.B. and Chattopadhyay, J.P. Effect of dietary saturated fatty acids on hormone-sensitive lipolysis in rat adipocytes. *J. Nutr.*, 116, 1088, 1986.

259. Matsuo, T., Sumida, H. and Suzuki, M. Beef tallow diet decreases beta-adrenergic receptor binding and lipolytic activities in different adipose tissues of rat. *Metabolism*, 44, 1271, 1995.

260. Awad, A.B. and Chattopadhyay, J.P. Effect of dietary saturated fatty acids on intra-cellular free fatty acids and kinetic properties of hormone-sensitive lipase of rat adipocytes. *J. Nutr.*, 116, 1095, 1986.

261. Larking, P.W. and Nye, E.R. The effect of dietary lipids on lipolysis in rat adipose tissue. *Br. J. Nutr.*, 33, 291, 1975.

262. Awad, A.B. and Zepp, E.A. Alteration of rat adipose tissue lipolytic response to norepinephrine by dietary fatty acid manipulation. *Biochem. Biophys. Res. Commun.*, 86, 138, 1979.

263. Fotovati, A., Hayashi, T. and Ito, T. Lipolytic effect of BRL 35 135, a beta3 agonist, and its interaction with dietary lipids on the accumulation of fats in rat body. *J. Nutr. Biochem.*, 12, 153, 2001.

264. Matsuo, T. and Suzuki, M. Beef tallow diet decreases lipoprotein lipase activities in brown adipose tissue, heart, and soleus muscle by reducing sympathetic activities in rats. *J. Nutr. Sci. Vitaminol.*, 40, 569, 1994.

265. Mersmann, H.J. et al. Beta-adrenergic receptor-mediated functions in porcine adipose tissue are not affected differently by saturated vs. unsaturated dietary fats. *J. Nutr.*, 122, 1952, 1992.

266. Awad, A.B., Bernardis, L.L. and Fink, C.S. Failure to demonstrate an effect of dietary fatty acid composition on body weight, body composition and parameters of lipid metabolism in mature rats. *J. Nutr.*, 120, 1277, 1990.

267. Portillo, M.P. et al. Modifications induced by dietary lipid source in adipose tissue phospholipid fatty acids and their consequences in lipid mobilization. *Br. J. Nutr.*, 82, 319, 1999.

268. Sweeney, J.H. Dietary factors that influence the dextrose tolerance test. *Arch. Intern. Med.*, 40, 818, 1927.

269. Himsworth, H.P. Dietetic factors influencing the glucose tolerance and activity of insulin. *J. Physiol.*, 81, 29, 1934.

270. Lichtenstein, A.H. and Schwab, U.S. Relationship of dietary fat to glucose metabolism. *Atherosclerosis*, 150, 227, 2000.

271. Blazquez, E. and Quijada, C.L. The effect of a high-fat diet on glucose, insulin sensitivity and plasma insulin in rats. *J. Endocrinol.*, 42, 489, 1968.

272. Harris, R.B. and Kor, H. Insulin insensitivity is rapidly reversed in rats by reducing dietary fat from 40 to 30% of energy. *J. Nutr.*, 122, 1811, 1992.

273. Yakubu, F. et al. Insulin action in rats is influenced by amount and composition of dietary fat. *Obes. Res.*, 1, 481, 1993.

274. Grundleger, M.L. and Thenen, S.W. Decreased insulin binding, glucose transport, and glucose metabolism in soleus muscle of rats fed a high fat diet. *Diabetes*, 31, 232, 1982.

275. Lavau, M. et al. Mechanism of insulin resistance in adipocytes of rats fed a high-fat diet. *J. Lipid Res.*, 20, 8, 1979.

276. Susini, C. and Lavau, M. In-vitro and in-vivo responsiveness of muscle and adipose tissue to insulin in rats rendered obese by a high-fat diet. *Diabetes*, 27, 114, 1978.

277. Hissin, P.J. et al. A possible mechanism of insulin resistance in the rat adipose cell with high-fat/low-carbohydrate feeding. Depletion of intracellular glucose transport systems. *Diabetes*, 31, 589, 1982.

278. Kraegen, E.W. et al. In vivo insulin resistance in individual peripheral tissues of the high fat fed rat: assessment by euglycaemic clamp plus deoxyglucose administration. *Diabetologia*, 29, 192, 1986.

279. Kraegen, E.W. et al. Development of muscle insulin resistance after liver insulin resistance in high-fat-fed rats. *Diabetes*, 40, 1397, 1991.

280. van Amelsvoort, J.M., van der Beek, A. and Stam, J.J. Dietary influence on the insulin function in the epididymal fat cell of the Wistar rat. III. Effect of the ratio carbohydrate to fat. *Ann. Nutr. Metab.*, 32, 160, 1988.

281. Oakes, N.D. et al. Mechanisms of liver and muscle insulin resistance induced by chronic high-fat feeding. *Diabetes*, 46, 1768, 1997.

282. Frangioudakis, G., Ye, J.M. and Cooney, G.J. Both saturated and n-6 polyunsaturated fat diets reduce phosphorylation of insulin receptor substrate-1 and protein kinase B in muscle during the initial stages of in vivo insulin stimulation. *Endocrinology*, 146, 5596, 2005.

283. Bisschop, P.H. et al. Dietary fat content alters insulin-mediated glucose metabolism in healthy men. *Am. J. Clin. Nutr.*, 73, 554, 2001.

284. Pagliassotti, M.J. et al. Developmental stage modifies diet-induced peripheral insulin resistance in rats. *Am. J. Physiol. Regul. Integr. Comp. Physiol.*, 278, R66, 2000.

285. Lopez-Miranda, J. et al. Dietary fat, genes and insulin sensitivity. *J. Mol. Med.*, 85, 209, 2007.

286. Roche, H.M., Phillips, C. and Gibney, M.J. The metabolic syndrome: the crossroads of diet and genetics. *Proc. Nutr. Soc.*, 64, 371, 2005.

287. Ip, C. et al. Insulin binding and insulin response of adipocytes from rats adapted to fat feeding. *J. Lipid Res.*, 17, 588, 1976.

288. Sun, J.V., Tepperman, H.M. and Tepperman, J. A comparison of insulin binding by liver plasma membranes of rats fed a high glucose diet or a high fat diet. *J. Lipid Res.*, 18, 533, 1977.

289. Nagy, K., Levy, J. and Grunberger, G. High-fat feeding induces tissue-specific alteration in proportion of activated insulin receptors in rats. *Acta Endocrinol.*, 122, 361, 1990.

290. Begum, N., Tepperman, H.M. and Tepperman, J. Insulin-induced internalization and replacement of insulin receptors in adipocytes of rats adapted to fat feeding. *Diabetes*, 34, 1272, 1985.

291. Salans, L.B. et al. Effects of dietary composition on glucose metabolism in rat adipose cells. *Am. J. Physiol.*, 240, E175, 1981.

292. Watarai, T. et al. Alteration of insulin-receptor kinase activity by high-fat feeding. *Diabetes*, 37, 1397, 1988.

293. Rosholt, M.N., King, P.A. and Horton, E.S. High-fat diet reduces glucose transporter responses to both insulin and exercise. *Am. J. Physiol.*, 266, R95, 1994.

294. Hunnicutt, J.W. et al. Saturated fatty acid–induced insulin resistance in rat adipocytes. *Diabetes*, 43, 540, 1994.

295. Steppan, C.M. and Lazar, M.A. The current biology of resistin. *J. Intern. Med.*, 255, 439, 2004.

296. Borst, S.E. The role of TNF-alpha in insulin resistance. *Endocrine*, 23, 177, 2004.

297. Borst, S.E. and Conover, C.F. High-fat induces increased tissue expression of TNFα. *Life Sci.*, 77, 2156, 2005.

298. Park, S.Y. et al. Unraveling the temporal pattern of diet-induced insulin resistance in individual organs and cardiac dysfunction in C57BL/6 mice. *Diabetes*, 54, 3530, 2005.

299. Pitombo, C. et al. Amelioration of diet-induced diabetes mellitus by removal of visceral fat. *J. Endocrinol.*, 191, 699, 2006.

300. Barnea, M. et al. A high-fat diet has a tissue-specific effect on adiponectin and related enzyme expression. *Obesity*, 14, 2145, 2006.

301. Bullen Jr., J.W. et al. Regulation of adiponectin and its receptor in response to development of diet-induced obesity in mice. *Am. J. Physiol. Endocrinol. Metab.*, 292, E1079, 2007.

302. Lovejoy, J.C. Dietary fatty acids and insulin resistance. *Curr. Atheroscler. Rep.*, 1, 215, 1999.

303. Haag, M. and Dippenaar, N.G. Dietary fats, fatty acids and insulin resistance: short review of a multifaceted connection. *Med. Sci. Monit.*, 11, RA359, 2005.

304. Riccardi, G., Giacco, R. and Rivellese, A.A. Dietary fat, insulin sensitivity and the metabolic syndrome. *Clin. Nutr.*, 23, 447, 2004.

305. Storlien, L.H. et al. Does dietary fat influence insulin action? *Ann. N.Y. Acad. Sci.*, 827, 287, 1997.

306. Vessby, B. et al. Substituting dietary saturated for monounsaturated fat impairs insulin sensitivity in healthy men and women: the KANWU study. *Diabetologia*, 44, 312, 2001.

307. Perez-Jimenez, F. et al. A Mediterranean and a high-carbohydrate diet improve glucose metabolism in healthy young persons. *Diabetologia*, 44, 2038, 2001.

308. van Amelsvoort, J. et al. Dietary influence on the insulin function in the epididymal fat cell of the Wistar rat: I. Effect of type of fat. *Ann. Nutr. Metab.*, 32, 138, 1988.

309. Field, C.J. et al. Dietary fat and the diabetic state alter insulin binding and the fatty acyl composition of the adipocyte plasma membrane. *Biochem. J.*, 253, 417, 1988.

310. Field, C.J. et al. The effect of dietary fat content and composition on adipocyte lipids in normal and diabetic states. *Int. J. Obes.*, 13, 747, 1989.

311. Field, C.J, Toyomizu, M. and Clandinin, M.T. Relationship between dietary fat, adipocyte membrane composition and insulin binding in the rat. *J. Nutr.*, 119, 1483, 1989.

312. Podolin, D.A. et al. Menhaden oil prevents but does not reverse sucrose-induced insulin resistance in rats. *Am. J. Physiol. Regul. Integr. Comp. Physiol.*, 274, R840, 1998.

313. Storlien, L.H. et al. Fish oil prevents insulin resistance induced by high-fat feeding in rats. *Science*, 237, 885, 1987.

314. Storlien, L.H. et al. Influence of dietary fat composition on development of insulin resistance in rats. Relationship to muscle triglyceride and omega-3 fatty acids in muscle phospholipid. *Diabetes*, 40, 280, 1991.

315. Lovejoy, J.C. et al. Effects of diets enriched in saturated (palmitic), monounsaturated (oleic), or trans (elaidic) fatty acids on insulin sensitivity and substrate oxidation in healthy adults. *Diabetes Care*, 25, 1283, 2002.

316. Kim, Y.B. et al. Gene expression of insulin signal-transduction pathway intermediates is lower in rats fed a beef tallow diet than in rats fed a safflower oil diet. *Metabolism*, 45, 1080, 1996.

317. Ghafoorunissa. et al. Dietary (n-3) long chain polyunsaturated fatty acids prevent sucrose-induced insulin resistance in rats. *J. Nutr.*, 135, 2634, 2005.

318. Haugaard, S.B. et al. Dietary intervention increases n-3 long-chain polyunsaturated fatty acids in skeletal muscle membrane phospholipids of obese subjects. Implication for insulin sensitivity. *Clin. Endocrinol.*, 64, 169, 2006.

319. Lombardo, Y.B. and Chicco, A.G. Effect of dietary polyunsaturated n-3 fatty acids on dyslipidemia and insulin resistance in rodents and humans. A review. *J. Nutr. Biochem.*, 17, 1, 2006.

320. Nettleton, J.A. and Katz, R. N-3 long-chain polyunsaturated fatty acids in type 2 diabetes: a review. *J. Am. Diet. Assoc.*, 105, 428, 2005.

321. Luo, J. et al. Dietary (n-3) polyunsaturated fatty acids improve adipocyte insulin action and glucose metabolism in insulin-resistant rats: relation to membrane fatty acids. *J. Nutr.*, 126, 1951, 1996.

322. Fickova, M. et al. Dietary (n-3) and (n-6) polyunsaturated fatty acids rapidly modify fatty acid composition and insulin effects in rat adipocytes. *J. Nutr.*, 128, 512, 1998.

323. Ezaki, O. et al. Effects of fish and safflower oil feeding on subcellular glucose transporter distributions in rat adipocytes. *Am. J. Physiol.*, 263, E94, 1992.

324. Malasanos, T.H. and Stacpoole, P.W. Biological effects of omega-3 fatty acids in diabetes mellitus. *Diabetes Care*, 14, 1160, 1991.

325. Mostad, I.L. et al. Effect of n-3 fatty acids in subjects with type-2 diabetes: reduction of insulin sensitivity and time-dependent alteration from carbohydrate to fat oxidation. *Am. J. Clin. Nutr.*, 84, 540, 2006.

326. Salmeron, J. et al. Dietary fat intake and risk of type 2 diabetes in women. *Am. J. Clin. Nutr.*, 73, 1019, 2001.

327. Ibrahim, A., Natrajan, S. and Ghafoorunissa, R. Dietary trans-fatty acids alter adipocyte plasma membrane fatty acid composition and insulin sensitivity in rats. *Metabolism*, 54, 240, 2005.

328. Riserus, U. Trans fatty acids and insulin resistance. *Atherosclerosis*, 7, 37, 2006.

329. Lemmonier, D. Effect of age, sex, and site on the cellularity of the adipose tissue in mice and rats rendered obese by a high fat diet. *J. Clin. Invest.*, 51, 2907, 1972.

330. Greenwood, M.R. and Hirsch, J. Postnatal development of adipocyte cellularity in the normal rat. *J. Lipid Res.*, 15, 474, 1974.

331. Johnson, P.R. et al. Adipose tissue hyperplasia and hyperinsulinemia in Zucker obese female rats: a developmental study. *Metabolism*, 27, 1941, 1978.

332. Obst, B. et al. Adipocyte size and number in dietary resistant and susceptible rats. *Am. J. Physiol.*, 240, E47, 1981.

333. Bourgeois, F., Alexiu, A. and Lemmonier, D. Dietary-induced obesity: effects of dietary fats on adipose cellularity in mice. *Br. J. Nutr.*, 49, 17, 1983.

334. Massiera, F. et al. Arachidonic acid and prostacyclin signaling promote adipose tissue development: a human health concern? *J. Lipid Res.*, 44, 271, 2003.

335. Bjorntorp, P., Karlsson, M. and Pettersson, P. Expansion of adipose tissue storage capacity at different ages in rats. *Metabolism*, 31, 366, 1982.

336. Faust, I.M. et al. Diet-induced adipocyte number increase in adult rats: a new model of obesity. *Am. J. Physiol.*, 235, E279, 1978.

337. Lemmonier, D. and Alexiu, A. Nutritional, genetic and hormonal aspects of adipose tissue cellularity. in *The Regulation of the Adipose Tissue Mass*, Vague, J and Boyer, J., Eds., Excerpta Medica, Amsterdam, 1974, p. 158.

338. Launay, M., Vodovar, N. and Raulin, J. Developpement du tissu adipeux: Nombre et taille des cellules en fonction de la valeur energetique et de l'insaturation des lipides du regime. *Bull. Soc. Chim. Biol.*, 50, 439, 1968.

339. Raulin, J. and Launay, M. Enrichissement en ADN et ARN du tissu adipeux epididymaire du rat administration de lipides insatures. *Nutr. Diet.*, 9, 208, 1967.

340. Raulin, J. et al. Remaniements structuraux du tissu adipeux et apport excessif d'acides gras essentiels. *Nutr. Metab.*, 13, 249, 1971.

341. Lemmonier, D., Alexiu, A. and Lanteaume, M.T. Effect of two dietary lipids on the cellularity of the rat adipose tissue. *J. Physiol. (Paris)*, 66, 729, 1973.

342. Shillabeer, G. and Lau, D.C.W. Regulation of new fat cell formation in rats: the role of dietary fats. *J. Lipid Res.*, 35, 592, 1994.

343. Kirtland, J., Gurr, M. and Widdowson, E.M. Body lipids of guinea pigs exposed to different dietary fats from mid-gestation to 3 months of age. I. The cellularity of adipose tissue. *Nutr. Metab.*, 20, 338, 1976.

344. Edwards, M. et al. Effect of dietary fat and aging on adipose tissue cellularity in mice differing in genetic predisposition to obesity. *Growth Dev. Aging*, 57, 45, 1993.

345. Kirtland, J. and Gurr, M.I. The effect of different dietary fats on cell size and number in rat epididymal fat pad. *Br. J. Nutr.*, 39, 19, 1978.

346. Hill, J.O. et al. Development of dietary obesity in rats: influence of amount and composition of dietary fat. *Int. J. Obes. Relat. Metab. Disord.*, 16, 321, 1992.

347. Hill, J.O. et al. Lipid accumulation and body fat distribution is influenced by type of dietary fat fed to rats. *Int. J. Obes.*, 17, 223, 1993.

348. Okuno, M. et al. Perilla oil prevents the excessive growth of visceral adipose tissue in rats by down-regulating adipocyte differentiation. *J. Nutr.*, 127, 1752, 1997.

349. Parrish, C., Pathy, D. and Angel, A. Dietary fish oils limit adipose tissue hypertrophy in rats. *Metabolism*, 39, 217, 1990.

350. Garaulet, M. et al. Relationship between fat cell size and number and fatty acid composition in adipose tissue from different fat depots in overweight/obese humans. *Int. J. Obes.*, 30, 899, 2006.

351. Skurnick-Minot, G. et al. Whole-body fat mass and insulin sensitivity in type 2 diabetic women: effect of n-3 polyunsaturated fatty acids. *Diabetes*, 53(Suppl. 2), A44, 2004.

352. Huber, J. et al. Prevention of high-fat diet-induced adipose tissue remodeling in obese diabetic mice by n-3 polyunsaturated fatty acids. *Int. J. Obes.*, 31, 1004, 2007.

353. Ailhaud, G. et al. Temporal changes in dietary fat: role of n-6 polyunsaturated fatty acids in excessive adipose tissue development and relationship to obesity. *Prog. Lipid Res.*, 45, 203, 2006.

354. Ruzickova, J. et al. Omega-3 PUFA of marine origin limit diet-induced obesity in mice by reducing cellularity of adipose tissue. *Lipids*, 39, 1177, 2004.

355. Klyde, B.J. and Hirsch, J. Increased cellular proliferation in adipose tissue of adult rats fed a high-fat diet. *J. Lipid Res.*, 20, 705, 1979.

356. Ellis, J., McDonald, R. and Stern, J.S. A diet high in fat stimulates adipocyte proliferation in older (22 months) rats. *Exp. Gerontol.*, 25, 141, 1990.

357. Bjorntorp, P. et al. Dietary and species influence on potential of plasma to stimulate differentiation and lipid accumulation in cultured adipocyte precursors. *J. Lipid Res.*, 26, 1444, 1985.

358. Shillabeer, G., Forden, J.M. and Lau, D.C.W. Induction of preadipocyte differentiation by mature fat cells in the rat. *J. Clin. Invest.*, 84, 381, 1989.

359. Lau, D.C.W. et al. Influence of paracrine factors on preadipocyte replication and differentiation. *Int. J. Obes.*, 14, 193, 1990.

360. Marques, B. et al. Insulin-like growth factor-I mediates high fat-diet induced adipogenesis in Osborne–Mendel rats. *Am. J. Physiol. Regul. Integr. Comp. Physiol.*, 278, R654, 2000.

361. Latimer, A. et al. Regional differences in adipose tissue insulin-like growth factor (IGF-I) content in high fat fed Osborne–Mendel rats. *FASEB J.*, 10, A218, 1996.
362. Crossno Jr., J.T. et al. Rosiglitazone promotes development of a novel adipocyte population from bone marrow-derived circulating progenitor cells. *J. Clin. Invest.*, 116, 3220, 2006.
363. Bartness, T.J. et al. Reversal of high fat diet-induced obesity in female rats. *Am. J. Physiol.*, 263, R790, 1992.
364. Hill, J.O. et al. Reversal of dietary obesity is influenced by its duration and severity. *Int. J. Obes.*, 13, 711, 1989.

25 Influence of Dietary Fat on the Development of Cancer

Howard Perry Glauert

CONTENTS

I. INTRODUCTION

Cancer is currently the second leading cause of death in the United States. It is estimated that over 1,400,000 people will be diagnosed with cancer in the United States in 2007 and that over 550,000 will die from it [1]. One of the primary mechanisms for reducing cancer deaths may be by altering the diet, and one proposed way is by decreasing the consumption of dietary fat. For example, the American Institute for Cancer Research states to "limit consumption of fatty foods, particularly those of animal origin" [2]. On the other hand, the American Cancer Society no longer specifically recommends lowering fat intake, and instead advises individuals to "consume a healthy diet with an emphasis on plant sources" [1].

In this chapter, the role of dietary fat on the development of human and experimental cancer is discussed. Because of the large number of studies published, reviews are cited where possible.

II. EPIDEMIOLOGICAL STUDIES

Numerous epidemiological studies have examined the effect of dietary fat on human cancer. Several correlational studies have noted an increase in the rates of colon, breast, and other cancers in areas where dietary fat consumption is high [3]. Additionally, studies with immigrant populations have identified dietary fat intake as a causative factor in the development of these cancers [3].

For colon cancer, epidemiological studies have not reached a clear consensus about the influence of dietary fat. Case-control studies overall have not found a positive association with dietary fat, although many have observed a positive association with meat intake [4]. Prospective epidemiological studies have conflicting results: some studies found a positive association [5–7], others saw no effect [8–22], and others saw an actual protective effect of high-fat intakes [23–25]. In several of these studies, the consumption of red meat was found to be significantly correlated with colon cancer risk, but independently of fat intake. The Women's Health Initiative intervention study was recently published [26]. In this study, 19,500 women lowered their fat intake by about 10% compared with 29,000 women who did not alter their diet, for a follow-up period that averaged 8 years. The intervention group had a relative risk of 1.08, which was not statistically significant, indicating that a diet lower in fat did not inhibit the development of colon cancer in this study.

Numerous epidemiological studies have attempted to identify factors which influence breast cancer risk in humans. Established breast cancer risk factors include age of first pregnancy, body build, age at menarche or menopause, and the amount of radiation received by the chest [27]. The effect of dietary fat has been studied in correlational, case-control, and prospective epidemiological studies. Studies examining international correlations between dietary fat intake and breast cancer risk, and migrant studies have reported a positive association between dietary fat intake and breast cancer risk [27]. Case-control studies generally have also supported a connection between total fat intake and breast cancer risk [4]. A combined analysis of 12 case-control studies found a significant positive association between breast cancer risk and saturated fat intake [28]. Most prospective studies, however, did not find any link between dietary fat intake and the development of breast cancer [29–45]. Furthermore, a combined analysis of seven of these prospective studies did not find any evidence of a link between dietary fat intake and breast cancer risk, even for women consuming <20% of their calories as dietary fat [46]. The Women's Health Initiative intervention study (described in the preceding paragraph) examined the effect of low-fat diets on the development of breast cancer [47]. Although dietary fat did not significantly affect the development of breast cancer, there was a relative risk of 0.91 in the low-fat intervention group.

For the pancreas, international comparisons do not show as strong of a trend as with colon or breast cancer [48]. Case-control studies which have examined total or saturated fat intakes do not show a clear trend; however, studies examining the consumption of meat or cholesterol tend to show a positive correlation [48]. In prospective studies, Nothlings et al. [49] and Michaud et al. [50,51] found that dietary fat did not influence the development of pancreatic cancer. However, Stolzenberg-Solomon et al. [52] found that dietary fat enhanced the development of pancreatic cancer. Five prospective studies have examined the relationship of meat consumption with pancreatic cancer; four of the five reported a positive correlation [49,50,53–55].

A number of case-control and prospective studies have examined the role of dietary fat in prostate cancer. Several case-control studies observed a positive correlation between the intake of total and saturated fat and the development of prostate cancer, although others, particularly the more recent studies, did not see an effect [4,56]. Four prospective studies have been performed, with two of these observing a positive association between dietary fat and prostate cancer incidence or mortality and two observing no effect [4,56].

Fewer studies have been conducted for other major forms of human cancer. For endometrial cancer, several but not all case-control studies have noted an association with dietary fat [57–59]. Case-control studies examining dietary fat and bladder cancer showed an association in some but not all studies; a prospective study did not observe a correlation between dietary fat intake and the development of bladder cancer [60,61]. Dietary fat has been found to be a risk factor for ovarian cancer in some epidemiological studies but not in others [62–67]. Lung cancer risk was not found to be significantly affected by dietary fat in two prospective studies, but several case-control studies have observed an association [68–70], although several investigators indicated that their results may have been affected by confounding from smoking. For testicular cancer, case-control studies have observed an association between high-fat diets and increased incidence [71,72]. Using case-control and cohort study designs, Granger et al. [73] found that increased dietary fat consumption protected against the development of skin cancer. Davies et al. [74], however, found that dietary fat did not influence basal cell carcinoma development. The development of esophageal cancer was found to be increased by dietary fat in a case-control study [75].

III. EXPERIMENTAL CARCINOGENESIS STUDIES

A. Skin Carcinogenesis

Mouse skin is one of the oldest and most widely used systems for studying chemical carcinogenesis, including multistage carcinogenesis. Two-stage carcinogenesis (initiation–promotion) was first observed in mouse skin and involves initiation by a subcarcinogenic dose of radiation or of a chemical such as a polycyclic aromatic hydrocarbon (PAH) followed by the long-term administration of croton oil or its active ingredient 12-O-tetradecanoylphorbol-13-acetate (TPA) [76]. More recently, transgenic skin carcinogenesis models have been developed [77,78].

Most studies examining dietary fat have studied complete carcinogenesis by PAH or UV light. Early studies demonstrated that high-fat diets enhanced skin carcinogenesis induced by tar [79] or PAH [80–86]. In studies where skin tumors were induced by UV light, Mathews-Roth and Krinsky [87] found that high-fat diets increased skin carcinogenesis, whereas Black et al. [88] found that high-fat diets did not increase skin carcinogenesis, but that feeding a saturated fat inhibited tumorigenesis.

The effect of fatty acids on the initiation and promotion of skin carcinogenesis has also been studied. Certain fatty acids—oleic acid and lauric acid—were found to have promoting activity when applied daily to mouse skin after a single application of 7,12-dimethylbenz(a)anthracene (DMBA); stearic acid and palmitic acid did not have any effect [89]. When diets varying in their fat content were fed during the promotion stages of DMBA-initiated, TPA-promoted mouse skin carcinogenesis, high-fat diets were found to enhance the promotion of skin carcinogenesis in some studies [90,91] but not in others [92,93]. High-fat diets also partially offset the tumor inhibitory effects of caloric restriction [94]. Locniskar et al. [95] found that substituting menhaden oil for corn oil or coconut oil did not affect skin tumor promotion by TPA. When benzoyl peroxide was used as the promoting agent, mice fed mainly coconut oil had the highest tumor incidence and mice fed corn oil had the lowest tumor incidence, with those fed mainly menhaden oil having intermediate tumor incidence [96]. In a study using mezerein as the promoting agent, high-fat diets did not increase the skin carcinogenesis [97]. High-fat diets were found not to affect or slightly inhibit initiation [90,98], and substituting coconut oil for corn oil did not influence UV-induced skin carcinogenesis [99].

B. Hepatocarcinogenesis

Many early studies of dietary fat and cancer used the liver as the target organ. In these studies, aromatic amines and azo dyes were frequently used to induce hepatocellular carcinomas. In later studies, effects of dietary fat on initiation and promotion in the liver were examined. In initiation–promotion protocols, the administration of a single subcarcinogenic dose of a carcinogen (such as diethylnitrosamine [DEN]) or DMBA along with a proliferative stimulus (such as partial hepatectomy) followed by the long-term feeding of chemicals, such as phenobarbital, 2,3,7,8-tetrachlorodibenzo-p-dioxin, or polyhalogenated biphenyls, leads to a high incidence of hepatocellular adenomas and carcinomas [100,101]. Transgenic mouse models of liver carcinogenesis have also been developed [102]. In addition, foci of putative preneoplastic hepatocytes appear before the development of gross tumors. These foci, known as altered hepatic foci or enzyme-altered foci, contain cells which exhibit qualitatively altered enzyme activities or alterations in one or more cell functions [100]. The enzymes most frequently studied include γ-glutamyl transpeptidase (GGT) and placental glutathione-*S*-transferase (PGST), which are not normally present in adult liver but often present in foci; and ATPase and glucose-6-phosphatase, which are normally present but frequently missing from foci [103,104]. Altered hepatic foci can also be identified on hematoxylin- and eosin-stained tissue [105,106]. The appearance of foci has been correlated with the later development of malignant neoplasms [107,108].

The first studies examined the effect of dietary fat on the induction of hepatocellular carcinomas by complete hepatocarcinogens. In the liver, increasing the fat content of the diet enhances the development of 2-acetylaminofluorene (AAF)-, *p*-dimethylaminoazobenzene (DAB)-, and aflatoxin B_1 (AFB)-induced tumors and GGT-positive foci in rats [109–113]. Furthermore, hepatocarcinogenesis by DAB is enhanced by feeding a diet which contains a greater proportion of polyunsaturated fatty acids [114,115]. In these studies, however, the diets were administered at the same time as the carcinogen injections, so that the stage of carcinogenesis which was affected could not be determined.

More recent studies have examined whether this enhancement of hepatocarcinogenesis is caused by an effect on the initiation of carcinogenesis, the promotion of carcinogenesis, or both. Misslbeck et al. [116] found that increasing the corn oil content of the diet after the administration of 10 doses of aflatoxin increased the number and size of GGT-positive foci, but Baldwin and Parker [117], using a similar protocol, found no effect of dietary corn oil. Glauert and Pitot [118] similarly found that increasing the safflower oil or palm oil content of the diet did not promote DEN-induced GGT-positive foci or greatly affect phenobarbital promotion of GGT-positive foci. The promotion of GGT-positive foci by dietary tryptophan is also not affected by dietary fat [119]. Newberne et al. [120] found that increasing dietary corn oil (but not beef fat) during and after the administration of AFB increased the incidence of hepatic tumors, but not when the diets were fed only after AFB administration. Baldwin and Parker [117] also found that increasing the corn oil content of the diet before and during AFB administration increased the number and volume of GGT-positive foci. When rats are fed diets high in polyunsaturated fatty acids (but not in saturated fatty acids) before receiving the hepatocarcinogen DEN, they develop more GGT-positive and ATPase-negative foci than rats fed low-fat diets [121]. Finally, the feeding of diets high in corn oil but not lard enhanced the initiation of PGST-positive foci induced by azoxymethane (AOM) [122]. The results of these studies suggest that the enhancement of hepatocarcinogenesis by dietary fat is primarily due to an effect on initiation, and that polyunsaturated fats have a greater effect than do saturated fats.

C. COLON CARCINOGENESIS

Studies in experimental animals have produced differing results. A variety of chemicals have been used to induce colon tumors, usually in rats or mice. These include 1,2-dimethylhydrazine (DMH) and its metabolites AOM and methylazoxymethanol (MAM); 3,2′-dimethyl-4-aminobiphenyl (DMAB); methylnitrosourea (MNU); and *N*-methyl-*N*′-nitro-*N*-nitrosoguanidine (MNNG) [123–126]. DMH and AOM have been used most frequently to study nutritional effects. Both can induce colon tumors by single [127–131] or multiple [132–136] injections. The Min mouse, which has a mutation in the mouse homolog of the adenomatous polyposis coli (APC) gene, develops colon tumors spontaneously and is used as a model of colon carcinogenesis [137]. In addition to tumors, putative preneoplastic lesions, aberrant crypt foci (ACF), are induced by colon carcinogens [138]. ACF, which are identified by fixing the colon in formalin and then staining with methylene blue, are stained darker and are larger than normal crypts [138]. Some but not all studies have shown that ACF correlate well with the later appearance of adenocarcinomas [139–142].

Animal studies examining the effect of dietary fat have used a variety of protocols, and the results obtained often have been dependent on the investigator's protocol. In these studies, rats or mice were subjected to multiple doses of a colon carcinogen, with the dietary fat content being varied (isocalorically) during, and frequently before or after, the carcinogen injections. Some of these studies found an enhancement when the dietary fat content of the diet was increased, but others saw no effect or even an inhibition of tumor development [134,143–151]. High-fat diets were found to influence the early stages of carcinogenesis more than the later stages [152]. In several studies where fat was found to enhance colon carcinogenesis, fat was either added to an unrefined (chow) diet or was substituted for carbohydrate on a weight basis, so that the ratio of calories to essential nutrients was altered; therefore, the effect could have been due to a lower consumption of essential nutrients rather than to an effect of fat [132,133,153–159]. In the Min model, high-fat diets were found to increase colon carcinogenesis in one study but not another [160,161]. Increasing the fat content of the diet has been found to increase the number of ACF induced by colon carcinogens in several but not all studies [151,162–172]. The type of fat (unsaturated vs. saturated) in the diet also produced conflicting results [149,168,173]. ω-3 Fatty acids can also influence colon carcinogenesis: feeding fish or flaxseed oil in place of corn oil, or eicosapentanoic acid in place of linoleic acid, decreases the development of DMH- or AOM-induced colon tumors, but adding menhaden oil to a low-fat diet does not affect colon carcinogenesis [151,174–181]. Olive oil, high in ω-9 fatty acids, was also found to inhibit colon carcinogenesis when substituted for polyunsaturated fatty acids [179].

D. Pancreatic Carcinogenesis

Dietary fat has been studied extensively in animal models. A common model is induction of pancreatic tumors by azaserine; however, azaserine produces tumors in acinar cells [182], whereas the primary site in humans is the ductal cell. Tumors can be produced in pancreatic ductal cells in hamsters, by the chemicals N-nitroso-bis-(2-oxypropyl)amine (BOP) and N-nitroso-bis-(2-hydroxypropyl) amine (BHP) [182]. A number of transgenic models have been developed [182,183]. A number of models used regulatory elements from the rat elastase gene, which targets acinar cells. These constructs produced acinar tumors or mixed acinar/ductal tumors [182,183]. Recently, a new model has been developed, which uses an oncogenic K-ras (KRAS[G12D]) inserted into the endogenous K-ras locus [184]. The gene has a Lox-STOP-Lox (LSL) construct-inserted upstream. These mice are interbred with mice containing the Cre recombinase downstream from a pancreatic-specific promoter, either PDX-1 or P48. The PDX-1-Cre;LSL-KRAS[G12D] mice develop pancreatic intraepithelial neoplasia (PanINs), which progress over time [184]. In addition, when these mice are crossed to mice containing p53 mutations or Ink4a/Arf deficiency, the rapid development of pancreatic adenocarcinomas is observed [185,186].

Dietary fat has been found to influence tumorigenesis in both rats and hamsters. In rats, feeding high-fat diets after, or during and after, the injection of azaserine enhances the development of pancreatic tumors and putative preneoplastic lesions [187–195]. Pancreatic carcinogenesis induced by N-nitroso(2-hydroxypropyl)(2-oxopropyl)amine in rats is also enhanced by feeding high-fat diets [196]. In several of these studies, the effect cannot be attributed unequivocally to dietary fat because fat was substituted for carbohydrate on a weight basis. In hamsters, BOP-induced pancreatic carcinogenesis is also increased by feeding high-fat diets [193–195,197–201]. Roebuck and colleagues [187,189,192] found that polyunsaturated fat, but not saturated fat, enhanced pancreatic carcinogenesis, and that a certain level of essential fatty acids are required for the enhancement of pancreatic carcinogenesis. Increased linoleic acid was also found to increase metastases to the liver in hamsters [202]. Appel et al. [203], however, found that increasing the linoleic acid content of the diet did not increase pancreatic carcinogenesis in either rats or hamsters. Birt et al. [200] found that feeding a saturated fat (beef tallow) enhanced pancreatic carcinogenesis in hamsters greater than a polyunsaturated fat (corn oil). Studies using fish oil have produced differing results, depending on the experimental protocol. Substituting fish oil for oils high in polyunsaturated fats decreases [204,205] or does not affect [206] the development of azaserine-induced preneoplastic lesions in rats. Adding fish oil to a diet containing adequate polyunsaturated fatty acids enhances azaserine-induced carcinogenesis in rats and BOP-induced carcinogenesis in hamsters [207–209].

However, Heukamp et al. [210] found that increasing dietary n-3 fatty acids inhibited the incidence but not the number of liver metastases in BOP-treated hamsters compared with hamsters fed a low-fat diet or a diet enriched in n-3, n-6, and n-9 fatty acids; the incidence of pancreatic adenocarcinomas did not differ among the diets.

Finally, it has been observed in 2 year carcinogenesis studies in which corn oil gavage has been used as the vehicle for the carcinogen that a higher incidence of pancreatic acinar cell adenomas is present in corn oil gavage-treated male Fischer-344 control rats than in untreated controls [211,212]. This association was not observed in female rats or in male or female B6C3F$_1$ mice.

E. Mammary Carcinogenesis

The effect of dietary fat on mammary carcinogenesis in experimental animals has been examined extensively: over 100 experiments have been conducted [213–215]. The primary model used is a rat model (usually the Sprague–Dawley strain) in which mammary tumors are induced by DMBA or MNU. Genetically engineered models have also been developed, in which the Erbb2 or simian virus 40 (SV40) T/t-antigens are overexpressed in mammary epithelial cells [216]. The use of these models is advantageous because tumor latency, tumor size, and tumor progression can easily be quantified by palpation of mammary tumors as they appear. Increasing the fat content of the diet clearly enhances the development of mammary tumors [213–215]. In the rat model, a high-fat diet increases tumorigenesis both when it is fed during and after carcinogen administration, and when it is fed only after carcinogen injection. Feeding a diet high in fish oil instead of a diet high in polyunsaturated fat decreases the incidence of DMBA-induced tumors in rats [213–215]. A meta-analysis of experimental animal studies found that n-6 fatty acids strongly enhanced carcinogenesis; saturated fatty acids were weaker at enhancing carcinogenesis; monounsaturated fatty acids had no effect; and n-3 fatty acids weakly (but nonsignificantly) inhibited carcinogenesis [215].

F. Other Sites

Dietary fat has also been studied for its effect on experimental carcinogenesis in other organs. In the lung, dietary fat enhanced benzo(a)pyrene (BP)- or BOP-induced carcinogenesis in hamsters [198,217], whereas in mice a high-fat diet did not affect spontaneous carcinogenesis in one study but the feeding of egg extracts enhanced it in another [85,218]. In the prostate, several studies have found that high-fat diets enhance the growth of transplantable prostate tumors, but that inconsistent effects are seen in chemically induced prostate carcinogenesis models [219–222].

IV. MECHANISMS BY WHICH DIETARY FAT MAY INFLUENCE CARCINOGENESIS

A. Membrane Fluidity

An important function of dietary fatty acids is their presence in membrane lipids. Altering the fatty acid content of the diet alters the composition of membrane lipids, particularly in certain tissues; feeding diets high in n-6 or n-3 fatty acids increases the concentrations of these fatty acids in membrane lipids [223]. The activities of membrane-bound enzymes are increased in membranes that are more fluid, that is, that have a higher content of polyunsaturated fatty acids [223]. The alteration by dietary fatty acids of the catalytic abilities of membrane-bound enzymes, such as cytochrome P-450, may play an important role in carcinogenesis.

B. Toxicity

One possible mechanism by which dietary fat may enhance carcinogenesis is by the toxicity of fatty acids or of metabolites that increase after the feeding of high-fat diets. Such toxicity would bring about a proliferative response in the tissue to replace lost cells. Cellular genes involved in cell proliferation, including cellular oncogenes, would likely be increased.

In the colon, toxicity may play a role in the enhancement of carcinogenesis by dietary fat. One hypothesis for the effect of dietary fat is that dietary fat increases the concentration of metabolites with carcinogenic or promoting activity in the fecal stream. Bile acids, particularly secondary bile acids, have promoting activity in the colon [224–226]; their concentration in the feces has been found to be increased by dietary fat in some but not all studies [132,227–230]. Bile acids function as detergents; therefore, high concentrations may be toxic to epithelial cells in the colon. Several studies have shown bile acids to induce apoptosis [231]. This may result in a compensatory increase in cell proliferation; most studies have found that increasing the concentration of bile acids in vivo or in vitro increases colon epithelial cell proliferation [232–242]. In addition, several studies found that bile acids induced DNA damage [231].

C. Eicosanoid Metabolism

Another mechanism by which dietary fat may influence carcinogenesis is by altering the synthesis of eicosanoids. Fatty acids that are consumed in the diet can be metabolized to a variety of other compounds, including longer and more unsaturated fatty acids, prostaglandins, leukotrienes, thromboxanes, hydroperoxyeicosatetranoic acids, and hydroxyeicosatetranoic acids [243]. Altering the type of fatty acid in the diet has been found to change the amounts and composition of the eicosanoids that are produced by the body [244,245]. *n*-3 Fatty acids antagonize the metabolism of arachidonic acid to eicosanoids, which may be a mechanism in their inhibition of carcinogenesis [244,245]. Specific eicosanoids bind to receptors and cause specific alterations in gene expression and cellular function [246–249]; some of which may be related to carcinogenesis. It has been found that inhibition of eicosanoid synthesis inhibits tumor promotion in several tissues [250–253].

D. Caloric Effect

The issue of whether the enhancing effect of fat in carcinogenesis is due to higher consumption of calories or more efficient utilization of energy has been examined in several tissues. The earliest study was conducted by Boutwell et al. [86] using the mouse skin carcinogenesis system; they attributed most of the enhancing effect of dietary fat to an increased consumption of calories. Birt et al. [90], however, found that the promotion of skin carcinogenesis was enhanced even though the high-fat diets were pair-fed. The greater caloric density of fat has also been proposed to play a role in colon tumorigenesis. Caloric restriction inhibits chemically induced colon carcinogenesis, even if the percentage of dietary fat in the diet is greatly increased [254,255]. In the pancreas, the enhancement by dietary fat appears to be an effect of dietary fat rather than of an increased consumption of calories, as pair-feeding does not inhibit the enhancing effect of dietary fat in hamsters [199]. Several studies have suggested that the enhancement of mammary carcinogenesis by dietary fat may be caused, at least in part, by an alteration in the efficiency of energy utilization [256–258]. Using a combined statistical analysis of over 100 animal experiments, Fay et al. [215] found that energy intake was not responsible for the enhancing effect of polyunsaturated fats on mammary carcinogenesis, although there was a slight (but not significant) effect. Finally, caloric restriction has been found to inhibit tumorigenesis in many tissues in experimental animals [259].

E. Effect on Tumor Initiation

Dietary fat may also affect the initiation stage of carcinogenesis. Since initiation involves the mutation of DNA, its alteration (by dietary fat or other agents) would mainly affect the structure of genes rather than their expression. In rat liver, dietary fat appears to enhance carcinogenesis primarily by an effect on initiation. In other tissues, many protocols have varied the levels of dietary fat during the time of carcinogen injections; therefore, dietary fat may be affecting some aspect of initiation in these studies. Higher levels of dietary fat may enhance initiation of carcinogenesis by several mechanisms, including alterations in absorption of the carcinogen from the gut, transport to the target organ, uptake by the target organ, metabolism by cytochrome P-450 or other drug-metabolizing enzymes to a form which can react with DNA, and DNA repair. Several of these processes occur in membranes, whose lipid composition can be altered by changing the amount or type of dietary fat [260,261]. Increasing the fat content of the diet increases cytochrome P-450 and related activities [262–265]. Western analysis has indicated that higher amounts of enzyme protein are present after feeding diets high in polyunsaturated fat [266]. Therefore, dietary fat may be affecting both gene expression and the surrounding matrix necessary for optimum enzyme activity. The metabolism of several chemicals, including hexobarbital, aniline, ethylmorphine, benzo(a) pyrene, and dimethylnitrosamine, is also enhanced by feeding diets high in polyunsaturated fatty acids [262,263,267–269].

F. Lipid Peroxidation

Another way in which dietary fat could affect carcinogenesis is through lipid peroxidation. Polyunsaturated fatty acids are susceptible to lipid peroxidation; therefore, diets high in polyunsaturated fat could result in increased consumption of oxidized lipids present in the diet or increased lipid peroxidation in the body. Lipid peroxidation could affect carcinogenesis in a number of ways. Several products of lipid peroxidation are very toxic [270] and could influence the carcinogenic process through toxicity as described earlier. Lipid peroxidation products have the potential to exert genotoxicity and therefore could bring about tumor initiation. One of the products of lipid peroxidation is malondialdehyde, which forms DNA adducts and is mutagenic [271]. Another potential product is the hydroxyl radical, which can form DNA adducts such as 8-hydroxyguanosine [272]. Finally, oxidation products could act on signal transduction pathways leading to altered cell proliferation or apoptosis [273].

G. Alteration of Specific Gene Expression

Altering the level of dietary fat changes the expression of many genes. Most of the genes studied, however, are related to carbohydrate or lipid metabolism and are not likely to play a role in carcinogenesis [274]. Dietary fat may alter signal transduction pathways that lead to altered cell proliferation or apoptosis. Fatty acids have been found to bind to several transcription factors, including the peroxisome proliferator-activated receptors (PPARs), PPARα, PPARβ, and PPARγ; the liver X receptors (LXRs), LXRα and LXRβ; and hepatocyte nuclear factor-4 (HNF-4) [275]. PPARα activators induce hepatic tumors, but only in specific rodent species [276]; PPARγ, however, has antineoplastic properties [277]. Several studies have examined effects on protein kinase C (PKC), which consists of at least 12 subtypes [275]. Increasing the fat content of the diet has been found to increase PKC activity in the colon, skin, and mammary gland [162,278–282].

Several papers have been published on the effect of dietary fat on the expression of oncogenes and tumor suppressor genes. In the colon, Guillem et al. [143] found that dietary fat did not affect the expression of either c-*myc* or c-H-*ras* oncogenes in normal or tumor tissue, whereas Singh et al. [283] found that dietary corn oil increased AOM-induced expression of *ras*. In contrast, fish oil was found to inhibit *ras* expression. In the mammary gland, DeWille et al. [284] found that high corn oil diets increased *ras* mRNA levels in mammary tumors from MMTV/v-Ha-*ras* transgenic mice. However, high-fat diets decreased the frequency of *ras* mutations in rat mammary gland tumors induced by the heterocyclic amine 2-amino-1-methyl-6-phenylimidazo[4,5-*b*]pyridine (PhIP) [285]. Substituting *n*-3 fatty acid-rich oils for corn oil decreased the expression of the H-*ras* oncogene in the mammary gland in one study but had no effect in the other [286,287].

V. SUMMARY AND CONCLUSIONS

Clearly, there is much variability in studies of dietary fat and cancer, both in epidemiological and experimental studies. In epidemiological studies, a relationship between dietary fat and breast cancer has been found in correlational and case-control studies, but prospective studies do not support a role for dietary fat. Prospective epidemiological studies examining the role of dietary fat in the development of colon, pancreatic, and prostate cancers have produced conflicting results. The Women's Health Initiative intervention studies did not show any significant effects for dietary fat in the development of either colon or breast cancer in women. In experimental studies, dietary fat generally enhances chemically induced skin, liver, pancreatic, and mammary carcinogenesis, whereas conflicting results have been seen in colon carcinogenesis. Dietary fat appears to act primarily during the promotional stage of carcinogenesis in all of these models except the liver, where the effect of dietary fat is primarily on initiation. Because of the variability seen in studies of dietary fat and cancer (particularly prospective epidemiological studies), recommendations for preventing human cancer should not include decreasing the fat content of the diet.

The mechanisms by which high-fat diets enhance experimental carcinogenesis are unclear, but probably involve several mechanisms, some of which are organ-specific. Nearly all of the mechanisms by which dietary fat may influence carcinogenesis involve alterations in genetic expression. The determination of genes which are turned on or off by dietary fat or its metabolic or oxidative products will likely provide answers as to the role of dietary fat in carcinogenesis.

REFERENCES

1. American Cancer Society, *Cancer Facts and Figures*, American Cancer Society, Atlanta, 2007.
2. World Cancer Research Fund/American Institute for Cancer Research, *Food, Nutrition and the Prevention of Cancer: A Global Perspective*, American Institute for Cancer Research, Washington, D.C., 1997.
3. Fraser, D., Nutrition and cancer: epidemiological aspects, *Public Health Rev*, 24, 113, 1996.
4. Kushi, L. and Giovannucci, E., Dietary fat and cancer, *Am J Med*, 113(Suppl 9B), 63S, 2002.
5. Willett, W.C. et al., Relation of meat, fat, and fiber intake to the risk of colon cancer in a prospective study among women, *N Engl J Med*, 323, 1664, 1990.
6. Giovannucci, E. et al., Relationship of diet to risk of colorectal adenoma in men, *J Natl Cancer Inst*, 84(2), 91, 1992.
7. Singh, P.N. and Fraser, G.E., Dietary risk factors for colon cancer in a low-risk population, *Am J Epidemiol*, 148(8), 761, 1998.
8. Garland, C. et al., Dietary vitamin D and calcium and risk of colorectal cancer: a 19-year prospective study in men, *Lancet*, 1, 307, 1985.
9. Phillips, R.L. and Snowdon, D.A., Dietary relationships with fatal colorectal cancer among Seventh-Day Adventists, *J Natl Cancer Inst*, 74, 307, 1985.
10. Thun, M.J. et al., Risk factors for fatal colon cancer in a large prospective study, *J Natl Cancer Inst*, 84(19), 1491, 1992.
11. Goldbohm, R.A. et al., A prospective cohort study on the relation between meat consumption and the risk of colon cancer, *Cancer Res*, 54(3), 718, 1994.
12. Giovannucci, E. et al., Intake of fat, meat, and fiber in relation to risk of colon cancer in men, *Cancer Res*, 54(9), 2390, 1994.
13. Bostick, R.M. et al., Sugar, meat, and fat intake, and non-dietary risk factors for colon cancer incidence in Iowa women (United States), *Cancer Causes Control*, 5, 38, 1994.
14. Gaard, M., Tretli, S., and Loken, E.B., Dietary factors and risk of colon cancer: a prospective study of 50,535 young Norwegian men and women, *Eur J Cancer Prev*, 5, 445, 1996.
15. Kato, I. et al., Prospective study of diet and female colorectal cancer: the New York University Women's Health Study, *Nutr Cancer*, 28(3), 276, 1997.
16. Pietinen, P. et al., Diet and risk of colorectal cancer in a cohort of Finnish men, *Cancer Causes Control*, 10(5), 387, 1999.

17. Jarvinen, R. et al., Prospective study on milk products, calcium and cancers of the colon and rectum, *Eur J Clin Nutr*, 55(11), 1000, 2001.

18. Terry, P. et al., Prospective study of major dietary patterns and colorectal cancer risk in women, *Am J Epidemiol*, 154(12), 1143, 2001.

19. Terry, P. et al., No association between fat and fatty acids intake and risk of colorectal cancer, *Cancer Epidemiol Biomarkers Prev*, 10(8), 913, 2001.

20. Flood, A. et al., Meat, fat, and their subtypes as risk factors for colorectal cancer in a prospective cohort of women, *Am J Epidemiol*, 158(1), 59, 2003.

21. Robertson, D.J. et al., Fat, fiber, meat and the risk of colorectal adenomas, *Am J Gastroenterol*, 100(12), 2789, 2005.

22. Oba, S. et al., The relationship between the consumption of meat, fat, and coffee and the risk of colon cancer: a prospective study in Japan, *Cancer Lett*, 244(2), 260, 2006.

23. Hirayama, T., A large-scale cohort study on the relationship between diet and selected cancers of digestive organs, in *Gastrointestinal Cancer: Endogenous Factors (Banbury Report 7)*, Bruce, W.R., Correa, P., Lipkin, M., Tannenbaum, S.R., and Wilkins, T.D., Eds., Cold Spring Harbor Laboratory, Cold Spring Harbor, NY, 1981, p. 409.

24. Stemmermann, G.N., Monura, A.M.Y., and Heilbrun, L.K., Dietary fat and the risk of colorectal cancer, *Cancer Res*, 44, 4633, 1984.

25. Chyou, P.H., Nomura, A.M., and Stemmermann, G.N., A prospective study of colon and rectal cancer among Hawaii Japanese men, *Ann Epidemiol*, 6, 276, 1996.

26. Beresford, S.A. et al., Low-fat dietary pattern and risk of colorectal cancer: the Women's Health Initiative Randomized Controlled Dietary Modification Trial, *JAMA*, 295(6), 643, 2006.

27. Kelsey, J.L. and Berkowitz, G.S., Breast cancer epidemiology, *Cancer Res*, 48, 5615, 1988.

28. Howe, G.R. et al., Dietary factors and risk of breast cancer: combined analysis of 12 case-control studies, *J Natl Cancer Inst*, 82(7), 561, 1990.

29. Jones, D.Y. et al., Dietary fat and breast cancer in the National Health and Nutrition Examination Survey I. Epidemiological followup study, *J Natl Cancer Inst*, 79, 465, 1987.

30. Willett, W.C. et al., Dietary fat and the risk of breast cancer, *N Engl J Med*, 316, 22, 1987.

31. Mills, P.K. et al., Dietary habits and breast cancer incidence among Seventh-day Adventists, *Cancer*, 64, 582, 1989.

32. Knekt, P. et al., Dietary fat and risk of breast cancer, *Am J Clin Nutr*, 52, 903, 1990.

33. Howe, G.R. et al., A cohort study of fat intake and risk of breast cancer, *J Natl Cancer Inst*, 83, 336, 1991.

34. Graham, S. et al., Diet in the epidemiology of postmenopausal breast cancer in the New York State cohort, *Am J Epidemiol*, 136(11), 1327, 1992.

35. Kushi, L.H. et al., Dietary fat and postmenopausal breast cancer, *J Natl Cancer Inst*, 84, 1092, 1992.

36. Willett, W.C. et al., Dietary fat and fiber in relation to risk of breast cancer—an 8-year follow-up, *JAMA*, 268(15), 2037, 1992.

37. van den Brandt, P.A. et al., A prospective cohort study on dietary fat and the risk of postmenopausal breast cancer, *Cancer Res*, 53, 75, 1993.

38. Toniolo, P. et al., Consumption of meat, animal products, protein, and fat and risk of breast cancer: a prospective cohort study in New York, *Epidemiology*, 5(4), 391, 1994.

39. Gaard, M., Tretli, S., and Loken, E.B., Dietary fat and the risk of breast cancer: a prospective study of 25,892 Norweigian women, *Int J Cancer*, 63, 13, 1995.

40. Holmes, M.D. et al., Association of dietary intake of fat and fatty acids with risk of breast cancer, *JAMA*, 281(10), 914, 1999.

41. Velie, E. et al., Dietary fat, fat subtypes, and breast cancer in postmenopausal women: a prospective cohort study, *J Natl Cancer Inst*, 92(10), 833, 2000.

42. Thiebaut, A.C. and Clavel-Chapelon, F., Fat consumption and breast cancer: preliminary results from the E3N-Epic cohort, *Bull Cancer*, 88(10), 954, 2001.

43. Terry, P. et al., A prospective study of major dietary patterns and the risk of breast cancer, *Cancer Epidemiol Biomarkers Prev*, 10(12), 1281, 2001.

44. Byrne, C., Rockett, H., and Holmes, M.D., Dietary fat, fat subtypes, and breast cancer risk: lack of an association among postmenopausal women with no history of benign breast disease, *Cancer Epidemiol Biomarkers Prev*, 11(3), 261, 2002.

45. Kim, E.H. et al., Dietary fat and risk of postmenopausal breast cancer in a 20-year follow-up, *Am J Epidemiol*, 164(10), 990, 2006.
46. Hunter, D.J. and Willett, W.C., Nutrition and breast cancer, *Cancer Causes Control*, 7, 56, 1996.
47. Prentice, R.L. et al., Low-fat dietary pattern and risk of invasive breast cancer: the Women's Health Initiative Randomized Controlled Dietary Modification Trial, *JAMA*, 295(6), 629, 2006.
48. Howe, G.R. and Burch, J.D., Nutrition and pancreatic cancer, *Cancer Causes Control*, 7, 69, 1996.
49. Nothlings, U. et al., Meat and fat intake as risk factors for pancreatic cancer: the multiethnic cohort study, *J Natl Cancer Inst*, 97(19), 1458, 2005.
50. Michaud, D.S. et al., Dietary meat, dairy products, fat, and cholesterol and pancreatic cancer risk in a prospective study, *Am J Epidemiol*, 157(12), 1115, 2003.
51. Michaud, D.S. et al., Dietary patterns and pancreatic cancer risk in men and women, *J Natl Cancer Inst*, 97(7), 518, 2005.
52. Stolzenberg-Solomon, R.Z. et al., Prospective study of diet and pancreatic cancer in male smokers, *Am J Epidemiol*, 155(9), 783, 2002.
53. Mills, P.K. et al., Dietary habits and past medical history as related to fatal pancreas cancer risk among adventists, *Cancer*, 61, 2578, 1988.
54. Hirayama, T., Epidemiology of pancreatic cancer in Japan, *Jpn J Clin Oncol*, 19, 208, 1989.
55. Zheng, W. et al., A cohort study of smoking, alcohol consumption, and dietary factors for pancreatic cancer (United States), *Cancer Causes Control*, 4, 477, 1993.
56. Wu, K. et al., Dietary patterns and risk of prostate cancer in U.S. men, *Cancer Epidemiol Biomarkers Prev*, 15(1), 167, 2006.
57. Hill, H.A. and Austin, H., Nutrition and endometrial cancer, *Cancer Causes Control*, 7, 19, 1996.
58. McCann, S.E. et al., Diet in the epidemiology of endometrial cancer in western New York (United States), *Cancer Causes Control*, 11(10), 965, 2000.
59. Salazar-Martinez, E. et al., Dietary factors and endometrial cancer risk. Results of a case-control study in Mexico, *Int J Gynecol Cancer*, 15(5), 938, 2005.
60. La Vecchia, C. and Negri, E., Nutrition and bladder cancer, *Cancer Causes Control*, 7, 95, 1996.
61. Radosavljevic, V. et al., Diet and bladder cancer: a case-control study, *Int Urol Nephrol*, 37(2), 283, 2005.
62. Kushi, L.H. et al., Prospective study of diet and ovarian cancer, *Am J Epidemiol*, 149(1), 21, 1999.
63. Pan, S.Y. et al., A case-control study of diet and the risk of ovarian cancer, *Cancer Epidemiol Biomarkers Prev*, 13(9), 1521, 2004.
64. Zhang, M. et al., Diet and ovarian cancer risk: a case-control study in China, *Br J Cancer*, 86(5), 712, 2002.
65. Bertone, E.R. et al., Dietary fat intake and ovarian cancer in a cohort of US women, *Am J Epidemiol*, 156(1), 22, 2002.
66. Risch, H.A. et al., Dietary fat intake and risk of epithelial ovarian cancer, *J Natl Cancer Inst*, 86(18), 1409, 1994.
67. Genkinger, J.M. et al., A pooled analysis of 12 cohort studies of dietary fat, cholesterol and egg intake and ovarian cancer, *Cancer Causes Control*, 17(3), 273, 2006.
68. Ziegler, R.G., Mayne, S.T., and Swanson, C.A., Nutrition and lung cancer, *Cancer Causes Control*, 7, 157, 1996.
69. Alavanja, M.C. et al., Lung cancer risk and red meat consumption among Iowa women, *Lung Cancer*, 34(1), 37, 2001.
70. Mohr, D.L. et al., Southern cooking and lung cancer, *Nutr Cancer*, 35(1), 34, 1999.
71. Bonner, M.R., McCann, S.E., and Moysich, K.B., Dietary factors and the risk of testicular cancer, *Nutr Cancer*, 44(1), 35, 2002.
72. Sigurdson, A.J. et al., A case-control study of diet and testicular carcinoma, *Nutr Cancer*, 34(1), 20, 1999.
73. Granger, R.H. et al., Association between dietary fat and skin cancer in an Australian population using case-control and cohort study designs, *BMC Cancer*, 6, 141, 2006.
74. Davies, T.W. et al., Diet and basal cell skin cancer: results from the EPIC-Norfolk cohort, *Br J Dermatol*, 146(6), 1017, 2002.
75. Wolfgarten, E. et al., Coincidence of nutritional habits and esophageal cancer in Germany, *Onkologie*, 24(6), 546, 2001.

76. Berenblum, I. and Shubik, P., The role of croton oil applications, associated with a single painting of a carcinogen, in tumour induction of the mouse's skin, *Br J Cancer*, 1, 379, 1947.

77. Greenhalgh, D.A., Wang, X.J., and Roop, D.R., Multistage epidermal carcinogenesis in transgenic mice: cooperativity and paradox, *J Investig Dermatol Symp Proc*, 1(2), 162, 1996.

78. Humble, M.C. et al., Biological, cellular, and molecular characteristics of an inducible transgenic skin tumor model: a review, *Oncogene*, 24(56), 8217, 2005.

79. Watson, A.F. and Mellanby, E., Tar cancer in mice. II. The condition of the skin when modified by external treatment or diet, as a factor in influencing the cancerous reaction, *Br J Exp Pathol*, 11, 311, 1930.

80. Lavik, P.S. and Baumann, C.A., Further studies on the tumor-promoting action of fat, *Cancer Res*, 3, 749, 1943.

81. Lavik, P.S. and Baumann, C.A., Dietary fat and tumor formation, *Cancer Res*, 1, 181, 1941.

82. Jacobi, H.P. and Baumann, C.A., The effect of fat on tumor formation, *Am J Cancer*, 39, 338, 1940.

83. Baumann, C.A., Jacobi, H.P., and Rusch, H.P., The effect of diet on experimental tumor production, *Am J Hyg*, 30A, 1, 1939.

84. Tannenbaum, A., The dependence of the genesis of induced skin tumors on the fat content of the diet during different stages of carcinogenesis, *Cancer Res*, 4, 683, 1944.

85. Tannenbaum, A., The genesis and growth of tumors. III. Effects of a high fat diet, *Cancer Res*, 2, 468, 1942.

86. Boutwell, R.K., Brush, M.K., and Rusch, H.P., The stimulating effect of dietary fat on carcinogenesis, *Cancer Res*, 9, 741, 1949.

87. Mathews-Roth, M.M. and Krinsky, N.I., Effect of dietary fat level on UV-B induced skin tumors, and anti-tumor action of beta-carotene, *Photochem Photobiol*, 40(5), 671, 1984.

88. Black, H.S. et al., Influence of dietary lipid upon ultraviolet-light carcinogenesis, *Nutr Cancer*, 5(2), 59, 1983.

89. Holsti, P., Tumor promoting effects of some long chain fatty acids in experimental skin carcinogenesis in the mouse, *Acta Pathol Microbiol Scand*, 46, 51, 1959.

90. Birt, D.F. et al., Dietary fat effects on the initiation and promotion of two-stage skin tumorigenesis in the SENCAR mouse, *Cancer Res*, 49, 4170, 1989.

91. Birt, D.F. et al., Acceleration of papilloma growth in mice fed high-fat diets during promotion of two-stage skin carcinogenesis, *Nutr Cancer*, 12(2), 161, 1989.

92. Locniskar, M. et al., The effect of the level of dietary corn oil on mouse skin carcinogenesis, *Nutr Cancer*, 16(1), 1, 1991.

93. Lo, H.H. et al., Effects of type and amount of dietary fat on mouse skin tumor promotion, *Nutr Cancer*, 22(1), 43, 1994.

94. Birt, D.F. et al., High-fat diet blocks the inhibition of skin carcinogenesis and reductions in protein kinase C by moderate energy restriction, *Mol Carcinog*, 16(2), 115, 1996.

95. Locniskar, M. et al., Lack of a protective effect of menhaden oil on skin tumor promotion by 12-*O*-tetradecanoylphorbol-13-acetate, *Carcinogenesis*, 11(9), 1641, 1990.

96. Locniskar, M. et al., The effect of dietary lipid on skin tumor promotion by benzoyl peroxide: comparison of fish, coconut and corn oil, *Carcinogenesis*, 12(6), 1023, 1991.

97. Birt, D.F. et al., Consumption of reduced-energy/low-fat diet or constant-energy/high-fat diet during mezerein treatment inhibited mouse skin tumor promotion, *Carcinogenesis*, 15(10), 2341, 1994.

98. Locniskar, M. et al., The effect of various dietary fats on skin tumor initiation, *Nutr Cancer*, 16(3–4), 189, 1991.

99. Berton, T.R. et al., Comparison of ultraviolet light-induced skin carcinogenesis and ornithine decarboxylase activity in sencar and hairless SKH-1 mice fed a constant level of dietary lipid varying in corn and coconut oil, *Nutr Cancer*, 26(3), 353, 1996.

100. Pitot, H.C. and Dragan, Y.P., Chemical induction of hepatic neoplasia, in *The Liver: Biology and Pathobiology*, Third Edition, Arias, I.M., Boyer, J.L., Fausto, N., Jakoby, W.B., Schachter, D.A., and Shafritz, D.A., Eds., Raven Press, New York, 1994, p. 1467.

101. Glauert, H.P., Robertson, L.W., and Silberhorn, E.M., PCBs and tumor promotion, in *PCBs: Recent Advances in Environmental Toxicology and Health Effects*, Robertson, L.W. and Hansen, L.G., Eds., University Press of Kentucky, Lexington, KY, 2001, p. 355.

102. Calvisi, D.F. and Thorgeirsson, S.S., Molecular mechanisms of hepatocarcinogenesis in transgenic mouse models of liver cancer, *Toxicol Pathol*, 33(1), 181, 2005.

103. Pitot, H.C., Glauert, H.P., and Hanigan, M., The significance of biochemical markers in the characterization of putative initiated cell populations in rodent liver, *Cancer Lett*, 29, 1, 1985.

104. Hendrich, S., Campbell, H.A., and Pitot, H.C., Quantitative stereological evaluation of four histochemical markers of altered foci in multistage hepatocarcinogenesis in the rat, *Carcinogenesis*, 8(9), 1245, 1987.

105. Bannasch, P. et al., Significance of sequential cellular changes inside and outside foci of altered hepatocytes during hepatocarcinogenesis, *Toxicol Pathol*, 17(4), 617, 1989.

106. Harada, T. et al., Morphological and stereological characterization of hepatic foci of cellular alteration in control Fischer 344 rats, *Toxicol Pathol*, 17(4), 579, 1989.

107. Emmelot, P. and Scherer, E., The first relevant cell stage in rat liver carcinogenesis. A quantitative approach, *Biochim Biophys Acta*, 605, 247, 1980.

108. Kunz, H.W. et al., Quantitative aspects of chemical carcinogenesis and tumor promotion in liver, *Environ Health Perspect*, 50, 113, 1983.

109. Kline, B.E. et al., Certain effects of dietary fats on the production of liver tumors in rats fed *p*-dimethylaminoazobenzene, *Cancer Res*, 6, 5, 1946.

110. Sugai, M. et al., The effect of heated fat on the carcinogenic activity of 2-acetylaminofluorene, *Cancer Res*, 22, 510, 1962.

111. McCay, P.B. et al., Interactions between dietary fats and antioxidants on DMBA-induced hyperplastic nodules and hepatomas, *J Environ Pathol Toxicol*, 3, 451, 1980.

112. Baldwin, S. and Parker, R.S., The effect of dietary fat and selenium on the development of preneoplastic lesions in rat liver, *Nutr Cancer*, 8(4), 273, 1986.

113. Hietanen, E. et al., Mechanisms of fat-related modulation of *N*-nitrosodiethylamine-induced tumors in rats: organ distribution, blood lipids, enzymes and pro-oxidant state, *Carcinogenesis*, 12(4), 591, 1991.

114. Miller, J.A. et al., The carcinogenicity of *p*-dimethylaminoazobenzene in diets containing hydrogenated coconut oil, *Cancer Res*, 4, 153, 1944.

115. Miller, J.A. et al., The effect of certain lipids on the carcinogenicity of *p*-dimethylaminoazobenzene, *Cancer Res*, 4, 756, 1944.

116. Misslbeck, N.G., Campbell, T.C., and Roe, D.A., Effect of ethanol consumed in combination with high or low fat diets on the post initiation phase of hepatocarcinogenesis, *J Nutr*, 114, 2311, 1984.

117. Baldwin, S. and Parker, R.S., Influence of dietary fat and selenium in initiation and promotion of aflatoxin B-1-induced preneoplastic foci in rat liver, *Carcinogenesis*, 8(1), 101, 1987.

118. Glauert, H.P. and Pitot, H.C., Effect of dietary fat on the promotion of diethylnitrosamine-induced hepatocarcinogenesis in female rats, *Proc Soc Exp Biol Med*, 181, 498, 1986.

119. Sidransky, H., Verney, E., and Wang, D., Effects of varying fat content of a high tryptophan diet on induction of gamma-glutamyltranspeptidase positive foci in the livers of rats treated with hepatocarcinogen, *Cancer Lett*, 31, 235, 1986.

120. Newberne, P.M., Weigert, J., and Kula, N., Effects of dietary fat on hepatic mixed-functions oxidases and hepatocellular carcinoma induced by aflatoxin B1 in rats, *Cancer Res*, 39, 3986, 1979.

121. Glauert, H.P. et al., Effect of dietary fat on the initiation of hepatocarcinogenesis by diethylnitrosamine or 2-acetylaminofluorene in rats, *Carcinogenesis*, 12(6), 991, 1991.

122. Rahman, K.M. et al., Effect of types and amount of dietary fat during the initiation phase of hepatocarcinogenesis, *Nutr Cancer*, 39(2), 220, 2001.

123. Druckrey, H., Production of colonic carcinomas by 1,2-dialkyhydrazines and azoalkanes, in *Carcinoma of the Colon and Antecedent Epithelium*, Burdette, W.J., Ed., Thomas, Springfield, IL, 1970, p. 267.

124. Reddy, B.S. and Ohmori, T., Effect of intestinal microflora and dietary fat on 3,2′-dimethyl-4-aminobiphenyl-induced colon carcinogenesis in F344 rats, *Cancer Res*, 41, 1363, 1981.

125. Nauss, K.M. et al., Lack of effect of dietary fat on *N*-nitrosomethylurea (NMU)-induced colon tumorigenesis in rats, *Carcinogenesis*, 5, 255, 1984.

126. Rogers, A.E. and Nauss, K.M., Rodent models for carcinoma of the colon, *Dig Dis Sci*, 30, 87S, 1985.

127. Ward, J.M., Dose response to a single injection of azoxymethane in rats, *Vet Pathol*, 12, 165, 1975.

128. Schiller, C.M., Curley, W.H., and McConnell, E.E., Induction of colon tumors by a single oral dose of 1,2-dimethlhydrazine, *Cancer Lett*, 11, 75, 1980.

129. Decaens, C. et al., Induction of rat intestinal carcinogenesis with single doses, low and high repeated doses of 1,2-dimethylhydrazine, *Carcinogenesis*, 10, 69, 1989.
130. Glauert, H.P. and Weeks, J.A., Dose- and time-response of colon carcinogenesis in Fischer-344 rats after a single dose of 1,2-dimethylhydrazine, *Toxicol Lett*, 48, 283, 1989.
131. Karkare, M.R., Clark, T.D., and Glauert, H.P., Effect of dietary calcium on colon carcinogenesis induced by a single injection of 1,2-dimethylhydrazine in rats, *J Nutr*, 121, 568, 1991.
132. Reddy, B.S., Weisburger, J.H., and Wynder, E.L., Effects of dietary fat level and dimethylhydrazine on fecal bile acid and neutral sterol excretion and colon carcinogenesis in rats, *J Natl Cancer Inst*, 52, 507, 1974.
133. Bull, A.W. et al., Promotion of azoxymethane-induced intestinal cancer by high fat diet in rats, *Cancer Res*, 39, 4956, 1979.
134. Glauert, H.P., Bennink, M.R., and Sander, C.H., Enhancement of 1,2-dimethylhydrazine-induced colon carcinogenesis in mice by dietary agar, *Food Cosmet Toxicol*, 19, 281, 1981.
135. Nauss, K.M., Locniskar, M., and Newberne, P.M., Effect of alterations in the quality and quantity of dietary fat and 1,2-dimethylhydrazine-induced colon carcinogenesis in rats, *Cancer Res*, 43, 4083, 1983.
136. Sakaguchi, M. et al., Effects of dietary saturated and unsaturated fatty acids on fecal bile acids and colon carcinogenesis induced by azoxymethane in rats, *Cancer Res*, 46, 61, 1986.
137. Thompson, M.B., The Min mouse: a genetic model for intestinal carcinogenesis, *Toxicol Pathol*, 25(3), 329, 1997.
138. Bird, R.P., Observation and quantification of aberrant crypts in the murine colon treated with a colon carcinogen: preliminary findings, *Cancer Lett*, 37, 147, 1987.
139. Hardman, W.E., Heitman, D.W., and Cameron, I.L., Suppression of the progression of 1,2 dimethylhydrazine (DMH) induced colon carcinogenesis by 20% dietary corn oil in rats supplemented with dietary pectin, *Proc Am Assoc Cancer Res*, 32(1), 131, 1991.
140. Pereira, M.A. et al., Use of azoxymethane-induced foci of aberrant crypts in rat colon to identify potential cancer chemopreventive agents, *Carcinogenesis*, 15(5), 1049, 1994.
141. Wargovich, M.J. et al., Aberrant crypts as a biomarker for colon cancer: evaluation of potential chemopreventive agents in the rat, *Cancer Epidemiol Biomarkers Prev*, 5, 355, 1996.
142. Alrawi, S.J. et al., Aberrant crypt foci, *Anticancer Res*, 26(1A), 107, 2006.
143. Guillem, J.G. et al., Changes in expression of oncogenes and endogenous retroviral-like sequences during colon carcinogenesis, *Cancer Res*, 48(14), 3964, 1988.
144. Wargovich, M.J. et al., Inhibition of the promotional phase of azoxymethane-induced colon carcinogenesis in the F344 rat by calcium lactate: effect of simulating two human nutrient density levels, *Cancer Lett*, 53, 17, 1990.
145. Zhao, L.P. et al., Quantitative review of studies of dietary fat and rat colon carcinoma, *Nutr Cancer*, 15, 169, 1991.
146. Clinton, S.K. et al., The combined effects of dietary fat, protein, and energy intake on azoxymethane-induced intestinal and renal carcinogenesis, *Cancer Res*, 52, 857, 1992.
147. Hardman, W.E. and Cameron, I.L., Site specific reduction of colon cancer incidence, without a concomitant reduction in cryptal cell proliferation, in 1,2-dimethylhydrazine treated rats by diets containing 10% pectin with 5% or 20% corn oil, *Carcinogenesis*, 16, 1425, 1995.
148. Rijnkels, J.M. et al., Modulation of dietary fat-enhanced colorectal carcinogenesis in *N*-methyl-*N'*-nitro-*N*-nitrosoguanidine-treated rats by a vegetables–fruit mixture, *Nutr Cancer*, 29(1), 90, 1997.
149. Takeshita, M. et al., Lack of promotion of colon carcinogenesis by high-oleic safflower oil, *Cancer*, 79 (8), 1487, 1997.
150. Wijnands, M.V. et al., A comparison of the effects of dietary cellulose and fermentable galacto-oligosaccharide, in a rat model of colorectal carcinogenesis: fermentable fibre confers greater protection than non-fermentable fibre in both high and low fat backgrounds, *Carcinogenesis*, 20(4), 651, 1999.
151. Rao, C.V. et al., Modulation of experimental colon tumorigenesis by types and amounts of dietary fatty acids, *Cancer Res*, 61(5), 1927, 2001.
152. Bird, R.P. et al., Inability of low- or high-fat diet to modulate late stages of colon carcinogenesis in Sprague–Dawley rats, *Cancer Res*, 56(13), 2896, 1996.
153. Reddy, B.S. et al., Effect of quality and quantity of dietary fat and dimethylhydrazine on colon carcinogenesis in rats, *Proc Soc Exp Biol Med*, 151, 237, 1976.

154. Reddy, B.S., Watanabe, K., and Weisburger, J.H., Effect of high-fat diet on colon carcinogenesis in F344 rats treated with 1,2-dimethylhydrazine, methylazoxymethanol acetate, or methylnitrosourea, *Cancer Res*, 37, 4156, 1977.

155. Nigro, N.D. et al., Effect of dietary fat on intestinal tumor formation by azoxymethane in rats, *J Natl Cancer Inst*, 54, 439, 1975.

156. Bull, A.W., Bronstein, J.C., and Nigro, N.D., The essential fatty acid requirement for azoxymethane-induced intestinal carcinogenesis in rats, *Lipids*, 24, 340, 1989.

157. Bansal, B.R., Rhoads, J.E.J., and Bansal, S.C., Effects of diet on colon carcinogenesis and the immune system in rats treated with 1,2-dimethylhydrazine, *Cancer Res*, 38, 3293, 1978.

158. Schmaehl, D., Habs, M., and Habs, H., Influence of a non-synthetic diet with a high fat content on the local occurrence of colonic carcinomas induced by *N*-nitroso-acetoxymethylmethylamine (AMMN) in Sprague–Dawley rats, *Hepatogastroenterology*, 30, 30, 1983.

159. Fujise, T. et al., Long term feeding of various fat diets modulates azoxymethane-induced colon carcinogenesis through Wnt/beta-catenin signaling in rats, *Am J Physiol Gastrointest Liver Physiol*, 292, G1150, 2007.

160. Wasan, H.S. et al., Dietary fat influences on polyp phenotype in multiple intestinal neoplasia mice, *Proc Natl Acad Sci U S A*, 94(7), 3308, 1997.

161. van Kranen, H.J. et al., Effects of dietary fat and a vegetable–fruit mixture on the development of intestinal neoplasia in the ApcMin mouse, *Carcinogenesis*, 19(9), 1597, 1998.

162. Lafave, L.M.Z., Kumarathasan, P., and Bird, R.P., Effect of dietary fat on colonic protein kinase C and induction of aberrant crypt foci, *Lipids*, 29(10), 693, 1994.

163. Kristiansen, E., Thorup, I., and Meyer, O., Influence of different diets on development of DMH-induced aberrant crypt foci and colon tumor incidence in Wistar rats, *Nutr Cancer*, 23, 151, 1995.

164. Koohestani, N. et al., Insulin resistance and promotion of aberrant crypt foci in the colons of rats on a high-fat diet, *Nutr Cancer*, 29(1), 69, 1997.

165. Morotomi, M. et al., Effects of a high-fat diet on azoxymethane-induced aberrant crypt foci and fecal biochemistry and microbial activity in rats, *Nutr Cancer*, 27, 84, 1997.

166. Hambly, R.J. et al., Influence of diets containing high and low risk factors for colon cancer on early stages of carcinogenesis in human flora-associated (HFA) rats, *Carcinogenesis*, 18(8), 1535, 1997.

167. Baijal, P.K., Fitzpatrick, D.W., and Bird, R.P., Comparative effects of secondary bile acids, deoxycholic and lithocholic acids, on aberrant crypt foci growth in the postinitiation phases of colon carcinogenesis, *Nutr Cancer*, 31(2), 81, 1998.

168. Parnaud, G. et al., Effect of meat (beef, chicken, and bacon) on rat colon carcinogenesis, *Nutr Cancer*, 32(3), 165, 1998.

169. Wan, G., Kato, N., and Watanabe, H., High fat diet elevates the activity of inducible nitric oxide synthase and 1,2-dimethylhydrazine-induced aberrant crypt foci in colon of rats, *Oncol Rep*, 7(2), 391, 2000.

170. Liu, Z. et al., High fat diet enhances colonic cell proliferation and carcinogenesis in rats by elevating serum leptin, *Int J Oncol*, 19(5), 1009, 2001.

171. Ju, J. et al., Effects of green tea and high-fat diet on arachidonic acid metabolism and aberrant crypt foci formation in an azoxymethane-induced colon carcinogenesis mouse model, *Nutr Cancer*, 46(2), 172, 2003.

172. Choi, S.Y. et al., Effects of quercetin and beta-carotene supplementation on azoxymethane-induced colon carcinogenesis and inflammatory responses in rats fed with high-fat diet rich in omega-6 fatty acids, *Biofactors*, 27(1–4), 137, 2006.

173. Sakaguchi, M. et al., Effect of dietary unsaturated and saturated fats on azoxymethane-induced colon carcinogenesis in rats, *Cancer Res*, 44, 1472, 1984.

174. Reddy, B.S. and Maruyama, H., Effect of fish oil on azoxymethane-induced colon carcinogenesis in male F344 rats, *Cancer Res*, 46, 3367, 1986.

175. Reddy, B.S. and Sugie, S., Effect of different levels of omega-3 and omega-6 fatty acids on azoxymethane-induced colon carcinogenesis in F344 rats, *Cancer Res*, 48, 6642, 1988.

176. Minoura, T. et al., Effect of dietary eicosapentaenoic acid on azoxymethane-induced colon carcinogenesis in rats, *Cancer Res*, 48, 4790, 1988.

177. Nelson, R.L. et al., A comparison of dietary fish oil and corn oil in experimental colorectal carcinogenesis, *Nutr Cancer*, 11, 215, 1988.

178. Latham, P., Lund, E.K., and Johnson, I.T., Dietary *n*-3 PUFA increases the apoptotic response to 1,2-dimethylhydrazine, reduces mitosis and suppresses the induction of carcinogenesis in the rat colon, *Carcinogenesis*, 20(4), 645, 1999.

179. Bartoli, R. et al., Effect of olive oil on early and late events of colon carcinogenesis in rats: modulation of arachidonic acid metabolism and local prostaglandin E(2) synthesis, *Gut*, 46(2), 191, 2000.

180. Dommels, Y.E. et al., Effects of high fat fish oil and high fat corn oil diets on initiation of AOM-induced colonic aberrant crypt foci in male F344 rats, *Food Chem Toxicol*, 41(12), 1739, 2003.

181. Dwivedi, C., Natarajan, K., and Matthees, D.P., Chemopreventive effects of dietary flaxseed oil on colon tumor development, *Nutr Cancer*, 51(1), 52, 2005.

182. Grippo, P.J. and Sandgren, E.P., Modeling pancreatic cancer in animals to address specific hypotheses, in *Methods in Molecular Medicine, Vol. 103: Pancreatic Cancer: Methods and Protocols*, Su, G., Ed., Humana Press, Totowa, NJ, 2005, p. 217.

183. Leach, S.D., Mouse models of pancreatic cancer: the fur is finally flying! *Cancer Cell*, 5(1), 7, 2004.

184. Hingorani, S.R. et al., Preinvasive and invasive ductal pancreatic cancer and its early detection in the mouse, *Cancer Cell*, 4(6), 437, 2003.

185. Hingorani, S.R. et al., Trp53R172H and KrasG12D cooperate to promote chromosomal instability and widely metastatic pancreatic ductal adenocarcinoma in mice, *Cancer Cell*, 7(5), 469, 2005.

186. Aguirre, A.J. et al., Activated Kras and Ink4a/Arf deficiency cooperate to produce metastatic pancreatic ductal adenocarcinoma, *Genes Dev*, 17(24), 3112, 2003.

187. Roebuck, B.D., Yager, J.D., and Longnecker, D.S., Dietary modulation of azaserine-induced pancreatic carcinogenesis in the rat, *Cancer Res*, 41, 888, 1981.

188. Roebuck, B.D. et al., Promotion by unsaturated fat of azaserine-induced pancreatic carcinogenesis in the rat, *Cancer Res*, 41, 3961, 1981.

189. Roebuck, B.D. et al., Carcinogen-induced lesions in the rat pancreas: effects of varying levels of essential fatty acid, *Cancer Res*, 45, 5252, 1985.

190. Roebuck, B.D. et al., Effects of dietary fats and soybean protein on azaserine-induced pancreatic carcinogenesis and plasma cholecystokinin in the rat, *Cancer Res*, 47, 1333, 1987.

191. O'Connor, T.P., Roebuck, B.D., and Campbell, T.C., Dietary intervention during the postdosing phase of 1-azaserine-induced preneoplastic lesions, *J Natl Cancer Inst*, 75, 955, 1985.

192. Roebuck, B.D., Effects of high levels of dietary fats on the growth of azaserine-induced foci in the rat pancreas, *Lipids*, 21, 281, 1986.

193. Woutersen, R.A. and van Garderen-Hoetmer, A., Inhibition of dietary fat promoted development of (pre) neoplastic lesions in exocrine pancreas of rats and hamsters by supplemental vitamins A, C and E, *Cancer Lett*, 41, 179, 1988.

194. Woutersen, R.A. et al., Modulation of dietary fat-promoted pancreatic carcinogenesis in rats and hamsters by chronic ethanol ingestion, *Carcinogenesis*, 10(3), 453, 1989.

195. Woutersen, R.A. et al., Modulation of dietary fat-promoted pancreatic carcinogenesis in rats and hamsters by chronic coffee ingestion, *Carcinogenesis*, 10(2), 311, 1989.

196. Longnecker, D.S., Roebuck, B.D., and Kuhlmann, E.T., Enhancement of pancreatic carcinogenesis by a dietary unsaturated fat in rats treated with saline or *N*-nitroso(2-hydroxypropyl)(2-oxopropyl)amine, *J Natl Cancer Inst*, 74(1), 219, 1985.

197. Birt, D.F., Salmasi, S., and Pour, P.M., Enhancement of experimental pancreatic cancer in Syrian golden hamsters by dietary fat, *J Natl Cancer Inst*, 67(6), 1327, 1981.

198. Birt, D.F. and Pour, P.M., Increased tumorigenesis induced by *N*-nitrosobis(2-oxopropyl)amine in Syrian golden hamsters fed high-fat diets, *J Natl Cancer Inst*, 70, 1135, 1983.

199. Birt, D.F. et al., Enhancement of pancreatic carcinogenesis in hamsters fed a high-fat diet ad libitum and at a controlled calorie intake, *Cancer Res*, 49, 5848, 1989.

200. Birt, D.F. et al., Comparison of the effects of dietary beef tallow and corn oil on pancreatic carcinogenesis in the hamster model, *Carcinogenesis*, 11(5), 745, 1990.

201. Herrington, M.K. et al., Effects of high-fat diet and cholecystokinin receptor blockade on promotion of pancreatic ductal cell tumors in the hamster, *Nutr Cancer*, 28, 219, 1997.

202. Wenger, F.A. et al., Does dietary alpha-linolenic acid promote liver metastases in pancreatic carcinoma initiated by BOP in Syrian hamster? *Ann Nutr Metab*, 43(2), 121, 1999.

203. Appel, M.J., van Garderen-Hoetmer, A., and Woutersen, R.A., Effects of dietary linoleic acid on pancreatic carcinogenesis in rats and hamsters, *Cancer Res*, 54, 2113, 1994.

204. O'Connor, T.P. et al., Effect of dietary intake of fish oil and fish protein on the development of l-azaserine-induced preneoplastic lesions in the rat pancreas, *J Natl Cancer Inst*, 75(5), 959, 1985.

205. O'Connor, T.P. et al., Effect of dietary omega-3 and omega-6 fatty acids on development of azaserine-induced preneoplastic lesions in rat pancreas, *J Natl Cancer Inst*, 81(11), 858, 1989.

206. Appel, M.J. and Woutersen, R.A., Modulation of growth and cell turnover of preneoplastic lesions and of prostaglandin levels in rat pancreas by dietary fish oil, *Carcinogenesis*, 15(10), 2107, 1994.

207. Appel, M.J. and Woutersen, R.A., Effects of dietary fish oil (MaxEPA) on *N*-nitrosobis(2-oxopropyl) amine (BOP)-induced pancreatic carcinogenesis in hamsters, *Cancer Lett*, 94, 179, 1995.

208. Appel, M.J. and Woutersen, R.A., Dietary fish oil (MaxEPA) enhances pancreatic carcinogenesis in azaserine-treated rats, *Br J Cancer*, 73, 36, 1996.

209. Appel, M.J. and Woutersen, R.A., Effects of a diet high in fish oil (MaxEPA) on the formation of micronucleated erythrocytes in blood and on the number of atypical acinar cell foci Induced in rat pancreas by azaserine, *Nutr Cancer*, 47(1), 57, 2003.

210. Heukamp, I. et al., Influence of different dietary fat intake on liver metastasis and hepatic lipid peroxidation in BOP-induced pancreatic cancer in Syrian hamsters, *Pancreatology*, 6(1–2), 96, 2006.

211. Eustis, S.L. and Boorman, G.A., Proliferative lesions of the exocrine pancreas: relationship to corn oil gavage in the National Toxicology Program, *J Natl Cancer Inst*, 75(6), 1067, 1985.

212. Haseman, J.K. et al., Neoplasms observed in untreated and corn oil gavage control groups of F344/N rats and (C57BL/6N X C3H/HeN)F1 (B6C3F1) mice, *J Natl Cancer Inst*, 75(5), 975, 1985.

213. Freedman, L.S., Clifford, C., and Messina, M., Analysis of dietary fat, calories, body weight, and the development of mammary tumors in rats and mice: a review, *Cancer Res*, 50, 5710, 1990.

214. Welsch, C.W., Review of the effects of dietary fat on experimental mammary gland tumorigenesis: role of lipid peroxidation, *Free Radic Biol Med*, 18(4), 757, 1995.

215. Fay, M.P. et al., Effect of different types and amounts of fat on the development of mammary tumors in rodents: a review, *Cancer Res*, 57(18), 3979, 1997.

216. Green, J.E. and Hudson, T., The promise of genetically engineered mice for cancer prevention studies, *Nat Rev Cancer*, 5(3), 184, 2005.

217. Beems, R.B. and van Beek, L., Modifying effect of dietary fat on benzo(a)pyrene-induced respiratory tract tumors in hamsters, *Carcinogenesis*, 5, 413, 1984.

218. Szepsenwol, J., Carcinogenic effect of ether extract of whole egg, alcohol extract of egg yolk, and powdered egg free of the ether extractable part in mice, *Proc Soc Exp Biol Med*, 116, 1136, 1964.

219. Zhou, J.R. and Blackburn, G.L., Bridging animal and human studies: what are the missing segments in dietary fat and prostate cancer? *Am J Clin Nutr*, 66(Suppl), 1572S, 1997.

220. Rose, D.P., Effects of dietary fatty acids on breast and prostate cancers: evidence from in vitro experiments and animal studies, *Am J Clin Nutr*, 66(6 Suppl), 1513S, 1997.

221. Mori, T. et al., Beef tallow, but not perilla or corn oil, promotion of rat prostate and intestinal carcinogenesis by 3,2′-dimethyl-4-aminobiphenyl, *Jpn J Cancer Res*, 92(10), 1026, 2001.

222. Leung, G. et al., No effect of a high-fat diet on promotion of sex hormone-induced prostate and mammary carcinogenesis in the Noble rat model, *Br J Nutr*, 88(4), 399, 2002.

223. Murphy, M.G., Dietary fatty acids and membrane protein function, *J Nutr Biochem*, 1, 68, 1990.

224. Reddy, B.S. et al., Promoting effect of sodium deoxycholate on colon adenocarcinomas in germ-free rats, *J Natl Cancer Inst*, 56, 441, 1976.

225. Reddy, B.S. et al., Promoting effect of bile acids in colon carcinogenesis in germ-free and conventional F344 rats, *Cancer Res*, 37, 3238, 1977.

226. Narisawa, T. et al., Promoting effect of bile acids on colon carcinogenesis after intrarectal instillation of *N*-methyl-*N*′-nitro-*N*-nitrosoguanidine in rats, *J Natl Cancer Inst*, 53, 1093, 1974.

227. Reddy, B.S. et al., Effect of type and amount of dietary fat and 1,2-dimethylhydrazine on biliary bile acids, fecal bile acids and neutral sterols in rats, *Cancer Res*, 37, 2132, 1977.

228. Reddy, B.S. et al., Effect of high-fat, high-beef diet and of mode of cooking of beef in the diet on fecal bacterial enzymes and fecal bile acids and neutral sterols, *J Nutr*, 110(9), 1880, 1980.

229. Glauert, H.P. and Bennink, M.R., Influence of diet or intrarectal bile acid injections on colon epithelial cell proliferation in rats previously injected with 1,2-dimethylhydrazine, *J Nutr*, 113, 475, 1983.

230. Gallaher, D.D. and Franz, P.M., Effects of corn oil and wheat brans on bile acid metabolism in rats, *J Nutr*, 120, 1320, 1990.

231. Bernstein, H. et al., Bile acids as carcinogens in human gastrointestinal cancers, *Mutat Res*, 589(1), 47, 2005.

232. Cohen, B.I. et al., Effect of cholic acid feeding on *N*-methyl-*N*-nitrosourea-induced colon tumors and cell kinetics in rats, *J Natl Cancer Inst*, 64, 573, 1980.

233. Deschner, E.E. and Raicht, R.F., Influence of bile on kinetic behavior of colonic epithelial cells of the rat, *Digestion*, 19, 322, 1979.

234. Deschner, E.E., Cohen, B.I., and Raicht, R.F., Acute and chronic effect of dietary cholic acid on colonic epithelial cell proliferation, *Digestion*, 21, 290, 1981.

235. Wargovich, M.J. et al., Calcium ameliorates the toxic effect of deoxycholic acid on colonic epithelium, *Carcinogenesis*, 4, 1205, 1983.

236. Skraastad, O. and Reichelt, K.L., An endogenous colon mitosis inhibitor and dietary calcium inhibit increased colonic cell proliferation induced by cholic acid, *Scand J Gastroenterol*, 23, 801, 1988.

237. Bartram, H.P. et al., Effects of calcium and deoxycholic acid on human colonic cell proliferation in vitro, *Ann Nutr Metab*, 41(5), 315, 1997.

238. Peiffer, L.P., Peters, D.J., and McGarrity, T.J., Differential effects of deoxycholic acid on proliferation of neoplastic and differentiated colonocytes in vitro, *Dig Dis Sci*, 42(11), 2234, 1997.

239. Ochsenkuhn, T. et al., Colonic mucosal proliferation is related to serum deoxycholic acid levels, *Cancer*, 85(8), 1664, 1999.

240. Milovic, V. et al., Effects of deoxycholate on human colon cancer cells: apoptosis or proliferation, *Eur J Clin Invest*, 32(1), 29, 2002.

241. Ochsenkuhn, T. et al., Does ursodeoxycholic acid change the proliferation of the colorectal mucosa? A randomized, placebo-controlled study, *Digestion*, 68(4), 209, 2003.

242. Cheng, K. and Raufman, J.P., Bile acid-induced proliferation of a human colon cancer cell line is mediated by transactivation of epidermal growth factor receptors, *Biochem Pharmacol*, 70(7), 1035, 2005.

243. Rosenthal, M.D., Fatty acid metabolism of isolated mammalian cells, *Prog Lipid Res*, 26, 87, 1987.

244. McEntee, M.F. and Whelan, J., Dietary polyunsaturated fatty acids and colorectal neoplasia, *Biomed Pharmacother*, 56(8), 380, 2002.

245. Whelan, J. and McEntee, M.F., Dietary (*n*-6) PUFA and intestinal tumorigenesis, *J Nutr*, 134(12 Suppl), 3421S, 2004.

246. Hanasaki, K. and Arita, H., Phospholipase A2 receptor: a regulator of biological functions of secretory phospholipase A2, *Prostaglandins Other Lipid Mediat*, 68–69, 71, 2002.

247. Tsuboi, K., Sugimoto, Y., and Ichikawa, A., Prostanoid receptor subtypes, *Prostaglandins Other Lipid Mediat*, 68–69, 535, 2002.

248. Kobayashi, T. and Narumiya, S., Function of prostanoid receptors: studies on knockout mice, *Prostaglandins Other Lipid Mediat*, 68–69, 557, 2002.

249. Toda, A., Yokomizo, T., and Shimizu, T., Leukotriene B4 receptors, *Prostaglandins Other Lipid Mediat*, 68–69, 575, 2002.

250. Fischer, S.M. et al., Events associated with mouse skin tumor promotion with respect to arachidonic acid metabolism: a comparison between SENCAR and NMRI mice, *Cancer Res*, 47, 3174, 1987.

251. Steele, V.E. et al., Potential use of lipoxygenase inhibitors for cancer chemoprevention, *Expert Opin Investig Drugs*, 9(9), 2121, 2000.

252. Richter, M. et al., Growth inhibition and induction of apoptosis in colorectal tumor cells by cyclooxygenase inhibitors, *Carcinogenesis*, 22(1), 17, 2001.

253. Mao, J.T. et al., Chemoprevention strategies with cyclooxygenase-2 inhibitors for lung cancer, *Clin Lung Cancer*, 7(1), 30, 2005.

254. Reddy, B.S., Wang, C.X., and Maruyama, H., Effect of restricted caloric intake on azoxymethane-induced colon tumor incidence in male F344 rats, *Cancer Res*, 47, 1226, 1987.

255. Klurfeld, D.M., Weber, M.M., and Kritchevsky, D., Inhibition of chemically induced mammary and colon tumor promotion by caloric restriction in rats fed increased dietary fat, *Cancer Res*, 47, 2759, 1987.

256. Thompson, H.J. et al., Effect of energy intake on the promotion of mammary carcinogenesis by dietary fat, *Nutr Cancer*, 7, 37, 1985.

257. Boissonneault, G.A., Elson, C.E., and Pariza, M.W., Net energy effects of dietary fat on chemically induced mammary carcinogenesis in F344 rats, *J Natl Cancer Inst*, 76(2), 335, 1986.

258. Welsch, C.W. et al., Enhancement of mammary carcinogenesis by high levels of dietary fat: a phenomenon dependent on ad libitum feeding, *J Natl Cancer Inst*, 82(20), 1615, 1990.

259. Boissonneault, G.A., Calories and carcinogenesis: modulation by growth factors, in *Nutrition, Toxicity, and Cancer*, Rowland, I.R., Ed., CRC Press, Boca Raton, FL, 1991.

260. Neelands, P.J. and Clandinin, M.T., Diet fat influences liver plasma-membrane lipid composition and glucagon-stimulated adenylate cyclase activity, *Biochem J*, 212, 573, 1983.

261. Baldwin, S. and Parker, R.S., Effects of dietary fat level and aflatoxin B-1 treatment on rat hepatic lipid composition, *Food Chem Toxicol*, 23(12), 1049, 1985.

262. Wade, A.E., Norred, W.P., and Evans, J.S., Lipids in drug detoxification, in *Nutrition and Drug Interrelations*, Hathcock, J.N. and Coon, J., Eds., Academic Press, New York, 1978, p. 475.

263. Hammer, C.T. and Wills, E.D., Dependence of the rate of metabolism of benzo(a)pyrene on the fatty acid composition of the liver endoplasmic reticulum and on dietary lipids, *Nutr Cancer*, 2, 113, 1980.

264. Cassanol, P. et al., The effect of dietary imbalances on the activation of benzo[a]pyrene by the metabolizing enzymes from rat liver, *Mutat Res*, 191, 67, 1987.

265. Rutten, A.A.J.J.L. and Flake, H.E., Influence of high dietary levels of fat on rat hepatic phase I and II biotransformation enzyme activities, *Nutr Rep Int*, 36(1), 109, 1987.

266. Kim, H.J., Choi, E.S., and Wade, A.E., Effect of dietary fat on the induction of hepatic microsomal cytochrome P450 isozymes by phenobarbital, *Biochem Pharmacol*, 39(9), 1423, 1990.

267. Lam, T.C.L. and Wade, A.E., Influence of dietary lipid on the metabolism of hexobarbital by the isolated, perfused rat liver, *Pharmacology*, 21, 64, 1980.

268. Lam, T.C.L. and Wade, A.E., Effect of dietary lipid on benzo(a)pyrene metabolism by perfused rat liver, *Drug Nutr Interact*, 1, 31, 1981.

269. Wade, A.E., Harley, W., and Bunce, O.R., The effects of dietary corn oil on the metabolism and mutagenic activation of *N*-nitrosodimethylamine (DMN) by hepatic microsomes from male and female rats, *Mutat Res*, 102, 113, 1982.

270. Chow, C.K., Biological effects of oxidized fatty acids, in *Fatty Acids in Foods and Their Health Implications*, Second Edition, Chow, C.K., Ed., Marcel Dekker, New York, 2000, p. 687.

271. Marnett, L.J., Oxy radicals, lipid peroxidation and DNA damage, *Toxicology*, 181–182, 219, 2002.

272. Poulsen, H.E., Oxidative DNA modifications, *Exp Toxicol Pathol*, 57(Suppl 1), 161, 2005.

273. West, J.D. and Marnett, L.J., Endogenous reactive intermediates as modulators of cell signaling and cell death, *Chem Res Toxicol*, 19(2), 173, 2006.

274. Hillgartner, F., Salati, L.M., and Goodridge, A.G., Physiological and molecular mechanisms involved in nutritional regulation of fatty acid synthesis, *Physiol Rev*, 75(1), 47, 1995.

275. Jump, D.B., Fatty acid regulation of gene transcription, *Crit Rev Clin Lab Sci*, 41(1), 41, 2004.

276. O'Brien, M.L., Spear, B.T., and Glauert, H.P., Role of oxidative stress in peroxisome proliferator-mediated carcinogenesis, *Crit Rev Toxicol*, 35(1), 61, 2005.

277. Wang, T. et al., Peroxisome proliferator-activated receptor gamma in malignant diseases, *Crit Rev Oncol Hematol*, 58(1), 1, 2006.

278. Rao, C.V. et al., Mechanisms in the chemoprevention of colon cancer: modulation of protein kinase C, tyrosine protein kinase and diacylglycerol kinase activities by 1,4-phenylenebis-(methylene)selenocyanate and impact of low-fat diet, *Int J Oncol*, 16(3), 519, 2000.

279. Hilakivi-Clarke, L. and Clarke, R., Timing of dietary fat exposure and mammary tumorigenesis: role of estrogen receptor and protein kinase C activity, *Mol Cell Biochem*, 188(1–2), 5, 1998.

280. Pajari, A.M., Rasilo, M.L., and Mutanen, M., Protein kinase C activation in rat colonic mucosa after diets differing in their fatty acid composition, *Cancer Lett*, 114, 101, 1997.

281. Birt, D.F., Dietary modulation of epidermal protein kinase C: mediation by diacylglycerol, *J Nutr*, 125(6 Suppl), S1673, 1995.

282. Reddy, B.S. et al., Effect of amount and types of dietary fat on intestinal bacterial 7 alpha-dehydroxylase and phosphatidylinositol-specific phospholipase C and colonic mucosal diacylglycerol kinase and PKC activities during different stages of colon tumor promotion, *Cancer Res*, 56(10), 2314, 1996.

283. Singh, J., Hamid, R., and Reddy, B.S., Dietary fat and colon cancer: modulating effect of types and amount of dietary fat on ras-p21 function during promotion and progression stages of colon cancer, *Cancer Res*, 57(2), 253, 1997.

284. DeWille, J.W. et al., Dietary fat promotes mammary tumorigenesis in MMTV/v-ha-ras transgenic mice, *Cancer Lett*, 69(1), 59, 1993.

285. Roberts-Thomson, S.J. and Snyderwine, E.G., Effect of dietary fat on codon 12 and 13 Ha-ras gene mutations in 2-amino-1-methyl-6-phenylimidazo[4,5-*b*]pyridine-induced rat mammary gland tumors, *Mol Carcinog*, 20(4), 348, 1997.

286. Karmali, R.A. et al., II. Effect of *n*-3 and *n*-6 fatty acids on mammary H-ras expression and PGE-2 levels in DMBA-treated rats, *Anticancer Res*, 9, 1169, 1989.

287. Ronai, Z., Lau, Y.Y., and Cohen, L.A., Dietary *n*-3 fatty acids do not affect induction of Ha-ras mutations in mammary glands of NMU-treated rats, *Mol Carcinog*, 4(2), 120, 1991.

26 Lipid-Based Synthetic Fat Substitutes

Casimir C. Akoh

CONTENTS

I. INTRODUCTION: WHY SYNTHETIC FAT SUBSTITUTES?

Fat is an important macromolecular component of plant and animal tissues. The various fats contribute to the physical and functional properties (solubility, viscosity, rheology, melting behavior, emulsification, body, creaminess, heat conduction, carrier of lipophilic vitamins, and flavorants) of most food products, affecting as well sensory (appearance, taste, mouthfeel, lubricity, flavor) and nutritional (satiety, calories, essential fatty acid source, health benefits) aspects of food. These properties are difficult to duplicate in food formulations without adding fats. The amount and type of fat present in a food determine the characteristics of that food and consumer acceptance.

Fat is still the number one nutritional concern for most people in developed countries. The present estimate of average fat calories consumed by most Americans is 35%–37%. Recommendation by the U.S. Senate Select Committee on Nutrition and Human Needs and the Surgeon General is that fat consumption be reduced to 30% of total calories of the diet [1]. Excessive intake of fat in the diet has been linked to certain diseases, such as heart disease, cancer, obesity, and possibly gallbladder disease [2,3]. Increased saturated fat intake is associated with high blood cholesterol and increased risk of coronary heart disease. It has been difficult for individuals to change their dietary habits to reduce or minimize fat intake while enjoying their favorite foods. This problem and the interest shown by consumers in alternative fats and foods low in calories or without calories led to the search by the food industry and scientific community for the "ideal" fat substitute [4,5].

A real fat substitute must be able to provide all the attributes of fats and replace the calories from fat on a 1:1 weight basis [2]. Fats are the most concentrated source of energy: a given amount of fat contains more than twice the calories (9 kcal/g) of other macronutrients such as proteins and carbohydrates (4 kcal/g). An ideal fat substitute must look and function like fat and be able to substantially reduce caloric contributions to food. In recent years, health-conscious consumers have shown interest in reducing calories from fat by modifying their diets, exercising, and eating healthier foods. Survey results show that consumers were ready for fat substitutes and replacers in their foods [6].

II. CLASSIFICATION OF FAT REPLACERS

Fat replacers are divided into two main groups: fat mimetics and fat substitutes. They are classified as carbohydrate-based, protein-based, and lipid-based fat replacers (Table 26.1) or combinations thereof. The protein-based and carbohydrate-based replacers are widely regarded as *fat mimetics*.

TABLE 26.1
Typical Examples of Fat Replacers

Compound	Class	Caloric Content (kcal/g)	Absorbability	Uses
Simplesse	Protein-based	1–4	Absorbable	FDA approved. Dairy products, dressings, spreads. Not stable for frying and baking
Maltodextrin	Carbohydrate-based	4–4.5	Absorbable	Low-fat table spreads, dressings, baked goods, desserts. Not stable for frying
Olestra/Olean	Lipid-based	0	Nonabsorbable	FDA approved for savory snacks. Stable to frying, baking, and cooking temperatures. Supplementation with fat-soluble vitamins required. Can be used for dairy products, spreads, and dressing but will require separate approval

The fat mimetics are either proteins or carbohydrates that have been physically or chemically processed to mimic the properties and functions of fats in food systems; they are not fats. They tend to adsorb a large amount of water, are not stable at frying temperature, and may produce food that is not microbiologically shelf stable; they contribute some calories (1–4 kcal/g) to the diet. The fat mimetics do not possess all the organoleptic, physical, chemical, and functional properties of fats and cannot replace calories from fat on a 1:1 weight basis. They cannot carry lipid-soluble flavor compounds because they cannot lower the vapor pressure of lipophilic flavor molecules, and most foods prepared with fat mimetics are often perceived by consumers as lacking in taste. Although fat mimetics need a delivery system such as an emulsifier to carry lipid-soluble flavors, they can carry water-soluble flavors. Some fat mimetics have mouthfeel and physical properties approximating those of triacylglycerols but are not suitable for frying operations because they can be denatured (protein-based substances) or caramelized (those based on carbohydrates). They can, however, be used for baking and retort cooking operations.

Fat substitutes are believed to be compounds that physically and chemically resemble triacylglycerols. They are stable to high-temperature cooking and frying operations, and in theory, can replace fats and oils on a 1:1 weight basis in foods. In the literature, the term "fat substitutes" has been used interchangeably with "fat replacers," but not with "fat mimetics," and this can be confusing. Several lipid-based synthetic low-calorie or zero-calorie fat substitutes belong to the fat substitute group. A good example is sucrose fatty acid polyester (SPE), which was originally developed as olestra and marketed as Olean by Procter & Gamble to replace edible fats and oils in the diet [7]. A number of other fat substitutes have been developed or are under development [8–18]. Among these, the carbohydrate and alkyl glycoside fatty acid polyesters and the structured lipids (SLs) have functional and physical properties resembling those of triacylglycerols, while contributing few to no calories to the diet [10–20].

Some of the lipid-based fat substitutes can be added to food products to replace the functional properties of fats, including frying (not possible with protein-based fat replacers such as Simplesse, and carbohydrate-based fat replacers such as maltodextrin), while reducing caloric contributions from fats and oils. Table 26.2 lists the applications and functions of some fat replacers. Because the ideal fat substitute does not exist, a systems approach to reduced-fat or low-fat food formulations has been proposed. Each type of food product will require a different approach to address the difficulties of formulating a counterpart that has reduced, no, or low fat. Simply put, a systems approach uses a combination of different ingredients that may or may not belong to either of the classes of fat replacers and requires a basic knowledge of ingredient technology to formulate desired products. The system may contain emulsifiers, fat substitutes or mimetics, fibers, water control ingredients, flavor, and bulking agents. Water or moisture control poses one of the greatest challenges in formulating reduced-fat snack and baked goods. In these systems, water is used to replace fat, to increase bulk, or for functionality. A detailed review of fat mimetics is outside the scope of this chapter, which concentrates on lipid-based fat substitutes.

III. TYPES OF LIPID-BASED SYNTHETIC FAT SUBSTITUTES

Lipid-based fat substitutes include carbohydrate fatty acid polyesters such as sucrose polyester, sorbitol polyester, raffinose polyester, stachyose polyester, and alkyl glycoside fatty acid polyesters. Others include Caprenin, Salatrim (short and long acyl triglyceride molecules, marketed as Benefat), structured lipids, medium-chain triacylglycerols (MCTs), mono- and diacylglycerols, esterified propoxylated glycerol (EPG), dialkyl dihexadecylmalonate (DDM), and trialkoxytricarballylate (TATCA), to name a few. The composition and sources or developers of the lipid-based fat substitutes are shown in Table 26.3.

Of all the lipid-based fat substitutes, only sorbitol, trehalose, raffinose, and stachyose polyesters have a chance to compete with Olestra as nondigestible zero-calorie fat substitutes. Others are either partially hydrolyzed or fully hydrolyzed and absorbed, thus contributing some calories to the diet.

TABLE 26.2

Applications and Functions of Some Fat Replacers

Specific Application	Fat Replacer	General Functions[a]
Baked goods	Lipid-based	Emulsification; cohesiveness; tenderizer, flavor carrier, shortening replacer, antistaling agent; prevention of retrogradation of starch; dough conditioner
	Carbohydrate-based	Moisture retention; retard staling
	Protein-based	Texturizer
Frying and cooking	Lipid-based	Texturizer; flavor, crispiness; heat conduction
Salad dressing	Lipid-based	Emulsification; mouthfeel; hold flavorants
	Carbohydrate-based	Increase viscosity; mouthfeel; texturizer
	Protein-based	Texturizer; mouthfeel
Frozen desserts	Lipid-based	Emulsification; texture
	Carbohydrate-based	Increase viscosity; texturizer, thickener
	Protein-based	Texturizer; stabilizer
Margarines, shortenings, spreads, and butter	Lipid-based	Spreadability; emulsification; flavor; plasticity
	Carbohydrate-based	Mouthfeel
	Protein-based	Texturizer
Confectionery	Lipid-based	Emulsification; texturizer
	Carbohydrate-based	Mouthfeel; texturizer
	Protein-based	Mouthfeel; texturizer
Processed meat products	Lipid-based	Emulsification; texturizer; mouthfeel
	Carbohydrate-based	Increase water-holding capacity; texturizer; mouthfeel
	Protein-based	Texturizer; mouthfeel; water holding
Dairy products	Lipid-based	Flavor; body; mouthfeel; texture; stabilizer; increases overrun
	Carbohydrate-based	Increase viscosity; thickener; gelling agent; stabilizer
	Protein-based	Stabilizer; emulsification
Soups, sauces, and gravies	Lipid-based	Mouthfeel; lubricity
	Carbohydrate-based	Thickener; mouthfeel; texturizer
	Protein-based	Texturizer
Snack products	Lipid-based	Emulsification; flavor
	Carbohydrate-based	Texturizer; formulation aid
	Protein-based	Texturizer

[a] Functions are in addition to serving as a fat replacer.

A. Strategies for Designing Lipid-Based Fat Substitutes

Several strategies were suggested for designing low-calorie or zero-calorie lipid-based synthetic fat substitutes [2,16]. The basic premise is to reengineer, redesign, chemically alter, or synthesize conventional fats and oils such that they retain the conventional functional and physical properties of fats and oils in foods but contribute few or no calories because of reduced susceptibility to hydrolysis and/or absorption in the lumen. Possible strategies, rationale, and examples are as follows:

1. Replace the glycerol moiety of the triacylglycerol with alternative alcohols (e.g., carbohydrates, polyols, neopentyl alcohol). This ensures steric protection of the ester bonds. Branching interferes with hydrolysis by pancreatic lipase. Examples are sucrose fatty acid esters (SFEs), sucrose polyesters, other carbohydrate fatty acid polyesters, alkyl glycoside fatty acid polyesters, and polyglycerol esters.

TABLE 26.3

Types of Lipid-Based Fat Substitutes

Name	Composition	Source/Developer
Olestra or Olean	Sucrose polyester of fatty acids (6–8 fatty acids)	Procter & Gamble (FDA-approved, 1996), Akoh and Swanson, Unilever
Caprenin	Caprocaprylobehenin-structured triacylglycerol C8:0, C10:0, C22:0	Procter & Gamble, GRAS requested[a]
Salatrim/Benefat	(C18:0, C2:0, C4:0)-structured triacylglycerol	Nabisco Foods Group/Cultor Food Science
EPG	Esterified propoxylated glycerols	ARCO Chemical Co. CPC International
DDM	Dialkyl dihexadecylmalonate	Frito-Lay, Inc.
TATCA	Trialkoxytricarballylate	CPC International
TAC	Trialkoxycitrate	CPC International
Alkyl glycoside polyesters	Alkyl glycosides + fatty acids	Akoh and Swanson, Curtice Burns, Inc.
Trehalose, raffinose, stachyose polyesters	Carbohydrate + fatty acids (all similar to olestra)	Akoh and Swanson, Curtice Burns, Inc.
Sorbestrin	Sorbitol or cyclic sorbitol + fatty acids	Cultor Food Science
PGE	Polglycerol esters–emulsifiers	Lonza, Inc.
Sucrose esters	Sucrose with 1–4 fatty acids as emulsifiers	Mitsubushi Chemical America, Inc., Crodesta
TGE	Trialkoxyglyceryl ether	CPC International
MCT	Medium-chain triacylglycerols (C6–C10 fatty acids)	ABITEC Corp., Stepan Co.
Phenylmethylpolysiloxane	Organic derivatives of silica	Dow Corning Corp.

[a] Developer has petitioned the FDA to obtain "Generally Recognized as Safe" status for the product.

2. Replace the long-chain fatty acids with alternative acids (to confer steric protection to the ester bonds). Examples are branched carboxylic esters of glycerol and structured lipids such as Caprenin and Salatrim/Benefat. Caprenin and Salatrim contain poorly or less absorbed long-chain saturated fatty acids and easily absorbed short- and/or medium-chain fatty acids esterified to the glycerol. The short- and medium-chain fatty acids have lower heats of combustion than long-chain fatty acids.

3. Reverse the ester linkage in triacylglycerols by replacing the glycerol moiety with a polycarboxylic acid, amino acid, or other polyfunctional acid and esterify with a long-chain alcohol. Examples include TATCA and trialkoxycitrate (TAC).

4. Reduce the ester linkage of the glycerol moiety to an ether linkage. This product is not a good substrate for lipases, which do not hydrolyze ether bonds as fast as ester bonds. Examples include diether monoesters of glycerol, triglyceryl ethers, and trialkoxyglyceryl ether (TGE).

5. Apply chemistry unrelated to triacylglycerol structure. A good example is the use of polymeric materials having physical and functional properties similar to those of conventional fats and oils such as phenylmethylsiloxane (PS) or silicone oil and paraffins.

6. Evaluate naturally occurring substances as potential low-calorie fat substitutes. Jojoba oil is an excellent example.

7. Use enzymes to synthesize reduced-calorie fat substitutes. Examples include sugar mono- and diesters, glycerophospholipids, mono- and diacylglycerols, and structured lipids.

8. Introduce oxypropylene group between glycerol and fatty acids to form propoxylated molecules. An example is EPG.

B. Olestra or Sucrose Polyester: Brief History of Development

SPE development dates back to the year 1880, when a derivative of sucrose was prepared by acetylation to produce sucrose octaacetate (SOAc) (i.e., sucrose containing eight acetate groups). Following this, other carbohydrate acetates were successfully prepared. In 1921, Hess and Messner [21] synthesized sucrose octapalmitate (sucrose esterified with eight molecules of palmitic acid, a long-chain fatty acid) and sucrose octastearate. In 1952, the concept of sucrose polyester production was initiated when the president of the Sugar Research Foundation, Henry B. Hass, asked Foster D. Snell to look into the possibility of "hanging a fat tail on sucrose" for use in detergents. The idea was that since sucrose is highly hydrophilic, a lipophilic tail on sucrose would result in a molecule that is amphiphilic (both water- and oil-loving), hence able to serve as an excellent surfactant. It was anticipated that production would be easy and the product biodegradable under aerobic and anaerobic conditions. It turned out that the chemical synthesis was not that easy without the use of solvents like dimethylformamide (DMF), dimethylsulfoxide (DMS), and dimethylpyrolidone (DMP) to solubilize sucrose and free fatty acids. This process, called the Hass–Snell process, was applicable only to the synthesis of sucrose mono- and diesters, otherwise called SFEs. These are digestible and good nonionic surfactants, as we shall see later in this chapter. By today's standard, the solvents used are not food grade; therefore, products made in them are unacceptable for human consumption.

The other concept was to find means of reducing fat-derived calories without resorting to dilution with, say, water, air, carbohydrates, and proteins. The aim was to somehow come up with a fatlike molecule that would significantly reduce fat calories by preventing their hydrolysis and absorption. This led to the discovery of a nondigestible and nonabsorbable fatlike molecule called SPE, now known as olestra (generic name) or Olean (brand name), by Mattson and Volpenhein [7] while working on the absorption of fats by infants.

Sucrose is a nonreducing disaccharide and the common table sugar. "Olestra" or "sucrose polyester" refers to sucrose esterified with six to eight fatty acids. SPEs become undigestible when the number of fatty acids esterified is >4. The structure of sucrose polyester is given in Figure 26.1. Procter & Gamble, which was granted the original patent for sucrose polyester in 1971, spent over $250 million over the last 25 years to develop this fat substitute. The original application for use of sucrose polyester as a food additive, filed with the Food and Drug Administration (FDA) in April 1987, was withdrawn and modified. But on January 24, 1996, the FDA approved olestra for limited use in savory snacks (namely, chips, curls, and crackers). Before the approval, sucrose polyester was evaluated in over 10 animal studies and in 25–30 clinical trials. The approval of olestra was not without controversy. The Center for Science in the Public Interest (CSPI), a Washington consumer

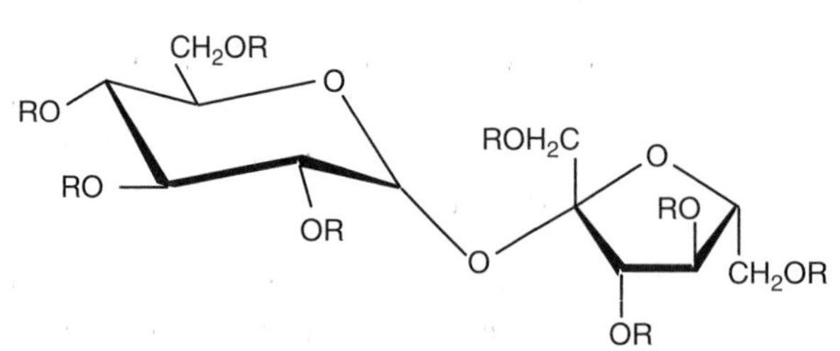

Sucrose polyester (olestra or Olean)
[α-Glucopyranosyl-(1 → 2)-β-fructofuranoside linkage]

where R = alkyl part of the acyl group of fatty acids $\left(R-\overset{O}{\overset{\|}{C}}- \right)$

FIGURE 26.1 Structure of sucrose polyester (Olean).

advocate group, believes that olestra deprives the body of some of the essential vitamins and carotenoid that may protect against cancer. In 1996, in Iowa, Wisconsin, and Colorado, Frito-Lay, a unit of PepsiCo, test-marketed chips made with the olestra.

1. Synthetic Approaches

SPE can be synthesized in the presence or absence of organic solvent. Direct esterification of sucrose with fatty acid is very difficult. The solvent-free process is widely used for the current production of sucrose polyester [18,22]. The synthesis may involve reactions of the following types:

1. *Transesterification.* Fatty acid methyl esters (FAMEs) and sucrose are reacted in the presence of potassium soaps to form a homogeneous melt followed by the addition of excess FAME and NaH at 130°C–150°C. In some cases, potassium carbonate is added to aid the reaction. The function of soap is to help solubilize sucrose and FAME. Methanol, a by-product of the transesterification reaction, is distilled off (Figure 26.2). The active catalyst is the sucrate ion generated with alkali metal hydrides. This is a two-stage transesterification process, and up to 8–9 h may be needed to achieve 90% yield of SPE.

2. *Interesterification (ester interchange).* This involves reacting a short-chain alkyl ester such as SOAc with FAME in the presence of sodium methoxide ($NaOCH_3$) or Na or K metal as catalyst [14]. The reaction requires extremely anhydrous conditions to prevent hydrolysis of formed product, catalyst inactivation, or explosion (when Na is in contact with water). The temperature of the reaction with Na catalyst is lower (105°C–130°C). Reaction times of 2–6 h and pressure of 0–5 mmHg are required to achieve >95% yield of SPE [14]. This is a simple ester–ester interchange reaction (Figure 26.3). The methyl acetate formed is trapped with a liquid nitrogen (−196°C).

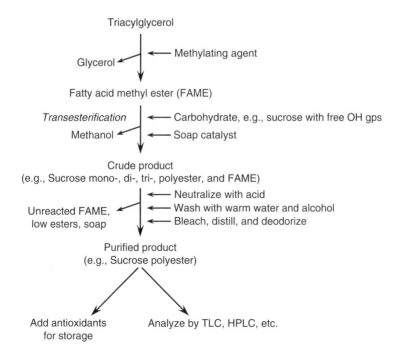

FIGURE 26.2 Synthetic scheme for olestra by transesterification and processing.

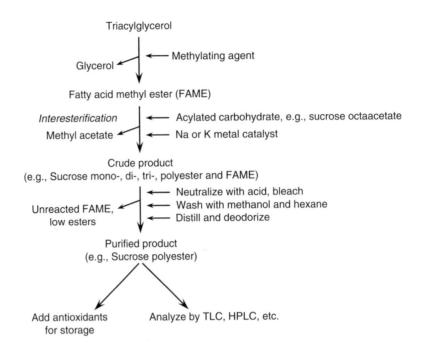

FIGURE 26.3 Synthetic scheme for olestra by interesterification (ester interchange) and processing.

Synthesis based on Na metal (potentially explosive and flammable compound) as catalyst and SOAc and FAME as substrates may not be suitable for industrial adaptation. Therefore, use of milder catalysts such as sodium methoxide, potassium soap, and potassium carbonate are encouraged. Shieh et al. [23] reported the optimized synthesis of SPE using potassium hydroxide (KOH) in methanol plus FAME to form soap (potassium soap) followed by the addition of potassium carbonate. The reaction time at 144°C was 11.5 h.

The triacylglycerol for the synthesis can come from vegetable oils and fats, alone or in combination. The type of product desired dictates the type of fatty acids needed for synthesis to achieve desired functionality. In most cases, the fatty acid profiles of finished products will resemble those of the triacylglycerol source [22]. It should be noted that the reactions above are random processes and therefore the specific position and type of fatty acid on the sucrose molecule will vary from product to product.

2. Analyses of Olestra (SPE)

No matter what synthetic approach is used, at the end of the reaction, the vessel is cooled down to ~50°C or less and the product neutralized with acid, washed, bleached, distilled (to remove unreacted substrates and sucrose ester with low degree of esterification), deodorized, and analyzed for extent of esterification [22]. Antioxidant(s) may or may not be added to prevent oxidation during storage (Figures 26.2 and 26.3). Olestra can be analyzed by any of the following techniques or combinations thereof:

1. Thin-layer chromatography (TLC), Iatroscan, and TLC-recording densitometry for degree of esterification [18]
2. Gas–liquid chromatography to determine the fatty acid composition of the product after transmethylation of olestra [10,18,24] or to determine olestra content after derivatization (silyl or acetate derivatives)
3. Column chromatography to separate products of small-scale synthesis [10,14]

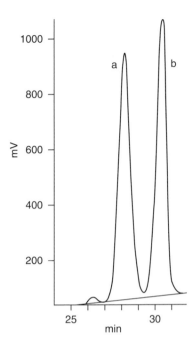

FIGURE 26.4 HPLC chromatographic analysis of a mixture of olestra and triacylglycerol blend on four gel permeation chromatography columns arranged in series: (a) olestra and (b) a triacylglycerol; injection volume, 200 µL, flow rate, 1.0 mL/min; a light-scattering mass detector was used. (From Chase, G.W., Akoh, C.C., and Eitenmiller, R.R., *J. Am. Oil Chem. Soc.*, 71, 1273, 1994. With permission.)

4. High-performance liquid chromatography (HPLC) to analyze the content of olestra in, say, salad dressing, cooking oil, margarine, and spreads whose ingredients include triacylglycerols and fat-soluble vitamins, or analysis by size-exclusion chromatography (SEC) for oxidized, polymerized, or heated olestra [25–27]
5. Fourier transform infrared spectroscopy (FTIR) for functional group and hydroxyl value determinations [14,18]
6. Nuclear magnetic resonance (NMR) spectroscopy for structural elucidation [14,18]
7. Supercritical fluid chromatography (SFC) of trimethylsilyl ether derivatives of olestra for separation and determination of degree of esterification [28]
8. Spectrophotometry for analysis of olestra content in foods [29]
9. Polarimetry measurements for degree of esterification [30]
10. Desorption mass spectrometry (MS) for characterization of olestra [31]
11. Hyphenated chromatography techniques, such as HPLC–FTIR, GC–MS, and HPLC–GC–FTIR, may also be useful

Figure 26.4 shows the result of HPLC separation of olestra or sucrose polyester and triacylglycerol blend on a gel permeation chromatography column [32].

C. Olestra-Type Fat Substitutes

1. Sorbitol Polyester

Sorbitol or glucitol is a sugar alcohol (polyol) made by hydrogenation or electrolytic reduction of glucose. Other polyols like xylitol, mannitol, and lactitol can equally be used for the synthesis of fat

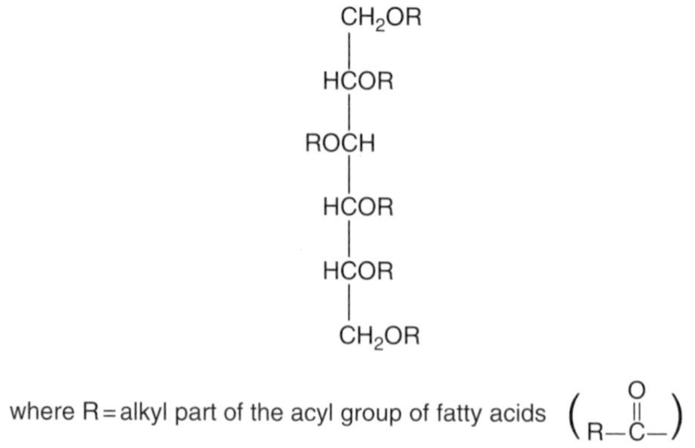

$$CH_2OR$$
$$|$$
$$HCOR$$
$$|$$
$$ROCH$$
$$|$$
$$HCOR$$
$$|$$
$$HCOR$$
$$|$$
$$CH_2OR$$

where R = alkyl part of the acyl group of fatty acids $\left(R-\overset{O}{\underset{||}{C}}- \right)$

FIGURE 26.5 Structure of sorbitol polyester.

substitutes. Up to six hydroxyl groups can be esterified with fatty acids to produce a nondigestible sorbitol polyester. The structure is shown in Figure 26.5. Sorbitol polyester can be synthesized by interesterification of sorbitol hexaacetate with FAME in the absence of organic solvent [13] or by transesterification reaction between sorbitol and FAME as described earlier.

Chung et al. [33] optimized the synthesis of sorbitol fatty acid polyesters. Procter & Gamble also worked on sorbitol polyesters but did not follow up on their development. The properties and applications in food have not been extensively studied as with olestra. However, sorbitol polyesters are proposed as zero-calorie fat substitutes. They are less viscous than sucrose polyester and slightly more viscous than vegetable oils [13]. Based on the known structure and limited metabolic studies, sorbitol polyesters may be used in place of olestra in food products. They are stable to high heat and taste and function like fats.

2. Sorbestrin

Sorbestrin is sorbitol or cyclic sorbitol containing 3–5 fatty acids esterified to the OH group. It was developed by Pfizer Inc. Sorbestrin contains 1.5 kcal/g and has a bland oil-like taste with cloud point between 15°C and 13°C. It is mainly a clear liquid and can serve as a reduced-calorie fat substitute. It is thermally stable and can withstand frying temperatures. It is intended for use in frying, baking, and salad dressings. Sorbestrin is not yet commercially available and will require food additive petition (FAP) and FDA approval before use. Akoh and Swanson [13] synthesized sorbitol hexaoleate as a low-calorie fat substitute. Pfizer noted that for humans, sorbestrin consumption in the future is estimated as 12 g/day. The structure of sorbestrin is shown in Figure 26.6.

3. Trehalose Polyester

Trehalose is a nonreducing disaccharide, the major sugar of insect hemolymph, fungi, and yeasts. It is made of two α-d-glucopyranose components and closely resembles sucrose in their physical properties. The nonreducing sugars are better substrates for the synthesis of fat substitutes than the reducing sugars. This is because the anomeric carbon atoms (C1) of nonreducing sugars are protected and not very susceptible to thermal degradation. The reducing sugars degrade and caramelize at the high temperatures required for transesterification. The use of acetyl derivatives of nonreducing sugars allows reactions to be carried out at reduced temperatures (<105°C–120°C). Unfortunately, this process is not applicable to acetylated reducing sugars such as glucose pentaacetate.

Trehalose octaacetate has been used to interesterify FAME catalyzed by Na metal to produce highly substituted trehalose polyester with properties similar to those of sucrose polyester [13].

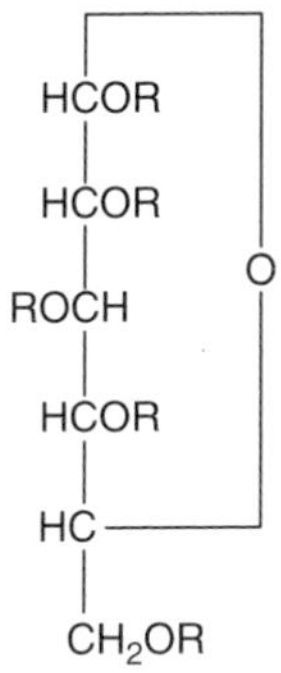

HCOR
|
HCOR
|
ROCH   O
|
HCOR
|
HC
|
CH₂OR

where R = alkyl part of the acyl group of fatty acids $\left(R-\overset{O}{\overset{\|}{C}}- \right)$

FIGURE 26.6 Structure of sorbestrin (cyclic form).

Therefore, trehalose polyester can be used as a zero-calorie fat substitute in place of sucrose polyester. The only problem is cost-effectiveness, since sucrose is a cheaper substrate than trehalose. The structure of trehalose polyester is depicted in Figure 26.7.

4. Raffinose Polyester

Raffinose, like sucrose, is a heterogeneous nonreducing sugar. Raffinose is made of galactose–glucose–fructose units. It has 11 hydroxyl groups that can be esterified with fatty acids. It is well established that as the degree of substitution of a carbohydrate with fatty acids increases, the susceptibility to hydrolysis and absorption decreases. Based on this fact, raffinose undecaacetate was used to interesterify FAME to produce raffinose polyester containing 10–11 fatty acids in 99% yield. The reaction was catalyzed by Na metal at 110°C for 2–3 h [10]. The product had consistency in the range of salad oils and sucrose polyesters, with some raffinose polyesters being slightly more viscous than sucrose polyesters and salad oils depending on the fatty acid composition and degree of substitution. The structure of raffinose polyester is shown in Figure 26.8.

Metabolic studies in mice indicate that raffinose polyester is not hydrolyzed by intestinal lipases and is not absorbed [11]. About 93%–98% of the raffinose polyester fed to the mice was recovered unchanged from the feces. Thus, raffinose polyester can serve as an excellent zero-calorie fat

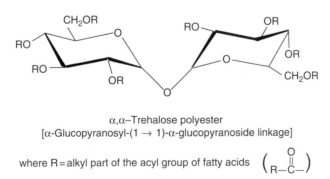

α,α–Trehalose polyester
[α-Glucopyranosyl-(1 → 1)-α-glucopyranoside linkage]

where R = alkyl part of the acyl group of fatty acids $\left(R-\overset{O}{\overset{\|}{C}}- \right)$

FIGURE 26.7 Structure of trehalose polyester.

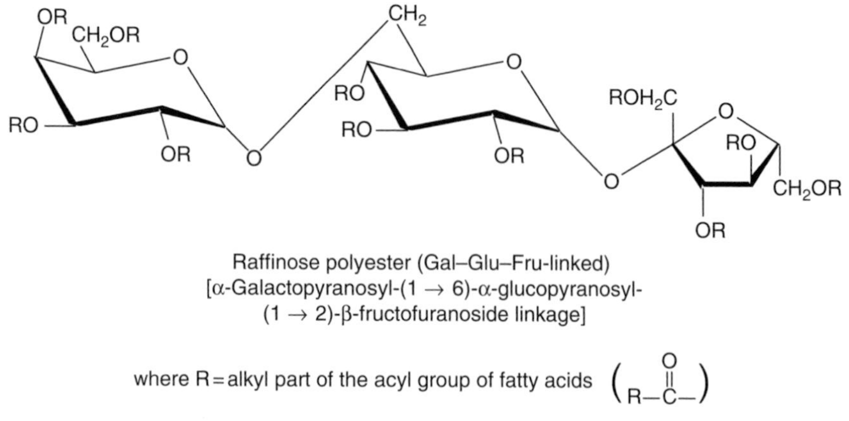

Raffinose polyester (Gal–Glu–Fru-linked)
[α-Galactopyranosyl-(1 → 6)-α-glucopyranosyl-
(1 → 2)-β-fructofuranoside linkage]

where R = alkyl part of the acyl group of fatty acids $\left(R-\overset{O}{\underset{\|}{C}}- \right)$

FIGURE 26.8 Structure of raffinose polyester.

substitute. Now that olestra has received FDA approval, more interest is expected on the possible use of raffinose polyester and others as alternative fat substitutes.

5. Stachyose Polyester

Stachyose, a nonreducing heterogeneous tetrasaccharide, is similar to raffinose and sucrose except that stachyose contains galactose–galactose–glucose–fructose. There are 14 available OH groups for esterification with long-chain fatty acids. Stachyose can be completely acetylated and used as the substrate for interesterification with FAME, catalyzed by Na metal as described for sucrose and raffinose acetates. The degree of substitution as determined by ^{13}C NMR spectroscopy was about 12 fatty acids per molecule of stachyose [12]. The higher the degree of acetylation of the substrate, the lower the melting point, and the lower the temperature requirement for the interesterification. The only requirement is that the temperature be high enough to melt the Na catalyst (>98°C).

The structure of stachyose polyester is given in Figure 26.9. It is conceivable that this molecule will be highly resistant to hydrolysis and absorption and able to serve as an excellent zero-calorie fat substitute. The drawback is the cost, since again sucrose is cheaper.

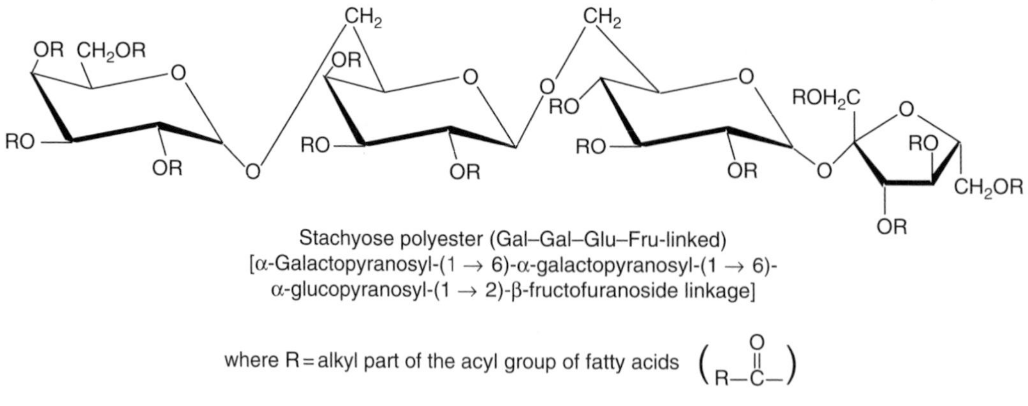

Stachyose polyester (Gal–Gal–Glu–Fru-linked)
[α-Galactopyranosyl-(1 → 6)-α-galactopyranosyl-(1 → 6)-
α-glucopyranosyl-(1 → 2)-β-fructofuranoside linkage]

where R = alkyl part of the acyl group of fatty acids $\left(R-\overset{O}{\underset{\|}{C}}- \right)$

FIGURE 26.9 Structure of stachyose polyester.

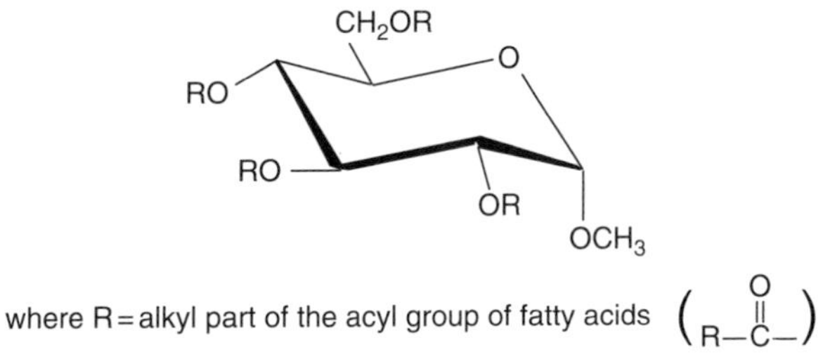

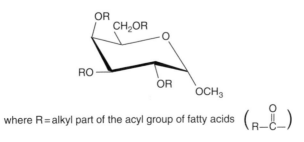

FIGURE 26.10 Structure of methyl glucoside polyester.

D. ALKYL GLYCOSIDE POLYESTERS

Alkyl glycosides are prepared by reacting the sugar (e.g., glucose) with a desired alcohol (methanol for methyl glucoside) in the presence of acid catalysts. This glycosylation is important to convert the reducing C1 anomeric centers to nonreducing, less reactive anomeric centers. The alkyl glycosides become substrates for transesterification with FAME in the presence of potassium soap or sodium methoxide as catalysts. Alternatively, the remaining hydroxyl groups can be acetylated and used for the interesterification reaction catalyzed by Na metal as described for sucrose polyester, to produce alkyl glycoside fatty acid polyesters [12].

The different types of alkyl glycoside polyesters synthesized by the latter process are shown in Figures 26.10 through 26.12. Up to 99% product yields have been reported. Physical properties are closer to those of vegetable oils than to sucrose polyester, partly because a maximum of only four fatty acids can be esterified. The alkyl glycosides are partially hydrolyzed and absorbed and can serve as reduced-calorie but not as zero-calorie fat substitutes.

E. SUCROSE FATTY ACID ESTERS

SFEs with a degree of substitution (DS) of 1–3 are highly hydrophilic, digestible, and absorbable; they are used as solubilization, wetting, dispersion, emulsifying (especially oil-in-water, O/W, and some water-in-oil, W/O, emulsions), and stabilization agents, and as antimicrobial and protective coatings for fruits [8,34,35]. They have a wide range of hydrophilic–lipophilic balance (HLB), namely, 1–16. They are tasteless, odorless, nontoxic, and biodegradable, and they can be used in food, cosmetic, and pharmaceutical applications. The structure of a sucrose monoester is shown in Figure 26.13.

1. Synthetic Approaches

Sucrose has eight hydroxyl groups that can be replaced or esterified with fatty acids. The problem for chemists was how to bring two immiscible molecules (sucrose and fatty acids or FAMEs) to

FIGURE 26.11 Structure of methyl galactoside polyester.

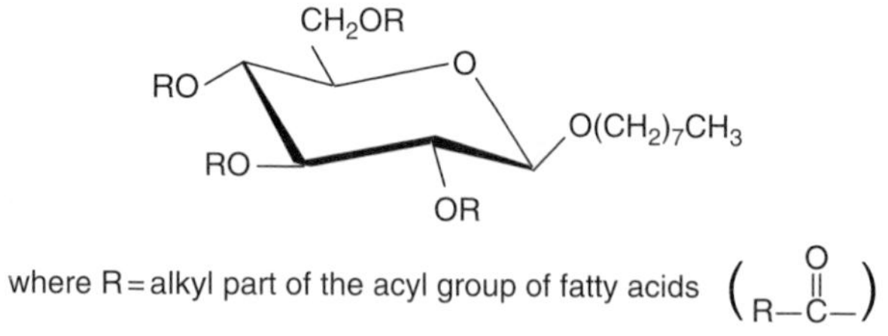

CH₂OR

RO

RO

OR

O(CH₂)₇CH₃

where R = alkyl part of the acyl group of fatty acids $\left(R-\overset{\overset{\displaystyle O}{\|}}{C}- \right)$

FIGURE 26.12 Structure of octyl β-glucoside polyester.

react in the presence of a suitable catalyst. A logical approach would be to dissolve them in a mutual solvent. To obtain a homogeneous solution of the reactants as required for synthesis of mono- and diesters of sucrose, a transesterification reaction was set up: that is, sucrose and free fatty acids were solubilized in mutual organic solvents such as DMF, DMS, and DMP. SFEs produced by this Hass–Snell process were not approved for use in food because of the potentially toxic solvents used [30,34]. Later, Feuge et al. [36] described a solvent-free interesterification between molten sucrose and FAMEs at 170°C–187°C in the presence of lithium, potassium, and sodium soaps as solubilizers and as catalysts. In both processes, the unreacted FAME and free fatty acids are removed by distillation. The higher the monoester content of the purified product, the better the SFEs are as emulsifiers.

Enzymatic synthesis of sugar fatty acid esters including SFE has not been very successful compared with chemical synthetic procedures. One of the major reasons is that the use of mutual organic solvents (suitable for the solubilization of both sugars and fatty acids) mentioned earlier induces low catalytic activity and rapid inactivation of the enzymes [37]. Moreover, the use of enzymatic procedures was shown to be limited to only the synthesis of sugar fatty acid esters with a DS below 2 [38–52]. There have been a few reports on the lipase-catalyzed synthesis of SFE in organic solvents [38,39]. The lipase-catalyzed synthesis of other kinds of sugar fatty acid esters using organic solvent systems and sugar substrates, such as fructose [40–43], glucose [44–49], mannose [45,50,51], galactose [45], and maltose [38,52], has been reported with mixed results. Because these sugar fatty acid esters that were prepared by the enzymatic procedures have low DS values of 1–2, they may be applicable as emulsifiers rather than as fat substitutes by the food industry.

2. Emulsification Properties

SFEs were approved for use in Japan in 1959 and in the United States in 1983 [34]. Approved uses include:

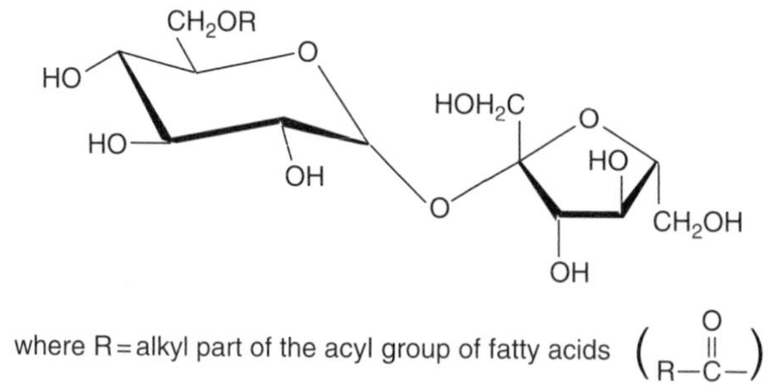

CH₂OR

HO

HO

OH

HOH₂C

O

HO

CH₂OH

O

OH

where R = alkyl part of the acyl group of fatty acids $\left(R-\overset{\overset{\displaystyle O}{\|}}{C}- \right)$

FIGURE 26.13 Structure of sucrose monoester, a sucrose ester.

TABLE 26.4
Some Properties of Sucrose Esters and Polyesters
Based on Degree of Substitution

Property	Degree of Substitution	Approximate HLB Value
Hydrophilic	1–4	5–16
Lipophilic	5–8	1–3
Digestibility		
Nondigestible	4–8	1–3
Digestible	1–3	5–16
Absorbability		
Nonabsorbable	4–8	1–3
Absorbable	1–3	5–16
Emulsification		
O/W emulsion	1–4	5–16
W/O emulsion	5–8	1–5
Antimicrobial	1–2	15–16

1. As emulsifiers in baked goods and baking mixes, dairy product analogs, frozen desserts and mixes, and whipping milk products
2. As texturizers in biscuit mixes
3. As components of protective coating for fresh apples, bananas, pears, pineapples, avocados, plantains, limes, melons, papaya, peaches, and plums to retard ripening and spoilage

The properties of SFE and olestra depend in part on the DS of sucrose with fatty acids, as shown in Table 26.4. SFE is used extensively in baked goods to improve the finished quality of frozen bread dough and sugar snap cookies. As a batter-aerating agent, SFE improves sponge and angel food cakes by increasing cake volume by 10%–20%. Other functional users of SFE are shown in Table 26.5. These esters have increased use in baked foods for several reasons:

1. They are easy to solubilize, can form good starch complexes, and can delay or control starch granule gelatinization. They prevent sticking to machinery and result in a softer crumb.
2. They are similar to the natural glycolipids found in wheat flour. Therefore, they promote expansion of gluten, stabilize aerated bubbles in batter, and make the manufacturing process easier.
3. They can inhibit crystal growth, such as ice crystals in frozen dough and sugar crystals in sweet baked goods.
4. They can increase loaf and cake volume and tenderness. They increase cookie spread factor and can affect crumb firmness and texture of sponge cakes.

However, Oh and Swanson [53] recently found that the addition of SFE with relatively high DS values as an emulsifier in chocolate products may prevent or retard the "bloom" formation, which plays an important role in the deterioration of quality and shelf life of these products.

TABLE 26.5

Some Functional Uses of Sucrose Fatty Acid Esters

Product[a]	Function
Bread	Increase loaf volume and maintain softness
Noodles	Prevent sticking of mixed dough
Cake	Increase cake volume; shorten whipping time
Crackers, cookies	Stabilize emulsion; prevent sticking to machinery; increase volume
Ice cream	Improve overrun by promoting stable emulsion, thus preventing excessive cohesion of fat during freezing
Whipping cream	Prevent water separation
Margarine and W/O emulsions	Emulsification; prevent spattering
Shortening	Stabilize emulsion; increase water-holding capacity
O/W Emulsions	Stabilize emulsions in a wide range of HLB values
Processed meat	Increase water-holding capacity of sausages; prevent separation of bolognas
Fruits	Coating to maintain freshness and extend shelf life
Drugs	Stabilize fat-soluble vitamins; lubricant; binder and filler
Cosmetics	Softness to skin; smoothness
Detergents	Cleaning agents for baby bottles; vegetables, and fruits
Antimicrobials	Prevent growth of microorganisms

[a] All used at no more than 10%.

3. Antimicrobial Properties

The antimicrobial properties of SFE depend on the type and chain length of the fatty acid esterified to sucrose. Medium-chain fatty acids such as lauric acid esterified to sucrose are better antimicrobial agents than sucrose esters of long-chain fatty acids. Sucrose monoesters are more potent than the di-, tri-, and polyesters against gram-positive bacteria. In general, SFEs exert more inhibitory action against gram-positive bacteria than against gram-negative bacteria. However, the polyesters are more active against gram-positive bacteria. The antimicrobial action of SFE is biostatic rather than biocidal in most cases. The antimicrobial properties of SFE received an excellent review from Marshall and Bullerman [54].

F. ESTERIFIED PROPOXYLATED GLYCEROL

EPG contains an oxypropylene group between the glycerol and fatty acids as follows:

$$EPG = P(OH)_{a+c}(EPO)_n(FE)_b,$$

where
P is the polyol with 2–8 primary OH groups (e.g., glucose)
C is 0–8 secondary and tertiary OH groups
EPO is C3–C6 epoxide
FE is the acyl group of fatty acids C8–C24
n is the minimum epoxylation index average number ($\geq a$)
$a + c$ is in the range of 3–8
$b \leq a + c$

EPG is synthesized from glycerol and propylene oxide, which is subsequently esterified with fatty acids to yield oil-like product. The physical properties of the finished product, like natural triacylglycerols, depend on the type of fatty acid esterified. EPGs resemble triacylglycerols and can be used to replace fat in most food applications [55]. Early results indicate that EPG is poorly hydrolyzed in animals fed the product for 30 days [9]. It was anticipated that up to 3–4 more years of research is needed before ARCO Chemical Company will be able to petition the FDA for food additive status, with use in foods as a fat substitute contemplated. This compound may also have applications in the cosmetic and pharmaceutical industries.

G. DIALKYL DIHEXADECYLMALONATE

DDM is synthesized from malonic acid, fatty acids, and hexadecane [56]. It is a low-calorie fat substitute suitable for high-temperature frying. DDM is absorbed, distributed, and eliminated through the liver. Frying test results indicate that there is no difference between potato and corn chips fried in a DDM–vegetable oil blend (60:40) and fried in conventional vegetable oil. Frito-Lay was expected to file a formal petition to the FDA for use of DDM as a fat substitute, on completion of further testing and safety studies.

H. JOJOBA OIL

Jojoba oil is composed of a mixture of linear esters of monounsaturated fatty acids and fatty alcohols containing 20–22 carbon atoms each (Figure 26.14). It is liquid at 10°C and above. Nestec Limited (Switzerland) started research on the possible use of jojoba oil as a low-calorie food ingredient in 1979. Results indicate that jojoba oil is not fully hydrolyzed by the pancreatic lipases because of their structure. About 40% of the oil was absorbed by rats fed the oil [16]. Safety and tolerance levels have been big issues. Indeed, rats consuming >16% of jojoba oil developed diarrhea and eventually died. Some levels of jojoba oil are stored in the liver. Possible use in control of obesity must be under careful medical supervision. No extreme adverse effects on liver, kidney, heart, or reproductive organ have been reported. Before jojoba could be marketed as a low-calorie oil, it would have to receive FDA approval [4].

I. POLYCARBOXYLIC ACID ESTERS AND ETHERS

Polycarboxylic acids containing 2–4 carboxylic acid groups esterified with saturated or unsaturated alcohols having straight or branched carbon chains consisting of 8–30 carbon atoms have been proposed as low-calorie edible oils [4,9,16,55,57]. Because the ester groups are reversed (fatty alcohol esterified onto a polycarboxylic acid backbone) from the corresponding esters present in triacylglycerols, the polycarboxylic acid esters are not susceptible to complete hydrolysis by lipases. Hamm [16] also reported that TATCA (Figure 26.15), TAC (Figure 26.16), TGE (Figure 26.17), and jojoba oil can serve as replacements for conventional edible fats and oils (Table 26.2). TATCA resembles triacylglycerol except that the glycerol backbone is replaced with tricarballylic acid and fatty acids with saturated and unsaturated alcohols.

TATCA is synthesized from tricarballylic acid and excess oleyl alcohol by solvent-free esterification at 135°C–150°C and a vacuum of 7 mmHg. TAC is synthesized by essentially the same procedure as the TATCA except that molecular sieving is not used to drive the reaction to the right (product formation). TATCA can replace vegetable oil in cooking and spreadable products such as

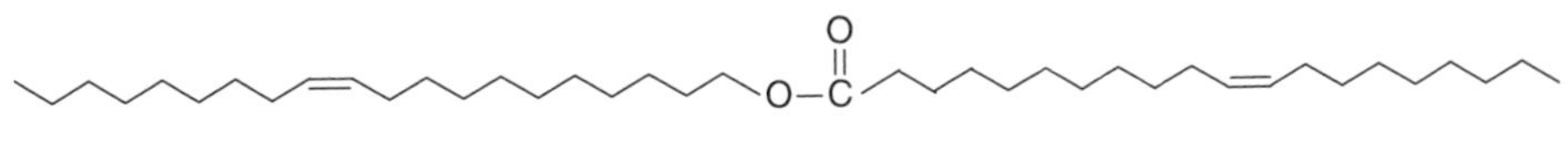

FIGURE 26.14 Structure of jojoba oil.

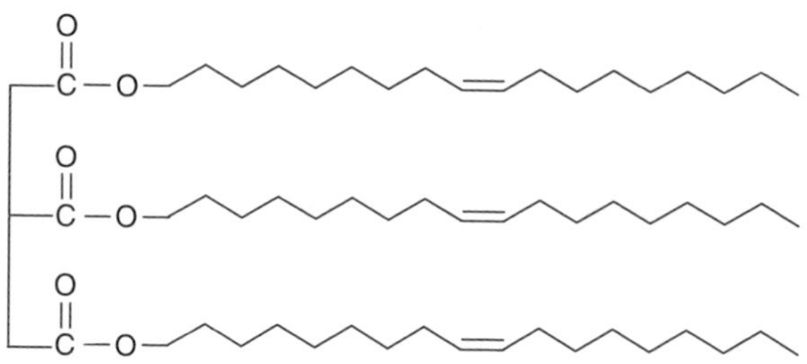

FIGURE 26.15 Structure of trialkoxytricarballylate.

margarine and mayonnaise. There are problems associated with the consumption of polycarboxylic acid esters. For example, in rat-feeding experiments, anal leakage, weakness, depression, and death were observed when TATCA and jojoba oil were fed at moderate to high dose levels (1–3 g) [16]. Weight gain data in rats indicated that both jojoba oil and TATCA are lower in calorie value than corn oil.

J. POLYGLYCEROL ESTERS

The process for preparation and purification of polyglycerol and polyglycerol esters was described by Babayan [58,59]. Esterification of glycerol with long-chain fatty acids reduces absorption by 31%–39% compared with corn oil, which is 98% absorbed. As the molecular weight of the polyglycerol portion increases, the hydrophilicity of the molecule also increases. But as the fatty acid chain length increases, hydrophilicity decreases. It was reported that male rats consuming 1 g/day of polyglycerol esters experienced weight gain comparable with that of rats fed lard. There were no abnormalities, except that diarrhea was observed in rats fed polyglycerol esters. Presently, polyglycerol esters are used as emulsifiers and dietetic acids such as in Weight Watchers ice cream. They can also be used in shortenings, margarines, bakery products, frozen desserts, ice cream, and confectioneries. The structure of a polyglycerol ester is shown in Figure 26.18.

K. POLYSILOXANE

Polysiloxane (PS) and phenylmethylpolysiloxane (a substituted polysiloxane) are organic derivatives of silica (SiO_2) with a linear polymeric structure (Figure 26.19). The polymeric molecules are inert, nontoxic, and varied in viscosity. Some have viscosity close to that of soybean oil. They are stable and not susceptible to oxidation as carbohydrate polyesters and vegetable oils. They are proposed as noncaloric, nonabsorbable liquid oil substitutes [3]. Studies in female obese Zucker rats resulted in weight reduction, and the animals did not compensate for the caloric dilution with PS by increasing their food intake [3].

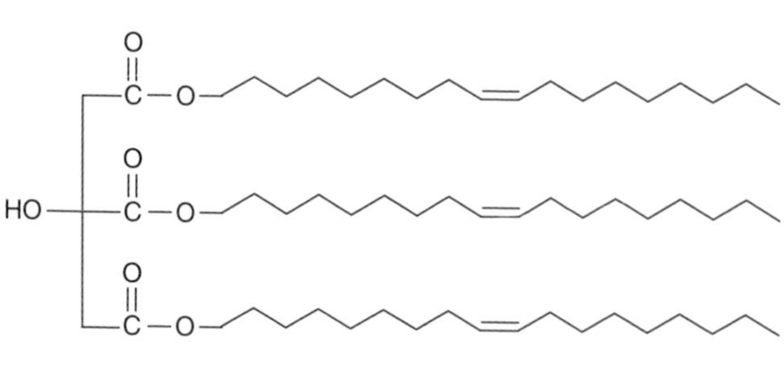

FIGURE 26.16 Structure of trialkoxycitrate.

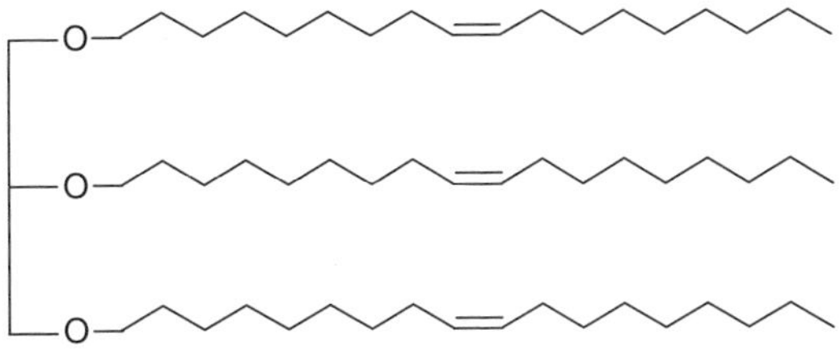

FIGURE 26.17 Structure of trialkoxyglyceryl ether.

IV. GENERAL PROPERTIES OF SYNTHETIC FAT SUBSTITUTES

Most of the physical properties studies have been conducted with SFE (DS < 4), which are digestible and absorbable. The surface-active properties of the mono- and diesters of sucrose were studied by Osipow et al. [30]. These esters were found to be good emulsifying agents and detergents with low toxicity. SFEs may be useful in cosmetic, pharmaceutical, and food applications. They are soluble in warm water, ethanol, methanol, and acetone. Other properties reported include foaming, stability, wetting, softening point, percent acyl radical (determined by saponification of the sucrose esters), and surface and interfacial tension reduction. More information on the physical properties of sugar ester emulsifiers such as solubility, melting and decomposition temperatures, surface tension, solubilizing ability, foaming, and emulsification ability is available [34]. The HLB of food-grade sucrose ester emulsifiers has been reported in two sources [34,60]. The HLB of sucrose esters and sucrose ester–glyceride blends as emulsifiers was reported to be dependent on (1) degree of substitution or esterification (number of OH groups esterified with fatty acids), (2) alkyl chain length in the ester group, and (3) the presence of acyl double or triple bonds (i.e., degree of unsaturation). Sucrose esters are synergistic with sucrose polyesters of low HLB values in stabilizing O/W emulsions.

The physical properties of the more completely substituted carbohydrate and alkyl glycoside fatty acid polyesters (HLB $\geq$ 4) have been reported [10,12–14,61] and are summarized in Table 26.6. The color, consistency, density, specific gravity, and refractive indices of these polyesters approximate those of commercial vegetable and salad oils. However, the apparent viscosities of some of the carbohydrate polyesters were significantly greater than the apparent viscosities of salad oil. The HLB values of the carbohydrate polyesters were between HLB 2 and 6, suggesting that they are capable of promoting W/O emulsions and might be useful as emulsifiers in butter, margarine, low-fat spreads, caramel, chocolate, candy, and shortening. The melting behavior of carbohydrate polyesters, such as sucrose and raffinose polyesters, has been reported [2,14,61]. The melting point decreases with an increase in unsaturation of fatty acids. Carbohydrate or alkyl glycoside polyester with a desirable melting point range and other physical properties is prepared by blending FAMEs of various degrees of saturation and unsaturation before synthesis of the saccharide polyester [14].

Lipid-based fat substitutes are subjected to oxidation just like conventional triacylglycerols. The addition of the antioxidant TBHQ was reported to greatly improve the stability of liquid

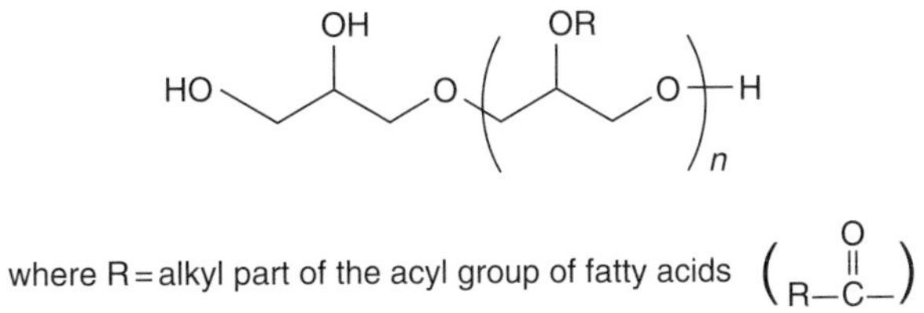

where R = alkyl part of the acyl group of fatty acids $\left(\begin{array}{c} O \\ \parallel \\ R-C- \end{array} \right)$

FIGURE 26.18 Structure of a polyglycerol ester.

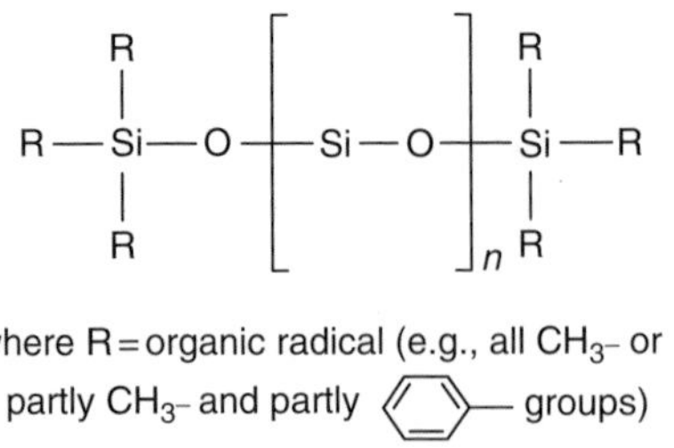

where R = organic radical (e.g., all CH_3– or
partly CH_3– and partly ⬡— groups)

FIGURE 26.19 Structure of polysiloxane.

carbohydrate fatty acid polyester fat substitutes and vegetable oils. The degree of added stability was greater in the fat substitutes and in the refined, bleached, and deodorized (RBD) soybean oil than in the crude soybean oil [62]. Part of the explanation was that RBD oils and fat substitutes made with them have lost some of the protective natural tocopherols, which must be restored to improve storage stability.

V. METABOLISM OF LIPID-BASED FAT SUBSTITUTES

Lipid-based fat substitutes, as exemplified by sucrose, trehalose, sorbitol, raffinose, stachyose, and sorbestrin, are not hydrolyzed by pancreatic lipase and consequently are not taken up by the intestinal mucosa [7,11,20,63]. However, alkyl glycoside fatty acid polyesters may be partially hydrolyzed and absorbed [19]. Mattson and Volpenhein [64] reported that as the number of ester groups increased from 4 to 8 in sucrose polyester, the rate of hydrolysis by lipase decreased. In other words, digestibility and absorbability of carbohydrate polyesters in rats and humans are inversely

TABLE 26.6
Comparison of General Properties of Fat Replacers

Property	Protein-Based	Carbohydrate-Based	Lipid-Based
Physical form	Powder	Powder/liquid	Powder, solid, liquid
Taste[a]	Bland	Bland	Bland
Caloric value, kcal/g	1–4	1–4.5	0–8.3
Melting point, °C[b]	Variable	Variable	Range variable depending on fatty acid, unsaturation, and degree of substitution
Color	White	White	Yellow to golden yellow
Odor	None	None	None
Soluble at 25°C in:	Water	Water, insoluble	Hexane, vegetable oils
Viscosity	Variable	Variable	Olestra more viscous than vegetable oils
Oxidative stability	Stable	Stable	Oil type of olestra not very stable; medium-chain length triacylglycerols very stable
Stability to frying temperature	Not stable	Not stable	Most are stable

[a] Bland is a general term applicable to most but not all fat replacers.
[b] Variable melting point; depends on the specific compound. Actual values may be obtained from the supplier.

related to the degree of substitution or esterification [20,64]. Because of the molecular size of the fat substitutes and steric hindrance, the lipases cannot get to the substrate, and subsequently the formation of the enzyme–substrate (ES) complex required for lipase action on carbohydrate polyesters is prevented [11]. Poor digestibility and absorbability of carbohydrate polyesters indicate that these substances supply few or zero calories; hence, they are called low-calorie fat substitutes. Olestra is an FDA-approved, zero-calorie fat substitute. Because olestra is the most studied and the only synthetic lipid-based fat substitute approved, it is the focus of most of the discussions.

A. Effect of Olestra on Serum Lipids and Weight Reduction

An early report by Crouse and Grundy [65] indicated that sucrose polyester (olestra) interferes with the absorption of cholesterol because cholesterol is soluble in the fat substitute. According to Grundy et al. [66], nondiabetic patients on a calorie-restricted diet plus SPE exhibited decreases in total cholesterol and low-density lipoprotein (LDL) cholesterol of 20% and 26%, respectively. Reduction in serum cholesterol was at the expense of LDL-cholesterol. Diabetic patients with hypertriacylglycerolemia caloric restriction showed a marked reduction in plasma triacylglycerols with or without sucrose polyester consumption. Caloric restriction apparently reduced cholesterol by reducing cholesterol synthesis. Sucrose polyester has little effect on the concentration of high-density lipoprotein (HDL)-cholesterol [63,67,68]. LDL-cholesterol is associated with the development of atherosclerotic plaques, whereas HDL-cholesterol is associated with the prevention of atherosclerotic plaques. Patients with high cholesterol levels are advised to lower their serum cholesterol levels by losing weight and reducing both total fat intake (especially saturated fats) and cholesterol intake. Sucrose polyester may also be useful in weight reduction [11,19,63]. Other carbohydrate polyesters can assist in the reduction of cholesterol levels [5,69–77]. As cholesterol enters the digestive tract, it dissolves into the carbohydrate polyester oil phase and is excreted along with the carbohydrate polyester. Mellies et al. [69] reported that in obese hypercholesterolemic outpatients, SPE induced significant reductions in LDL-cholesterol beyond the effects of weight reduction. Glueck et al. [70] found that substitution of SPE for dietary fats in hypocaloric diets in obese women heterozygous for familial hypercholesterolemia resulted in 23% reduction of LDL-cholesterol and in weight loss. Mattson and Jandacek [74] reported that [4-^{14}C] cholesterol injected into rats that were subsequently fed SPE diets appeared in the feces of SPE-fed animals. Cholesterol absorption was reduced when 7 g of SPE was fed twice a day to 20 normocholesterolemic male inpatients in a double-blind crossover trial [72]. It should be noted that Procter & Gamble did not pursue the approval of olestra as a cholesterol-reducing agent.

B. Effect of Olestra on Macronutrients and Fat-Soluble Vitamins

Olestra does not affect the availability of macronutrients such as carbohydrates, triacylglycerols, and proteins or that of micronutrients such as water-soluble vitamins; olestra consumption was, however, found to increase the intake of carbohydrate but without affecting total daily energy intake or usual patterns of hunger and fullness. SPEs may affect the availability of some fat-soluble vitamins such as vitamins A, D, E, and K, β-carotene, and lycopene [78], and cholesterol. In heavy snack eaters, olestra can reduce vitamin A by 5% and vitamin E by about 3%. Addition of retinyl palmitate and tocopheryl acetate can offset the losses of vitamins A and E, respectively. The FDA requires that foods containing olestra be supplemented with fat-soluble vitamins to replace the amount normally found in vegetable oil and to compensate for olestra's effect on vitamin absorption (Table 26.7).

Mellies et al. [68] reported earlier that plasma vitamin E was significantly reduced when 40 g/day of sucrose polyester was consumed by hypercholesterolemic outpatients compared with patients consuming a placebo. Crouse and Grundy [65] reported that vitamin E concentrations

TABLE 26.7

Compensation Levels of Fat-Soluble Vitamins in Olestra-Containing Snacks

Vitamin	Suggested Supplementation Level[a]
A	51 retinol eq/g olestra as retinyl palmitate or retinyl acetate ($= 170$ IU/g olestra or $0.34 \times$ RDA/10 g olestra)
D	12 IU vitamin D/g olestra $= (0.3 \times$ RDA/10 g olestra)
E	1.9 mg α-tocopherol eq/g olestra $= (0.94 \times$ RDA/10 g olestra)
K	8 μg vitamin K_1/g olestra $= (1.0 \times$ RDA/10 g olestra)

Source: Adapted from *Federal Register*, Food Additives Permitted for Direct Addition to Food for Human Consumption: Olestra, Final Rule. 21 CFR Part 172, January 30, 1996: Vol. 61(20), 3118.

[a] IU, international unit; RDA, recommended daily allowance. The suggested levels are to compensate for amounts that are not absorbed from the diet because of olestra action.

decreased by 24% and vitamin A was not affected. Fallat et al. [79] reported that both vitamins A and E were significantly reduced when SPE was added to patients' diet at 50 g/week of diet. In a later study, Mellies et al. [69] reported no decrease in vitamin A, but a 23% decrease in vitamin E. Daily consumption of 3 g of SPE led to significant reductions in plasma concentrations of β-carotene and lycopene in humans [78].

There is no clear consensus on the loss of vitamin A and carotenoids when consumed with olestra, and this may warrant further investigation. Vitamin D decreased in obese, hypercholesterolemic outpatients fed low-fat diets with or without SPE or conventional fat placebo [69]. Miller et al. [80] supplemented both vitamins A and E in the diets of beagle dogs fed for 20 months with SPE at 10% and found no effect on the level of vitamins D and K, but concluded that with supplementation, vitamins A and E status remained sufficient in all groups. Another study suggested that vitamin K was not affected by consumption of SPE [68]. Nutritional implications of fat substitute consumption and impact on fat intake were reported recently [9,81–84]. Potential nutritional benefits of nondigestible lipid-based fat substitutes are shown in Table 26.8.

TABLE 26.8

Some Nutritional Uses of Nondigestible Lipid-Based Fat Substitutes

Replace saturated fat in the diet
Contribute reduced to zero calories to the diet
Reduce total fat intake in the diet
Reduce total cholesterol level
Reduce serum and plasma triacylglycerol levels
Reduce LDL-cholesterol level
Maintain ideal body weight or promote weight loss
Maintain HDL-cholesterol level
Reduce coronary heart disease risk factors
Reduce risk to certain kinds of cancer

Carbohydrate polyesters in general, if taken in moderation, have the potential to help consumers lower their total fat intake, to reduce their saturated fat and cholesterol levels, to lose weight, and to achieve a healthier lifestyle. Again, they may benefit persons at high risk for coronary heart disease, colon cancer, and obesity [4,63,77].

C. SIDE EFFECTS AND LIMITATIONS

Anal leakage or oil loss resulted from consumption of large amounts of liquid olestra at the early stages of olestra development and studies. This can be prevented by addition of saturated fatty acids and their salts [75]. It was suggested that oil loss can also be prevented by controlling the rheology (viscosity) and stiffness of carbohydrate polyesters by simply adjusting the fatty acid composition of the fats and oils used in their synthesis [11,19,67,77]. Other gastrointestinal symptoms due to consumption of olestra are flatulence, soft stools, fecal urgency, diarrhea, and increased bowel movements [5,68,69]. Indeed, it was recommended that soft stool can be prevented by using olestra with semisolid consistency or by increasing the viscosity of the olestra or product [67].

The most recently publicized problems with consuming olestra are its effects on serum carotenoid (antioxidant vitamin) levels, fat-soluble vitamins, gastrointestinal tract (oil loss, loose stools), and oral contraceptives or lipophilic drugs (Table 26.9). However, the role of carotenoids in health is still under investigation and not fully understood. Also not known is the effect of high levels of olestra consumption on consumers who may indulge in an effort to reduce or control their weight. To protect consumers, the FDA requires that products containing olestra bear labels [67] stating: "This product contains Olestra. Olestra may cause abdominal cramping and loose stools. Olestra inhibits the absorption of some vitamins and other nutrients. Vitamins A, D, E, and K have been added." Vitamins A, D, E, and K are required to be added as shown (Table 26.7) to protect consumers from depleting their essential vitamins [67]. In other words, the added vitamins will compensate for the amounts that are not absorbed due to olestra's interference with their absorption when olestra is eaten at the same time as foods containing the vitamins.

Olestra has the potential to benefit some segments of the population. From the available data, this substance is considered safe and nontoxic: because olestra is not digested or absorbed, no major component of it is available to produce unsafe or toxic effects [67]. Research has shown that olestra is not genotoxic and does not affect reproduction or cause birth defects when consumed at certain levels. The long-term effects on humans of olestra use are open to debate. The published estimated

---

TABLE 26.9
Limitations to Olestra Consumption

Anal leakage or "oil loss" or fecal urgency

Decrease or loss of vitamins E and A from other foods
 (FDA-recommended compensation [67])[a]

Soft stool

Diarrhea

Increased bowel movement

Loss of carotenoids [78]; 60% decrease in β-carotene
 (supplementation under debate)

May affect drug absorption (coumarin, oral contraceptives—
 magnitude of effect small)

Inflammatory bowel disease

[a] FDA recommendations informing the consumer, through labeling, of some of these limitations, are given in Ref. [67].

---

daily intake (EDI) of olestra is 7 g/person for chronic consumption by the 90th percentile snack eaters and 20 g/person for short-term consumption [67].

Foods made with olestra compared with their commercially available counterparts taste the same, have the same flavor, texture, and appearance, and perhaps are less oily than the commercial product [67]. Recent studies showed that olestra was not associated with increased incidence or severity of gastrointestinal symptoms [85,86].

VI. SAFETY AND REGULATORY UPDATES

For FDA approval of any new food additive, the petitioner must conduct toxicological, clinical, gastrointestinal, and nutritional testing to prove that the compound is safe and submit a FAP. Since the discovery of SPE (olestra) in 1968 by Mattson and Volpenhein, extensive studies have been carried out in seven different species of animals, including long-term studies in monkeys and over 60 clinical trials involving over 8000 men, women, and children in several universities and medical research centers. The results show that olestra is safe: it is not absorbed or metabolized; it is nontoxic, nonmutagenic, and noncarcinogenic; and it does not affect reproduction or cause birth defects [64,75,78,87–91]. Olestra is biodegradable in sludge-amended soils and does not adversely affect the physical properties of the soil. It is not toxic to plants and animals. Gastrointestinal testing revealed that olestra does not affect gastrointestinal morphology, transit rate, bile acid physiology, bowel movement, pancreatic response, or intestinal microflora to any appreciable extent. It does not significantly affect the absorption or efficacy of orally dosed lipid-soluble drugs such as diazepam (tranquilizer), propranolol (cardiovascular agent), aspirin (analgesic), ethinyl estradiol (oral contraceptive), and norethindrone (oral contraceptive) [67].

The results reported by Miller et al. [80] indicated that sucrose polyester was not toxic when fed at 10% level to dogs in a 20 month study. No SPE was detected in the liver, heart, kidney, spleen, lymph nodes, and adipose tissues of 26 monkeys fed SPE at 0%, 2%, 4%, or 6% on the diets for 2 months. In addition, no SPE was detected in liver of these monkeys after 2 years of 9% SPE feeding. Daily consumption of 18 g of SPE did not affect the absorption or efficacy of the highly lipophilic oral contraceptives containing 300 μg of norgestrel and 30 μg of ethinyl estradiol (Lo/Oral-28) after 28 days [92]. SPE is not genotoxic in the *Salmonella*/mammalian microsome test, the L5178Y thymidine kinate (TK +/−) mouse lymphoma assay, an unscheduled DNA synthesis array in primary rat hepatocytes, or an in vitro cytogenetic assay in Chinese hamster ovary cells [90]. When heated sucrose polyester–vegetable oil blends were fed at 5% level to rats for 91 days, there were no toxic effects detected on hematological and histomorphology evaluation [88]. Excellent reviews on the regulatory aspects, labeling, and safety requirements of fat substitutes are available [2,9,89,93].

The current FDA approval for olestra is only for use in ready-to-eat savory or salty snacks and not sweet snacks [67]. It can be used up to 100% replacement for conventional fats in flavored and unflavored extruded snacks and crackers. Olestra can be used in frying, dough conditioners, oil sprays, and flavors in savory snacks. It cannot be used in other foods, including fat-free cakes and french fries. A separate petition would have to be filed for each additional use in the future, and each such product would be subjected to the same regulatory approval process. The major factors considered by the FDA and before the ruling on olestra use in savory snacks are summarized in Table 26.10.

VII. APPLICATIONS OF SUCROSE ESTERS

SFEs are sugar esters with a lower degree of substitution or esterification (DS 1–3). SFEs are FDA-approved and are currently marketed by Mitsubishi Chemical America, Inc. (White Plains, NY) and Crodesta, Inc. (New York). They are digestible and absorbable and can be used as emulsifiers and as lubricating, wetting or dispersion, solubilization, stabilization, anticaking, antimicrobial, or

TABLE 26.10

FDA Determination Leading to the Approval of Olestra

Item/Property	Effect as Petitioned	FDA Ruling
Stability	Stable as triacylglycerols; less polymer formation; higher free fatty acid formed	No difference in by-product formation during frying and backing
Absorption	Small amounts of penta- and lower esters are absorbed and metabolized to fatty acids and sucrose	Essentially not absorbed
Genetic toxicity	No evidence of genetic toxicity or mutagenicity from heated olestra	Not genotoxic
Teratogenicity	Not teratogenic or embryotoxic	Not teratogenic
Carcinogenicity	Not adverse treatment-related effect	No association to adenomas, pituitary, leukemia, basophilic foci with olestra treatment
Drug absorption	No effect on lipophilic and nonlipophilic drug absorption	Magnitude of effect on drug absorption (including Coumadin, oral contraceptives) is not significant
Water-soluble nutrients	No adverse effect	No significant effect; no supplementation required
Fat-soluble vitamins	No significant decrease in vitamin A; dose-related significant decrease in vitamin E; small decrease in vitamin D; dose–response decrease in vitamin K_1 60% decrease in β-carotene	Not significant as analyzed; vitamin E decrease; serum vitamin D decrease; potential for vitamin K decrease on long-term olestra use Carotenoids' effect not conclusive; supplementation for vitamins A, D, E, K recommended
Gastrointestinal tract	GI symptoms not clinically significant	Dose–response for diarrhea/loose stools and fecal urgency; recommend informing consumer through labeling
	Oil loss if large liquid olestra is consumed	Low stiffness leads to oil loss, not a health hazard
	Increased bowel movement, soft stools	Symptoms not the same as in high-fiber diets
	Inflammatory bowel disease affected little	Not highly exacerbated at 20 g/day
	No significant health effects on young children	Data from adults can be extrapolated to children
Intestinal microflora	Olestra is not metabolized by GI microflora; microbial counts, short-chain fatty acid, breath hydrogen not greatly affected	No metabolism by microflora; no differences as observed with dietary changes
Bile acid metabolism	No effect on bile acid synthesis or excretion	No major changes in bile acid metabolism

viscosity-reducing agents, in baking or as coatings for bananas, apples, pears, pineapples, and so on. Alkyl glycoside mono- and diesters can be used for similar applications. Table 26.5 shows some functional uses of SFEs. The concentrations of SFE added to foods may range from 0.005% to 10%.

VIII. APPLICATIONS OF SYNTHETIC FAT SUBSTITUTES

Carbohydrate polyesters such as sucrose polyester have the potential to lower cholesterol levels in certain lipid disorders [66,68,69]. Daily substitution of fat with 30 g of SPE will reduce caloric intake by 270 cal. Grundy et al. [66] reported that SPEs were tolerated by diabetic patients at a maximum of 90 g/day when total caloric intake was reduced to 1000 kcal/day. Thus, SPEs and other carbohydrate polyesters may be used for therapeutic weight reduction [63]. They have the benefit of adding bulk to the diet without adding calories. Consumption of carbohydrate polyesters may benefit persons at high risk for coronary heart disease, colon cancer, and obesity [63,81]. Table 26.8 lists some nutritional uses and applications of nondigestible fat substitutes.

Carbohydrate fatty acid polyesters are poor surfactants for O/W emulsions when used alone, but when blended with other emulsifiers or SFEs, they exhibit synergistic effects [2]. Food made with SPE has 10%–50% fewer calories than food made with regular oils and shortenings. For example, an order of french fries prepared with 100% SPE contains about 144 cal, 52.7% less than the same item with conventional oil. Thus, SPEs can be used for cooking, frying, and baking. They are useful in dairy products, spreads, dressings, margarines, sauces, confectioneries, and dessert manufacture.

IX. PERSPECTIVES

With the approval of olestra, it was expected that studies by other competitors will be accelerated and alternative compounds will be developed to challenge the olestra market. This has not happened. Other approval requests for uses of olestra are expected to be made if the test market results for olestra-made snacks are favorable. However, if more problems arise from the test marketing, more opposition to requested approvals is expected. Approval for a new food additive is not easy: the additive petition for olestra/Olean contained over 150,000 pages that included safety data on more than 100 laboratory studies on 7 animal species and more than 60 clinical trials involving over 8000 men, women, and children. Therefore, any new food additive or new use of an approved ingredient must undergo the same FDA petition and approval process, and studies must include toxicologic, preclinical, and clinical data showing that the substance is safe and efficacious.

The consumers were willing to try low-fat or reduced-calorie foods [6]. The market for fat substitutes is likely to increase. The labeling requirement for fat was revised to include nondigestible fat substitutes and in the future may include reduced-calorie structured lipids. Foods containing fat substitutes will be so labeled to inform the consumer of potential benefits and health implications. Obviously, more research is still needed in this area of consumer interest. The FDA and the government must be willing to support more research through independent investigators in the interest of the consumer. The need for more studies on the long-term effect of fat substitutes on the immune system, tolerance levels/dosage, essential fatty acid availability, degradation in the soil, physical properties, applications, stability, and analysis in food matrices should be emphasized.

Reduced-calorie fats and fat substitutes are likely to offer the public new strategies and more choices for reducing fat consumption. The specialty or structured lipids may compete with the zero-calorie fats or may come to be preferred because they can be designed for optimal nutrition, health, and disease prevention or to contain fewer calories than conventional fats and oils (see Chapters 31 and 32). A partial replacement and systems approach, rather than total replacement of dietary fats with olestra-type materials, and consumption in moderation, should be encouraged. Total elimination of fat in the diet is not nutritionally prudent. A systems approach to formulating reduced-, low-, or no-calorie foods seems appropriate at this time, barring the absence of a "magic bullet" or an ideal fat substitute.

REFERENCES

1. Select Committee on Nutrition and Human Needs. *U.S. Senate Dietary Goals for the United States.* U.S. Government Printing Office, Washington, D.C., 1977.
2. C.C. Akoh and B.G. Swanson, eds. *Carbohydrate Polyesters as Fat Substitutes.* Dekker, New York, NY, 1994, 269 p.
3. E.F. Bracco, N. Baba, and S.A. Hashim. Polysiloxane: Potential noncaloric fat substitute: Effects on body composition of obese Zucker rats. *Am. J. Clin. Nutr.* 46:784–789 (1987).
4. B.F. Haumann. Getting the fat out—Researchers seek substitutes for full-fat fat. *J. Am. Oil Chem. Soc.* 63:278–288 (1986).

5. R.G. LaBarge. The search for a low-caloric oil. *Food Technol.* 42(1):84–90 (1988).

6. C.M. Bruhn, A. Cotter, K. Diaz-Knauf, J. Sutherlin, E. West, N. Wightman, E. Williamson, and M. Yaffee. Consumer attitudes and market potential for foods using fat substitutes. *Food Technol.* 46(4):81–86 (1992).

7. F.H. Mattson and R.A. Volpenhein. Rate and extent of absorption of the fatty acids of fully esterified glycerol, erythritol, xylitol and sucrose as measured in thoracic duct cannulated rats. *J. Nutr.* 102:1177–1180 (1972).

8. K.A. Harrigan and W.M. Breene. Fat substitutes: Sucrose esters and Simplesse. *Cereal Foods World* 34:261–267 (1989).

9. S.N. Gershoff. Nutrition evaluation of dietary fat substitutes. *Nutr. Rev.* 53:305–313 (1995).

10. C.C. Akoh and B.G. Swanson. One-stage synthesis of raffinose fatty acid polyesters. *J. Food Sci.* 52:1570–1576 (1987).

11. C.C. Akoh and B.G. Swanson. Preliminary raffinose polyester and methyl glucoside polyester feeding trials with mice. *Nutr. Rep. Int.* 39:659–666 (1989).

12. C.C. Akoh and B.G. Swanson. Synthesis and properties of alkyl glycoside and stachyose fatty acid polyesters. *J. Am. Oil Chem. Soc.* 66:1295–1301 (1989).

13. C.C. Akoh and B.G. Swanson. Preparation of trehalose and sorbitol fatty acid polyesters by interesterification. *J. Am. Oil Chem. Soc.* 66:1581–1587 (1989).

14. C.C. Akoh and B.G. Swanson. Optimized synthesis of sucrose polyesters: Comparison of physical properties of sucrose polyesters, raffinose polyesters and salad oils. *J. Food Sci.* 55:236–243 (1990).

15. R.E. Smith, J.W. Finley, and G.A. Leveille. Overview of Salatrim, a family of low-calorie fats. *J. Agric. Food Chem.* 42:432–434 (1994).

16. D.J. Hamm. Preparation and evaluation of trialkoxytricarballylate, trialkoxycitrate, trialkoxyglycerylether, jojoba oil and sucrose polyester as low calories replacements of edible fats and oils. *J. Food Sci.* 49:419–428 (1984).

17. C.J. Glueck, P.A. Streicher, E.K. Illig, and K.D. Weber. Dietary fat substitutes. *Nutr. Res.* 14:1605–1619 (1994).

18. G.P. Rizzi and H.M. Taylor. A solvent-free synthesis of sucrose polyesters. *J. Am. Oil Chem. Soc.* 55:398–401 (1978).

19. C.C. Akoh and B.G. Swanson. Absorbability and weight gain by mice fed methyl glucoside fatty acid polyesters: Potential fat substitutes. *J. Nutr. Biochem.* 2:652–655 (1991).

20. F.H. Mattson and G.A. Nolen. Absorbability by rats of compounds containing from one to eight ester groups. *J. Nutr.* 102:1171–1175 (1972).

21. K. Hess and E. Messner. Über die synthese von fettsanrederivaten der Zuckerarten. *Chem. Ber.* 54:499–523 (1921).

22. C.C. Akoh. Synthesis of carbohydrate fatty acid polyesters. In: *Carbohydrate Polyesters as Fat Substitutes* (C.C. Akoh and B.G. Swanson, eds.). Dekker, New York, NY, 1994, pp. 9–35.

23. C.J. Shieh, P.E. Koehler, and C.C. Akoh. Optimization of sucrose polyester synthesis using response surface methodology. *J. Food Sci.* 61:97–100 (1996).

24. R.J. Jandacek and M.R. Webb. Physical properties of pure sucrose octaesters. *Chem. Phys. Lipid* 22:163–176 (1978).

25. G.W. Chase, C.C. Akoh, and R.R. Eitenmiller. Liquid chromatographic analysis of sucrose polyester in salad dressing by evaporative light scattering mass detection. *J. AOAC Int.* 78:1324–1327 (1995).

26. G.W. Chase, C.C. Akoh, R.R. Eitenmiller, and W.O. Landen. Liquid chromatographic method for the concurrent analysis of sucrose polyester, vitamin A palmitate, and β-carotene in margarine. *J. Liquid Chromatogr.* 18:3129–3138 (1995).

27. D.R. Gardner and R.A. Sanders. Isolation and characterization of polymers in heated olestra and olestra/triglyceride blend. *J. Am. Oil Chem. Soc.* 67:788–796 (1990).

28. T.T. Boutte and B.G. Swanson. Supercritical fluid extraction and chromatography of sucrose and methyl glucose polyester. In: *Carbohydrate Polyesters as Fat Substitutes* (C.C. Akoh and B.G. Swanson, eds.). Dekker, New York, NY, 1994, pp. 65–93.

29. M.A. Drake, C.W. Nagel, and B.G. Swanson. Sucrose polyester content in foods by a colorimetric method. *J. Food Sci.* 59:655–656 (1994).

30. L. Osipow, F.D. Snell, W.C. York, and A. Finchler. Methods of preparation—Fatty acid esters of sucrose. *Ind. Eng. Chem.* 48:1459–1462 (1956).

31. R.A. Sanders, D.R. Gardner, M.P. Lacey, and T. Keough. Desorption mass spectrometry of Olestra. *J. Am. Oil Chem. Soc.* 69:760–771 (1992).

32. G.W. Chase, C.C. Akoh, and R.R. Eitenmiller. Evaporative light scattering mass detection for high performance liquid chromatographic analysis of sucrose polyester blends in cooking oils. *J. Am. Oil Chem. Soc.* 71:1273–1276 (1994).

33. H.Y. Chung, J. Park, J.H. Kim, and U.Y. Kong. Preparation of sorbitol fatty acid polyesters, potential fat substitutes: Optimization of reaction conditions by response surface methodology. *J. Am. Oil Chem. Soc.* 73:637–643 (1996).

34. Mitsubishi-Kasei Food Corporation. *Ryoto Sugar Ester Technical Information: Nonionic Surfactant/Sucrose Fatty Acid Ester/Food Additive.* M-KFC, Tokyo, Japan, 1987, 20 p.

35. S.R. Drake, J.K. Fellman, and J.W. Nelson. Postharvest use of sucrose polyesters for extending the shelf-life of stored "Golden Delicious" apples. *J. Food Sci.* 52:1283–1285 (1987).

36. R.O. Feuge, H.J. Zeringue, T.J. Weiss, and M. Brown. Preparation of sucrose esters by interesterification. *J. Am. Oil Chem. Soc.* 47:56–60 (1970).

37. E.N. Vulfson. Enzymatic synthesis of surfactants. In: *Novel Surfactants: Preparation, Applications, and Biodegradability; Surfactant Science Series (Volume 74)* (K. Holmberg, ed.). Dekker, New York, NY, 1988, pp. 279–300.

38. M. Ferrer, M.A. Cruces, F.J. Plou, M. Bernabe, and A. Ballesteros. A simple procedure for the regioselective synthesis of fatty acid esters of maltose, leucrose, maltotriose and *n*-dodecyl maltosides. *Tetrahedron* 56:4053–4061 (2000).

39. M.A. Ku and Y.D. Hang. Enzymatic synthesis of esters in organic medium with lipase from *Byssochlamys fulva*. *Biotechnol. Lett.* 17:1081–1084 (1995).

40. S. Sabeder, M. Habulin, and Z. Knez. Lipase-catalyzed synthesis of fatty acid fructose esters. *J. Food Eng.* 77:880–886 (2006).

41. F. Chamouleau, D. Coulon, M. Girardin, and M. Ghoul. Influence of water activity and water content on sugar esters lipase-catalyzed synthesis in organic media. *J. Mol. Catal., B Enzym.* 11:949–954 (2001).

42. S. Soultani, J.M. Engasser, and M. Ghoul. Effect of acyl donor chain length and sugar/acyl donor molar ratio on enzymatic synthesis of fatty acid fructose esters. *J. Mol. Catal., B Enzym.* 11:725–731 (2001).

43. A. Schlotterbeck, S. Lang, V. Wray, and F. Wagner. Lipase-catalyzed monoacylation of fructose. *Biotechnol. Lett.* 15:61–64 (1993).

44. D.S. Kelkar, A.R. Kumar, and S.S. Zinjarde. Hydrocarbon emulsification and enhanced crude oil degradation by lauroyl glucose ester. *Bioresour. Technol.* 98:1505–1508 (2007).

45. J. Chen, Y. Kimura, and S. Adachi. Continuous synthesis of 6-*O*-linoleoyl hexose using a packed-bed reactor system with immobilized lipase. *Biochem. Eng. J.* 22:145–149 (2005).

46. P. Degn and W. Zimmermann. Optimization of carbohydrate fatty acid ester synthesis in organic media by a lipase from *Candida antarctica*. *Biotechnol. Bioeng.* 74:483–491 (2001).

47. Y. Watanabe, Y. Miyawaki, S. Adachi, K. Nakanishi, and R. Matsuno. Synthesis of lauroyl saccharides through lipase-catalyzed condensation in microaqueous water-miscible solvents. *J. Mol. Catal., B Enzym.* 10:241–247 (2000).

48. J.A. Arcos, M. Bernabe, and C. Otero. Quantitative enzymatic production of 6-*O*-acylglucose esters. *Biotechnol. Bioeng.* 57:505–509 (1998).

49. I. Ikeda and A.M. Klibanov. Lipase-catalyzed acylation of sugars solubilized in hydrophobic solvents by complexation. *Biotechnol. Bioeng.* 42:788–791 (1993).

50. Y. Watanabe, Y. Miyawaki, S. Adachi, K. Nakanishi, and R. Matsuno. Equilibrium constant for lipase-catalyzed condensation of mannose and lauric acid in water-miscible organic solvents. *Enzyme Microb. Technol.* 29:494–498 (2001).

51. Y. Watanabe, Y. Miyawaki, S. Adachi, K. Nakanishi, and R. Matsuno. Continuous production of acyl mannoses by immobilized lipase using a packed-bed reactor and their surfactant properties. *Biochem. Eng. J.* 8:213–216 (2001).

52. N.R. Pedersen, R. Wimmer, J. Emmersen, P. Degn, and L.H. Pedersen. Effect of fatty acid chain length on initial reaction rates and regioselectivity of lipase-catalysed esterification of disaccharides. *Carbohydr. Res.* 337:1179–1184 (2002).

53. J.H. Oh and B.G. Swanson. Polymorphic transitions of cocoa butter affected by high hydrostatic pressure and sucrose polyesters. *J. Am. Oil Chem. Soc.* 83:1007–1014 (2006).

54. D.L. Marshall and L.B. Bullerman. Antimicrobial properties of sucrose fatty acid esters. In: *Carbohydrate Polyesters as Fat Substitutes* (C.C. Akoh and B.G. Swanson, eds.). Dekker, New York, NY, 1994, pp. 149–167.

55. W.E. Artz and S.L. Hansen. Other fat substitutes. In: *Carbohydrate Polyesters as Fat Substitutes* (C.C. Akoh and B.G. Swanson, eds.). Dekker, New York, NY, 1994, pp. 197–236.

56. S.A. Schlicker and C. Regan. Innovations in reduced-calorie foods: A review of fat and sugar replacement technologies. *Top. Clin. Nutr.* 6:50–60 (1990).

57. J.F. White and M.R. Pollard. Non-digestible fat substitutes of low-caloric value. U.S. Patent 4,861,613 (1989).

58. V.K. Babayan. Polyglycerol esters: Unique additives for the bakery industry. *Cereal Foods World* 27:510–512 (1982).

59. V.K. Babayan. Polyglycerols and polyglycerol esters in nutrition, health and disease. *J. Environ. Pathol. Toxicol. Oncol.* 6:15–24 (1986).

60. R.K. Gupta, K. James, and F.J. Smith. Sucrose esters and sucrose ester/glyceride blends as emulsifiers. *J. Am. Oil Chem. Soc.* 60:862–869 (1983).

61. C.J. Shieh, C.C. Akoh, and P.E. Koehler. Formulation and optimization of sucrose polyester physical properties by mixture response surface methodology. *J. Am. Oil Chem. Soc.* 73:455–460 (1996).

62. C.C. Akoh. Oxidative stability of fat substitutes and vegetable oils by the oxidative stability index method. *J. Am. Oil Chem. Soc.* 71:211–216 (1994).

63. B.M. Grossman, C.C. Akoh, J.K. Hobbs, and R.J. Martin. Effects of a fat substitute, sucrose polyester, on food intake, body composition and serum factors in lean and obese Zucker rats. *Obes. Res.* 2:271–278 (1994).

64. F.H. Mattson and R.A. Volpenhein. Hydrolysis of fully esterified alcohols containing from one to eight hydroxyl groups by the lipolytic enzymes of rat pancreatic juice. *J. Lipid Res.* 13:325–328 (1972).

65. J.R. Crouse and S.M. Grundy. Effects of sucrose polyester on cholesterol metabolism in man. *Metabolism* 28:994–1000 (1979).

66. S.M. Grundy, J.V. Anastasia, Y.A. Kesaniemi, and J. Abrams. Influence of sucrose polyester on plasma lipoproteins, and cholesterol metabolism in obese patients with and without diabetes mellitus. *Am. J. Clin. Nutr.* 44:620–629 (1986).

67. *Federal Register*. Food Additives Permitted for Direct Addition to Food for Human Consumption: Olestra, Final Rule. 21 CFR Part 172, January 30, 1996: Vol. 61(20):3118–3173.

68. M.J. Mellies, R.J. Jandacek, J.D. Taulbee, M.B. Tewksbury, G. Lamkin, L. Baehler, P. King, D. Boggs, S. Goldman, A. Gouge, R. Tsang, and C.J. Glueck. A double-blind, placebo-controlled study of sucrose polyester in hypercholesterolemic outpatients. *Am. J. Clin. Nutr.* 37:339–346 (1983).

69. M.J. Mellies, C. Vitale, R.J. Jandacek, G.E. Lamkin, and C.J. Glueck. The substitution of sucrose polyester for dietary fat in obese, hypercholesterolemic outpatients. *Am. J. Clin. Nutr.* 41:1–12 (1985).

70. C.J. Glueck, R. Jandacek, E. Hogg, C. Allen, L. Baehler, and M. Tewksbury. Sucrose polyester: Substitution for dietary fats in hypocaloric diets in the treatment of familial hypercholesterolemia. *Am. J. Clin. Nutr.* 37:347–354 (1983).

71. C.J. Glueck, F.H. Mattson, and R.J. Jandacek. The lowering of plasma cholesterol by sucrose polyester in subjects consuming diets with 800, 300, or less than 50 mg of cholesterol per day. *Am. J. Clin. Nutr.* 32:1636–1644 (1979).

72. R.J. Jandacek, M.M. Ramirez, and J.R. Crouse. Effects of partial replacement of dietary fat by Olestra on dietary cholesterol absorption in man. *Metabolism* 39:848–852 (1990).

73. M.R. Adams, M.R. McMahan, F.H. Mattson, and T.B. Clarkson. The long-term effects of dietary sucrose polyester on African green monkeys. *Proc. Soc. Exp. Biol. Med.* 167:346–353 (1981).

74. F.H. Mattson and R.J. Jandacek. The effect of a non-absorbable fat on the turnover of plasma cholesterol in the rat. *Lipids* 20:273–277 (1985).

75. R.J. Jandacek. Developing a fat substitute. *Chemtechnology* 7:398–402 (1991).

76. R.J. Jandacek. The development of Olestra, a noncaloric substitute for dietary fat. *J. Chem. Educ.* 68:476–479 (1991).

77. C.A. Bernhardt. Olestra—A noncaloric fat replacement. *Food Technol. Int. Eur.* 176–178 (1988).

78. J.A. Westrate and K.H. van het Hof. Sucrose polyester and plasma carotenoid concentrations in healthy subjects. *Am. J. Clin. Nutr.* 62:591–597 (1995).

79. R.W. Fallat, C.J. Glueck, R. Lutner, and F.H. Mattson. Short term study of sucrose polyester a non-absorbable fat-like material as a dietary agent for lowering plasma cholesterol. *Am. J. Clin. Nutr.* 29:1204–1215 (1976).
80. K.W. Miller, F.E. Wood, S.B. Stuard, and C.L. Alden. A 20-month Olestra feeding study in dogs. *Food Chem. Toxicol.* 29:427–435 (1991).
81. R.B. Toma, D.J. Curtis, and C. Sobotor. Sucrose polyester: Its metabolic role and possible future applications. *Food Technol.* 42(1):93–95 (1988).
82. C.A. Hassel. Nutritional implications of fat substitutes. *Cereal Foods World* 38:142–144 (1993).
83. J. Giese. Fats, oils, and fat replacers. *Food Technol.* 50(4):78–84 (1996).
84. G.D. Miller and S.M. Groziak. Impact of fat substitutes on fat intake. *Lipids* 31:S293–S296 (1996).
85. L.J. Cheskin, R. Miday, N. Zorich, and T. Filloon. Gastrointestinal symptoms following consumption of olestra or regular triglyceride potato chips. *JAMA* 279:150–152 (1998).
86. R.J. Jandacek, J.J. Kester, A.J. Papa, T.J. Wehmeier, and P.Y.T. Lin. Olestra formulation and the gastrointestinal track. *Lipids* 34:771–783 (1999).
87. F.E. Wood, W.J. Tierney, A.L. Knezevich, H.F. Bolte, J.K. Maurer, and R.D. Bruce. Chronic toxicity and carcinogenicity studies of Olestra in Fischer 344 rats. *Food Chem. Toxicol.* 29:223–230 (1991).
88. K.W. Miller and P.H. Long. A 91-day feeding study in rats with heated Olestra/vegetable oil blends. *Food Chem. Toxicol.* 28:307–315 (1990).
89. C.M. Bergholz. Safety evaluation of Olestra, a nonabsorbed, fatlike fat replacement. *Crit. Rev. Food Sci. Nutr.* 32:141–146 (1992).
90. K.L. Skare, J.A. Skare, and E.D. Thompson. Evaluation of Olestra in short-term genotoxicity assays. *Food Chem. Toxicol.* 28:69–73 (1990).
91. F.E. Wood, B.R. DeMark, E.J. Hollenbach, M.C. Sargent, and K.C. Triebwasser. Analysis of liver tissue for Olestra following long-term feeding to rats and monkeys. *Food Chem. Toxicol.* 29:231–236 (1991).
92. K.W. Miller, D.S. Williams, S.B. Carter, M.B. Jones, and D.R. Mishell. The effect of Olestra on systemic levels of oral contraceptives. *Clin. Pharmacol. Ther.* 48:34–40 (1990).
93. J.E. Vanderveen and W.H. Glinsmann. Fat substitutes: A regulatory perspective. *Annu. Rev. Nutr.* 12:473–487 (1992).

27 Food Applications of Lipids

Frank D. Gunstone

CONTENTS

I. FATS AVAILABLE FOR FOOD APPLICATIONS

A. INTRODUCTION

The annual production of oils and fats now exceeds 150 million tons and is on a rising curve. Table 27.1 contains data for four major vegetable oils, some minor vegetable oils, and the animal fats. These show the increasing dominance of vegetable oils and particularly of palm oil, soybean oil, rapeseed/canola oil, and sunflower oil. Until the recent rapid increase in the production of biodiesel, global oil and fat production was split between food use, animal feed, and the oleochemical industry in a ratio of about 80:6:14, but this ratio is changing and it has been predicted that by 2020 the distribution will change to 68:6:26 [5]. The fast-growing production of biodiesel is likely to put upward pressure on oil/fat prices.

Oil World [2,6] provides regular statistics on 17 oils and fats, a list that does not include cocoa butter (Section VI), rice bran oil, and many specialty oils. These 17 oils and fats are detailed in the footnote of Table 27.1. The 17 oils vary in their fatty acid and triacylglycerol composition and have differing physical, chemical, and nutritional properties. The food technologist has to work with these resources to produce a wide range of products with optimized properties. Similar products can frequently be obtained with more than one selection of oils and fats, as is apparent in the fact that similar food products available in different regions of the world are made from different blends of natural or processed oils. What happens when the natural products do not provide the necessary range of properties? Lipid scientists and technologists have devised several procedures to extend the

TABLE 27.1

Production of 17 Commodity Oils and Fats (Million Metric Tons) in 1986/1990 and 1996/2000 (Average Annual Values) and in More Recent Years

	1986/1990	1996/2000	2003/2004	2004/2005	2005/2006	2006/2007[a]
Oils and fats [17]	75.7	103.4	130.3	138.4	145.8	152.9
Soybean	15.3	22.8	30.9	32.9	34.8	36.6
Palm	9.2	17.9	29.9	33.3	35.2	37.6
Rapeseed	7.5	12.6	14.4	15.7	17.7	18.6
Sunflower	7.2	9.1	9.6	9.4	10.5	10.8
Lauric oils [2]	4.3	5.4	6.6	7.1	7.4	7.7
Other vegetable oils [7]	12.3	14.9	16.1	16.6	16.3	17.0
Animal fats [4]	19.9	20.7	22.9	23.4	23.9	24.5

Sources: The figures are taken from *Oil World 2020*, ISTA Mielke GmbH, Hamburg, Germany, 1999; *Oil World Annual 2006*, ISTA Mielke GmbH, Hamburg, Germany, 2006.

Lauric oils: palm kernel and coconut; other vegetable oils: cottonseed, groundnut, sesame, corn, olive, linseed, and castor; animal fats: butter, lard, tallow, and fish oil.

Averages of 5 year periods are cited to avoid unusually high or low figures that occur in some years.

[a] Figures for 2006/2007 are forecasts.

range of natural fats. These are listed in Table 27.2. Some procedures are discussed in greater detail in other chapters in this book, and they have been described elsewhere by the present author [4,7,8].

B. Fatty Acid Composition of Major Oils

Palmitic (16:0), oleic (18:1), and linoleic acids (18:2) are the dominant fatty acids in vegetable oils. Among commodity oils, these three acids make up 85%–90% of the total [3]. Many oils and fats differ only in the relative proportion of these three acids. However, other fatty acids are present and may be important. Stearic acid (18:0) is a minor component in virtually all oils; linolenic acid (18:3) becomes important in some (soybean, rapeseed, linseed, and several minor oils); and the so-called lauric oils are rich in lauric (12:0) and myristic acids (14:0) (Table 27.3). Of the processes listed in

TABLE 27.2

Methods Employed to Extend the Usefulness and Improve the Properties of Natural Oils

Blending	Mixing of Two or More Oils
Fractionation	Separation of oils into two or more fractions
Partial hydrogenation	Saturation of some double bonds accompanied by double-bond isomerization
Interesterification with chemical or enzymatic catalyst	Reorganization of fatty acids among triacylglycerol molecules
Domestication of wild crops	Conversion of wild crops to crops that can be cultivated commercially
Seed breeding by traditional methods	Cross breeding using irradiation or mutagenesis if necessary
Seed breeding by genetic modification	Crossing between species
Lipids from microorganisms	Production under fermentation conditions

Details of many of these methods are given in Ref. [7].

TABLE 27.3

Typical Fatty Acid Composition (% wt) of Four Major Vegetable Oils, Lauric Oils, Animal Fats, and Hydrogenated and Fractionated Oils

	16:0	18:0	18:1	18:2	Other Acids
Soybean	11	4	23	53	18:3 8%
Soybean (hydrog)[a]	11	4–14	40–75	0–40	18:3 0%–3%
Palm	44	4	39	11	
Palm olein	41	4	41	12	
Palm stearin	47–74	4–6	16–37	3–10	
Rapeseed	4	2	62	22	18:3 10%
Sunflower	6	5	20	60	
NuSun[b]	4	5	65	26	
Sunola	4	5	81	8	
Olive	10	2	78	7	
Palm kernel	8	2	15	2	8:0 3%, 10:0 3%, 12:0 48%, 14:0 16%
Coconut	9	3	6	2	8:0 7%, 10:0 7%, 12:0 48%, 14:0 18%
Butter	26	11	28	2	4:0–12:0 13%, 14:0 12%, 16:1 3%
Lard	26	11	44	11	14:0 2%, 16:1 5%
Tallow	27	7	48	2	14:0 3%, 16:1 11%

High-oleic varieties of the major vegetable oils are becoming available as niche products.

[a] Hydrogenated soybean oil normally includes 9%–40% of 18:1 *trans* isomers.
[b] NuSun is a sunflower oil with the levels of oleic acid shown in the table.

Table 27.2, it is important to note that blending and fractionation lead to changes in the proportions of fatty acids and that interesterification does not change fatty acid composition but modifies the ways in which the acids are incorporated into triacylglycerols. However, (partial) hydrogenation results in the formation of novel fatty acids with structures different from those present in the native oils. New acids are formed with double bonds of changed configuration and changed position in the carbon chain.

C. AVAILABILITY OF OILS AND FATS IN DIFFERENT REGIONS OF THE WORLD

In the days before extensive world trade in perishable goods, foods were based on local availability and food traditions became established. These are still apparent throughout the world. For example, communities living near the seashore or a lakeside often consume more fish in their diet than other communities not so favorably placed with respect to these healthy foods. In addition, countries and regions that produce large quantities of a particular oil generally consume a large amount of that material. This is very apparent in certain developing countries but it is also true of the developed world. It has been reported that 80% of the fat consumed in the United States in 2004 was soybean oil (before or after some degree of hydrogenation) [9]. Some relevant data are presented in Table 27.4 but these need careful interpretation. "Disappearance" for a country or region represents locally produced oil *plus* imported oil *minus* exported oil with a small adjustment for changes in stock levels at the beginning and end of the reporting season. The figures include food and nonfood uses and make no allowance for waste or use of commodities outside the 17 listed in the reports. As already indicated, on a worldwide basis about 20% (or more) of all oils and fats are used for nonfood purposes. This proportion will be lower in many developing countries but higher in areas where the oleochemical industry is concentrated, such as North America, Western Europe, Japan, and increasingly in Southeast Asia. It is also relevant for this consideration that oils used extensively for

TABLE 27.4

Annual Average Disappearance of Oils and Fats in 2005/2006 in Selected Countries in Percentage Terms

	Can	United States	Arg	Brz	EU	China	India	Indon	Japan	Mal	Phil
Soya	23	52	40	66	11	28	20	0	22	2	6
Palm	2	3	0	4	18	19	22	84	17	61	38
Rape	32	6	0	1	25	18	17	0	35	1	0
Sun	5	1	30	1	10	1	4	0	1	1	0
PKO	2	1	1	1	3	1	1	10	2	31	1
Coco	1	3	0	0	3	1	3	4	2	3	37
Olive	2	1	0	1	8	0	0	0	1	0	0
Cotton	1	3	0	4	0	6	6	0	0	0	0
Peanut	0	1	1	0	0	8	9	0	0	0	0
Corn	4	5	1	1	1	0	0	0	4	0	0
Butter	5	4	6	2	7	0	14	0	3	0	2
Lard	9	3	3	8	7	13	0	0	2	0	11
Tallow	11	17	16	11	4	4	1	0	4	0	4
Other	2	1	1	1	2	2	2	0	6	0	1

Countries: Canada, Argentina, Brazil, European Union, Indonesia, Malaysia, and Philippines.

oleochemical purposes include the lauric oils (coconut and palm kernel), tallow, and, to a lesser extent, palm oil and palm stearin. Rapeseed oil is used extensively in Europe to produce biodiesel. In the United States, this commodity is based mainly on soybean oil and Malaysia and Indonesia plan to use large volumes of palm oil for this purpose.

The figures in Table 27.4 show that there are several countries in which an oil and fat grown locally is also the major fat consumed (for all purposes) at a level close to or higher than 50% of total disappearance, Disappearance of soybean oil is 66% in Brazil and 52% in the United States and disappearance of palm oil is 84% in Indonesia and 61% in Malaysia. The situation is different in countries that import large amounts of seed for local crushing or import oils and fats and generally use a wider range of oils. This is apparent in the figures in Table 27.4 for EU-25, Japan, China, and India. EU-25 is an association of states covering a wide geographic region, and the figures in Table 27.5 reveal that there are significant differences in oil and fat disappearance between individual European countries and particularly between those in the north and the south. Rapeseed oil, sunflower oil, and olive oil are much used in the European countries in which they are produced. Butter is still a significant fat in Europe except in the Mediterranean countries producing olive oil. Figures for disappearance cover all·uses though food use dominates. The high consumption of rapeseed oil in Germany is an indication of the large amount of this oil being used to produce biodiesel.

With the free movement of oilseeds and of oils and fats, trading nations have a wide choice among these commodities. With the development of high-quality refined oils, many of them become interchangeable though attention must be paid to fatty acid and triacylglycerol composition; physical, nutritional, and chemical properties; price; and customer concerns. Food producers monitor their supplies to make best use of the cheapest oils, and when customers express concerns, such as those over oils from genetically modified sources (mainly soybean oil) or oils containing *trans* acids or ω-3 acids, producers switch to recipes with appropriate levels of these materials. The message here is that while the major producing countries make extensive use of local products, there are regions of the world in which flexibility of supply is possible and is cherished by producers and consumers alike.

TABLE 27.5

Disappearance of Oils and Fats during 2005/2006 in Selected EU-25 Countries in Percentage Terms

	EU-25	Fr	Ger	It	Pol	Sp	UK
Soybean	11	6	7	15	9	9	9
Palm	18	14	12	14	10	10	36
Rapeseed	25	27	45	7	35	2	25
Sunflower	10	17	5	10	3	24	5
Palm kernel	3	1	5	1	1	1	3
Coconut	3	2	6	2	1	2	1
Olive	8	4	1	27	1	27	2
Cottonseed	0	0	0	1	0	2	0
Groundnut	0	1	0	2	0	0	0
Corn	1	1	0	2	0	0	1
Butter	7	18	8	5	10	1	9
Lard	7	4	7	7	24	13	1
Tallow	4	0	1	5	5	7	6
Other	2	4	2	2	1	2	2

Countries: France, Germany, Italy, Poland, Spain, and United Kingdom.

II. FRYING OILS AND FATS

The use of oils and fats as a frying medium in both shallow and in deep-frying mode is an important component in the overall picture of food applications. Between 20 and 40 million tons of oils and fats is used in this way. This represents a major share of the 110–120 million tons used for dietary purposes. Of course, it must be remembered that while some of the frying oil is consumed along with the fried food, much is thrown away (shallow pan frying) or ultimately finds other uses as spent frying oil. The importance of frying is reflected by the fact that this topic has a chapter to itself, and this matter will not be pursued here except to provide some general references that may have been overlooked in Chapter 7 [10–15].

III. SPREADS: BUTTER, GHEE, MARGARINE, VANASPATI

A. BUTTER

Butter from cows' milk fat has been used primarily as a spread, but also for baking and frying, for many centuries. With the development of good-quality margarine and other spreads, butter has become less popular. The disadvantages associated with butter are its relatively high price, its poor spreadability (especially from the refrigerator), and its poor health profile resulting from its high fat content, its high level of saturated acids and of cholesterol, and the presence of *trans* unsaturated fatty acids. Its advantages are its "wholly natural" profile and its superb flavor. The perceived superiority of butter is reflected in the fact that spread manufacturers seek to have products with the flavor and appearance of butter and by their use of names like "I Can't Believe It's Not Butter" for one brand of spread. Although the name *butter* is jealously guarded and it is not permissible to take anything away or to add anything to a product that is to be called butter, nevertheless ways of overcoming the disadvantages are reported. Refs. [1,16–19] provide general information on butter.

Butter is a water-in-oil emulsion consisting of fat (80%–82%) and an aqueous phase (18%–20%). The legal limit for water is 16% and the aqueous phase also contains salt and milk-solids-not-fat. It is made from cow's milk (3%–4% fat), which is converted first to cream (30%–45% fat) by

TABLE 27.6

Production and Disappearance (Million Metric Tons) of Butter on a Fat Basis for Selected Countries in the Period 2001/2005

Country	Production					Disappearance				
	2001	2002	2003	2004	2005	2001	2002	2003	2004	2005
World	6.13	6.33	6.39	6.48	6.67	6.12	6.27	6.36	6.54	6.65
EU-25	1.81	1.86	1.86	1.83	1.86	1.75	1.71	1.70	1.70	1.72
CIS	0.40	0.38	0.40	0.41	0.43	0.45	0.47	0.49	0.45	0.46
United States	0.46	0.50	0.46	0.46	0.50	0.48	0.49	0.51	0.53	0.53
India	1.53	1.64	1.69	1.78	1.86	1.53	1.65	1.69	1.78	1.86
Pakistan	0.48	0.49	0.49	0.50	0.51	0.48	0.49	0.50	0.50	0.51
New Zealand	0.33	0.33	0.35	0.34	0.33	0.03	0.03	0.03	0.03	−0.03

Source: From *Oil World Annual 2006*, ISTA Mielke GmbH, Hamburg, Germany, 2006.

CIS, Commonwealth of Independent States (Former Soviet Union).

centrifuging and then to butter by churning and kneading. During churning, there is a phase inversion from an oil-in-water to a water-in-oil emulsion. Details of annual production and disappearance are given in Table 27.6. Production of butter, which peaked in the 1980s at 6.0–6.4 MMT, fell to below 6 MMT but is now 6.7 MMT and is predicted to rise steadily to 7.8 MMT in the next 15 years. The biggest consumers are in the Indian subcontinent (India and Pakistan) and in EU-25 within which France has a consumption equivalent to >18% of total oil and fat disappearance for that country (Table 27.5).

In many countries, the composition of milk fat changes between summer and winter because of changing dietary intake (Table 27.7). The fat is mainly triacylglycerols (97%–98%) along with some free acids, monoacylglycerols, and diacylglycerols. In addition, cholesterol (0.2%–0.4%), phospholipids (0.2%–1.0%), and traces of carotenoids, squalene, and vitamins A and D are present.

Following several thorough examinations, cows' milk fat is known to contain >500 different fatty acids. Most are present only at very low levels but some of these, such as the lactones which provide important flavor notes, are important. Among the many fatty acids, some of them are as follows (see also Table 27.7):

- Saturated acids are in the range 4:0–18:0 including some odd-chain members.
- Low levels of iso-[Me_2CH-], anteiso-[$MeCH_2CH(Me)-$], and other branched chain acids.
- A significant level of monoene acids (28%–31%) that is mainly oleic but includes other isomers, among which are some with *trans* configuration.

TABLE 27.7

Major Fatty Acids (% wt) in Cows' Milk Fat

Month	4:0	6:0	8:0	10:0	12:0	14:0	16:0	18:0	20:0	18:1	18:2	Other
June	4.2	2.5	2.3	2.2	2.4	9.0	22.1	14.3	2.6	30.4	1.2	6.8
December	3.5	2.2	1.1	2.6	2.8	10.6	26.0	11.6	2.3	24.8	2.8	9.7
Average	3.6	2.2	1.2	2.8	2.8	10.1	25.0	12.1	2.1	27.1	2.4	8.6

These figures relate to the Northern hemisphere so that June and December represent summer and winter seasons, respectively.

- *Trans* acids, produced by biohydrogenation of dietary lipids, are significant components of all ruminant milk fats (about 4%–8%). They are mainly C_{16} and C_{18} monoene acids of which vaccenic acid (11t-18:1) is the major component.
- Very low levels of polyene fatty acids and even those cited as linoleic or linolenic are not entirely the all-cis isomers.
- Trace amounts of oxo (keto) and hydroxy acids and lactones of which the latter are important flavor components.

The fatty acid composition of milk depends on the diet of the cow, so that in many countries there is a difference in composition between winter (fed indoors) and summer (pasture fed). Composition can be further modified by changes to the diet. An exciting development in lipid science in recent years has been the recognition of the importance of octadecadienoic acids with conjugated unsaturation (conjugated linoleic acid [CLA]). These are produced by ruminants and appear in low but significant levels in the milk and meat of these animals. This topic is important enough to merit a chapter of its own in this volume (Chapter 23).

The level of butyric acid at around 4% by weight may seem insignificant, but it should be recognized that this is equivalent to about 8.5% mol. Since this acid is likely to occur only once in any triacylglycerol molecule (at the *sn*-3 position), a quarter of all the triacylglycerol molecules in butter contain butyric acid.

The C_4–C_{14} acids and about one-half of the C_{16} saturated acids are produced by de novo synthesis in the mammary gland, whereas the rest of the C_{16} and the saturated and unsaturated C_{18} acids are derived from dietary sources or by mobilization of body fat reserves during early lactation. It follows that only this half of milk fat fatty acids is subjected to change by modification of dietary intake. Furthermore, since in ruminants (free) unsaturated acids are subjected to biohydrogenation in the rumen, it is necessary to protect such acids during their passage through the rumen if these are to be incorporated unchanged in the milk fat. This was first achieved through coating the oil/fat (soybean, linseed, rapeseed/canola), but two other methods are now more commonly employed. In the first, calcium salts are used as lipid source. These remain as (unreactive) salts in the rumen but are converted to acids in the more acidic conditions of the abomasum and enter the duodenum as fatty acids available for digestion. Alternatively, the lipid is hardened to the point where it remains solid in the rumen but melts in the abomasum. The resulting changes in the milk fat may seem small in terms of fatty acid composition, but are slightly greater in their effect on triacylglycerol composition and may be sufficient to allow the butter to spread directly from the refrigerator. It is important that the supplement contains appropriate proportions of *n*-9, *n*-6, and *n*-3 unsaturated acids and that it is over 75% protected from metabolism in the rumen [20,21].

In times of oversupply, there is an interest in extending the range of applications of milk fat by fractionation. However, the triacylglycerol composition of milk fat is so complex (no individual triacylglycerol exceeds 5%) that differences between crystallized fractions are not so marked as with simpler vegetable oils, such as palm oil.

Nevertheless, useful separations have been achieved producing fractions that are harder and softer than the original milk fat. The lower melting (softer) fractions are employed to make spreadable butter and the harder fractions find pastry applications. Anhydrous milk fat itself is used to make cakes. Mixed with the oleic fraction, it is used in cookies, biscuits, and butter cream; mixed with the stearic fraction, it is used in fermented pastries and puff pastry; and the oleic fraction on its own is used in ice cream cones, waffles, butter sponges, and in chocolate for ice cream bars [22,23].

In Europe, butters with reduced fat levels (and therefore reduced caloric values) are designated as butter (80%–90% fat), three-quarters fat butter (60%–62% fat), half fat butter (39%–41% fat), and dairy fat spreads (other fat levels). In the United States, "light butter" must contain less than half of the normal level of fat and "reduced butter" less than one-quarter of the normal level.

Products are available in many countries that are blends of butter and vegetable oil—generally soybean oil. These cannot be called butter but are given an appropriate name that the consumer comes to think of as "spreadable butter." Spreadable butters developed in New Zealand are made by fractionation of butter followed by recombination of appropriate fractions.

B. GHEE

In India, milk fat is consumed partly as butter but also as ghee, though the latter is declining and is now probably below one-quarter of the combined total. Nevertheless, demand is growing in other countries, probably reflecting the migration of Indian citizens. Ghee is a concentrate of butterfat with >99% milk fat and <0.2% moisture. It has a shelf life of 6–8 months, even at ambient tropical temperatures. Butter or cream is converted to ghee by controlled heating to reduce the water content to <0.2%. In other procedures, the aqueous fraction is allowed to separate and some of it is run off before residual moisture is removed by heating. Ghee has a cooked caramelized flavor varying slightly with the method of preparation [16,19,24]. The vegetable oil–based alternative to ghee is called vanaspati (Section III.D).

C. MARGARINE AND SPREADS

Margarine has been produced for over 100 years. During the 1860s, large sections of the European population migrated from country to town and changed from rural to urban occupations. At the same time, there was a rapid increase in population in Europe and a general recession in agriculture leading to a shortage of butter, especially for the growing urban population. As a consequence, the price rose beyond the reach of many poor people. The situation was so bad in France that the government offered a prize for the best proposal for a butter substitute that would be cheaper and would also keep better.

The prize went to the French chemist, Hippolyte Mège Mouriés, who patented his product in France and Britain in 1869. His process required the softer component from fractionated tallow, skimmed milk, and macerated cows' udder. The product was described as mixed glycerol esters of oleic and margaric acids and was therefore called oleomargarine. Margaric acid was thought to be heptadecanoic acid (17:0), but it was actually a eutectic mixture of palmitic (16:0) and stearic (18:0) acids. Even, this early process involved fractionation and enzymes. (Both margaric and margarine should be pronounced with a hard *g* as in Margaret. All three words come from the Greek word for "pearl"—*margarites*.)

Margarine was long considered as a cheap and inferior substitute for butter. In several countries, regulations were passed that prohibited the addition of coloring matter so that white margarine would compare even less favorably with the more familiar yellow butter. Now the situation is different. These impediments have largely disappeared and margarine is widely accepted as having several advantages over butter. It is a more flexible product that can be varied for different markets and modified to meet new nutritional demands, such as desirable levels of cholesterol, phytosterols, saturated or *trans* acids, and fat content, as well as the statutory levels of certain vitamins. General information on margarine is available in Refs. [1,16,17,19]. Table margarine is made from appropriate oils and fats (soybean, rapeseed/canola, sunflower, cottonseed, palm, palm kernel, coconut), which may have been fractionated, blended, hydrogenated in varying degrees, and/or interesterified. Fish oil (hydrogenated or not) may also be included. Other ingredients include surface-active agents, proteins, salt, and water along with preservatives, flavors, and vitamins.

Margarine production involves three basic steps: emulsification of the oil and aqueous phases, crystallization of the fat phase, and plasticification of the crystallized emulsion. Water-in-oil emulsions are cooled in scraped-wall heat exchangers during which time fat crystallization is initiated, a process known as *nucleation*, during which the emulsion drop size is reduced. There follows a maturing stage in working units during which crystallization approaches equilibrium, though crystallization may continue even after the product has been packed. The lipid in a margarine is part solid (fat) and part liquid (oil), and the proportion of these two varies with temperature. The solid/liquid ratio at different temperatures is of paramount importance in relation to the physical nature of the product.

Individual crystals are between 0.1 and several micrometers in size and form clusters or aggregates of 10–30 μm. One gram of fat phase may contain up to 10^{12} individual crystals. The aqueous phase is present in droplets, generally 2–4 μm in diameter, stabilized by a coating of fat crystals.

Margarine taken from the refrigerator at 4°C should spread easily. For this to happen, the proportion of solids should be 30%–40% at that temperature and should not exceed the higher value. For the sample to "stand up" at room temperature (and not collapse to an oily liquid), it should still have 10%–20% solids at 10°C. Finally, as it melts completely in the mouth and does not have a waxy mouthfeel, the solid content at 35°C should be <3%. These are important parameters that can be attained with many different fat blends. Formulations have to be changed slightly to make the product suitable for use in hot climates.

Fats usually crystallize in two different forms, known as β′ and β. Of these, the β-form is thermodynamically more stable and will therefore be formed in many fats and fat blends. But sometimes, the fat remains in the slightly less-stable β′-form. For margarines and other spreads, the β′-form is preferred because the crystals are smaller, are able to entrap more liquid to give firm products with good texture and mouthfeel, and impart a high gloss to the product. The β-crystals, on the other hand, start small but tend to agglomerate and can trap less liquid. It is therefore desirable to choose a blend of oils that crystallize in the β′-form.

Margarines and shortenings made from rapeseed/canola, sunflower, and soybean oil after partial hydrogenation tend to develop β-crystals. This can be inhibited or prevented by the incorporation of some cottonseed oil, hydrogenated palm oil or palm olein, tallow, modified lard, or hydrogenated fish oil, all of which stabilize crystals in the β′-form. The canola, sunflower, and soybean oils share very high levels of C_{18} acids, whereas the remainder have appreciable levels of C_{16} acids (or other chain length in the case of fish oil) along with the C_{18} acids and thus contain more triacylglycerols with acids of mixed chain length.

To make spreads (and shortenings) from readily available liquid vegetable oils, it is necessary to "harden" them. This requires that the proportion of solid triacylglycerols be increased and for most of the last 100 years this has been achieved by partial hydrogenation that converts linoleic acid to saturated acids and to monoene acids rich in *trans* isomers. Since the latter have higher melting point than their cis isomers, this was seen as an additional route to solid compounds. However, during the later years of the twentieth century, it was shown by researchers in Europe that *trans* acids have cholesterol-raising powers even greater than the saturated acids [25]. One country in Europe (Denmark) prohibited the use of fats and fatty ingredients with *trans* content >2% and food producers in other European countries developed recipes with lower levels of *trans* acids. This could result in more saturated acids but the combined content of saturated and *trans* acids was lowered. Changes in the United States were spurred on by legislation, operative from January 2006, requiring separate labeling of *trans* acids (excluding CLA). A product can be labeled as *trans*-free if the content of *trans* acids is <0.5 g per serving and many food companies in the United States changed their recipes to deliver *trans* acids below this limit. They are then able to claim *trans*-free products. This has been achieved, in part, by optimizing the partial hydrogenation procedure to minimize (but not eliminate) *trans* acids and also by a new approach in which unhydrogenated oil is blended with hard stock and the mixture subjected to interesterification. The hardstock may be a lauric oil, palmstearin, or a fully hydrogenated oil. Since the last contains very little unsaturated acid, *trans* acids must be virtually absent. A problem may remain if it is necessary to declare the presence of hydrogenated vegetable oil because of the unwarranted perception that this is undesirable. In some quarters, it is feared that one day chemical interesterification may also be perceived to be unacceptable because of the use of "chemicals." If that happens, suppliers will have to use

TABLE 27.8
Approximate Fatty Acid Composition of Spreading Fats (%)

Fat	Saturated	Monoene[a]	Polyene
Butter	63–70	28–31	1–3
Soft (tub)	17–19	35–52	29–48

[a] Soft margarines traditionally contain 10%–18% *trans* acids but products with lower levels of *trans* acids are now produced.

enzymatic interesterification. This has some advantages but is still costlier than chemical interesterification [26–30].

Listing all of the formulations used to make margarines is impossible, and the following list is merely indicative (in the following blends, "hydrogenated" means "partially hydrogenated"). These blends are now modified along the lines indicated in the previous paragraph.

- Blends of hydrogenated soybean oils with unhydrogenated soybean oil
- Blends of canola oil, hydrogenated canola oil, and either hydrogenated palm oil or palm stearin
- Blends of various hydrogenated cottonseed oils
- Blends of edible tallow with vegetable oils (soybean, coconut)
- Blends of palm oil with hydrogenated palm oil and a liquid oil (rapeseed, sunflower, soybean, cottonseed, olive)

For hot climates a harder formulation is required, as in the following examples from Malaysia:

- Palm oil (60%), palm kernel oil (30%), and palm stearin (10%)
- Palm stearin (45%), palm kernel oil (40%), and liquid oil (15%)

Table 27.8 gives details of the fatty acid composition of butter and of soft tub margarine and Table 27.9 provides information on production levels of margarine in the period 2000–2005.

Margarine is expected to have a shelf life of about 12 weeks. With good ingredients and the absence of pro-oxidants (e.g., copper), oxidative deterioration is not likely to be a problem.

TABLE 27.9
Production of Margarine (Million Metric Tons) Expressed in Terms of Normal Fat Levels and Including Vanaspati in the Period 2000–2005

Country	2000	2001	2002	2003	2004	2005
World	9.78	9.75	9.85	9.66	9.74	9.83
EU-25	2.64	2.60	2.61	2.49	2.46	2.41
Pakistan	1.50	1.53	1.55	1.57	1.58	1.61
CIS	0.83	0.93	0.94	1.03	1.12	1.23
India	1.36	1.40	1.43	1.21	1.22	1.10
Turkey	0.48	0.49	0.51	0.54	0.54	0.52
The United States	1.04	0.82	0.77	0.73	0.71	0.56
Brazil	0.48	0.49	0.49	0.49	0.49	0.49
Other	1.45	1.49	1.55	1.60	1.62	1.93

The reduction in Indian production results from increased imports from Sri Lanka and Nepal following new tariff agreements.

TABLE 27.10
Presence of *Trans* Fatty Acids in Margarines

Country (Year of Publication)	Range (%)	Mean (%)	References[a]
Germany (1997)	0.2–5	1.5	[20]
Belgium, Hungary, and Britain (1996)	1–24	9.7	[21]
Denmark (1998), soft		0.4	[22]
Denmark (1998), hard		4.1	[22]
Canada (1998), tub	1–46[b]	18.8	[23]
Canada (1998), hard	16–44	34	[23]
Hawaii (2001), cup	1–19	12.1	[24]
Hawaii (2001), carton	18–27	23.4	[24]

[a] See also Ref. [26].
[b] Mainly 15%–20%.

However, care must be taken to avoid microbiological contamination in the aqueous phase. This is achieved by hygienic practices during manufacture, the addition of some salt (8%–10% in the aqueous phase, corresponding to slightly >1% in the margarine), control of pH of any cultured milk that may be used, and careful attention to droplet size in the emulsion.

The levels of total *trans* acids (mainly 18:1 but also some 18:2 and 18:3) in margarines from various countries are listed in Table 27.10 [31–35]. Levels have declined over the last 10 years and are now noticeably lower. However, spreads are not the only source of dietary *trans* fatty acids. Such acids are also obtained from dairy produce and from baked goods made with partially hydrogenated vegetable fats. Ratnayake et al. [34] reported in 1998 that with a *trans* fatty acid consumption of about 8.4 g/day in Canada, only about 0.96 g (11%) comes from the consumption of margarine. Wolff et al. [36] in 2000 drew attention to the very different profile of *trans* monoene fatty acids consumed in France and in Germany compared with consumption in North America. These differences reflect the differing nature of *trans* acids from dairy produce on the one hand and industrially hydrogenated vegetable oils on the other.

Margarines are now available with added phytosterols, which, it is claimed, are capable of reducing blood cholesterol levels. The phytosterols, added at around 8% level, are obtained from tall oil and added to spreads as hydrogenated sterol esters (stanols) or from soybean oil and added as unsaturated sterol esters. Spreads are suitable foods for phytosterol addition because they are used widely and regularly, and are unlikely to be overconsumed. Intake of phytosterols is normally 200–400 mg/day, though higher for vegetarians, but the intake of 1.6–3.3 g/day, recommended by those offering this special margarine, is markedly higher. Normally about 50% of ingested cholesterol is absorbed but with an adequate intake of phytosterols, which are absorbed only at 5% level, absorption of cholesterol falls to about 20%.

Spreads with reduced levels of fat (40% or less) are popular with consumers (as an alternative to discipline in the amount of normal spread consumed). These spreads contain more water than the full-fat spreads and require emulsifiers (monoacylglycerols or polyglycerol esters). It is also usual to add thickeners, such as gelatin, sodium alginate, pectin, and carrageenan, to the aqueous phase. Industrial margarines are used mainly for bakery products and are discussed in Section IV.

D. VANASPATI

The production of vanaspati in 1998 [24] was 4.7 MMT (mainly in Pakistan 1.4, India 1.0, Iran 0.5, Egypt 0.4, and Turkey 0.4). Vanaspati can be considered as vegetable ghee. It is used mainly for frying and for the preparation of sauces, sweets, and desserts. Traditionally, vanaspati was a blend of hydrogenated seed oils (cottonseed, groundnut, soybean, rapeseed/canola, and palm), but increasingly palm oil has become a significant component. The product should melt between 31°C and 41°C, though generally it is close to 37°C in India and 36 ± 2°C in Pakistan. A wide range of oils is used, including soybean, rapeseed, sunflower, cottonseed, palm olein, and palm oil. Because of the method of production involving hydrogenation, vanaspati contains high levels of acids with *trans* unsaturation (>50% in India and about 27% in Pakistan). However with increasing use of palm oil in vanaspati, the need for hydrogenation is reduced with a consequent fall in the level of *trans* acids. Figures around 3% have been reported in Pakistan [16,19,24].

IV. BAKING FATS, DOUGHS, SHORTENINGS

The use of oils and fats in baking processes ranks with frying and spreads as a major food use of these materials. The products range from breads and layered doughs to cakes, biscuits (cookies) and biscuit fillings, piecrusts, short pastry, and puff pastry. The fats used to produce this wide range of baked goods vary in their properties and particularly in their melting behavior and plasticity. It is possible to attain appropriate properties with different blends of oils, and preferred mixtures vary in different regions of the world. In addition to the desired physical properties, it is necessary to meet two further requirements. One is oxidative stability related to the shelf life of the baked goods. The other is the need to respond to current nutritional demands. A good baked item is tasty, has good texture, has a reasonable shelf life in terms of rancidity and palatability and texture, and is a healthy food. Sometimes, the pressure for appropriate physical properties and nutritional requirements work in opposite directions and a compromise has to be made. As already discussed with the spreads, a plastic fat containing solid and liquid components must have some solid triacylglycerols, which implies a certain level of saturated acids or of acids with *trans* unsaturation despite the nutritional concerns associated with these compounds.

Fats used to make doughs of various kinds are almost entirely plastic fats, that is, mixtures of solid and liquid components that appear solid at certain temperatures and that deform when a pressure is applied. Fats exert their influence by interaction with the flour and (sometimes) sugar, which are the other major components of a baked product.

Going back to the important physical properties, the solid fat ratio at various temperatures is now usually measured by pulsed ^{1}H nuclear magnetic resonance. Plasticity depends on the solid components being in the correct polymorphic form (see Section III.C). Tests have also been devised to determine the extent of oxidation and to assess shelf life.

Baking fats may include butter or margarine; both of them have >80% fat and also contain an aqueous phase, or they may be shortenings with 100% fat. These are so described as they give pastry the crispness and flakiness that is suitable for its edible purpose. Industrial margarine has the fat/water ratio required of margarine but differs from margarine spread in that it has fat components that produce the physical properties required by its final end use. Changes in the composition of fat in margarine spread designed to increase their nutritional value have not always carried through to the baking fats, which are often richer in saturated fatty acids and/or acids with *trans* unsaturation (see Section III.C). But there seems little doubt that the appropriate changes will come [19,27–29]. Baked goods contain what is described as "hidden" fat, and it is easy to forget the presence of fat when delicious pastries, cakes, and biscuits are eaten.

The prime function of fat in a cake is to assist in aeration and to modify the texture of the product. The first stage in making a cake is to produce a batter containing a fine dispersion of air bubbles largely stabilized by fat crystals. During baking, the fat melts and the water-in-oil emulsion inverts with the air being trapped in the aqueous phase. As baking continues, the starch is hydrated and gelatinized, the protein starts to coagulate, and the air cells expand through the presence of steam and carbon dioxide (produced from baking powder).

In short pastry, aeration is only of secondary importance. The fat needs to be fairly firm and should be distributed throughout the dough as a thin film; lard, beef tallow olein, and hardened vegetable oils may be employed. Sometimes butter or margarine is used.

In puff pastry (piecrusts, Danish pastries, croissants), fat acts as a barrier separating the layers of dough from one another. Liberation of gas or steam during baking produces a layer structure. This requires a fat of higher melting point than normal (about 42°C) with a higher solid fat content achieved through an appropriate degree of hydrogenation. Small amounts of fat (2%–5%) are added to bread dough. Additional information is available in Refs. [37–39].

V. SALAD OILS, MAYONNAISE AND SALAD CREAM, FRENCH DRESSINGS

Salad oils, used to make mayonnaise and salad cream, should be oxidatively stable and free of solids even when stored in a refrigerator at about 4°C. Several vegetable oils may be used. Those containing linolenic acid (soybean oil, rapeseed/canola oil) are usually lightly hydrogenated (brush hydrogenation) to enhance oxidative stability. All oils are generally winterized to remove high-melting glycerides that would crystallize, as well as waxes present in some solvent-extracted oils. The latter lead to a haze in the oil when it is cooled. Salad oils must pass a "cold test," which requires that the oil remain clear for 5.5 h at refrigeration temperature. After appropriate treatment, soybean, rapeseed/canola, corn, and sunflower oils are used to produce mayonnaise.

Mayonnaise is an oil-in-water emulsion containing between 65% (legal minimum) and 80% of oil. The aqueous phase contains vinegar, citric acid, and egg yolk. The latter contains lecithin, which serves as an emulsifying agent. Lemon and/or lime juice, salt, syrups, seasonings, spices, and antioxidants are optional constituents. These components may be mixed together at temperatures not exceeding 5°C (cold process) or at temperatures around 70°C (hot process). A typical mayonnaise contains vegetable oil (75%–80% by weight), vinegar (9.4%–10.8%), egg yolk (7.0%–9.0%), and small amounts of sugar, salt, mustard, and pepper [38,40]. "Light" mayonnaise contains about 30%–40% of oil, and in low-calorie dressings the level is 3%–10%.

Salad creams are similar but contain much less oil (30%–40%) along with cooked starch materials, emulsifiers, and gums to provide stability and thickness. They are cheaper than mayonnaise.

French dressings are temporary emulsions of oil, vinegar or lemon juice, and seasonings. As the emulsions are not stable, the dressings should be shaken before use. A nonseparating product can be made by the addition of egg yolk or other emulsifying agent. Additional information is available in Refs. [13,19,41,42].

VI. CHOCOLATE AND CONFECTIONERY FATS

Chocolate is an important fat-containing food based mainly, but not always entirely, on cocoa butter. Confectionery fats are materials with similar physical/functional properties to cocoa butter. Legal definitions of chocolate limit the amount of fat other than cocoa butter that may be used. The incorporation of milk fat into chocolate, the limited use of other fats, and the complete replacement of cocoa butter are discussed later in this section. The most recent books on this topic are those published by Beckett [43] and Timms [44].

Annual production of cocoa beans (2.9 MMT), cocoa butter (1.5 MMT), and chocolate (5 MMT) is reported to be at about the levels indicated in parentheses. A second source gives a figure of about 1.7 MMT for cocoa butter. Cocoa beans contain 50%–55% fat. Production figures for cocoa butter are not included in the statistics generally cited for oil and fat production (Section I.C).

Harvested pods are broken open and left in heaps on the ground for about a week during which time the sugars ferment. The beans are then sun-dried and are ready for transportation and storage. To recover the important components, the beans are roasted (~150°C), shells are separated from the cocoa nib, and the latter is ground to produce cocoa mass. When this is pressed, it yields cocoa butter and cocoa powder, which still contains some fat. Typically, 100 g of beans produce 40 g of cocoa butter by pressing, expelling, or solvent extraction, 40 g of cocoa powder, and 20 g of waste material (shell, moisture, dirt, etc.). Cocoa powder is the residue after extraction and still contains 10%–24% fat. Increasingly, the beans are processed in the country where they grow and cocoa liquor, cocoa powder, and cocoa butter (usually in 25 kg parcels) are exported to the chocolate-producing countries [45].

Cocoa butter is a solid fat (mp 32°C–35°C) obtained from the cocoa bean (*Theobroma cacao*) along with cocoa powder. Both the butter and the powder are important ingredients in chocolate. Cocoa butter is in high demand because its characteristic melting behavior gives it properties that are significant in chocolate. At ambient temperatures it is hard and brittle, giving chocolate its characteristic snap, but also it has a steep melting curve that allows for complete melting at mouth temperature. This gives a cooling sensation and a smooth creamy texture. Typical figures for Ghanaian cocoa butter given in Table 27.11 show that the content of solid falls from 45% to 1% between 30°C and 35°C. The hardness of cocoa butter is related to its solid fat content at 20°C and 25°C. This melting behavior is linked to the chemical composition of cocoa butter. The fat is rich in palmitic (24%–30%), stearic (30%–36%), and oleic acids (32%–39%), and its major triacylglycerols are of the kind SOS where S represents saturated acyl chains in the 1- and 3-positions and O represents an oleyl chain in the 2-position. There are three major components: POP, POSt, and StOSt (P = palmitic acid and St = stearic acid). Cocoa butter has a high content of saturated acids,

TABLE 27.11

Composition and Properties of Cocoa Butter from Different Countries

Factor	Ghana	Ivory Coast	Brazil	Malaysia
Iodine value	35.8	36.3	40.7	34.2
Melting point (°C)	32.2	32.0	32.0	34.3
Diacylglycerols	1.9	2.1	2.0	1.8
Free acid (%)	1.53	2.28	1.24	1.21
Component acids				
Palmitic	24.8	25.4	23.7	24.8
Stearic	37.1	35.0	32.9	37.1
Oleic	33.1	34.1	37.4	33.2
Linoleic	2.6	3.3	4.0	2.6
Arachidic	1.1	1.0	1.0	1.1
Component triacylglycerols				
Trisaturated	0.7	0.6	Trace	1.3
Monounsaturated	84.0	82.6	71.9	87.5
POP	15.3	15.2	13.6	15.1
POSt	40.1	39.0	33.7	40.4
StOSt	27.5	27.1	23.8	31.0
Diunsaturated	14.0	15.5	24.1	10.9
Polyunsaturated	1.3	1.3	4.0	0.3
Solid content (pulsed NMR)–tempering 40 h at 26°C				
20°C (%)	76.0	75.1	62.6	82.6
25°C (%)	69.6	66.7	53.3	77.1
30°C (%)	45.0	42.8	23.3	57.7
35°C (%)	1.1	0.0	1.0	2.6

Source: Adapted from Shukla, V.K.S., *INFORM*, 8, 152, 1997.

The original paper contains more details along with information on cocoa butter from India, Nigeria, and Sri Lanka.

which raises health concerns; however, it has been argued that much of this is stearic acid, which is not considered to be cholesterolemic. Cocoa butter is also a rich source of flavonoids, which have powerful antioxidant activity [46].

Cocoa is grown mainly in West Africa, Southeast Asia, and South and Central America. The composition of cocoa butter from these different sources varies slightly, as shown in Table 27.11 for cocoa butter from Ghana, Ivory Coast, Brazil, and Malaysia [47]. There are small differences in fatty acid composition that are reflected in the iodine value and the melting point, but more significantly in the triacylglycerol composition and the melting profile. The content of the important SOS triacylglycerols varies between 87.5% in Malaysian and 71.9% in Brazilian cocoa butter, with the African samples midway between these extremes. There is some evidence that the cocoa butters of different geographic origin are becoming more alike [45].

The crystal structure of cocoa butter has been studied intensively because of its importance to our understanding of the nature of chocolate. The solid fat has six crystalline forms designated I–VI. Some crystals show double-chain length (D) and some triple-chain length (L). The six forms have the following melting points (°C) and D/T structure: I (17.3, D), II (23.3, D), III (25.5, D), IV (27.3, D), V (33.8, T), and VI (36.3, T). Of these, form V is the one preferred for chocolate. This crystalline form gives good molding characteristics and has a stable gloss and favorable snap at room temperature. Procedures to promote this form are necessary, and its change to form VI must be inhibited. Form V is usually obtained as a result of extensive tempering, that is, putting molten chocolate through a series of cooling and heating processes that have been found to optimize production of the appropriate polymorph. Alternatively, molten chocolate can be seeded with cocoa butter already in form V.

Transition from form V to the more stable form VI leads to the appearance of white crystals of fat on the surface of the chocolate. This phenomenon is termed "bloom." It is promoted by fluctuations in temperature during storage and by migration of liquid oils from nut centers. It is a harmless change but is considered undesirable because it may be mistaken for microbiological contamination. Bloom can be inhibited by the addition of a little 2-oleo 1,3-dibehenin (BOB) to the cocoa butter. This phenomenon is discussed in more detail by Smith [48], Padley [49], Hammond and Gedney [50], and by Lonchampt and Hartel [51].

The simplest plain chocolate contains sugar and cocoa liquor, with cocoa butter the only fat present. Padley [49] reports that a typical plain chocolate has cocoa mass (~40%, which contains some cocoa butter), sugar (~48%), added cocoa butter (~12%), and small amounts of lecithin and other materials. In some European countries, it is permissible to replace cocoa butter with up to 5% of another fat with similar fatty acid and triacylglycerol composition taken from a prescribed list of tropical fats [52,53]. The permitted tropical fats come from palm, illipe, shea, sal, kokum, and mango and may be used in a fractionated form (Table 27.12).

Milk chocolate contains between 3.5% and 9% of milk fat, and white chocolate is based on sugar and cocoa liquor and cocoa butter (without cocoa mass). If the latter is not entirely refined, it will retain some of the flavor normally associated with chocolate. Chocolate normally contains up to

TABLE 27.12
Typical Triacylglycerol Composition of Cocoa Butter and Some Materials Used as Permitted Partial Replacers

	Cocoa Butter	Palm Midfraction	Illipe	Shea Stearin	Sal Stearin	Mango Kernel Stearin
POP	16	43	7	1	2	3
POSt	38	8	31	10	13	15
StOSt	23	1	50	66	64	65
Total	77	52	88	77	60	83

0.4% of lecithin, usually from soybeans. This aids the processing of the chocolate by reducing the viscosity of molten chocolate. Polyglycerol ricinoleate is sometimes added to optimize viscosity.

Cocoa butter alternatives (CBA) is a general name covering cocoa butter equivalents (CBE), cocoa butter improvers (CBI), cocoa butter replacers (CBR), and cocoa butter substitutes (CBS) [48–51].

CBE have the same general chemical composition and hence the same physical properties as cocoa butter and include the tropical oils described earlier and sometimes designated as hard butters. These can be blended to give mixtures of POP, POSt, and StOSt very similar to cocoa butter and fully miscible with it. The level at which cocoa butter can be replaced by a CBE is limited in some countries on a legal basis and not on a functional basis. CBE which must be compatible with cocoa butter by virtue of their similar fatty acid and triacylglycerol composition, have a melting range equivalent to that of cocoa butter, yield the β-polymorph when processed and tempered in the same way as cocoa butter, and give a product that is at least as good as cocoa butter with respect to bloom. The market for CBE in those European countries where their use in chocolate is permitted is estimated to be 20,000–25,000 t, but it could rise to three times this level if all European Community countries accepted their use as legal.

CBR are usually based on vegetable oils (soybean, cottonseed, palm) that have been fractionated and partially hydrogenated. They contain *trans* unsaturated acids at levels up to 60% and have a different triacylglycerol composition from cocoa butter. They do not require tempering but should be compatible with cocoa butter.

CBS are usually based on lauric fats. They share some of the physical properties of cocoa butter but have a different composition. Coatings based on CBS fats do not require to be tempered but are used in the molten state for enrobing. They give a superior gloss and have very sharp melting characteristics. Further information is given in Refs. [19,52–57].

Chocolate spreads are increasing in popularity. These are made to be stored at room temperature but are often kept in a refrigerator and must spread at this temperature. They contain about 30% fat and like other spreads contain an appropriate mixture of solid and liquid fats with the former preferably in the β'-form [54].

VII. ICE CREAM

The annual production of ice cream in the United States in 2002 was reported to be about 54 million hectoliters (i.e., 5400 million liters), suggesting that the global figure is at least twice this level. This quantity of ice cream will contain around 0.8–1.0 MMT of fat, which will be mainly from milk fat but include a range of vegetable fats such as sunflower, groundnut, palm, palm kernel, and coconut.

Ice cream contains water (60%–70%) and total solids (30%–40%), with the latter including fat (5%–12%), milk solids other than fat (10%–12%), sucrose (12%–14%), glucose solids (2%–4%), emulsifier (0.2%–0.5%), and stabilizer (0.1%–0.3%). Legal requirements for fat vary from country to country as does the possibility of replacing some or all of the dairy fat with vegetable fat.

Fat in ice cream contributes to structure. It stabilizes the aerated foam, improves melting resistance, imparts creaminess, and contributes to taste. The most important properties are melting characteristics, solid-to-liquid ratio at various temperatures, and its taste profile.

Production of ice cream occurs through nine stages: selection and weighing of ingredients, mixing of these in an appropriate sequence at 20°C–35°C, pasteurization at 70°C–75°C or sterilization at 95°C, homogenization at 75°C, cooling to <5°C, ageing at 5°C for at least 4 h, freezing at −5°C to −10°C, hardening at −25°C to −35°C, and storage at −18°C to −20°C [19,58,59].

VIII. INCORPORATION OF VEGETABLE OILS INTO DAIRY PRODUCTS

Vegetable oils may be incorporated into dairy products as a replacement for dairy fat. This happens when local supplies of milk fat are inadequate as in some tropical countries, where the climate is not suitable for large-scale dairy farming and also for consumers concerned about the saturated acids and cholesterol present in milk fat. In addition, it is possible to produce milk fat replacements in a more convenient form as, for example, in long-life cream. The possible use of vegetable fat in ice cream has already been discussed in Section VII.

The so-called filled milk is made from skim milk powder reconstituted with an appropriate vegetable oil. This should be free of linolenic acid, have a low content of linoleic acid, and contain antioxidant so that it is oxidatively stable. Palm oil, palm kernel oil, and coconut oil are most frequently used, and these may be partially hydrogenated to provide further stability against oxidation.

Nondairy coffee whiteners, available in powder or liquid form, generally contain 35%–45% fat, which is usually partially hydrogenated palm kernel oil.

Cheeses have been developed based on vegetable fat rather than dairy fat. Several formulations have been described incorporating soybean oil with or without hydrogenation, palm oil, rapeseed oil, lauric oils, and high-oleic sunflower oil. Attempts have been made to incorporate into these products some of the short-chain acids that are characteristic of milk fat and give cheese some of its characteristic flavor [60–63].

Nondairy whipping creams made with hardened palm kernel oil and coconut oil (each about 17%) are convenient because they have a long shelf life at ambient temperature. They are popular in Britain. First produced for the bakery and catering market with high overrun and good shape retention, they are now supplied to the retail market for domestic use. Pouring creams, containing about 9% of each of the two lauric oils, are also available. Both creams also contain buttermilk powder (7%), guar gum (0.10%–0.15%), emulsifying agent (0.30%–0.35%), β-carotene (0.25%), and water.

IX. EDIBLE COATINGS AND SPRAY PROCESSING

Foods are sometimes coated with thin layers of edible material to extend shelf life by minimizing moisture loss, to provide gloss for aesthetic reasons, and to reduce the complexity and cost of packaging. The thin layers may be carbohydrate, protein, lipid, or some combination of these. The lipids most commonly used are waxes (candelilla, carnauba, or rice bran), appropriate triacylglycerols, or acetylated monoacylglycerols. The latter are capable of producing flexible films at temperatures below those appropriate for the waxes even though they are poorer moisture barriers. The foods most frequently coated are citrus fruits (oranges and lemons), deciduous fruits (apples), vegetables (cucumbers, tomatoes, potatoes), candies and confectioneries, nuts, raisins, cheeses, and starch-based products (cereals, doughnuts, and ice cream cones and wafers).

Vegetable oils used to coat food products must be liquid at room temperature and must have high oxidative stability. They serve as a moisture barrier, a flavor carrier, a lubricant, or a release agent, as an antidust or anticake agent, and as a gloss enhancer. They are used at low levels and are sprayed onto large exposed surfaces of products during roasting, frying, or handling. Traditionally, they are made from commodity oils like soybean or cottonseed. These are cheap but require elaborate processing (partial hydrogenation and fractionation) to develop the required physical state and chemical stability. New high-oleic oils may also be used. These are costlier but they bring added value in terms of their superior nutritional properties, resulting from lower levels of *trans* acids and saturated acids and in the reduced need for processing. Lauric oils, such as coconut oil, palm kernel oil, are used to spray cracker-type biscuits to provide an attractive appearance, maintain crispness by acting as a barrier to moisture, and improve eating quality [64,68].

X. EMULSIFYING AGENTS

Fatty acids and their derivatives are amphiphilic. This means that their molecules have hydrophilic (lipophobic) and lipophilic (hydrophobic) regions. If these are appropriately balanced, then the molecules can exist in a physically stable form between aqueous and fatty substances. They can therefore be used to stabilize both oil-in-water and water-in-oil emulsions and are important components of many of the fat-based products that have been described in the earlier sections of this chapter. Applications of emulsifiers in foods include film coatings, stabilizing and destabilizing emulsions, modification of fat crystallization, dough strengthening, crumb softening, and texturization of starch-based foods. Krog [65] estimates that production of food emulsifiers is about 250,000 t of which about 75% is monoacylglycerols or compounds derived from these.

Monoacylglycerols are most often made by glycerolysis of natural triacylglycerol mixtures in the presence of an alkaline catalyst (180°C–230°C, 1 h). Fat and glycerol (30% by weight) will give a mixture of monoacylglycerols (around 58%, mainly the 1-isomer), diacylglycerols (about 36%), and triacylglycerols (about 6%). This mixture can be used in this form or it can be subjected to high-vacuum thin-film molecular distillation to give a monoacylglycerol product (around 95% and at least 90% of the 1-monoester) with only low levels of diacylglycerols, triacylglycerols, and free acids. Attempts are made to develop an enzyme-catalyzed glycerolysis reaction that occurs under milder reaction conditions. The oils most commonly used include lard, tallow, soybean, cottonseed, sunflower, palm, and palm kernel oil—all in hydrogenated or nonhydrogenated form. Glycerol monostearate (GMS) is a commonly used product of this type.

The properties desired in a monoacylglycerol for some specific use may be improved by acylation of one of the free hydroxyl groups by reaction with acid (lactic, citric) or acid anhydride (acetic, succinic, diacetyltartaric). For the most part, these have the structures shown:

$$CH_3(CH_2)_n COOCHCH(OH)CHOCOR,$$

where R is:

CH$_3$ (acetate)
CH(OH)COOH (lactate)
CH$_2$CH$_2$COOH (succinate)
CHOAcCHOAcCOOH (diacetyltartrate)
CH$_2$C(OH,COOH)CH$_2$COOH (citrate)

Propylene glycol (CH$_3$CHOHCH$_2$OH) also reacts with fatty acids to give mixtures of mono- (about 55%, mainly 1-acyl) and diacyl esters (about 45%). A 90% monoacyl fraction can be obtained by molecular distillation.

Other compounds include the partial esters of polyglycerols (a polyether with 2–10 glycerol units but mainly 2–4 units), sorbitan and its polyethylene oxide derivatives, the 6-monoacylate sucrose, and stearoyl lactate, usually as the sodium or calcium salt [65–67].

REFERENCES

1. E. Flack. Butter, margarine, spreads, and baking fats. In: *Lipid Technologies and Applications* (F.D. Gunstone and F.B. Padley, eds.). Marcel Dekker, New York, 1997, p. 305.
2. *Oil World 2020*. ISTA Mielke GmbH, Hamburg, Germany, 1999.
3. F.D. Gunstone. Fatty acid production for human consumption. *INFORM*, 16:736 (2005).
4. F.D. Gunstone. Movements towards tailor-made fats. *Prog. Lipid Res.*, 37:277 (1998).
5. F.D. Gunstone. Will oil and fat supply meet oil and fat demand in 2020? *INFORM*, 17:541 (2006).
6. *Oil World Annual 2006*. ISTA Mielke GmbH, Hamburg, Germany, 2006.
7. F.D. Gunstone. Introduction: Modifying lipids—why and how? In: *Modifying Lipids for Use in Food* (F.D. Gunstone, ed.). Woodhead Publishing Ltd., Cambridge, England, 2006, pp. 1–8. (Other chapters in this book describe the modifying processes in greater detail.)
8. F.D. Gunstone. Procedures used for lipid modification. In: *Structured and Modified Lipids* (F.D. Gunstone, ed.). Marcel Dekker, New York, 2001, pp. 11–35.

9. www.soystats.com

10. C. Gertz, et al. Review articles, research papers, and abstracts of the 3rd International Symposium on Deep Fat Frying—Optimal operation. *Eur. J. Lipid Sci. Technol.*, 102:505 (2001).

11. E.G. Perkins and M.D. Erickson (eds.). *Deep Frying: Chemistry, Nutrition, and Practical Applications*. AOCS Press, Champaign, IL, 1996.

12. D. Boskou and I. Elmadfa (eds.). *Frying of Food—Oxidation, Nutrient and Non-Nutrient Antioxidants, Biologically Active Compounds and High Temperatures*. Technomic, Lancaster, PA, 1999.

13. T.L. Mounts. Frying oils and salad oils. In: *Lipid Technologies and Applications* (F.D. Gunstone and F.B. Padley, eds.). Marcel Dekker, New York, 1997, p. 433.

14. J.B. Rossell. Manufacture and use of fats in frying. In: *LFRA Oils and Fats Handbook Series, Vol. 1, Vegetable Oils and Fats* (B. Rossell, ed.). Leatherhead, UK, 1999.

15. C.G. Gertz. Developments in frying oils, In: *Modifying Lipids for Use in Food* (F.D. Gunstone, ed.). Woodhead Publishing Ltd., Cambridge, England, 2006, p. 517.

16. K.S. Rajah. Manufacture of marketable yellow fats: Butter, margarine spread, ghee and vanaspati. In: *LFRA Oils and Fats Handbook Series, Vol. 1, Vegetable Oils and Fats* (B. Rossell ed.). Leatherhead, UK, 1999.

17. D. Hettinga. Butter. In: *Bailey's Industrial Oil and Fat Products*, 6th edition, Vol. 2 (F. Shahidi, ed.). Wiley, New York, 2005, p. 1.

18. W.F.J. De Greyt and M.J. Kellens. Improvements of the nutritional and physiochemical properties of milk fat. In: *Structured and Modified Lipids* (F.D. Gunstone, ed.). Marcel Dekker, New York, 2001, p. 285.

19. A.J. Dijkstra, Food uses of oils and fats. In: *The Lipid Handbook*, 3rd edition (F.D. Gunstone, J.L. Harwood, and A.J. Dijkstra, eds.). Taylor & Francis, Boca Raton, FL, USA, 2007, p. 336.

20. S.K. Gulati, J.R. Ashes, and T.W. Scott. Dietary induced changes in the physical and nutritional characteristics of butter fat. *Lipid Technol.*, 10:10 (1999).

21. W.J.P. Harris. The use of fats in animal feeds. *Lipid Technol.*, 7:130 (1995).

22. W. De Greyt and A. Huyghebaert. Food and non-food applications of milk fat. *Lipid Technol.*, 5:138 (1995).

23. K.E. Keylegian. Contemporary issues in milk fat technology. *Lipid Technol.*, 11:132 (1999).

24. K.T. Achaya. Ghee, vanaspati, and special fats in India. In: *Lipid Technologies and Applications* (F.D. Gunstone and F.G. Padley, eds.). Marcel Dekker, New York, 1997, p. 369.

25. R.P. Mensink and M.B. Katan. Effect of dietary *trans* fatty acids on high density and low density lipoprotein cholesterol levels in healthy subjects. *N. Engl. J. Med.*, 323:439 (1990).

26. H. Matsuzaki, T. Okamoto, M. Aoyama, T. Maruyama, I. Niiya, T. Yanagita, and M. Sugano. *Trans* fatty acids marketed in eleven countries. *J. Oleo Sci.*, 51:555 (2002).

27. G.R. List. Decreasing *trans* and saturated fatty acid content in food oils. *Food Technol.*, 58:23 (2004).

28. G.R. List. Processing and reformulation for nutrition labelling of *trans* fatty acids. *Lipid Technol.*, 16:173 (2004).

29. G.R. List and J.W. King. Hydrogenation of lipids for use in food. In: *Modifying Lipids for Use in Food* (F.D. Gunstone, ed.). Woodhead Publishing Ltd., Cambridge, England, 2006, p. 173.

30. E. Floter and G. van Duijn. *Trans*-free fats for use in food. In: *Modifying Lipids for Use in Food* (F.D. Gunstone, ed.). Woodhead Publishing Ltd., Cambridge, England, 2006, p. 429.

31. J. Fritsche and H. Steinhart. *Trans* fatty acid content in German margarines. *Fett/Lipid*, 99:214 (1997).

32. W. De Greyt, O. Radanyi, M. Kellens, and A. Huyghebaert. Contribution to *trans*-fatty acids from vegetable oils and margarines to the Belgian diet. *Fett/Lipid*, 98:30 (1996).

33. L. Ovesen, T. Leth, and K. Hansen. Fatty acid composition and contents of *trans* mono-unsaturated fatty acids in frying fats, and in margarines and shortenings marketed in Denmark. *J. Am. Oil Chem. Soc.*, 75:1079 (1998).

34. W.M.N. Ratnayake, G. Pelletier, R. Hollywood, S. Bacler, and D. Leyte. *Trans* fatty acids in Canadian margarines: Recent trends. *J. Am. Oil Chem. Soc.*, 75:1587 (1998).

35. T. Okamoto, H. Matsuzaki, T. Maruyama, I. Niiya, and M. Sugano. *Trans* fatty acid contents of margarines and baked confectioneries produced in the United States. *J. Oleo Sci.*, 50:137 (2001).

36. R.L. Wolff, N.A. Combe, F. Destaillets, C. Boue, D. Precht, J. Molkentin, and B. Entressangles. Follow-up of the $\Delta 4$ to $\Delta 16$ *trans*-18:1 isomer profile and content in French processed foods containing partially hydrogenated vegetable oils during the period 1995–1999. Analytical and nutritional implications. *Lipids*, 35:815 (2000).

37. J. Podmore. Manufacture and use of fats in dough. In: *LFRA Oils and Fats Handbook Series, Vol. 1, Vegetable Oils and Fats* (B. Rossell, ed.). Leatherhead Food RA, Leatherhead UK, 1999, p. D1.

38. D.J. Metzroth. Shortenings: Science and technology. In: *Bailey's Industrial Oil and Fat Products*, 6th edition, Vol. 4 (F. Shahidi, ed.). Wiley, New York, 2005, p. 83.

39. R.D. O'Brien. Shortening: Types and formulations. In: *Bailey's Industrial Oil and Fat Products*, 6th edition, Vol. 4 (F. Shahidi, ed.). Wiley, New York, 2005, p. 125.

40. S.E. Hill and R.G. Krishnamurty. Cooking oils, salad oils, and dressings. In: *Bailey's Industrial Oil and Fat Products*, 6th edition, Vol. 4 (F. Shahidi, ed.). Wiley, New York, 2005, p. 175.

41. J. Podmore. Applications of edible oils. In: *Edible Oil Processing* (W. Hamm and R.J. Hamilton, eds.). Sheffield Academic Press, Sheffield, 2000, p. 205.

42. P. Aikens. Manufacture and use of fats in salad oils and mayonnaise. In: *LFRA Oils and Fats Handbook Series, Vol. 1, Vegetable Oils and Fats* (B. Rossell, ed.). Leatherhead Food RA, Leatherhead UK, 1999, p. S1.

43. S. Beckett. *The Science of Chocolate*. RSC, Cambridge, England, 2000.

44. R.E. Timms. *Confectionery Fats Handbook—Properties, Production, and Application*. The Oily Press, Bridgwater, England, 2003.

45. R.E. Timms and I.M. Stewart. Cocoa butter, a unique vegetable fat. *Lipid Technol. Newslett.*, 5:101 (1999).

46. T. Krawczyk. Chocolate. *INFORM*, 11:1265 (2000).

47. V.K.S. Shukla. Chocolate—the chemistry of pleasure. *INFORM*, 8:152 (1997).

48. K.W. Smith. Cocoa butter and cocoa butter equivalents. In: *Structured and Modified Lipids* (F.D. Gunstone, ed.). Marcel Dekker, New York, 2001, p. 401.

49. F.B. Padley. Chocolate and confectionery fats. In: *Lipid Technologies and Applications* (F.D. Gunstone and F.B. Padley, eds.). Marcel Dekker, New York, 1997, p. 391.

50. E. Hammond and S. Gedney. Fat bloom. www.britanniafood.com/guest contributions

51. P. Lonchampt and R.W. Hartel. Fat bloom in chocolate and compound coatings. *Eur. J. Lipid Sci. Technol.*, 106:241 (2004).

52. G. Talbot. Manufacture and use of fats in non-chocolate confectionery. In: *LFRA Oils and Fats Handbook Series, Vol. 1, Vegetable Oils and Fats* (B. Rossell, ed.). Leatherhead Food RA, Leatherhead UK, 1999, p. Co-1.

53. B. Eagle. The chocolate directive. www.britanniafood.com/guestcontributions

54. S. Norberg. Chocolate and confectionery fats. In: *Modifying Lipids for Use in Food* (F.D. Gunstone, ed.). Woodhead Publishing Ltd., Cambridge, England, 2006, p. 488.

55. J. Birkett. Manufacture and use of fats in chocolate. In: *LFRA Oils and Fats Handbook Series, Vol. 1, Vegetable Oils and Fats* (B. Rossell, ed.). Leatherhead Food RA, Leatherhead UK, 1999, Chap. 1.

56. V.J.S. Shukla. Confectionery lipids. In: *Bailey's Industrial Oil and Fat Products*, 6th edition, Vol. 4 (F. Shahidi, ed.). Wiley, New York, 2005, p. 125.

57. W. Hamm and R.E. Timms (eds.). Production and application of confectionery fats. Paper presented at a meeting in October 1966, P.J. Barnes and Associates, Bridgwater, UK, 1997.

58. E. Flack. Manufacture and use of fats in ice cream. In: *LFRA Oils and Fats Handbook Series, Vol. 1, Vegetable Oils and Fats* (B. Rossell, ed.). Leatherhead Food RA, Leatherhead UK, 1999.

59. H.D. Goff. Ice cream. In: *Lipid Technologies and Applications* (F.D. Gunstone and F.B. Padley, eds.). Marcel Dekker, New York, 1997, p. 329.

60. L. Yu and E.G. Hammond. The modification and analysis of vegetable oil for cheese making. *J. Am. Oil Chem. Soc.*, 77:911 (2000).

61. L. Yu and E.G. Hammond. Production and characterisation of a Swiss cheese-like product from modified vegetable oils. *J. Am. Oil Chem. Soc.*, 77:917 (2000).

62. E. Hammond. Filled and artificial dairy products and altered milk fats. In: *Modifying Lipids for Use in Food* (F.D. Gunstone, ed.). Woodhead Publishing Ltd., Cambridge, England, 2006, p. 462.

63. I.J. Campbell and M.G. Jones. Cream alternatives. In: *Lipid Technologies and Applications* (F.D. Gunstone and F.B. Padley, eds.). Marcel Dekker, New York, 1997, p. 355.

64. Hewin International. *Edible Oils and Fats in Western European, North American and Asian Markets*. John Wiley & Sons, New York, 2000.

65. N. Krog. Food emulsifiers. In: *Lipid Technologies and Applications* (F.D. Gunstone and F.B. Padley, eds.). Marcel Dekker, New York, 1997, p. 521.

66. C.E. Stauffer. Emulsifiers for the food industry. In: *Bailey's Industrial Oil and Fat Products,* 6th edition, Vol. 4 (F. Shahidi, ed.). Wiley, New York, 2005, p. 229.

67. A.J. Dijkstra, Food grade emulsifiers. In: *The Lipid Handbook*, 3rd edition (F.D. Gunstone, J.L. Harwood, and A.J. Dijkstra, eds.). Taylor & Francis, Boca Raton, FL, USA, 2007, p. 318.

68. T.L. Shellhammer and J.M. Krochta. Edible coatings and film barriers. In: *Lipid Technologies and Applications* (F.D. Gunstone and F.B. Padley, eds.). Marcel Dekker, New York, 1997, p. 453.

Part V

Biotechnology and Biochemistry

28 Lipid Biotechnology

Nikolaus Weber and Kumar D. Mukherjee

CONTENTS

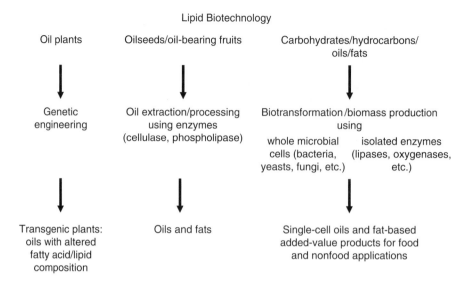

SCHEME 28.1 Major areas of lipid biotechnology.

I. INTRODUCTION

Lipid biotechnology broadly covers the areas outlined in Scheme 28.1. Genetic engineering of oilseeds and oil-bearing fruits for improved agronomic properties and altered fatty acid and lipid composition is a rapidly expanding area of lipid biotechnology [1]. Plant genetic engineering of edible oilseed crops is covered in Chapter 32 of this book.

This chapter will initially cover the use of microorganisms, such as microalgae, yeasts, molds, and bacteria, for the production of oils and fats containing triacylglycerols and other lipids of commercial interest from nonlipidic and lipid-containing carbon sources. This will be followed by biotransformation of fats, oils, and their constituent fatty acids using whole microbial cells or enzymes isolated from various organisms for the production of fat-based added-value products, as outlined in Scheme 28.2. A bulk of this section will be devoted to the application of triacylglycerol lipases and phospholipases for the preparation of specialty products. Finally, several examples will be given on the applications of enzymes in the processing of oilseeds and other oil-bearing materials as well as the use of enzymes in the processing of fats and oils.

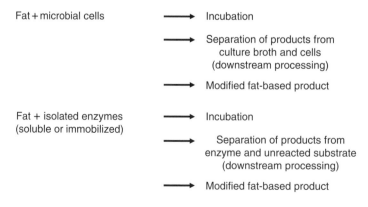

SCHEME 28.2 General principles of modification of fats and other lipids by biotransformation using whole microbial cells or enzymes isolated from various organisms.

II. MICROBIAL PRODUCTION OF FATS AND OTHER LIPIDS

A. Fats and Lipids from Biomass of Microorganisms Using Nonlipidic or Lipid-Containing Carbon Sources—Single-Cell Oils

Some microorganisms, such as eukaryotic yeasts, molds, and algae, are known to produce triacylglycerols in their biomass (single-cell oils, SCOs) similar to plant oils, whereas prokaryotic organisms, such as bacteria, produce more specific lipids, for example, wax esters, polyesters, poly-β-hydroxybutyrate, etc. This section will cover a few selected microorganisms that have some potentials of application for the production of triacylglycerols and other lipids from nonlipidic and lipid-containing carbon sources. For a more comprehensive review on this subject, the reader is referred to Refs. [2–7].

1. Oils and Fats

Lipid content and the levels of major constituent fatty acids in the biomass of a few selected species of lipid-accumulating microorganisms are given in Table 28.1 for microalgae, Table 28.2 for yeasts, and Table 28.3 for molds (fungi).

Some microalgae, as opposed to macroalgae (seaweeds, phytoplankton, etc.), are known to produce substantial amounts of lipids with widely varying fatty acid composition in their biomass (Table 28.1). Due to relatively high cost of growing algae, these organisms are being considered for the production of specialty lipids to be discussed later, rather than common oils and fats for food uses.

Biomasses of oleaginous yeasts, listed in Table 28.2, contain high levels of lipids, predominantly triacylglycerols, having fatty acid composition resembling several edible fats and oils [8]. Oleaginous yeasts can grow well on a wide variety of substrates, such as pure sugars (e.g., glucose, sucrose, and fructose), mixed sugars contained in molasses, lactose contained in whey, ethanol, and so forth. Whey is a soluble liquor by-product of the cheese-making industry that contains up to 5% w/v lactose in addition to noncoagulated protein. The protein can be separated from lactose by ultrafiltration, and the resulting permeate-containing lactose serves as an excellent substrate for the oleaginous yeast, *Candida curvata* D (syn. *Apiotrichum curvatum* ATCC 20509). The fat in the resulting biomass has fatty acid composition and triglyceride structure resembling the natural cocoa butter [5] that fetches a high price. Use of *C. curvata* D for the production of cocoa butter substitute from whey permeate is considered as a commercial process.

TABLE 28.1
Lipid Content and Levels of Major Constituent Fatty Acids in Biomass of Lipid-Accumulating Microalgae

Organism	Lipid Content (% w/w of Biomass)	Fatty Acid (% w/w of Total)[a]				
		16:0	16:1ω9	18:1	18:2	18:3ω3
Chlorella vulgaris	39	16	2	58	9	14
Botryococcus braunii	53–70	12	2	59	4	
Scenedesmus acutus	26	15	1	8	20	30
Navicula pelliculosa	22–32	21	57	5	2	

Source: From Ratledge, C. in *Microbial Lipids*, Vol. 2, Ratledge, C. and Wilkinson, S.G., eds, Academic Press, London, 1989, 567.

[a] Fatty acids are designated by number of C-atoms (left of colon) and number of *cis* double bonds (right of colon).

TABLE 28.2

Lipid Content and Levels of Major Constituent Fatty Acids in Biomass of Lipid-Accumulating Yeasts

Organism	Carbon Source	Lipid Content (% w/w of Biomass)	Fatty Acid (% w/w of Total)[a]				
			16:0	16:1	18:0	18:1	18:2
Candida curvata D (syn. *Apiotrichum curvatum*)	Whey (lactose)	58	32		15	44	8
Cryptococcus albidus	Glucose, ethanol	65	16	1	3	56	
Lipomyces lipofer	Ethanol, glucose	64	37	4	7	48	3
Lipomyces starkeyi	Ethanol, glucose	63	34	6	5	51	3
Rhodosporidium toruloides	Glucose, mixed sugars	66	18	3	3	66	0
Trichosporon pullulans	Ethanol, glucose	65	15		2	57	24
Rhodotorula glutinis	Glucose, mixed sugars	72	37	1	3	47	8
Yarrowia lipolytica (syn. *Candida lipolytica*)	Glucose	32–36	11	6	1	28	51

Source: From Ratledge, C. in *Microbial Lipids*, Vol. 2, Ratledge, C. and Wilkinson, S.G., eds, Academic Press, London, 1989, 567.

[a] Fatty acids are designated by number of C-atoms (left of colon) and number of *cis* double bonds (right to colon).

Some oleaginous molds synthesize high proportions of lipids, predominantly triacylglycerols, in their biomass using glucose as carbon source (Table 28.3). Fatty acid composition of the lipids of different molds varies widely [6]. Although some molds, for example, *Aspergillus terreus* and *Tolyposporium ehrenbergii*, produce SCOs having fatty acid composition similar to that of edible plant oils, others produce substantial proportions of less common fatty acids (Table 28.3). For example, *Claviceps purpurea* produces large proportions of ricinoleic acid, and some *Mucor* species produce substantial proportions of γ-linolenic acid (GLA) (Table 28.3). Use of such microorganisms for the production of specific polyunsaturated fatty acids will be discussed in Section II.B.3.

TABLE 28.3

Lipid Content and Levels of Major Constituent Fatty Acids in Biomass of Lipid-Accumulating Molds Grown on Glucose as Carbon Source

Organism	Lipid Content (% w/w of Biomass)	Fatty Acid (% w/w of Total)[a]					
		16:0	16:1	18:0	18:1	18:2	18:3
Aspergillus terreus	57	23			14	40	21
Claviceps purpurea[b]	31–60	23	6	2	19	8	
Mucor ramannianus	56	19	3	4	28	14	31[c]
Tolyposporium ehrenbergii	41	7	1	5	81	2	

Source: From Ratledge, C. in *Microbial Lipids*, Vol. 2, Ratledge, C. and Wilkinson, S.G., eds, Academic Press, London, 1989, 567.

[a] Fatty acids are designated by number of C-atoms (left of colon) and number of *cis* double bonds (right of colon).
[b] Contains 42% 12-hydroxy-*cis*-9-octadecenoic (ricinoleic) acid.
[c] 18:3ω6 (γ-linolenic acid).

Bacteria that have been reported to produce substantial proportions of triacylglycerols in their biomass include Actinomycetales, especially mycobacteria, corynebacteria, and *Nocardia* spp. [9]. The biomass of an *Arthrobacter* sp. contains as much as 80% lipids of which about 90% are triacylglycerols [10]. However, cell yields of such organisms are rather low, and the presence of potentially toxic lipids in their biomass restricts the usefulness of such microorganisms as producer of triglyceride-rich SCOs.

2. Polyunsaturated Fatty Acids

Polyunsaturated fatty acids of the ω3 and ω6 series, which are precursors of eicosanoids, are of great interest in specialty products, such as nutraceuticals, due to their ability to modulate the metabolism of eicosanoids, which play a great role in health and disease (see Chapter 21 in Part IV of this book).

Several species of microalgae are known to produce large proportions of ω3 and ω6 polyunsaturated fatty acids and eicosanoids in their biomass [11–13] (Table 28.4). Thus, a docosahexaenoic acid-containing single-cell oil (DHASCO) containing about 40% docosahexaenoic acid (DHA) (22:6ω3) is produced commercially from the biomass of the heterotrophic, nonphotosynthetic marine alga, *Crypthecodinium cohnii* [14,15].

Selected fungi that have been reported to produce ω3 and ω6 polyunsaturated fatty acids in their biomass are listed in Table 28.5. Especially, *M. javanicus* [5], *Mortierella isabellina*, and other *Mortierella* species [8,30,31] have been used for the commercial production of oils containing GLA (18:3ω6) as substitute for rather expensive plant oils, such as those from seeds of borage (*Borago officinalis*), evening primrose (*Oenothera biennis*), black currant, and other *Ribes* spp. that also contain 10%–25% GLA. Typically, a commercially produced oil from the biomass of a *Mortierella* sp. contains as major constituent of fatty acids such as palmitic (27%), stearic (6%), oleic (44%), linoleic (12%), and γ-linolenic acids (8%) [30]. Optimization of culture conditions of a *M. ramanniana* species has yielded an oil containing 18% GLA [32]. Similarly, commercially produced arachidonic acid-containing single-cell oil (ARASCO) from the biomass of *M. alpina* contains as much as 40% arachidonic acid (20:4ω6) [33]. Recently, culture conditions have been optimized for the production of an oil containing about 40% dihomo-γ-linolenic acid (20:3ω6) from the biomass of *M. alpina* [34].

A few species of marine bacteria have also been found to produce substantial proportions of polyunsaturated fatty acids, such as eicosapentaenoic acid [35,36] and DHA [37] in the lipids of their biomass.

TABLE 28.4
Levels of Major Constituent Polyunsaturated Fatty Acids in Biomass of Lipid-Accumulating Microalgae

Organism	Major Polyunsaturated Fatty Acid	
	Fatty Acid[a]	% w/w of Total Fatty Acids
Spirulina platensis	18:3ω6	21
Dunaliella tertiolecta	18:3ω6	32
Porphyridium cruentum	20:4ω6	60
Chlorella minutissima	20:5ω3	45
Navicula saprophilla	20:5ω3	22

Source: From Ratledge, C. in *Microbial Lipids*, Vol. 2, Ratledge, C. and Wilkinson, S.G., eds, Academic Press, London, 1989, 567.

[a] Fatty acids are designated by number of C-atoms (left of colon) and number of *cis* double bonds (right of colon).

TABLE 28.5

Levels of Major Constituent Polyunsaturated Fatty Acids in Biomass of Lipid-Accumulating Molds

Organism	Fatty Acid[a]	Major Polyunsaturated Fatty Acid	References
		% w/w of Total Fatty Acids	
Mortierella isabellina	18:3ω6	8	[2]
Mucor javanicus	18:3ω6	15–18	[2]
Mucor ambiguous	18:3ω6	11–14	[16]
Mortierella rammaniana	18:3ω6	26	[17]
Mortierella alpina	20:3ω6	23	[18]
	20:4ω6	11	
Mortierella alpina	20:4ω6	69–79	[19]
Mortierella elongata	20:4ω6	15	[20]
Mortierella alpina	20:4ω6	31	[21]
Mortierella alpina	20:4ω6	43–66	[22]
Mortierella alpina	20:4ω6	32–57	[23]
Mortierella alpina	20:5ω3	15	[24]
Pythium irregulare	20:5ω3	25	[25]
Pythium irregulare	20:5ω3	24	[26]
Thraustochytrium aureum	20:6ω3	40–50	[27]
Thraustochytrium aureum	20:6ω3	29	[28]
Thraustochytrium aureum	20:6ω3	40	[29]

[a] Fatty acids are designated by number of C-atoms (left of colon) and number of *cis* double bonds (right of colon).

Enzymes from transgenic organisms for the production of long-chain polyunsaturated fatty acid–enriched oils [38], microbial polyunsaturated fatty acid production [39], and processing technologies and applications for SCOs as sources of nutraceutical and specialty lipids [40] have been recently reviewed.

3. Wax Esters

Several reports are known on the production of wax esters by *Acinetobacter* sp. H01-N grown on a wide variety of *n*-alkanes of chain lengths ranging from C_{16} to C_{20} [41,42]. In general, the chain lengths of the waxes formed depend on the chain length of the alkane [41], for example,

$$C_n - \text{alkane} \rightarrow C_{2n} + C_{2n-2} + C_{2n-4} \text{ wax esters}$$

In addition to saturated wax esters, those containing up to two double bonds are also formed from *n*-alkanes and carbon sources, such as ethanol, acetic acid, propanol, and propionic acid [42]. Typical composition of wax esters produced by *Acinetobacter* sp. H01-N from various carbon sources is summarized in Table 28.6.

The microalga *Euglena gracilis* (ATCC 12716), grown on yeast malt extract, synthesizes wax esters having saturated even carbon numbered fatty acids and alcohols ranging from C_{12} to C_{18} as the major constituents [43].

4. Biosurfactants

Yeasts, molds, and bacteria are known to produce a wide variety of extracellular glycolipids and other surface-active substances, which are termed biosurfactants [4,44,45]. Table 28.7

TABLE 28.6

Wax Esters Formed by *Acinetobacter* sp. H01-N
from Various Substrates

Substrate	Wax Esters[a]	Constituents of Wax Esters[a]	
		Fatty Acids	Fatty Alcohols
n-Hexadecane	32:0 + 32:1 + 32:2	16:0 + 16:1	16:0 + 16:1
	36:0 + 38:0 + 40:0	16:0 + 16:1	
n-Eicosane	36:1 + 38:1 + 40:1	18:0 + 18:1	20:0 + 20:1
	36:2 + 38:2 + 40:2	20:0 + 20:1	
Ethanol and acetic acid	32:0 + 34:0 + 36:0	16:0 + 16:1	16:0 + 16:1
	32:1 + 34:1 + 36:1	18:0 + 18:1	18:0 + 18:1
	32:2 + 34:2 + 36:2		

Source: From Neidelman, S.L. and Geigert, J., *J. Am. Oil Chem. Soc.*, 61, 290, 1984.

[a] Wax esters, fatty acids, and fatty alcohols are designated by number of C-atoms (left of colon) and number of *cis* double bonds (right of colon).

summarizes the microorganisms and the carbon sources used for the production of several important biosurfactants.

Two common forms of sophorolipids formed from glucose and palm oil are composed of a glycoside of the disaccharide sophorose in which 17-hydroxystearic acid is bound glycosidically at the C-1 hydroxy group or as a lactone in which 17-hydroxystearic acid is bound to sophorose both glycosidically at the C-1 hydroxy group and as an ester at the carboxy end (Figure 28.1). Sophorolipids from some microorganisms contain 13-hydroxydocosanoic acid as the constituent hydroxy acid [4,46–49].

A few common forms of rhamnolipids (Figure 28.2) obtained from microorganisms grown on soybean oil as carbon source consist of rhamnose bound glycosidically to 3-hydroxy fatty acids and their estolides [50].

Cellobiose lipids produced microbially from coconut oil as carbon source (Figure 28.3) contain fatty acids or 3-hydroxy fatty acids esterified at the 2-position of cellobiose and 15,16-dihydroxy-palmitic acid bound glycosidically to cellobiose [53].

Some other biosurfactants include emulsan, a lipopolysaccharide produced by an *Acinetobacter* strain [54], surfactin, produced by *Bacillus subtilis*, which is a heptapeptide linked with β-hydroxy-myristic acid by an ester and an amide bond [55] and penta- and disaccharide lipids produced by a

TABLE 28.7

Microbial Glycolipids as Biosurfactants

Glycolipid	Microorganism	Carbon Source	References
Sophorolipids	*Torulopsis bombicola*	Glucose/palm oil	[46]
		Glucose/safflower oil	[47]
		Oleic acid	[48]
	Candida apicola	Glucose	[49]
Rhamnolipids	*Pseudomonas* sp.	Alkanes/glycerol	[50]
Trehalose lipids	*Rhodococcus erythropolis*	Alkanes	[51]
	Arthrobacter sp.	Alkanes	[52]
Cellobiose lipids	*Ustilago maydis*	Coconut oil	[53]

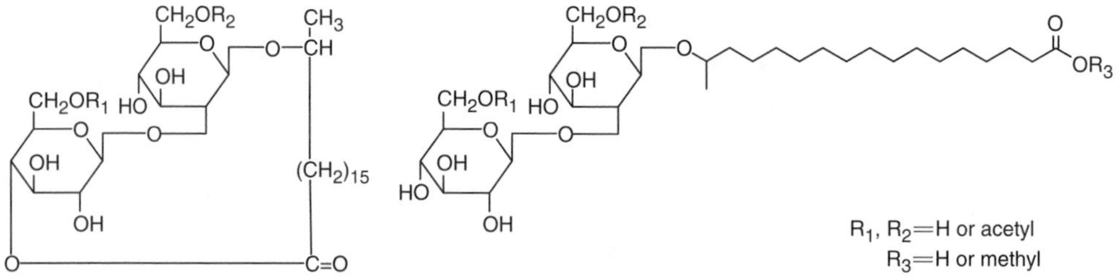

FIGURE 28.1 Common forms of microbial sophorolipids. (According to Boulton, C.A., *Microbial Lipids*, Vol. 2, Ratledge, C. and Wilkinson, S.G., eds, Academic Press, London, 1989. With permission.)

N. corynebacteroides strain [56]. Glycolipids containing sophorose and ω- or (ω-1)-hydroxy C_{16}/C_{18} acids, bound as glycosides or esters, have been produced by the yeast *Candida bombicola* grown on glucose and 2-dodecanol [57]. Production of biosurfactants by fermentation of fats and oils and their coproducts has been discussed in a recent review [58].

B. BIOTRANSFORMATION OF FATS AND LIPIDS USING WHOLE MICROBIAL CELLS

Bioconversions of fats using whole microbial cells is a promising approach for large-scale economical production of fat-based products. The following selected examples show the applications of microbial biotransformations for the preparation of specific products from fats, fatty acids, and their derivatives.

1. Wax Esters

Cells of the alga *E. gracilis* [59,60] and other algae [61], as well as microorganisms such as *Corynebacterium* sp. [62], have been shown to catalyze the esterification of fatty acids with long-chain alcohols yielding wax esters (Figure 28.4).

2. Hydroxy Acids and Other Oxygenated Fatty Acids

Numerous examples of microbial oxidation of unsaturated fatty acids for the preparation of hydroxy fatty acids have become known (see Ref. [63] for review). The microbial enzymes involved in

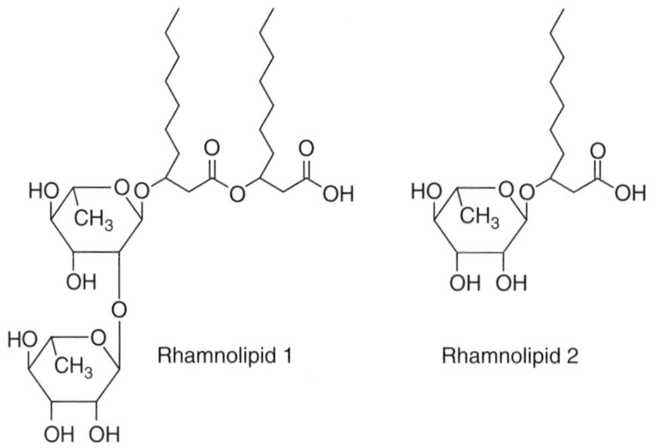

FIGURE 28.2 Common forms of microbial rhamnolipids. (According to Boulton, C.A., *Microbial Lipids*, Vol. 2, Ratledge, C. and Wilkinson, S.G., eds, Academic Press, London, 1989. With permission.)

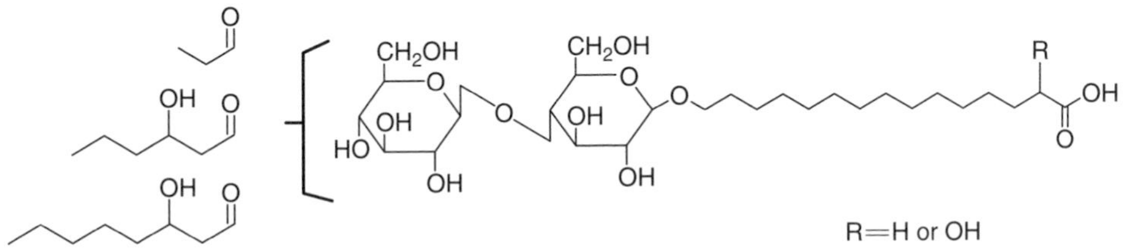

FIGURE 28.3 Common forms of microbial cellobiose lipids. (According to Boulton, C.A., *Microbial Lipids*, Vol. 2, Ratledge, C. and Wilkinson, S.G., eds, Academic Press, London, 1989. With permission.)

oxidation of unsaturated fatty acids to hydroxy acids via hydration (Figure 28.5) are termed monooxygenases, which catalyze the incorporation of one atom of molecular oxygen into a substrate, while the other atom is reduced to water [64]. Dioxygenases catalyze the incorporation of both atoms of molecular oxygen into the substrate.

Since the initial reports by Wallen et al. [65] on the hydration of the olefinic bond of oleic acid to 10-hydroxystearic acid by *Pseudomonas* sp., hydration of oleic acid catalyzed by *Nocardia* sp. [66–68], *Corynebacterium* sp. [69], and *Micrococcus* sp. [70] has become known.

Anaerobic microbial hydration of *cis*-9 olefinic bond has been reported to yield 10-hydroxy-12-*cis*-octadecenoic acid from linoleic acid, 10-hydroxy-12,15-*cis,cis*-octadecadienoic acid from α-linolenic acid, and 10,12-dihydroxystearic acid from ricinoleic acid, as shown in Figure 28.6 [71]. Recently, hydration catalyzed by *N. cholesterolicum* has been shown to produce 10-hydroxy-12-*cis*-octadecenoic acid from linoleic acid in 71% yield and 10-hydroxy-12,15-*cis,cis*-octadecadienoic acid from α-linolenic acid in 77% yield [72].

A cyanobacterium, *Phormidium tenue*, converts linoleic acid via 9-hydroperoxy and 13-hydroperoxyoctadecadienoic acids to the corresponding hydroxyoctadecadienoic acids [73]. Moderate proportions of 15-hydroxy-, 16-hydroxy-, and 17-hydroxy-*cis*-9-octadecenoic acids have been prepared from oleic acid by *B. megaterium* and *B. pumilus* [74].

Strains of *Candida tropicalis* have been engineered with enhanced ω-hydroxylase activity to produce ω-hydroxylauric acid from lauric acid, as shown in Figure 28.7 [75].

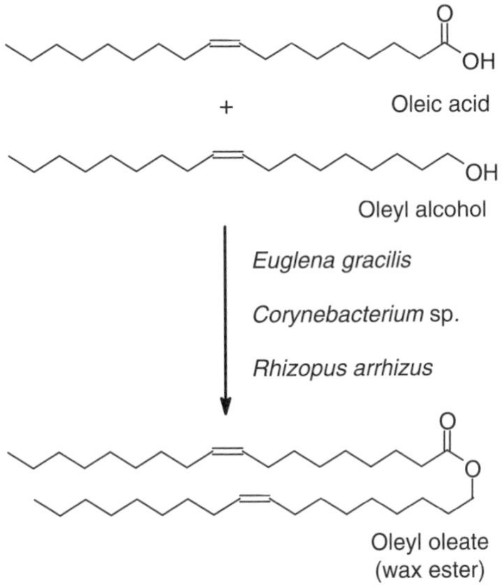

Oleic acid

+

Oleyl alcohol

Euglena gracilis

Corynebacterium sp.

Rhizopus arrhizus

Oleyl oleate
(wax ester)

FIGURE 28.4 Microbial production of wax esters.

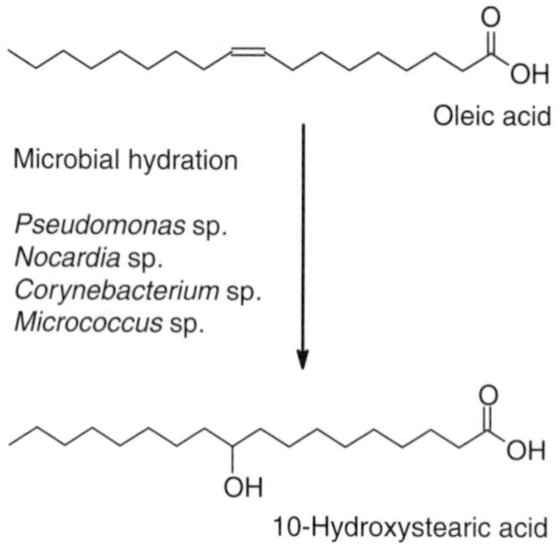

FIGURE 28.5 Microbial production of 10-hydroxystearic acid by hydration of oleic acid.

A bacterial isolate from *Pseudomonas aeruginosa* has been found to convert oleic acid to 7,10-dihydroxy-*cis*-8-octadecenoic acid (Figure 28.8), possibly via hydration and hydroxylation [76].

A *Flavobacterium* sp. has been shown to convert oleic acid to 10-ketostearic acid as the main product (Figure 28.9) in addition to 10-hydroxystearic acid [77]. Moreover, this species converts linoleic acid to 10-hydroxy-12-*cis*-octadecenoic acid [78], and α-linolenic acid and GLA to

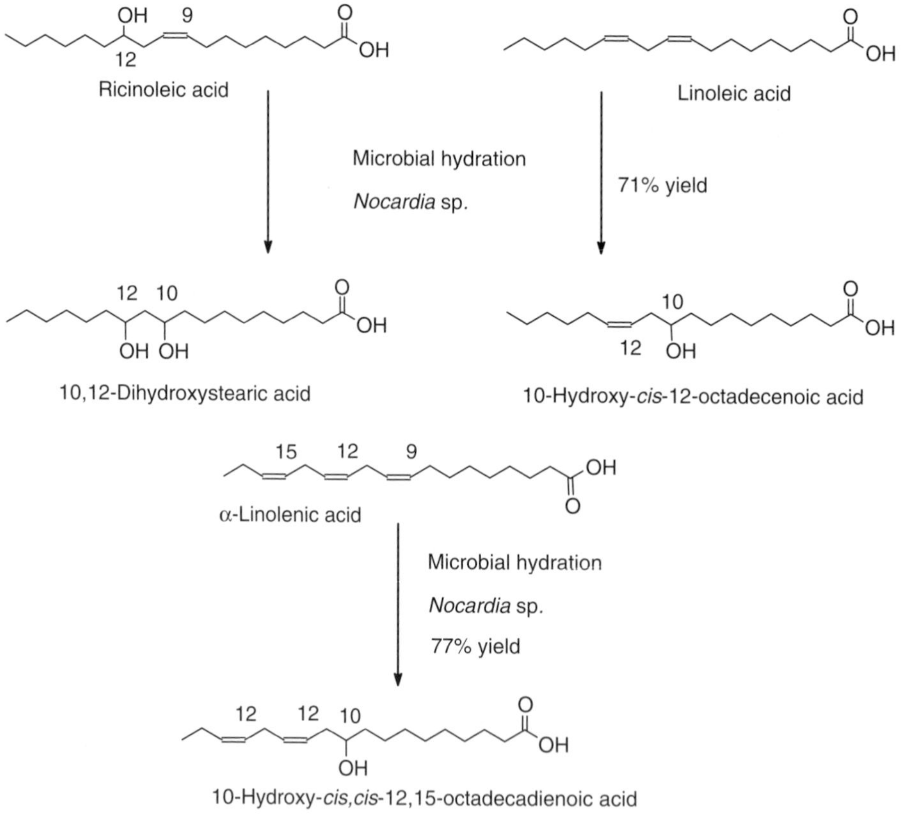

FIGURE 28.6 Microbial production of hydroxy acids by hydration of ricinoleic, linoleic, and linolenic acids.

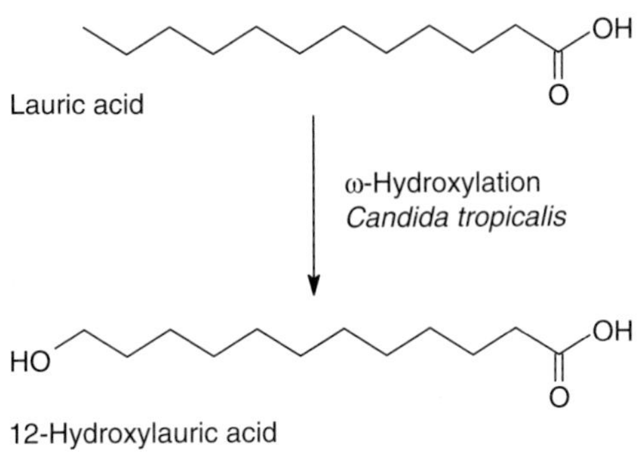

FIGURE 28.7 Microbial production of hydroxy acids by ω-hydroxylation.

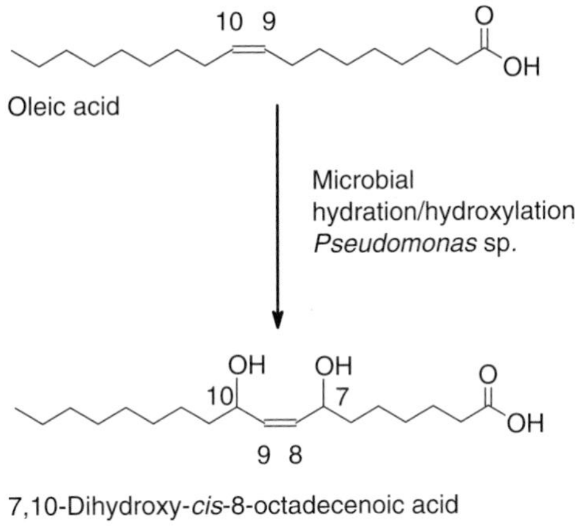

FIGURE 28.8 Microbial production of dihydroxy acids via hydration/hydroxylation.

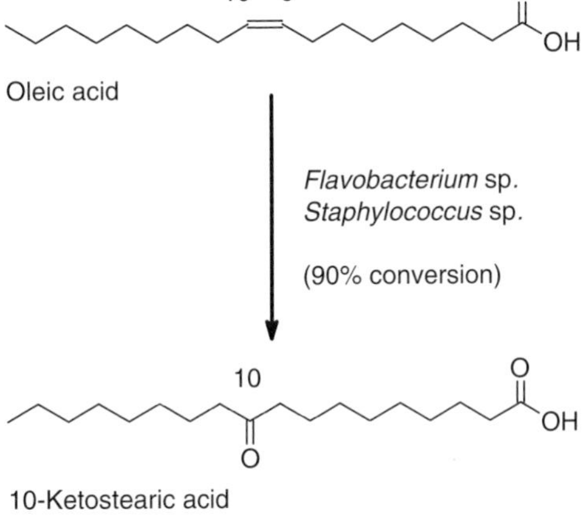

FIGURE 28.9 Microbial production of keto acids.

10-hydroxy-12,15-*cis,cis*-octadecadienoic acid and 10-hydroxy-6,12-*cis,cis*-octadecadienoic acid, respectively [79]. A *Staphylococcus* sp. has also been found to convert oleic acid to 10-ketostearic acid with 90% conversion and 85% yield [80].

Other examples of microbial biotransformation for the preparation of oxygenated fatty acids include the conversion of oleic acid to 7-hydroxy-17-oxo-9-*cis*-octadecenoic acid by a *Bacillus* strain [81] and to 10-hydroxy-8-*trans*-octadecenoic acid [82] and 7,10-dihydroxy-8-*trans*-octadecenoic acid [83] by *Pseudomonas* sp., formation of 3-*R*-hydroxy-polyunsaturated fatty acids from fatty acids containing a *cis*-5,*cis*-8-diene system by the yeast *Dipodascopsis uninucleata* [84], and the biotransformation of linoleic acid to 8-*R*-hydroxy-*cis*-9,*cis*-12-octadecadienoic acid by the fungus *Leptomitus lacteus* [85]. Furthermore, using *P. aeruginosa*, ricinoleic acid has been converted to 7,10,12-trihydroxy-8-*trans*-octadecenoic acid [83] via intermediate formation of 10,12-dihydroxy-8-*trans*-octadecenoic acid [86].

Numerous examples for the production of oxygenated fatty acids via hydroxylation of unsaturated fatty acids by microorganisms have been given in a recent review [87]. Moreover, oxygenation reactions, for example, epoxydation and hydroxylation, of fatty acids and their derivatives by microbial fermentation have been subjects of a recent review [88].

3. Polyunsaturated Fatty Acids

Fungi of the *Mortierella* sp. are capable of converting linoleic acid into GLA [89], and an oil containing α-linolenic acid into an oil containing eicosapentaenoic acid, as shown in Figure 28.10 [90]. Various microorganisms that are capable of converting linoleic and linolenic acids to higher polyunsaturated fatty acids, such as GLA, arachidonic acid, eicosapentaenoic acid, and DHA, have been described in a recent review [38].

4. Dicarboxylic Acids

Species of yeast, *C. tropicalis*, catalyzes the conversion of oleic acid to *cis*-9-octadecene-1,18-dioic acid [91] (Figure 28.11) and elaidic acid to *trans*-9-octadecene-1,18-dioic acid [92]. An industrial strain of *C. tropicalis* has been engineered to convert methyl myristate, methyl palmitate, methyl stearate, oleic acid, and erucic acid to the corresponding α,ω-alkanedicarboxylic acids, as outlined in Figure 28.11 [75].

A mutant of *C. tropicalis* efficiently converts linoleic acid into *cis*-6,*cis*-9-octadecadiene-1,18-dioic acid, 3-hydroxy-*cis*-9,*cis*-12-octadecadiene-1,18-dioic acid, and 3-hydroxy-*cis*-5,

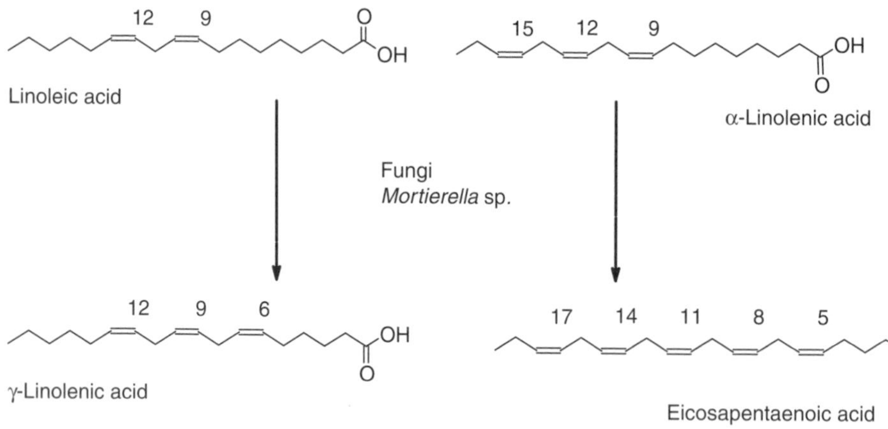

FIGURE 28.10 Microbial production of polyunsaturated fatty acids.

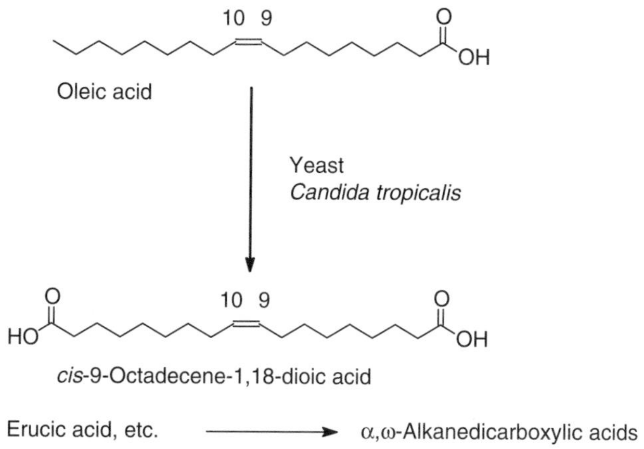

FIGURE 28.11 Microbial production of dicarboxylic acids.

cis-8-octadecadiene-1,18-dioic acid [93]. *C. cloacae* cells oxidize long-chain fatty acids to the corresponding α,ω-dicarboxylic acids [94]. Microbial production of α,ω-dicarboxylic acids has been recently reviewed [87].

5. Ketones and Lactones

Fatty acids and their esters (triacylglycerols) have been converted to methylalkyl-ketones having an alkyl chain one carbon shorter than the substrate using filamentous fungus *Penicillium roquefortii* [95] and the fungus *Trichoderma* sp. [96]. A black yeast, *Aureobasidium* sp., produces from lauric acid-rich palm kernel oil *n*-undecane-2-one and *n*-undecane-2-ol in good yields [97]. A fungus, *Fusarium avenaceum* strain, produces monohydroxynonane-2-ones from tricaprin [98]. These reactions are outlined in Figure 28.12.

The fungi *Mucor javanicus* and *M. miehei* catalyze the lactonization of ω-hydroxy fatty acids, for example, 15-hydroxypentadecanoic and 16-hydroxyhexadecanoic acids, to the corresponding macrocyclic mono- and oligolactones [99], as shown in Figure 28.13. 10-Hydroxystearic acid, prepared by microbial oxidation of oleic acid, has been converted to γ-dodecalactone by baker's yeast, possibly via β-oxidation [100].

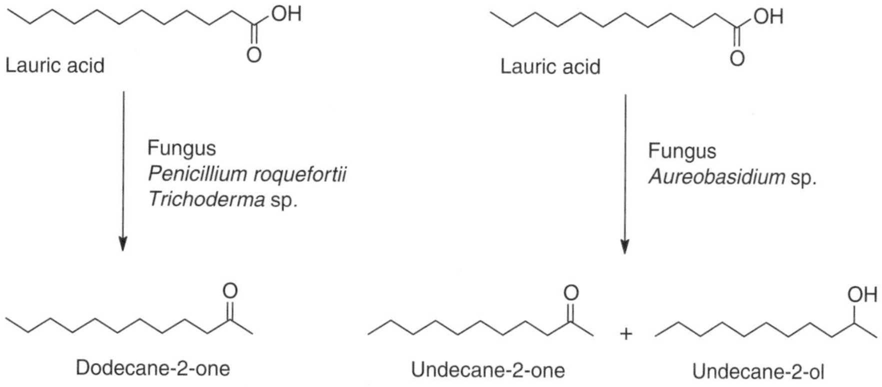

FIGURE 28.12 Microbial production of ketones.

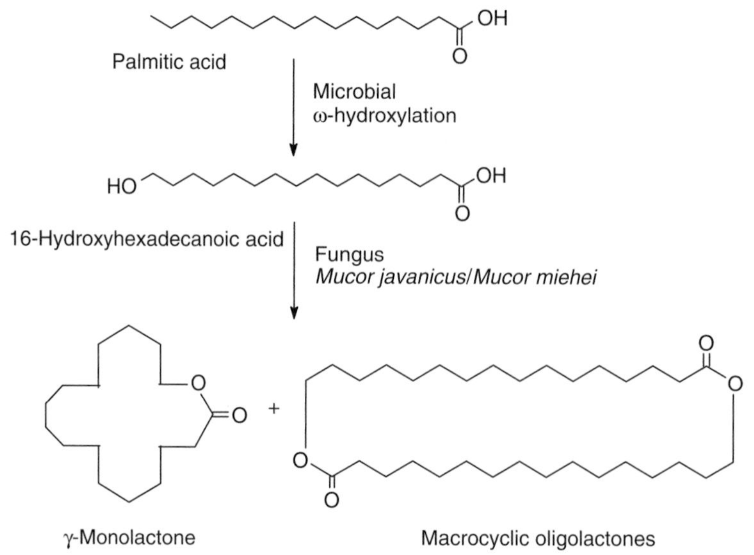

FIGURE 28.13 Microbial production of lactones.

III.   MODIFICATION OF FATS AND OTHER LIPIDS USING ISOLATED ENZYMES

Enzymes of lipid metabolism, isolated from microorganisms, plants, and animal tissues, are highly suitable for biotransformations of fats and other lipids because many of these enzymatic reactions do not require cofactors, and a large number of such enzyme preparations are commercially available. This section will cover the applications of lipolytic enzymes, such as triacylglycerol lipases and phospholipases, as well as the use of oxygenases, lipoxygenases, and epoxide hydrolases in the biomodification of fats and other lipids.

A.   Triacylglycerol Lipases

General properties of triacylglycerol lipases (triacylglycerol acylhydrolases EC 3.1.1.3) are covered in Chapter 29, and the specific applications of triacylglycerol lipases in interesterification reactions are dealt with in Chapters 30 and 31.

Triacylglycerol lipases not only catalyze the hydrolysis; in reaction media with low water content or in the presence of a less polar organic solvent they also catalyze the reverse reaction of lipolysis, that is, esterification of a fatty acid with an alcohol. Moreover, in media with low water content, lipases catalyze a wide variety of interesterification and transesterification reactions, for example, between triacylglycerols or between a triacylglycerol and a fatty acid (acidolysis) or an alcohol (alcoholysis) or glycerol (glycerolysis) [101–104]. The ability of lipases to catalyze the hydrolysis, esterification, and transesterification of lipids has been exploited in commercial processes for the modification of fats and other lipids.

The substrate specificities and regioselectivities in the hydrolysis of triacylglycerols, catalyzed by some common and commercially available lipases, are summarized in Table 28.8 [105–109]. It is evident that the triacylglycerol lipases have widely varying substrate specificities preferring substrates with long- and medium-chain fatty acids over the short chain ones and vice versa. Moreover, specificity of lipases for the fatty acids esterified at the *sn*-1, *sn*-2, and *sn*-3 positions of the glycerol backbone vary widely, ranging from nonspecificity for either of the three *sn*-1, *sn*-2, and *sn*-3 positions to strong *sn*-1,3 or *sn*-3 specificity.

TABLE 28.8

Specificity of Triacylglycerol Lipases from Different Sources

Source of Lipase	Fatty Acid Specificity[a]	Positional Specificity
Micoorganisms		
Aspergillus niger	S, M, L	*sn*-1,3 ≫ *sn*-2
Candida antarctica	S > M, L	*sn*-3
Candida rugosa (syn. *C. cylindracea*)	S, L > M	*sn*-1,2,3
Chromobacterium viscosum	S, M, L	*sn*-1,2,3
Rhizomucor miehei	S > M, L	*sn*-1,3 ≫ *sn*-2
Penicillium roquefortii	S, M ≫ L	*sn*-1,3
Pseudomonas aeruginosa[b]	S, M, L	*sn*-1
Pseudomonas fluorescens	S, L > M	*sn*-1,2,3
Rhizopus delemar	S, M, L	*sn*-1,2,3
Rhizopus oryzae	M, L > S	*sn*-1,3 ≫ *sn*-2
Plants		
Rapeseed (*Brassica napus*)[c]	S > M, L	*sn*-1,3 > *sn*-2
Papaya (*Carica papaya*) latex[d]		*sn*-3
Animal tissues		
Porcine pancreatic	S > M, L	*sn*-1,3
Rabbit gastric[b]	S, M, L	*sn*-3

Source: Adapted from Godfrey, T., *Lipid Technol.*, 7, 58, 1995.

[a] S, short chain; M, medium chain; L, long chain.
[b] Data from Villeneuve, P., Pina, M., Montet, D., and Graille, J., *Chem. Phys. Lipids*, 76, 109, 1995.
[c] Data from Hills, M.J. and Mukherjee, K.D., *Appl. Biochem. Biotechnol.*, 26, 1, 1990.
[d] Data from Villeneuve, P., Pina, M., Montet, D., and Graille, J., *Lipids*, 72, 753, 1995.

In general, the substrate specificity and positional specificity of triacylglycerol lipases observed in the hydrolysis reactions are also maintained in the reverse reaction of hydrolysis, that is, esterification (Figure 28.14), or in interesterification and transesterification reactions (Figure 28.15). These properties of triacylglycerol lipases permit their use as biocatalysts for the preparation of specific lipid products of definite composition and structure that often cannot be obtained by reactions carried out using chemical catalysts [101]. This section outlines some current commercial applications and potentially interesting uses of lipase-catalyzed reactions [110–113] for the production of specialty products from oils and fats.

1. Structured Triacylglycerols

a. Cocoa Butter Substitutes
Some typical applications of lipase-catalyzed interesterification reactions include the preparation, from inexpensive starting materials, of products resembling cocoa butter in their triacylglycerol structure and physical properties. Commercial processes for the preparation of cocoa butter substitutes involve interesterification of palm oil midfraction with stearic acid or ethyl stearate using *sn*-1,3-specific lipases, as shown in Figure 28.16 [103,114–121].

b. Human Milk Fat Replacers
The triacylglycerols of human milk contain the palmitic acid esterified predominantly at the *sn*-2 position. Structured triacylglycerols resembling triacylglycerols of human milk are produced by transesterification of tripalmitin, derived from palm oil, with oleic acid or polyunsaturated fatty

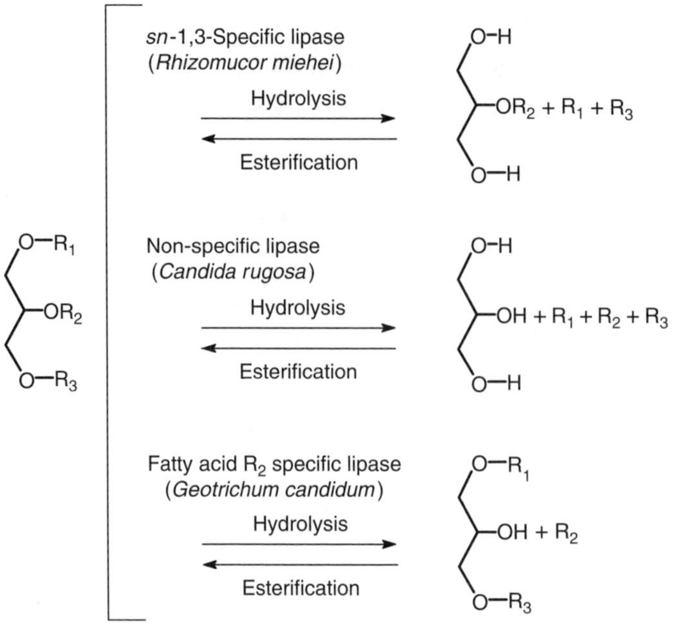

FIGURE 28.14 Specificity of triacylglycerol lipases in hydrolysis and esterification: R_1, R_2, R_3, fatty acids/acyl moieties.

acids, obtained from plant oils, using sn-1,3-specific lipases as biocatalyst, as outlined in Figure 28.17 [118–120,122,123]. Such triacylglycerols are used in infant food formulations.

c. Nutraceuticals

Possible applications of interesterification reactions catalyzed by sn-1,3-specific lipases include, for example, the preparation of structured triacylglycerols for use in specific dietetic products

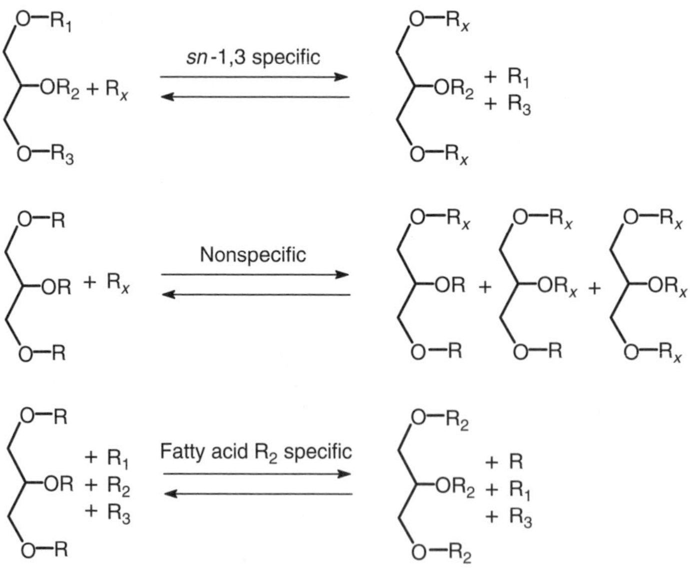

FIGURE 28.15 Specificity of triacylglycerol lipases in interesterification and transesterification: R_1, R_2, R_3, fatty acids/acyl moieties.

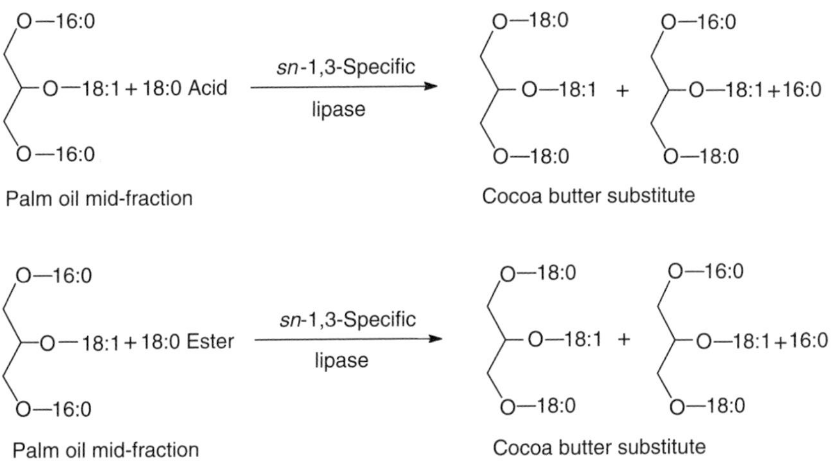

FIGURE 28.16 Lipase-catalyzed transesterification for the production of cocoa butter substitutes.

(nutraceuticals). Thus, interesterification of a common plant oil, such as sunflower oil, with a medium-chain fatty acid using an *sn*-1,3-specific lipase would yield triacylglycerols containing medium-chain acyl moieties at the *sn*-1,3 positions and long-chain acyl moieties at the *sn*-2 position, as shown in Figure 28.18 [124–128]. Such products, which do not occur in nature and are difficult to prepare by chemical synthesis, may find interesting dietetic applications [129].

Evidence has accumulated lately that nutritional properties of triacylglycerols can be altered beneficially by structuring such lipids, for example, by inserting certain fatty acyl moieties at specific positions of the glycerol backbone to yield structured triacylglycerols [119,120,130–141]. Especially, structured triacylglycerols, prepared by lipase-catalyzed transesterification in which the physiologically active ω3 or ω6 polyunsaturated fatty acids, such as DHA and GLA (eicosanoid precursors), are esterified at specific positions of glycerol backbone, as shown in Figure 28.19 [142–145]. They are envisaged to exhibit interesting biological properties [134] that might enable their use in specific nutraceutical products and infant feed (Figure 28.19).

Structured triacylglycerols containing two molecules of caprylic acid and one molecule of erucic acid have been prepared by lipase-catalyzed esterification of caprylic acid to monoerucin. The resulting triacylglycerols yield on subsequent hydrogenation of the erucoyl moieties to behenoyl moieties products resembling Caprenin, a commercially available low-calorie triglyceride [146]. Structured triacylglycerols of the type monoglyceride diacetates and diglyceride monoacetates are prepared using *sn*-1,3-specific lipase by transesterification of plant oils with triacetin [147] or an alkyl acetate, for example, ethyl acetate [148], as shown in Figure 28.20.

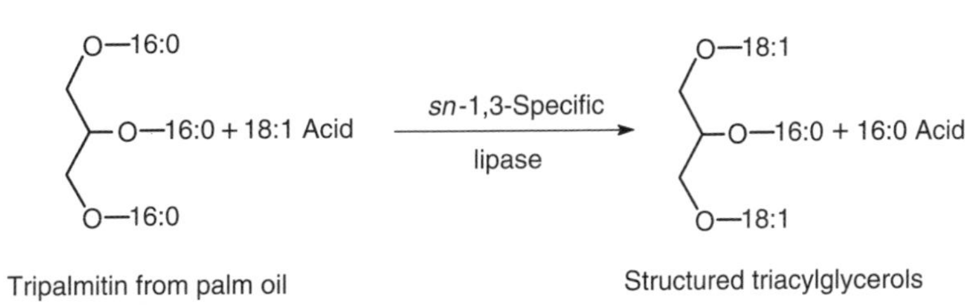

FIGURE 28.17 Preparation of structured triacylglycerols for use as human milk fat replacers by lipase-catalyzed transesterification.

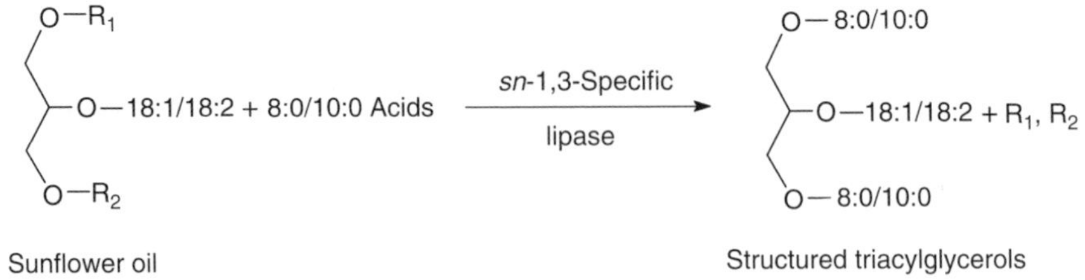

FIGURE 28.18 Preparation of structured triacylglycerols for use as nutraceuticals by lipase-catalyzed transesterification.

2. Bioesters, Long-Chain Esters, and Flavor Esters

Esterification reactions catalyzed by a nonspecific lipase B from *C. antarctica* (Figure 28.21) is used commercially to produce a wide variety of fatty acid esters, the bioesters, such as isopropyl myristate, isopropyl palmitate, octyl palmitate, octyl stearate, decyl oleate, and cetyl palmitate, which are used in personal-care products [149].

Short-chain esters have numerous applications in food industries as flavoring components. Lipase-catalyzed esterification and interesterification for the synthesis of these esters have received considerable attention [150–158]. Applications of lipases for the preparation of esters have been recently reviewed [159–161].

3. Wax Esters and Steryl Esters

Esterification of mixtures of long-chain and very long chain monounsaturated fatty acids (VLCMFAs) with the corresponding mixtures of alcohols using a lipase from *Rhizomucor miehei* (Lipozyme) as catalyst provides wax esters (Figure 28.22) in almost theoretical yields [162].

The high rates of interesterification of triacylglycerols with a long-chain alcohol [102] indicate that alcoholysis reactions should be useful for the production of wax esters of good commercial value. Using lipozyme as biocatalyst, wax esters resembling jojoba oil are obtained in high yield by alcoholysis of seed oils from *Sinapis alba*, *Lunaria annua* [162], or *Crambe abyssinica* [163], which contain large proportions of VLCMFAs esterified at the *sn*-1 and *sn*-3 positions, with very long-chain alcohols derived from these oils, as shown in Figure 28.23. Similarly, long-chain wax esters resembling jojoba oil were obtained in high yields when fatty acids obtained from seed oils of crambe (*C. abyssinica*) and camelina (*Camelina sativa*) were esterified with oleyl alcohol or the alcohols derived from crambe and camelina oils. Novozym 435 (immobilized lipase B from *Candida antarctica*) or papaya (*Carica papaya*) latex lipase were used as biocatalysts, and vacuum was applied to remove the water formed [164]. Further examples of lipase-catalyzed preparation of wax esters via esterification [165–170] and interesterification [167–169,171–174] are known.

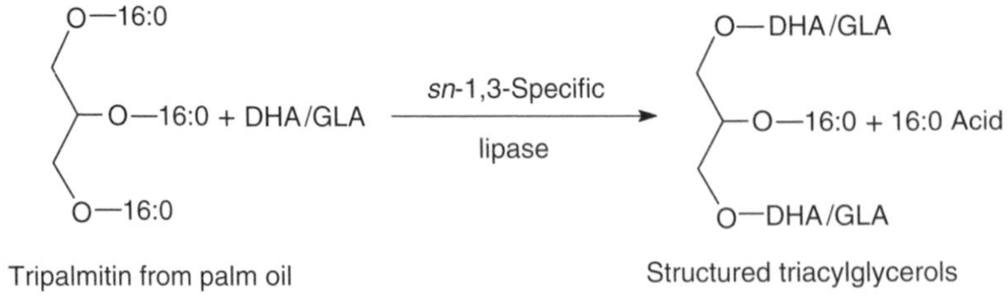

FIGURE 28.19 Preparation of structured triacylglycerols for use in infant foods and nutraceuticals by lipase-catalyzed transesterification.

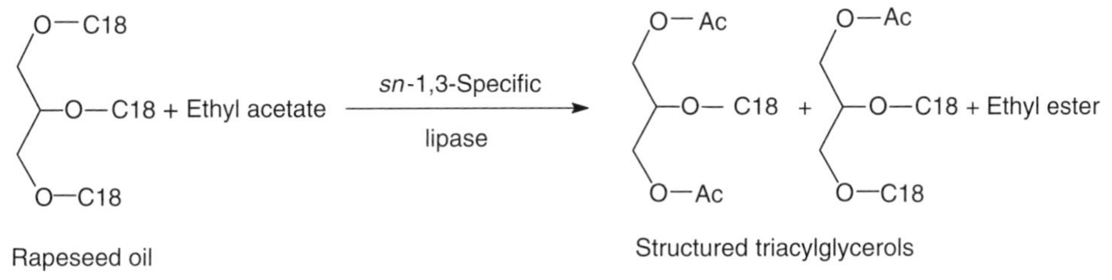

FIGURE 28.20 Preparation of structured triacylglycerols of the monoglyceride diacetate and diglyceride monoacetate types by lipase-catalyzed transesterification: Ac, acetyl moieties.

Unusual wax esters have also been obtained in good yields by lipase-catalyzed reactions, such as esterification of decanol with fatty acids, for example, 9(10)-hydroxymethyl-octadec-10-enoic acid, and transesterification of octanol with methyl esters of 9,10-epoxy- or 9-oxodecanoic acids [175].

Lipozyme has been shown to catalyze the esterification of a great variety of carboxylic acids, including short-chain, long-chain, and branched-chain acids to different types of alcohols, ranging from short-chain and long-chain alkanols to cyclic alcohols [176], giving almost theoretical yields if the reaction water formed by esterification is efficiently removed [115,176].

Esterification catalyzed by immobilized lipases from *R. miehei* [115,176,177] and *Candida rugosa* [178] as well as surfactant-coated microbial lipases [179] has been carried out for the preparation of a wide variety of alkyl esters of fatty acids in high yields. Moreover, lipase-catalyzed transesterification (alcoholysis) of triacylglycerols with an alcohol, such as *n*-butanol [180], ethanol, or iso-propanol [181], provides alkyl esters in high yields, whereby the use of silica gel as an adsorbent for glycerol formed by the reaction greatly enhances the yield [180].

Transesterification (alcoholysis) of low-erucic rapeseed oil with 2-ethyl-1-hexanol, catalyzed by lipase from *C. rugosa*, provides 2-ethyl-1-hexyl esters of rapeseed fatty acids in high yields that can serve as a solvent for printing ink [182].

Butyl esters of fatty acids are useful as lubricants, hydraulic fluids, biodiesel additives, and plasticizers for polyvinylchloride. Butyl oleate has been obtained in high yields by lipase-catalyzed esterification of oleic acid with *n*-butanol [182].

Long-chain alkyl esters of ricinoleic acid have been prepared in high yields by lipase-catalyzed reactions, such as esterification of castor oil fatty acids with a long-chain alcohol or transesterification of castor oil triacylglycerols with a long-chain alcohol [183].

Medium- and long-chain dialkyl thia-alkanedioate antioxidants such as dialkyl 3,3'-thiodipropionates (dioctyl-, didodecyl-, dihexadecyl-, and dioleyl-3,3'-thiodipropionate) have been prepared in high yield by lipase-catalyzed esterification and transesterification of 3,3'-thiodipropionic acid (4-thiaheptane-1,7-dioic acid) and its dimethyl ester, respectively, with the corresponding medium- or long-chain 1-alkanols, that is, 1-octanol, 1-dodecanol, 1-hexadecanol, and *cis*-9-octadecen-1-ol, in vacuo at moderate temperatures without solvents [184] (Figure 28.24). Immobilized lipase B from *C. antarctica* (Novozym 435) is the most active biocatalyst for the preparation of such medium- and long-chain dialkyl 3,3'-thiodipropionates, whereas the immobilized lipases from *R. miehei* (Lipozyme RM IM) and *Thermomyces lanuginosus* (Lipozyme TL IM) are by far less active. Similarly, dihexadecyl

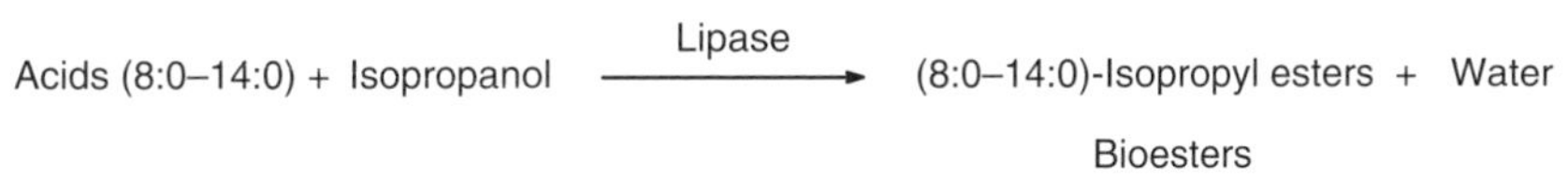

FIGURE 28.21 Lipase-catalyzed synthesis of bioesters.

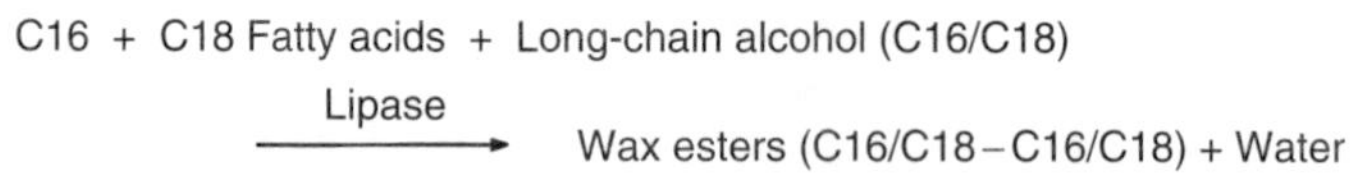

C16 + C18 Fatty acids + Long-chain alcohol (C16/C18)

Lipase
$\xrightarrow{\hspace{2cm}}$ Wax esters (C16/C18 – C16/C18) + Water

FIGURE 28.22 Preparation of wax esters by lipase-catalyzed esterification of fatty acids with long-chain alcohols.

2,2'-thiodiacetate has been prepared in moderate yield using 2,2'-thiodiacetic acid or diethyl 2,2'-thiodiacetate and 1-hexadecanol as the starting materials and Novozym 435 as the biocatalyst [184].

Steryl esters of polyunsaturated fatty acids have been obtained by esterification catalyzed by lipase from *Pseudomonas* sp., but at rather low reaction rates [185]. Recently, fatty acyl esters of phytosterols and phytostanols have been obtained at high rates and in near-quantitative yields by esterification of the sterols with the fatty acids or their transesterification with alkyl esters of fatty acids, both under vacuum using *C. rugosa* lipase as biocatalyst [186–190] (Figure 28.25). Such steryl esters are being used as blood cholesterol-lowering food supplements added to margarines, vegetable oils, mayonnaise, and yoghurts [191–195].

4. Monoacylglycerols and Diacylglycerols

Lipase-catalyzed partial hydrolysis of oils [196,197] and esterification of fatty acids with glycerol [198–207] have been carried out for the production of monoacylglycerols (Figure 28.26). Lipase-catalyzed esterification of glycerol with fatty acids under vacuum provides symmetrical 1,3-diacylglycerols in good yields [208]. Moreover, lipase-catalyzed transesterification of glycerol with an alkyl ester of a fatty acid [203,204] or of triacylglycerols with an alcohol, such as ethanol [209] or *n*-butanol [210], provides good yields of monoacylglycerols.

Interesterification (glycerolysis) of triacylglycerols with glycerol, catalyzed by lipases, as shown in Figure 28.27, has been by far most successful for the preparation of monoacylglycerols [211–220].

Monoacylglycerols of less common fatty acids, such as 9(10)-acetonyloctadecanoic acid, have been obtained in high yields by one-pot reaction of this acid with glycerol and lipase in the presence of phenylboronic acid as a solubilizing agent [175].

For the preparation of diacylglycerol (DAG) oil for food use common and highly unsaturated fatty acids have been esterified to glycerol, catalyzed by lipase from *R. miehei* [221,222]. Interesterification of rapeseed oil triacylglycerols with commercial preparations of monoacylglycerols, such as Monomuls 90-O18, Mulgaprime 90, and Nutrisoft 55, catalyzed by immobilized lipase from *R. miehei* (Lipozyme RM IM) in vacuo at 60°C, has led to extensive (60% to up to 75%) formation of DAGs [223]. Esterification of rapeseed oil fatty acids with Nutrisoft, catalyzed by Lipozyme RM in vacuo at 60°C, also provides DAGs in yields of 60%–70%. Esterification of rapeseed oil fatty acids with glycerol in vacuo at 60°C, catalyzed by Lipozyme RM and lipases from *T. lanuginosus* (Lipozyme TL IM) and *C. antarctica* (lipase B, Novozym 435), also yields DAGs, however, to a lower extent (40%–45%). Glycerolysis of rapeseed oil triacylglycerols with glycerol in vacuo at

O—18:1/22:1

|—O—C18 + 22:1 Alcohol $\xrightarrow[\text{lipase}]{\textit{sn}\text{-1,3-Specific}}$ |—O—C18 + 18:1/22:1 — 22:1 Ester

O—18:1/22:1   O—H   O—H

High-erucic oil   Jojoba wax substitute

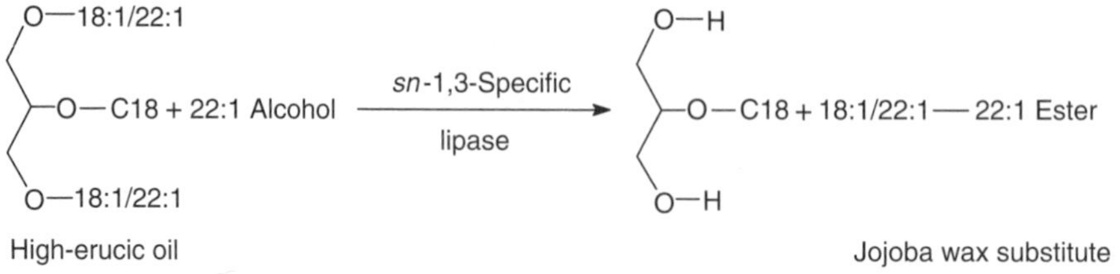

FIGURE 28.23 Preparation of wax esters resembling jojoba oil by alcoholysis of triacylglycerols of oils high in erucic acid with very long chain alcohols, catalyzed by *sn*-1,3-specific triacylglycerol lipases.

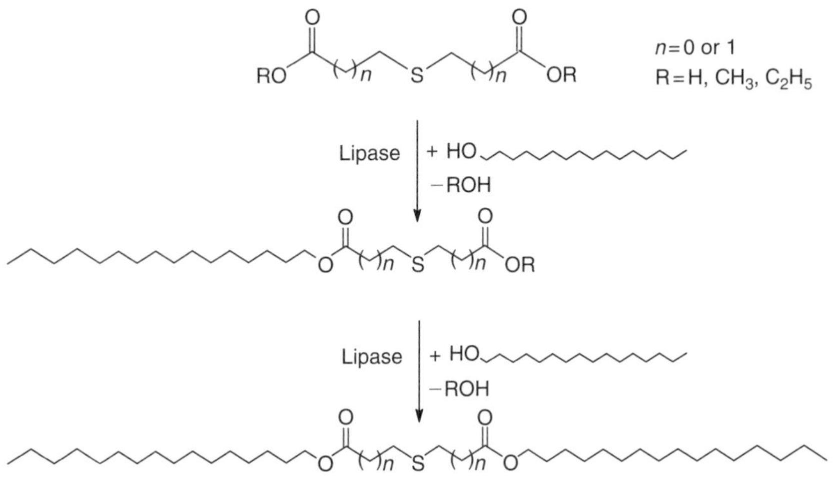

$n = 0$ or 1
$R = H$, CH_3, C_2H_5

FIGURE 28.24 Reaction scheme of the successive lipase-catalyzed esterification or transesterification of thia-alkanedioates such as 3,3′-thiodipropionic acid ($n = 1$; $R = H$) and dimethyl 3,3′-thiodipropionate ($n = 1$; $R = $ methyl) as well as 2,2′-thiodiacetic acid ($n = 0$; $R = H$) and diethyl 2,2′-thiodiacetate ($n = 0$; $R = $ ethyl) with 1-hexadecanol yielding, for example, dihexadecyl 3,3′-thiodipropionate and dihexadecyl 2,2′-thiodiaceate.

60°C, catalyzed by Lipozyme TL and Novozym 435, provides DAGs to the extent of 50% and less. The products of esterification of rapeseed oil fatty acids with Monomuls and glycerol yield upon short-path vacuum distillation residues (DAG oils) containing 66%–70% DAGs [223].

It has been found that DAG, particularly *sn*-1,3-diacylglycerols, may have beneficial effects with regard to the prevention of obesity [221] and lipemia [224], despite having similar energy value and digestibility as known for triacylglycerols [225]. In Japan, DAG have been granted the permission to be termed as a "Food for Specified Health Use" (FOSHU) since 1999 and a cooking oil containing around 80% DAG (DAG oil) as well as mayonnaise are on the market [226]. In the United States, DAG oil has received the status "Generally Recognized as Safe" (GRAS) by the U.S. Food and Drug Administration [226]. Enzymatic production of DAGs [227] and their beneficial physiological functions have been recently reviewed [227–230].

5. Lipophilic Phenolics

The beneficial effects of hydroxycinnamic acids and other plant phenolics on health have been attributed to their antioxidant capacity, particularly against oxidative attacks by their radical-scavenging activity. Both antioxidant capacity and biological availability of several of these

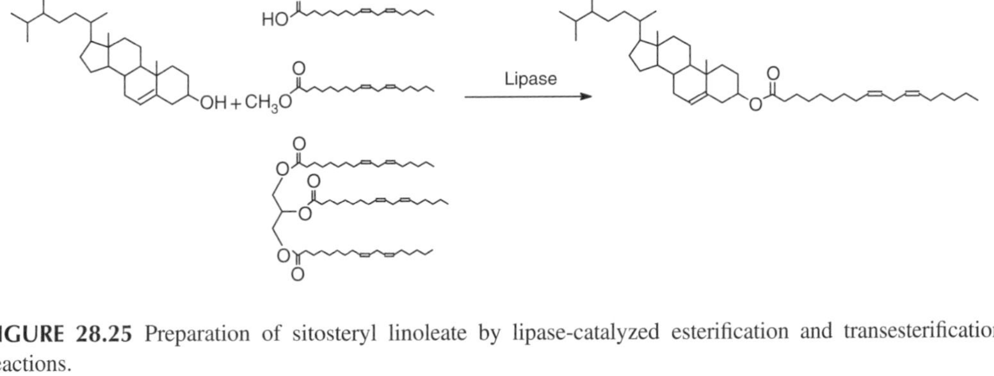

FIGURE 28.25 Preparation of sitosteryl linoleate by lipase-catalyzed esterification and transesterification reactions.

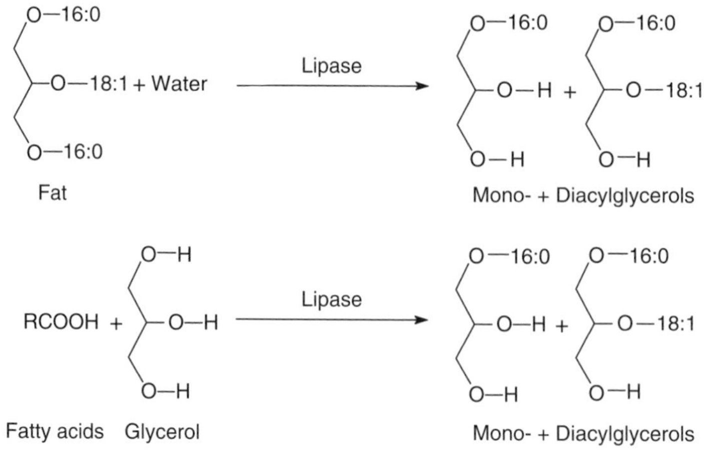

FIGURE 28.26 Preparation of monoacylglycerols by lipase-catalyzed partial hydrolysis of fats (*top*) and esterification of fatty acids with glycerol (*bottom*).

compounds may be further improved by increasing their lipophilicity, and the range of applications of such lipophilic antioxidants may be extended by their possible use as additives for food and technical applications. In addition, similar compounds with inverse chemical structure such as hydroxycinnamyl alkanoates have been isolated from apple fruits. Such retro compounds also have antioxidant properties as they are known for alkyl hydroxycinnamates.

Various medium- or long-chain alkyl cinnamates and hydroxycinnamates, including oleyl *p*-coumarate as well as palmityl and oleyl ferulates, have been prepared in high yield by lipase-catalyzed transesterification of an equimolar mixture of a short-chain alkyl cinnamate and a fatty alcohol such as lauryl, palmityl, and oleyl alcohols under partial vacuum at moderate temperature in the absence of solvents and drying agents in direct contact with the reaction mixture (Figure 28.28). This method, developed for the preparation of lipophilic phenolics, is far superior to lipase-catalyzed syntheses described by others [231]. Immobilized lipase B from *C. antarctica* (Novozym 435) is the most effective biocatalyst for the various transesterification reactions. Transesterification activity of this enzyme is up to 56-fold higher than esterification activity for the preparation of medium- and long-chain alkyl ferulates [232,233]. The relative transesterification activities found for *C. antarctica* lipase are of the following order: hydrocinnamate > cinnamate > 4-hydroxyhydrocinnamate > 3-methoxycinnamate > 2-methoxycinnamate ≈ 4-methoxycinnamate ≈ 3-hydroxycinnamate > hydrocaffeate ≈ 4-hydroxycinnamate > ferulate > 2-hydroxycinnamate > caffeate ≈ sinapate. With respect to the position of the hydroxy substituents at the phenyl moiety, the transesterification activity of Novozym 435 increases in the order *meta* > *para* > *ortho*. Compounds with inverse chemical structure, that is, 3-phenylpropyl alkanoates such as 3-(4-hydroxyphenyl)

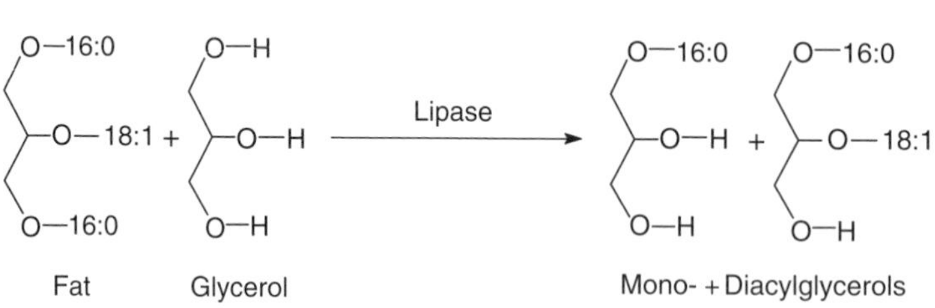

FIGURE 28.27 Preparation of mono- and diacylglycerols by lipase-catalyzed interesterification of fats with glycerol (glycerolysis).

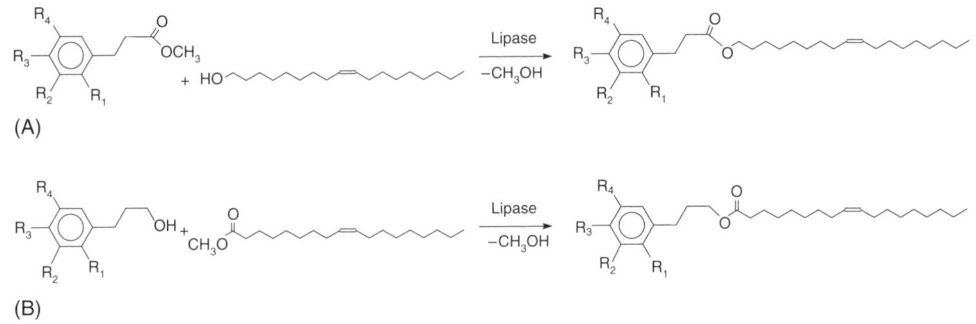

(A)

(B)

FIGURE 28.28 Transesterification of (A) cinnamoyl methyl esters with oleyl alcohol and (B) methyl oleate with dihydrocinnamyl alcohols (R_1, R_2, R_3, R_4 = H, OH, or OCH_3). (According to Vosmann, K., Weitkamp, P., and Weber, N., *J. Agric. Food Chem.*, 54, 2969, 2006; Weitkamp, P., Vosmann, K., and Weber, N., *J. Agric. Food Chem.*, 54, 7062, 2006.)

propyl oleate and 3-(3,4-dimethoxyphenyl)propyl oleate, have been obtained by Novozym 435-catalyzed transesterification of fatty acid methyl esters with the corresponding 3-phenylpropan-1-ols in high yield, as well (Figure 28.28).

6. Lactones and Estolides

Musk lactones are used in the fragrance industry. Hexadecanolide, a musk monolactone, has been obtained in good yields by intramolecular lactonization of 16-hydroxyhexadecanoic acid, catalyzed by immobilized lipase from *C. antarctica*, whereby oligolactones are not formed by intermolecular lactonization [234].

Reaction of lesquerolic (14-hydroxy-11-eicosenoic) acid with oleic acid, catalyzed by lipase from *C. rugosa*, produces mainly monoestolides containing one molecule each of the hydroxy acid and oleic acid per molecule, whereas the corresponding reaction, catalyzed by *Pseudomonas* sp. lipase, produces substantial proportions of monoestolides containing two molecules of lesquerolic acid per molecule besides diestolides [235]. Properties of mono- and polyestolides, synthesized chemically, can be substantially improved by esterification of the estolides with fatty alcohols or α-, ω-diols, catalyzed by lipase from *R. miehei* [235].

7. Fatty Acid Esters of Sugars, Alkylglycosides, and Other Hydroxy Compounds

Lipase-catalyzed esterification has been carried out for the synthesis of fatty acyl esters of carbohydrates that can be used as emulsifiers. Esters of monosaccharides and disaccharides [236–247] as well as those of sugar alcohols [237,238,240,241,248] and other polyols [249] have been prepared in good yields and excellent regioselectivity by esterification catalyzed by microbial lipases (Figure 28.29).

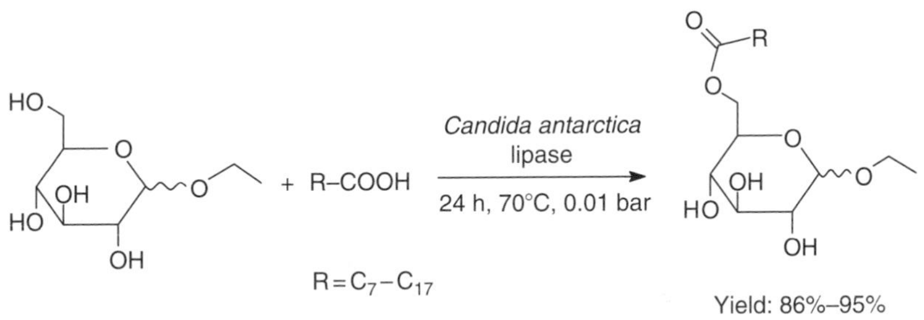

Candida antarctica lipase
24 h, 70°C, 0.01 bar

+ R–COOH

$R = C_7 - C_{17}$

Yield: 86%–95%

FIGURE 28.29 Preparation of sugar esters of fatty acids by lipase-catalyzed esterification.

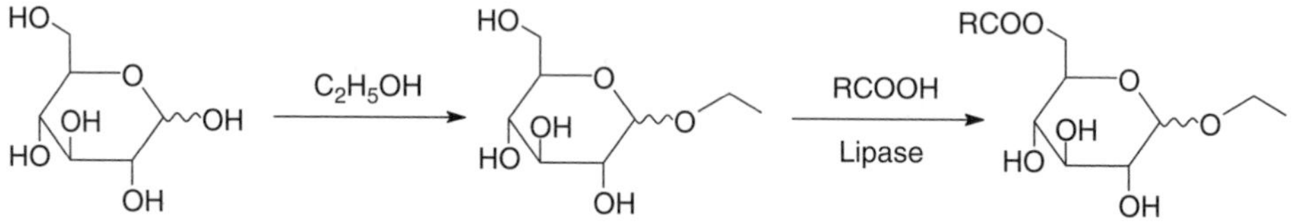

FIGURE 28.30 Preparation of esters of alkylglycosides by lipase-catalyzed esterification.

Transesterification of sugars with short-chain alkyl esters of fatty acids also provides sugar esters in good yields [236,250].

Recently, lipase-catalyzed esterification and transesterification reactions using less toxic solvents, such as *tert*-butanol [251,252] or acetone [253,254], have been used successfully for the preparation of sugar esters.

Triacylglycerols, contained in common fats and oils, as well as wax esters of jojoba oil, have been transesterified with various sugar alcohols in pyridine using lipases to yield primary monoesters of sugar alcohols having excellent surfactant properties [255]. Transesterification [256,257] and esterification [258,259] reactions, catalyzed by lipases, have also been applied for the preparation of fatty acid esters of alkyl glycosides (Figure 28.30).

Fatty acid esters of polyols are useful as surfactants. Polyglycerol fatty acid esters have been prepared in good yields by transesterification of polyglycerol, adsorbed on silica gel, with methyl esters of fatty acids [260].

Propylene glycol monoesters, suitable as emulsifiers, have been prepared in good yields by reacting a fatty acid anhydride with 1,2-propanediol in the presence of lipase from a *Pseudomonas* sp. [261]. Esterification of eicosapentaenoic acid and DHA with 1,2-propanediol in the presence of lipase from *R. miehei* provides propylene glycol monoester emulsifiers that are potentially beneficial to health [262].

Polyethylene glycol esters of fatty acids, widely used as nonionic surfactants, have been prepared in essentially quantitative yields by esterification of oleic acid with polyethylene glycol, catalyzed by lipase from *R. miehei* [263].

Medium- and long-chain alcohols (C_8 to C_{16}) have been efficiently esterified with lactic acid and glycolic acid using lipase B from *C. antarctica* (Novozym 435) as biocatalyst [264]. Using the same lipase ethyl lactate has been transesterified with *n*-octyl-β-D-glucopyranoside to obtain *n*-octyl-β-D-glucopyranosyl lactate in high yield [265].

Ascorbyl palmitate, used as antioxidant in foods and cosmetics, has been prepared in good yields by esterification of ascorbic acid with palmitic acid using lipase from *B. stearothermophilus* SB 1 [266] and *C. antarctica* (Novozym 435) [267]. Esterification of cinnamic acid with 1-octanol using Novozym 435 also provides the octyl ester in moderate yields [267]. Similarly, using lipase B from *C. antarctica* (Chirazyme L2), 6-*O*-palmitoyl-L-ascorbic acid and 6-*O*-eicosapentaenoyl-L-ascorbic acid have been prepared, respectively, via transesterification with vinyl palmitate [268] and condensation with eicosapentaenoic acid [269]. Transesterification of L-methyl lactate with ascorbic acid or retinol using Novozym 435 gives high yields of ascorbyl-L-lactate and retinyl-L-lactate, respectively [270].

8. Amides

Reaction of a triacylglycerol mixture, such as soybean oil, with lysine, catalyzed by lipase from *R. miehei* yields acyl amides, that is, *N*-ε-acyllysines [271]. Reaction of ethyl octanoate with ammonia (ammonolysis), catalyzed by lipase from *C. antarctica*, provides octanamide in near-quantitative yield [272]. One-pot enzymatic synthesis of octanamide in high yield via esterification of octanoic acid with ethanol, followed by ammonolysis of the resulting ethyl octanoate, both reactions being conducted using the lipase from *C. antarctica* has also been reported [272]. Lipase-catalyzed

Fat + Water $\xrightarrow{\text{Lipase}}$ Fatty acids + Glycerol

FIGURE 28.31 Preparation of fatty acids and glycerol by lipase-catalyzed hydrolysis of fats.

direct amidation of carboxylic acids by ammonia and ammonium salts has been reported [273], and various applications of lipase-catalyzed aminolysis and ammonolysis have been recently reviewed [274].

9. Fatty Acids

Lipase-catalyzed hydrolysis can be applied for the production of fatty acids from fats using, for example, a nonspecific lipase preparation from *C. rugosa* (syn. *C. cylindracea*) [275–283], *Pseudomonas* sp. [284–288], *Aspergillus* sp. [289], *T. lanuginosus* [290], and *Chromobacterium viscosum* [291] as an alternative mild process compared with drastic steam splitting (Figure 28.31). Specifically, fatty acids have been obtained by lipase-catalyzed hydrolysis of technically important fats, such as animal fats [281], castor oil [288], palm stearin [287], and the thermally labile high-α-linolenic acid-containing perilla oil [283]. More recently, enzymatic presplitting of oils before steam splitting is considered as an economically feasible alternative to complete enzymatic hydrolysis or steam splitting alone [292].

a. Fatty Acid Concentrates

Triacylglycerol lipases from various organisms have one common feature in their selectivity toward groups of fatty acids/acyl moieties having olefinic bonds at definite positions [293] or geometric configuration [294]. Thus, several lipases from microorganisms, plants, and animal tissues discriminate against fatty acids/acyl moieties having a *cis*-4, *cis*-6, or a *cis*-8 double bond as substrates in hydrolysis, esterification, and interesterification reactions, as summarized in Table 28.9.

Data presented in Figure 28.32 show for example the substrate specificity in esterification of a wide variety of fatty acids with *n*-butanol using the latex from papaya plant (*Carica papaya*) as biocatalyst [301]. In these studies, a mixture of the fatty acid examined and the reference standard, myristic acid, at equal molar concentrations in *n*-hexane was reacted with *n*-butanol using the above

TABLE 28.9
Specificity of Triacylglycerol Lipases from Different Sources toward Various Fatty Acids/Acyl Moieties

Source of Lipase	Discrimination against[a]	References
Microorganisms		
Candida rugosa (syn. *C. cylindracea*)	All-*cis*-4,7,10,13,16,19-DHA, petroselinic (*cis*-6-	[295–297]
Penicillium cyclopium	octadecenoic) acid, GLA (all-*cis*-6,9,12-octadecatrienoic)	
Penicillium sp. (lipase G)	acid, stearidonic (all-*cis*-6,9,12,15-octadecatetraenoic) acid,	
Rhizomucor miehei	dihomo-γ-linolenic (all-*cis*-6,9,12-eicosatrienoic) acid	
Rhizopus arrhizus		
Plants		
Rape (*Brassica napus*) seedlings	DHA, petroselinic, GLA, stearidonic, and dihomo-γ-	[298–300]
Papaya (*Carica papaya*) latex	linolenic acids	[301]
Animal tissues		
Porcine pancreas	DHA, petroselinic acid, GLA, and stearidonic acid	[295]

[a] DHA, docosahexaenoic acid; GLA, γ-linolenic acid.

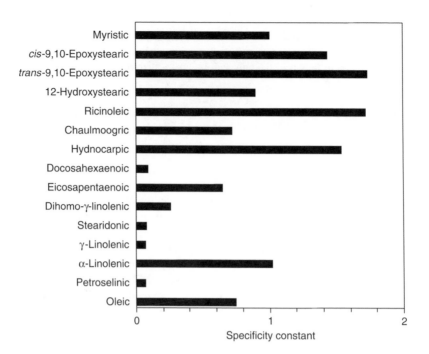

FIGURE 28.32 Specificity constants in the esterification of mixtures of myristic acid (reference standard) and individual fatty acids with *n*-butanol in hexane using papaya latex as biocatalyst. (According to Mukherjee, K.D. and Kiewitt, I., *J. Agric. Food Chem.*, 44, 1948, 1996. With permission.)

biocatalyst and the course of formation of butyl esters under competitive conditions was followed. The competitive factor α was determined according to Rangheard et al. [297] from the concentrations of the two substrates (Ac1X and Ac2X) at time X by the equation:

$$\alpha = \frac{VAc1X/KAc1X}{VAc2X/KAc2X},$$

where
  V is maximal velocity
  K is the Michaelis constant

The competitive factor α was calculated from the substrate concentrations Ac1X0 and Ac2X0 at time zero as follows:

$$\alpha = \frac{\text{Log}[\text{Ac1}X0/\text{Ac1}X]}{\text{Log}[\text{Ac2}X0/\text{Ac2}X]}.$$

From the competitive factor specificity constant was calculated as $1/\alpha$ with reference to the specificity constant of myristic acid taken as 1.00. The higher the specificity constant of a fatty acid, the greater is the specificity of the biocatalyst for that particular fatty acid. The data presented in Figure 28.32 show that the fatty acids having a *cis*-4, *cis*-6, or a *cis*-8 double bond are poor substrates in esterification reactions as compared with those having a *cis*-5 or *cis*-9 double bond or fatty acids having hydroxy, epoxy, or cyclopentenyl groups.

The above substrate specificities have been utilized for the enrichment of definite fatty acids or their derivatives from mixtures via kinetic resolution [302,303], for example, by selective hydrolysis, as shown in Scheme 28.3, and the following examples.

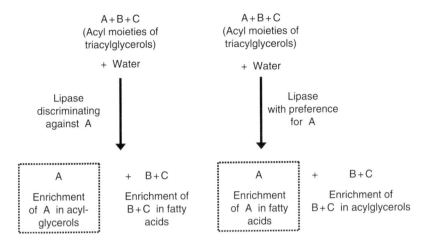

SCHEME 28.3 Principle of kinetic resolution via lipase-catalyzed hydrolysis.

Despite relatively high prices of lipase preparations, lipase-catalyzed hydrolysis could be economically attractive for the preparation of specific products of high commercial value, such as polyunsaturated (ω3) fatty acid concentrates via selective hydrolysis of marine oils, catalyzed by fatty acid-specific lipases that enable the enrichment of docosahexaenoic 22:6ω3 and eicosapentae-noic 20:5ω3 acids in the unhydrolyzed acylglycerols, as outlined in Figure 28.33 [302,304–308]. Such polyunsaturated fatty acids that are interesting as dietetic products [309] cannot be obtained by conventional steam splitting without substantial decomposition.

In a commercial process, fish oil is partially hydrolyzed by *Candida rugosa* lipase to yield an acylglycerol fraction enriched in 20:5ω3, and especially in 22:6ω3. The acylglycerol fraction is subsequently isolated by evaporation and converted to triacylglycerols via hydrolysis and reesterification, both catalyzed by *R. miehei* lipase [310]. Using *Rhizopus delemar* lipase, selective esterification of tuna oil fatty acids with lauryl alcohol, extraction of the unreacted fatty acids, and their repeated esterification with lauryl alcohol have resulted in an unesterified fatty acid fraction containing 91% 22:6 ω3 [311]. Selective interesterification of tuna oil triacylglycerols with ethanol using *Rhizomucor miehei* lipase as biocatalyst yields an acylglycerol fraction containing 49% 22:6ω3,

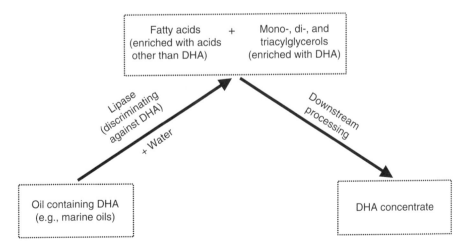

FIGURE 28.33 Preparation of concentrates of docosahexaenoic acid via lipase-catalyzed selective hydrolysis of marine oils.

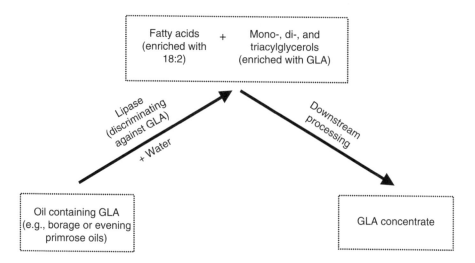

FIGURE 28.34 Preparation of concentrates of γ-linolenic acid (GLA) via lipase-catalyzed selective hydrolysis of borage oil or evening primrose oil.

whereas selective esterification of tuna oil fatty acids with ethanol yields an unesterified fatty acid fraction containing 74% 22:6ω3 [312].

Fatty acids generated by lipase-catalyzed hydrolysis of a commercial SCO from *Mortierella alpina* have been subjected to selective esterification with lauryl alcohol, catalyzed by lipase from *C. rugosa*. This leads to an increase in the arachidonic acid content from 25% in the starting fatty acid mixture to over 50% in the fatty acids that remained unesterified [313].

In addition, GLA, a constituent of certain seed oils such as borage oil and evening primrose oil, can be prepared as a concentrate together with linoleic acid by lipase-catalyzed selective hydrolysis (Figure 28.34) under mild conditions, and typical data obtained with lipase from *C. rugosa* are shown in Figure 28.35 [314]. Very recently, lipase-catalyzed selective hydrolysis of a microbial oil from *Mortierella alpina* has been employed to enrich arachidonic acid in the acylglycerols [315].

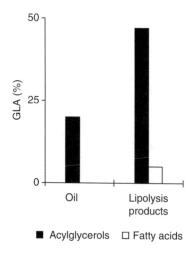

FIGURE 28.35 Enrichment of γ-linolenic acid (GLA) from borage oil by selective hydrolysis catalyzed by triacylglycerol lipase from *Candida cylindracea* (syn. *C. rugosa*): reaction temperature, 20°C; reaction time, 2 h; degree of hydrolysis, 89%. (According to Watanabe, Y., Minemoto, Y., Adachi, S., Nakanishi, K., Shimada, Y., and Matsuno, R., *Biotechnol. Lett.*, 22, 637, 2000.)

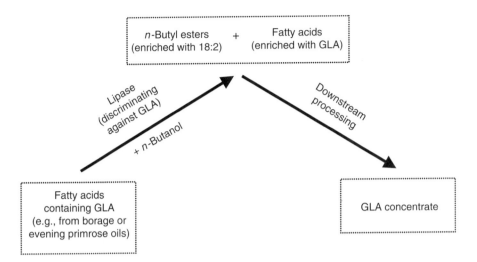

FIGURE 28.36 Preparation of concentrates of γ-linolenic acid (GLA) via lipase-catalyzed selective esterification of fatty acids from borage oil or evening primrose oil with *n*-butanol.

The ability of lipase preparations from plants and microorganisms to discriminate against fatty acids/acyl moieties having *cis*-4, *cis*-6, or *cis*-8 double bonds (Table 28.9) has been utilized for the enrichment of GLA from fatty acid mixtures, derived from plant and microbial oils, by selective esterification of the fatty acids, other than GLA, with *n*-butanol, as outlined in Figure 28.36 [316–321]. Similarly, lipase-catalyzed esterification has been applied to enrich DHA from fatty acid mixtures, derived from marine oils [317,321]. Such concentrates might find nutraceutical applications in capsules.

Typical data on enrichment of GLA via lipase-catalyzed selective esterification of fatty acids from borage oil with *n*-butanol are given in Figure 28.37.

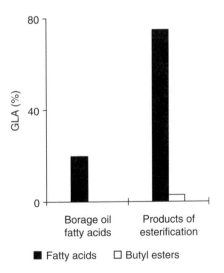

FIGURE 28.37 Enrichment of γ-linolenic acid (GLA) from borage oil fatty acids by selective esterification with *n*-butanol, catalyzed by triacylglycerol lipase from *Rhizomucor miehei* (Lipozyme): reaction temperature, 60°C; reaction time, 2 h; degree of esterification, 91%. (According to Syed Rahmatullah, M.S.K., Shukla, V.K.S., and Mukherjee, K.D., *J. Am. Oil Chem. Soc.*, 71, 563, 1994.)

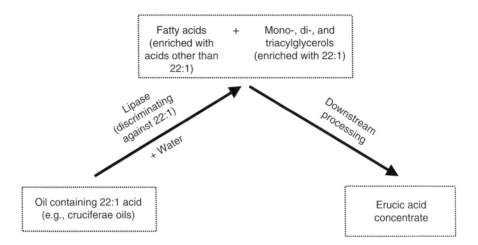

FIGURE 28.38 Enrichment of very long chain monounsaturated fatty acids from high erucic acid seed oils via lipase-catalyzed selective hydrolysis.

Conjugated linoleic acids (CLA), predominantly a mixture of *cis*-9,*trans*-11-octadecadienoic and *trans*-10,*cis*-12-octadecadienoic acids, have gained some interest as beneficial supplements for foods and feeds because of their potentially anticarcinogenic and immunological properties. However, it is not known which of the two major isomers is physiologically more active. With this background, a mixture containing the two CLA isomers in equal amounts was subjected to esterification with dodecanol using lipase from *Geotrichum candidum* as catalyst [322]. This resulted in selective esterification of the *cis*-9,*trans*-11 isomer and enrichment of the *trans*-10,*cis*-12 isomer in the unesterified fatty acid fraction. Separation of the two fractions by molecular distillation yielded an ester fraction containing 91% *cis*-9,*trans*-11 isomer and a fatty acid fraction containing 82% *trans*-10,*cis*-12 isomer. Nutritional implications of CLA have been recently reviewed [323,324].

Selective hydrolysis of high-erucic oils, catalyzed by lipases from *G. candidum* [325,326] and *C. rugosa* [326,327], leads to enrichment of erucic acid in the unhydrolyzed acylglycerols, as outlined in Figure 28.38. The DAGs formed by hydrolysis using the lipase from *C. rugosa* contain as much as 95% of erucic acid [326]. In the esterification of individual fatty acids with *n*-butanol, catalyzed by lipase from *G. candidum*, erucic acid is discriminated against [328,329].

The selectivity of commercial lipases toward VLCMFAs in hydrolysis and transesterification reactions has been determined using high-erucic oils from white mustard (*Sinapis alba*), oriental mustard (*Brassica juncea*), and honesty (*L. annua*) seeds [148]. The lipases from *C. rugosa* and *G. candidum* selectively cleave the C_{18} fatty acids from the triacylglycerols, which results in enrichment of these fatty acids from about 40% in the starting oil to approximately 60% and 89%, respectively, in the fatty acid fraction. Concomitantly, the level of erucic acid and the other VLCMFA is raised in the acylglycerol fraction from 51% in the starting oil to about 80% and 72%, respectively, as shown in Figure 28.39.

The *sn*-1,3-specific lipases from porcine pancreas, *Chromobacterium viscosum*, *Rhizopus arrhizus*, and *Rhizomucor miehei* selectively cleave the VLCMFA, esterified almost exclusively at the *sn*-1,3 positions of the high-erucic triacylglycerols from Cruciferae, which results in enrichment of the VLCMFA in the fatty acid fraction to 65%–75%, whereas the C_{18} fatty acids are enriched in the acylglycerol fraction [148].

Selective hydrolysis of erucic acid- and nervonic acid-rich triacylglycerols of *L. annua*, catalyzed by the lipase from *Candida rugosa*, leads to preferential cleavage of the C_{18} fatty acids, resulting in their enrichment from 36% in the starting oil to 79% in the fatty acids; concomitantly,

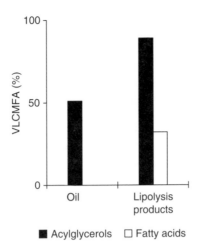

FIGURE 28.39 Enrichment of very long chain monounsaturated fatty acids (VLCMFA = eicosenoic + erucic + nervonic) from white mustard seed oil by selective hydrolysis catalyzed by triacylglycerol lipase from *Candida cylindracea* (syn. *C. rugosa*): reaction temperature, 20°C; reaction time, 1.25 h; degree of hydrolysis, 49%. (According to Mukherjee, K.D. and Kiewitt, I., *Appl. Microbiol. Biotechnol.*, 44, 557, 1996.)

the VLCMFA are enriched in the di- and triacylglycerols. The DAGs, the major (55%) products of lipolysis, are almost exclusively (>99%) composed of VLCMFA [148].

Lipase-catalyzed selective hydrolysis of triacylglycerols from meadowfoam oil or selective esterification of meadowfoam fatty acids has been reported for the enrichment of *cis*-5-eicosenoic acid in the acylglycerols and unesterified fatty acids, respectively [330].

Fennel (*Foeniculum vulgare*) oil has been selectively hydrolyzed using a lipase from *Rhizopus arrhizus* for the enrichment of petroselinic acid in unhydrolyzed acylglycerols [331]. The acylglycerols were separated from the fatty acids using an ion exchange resin and subsequently hydrolyzed by the lipase from *C. rugosa* to yield a fatty acid concentrate containing 96% of petroselinic acid. Highly purified concentrate of petroselinic acid has also been prepared from fatty acids of coriander oil via selective esterification with *n*-butanol, catalyzed by lipase from germinating rapeseed (Figure 28.40) [321].

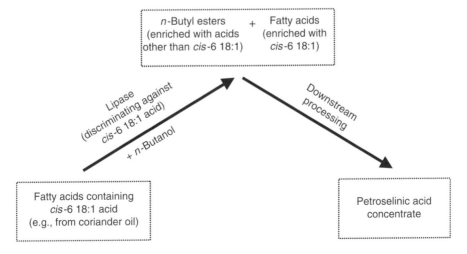

FIGURE 28.40 Preparation of concentrates of petroselinic acid via lipase-catalyzed selective esterification of fatty acids from coriander oil with *n*-butanol.

Fatty acid concentrates containing 85% hydroxy acids, such as lesquerolic (14-hydroxy-*cis*-11-eicosenoic) and auricolic (14-hydroxy-*cis*-11,*cis*-17-eicosadienoic) acids, have been prepared from lesquerella oil by their selective cleavage catalyzed by lipase from *R. arrhizus* [332].

Fatty acids of *Biota orientalis* seed oil have been selectively esterified with *n*-butanol using lipase from *C. rugosa* to enrich *cis*-5-polyunsaturated fatty acids, for example, all-*cis*-5,11,14-octadecatrienoic and all-*cis*-5,11,14,17-octadecatetraenoic acids. The level of total *cis*-5-polyunsaturated fatty acids is raised from about 15% in the starting material to about 73% in the unesterified fatty acids [333]. Selective hydrolysis of the seed oil of *B. orientalis* by the lipase from *C. rugosa* leads to enrichment of the *cis*-5-polyunsaturated fatty acids in the acylglycerols to about 41% [333]. Apparently, fatty acids/acyl moieties having as *cis*-5 double bond are also discriminated against by some lipases. A few comprehensive reviews [293,334–336] cover the applications of lipase-catalyzed reactions for the enrichment of fatty acids via kinetic resolution.

10. Phospholipids

Interestingly, not only phospholipases A_1 and A_2, as described in Section III.B, but also some triacylglycerol lipases cleave the fatty acids from the *sn*-1 and *sn*-2 positions of diacylglycerophospholipids and also catalyze ester exchange reactions to modify the composition of the acyl moieties at the *sn*-1 and *sn*-2 positions of glycerophospholipids. Interestingly, triacylglycerol lipases, such as those from *R. arrhizus* and *Rhizomucor miehei*, have recently been found also to be able to catalyze the acyl exchange of galactolipids, for example, via acidolysis of heptadecanoic acid with digalactosyldiacylglycerols (DGDG) [337]. Modification of phospholipids using triacylglycerol lipases will be covered in Section III.B.

B. Phospholipases

Figure 28.41 shows the reactions catalyzed by phospholipases (see also Chapter 29). Phospholipases A_1 and A_2 hydrolyze the acyl moieties from the *sn*-1 and *sn*-2 positions, respectively, of glycerophospholipids, such as diacylglycerophosphocholines. Phospholipase C cleaves the polar head groups, such as phosphocholine or phosphoethanolamine residues, esterified at the *sn*-3 position of these phospholipids yielding DAGs. Phospholipase D cleaves the bases or alcohols, such as choline, ethanolamine, or glycerol, from these phospholipids yielding phosphatidic acids (PA). Under certain conditions, such as in the presence of less polar organic solvents at low water content, most of the above reactions can be reversed to modify the composition of the acyl moieties or the head groups of the phospholipids. A recent review covers the industrial applications of phospholipases [338].

Phospholipids, such as commercial soya lecithin or egg lecithin, are widely used for their emulsifying and other functional properties in food, cosmetic, and pharmaceutical products. The composition of phospholipids can be altered to modify their properties by chemical reactions [339] or enzymatic reactions catalyzed by phospholipases [101,132,340].

1. Phospholipids Modified by Phospholipases A_1 and A_2 and Triacylglycerol Lipases

The following examples show the various possibilities of modifying the composition of acyl moieties of phospholipids by interesterification reactions catalyzed by phospholipase A_1, phospholipase A_2, or triacylglycerol lipase. Such modified phospholipids may find interesting biomedical applications.

Selective hydrolysis of diacylglycerophospholipids, catalyzed by phospholipase A_2 or A_1, yields 1-acyl- or 2-acyl-lysoglycerophospholipids, respectively [132] (Figure 28.41). Phospholipases A_1 and A_2 as well as regiospecific or nonregiospecific triacylglycerol lipases have been found to cleave the fatty acids from the *sn*-1 and *sn*-2 positions of diacylglycerophospholipids to yield *sn*-1- or *sn*-2-lysoglycerophospholipids with interesting functional properties [341–346]. Studies on hydrolysis of

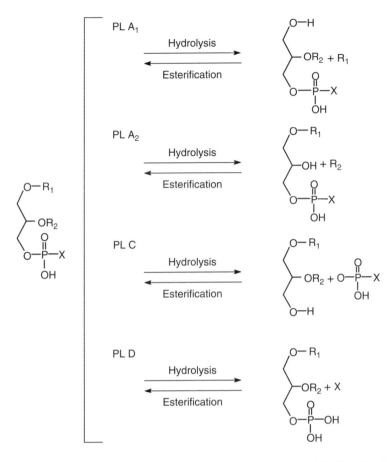

FIGURE 28.41 Reactions catalyzed by phospholipases (PL) A_1, A_2, C, and D: R_1, R_2, fatty acids/acyl moieties; X, alcohol (e.g., choline, ethanolamine, etc.).

soybean phospholipids have revealed that fungal triacylglycerol lipase preparations that also contain phospholipase A_1 and A_2 activities as well as lysophospholipase activity are more efficient in the cleavage of fatty acids than fungal and mammalian enzyme preparations that have only phospholipase A_1 and A_2 activities [347]. Moreover, fungal preparations of both triacylglycerol lipases as well as phospholipase A_1 cleave in the course of time the fatty acids esterified at both *sn*-1 and *sn*-2 positions of diacylglycerophospholipids yielding completely deacylated products, for example, glycerophosphocholine [347]. However, phospholipase A_2 preparations of fungal as well as mammalian pancreatic origin yield primarily *sn*-1-acyl-lysophospholipid by selective partial deacylation at the *sn*-2 position [347].

Esterification of *sn*-1-acyl-lysoglycerophosphocholines with eicosapentaenoic acid and DHA, catalyzed by porcine pancreatic phospholipase A_2 in a microemulsion system containing small amounts of water, has been carried out to prepare diacylglycerophosphocholines containing well over 30% ω3 polyunsaturated fatty acids (ω3 PUFAs), as outlined in Figure 28.42 [348]. Transesterification of phosphatidylcholine (PC) with ethyl eicosapentaenoate, catalyzed by porcine pancreatic phospholipase A_2 in the presence of toluene, has led to about 14% incorporation of eicosapentaenoyl moieties into PC [349].

Long-chain PUFAs from fish oil, dissolved in propane [350] or isooctane [351] have been esterified to *sn*-1-acyl-lysophosphatidylcholine to an extent of about 20%–25% using porcine pancreatic phospholipase A_2 as biocatalyst. Similarly, phospholipase A_2-mediated esterification of

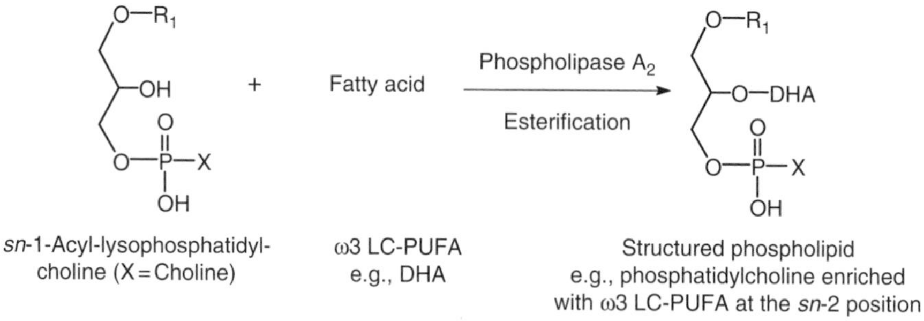

FIGURE 28.42 Preparation of structured phospholipids by esterification of lysophospholipids with ω3 polyunsaturated fatty acids (ω3 PUFA) catalyzed by phospholipase A_2.

eicosapentaenoic acid to lysophosphatidylcholine in the presence of formamide has been used to prepare therapeutic phospholipids in yields of about 60% [352]. 1-Ricinoleoyl-2-acyl-*sn*-glycero-3-phosphocholine has been prepared from soya and egg PC using phospholipase A_1 [353].

Lysophosphatidic acid has been prepared in a yield of 32% by direct solvent-free esterification of fatty acids to *sn*-glycerol-3-phosphate, catalyzed by triacylglycerol lipase from *R. miehei* [354]. Immobilized *sn*-1,3-specific triacylglycerol lipase from *Rhizopus arrhizus* has been found to efficiently catalyze the transesterification of DL-glycero-3-phosphate with vinyl laurate yielding lysophosphatidic acids (1-acyl-*rac*-glycero-3-phosphate) and PAs (1,2-diacyl-*rac*-glycero-3-phosphate) in a total conversion of >95% [355]. The conversions were lower (55%) with oleic acid as acyl donor in the corresponding esterification reaction [355].

Transesterification of L-α-glycerophosphocholine with vinyl esters of fatty acids such as vinyl laurate, catalyzed by *C. antarctica* lipase B (Novozym 435) in the presence of *tert*-butanol, gives predominantly 1-acyl-lysophosphatidylcholine with high (>95%) conversion [356]. Similarly, esterification of fatty acids with L-α-glycerophosphocholine, catalyzed by *Rhizomucor miehei* lipase (Lipozyme RM IM) in the presence of dimethylformamide, produces 1-acyl-lysophosphatidylcholine with high (90%) conversion [357].

Transesterification reactions, such as acidolysis, that is, exchange of the constituent fatty acids of diacylglycerophosphocholines, have been carried out against other fatty acids added as reaction partners using *sn*-1,3-specific triacylglycerol lipase from *Rhizopus delemar* as biocatalyst [358] (Figure 28.41). Similarly, transesterification of diacylglycerophospholipids, catalyzed by *sn*-1,3-specific or nonspecific triacylglycerol lipases, has been applied to modify the fatty acid composition of diacylglycerophospholipids, specifically at the *sn*-1 position or at both *sn*-1 and *sn*-2 positions [346,347,359–364]. In transesterification of PC with a fatty acid, catalyzed by triacylglycerol lipases from *R. delemar*, *Rhizomucor miehei* [360], or *Rhizopus arrhizus* [361,362], acyl exchange occurs almost exclusively at the *sn*-1 position. Polyunsaturated fatty acids, especially ω3 PUFA, have been incorporated into phospholipids by transesterification catalyzed by *sn*-1,3-specific triacylglycerol lipases from *R. delemar* [363] and *Rhizomucor miehei* [364], as outlined in Figure 28.43.

Transesterification of eicosapentaenoic acid with PC from soybean, catalyzed by an *sn*-1,3-specific triacylglycerol lipase from *R. miehei* in the presence of a combination of water and propylene glycol, yields a therapeutically beneficial phospholipid [352]. Moreover, transesterification, such as alcoholysis of a phospholipid with an alcohol, such as ethanol, isopropanol, or *n*-butanol, catalyzed by an *sn*-1,3-specific lipase from *R. miehei*, has been employed for the preparation of lysophospholipids (Figure 28.44) [365].

Preparation of polyunsaturated phospholipids via reactions catalyzed by phospholipase A_2 and triacylglycerol lipases and functional properties of such products have been reviewed recently [366,367].

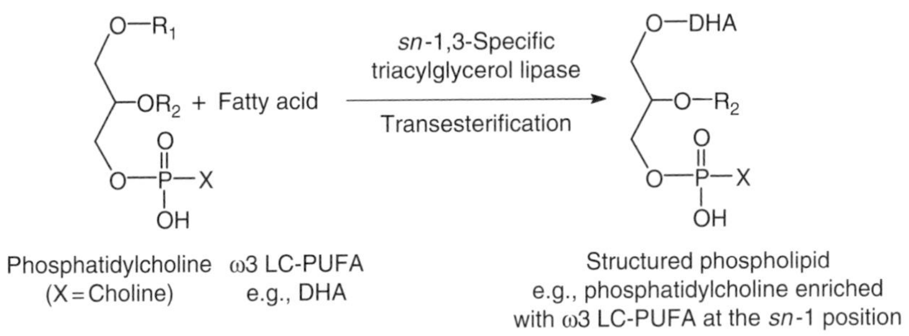

FIGURE 28.43 Preparation of structured phospholipids by transesterification of diacylglycerophospholipids with ω3 polyunsaturated fatty acids catalyzed by triacylglycerol lipases.

2. Phospholipids Modified by Phospholipase D

The following examples show the various possibilities of modifying the composition of polar head groups of phospholipids by phospholipase D-catalyzed reactions. Such modified phospholipids may find interesting biomedical applications.

Transphosphatidylation (base exchange) reactions of phospholipids, catalyzed by phospholipase D (Figure 28.41), can be utilized for the preparation of specific phospholipids. For example, phospholipase D-catalyzed transphosphatidylation reaction of egg lecithin (predominantly PCs and phosphatidylethanolamines [PEs]) with glycerol yields phosphatidylglycerols with the simultaneous formation of choline and ethanolamine (Figure 28.45) [368,369].

Phosphatidylglycerols may find biomedical applications as physiologically active pulmonary surfactant [370]. Efficient methods have been described for the preparation of phosphatidylglycerols from PCs and glycerol by transphosphatidylation catalyzed by phospholipase D [371–373]. Transphosphatidylation reactions catalyzed by phospholipase D have also been carried out to convert ethanolamine plasmalogens to their dimethylethanolamine or choline analogs [374] and to obtain PEs [375] or phosphatidylserines [376,377] from PCs (Figure 28.46). PC content of commercial lecithins has been increased by transphosphatidylation of lecithins with choline chloride catalyzed by phospholipase D [378].

Another example of phospholipase D-catalyzed transphosphatidylation reaction is the synthesis of structural analogs of platelet-activating factor (PAF) (1-O-alkyl-2-acetyl-sn-glycero-3-phosphocholine), by the replacement of choline by primary cyclic alcohols [379]. Moreover, transphosphatidylation of pure glycerophospholipids or commercial products, for example, soy lecithin or egg lecithin, with a sugar such as glucose, using phospholipase D from an Actinomadura sp. as biocatalyst and diethyl ether or tert-butanol as solvent, affords the phosphatidylglucose or other phosphatidylsaccharides in yields as high as 85% (Figure 28.47) [380].

Transphosphatidylation of PC with 1-monolauroyl-rac-glycerol, catalyzed by phospholipase D from Streptomyces sp., yields 1-lauroyl-lysophosphatidylglycerol, which has been subsequently cleaved by phospholipase C from Bacillus cereus to yield 1-lauroyl-rac-lysoglycerophosphate [381]. Similarly, transphosphatidylation of PC with 1-lauroyl-dihydroxyacetone, catalyzed by phospholipase D, yields 1-lauroyl-phosphatidyldihydroxyacetone, which has been subsequently

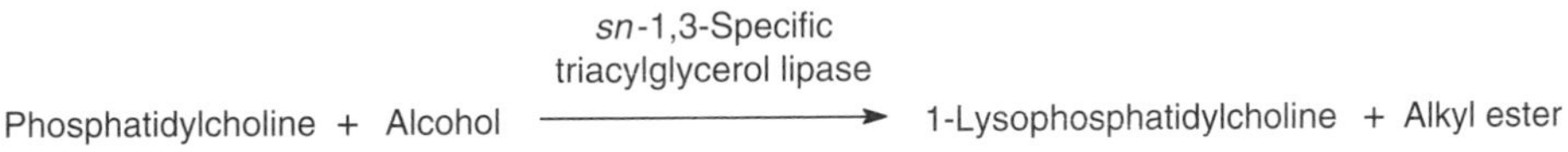

FIGURE 28.44 Transesterification of phosphatidylcholines with an alcohol (alcoholysis) catalyzed by triacylglycerol lipases for the preparation of lysophosphatidylcholines.

$$\text{Phosphatidylcholine} + \text{Glycerol} \xrightarrow{\text{Phospolipase D}} \text{Phosphatidylglycerol} + \text{Choline}$$

FIGURE 28.45 Transphosphatidylation of phosphatidylcholines with glycerol catalyzed by phospholipase D for the preparation of phosphatidylglycerols.

cleaved by phospholipase C to yield 1-lauroyl-dihydroxyacetonephosphate [381]. Preparation of polyunsaturated phospholipids via exchange of polar head groups catalyzed by phospholipase D and functional properties of such products have been reviewed recently [366].

C. OTHER ENZYMES

1. Lipoxygenases

Lipoxygenase from soybean converts linoleic acid or other compounds having a *cis,cis*-1,4-pentadiene system to conjugated hydroperoxides (Figure 28.48) [382]. Soybean lipoxygenase has been used, for example, to prepare 9-hydroperoxy-γ-linolenic acid from GLA [383], hydroperoxides of acylglycerols and phospholipids [384,385], and dimers of linoleic acid [386].

Recently, lipoxygenase immobilized in a packed-bed column reactor has been used to convert linoleic acid to hydroperoxy-octadecadienoic acid [387]. Furthermore, a double-fed batch reactor fed with a mixture of linoleic acid together with a crude lipoxygenase extract from defatted soybean flour has been used to obtain hydroperoxy-octadecadienoic acid, which is subsequently reduced in situ by cysteine, also contained in the reactor, to give 13(*S*)-hydroxy-9-*cis*,12-*cis*-octadecadienoic acid in high yield [388]. Reactions catalyzed by lipoxygenases have been recently reviewed [88].

2. Oxygenases

Crude enzyme preparations from a variety of organisms have been shown to exhibit interesting activities that can be utilized for the biotransformation of fats and other lipids. For example, enzyme preparation from plants have been shown to catalyze ω-hydroxylation (Figure 28.49) [389] and epoxidation (Figure 28.50) [390] of fatty acids. Recently, a peroxygenase isolated from oat (*Avena sativa*) has been immobilized on synthetic membranes and employed for the epoxidation of oleic acid using hydrogen peroxide or organic hydroperoxides as oxidants [391].

Enzyme preparations containing alcohol oxidase, isolated from the yeast *C. tropicalis*, catalyze the oxidation of long-chain alcohols, diols, and ω-hydroxy fatty acids to the corresponding aldehydes (Figure 28.51) [392]. Alcohol oxidase preparations, isolated from the yeast *C. maltosa*, catalyze the oxidation of 1-alkanols and 2-alkanols to the corresponding aldehydes and ketones,

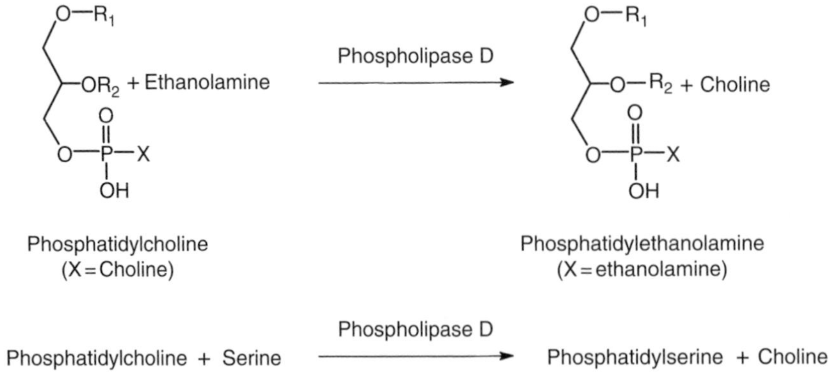

Phosphatidylcholine
(X = Choline)

Phosphatidylethanolamine
(X = ethanolamine)

$$\text{Phosphatidylcholine} + \text{Serine} \xrightarrow{\text{Phospholipase D}} \text{Phosphatidylserine} + \text{Choline}$$

FIGURE 28.46 Transphosphatidylation reactions catalyzed by phospholipase D.

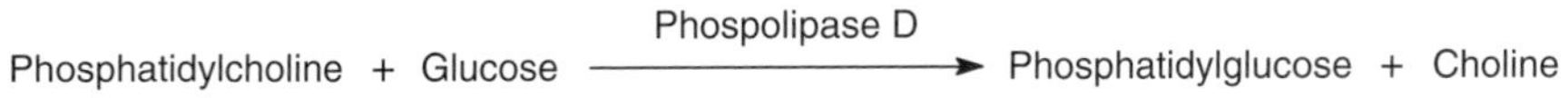

FIGURE 28.47 Transphosphatidylation of phosphatidylcholines with glucose catalyzed by phospholipase D for the preparation of phosphatidylglucose.

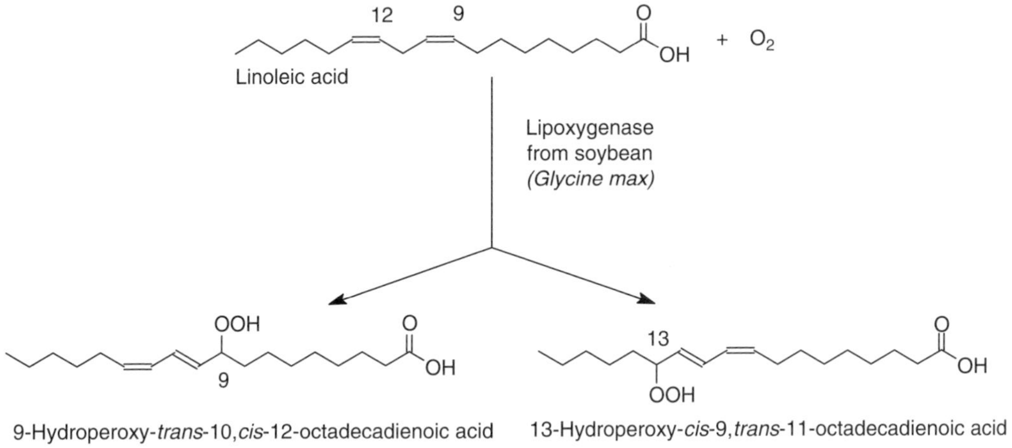

FIGURE 28.48 Preparation of hydroperoxides from linoleic acid by soybean lipoxygenase.

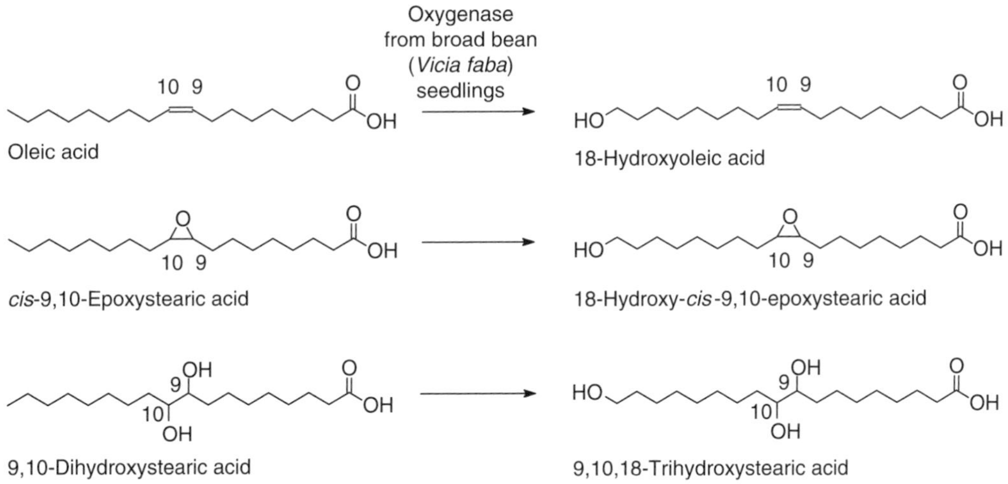

FIGURE 28.49 Enzymatic production of hydroxy acids by ω-hydroxylation.

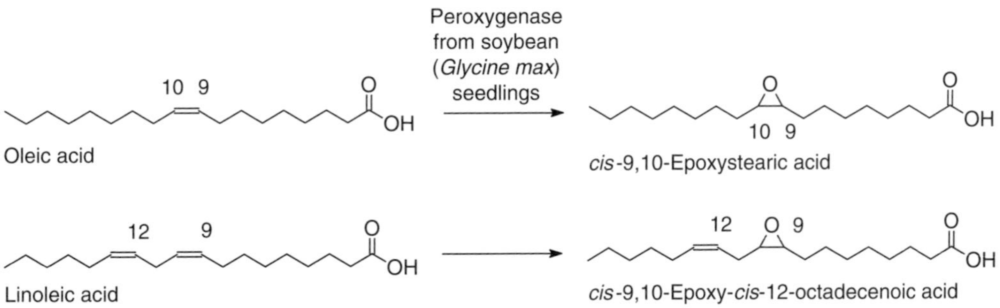

FIGURE 28.50 Enzymatic production of epoxy acids.

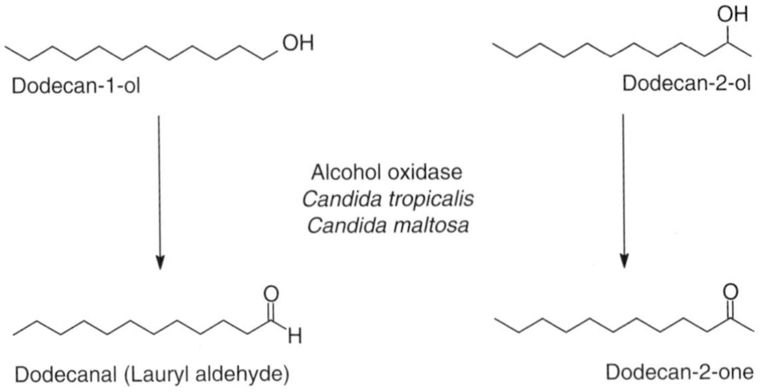

FIGURE 28.51 Enzymatic production of long-chain aldehydes and ketones.

respectively [393]. Reactions, catalyzed by various oxygenases, have been described in a recent review [88].

3. Epoxide Hydrolases

Expoxide hydrolases (EC 3.3.2.3) are ubiquitous in nature [394]. They catalyze the hydrolysis of epoxides to vicinal diols. In particular, epoxide hydrolases from higher plants are well characterized [395–399]. Epoxide hydrolase from soybean seedlings catalyzes the hydration of *cis*-9,10-epoxy-stearic acid to *threo*-9,10-dihydroxystearic acid (Figure 28.52) [395]. The two positional isomers of linoleic acid monoepoxides are hydrated to their corresponding *vic*-diols by soybean fatty acid epoxide hydrolase (Figure 28.53) [396].

This epoxide hydrolase has been found to be highly enantioselective with strong preference for the enantiomers *cis*-9*R*,10*S*-epoxy-*cis*-12-octadecenoic acid and *cis*-12*R*-,13*S*-epoxy-*cis*-9-octadecenoic acid [396–398].

Strong enantioselection has also been reported for rabbit liver microsomal epoxide hydrolase [400]. Cloning and expression of soluble epoxide hydrolase from potato [401] and purification as well as immobilization of epoxide hydrolase from rat liver [402] have been reported. Commercial avail-ability of such enzyme preparations is a prerequisite for their application in biotransformation of fats for the preparation of products via hydration of epoxides, for example, hydroxylated fatty acids [403].

IV. USE OF ENZYMES IN TECHNOLOGY OF OILSEEDS, OILS, AND FATS

Enzymes are gaining importance as processing aids in the technology of oilseeds, oils, and fats. Use of enzymes to facilitate the recovery of oils from oilseeds and other oil-bearing materials has become known lately. Very recently, an enzymatic process was developed for degumming of oils and fats in large-scale commercial operations.

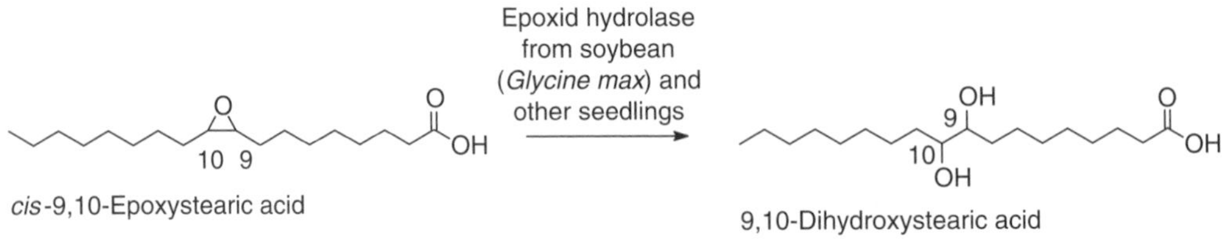

FIGURE 28.52 Preparation of dihydroxy fatty acids by epoxide hydrolase.

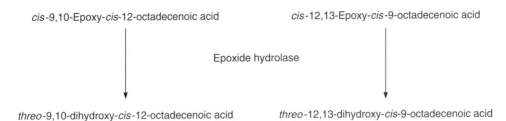

cis-9,10-Epoxy-cis-12-octadecenoic acid   cis-12,13-Epoxy-cis-9-octadecenoic acid

Epoxide hydrolase

threo-9,10-dihydroxy-cis-12-octadecenoic acid   threo-12,13-dihydroxy-cis-9-octadecenoic acid

FIGURE 28.53 Hydration of linoleic acid monoepoxides to *vic*-diols by epoxide hydrolase.

A. ENZYMES FOR PRETREATMENT OF OILSEEDS BEFORE OIL EXTRACTION

Conventional techniques for the recovery of oils from seeds and fruits involve grinding and conditioning by heat and moisture to disintegrate the oil-bearing cells, followed by mechanical pressing in hydraulic presses or expellers or extraction by organic solvents, such as hexane (see Chapter 8). Fats from animal tissues are frequently recovered by rendering, that is, heat treatment with live steam. Lately, aqueous extraction processes have become known in which the seeds or fruits are ground with water to disrupt the oil-bearing cells, followed by centrifugation to separate the oil from the solids and the aqueous phase [404].

Extensive mechanical rupturing of the cells of oil-bearing seeds is the prerequisite for efficient oil extraction [405]. Several reports have suggested the use of enzymes for the rupture of the plant cell walls to release the oil contained in the cell before recovery of the oil by mechanical pressing, solvent extraction, or aqueous extraction. Plant cell walls are generally composed of unlignified cellulose fibers to which strands of hemicellulose are attached. The cellulose fibers are often embedded in a matrix of pectic substances linked to structural protein [406]. Since substantial differences are observed in the polysaccharide composition of the cell walls of different plant species (Table 28.10), different combinations of cell wall-degrading enzymes (carbohydrases and proteases) have to be used for individual seeds or fruits [117,406,407]. Enzymes used in cocktails for cell wall degradation include amylase, cellulase, polygalacturonase, pectinase, hemicellulase, galactomannase, and proteases [117,407].

Enzyme pretreatment followed by mechanical expelling for improved oil recovery has been used for rapeseed [117,408] and soybean [409]. Typically, treatment of flaked rapeseed with commercial enzyme preparations (SP 249, Novo Nordisk Biochem North America, Inc., Franklinton, NC,

TABLE 28.10
Approximate Composition (%) of Cell Wall Polysaccharides of Some Oil-Bearing Materials

Polysaccharide	Rapeseed	Coconut	Corn Germ
Pectic substances	39	—	<1
Mannans	—	61	—
Galactomannans	—	26	—
Arabinogalactans	8	Some	—
Celluloses	22	13	39
Hemicelluloses			
Xyloglucans	29	—	—
Arabinoxylans	—	—	40
Others	2	—	10

Source: From Christensen, M., *Int. News Fats Oils Relat. Mater.*, 2, 984, 1991.

United States, and Olease, Biocon [US] Inc., Lexington, KY, United States) at 30% moisture and 50°C for 6 h followed by drying and expelling gave 90%–93% recovery of oil as compared with 72% recovery of the controls not treated with the enzymes [408]. Enzyme-assisted expeller process is now a commercial process for partial oil recovery from rapeseed [410]. In the commercial process the oil release is substantially enhanced and the resulting rapeseed cake with a superior nutritional value is preferred in a number of animal feeds. Enzyme-assisted pressing for the production of virgin grade olive oil has also been reported [117,407].

Enzyme pretreatment followed by solvent extraction for enhanced oil recovery has been used for melon seeds [411] and rapeseed [412]. Incubation of autoclaved and moistened rapeseed flakes (30% moisture) with carbohydrases for 12 h, followed by drying to 4% moisture and extraction with hexane resulted in 4.0%–4.7% enhancement of oil extraction with the different enzyme preparations in the following order: mixed activity enzyme > β-glucanase > pectinase > hemicellulase > cellulase [412]. A disadvantage of the process is a rather long incubation time.

Since the earlier publication of Lanzani et al. [413] on rapeseed, sesame seed, sunflower seed, soybean, and peanut, several reports have appeared on enzyme pretreatment followed by aqueous extraction for enhanced oil recovery from coconut [406,414–417], corn germ [406,418], avocado [419,420], olives [406,421,422], mustard seed and rice bran [423], *Jatropha curca* seeds [424], and cocoa beans [425]. Enzyme-assisted aqueous extraction has been extensively studied on rapeseed using the mixed enzyme preparation SP-311 (Novo) and found to yield an oil with low phospholipid content (phosphorus < 3 ppm) that can be subjected to physical refining without complex degumming procedures [406,410]. The protein recovered by aqueous enzymatic extraction has higher nutritive/market value than the rapeseed protein obtained by conventional pressing followed by solvent extraction. However, the oil yield in aqueous enzymatic processing is somewhat lower (90%–92%) compared with the conventional process (about 97.5%) [410].

Addition of plant proteases, such as papain or bromelain, to macerated fish followed by incubation at 65°C for 30 min and at 50°C–55°C for 1–2 h has been shown to improve the recovery of fish oils [426]. Thermostable microbial proteases operating at higher temperatures (70°C–75°C) appear to be more suitable for this purpose because they reduce the risk of microbial growth and contamination [117].

B. Enzymatic Degumming of Oils

Degumming is an important processing step during the refining of vegetable oils, such as soybean and rapeseed (canola). Untreated vegetable oils contain varying proportions of phospholipids (phosphatides), such as PC, phosphatidylinositols (PI), PE, and PA, which impair the quality and stability of oils if they are not removed. Especially for vegetable oils that are subjected to physical refining, it is very important to eliminate the phospholipids before treatment of the oil with steam under vacuum for the removal of free (unesterified) fatty acids.

In the current industrial practice of degumming [427], the raw oils are treated with water or aqueous solutions of phosphoric acid, citric acid, or an alkali in order to hydrate the phospholipids that subsequently flocculate and thus can be removed by centrifugation (see also Chapter 8). Among the different classes of phospholipids PC and PI are readily hydratable, both PE and PA are hydratable when combined with potassium and unhydratable when complexed with divalent metal ions. Moreover, PA occurs as a partially dissociated acid and is nonhydratable [427,428]. Phospholipase D plays an important role in the formation of nonhydratable PA. Especially during the processing of soybeans phospholipase D can remain fairly active even at a temperature of 65°C in the presence of water-saturated hexane, thereby catalyzing the hydrolysis of hydratable phospholipids, for example, PC to nonhydratable PA; however, in aqueous media this enzyme is readily deactivated by heat [429].

Recently, Lurgi AG (Frankfurt, Germany) developed an enzymatic process (EnzyMax) for degumming of oils that involves treatment of the raw oil with phospholipase A_2 that cleaves the

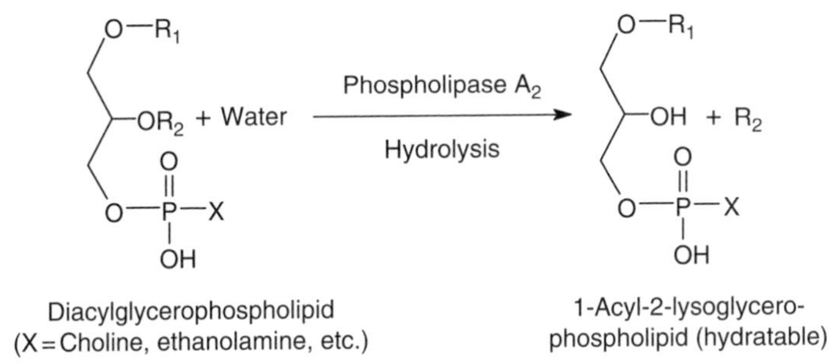

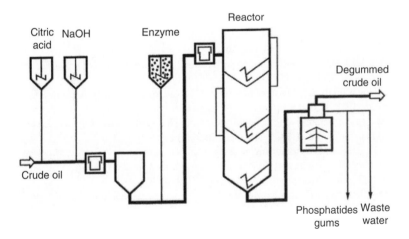

FIGURE 28.54 Principle of enzymatic degumming of oils and fats using phospholipase A_2 for converting diacylglycerophospholipids to easily hydratable lysophospholipids.

fatty acids esterified at the *sn*-2 position of the diacylglycerophospholipids yielding *sn*-1-acyl-lysoglycerophospholipids, including *sn*-1-acyl-lyso-PE and *sn*-1-acyl-lyso-PA (Figure 28.54), that can be readily hydrated and removed [428,429].

Figure 28.55 shows the flow sheet of the commercial EnzyMax process for degumming of oils. The enzyme used is a commercially available phospholipase A_2 (lecitase activity 10,000 IU/mL) isolated from porcine pancreas. Recently, a microbial phospholipase A_2 that is more economical than the porcine enzyme has been used successfully for degumming [430]. The process consists of three stages. In the first stage, solutions of citric acid (50% w/v) and sodium hydroxide (3% w/v) are dispersed with the crude oil or water-degummed oil and the pH adjusted to approximately 5.0. The resulting mixture, maintained at a temperature of 60°C, is intimately mixed with the enzyme solution (enzyme concentrate diluted with water to 0.2% v/v) and passed through two or more enzyme reactors. The residence time of the mixture in the reactors varies from 1 to 6 h, depending on the phosphatide content of the starting material and the quality requirement on the degummed oil. The effluent from the enzyme reactors is finally admitted to a separator to obtain the streams of degummed oil and aqueous phase containing the phosphatide sludge containing the still active enzyme that is generally reused several times by recycling into the process.

FIGURE 28.55 Flow sheet of the commercial EnzyMax process for degumming oils using phospholipase A_2. (Adapted from Buchold, H., *Proceedings of Latin American Congress and Exhibit on Fats and Oils Processing*, Barrera-Arellano, D., Regitano d'Arce, M.A.B., and Gonçalves, L.Ap.G., eds, Sociedade Brasileira de Óleos e Gorduras, UNICAMP, Campinas, 1995.)

Operational data from commercial plants for degumming of 540 ton/day rapeseed oil and 400 ton/day soybean oil show a likely average consumption of 10 g enzyme concentrate/ton of oil and residual phosphorus content of the degummed oil less than 10 ppm. The enzymatically degummed oil can be subjected to physical refining after bleaching. Due to low phosphorus content and absence of soaps, enzymatically degummed oils, as compared with those obtained by other degumming processes, can be bleached with standard bleaching earths instead of high-activity and acidic bleaching earths. Additional economic advantages of enzymatic degumming as compared with chemical neutralization and soapstock splitting are due to lower oil loss, less consumption of water, and very little effluent wastes of environmental concern.

V. PERSPECTIVES

Lipid biotechnology is a very young discipline. Although breathtaking progress has been made in genetic engineering for alteration of composition of oilseeds and oils (see Chapter 30), the commercial applications of biotechnology for the production and modification of fats and oils using whole microbial cells and isolated enzymes are restricted so far to a few areas, such as enzymatic production of cocoa butter substitutes, human milk fat replacers, and bioester as well as enzymatic degumming of fats and oils. Factors limiting the application of biotechnology in the area of lipids include the productivity and the cost of downstream processing in microbial processes and the price as well as the reuse properties of enzymes, such as lipases. In this regard, optimism is raised in the current literature, which contains numerous research activities aimed at development of novel lipid-based specialty products as well as mass fat products, such as feedstocks for margarines using biotechnology. With increasing production and use of enzymes in detergents, one might foresee the commercial bulk-scale availability of such biocatalysts at competitive prices that should facilitate their applications in wide areas of lipid biotechnology.

REFERENCES

1. Murphy, D.J., Engineering oil production in rapeseed and other oil crops, *Trends Biotechnol.*, 14, 206, 1996.
2. Ratledge, C., Biotechnology of oils and fats, in *Microbial Lipids*, Vol. 2, Ratledge, C. and Wilkinson, S.G., Eds., Academic Press, London, 1989, p. 567.
3. Cobelas, M.A. and Lechado, J.Z., Lipids in microalgae. A review. I. Biochemistry, *Grasas Aceites*, 40, 118, 1989.
4. Boulton, C.A., Extracellular microbial lipids, in *Microbial Lipids*, Vol. 2, Ratledge, C. and Wilkinson, S.G., Eds., Academic Press, London, 1989, p. 669.
5. Ratledge, C., Microbial oils and fats: perspectives and prospects, in *Fats for the Future*, Cambie, R.C., Ed., Ellis Horwood, Chichester, 1989, p. 153.
6. Lösel, D., Fungal lipids, in *Microbial Lipids*, Vol. 1, Ratledge, C. and Wilkinson, S.G., Eds., Academic Press, London, 1989, p. 699.
7. Kayama, M., Araki, S., and Sato, S., Lipids of marine plants, in *Marine Biogenic Lipids, Fats, and Oils*, Vol. 2, Ackman, R.A., Ed., CRC Press, Boca Raton, FL, 1989, p. 3.
8. Leman, J., Oleaginous microorganisms: an assessment of the potential, *Adv. Appl. Microbiol.*, 43, 195, 1997.
9. Brennan, J., Mycobacterium and other actinomycetes, in *Microbial Lipids*, Vol. 1, Ratledge, C. and Wilkinson, S.G., Eds., Academic Press, London, 1989, p. 203.
10. Wayman, M., Jenkins, A.D., and Kormendy, A.G., Bacterial production of oils and fats, in *Biotechnology for the Oils and Fats Industry*, Ratledge, C., Dawson, P., and Rattray, J., Eds., American Oil Chemists' Society, Champaign, IL, 1984, p. 129.
11. Wood, B.J.B., Lipids in algae and protozoa, in *Microbial Lipids*, Vol. 1, Ratledge, C. and Wilkinson, S.G., Eds., Academic Press, London, 1989, p. 807.
12. Gerwick, C.H. and Bernait M.W., Eicosanoids and related compounds from marine algae, in *Marine Biotechnology*, Vol. 1, *Pharmaceutical and Bioactive Natural Products*, Attaway, D.H. and Zaborsky, O.R., Eds., Plenum Press, New York, NY, 1993, p. 101.

13. Molina Grima, E., Sánchez Pérez, J.A., Garcia Camacho, F., Robles Medina, A., Giménez Giménez, A., and López Alonso, A.D., The production of polyunsaturated fatty acids by microalgae: from strain selection to product purification, *Process Biochem.*, 30, 711, 1995.

14. Kyle, D.J., Production and use of a single cell oil which is highly enriched in docosahexaenoic acid, *Lipid Technol.*, 8, 107, 1996.

15. Mukherjee, K.D., Production and use of microbial oils, *Int. News Fats Oils Relat. Mater.*, 10, 308, 1999.

16. Fukuda, H. and Morikawa, H., Enhancement of γ-linolenic acid production by *Mucor ambiguus* with nonionic surfactants, *Appl. Microbiol. Biotechnol.*, 27, 15, 1987.

17. Hansson, L. and Dostálek, M., Effect of culture conditions on mycelial growth and production of γ-linolenic acid by the fungus *Mortierella ramanniana*, *Appl. Microbiol. Biotechnol.*, 28, 240, 1988.

18. Shimizu, M., Akimoto, K., Kawashima, H., Shinmen, Y., and Yamada, H., Production of di-homo-γ-linolenic acid by *Mortierella alpina* 1S-4, *J. Am. Oil Chem. Soc.*, 66, 237, 1989.

19. Totani, N. and Oba, K., The filamentous fungus *Mortierella alpina* high in arachidonic acid, *Lipids*, 66, 1060, 1987.

20. Yamada, H., Shimizu, S., and Shinmen, Y., Production of arachidonic acid by *Mortierella elongata* 1S-5, *Agric. Biol. Chem.*, 51, 785, 1987.

21. Shinmen, Y., Shimizu, S., Akimoto, K., Kawashima, H., and Yamada, H., Production of arachidonic acid by *Mortierella* fungi. Selection of a potent producer and optimization of culture conditions for large-scale production, *Appl. Microbiol. Biotechnol.*, 31, 11, 1989.

22. Bajpai, P.K., Bajpai, P., and Ward, O.P., Arachidonic acid production by fungi, *Appl. Environ. Microbiol.*, 57, 1255, 1991.

23. Lindberg, A.-M. and Molin, G., Effect of temperature and glucose supply on the production of polyunsaturated fatty acids by the fungus *Mortierella alpina* CBS 343.66 in fermentor cultures, *Appl. Microbiol. Biotechnol.*, 39, 450, 1993.

24. Shimizu, S., Shinmen, Y., Kawashima, H., Akimoto K., and Yamada, H., Fungal mycelia as a novel source of eicosapentaenoic acid. Activation of enzyme(s) involved in eicosapentaenoic acid production at low temperature, *Biochem. Biophys. Res. Commun.*, 150, 335, 1988.

25. O'Brien, D.J., Kurantz, M.J., and Kwoczak, R., Production of eicosapentaenoic acid by filamentous fungus *Pythium irregulare*, *Appl. Microbiol. Biotechnol.*, 40, 211, 1993.

26. O'Brien, D.J. and Senske, G., Recovery of eiocosapentaenoic acid from fungal mycelia by solvent extraction, *J. Am. Oil Chem. Soc.*, 71, 947, 1994.

27. Bajpai, P.K., Bajpai, P., and Ward, O.P., Optimization of production of docosahexaenoic acid (DHA) by *Thraustochytrium aureum* ATCC 34304, *J. Am. Oil Chem. Soc.*, 68, 509, 1991.

28. Kendrick, A. and Ratledge, C., Lipids of selected moulds grown for production of n-3 and n-6 polyunsaturated fatty acids, *Lipids*, 27, 15, 1992.

29. Suzuki, O., Iida, I., Nakahara, T., Yokochi, T., Kamisaka, Y., Yagi, H., and Yamaoka, M., Improvement of docosahexaenoic acid production in a culture of *Thraustochytrium aureum* by medium optimization, *J. Ferment. Bioeng.*, 81, 76, 1996.

30. Suzuki, O., Production of γ-linolenic acid by fungi and its industrialization, in *Proceedings of the World Conference on Biotechnology for the Fats and Oils Industry*, Applewhite, T.H., Ed., American Oil Chemists' Society, Champaign, IL, 1988, p. 110.

31. Certik, M. and Shimizu, S., Biosynthesis and regulation of microbial polyunsaturated fatty acid production, *J. Biosci. Bioeng.*, 87, 1, 1999.

32. Hiruta, O., Futamura, T., Takebe, H., Satoh, A., Kamisaka, Y., Yokochi, T., Nakahara, T., and Suzuki, O., Optimization and scale-up of γ-linolenic acid production by *Mortierella ramanniana* MM 15-1, a high γ-linolenic acid producing mutant, *J. Ferment. Bioeng.*, 82, 366, 1996.

33. Kyle, D.J., Production and use of a single cell oil highly enriched in arachidonic acid, *Lipid Technol.*, 9, 116, 1997.

34. Kawashima, H., Akimoto, K., Higashiyama, K., Fujikawa, S., and Shimizu, S., Industrial production of dihomo-γ-linolenic acid by a Δ5 desaturase-defective mutant of *Mortierella alpina* 1S-4 fungus, *J. Am. Oil Chem. Soc.*, 77, 1135, 2000.

35. Yazawa, K., Araki, K., Okazaki, N., Watanabe, K., Ishikawa, C., Inoue, A., Numano, N., and Kondo K., Production of eicosapentaenoic acid by marine bacteria, *J. Biochem.*, 103, 5, 1988.

36. Akimoto, M., Ishii, T., Yamagaki, K., Ohtaguchi, K., Koide, K., and Yazawa, K., Production of eicosapentaenoic acid by a bacterium isolated from mackarel intestines, *J. Am. Oil Chem. Soc.*, 67, 911, 1990.

37. Yano, Y., Nakayama, A., Saito, H., and Ishihara, K., Production of docosahexaenoic acid by marine bacteria isolated from deep sea fish, *Lipids*, 29, 527, 1994.
38. Huang, Y.-S., Pereira, S.L., and Leonard, A.E., Enzymes for the transgenic production of long-chain polyunsaturated fatty acid-enriched oils, in *Handbook of Industrial Biocatalysis*, Hou, C.T., Ed., Taylor & Francis, Boca Raton, FL, 2005, Chap. 1.
39. Nakahara, T., Microbial polyunsaturated fatty acid production, in *Handbook of Industrial Biocatalysis*, Hou, C.T., Ed., Taylor & Francis, Boca Raton, FL, 2005, Chap. 17.
40. Senanayake, S.P.J.N. and Fichtali, J., Single-cell oils as sources of nutraceutical and specialty lipids: processing technologies and application, in *Nutraceutical and Specialty Lipids and Their Co-Products*, Shahidi, F., Ed., Taylor & Francis, Boca Raton, FL, 2005, Chap. 16.
41. Dewitt, S., Ervin, J.L., Howes-Orchison, D., Dalietos, D., Neidelman, S.L., and Geigert, J., Saturated and unsaturated wax esters produced by *Acinetobacter* sp. H01-N grown on C_{16}–C_{20} *n*-alkanes, *J. Am. Oil Chem. Soc.*, 59, 69, 1982.
42. Neidelman, S.L. and Geigert, J., Biotechnology and oleochemicals: changing patterns, *J. Am. Oil Chem. Soc.*, 61, 290, 1984.
43. Koritala, S., Microbiological synthesis of wax esters by *Euglena gracilis*, *J. Am. Oil Chem. Soc.*, 66, 133, 1989.
44. Wagner, F., Strategies of biosurfactant production, *Fat Sci. Technol.*, 89, 586, 1987.
45. Banat, I.M., Makkar, R.S., and Cameotra, S.S., Potential commercial applications of microbial surfactants, *Appl. Microbiol. Biotechnol.*, 53, 495, 2000.
46. Inoue, S., Biosurfactants in cosmetic applications, in *Proceedings of the World Conference on Biotechnology for the Fats and Oils Industry*, Applewhite, T.H., Ed., American Oil Chemists' Society, Champaign, IL, 1988, p. 206.
47. Zhou, Q.H., Klekner, V., and Kosaric, N., Production of sophorose lipids by *Torulopsis bombicola* from safflower oil and glucose, *J. Am. Oil Chem. Soc.*, 69, 89, 1992.
48. Asmer, H.-J., Lang, S., Wagner, F., and Wray, V., Microbial production, structure elucidation and bioconversion of sophorose lipids, *J. Am. Oil Chem. Soc.*, 65, 1460, 1988.
49. Hommel, R.K., Weber, L., Weis, A., Himmelreich, U., Rilke, O., and Kleber H.-P., Production of sophorose lipid by *Candida (Torulopsis) apicola* grown on glucose, *J. Biotechnol.*, 33,147, 1994.
50. Syldatk, C., Lang, S., Matulovic, U., and Wagner, F., Production of four interfacial active rhamnolipids from *n*-alkanes or glycerol by resting cells of *Pseudomonas* species DSM 2874, *Z. Naturforsch.*, 40c, 61, 1985.
51. Ristau, E. and Wagner, F., Formation of novel anionic trehalose-tetraesters from *Rhodococcus erythropolis* under growth limiting conditions, *Biotechnol. Lett.*, 5, 95, 1983.
52. Passeri, A., Lang, S., Wagner, F., and Wray, V., Marine biosurfactants. II. Production and characterization of an anionic trehalose tetraester from the marine bacterium *Arthrobacter* sp. EK1, *Z. Naturforsch.*, 46c, 204, 1991.
53. Frautz, B., Lang, S., and Wagner F., Formation of cellobiose lipids by growing and resting cells of *Ustilago maydis*, *Biotechnol. Lett.*, 8, 757, 1986.
54. Shabtai, Y. and Wang, D.I.C., Production of Emulsan in a fermentation process using soybean oil (SBO) in a carbon-nitrogen coordinated feed, *Biotechnol. Bioeng.*, 35, 753, 1990.
55. Kakinuma, A., Ouchida, A., Shima, T., Sugino, H., Isono, M., Tamura, G., and Arima, K., Confirmation of the structure of surfactin by mass spectrometry, *Agric. Biol. Chem.*, 33, 1669, 1969.
56. Powalla, M., Lang, S., and Wray, V., Penta- and disaccharide lipid formation by *Nocardia corynebacteroides* grown on *n*-alkanes, *Appl. Microbiol. Biotechnol.*, 31, 473, 1989.
57. Brakemeier, A., Wullbrandt, D., and Lang, S., *Candida bombicola*: production of novel alkyl glycosides based on glucose/2-dodecanol, *Appl. Microbiol. Biotechnol.*, 50, 161, 1998.
58. Solaiman, D.K.Y., Ashby, R.D., and Foglia, T.A., Production of biosurfactants by fermentation of fats and oils and their coproducts, in *Handbook of Industrial Biocatalysis*, Hou, C.T., Ed., Taylor & Francis, Boca Raton, FL, 2005, Chap. 14.
59. Inui, H., Ohya, O., Miyatake, K., Nakano, Y., and Kitaoka, S., Assimilation and metabolism of fatty alcohols in *Euglena gracilis*, *Biochim. Biophys. Acta*, 875, 543, 1986.
60. Tani, Y., Okumura, M., and Ii, S., Liquid wax ester production by *Euglena gracilis*, *Agric. Biol. Chem.*, 51, 225, 1987.
61. Weete, J.D., Algal and fungal waxes, in *Chemistry and Biochemistry of Natural Waxes*, Kolattukudy, P.E., Ed., Elsevier, Amsterdam, 1976, p. 349.

62. Seo, C.W., Yamada, Y., and Okada, H., Synthesis of fatty acid esters by *Corynebacterium*, *Agric. Biol. Chem.*, 46, 405, 1982.

63. Hou, C.T., Microbial oxidation of unsaturated fatty acids, *Adv. Appl. Microbiol.*, 41, 1, 1995.

64. Faber, K., *Biotransformations in Organic Chemistry: A Textbook*, 3rd edn, Springer-Verlag, Berlin, 1997, p. 206.

65. Wallen, L.L., Benedict, R.G., and Jackson, R.W., The microbiological production of 10-hydroxystearic acid from oleic acid, *Arch. Biochem. Biophys.*, 99, 249, 1962.

66. Litchfield, J.H. and Pierce G.E., Microbiological synthesis of hydroxy-fatty acids and keto-fatty acids, *U.S. Patent*, 4, 582, 804, 1986.

67. Koritala, S., Hosie, L., Hou, C.T., Hesseltine, C.W., and Bagby, M.O., Microbial conversion of oleic acid to 10-hydroxystearic acid, *Appl. Microbiol. Biotechnol.*, 32, 299, 1989.

68. El Sharkawy, S.H., Yang, W., Dostal, L., and Rosazza, J.P.N., Microbial oxidation of oleic acid, *Appl. Environ. Microbiol.*, 58, 2116, 1992.

69. Seo, C.W., Yamada, Y., Takada, N., and Okada, H., Hydration of squalene and oleic acid by *Corynebacterium* sp. S-401, *Agric. Biol. Chem.*, 45, 2025, 1981.

70. Blank, W., Takayanagi, H., Kido, T., Meussdörffer, F., Esaki, N., and Soda, K., Transformation of oleic acid and its esters by *Sarcina lutea*, *Agric. Biol. Chem.*, 55, 2651, 1991.

71. Wallen, L.L., Davis, E.N., Wu, Y.V., and Rohwedder, W.K., Stereospecific hydration of unsaturated fatty acids by bacteria, *Lipids*, 6, 745, 1972.

72. Koritala, S. and Bagby M.O., Microbial conversion of linoleic and linolenic acids to unsaturated hydroxy fatty acids, *J. Am. Oil Chem. Soc.*, 69, 575, 1992.

73. Murakami, N., Morimoto, T., Shirahashi, H., Ueda, T., Nagai, S.-I., Sakakibara, J., and Yamada, N., Bioreduction of hydroperoxy fatty acid by cyanobacterium. *Phormidium tenue*, *Bioorg. Med. Chem. Lett.*, 2, 149, 1992.

74. Lanser, A.C., Plattner, R.D., and Bagby, M.O., Production of 15-, 16-, and 17-hydroxy-9-octadecenoic acids by bioconversion of oleic acid with *Bacillus pumilus*, *J. Am. Oil Chem. Soc.*, 69, 363, 1992.

75. Picataggio, S., Rohrer, T., Deanda, K., Lanning, D., Reynolds, R., Mielenz, J., and Eirich, L.D., Metabolic engineering of *Candida tropicalis* for the production of long-chain dicarboxylic acids, *Biotechnology*, 10, 894, 1992.

76. Hou, C.T., Bagby, M.O., Plattner, R.D., and Koritala, S., A novel compound, 7,10-dihydroxy-8(E)-octadecenoic acid from oleic acid by bioconversion, *J. Am. Oil Chem. Soc.*, 68, 99, 1991.

77. Hou, C.T., Production of 10-ketostearic acid from oleic acid by *Flavobacterium* sp. strain DS5 (NRRL B-14859), *Appl. Environ. Microbiol.*, 60, 3760, 1994.

78. Hou, C.T., Conversion of linoleic acid to 10-hydroxy-12(Z)-octadecenoic acid by *Flavobacterium* sp. (NRRL B-14859), *J. Am. Oil Chem. Soc.*, 71, 975, 1994.

79. Hou, C.T., Production of hydroxy fatty acids from unsaturated fatty acids by *Flavobacterium* sp. DS5 hydratase, a C-10 positional- and *cis* unsaturation-specific enzyme, *J. Am. Oil Chem. Soc.*, 72, 1265, 1995.

80. Lanser, A.C., Conversion of oleic acid to 10-ketostearic acid by a *Staphylococcus* species, *J. Am. Oil Chem. Soc.*, 70, 543, 1993.

81. Lanser, A.C. and Manthey, L.K., Bioconversion of oleic acid by *Bacillus* strain NRRL BD-447: identification of 7-hydroxy-17-oxo-9-*cis*-octadecenoic acid, *J. Am. Oil Chem. Soc.*, 76, 1023, 1999.

82. Bastida, J., de Andrés, C., Culleré, J., Busquets, M., and Manresa. A., Biotransformation of oleic acid into 10-hydroxy-8-E-octadecenoic acid by *Pseudomonas* sp. 42A2, *Biotechnol. Lett.*, 21, 1031, 1999.

83. Kuo, T.M., Manthey, L.K., and Hou, C.T., Fatty acid bioconversions by *Pseudomonas aeruginosa* PR3, *J. Am. Oil Chem. Soc.*, 75, 875, 1998.

84. Venter, P., Kock, J.L.F., Sravan Kumar, G., Botha, A., Coetzee, D.J., Botes, P.J., Bhatt, R.K., Falck, J.R., Schewe, T., and Nigam, S., Production of 3-*R*-hydroxy-polyenoic fatty acids by the yeast *Dipodascopis uninucleata*, *Lipids*, 32, 1277, 1997.

85. Fox, S.R., Akpinar, A., Prabhune, A.A., Friend, J., and Ratledge, C., The biosynthesis of oxylipins of linoleic acid by the sewage fungus *Leptomitus lacteus*, including the identification of 8*R*-Hydroxy-9*Z*,12*Z*-octadecadienoic acid, *Lipids*, 35, 23, 2000.

86. Kim, H., Kuo, T.M., and Hou, C.T., Production of 10,12-dihydroxy-8(E)-octadecenoic acid, an intermediate in the conversion of ricinoleic acid to 7,10,12-trihydroxy-8(E)-octadecenoic acid by *Pseudomonas aeruginosa* PR3, *J. Ind. Microbiol.*, 24, 167, 2000.

87. Hou, C.T. and Hosokawa, M., Production of value-added industrial products from vegetable oils: oxygenated fatty acids, in *Handbook of Industrial Biocatalysis*, Hou, C.T., Ed., Taylor & Francis, Boca Raton, FL, 2005, Chap. 7.

88. Maurer, S.C. and Schmid, R.D., Biocatalysts for the epoxidation and hydroxylation of fatty acids and fatty alcohols, in *Handbook of Industrial Biocatalysis*, Hou, C.T., Ed., Taylor & Francis, Boca Raton, IL, 2005, Chap. 4.

89. Kamisaka, Y., Yokochi, T., Nakahara, T., and Suzuki, O., Incorporation of linoleic acid and its conversion to γ-linolenic acid in fungi, *Lipids*, 25, 54, 1990.

90. Shimizu, S., Kawashima, H., Akimoto, K., Shinmen, Y., and Yamada, H., Microbial conversion of an oil containing α-linolenic acid to an oil containing eicosapentaenoic acid, *J. Am. Oil Chem. Soc.*, 66, 342, 1989.

91. Yi, Z.-H. and Rehm, H.-J., Identification and production of Δ9-*cis*-1,18-octadecenedioic acid by *Candida tropicalis*, *Appl. Microbiol. Biotechnol.*, 30, 327, 1989.

92. Yi, Z.-H. and Rehm, H.-J., Bioconversion of elaidic acid to Δ9-*trans*-1,18-octadecenedioic acid by *Candida tropicalis*, *Appl. Microbiol. Biotechnol.*, 29, 305, 1988.

93. Fabritius, D., Schäfer, H.-J., and Steinbüchel, A., Biotransformation of linoleic acid with the *Candida tropicalis* M25 mutant, *Appl. Microbiol. Biotechnol.*, 48, 83, 1997.

94. Green, K.D., Turner, M.K., and Woodley, J.M., *Candida cloacae* oxidation of long-chain fatty acids to dioic acids, *Enzyme Microb. Technol.*, 27, 205, 2000.

95. Creuly, C., Larroche, C., and Gros, J.-B., Bioconversion of fatty acids into methyl ketones by spores of *Penicillium roquefortii* in a water-organic solvent, two-phase system, *Enzyme Microb. Technol.*, 14, 669, 1992.

96. Yagi, T., Kawaguchi, M., Hatano, T., Fukui, F., and Fukui, S., Production of ketoalkanes from fatty acid esters by a fungus, *Trichoderma*, *J. Ferment. Bioeng.*, 68, 188, 1989.

97. Yagi, T., Kawaguchi, M., Hatano, T., Hatano, A., Nakanishi, T., Fukui, F., and Fukui. S., Formation of *n*-alkane-2-ones and *n*-alkane-2-ols from triglycerides by a black yeast, *Aureobasidium*, *J. Ferment. Bioeng.*, 71, 93, 1991.

98. Yagi, T., Hatano, A., Hatano, T., Fukui, F., and Fukui, S., Formation of monohydroxy-*n*-nonane-2-ones from tricaprin by *Fusarium avenaceum* f. sp. *fabae* IFO 7158, *J. Ferment. Bioeng.*, 71, 176, 1991.

99. Antczak, U., Góra, J., Antczak, T., and Galas, E., Enzymatic lactonization of 15-hydroxypentadecanoic and 16-hydroxyhexadecanoic acids to macrocyclic lactones, *Enzyme Microb. Technol.*, 13, 589, 1991.

100. Gocho, S., Tabogami, N., Inagaki, M., Kawabata, C., and Komai, T., Biotransformation of oleic acid to optically active γ-dodecalactone, *Biosci. Biotech. Biochem.*, 59, 1571, 1995.

101. Mukherjee, K.D., Lipase-catalyzed reactions for modification of fats and other lipids, *Biocatalysis*, 3, 277, 1990.

102. Schuch, R. and Mukherjee, K.D., Interesterification of lipids using an immobilized *sn*-1,3-specific triacylglycerol lipase, *J. Agric. Food Chem.*, 35, 1005, 1987.

103. Macrae, A.R., Lipase-catalyzed interesterification of oils and fats, *J. Am. Oil Chem. Soc.*, 60, 243A, 1983.

104. McNeill, G.P., Enzymic processes, in *Lipid Synthesis and Manufacture*, Gunstone, F.D., Ed., Academic Press, Sheffield, 1998, p. 288.

105. Godfrey, T., Lipases for industrial use, *Lipid Technol.*, 7, 58, 1995.

106. Villeneuve, P., Pina, M., Montet, D., and Graille, J., Determination of lipase specificities through the use of chiral triglycerides and their racemics, *Chem. Phys. Lipids*, 76, 109, 1995.

107. Hills, M.J. and Mukherjee, K.D., Triacylglycerol lipase from rape (*Brassica napus* L.) suitable for biotechnological purposes, *Appl. Biochem. Biotechnol.*, 26, 1, 1990.

108. Villeneuve, P., Pina, M., Montet, D., and Graille, J., *Carica papaya* latex lipase: *sn*-3 stereoselectivity or short-chain selectivity? Model chiral triglycerides are removing ambiguity, *Lipids*, 72, 753, 1995.

109. Villeneuve, P. and Foglia, T., Lipase specificities: potential application in lipid bioconversions, *Int. News Fats Oils Relat. Mater.*, 8, 640, 1997.

110. Gupta, R., Rathi, P., and Bradoo, S., Lipase mediated upgradation of dietary fats and oils, *Crit. Rev. Food Sci. Nutr.*, 43, 635, 2003.

111. Gupta, R., Gupta, N., and Rathi, P., Bacterial lipases: an overview of production, purification and biochemical properties, *Appl. Microbiol. Biotechnol.*, 64, 763, 2004.

112. Bornscheuer, U.T., Bessler, C., Srinivas, R., and Krishna, S.H., Optimizing lipases and related enzymes for efficient application, *Trends Biotechnol.*, 20, 433, 2002.

113. Otero, C., Berrendero, M.A., Cardenas, F., Alvarez, E., and Elson, S.W., General characterization of noncommercial microbial lipases in hydrolytic and synthetic reactions, *Appl. Biochem. Biotechnol.*, 120, 209, 2005.

114. Bloomer, S., Adlercreutz, P., and Mattiasson, B., Triglyceride interesterification by lipases. Cocoa butter equivalents from fraction of palm oil, *J. Am. Oil Chem. Soc.*, 67, 519, 1990.

115. Bloomer, S., Adlercreutz, P., and Mattiasson, B., Kilogram-scale ester synthesis of acyl donor and use in lipase-catalyzed interesterifications, *J. Am. Oil Chem. Soc.*, 69, 966, 1992.

116. Kawahara, Y., Progress in fats, oils food technology, *Int. News Fats Oils Relat. Mater.*, 4, 663, 1993.

117. Owusu-Ansah, Y.J., Enzymes in lipid technology and cocoa butter substitutes, in *Technological Advances in Improved and Alternative Sources of Lipids*, Kamel, B.A. and Kakuda, Y., Eds., Blackie Academic and Professional, London, 1994, p. 361.

118. Quinlan, P. and Moore, S., Modification of triglycerides by lipases: process technology and its application to the production of nutritionally improved fats, *Int. News Fats Oils Relat. Mater.*, 4, 580, 1993.

119. Bornscheuer, U.T., Adamczak, M., and Suomanou, M.M., Lipase-catalyzed synthesis of modified lipids, in *Lipids for Functional Foods and Nutraceuticals*, Gunstone, F.D., Ed., The Oily Press, Bridgwater, 2003, Chap. 6.

120. Sellappan, S. and Akoh, C.C., Application of lipases in modification of food lipids, in *Handbook of Industrial Biocatalysis*, Hou, C.T., Ed., Taylor & Francis, Boca Raton, FL, 2005, Chap. 9.

121. Yamada, K., Ibuki, M., and McBrayer, T., Cocoa butter, cocoa butter equivalents, and cocoa butter replacers, in *Healthful Lipids*, Akoh, C.C. and Lai, O.-M., Eds., AOCS Press, Champaign, IL, 2005, Chap. 26.

122. Mukherjee, K.D. and Kiewitt, I., Structured triacylglycerols resembling human milk fat by transesterification catalyzed by papaya (*Carcia papaya*) latex, *Biotechnol. Lett.*, 20, 613, 1998.

123. Weber, N. and Mukherjee, K.D., Lipids in infant formulas and human milk fat substitutes, in *Healthful Lipids*, Akoh, C.C. and Lai, O.-M., Eds., AOCS Press, Champaign, IL, 2005, Chap. 25.

124. Hansen, T.T., Specific esterification with immobilized lipase, in *Proceedings of International Symposium on New Aspects of Dietary Lipids: Benefits, Hazards, and Use*, Swedish Institute of Food Research (SIK), Göteborg, 1990, p. 171.

125. Xu, X., Production of specific-structured triacylglycerols by lipase-catalyzed reactions: a review, *Eur. J. Lipid Sci. Technol.*, 102, 287, 2000.

126. Foglia, T.A. and Villeneuve, P., *Carica papaya* latex-catalyzed synthesis of structured triacylglycerols, *J. Am. Oil Chem. Soc.*, 74, 1447, 1997.

127. Mangos, T.J., Jones, K.C., and Foglia, T.A., Lipase-catalyzed synthesis of structured low-calorie triacylglycerols, *J. Am. Oil Chem. Soc.*, 761, 127, 1999.

128. Lee, K.-T. and Foglia, T.A., Synthesis, purification, and characterization of structured lipids produced from chicken fat, *J. Am. Oil Chem. Soc.*, 77, 1027, 2000.

129. Jandacek, R.J., Whiteside, J.A., Holcombe, B.N., Volpenhein, R.A., and Taulbee, J.D., The rapid hydrolysis and efficient absorption of triglycerides with octanoic acid in the 1 and 3 positions and long chain fatty acid in the 2 position, *Am. J. Clin. Nutr.*, 45, 940, 1987.

130. Bracco, U., Effect of triglyceride structure on fat absorption, *Am. J. Clin. Nutr.*, 60, 1002S, 1994.

131. de Fouw, N.J., Kivits, G.A.A., Quinlan, P.T., and van Nielen, W.G.L., Absorption of isomeric, palmitic acid-containing triacylglycerols resembling human milk fat in the adult rat, *Lipids*, 29, 765, 1994.

132. Jensen, R.G., Gerrior, S.A., Hagerty, M.M., and Macmahon, K.E., Preparation of acylglycerols and phospholipids with the aid of lipolytic enzymes, *J. Am. Oil Chem. Soc.*, 55, 422, 1978.

133. Lien, E.L., The role of fatty acid composition and positional distribution on fat absorption in infants, *J. Pediatr.*, 125, S62, 1994.

134. Christensen, M.S., Hoy, C.-E., Becker, C.C., and Redgrave, T.G., Intestinal absorption and lymphatic transport of eicosapentaenoic (EPA), docosahexaenoic (DHA), and decanoic acids: dependence on intramolecular triacylglycerol structure, *Am. J. Clin. Nutr.*, 61, 56, 1995.

135. Yamane, T., Enzymatic synthesis of symmetrical triacylglycerols containing polyunsaturated fatty acids, in *Healthful Lipids*, Akoh, C.C. and Lai, O.-M., Eds., AOCS Press, Champaign, IL, 2005, Chap. 18.

136. Akoh, C.C., Structured and specialty lipids, in *Healthful Lipids*, Akoh, C.C. and Lai, O.-M., AOCS Press, Champaign, IL, 2005, Chap. 24.

137. Haraldsson, G.G., Structured triacylglycerols comprising omega-3 polyunsaturated fatty acids, in *Handbook of Industrial Biocatalysis*, Hou, C.T., Ed., Taylor & Francis, Boca Raton, FL, 2005, Chap. 18.

138. Lee, K.-T., Foglia, T.A., and Lee, J.-H., Low-calorie fat substitutes: synthesis and analysis, in *Handbook of Industrial Biocatalysis*, Hou, C.T., Ed., Taylor & Francis, Boca Raton, FL, 2005, Chap. 16.

139. Christophe, A.B., Structure-related effects on absorption and metabolism of nutraceutical and specialty lipids, in *Nutraceutical and Specialty Lipids and Their Co-Products*, Shahidi, F., Ed., Taylor & Francis, Boca Raton, FL, 2005, Chap. 23.

140. Gandhi, N.N. and Mukherjee, K.D., Synthesis of designer lipids using papaya (*Carica papaya*) latex lipase, *J. Mol. Catal. B Enzym.*, 11, 271, 2001.

141. De Maria, P.D., Sinistera, J.V., Tsai, S.W., and Alcantara, A.R., *Carica papaya* lipase (CPL): an emerging and versatile biocatalyst, *Biotechnol. Adv.*, 24, 493, 2006.

142. Akoh, C.C., Jennings, B.H., and Lillard, D.A., Enzymatic modification of trilinolein: incorporation of n-3 polyunsaturated fatty acids, *J. Am. Oil Chem. Soc.*, 72, 1317, 1994.

143. Shimada, Y., Suenaga, M., Sugihara, A., Nakai, S., and Tominaga, Y., Continuous production of structured lipid containing γ-linolenic and caprylic acids by immobilized *Rhizopus delemar* lipase, *J. Am. Oil Chem. Soc.*, 76, 189, 1999.

144. Jennings, B.H. and Akoh, C.C., Enzymatic modification of triacylglycerols of high eicosapentaenoic and docosahexaenoic acids content to produce structured lipids, *J. Am. Oil Chem. Soc.*, 76, 1113, 1999.

145. Irimescu, R., Yasui, M., Iwasaki, Y., Shimidzu, N., and Yamane, T., Enzymatic synthesis of 1,3-dicapryloyl-2-eicosapentaenoylglycerol, *J. Am. Oil Chem. Soc.*, 77, 501, 2000.

146. McNeill, G.P. and Sonnet, P.E., Low-calorie triglyceride synthesis by lipase-catalyzed esterification of monoglycerides, *J. Am. Oil Chem. Soc.*, 72, 1301, 1995.

147. Kuo, S.-J. and Parkin, K.L., Acetylacylglycerol formation by lipase in microaqueous milieu: effects of acetyl group donor and environmental factors, *J. Agric. Food Chem.*, 43, 1775, 1995.

148. Mukherjee, K.D. and Kiewitt, I., Enrichment of very-long-chain mono-unsaturated fatty acids by lipase-catalysed hydrolysis and transesterification, *Appl. Microbiol. Biotechnol.*, 44, 557, 1996.

149. Macrae, A.R., Roehl, E.-L., and Brand, H.M., Bio-esters, *Seifen Öle Fette Wachse*, 116, 201, 1990.

150. Claon, P.A. and Akoh, C.C., Enzymatic synthesis of geraniol and citronellol esters by direct esterification in *n*-hexane, *Biotechnol. Lett.*, 15, 1211, 1993.

151. Gillies, B., Yamazaki, H., and Armstrong, D.W., Production of flavor esters by immobilized lipase, *Biotechnol. Lett.*, 9, 709, 1987.

152. Langrand, G., Rondot, N., Triantaphylides, C., and Baratti, J., Short-chain flavor esters synthesis by microbial lipase, *Biotechnol. Lett.*, 12, 581, 1990.

153. Welsh, F.W. and Williams, R.E., Lipase-mediated production of ethyl butyrate in non-aqueous systems, *Enzyme Microb. Technol.*, 12, 743, 1990.

154. Kim, J., Altreuter, D.H., Clark, D.S., and Dordick, J.S., Rapid synthesis of fatty acid esters for use as potential food flavors, *J. Am. Oil Chem. Soc.*, 75, 1109, 1998.

155. De, B.K., Chatterjee, T., and Bhattacharyya, D.K., Synthesis of geranyl and citronellyl esters of coconut oil fatty acids through alcoholysis by *Rhizomucor miehei* lipase catalysis, *J. Am. Oil Chem. Soc.*, 76, 1501, 1999.

156. Shimada, Y., Hirota, Y., Baba, T., Kato, S., Sugihara, A., Moriyama, S., Tominaga, Y., and Terai, T., Enzymatic synthesis of L-menthyl esters in organic solvent-free system, *J. Am. Oil Chem. Soc.*, 76, 1139, 1999.

157. Gandhi, N.N. and Mukherjee, K.D., Specificity of papaya lipase in esterification with respect to the chemical structure of substrates, *J. Agric. Food Chem.*, 48, 566, 2000.

158. Molinari, F., Marianelli, G., and Aragozzini, F., Production of flavour esters by *Rhizopus oryzae*, *Appl. Microbiol. Biotechnol.*, 43, 967, 1995.

159. Krishna, S.H. and Karanth, N.G., Lipases and lipase-catalyzed esterification reactions in nonaqueous media, *Catal. Rev.-Sci. Eng.*, 44, 499, 2002.

160. De Maria, P.D., Carboni-Oerlemans, C., Tuin, B., Bargeman, G., van der Meer, A., and van Gemert, R., Biotechnological applications of *Candida antarctica* lipase A: state-of-the-art, *J. Mol. Catal. B: Enzym.*, 37, 36, 2005.

161. Hasan, F., Shah, A.A., and Hameed, A., Industrial applications of microbial lipases, *Enzyme Microb. Technol.*, 39, 235, 2006.

162. Mukherjee, K.D. and Kiewitt, I., Preparation of esters resembling natural waxes by lipase-catalyzed reactions, *J. Agric. Food Chem.*, 36, 1333, 1988.

163. Steinke, G., Kirchhoff, R., and Mukherjee, K.D., Lipase-catalyzed alcoholysis of crambe oil and camelina oil for the preparation of long-chain esters, *J. Am. Oil Chem. Soc.*, 77, 361, 2000.

164. Steinke, G., Weitkamp, P., Klein, E., and Mukherjee, K.D., High-yield preparation of wax esters via lipase-catalyzed esterification using fatty acids and alcohols from crambe and camelina oils, *J. Agric. Food Chem.*, 49, 647, 2001.

165. Bloomer, S., Adlercreutz, P., and Mattiasson, B., Facile synthesis of fatty acid esters in high yields, *Enzyme Microb. Technol.*, 14, 546, 1992.

166. Habulin, M., Krmelj, V., and Knez, Z., Synthesis of oleic acid esters catalyzed by immobilized lipase, *J. Agric. Food Chem.*, 44, 338, 1996.

167. Hayes, D.G. and Kleiman, R., Lipase-catalyzed synthesis of lesquerolic acid wax and diol esters and their properties, *J. Am. Oil Chem. Soc.*, 73, 1385, 1996.

168. Trani, M., Ergan, F., and André, G., Lipase-catalyzed production of wax esters, *J. Am. Oil Chem. Soc.*, 68, 20, 1991.

169. Ucciani, E., Schmitt-Rozieres, M., Debal, A., and Comeau, L.C., Enzymatic synthesis of some wax-esters, *Fett/Lipid*, 98, 206, 1996.

170. Wehtje, E., Costes, D., and Adlercreutz, P., Continuous lipase-catalyzed production of wax ester using silicone tubing, *J. Am. Oil Chem. Soc.*, 76, 1489, 1999.

171. De, B.K., Bhattacharyya, D.K., and Bandhu, C., Enzymatic synthesis of fatty alcohol esters by alcoholysis, *J. Am. Oil Chem. Soc.*, 76, 451, 1999.

172. Decagny, B., Jan, S., Vuillemard, J.C., Sarazin, C., Séguin, J.P., Gosselin, C., Barbotin, J.N., and Ergan, F., Synthesis of wax ester through triolein alcoholysis: choice of the lipase and study of the mechanism, *Enzyme Microb. Technol.*, 22, 578, 1998.

173. Goma-Doncescu, N. and Legoy, M.D., An original transesterification route for fatty acid ester production from vegetable oils in a solvent-free system, *J. Am. Oil Chem. Soc.*, 74, 1137, 1997.

174. Hallberg, M.L., Wang, D., and Härröd, M., Enzymatic synthesis of wax esters from rapeseed fatty acid methyl esters and a fatty alcohol, *J. Am. Oil Chem. Soc.*, 76, 183, 1999.

175. Multzsch, R., Lokotsch, W., Steffen, B., Lang, S., Metzger, J.O., Schäfer, H.J., Warwel, S., and Wagner, F., Enzymatic production and physicochemical characterization of uncommon wax esters and monoglycerides, *J. Am. Oil Chem. Soc.*, 71, 721, 1994.

176. Miller, C., Austin, H., Posorske, L., and Gonzlez, J., Characteristics of an immobilized lipase for the commercial synthesis of esters, *J. Am. Oil Chem. Soc.*, 65, 927, 1988.

177. Rocha, J.M.S., Gil, M.H., and Garcia, F.A.P., Synthesis of *n*-octyl oleate with lipase from *Mucor miehei* immobilized onto polyethylene based graft copolymers, *Biocatalysis*, 9, 157, 1994.

178. Basri, M., Ampon, K., Wan Yunus, W.M.Z., Razak, C.N.A., and Salleh, A.B., Enzymic synthesis of fatty esters by hydrophobic lipase derivatives immobilized on organic polymer beads, *J. Am. Oil Chem. Soc.*, 72, 407, 1995.

179. Goto, M., Kamiya, N., Miyata, M., and Nakashio, F., Enzymatic esterification by surfactant-coated lipase in organic media, *Biotechnol. Prog.*, 10, 263, 1994.

180. Stevenson, D.E., Stanley, R.A., and Furneaux, R.H., Near-quantitative production of fatty acid alkyl esters by lipase-catalyzed alcoholysis of fats and oils with adsorption of glycerol by silica gel, *Enzyme Microb. Technol.*, 16, 478, 1994.

181. Shaw, J.-F., Wang, D.-L., and Wang, Y.J., Lipase-catalysed ethanolysis and isopropanolysis of triglycerides with long-chain fatty acids, *Enzyme Microb. Technol.*, 13, 544, 1991.

182. Linko, Y.-Y., Lämsä, M., Huhtala, A., and Rantanen, O., Lipase biocatalysis in the production of esters, *J. Am. Oil Chem. Soc.*, 72, 1293, 1995.

183. Ghoshray, S. and Bhattacharyya, D.K., Enzymatic preparation of ricinoleic acid esters of long-chain monohydric alcohols and properties of the esters, *J. Am. Oil Chem Soc.*, 69, 85, 1992.

184. Weber, N., Klein, E., and Vosmann, K., Dialkyl 3,3'-thiodipropionate and 2,2'-thiodiacetate antioxidants by lipase-catalyzed esterification and transesterification, *J. Agric. Food Chem.*, 54, 2957, 2006.

185. Yamada, Y., Hirota, Y., Baba, T., Sugihara, A., Moriyama, S., Tominaga, Y., and Terai, T., Enzymatic synthesis of steryl esters of polyunsaturated fatty acids, *J. Am. Oil Chem. Soc.*, 76, 713, 1999.

186. Weber, N., Weitkamp, P., and Mukherjee, K.D., Fatty acid steryl, stanyl and steroid esters by esterification and transesterification in vacuo using *Candida rugosa* lipase as catalyst, *J. Agric. Food Chem.*, 49, 67, 2001.

187. Weber, N., Weitkamp, P., and Mukherjee, K.D., Steryl and stanyl esters of fatty acids by solvent-free esterification and transesterification in vacuo using lipases from *Rhizomucor miehei*, *Candida antarctica* and *Carica papaya*, *J. Agric. Food Chem.*, 49, 5210, 2001.

188. Weber, N., Weitkamp, P., and Mukherjee, K.D., Cholesterol-lowering food additives: lipase-catalyzed preparation of phytosterol and phytostanol esters, *Food Res. Int.*, 35, 177, 2002.

189. Weber, N., Weitkamp, P., and Mukherjee, K.D., *Thermomyces lanuginosus* lipase-catalyzed transesterification of sterols by methyl oleate, *Eur. J. Lipid Sci. Technol.*, 105, 624, 2003.

190. Weber, N., Weitkamp, P., and Mukherjee, K.D., Steryl and stanyl esters by solvent-free lipase-catalyzed esterification and transesterification in vacuo, *Fresenius Environ. Bull.*, 12, 517, 2003.

191. Wester, I., Cholesterol-lowering effect of plant sterols, *Eur. J. Lipid Sci. Technol.*, 102, 37, 2000.

192. Salo, P., Wester, I., and Hopia, A., Phytosterols, in *Lipids for Functional Foods and Nutraceuticals*, Gunstone, F.D., Ed., The Oily Press, Bridgwater, 2003, Chap. 7.

193. Weber, N. and Mukherjee, K.D., Plant sterols and steryl esters in functional foods and nutraceuticals, in *Nutraceutical and Specialty Lipids and Their Co-Products*, Shahidi, F., Ed., Taylor & Francis, Boca Raton, FL, 2005, Chap. 27.

194. Moreau, R., Phytosterols and phytosterol esters, in *Healthful Lipids*, Akoh, C.C. and Lai, O.-M., Eds., AOCS Press, Champaign, IL, 2005, Chap. 15.

195. Salo, P., Hopia, N., Ekblom, J., Lahtinen, R., and Laasko, P., Plant stanol ester as a cholesterol-lowering ingredient of Benecol® foods, in *Healthful Lipids*, Akoh, C.C. and Lai, O.-M., Eds., AOCS Press, Champaign, IL, 2005, Chap. 29.

196. Holmberg, K. and Österberg, E., Enzymatic preparation of monoglycerides in micro-emulsion, *J. Am. Oil Chem. Soc.*, 65, 1544, 1988.

197. Plou, F.J., Barandiarán, M., Calvo, M.V., Ballesteros, A., and Pastor, E., High-yield production of mono- and di-oleoylglycerol by lipase-catalyzed hydrolysis of triolein, *Enzyme Microb. Technol.*, 18, 66, 1996.

198. Akoh, C.C., Cooper, C., and Nwosu, C.V., Lipase G-catalyzed synthesis of monoglycerides in organic solvent and analysis by HPLC, *J. Am. Oil Chem. Soc.*, 69, 257, 1992.

199. Berger, M. and Schneider, M.P., Enzymatic esterification of glycerol II. Lipase-catalyzed synthesis of regioisomerically pure 1(3)-*rac*-monoacylglycerols, *J. Am. Oil Chem. Soc.*, 69, 961, 1992.

200. Steffen, B., Ziemann, A., Lang, S., and Wagner, F., Enzymatic monoacylation of trihydroxy compounds, *Biotechnol. Lett.*, 14, 773, 1992.

201. Janssen, A.E.M., van der Padt, A., and van't Riet, K., Solvent effects on lipase-catalyzed esterification of glycerol and fatty acids, *Biotechnol. Bioeng.*, 42, 953, 1993.

202. Kwon, S.J., Han, J.J., and Rhee, J.S., Production and in situ separation of mono- or diacylglycerol catalyzed by lipases in *n*-hexane, *Enzyme Microb. Technol.*, 17, 700, 1995.

203. Pastor, E., Otero, C., and Ballesteros, A., Synthesis of mono- and dioleoylglycerols using an immobilized lipase, *Appl. Biochem. Biotechnol.*, 50, 251, 1995.

204. Pastor, E., Otero, C., and Ballesteros, A., Enzymatic preparation of mono- and distearin by glycerolysis of ethyl stearate and direct esterification of glycerol in the presence of a lipase from *Candida antarctica* (Novozym 435), *Biocatal. Biotransformation*, 12, 147, 1995.

205. Schuch, R. and Mukherjee, K.D., Lipase-catalyzed reactions of fatty acids with glycerol and acylglycerols, *Appl. Microbiol. Biotechnol.*, 30, 332, 1989.

206. Singh, C.P., Shah, D.O., and Holmberg, K., Synthesis of mono- and diglycerides in water-in-oil microemulsions, *J. Am. Oil Chem. Soc.*, 71, 583, 1994.

207. Arcos, J.A., Garcia, H.S., and Hill Jr., C.G., Continuous enzymatic esterification of glycerol with (poly) unsaturated fatty acids in a packed-bed reactor, *Biotechnol. Bioeng.*, 68, 563, 2000.

208. Rosu, R., Yasui, M., Iwasaki, Y., and Yamane, T., Enzymatic synthesis of symmetrical 1,3-diacylglycerols by direct esterification of glycerol in solvent-free system, *J. Am. Oil Chem. Soc.*, 76, 839, 1999.

209. Millqvist, A., Adlercreutz, P., and Mattiasson, B., Lipase-catalyzed alcoholysis of triglycerides for the preparation of 2-monoglycerides, *Enzyme Microb. Technol.*, 16, 1042, 1994.

210. Mukesh, D., Iyer, R.S., Wagh, J., Mokashi, A.A., Banerji, A.A., Newadkar, R.V., and Bevinakatti, H.S., Lipase catalysed transesterification of castor oil, *Biotechnol. Lett.*, 15, 251, 1993.

211. Bornscheuer, U.T., Lipase-catalyzed synthesis of monoacylglycerols, *Enzyme Microb. Technol.*, 17, 578, 1995.

212. Bornscheuer, U.T., Stamatis, H., Xenakis, A., Yamane, T., and Kolisis, F.N., A comparison of different strategies for lipase-catalyzed synthesis of partial glycerides, *Biotechnol. Lett.*, 16, 697, 1994.

213. Bornscheuer, U.T. and Yamane, T., Activity and stability of lipase in the solid-phase glycerolysis of triolein, *Enzyme Microb. Technol.*, 16, 864, 1995.

214. Chang, P.S., Rhee, J.S., and Kim, J.-J., Continuous glycerolysis of olive oil by *Chromobacterium viscosum* lipase, *Biotechnol. Bioeng.*, 38, 1159, 1991.

215. McNeill, G.P., Shimizu, S., and Yamane, T., High-yield enzymatic glycerolysis of fats and oils, *J. Am. Oil Chem. Soc.*, 68, 1, 1991.

216. McNeill, G.P. and Yamane, T., Further improvement in the yield of monoglycerides during enzymatic glycerolysis of fats and oils, *J. Am. Oil Chem. Soc.*, 67, 771, 1991.

217. Myrnes, B., Barstad, H., Olsen, R.L., and Elvevoll, E.O., Solvent-free enzymatic glycerolysis of marine oils, *J. Am. Oil Chem. Soc.*, 72, 1339, 1995.

218. Wang, L.-L., Yang, B.-K., Parkin, K.L., and Johnson, E.A., Inhibition of *Listeria monocytogenes* by monoacylglycerols synthesized from coconut oil and milkfat by lipase-catalyzed glycerolysis, *J. Agric. Food Chem.*, 41, 1000, 1993.

219. Yang, B. and Parkin, K., Monoacylglycerol production from butteroil by glycerolysis with a gel-entrapped microbial lipase in microaqueous media, *J. Food Sci.*, 59, 47, 1994.

220. Edmundo, C., Valérie, D., Didier, C., and Alain, M., Efficient lipase-catalyzed production of tailor-made emulsifiers using solvent engineering coupled to extractive processing, *J. Am. Oil Chem. Soc.*, 75, 309, 1998.

221. Nagao, T., Watanabe, H., Goto, N., Onizawa, K., Taguchi, H., Matsuo, N., Yasukawa, T., Tsushima, R., Shimasaki, H., and Itakura, H., Dietary diacylglycerol suppresses accumulation of body fat compared to triacylglycerol in men in a double-blind controlled trial, *J. Nutr.*, 130, 792, 2000.

222. Watanabe, T., Shimizu, M., Sugiura, M., Sato, M., Kohori, J., Yamada, N., and Nakanishi, K., Optimization of reaction conditions for the production of DAG using immobilized 1,3-regiospecific lipase Lipozyme RM IM, *J. Am. Oil Chem. Soc.*, 80, 1201, 2003.

223. Weber, N. and Mukherjee, K.D., Solvent-free lipase-catalyzed preparation of diacylglycerols, *J. Agric. Food Chem.*, 52, 5347, 2004.

224. Yamamoto, K., Asakawa, H., Tokunaga, K., Watanabe, H., Matsuo, N., Tokimitsu, I., and Yagi, N., Long-term ingestion of dietary diacylglycerol lowers serum triacylglycerol in type II diabetic patients with hypertriglyceridemia, *J. Nutr.*, 131, 3204, 2001.

225. Sakaguchi, H., Marketing a healthy oil, *Oil Fats Internat.*, 18, 2001.

226. Watkins, C., Time for an oil change, *Int. News Fats Oils Relat. Mat.*, 14, 70, 2003.

227. Yamada, N., Matsuo, N., Watanabe, T., and Yanagita, T., Enzymatic production of diacylglycerol and its beneficial physiological functions, in *Handbook of Industrial Biocatalysis*, Hou, C.T., Ed., Taylor & Francis, Boca Raton, FL, 2005, Chap. 11.

228. Watanabe, H. and Matsuo, N., Diacylglycerols, in *Lipids for Functional Foods and Nutraceuticals*, Gunstone, F.D., Ed., The Oily Press, Bridgwater, 2003, Chap. 5.

229. Matsuo, N., Nutritional characteristics of diacylglycerol oil and its health benefits, in *Healthful Lipids*, Akoh, C.C. and Lai, O.-M., Eds., AOCS Press, Champaign, IL, 2005, Chap. 28.

230. Flickinger, B.D., Diacylglycerols (DAGs) and their mode of action, in *Nutraceutical and Specialty Lipids and Their Co-Products*, Shahidi, F., Ed., Taylor & Francis, Boca Raton, FL, 2005, Chap. 11.

231. Figueroa-Espinoza, M.-C. and Villeneuve, P., Phenolic acids enzymatic lipophilization, *J. Agric. Food Chem.*, 53, 2779, 2005.

232. Vosmann, K., Weitkamp, P., and Weber, N., Solvent-free lipase-catalyzed preparation of long-chain alkyl phenylpropanoates and phenylpropyl alkanoates, *J. Agric. Food Chem.*, 54, 2969, 2006.

233. Weitkamp, P., Vosmann, K., and Weber, N., Highly efficient preparation of lipophilic hydroxycinnamates by solvent-free lipase-catalyzed transesterification, *J. Agric. Food Chem.*, 54, 7062, 2006.

234. Robinson, G.K., Alston, M.J., Knowles, C.J., Cheetham, P.S.J., and Motion, K.R., An investigation into the factors influencing lipase-catalyzed intramolecular lactonization in microaqueous systems, *Enzyme Microb. Technol.*, 16, 855, 1994.

235. Hayes, D.G. and Kleiman, R.R., Lipase-catalyzed synthesis and properties of estolides and their esters, *J. Am. Oil Chem. Soc.*, 17, 183, 1995.

236. Couldon, D., Girardin, M., Rovel, B., and Ghoul, M., Comparison of direct esterification and transesterification of fructose by *Candida antarctica* lipase, *Biotechnol. Lett.*, 17, 183, 1995.

237. Ducret, A., Giroux, A., Trani, M., and Lottie, R., Enzymatic preparation of biosurfactants from sugars or sugar alcohols and fatty acids in organic media under reduced pressure, *Biotechnol. Bioeng.*, 48, 214, 1995.

238. Ducret, A., Giroux, A., Trani, M., and Lortie, R., Characterization of enzymatically prepared biosurfactants, *J. Am. Oil Chem. Soc.*, 73, 109, 1996.

239. Fregapane, G., Sarney, D.B., Greenberg, S.G., Knight, D.J., and Vulfson, E.N., Enzymatic synthesis of monosaccharide fatty acid esters and their comparison with conventional products, *J. Am. Oil Chem. Soc.*, 71, 87, 1994.

240. Janssen, A.E.M., Klabbers, C., Franssen, M.C.R., and van't Riet, K., Enzymatic synthesis of carbohydrate esters in 2-pyrrolidone, *Enzyme Microb. Technol.*, 13, 565, 1991.

241. Janssen, A.E.M., Lefferts, A.G., and van't Riet, K., Enzymatic synthesis of carbohydrate esters in aqueous media, *Biotechnol. Lett.*, 12, 711, 1990.

242. Khaled, N., Montet, D., Pina, M., and Graille, J., Fructose oleate synthesis in a fixed catalyst bed reactor, *Biotechnol. Lett.*, 13, 167, 1991.

243. Ku, M.A. and Hang, Y.D., Enzymatic synthesis of esters in organic medium with lipase from *Byssochlamys fulva*, *Biotechnol. Lett.*, 17, 1081, 1995.

244. Oguntimein, G.B., Erdmann, H., and Schmid, R.D., Lipase catalysed synthesis of sugar ester in organic solvents, *Biotechnol. Lett.*, 15, 175, 1993.

245. Sarney, D.B., Kapeller, H., Fregapane, G., and Vulfson, E.N., Chemo-enzymatic synthesis of disaccharide fatty acid esters, *J. Am. Oil Chem. Soc.*, 71, 711, 1994.

246. Scheckermann, C., Schlotterbeck, A., Schmidt, M., Wray, V., and Lang, S., Enzymatic monoacylation of fructose by two procedures, *Enzyme Microb. Technol.*, 17, 157, 1995.

247. Seino, H., Uchibori, T., Nishitani, T., and Inamasu, S., Enzymatic synthesis of carbohydrate esters of fatty acids. I. Esterification of sucrose, glucose, fructose and sorbitol, *J. Am. Oil Chem. Soc.*, 61, 1761, 1984.

248. Mukesh, D., Sheth, D., Mokashi, A., Wagh, J., Tilak, J.M., Banerji, A.A., and Thakkar, K.R., Lipase catalysed esterification of isosorbide and sorbitol, *Biotechnol. Lett.*, 15, 1243, 1993.

249. Hayes, D.G. and Gulari, E., Formation of polyol-fatty acid esters by lipases in reverse miceller media, *Biotechnol. Bioeng.*, 40, 110, 1992.

250. Soedjak, H.S. and Spradlin, J.E., Enzymatic transesterification of sugars in anhydrous pyridine, *Biocatalysis*, 11, 241, 1994.

251. Zhang, X. and Hayes D.G., Increased rate of lipase-catalyzed saccharide-fatty acid esterification by control of reaction medium, *J. Am. Oil Chem. Soc.*, 76, 1495, 1999.

252. Degn, P., Pedersen, L.H., Duus, J.Ø., and Zimmermann, W., Lipase-catalysed synthesis of glucose fatty acid esters in *tert*-butanol, *Biotechnol. Lett.*, 21, 275, 1999.

253. Arcos, J.A., Bernabé, M., and Otero, C., Different strategies for selective monoacylation of hexoaldoses in acetone, *J. Surfactants Detergents*, 1, 345, 1998.

254. Yan, Y., Bornscheuer, U.T., Cao, L., and Schmid, R.D., Lipase-catalyzed solid-phase synthesis of sugar fatty acid esters, Removal of byproducts by azeotropic distillation, *Enzyme Microb. Technol.*, 25, 725, 1999.

255. Chopineau, J., McCafferty, F.D., Therisod, M., and Klibanov, A., Production of biosurfactants from sugar alcohols and vegetable oil catalyzed by lipases in nonaqueous medium, *Biotechnol. Bioeng.*, 31, 208, 1988.

256. Akoh, C.C. and Mutua, L.N., Synthesis of alkyl glycoside fatty acid esters: effect of reaction parameters and the incorporation of n-3 polyunsaturated fatty acids, *Enzyme Microb. Technol.*, 16, 115, 1994.

257. Mutua, L.N. and Akoh, C.C., Synthesis of alkyl glycoside fatty acid esters in non-aqueous media by *Candida* sp. lipase, *J. Am. Oil Chem. Soc.*, 70, 43, 1993.

258. de Goede, A.T.J.W., van Oosterom, M., van Deurzen, M.P.J., Sheldon, R.J., van Bekkum, H., and van Rantwijk, F., Selective lipase-catalyzed esterification of alkyl glycosides, *Biocatalysis*, 9, 145, 1994.

259. Nakano, H., Kiku, Y., Ando, K., Kawashima, Y., Kitahata, S., Tominaga, Y., and Takenishi, S., Lipase-catalyzed esterification of mono- and oligoglycosides with various fatty acids, *J. Ferment. Bioeng.*, 78, 70, 1994.

260. Charlemagne, D. and Legoy, M.D., Enzymatic synthesis of polyglycerol-fatty acid esters in a solvent-free system, *J. Am. Oil Chem. Soc.*, 72, 61, 1995.

261. Shaw, J.-F. and Lo, S., Production of propylene glycol fatty acid monoesters by lipase-catalyzed reactions in organic solvents, *J. Am. Oil Chem. Soc.*, 71, 715, 1994.

262. Liu, K.-J. and Shaw, J.-F., Synthesis of propylene glycol monoesters of docosahexaenoic acid and eicosapentaenoic acid by lipase-catalyzed esterification in organic solvents, *J. Am. Oil Chem. Soc.*, 72, 1271, 1995.

263. Janssen, G.J. and Haas, M., Lipase-catalyzed synthesis of oleic acid esters of polyethylene glycol 400, *Biotechnol. Lett.*, 16, 163, 1994.

264. Torres, C. and Otero, C., Enzymatic synthesis of lactate and glycolate esters of fatty alcohols, Part I, *Enzyme Microb. Technol.*, 25, 745, 1999.

265. Torres, C., Bernabé, M., and Otero, C., Enzymatic synthesis of lactic acid derivatives with emulsifying properties, *Biotechnol. Lett.*, 22, 331, 2000.

266. Bradoo, S., Saxena, R.K., and Gupta, R., High yields of ascorbyl palmitate by thermostable lipase-mediated esterification, *J. Am. Oil Chem. Soc.*, 76, 1291, 1999.

267. Stamatis, H., Sereti, V., and Kolisis, F.N., Studies on the enzymatic synthesis of lipophilic derivatives of natural antioxidants, *J. Am. Oil Chem. Soc.*, 76, 1505, 1999.

268. Yan, Y., Bornscheuer, U.T., and Schmid, R.D., Lipase-catalyzed synthesis of vitamin C fatty acid esters, *Biotechnol. Lett.*, 21, 1051, 1999.

269. Watanabe, Y., Minemoto, Y., Adachi, S., Nakanishi, K., Shimada, Y., and Matsuno, R., Lipase-catalyzed synthesis of 6-*O*-eicosapentaenoyl L-ascorbate in acetone and its autoxidation, *Biotechnol. Lett.*, 22, 637, 2000.

270. Maugard, T., Tudella, J., and Legoy, M.D., Study of vitamin ester synthesis by lipase-catalyzed transesterification in organic media, *Biotechnol. Prog.*, 16, 358, 2000.

271. Montet, D., Servat, F., Pina, M., Graille, J., Galzy, P., Arnaud, A., Ledon, H., and Marcou, L., Enzymatic synthesis of *N*-ε-acyllysines, *J. Am. Oil Chem. Soc.*, 68, 6, 1990.

272. de Zoete, M.C., Kock-van Dalen, A.C., van Rantwijk, F., and Sheldon, R.A., A new enzymatic reaction: enzyme catalyzed ammonolysis of carboxylic esters, *Biocatalysis*, 10, 307, 1994.

273. Litjens, M.J.J., Straathof, A.J.J., Jongejan, J.A., and Heijnen, J.J., Exploration of lipase-catalyzed direct amidation of free carboxylic acids with ammonia in organic solvents, *Tetrahedron*, 55, 12411, 1999.

274. Gotor, V., Non-conventional hydrolase chemistry: amide and carbamate bond formation catalyzed by lipases, *Bioorg. Med. Chem.*, 7, 2189, 1999.

275. Bühler, M. and Wandrey, C., Continuous use of lipases in fat hydrolysis, *Fat Sci. Technol.*, 89, 598, 1987.

276. Hoq, M.M., Yamane, T., Shimizu, S., Funada, T., and Ishida, S., Continuous hydrolysis of olive oil by lipase in microporous hydrophobic membrane bioreactor, *J. Am. Oil Chem. Soc.*, 62, 1016, 1988.

277. Linfield, W.M., Barauskas, R.A., Sivieri, L., Serota, S., and Stevenson Sr., R.W., Enzymatic fat hydrolysis and synthesis, *J. Am. Oil Chem. Soc.*, 61, 191, 1984.

278. Linfield, W.M., O'Brien, D.J., Serota, S., and Barauskas, R.A., Lipid-lipase interactions. I. Fat splitting with lipase from *Candida rugosa*, *J. Am. Oil Chem. Soc.*, 61, 1067, 1984.

279. Marangoni, A.G., *Candida* and *Pseudomonas* lipase-catalyzed hydrolysis of butteroil in the absence of organic solvents, *J. Food Sci.*, 59, 1096, 1994.

280. Pronk, W., Kerkhof, P.J.A.M., van Helden, C., and van't Riet, K., The hydrolysis of triglycerides by immobilized lipase in a hydrophilic membrane reactor, *Biotechnol. Bioeng.*, 32, 512, 1988.

281. Virto, M.D., Agud, I., Moneto, S., Blanco, A., Solozabal, R., Lascaray, J.M., Llama, M.J., Serra, J.L., Landeta, L.C., and Renobales, M.D., Hydrolysis of animal fats by immobilized *Candida rugosa* lipase, *Enzyme Microb. Technol.*, 16, 61, 1994.

282. Wang, Y.J., Sheu, J.Y., Wang, F.F., and Shaw, J.F., Lipase-catalyzed oil hydrolysis in the absence of added emulsifier, *Biotechnol. Bioeng.*, 31, 628, 1988.

283. Watanabe, T., Suzuki, Y., Sagesaka, Y., and Kohashi, M., Immobilization of lipases on polyethylene and application to perilla oil hydrolysis for production of α-linolenic acid, *J. Nutr. Sci. Vitaminol.*, 41, 307, 1995.

284. Kosugi, Y., Suzuki, H., and Funuda, T., Hydrolysis of beef tallow by lipase from *Pseudomonas* sp., *Biotechnol. Bioeng.*, 31, 349, 1988.

285. Kosugi, Y., Tanaka, H., and Tomizuka, N., Continuous hydrolysis of oil by immobilized lipase in a countercurrent reactor, *Biotechnol. Bioeng.*, 36, 617, 1990.

286. Kosugi, Y. and Tomizuka, N., Continuous lipolysis reactor with a loop connecting an immobilized lipase column and an oil–water separator, *J. Am. Oil Chem. Soc.*, 72, 1329, 1995.

287. Tanigaki, M., Sakata, M., Takaya, H., and Mimura, K., Hydrolysis of palm stearin oil by a thermostable lipase in a draft tube-type reactor, *J. Ferment. Bioeng.*, 80, 340, 1995.

288. Yamamoto, K. and Fujiwara, N., The hydrolysis of castor oil using a lipase from *Pseudomonas* sp. f-B-24: positional and substrate specificity of the enzyme and optimum reaction conditions, *Biosci. Biotech. Biochem.*, 59, 1262, 1995.

289. Fu, X., Zhu, X., Gao, K., and Duan, J., Oil and fat hydrolysis with lipase from *Aspergillus* sp., *J. Am. Oil Chem. Soc.*, 72, 527, 1995.

290. Taylor, F., Panzer, C.C., Craig Jr., J.C., and O'Brien, D.J., Continuous hydrolysis of tallow with immobilized lipase in a microporous membrane, *Biotechnol. Bioeng.*, 28, 1318, 1986.

291. Prazeres, D.M.F., Garcia, F.A.P., and Cabral, J.M.S., An ultrafiltration membrane bioreactor for the lipolysis of olive oil in reversed micellar media, *Biotechnol. Bioeng.*, 41, 761, 1993.

292. Anderson, K.W., Enzyme pre-splitting, *Int. News Fats Oils Relat. Mater.*, 7, 81, 1996.

293. Mukherjee, K.D., Fractionation of fatty acids and other lipids using lipases, in *Enzymes in Lipid Modification*, Bornscheuer, U.T., Ed., Wiley-VCH, Weinheim, 2000, p. 23.

294. Borgdorf, R. and Warwel, S., Substrate selectivity of various lipases in the esterification of *cis*- and *trans*-9-octadecenoic acid, *Appl. Microbiol. Biotechnol.*, 51, 480, 1999.

295. Mukherjee, K.D., Kiewitt, I., and Hills, M.J., Substrate specificity of lipases in view of kinetic resolution of unsaturated fatty acids, *Appl. Microbiol. Biotechnol.*, 40, 489, 1993.

296. Osterberg, E., Blomstrom, A.-C., and Holmberg, K., Lipase catalyzed transesterification of unsaturated lipids in a microemulsion, *J. Am. Oil Chem. Soc.*, 66, 1330, 1989.

297. Rangheard, M.-S., Langrand, G., Triantaphylides, C., and Baratti, J., Multicompetetive enzymatic reactions in organic media: a simple test for the determination of lipase fatty acid specificity, *Biochim. Biophys. Acta*, 1004, 20, 1989.

298. Hills, M.J., Kiewitt, I., and Mukherjee, K.D., Lipase from *Brassica napus* L. discriminates against *cis*-4 and *cis*-6 unsaturated fatty acids and secondary and tertiary alcohols, *Biochim. Biophys. Acta*, 1042, 237, 1990.

299. Jachmanián, I. and Mukherjee, K.D., Germinating rapeseed as biocatalyst: hydrolysis of oils containing common and unusual fatty acids, *J. Agric. Food Chem.*, 43, 2997, 1995.

300. Jachmanián, I., Schulte, E., and Mukherjee, K.D., Substrate selectivity in esterification of less common fatty acids catalysed by lipases from different sources, *Appl. Microbiol. Biotechnol.*, 44, 563, 1996.

301. Mukherjee, K.D. and Kiewitt, I., Specificity of *Carica papaya* latex as biocatalyst in the esterification of fatty acids with 1-butanol, *J. Agric. Food Chem.*, 44, 1948, 1996.

302. Mukherjee, K.D., Fractionation of fatty acids and other lipids via lipase-catalyzed reactions, *J. Franc. Oleagineux Corps Gras Lipides*, 2, 365, 1995.

303. Mukherjee, K.D., Lipase-catalyzed reactions for the fractionation of fatty acids, in *Engineering of/with Lipases*, Malcata, F.X., Ed., Kluwer Academic, Dordrecht, 1996, p. 51.

304. Hoshino, T., Yamane, T., and Shimizu, S., Selective hydrolysis of fish oil by lipase to concentrate n-3 polyunsaturated fatty acids, *Agric. Biol. Chem.*, 54, 1459, 1990.

305. Maehr, H., Zenchoff, G., and Coffen, D.L., Enzymatic enhancement of n-3 fatty acid content in fish oils, *J. Am. Oil Chem. Soc.*, 71, 463, 1994.

306. Shimada, Y., Maruyama, K., Okazaki, S., Nakamura, M., Sugihara, A., and Tominaga, Y., Enrichment of polyunsaturated fatty acids with *Geotrichum candidum* lipase, *J. Am. Oil Chem. Soc.*, 71, 951, 1994.

307. Tanaka, Y., Hirano, J., and Funada, T., Concentration of docosahexaenoic acid in glyceride by hydrolysis of fish oil with *Candida cylindracea* lipase, *J. Am. Oil Chem. Soc.*, 69, 1210, 1992.

308. Yadwad, V.B., Ward, O.P., and Noronha, L.C., Application of lipase to concentrate the docosahexaenoic acid (DHA) fraction of fish oil, *Biotechnol. Bioeng.*, 38, 956, 1991.

309. Lawson, L.D. and Hughes, B.G., Human absorption of fish oil fatty acids as triacylglycerols, free acids, or ethyl esters, *Biochim. Biophys. Res. Commun.*, 152, 328, 1988.

310. Moore, S.R. and McNeill, G.P., Production of triglycerides enriched in long-chain n-3 polyunsaturated fatty acids from fish oil, *J. Am. Oil Chem. Soc.*, 73, 1409, 1996.

311. Shimada, Y., Maruyama, K., Sugihara, A.S., and Tominaga, Y., Purification of docosahexaenoic acid from tuna oil by a two-step enzymatic method: hydrolysis and selective esterification, *J. Am. Oil Chem. Soc.*, 74, 1441, 1997.

312. Haraldsson, G.G. and Kristinsson, B., Separation of eicosapentaenoic acid and docosahexaenoic acid in fish oil by kinetic resolution using lipase, *J. Am. Oil Chem. Soc.*, 75, 1551, 1998.

313. Shimida, Y., Sugihara, A., Minamigawa, Y., Higashiyama, K., Akimoto, K., Fujikawa, S., Komemushi, S., and Tominaga, Y., Enzymatic enrichment of arachidonic acid from *Mortierella* single-cell oil, *J. Am. Oil Chem. Soc.*, 75, 1213, 1998.

314. Syed Rahmatullah, M.S.K., Shukla, V.K.S., and Mukherjee, K.D., Enrichment of γ-linolenic acid from evening primrose oil and borage oil via lipase-catalyzed hydrolysis, *J. Am. Oil Chem. Soc.*, 71, 569, 1994.

315. Shimada, Y., Sugihara, A., Maruyama, K., Nagao, T., Nakayama, S., Nakano, H., and Tominaga, Y., Enrichment of arachidonic acid: selective hydrolysis of a single-cell oil from *Mortierella* with *Candida cylindracea* lipase, *J. Am. Oil Chem. Soc.*, 71, 1323, 1995.

316. Hills, M.J., Kiewitt, I., and Mukherjee, K.D., Enzymatic fractionation of evening primrose oil by rape lipase: enrichment of γ-linolenic acid, *Biotechnol. Lett.*, 11, 629, 1989.

317. Hills, M.J., Kiewitt, I., and Mukherjee, K.D., Enzymatic fractionation of fatty acids: enrichment of γ-linolenic acid and docasahexaenoic acid by selective esterification catalyzed by lipases, *J. Am. Oil Chem. Soc.*, 67, 561, 1990.

318. Langholz, P., Anderson, P., Forskov, T., and Schmidtsdorff, W., Application of a specificity of *Mucor miehei* lipase to concentrate docosahexaenoic acid (DHA), *J. Am. Oil Chem. Soc.*, 66, 1120, 1989.

319. Mukherjee, K.D. and Kiewitt, I., Enrichment of γ-linolenic acid from fungal oil by lipase-catalysed reactions, *Appl. Microbiol. Biotechnol.*, 35, 579, 1991.

320. Syed Rahmatullah, M.S.K., Shukla, V.K.S., and Mukherjee, K.D., γ-Linolenic acid concentrates from borage and evening primrose oil fatty acids via lipase-catalyzed esterification, *J. Am. Oil Chem. Soc.*, 71, 563, 1994.

321. Jachmanián, I. and Mukherjee, K.D., Esterification and interesterification reactions catalyzed by acetone powder from germinating rapeseed, *J. Am. Oil Chem. Soc.*, 73, 1527, 1996.

322. McNeill, G.P., Rawlins, C., and Peilow, A.C., Enzymatic enrichment of conjugated linoleic acid isomers and incorporation into triglycerides, *J. Am. Oil Chem. Soc.*, 76, 1265, 1999.

323. O'Shea, M., van der Zee, M., and Mohede, I., CLA sources and human studies, in *Healthful Lipids*, Akoh, C.C. and Lai, O.-M., Eds., AOCS Press, Champaign, IL, 2005, Chap. 12.

324. Watkins, B.A. and Li, Y., Conjugated linoleic acids (CLAs): food, nutrition, and health, in *Nutraceutical and Specialty Lipids and Their Co-Products*, Shahidi, F., Ed., Taylor & Francis, Boca Raton, FL, 2005, Chap. 12.

325. Baillargeon, M.W. and Sonnet, P.E., Selective hydrolysis by *Geotrichum candidum* NRRL Y-553 lipase, *Biotechnol. Lett.*, 13, 871, 1991.

326. McNeill, G.P. and Sonnet, P.E., Isolation of erucic acid from rapeseed oil by lipase-catalyzed hydrolysis, *J. Am. Oil Chem. Soc.*, 72, 213, 1995.

327. Kaimal, T.N.B., Prasad, R.B.N., and Chandrasekhara Rao, T., A novel lipase hydrolysis method to concentrate erucic acid glycerides in cruciferae oils, *Biotechnol. Lett.*, 15, 353, 1993.

328. Sonnet, P.E., Foglia, T.A., and Baillargeon, M.A., Fatty acid selectivity of lipases of *Geotrichum candidum*, *J. Am. Oil Chem. Soc.*, 70, 1043, 1993.

329. Sonnet, P.E., Foglia, T.A., and Feairheller, S.H., Fatty acid selectivity of lipases: erucic acid from rapeseed oil, *J. Am. Oil Chem. Soc.*, 70, 387, 1993.

330. Hayes, D.G. and Kleiman, R.R., The isolation and recovery of fatty acids with Δ5 unsaturation from meadowfoam oil by lipase-catalyzed hydrolysis and esterification, *J. Am. Oil Chem. Soc.*, 70, 555, 1993.

331. Mbayhoudel, K. and Comeau, L.-C., Obtention sélective de l'acide pétrosélinique à partir de l'huile de fenouil par hydrolyse enzymatique, *Rev. Franc. Corps Gras*, 36, 427, 1989.

332. Hayes, D.G. and Kleiman, R.R., Recovery of hydroxy fatty acids from lesquerella oil with lipases, *J. Am. Oil Chem. Soc.*, 69, 982, 1992.

333. Lie Ken Jie, M.S.F. and Syed Rahmatullah, M.S.K., Enzymatic enrichment of C20 *cis*-5 polyunsaturated fatty acids from *Biota orientalis* seed oil, *J. Am. Oil Chem. Soc.*, 72, 245, 1995.

334. Mukherjee, K.D., Lipase-catalyzed kinetic resolution for the fractionation of fatty acids and other lipids, in *Handbook of Industrial Biocatalysis*, Hou, C.T., Ed., Taylor & Francis, Boca Raton, FL, 2005, Chap. 5.

335. Shimada, Y., Nagao, T., and Watanabe, Y., Application of multistep reactions with lipases to the oil and fat industry, in *Nutraceutical and Specialty Lipids and Their Co-Products*, Shahidi, F., Ed., Taylor & Francis, Boca Raton, FL, 2005, Chap. 23.

336. Shimada, Y., Nagao, T., and Watanabe, Y., Application of lipases to industrial-scale purification of oil- and fat-related compounds, in *Handbook of Industrial Biocatalysis*, Hou, C.T., Ed., Taylor & Francis, Boca Raton, FL, 2005, Chap. 8.

337. Persson, M., Svensson, I., and Adlercreutz, P., Enzymatic fatty acid exchange in digalactosyldiacylglycerol, *Chem. Phys. Lipids*, 104, 13, 2000.

338. De Maria, L., Vind, J., Oxenbøll, K.M., Svendsen, A., and Patkar, S., Phospholipases and their industrial applications, *Appl. Microbiol. Biotechnol.*, 74, 290, 2007.

339. Ziegelitz, R., Lecithin processing possibilities, *Int. News Fats Oils Relat. Mater.*, 6, 1224, 1995.
340. Vulfson, E.N., Industrial applications of lipases, in *Lipases: Their Structure, Biochemistry and Application*, Woolley, P. and Petersen, S.B., Eds., Cambridge University Press, Cambridge, 1994, 271.
341. Morimoto, T., Murakami, N., Nagatsu, A., and Sakakibara, J., Regiospecific deacylation of glycerophospholipids by use of *Mucor javanicus* lipase, *Tetrahedron Lett.*, 34, 2487, 1993.
342. Aura, A.-M., Forssell, P., Mustranta, A., Suortti, T., and Poutanen, K., Enzymatic hydrolysis of oat and soya lecithin: effects on functional properties, *J. Am. Oil Chem. Soc.*, 71, 887, 1994.
343. Haas, M.J., Cichowicz, D.J., Phillips, J., and Moreau, R., The hydrolysis of phosphatidylcholine by an immobilized lipase: optimization of hydrolysis in organic solvents, *J. Am. Oil Chem. Soc.*, 70, 111, 1993.
344. Haas, M.J., Scott, K., Jun, W., and Janssen, G., Enzymatic phosphatidylcholine hydrolysis in organic solvents: an examination of selected commercially available lipases, *J. Am. Oil Chem. Soc.*, 71, 483, 1994.
345. Haas, M.J., Cichowicz, D.J., Jun, W., and Scott, K., The enzymatic hydrolysis of triglyceride-phospholipid mixtures in an organic solvent, *J. Am. Oil Chem. Soc.*, 72, 519, 1995.
346. Mustranta, A., Forssell, P., Aura, A.-M., Suorti, T., and Poutanen, K., Modification of phospholipids with lipases and phospholipases, *Biocatalysis*, 9, 194, 1994.
347. Mustranta, A., Forssell, P., and Poutanen, K., Comparison of lipases and phospholipases in the hydrolysis of phospholipids, *Process Biochem.*, 30, 393, 1995.
348. Na, A., Eriksson, C., Eriksson, S.-G., Österberg, E., and Holmberg, K., Synthesis of phosphatidylcholine with (n-3) fatty acids by phospholipase A2 in microemulsion, *J. Am. Oil Chem. Soc.*, 67, 766, 1990.
349. Park, C.W., Kwon, S.J., Han, J.J., and Rhee, J.S., Transesterification of phosphatidylcholine with eicosapentaenoic acid ethyl ester using phospholipase A_2 in organic solvent, *Biotechnol. Lett.*, 22, 147, 2000.
350. Härröd, M. and Elfman, I., Enzymatic synthesis of phosphatidylcholine with fatty acids, isooctane, carbon dioxide, and propane as solvents, *J. Am. Oil Chem. Soc.*, 72, 641, 1995.
351. Lilja-Hallberg, M. and Härröd, M., Enzymatic and non-enzymatic esterification of long polyunsaturated fatty acids and lysophosphatidylcholine in isooctane, *Biocatal. Biotransformation*, 12, 55, 1995.
352. Hosokawa, M., Takahashi, K., Kikuchi, Y., and Hatano, M., Preparation of therapeutic phospholipids through porcine pancreatic phospholipase A2-mediated esterification and Lipozyme-mediated acidolysis, *J. Am. Oil Chem. Soc.*, 72, 1287, 1995.
353. Vijeeta, T., Reddy, J.R.C., Rao, B.V.S.K., Karuna, M.S.L., and Prasad, R.B.N., Phospholipase-mediated preparation of 1-ricinoleoyl-2-acyl-*sn*-glycero-3-phosphocholine from soya and egg phosphatidylcholine, *Biotechnol. Lett.*, 26, 1077, 2004.
354. Han, J.J. and Rhee, J.S., Lipase-catalyzed synthesis of lysophosphatidic acid in a solvent free system, *Biotechnol. Lett.*, 17, 531, 1995.
355. Virto, C., Svensson, I., and Adlercreutz, P., Enzymatic synthesis of lysophosphatidic acid and phosphatidic acid, *Enzyme Microb. Technol.*, 24, 651, 1999.
356. Virto, C. and Adlercreutz, P., Lysophosphatidylcholine synthesis with *Candida antarctica* lipase B (Novozym 435), *Enzyme Microb. Technol.*, 26, 630, 2000.
357. Kim, J. and Kim, B.-G., Lipase-catalyzed synthesis of lysophosphatidylcholine using organic cosolvent for in situ water activity control, *J. Am. Oil Chem. Soc.*, 77, 791, 2000.
358. Brockerhoff, H., Schmidt, P.C., Fong, J.W., and Tirri, L.J., Introduction of labeled fatty acid in position 1 of phosphoglycerides, *Lipids*, 11, 421, 1974.
359. Yagi, T., Nakanishi, T., Yoshizawa, Y., and Fukui, F., The enzymatic acyl exchange of phospholipids with lipases, *J. Ferment. Bioeng.*, 69, 23,1990.
360. Svensson, I., Adlercreutz, P., and Mattiasson, B., Interesterification of phosphatidylcholine with lipases in organic media, *Appl. Microbiol. Biotechnol.*, 33, 255, 1990.
361. Svensson, I., Adlercreutz, P., and Mattiasson, B., Lipase-catalyzed transesterification of phosphatidylcholine at controlled water activity, *J. Am. Oil Chem. Soc.*, 69, 986, 1992.
362. Svensson, I., Adlercreutz, P., Mattiasson, B., Miesiz, Y., and Larsson, K., Phase behaviour of aqueous systems of enzymatically modified phosphatidylcholines with one hexadecyl and one hexyl or octyl chain, *Chem. Phys. Lipids*, 66, 195, 1993.
363. Totani, Y. and Hara, S., Preparation of polyunsaturated phospholipids by lipase-catalyzed transesterification, *J. Am. Oil Chem. Soc.*, 68, 848, 1991.
364. Mutua, L.N. and Akoh, C.C., Lipase-catalyzed modification of phospholipids: incorporation of n-3 fatty acids into biosurfactants, *J. Am. Oil Chem. Soc.*, 70, 125, 1993.

365. Sarney, D.B., Fregapane, G., and Vulfson, E.N., Lipase-catalyzed synthesis of lysophospholipids in a continuous bioreactor, *J. Am. Oil Chem. Soc.*, 71, 93, 1994.

366. Hosokawa, M. and Takahashi, K., Preparation of polyunsaturated phospholipids and their functional properties, in *Handbook of Industrial Biocatalysis*, Hou, C.T., Ed., Taylor & Francis, Boca Raton, FL, 2005, Chap. 13.

367. Joshi, A., Paratkar, S.G., and Thorat, B.N., Modification of lecithin by physical, chemical and enzymatic methods, *Eur. J. Lipid Sci. Technol.*, 108, 363, 2006.

368. Joutti, A. and Renkonen, O., The structure of phosphatidyl glycerol prepared by phospholipase D-catalyzed transphosphatidylation from egg lecithin and glycerol, *Chem. Phys. Lipids*, 17, 264, 1976.

369. Kanfer, J.N., The base exchange enzymes and phospholipase D of mammalian tissue, *Can. J. Biochem.*, 58, 1370, 1980.

370. Tanaka, Y., Takei, T., Aiba, T., Masuda, K., Kiuchi, A., and Fujiwara, T., Development of synthetic lung surfactant, *J. Lipid Res.*, 27, 475, 1986.

371. Juneja, L.R., Hibi, N., Inagaki, N., Yamane, T., and Shimizu, S., Comparative study on conversion of phosphatidylcholine to phosphatidylglycerol by cabbage phospholipase D in micelle and emulsion systems, *Enzyme Microb. Technol.*, 9, 350, 1987.

372. Lee, S.Y., Hibi, N., Yamane, T., and Shimizu, S., Phosphatidylglycerol synthesis by phospholipase D in a microporous membrane bioreactor, *J. Ferment. Technol.*, 63, 37, 1985.

373. Juneja, L.R., Hibi, N., Yamane, T., and Shimizu, S., Repeated batch and continuous operations for phosphatidylglycerol synthesis from phosphatidylcholine with immobilized phospholipase D, *Appl. Microbiol. Biotechnol.*, 27, 146, 1987.

374. Achterberg, V., Fricke, H., and Gercken, G., Conversion of radiolabelled ethanolamine plasmalogen into dimethylethanolamine and choline analogue by transphosphatidylation by phospholipase D from cabbage, *Chem. Phys. Lipids*, 41, 349, 1986.

375. Juneja, L.R., Kazuoka, T., Yamane, T., and Shimizu, S., Kinetic evaluation of conversion of phosphatidylcholine to phosphatidylethanolamine by phospholipase D from different sources, *Biochim. Biophys. Acta*, 960, 334, 1988.

376. Juneja, L.R., Kazuoka, T., Goto, N., Yamane, T., and Shimizu, S., Conversion of phosphatidylcholine to phosphatidylserine by various phospholipases D in the presence of L- or D-serine, *Biochim. Biophys. Acta*, 1003, 277, 1989.

377. Juneja, L.R., Shimizu, S., and Yamane, T., Increasing productivity by removing choline in conversion of phosphatidylcholine to phosphatidylserine by phospholipase D, *J. Ferment. Bioeng.*, 73, 357, 1992.

378. Juneja, L.R., Shimizu, S., and Yamane, T., Enzymatic method of increasing phosphatidylcholine content of lecithin, *J. Am. Oil Chem. Soc.*, 66, 714, 1983.

379. Testet-Lamant, V., Archaimbault, B., Durand, J., and Rigaud, M., Enzymatic synthesis of structural analogs of PAF-acether by phospholipase D-catalysed transphosphatidylation, *Biochim. Biophys. Acta*, 1123, 347, 1992.

380. Kokusho, Y., Tsunoda, A., Kato, S., Machida, H., and Iwasaki, S., Production of various phosphatidyl-saccharides by phospholipase D from *Actinomadura* sp. strain no. 362, *Biosci. Biotech. Biochem.*, 57, 1302, 1993.

381. Virto, C. and Adlercreutz, P., Two-enzyme system for the synthesis of 1-lauroyl-*rac*-glycerophosphate (lysophosphatidic acid) and 1-lauroyl-dihydroxyacetonephosphate, *Chem. Phys. Lipids*, 104, 175, 2000.

382. Siedow, J.N., Plant lipoxygenase: structure and function, *Annu. Rev. Plant Physiol. Plant Mol. Biol.*, 42, 145, 1992.

383. Hiruta, O., Nakahara, T., Yokochi, T., Kamisaka, Y., and Suzuki, O., Production of 9-hydroperoxy-γ-linolenic acid by soybean lipoxygenase in a two-phase system, *J. Am. Oil Chem. Soc.*, 65, 1911, 1988.

384. Luquet, M.P., Pourplanche, C., Podevin, M., Thompson, G., and Larreta-Garde, V., A new possibility for the direct use of soybean lipoxygenase on concentrated triglycerides, *Enzyme Microb. Technol.*, 15, 842, 1993.

385. Piazza, G. and Nuñez, A., Oxidation of acylglycerols and phosphoglycerides by soybean lipoxygenase, *J. Am. Oil Chem. Soc.*, 72, 463, 1995.

386. van der Heijdt, L.M., van der Lecq, F., Lachmansingh, A., Versluis, K., van der Kerk-van Hoof, A., Veldink, G.A., and Vliegenthart, J.F.G., Formation of octadecadienoate dimers by soybean lipoxygenases, *Lipids*, 28, 779, 1993.

387. Hsu, A.-F., Wu, E., Shen, S., Foglia, T.A., and Jones, K., Immobilized lipoxygenase in a packed-bed column bioreactor: continuous oxygenation of linoleic acid, *Biotechnol. Appl. Biochem.*, 30, 245, 1999.

388. Elshof, M.B.W., Veldink, G.A., and Vliegenthart, J.F.G., Biocatalytic hydroxylation of linoleic acid in a double-fed batch system with lipoxygenase and cysteine, *Fett/Lipid*, 100, 246, 1998.

389. Pinot, F., Jalaün, J.P., Bosch, H., Lesot, A., Mioskowski, C., and Durst, F., ω-Hydroxylation of Z9-octadecenoic, Z9,10-epoxystearic and 9,10-dihydroxystearic acids by microsomal cytochrome P450 systems from *Vicia faba*, *Biochem. Biophys. Res. Commun.*, 184, 183, 1992.

390. Blée, E. and Schuber, F., Efficient epoxydation of unsaturated fatty acids by a hydroperoxide-dependent oxygenase, *J. Biol. Chem.*, 265, 12887, 1990.

391. Piazza, G.J., Foglia, T.A., and Nuñez, A., Epoxidation of fatty acids with membrane-supported peroxigenase, *Biotechnol. Lett.*, 22, 217, 2000.

392. Kemp, G.D., Dickinson, F.M., and Ratledge, C., Activity and substrate specificity of fatty alcohol oxidase of *Candida tropicalis* in organic solvents, *Appl. Microbiol. Biotechnol.*, 34, 441, 1991.

393. Mauersberger, S., Drechsler, H., Oehme, G., and Müller, H.-G., Substrate specificity and stereoselectivity of fatty alcohol oxidase from the yeast *Candida maltosa*, *Appl. Microbiol. Biotechnol.*, 37, 66, 1992.

394. Thomas, H. and Oesch, F., Functions of epoxide hydrolases, *ISI Atlas Sci.: Biochem.*, 1, 287, 1988.

395. Blée, E. and Schuber, F., Occurrence of fatty acid epoxide hydrolases in soybean (*Glycine max*), *Biochem. J.*, 282, 711, 1992.

396. Blée, E. and Schuber, F., Enantioselectivity of the hydrolysis of linoleic acid monoepoxides catalyzed by soybean fatty acid epoxide hydrolase, *Biochem. Biophys. Res. Commun.*, 187, 171, 1992.

397. Blée, E. and Schuber, F., Regio- and enantioselectivity of soybean fatty acid epoxide hydrolase, *J. Biol. Chem.*, 267, 11881, 1992.

398. Blée, E. and Schuber, F., Stereocontrolled hydrolysis of the linoleic acid monoepoxide regioisomers catalyzed by soybean epoxide hydrolase, *Eur. J. Biochem.*, 230, 229, 1995.

399. Stark, A., Lundholm, A.K., and Meijer, J., Comparison of fatty acid epoxide hydrolase activity in seeds from different plant species, *Phytochemistry*, 38, 31, 1995.

400. Bellucci, G., Chiappe, C., Conti, L., Marioni, F., and Pierini, G., Substrate enantioselection in the microsomal epoxide hydrolase catalyzed hydrolysis of monosubstituted oxiranes. Effects of branching of alkyl chains, *J. Org. Chem.*, 54, 5978, 1989.

401. Stapleton, A., Beetham, J.K., Pinot, F., Garbarino, J.E., Rockhold, D.R., Friedman, M., Hammock, B.D., and Belknap, W.R., Cloning and expression of soluble epoxide hydrolase from potato, *Plant J.*, 6, 251, 1994.

402. Ibrahim, M., Hubert, P., Dellacherie, E., Magdalou, J., and Siest, G., Immobilization of epoxide hydrolase purified from rat liver microsomes, *Biotechnol. Lett.*, 6, 771, 1984.

403. Weber, N., Vosmann, K., Fehling, E., Mukherjee, K.D., and Bergenthal, D., Analysis of hydroxylated fatty acids from plant oils, *J. Am. Oil Chem. Soc.*, 72, 361, 1995.

404. Cater, M., Rhee, K.C., Hagenmaier, R.D., and Mattil, K.F., Aqueous extraction—an alternative to oil milling, *J. Am. Oil Chem. Soc.*, 51, 137, 1974.

405. Diosady, L.L., Rubin, L.J., Ting, N., and Trass, O., Rapid extraction of canola oil, *J. Am. Oil Chem. Soc.*, 60, 1658, 1983.

406. Christensen, M., Extraction by aqueous enzymatic processes, *Int. News Fats Oils Relat. Mater.*, 2, 984, 1991.

407. Rosenthal, A., Pyle, D.L., and Niranjan, K., Aqueous and enzymatic processes for edible oil extraction, *Enzyme Microb. Technol.*, 19, 402, 1996.

408. Sosulski, K. and Sosulski, F.W., Enzyme-aided vs. two-stage processing of canola: technology, product quality and cost evaluation, *J. Am. Oil Chem. Soc.*, 70, 825, 1993.

409. Smith, D.D., Agrawal, Y.C., Sarkar, B.C., and Singh, B.P.N., Enzymatic hydrolysis pretreatment for mechanical expelling of soybeans, *J. Am. Oil Chem. Soc.*, 70, 885, 1993.

410. Carr, R. and Mikle, J., Recent applications of canola processing, in *Proceedings of Latin American Congress and Exhibit on Fats and Oils Processing*, Barrera-Arellano, D., Regitano d'Arce, M.A.B., and Gonçalves, L.Ap.G., Eds., Sociedade Brasileira de Óleos e Gorduras, UNICAMP, Campinas, 1995, p. 9.

411. Fullbrook, P.D., The use of enzymes in the processing of oilseeds, *J. Am. Oil Chem. Soc.*, 60, 476, 1983.

412. Sosulski, K., Sosulski, F.W., and Coxworth, E., Carbohydrase hydrolysis of canola to enhance oil extraction with hexane, *J. Am. Oil Chem. Soc.*, 65, 357, 1988.

413. Lanzani, A., Petrini, M.C., Cozzoli, O., Gallavresi, P., Carola, C., and Jacini, G., On the use of enzymes for vegetable oil extraction. A preliminary report, *Riv. Ital. Sostanze Grasse*, LII, 226, 1975.

414. McGlone, O.C., Canales, L.A., and Carter, J.V., Coconut extraction by a new enzymatic process, *J. Food Sci.*, 51, 695, 1986.

415. Barrios, V.A., Olmos, D.A., Noyola, R.A., and Lopez-Munguia, C.A., Optimization of an enzymatic process for coconut oil extraction, *Oleagineux*, 45, 35, 1990.

416. Che Man, Y.B., Suhardiyono, A.B., Asbi, A.B., Azudin, M.N., and Wei, L.S., Aqueous enzymatic extraction of coconut oil, *J. Am. Oil Chem. Soc.*, 73, 683, 1996.

417. Che Man, Y.B., Abdul Karim, M.I.B., and Teng, C.T., Extraction of coconut oil with *Lactobacillus plantarum* 1041 1 AM, *J. Am. Oil Chem. Soc.*, 74, 1115, 1997.

418. Bocevska, M., Karlovic, D., Turkulov, J., and Pericin, D., Quality of corn germ oil obtained by aqueous enzymatic extraction, *J. Am. Oil Chem. Soc.*, 70, 1273, 1993.

419. Buenrosto, M. and Lopez-Menguia, C.A., Enzymatic extraction of avocado oil, *Biotechnol. Lett.*, 8, 505, 1986.

420. Freitas, S.P., Lago, R.C.A., Jablonka, F.H., and Hartmann, L., Extraction aqueuse enzymatique de l'huile d'avocat à partir de la pulpe fraîche, *Rev. Franc. Corps Gras*, 40, 365, 1993.

421. Ranalli, A. and De Mattia, G., Characterization of olive oil produced with a new enzyme processing aid, *J. Am. Oil Chem. Soc.*, 74, 1105, 1997.

422. Obergföll, H.-M., The use of enzymes in the extraction of olive oil, *Oleagineux Corps Gras Lipides*, 4, 35, 1997.

423. Sengupta, R. and Bhattacharyya, D.K., Enzymatic extraction of mustard seed and rice bran, *J. Am. Oil Chem. Soc.*, 73, 687, 1996.

424. Winkler, E., Foidl, N., Gübitz, G.M., Staubmann, R., and Steiner, W., Enzyme-supported oil extraction from *Jatropha curca* seeds, *Appl. Biochem. Biotechnol.*, 63–5, 449, 1997.

425. Tano-Debrah, K. and Ohta, Y., Application of enzyme-assisted aqueous fat extraction to cocoa fat, *J. Am. Oil Chem. Soc.*, 72, 1409, 1995.

426. Godfrey, T., Edible oils, in *Industrial Enzymology. The Application of Enzymes in Industry*, Godfrey, T. and Reichelt, J., Eds., Nature Press, New York, NY, 1983, p. 424.

427. Dijkstra, A.J., Degumming, refining, washing and drying of fats and oils, in *Proceedings of the World Conference on Oilseed Technology and Utilization*, Applewhite, T.H., Ed., American Oil Chemists' Society, Champaign, IL, 1992, p. 138.

428. Buchold, H., EnzyMax—a state of the art degumming process and its application in the oil industry, in *Proceedings of Latin American Congress and Exhibit on Fats and Oils Processing*, Barrera-Arellano, D., Regitano d'Arce, M.A.B., and Gonçalves, L.Ap.G., Eds., Sociedade Brasileira de Óleos e Gorduras, UNICAMP, Campinas, 1995, p. 29.

429. Simpson, T.A., Phospholipase activity in hexane, *J. Am. Oil Chem. Soc.*, 68, 176, 1995.

430. Dahlke, K., Buchold, H., Münch, E.-W., and Paultiz, B., First experiences with enzymatic oil refining, *Int. News Fats Oils Relat. Mater.*, 6, 1284, 1995.

29 Microbial Lipases

John D. Weete, Oi-Ming Lai, and Casimir C. Akoh

CONTENTS

I. INTRODUCTION

Lipases (triacylglycerol acylhydrolases, EC 3.1.1.3) are enzymes that catalyze the reversible hydrolysis of triacylglycerols (TAGs) under natural conditions. Widely distributed in animals, plants, and microbes, lipases differ from other esterases and are unique in that their activity is greatest against water-insoluble substrates and is enhanced at the substrate (oil)–water interface; that is, they exhibit "interfacial activation." Optimum activities are obtained in systems such as emulsions, where high surface areas of the substrate can be obtained. Lipases are active not only in normal-phase emulsions where the substrate is emulsified into an aqueous system (oil-in-water), but also they are also active, often more active, in invert (water-in-oil) emulsions and in reverse micelle systems containing an organic solvent solution of the substrate. Furthermore, lipases are exceedingly versatile in that they can also catalyze transesterification reactions and the stereospecific synthesis of esters, and they can act on a broad range of substrates.

Major advances have been made in our understanding of lipolytic enzymes over the past few years through solving the crystal structures of lipases from several sources. Knowledge of their structures has given insight into the mechanism of action, interfacial activation, specificity, and the nature of the active site. The high current interest in lipases, shown by the large volume of recent scientific literature on the topic, is driven by the great potential in a diversity of commercial applications for these enzymes. This chapter gives an overview of lipases and their properties with a focus on the topics of current research interest such as lipase structure and its implications in interfacial activation and selectivity. The emphasis is on microbial lipases, particularly from fungi, with references to human pancreatic lipase for comparison.

II. LIPASES FROM DIFFERENT SOURCES

A. ANIMALS

Lipases from various organs and tissues of several mammalian species have been investigated, but human and other pancreatic lipases are the most thoroughly studied. Pancreatic lipases are secreted into the duodenum and active on dietary TAGs. They are a class of structurally similar 50 kDa glycoproteins that are characterized by their specificity toward TAGs with little or no activity toward phospholipids, activation at the oil (substrate)–water interface and by colipase, and inhibition at micellar concentrations of bile salts. Pancreatic lipase can catalyze the complete breakdown of TAGs to free fatty acids (FFA) and glycerol. Nonhuman pancreatic lipases may differ from corresponding human lipases; for example, guinea pig pancreatic lipase differs in that it exhibits phospholipase A activity [1].

Other mammalian lipases have been studied and, while similar, they exhibit some characteristics that differ from those of pancreatic lipase. In addition to pancreatic lipase, fat digestion is aided by a series of lingual, pharyngeal, and gastric lipases that may be responsible for up to 50% of dietary fat breakdown [2]. Hence, in newborns, especially the premature and patients with cystic fibrosis where pancreatic lipase and bile acid levels are low, gastric lipolysis remains normal and a major portion of the fat digestion occurs without the presence of pancreatic lipase. Highly stable at low pH, these enzymes are activated by bile salts and show a preference for the sn-3 position of the substrate. Human lipoprotein lipase, which functions in hydrolyzing TAGs in chylomicrons and very low-density lipoproteins, shows many similarities to pancreatic lipase (e.g., high sequence homology, presence of the lid). However, this lipase is not responsive to colipase but instead requires apolipoprotein C-II (apoC-II) for activity, functions as a dimer, and is activated by heparin [3–5]. Hepatic lipase is confined to the liver, where it is also involved in the metabolism of lipoproteins, but not activated by apoC-II [6]. Human milk lipase, which functions in the digestion of milk fat ingested by infants, is activated by bile salt [7]. It is a glycoprotein, containing 10 wt% of carbohydrate [8], has an isoelectric point of 4 [9], and contains 12 and 10 mol% of proline and glycine, respectively in its amino acid composition [8,9]. The amount of this lipase secreted in milk is ~0.2 mg/mL or 1% of the milk protein [9].

B. PLANTS

Plant lipases have not received the same attention as those from other sources, but they have been reviewed by Mukherjee and Mills [10]. Oilseed lipases have been of greatest interest among the plant lipases, and those from a variety of plant species show differences in their substrate specificity, pH optima, reactivity toward sulfhydryl reagents, hydrophobicity, and subcellular location [11]. In wheat, rice, cereals, and barley, for example, most of the lipolytic activities are found in the embryo (germ) and aleurone layer (the bran) of the grains. These lipases are relatively specific for the native TAGs of the species from which they were isolated. They are absent from the ungerminated seed and formed during germination to hydrolyze the reserve TAGs stored in lipid bodies. This provides energy and carbon skeletons during embryonic growth.

C. MICROBES

There is substantial current interest in developing microbial lipases for use in biomedical and industrial applications because of their versatility and availability; in addition, they can be produced less expensively than corresponding mammalian enzymes. Animal and plant lipases are generally less thermostable than microbial extracellular lipases [12]. Hou and Johnston [13] screened 1229 bacteria, yeasts, actinomycetes, and fungi and found that about 25% were lipase-positive. Microbial lipases that have received the greatest attention are inducible, extracellular enzymes having properties that are generally similar to those of human pancreatic lipase, despite differences in detail. At least some of the microbes produce a mixture of extracellular lipases formed from multiple genes [14–18] and some lipases vary by degrees of glycosylation. Fungal lipases typically exist as monomers with molecular masses ranging from about 30 to 60 kDa. They vary in specificity, specific activity, temperature stability, and other properties; however, dimeric lipases have been reported (Table 29.1) [26,29]. The characteristics of fungal lipases have been reviewed by Antonian [31], whereas the versatility of fungal lipases (from genera such as *Candida*, *Geotrichum*, *Rhizopus*, and *Thermomyces*) in biotechnology was illustrated extensively by Gandhi [32], Benjamin and Pandey [33], and Pandey et al. [34]. Lipases from *Rhizomucor*, *Rhizopus*, and *Candida* are currently used commercially. Examples of such commercially available fungal lipases include Lipozyme RMIM from *Rhizomucor miehei* (Novozymes A/S, Bagsvaerd, Denmark), Lipase A "Amano" (Amano Enzyme Inc., Nagoya, Japan) and Lipolase 100T (Novozymes A/S, Bagsvaerd, Denmark) from *Aspergillus niger*, Lipomod and Lipase F-AP 15 from *Rhizopus oryzae* (Novozymes A/S, Bagsvaerd, Denmark), Lipase AYS "Amano"(Amano Enzyme Inc., Nagoya, Japan), and Lipase MY (Meito Sangyo, Tokyo, Japan) from *Candida rugosa*, Novozyme 435 from *Candida antarctica* (Novozymes A/S, Bagsvaerd, Denmark), and so on.

III.   CLASSIFICATION OF MICROBIAL LIPASES

Microbial lipases can be classified based on a comparison of the amino acid sequences as well as some fundamental biological properties [35]. Such classification allows one to predict (1) important structural features such as catalytic site residues or the presence of disulfide bonds, (2) types of secretion mechanisms and requirement for lipase-specific foldase, and (3) the potential relationship to other enzyme families. The classification will contribute to a faster identification and easier characterization of novel bacterial lipolytic enzymes. Microbial lipases have been classified into eight different families (Table 29.2).

Family I is the largest and consists of six subfamilies. Families I.1 and I.2 consist of lipases from the *Pseudomonas* family. These lipases usually show pronounced differences in regio- and enantioselectivity despite a high degree (>40%) of amino acid sequence homology [36]. Lipases from subfamily I.1, also known as true lipases have molecular masses in the range of 30–32 kDa and display higher sequence similarity to *Pseudomonas aeruginosa* lipase. Lipases from subfamily I.2 have slightly larger size (33 kDa). All lipases in subfamily I.1 and I.2 require lipase-specific chaperon or lipase-specific foldase (Lif) for expression. Lipases from subfamily I.3 secrete lipases with higher molecular mass than the previous two, lack in the N-terminal signal peptide and Cys residue. Most of the Gram-positive bacteria such as those from *Bacillus* and *Staphylococcus* family

TABLE 29.1
Properties of Some Recently Characterized Extracellular Microbial Lipases

Organism	Molecular Weight (kDa)	pH Optima	Optimum Temperature (°C)	Specific Activity (U/mg)	Specificity	References
Neurospora sp. TT-241	55	6.5	45	8203	No specificity but preferred 1- and 3-positions	[19]
Botrytis cinerea	60	6.0	38	2574	Highest activity with oleic acid esters	[20]
Pseudomonas fluorescens AK102	33	8–10	55	6200	Nonspecific for fatty acids	[21]
Rhizopus niveus 1f04759						[22]
Lipase I	34	6.0–6.5	35	4966	—	
Lipase II	30	6.0	40	6198	—	
Candida parapsilosis	160	6.5	45	—	High specificity for long-chain fatty acids, particularly polyunsaturated fatty acids	[23]
Propionibacterium acidipropionici	6–8	7.0	30	—	Preferred substrates with high saturated fatty acids	[24]
Neurospora crassa	54	7.0	30	44	Preferred triglycerides with C16 and C18 acyl chains	[25]
Pythium ultimum #144	270	8.0	30	63	1,3-Specific, preferring substrates with higher unsaturation	[26]
Rhizopus delemar ATCC 34612	30.3	8.0–8.5	30	7638	—	[27]
Fusarium heterosporium	31	5.5–6.0	45–50	2010	1,3-Specific; preferred triacylglycerols with C6–C12 fatty acids	[28]
Penicillium roquefortii	25	6.0–7.0	30	4063	Preferred triacylglycerols with C4–C6 fatty acids	[29]
Penicillium sp. uzim–4	27	7.0	25	1001	1,3-Specific, discriminates against diglycerides and active at low surface pressures	[30]

TABLE 29.2

Classification of Microbial Lipolytic Enzymes

Family	Sub-family	Enzyme-Producing Strain	Accession Number	Similarity (%) Family	Similarity (%) Subfamily	Properties
I	1	*Pseudomonas aeruginosa*[a]	D50587	100		True lipases
		Pseudomonas fluorescens C9	AF031226	95		
		Vibrio cholera	X16945	57		
		Acinetobacter calcoaceticus	X80800	43		
		Pseudomonas fragi	X14033	40		
		Pseudomonas wisconsinensis	U88907	39		
		Proteus vulgaris	U33845	38		
	2	*Bulkholderia glumae*[a]	X70354	35	100	
		Chromobacterium viscosum[a]	Q05489	35	100	
		Bulkholderia cepacia[a]	M58494	33	78	
		Pseudomonas luteola	AF050153	33	77	
	3	*P. fluorescens* SIK W1	D11455	14	100	
		Serratia marcescens	D13253	15	51	
	4	*Bacillus subtilis*	M74010	16	100	
		Bacillus pumilus	A34992	13	80	
		Bacillus licheniformis	U35855	13	80	
	5	*Geobacillus stearothermophilus*	U78785	15	100	
		Geobacillus thermocatenulatus	X95309	14	94	
		Geobacillus thermoleovorans	AF134840	14	92	
	6	*Staphylococcus aureus*	M12715	14	100	Phospholipases
		Staphylococcus haemolyticus	AF096928	15	45	
		Staphylococcus epidermidis	AF090142	13	44	
		Staphylococcus hyicus	X02844	15	36	
		Staphylococcus xylosus	AF208229	14	36	
		Staphylococcus warneri	AF208033	12	36	
	7	*Propionibacterium acnes*	X99255	14	100	
		Streptomyces cinnamoneus	U80063	14	50	
II (GDSL)		*Aeromonas hydrophila*	P10480	100		Secreted acyltransferase
		Streptomyces scabies[a]	M57579	36		Secreted esterase
		P. aeruginosa	AF005091	35		Outer membrane-bound esterase
		Salmonella typhimurium	AF047014	28		Outer membrane-bound esterase
		Photorhabdus luminescens	X66379	28		Secreted esterase
III		*Streptomyces exfoliates*[a]	M86351	100		Extracellular lipase
		Streptomyces albus	U03114	82		Extracellular lipase
		Moraxella sp.	X53053	33		Extracellular esterase 1
IV (HSL)		*Alicyclobacillus acidocaldarius*	X62835	100		Esterase
		Pseudomonas sp. B11-1	AF034088	54		Lipase
		Archaeoglobus fulgidus	AE000985	48		Carboxylesterase
		Alcaligenes eutrophus	L36817	40		Putative lipase
		Escherichia coli	AE000153	36		Carboxylesterase
		Moraxella sp.	X53868	25		Extracellular esterase 2
V		*Pseudomonas oleovorans*	M58445	100		PHA-depolymerase
		Haemophilus influenzae	U32704	41		Putative esterase
		Psychrobacter immobilis	X67712	34		Extracellular esterase
		Moraxella sp.	X53869	34		Extracellular esterase 3
		Sulfolobus acidocaldarius	AF071233	32		Esterase
		Acetobacter pasteurianus	AB013096	20		Esterase

(*continued*)

TABLE 29.2 (continued)
Classification of Microbial Lipolytic Enzymes

Family	Sub-family	Enzyme-Producing Strain	Accession Number	Similarity (%) Family Subfamily		Properties
VI		*Synechocystis* sp.	D90904	100		Carboxylesterase
		Spirulina platensis	S70419	50		
		P. fluorescens[a]	S79600	24		
		Rickettsia prowazekii	Y11778	20		
		Chlamydia trachomatis	AE001287	16		
VII		*Arthrobacter oxydans*	Q01470	100		Carbamate hydrolase
		B. subtilis	P37967	48		*p*-Nitrobenzyl esterase
		Streptomyces coelicolor	CAA22794	45		Putative carboxylesterase
VIII		*Arthrobacter globiformis*	AAA99492	100		Stereoselective esterase
		Streptomyces chrysomallus	CAA78842	43		Cell-bound esterase
		P. fluorescens SIK WI	AAC60471	40		Esterase III

Source: From Arpigny, J.L. and Jaeger, K.-E. *Biochem. J.*, 343, 1999, 177. With permission.

GDSL, Gly–Asp–Ser–(Leu); HSL, hormone-sensitive lipase.

[a] Lipolytic enzyme with known three-dimensional structure.

are classified in Groups I.4–I.6. For example, *Bacillus subtilis*, *Bacillus pumilus*, *Geobacillus stearothermophilus*, *Geobacillus thermocatenulatus*, *Staphylococcus aureus*, and *Staphylococcus hyicus* were classified in these subfamilies. The subfamily I.7 consisted of lipolytic enzymes from *Propionibacterium acnes* and *Streptomyces cinnamoneus*, which show significant similarity to each other. The central region of these proteins (residues 50–150) is ~50% similar to lipases from *Bacillus subtilis* and from subfamily I.2. No similarity was observed between the *S. cinnamoneus* lipases and other *Streptomyces* lipases known so far [35].

Family II exhibits a Gly–Asp–Ser–(Leu) [GDSL] motif containing the active site serine residues and not the conventional pentapeptide Gly–Xaa–Ser–Xaa–Gly, whereas Family III enzymes display the canonical fold of α/β-hydrolases and contain a typical catalytic triad usually formed by Ser, His, and Asp residues. This triad is functionally (but not structurally) identical with that of trypsin and subtilisin.

Several bacterial enzymes in Family IV show similar amino acid sequence to the mammalian hormone-sensitive lipase (HSL) [37]. It was once thought that the relatively high activity of enzymes in the HSL family at temperatures below 15°C was because of the sequence blocks that are highly conserved in these enzymes. However, the distinct sequence similarity between esterases from psychrophilic (*Moraxella* sp., *Psychrobacter immobilis*), mesophilic (*Escherichia coli*, *Alcaligenes eutrophus*), and thermophilic (*Alicyclobacillus acidocaldarius*, *Archeoglobus fulgidus*) indicates that the microorganisms' temperature adaptation is not responsible for the extensive conserved sequence motifs of these enzymes.

Enzymes grouped in Family V originate from mesophilic bacteria (*Pseudomonas oleororans*, *Haemophilus influenzae*, *Acetobacter pasteuriannus*) as well as from cold-adapted (*Sulfolobus acidocaldarius*) organisms. They show similarity to bacterial nonlipolytic enzymes such as epoxide hydrolases, dehalogenases, and haloperoxides, which also possess the α/β-hydrolase fold and a catalytic triad [38].

The enzymes classified in Family VI are among the smallest esterases known with molecular masses in the range of 23–26 kDa, active in dimer form, has the α/β-hydrolase fold and the typical Ser–Asp–His catalytic triad. Very little information is reported on the other enzymes in this family.

Bacterial esterases in Family VII have large molecular masses (55 kDa) and share similar amino acid sequence homology with eukaryotic acetylcholine esterases and intestine/liver carboxylesterases [35]. The esterase from *Arthrobacter oxydans* is active against the phenylcarbamate herbicides by hydrolyzing the central carbamate bond [39]. It is plasmid-encoded and has the potential to be transmitted to other strains or species.

The enzymes in the last family, Family VIII, show striking similarity to several class C β-lactamases, and has about 150 residues that are similar to *Enterobacter cloacae ampC* gene product, suggesting that the esterases in this family have an active site similar to that found in class C β-lactamases, which involves a Ser–Xaa–Xaa–Lys motif in the N terminus of both enzymes. However, Kim et al. [40] reported that the consensus sequence Gly–Xaa–Ser–Xaa–Gly in *Pseudomonas fluorescens* esterase is involved in the active site of the enzyme. This motif, which is not conserved in *Arthrobacter globiformis*, is also found in the *Streptomyces chrysomallus* esterase but lies near the C terminus and no histidine residue follows it in the sequence, implying that the order of the catalytic sequence (Ser–Asp–His) found in the entire superfamily of lipases and esterases is not conserved in this case. Obviously, more structural information is required to describe the catalytic mechanism of Family VIII esterases [35].

The attempt by Arpigny and Jaeger [35] in classifying bacterial esterases and lipases based on information available from protein and nucleotide databases showed that lipolytic enzymes display a wide diversity of properties and relatedness to other protein families, despite a highly conserved tertiary fold and sequence similarities. As more structural and kinetic information is made available from the continuing genome-sequencing projects, it is hoped that such classification will serve as a basis for a more complete and evolving bacterial lipolytic enzymes' classification.

IV. PRODUCTION, ISOLATION, AND PURIFICATION

A. LIPASE PRODUCTION

Microbial extracellular lipase production for laboratory study is typically carried out under liquid shake culture conditions or in small fermentors. Conditions for optimum lipase production seem to be variable depending on the species. Typically, TAGs (olive or soybean oil are commonly used) are placed in the culture medium to induce lipase production [26,41,42], but fatty acids (FA) may also induce lipase production [43–45]. However, lipase production has been studied in the absence of lipids (e.g., with sugars as the carbon source) [46,47]. Chang et al. [48] reported that Tween 80 and Tween 20 in the culture medium promoted lipase production and a change in the multiple forms of lipase produced by *C. rugosa*. The following is a specific example of conditions for the maximal production of lipases by *Geotrichum candidum*: growth for 24 h in liquid medium containing 1% soybean oil, 5% peptone, 0.1% $NaNO_3$, and 0.1% $MgSO_4$ at pH 7.0, 30°C, and shaking at 300 rpm [14]. In short, lipase production is influenced by both nutritional as well as physical factors and these factors are described as follows.

1. Effect of Nutritional Factors

a. *Nitrogen Sources*
Nitrogen source is required in large quantities because it amounts to ~10% of the dry weight of bacteria. It can be supplied from organic or inorganic sources in the form of ammonia, nitrate, nitrogen-containing compounds, and molecular nitrogen. Lin et al. [49] reported an extracellular alkaline lipase produced by *Pseudomonas alcaligenes* F-111 in a medium that contained peptone (1.5%, v/v) and yeast extract (0.5%, v/v), whereas Chander et al. [50] and Salleh et al. [51] showed enhanced *Aspergillus wentii* and *R. oryzae* lipase production using peptone as the nitrogen source. Organic nitrogen source such as corn-steep liquor has been found to increase the lipase production in *Humicola lanuginosa* No. 3 [52], *Syncephalastrum racemosum* [53], and *R. oryzae* [54].

Researchers have also used inorganic nitrogen sources such as ammonium nitrate and sodium nitrate in their effort to maximize lipase production in *Candida* sp. and *Yarrowia* sp. [55] and ammonium sulfate in the production of *Saccharomyces lipolytica* lipase [56].

b. Carbon Sources

Carbon sources are important substrates for the energy production in microorganisms. According to Chopra and Chander [53], fructose 1% (w/v) is a good promoter to *Syncephalastrum racemosum* lipase production but not raffinose, galactose, maltose, lactose, mannitol, and glucose. However, in certain cases, it was observed that media supplemented with carbohydrates promoted good growth of the microorganism but not lipase activity [57–59]. Nahas [58] reported that the presence of monosaccharides or dissacharides or glycerol suppressed *Rhizopus oligosporus* lipase production in the growth medium. Similar observations were made by Macrae [59] on *G. candidum* lipase using glucose as the carbon source. Lipase production occurred only after glucose has been exhausted from the medium and growth almost ceased.

c. Lipid Sources

As mentioned in Section IV.A earlier, TAGs and fatty acids both can be used to induce lipase production. Olive oil, tea oil, and oleic acid significantly enhanced the activities of *Rhizopus chinensis* lipase [60], whereas rapeseed and corn oil at 3% were found to be the best in promoting growth of *R. oryzae* and 2% was the optimal for lipase production [54]. Another interesting observation was made on *H. lanuginosa* lipase by Che Omar et al. [61]. Oleic and elaidic acid (a stereoisomer of oleic acid) were found to have different effect on lipase production in *H. lanuginosa* with oleic acid exhibiting a higher stimulating effect than elaidic acid.

d. Surfactants

Lipase–substrate–detergent interaction can be complex due to many parameters that affect the interaction such as micelle formation, concentration of free and micellar substrate, their availability to enzyme, enzyme denaturation/inactivation by detergent, degree and mode of enzyme activation by the hydrophobic interactions, and the structure of the enzyme at the water–oil interface. Cationic surface active agent such as hexadecyltrimethylammonium bromide (Cetrimide) stimulated production of *Flavobacterium odoratum* lipase 175% higher relative to the control (absence of detergent) compared with Tween 80 which increased the lipase activity to 170% of the control [62]. In contrast, Tweens and sulfonates were found to inactivate *Pseudomonas* sp. lipase completely in 90 min [63] while *Pseudomonas* sp. KWI-56 lipase was markedly inhibited by ionic surfactants such as sodium dodecylsulfate (SDS) and sodium lauryl benzenesulfonate (LBS) [64].

e. Metal Ions

Metal ions play an important role in about one-third of enzymes. They modify electron flow in a substrate or enzyme, thus effectively controlling an enzyme-catalyzed reaction. They also serve to bind and orient substrate with respect to functional groups in the active site, and provide a site for redox activity if the metal has several valence states [65]. When magnesium ions were added into the medium, lipase production in *Pseudomonas pseudoalcaligenes* F-111 [49] and *P. aeruginosa* [66] was enhanced and in *Bacillus* sp., the production of lipase was enhanced several fold when magnesium, iron, and calcium ions were added into the production media [67].

2. Effect of Physical Factors

The growth and lipase production in microorganisms are significantly affected by certain physical parameters such as incubation temperature, initial pH of medium, inoculum size, and agitation speed.

a. Incubation Temperature

Temperature is a critical parameter that has to be controlled if maximum growth and lipase production have to be achieved and it varies according to the microorganisms. *Pseudomonas* sp. lipases exhibit maximum production at 45°C, with substantial activity between 20°C and 60°C and a little activity at 80°C [63]. On the other hand, *Pseudomonas aurantiogriseum* exhibited highest lipase production at 29°C and at 26°C and 32°C, the lipase production decreased by 1–2 orders of magnitude. This result showed the importance of temperature control during fermentation, since relatively small variations can greatly influence the productivity [68]. For some microorganisms, lipase production is growth-associated, as in the case of *Acinetobacter radioresisten*, which showed the highest yield between 20°C and 30°C that coincided with the exponential growth of the microorganism [69].

b.   Initial pH of Medium

The initial pH of the medium contributes to efficient lipase production by microorganisms as reported in many published works. Optimum pH of *Pseudomonas* sp. lipase with olive oil as substrate was pH 9, a characteristic shared by many lipases from the *Pseudomonas* genus [63]. Such lipases have great potential in the detergent industry, since it has the capability to function at alkaline pH.

c.   Inoculum Size

The effect of inoculum size on lipase production can be demonstrated experimentally by using different inoculum sizes. The lipase production rate by *Rhizopus delemar* is highest at low inoculum sizes, since the rate of oxygen transfer in the flask can be rate limiting at high inoculum size due to high oxygen consumption uptake by the microorganisms [70]. In addition, greater inoculum size can lead to an impaired nutrient supply and thereby decreasing production capacity [71].

d.   Shaking Conditions

Shaking of growth culture during cultivation also gave contradictory effect on different micro-organisms. Lipase production increases with the shaking condition literally. Shaking enhanced production of both intracellular and extracellular lipases by *R. oryzae* [51]. For extracellular lipase production, 100 rpm was the optimal shaking rate, whereas for intracellular lipases, 150 rpm was the best. In shaken cultures, lipase production was enhanced by 70% for *Penicillium chrysogenum* [72], 50% for *Aspergillus wentii* [50], while the stationary culture experienced decreased growth and poor lipase production [73]. In contrast, when *S. racemosum* was grown in the nutrient medium in static condition, the enzyme production per unit growth was comparatively high [53].

B.   ISOLATION AND PURIFICATION

Extracellular lipases have been isolated, purified, and characterized from numerous microbial species. Some examples of recently characterized lipases are illustrated in Table 29.1. The specific isolation and purification methodology differs in detail from study to study, but it generally involves the ammonium sulfate precipitation of proteins from the culture medium after removal of cells or mycelium, and then fractionating the proteins through a series of ion exchange, affinity, and gel filtration columns. About 80% of purification schemes attempted and reported so far in the literature have used a precipitation step, with 60% of these using ammonium sulfate and 35% using ethanol, acetone, or an acid (usually hydrochloric) [74]. The precipitation is usually used as an initial step to purification followed by chromatography separation. Increase in lipase activity depends on the concentration of ammonium sulfate solution used. This step is less affected by interfering non-protein materials than chromatographic methods. Chromatographic separation involves passing a solution of enzymes through a medium that shows selective absorption for different solutes. Most of the time, a combination of chromatographic steps has proved to be efficient in purifying enzymes to homogeneity. The most common column used in the first chromatography step was the ion exchange chromatography. Gel filtration chromatography, which separates protein based on molecular size, is usually used after ion exchange chromatography. Affinity chromatography is an efficient technique used to purify enzymes based on ligand binding, but only 27% of studies reported had used affinity chromatography as a purification step. This is due to the high cost of the resins used. Most researchers preferred the less expensive resins such as those used in ion exchange and gel filtration chromatography. Several recent publications can be consulted for specific procedures for isolating lipases [20,21,26–29,65,75–78]. The following is given as an example of a lipase from *Neurospora* sp. TT-241 that was isolated and purified 371-fold [19].

1. *Ammonium sulfate fractionation.* Add solid ammonium sulfate to the clarified aqueous extract (or culture medium) up to 60% saturation. The resulting precipitate is collected by centrifugation at 12,000 × *g* for 30 min and dissolved in a minimal volume of buffer A (50 mM phosphate buffer, pH 7.0). The enzyme solution is dialyzed overnight against a 50-fold excess of the same buffer.

2. *Sephadex G-100 gel filtration chromatography.* The enzyme solution (70 mL) is applied to a Sephadex G-100 column (4 cm × 120 cm), eluted with buffer A at a flow rate of 40 mL/h, and 8 mL fractions are collected.

3. *Toyopearl phenyl-650M column chromatography.* Add ammonium sulfate to the pooled active fractions from gel filtration to a final concentration of 1 M. The enzyme solution is then applied to a Toyopearl phenyl-650M column (2.0 cm × 18 cm) that has been preequilibrated with buffer A containing 1 M ammonium sulfate. The column is washed with 300 mL of buffer A containing 1 M ammonium sulfate and then eluted with 1 L of a linear gradient from 1 to 0 M ammonium sulfate in buffer A at a flow rate of 60 mL/h; 4 mL fractions are collected.

4. *Ultrogel-HA hydroxyapatite column chromatography.* The pooled active fractions from the Toyopearl column are further purified by passage through an Ultrogel-HA column (3.0 cm × 9 cm), preequilibrated with 10 mM phosphate buffer (pH 7.0). After being washed with 10 mM phosphate buffer, the column is eluted with a 1 L linear gradient from 10 to 500 mM phosphate buffer at a flow rate of 50 mL/h, and 5 mL fractions are collected. The active fractions are pooled and stored at −20°C.

These procedures are carried out at 4°C. The purity of lipase active fractions is determined by SDS polyacrylamide gel electrophoresis.

V. ASSAY OF LIPASES

A. REVIEW OF EMULSION SYSTEMS

Lipases are assayed in emulsions where high substrate surface areas can be achieved. Since, in addition to high interfacial area, the chemical and physical environments at the substrate–water interface are important to the adsorption and activity of lipases, it is useful to understand the nature of the different types of emulsions used for lipase assay or in various applications. Emulsions are mixtures of two immiscible liquids (e.g., oil and water); one of the components is dispersed as very small droplets, or particles, and the mixture is stabilized by a surface active agent, or surfactant. Emulsions are classified as macro- or microemulsions where the dispersed particles are either greater or smaller than 1 μm in diameter, respectively. Macroemulsions are turbid, milky in color, and thermodynamically unstable (i.e., they will ultimately separate into the two liquid phases). On the other hand, microemulsions are homogeneous and stable.

Depending on the conditions of formation, and particularly the nature of the surfactant, macro-emulsions may be normal phase (oil-in-water) or reversed phase (water-in-oil, invert). In the former, the oil is emulsified into the aqueous phase, and the surfactant forms a monolayer film around the dispersed oil droplets, whereby the hydrophobic moiety of the surfactant extends into the oil and the polar moiety is at the droplet surface (Figure 29.1). In reversed-phase emulsions, the aqueous phase is dispersed in the oil with the orientation of the surfactant molecules reversed (Figure 29.1).

Amphiphilic (surfactant) molecules undergo self-organization into spheroidal particles when dissolved in certain organic solvents such as isooctane with the polar head groups oriented inward

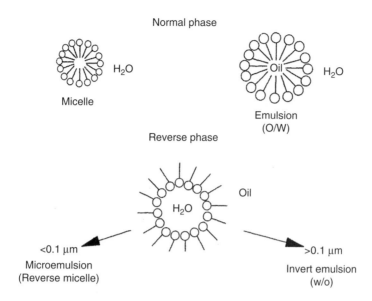

FIGURE 29.1 Diagrammatic representation of globules (particles) in various types of emulsions. In the ball-and-stick model of amphiphilic (surfactant) molecules, the ball represents the polar head group and the stick the hydrophobic region of the molecule.

and hydrophobic tails outward. These particles are referred to as reverse micelles and are <1 µm in diameter (Figure 29.1). Water can be "solubilized" in the organic solvent by becoming entrapped in the particles at up to several dozens of molecules per molecule of surfactant. Typically, reverse micelles are formed when the molar ratio of water to surfactant is <15, which is expressed as water activity, w_o (i.e., molarity of water to molarity of surfactant) or R. Enzymes such as lipases can be entrapped within the aqueous phase particles of the invert emulsions or reverse micelles of microemulsions, where they retain their activity. In the latter case, the enzymes are isolated from the organic solvent. Luisi et al. [79] and Sanchez-Ferrer and Garcia-Carmona [80] can be consulted for more information on microemulsions.

B. NORMAL-PHASE EMULSIONS

Most lipase assays for hydrolytic reactions are carried out in normal-phase emulsions whereby the water-insoluble oil substrate is emulsified by sonication into a buffered aqueous enzyme preparation. The emulsion is stabilized with an emulsifier and the aqueous phase often contains Ca^{2+}. The specific ingredients, buffer types, pH, and relative amounts of components vary widely with the author and specific lipase being assayed. An example of such an assay is as follows: 2 mL 0.2 M Tris maleate–NaOH buffer (pH 8.2), 1 mL 0.03 M Ca_2Cl, 5 mL distilled water, 1 mL olive oil, and 1 mL enzyme solution. The emulsion is incubated at 30°C for 60 min or an appropriate time depending on lipase activity, and the reaction is stopped with 20 mL of acetone ethanol (1:1). Common emulsifiers used in lipase assays include polyvinyl alcohol and gum arabic at 1%–2% by volume of assay mixture. Hydrolytic activity is most often determined by titration of the fatty acid products of the reaction with NaOH. Other methods and assay conditions have been reviewed by Jensen [81].

C. INVERT EMULSIONS

Using lipases from several fungi, Mozaffar and Weete [82] reported a reaction mixture containing 5 mL olive oil, 0.1 mL 520 mM taurocholic acid in 50 mM sodium phosphate buffer (pH 7.5), and 0.1 mL enzyme preparation. Final concentrations of taurocholic acid and water in the 5.2 mL of reaction mixture were 10 mM and 4%. The mixture was emulsified by vortexing for about 30 s and incubated at 45°C without shaking for 30 min, whereon the reaction was stopped as described earlier. Up to 90% of the substrate was hydrolyzed under these conditions by *C. rugosa* lipase when the incubation time was extended to 48 h with periodic additions of buffer (0.2 mL) during the incubation.

D. MICROEMULSION: REVERSE MICELLES

1. Hydrolysis

The desired amount of concentrated buffered lipase solution is poured into 5 mL of 50 mM aerosol optical thickness (AOT) [aerosol-OT, bis (2-ethylhexyl) sodium sulfosuccinate]–isooctane solution containing 10% v/v of the substrate. The amount of enzyme solution depends on the R value (e.g., 10.5). The reaction is initiated by vortexing until clear, and the mixture is incubated at 30°C for 15 min; then 0.4 mL is added to 4.6 mL benzene and 1.0 mL cupric acetate–pyridine solution, and the reaction is stopped by vortexing [83]. Fatty acids liberated by the hydrolytic reaction are determined according to Lowry and Tinsley [84]. Other examples of studies involving hydrolysis by lipases in organic solvent–reverse micelle systems are cited in Table 29.3 [83,85–126].

2. Transesterification

Using a lipase from *R. delemar*, Osterberg et al. [27] used the following system for transesterifing TAGs with stearic acid: isooctane (91.65 wt%) was mixed with AOT at 100–200 mM, aqueous 0.066 M phosphate buffer (pH 6) (1.0%), and substrate (5.0%). The enzyme in the buffer was used at 1.5 U/mg substrate. The reaction was carried out at 35°C under nitrogen with magnetic stirring, and was stopped by raising the temperature to 100°C and holding at that temperature for 10 min. Other examples of studies involving transesterification and synthetic reactions in organic solvent–reverse micelle systems are cited in Table 29.3.

TABLE 29.3

Representative Applications of Microbial Lipases Using Organic Solvent Reverse Micelle Systems

Reaction Type	Topic	References[a]
Synthesis	Wax ester synthesis in a membrane reactor with lipase–surfactant complex in hexane	[85]
	Formation of polyol–fatty acid esters	[86]
	Esterification reactions catalyzed by lipases in microemulsions: role of enzyme localization in relation to selectivity	[87]
	Esterification of oleic acid with glycerol in monolayer and microemulsion systems	[88]
	Synthesis of sugar esters	[89]
	Polyunsaturated fatty acid glyceride synthesis	[90]
	Monoacylation of fructose	[91]
	Double enantioselective esterification of racemic acids and alcohols	[92]
	Esterification of oleic acid and methanol in hexane	[93]
	Esterification of glycerol: synthesis of regioisomerically pure 1,3-*sn*-diacylglycerols and monoacylglycerols	[94,95]
	Synthesis of DHA-rich triglycerides	[96]
	Synthesis of acylated glucose	[97]
	Synthesis of mono- and diglycerides	[98]
	Concentration of EPA by selective esterification	[99]
	Esterification of lauric acid	[100]

TABLE 29.3 (continued)

Representative Applications of Microbial Lipases Using Organic Solvent Reverse Micelle Systems

Reaction Type	Topic	References[a]
Interesterification	Interesterification of butterfat	[101]
	Interesterification of triglycerides	[102]
	Enzymatic interesterification of triolein in canola lecithin–hexane reverse micelles	[103]
	Alteration of melting point of tallow–rapeseed oil	[104]
	Interesterification of phosphatidylcholine	[105]
	Incorporation of long-chain fatty acids into medium-chain triglycerides	[106]
	Kilogram-scale ester synthesis	[107]
	Incorporation of n-3 polyunsaturated fatty acids into vegetable oils	[108]
	Incorporation of EPA and DHA into groundnut oil	[109]
	Interesterification of milk fat with oleic acid	[110]
	Incorporation of exogenous DHA into bacterial phospholipids	[111]
	Interesterification of triglycerides and fatty acids by surfactant-modified lipase	[112]
	Transesterification of rapeseed oil and 2-ethylhexanol	[113]
	Transesterification of cocoa butter	[114]
	Modification of phospholipids	[115]
	Diacylglycerol formation	[116]
	Transesterification of palm oil	[117]
	Transesterification of high oleic sunflower oil	[118]
	Transesterification of sunflower oil	[119]
	Transesterification of trilinolein and trilinolenin with selected phenolic acids	[120]
Hydrolysis	Hydrolysis of phosphatidylcholine by an immobolized lipase	[121]
	Palm kernel olein hydrolysis	[122]
	Olive oil hydrolysis	[83,123]
	Production of polyunsaturated fatty acid-enriched fish oil	[124]
	Enrichment of γ-linolenic acid from evening primrose oil and borage	[125,126]

[a] The references cited are representative; they are not intended to be inclusive.

VI. PROPERTIES AND REACTIONS

A. STRUCTURE

Over the past several years, the crystal structures of several mammalian and microbial lipases have been determined. Generally, lipases are α/β-proteins with a central core of a mixed β-sheet containing the catalytic triad composed of Ser⋯His⋯Asp, and a surface loop restricting access of the substrate to the active site.

1. Animal Lipases

Human pancreatic lipase is folded into two domains, a larger N-terminal domain (comprising residues 1–335) and a smaller C-terminal domain (residues 336–449). The core of the N domain is formed by a nine-stranded, β-pleated sheet in which most of the strands run parallel to one another. Seven α-helical segments of varying length occur in the strand connections, and six of them pack against the two faces of the core sheet. The C domain is formed by two layers of antiparallel sheets, the strands of which are connected by loops of varying length. The N domain contains the active site, a glycosylation site, a Ca^{2+}-binding site, and possibly a heparin-binding site. The active site is buried beneath a short amphipathic α-helical surface loop, termed the "flap" or "lid" [127,128]. Colipase (see later) binds exclusively to the C domain of the protein through hydrophobic interactions and ion pairing [129], and the binding does not induce a conformational change [130–132]. Calcium ions activate and reduce the lag phase of human pancreatic lipase, particularly for mixed bile acid–lipase complexes [133].

The structures of pancreatic lipase from other systems have been recently reported, including guinea pig [1] and horse [134]. Overall, the nonhuman pancreatic lipases are structurally similar to that from humans but may differ in detail. For example, the guinea pig pancreatic lipase does not possess the lid that is typical of other lipases [1]. In addition, unlike the pancreatic lipases from humans and pigs, the horse enzyme is not glycosylated [134].

2. Fungal Lipases

The crystal structures of several fungal lipases have also been determined: for example, *Rhizomucor miehei* [135,136], *G. candidum* [137,138], *C. rugosa* [139], *C. antarctica* [140], and *H. lanuginosa* [141]. Although there are no obvious sequence similarities between pancreatic and fungal lipases, except for the Gly–X–Ser–X–Gly consensus sequence in the active site region [31,134], there are structural similarities. Fungal lipases are α/β-proteins and have similar topologies based on a large central mixed β-pleated sheet pattern, mainly parallel, but the connectivities between strands vary. The two lipases (see later) of *G. candidum* have 554 amino acids and share about 85% sequence homology [16,142]. In addition, many fungal lipases possess the serine protease catalytic triad Ser$\cdots$His$\cdots$Asp in their active sites; however, glutamic acid is substituted for aspartic acid in the lipases from *G. candidum* [137] and *C. rugosa* [139]. Lipase II gene from *G. candidum* has been cloned [16,143], and the results of probing the active site by site-directed mutagenesis are consistent with x-ray crystallography data in that the Ser$\cdots$His$\cdots$Glu is the active site [144].

Fungal lipases also possess a lid that prevents access of the substrate to the active site, form a functional oxyanion hole, and have an interfacial-binding site (Gly–X–Ser–X–Gly), but the motion of the lid differs between fungal and pancreatic lipases. Lid rearrangements by human pancreatic lipase [131,132], *C. rugosa* [139], and *G. candidum* [137] involve more than one loop. The lid is closed and covers the active site in *G. candidum* and is open in *C. rugosa*; otherwise, lipases from these two sources are very similar in sequence and structure. The three-dimensional structure of the extracellular lipase from *Rhizopus delemar* based on x-ray crystallographic coordinates is shown in Figure 29.2, with the lid partially open exposing the active site residues Ser[145], Asp[204], and His[257] (Figure 29.2A) and residues 86–92 in the partially closed conformation (Figure 29.2B) [144].

3. Bacterial Lipases

Lipases from several *Pseudomonas* species have been crystallized, and preliminary x-ray crystallographic analyses of such species as *Pseudomonas cepacia* [145], *Pseudomonas glumae* [146], and *P. fluorescens* [147] have been conducted. Their molecular masses, which are in 30–35 kDa range, show extensive sequence homology to one another but little to those of other lipases; the common G–X–S–X–G sequence in the active site region is conserved, and activity appears at the substrate–water interface. On the other hand, lipases from *Bacillus* species have molecular weights at about 19 kDa, have the sequence A–X–S–X–G instead of the characteristic sequence, and do not exhibit interfacial activation [148].

B. INTERFACIAL ACTIVATION AND THE HYDROLYTIC REACTION

In 1958, Sarda and Desnuelle [149] showed that pancreatic lipase does not exhibit normal Michaelis–Menten kinetics with respect to substrate concentration. Lipases are inactive in aqueous media with the substrate present in its monomeric form, but there is a sharp increase in activity when the

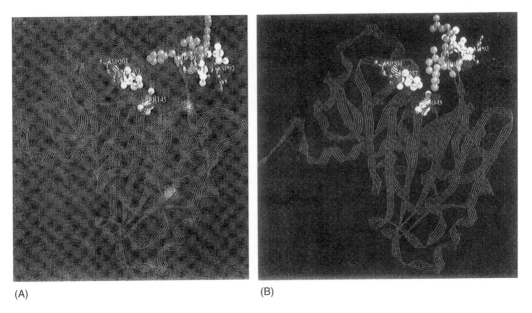

(A)   (B)

FIGURE 29.2 Three-dimensional representation of the extracellular lipase from the fungus *Rhizopus delemar* showing the lid partially open (A) and closed (B). (From Mutua, L.N. and Akoh, C.C., *J. Am. Oil Chem. Soc.* 70, 125, 1993.) (Plates courtesy of Dr. Michael Haas, U.S. Department of Agriculture.)

substrate exceeds the critical micelle concentration. The inactive enzyme must first adsorb to the surface of the bulk substrate, which initiates interfacial activation (Figure 29.3). Interfacial enzyme kinetics in lipolysis was reviewed by Verger and de Haas [150].

Solving the crystal structures of lipases has given insight into the mechanism of interfacial activation [151]. It is believed that the preferred conformation in aqueous solution is with the lid covering the active site, thus denying access to the substrate. Adsorption of the lipase to the interface involves a conformational change in the enzyme whereas the lid, which covers a cavity containing the active site and is held in place by mostly hydrophobic and some hydrogen bonds, undergoes reorientation. This is accompanied by additional conformational changes that expose the active site and a larger hydrophobic site, and allow access of the substrate to the active site. The working hypothesis for the mechanism of interfacial activation is based on the three-dimensional structural

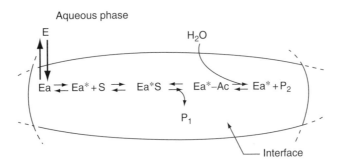

FIGURE 29.3 Schematic representation of the adsorption and hydrolytic activities of lipases at the oil (below the plane) and aqueous (above the plane) interface: E, enzyme in the aqueous phase; Ea, enzyme adsorbed at the interface; Ea*, activated enzyme; S, Substrate; Ea*S, enzyme–substrate complex; Ea*–Ac, acylenzyme; P_1, product (diacylglycerol); P_2, product (fatty acid).

analyses of lipases from *Rhizomucor miehei* [135,152], human pancreatic lipase [128], *G. candidum* [137,138], and *C. rugosa* [139]. The two latter lipases, for example, are globular, single-domain proteins built around an 11-stranded-mixed β-sheet with domains of 45 Å × 60 Å × 65 Å [153].

Activation of *R. miehei* lipase involves the movement of a 15 amino acid long lid in a hinge-type, rigid-body motion that transports some of the atoms of a short α-helix by more than 12 Å [152]. This, combined with another hinge movement, results in the exposure of a hydrophobic area representing 8% of the total molecular surface. In *C. rugosa*, comparison of open [139] and closed [154] conformation indicates that activation of the lipase requires the movement and refolding, including a *cis*-to-*trans* isomerization of a proline residue, of a single surface loop to expose a large hydrophobic surface where the substrate likely interacts. Lid reorganization contributes to the formation of a catalytically competent oxyanion hole and creation of a fully functional active site [153]. The scissile fatty acyl chain is bound in a narrow, hydrophobic tunnel, where modeling studies suggest that the substrate must adopt a tuning fork conformation [153,155]. There is a tryptophan residue at the tip of the lid (Trp^{89}) in the lipase from *H. lanuginosa* that plays an important role in hydrolytic activity [156]. When Trp^{89} is substituted with other amino acids, activity drops substantially and variably depending on the substituent amino acid.

Visualizing how the hydrophobic substrate molecule, which is buried in the oil surface, gains access to the active site, which is buried within the water-soluble lipase molecule is difficult. Blow [157] has suggested that on activation, a "hydrophobic seal" forms at the interface that allows the substrate to enter the active site without interacting with the bulk water; that is, the enzyme partially withdraws the substrate molecule from the bulk oil with at least some of the acyl chains projecting into the lipid.

Cygler et al. [153] have postulated that there are two tetrahedral intermediates in the lipase-catalyzed hydrolysis of esters. Formation of a noncovalent Michaelis complex between the lipase and TAG is followed by formation of a tetrahedral, hemiacetal intermediate resulting from a nucleophilic attack by the serine O^γ. The oxyanion resulting from formation of the enzyme–substrate complex is stabilized by the amide groups of the oxyanion hole (e.g., Gly^{123} in *C. rugosa*) and α-helix following the active site serine [153]. Intermediate formation is followed by cleavage of the substrate ester bond, breakdown of the tetrahedral intermediate to the acyl enzyme, and protonation and dissociation of the diacylglycerol (DAG). The serine ester of the acylated enzyme is attacked by an activated water molecule to form a second tetrahedral intermediate, which is cleaved to give rise to the protonated enzyme (serine residue) and fatty acid.

C. ACTIVATION/INHIBITION

Calcium may stimulate lipase-catalyzed hydrolytic activity by (1) binding to the enzyme resulting in a change in conformation, (2) facilitating adsorption of the lipase to the substrate–water interface, and/or (3) removing from the interface fatty acid products of hydrolysis that may reduce end-product inhibition of the reaction. Activation of human pancreatic lipase by calcium is complex and variable depending on the substrate and presence or absence of bile acids [133]. Calcium effects on microbial lipase activity may be variable depending on the enzyme source and assay conditions. For example, stimulation of *C. rugosa* lipase activity was attributed to the formation of calcium salts of fatty acid products in a normal-phase emulsion, with olive oil as the substrate but not tributyrin; however, calcium had no effect in an invert emulsion [158]. In a nonemulsion system (i.e., without emulsifier), calcium had no effect on *C. rugosa* lipase activity with olive oil as the substrate but tended to offset the inhibitory effects of bile acid (Mozaffar and Weete, unpublished).

Human pancreatic lipase is inhibited by the bile salts, and the inhibition can be overcome by 10 kDa protein colipase. Bile salt coating of the substrate micelles creates a negatively charged surface that is believed to inhibit adsorption of the bile salt–lipase complex to the interface [159]. Colipase overcomes the inhibitory effect of the bile salts through formation of a 1:1 complex with lipase that facilitates adsorption at bile salt-coated interfaces [130,159]. Naka and Nakamura [160] found that although the bile salt sodium taurodeoxycholate inhibited pancreatic lipase activity when tributyrin was the substrate, a result that has been widely reported and cited by others, and colipase could reverse the inhibition, the bile salt actually stimulated hydrolytic activity when triolein was the substrate. This was attributed to the fact that triolein is a more natural substrate for the lipase than tributyrin. On the other hand, when sodium taurocholate was added to an emulsion assay of the lipase from *C. rugosa* with olive oil as the substrate, activity was progressively inhibited from 0.1 to 0.8 mM concentration of the bile salt [161]. Relatively high activity at the lowest concentration was attributed to the role of the bile salt in the stabilization of oil particles in the emulsion and providing high interfacial area for adsorption of the lipase, and inhibition was due to interaction with the enzyme such that adsorption was reduced.

A variety of substances have been shown to inhibit lipase activity; examples include anionic surfactants, certain proteins, metal ions, boronic acids, phosphorus-containing compounds such as diethyl *p*-nitrophenyl phosphate, phenylmethyl sulfonylfluoride, certain carbamates, β-lactones, and diisopropylfluorophosphate [162].

D. Selectivity

Lipases can be separated into three groups according to specificity [163,164]. The first group shows no marked specificity with respect to the position of the acyl group on the glycerol molecule, or to the specific nature of the fatty acid component of the substrate. Complete breakdown of the substrate to glycerol and fatty acids occurs with nonspecific lipases. Examples of such lipases are those from *C. (cylindracae) rugosa*, *Corynebacterium acnes*, and *Staphylococcus aureus*. The second group attacks the ester bonds specifically at the 1- and 3-positions of the substrate, with mixtures of DAG and monoacylglycerol (MAG) as products. Because of the instability of intermediate 1,2-DAG, 2, 3-DAG, and 2-MAG (i.e., migration of the fatty acid from the 2-position to the 1- or 3-position), these lipases may catalyze the complete breakdown of the substrates. The positional specificity of lipases on *sn*-1 and *sn*-3 positions of the TAGs is due to the steric hindrance conflict that prevents the fatty acid located at the *sn*-2 from binding to the active sites [165,166]. Most microbial lipases fall into this group; examples include those from *Aspergillus niger*, *Rhizopus delemar* (*oryzae*), *Rhizomucor miehei*, and *Mucor javanicus*. Members of the third group of lipases show preference for a specific fatty acid or chain length range, and are less common. The most widely studied lipase in this regard is that from *G. candidum*, which shows specificity for long-chain fatty acids with a *cis* double bond in the C9 position (see later) [167]. Other lipases that show some preference for specific fatty acids are those from *Candida rugosa* (C18:1 *cis*-9), strains of *A. niger* (C10 and C12 or C18:1 *cis*-9), *Mucor miehei* (C12), *Rhizopus arrhizus* (C8, C10) [164], *H. lanuginosa* #3 [52], and human gastric lipase (C8 and C10) [168]. A lipase from *Penicillium camembertii* hydrolyzes only MAG and DAG [169]. The positional and FA selectivity of some fungal lipases are given in Table 29.4.

Many new bacterial lipolytic enzymes have been studied and published but there has been very few attempts made to organize all the information published. Usually, lipolytic enzymes are characterized by their ability to catalyze a broad range of reactions and the different assay methods used such as pH-stat, monolayer technique, and hydrolysis of *p*-nitrophenyl esters prevented direct comparison of the results obtained. To standardize the methods used, comparative studies have been conducted [182] but these studies only focused on a small number of bacterial lipases [183]. However, through the elucidation of many gene sequences and the resolution of numerous crystal structures, efforts have been made to classify the lipolytic enzymes. The work of Arpigny and Jaeger [35] as described in Section III is an example of such efforts. Arpigny and Jaeger [35] attempted to classify the lipolytic enzymes based on conserved sequence motifs and to relate them to the three-dimensional structural elements involved in substrate recognition and catalysis. Shimada [184] categorized industrial lipases into five groups based on the homologies of their primary structures deduced from the nucleotide sequences (Table 29.5). According to him, homology of the primary

TABLE 29.4

Positional and Fatty Acid Selectivity of Some Fungal Lipases

Organism	Positional Selectivity	Fatty Acid Selectivity	References
Aspergillus niger	1,3 ≫ 2	10:0; 12:0	[164]
Candida rugosa	Nonspecific	18 (*cis*-9)	[164]
Candida antarctica A	2-Specific	—	[170]
Geotrichum candidum	Nonspecific	18:1 (*cis*-9)	[171]
		10:0 and 18:0	[172]
		cis-9 Unsaturated acids	[76]
		double bond *cis*-9	[18]
	Double bond *cis*-9		
	2-Specific and nonspecific	Triolein and tricaprylin	[173]
		cis-Δ^9 Fatty acid	[174]
		Triolein, tricaprylin, and methyl oleate	[175]
		Triolein and methyl oleate inside ester bond of triolein	[176]
Humicola lanuginosa	1,3-Specific	12:0	[52]
Mucor javanicus	1,3 > 2	MCFA, LCFA > SCFA	[177]
Mucor miehei	1,3-Specific	12:0	[178]
		PUFA	[179]
Neurospora crassa	1,3-Specific	16:0, 18:0	[25]
Rhizomucor miehei	1 > 3 ≫ 2	SCFA > MCFA, LCFA	[177]
Rhizopus arrhizus	1,3-Specific	16:0, 18:0	[180]
		18:1 (*cis*-6)	[181]
		SCFA, MCFA > LCFA	[177]
Rhizopus delemar	1,3 ≫ 2	MCFA, LCFA ≫ SCFA	[177]
Penicillium camembertii	1,3-Specific	MAG, DAG > TAG	[177]

Abbreviations: PUFA, polyunsaturated fatty acids; SCFA, short-chain fatty acid; MCFA, medium-chain fatty acids; LCFA, long-chain fatty acid; MAG, monoacylglycerols; DAG, diacylglycerols; TAG, triacylglycerols.

structure shows similarity of tertiary structure, and lipases with similar tertiary structure are assumed to have similar properties. It is interesting to note that although Shimada had taken a simpler approach to categorizing the lipolytic enzymes, there were some similarities in his classification with that of Arpigny and Jaeger [35].

Negative selectivity has also been observed where the lipase from *Candida cylindracea* (*rugosa*) selected against TAG molecules containing docosahexanoic acid (DHA) [185]; *G. candidum* lipase preparations have shown similar discrimination for γ-linolenate (GLA) in borage oil [186], and erucic acid from rapeseed oil [187]. Lipases from *Brassica napus* and *M. miehei* were also found to discriminate against polyunsaturated acids, such as GLA and DHA [179].

Lipases that are specific to FA for their chain lengths and unsaturation in catalyzing hydrolysis reaction of TAG have been used to develop concentrated forms of FFA including polyunsaturated fatty acid (PUFA) and GLA. Shimada et al. [188] used lipase of *G. candidum* to produce PUFA-rich oil from tuna oil. The principle involved is that if there is a lipase that does not digest PUFA esters, then the PUFA will be concentrated in glycerides by hydrolyzing PUFA-containing oil with that lipase. Thus, by using the selective specificity of a lipase, certain FA can be enriched in glycerides. The work done by Shimada et al. [188] proved that *G. candidum* lipase hydrolyzed palmitic and oleic acids more effectively than *Fusarium heterosporum* and *C. cylindracea* lipases, and less effectively for eicosapentanoic acid (EPA) and DHA esters. Therefore, *G. candidum* lipase enriched EPA and DHA as its reactivities on DHA and EPA esters were very low. The substrate selectivity and

TABLE 29.5
Classification of Industrial Lipases Based on Their Primary Structure

Group	Microorganism	Property
Bacteria		
Group 1	*Bulkholderia cepacia*	Positional specificity: nonspecific or 1,3-position preferential
	Bulkholderia glumae	
	Pseudomonas aeruginosa	Acts somewhat on PUFA
Group 2	*Pseudomonas fluorescens*	Positional specificity: nonspecific or 1,3-position preferential
	Serratia marcescens	Acts somewhat on PUFA
Yeast		
Group 3	*Candida rugosa*	Positional specificity: nonspecific
	Geotrichum candidum	Act very weakly on C20 FA and PUFA
		FA specificity: relatively strict
		Acts on sterol and L-menthol
		Hydrolysis activity: strong
Group 4	*Candida antarctica*	Positional specificity: 1,3-position preferential, 1,3-position specific in a reaction
		Acts strongly on PUFA and short-chained alcohols
Fungi		
Group 5	*Rhizomucor miehei*	Positional specificity: 1,3-position specific
	Rhizopus oryzae	
	Thermomyces lanuginose	Acts weakly on PUFA
	Fusarium heterosporum	Acts strongly on C8–C24 saturated and monoenoic FA
	Penicillium camembertii	Positional specificity: 1,3-position specific
		Does not act on TAG

Source: From Shimada, Y. in *Handbook of Functional Lipids*, C.C. Akoh, ed., CRC Press, Boca Raton, FL, 2006, 437–455. With permission.

regioselectivity of lipase can be exploited advantageously for use in structural determination of TAG, synthesis of a specific and defined set of MAG and/or DAG [108], and for the preparation of FA.

In addition, there have been numerous studies on the enantioselectivity catalyzed by lipases and esterases for the formation of optically pure chiral compounds. The fungal lipase from *C. rugosa* has been shown to hydrolyze octyl-2-chloropropionate with high stereoselectivity on a large scale, yielding 46% (R) acid and 45% of (S) ester [189]. The formation of optically active synthons and the ready availability of such materials for synthesis are important because optically pure end products are often prerequisite for incorporation in ethical drug formulations and important agricultural aids in the future [190].

In some instances, it is possible to couple FA selectivity to positional selectivity, for example, when a natural TAG has a high proportion of a particular FA in the primary positions. In one case, *R. delemar* lipase was employed in a biomembrane reactor to remove erucic acid from the primary positions of crambe oil [191]. Similarly, lesquerolic acid (14-hydroxy-Z-11-eicosanoic acid) has been removed from lesquerella oil by means of the positionally selective lipase of *R. arrhizus* [192]. Contrastingly, a lipase of *C. rugosa* was reported as positionally nonselective but has sufficient FA selectivity that it could be used to prepare 1,3-dierucin acid from rapeseed oil, that is, the lipase is positionally nonselective but favors hydrolysis or acid residues shorter than that of erucic acid, most of which occur at the 2-position [193]. Likewise, the lipases of *G. candidum* and *C. rugosa* are useful in the production of oil containing high concentrations of PUFA. *G. candidum* lipase can enrich DHA and EPA [194] and *C. rugosa* lipase can enrich DHA, but not EPA [194,195].

Lipase selectivity/specificity may be due to structural features of the substrate (e.g., fatty acid chain length, unsaturation, stereochemistry), physicochemical factors at the interface, and/or differences in the binding sites of the enzyme. Stereoselectivity of enzymes can be influenced by temperature and hydrophobicity of the solvent. Recently, Rogalska et al. [196] showed that the enantioselectivity of lipases from *Rhizomucor miehei, C. antarctica* B, lipoprotein lipase, and human gastric lipase toward monolayers of racemic dicaprin was enhanced at low surface pressures, while catalytic activity decreased.

Variations in specificity of lipase preparations from different fungi, different strains of the same species, and the same strain cultured under different conditions may be due to the production of multiple isoforms with differing specificities [174]. The lipase(s) of *G. candidum* have been of particular interest because those from some strains exhibit a relatively high preference for ester bonds involving fatty acids with a *cis*-9 double bond (e.g., oleic acid). Purification of *G. candidum* lipases and isolation of isozymes have been carried out [197–201]. *Geotrichum* species and strains produce at least two glycosylated lipases, most often designated lipases I and II, or less often A and B, which are coded for by two genes [16,17,142,202]. Additional isoforms may be produced that differ by degrees of glycosylation. There are conflicting reports on the specificities of the two lipases. For example, of the two lipases produced by *G. candidum* CMICC 335426, lipase B showed high specificity for the *cis*-9 unsaturates than lipase A of this strain and, according to Sidebottom et al. [18] lipases I and II of ATCC 34614 showed no preference. Bertolini et al. [202] cloned lipases I and II from *G. candidum* ATCC 34614 into *Saccharomyces cerevisiae*, then isolated and purified the two lipases from this yeast and determined their substrate specificities. Lipase I showed higher specificity than lipase II for long-chain unsaturated fatty acyl chains with *cis*-9 double bond, and lipase II showed a preference for substrates having short acyl chains (C_8–C_{14}). These investigators also showed that sequence variation in the N-terminal amino acids of these lipases, or the lid, does not contribute to variation in substrate preference. Not all strains of *G. candidum* exhibit the preference for unsaturated fatty acids [11]. The specificities of isoforms of lipases from *Geotrichum* are shown in Table 29.6 [18,203–206].

Cygler et al. [153] suggested that the basis for the selectivity differences between lipases I and II from *G. candidum* involves key amino acid residues along the internal cavity presumed to be the binding site of the scissile acyl chain and that selectivity does not involve the lid. Studies with a cloned lipase from *Rhizopus delemar* [207] appear to support this suggestion. For example, substitution of certain amino acids (e.g., Phe[95] → Asp) through site-directed mutagenesis in the substrate-binding region resulted in an almost twofold increase in the preference for tricaprylin relative to tributyrin and triolein [208]. In addition, substitution of Thr[83] is believed to be involved in oxyanion binding with Ala-eliminated activity.

At least five genes encoding lipases have been found in *C. cylindracea* (*rugosa*), but the individual isoforms have not been purified and had their specificities determined [209]. Based on gene sequences, the lipase genes of *C. cylindracea* and *G. candidum* appear to belong to the cholinesterase family.

E. Immobilization

One of the limitations to the industrial uses of lipases is cost-effectiveness. This can be improved by reuse of the lipase which can be accomplished by immobilizing the enzyme on an inert support. Immobilized lipases have been studied for hydrolytic [15,121,210–215], synthetic [216–220], and inter-/transesterification [221–224] reactions in both aqueous and organic media. Materials tested as solid supports for lipases include ion exchange resins (DEAE-Sephadex A50 or Amberlite IRA94) [214] and others [211], adsorbants such as silica gel [220], microporous polypropylene [210], and nylon [225]. In other cases, lipases have been immobilized by entrapping them in gels of photo-cross-linkable resins (ENT and ENTP) [212]. Immobilization using methods described in the references cited earlier does not involve chemical modification of the enzyme. Recently, Braun et al. [218] achieved immobilization of *C. rugosa* lipase on nylon by covalent attachment to the

TABLE 29.6
Selectivities of Multiple Extracellular Lipases from *Geotrichum* Species/Strains and *Mucor miehei*

Fungus/Strain	Molecular Mass (kDa)	Selectivity	References
Hydrolysis			
Geotrichum candidum ATCC 34614			
Lipase I	50.1	No preference for C18:1	[18]
Lipase II	55.5	No preference for C18:1	
G. candidum CMICC 335426			
Lipase A	53.7	Preference for C16:0 relative to C18:1Δ^9	
Lipase B	48.9	Selectivity for fatty acids with *cis*-9 double bond	
G. candidum ATCC 66592			
Lipase I	61	Hydrolyzed C16:0 methyl ester at 60% of initial velocity to that of C18:1 methyl ester	[203]
Lipase II	57	Hydrolyzed C16:0 methyl ester at initial velocity that was only 7% of that of C18:1 methyl ester	
G. candidum ATCC 34614			
Lipase I	64	Showed high preference for triolein and TAG with C8	[204]
Lipase II	66	Same as lipase I	
Geotrichum sp. F0401B			
Lipase A	62	Incompletely 1,3-specific toward triolein but able to hydrolyze at 2-position at slower rate	[205]
Lipase B	—	Nonspecific positional specificity (presumed mixture of A and C)	
Lipase C	58	Incompletely 2-specific	
Synthesis			
G. candidum NRRL Y-553		Reactivities slow for γ-linolenic and ricinoleic acids relative to oleic acid; oleic acid esterifies 2.5 times faster than C16:0 to 1-butanol, but 50 times faster with 2-methyl-1-propanol or cyclopentanol	[206]

support after conversion of carbohydrate groups on the enzyme to dialdehydes. More recently, Goto et al. [100] proposed a novel method of immobilizing enzymes onto porous hollow-fiber membrane modified using radiation-induced graft polymerization. An ion exchange group containing the grafted hollow-fiber membrane extends from the pore surface toward the pore interior and through mutual repulsion, captures the enzymes via electrostatic interaction.

Besides the use of immobilized purified enzymes, there are studies reported on the use of whole cells, or the so-called naturally immobilized enzymes. Cells of microorganisms like bacteria, yeast, and filamentous fungi produce their own enzymes within the confines of the peripheral plasma membrane, the so-called intracellular or naturally bound enzymes. Such naturally immobilized enzymes can be used directly without the laborious procedures of extraction, purification, and immobilization and thus are inexpensive and time saving. It can be used directly after physical separation from the growth medium and after drying if the condition requires low moisture content. Since extraction from the source cells is not required, loss of catalytic activity can be prevented. Such enzymes also do not require the addition of cofactors. Studies on the use of mycelium-bound lipase (MBL) have been reported [226–231].

Whole cell-immobilized enzymes can be obtained in alternative forms of growing cultures, resting cultures, spore cultures, or immobilized cultures [190]. Growing cultures are based on conventional fermentation practice where substrate to be biotransformed is added to the batch-grown cultures, and the incubation is continued until all of the added substrate disappears and/or the bioconversion cease. In this technique, it is crucial to avoid overmetabolism, which serves to erode the yield of desired products. The use of resting cells minimizes these problems. Resting cells are nongrowing live cells obtained by removing growing cells from liquid growing medium at a time in the growth phase when the potential of the cells to undertake the desired biotransformation is optimal. Spore cultures, even though metabolically inert, like resting cells can undertake desired reactions when suspended in dilute buffer.

Immobilized culture of microorganism has been much studied and reviewed [232–235]. Immobilized microbial cells are attractive in that they are likely to remain operationally active for very much longer than an actively growing culture, are easily removed from the reaction mixture, and can be reused repeatedly [236]. Long et al. [237] demonstrated that the MBL of *Aspergillus flavus* can be stabilized by cross-linking with glutaraldehyde, methylglyoxal, and ethylenediamine. But the activity and stability of the MBL was affected by the concentration of the bifunctional reagent used and exposure times. The potential for enhancement of enzyme stability by immobilization and its limitations has been discussed in length by Klibanov [238].

In naturally bound enzymes, fungal mycelia are used as a direct source of the enzyme and thereby eliminate the need for isolation and external immobilization procedures in cases where the lipase is already bound to the cell wall. Batches of mycelia are grown in shake-flask culture, harvested, freeze-dried, defatted, ground into fine particles, and stored in vacuo at room temperature until required for use [239]. There have been reports of usage of microbial cells having lipase in microaqueous systems in bioreactor. The microbial cells used can be wet or in semidried forms and suspended in water-immiscible organic solvent [240], packed in a column [241], or immobilized by entrapping in gels having an appropriate hydrophobicity–hydrophilicity balance [242]. It has also been reported that an extracellular thermostable lipase produced by *Pseudomonas mephitica* var. *lipolytica* retains its activity in the supernatant fluid separated by centrifugation from a mixture obtained when fully grown cells are homogenized with small glass beads (under cooling provided by solid carbon dioxide) and disrupted with a sonic disintegrator [243]. Binding of lipases to whole dried cells or the so-called MBL has been successfully employed for exogenous lipases produced by fungal species [244], *R. arrhizus* [241], *A. flavus* [227], *Rhizomucor meihei* [229], or by bacteria [243]. For example of application in industry scale, the sugar beet industry uses immobilized α-galactosidase to hydrolyze the indigestible sugar raffinose into sucrose and galactose. This enzyme is fixed within pellets of the fungus *Mortierella vinaceae*, which are formed under particular growth conditions, and the untreated cells can be used directly as biocatalyst [245].

Szczesna-Antczak et al. [246] compared the stability and performance of *Mucor circinelloides* lipase either immobilized in situ in the mycelium pellets or isolated from mycelium and immobilized on solid carriers (cellulose palmitate and octate, diatomaceous earth, modified porous glass) and found that lipase isolated and immobilized on solid supports gave lower activity in the synthesis of oleic and caprylic esters of propanol and sucrose, and weaker hydrolytic activity compared with the MBL preparation. They explained that the higher activity of MBL was due to the more favorable spatial arrangement of the lipase molecules anchored in the cell membranes and this facilitated the binding of substrates in the catalytic site of the enzyme. Another reason for the lower synthetic and hydrolytic activity of the purified immobilized lipase was due to the physicochemical properties of the carriers, which were different from the natural matrix, thus changing the enzyme's microenvironment with a negative impact on its catalytic activity. The activity and operational stability of *M. circinelloides* lipase derived by defatting and dehydrating the mycelium pellets was superior and enhanced after entrapment of the MBL in polyvinyl pyrrolidone containing chitosan beads solidified with hexametapolyphosphate [246].

VII. INDUSTRIAL APPLICATIONS

A. General Uses

The current industrial enzyme market is estimated over $600 million, with lipases representing about 4% of the worldwide market. The three major industrial enzyme companies worldwide are Novo Nordisk (>50%), Genencor (35%), and Solvay. Lipases are currently used, or have the potential for use, in a wide range of applications: in the dairy industry for cheese flavor enhancement, acceleration of cheese ripening, and lipolysis of butterfat and cream; in the oleochemical industry for hydrolysis, glycerolysis, and alcoholysis of fats and oils; and for the synthesis of structured triglycerides, surfactants, ingredients of personal care products, pharmaceuticals, agrochemicals, and polymers [247,248]. Review of lipase applications has been enormous. Industrial applications of microbial lipases have been discussed by Seitz [249]; hydrolysis of glycerides by lipases has been reported by Nielson [250]; the present and future applications of lipases have been reviewed by Macrae and Hammond [251]; applications of lipase-catalyzed hydrolysis have been listed by Iwai and Tsujisaka [252]; the importance of biotechnology in relation to the fats and oils industry has been reviewed by Macrae [253]; various aspects of enzymes that are useful in the lipid industry including sources, properties, reaction catalysis, some applications, and engineering aspects have been discussed by Yamane [12]; and some applications of lipase in industry have been discussed by Godfrey [254] and Gandhi [32]. Many of these processes have been patented extensively since the early 1900s. Lai et al. [255] reviewed the patent literature on lipid technology.

The Colgate–Emery process, currently used in the steam fat-splitting of TAGs, requires 240°C–260°C and 700 psi, has energy costs, and results in an impure product requiring redistillation to remove impurities and degradation products. In addition, this process is not suitable for highly unsaturated TAGs [256]. Lipase-catalyzed reactions offer several benefits over chemical reactions, including stereospecificity, milder reaction conditions (room temperature, atmospheric pressure), cleaner products, and reduced waste materials [114,257,258].

The largest current use of industrial enzymes is in laundry detergents, where they combine environmental friendliness and biodegradability with a low energy requirement and efficiency at low concentrations. The current U.S. market share of enzymatic laundry detergents is approaching 80%, and the U.S. detergent enzyme market is about $140 million. Essentially four types of enzymes are used in detergents: proteases, amylases, lipases, and cellulases. These enzymes perform multiple functions (e.g., stain removal, antiredeposition, whiteness/brightness retention, and fabric softening). Proteases were the first and are the most widely used enzymes in detergent formulations. Lipases are relatively new introductions to detergents, where they attack oily and greasy soils and contribute to making the detergents particularly effective at lower wash temperatures. However, a current limitation is that most lipases are unstable in alkaline conditions in the presence of anionic surfactants used in laundry detergents [21]. However, some lipases may be relatively resistant to certain surfactants [259].

B. New Lipases/Modification of Known Lipases

Early studies with fungal lipases focused on the isolation and characterization of extracellular lipases from various species. Some of the thoroughly studied fungal lipases include those from C. (cylindracea) rugosa, R. miehei, Penicillium camembertii, H. lanuginosa, C. antarctica B, Rhizopus delemar, and G. candidum, all of which are commercially available. Many of these lipases have relatively high specific activities: 3485 U/mg for M. miehei lipase A [15] and 7638 U/mg for R. delemar [27] (see Table 29.1 for other examples). Nevertheless, the research and development concerning lipase application in the lipid industry is not as dynamic as those in the area of carbohydrate, protein, and amino acids. Currently, lipase technology is mainly restricted to those operations where the cost of the product is high, making the enzyme cost low in relation, that is, this technology is applicable to high-value fine chemicals. Apart from this, lipase employment may be

TABLE 29.7
Properties of Cloned Lipase Genes

Source	Host	Vectors	Recombinant Plasmid	Size of Insert (bp)	References
Pseudomonas sp. strain KB700A	*Escherichia coli* TG1	pUC18	KB-lip	1422	[262]
Pseudomonas sp. strain B11-1	*E. coli* C600	pUC118	pPL2-1	924	[264]
Pseudomonas fluorescens No. 33	*E. coli* JM109	pUC19	pSHL2	1434	[264]
P. fluorescens SIK W1	*E. coli* JM83	pUC19	pJH92	1600	[265]
Pseudomonas KWI-56	*E. coli* HB101	pUC19	pLP64	1092	[64]
Staphylococcus haemolyticus L62	*E. coli* XL1 Blue	pBluescript II SK (+)	pSHL	2136	[266]
Bacillus stearothermophilus L1	*E. coli* RR1	pUC19	pLIP1	1254	[267]

attractive in processes that involve thermolabile substrates or products, such as phospholipids, or that may entail a number of side reactions, such as oxidation, racemization, and dehydration, and in processes, where high enantio- and/or regioselectivity is required. The latter is a problem frequently faced with the high temperatures and/or mineral acid catalysts, which are highly nonspecific with respect to the types of reactions catalyzed. Lipases that have received the most attention are mainly those having relatively high activities or certain properties that make them commercially attractive. Other than additional strains of known lipase producers, there seems to be no pattern among the fungi or yeasts from a taxonomic point of view that would direct future studies on where to find prolific lipase producers or lipases with specific properties.

Lipases have been modified using either chemical or molecular approaches to alter their properties and to identify structure–activity relationships. For example, lipases have been chemically modified with polyethylene glycol to render them more soluble in organic media. Recently, Kodera et al. [260] produced amphipathic chain-shaped and copolymer derivatives of lipases from *Pseudomonas fragi* or *P. cepacia* that were soluble in aqueous and hydrophobic media and exhibited catalytic activities for esterification and transesterification reactions, as well as for hydrolysis. The modified lipase showed preference for the *R* isomer of secondary alcohols in esterification reactions.

Molecular approaches have been used to increase the production of a lipase from the fungus, *R. delemar* [207]. The gene for this lipase codes for a preproenzyme that is posttranslationally modified to the mature enzyme. A cloned cDNA for the precursor polypeptide of the lipase [261] was altered by site-directed mutagenesis to produce fragments that code for the proenzyme and mature enzyme [207]. When inserted into *Escherichia coli* BL21 (DE3), the quantities of lipase from a 1 L culture exceeded those obtained from the fungal culture by 100-fold. Other examples of gene modification of lipases are given in Section VI.D, while Table 29.7 [64,262–267] lists some of the reported properties of cloned lipase genes.

C. PRODUCTION SYNTHESIS/MODIFICATION

There are many examples of uses for lipases in product synthesis/modification. One of the major areas of interest is in the use of lipase-catalyzed interesterification to improve the nutritional value, or alter the physical properties, of vegetable or fish oils. This is achieved, for example, by increasing the content of DHA or EPA of these oils. These long-chain ω-3 (*n*-3) fatty acids have been

TABLE 29.8
Production of Polyunsaturated Fatty Acid-Enriched Acylglycerols Using Lipase-Catalyzed Esterification

Lipase	Fatty Acid-Incorporated	Product	References
Candida antarctica lipase B, *Mucor miehei, Pseudomonas cepacia, Penicillium camembertii*	EPA, DHA	Production of EPA and DHA TAG	[273]
M. miehei	Conjugated linoleic acid, linoleic and oleic acids	Production of PUFA-enriched MAG, DAG, and TAG	[274]
C. antarctica lipase B and *M. miehei*	EPA, DHA	Production of EPA and DHA TAG	[275]
C. antarctica lipase B	PUFA from cod liver oil and microalgae oils	Production of PUFA-enriched TAG	[276]
Chromobacterium viscosum, M. miehei, Pseudomonas sp., *Candida rugosa, Rhizopus niveus, Aspergillus niger,* and *Rhizopus oryzae*	ω-3 Fatty acids	Production of acylglycerols containing ω-3 fatty acids	[277]
Novozyme 435	*n*-3 PUFA from menhaden oil	Enrichment of *n*-3 PUFA in hazelnut oil	[278]
Pseudomonas sp.	EPA	EPA was found to be incorporated mainly in the *sn*-1,3 positions of the TAG molecules and suitable for applications where quick energy release and EPA supplementation are required	[279]
Novozyme 435, Lipozyme RMIM, *Pseudomonas* sp., *A. niger, C. rugosa*	EPA and DHA	Highest DHA and EPA incorporation into high-laurate canola oil was 37.3% and 61.6%, respectively	[280]
Mucor miehei (Lipozyme IM 60), *Pseudomonas* sp. (PS-30)	EPA and DHA	Production of high EPA and DHA acylglycerols	[269]

incorporated into several vegetable oils using a lipase from *M. miehei* [109,268], medium-chain triglycerides [106], cod liver oil [269], borage, and evening primrose oil [270]. Halldorsson et al. [271] have also used lipase-catalyzed glycerolysis to separate DHA and EPA in fish oil. The *n*-3 fatty acid content of menhaden and anchovy oils [272] and tuna oil [185,188] has also been increased by lipase-catalyzed interesterification. Another fatty acid of interest is GLA, which is applicable in a wide range of clinical disorders. GLA has been enriched in evening primrose and borage oils by several fungal lipases [125,126]. Table 29.8 summarizes some of the work done on the production of PUFA-enriched acylglycerols by lipase-catalyzed esterification.

Other research involving synthesis/modification includes the synthesis of mono- and diglycerides [98,116,281–284] including regioisomerically pure products [94,95], synthesis of acetylated glucose [97], modification of phospholipids into biosurfactants [115], hydrolysis of phosphatidylcholine [121], and production of high-value specialty fats such as cocoa butter substitutes or hardened vegetable oils with butterfat properties [257,285,286]. The production of high-value fats takes advantage of the 1,3-specificity of lipases that could not be achieved by chemical synthesis [114]. Some recent examples of research involving synthesis/modification by lipases were given in Tables 29.3 and 29.9.

TABLE 29.9
Lipase-Catalyzed Production of Structured Lipids

Lipase	Substrate	Product	References
Lipozyme IM	Canola oil and caprylic acid	Products containing 40.1% caprylic acid with 3.4% caprylic acid in sn-2 position	[287]
Lipozyme IM	Menhaden oil	Products containing 40% caprylic acid and 30% EPA and DHA	[288]
Lipozyme IM	Canola oil and caprylic acid (acidolysis); Lipozyme IM-hydrolyzed canola oil and caprylic acid (esterification)	59.9% New TAG with acidolysis and 82.8% new TAG with esterification	[289]
Lipozyme IM	MCT and oleic acid	58% Di-incorporated specific structured lipid	[290]
Lipozyme IM	Sunflower, safflower, borage, linseed oils, and capric and caprylic acids	35%–47% Incorporation, 40% dimedium-chain fatty acid-incorporated TAG	[291]
Rhizopus arrhizus immobilized on Celite	Palm oil midfraction and stearic acid	50%–55% Incorporation	[292]
Carica papaya lipase	Tripalmitin + alkyl esters of caprylic acid	Transesterification with n-butyl and n-propyl was faster than when ethyl and methyl caprylates were used	[293]
Rhizomucor miehei	Palm stearin and palm kernel olein	Margarine prepared from the transesterified blends had acceptable PV levels although slight posthardening was observed after 3 month storage	[294]
Lipozyme IM 60	Palm stearin and palm kernel olein	Production of low melting TAG caused a sharp drop in SFC which shifted from the range of 15°C–20°C to 10°C–15°C	[295]

Lipase	Substrate	Result	Reference
R. miehei, Aspergillus niger, Rhizopus javanicus, Rhizopus niveus, Alcaligenes sp., *Candida rugosa*, and *Pseudomonas* sp.	Palm stearin + anhydrous milk fat	Highest degree and rate of transesterification was obtained when *Pseudomonas* sp. lipase was used followed by *R. miehei* lipase	[296]
Lipozyme IM 20	Glycerol and triolein	Monoolein synthesis enhanced from 10.6 mol% in *n*-hexane to 64 mol% in 2-methyl-2-butanol	[297]
Mixtures of *R. miehei* and *Alcaligenes* sp.; mixtures of *Pseudomonas camembertii* and *Alcaligenes* sp.	Glycerol and conjugated linoleic acid (CLA)	TAG containing CLA reached 82%–83% after 47 h using 1 wt% lipases	[298]
Lipozyme TLIM, Lipozyme RMIM, Novozyme 435	Menhaden oil and pinolenic acid	Incorporation of pinolenic acid was 19.4 mol% for Novozyme 435, 16.1, and 13.6 mol% for Lipozyme TLIM and Lipozyme RMIM, respectively	[299]
Lipozyme RMIM	Stearic acid and blends of palm olein and palm kernel oil	An incorporation of 42% stearic acid into blends of palm olein and palm kernel oil can be achieved	[300]
P. camembertii	CLA and glycerol	A high degree of esterification (84%) was achieved but produced equal amounts of MAG and DAG	[301]
Lipozyme TLIM and Novozyme 435	Phosphatidylcholine (PC), palmitic and stearic acids	58.6% and 57.1% palmitic acid was incorporated using Lipozyme TLIM and 56% and 61% using Novozyme 435 in egg and soybean PC from an initial content of 37.4% and 16.8%, respectively. Stearic acid incorporation was 44.7% and 46.3% with Lipozyme TLIM and 37.2% and 55.8% using Novozyme 435	[302]

VIII. SUMMARY

Lipases are exceedingly interesting enzymes because the relationship between their structure and activity presents an intellectual challenge and because their versatility offers a broad range of possible industrial applications. However, interest in the lipases has begun to move from academic curiosity to full commercialization in terms of the availability of lipases and their industrial use. For example, 50 t/year of the chiral intermediate methyl methoxyphenyl glycidate is produced based on a lipase-catalyzed process [218].

Although lipases have high potential for a variety of industrial applications, their use at the present time is limited by several factors, such as lack of cost-effective systems or processes for producing sufficient enzyme, heterogeneity of available preparations, and absence of lipases with properties required for certain applications [207]. As with proteases, protein engineering can be applied to lipases to target numerous specific characteristics. Alteration of amino acid sequences will result in variants with modified specific activity, increased k_{cat}, altered pH and thermal activity profiles, increased stability (with respect to temperature, pH, and chemical agents such as oxidants and proteases), and show altered pI, surface hydrophobicity, and substrate specificity [303,304]. Currently, lipase genes from fungal sources (e.g., *G. candidum* and *C. rugosa*) are cloned and subjected to site-directed mutation to gain insight into structure–activity relationships, mainly with respect to selectivity, on which to base protein engineering strategies. Despite the enormous progress that has been made in this regard, the molecular basis for selectivity is still not well understood.

REFERENCES

1. A. Hjorth, E. Carriere, C. Cudrey, H. Woldike, E. Boel, D.M. Lawson, F. Ferrato, C. Cambillau, C.G. Dobson, L. Thim, and R. Verger. A structural domain (the lid) found in pancreatic lipases is absent in the guinea pig (phospholipase) lipase. *Biochemistry* 32:4702 (1993).
2. F. Carrere, Y. Gargouri, H. Moreau, S. Ransae, E. Rogalska, and R. Verger. Gastric lipases: Cellular, biochemical and kinetic aspects. In: *Lipases* (P. Woolley and S.B. Petersen, eds.). Cambridge University Press, Cambridge, 1994, p. 181.
3. K.A. Dugi, H.L. Dechek, G.D. Tally, H.B. Brewer, and S. Santamarina-Fojo. Human lipoprotein lipase: The loop covering the catalytic site is essential for interaction with substrates. *J. Biol. Chem.* 267:25086 (1992).
4. F. Faustinella, L.C. Smith, and L. Chan. Functional topology of surface loop shielding the catalytic center in lipoprotein lipase. *Biochemistry* 31:7219 (1992).
5. H.A. van Tilbeurgh, J.-M. Lalouel Roussel, and C. Cambillau. Lipoprotein lipase. *J. Biol. Chem.* 269:4626 (1994).
6. S.B. Clark and H.L. Laboda. Triolein–phosphatidylcholine cholesterol emulsions as substrates for lipoprotein and hepatic lipases. *Lipids* 26:68 (1991).
7. T. Baba, D. Downs, K.W. Jackson, J. Tang, and C.-S. Wang. Structure of human milk activated lipase. *Biochemistry* 30:500 (1991).
8. C.-S. Wang. Human milk bile salt-activated lipase. Further characterization and kinetic studies. *J. Biol. Chem.* 256:10198 (1981).
9. L. Bläckberg and O. Hernell. The bile-salt-stimulated lipase in human milk: Purification and characterization. *Eur. J. Biochem.* 116:221 (1981).
10. K.D. Mukherjee and M.J. Mills. Lipases from plants. In: *Lipases* (P. Woolley and S.B. Petersen, eds.). Cambridge University Press, Cambridge, 1994, p. 49.
11. A.H.C. Huang, Y.H. Lin, and S.M. Wang. Characteristics and biosynthesis of seed lipases in maize and other plant species. *J. Am. Oil Chem. Soc.* 65:897 (1988).
12. T. Yamane. Enzyme technology for the lipids industry: An engineering overview. *J. Am. Oil Chem. Soc.* 64(12):1657 (1987).
13. C.T. Hou and T.M. Johnston. Screening of lipase activities with cultures from the agricultural research culture collection. *J. Am. Oil Chem. Soc.* 69:1088 (1992).

14. M.W. Baillargeon, R.G. Brisline, and P.E. Sonnet. Evaluation of strains of *Geotrichum candidum* for lipase production and fatty acid specificity. *Appl. Microbiol. Biotechnol.* 30:92 (1989).

15. B. Huge-Jensen, D.R. Galluzzo, and R.G. Jensen. Studies on free and immobilized lipases from *Mucor miehei*. *J. Am. Oil Chem. Soc.* 65:905 (1988).

16. Y. Shimada, A. Sugichara, T. Tizumi, and Y. Tominaga. cDNA cloning and characterization of *Geotrichum candidum* lipase II. *J. Biochem.* 107:703 (1990).

17. M.C. Bertolini, L. Laramee, D.Y. Thomas, M. Cygler, J.D. Schrag, and T. Vernet. Polymorphism in the lipase genes of *Geotrichum candidum* strains. *Eur. J. Biochem.* 219:119 (1994).

18. C.M. Sidebottom, E. Charton, P.P.J. Dunn, G. Mycock, C. Davies, L. Sutton, A.R. Macrae, and A.R. Slabas. *Geotrichum candidum* produces several lipases with markedly different substrate specificities. *Eur. J. Biochem.* 202:485 (1991).

19. S.-F. Lin, J.-C. Lee, and C.-M. Chion. Purification and characterization of a lipase from *Neurospora* sp. TT-241. *J. Am. Oil Chem. Soc.* 73:739 (1996).

20. P. Commenil, L. Belingheri, M. Sancholle, and B. Dehorter. Purification and properties of an extracellular lipase from the fungus *Botrytris cenerea*. *Lipids* 30:351 (1995).

21. Y. Kojima, M. Yokoe, and T. Mase. Purification and characterization of an alkaline lipase from *Pseudomonas fluorescens*. *Biosci. Biotechnol. Biochem.* 58:1564 (1994).

22. M. Kohno, W. Kugimiya, Y. Hashemoto, and Y. Moreta. Purification, characterization, and crystallization of two types of lipase from *Rhizopus niveus*. *Biosci. Biotechnol. Biochem.* 58:1007 (1994).

23. A. Riaublanc, R. Ratomahenina, P. Galzy, and M. Nicolas. Peculiar properties of lipase from *Candida parapsilosis* (Ashland) Langeron and Talice. *J. Am. Oil Chem. Soc.* 70:497 (1993).

24. R. Sarada and R. Joseph. Purification and properties of lipase from the anaerobe *Propionibactrium acidipropionici*. *J. Am. Oil Chem. Soc.* 69:974 (1992).

25. N. Kundu, J. Basu, M. Guchhait, and P. Chakrabarti. Purification and characterization of an extracellular lipase from the conidia of *Neurospora crassa*. *J. Gen. Microbiol.* 133:149 (1987).

26. Z. Mozaffar and J.D. Weete. Purification of properties of an extracellular lipase from *Pythium ultimum*. *Lipids* 28:377 (1993).

27. M.J. Haas, D.J. Cichowicz, and D.G. Bailey. Purification and characterization of an extracellular lipase from the fungus *Rhizopus delemar*. *Lipids* 27:571 (1992).

28. Y. Shimada, C. Koga, A. Sugihara, T. Nagao, N. Takada, S. Tsunasawa, and Y. Tominaga. Purification and characterization of a novel solvent-tolerant lipase from *Fusarium heterosporum*. *J. Ferment. Bioeng.* 75:349 (1993).

29. T. Mase, Y. Matsumiya, and A. Matsuura. Purification and characterization of *Penicillium roquefortii* 1AM 7268 lipase. *Biosci. Biotech. Biochem.* 59:329 (1995).

30. K. Gulomova, E. Ziomek, J.D. Schrag, K. Davranov, and M. Cygler. Purification and properties of a *Penicillum* sp. which discriminates against diglycerides. *Lipids* 31:379 (1996).

31. E. Antonian. Recent advances in the purification, characterization and structure determination of lipases. *Lipids* 23:1101 (1988).

32. N.N. Gandhi. Applications of lipase. *J. Am. Oil Chem. Soc.* 74:621 (1997).

33. S. Benjamin and A. Pandey. Mixed-solid substrate fermentation. A novel process for enhanced lipase production by *Candida rugosa*. *Yeast* 14:1069 (1998).

34. A. Pandey, S. Benjamin, C.R. Soccol, P. Nigam, M. Krieger, and V.T. Soccol. The realm of microbial lipases in biotechnology. *Biotechnol. Appl. Biochem.* 29:119 (1999).

35. J.L. Arpigny and K.-E. Jaeger. Bacterial lipolytic enzymes: Classification and properties. *Biochem. J.* 343:177 (1999).

36. E.J. Gilbert. *Pseudomonas* lipases: Biochemical properties and molecular cloning. *Enzyme Microb. Technol.* 15:634 (1993).

37. J.A. Contreras, M. Karlsson, T. Osterlund, H. Laurell, A. Svenson, and C. Holm. Hormone-sensitive lipase is structurally related to acetylcholinesterase, bile salt-stimulated lipase, and several fungal lipases. Building of a three-dimensional model for the catalytic domain of hormone-sensitive lipase. *J. Biol. Chem.* 271:31426 (1996).

38. E. Misawa, C.K. Chion, I.V. Archer, M.P. Woodland, N.Y. Zhou, S.F. Carter, D.A. Widdowson, and D.J. Leak. Characterisation of a catabolic epoxide hydrolase from a *Corynebacterium* sp. *Eur. J. Biochem.* 253:173 (1998).

39. H.D. Pohlenz, W. Boidol, I. Schuttke, and W.R. Streiber. Purification and properties of an *Arthrobacter oxydans* P52 carbamate hydrolase specific for the herbicide phenmedipham and nucleotide sequence of the corresponding gene. *J. Bacteriol.* 174:6600 (1992).

40. Y.S. Kim, H.B. Lee, K.D. Choi, S. Park, and O.J. Yoo. Cloning of *Pseudomonas* fluorescens carboxylesterase gene and characterization of its products expressed in *Escherichia coli*. *Biosci. Biotechnol. Biochem.* 58:111 (1994).

41. M.W. Akhtar, A.Q. Mirza, and M.I.D. Chuigh Tai. Lipase induction in *Mucor heimalis*. *Appl. Environ. Microbiol.* 40:257 (1980).

42. K. Ohnishi, Y. Yoshida, and J. Sekiguchi. Lipase production of *Aspergillus oryzae*. *J. Ferment. Bioeng.* 77:490 (1994).

43. N. Obradors, J.L. Montesinos, F. Valero, F.J. Lafuente, and C. Sola. Effects of different fatty acids in lipase production by *Candida rugosa*. *Biotechnol. Lett.* 15:357 (1993).

44. D.D. Hegedus and G.G. Khachatourians. Production of an extracellular lipase by *Beauveria bassiana*. *Biotechnol. Lett.* 10:637 (1988).

45. T. Sugiura, Y. Ota, and Y. Minoda. Effects of fatty acids, lipase activator, phospholipids, and related substances on the lipase production by *Candida paralipolytica*. *Agric. Biol. Chem.* 39:1689 (1975).

46. A.K. Chopra and H. Chander. Factors affecting lipase production in *Syncephalastrum racemosum*. *J. Appl. Bacteriol.* 54:163 (1983).

47. R. Sharma, Y. Chisti, and U.C. Banerjee. Production, purification, characterization and applications of lipases. *Biotechnol. Adv.* 19:627 (2001).

48. R.-C. Chang, S.-J. Chou, and J.-F. Shaw. Multiple forms and functions of *Candida rugosa* lipase. *Biotechnol. Appl. Biochem.* 19:93 (1994).

49. S.F. Lin, C.M. Chiou, C.M. Yeh, and Y.C. Tsai. Purification and partial characterization of an alkaline lipase from *Pseudomonas pseudoalcaligenes* F-111. *Appl. Environ. Microb.* 62:1093 (1996).

50. H. Chander, V.K. Batish, S.S. Sanabhadti, and R.A. Srinivasan. Factors affecting lipase production in *Aspergillus wentii*. *J. Food Sci.* 45:598 (1980).

51. A.B. Salleh, R. Musani, M. Basri, K. Ampon, W.M.Z. Yunus, and C.N.A. Razak. Extra and intracellular lipases from a thermophilic *Rhizopus oryzae* and factors affecting their production. *Can. J. Microbiol.* 39:978 (1993).

52. I. Che Omar, M. Hayashi, and S. Nagai. Purification and some properties of a thermostable lipase from *Humicola lanuginosa* No. 3. *Agric. Biol. Chem.* 51:37 (1987).

53. A.K. Chopra and H. Chander. Factors affecting lipase production in *Syncephalastrum racemosum*. *J. Appl. Bacteriol.* 54:163 (1983).

54. M. Essamri, V. Deyris, and L. Comeau. Optimization of lipase production by *Rhizopus oryzae* and study on the stability of lipase activity in organic solvents. *J. Biotechnol.* 60:97 (1998).

55. C. Novotny, L. Dolezalova, P. Musil, and M. Noval. The production of lipase by some *Candida* and *Yarrowia* yeast. *J. Basic Microbiol.* 28:221 (1988).

56. K.H. Tan and C.O. Gill. Batch growth of *Saccharomyces lipolytica* on animal fats. *Appl. Microbiol. Biotechnol.* 21:292 (1985).

57. E. Dalmau, J.L. Montesinos, M. Lotti, and C. Casas. Effect of different carbon sources on lipase production by *Candida rugosa*. *Enzyme Microb. Technol.* 26:657 (2000).

58. E. Nahas. Control of lipase production by *Rhizopus oligosporus* under various growth conditions. *J. Gen. Microbiol.* 134:227 (1988).

59. A.R. Macrae. Lipase-catalyzed interesterification of oils and fats. *J. Am. Oil Chem. Soc.* 62:284 (1983).

60. T. Nakashima, H. Fukuda, S. Kyotani, and H. Morikawa. Culture conditions for intracellular lipase production by *Rhizopus chinensis* and its immobilization within biomass support particles. *J. Ferment. Technol.* 66:441 (1988).

61. I. Che Omar, N. Nishio, and S. Nagai. Production of a thermostable lipase by *Humicola lanuginosa* grown on sorbitol–corn steep liquor medium. *Agric. Biol. Chem.* 51:2145 (1987).

62. R.B. Labuschagne, A.V. Tonder, and D. Litthauer. *Flavobacterium odoratum* lipase: Isolation and characterization. *Enzyme Microb. Technol.* 21:52 (1997).

63. X.G. Gao, S.G. Cao, and K.C. Zhang. Production, properties and application of non-aqueous enzymatic catalysis of lipase from a newly isolated *Pseudomonas* strain. *Enzyme Microb. Technol.* 27:74 (2000).

64. T. Iizumi, K. Nakamura, and T. Fukase. Purification and characterization of a thermostable lipase from a newly isolated *Pseudomonas* sp. KWI-56. *Agric. Biol. Chem.* 54:1253 (1990).

65. J.P. Glusker, A.K. Katz, and C.W. Bock. Metal ions in biological systems. *Rigaku J.* 16:8 (1999).

66. C. Sharon, S. Rurugoh, T. Yamakido, H. Ogawa, and Y. Kato. Purification and characterization of a lipase from *Pseudomonas aeruginosa* KKA-5 and its role in castor oil hydrolysis. *J. Ind. Microbiol. Biotechnol.* 20:304 (1998).

67. P.H. Janssen, C.R. Monk, and H.W. Morgan. A thermophilic, lipolytic *Bacillus* sp. and continuous assay of its *p*-nitrophenyl-palmitate esterase activity. *FEMS Microbiol. Lett.* 120:195 (1994).

68. V.M.G. Lima, N. Krieger, M.I.M. Sarquis, D.A. Mitchell, L.P. Ramos, and J.D. Fontana. Effect of nitrogen and carbon sources on lipase production by *Penicillium aurantiogriseum*. *Food Technol. Biotechnol.* 41:105 (2003).

69. S.J. Chen, C.Y. Cheng, and T.L. Chen. Production of an alkaline lipase by *Acinetobacter radioresisten*. *J. Ferment. Bioeng.* 86:308 (1998).

70. M.L.F. Giuseppin. Effects of dissolved oxygen concentration on lipase production by *Rhizopus delemar*. *Appl. Microb. Biotechnol.* 20:161 (1984).

71. B. Kopp. Long term alkaloid production by immobilized cells of *Claviceps purpurea*. *Methods Enzymol.* 136:317 (1987).

72. H. Chander, S. Sanabhadti, J. Elias, and B. Ranganathan. Factors affecting lipase production by *Penicillium chrysogenum*. *J. Food Sci.* 42:1677 (1977).

73. H. Chander, V.K. Batish, D.R. Ghodekar, and R.A. Srinivasan. Factors affecting lipase production in *Rhizopus nigricans*. *J. Dairy Sci.* 64:193 (1981).

74. R.K. Saxena, W.S. Davidson, A. Sheoran, and B. Giri. Purification strategies for microbial lipases. *J. Microb. Methods Rev.* 52:1 (2003).

75. K. Torossian and S.W. Bell. Purification and characterization of an acid resistant triacylglycerol lipase from *Aspergillus niger*. *Biotechnol. Appl. Biochem.* 13:205 (1991).

76. M.W. Baillargeon and S.G. McCarthy. *Geotrichum candidum* NRRL-Y-553: Purification, characterization and fatty acid specificity. *Lipids* 26:831 (1991).

77. H. Dong, S. Gao, S.P. Han, and S.G. Cao. Purification and characterization of a *Pseudomonas* sp. lipase and its properties in non-aqueous media. *Biotechnol. Appl. Biochem.* 30:251 (1999).

78. H. Ogino, S. Nakagawa, K. Shinya, T. Muto, N. Fujimura, M. Yasuda, and H. Ishikawa. Purification and characterization of organic solvent-stable lipase from organic solvent-tolerant *Pseudomonas aeruginosa* LST-03. *J. Biosci. Bioeng.* 89:451 (2000).

79. P.L. Luisi, M. Giomini, M.P. Pileni, and B.H. Robinson. Reverse micelles as hosts for proteins and small molecules. *Biochim. Biophys. Acta* 947:209 (1988).

80. A. Sanchoz-Ferrer and F. Garcia-Carmona. Biocatalysis in reverse self-assembling structures: Reverse micelles and reverse vesicles. *Enzyme Microb. Technol.* 16:409 (1994).

81. R.D. Jensen. Detection and determination of lipase (acylglycerol hydrolase) activity from various sources. *Lipids* 18:650 (1983).

82. Z. Mozaffar and J.D. Weete. Invert emulsion as a medium for fungal lipase activity. *J. Am. Oil Chem. Soc.* 72:1361 (1995).

83. D. Han and J.S. Rhee. Characteristics of lipase-catalyzed hydrolysis of olive oil in AOT-isooctane reversed micelles. *Biotechnol. Bioeng.* 28:1250 (1986).

84. R.R. Lowry and I.J. Tinsley. A derivatization method for determination of brominated fatty acid. *J. Am. Oil Chem. Soc.* 53:470 (1976).

85. Y. Isono, H. Nabetani, and M. Nakajima. Wax ester synthesis in a membrane reactor with lipase–surfactant complex in hexane. *J. Am. Oil Chem. Soc.* 72:887 (1995).

86. D.G. Hayes and E. Gulari. Formation of polyol–fatty acid esters by lipases in reverse micellar media. *Biotechnol. Bioeng.* 40:110 (1992).

87. H. Stamatis, A. Xenakis, M. Provelegiou, and F.N. Kolisis. Esterification reactions catalyzed by lipase in microemulsions: The role of enzyme location in relation to its selectivity. *Biotechnol. Bioeng.* 42:103 (1993).

88. C.P. Singh, P. Skagerlind, K. Holmberg, and D.D. Shah. A comparison between lipase-catalyzed esterification of oleic acid with glycerol in monolayer and microemulsion systems. *J. Am. Oil Chem. Soc.* 71:1405 (1994).

89. G.B. Oguntimein, H. Erdmann, and R.D. Schmid. Lipase catalyzed synthesis of sugar ester in organic solvents. *Biotechnol. Lett.* 15:175 (1993).

90. K.K. Osada, K. Takahashi, and M. Hatano. Polyunsaturated fatty glyceride synthesis by microbial lipases. *J. Am. Oil Chem. Soc.* 67:921 (1990).

91. A. Schlotterbeck, S. Lang, V. Wray, and F. Wagner. Lipase-catalyzed monoacylation of fructose. *Biotechnol. Lett.* 15:61 (1993).

92. P.-Y. Chen, S.-H. Wu, and K.-T. Wang. Double enantioselective esterification of racemic acids and alcohols by lipase from *Candida cylindracea*. *Biotechnol. Lett.* 15:181 (1993).

93. S. Ramamurthi and A.R. McCurdy. Lipase-catalyzed esterification of oleic acid and methanol in hexane—A kinetic study. *J. Am. Oil Chem. Soc.* 71:927 (1994).

94. M. Berger, K. Laumen, and M.P. Schneider. Enzymatic esterification of glycerol: I. Lipase-catalyzed synthesis of regioisomerically pure 1,3-*sn* diacylglycerols. *J. Am. Oil Chem. Soc.* 69:955 (1992).

95. M. Berger and M.P. Schneider. Enzymatic esterification of glycerol II. Lipase-catalyzed synthesis of regioisomerically pure 1 (3)-*rac*-monoacylglycerols. *J. Am. Oil Chem. Soc.* 69:961 (1992).

96. Y. Tanaka, J. Hirano, and T. Fundada. Synthesis of docosahexaenoic acid-rich triglyceride with immobilized *Chromobacterium viscosum* lipase. *J. Am. Oil Chem. Soc.* 71:331 (1994).

97. C.C. Akoh. Enzymatic synthesis of acetylated glucose fatty acid esters in organic solvent. *J. Am. Oil Chem. Soc.* 71:319 (1994).

98. C.P. Singh, D.O. Shah, and K. Holmberg. Synthesis of mono- and diglycerides in water-in-oil microemulsions. *J. Am. Oil Chem. Soc.* 71:583 (1994).

99. A.R. Fajardo, L.E. Cerdan, A.R. Medina, M.M.M. Martinez, E.H. Pena, and E.M. Grima. Concentration of eicosapentaenoic acid by selective esterification using lipases. *J. Am. Oil Chem. Soc.* 83:215 (2006).

100. M. Goto, H. Kawahita, K. Uezu, S. Tsuneda, K. Saito, M. Goto, M. Tamada, and T. Sugo. Esterification of lauric acid using lipase immobilized in the micropores of a hollow-fiber membrane. *J. Am. Oil Chem. Soc.* 83:209 (2006).

101. M. Safari, S. Kermasha, L. Lamboursain, and J.D. Sheppard. Interestification of butterfat by lipase from *Rhizopus niveus* in reverse micellar systems. *Biosci. Biotechnol. Biochem.* 58:1553 (1994).

102. M. Goto, N. Kamiya, and F. Nakashio. Enzymatic interesterification of triglyceride with surfactant coated lipase in organic media. *Biotechnol. Bioeng.* 45:27 (1995).

103. A.G. Marangoni, R. McCurdy, and E.D. Brown. Enzymatic interestification of triolein with tripalmitin in canola lecithin-hexane reverse-micelles. *J. Am. Oil Chem. Soc.* 70:737 (1993).

104. P. Forssel, R. Kervinen, M. Lappi, P. Linko, T. Suortti, and K. Poutanen. Effect of enzymatic interesterification on the melting point of tallow-rapeseed oil (LEAR) mixture. *J. Am. Oil Chem. Soc.* 69:126 (1992).

105. I. Svensson, P. Adlercreutz, and B. Mattiasson. Lipase-catalyzed transesterification of phosphatidylcholine at controlled water activity. *J. Am. Oil Chem. Soc.* 69:986 (1992).

106. A. Shishikura, K. Fujimoto, T. Suzuki, and K. Arai. Improved lipase-catalyzed incorporation of long-chain fatty acids into medium-chain triglycerides assisted by supercritical CO_2 extraction. *J. Am. Oil Chem. Soc.* 71:961 (1994).

107. S. Bloomer, P. Adlercreutz, and B. Mattiasson. Kilogram-scale ester synthesis of acyl donor and use in lipase-catalyzed interesterifications. *J. Am. Oil Chem. Soc.* 169:966 (1992).

108. K. Huang and C.C. Akoh. Lipase-catalyzed incorporation of *n*-3 polyunsaturated fatty acids into vegetable oils. *J. Am. Oil Chem. Soc.* 71:1277 (1994).

109. R. Sridhar and G. Lakshminarayana. Incorporation of EPA and DHA into groundnut oil by lipase-catalyzed ester exchange. *J. Am. Oil Chem. Soc.* 69:1041 (1992).

110. T. Oba and B. Witholt. Interesterification of milk fat with oleic acid catalyzed by immobilized *Rhizopus oryzae* lipase. *J. Dairy Sci.* 77:1790 (1994).

111. K. Watanake, C. Ishikawa, H. Inoue, D. Cenhua, K. Yazawa, and K. Kondo. Incorporation of exogenous decosahexaenoic acid into various bacterial phospholipids. *J. Am. Oil Chem. Soc.* 71:325 (1994).

112. S. Basheer, K. Mogi, and M. Nakajima. Surfactant modified lipase for the catalysis of the interesterification of triglycerides and fatty acids. *Biotechnol. Bioeng.* 45:187 (1995).

113. Y.-Y. Linko, M. Lamsa, A. Huhtala, and P. Linko. Lipase-catalyzed transesterification of rapeseed oil and 2-ethyl-1-hexanol. *J. Am. Oil Chem. Soc.* 71:1411 (1994).

114. S.H. Goh, S.K. Yeong, and C.W. Wang. Transesterification of cocoa butter by fungal lipases: Effect of solvent on 1,3-specificity. *J. Am. Oil Chem. Soc.* 70:567 (1993).

115. L.N. Mutua and C.C. Akoh. Lipase catalyzed modification of phospholipids: Incorporation of *n*-3 fatty acids into biosurfactants. *J. Am. Oil Chem. Soc.* 70:125 (1993).

116. T. Yamane, S.T. Kang, K. Kawahara, and Y. Koizumi. High yield diacylglycerol formation by solid phase enzymatic glycerolysis of hydrogenated beef tallow. *J. Am. Oil Chem. Soc.* 71:339 (1994).

117. H.M. Ghazali, S. Hamidah, and Y.B. Cheman. Enzymatic transesterification of palm olein with non-specific and 1,3-specific lipases. *J. Am. Oil Chem. Soc.* 72:633 (1995).

118. V. Dossat, D. Combes, and A. Marty. Lipase-catalyzed transesterification of high oleic sunflower oil. *Enzyme Microb. Technol.* 30:90 (2002).

119. A.C. Oliveira and M.F. Rosa. Enzymatic transesterification of sunflower oil in an aqueous-oil biphasic system. *J. Am. Oil Chem. Soc.* 83:21 (2006).

120. K. Sabally, S. Karboune, R. St-Louis, and S. Kermasha. Lipase-catalyzed transesterification of trilinolein and trilinolenin with selected phenolic acids. *J. Am. Oil Chem. Soc.* 83:101 (2006).

121. M.J. Haas, D.J. Cichowicz, J. Phillips, and R. Moreau. The hydrolysis of phosphatidylcholine by an immobilized lipase: Optimization of hydrolysis in organic solvents. *J. Am. Oil Chem. Soc.* 70:111 (1993).

122. T. Kim and K. Chung. Some characteristics of palm kernel olein hydrolysis by *Rhizopus arrhizus* lipase in reversed micelle of AOT in isooctane, and additive effect. *Enzyme Microb. Technol.* 11:528 (1989).

123. D. Han and J.S. Rhee. Batchwise hydrolysis of olive oil by lipase in AOT-isooctane reverse-micelles. *Biotechnol. Lett.* 7:651 (1985).

124. T. Yamane, T. Suzuki, Y. Sahashi, L. Vikersveen, and T. Hoshino. Production of *n*-3 polyunsaturated fatty acid-enriched fish oil by lipase-catalyzed acidolysis without solvent. *J. Am. Oil Chem. Soc.* 69:1104 (1992).

125. M.S.K. Rahmatullah, V.K.S. Shukla, and K.D. Mukherjee. Enrichment of GLA from borage and evening primrose oils via lipase-catalyzed hydrolysis. *J. Am. Oil Chem. Soc.* 71:569 (1994).

126. M.S.K. Rahmatullah, V.K.S. Shukla, and K.D. Mukherjee. GLA concentrates from borage and evening primrose oil fatty acids via lipase catalyzed esterification. *J. Am. Oil Chem. Soc.* 71:563 (1994).

127. F.R. Winkler and K. Gubernator. Structure and mechanisms of action of human pancreatic lipase. In: *Lipases* (P. Woolley and S.B. Petersen, eds.). Cambridge University Press, Cambridge, 1994, p. 139.

128. F.K. Winkler, A. D'Arcy, and W. Hunziker. Structure of human pancreatic lipase. *Nature* 343:771 (1990).

129. C. Erlandson, J.A. Barrowman, and B. Borgstrom. Chemical modification of pancreatic colipase. *Biochim. Biophys. Acta* 489:150 (1977).

130. H. van Tilbeurgh, L. Sanda, R. Verger, and C. Cambillau. Structure of the pancreatic lipase–colipase complex. *Nature* 359:159 (1992).

131. H. van Tilbeurgh, M.P. Egloff, C. Martinez, N. Rugani, R. Verger, and C. Cambillau. Interfacial activation of the lipase–procolipase complex by mixed micelles revealed by X-ray crystallography. *Nature* 362:814 (1993).

132. H. van Tilbeurgh, Y. Gargouri, C. DeZan, M.-P. Egloff, M.-P. Nesa, N. Rugame, L. Sarda, R. Verger, and C. Cambillau. Crystallization of pancreatic prolipase and of its complex with pancreatic lipase. *J. Mol. Biol.* 229:552 (1993).

133. F.J. Alvarez and V.J. Stella. The role of calcium ions and bile acids on the pancreatic lipase-catalyzed hydrolysis of triglyceride emulsions stabilized with lecithin. *Pharm. Res.* 6:449 (1989).

134. Y. Bourne, C. Martinez, B. Kerfelec, D. Lombardo, C. Chapus, and C. Cambillau. Horse pancreatic lipase. *J. Mol. Biol.* 238:709 (1994).

135. L. Brady, A.M. Brzozowski, Z.S. Derewenda, E. Dodson, G. Dodson, S. Tolley, J.P. Tuskenburg, L. Christiansen, B. Huge-Jensen, L. Novskov, L. Thim, and U. Menge. Serine protease triad forms the catalytic center of a triacylglycerol lipase. *Nature* 343:767 (1990).

136. Z.S. Derewenda, U. Derewenda, and G.G. Dodson. The crystal and molecular structure of the *Rhizomucor miehei* triacylglycerol lipase at 1.9 Å resolution. *J. Mol. Biol.* 227:818 (1992).

137. J.D. Schrag and M. Cygler. 1.8 Å refined structure of the lipase from *Geotrichum candidum*. *J. Mol. Biol.* 230:575 (1993).

138. J.D. Schrag, Y. Li, S. Wu, and M. Cygler. Ser–His–Glu triad forms the catalytic site of the lipase from *Geotrichum candidum*. *Nature* 351:761 (1991).

139. P. Grochulski, Y. Li, J.D. Schrag, F. Bouthellier, P. Smith, D. Harrison, B. Rubin, and M. Cygler. Insights into interfacial activation from an open structure of *Candida rugosa* lipase. *J. Biol. Chem.* 268:12843 (1993).

140. J. Uppenberg, S. Atkar, T. Bergfors, and T.A. Jones. Crystallization and preliminary x-ray studies of lipase from *Candida antarctica*. *J. Mol. Biol.* 235:790 (1994).

141. D.M. Lawson, A.M. Brzozowski, G.G. Dodson, R.E. Hubbard, B. Huge-Jensen, E. Boel, and Z.S. Derewenda. Three-dimensional structures of two lipases from filamentous fungi. In: *Lipases* (P. Woolley and S.B. Petersen, eds.). Cambridge University Press, Cambridge, 1994, pp. 77–94.

142. Y. Shimada, A. Sugihara, Y. Tominaga, T. Iizumi, and S. Tsunasawa. cDNA molecular cloning of *Geotrichum candidum* lipase. *J. Biochem.* 106:383 (1989).

143. T. Vernet, E. Ziomek, A. Recktenwald, J.D. Schrag, C. de Montigny, D.C. Tessier, D.Y. Thomas, and M. Cygler. Cloning and expression of *Geotrichum candidum* lipase II gene in yeast. *J Biol. Chem.* 268:26212 (1993).

144. U. Derewenda, L. Swenson, Y. Wei, R. Green, P.M. Kobos, R. Joerger, M.J. Haas, and Z.S. Derewenda. Conformational lability of lipases observed in the absence of an oil–water interface: Crystallographic studies of enzymes from the fungi *Humicola lanuginosa* and *Rhizopus delemar*. *J. Lipid Res.* 35:524 (1994).

145. K.K. Kin, Y.H. Kwang, H.S. Jeon, S. Kim, R.M. Sweet, C.H. Yang, and S.W. Suh. Crystallization and preliminary x-ray crystallographic analysis of lipase from *Pseudomonas cepacia*. *J. Mol. Biol.* 227:1258 (1992).

146. A. Cleasly, E. Garman, M.R. Egmond, and M. Batenburg. Crystallization and preliminary x-ray study of a lipase from *Pseudomonas glumae*. *J. Mol. Biol.* 224:281 (1992).

147. S. Larson, J. Day, A. Greenwood, J. Oliver, J. Rubingh, and A. McPherson. Preliminary investigation of crystals of the neutral lipase from *Pseudomonas fluorescens*. *J. Mol. Biol.* 222:21 (1991).

148. S. Ransae, M. Blaanin, E. Lesuisse, K. Schanck, C. Colson, and B.W. Dijkstra. Crystallization and preliminary x-ray analysis of a lipase from *Bacillus subtilis*. *J. Mol. Biol.* 238:857 (1994).

149. L. Sarda and P. Desnuelle. Action de la lipase pancréatique sur les esters en emulsion. *Biochim. Biophys. Acta* 30:513 (1958).

150. R. Verger and G.H. de Haas. Interfacial enzyme kinetics of lipolysis. *Annu. Rev. Biophys. Bioeng.* 5:77 (1996).

151. A.M. Brzozowski, Z.S. Derewenda, G. Dobson, D.M. Lawson, J.P. Turkenburg, F. Bjorkling, B. Huge-Jensen, S.A. Patkan, and L. Thim. A model for interfacial activation in lipases from the structure of a fungal lipase–inhibitor complex. *Nature* 351:491 (1991).

152. U. Derewenda, A.M. Brzozowski, D.M. Lawson, and Z.S. Derewenda. Catalysis at the interface: The anatomy of a conformational change in a triglyceride lipase. *Biochemistry* 31:1532 (1992).

153. M. Cygler, P. Grochulski, and J.D. Schrag. Structural determinants defining common selectivity of lipases toward primary alcohols. *Can. J. Microbiol.* 41:289 (1995).

154. P. Grochulski, J.D. Schrag, and M. Cygler. Two conformational states of *Candida rugosa* lipase. *Protein Sci.* 3:82 (1994).

155. P. Grochulski, F. Bouthellier, R.J. Kazlauskas, A.N. Serreqi, J.D. Schrag, E. Ziomek, and M. Cygler. Analogs of reaction intermediates identify a unique substrate binding site in *Candida rugosa* lipase. *Biochemistry* 33:3494 (1994).

156. M. Holmquist, M. Martinelle, J.G. Clausen, S. Patkar, A. Svendsen, and K. Hult. TRP89 in the lid of *Humicola lanuginosa* lipase is important for efficient hydrolysis of tributyrin. *Lipids* 29:599 (1994).

157. D. Blow. Lipases at the interface. *Nature* 351:444 (1991).

158. Y.J. Wang, J.Y. Sheu, F.F. Wang, and J.-F. Shaw. Lipase-catalyzed oil hydrolysis in the absence of added emulsifier. *Biotechnol. Bioeng.* 31:628 (1987).

159. W.E. Momsen and H.L. Brockman. Inhibition of pancreatic lipase B activity by taurodeoxycholate and its reversal by colipase. *J. Biol. Chem.* 251:384 (1976).

160. Y. Naka and T. Nakamura. Effects of colipase, bile salts, Na^+ and Ca^{2+} on pancreatic lipase activity. *Biosci. Biotechnol. Biochem.* 58:2121 (1994).

161. Z. Mozaffar, J.D. Weete, and R. Dute. Influence of surfactants on an extracellular lipase from *Pythium ultimum*. *J. Am. Oil Chem. Soc.* 71:75 (1994).

162. S. Patkar and F. Bjorkling. Lipase inhibitors. In: *Lipases* (P. Woolley and S.B. Petersen, eds.). Cambridge University Press, Cambridge, 1994, p. 207.

163. H.R. Macrae. Lipase-catalyzed interesterification of oils and fats. *J. Am. Oil Chem. Soc.* 60:291 (1983).

164. P.E. Sonnet. Lipase selectivities. *J. Am. Oil Chem. Soc.* 65:900 (1988).

165. P. Stadler, A. Kovac, L. Haalck, F. Spener, and F. Paltauf. Stereoselectivity of microbial lipases: The substitution at position *sn*-2 of triacylglycerol analogs influence the stereoselectivity of different microbial lipases. *Eur. J. Biochem.* 227:335 (1995).

166. A. Kovac, P. Stadler, L. Haalck, F. Spener, and F. Paltauf. Hydrolysis and esterification of acylglycerols and analogs in aqueous medium catalysed by microbial lipases. *Biochim. Biophys. Acta* 1301:57 (1996).

167. R.G. Jensen, J. Sampugna, J.G. Guinn, D.L. Carpenter, and T.A. Marks. Specificity of a lipase from *Geotrichum candidum* for *cis*-octadecenoic acid. *J. Am. Chem. Soc.* 42:1029 (1965).
168. R.E. Jensen, F.A. deJong, L.G. Lambert-Davis, and M. Hamosh. Fatty acid and positional selectivities of gastric lipases from premature human infants. *Lipids* 29:433 (1994).
169. S. Yamaguchi and T. Mase. Purification and characterization of mono and diacylglycerol lipase isolated from *Penicillium camembertii* U-150. *Appl. Microbiol. Biotechnol.* 34:720 (1991).
170. P. Eigtved. Enzymes and lipid modification. In: *Advances in Applied Lipid Research* (F.B. Padley, ed.), Vol. 1. JAI Press Ltd., London, 1992, pp. 1–64.
171. R.G. Jensen. Characteristics of the lipase from the mold *Geotrichum candidum*: A review. *Lipids* 9:149 (1974).
172. M.K. Tahoun, E. Mostafa, R. Mashaly, and S. Abou-Donia. Lipase induction in *Geotrichum candidum*. *Milchwissenschaft* 37:86 (1982).
173. A. Sugihara, Y. Shimada, and Y. Tominaga. A novel *Geotrichum candidum* lipase with some preference for the 2-position on a triglyceride molecule. *Appl. Microbiol. Biotechnol.* 35:738 (1991).
174. E. Charton, C. Davies, and A. Macrae. Use of specific polyclonal antibodies to detect heterogeneous lipases from *Geotrichum candidum*. *Biochim. Biophys. Acta* 1127:191 (1992).
175. A. Sugihara, S. Hata, Y. Shimada, K. Goto, S. Tsunasawa, and Y. Tominaga. Characterisation of *Geotrichum candidum* lipase III with some preference for the inside ester bond of triglyceride. *Appl. Microbiol. Biotechnol.* 40:279 (1993).
176. A. Sugihara, Y. Shimada, M. Nakamura, T. Nagao, and Y. Tominaga. Positional and fatty acid specificities of *Geotrichum candidum* lipases. *Protein Eng.* 8:585 (1994).
177. X. Xu. Production of specific-structured triacylglycerols by lipase-catalyzed reactions: A review. *Eur. J. Lipid Sci. Technol.* 102:287 (2000).
178. P.E. Sonnet, T.A. Foglia, and S.H. Feairheller. Fatty acid selectivity of lipases: Erucic acid from rapeseed oil. *J. Am. Oil Chem. Soc.* 70:387 (1993).
179. M.J. Hills, I. Kiewitt, and K.D. Mukherjee. Enzymatic fractionation of fatty acids: Enrichment of γ-linolenic acid and docosahexaenoic acid by selective esterification catalysed by lipases. *J. Am. Oil Chem. Soc.* 67:561 (1990).
180. G. Benzonana and S. Esposito. On the positional and chain specificities of *Candida cylindracea* lipase. *Biochim. Biophys. Acta* 231:15 (1971).
181. K. Mbayhoudel and L.C. Comeau. Obtention selectice de l'acide petroselinic apartir de l'huile de fenouil par hydrolyse enzymatique. *Rev. Fr. Corps. Gras.* 36:427 (1989).
182. B.W. Baillargeon and S.G. McCarthy. *Geotrichum candidum* NRRL Y-553 lipase: Purification, characterization and fatty acid specificity. *Lipids* 26:831 (1991).
183. A. Svendsen, K. Borch, M. Barfoed, T.B. Nielsen, E. Gormsen, and S.A. Patkar. Biochemical properties of cloned lipases from the *Pseudomonas* family. *Biochim. Biophys. Acta* 1259:9 (1995).
184. Y. Shimada. Enzymatic modification of lipids for functional foods and nutraceuticals. In: *Handbook of Functional Lipids* (C.C. Akoh, ed.). CRC Press, Boca Raton, FL, 2006, pp. 437–455.
185. Y. Tanaka, T. Funada, J. Hirano, and R. Hashizuma. Triglyceride specificity of *Candida cylindracea* lipase: Effect of DHA on resistance of triglyceride to lipase. *J. Am. Oil Chem. Soc.* 70:1031 (1993).
186. T.A. Foglia and P.E. Sonnet. Fatty acid selectivity of lipases: Gamma-linolenic acid from borage oil. *J. Am. Oil Chem. Soc.* 72:417 (1995).
187. P.E. Sonnet, T.A. Foglia, and S.H. Feairheller. Fatty acid selectivity of lipases: Erucic acid from rape seed oil. *J. Am. Oil Chem. Soc.* 70:387 (1993).
188. Y. Shimada, K. Maruyama, S. Okazaki, M. Nakamura, A. Sugihara, and Y. Tominaga. Enrichment of polyunsaturated fatty acids with *Geotrichum candidum* lipase. *J. Am. Oil Chem. Soc.* 71:951 (1994).
189. E. Santaniello, P. Ferraboschi, P. Grisenti, and A. Manzocchi. The biocatalytic approach to the preparation of enantiomerically pure chiral building blocks. *Chem. Rev.* 92:1071 (1992).
190. S.M. Roberts, N.J. Turner, A.J. Willetts, and M.K. Turner. *Introduction to Biocatalysts Using Enzymes and Microorganisms*. Cambridge University Press, Cambridge, UK, 1995, pp. 34–97.
191. J.T.P. Derksen and F.P. Cuperus. Lipase-catalysed hydrolysis of new vegetable oils in a membrane bioreactor. *INFORM* 3:550 (1992).
192. D.G. Hayes and R. Kleiman. Recovery of hydroxy fatty acids from lesquerella oil with lipases. *J. Am. Oil Chem. Soc.* 69:982 (1992).
193. M. Trani, R. Lortie, and F. Ergan. Synthesis of trierucin from HEAR oil. *INFORM* 3:482 (1992).

194. Y. Tanaka, J. Hirano, and T. Funada. Concentration of docosahexaenoic acid in glyceride hydrolysis of fish oil with *Candida cylindracea* lipase. *J. Am. Oil Chem. Soc.* 69:1210 (1992).

195. Y. Shimada, K. Maruyama, S. Okazaki, M. Nakamura, A. Sugihara, and Y. Tominaga. Enrichment of polyunsaturated fatty acids with *Geotrichum candidum* lipase. *J. Am. Oil Chem. Soc.* 71:951 (1994).

196. E. Rogalska, S. Ransac, and R. Verger. Controlling lipase stereoselectivity via the surface pressure. *J. Biol. Chem.* 268:792 (1993).

197. Y. Tsujisaka, M. Iwai, and Y. Tominaga. Purification, crystallisation and some properties of lipase from *Geotrichum candidum* Link. *Agric. Biol. Chem.* 37:1457 (1973).

198. J. Kroll, C. Franzke, and S. Genz. Studies on the glyceride structure of fats. 5. Isolation and properties of lipase from *Geotrichum candidum*. *Pharmazie* 28:263 (1973).

199. E. Vandamme, K.H. Schanck-Brodrueck, C. Colson, and J.D.V. Honotier (S.A. Labofina) Eur. Pat. Appl. EP 243338 (Cl. C12N15100), October 28, 1987, GB Appl. 86/10,230, April 25, 1986, p. 19.

200. T. Jacobsen, J. Olsen, and K. Allerman. Production, partial purification and immunochemical characterization of multiple forms of lipase from *Geotrichum candidum*. *Enzyme Microb. Technol.* 11:90 (1989).

201. K. Veeraragavan, T. Colpitts, and B.F. Gibbs. Purification and characterization of two distinct lipases from *Geotrichum candidum*. *Biochim. Biophys. Acta* 1044:26 (1990).

202. M.C. Bertolini, J.D. Schrag, M. Cygler, E. Ziomek, D.Y. Thomas, and T. Vernet. Expression and characterization of *Geotrichum candidum* lipase I gene. *Eur. J. Biochem.* 228:863 (1995).

203. T. Jacobson and O.M. Paulsen. Separation and characterization of 61- and 57-kDa lipase from *Geotrichum candidum* ATCC 66592. *Can. J. Microbiol.* 38:75 (1992).

204. A. Sugihara, Y. Shimada, and Y. Tominaga. Separation and characterization of two molecular forms of *Geotrichum candidum* lipase. *J. Biochem.* 107:426 (1990).

205. T. Asahara, M. Matori, M. Ikemoto, and Y. Ota. Production of two types of lipases with opposite positional specificity by *Geotrichum* sp. F0401B. *Biosci. Biotechnol. Biochem.* 57:390 (1993).

206. P.E. Sonnet, T.A. Foglia, and M.W. Baillargeon. Fatty acid selectivity: The selectivity of lipases of *Geotrichum candidum*. *J. Am. Oil Chem. Soc.* 70:1043 (1993).

207. R.D. Joerger and M.J. Haas. Overexpression of a *Rhizopus delemar* lipase gene in *E. coli*. *Lipids* 28:81 (1993).

208. R.D. Joerger and M.J. Haas. Alteration of chain length selectivity of a *Rhizopus delemar* lipase. *Lipids* 29:377 (1994).

209. M. Lotti, R. Grandor, F. Fusetti, S. Longhi, S. Brocca, A. Tramontano, and L. Alberghina. Cloning and analysis of *Candida cylindracea* lipase sequences. *Gene* 124:45 (1993).

210. M.D. Virto, I. Agud, S. Montero, A. Blanco, R. Solozabal, J.M. Lascaray, M.J. Llama, J.L. Serra, L.C. Landeta, and M. de Renobales. Hydrolysis of animal fats by immobilized *Candida rugosa* lipase. *Enzyme Microb. Technol.* 16:61 (1994).

211. Y. Kosugi and H. Suzuki. Functional immobilization of lipase eliminating lipolysis product inhibition. *Biotechnol. Bioeng.* 40:369 (1992).

212. B.K. Yang and J.P. Chen. Gel matrix influence on hydrolysis of triglycerides by immobilized lipases. *Food Sci.* 59:424 (1994).

213. F. Taylor, M. Kurantz, and J.C. Craig, Jr. Kinetics of continuous hydrolysis of tallow in a multi-layered flatplate immobilized lipase reactor. *J. Am. Oil Chem. Soc.* 69:591 (1992).

214. D. Yang and J.S. Rhee. Continuous hydrolysis of olive oil by immobilized lipase in organic solvent. *Biotechnol. Bioeng.* 40:748 (1992).

215. Y. Kimura, A. Tanaka, K. Sonomoto, T. Nihira, and S. Fukui. Application of immobilized lipase to hydrolysis of triacylglyceride. *Eur. J. Appl. Microbiol. Biotechnol.* 17:107 (1983).

216. C. Miller, H. Austin, L. Posorske, and J. Gonzalez. Characteristics of an immobilized lipase for the commercial synthesis of esters. *J. Am. Oil Chem. Soc.* 65:927 (1988).

217. K. Takahashi, Y. Saito, and Y. Inada. Lipase made active in hydrophobic media. *J. Am. Oil Chem. Soc.* 65:911 (1988).

218. B. Braun, E. Klein, and J.L. Lopez. Immobilization of *Candida rugosa* lipase to nylon fibers using its carbohydrate groups as the chemical link. *Biotechnol. Bioeng.* 51:327 (1996).

219. Y. Dudal and R. Lortie. Influence of water activity on the synthesis of triolein catalyzed by immobilized *Mucor miehei* lipase. *Biotechnol. Bioeng.* 45:129 (1995).

220. P.E. Sonnet, G.P. McNeill, and W. Jun. Lipase of *Geotrichum candidum* immobilized on silica gel. *J. Am. Oil Chem. Soc.* 71:1421 (1994).

221. Y. Kosugi and N. Azuma. Synthesis of triacylglycerol from polyunsaturated fatty acid by immobilized lipase. *J. Am. Oil Chem. Soc.* 71:1397 (1994).

222. Y. Kosugi, T. Kunieda, and N. Azuma. Continual conversion of free fatty acid in rice bran oil to triacylglycerol by immobilized lipase. *J. Am. Oil Chem. Soc.* 71:445 (1994).

223. L.H. Posorske, G.K. LeFebvre, C.A. Miller, T.T. Hansen, and B.L. Glenvig. Process considerations of continuous fat modification with an immobilized lipase. *J. Am. Oil Chem. Soc.* 65:922 (1988).

224. S.-W. Cho and J.S. Rhee. Immobilization of lipase for effective interesterification of fats and oils in organic solvent. *Biotechnol. Bioeng.* 41:204 (1993).

225. C. Brady, L. Metcalfe, D. Slaboszewski, and D. Frank. Lipase immobilized on a hydrophobic, microporous support. *J. Am. Oil Chem. Soc.* 65:917 (1988).

226. K. Long, H.M. Ghazali, A. Ariff, K. Ampon, and C. Bucke. Mycelium-bound lipase from a locally isolated strain of *Aspergillus flavus* link: Pattern and factors involved in its production. *J. Chem. Technol. Biotechnol.* 67:157 (1996).

227. K. Long, H.M. Ghazali, A. Ariff, and C. Bucke. Acidolysis of several vegetable oils by mycelium-bound lipase of *Aspergillus flavus* Link. *J. Am. Oil Chem. Soc.* 74(9):1121 (1997).

228. M.Y.B. Liew, H.M. Ghazali, K. Long, O.M. Lai, and A.M. Yazid. Production and transesterification activity of mycelium-bound lipase from *Rhizomucor miehei*. *Asia Pac. J. Mol. Biol. Biotechnol.* 8:57 (2001).

229. M.Y.B. Liew, H.M. Ghazali, A.M. Yazid, O.M. Lai, M.C. Chow, M.S.A. Yusoff, and K. Long. Rheological properties of ice cream emulsion prepared from lipase-catalysed transesterified palm kernel olein: Anhydrous milkfat mixture. *J. Food Lipids* 8:131 (2001).

230. M.Y.B. Liew, H.M Ghazali, K. Long, O.M. Lai, and A.M. Yazid. Physical properties of palm kernel olein-anhydrous milk fat mixtures transesterified using mycelium-bound lipase from *Rhizomucor miehei*. *Food Chem.* 72:447 (2001).

231. T. Antczak, J. Graczyk, M. Szcesna-Antczak, and S. Bielecki. Activation of *Mucor circinelloides* lipase in organic medium. *J. Mol. Catal. B Enzym.* 19:287 (2002).

232. S.Y. Lee and J.S. Rhee. Production and partial purification of a lipase from *Pseudomonas putida* 3SK. *Enzyme Microb. Technol.* 15:617 (1993).

233. S.Y. Lee and J.S. Rhee. Hydrolysis of triglyceride by the whole cell of *Pseudomonas putida* 3SK in two-phase batch and continuous reactor system. *Biotechnol. Bioeng.* 44:437 (1994).

234. K. Bao, M. Kaieda, T. Matsumoto, A. Kondo, and H. Fukuda. Whole cell biocatalyst for biodiesel fuel production utilizing *Rhizopus oryzae* cells immobilized within biomass support particles. *Biochem. Eng.* 8:39 (2001).

235. T. Matsuda, Y. Nakajima, T. Harada, and K. Nakamura. Asymmetric reduction of simple aliphatic ketones with dried cells of *Geotrichum candidum*. *Tetrahedron Asymmetry* 13:971 (2002).

236. S. Fukui and A. Tanaka. Immobilised microbial cells. *Annu. Rev. Microbiol.* 36:145 (1982).

237. K. Long, H.M. Ghazali, A. Ariff, K. Ampon, and C. Bucke. In-situ crosslinking of *Aspergillus flavus* lipase: Improvement of activity, stability and properties. *Biotechnol. Lett.* 18:1169 (1996).

238. A.M. Klibanov. Enzyme stabilization by immobilization. *Anal. Biochem.* 93:1 (1979).

239. J.A. Blain, J.D.E. Patterson, and C.E.L. Shaw. The nature of mycelial lipolytic enzymes in filamentous fungi. *FEMS Microbiol. Lett.* 3:85 (1978).

240. B.C. Buckland, P. Dunnill, and M.D. Lilly. The enzymatic transformation of water-insoluble reactants in nonaqueous solvents conversion of cholesterol to cholest-4-ene-3-one by a *Nocardia* sp. *Biotechnol. Bioeng.* 17:815 (1975).

241. G. Bell, J.R. Todd, J.A. Blain, J.O.E Patterson, and C.E.L. Shaw. Hydrolysis of triglycerides by solid phase lipolytic enzyme of *Rhizopus arrhizus* in continuous reactor system. *Biotechnol. Bioeng.* 23:1703 (1981).

242. S. Fukui, A. Tanaka, and T. Iida. Immobilization of biocatalysts for bioprocesses in organic solvent media. In: *Biocatalysis in Organic Media* (C. Laane, J. Tramper, and M.D. Lilly, eds.). Elsevier, Amsterdam, 1986, p. 21.

243. Y. Kosugi and H. Suzuki. Fixation of cell-bound lipase and properties of the fixed lipases as an immobilised lipase. *J. Ferment. Technol.* 51:895 (1973).

244. J.D.E. Patterson, J.A. Blain, C.E.L. Shaw, R. Todd, and G. Bell. Synthesis of glycerides and esters by fungal cell-bound enzymes in continuous reactor system. *Biotechnol. Lett.* 1:211 (1979).

245. R.L. Antrim, W. Colilla, and B.J. Schnyder. Glucose isomerase production of high-fructose syrup. In: *Applied Biochemistry and Bioengineering, Vol. 2, Enzyme Technology* (L.B. Wingard, Jr., E. Katchalski-Katzir, and L. Goldstein, eds.). Academic Press, London and New York, 1972, p. 98.

246. M. Szczesna-Antczak, T. Antczak, M. Rzyska, Z. Modrzejewska, J. Patura, H. Kalinowska, and S. Bielecki. Stabilisation of an intracellular *Mucor circinelloides* lipase for application in non-aqueous media. *J. Mol. Catal. B Enzym.* 29:163 (2004).

247. F. Bjorkling, S.E. Godtfredsen, and O. Kirk. The future impact of industrial lipases. *Trends Biotechnol.* 9:360 (1991).

248. E.N. Vulfson. Industrial applications of lipases. In: *Lipases* (P. Woolley and S.B. Petersen, eds.). Cambridge University Press, Cambridge, 1994, p. 271.

249. E.W. Seitz. Industrial applications of microbial lipases—A review. *J. Am. Oil Chem. Soc.* 51:12 (1974).

250. T. Nielson. Industrial application possibilities of lipase. *Fat Sci. Technol.* 87:15 (1985).

251. A.R. Macrae and R.C. Hammond. Present and future application of lipases. *Biotechnol. Genet. Eng. Rev.* 3:193 (1983).

252. M. Iwai and Y. Tsujisaka. Fungal lipases. In: *Lipases* (B. Borgstrom and H.L. Brockman, eds.). Elsevier Science, Amsterdam, 1984, p. 443.

253. A.R. Macrae. Biotechnology in the fats and oils industry. In: *Proceeding of the World Conference on Emerging Technologies in the Fats and Oils Industry* (A.R. Baldwin, ed.). American Oil Chemists' Society, Champaign, 1986, pp. 7–13.

254. T. Godfrey. Lipases for industrial use. *Lipid Technol.* 7:58 (1995).

255. O.M. Lai, S.K. Lo, and C.C. Akoh. Patent review on lipid technology. In: *Healthful Lipids* (C.C. Akoh and O.M. Lai, eds.). AOCS Press, Champaign, IL, USA, 2005, p. 433.

256. X. Fu, X. Zhu, K. Gao, and J. Duan. Oil and fat hydrolysis with lipase from *Aspergillus* sp. *J. Am. Oil Chem. Soc.* 72:527 (1995).

257. J.B.M. Rattray. Biotechnology and the fats and oils industry—An overview. *J. Am. Oil Chem. Soc.* 61:1701 (1984).

258. M. Mittlebach. Lipase catalyzed alcoholysis of sunflower oil. *J. Am. Oil Chem. Soc.* 67:168 (1990).

259. J. Xia, X. Chen, and I.A. Nnanna. Activity and stability of *Penicillium cyclopium* lipase in surfactant and detergent solution. *J. Am. Oil Chem. Soc.* 73:115 (1996).

260. Y. Kodera, H. Nishimura, A. Matsushima, M. Hiroto, and Y. Inada. Lipase made active in hydrophobic media by coupling with polyethylene glycol. *J. Am. Oil Chem. Soc.* 71:335 (1994).

261. M.J. Haas, J. Allen, and T.R. Berka. Cloning, expression and characterization of a cDNA encoding a lipase from *Rhizopus delemar. Gene* 109:107 (1991).

262. N. Rashid, Y. Shimada, S. Ezaki, H. Atomi, and T. Imanaka. Low temperature lipase from psychrotropic *Pseudomonas* sp. strain KB700A. *Appl. Environ. Microbiol.* 67:4064 (2001).

263. D.W. Choo, T. Kurihara, T. Suzuki, K. Soda, and N. Esaki. A cold adapted lipase of an Alaskan psychrotroph, *Pseudomonas* sp. strain B11-1: Gene cloning and enzyme purification and characterization. *Appl. Environ. Microbiol.* 64:486 (1998).

264. H. Kimura, S. Hirose, H. Sakurai, K. Mikawa, F. Tomita, and K. Shimazaki. Molecular cloning and analysis of a lipase gene from *Pseudomonas fluorescens* No. 33. *Biosci. Biotechnol. Biochem.* 62:2233 (1998).

265. G.H. Chung, Y.P. Lee, O.J. Yoo, and J.S. Rhee. Overexpression of a thermostable lipase gene from *Pseudomonas fluorescens* in *Escherichia coli. Appl. Environ. Biotechnol.* 35:237 (1991).

266. B.C. Oh, H.K. Kim, J.K. Lee, S.C. Kang, and T.K. Oh. *Staphylococcus haemolyticus* lipase: Biochemical properties, substrate specificity and gene cloning. *FEMS Microbiol. Lett.* 179:385 (1999).

267. H.K. Kim, S.Y. Park, J.K. Lee, and T.K. Oh. Gene cloning and characterization of thermostable lipase from *Bacillus stearothermophilus* LI. *Biosci. Biotechnol. Biochem.* 62:66 (1998).

268. Z.-Y. Li and O.P. Ward. Enzyme catalyzed production of vegetable oils containing omega-3 polyunsaturated fatty acid. *Biotechnol. Lett.* 15:185 (1993).

269. Z.-Y. Li and O.P. Ward. Lipase catalyzed alcoholysis to concentrate the *n*-3 polyunsaturated fatty acid of cod liver oil. *Enzyme Microb. Technol.* 15:601 (1993).

270. S.P.J.N. Senanayake and F. Shahidi. Lipase-catalysed incorporation of docosahexaenoic acid (DHA) into borage oil: Optimization using response surface methodology. *Food Chem.* 77:115 (2002).

271. A. Halldorsson, B. Kristinsson, C. Glynn, and G.G. Haraldsson. Separation of EPA and DHA in fish oil by lipase-catalysed esterification with glycerol. *J. Am. Oil Chem. Soc.* 80:915 (2003).

272. H. Maehr, G. Zenchoff, and D.L. Cohen. Enzymatic enhancement of *n*-3 fatty acid content in fish oils. *J. Am. Oil Chem. Soc.* 71:463 (1994).

273. C.F. Torres, H.S. Garcia, J.J. Ries, and C.G. Hill, Jr. Esterification of glycerol with conjugated linoleic acid and long-chain fatty acids from fish oil. *J. Am. Oil Chem. Soc.* 78:1093 (2001).

274. J.A. Arcos, H.S. Garcia, and C.G. Hill, Jr. Continuous enzymatic esterification of glycerol with (poly) unsaturated fatty acids in a packed-bed reactor. *Biotechnol. Bioeng.* 68:563 (2000).

275. G.G. Haraldsson, A. Halldorsson, and E. Kulas. Chemoenzymatic synthesis of saturated triacylglycerols containing eicosapentaenoic and docosahexaenoic acids. *J. Am. Oil Chem. Soc.* 77:1139 (2000).

276. A.R. Medina, L.E. Cerdan, A.G. Gimenez, B.C. Paez, M.J.I. Gonzalez, and E.M. Grima. Lipase-catalyzed esterification of glycerol and polyunsaturated fatty acids from fish and microalgae oils. *J. Biotechnol.* 70:379 (1999).

277. Y. He and F. Shahidi. Enzymatic esterification of ω-3 fatty acid concentrates from seal blubber oil with glycerol. *J. Am. Oil Chem. Soc.* 74:1133 (1997).

278. A. Can and B. Ozcelik. Enrichment of hazelnut oil with long-chain *n*-3 PUFA by lipase-catalyzed acidolysis: Optimization by response surface methodology. *J. Am. Oil Chem. Soc.* 82:27 (2005).

279. F. Hamam, J. Daun, and F. Shahidi. Lipase-assisted acidolysis of high-laurate canola oil with eicosapentaenoic acid. *J. Am. Oil Chem. Soc.* 82:875 (2005).

280. F. Hamam and F. Shahidi. Structured lipids from high-laurate canola oil and long-chain omega-3 fatty acids. *J. Am. Oil Chem. Soc.* 82:731 (2005).

281. S.K. Lo, B.S. Baharin, C.P. Tan, and O.M. Lai. Enzyme-catalyzed production and chemical composition of diacylglycerols from corn oil deodoriser distillates. *Food Biotechnol.* 18:265 (2004).

282. S.K. Lo, B.S. Baharin, C.P. Tan, and O.M. Lai. Diacylglycerols from palm oil deodorizer distillate: Part 1. Synthesis by lipase-catalyzed esterification. *Food Sci. Technol. Int.* 10:149 (2004).

283. S.K. Lo, B.S. Baharin, C.P. Tan, and O.M. Lai. Diacylglycerols from palm oil deodorizer distillate: Part 2. Physical and chemical characterization. *Food Sci. Technol. Int.* 10:157 (2004).

284. S.K. Lo, B.S. Baharin, C.P. Tan, and O.M. Lai. Analysis of 1,2- and 1,3-positional isomers of diacylglycerols from vegetable oils by reversed-phase high-performance liquid chromatography. *J. Chromatogr. Sci.* 42:145 (2004).

285. F.C. Huang and Y.H. Ju. Interesterification of palm midfraction and stearic acid with *Rhizopus arrhizus* lipase immobilized on polypropylene. *J. Chin. I. Ch. E.* 28:73 (1997).

286. R. Sridhar, G. Lakshminarayana, and T.N.B. Kaimal. Modification of selected Indian vegetable fats into cocoa butter substitutes by lipase-catalyzed ester interchange. *J. Am. Oil Chem. Soc.* 68:726 (1991).

287. X. Xu, L.B. Fomuso, and C.C. Akoh. Synthesis of structured triacylglycerols by lipase-catalyzed acidolysis in a packed-bed bioreactor. *J. Agric. Food Chem.* 48:3 (2000).

288. X. Xu, L.B. Fomuso, and C.C. Akoh. Modification of menhaden oil by enzymatic acidolysis to produce structured lipids: Optimization by response surface design in a packed bed reactor. *J. Am. Oil Chem. Soc.* 77:171 (2000).

289. W.M. Willis and A.G. Marangoni. Assessment of lipase- and chemically-catalyzed lipid modification strategies for the production of structured lipids. *J. Am. Oil Chem. Soc.* 76:443 (1999).

290. X. Xu, H. Mu, C.E. Hoy, and J. Adler-Nissen. Production of specifically structured lipids by enzymatic interesterification in a pilot enzyme bed reactor: Process optimization by response surface methodology. *Fett/Lipid* 101:207 (1999).

291. H. Mu, X. Xu, and C.E. Hoy. Production of specific-structured triacylglycerols by lipase-catalyzed interesterification on a laboratory-scale continuous reactor. *J. Am. Oil Chem. Soc.* 75:1187 (1998).

292. L. Mojovic, S. Siler-Marinkovic, G. Kukic, B. Bugarski, and G. Vunjak-Nokakovic. *Rhizopus arrhizus* lipase-catalyzed interesterification of palm oil midfraction in a gas-lift reactor. *Enzyme Microb. Technol.* 16:159 (1994).

293. N.N. Gandhi and K.D. Mukherjee. Reactivity of medium-chain substrates in the interesterification of tripalmitin catalyzed by papaya lipase. *J. Am. Oil Chem. Soc.* 78:965 (2002).

294. O.M. Lai, H.M. Ghazali, F. Cho, and C.L. Chong. Physical and textural properties of an experimental table margarine prepared from lipase-catalysed transesterified palm stearin: Palm kernel olein mixture during storage. *Food Chem.* 71:173 (2000).

295. B.S. Chu, H.M. Ghazali, O.M. Lai, Y.B. Che Man, and S. Yusof. Physical and chemical properties of a lipase-transesterified palm stearin/palm kernel olein blend and its isopropanol-solid and high melting triacylglycerol fractions. *Food Chem.* 76:155 (2002).

296. O.M. Lai, H.M. Ghazali, F. Cho, and C.L. Chong. Enzymatic transesterification of palm stearin: Anhydrous milkfat mixtures using 1,3-specific and non-specific lipases. *Food Chem.* 70:221 (2000).

297. X. Rendón, A. López-Munguía, and E. Castillo. Solvent engineering applied to lipase-catalyzed glycerolysis of triolein. *J. Am. Oil Chem. Soc.* 78:1061 (2001).

298. T. Hirose, Y. Yamauchi-Sato, Y. Arai, and S. Negishi. Synthesis of triacylglycerol containing conjugated linoleic acid by esterification using two blended lipases. *J. Am. Oil Chem. Soc.* 83:35 (2006).

299. I.-H. Kim and C.G. Hill, Jr. Lipase-catalyzed acidolysis of menhaden oil with pinolenic acid. *J. Am. Oil Chem. Soc.* 83:109 (2006).

300. S.E. Lumor and C.C. Akoh. Enzymatic incorporation of stearic acid into a blend of palm olein and palm kernel oil: Optimization by response surface methodology. *J. Am. Oil Chem. Soc.* 82:421 (2005).

301. Y. Watanabe, Y. Yamauchi-Sato, T. Nagao, S. Negishi, T. Terai, T. Kobayashi, and Y. Shimada. Production of MAG of CLA by esterification with dehydration at ordinary temperature using *Penicillium camembertii* lipase. *J. Am. Oil Chem. Soc.* 82:619 (2005).

302. J.R.C. Reddy, T. Vijeeta, M.S.L. Karuna, B.V.S.K. Rao, and R.B.N. Prasad. Lipase-catalyzed preparation of palmitic and stearic acid-rich phosphatidylcholine. *J. Am. Oil Chem. Soc.* 82:727 (2005).

303. R. Bott, J.W. Shield, and A.J. Poulose. Protein engineering of lipases. In: *Lipases* (P. Woolley and S.B. Petersen, eds.). Cambridge University Press, Cambridge, 1994, p. 337.

304. Svendsen Engineered lipases for practical use. *INFORM* 5:619 (1994).

Enzymatic Interesterification

Wendy M. Willis and Alejandro G. Marangoni

CONTENTS

I. INTRODUCTION

The development of methods to improve the nutritional and functional properties of fats and oils is of great interest to food processors. The molecular weight, unsaturation, and positional distribution of fatty acid residues on the glycerol backbone of triacylglycerols are the principal factors determining the physical properties of fats and oils [1,2]. Chemical interesterification produces a complete positional randomization of acyl groups in triacylglycerols. It is used in the manufacture of shortenings, margarines, and spreads to improve their textural properties, modify melting behavior, and enhance stability [3,4]. Interest in interesterification from a nutritional and functional standpoint is increasing since it can be used to produce margarines with no *trans* unsaturated fatty acids, synthesize cocoa butter substitutes, and improve the nutritional quality of some fats and oils [5]. Recently, research efforts have been directed to substituting some chemical interesterification applications with enzymatic interesterification because of the inherent advantages associated with the enzymatic process. Enzymatic reactions are more specific, require less severe reaction conditions, and produce less waste. In addition, when immobilized, enzymes can be reused, thereby making them economically attractive [6]. Interesterification, whether chemical or enzymatic, is the exchange of acyl groups between an ester and an acid (acidolysis), an ester and an alcohol (alcoholysis), an ester and an ester (transesterification) [7].

The major components of fats and oils are triacylglycerols, the composition of which is specific to the origin of each fat or oil. The physical properties of various fats and oils are different because of the structure and distribution of fatty acids in the triacylglycerols [8]. In natural fats, acyl groups are distributed in a nonrandom fashion. During chemical or enzymatic interesterification, acyl groups are redistributed first intramolecularly, then intermolecularly until a random distribution is achieved. With enzymatic interesterification, more control of final product composition is possible, and glyceride mixtures that cannot be obtained using chemical interesterification can be produced [9,10]. At present, randomization of acyl group distribution using chemical interesterification is used to produce changes in crystal structure, solid fat content, and melting point of fats. Interesterification using lipase with particular specificities is used to produce high-value specialty fats, such as cocoa butter substitutes and confectionary fats [5].

Enzymatic interesterification is accomplished using lipases, which are enzymes obtained predominantly from bacterial yeast, and fungal sources. Extracellular microbial lipases are produced by microorganisms and released into their growth environment to digest lipid materials [9]. Lipases are defined as glycerol ester hydrolases (EC 3.1.1.3) because they catalyze the hydrolysis of carboxyl ester bonds in acylglycerols. Depending on the degree of hydrolysis, free fatty acids, monoacylglycerols, diacylglycerols, and glycerol are produced. Lipases are differentiated from esterases in that they act only on insoluble substrates. Long-chain triacylglycerols, the natural substrates of lipases, are insoluble in water, forming aggregates or dispersions in aqueous media. Lipases have a high affinity for hydrophobic surfaces and can be completely adsorbed from aqueous solution by emulsified long-chain triacylglycerols [11]. In the presence of excess water, lipases catalyze the hydrolysis of long-chain triacylglycerols, but under water-limiting conditions, the reverse reaction, ester synthesis, can be achieved [8,12]. Enzymatic interesterification systems are composed of a continuous water-immiscible phase, containing the lipid substrate, and an aqueous phase containing the lipase. Lipase-catalyzed interesterifications have been extensively studied in systems using organic solvents. However, if such a process is to be used in the food industry, solvent-free systems must be developed. Hence, the emphasis of this chapter will be on enzymatic interesterification performed in solvent-free systems.

A. Transesterification

As previously defined, transesterification is the exchange of acyl groups between two esters, namely, two triacylglycerols (Figure 30.1). Transesterification is used predominantly to alter the

$$R_1 - \overset{\overset{\displaystyle O}{\|}}{C} - O - R_2 + R_3 - \overset{\overset{\displaystyle O}{\|}}{C} - O - R_4 \rightleftharpoons R_1 - \overset{\overset{\displaystyle O}{\|}}{C} - O - R_4 + R_3 - \overset{\overset{\displaystyle O}{\|}}{C} - O - R_2$$

FIGURE 30.1 Lipase-catalyzed transesterification between two different triacylglycerols.

physical properties of individual fats and oils or fat–oil blends by altering the positional distribution of fatty acids in the triacylglycerols. Transesterification of butter using a nonspecific lipase has been reported to improve the plasticity of the fat [13]. Kalo et al. [14] found that transesterification of butterfat with a positionally nonspecific lipase at 40°C increased the level of saturated C48 to C54 triacylglycerols, monoene C38 and C46 to C52 triacylglycerols, and diene C40 to C54 triacylglycerols. These authors also found that the diacylglycerol content increased by 45% whereas the free fatty acid content doubled. Overall, lipase-catalyzed transesterification of butterfat at 40°C produced an increase in the solid fat content below 15°C and a decrease in the solid fat content above 15°C (Figure 30.2).

In another study, lipase-catalyzed transesterification of butter increased the relative proportion of C36 and C40–C48 saturated triacylglycerols, as well as triunsaturated triacylglycerols [15]. The resulting product had a 114% greater solid fat content at 20°C than the starting butter, with the solid fat content increasing from 22% to 46%. In general, lipase-catalyzed transesterification produces fat with a slightly lower solid fat content compared with chemical interesterification. This is attributed to contamination by monoacylglycerols, diacylglycerols, and free fatty acids, which are produced in the early stages of transesterification [8,13]. Kalo et al. [13] compared lipase-catalyzed transesterification with chemical interesterification of butter. He found that the solid fat content of butter increased from 41.2% to 42.2% at 20°C using lipase-catalyzed transesterification, whereas chemical interesterification produced butter with a solid fat content of 57.8% at 20°C. Transesterification has also been used to improve the textural properties of tallow and rapeseed oil mixtures as well as in the development of cocoa butter equivalents [16,17]. Forssell et al. [18] found that transesterification of a tallow and rapeseed oil blend decreased the solid fat content and melting point. The extent of melting point reduction was dependent on the mass fraction of the two lipid components. With a mass fraction of tallow to rapeseed oil of 0.8, the melting point was reduced by 6°C, whereas a mass

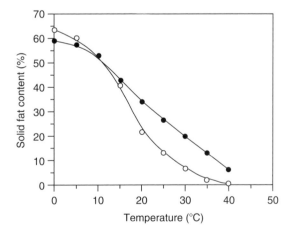

FIGURE 30.2 Solid fat content versus temperature profiles for native and enzymatically interesterified butterfat in the absence of solvent using lipase from *Pseudomonas fluorescens*. Nontransesterified butterfat (○); transesterified butterfat (●). (From Kalo, P., Parviainen, P., Vaara, K., Ali-Yrrkö, S., and Antila, M., *Milchwissenschaft*, 41, 82, 1986. With permission.)

fraction of 0.5 produced a 12°C decrease in melting point. A decrease in the solid fat content has also been observed on transesterification between palm oil and canola oil, due to a decrease in the level of triunsaturated triacylglycerols [19].

The attractiveness of cocoa butter to the chocolate and confectionary industry is based on the limited diversity of triacylglycerols in this fat, which gives it a unique, narrow melting range of 29°C–43°C. Chocolate can contain 30% cocoa butter, meaning that this fat determines the crystallization and melting properties of the chocolate. At 26°C cocoa butter is hard and brittle, but when eaten it melts completely in the mouth with a smooth, cool sensation. The major triacylglycerols in cocoa butter are 1-palmitoyl-2-oleoyl-3-stearoylglycerol (POS), 1,3-dipalmitoyl-2-oleoylglycerol (POP), and 1,3-distearoyl-2-oleoylglycerol (SOS) with levels of 41%–52%, 16%, and 18%–27%, respectively [8,17]. The main disadvantage of using cocoa butter in chocolate and confections is its high cost. A cocoa butter equivalent can be made from inexpensive fats and oils by interesterification. By transesterifying fully hydrogenated cottonseed and olive oil, Chang et al. [17] were able to produce a cocoa butter substitute with similar POS levels and slightly higher SOS levels than those found in cocoa butter. The melting range of the transesterified product was 29°C–49°C, compared with 29°C–43°C for cocoa butter. In order to remove the desired triacylglycerol product from the other triacylglycerols, trisaturated triacylglycerols were removed by crystallization in acetone. High-oleic sunflower oil and palm oil fraction have also been transesterified to obtain cocoa butter substitutes [5].

B. ACIDOLYSIS

Acidolysis, the transfer of an acyl group between an acid and an ester, is an effective means of incorporating novel free fatty acids into triacylglycerols (Figure 30.3). Acidolysis has been used to incorporate free acid or ethyl ester forms of eicosapentanoic acid (EPA) and docosahexanoic acid (DHA) into vegetable and fish oils to improve their nutritional properties. The nutritional benefits of consuming polyunsaturated fatty acids (PUFAs), such as EPA and DHA, derived from fish oils have been proven. When consumed, EPA reduces the risk of cardiovascular disease by reducing the tendency to form blood clots, whereas DHA consumption is required for proper nervous system and visual functions, due to its accumulation in the brain and retina [20,21]. Concentrations of EPA and DHA in fish oils to levels approaching 30% can be achieved using molecular distillation, winterization, and solvent crystallization. However, performing an acidolysis reaction between cod liver oil and free EPA and DHA, Yamane et al. [22] were able to increase the EPA content in the oil from 8.6% to 25% and the DHA content from 12.7% to 40% using immobilized lipase from *Mucor miehei*. Using ethyl esters of EPA, fish oil has been enriched by interesterification to contain 40% EPA and 25% DHA (wt%) [23]. During acidolysis in a fixed bed reactor, Yamane et al. [24] increased the PUFA content of cod liver oil by reducing the temperature to between −10°C and −20°C in the product reservoir. This led to crystallization and removal of more saturated fatty acids present in the fish oil. Lipases with strong specificities against EPA or DHA have also been used to enrich their content in fish oils [25]. Future development in lipase-catalyzed interesterification using EPA and DHA is directed to improving the nutritional quality of vegetable oils by enrichment with these fish oil–derived fatty acids. Acidolysis has also been used by Oba and Witholt [26] to incorporate oleic acid into milk fat. This process led to an increase in the level of unsaturated fatty acids in butter without losses in the characteristic flavor of butter. Acidolysis of milk fat with oleic acid was also found to decrease the crystallization temperature and lower the melting range of the milk lipids.

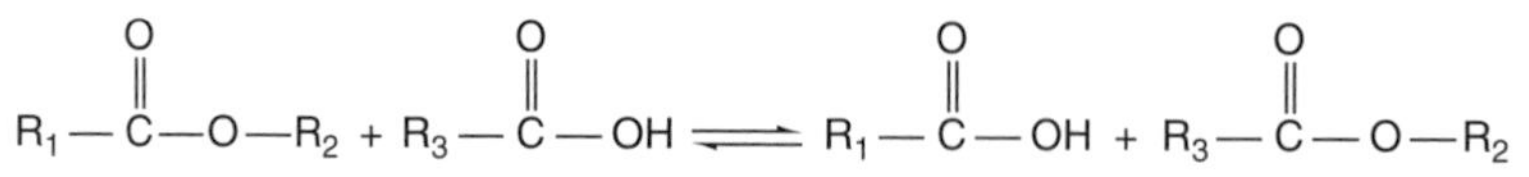

$$R_1 - \overset{\overset{\textstyle O}{\|}}{C} - O - R_2 + R_3 - \overset{\overset{\textstyle O}{\|}}{C} - OH \rightleftharpoons R_1 - \overset{\overset{\textstyle O}{\|}}{C} - OH + R_3 - \overset{\overset{\textstyle O}{\|}}{C} - O - R_2$$

FIGURE 30.3 Lipase-catalyzed acidolysis reaction between an acylglycerol and an acid.

Along with the enrichment of oils, acidolysis using EPA and DHA has also been useful in the synthesis of structured lipids. Structured lipids are composed of medium- and long-chain fatty acids, which meet the nutritional needs of hospital patients and those with special dietary needs. When consumed, medium-chain fatty acids, such as capric and caproic acids, are not incorporated into chylomicrons and are therefore not likely to be stored, but will be used for energy. They are readily oxidized in the liver and constitute a highly concentrated source of energy for premature babies and patients with fat malabsorption disease. Medium-chain fatty acids also possess a nutritional advantage compared with other fatty acids in that they are nontumor-producing forms of fat [27]. Long-chain fatty acids are also required by the body, especially in the form of PUFAs in the form of ω-3 and ω-6 fatty acids, which have been associated with reduced risk of platelet aggregation and cardiovascular disease, and the lowering of cholesterol [27,28]. When PUFAs are present in the sn-2 position and medium-chain fatty acids are present in the sn-1,3 positions, they are rapidly hydrolyzed by pancreatic lipase, absorbed, and oxidized for energy, whereas essential PUFAs are absorbed as 2-monoacyglycerols. Therefore, structuring triacylglycerols with medium-chain fatty acids and PUFAs can dramatically improve the nutritional properties of triacylglycerols [29]. Producing a triacylglycerol rich in EPA or DHA at the sn-2 position, with medium-chain fatty acids in the sn-1 or sn-3 positions, would provide maximal benefit, especially for intravenous use in hospitals [30]. Structured lipids that are reduced in caloric content have also been developed by esterifying long-chain monoacylglycerols containing behenic acid with capric acid. The produced triacylglycerols contain half the calories relative to natural triacylglycerols due to the incomplete absorption of behenic acid during digestion [31].

Acidolysis is also a common method for production of cocoa butter substitutes. The most common method is acidolysis of palm oil midfraction, which contains predominantly POP with stearic acid to increase the level of stearate in the lipid [32]. Chong et al. [33] also incorporated stearic acid into palmolein to produce 25% cocoa butter-like triacylglycerols.

C. ALCOHOLYSIS

As previously mentioned, alcoholysis is the esterification reaction between an alcohol and an ester (Figure 30.4). Alcoholysis has been used in the production of methyl esters from esterification of triacylglycerols and methanol with yields of up to 53% [34]). During alcoholysis, hydrolysis of triacylglycerols to produce diacylglycerols and monoacylglycerols can occur, in some cases reaching levels as high as 11%, although the presence of small amounts of alcohol can inhibit hydrolysis. The main use of alcoholysis is in the performance of glycerolysis reactions.

Glycerolysis is the exchange of acyl groups between glycerol and a triacylglycerol to produce monoacylglycerols, diacylglycerols, and triacylglycerols. There are several ways to produce monoacylglycerols, which are of great importance in the food industry as surface-active agents and emulsifiers. Monoacylglycerols can be produced by ester exchange between triacylglycerols and glycerols, or by free fatty acids and glycerol, although only the former reaction is termed glycerolysis (Figure 30.5). Glycerolysis is usually performed using nonspecific lipases, giving a wide range of reaction products (Figure 30.6).

High yields in lipase-catalyzed monoacylglycerol synthesis are achieved by temperature-induced crystallization of newly formed monoacylglycerols from the reaction mixture. This pushes the equilibrium of the reaction toward increased monoacylglycerol production. During glycerolysis, lipids containing saturated monoacylglycerols in the reaction product mixture crystallize at lower

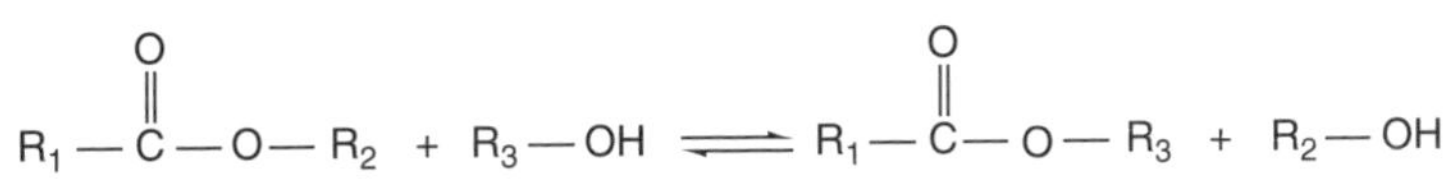

$$R_1 - \overset{\overset{\displaystyle O}{\|}}{C} - O - R_2 \ + \ R_3 - OH \ \rightleftharpoons \ R_1 - \overset{\overset{\displaystyle O}{\|}}{C} - O - R_3 \ + \ R_2 - OH$$

FIGURE 30.4 Lipase-catalyzed alcoholysis reaction between an acylglycerol and an alcohol.

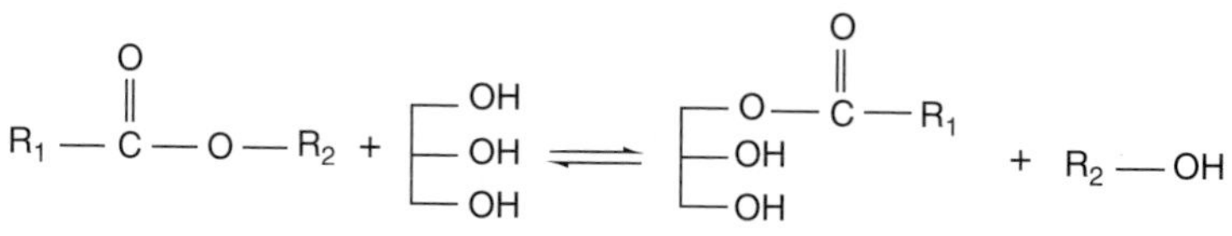

$$R_1 - \overset{\overset{\displaystyle O}{\|}}{C} - O - R_2 \;+\; \left[\begin{array}{l} -OH \\ -OH \\ -OH \end{array}\right] \;\rightleftharpoons\; \left[\begin{array}{l} -O-\overset{\overset{\displaystyle O}{\|}}{C}-R_1 \\ -OH \\ -OH \end{array}\right] \;+\; R_2-OH$$

FIGURE 30.5 Lipase-catalyzed glycerolysis reaction between glycerol and a triacylglycerol to produce monoacylglycerols.

temperatures than unsaturated monoacylglycerols [35]. *Pseudomonas fluorescens* and *Chromobacterium viscosum* have been shown to have high glycerolysis activity [36]. In glycerolysis reactions, T_c is defined as the critical temperature below which monoacylglycerols formed by glycerolysis crystallize out of the reaction mixture. Removal of monoacylglycerols from the reaction mixture pushes the equilibrium of the reaction toward increased monoacylglycerol production. Vegetable oils with low melting points due to the presence of long-chain unsaturated fatty acids have a much lower T_c than animal fats. The T_c for vegetable oils ranges from 5°C to 10°C, whereas it is between 30°C and 46°C for animal fats. By reducing the temperature below T_c, yields of monoacylglycerols can be increased from 30% up to yields as high as 90% [35,36]. Water content can also have an effect on glycerolysis since the reaction is an esterification. McNeill et al. [36] found that increasing the water content from 0.5% to 5.7% increased the production of monoacylglycerols, whereas higher levels of water did not increase the rate of reaction further. The main problem with lipase-catalyzed glycerolysis is the long reaction time in the order of 4–5 days required to produce high yields [36].

II.   LIPASES

A.   THREE-DIMENSIONAL STRUCTURE

While lipases can be derived from animal, bacterial, and fungal sources, they all tend to have similar three-dimensional structures. In the period from 1990 to 1995, crystallographers solved the high-resolution structures of 11 different lipases and esterases including 4 fungal lipases, 1 bacterial lipase, and human pancreatic lipase [12]. Comparison of the amino acid sequences has shown large differences between most lipases, yet all have been found to fold in similar ways and have similar catalytic sites. The characteristic patterns found in all lipases studied so far have included

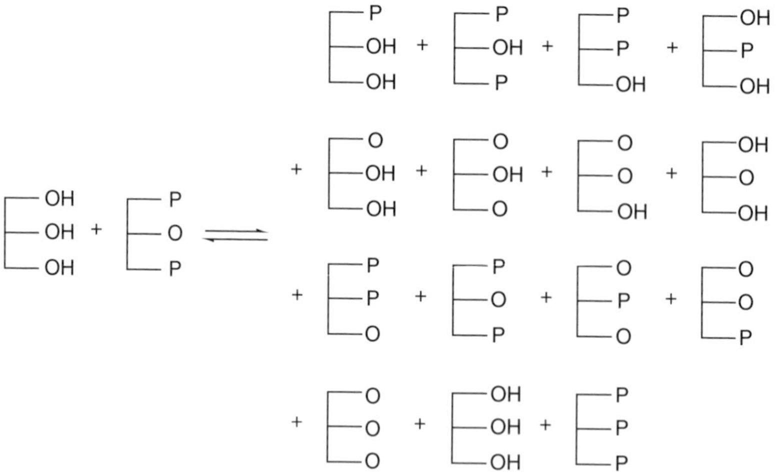

FIGURE 30.6 Products of a nonspecific lipase-catalyzed glycerolysis reaction between glycerol and 1,3-dipalmitoyl-2-oleoylglycerol.

α/β-structures with a mixed central β-sheet containing the catalytic residues. In general, a lipase is a polypeptide chain folded into two domains: the C-terminal domain and the N-terminal domain. The N-terminal domain contains the active site with a hydrophobic tunnel from the catalytic serine to the surface that can accommodate a long fatty acid chain.

In solution, a helical segment covers the active site of lipase, but in the presence of lipids or organic solvent, there is a conformational change in which the lid opens, exposing the hydrophobic core containing the active site. The structure of the lid differs for lipases in the number and position of the surface loops. For example, human pancreatic lipase has one α-helix (residues 237–261) in the loop covering the active site pocket [37,38]. The fact that the α-helix in the lid is amphipathic is very important in terms of the ability of the lipase to bind to lipid at the interface. If the amphiphilic properties of the loop are reduced, the activity of the enzyme is decreased [39]. The outside of the loop is relatively hydrophilic, whereas the side facing the catalytic site is hydrophobic. On association with the interface, the lid folds back, revealing its hydrophobic side which leads to increased interactions with the lipid at the interface [40]. The substrate can then enter the hydrophobic tunnel containing the active site.

B. Active Site

Koshland's modern induced fit hypothesis states that the active site does not have to be a preexisting rigid cavity, but instead can be a precise spatial arrangement of several amino acid residues that are held in the correct orientation by the other amino acids in the enzyme molecule [41]. The main component of the catalytic site is an α/β-hydrolase fold that contains a core of predominantly parallel β-sheets surrounded by α-helices. The folding determines the positioning of the catalytic triad composed of serine, histidine, and either glutamic acid or aspartic acid along with several oxyanion-stabilizing residues. The nucleophilic serine rests between a β-strand and an α-helix, whereas histidine and aspartic acid or glutamic acid rest on one side of the serine [12].

The importance of the serine residue for the catalytic activity of lipase has been demonstrated using site-directed mutagenesis. Substitution of Ser153 in human pancreatic lipase produces a drastic decrease in the catalytic activity of the enzyme, but has no effect on the ability of the enzyme to bind to micelles. As well, the presence of a highly hydrophobic sequence of amino acid residues has been verified in the vicinity of the active site, which is important in the interaction of the enzyme with the interface [42]. The chemical properties of the groups within the catalytic triad are consistent with a hydrophobic environment [11]. The process of opening the lid covering the active site causes the oxyanion hole to move into proper positioning for interaction with the substrate. For example, lipase from *Rhizomucor miehei* has a serine side chain at position 82 that assumes a favorable conformation for oxyanion interactions only after the lid has moved away from the active site [43]. During binding of the substrate with the enzyme, an ester binds in the active site, so that the alcohol portion of the substrate rests on a floor formed by the end of the β-strand while the acyl chain arranges itself in the hydrophobic pocket and tunnel region [42] (Figure 30.7).

In the lipase from *M. miehei*, the substrate-binding region is seven-carbon long. When longer chains are encountered, the rest of the carbons in the chain hang outside the hydrophobic tunnel [44]. When the lipase approaches the interface and the lid is folded back, an oxyanion-stabilizing residue is placed in proper orientation [12]. During hydrolysis, the tetrahedral intermediate is stabilized by hydrogen bonds with backbone amide groups of oxyanion-stabilizing residues. One stabilizing residue in the oxyanion hole is the amino acid following the catalytic serine, whereas the other one comes from a separate loop [37].

C. Activation by Interfaces

As previously stated, an advantage of enzyme-catalyzed interesterification in comparison with chemical methods is that it can operate effectively under relatively mild conditions. Enzyme-catalyzed reactions can increase the rate of a reaction by 10^6–10^{15} times even at 25°C [41].

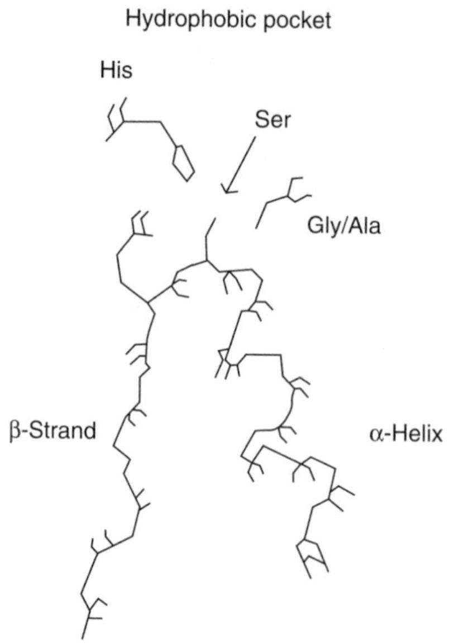

FIGURE 30.7 Crystal structure and location of catalytic residues of the active site of *Candida rugosa* lipase. (Adapted from Kaslauskas, A.J., *Trends Biotechnol.*, 12, 464, 1994.)

The kinetics of lipase-catalyzed interesterification can get complicated due to many factors that can affect the reaction. Activation by interfaces as well as participation of multiple substrates in the interesterification reaction must all be considered when describing the action of lipases at interfaces.

The natural substrates of lipases, long-chain triacylglycerols, are uncharged and insoluble in water and as such form two phases in aqueous solutions. The property of being active at lipid–water interfaces is unique to lipases. At low concentrations of lipids, termed *monomeric solutions*, the lipids are dissolved in aqueous phase. The maximal concentration of monomers in aqueous solution is the solubility limit or critical micelle concentration, after which triacylglycerols form emulsions. For example, the critical micelle concentration for triacetin in aqueous solution is 0.33 M, whereas for long-chain triacylglycerols, it can be as low as 1.0 μM [12,45]. It has been shown that lipases display almost no activity toward monomeric solutions of lipids, whereas the lipids are dissolved and do not form interfaces. Once the level of lipids exceeds the critical micellar concentration, the reaction rate increases dramatically by a factor of 10^3–10^4 in some cases, depending on the quality of the interface (Figure 30.8). Lipases have been found to act at several interfaces, including emulsions, bilayers, and micelles [46]. Action of lipases at the lipid–water interface is believed to follow two successive equilibria involving penetration of lipase into the interface, followed by the formation of the enzyme–substrate complex (Figure 30.9).

Initially, the enzyme penetrates the interface and undergoes a conformational change, folding back the lid and thereby increasing the hydrophobic surface area of the lipase making contact with the interface. The enzyme adsorbs to the interface following a Langmuir adsorption isotherm. Once adsorption has taken place, the enzyme is in its catalytically active form, meaning that interfacial activation has taken place. The lipid substrate can then fit into the active site and be transformed into product. The product is believed to be water soluble and leave the interface rapidly by diffusion into the surrounding solution. Several mechanisms have been proposed to explain interfacial activation of lipases. The first theory relates interfacial activation to a conformational change of the enzyme, where the lid moves to make the active site available to substrate molecules at the interface. The second theory points to changes in the concentration and organization of substrate molecules at the interface to cause activation of the lipase. In the presence of a nonsubstrate lipid interface, a lipase will not be active, but once the concentration of substrate in the interface exceeds that of

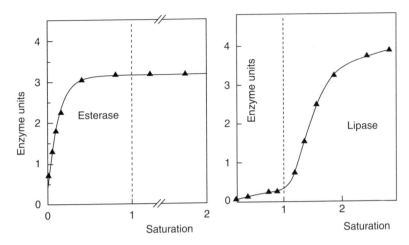

FIGURE 30.8 Comparison of the effect of substrate concentration on lipase and esterase activity at monomeric and saturation levels (beyond vertical dashed lines). (From Sarda, L. and Desnuelle, P., *Biochim. Biophys. Acta,* 30, 513, 1958. With permission.)

nonsubstrate lipids to become the continuous phase, lipase activity increases. There are several other theories as to why lipase activity is increased at an interface. One theory states that the higher substrate concentration at the interface produces more frequent collisions between the lipase and substrate than in monomeric solutions. Other theories involve decreased energy of activation induced by substrate aggregation, reduced hydration of the substrate, and progressive lipid-induced lipase aggregation at the interface [46].

In considering the action of lipases at interfaces, several factors have to be considered, including the reversibility of adsorption, the possibility of inactivation, and the quality of the interface. In general, lipases are considered to be reversibly adsorbed at interfaces, since by increasing surface pressure, lipases have been found to desorb from the interface [46]. The quality of the interface can affect the activity of lipases. Any factor that affects the affinity of the enzyme for the interface as well as packing and orientation of the molecules at the interface can affect activity [11].

D. Problem of Substrate Concentration

Since long-chain triacylglycerols are insoluble in water and form aggregates, lipase-catalyzed interesterification cannot be strictly governed by the Henri–Michaelis rule relating the rate of the reaction to the molar concentration of substrate in solution. In interesterification reactions, the insoluble substrate is in large excess as the continuous solvent phase, making it difficult to define its concentration in the reaction mixture. Since the substrate is insoluble, only the concentration of the substrate present at the interface, which is available to the lipase, is considered. Lipase activity is controlled by the concentration of micellar substrates at the interface and is independent of the molar concentration of the substrate [47]. In contrast, esterases, acting only on water-soluble

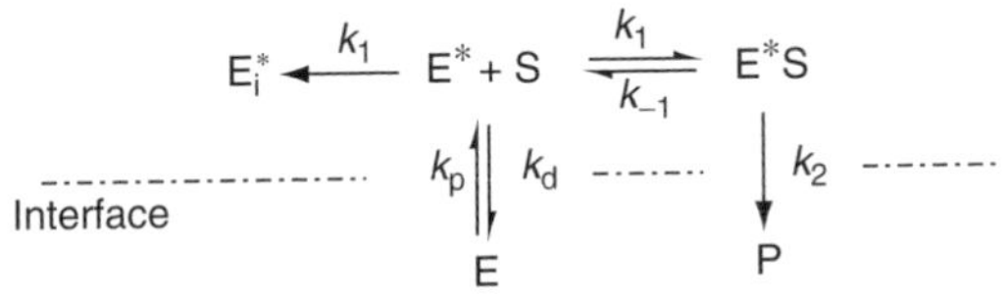

$$E_i^* \xleftarrow{\;k_1\;} E^* + S \; \underset{k_{-1}}{\overset{k_1}{\rightleftharpoons}} \; E^*S$$

$$\text{Interface} \quad \cdots\cdots \quad k_p \Big\Updownarrow k_d \;\cdots\cdots\; \Big\downarrow k_2 \;\cdots\cdots$$

$$E \qquad\qquad P$$

FIGURE 30.9 Model for activation and action of lipases at interfaces.

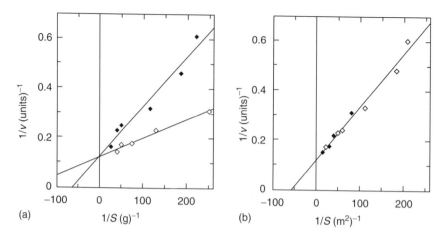

FIGURE 30.10 Lineweaver–Burk plot of lipase activity as a function of (a) mass of substrate at the interface and (b) area occupied by substrate at the interface. The comparison was made with assays containing a coarse emulsion (◆) and a fine emulsion (◇). (Adapted from Benzonana, G. and Desnuelle, P., *Biochimie*, 105, 121, 1965.)

substrates, have a Michaelis–Menten dependence on substrate concentration [42,48]. The dependence of lipase activity on the surface area of the interface as a measure of substrate concentration was proven by Benzonana and Desnuelle [49], who measured the rates of hydrolysis in coarse and fine emulsions (Figure 30.10). When the rate of the reaction for the two emulsions was plotted as a function of substrate mass, the same maximal rates of hydrolysis were obtained; however, there was a difference in the values for K_m. In contrast, when initial velocities were plotted as a function of interfacial area, the values of K_m and V_{max} were constant for both fine and coarse emulsions, indicating that the concentration of substrate at the interface (mol/m^3) directly determines the rate of the reaction [11,12,46,50]. Reaction rates have also been shown to be a function of the emulsion concentration. If a stock emulsion is diluted to different concentrations, a progress curve of rate versus concentration will be obtained, which on plotting gives a straight line in the Lineweaver–Burk plot. A relative K_m can be obtained from this plot, but to obtain the absolute value for K_m, the area of the interface must be known [11]. It is very difficult to obtain an accurate assessment of the interfacial area due to several factors. In free enzyme solutions, the size distribution of emulsion droplets and the degree of adsorption of enzyme to the interface must be known. It is difficult to estimate the surface area of the interface due to size heterogeneity and the possibility of coalescence of emulsion droplets. With immobilized enzyme, the size distribution and surface area of support particles and pores must be determined, as well as the degree of loading of the lipase [47]. Due to the difficulty in measuring these factors accurately, only relative K_m values are determined.

E.   KINETICS AND MECHANISM OF ACTION

Interesterification is a multisubstrate reaction, with the main substrates being glycerides, fatty acids, and water. This reaction can be considered a special case of chemical group transfer, involving sequential hydrolysis and esterification reactions [51]. Lipase-catalyzed interesterification follows a Ping-Pong Bi–Bi reaction for multisubstrate reactions [50,51]). The actual mechanism of acylation and deacylation of the glyceride in the active site is shown in Figure 30.11. During acylation, a covalent acyl–enzyme complex is formed by nucleophilic attack of the active site serine on the carbonyl carbon of the substrate. The serine is made a stronger nucleophile by the presence of histidine and aspartic acid residues. The histidine imidazole ring becomes protonated and positively charged, stabilized by the negative charge of the active site aspartic acid or glutamic acid residues.

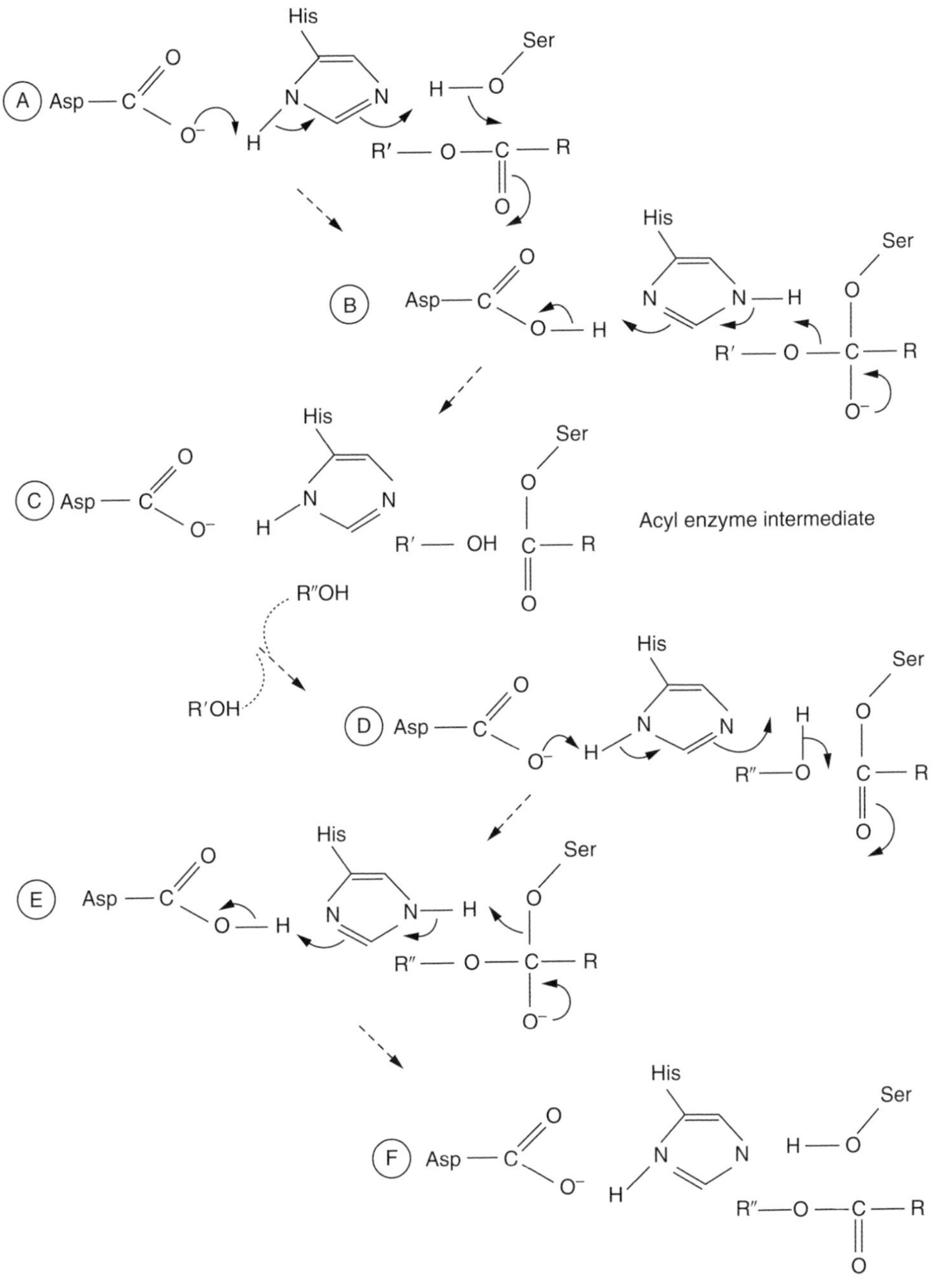

FIGURE 30.11 Catalytic mechanism for lipase-catalyzed interesterification, showing the catalytic site containing Asp/Glu, His, and Ser residues. (Adapted from Marangoni, A.G. and Rousseau, D., *Trends Food Sci. Technol.*, 6, 329, 1995.)

A tetrahedral intermediate is subsequently formed, stabilized by two hydrogen bonds formed with oxyanion-stabilizing residues [12]. A break in the carbon–oxygen bond of the ester causes release of the alcohol. During the reaction, the acylglycerol is associated with the catalytic triad through covalent bonds. Histidine hydrogen bonds with both serine and oxygen of the leaving alcohol. Nucleophilic attack by water or an alcohol causes the addition of a hydroxyl group to the carbonyl

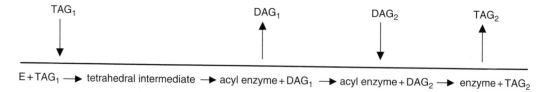

$$E + TAG_1 \longrightarrow \text{tetrahedral intermediate} \longrightarrow \text{acyl enzyme} + DAG_1 \longrightarrow \text{acyl enzyme} + DAG_2 \longrightarrow \text{enzyme} + TAG_2$$

FIGURE 30.12 The Ping-Pong Bi–Bi mechanism for lipase-catalyzed transesterification, with the transfer of an acyl group from one triacylglycerol (TAG_1) to a diacylglycerol (DAG_2) to form a new triacylglycerol (TAG_4).

carbon, producing a tetrahedral intermediate, which will rearrange, releasing the altered acylglycerol and regenerating the active site serine [42,52].

The first stage of interesterification involves hydrolysis of triacylglycerols with consumption of water to produce diacylglycerols, monoacylglycerols, and free fatty acids. Accumulation of hydrolysis products will continue during interesterification until an equilibrium is established [51]. Since lipases are involved in multisubstrate, multiproduct reactions, more complex kinetic mechanisms are required. Interesterification involves acylation and deacylation reactions, either of which can be the rate-limiting step [50,53]. The basic mechanism for a Ping-Pong Bi–Bi reaction using multiple substrates is shown in Figure 30.12.

Under steady-state conditions,

$$\frac{v}{V_{max}} = \frac{[AX][BX]}{K_{mBX}[AX] + K_{mAX}[B] + [A][B]},$$

where AX is the first substrate and BX is the second substrate [41]. It is difficult to study the kinetics of Ping-Pong Bi–Bi mechanisms due to the presence of two substrates. In order to study the kinetics, one substrate concentration is usually held constant whereas the other one is altered. In the case of lipase-catalyzed interesterification under aqueous conditions, there is the additional difficulty that the lipid substrate is also the reaction medium, which is in excess compared with other components. Even with measurable amounts of lipid substrate, it is difficult to develop rate equations since all species involved have to considered [50].

F. SPECIFICITY

The main advantage of lipases that differentiates enzymatic interesterification from chemical interesterification is their specificity. The fatty acid specificity of lipases has been exploited to produce structured lipids for medical foods and to enrich lipids with specific fatty acids to improve the nutritional properties of fats and oils. There are three main types of lipase specificities: positional, substrate, and stereo. Positional and fatty acid specificities are usually determined by partial hydrolysis of synthetic triacylglycerols and separation by thin-layer chromatography with subsequent extraction and analysis of the products. Other methods include conversion of the fatty acids produced during hydrolysis to methyl esters for gas chromatographic analysis [54].

1. Nonspecific Lipases

Certain lipases show no positional or fatty acid specificity during interesterification. Interesterification with these lipases after extended reaction times gives complete randomization of all fatty acids in all positions and gives the same products as chemical interesterification (Figure 30.13). Examples of nonspecific lipases include lipases derived from *Candida cylindraceae*, *Corynebacterium acnes*, and *Staphylococcus aureus* [9,10].

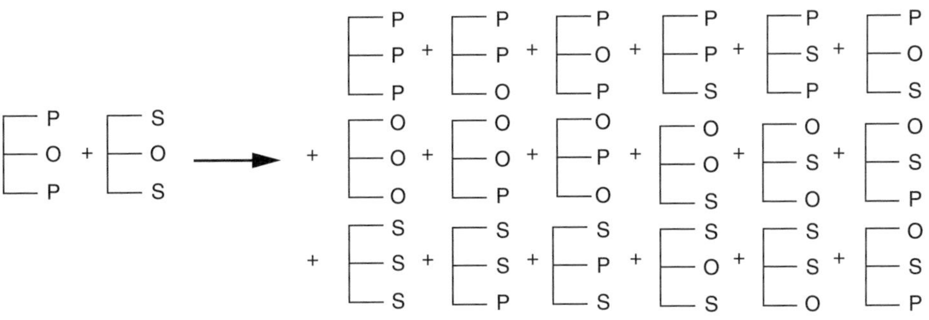

$$
\begin{bmatrix} P \\ O \\ P \end{bmatrix} + \begin{bmatrix} S \\ O \\ S \end{bmatrix} \longrightarrow \quad + \begin{bmatrix} P \\ P \\ P \end{bmatrix} + \begin{bmatrix} P \\ P \\ O \end{bmatrix} + \begin{bmatrix} P \\ O \\ P \end{bmatrix} + \begin{bmatrix} P \\ P \\ S \end{bmatrix} + \begin{bmatrix} P \\ S \\ P \end{bmatrix} + \begin{bmatrix} P \\ O \\ S \end{bmatrix}
$$

$$
+ \begin{bmatrix} O \\ O \\ O \end{bmatrix} + \begin{bmatrix} O \\ O \\ P \end{bmatrix} + \begin{bmatrix} O \\ P \\ O \end{bmatrix} + \begin{bmatrix} O \\ O \\ S \end{bmatrix} + \begin{bmatrix} O \\ S \\ O \end{bmatrix} + \begin{bmatrix} O \\ S \\ P \end{bmatrix}
$$

$$
+ \begin{bmatrix} S \\ S \\ S \end{bmatrix} + \begin{bmatrix} S \\ S \\ P \end{bmatrix} + \begin{bmatrix} S \\ P \\ S \end{bmatrix} + \begin{bmatrix} S \\ O \\ S \end{bmatrix} + \begin{bmatrix} S \\ S \\ O \end{bmatrix} + \begin{bmatrix} O \\ S \\ P \end{bmatrix}
$$

FIGURE 30.13 Triacylglycerol products from the transesterification of two triacylglycerols, 1,3-dipalmitoyl-2-oleoylglycerol and l,3-distearoyl-2-oleoylglycerol, using either a nonspecific lipase or chemical esterification.

2. Positional Specificity

Positional specificity, that is, specificity toward ester bonds in positions *sn*-1,3 of the triacylglycerol, results from an inability of lipases to act on position *sn*-2 on the triacylglycerol, due to steric hindrance (Figure 30.14). Steric hindrance prevents the fatty acid in position *sn*-2 from entering the active site [9,55]. An interesterification reaction using a 1,3-specific lipase will initially produce a mixture of triacylglycerols, 1,2- and 2,3-diacylglycerols, and free fatty acids [55]. After prolonged reaction periods, acyl migration can occur with the formation of 1,3-diacylglycerols, which allows some randomization of the fatty acids existing at the middle position of the triacylglycerols. In comparison with chemical interesterification, 1,3-specific lipase-catalyzed interesterification of oils with a high degree of unsaturation in the *sn*-2 position of the triacylglycerols will decrease the saturated to unsaturated fatty acid level [56]. Lipases that are 1,3-specific include those from *Aspergillus niger*, *M. miehei*, *Rhizopus arrhizus*, and *Rhizopus delemar* [9]. The specificity of individual lipases can change due to microenvironmental effects on the reactivity of functional groups or substrate molecules [57]. For example, lipase from *Pseudomonas fragi* is known to be 1,3-specific but has also produced random interesterification, possibly due to a microemulsion environment. As of yet, lipases that are specific toward fatty acids in the *sn*-2 position have been difficult to identify. Under aqueous conditions, one such lipase from *Candida parapsilosis* hydrolyzes the *sn*-2 position more rapidly than either of the *sn*-1 and *sn*-3 positions, and is also specific toward long-chain PUFAs [58].

The differences in the nutrition of chemically interesterified fats and oils compared with enzymatically interesterified samples can be linked to the positional specificity exhibited by some lipases. In fish oils and some vegetables oils that contain high degrees of essential PUFAs, these fatty acids are usually found in greater quantities in the *sn*-2 position. In the intestines, 2-monoacylglycerols are more easily absorbed than *sn*-1 or *sn*-3 monoacylglycerols. Using a 1,3-specific lipase, the fatty acid composition of positions 1 and 3 can be changed to meet the targeted structural requirements while retaining the nutritionally beneficial essential fatty acids in position 2. Using random chemical interesterification, retention and improvement in beneficial fatty acid content cannot be accomplished due to the complete randomization of the fatty acids in the triacylglycerols [59].

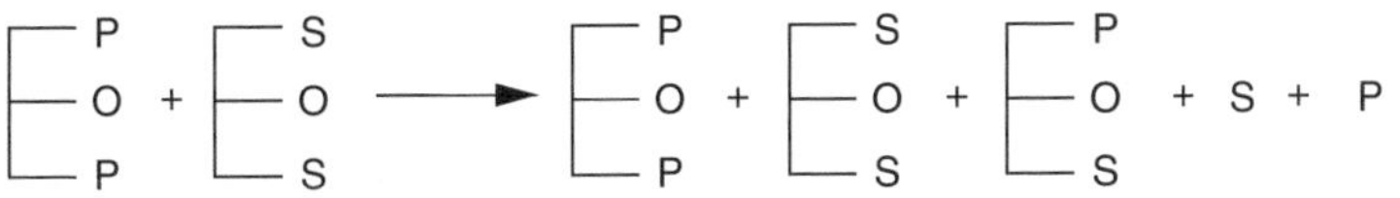

$$
\begin{bmatrix} P \\ O \\ P \end{bmatrix} + \begin{bmatrix} S \\ O \\ S \end{bmatrix} \longrightarrow \begin{bmatrix} P \\ O \\ P \end{bmatrix} + \begin{bmatrix} S \\ O \\ S \end{bmatrix} + \begin{bmatrix} P \\ O \\ S \end{bmatrix} + S + P
$$

FIGURE 30.14 Transesterification products of 1,3-dipalmitoyl-2-oleoylglycerol and 1,3-distearoyl-2-oleoyl-glycerol using a 1,3-specific lipase.

3. Stereospecificity

In triacylglycerols, the *sn*-1 and *sn*-3 positions are sterically distinct. Very few lipases differentiate between the two primary esters at the *sn*-1 and *sn*-3 positions, but when they do, the lipases possess stereospecificity. In reactions where the lipase is stereospecific, positions 1 and 3 are hydrolyzed at different rates. Stereospecificity is determined by the source of the lipase and the acyl groups, and can also depend on the lipid density at the interface, where an increase in substrate concentration can decrease specificity due to steric hindrance. Differences in chain length can also affect the specificity of the lipase [12]. Lipase from *Pseudomonas* species and porcine pancreatic lipase have shown stereoselectivity when certain acyl groups are hydrolyzed [60].

4. Fatty Acid Specificity

Many lipases are specific toward particular fatty acid substrates. Most lipases from microbial sources show little fatty acid specificity, with the exception of lipase from *Geotrichum candidum*, which is specific toward long-chain fatty acids containing *cis*-9 double bonds [9]. Lipases can also demonstrate fatty acid chain length specificity, with some being specific toward long-chain fatty acids and others being specific toward medium- and short-chain fatty acids. For example, porcine pancreatic lipase is specific toward short-chain fatty acids, whereas lipase from *Penicillium cyclopium* is specific toward long-chain fatty acids. As well, lipases from *A. niger* and *Aspergillus delemar* are specific toward both medium- and short-chain fatty acids [11,61]. Other lipases have been found to be specific toward fatty acids of varying lengths. Marangoni [62] found that in the hydrolysis of butter oil, lipase from *Candida rugosa* showed specificity toward butyric acid compared with *Pseudomonas fluorescens* lipase. With interesterification reactions in organic media, lipases can also be specific toward certain alcohol species. A large group of lipases from sources such as *C. cylindraceae*, *M. miehei*, and *R. arrhizus* have been found to be strongly specific against fatty acids containing the first double bond from the carboxyl end at an even-numbered carbon, such as *cis*-4, *cis*-6, and *cis*-8, resulting in slower esterification of these fatty acids in comparison with other unsaturated and saturated fatty acids. Fatty acid specificity by certain lipases can be used in the production of short-chain fatty acids for use as dairy flavors and in the concentration of EPA and DHA in fish oils by lipases with lower activity toward these fatty acids.

III. REACTION SYSTEMS

A. Enzymatic Interesterification in Microaqueous Organic Solvent Systems

Since the main substrates of lipases are long-chain triacylglycerols, which are insoluble in water, many experiments have been conducted in the presence of organic solvents. Organic solvents allow the fat or oil to be solubilized and convert two-phase systems to one-phase systems [63]. Stability can be improved by covalent attachment of polyethylene glycol (PEG) to free amino groups of the lipase, giving lipases amphiphilic properties and allowing their dissolution in organic solvents [64]. It has been reported that the thermal stability of lipases can be improved in microaqueous organic solvent systems since the lack of water prevents unfolding of the lipase at high temperatures [65]. Elliott and Parkin [65] found that porcine pancreatic lipase had optimal activity at 50°C in an emulsion, whereas the optimum increased to 70°C in a microaqueous organic solvent system using hexane. Lipase activity in organic solvents depends on the nature and concentration of the substrate and source of enzyme [63]. The specific organic solvent used can dramatically affect the activity of the lipase [66]. Lipases are more active in *n*-hexane and isooctane than other solvents, such as toluene, ethyl acetate, and acetylnitrile [28,44]. The polarity of solvents can be described by P, the partition coefficient of a solvent between water and octanol. This is an indication of the hydrophobicity of the solvent. No lipase activity is observed in solvents with a value for $\log P < 2$ [67,68]. The hydrophobicity of the solvent can also affect the degree of acyl migration during interesterification using a 1,3-specific lipase. Hexane tends to promote acyl migration due to the low solubility of free fatty acids and partial glycerides in hexane, which forces them into the microaqueous region around the lipase, providing optimum conditions for acyl migration. In contrast, the use of diethyl ether, in which free fatty acids and partial glycerides are more soluble, removes the products from the microaqueous environment and reduces the risk of acyl migration [6]. Since the choice of organic solvents based on minimization of acyl migration may conflict with maximization of interesterification, acyl migration is usually minimized simply by reducing reaction times. Lipases can be made more active and soluble in organic solvent systems by attachment of an amphiphilic group such as PEG. PEG reacts with the N-terminal or lysine amino groups, rendering the lipase more soluble in organic solvents [69]. The activity of lipases in organic solvent depends on the solubility of the solvent in water. Lipases are only active in water-immiscible solvents, since water-miscible organic solvents extract the water of hydration layer from the vicinity of the enzyme, thereby inactivating them [44]. Since the success of an interesterification reaction depends on the concentration of water in the system, the hydration state of the lipase plays a key role because a minimal amount of water is needed to maintain the enzyme in its active form. The use of hydrophobic solvents limits the flexibility of the enzyme, preventing it from assuming its most active conformation. Therefore, if organic solvents are used, the enzyme must be in its active conformation before the addition of the organic solvent. This can be accomplished by exposing the enzyme to an inhibitor or substrate, then drying it in its active conformation [12,70]. The advantage of using organic solvents in lipase-catalyzed interesterification reactions is that the water content can be carefully controlled. A water content higher than 1% can produce high degrees of hydrolysis, whereas water levels lower than 0.01% can prevent full hydration of the lipase and reduce the initial rate of hydrolysis [1]. Therefore, water levels between these two extremes are necessary to maximize the effectiveness of enzymatic interesterification in organic solvents. In microaqueous organic solvent systems, the effect of pH on lipase activity is complex because water levels are so low. It has been proven that enzymes in organic solvent systems have a memory of the pH of the last aqueous environment in which they were. Elliott and Parkin [65] found that porcine pancreatic lipase has an optimum activity in hexane after being exposed to pH values between 6.5 and 7.0. At pH 8.5, the decrease in activity was attributed to a change in the ionization state of the histidine in the active site.

A common form of organic solvent system used in lipase-catalyzed interesterification is that of reverse micelles. Reverse micelles, or micoremulsions, are defined as nanometer-sized water droplets dispersed in organic media with surfactants stabilizing the interface [29,71]. A common surfactant used is an anionic double-tailed surfactant called sodium-bis(2-ethylhexyl)sulfosuccinate (AOT). Reverse micelles are used in interesterification reactions because they increase the interfacial area and improve the interaction between lipase substrates [29]. As well, the use of microemulsions makes it possible to use polar and nonpolar reagents in the same reaction mixture [72]. Reverse micelles can be formed by gently agitating a mixture of AOT, lipid substrate, organic solvent, and lipase dissolved in buffer until the solution becomes clear. The lipase is trapped in an aqueous medium in the core of the micelle, avoiding direct contact with the organic medium [61]. Lecithin has been used to promote the formation of reverse micelles and to protect the lipase from nonpolar solvents [73,74]. At ionic strengths higher than 1 M, activity is decreased due to decreased solubility and activity of the lipase. The water content required for microemulsion systems is dependent on the desired reaction, although some level of water is necessary to hydrate the enzyme. For example, Holmberg et al. [75] found that 0.5% water was the optimum for the production of monoacylglycerols from palm oil in a microemulsion. The composition of the substrate can also affect the rate of interesterification in reverse micelles. Substrates with more amphiphilic properties are better because they can partition to the interface. More polar substrates tend to stay in the water phase and interact less with the interface [76]. The disadvantages of reverse micelle systems are that lipase activity is decreased rapidly, and the system can alter lipase specificity [73,76,77]. Reverse micelles can also be used with immobilized lipases, where the reverse micelle is formed around the support and immobilized lipase. This method has been used with hexane to produce cocoa butter equivalents [73]. Although they have been used in experimental form to produce triacylglycerols from diacylglycerols and oleic acid [78], as well as triacylglycerols suitable for use as cocoa butter substitutes [74], reverse micelles are not used in industrial enzymatic interesterification applications.

IV. IMMOBILIZATION

Immobilization of lipases has become increasingly popular for both hydrolysis and synthesis reactions. The advantages of immobilized enzyme systems compared with free enzyme systems include reusability, rapid termination of reactions, lowered cost, controlled product formation, and ease of separation of the enzyme from the reactants and products. In addition, immobilization of different lipases can affect their selectivity and chemical and physical properties. Immobilization also provides the possibility of achieving both purification of the lipase from an impure extract and immobilization simultaneously, with minimal inactivation of the lipase [79]. Methods for immobilization of enzymes include chemical forms, such as covalent bonding, and physical forms, such as adsorption and entrapment in a gel matrix or microcapsules [7,80].

The easiest and most common type of immobilization used in interesterification reactions is adsorption, which involves contacting an aqueous solution of the lipase with an organic or inorganic surface-active adsorbent. The objective of immobilization is to maximize the level of enzyme loading per unit volume of support. The process of adsorption can be accomplished through ion exchange or through hydrophobic or hydrophilic interactions and van der Waals interactions [81]. After a short period of mixing of the free enzyme and support, the immobilized enzyme is washed to remove any free enzyme that is left, after which the product is dried [79]. The same adsorption process can be accomplished by precipitating an aqueous lipase solution onto the support using acetone, ethanol, or methanol, then drying as previously described [9,81]. Although desorption can occur, most immobilized lipase preparations are stable in aqueous solutions for several weeks. The preparations are stable because as the lipase adsorbs to the support, it unfolds slightly, allowing several points of interaction between the lipase and support. In order for desorption to occur, simultaneous loss of interactions at all contact sites must occur, which is unlikely [82].

The degree of immobilization depends on several conditions, including pH, temperature, solvent type, ionic strength, and protein and adsorbent concentrations. The choice of carrier is dependent on its mechanical strength, loading capacity, cost, chemical durability, functionality, and hydrophobic or hydrophilic character [83]. In general, lipases retain the highest degree of activity when immobilized on hydrophobic supports, where desorption of lipase from the support after immobilization is negligible, and improved activity has been attributed to increased concentrations of hydrophobic substrate at the interface [7,50]. The disadvantages of using hydrophilic supports include high losses of activity due to changes in conformation of the lipase, steric hindrance, and prevention of access of hydrophobic substrates [7]. Common hydrophobic supports include polyethylene, polypropylene, styrene, and acrylic polymers, whereas hydrophilic supports include Duolite, Celite, silica gel, activated carbon, clay, and Sepharose [7]. The effectiveness of the immobilization process is influenced by the internal structure of the support. If a support with narrow pores is used, most of the enzyme will be immobilized on the surface of the support, which prevents the occurrence of internal mass transfer limitations. If a support containing larger pore sizes is used, such as Spherosil DEA, with an average diameter of 1480 Å, some lipase will be immobilized inside the pores, which can prevent access of the substrate to some of the lipase. This is due to preferential filling of pores and crevices by the lipase during immobilization [84,85]. The activity of lipases tends to decrease on immobilization, with activity being reduced by 20%–100% [79,81]. The activity of an immobilized enzyme relative to the free form can be compared by an effectiveness value, which is defined as the activity of immobilized enzyme divided by the activity of an equal amount of free enzyme determined under the same operating conditions. The effectiveness value can be used as a guide to the degree of inactivation of the enzyme caused by immobilization. For values close to 1.0, very little enzyme activity has been lost on immobilization, whereas values much lower than 1 indicate high degrees of enzyme inactivation [80].

The performance of an immobilized lipase can also be affected by handling and reaction conditions. Freeze drying of the immobilized enzyme before interesterification to substantially reduce the moisture content has been reported to dramatically improve activity. Molecular sieves can also be added to reaction systems to reduce the amount of water that accumulates during the reaction, which would in turn reduce the degree of hydrolysis [4]. The main disadvantage associated with adsorption as an immobilization method is that changes in pH, ionic strength, or temperature can cause desorption of lipase that has been adsorbed by ion exchange. Lipases adsorbed through hydrophobic or hydrophilic interactions can be desorbed by changes in temperature or substrate concentration [79].

A. FACTORS AFFECTING IMMOBILIZED LIPASE ACTIVITY

Immobilization can have an impact on the activity of lipases through steric, mass transfer, and electrostatic effects. During immobilization, the enzyme conformation can be affected and parts of the enzyme can be made inaccessible to the substrate due to steric hindrance.

1. Mass Transfer Effects

The kinetics of lipase-catalyzed interesterification can be affected by mass transfer limitations. The substrate must diffuse through the fluid boundary layer at the surface of the support into the pore structure of the support and react with the lipase. Once products have been released by the lipase, they must diffuse back out of the pore structure and away from the surface of the support. Mass transfer limitations fall into two categories: internal and external mass transfers. Internal mass transfer is the transport of substrate and product within the porous matrix of the support; it is affected by the size, depth, and smoothness of the pores. Internal mass transfer is diffusion-limited only. When the rate of diffusion inward is slower than the rate of conversion of substrate to product, the reaction is diffusion-limited, as there is not enough substrate available for the amount of enzyme present [86]. A diffusion coefficient for internal mass transfer in immobilized enzyme systems compared with free enzyme systems is defined as

$$D_e = \frac{D\psi}{\tau},$$

where
   D_e is the effective diffusion coefficient inside the support particles
   D is the diffusion coefficient in free solution
   ψ is the porosity of the particles
   τ is the tortuosity factor, defined as the distance of the path length traveled by molecules between two points in a particle

The effective diffusion coefficient varies inversely with the molecular weight of the substrate [80]. Internal diffusional limitations can be recognized if the activity increases when the support particles are crushed, since crushing would decrease the length of the pathway that the substrate would have to travel to reach the enzyme. The Thiele modulus, φ, can be used to evaluate the extent of internal mass transfer limitations:

$$\phi = L\lambda = L\left(\frac{V_{max}}{K_m D_e}\right)^{1/2},$$

where L is the half-thickness of the support particles. Internal mass transfer limitations can also be identified by measuring the initial velocity of the reaction at increasing enzyme concentrations.

If the rate of the reaction remains constant at increasing enzyme concentrations (amount of enzyme per gram of support), the reaction is mass transfer-limited. If the rate of reaction increases linearly with increasing enzyme concentration, the reaction is kinetically limited. Internal diffusion limitations can be reduced by decreasing the support particle size, increasing pore size and smoothness, using low-molecular-weight substrates, and using high substrate concentrations [80]. The difficulty with using smaller support particles in fixed bed reactors where internal mass transfer limitations are high is that it tends to increase the back pressure of the system [84].

External mass transfer limitations are the resistance to transport between the bulk solution and a poorly mixed fluid layer surrounding each support particle. External mass transfer can occur in packed bed and membrane reactors and is affected by both convection and diffusion [84]. If the reaction is faster than the rate of diffusion of substrate to the surface or product from the surface, this can affect the availability of substrate for lipase catalysis. If inadequate substrate quantities reach the enzyme, the rate of reaction will be lower than that of free enzyme. An increasing external mass transfer coefficient can be identified during kinetic analysis by an increasing slope of a Lineweaver–Burk plot [87]. In stirred reaction systems, external mass transfer limitations have been eliminated when there is no increase in the reaction rate with increasing rates of stirring. External mass transfer limitations can be reduced in packed bed reactors by increasing the flow rate, reducing the viscosity of the substrate, and increasing substrate concentration [80]. Changing the height-to-diameter ratio of a fixed bed reactor can also reduce external mass transfer limitation as it increases the linear velocity of the substrates.

2. Nernst Layer and Diffusion Layer

Immobilized lipases are surrounded by two different layers, which can create differences in substrate concentration between them and the bulk phase. The Nernst layer is a thin layer located directly next to the surface of the support. In the case of hydrophobic supports and hydrophobic substrates, such as triacylglycerols, the concentration of substrates in the Nernst layer is more than in the bulk solution since the hydrophobic substrate tends to partition toward the hydrophobic support material. Another layer surrounding the support particles is a diffusion or boundary layer. A concentration gradient is established between the diffusion layer and the bulk phase as substrate is converted to product by the lipase. The product concentration in the diffusion layer is higher than in the bulk phase as it must diffuse from the surface of the support into the bulk phase. Consequently, due to the higher product concentration in the diffusion layer, the substrate concentration is lower than in the bulk phase, producing concentration gradient with more substrate diffusing toward the support and immobilized lipase. Differences in substrate concentration between the Nernst layer and/or the boundary layer and the bulk phase can affect the determination of K_m since substrate concentration will be measured in the bulk layer, which may not be the concentration of substrate closer to the lipase. With a lower substrate concentration at the support in comparison with the bulk phase, the apparent K_m will appear higher and the activity will appear lower than its actual values. The opposite will occur with a higher substrate concentration at the interface.

A third factor that can affect the activity of immobilized lipase is electrostatic effects. If the support and substrate possess the same charge then they will experience repulsion, whereas if they have opposite charge they will be attracted. This factor can have an effect on the apparent K_m. As well, electrostatic effects can have an impact on other components in the reaction. For example, if the support was anionic, the local concentration of hydrogen ions would be higher in the vicinity of the immobilized lipase, which would cause a decrease in the pH around the enzyme.

Combining the electrostatic effects and the effect of the Nernst layer, the value of the apparent K_m can be modified as follows [88]:

$$K_m' = \left(K_m + \frac{x}{D} V_{max} \right) \frac{RT}{RT - xzFV},$$

where
  K'_m is the apparent K_m of the lipase
  x is the thickness of Nernst layer
  R is the universal gas constant
  T is the absolute temperature
  z is the valence of the substrate
  F is Faraday's constant
  V is the magnitude of the electric field around the enzyme support
  D is the diffusion coefficient of the substrate

If the thickness of the Nernst layer decreases, then the ratio x/D would decrease and K'_m would decrease, approaching K_m.

B. STABILITY OF IMMOBILIZED ENZYMES

The stability of immobilized enzymes depends on the method of immobilization and the susceptibility of the enzyme to inactivation. Inactivation can be caused by contaminants and changes in temperature, pH, and ionic strength. High shear, microbial contamination, fouling, and breakage of support particles have also been found to inactivate immobilized enzymes. Depending on the strength of the immobilization method, the enzyme can also be desorbed from the support. The stability of immobilized enzymes is evaluated by determining the half-life of the enzyme under the reaction conditions. In diffusion-limited systems, there is a linear decay in enzyme activity in time, as enzymes on the surface of the support are inactivated and the substrate diffuses further into the pores to reach enzyme molecules that have not been inactivated. In systems free of diffusional limitations, enzyme inactivation follows a first-order decay:

$$\ln \frac{N}{N_0} = -\lambda t,$$

where
  N_0 is the initial enzyme activity
  N is the activity at time t
  λ is the decay constant
  Using λ, the half-life of the immobilized lipase can be determined as follows:

$$\text{Half-life} = \frac{0.693}{\lambda}.$$

The half-lives of lipases in interesterification systems have been reported to range from 7 min to 7 months, with the large variability attributed to the source of lipases and different reaction conditions [50]. As previously stated, the half-life of the immobilized enzyme can be used to determine the productivity of the system. In order to avoid losses in productivity as the activity of the immobilized lipase decreases, the temperature can be raised to increase the reaction rate or, in fixed bed reactor systems, the flow rate can be reduced [80]. While these measures can improve the conversion rate, they can also increase the rate of enzyme inactivation in the case of temperature increases, or decrease the throughput in the case of reduced flow rate.

C. IMMOBILIZED ENZYME KINETICS

The previous discussion on the kinetics of lipase action was developed for soluble lipases acting on insoluble substrate, but assuming that diffusional and mass transfer effects are not rate-limiting, the same theories can be applied to immobilized lipases. When using immobilized lipases, the level of substrate in comparison with the level of enzyme must be considered. In general, there is a low average concentration of substrate in direct contact with the immobilized lipase due to high conversion rates, producing first-order, mixed first- and zero-order, or zero-order kinetics as opposed to zero-order Michaelis–Menten kinetics [80]. The rate of the reaction, v, is proportional to the substrate concentration at the interface where

$$v = \frac{V_{max}[S]}{K_m + [S]}.$$

The kinetics of immobilized lipases are also affected by the type of reactor used, since reactors differ in the amount of immobilized lipase used and in the method of substrate delivery, product removal, and degree of mixing.

V. ENZYMATIC INTERESTERIFICATION REACTORS

Reactors designed for immobilized enzyme reactions differ from one another based on several criteria. Reactors can be batch or flow-through systems and can differ in the degree of mixing involved during the reaction. For all reactor systems, the productivity of the system is defined as the volumetric activity × the operational stability of the immobilized enzyme, with units of kilograms of product per liter of reactor volume per year. The volumetric activity is determined as the mass of product obtained per liter of reactor per hour, whereas the operational stability is the half-life of the immobilized enzyme [80]. The most common reactor systems used include fixed bed, batch, continuous stirred tank, and membrane reactors.

A. Fixed Bed Reactor

A fixed bed reactor is a form of continuous flow reactor, where the immobilized enzyme is packed in a column or as a flat bed, and the substrate and product streams are pumped in and out of the reactor at the same rate. The main advantages of fixed bed reactors are their easy application to large-scale production, high efficiency, low cost, and ease of operation. A fixed bed reactor also provides more surface area per unit volume than a membrane reactor system [7]. A model fixed bed reactor for interesterification would consist of two columns in series: one for the reaction and a precolumn for fat-conditioning steps such as incorporation of water. Reservoirs attached to the columns would contain the feed streams and product streams. A pump would be required to keep the flow rate through the system constant, and the system would have to be water-jacketed to keep the reaction temperature constant (Figure 30.15).

Since water is required in minimal amounts for hydration of the enzyme during the reaction, the oil is first passed through a precolumn containing water-saturated silica or molecular sieves, which would allow the oil to become saturated with sufficient water to allow progression of the interesterification reaction without increasing the rate of hydrolysis. Interesterification in a fixed bed reactor can lead to increases in product formation through increased residence time in the reactor. Complete conversion to products will never be achieved, and with an increase in product levels, a loss in productivity will occur [89]. Using a fixed bed reactor with a silica precolumn for water saturation of the oil phase, Posorske et al. [89] produced a cocoa butter substitute from palm stearin and coconut oil. These authors found that decreasing the flow rates to increase the total product concentration caused a decrease in productivity. Decreasing the flow rates to increase product levels from 20% to 29% leads to a significant decrease in productivity. Fixed bed reactors are more efficient than batch reactors but are prone to fouling and compression. Dissolution of the oil in an organic solvent to reduce viscosity for flow through the packed bed may be required [89]. In addition, the substrate has to be treated to remove any particulates, inhibitors, and poisons that can build up over time and inactivate the lipase [8]. Macrae [9] found that after treatment of palm oil

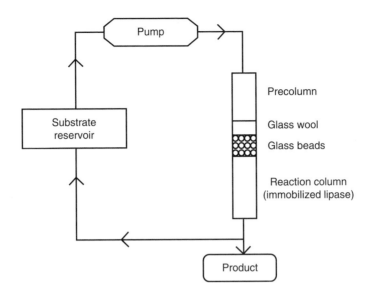

FIGURE 30.15 Fixed bed reactor for immobilized lipase–catalyzed interesterification.

midfraction and stearic acid to remove particulates, inhibitors, and poisons, acidolysis reached completion after 400 h and there was not appreciable loss in lipase activity even after 600 h of operation. Wisdom et al. [90] performed a pilot scale reaction using a 2.9 L fixed bed reactor to esterify shea oleine with stearic acid. It was found that with high-quality substrates, only a small loss of activity was exhibited after 3 days with the production of 50 kg of product. However, when a lower grade shea oil was used, there was rapid inactivation of the lipase.

The kinetics of a packed bed reactor are assumed to be the same as for a soluble lipase, where

$$\frac{dS}{dt} = \frac{V_{max}[S]}{K'_m + [S]}.$$

This can be rearranged and integrated to

$$[S_0]X = K'_m \ln(1 - X) + \frac{k_{cat}E_\tau}{Q},$$

where

[S_0] is the initial substrate concentration

X is the fraction of substrate that has been converted to product at any given time ($1-[S]/[S_0]$)

Q is the volumetric flow rate

E_τ is the total number of moles of enzyme present in the packed bed [80,91]

The residence time, τ, is based on the porosity of the packed bed and is defined as [92]

$$\tau = V_{tot} \frac{P}{Q},$$

where

V_{tot} is the volume of the reactor

P is the porosity of the bed

Q is the flow rate of the substrate

The porosity of the bed in a fixed bed reactor can produce internal transfer limitations. Ison et al. [84] studied the effects of pore size on lipase activity in a fixed bed reactor using Spherosil with a mean pore size of 1480 Å and Duolite with a mean pore size of 190 Å. The larger pore size of the Spherosil was found to produce a decrease in lipase activity. This loss in activity was due to the higher degree of enzyme loading during immobilization, making some of the lipase inaccessible to substrate. With the smaller pore size of Duolite, the lipase was immobilized only on the surface of the support, eliminating internal mass transfer limitations.

B. STIRRED BATCH REACTOR

A stirred batch reactor is a common system used in laboratory experiments with lipase-catalyzed interesterification due to its simplicity and low cost. No addition and removal of reactants and products are performed except at the initial and final stages of the reaction (Figure 30.16). The equation to characterize the kinetics of a stirred batch reactor is

$$[S_0]X - K'_m \ln(1 - X) = \frac{k_{cat}E_\tau t}{V},$$

where
  $[S_0]$ is the initial substrate concentration
  X is the fraction of substrate converted to product at any given time $(1-[S]/[S_0])$
  t is the reaction time
  E_τ is the total number of moles of enzyme present in the reactor
  V is the volume of the reactor

Kurashige [93] found that a batch reactor was useful in reducing the diacylglycerol content in palm oil by converting existing diacylglycerols and free fatty acids to triacylglycerols. Using lipase coadsorbed with lecithin on Celite under vacuum to keep the water content below 150 ppm, the author was able to increase the triacylglycerol content from 85% to 95% in 6 h. The rate of conversion in a stirred batch reactor decreases over time since there is a high initial level of substrate, which is reduced over time, with conversion to product. In order to maintain the same rate of conversion throughout the reaction, it would be necessary to add more immobilized enzyme to the reaction mixture [80]. A stirred batch reactor has the advantage of being relatively easy to

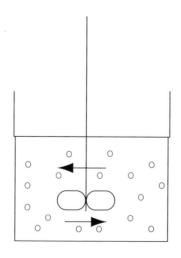

FIGURE 30.16 Stirred batch reactor for immobilized or free lipase–catalyzed interesterification.

build and free enzymes can be used, but it has the disadvantage that, unless immobilized, the enzyme cannot be reused. As well, a larger system or longer reaction times are required to achieve equivalent degrees of conversion in comparison with other systems, and side reactions can be significant [63]. Macrae [9] used a batch reactor to produce cocoa butter equivalents from the interesterification of palm oil midfraction and stearic acid. While product yields were high, by-products such as diacylglycerols and free fatty acids were formed. Therefore, it was necessary to isolate the desired triacylglycerol products using fat fractionation techniques.

C. CONTINUOUS STIRRED TANK REACTOR

A continuous stirred tank reactor combines components of both fixed bed and batch reactors. It is an agitated tank in which reactants and products are added and removed at the same rate, while providing continuous stirring to eliminate mass transfer limitations encountered in fixed bed reactors (Figure 30.17). Stirring also prevents the formation of temperature and concentration gradients between substrates or products. A continuous stirred tank reactor can be in the form of a tank with stirring from the top or bottom, or a column with stirring accomplished by propellers attached to the sides of the column [68]. The kinetics for a continuous stirred tank reactor, developed by Lilly and Sharp [94], first encompass the substrate balance in the system as

$$Q[S_i] - Q[S_0] = \frac{dS}{dt} V,$$

where
Q is the flow rate
$[S_i]$ is the initial substrate concentration entering the reactor
$[S_0]$ is the substrate concentration leaving the reactor
V is the steady-state liquid volume in the tank

Rearrangement gives modified equation here.

$$[S_0]X + K'_m \left(\frac{X}{1-X} \right) = \frac{k_{cat} E_\tau}{Q},$$

where
$[S_0]$ is the initial substrate concentration
X is the amount of substrate converted to product at any particular time $(1-[S]/[S_0])$
Q is the flow rate
E_τ is the total number of moles of enzyme present in the reactor

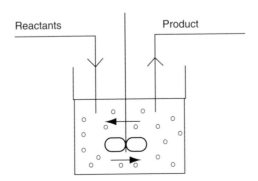

FIGURE 30.17 Continuous stirred tank reactor for immobilized lipase–catalyzed interesterification.

The main disadvantages of continuous stirred tank reactors are the higher power costs associated with continuous stirring, the possibility of breaking up support particles with agitation, and the requirement for a screen or filter at the outlet to prevent losses of the immobilized lipase [7,80].

D. MEMBRANE REACTORS

Immobilization of enzymes onto semipermeable membranes is an attractive alternative for lipase-catalyzed interesterification reactions. Membrane reactors involve two-phase systems, where the interface of two phases is at a membrane. The advantages of membrane systems are reduced pressure drops, reduced fluid channeling, high effective diffusivity, high chemical stability, and a high membrane surface area to volume ratio [95]. Membranes are commonly produced in the form of a bundle of hollow fibers and can be hydrophilic of hydrophobic in nature. Materials used in membrane systems are polypropylene, polyethylene, nylon, acrylic resin, and polyvinyl chloride. In a membrane such as microporous polypropylene, the pores have dimensions of 0.075 by 0.15 μm and the fibers have an internal diameter of 400 μm, providing 18 m^2 of surface area per gram of membrane [82]. With a hydrophilic membrane such as cellulose, the oil phase circulates through the inner fiber side whereas the aqueous components circulate on the shell side [63]. Immobilization of lipase can be accomplished by submerging the fibers in ethanol, rinsing them in buffer, then submerging them in lipase solution [82]. Another method involves dispersing the enzyme in the oil phase and using ultrafiltration to deposit the lipase on the inner fiber side. One of the substrates can diffuse through the membrane toward the interface where the enzyme is immobilized, van der Padt et al. [63] used hollow fibers made from cellulose to perform gylcerolysis of decanoic acid. Using a hydrophilic membrane bioreactor, the lipase activity was similar to the activity in emulsion systems. The hydrophilic membrane was found to be more effective for glycerolysis since the lipase was immobilized on the oil phase side, with the membrane preventing it from diffusing into the glycerol phase and being lost. Hoq et al. [96] used a hydrophobic polypropylene membrane to esterify oleic acid and glycerol. The lipase was adsorbed on the glycerol side, resulting in the loss of some enzyme in this phase. Therefore, use of a hydrophobic membrane would require the addition of more lipase to prevent losses in activity [7,64]. Membrane reactors have been used in glycerolysis and acidolysis reactions and have an advantage over more conventional stirred tank reactors in that the reaction and separation of substrates and product can be accomplished in one system. Having the substrates and products separated during the reaction is especially useful during the esterification reaction where water is produced. Hoq et al. [96,97] found that during esterification of oleic acid and glycerol, the excess water produced could be removed by passing the oleic acid stream through molecular sieves, thereby preventing losses in productivity from hydrolysis.

E. FLUIDIZED BED REACTOR

Fluidized bed reactors are reactors in which the immobilized enzyme and support are kept suspended by the upward flow of substrate or gas at high flow rates [80] (Figure 30.18). The advantages of fluidized bed reactors are that channeling problems are eliminated, there is less change in pressure at high flow rates and less coalescence of emulsion droplets. In addition, particulates do not have to be removed from the oil and there are no concentration gradients [7]. The main disadvantage of fluidized bed reactors is that small concentrations of enzyme can be used since a large void volume is required to keep the enzyme and support suspended. Mojovic et al. [98] used a gas lift reactor to produce a cocoa butter equivalent by interesterifying palm oil midfraction. These authors immobilized lipase encapsulated in lecithin reverse micelles in hexane; the reaction in the gas lift reactor was more efficient than in a stirred batch reactor. Equilibrium was reached 25% earlier and productivity was 2.8 times higher in the gas lift reactor.

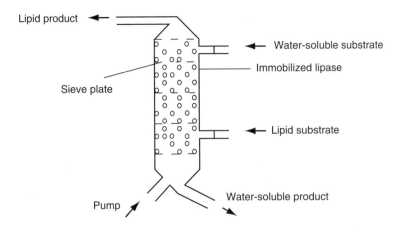

FIGURE 30.18 Fluidized bed reactor for immobilized lipase–catalyzed interesterification.

VI. FACTORS AFFECTING LIPASE ACTIVITY DURING INTERESTERIFICATION

In considering all of the factors involved in enzymatic interesterification, all components of the system must be examined; namely pH, water content, temperature, substrate composition, product composition, and lipase content.

A. pH

Lipases are only catalytically active at certain pHs, depending on their origin and the ionization state of residues in their active sites. While lipases contain basic, neutral, and acidic residues, the residues in the catalytic site are only active in one particular ionization state. The pH optima for most lipases lies between 7 and 9, although lipases can be active over a wide range of acid and alkaline pHs, from about pH 4 to 10 [50,99]. For example, the optimum pH for lipase from *Pseudomonas* species is around 8.5, whereas fungal lipases from *A. niger* and *R. delemar* are acidic lipases [100]. The effect of immobilization on the pH optimum of lipases is dependent on the partitioning of protons between the bulk phase and the microenvironment around the support and the restriction of proton diffusion by the support. If the lipase is immobilized on a polyanion matrix, the concentration of protons in the immediate vicinity of the support will be higher than in the bulk phase, thereby reducing the pH around the enzyme in comparison with the pH of the bulk phase. Since there is a difference in the perceived pH of the solution as measured by the pH of the bulk phase, the lipase would exhibit a shift in pH optimum toward a more basic pH. For instance, for a free lipase that has a pH optimum of 8.0, when immobilized on a polyanionic matrix, with the bulk solution at pH 8.0, the pH in the immediate vicinity of the lipase might be only 7.0. Therefore, while the reaction pH is 8.0, the lipase is operating at pH 7.0, which is below its optimum. The pH of the bulk solution would have to be increased to pH 9.0 to get the pH around the lipase to its optimum of 8.0. This phenomenon is only seen in solutions with ionized support and low ionic strength systems [101]. If protons are produced in the course of interesterification, the hydrogen ion concentration in the Nernst layer can be higher than in the bulk phase, thereby decreasing the pH in the vicinity of the lipase. Running an interesterification reaction with lipases at a pH well removed from the optimum can lead to rapid inactivation of the enzyme.

B. Temperature

In general, increasing the temperature increases the rate of interesterification, but very high temperatures can reduce the reaction rates due to irreversible denaturation of the enzyme. Animal and plant lipases are usually less thermostable than extracellular microbial lipases [99]. In a solvent-free system, the temperature must be high enough to keep the substrate in the liquid state [84,102]. Temperatures do not need to be as high in systems containing organic solvents since they easily solubilize hydrophobic substrates. However, for food industry applications, where organic solvents are avoided, the reaction temperatures are usually higher. Sometimes, the temperature has to be increased to as high as 60°C to liquify the substrate. Such high temperatures can seriously reduce the half-life of the lipase, although immobilization has been found to improve the stability of lipases under high temperature conditions. Immobilization fixes the enzyme in one conformation, which reduces the susceptibility of the enzyme to denaturation by heat. The optimal temperature for most immobilized lipases falls within the range of 30°C–62°C, whereas it tends to be slightly lower for free lipases [50]. Immobilized lipases are more stable to thermal deactivation because immobilization restricts movement and can reduce the degree of unfolding and denaturation. Hansen and Eigtved [103] found that even at a temperature of 60°C, immobilized lipase from *M. miehei* has a half-life of 1600 h.

C. WATER CONTENT AND WATER ACTIVITY

The activity of lipases at different water contents or water activity is dependent on the source of the enzyme. Lipases from molds seem to be more tolerant to low water activity than bacterial lipases. The optimal water content for interesterification by different lipases ranges from 0.04% to 11% (w/v), although most reactions require water contents of <1% for effective interesterification [15,50,104]. The water content in a reaction system is the determining factor as to whether the reaction equilibrium will be toward hydrolysis or ester synthesis. Ester synthesis depends on low water activity. Too low a water activity prevents all reactions from occurring because lipases need a certain amount of water to remain hydrated, which is essential for enzymatic activity [34,105]. As stated previously, lipases tend to retain the greatest degree of original activity when immobilized on hydrophobic supports. When the immobilized lipase is contacted with an oil-in-water emulsion, the oil phase tends to associate with and permeate the hydrophobic support, so that there is no aqueous shell surrounding the enzyme and support. It can be assumed that there is an ordered hydrophobic network of lipid molecules surrounding the support. Any water that reaches the enzyme for participation in hydrolysis and interesterification reactions must diffuse there from the bulk emulsion phase. Therefore, to avoid diffusional limitations, the oil phase must be well saturated with water [50]. Too much water can inhibit interesterification, probably due to decreased access of hydrophobic substrates to the immobilized enzyme. Abraham et al. [106] found that in a solvent-free system, interesterification dominated hydrolysis up to a water-to-lipase ratio of 0.9, after which hydrolysis became the predominant reaction. During interesterification, the reaction equilibrium can be forced away from ester synthesis due to accumulation of water, 1 mol of which is produced for every mole of ester synthesized during the reaction. The equilibrium can be pushed back toward ester synthesis by continuous removal of water produced during the reaction. Water activity can be kept constant by having a reaction vessel with a saturated salt solution in contact with the reaction mixture via the gas phase to continuously remove the water produced in the course of interesterification. Another method of water activity control that has proven useful with interesterification reactions is the use of silicone tubing containing the salt solution, immersed in the reaction vessel. Water vapor can be transferred out of the reaction system across the tubing wall and into the salt solution [107]. A very simple method for water removal involves adding molecular sieves near the end of the reaction, or running the reaction under a vacuum so that the water produced is continuously removed, while still allowing the lipase to retain its water of hydration [44,93,108]. Kurashige [93] ran an effective interesterification reaction with <150 ppm water maintained by running the reaction under vacuum.

D. ENZYME PURITY AND PRESENCE OF OTHER PROTEINS

During immobilization, adsorption of protein to surface-active supports is not limited solely to lipases. Other protein sources in the lipase solution can be adsorbed, and this can have an effect on the loading and activity of the immobilized enzyme. Use of a pure lipase solution for immobilization has been found to reduce activity of the lipase, whereas the presence of other proteins on the support can increase the activity of the immobilized lipase [90]. Nonprotein sources of contamination during immobilization are usually not a problem because the lipase is preferentially adsorbed to the support.

E.   SUBSTRATE COMPOSITION AND STERIC HINDRANCE

The composition of the substrate can have an effect on the rate of hydrolysis and interesterification by lipase. The presence of a hydroxyl group in the *sn*-2 position has a negative inductive effect, so that triacylglycerols are hydrolyzed at a faster rate than diacylglycerols, which are hydrolyzed at a faster rate than monoacylglycerols [11]. While the nucleophilicity of substrate is important to the rate of reaction, steric hindrance can have a much greater negative effect. If the composition of the substrate is such that it impedes access of the substrate to the active site, any improvements in the nucleophilicity will not improve the activity [109].

The conformation of the substrate can also have an effect on the rate of reaction. The hydrophobic tunnel in the lipase accepts aliphatic chains and aromatic rings more easily than branched structures [11,44]. For example, using carboxylic acids of differing chain lengths, Miller et al. [44] found that increasing the acyl group chain length up to seven carbons increased the esterification rate for lipase from *M. miehei*.

Oxidation of substrates, especially PUFAs, is possible and can cause inhibition and a decrease in activity of lipases, especially in reactions containing organic solvents. Inhibition is seen at hydroperoxide levels greater than 5 mequiv/kg oil and is attributed to the breakdown of hydroperoxides to free radicals [110]. Therefore, before running interesterification reactions, especially in flow-through systems such as fixed bed reactors which are more susceptible to poisoning and inactivation, oils containing high levels of PUFAs must be highly refined [89].

F.   SURFACE-ACTIVE AGENTS

The presence of surface-active agents used during the immobilization process can improve lipase activity during interesterification. The addition of lecithin or sugar esters as surface-active agents during the immobilization process can increase activity 10-fold when the preparation is used under microaqueous conditions [19]. In contrast, using surface-active agents to form an emulsion can dramatically decrease the rate of interesterification because they prevent contact between the lipase and substrate [111]. Adsorption at the interface can be inhibited by the presence of other non-substrate molecules, such as proteins. The presence of proteins other than lipase at the interface reduces the ability of the lipase to bind to the interface. Addition of protein in the presence of lipase can cause desorption of lipase from the interface.

Phospholipids, such as phosphatidylcholine, phosphatidylethanolamine, and phosphatidylinositol, can be found as minor components in oil, in quantities of 0.1%–3.2%. The presence of phospholipids can have a negative effect on lipase activity. The initial rate of reaction can be decreased due to initial competition between phosphatidylcholine and the triacylglycerols for the active site of the lipase. Phosphatidylethanolamine seems to have the most inhibitory effect on lipase action possibly due to the presence of the amine group. Due to their effects, the phospholipid content of oils must be <500 ppm to prolong the half-life of immobilized lipases during interesterification [112].

G.   PRODUCT ACCUMULATION

During interesterification of two triacylglycerols, the production of monoacylglycerols and diacylglycerols can lead to an increase in the rate of reaction, whereas the presence of high levels of free fatty acids can inhibit the initial hydrolysis of triacylglycerols [51]. In lipase-catalyzed interesterification, where hydrolysis is extensive, or in acidolysis reactions, the level of free fatty acids can have an impact on the rate of the reaction. During acidolysis of butter oil with undecanoic acid, Elliott and Parkin [65] reported that concentrations of undecanoic acid greater than 250 mM decreased the activity of porcine pancreatic lipase. Inhibition of lipase activity by free fatty acids agrees with the Michaelis–Menten model for uncompetitive inhibition by a substrate [65]:

$$v = \frac{V_{\max}[S_0]}{[S_0]\left(1 + \frac{[S_0]}{K_i}\right) + K_m},$$

where

[S_0] is the initial free fatty acid concentration
K_i is the inhibition constant
K_m is the Michaelis constant

The loss of activity by lipase in the presence of high concentrations of free fatty acids has been attributed to several factors. High levels of free fatty acids would produce high levels of free or ionized carboxylic acid groups, which would acidify the microaqueous phase surrounding the lipase or cause desorption of water from the interface. In addition, with short- and medium-chain fatty acids, there could be partitioning of fatty acids away from the interface into the surrounding water shell due to their increased solubility in water. This would limit access by the substrate to the interface [113]. Kuo and Parkin [113] found that there was less inhibition when longer chain fatty acids, such as C13:0 and C17:0, were used during acidolysis, compared with C5:0 and C9:0. The decrease in lipase activity was attributed to both increased solubility of the short-chain fatty acids and acidification of the aqueous phase.

VII. CONCLUSIONS

Despite the benefits of using lipase-catalyzed interesterification, it is unlikely that it will replace chemical interesterification in the future. This is due to the higher cost associated with enzymatic interesterification and the low cost of products, such as margarines and shortenings, that are currently produced using chemical interesterification. The main attraction of lipase-catalyzed inter-esterification reactions is in the specificities of individual lipases and their application to the development of novel fats and oils that cannot be produced by chemical means. Future applications will involve continued development of reduced-calorie products, enriched lipids, and structured lipids. In addition, research will continue in the area of the characterization of fatty acid specificities of new lipases particularly in the identification of 2-specific lipases. In order for any of these new applications to be useful in the food industry, scale-up studies simulating industrial processes are necessary.

REFERENCES

1. H.L. Goderis, G. Ampe, M.P. Feyton, B.L. Fouwé, W.M. Guffens, S.M. Van Cauwenbergh, and P.P. Tobback. Lipase-catalyzed ester exchange reactions in organic media with controlled humidity. *Biotechnol. Bioeng.* 30:258–266 (1987).
2. R.W. Stevenson, F.E. Luddy, and H.L. Rothbart. Enzymatic acyl exchange to vary saturation in diglycerides and triglycerides. *J. Am. Oil Chem. Soc.* 56:676–680 (1979).
3. W.W. Nawar. Chemistry. In: *Bailey's Industrial Oil and Fat Products, Vol. 1, Edible Oil and Fat Products, General Applications*, 5th edition (Y.H. Hui, ed.). John Wiley & Sons, New York, 1996, p. 409.
4. H.M. Ghazali, S. Hamidah, and Y.B. Che Man. Enzymatic transesterification of palm olein with nonspecific and 1,3 specific lipases. *J. Am. Oil Chem. Soc.* 72(6):633–639 (1995).
5. B. Fitch Haumann. Tools: Hydrogenation, interesterification. *INFORM* 5(6):668–678 (1994).

6. S.H. Goh, S.K. Yeong, and C.W. Wang. Transesterification of cocoa butter by fungal lipases: Effect of solvent on 1,3-specificity. *J. Am. Oil Chem. Soc.* 70(6):567–570 (1993).

7. F.X. Malcata, H.R. Reyes, H.S. Garcia, C.G. Hill Jr., and C.H. Amundson. Immobilized lipase reactors for modification of fats and oils-a review. *J. Am. Oil Chem. Soc.* 67(12):890–910 (1990).

8. A.R. Macrae. Interesterification of fats and oils. In: *Biocatalysts in Organic Syntheses* (J. Tramper, H.C. van der Plas, and P. Linko, eds.). *Proceedings of an International Symposium held at Noorwijkerhout*, The Netherlands (April 14–17, 1985), Elsevier Science Publishers, Amsterdam, 1985, pp. 195–208.

9. A.R. Macrae. Lipase-catalyzed interesterification of oils and fats. *J. Am. Oil Chem. Soc.* 60(2):291–294 (1983).

10. F.D. Gunstone. Chemical properties. In: *The Lipid Handbook*, 2nd edition (F.D. Gunstone, J.L. Harwood, and F.D. Padley, eds.). Chapman & Hall, London, 1994, pp. 594–595.

11. P. Desnuelle. The lipases. In: *The Enzymes*, Vol. 7, 3rd edition (P.D. Boyer, ed.). Academic Press, New York, 1972, pp. 575–616.

12. K.-E. Jaeger, S. Ransac, B.W. Dijkstra, C. Colson, M. van Heuvel, and O. Misset. Bacterial lipases. *FEMS Microbiol. Rev.* 15:29–63 (1994).

13. P. Kalo, P. Parviainen, K. Vaara, S. Ali-Yrrkö, and M. Antila. Changes in the triglyceride composition of butter fat induced by lipase and sodium methoxide catalysed interesterification reactions. *Milchwissenschaft* 41(2):82–85 (1986).

14. P. Kalo, H. Huotari, and M. Antila. *Pseudomonas fluorescens* lipase-catalysed inter-esterification of butter fat in the absence of solvent. *Milchwissenschaft* 45(5):281–285 (1990).

15. S. Bornaz, J. Fanni, and M. Parmentier. Limit of the solid fat content modification of butter. *J. Am. Oil Chem. Soc.* 71(12):1373–1380 (1994).

16. T.A. Foglia, K. Petruso, and S.H. Feairheller. Enzymatic interesterification of tallow–sunflower oil mixtures. *J. Am. Oil Chem. Soc.* 70(3):281–285 (1993).

17. M.K. Chang, G. Abraham, and V.T. John. Production of cocoa butter like fat from interesterification of vegetable oils. *J. Am. Oil Chem. Soc.* 67(11):832–834 (1990).

18. P. Forssell, R. Kervinen, M. Lappi, P. Linko, T. Suortti, and K. Poutanen. Effect of enzymatic interesterification on the melting point of tallow–rapeseed oil (LEAR) mixture. *J. Am. Oil Chem. Soc.* 69(2):126–129 (1992).

19. J. Kurashige, N. Matsuzaki, and H. Takahashi. Enzymatic modification of canola/palm oil mixtures: Effects on the fluidity of the mixture. *J. Am. Oil Chem. Soc.* 70(9):849–852 (1993).

20. Y. Tanaka, T. Funada, J. Hirano, and R. Hashizume. Triacylglycerol specificity of *Candida cylindraceae* lipase: Effects of docosahexaenoic acid on resistance of triacylglcerol to lipase. *J. Am. Oil Chem. Soc.* 70:1031–1034 (1993).

21. R.R. Brenner. Nutritional and hormonal factors influencing desaturation of essential fatty acids. *Prog. Lipid Res.* 20:41–47 (1982).

22. T. Yamane, T. Suzuki, Y. Sahashi, L. Vikersveen, and T. Hoshino. Production of *n*-3 polyunsaturated fatty acid-enriched fish oil by lipase-catalyzed acidolysis without solvent. *J. Am. Oil Chem. Soc.* 69(11):1104–1107 (1992).

23. G.G. Haraldsson, P.A. Höskulsson, S.Th. Sigurdsson, F. Thorsteinsson, and S. Gudbjarnason. Modification of the nutritional properties of fats using lipase catalyzed directed interesterification. *Tetrahedron Lett.* 30(13):1671–1674 (1995).

24. T. Yamane, T. Suzuki, and T. Hoshino. Increasing *n*-3 polyunsaturated fatty acid content of fish oil by temperature control of lipase-catalyzed acidolysis. *J. Am. Oil Chem. Soc.* 70(12):1285–1287 (1993).

25. Y.-Y. Linko and K. Hayakawa. Docosahexaenoic acid: A valuable nutraceutical? *Trends Food Sci. Technol.* 7:59–63 (1996).

26. T. Oba and B. Witholt. Interesterification of milk fat with oleic acid catalyzed by immobilized *Rhizopus arrhizus* lipase. *J. Dairy Sci.* 77:1790–1797 (1994).

27. J.P. Kennedy. Structured lipids: Fats of the future. *Food Technol.* 45:76–83 (1991).

28. C.C. Akoh, B.H. Jennings, and D.A. Lillard. Enzymatic modification of trilinolein: Incorporation of *n*-3 polyunsaturated fatty acids. *J. Am. Oil Chem. Soc.* 71(11):1317–1321 (1995).

29. P. Quinlan and S. Moore. Modification of triglycerides by lipase: Process technology and its application to the production of nutritionally improved fats. *INFORM* 4(5):580–585 (1993).

30. F. Bjorkling, S.E. Godtfredsen, and O. Kirk. The future impact of industrial lipase. *Trends Biotechnol.* 9:360–363 (1991).

31. G.P. McNeill and P.E. Sonnet. Low-calorie triglyceride synthesis by lipase-catalyzed esterification of monoglycerides. *J. Am. Oil Chem. Soc.* 72(11):1301–1307 (1995).

32. S. Bloomer, P. Adlercreutz, and B. Mattiasson. Triglyceride interesterification by lipases. 1. Cocoa butter equivalents from a fraction of palm oil. *J. Am. Oil Chem. Soc.* 67(8):519–524 (1990).

33. C.N. Chong, Y.M. Hoh, and C.W. Wang. Fractionation procedures for obtaining cocoa butter-like fat from enzymatically interesterified palm olein. *J. Am. Oil Chem. Soc.* 69(2):137–140 (1992).

34. D. Briand, E. Dubreucq, and P. Galzy. Enzymatic fatty esters synthesis in aqueous medium with lipase from *Candida parapsilosis* (Ashford) Langeron and Talice. *Biotechnol. Lett.* 16(8):813–818 (1994).

35. G.P. McNeill, D. Borowitz, and R.G. Berger. Selective distribution of saturated fatty acids into the monoglyceride fraction during enzymatic glycerolysis. *J. Am. Oil Chem. Soc.* 69(11):1098–1103 (1992).

36. G.P. McNeill, S. Shimizu, and T. Yamane. High-yield glycerolysis of fats and oils. *J. Am. Oil Chem. Soc.* 68(1):1–5 (1991).

37. A.J. Kaslauskas. Elucidating structure–mechanism relationships in lipases: Prospects for predicting and engineering catalytic properties, *Trends Biotechnol.* 12:464–472 (1994).

38. F.K. Winkler, A. D'Arcy, and W. Hunziker. Structure of human pancreatic lipase. *Nature* 343:771–773 (1990).

39. K.A. Dugi, H.L. Dichek, G.D. Talley, H.B. Brewer, and S. Santamarina-Fojo. Human lipoprotein lipase: The loop covering the active site is essential for interaction with lipid substrates. *J. Biol. Chem.* 267(35):25086–25091 (1992).

40. M.-P. Egloff, L. Sarda, R. Verger, C. Cambillau, and I.I. van Tilbeurgh. Crystallographic study of the structure of colipase and of the interaction with pancreatic lipase. *Protein Sci.* 4:44–57 (1988).

41. I.H. Segel. *Biochemical Calculations: How to Solve Mathematical Problems in General Biochemistry*, 2nd edition. John Wiley & Sons, New York, 1976, pp. 300–302.

42. D.S. Wong. Lipolytic enzymes. In: *Food Enzymes: Structure and Mechanism*, Chapman & Hall, New York, 1994, pp. 170–211.

43. A.M. Brzozowski, U. Derewenda, Z.S. Derewenda, G.G. Dodson, D.M. Lawson, J.P. Turkenburg, F. Bjorkling, B. Huge-Jensen, S.A. Patkar, and L. Thim. A model for interfacial activation in lipases form the structure of a fungal lipase–inhibitor complex. *Nature* 351:491–494 (1991).

44. C. Miller, H. Austin, L. Porsorske, and J. Gonziez. Characteristics of an immobilized lipase for the commercial synthesis of esters. *J. Am. Oil Chem. Soc.* 65(6):927–935 (1988).

45. L. Sarda and P. Desnuelle. Action de la lipase pancreatique sur les esters en émulsion. *Biochim. Biophys. Acta* 30:513–521 (1958).

46. R. Verger. Enzyme kinetics of lipolysis. In: *Enzyme Kinetics and Mechanism* (D.L. Purich, ed.). Academic Press, New York, 1980, pp. 340–391.

47. H. Brockerhoff and P.G. Jensen. *Lipolytic Enzymes.* Academic Press, London, 1974, pp. 13–21.

48. V.K. Antonov, V.L. Dyakov, A.A. Mishin, and T.V. Rotanova. Catalytic activity and association of pancreatic lipase. *Biochimie* 70:1235–1244 (1988).

49. G. Benzonana and P. Desnuelle. Etude anetique de l'action de la lipase pancreatique sur des triglycerides en emulsion. Essaie d'une enzomologie en milieu heterogene. *Biochimie* 105:121–136 (1965).

50. F.X. Malcata, H.R. Reyes, H.S. Garcia, C.G. Hill Jr., and C.H. Amundson. Kinetics and mechanisms catalyzed by immobilized lipases. *Enzyme Microb. Technol.* 14:426–446 (1992).

51. H.R. Reyes and C.G. Hill Jr. Kinetic modeling of interesterification reactions catalyzed by immobilized lipase. *Biotechnol. Bioeng.* 43:171–182 (1994).

52. A.G. Marangoni and D. Rousseau. Engineering triacylglycerols: The role of interesterification. *Trends Food Sci. Technol.* 6:329–335 (1995).

53. D.A. Miller, J.M. Prausnitz, and H.W. Blanch. Kinetics of lipase catalysed interesterification of triglycerides in cyclohexane. *Enzyme Microb. Technol.* 13:98–103 (1991).

54. P.E. Sonnet and J.A. Gazzillo. Evaluation of lipase selectivity for hydrolysis. *J. Am. Oil Chem. Soc.* 68(1):11–15 (1991).

55. A.R. Macrae and P. How. Rearrangement process. U.S. Patent 4,719,178 (1988).

56. S. Sil Roy and D.K. Bhattacharyya. Distinction between enzymically and chemically catalyzed interesterification. *J. Am. Oil Chem. Soc.* 70(12):1293–1294 (1993).

57. F. Pabai, S. Kermasha, and A. Morin. Lipase from *Pseudomonas fragi* CRDA 323: Partial purification, characterization and interesterification of butter fat. *Appl. Microbiol. Biotechnol.* 43:42–51 (1995).

58. A. Riaublanc, R. Ratomahenina, P. Galzy, and M. Nicolas. Peculiar properties from *Candida parapsilosis* (Ashford) Langeron Talice. *J. Am. Oil Chem. Soc.* 70(5):497–500 (1993).

59. S. Ray and D.K. Bhattacharyya. Comparative nutritional study of enzymatically and chemically interesterified palm oil products. *J. Am. Oil Chem. Soc.* 72(3):327–330 (1995).

60. H. Uzawa, T. Noguchi, Y. Nishida, H. Ohrui, and H. Meguro. Determination of the lipase stereoselectivities using circular dichroism (CD); lipase produce chiral di-*O*-acylglycerols from achiral tri-*O*-acylgylcerols. *Biochim. Biophys. Acta* 1168:253–260 (1993).

61. H. Stamatis, A. Xenakis, M. Provelegiou, and F.N. Kolisis. Esterification reactions catalyzed by lipase in microemulsions: The role of enzyme location in relation to its selectivity. *Biotechnol. Bioeng.* 42:103–110 (1993).

62. A.G. Marangoni. *Candida* and *Pseudomonoas* lipase-catalyzed hydrolysis of butteroil in the absence of organic solvents. *J. Food Sci.* 59(5):1096–1099 (1994).

63. A. van der Padt, M.J. Edema, J.J.W. Sewalt, and K. van't Riet. Enzymatic acylglycerol synthesis in a membrane bioreactor. *J. Am. Oil Chem. Soc.* 67(6):347–352 (1990).

64. M. Murakami, Y. Kawasaki, M. Kawanari, and H. Okai. Transesterification of oil by fatty acid-modified lipase. *J. Am. Oil Chem. Soc.* 70(6):571–574 (1993).

65. J.M. Elliott and K.L. Parkin. Lipase mediated acyl-exchange reactions with butteroil in anhydrous media. *J. Am. Oil Chem. Soc.* 68(3):171–175 (1991).

66. E. Santaniello, P. Ferraboschi, and P. Grisenti. Lipase-catalyzed transesterification in organic solvents: Applications to the preparation of enantiomerically pure compounds. *Enzyme Microb. Technol.* 15:367–382 (1993).

67. R.H. Valivety, G.A. Johnston, C.J. Suckling, and P.J. Hailing. Solvent effects on biocatalysis in organic systems: Equilibrium position and rates of lipase catalyzed esterification. *Biotechnol. Bioeng.* 38:1137–1143 (1991).

68. G. Carta, J.L. Gainer, and A.H. Benton. Enzymatic synthesis of esters using an immobilized lipase. *Biotechnol. Bioeng.* 37:1004–1009 (1991).

69. K. Takahashi, Y. Saito, and Y. Inada. Lipase made active in hydrophobic media. *J. Am. Oil Chem. Soc.* 65(6):911–916.

70. F. Monot, E. Paccard, F. Borzeix, M. Badin, and J.-P. Vandecasteele. Effect of lipase conditioning on its activity in organic media. *Appl. Microb. Biotechnol.* 39:483–486 (1993).

71. H. Stamatis, A. Xenakis, U. Menge, and F.N. Kolisis. Esterification reactions in water-in-oil microemulsions. *Biotechnol. Bioeng.* 42:931–937 (1993).

72. P.D.I. Fletcher, R.B. Freedman, B.H. Robinson, G.D. Rees, and R. Schomacker. Lipase-catalyzed ester synthesis in oil-continuous microemulsions. *Biochim. Biophys. Acta* 912:278–282 (1987).

73. L. Mojovic, S. Siler-Marinkovic, G. Kukíc, and G. Vunjak-Novakovíc. *Rhizopus arrhizus* lipase-catalyzed interesterification of the midfraction of palm oil to a cocoa butter equivalent fat. *Enzyme Microb. Technol.* 15:438–443 (1993).

74. A.G. Marangoni, R.D. McCurdy, and E.D. Brown. Enzymatic interesterification of triolein with tripalmitin in canola lecithin–hexane reverse micelles. *J. Am. Oil Chem. Soc.* 70(8):737–744 (1993).

75. K. Holmberg, B. Larsson, and M.-B. Stark. Enzymatic glycerolysis of a triacylglyceride in aqueous and nonaqueous microemulsions. *J. Am. Oil Chem. Soc.* 66(12):1796–1800 (1989).

76. D.G. Hayes and E. Gulari. Esterification reactions of lipase in reverse micelles. *Biotechnol. Bioeng.* 35:793–801.

77. M. Safari and S. Kermasha. Interesterification of butterfat by commercial microbial lipases in a cosurfactant-free microemulsion system. *J. Am. Oil Chem. Soc.* 71(9):969–973 (1994).

78. F. Ergan, M. Trani, and G. Andre. Use of lipases in multiphasic systems solely composed of substrates. *J. Am. Oil Chem. Soc.* 68(6):412–417 (1991).

79. O.R. Zaborsky. *Immobilized Enzymes*. CRC Press, Boca Raton, FL, 1973, p. 75, pp. 119–123.

80. P.S.J. Cheetham. The application of immobilized enzymes and cells and biochemical reactors in biotechnology: Principles of enzyme engineering. *Principles of Biotechnology*, 2nd edition (A. Wiseman, ed.). Chapman & Hall, New York, 1988, pp. 164–201.

81. A. Mustranta, P. Forssell, and K. Poutanen. Applications of immobilized lipases to transesterification and esterification reactions in non-aqueous systems. *Enzyme Microb. Technol.* 15:133–139 (1993).

82. F.X. Malcata, H.S. Garcia, C.G. Hill Jr., and C.H. Amundson. Hydrolysis of butteroil by immobilized lipase using a hollow-fiber reactor: Part I. Lipase adsorption studies. *Biotechnol. Bioeng.* 39:647–657 (1992).

83. I. Karube, Y. Yugeta, and S. Suzuki. Electric field control of lipase membrane activity. *Biotechnol. Bioeng.* 19:1493–1501 (1977).

84. A.P. Ison, A.R. Macrae, C.G. Smith, and J. Bosley. Mass transfer effects in solvent-free fat interesterification reactions: Influence on catalyst design. *Biotechnol. Bioeng.* 43:122–130 (1994).

85. R.A. Wisdom, P. Dunnill, and M.D. Lilly. Enzymatic interesterification of fats: Factors influencing the choice of support for immobilized lipase. *Enzyme Microb. Technol.* 6:443–446 (1984).

86. G. Abraham. Mass transfer in bioreactors. In: *Proceedings of the World Conference on Biotechnology for the Fats and Oils Industry* (T.H. Applewhite, ed.). American Oil Chemist's Society, Champaign, IL, 1988, pp. 219–225.

87. B.K. Hamilton, C.R. Gardner, and C.K. Colton. Basic concepts in the effects of mass transfer on immobilized enzyme kinetics. In: *Immobilized Enzymes in Food and Microbial Processes* (A.C. Olson and C.L. Cooney, eds.). Plenum Press, New York, 1974, pp. 205–219.

88. W.E. Hornby, M.D. Lilley, and E.M. Crook. Some changes in the reactivity of enzymes resulting from their chemical attachment of water-insoluble derivatives of cellulose. *Biochem. J.* 107:669–674 (1968).

89. L.H. Posorske, G.K. LeFebvre, C.A. Miller, T.T. Hansen, and B.L. Glenvig. Process considerations of continuous fat modification with immobilized lipase. *J. Am. Oil Chem. Soc.* 65:922–926 (1988).

90. R.A. Wisdom, P. Dunnill, and M.D. Lilly. Enzymic interesterification of fats: Laboratory and pilot-scale studies with immobilized lipase from *Rhizopus arrhizus*. *Biotechnol. Bioeng.* 29:1081–1085 (1987).

91. W.E. Hornby, M.D. Lilley, and E.M. Crook. Some changes in the reactivity of enzymes resulting from their chemical attachment of water-insoluble derivatives of cellulose. *J. Biochem.* 107:669–674 (1968).

92. P. Forssell, P. Parovuori, P. Linko, and K. Poutanen. Enzymatic transesterification of rapeseed oil and lauric acid in a continuous reactor. *J. Am. Oil Chem. Soc.* 70(11):1105–1109 (1993).

93. J. Kurashige. Enzymatic conversion of diglycerides to triglycerides in palm oils. In: *Proceedings of the World Conference on Biotechnology for the Fats and Oils Industry* (T.H. Applewhite, ed.). American Oil Chemist's Society, Champaign, IL, 1988, pp. 219–225.

94. M.D. Lilly and A.K. Sharp. The kinetics of enzymes attached to water-insoluble polymers. *Chem. Eng.* 215:CE12–CE18 (1968).

95. J. Kloosterman IV, P.D. van Wassenaar, and W.J. Bel. Modification of fats and oils in membrane bioreactors. In: *Proceedings of the World Conference on Biotechnology for the Fats and Oils Industry* (T.H. Applewhite, ed.). American Oil Chemist's Society, Champaign, IL, 1988, pp. 219–225.

96. M.M. Hoq, H. Tagami, T. Yamane, and S. Shimizu. Some characteristics of continuous glyceride synthesis by lipase in a microporous hydrophobic membrane bioreactor. *Agric. Biol. Chem.* 49:335–342 (1985).

97. M.M. Hoq, T. Yamane, and S. Shimizu. Continuous synthesis of glycerides by lipase in a microporous membrane bioreactor. *J. Am. Oil Chem. Soc.* 61:776–781 (1984).

98. L. Mojovic, S. Šiler-Marinkovic, G. Kukíc, B. Bugarski, and G. Vunjak-Novakovic. *Rhizopus arrhizus* lipase-catalyzed interesterification of palm oil in a gas-lift reactor. *Enzyme Microb. Technol.* 16:159–162 (1994).

99. T. Yamane. Enzyme technology for the lipids industry: An engineering overview. *J. Am. Oil Chem. Soc.* 4(12):1657–1662 (1987).

100. M. Iwai and Y. Tsujisaka. The purification and properties of three kinds of lipase from *Rhizopus delemar*. *Agric. Biol. Chem.* 38:241–247 (1974).

101. M.D. Trevan. *Immobilized Enzymes*. John Wiley & Sons, New York, 1980, pp. 16–26.

102. P. Forssell and K. Poutanen. Continuous enzymatic transesterification of rapeseed oil and lauric acid in a solvent-free system. In: *Biocatalysis in Non-Conventional Media* (J. Tramper et al., ed.). Elsevier Science Publishers, Amsterdam, 1992, pp. 491–495.

103. T.T. Hansen and P. Eigtved. A new immobilized lipase for oil and fat modifications. In: *Proceedings of the World Conference on Emerging Technologies in the Fats and Oils Industry* (A.R. Baldwin, ed.). Held at Cannes, France (November 3–8, 1985), American Oil Chemist's Society, Champaign, IL, 1986, pp. 365–369.

104. Z.-Y. Li and O.P. Ward. Lipase-catalyzed esterification of glycerol and *n*-3 polyunsaturated fatty acid concentrate in organic solvent. *J. Am. Oil Chem. Soc.* 70(8):746–748 (1995).

105. I. Svesson, E. Wehtje, P. Adlercreutz, and B. Mattiasson. Effects of water activity on reaction rates and equilibrium positions in enzymatic esterifications. *Biotechnol. Bioeng.* 44:549–556 (1994).

106. G. Abraham, M.A. Murray, and V.T. John. Interesterification selectivity in lipase catalyzed reactions of low molecular weight triglycerides. *Biotechnol. Lett.* 10(8):555–558 (1998).

107. I. Svensson, E. Wehtje, P. Adlercreutz, and B. Mattiasson. Effects of water activity on reaction rates and equilibrium positions in enzymatic esterifications. *Biotechnol. Bioeng.* 44:549–556 (1994).

108. P.E. Sonnet, G.P. McNeill, and W. Jun. Lipase of *Geotrichum candidum* immobilized on silica gel. *J. Am. Oil Chem. Soc.* 71(12):1421–1423 (1994).

109. H.S. Bevinakatti and A.A. Banerji. Lipase catalysis: Factors governing transesterification. *Biotechnol. Lett.* 10(6):397–398 (1988).

110. S. Bech Pedersen and G. Holmer. Studies of the fatty acid specificity of the lipase from *Rhizomucor miehei* toward 20:1*n*-9, 20:5*n*-3, 22:1*n*-9 and 22:6*n*-3. *J. Am. Oil Chem. Soc.* 72(2):239–243 (1995).

111. D. Briand, E. Dubreucq, and P. Galzy. Factors affecting the acyl transfer activity of the lipase from *Candida parapsilosis* in aqueous media. *J. Am. Oil Chem. Soc.* 72(11):1367–1373 (1995).

112. Y. Wang and M.H. Gordon. Effect of phospholipids in enzyme-catalyzed transesterification of oils. *J. Am. Oil Chem. Soc.* 68(8):588–590 (1991).

113. S.-I. Kuo and K.L. Parkin. Substrate preference for lipase-mediated acyl-exchange reactions with butteroil are concentration-dependent. *J. Am. Oil Chem. Soc.* 70(4):393–399 (1993).

31 Structured Lipids

Casimir C. Akoh and Byung Hee Kim

CONTENTS

I. INTRODUCTION

A. WHAT ARE STRUCTURED LIPIDS?

In a broad sense, structured lipids (SLs) are lipids that have been chemically or enzymatically modified from their natural biosynthetic form. In this definition of SLs, the scope of lipids includes triacylglycerols (TAGs) (the most common types of food lipids) as well as other types of acylglycerols, such as diacylglycerols, monoacylglycerols, and glycerophospholipids (phospholipids). The term modified means any alteration in the structure of the naturally occurring lipids. This definition includes the topics covered in Chapters 26–28 and 32. In a narrower sense and in many cases, SLs are specifically defined as TAGs that have been modified by incorporation of new fatty acids, restructured to change the positions of fatty acids, or the fatty acid profile, from the natural state, or

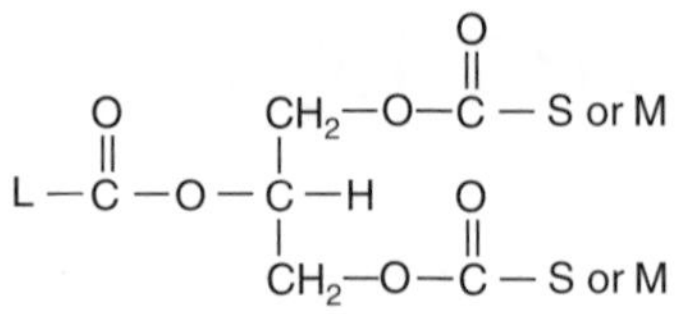

$$
\begin{array}{c}
\text{O}\\
\parallel\\
\text{CH}_2-\text{O}-\text{C}-\text{S or M}
\end{array}
$$

$$
\begin{array}{ccc}
\text{O} & & \text{CH}_2-\text{O}-\overset{\displaystyle\overset{\text{O}}{\parallel}}{\text{C}}-\text{S or M}\\
\parallel & & |\\
\text{L}-\text{C}-\text{O}-\text{C}-\text{H} & & \\
& & |\qquad\quad\text{O}\\
& & \text{CH}_2-\text{O}-\overset{\displaystyle\overset{\text{O}}{\parallel}}{\text{C}}-\text{S or M}
\end{array}
$$

FIGURE 31.1 General structure of structured lipids: S, L, and M: short-, medium-, and long-chain fatty acid, respectively; the positions of S, L, and M are interchangeable.

synthesized to yield novel TAGs. The fatty acid profiles of conventional TAGs are genetically defined and unique to each plant or animal species. In this chapter, SLs preferentially refer to TAGs containing mixtures of fatty acids (short chain and/or medium chain, plus long chain) esterified to the glycerol moiety, preferably in the same glycerol molecule. Figure 31.1 shows the general structure of SLs; their potency increases if each glycerol moiety contains both short- (SCFAs) or medium-chain fatty acids (MCFAs) and long-chain fatty acids (LCFAs). SLs combine the unique characteristics of component fatty acids such as melting behavior, digestion, absorption, and metabolism to enhance their use in foods, nutrition, and therapeutics. Individuals unable to metabolize certain dietary fats or with pancreatic insufficiency may benefit from the consumption of SLs.

SLs are often referred to as a new generation of fats that can be considered as nutraceuticals: food or parts of food that provide medical or health benefits, including the potential for the prevention and treatment of diseases [1]. Sometimes, they are referred to as functional foods or in the present context, as functional lipids. Functional foods is a term used to broadly describe foods that provide specific health benefits beyond basic nutrition. Medical foods (medical lipids) are foods (lipids) developed for use under medical supervision to treat or manage particular disease or nutritional deficiency states. Other terms used to describe functional foods are physiologic functional foods, pharmafoods, and nutritional foods. The nomenclature is still confusing and needs to be worked out by scientists in this field. SLs can be designed for use as medical or functional lipids, and as nutraceuticals.

B. RATIONALE FOR STRUCTURED LIPID DEVELOPMENT

Over the past several decades, long-chain triacylglycerols (LCTs), predominantly soybean and safflower oils, have been the standard lipids used in making fat emulsions for total parenteral nutrition (TPN) and enteral administration. The emulsion provides energy and serves as a source of essential fatty acids (EFAs). However, LCFAs are metabolized slowly in the body. It was then proposed that medium-chain triacylglycerols (MCTs) may be better than LCTs because the former are readily metabolized for quick energy. MCTs are not dependent on carnitine for transport into the mitochondria. They have higher plasma clearance, higher oxidation rate, improved nitrogen-sparing action, and less tendency to be deposited in the adipose tissue or to accumulate in the reticuloendothelial system (RES). One major disadvantage of using MCT emulsions is the lack of EFAs (18:2n-6). In addition, large doses of MCTs can lead to the accumulation of ketone bodies, a condition known as metabolic acidosis or ketonemia. It was suggested that combining MCTs and LCTs in the preparation of fat emulsions enables utilization of the benefits of both TAGs and may be theoretically better than pure LCT emulsions. An emulsion of MCTs and LCTs is called a physical mixture; however, a physical mixture is not equivalent to an SL. When MCTs and LCTs are chemically interesterified, the randomized product is called an SL. SLs are expected to be rapidly cleared and metabolized compared with LCTs.

For an SL to be beneficial, a minimum amount of LCFA is needed to meet EFA requirements. With the SL, LCFAs, MCFAs, and SCFAs can be delivered without the associated adverse effects of pure MCT emulsions. This is especially important when intravenous administration is considered

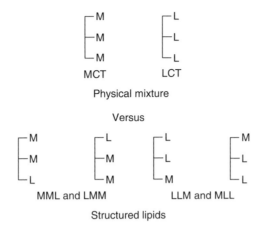

FIGURE 31.2 Structure of a physical mixture of medium-chain triacylglycerol and long-chain triacylglycerol, and structured lipid molecular species: M, medium-chain fatty acid; L, long-chain fatty acid. Note that physical mixture is not equivalent to structured lipid.

[2,3]. TAGs containing specific balances of medium-chain, *n*-3, *n*-6, *n*-9, and saturated fatty acids can be synthesized to reduce serum low density lipoprotein (LDL) cholesterol and TAG levels, prevent thrombosis, improve immune function, lessen the incidence of cancer, and improve nitrogen balance [1,4]. Although physical mixtures of TAGs have been administered to patients, an SL emulsion is more attractive due to the modified absorption rates of the SL molecule. Figure 31.2 shows the difference between a physical mixture of two TAGs and SL pairs of molecular species.

SLs can be manipulated to improve their physical characteristics such as melting points. SLs are texturally important in the manufacture of plastic fats such as margarines, modified butters, and shortenings. Caprenin, an SL produced by Procter & Gamble Company (Cincinnati, OH), consists of C8:0–C10:0–C22:0; it has the physical properties of cocoa butter but only about half the calories. Benefat, originally produced as Salatrim (see Section II.B.2), consists of SCFAs (C2:0–C4:0) and LCFAs (C18:0). Both products can be used as cocoa butter substitutes. Currently, they are manufactured through a chemical transesterification process. Due to the low caloric value of the SCFAs and the partial absorption of stearic acid on Salatrim, this product has strong potential for use as a low-calorie fat substitute in the future. The caloric content of Caprenin and Benefat is about 5 kcal/g (versus 9 kcal/g for a regular TAG). These SLs can also be manipulated for nutritive and therapeutic purposes, targeting specific diseases and metabolic conditions [4]. In the construction of SLs for nutritive and therapeutic use, it is important that the function and metabolism of various fatty acids be considered. This chapter focuses mainly on SLs and MCTs, emphasizing the use of enzymes for SL synthesis as an alternative to chemical processing.

II. PRODUCTION OF STRUCTURED LIPIDS

A. SOURCES OF FATTY ACIDS FOR STRUCTURED LIPID SYNTHESIS

SLs have been developed to optimize the benefit of fat substrate mixtures [5]. A variety of fatty acids are used in the synthesis of SLs, taking advantage of the functions and properties of each to obtain maximum benefits from a given SL. These fatty acids include SCFAs, MCFAs, polyunsaturated fatty acids (PUFAs), saturated LCFAs, and monounsaturated fatty acids. Table 31.1 gives the suggested levels of some of these fatty acids in SLs intended for clinical applications. The component fatty acids and their position in the TAG molecule determine the functional and physical properties, the metabolic fate, and the health benefits of the SL. It is therefore appropriate to review the function and metabolism of the component fatty acids.

TABLE 31.1

Suggested Optimum Levels of Fatty Acids for Structured Lipids in Clinical Nutrition

Fatty Acid	Levels and Function
n-3	2%–5% to enhance immune function, reduce blood clotting, lower serum triacylglycerols, and reduce risk of coronary heart disease
n-6	3%–4% to satisfy essential fatty acid requirement in the diet
n-9	Monounsaturated fatty acid (18:1*n*-9) for the balance of long-chain fatty acid
SCFA and MCFA[a]	30%–65% for quick energy and rapid absorption, especially for immature neonates, hospitalized patients, and individuals with lipid malabsorption disorders

Source: Modified from Kennedy, J.P., *Food Technol.*, 45, 76, 1991.

[a] Structured lipid containing short-chain fatty acids (SCFAs) and medium-chain fatty acids (MCFAs) as the main component.

1. Short-Chain Fatty Acids

The SCFAs range from C2:0 to C6:0. They occur ubiquitously in the gastrointestinal tract of mammals, where they are the end products of microbial digestion of carbohydrates [6]. In the human diet, SCFAs are usually taken in during consumption of bovine milk, which has a TAG mixture containing approximately 5%–10% butyric acid and 3%–5% caproic acid [7,8]. Butyric acid is found in butterfat, where it is present at about 30% of the TAG [9]. SCFAs, also known as volatile fatty acids, are more rapidly absorbed in the stomach than MCFAs because of their higher water solubility, smaller molecular size, and shorter chain length. Being hydrophilic, SCFAs have rates and mechanisms of absorption that are clearly distinguishable from those of lipophilic LCFAs [10]. SCFAs are mainly esterified to the *sn*-3 position in the milk of cows, sheep, and goats [7]. Under normal conditions, the end products of all carbohydrate digestion are the three major straight-chain SCFAs: acetate, propionate, and butyrate [11,12]; the longer SCFAs are generally found in smaller proportions except with diets containing high levels of sugar [13]. Microbial proteolysis followed by deamination also produces SCFA.

Using synthetic TAGs, Jensen et al. [14] have shown that human pancreatic gastric lipase can preferentially hydrolyze *sn*-3 esters over *sn*-1 esters in the ratio of 2:1. This enzyme has also shown some hydrolytic specificity for SCTs and MCTs, although later studies [15] reported in vitro optimal conditions for the hydrolysis of LCFAs by gastric lipase. Pancreatic lipase has been reported to attack only the primary ester group of TAG, independent of the nature of fatty acid attached [16]. Therefore, due to the positional and chain length specificity of the lipase, SCFAs attached to the *sn*-3 position of TAGs are likely to be completely hydrolyzed in the lumen of the stomach and small intestine. SCFAs are useful ingredients in the synthesis of low-calorie SLs such as Benefat because from heats of combustion, SCFAs are lower in caloric value than MCFAs and LCFAs. Examples of caloric values of SCFAs are as follows: acetic acid, 3.5 kcal; propionic acid, 5.0 kcal; butyric acid, 6.0 kcal; and caproic acid, 7.5 kcal.

2. Medium-Chain Fatty Acids and Triacylglycerols

MCTs contain C6:0 to C12:0 fatty acids esterified to glycerol backbone. MCTs serve as an excellent source of MCFAs for SL synthesis. MCTs are used for making lipid emulsions either alone or by blending with LCTs for parenteral and enteral nutrition. The MCT structure is given in Figure 31.3. MCTs are liquid or solid at room temperature, and their melting points depend on the fatty acid composition. MCTs are used as carriers for colors, flavors, vitamins, and pharmaceuticals [17]. MCFAs are commonly found in kernel oils or lauric fats; for example, coconut oil contains

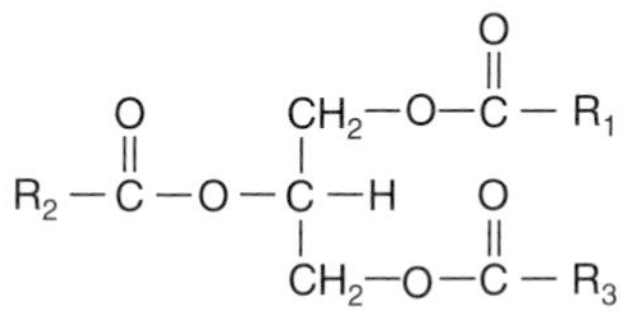

$$R_2 - \underset{\underset{O}{\|}}{C} - O - \underset{\underset{\underset{CH_2-O-\overset{\overset{O}{\|}}{C} - R_3}{|}}{\overset{\overset{CH_2-O-\overset{\overset{O}{\|}}{C} - R_1}{|}}{C}} - H$$

FIGURE 31.3 General structure of medium-chain triacylglycerols: R, alkyl group of MCFAs C6:0 to C12:0.

10%–15% C8:0 to C10:0 acid and is a raw material for MCT preparation [3]. MCT is synthesized chemically by direct esterification of MCFA and glycerol at high temperature and pressure, followed by alkali washing, steam refining, molecular distillation, and further purification. Enzymatically, MCTs have been synthesized with immobilized *Mucor miehei* lipase in a solvent-free system [18]. MCFAs have a viscosity of about 25–31 cP at 20°C and a bland odor and taste; as a result of the saturation of the fatty acids, they are extremely stable to oxidation [3]. MCTs have a caloric value of 8.3 kcal compared with 9 kcal for LCTs. This characteristic has made MCTs attractive for use in low-calorie desserts. MCTs may be used in reduced-calorie foods such as salad dressings, baked goods, and frozen dinners [17].

MCTs have several health benefits when consumed in mixtures containing LCTs. Toxicological studies on dogs have shown that consuming 100% MCT emulsions leads to the development of adverse effects in dogs, which include shaking of the head and vomiting and defecation, progressing to a coma [19]. It was theorized that these symptoms arose from elevated plasma concentration of MCFA or octonoate [19]. Some advantages of MCFA/MCT consumption include the following: (1) MCFAs are more readily oxidized than LCFAs; (2) carnitine is not required for MCT transport into the mitochondria, thus making MCT an ideal substrate for infants and stressed adults [20]; (3) MCFAs do not require chylomicron formation; and (4) MCFAs are transported back to the liver directly by the portal system. Absorption of SLs is discussed later in this chapter.

MCTs are not readily reesterified into TAGs and have more than twice the caloric density of proteins and carbohydrates, yet can be absorbed and metabolized as rapidly as glucose, whereas LCTs are metabolized more slowly [3]. Feeding diets containing 20% and 30% lipid concentrations in weight maintenance studies indicate that MCTs may be useful in the control of obesity [21]. MCTs appear to give satiety and satisfaction to some patients. Thermogenesis of MCT may be a factor in its very low tendency to deposit as depot fat [3].

Some reports suggest that MCTs can lower both serum cholesterol and tissue cholesterol in animals and man, even more significantly than conventional polyunsaturated oils [22]. However, a study by Cater et al. [23] showed that MCTs indeed raised plasma total cholesterol and TAG levels in mildly hypercholesterolemic men fed MCT, palm oil, or high oleic acid sunflower oil diets. A suggested mechanism for the cholesterol-raising ability of MCTs is as follows: acetyl CoA, which is the end product of MCT oxidation, is resynthesized into LCFAs; the LCFAs then mix with the hepatic LCFA pool; and the newly synthesized LCFA may then behave like dietary LCFA. In addition, the C8:0 may serve as precursor for de novo synthesis of LCFAs such as C14:0 and C16:0, which were detected in the plasma TAG [23]. There were no differences in the high-density lipoprotein (HDL) cholesterol concentrations among the subjects.

Evidence is pointing against the advisability of using MCTs in weight control because the level of MCTs (50%) required to achieve positive reduction is unlikely in human diet [24]. An SL containing MCFA and linoleic acid bound in the TAG is more effective for cystic fibrosis patients than safflower oil, which has about twice as much linoleic acid as the SL [25]. It appears that mobility, solubility, and ease of metabolism of MCFAs were responsible for the health benefits of the SL in these cases. In the SL, MCFAs provide not only a source of dense calories but also potentially fulfill a therapeutic purpose.

3. Omega-6 Fatty Acids

A common *n*-6 fatty acid is linoleic acid (18:2*n*-6). Linoleic acid is mainly found in most vegetable oils and in the seeds of most plants except coconut, cocoa, and palm nuts. Linoleic acids have a reducing effect on plasma cholesterol and an inhibitory effect on arterial thrombus formation [26]. The *n*-6 fatty acids cannot be synthesized by humans and mammals and are therefore considered EFAs. The inability of some animals to produce 18:2*n*-6 is attributed to the lack of a Δ12 desaturase, required to introduce a second double bond in oleic acid. Linoleic acid can be desaturated further, and elongated to arachidonic acid (20:4*n*-6), which is a precursor for eicosanoid formation, as shown in Figure 31.4.

Essentiality of fatty acids was reported by Burr and Burr in 1973 [27]. It is suggested that 1%–2% intake of linoleic acid in the diet is sufficient to prevent biochemical and clinical deficiency in infants. Adults consume enough 18:2*n*-6 in the diet, and deficiency is not a problem. The absence of linoleic acid in the diet is characterized by scaly dermatitis, excessive water loss via the skin, impaired growth and reproduction, and poor wound healing [28]. Nutritionists have suggested a 3%–4% content of *n*-6 fatty acids in SLs to fulfill the EFA requirements of SLs [1].

4. Omega-3 Fatty Acids

Omega-3 fatty acids are also known as EFAs because humans, like all mammals, cannot synthesize them and therefore must obtain them from their diets. The *n*-3 fatty acids are represented by linolenic acid (18:3*n*-3), which is commonly found in soybean and linseed oils and in the chloroplast of green leafy plants. Other polyunsaturated *n*-3 fatty acids (*n*-3 PUFAs) of interest in SL synthesis are eicosapentaenoic acid, 20:5*n*-3 (EPA), and docosahexaenoic acid, 22:6*n*-3 (DHA), which are commonly found in fish oils, particularly fatty fish. Children without enough *n*-3 PUFAs in their diet may suffer from neurological and visual disturbances, dermatitis, and growth retardation [29]. Therefore, *n*-3 PUFAs such as DHA must be included in their diet and in SL design.

SLs containing *n*-3 PUFAs and MCFAs have been synthesized chemically by hydrolysis and random esterification of fish oil and MCTs. They have been shown to inhibit tumor growth and to improve nitrogen balance in Yoshida sarcoma-bearing rats [30]. We have successfully used lipases as biocatalysts to synthesize position-specific SLs containing *n*-3 PUFAs with an ability to improve immune function and reduce serum cholesterol concentrations [31,32]. EPA is important in preventing heart attacks primarily due to its antithrombotic effect [33]. It was also shown to increase

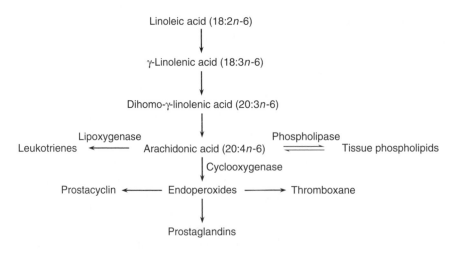

FIGURE 31.4 Pathway for eicosanoid biosynthesis.

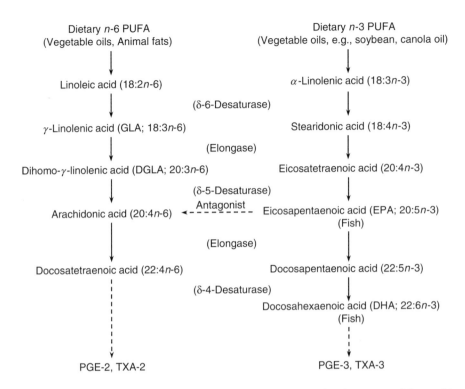

FIGURE 31.5 Pathways leading to the metabolism of dietary *n*-6 and *n*-3 polyunsaturated fatty acids.

bleeding time and to lower serum cholesterol concentrations [33]. Studies with nonhuman primates and human newborns suggest that DHA is essential for the normal functioning of the retina and brain, particularly in premature infants [34]. Other studies have shown that *n*-3 fatty acids can decrease the number and size of tumors and increase the time elapsed before the appearance of tumors [35].

The *n*-3 fatty acids are essential in growth and development throughout the life cycle of humans and therefore should be included in the diet. Nutritional experts consider a level of 2%–5% of *n*-3 fatty acids optimum in enhancing immune function in SL, as shown in Table 31.1. PUFAs of the *n*-3 series are antagonists of the arachidonic acid (20:4*n*-6) cascade (Figure 31.5). The mode of action of fish oil *n*-3 PUFAs on functions mediated by *n*-6 PUFAs is summarized in Table 31.2 [36].

TABLE 31.2

Mode of Action of *n*-3 PUFAs on Functions Mediated by *n*-6 PUFAs

Impair uptake of *n*-6 polyunsaturated fatty acid (PUFA).

Inhibit desaturases, especially δ-6-desaturase.

Compete with *n*-6 PUFAs for acyltransferases.

Displace arachidonic acid (20:4*n*-6) from specific phospholipid pools.

Dilute pools of free 20:4*n*-6.

Competitively inhibit cyclooxygenase and lipoxygenase.

Form eicosanoid analogs with less activity or competitively bind to eicosanoid sites.

Alter membrane properties and associated enzyme and receptor functions.

Source: Adapted from Kinsella, J.E. in *Omega-3 Fatty Acids in Health and Disease*, R.S. Lees and M. Karel, eds, Dekker, New York, 1990.

The n-3 PUFAs inhibit tissue eicosanoid biosynthesis by preventing the action of δ-6 desaturase and cyclooxygenase/lipoxygenase enzymes responsible for the conversion of 18:2n-6 to 20:4n-6 and 20:4n-6 to eicosanoids, respectively. The amount of 18:2n-6 determines the 20:4n-6 content of tissue phospholipid pools and affects eicosanoid production. Eicosanoids are divided into prostanoids (prostaglandins, prostacyclins, and thromboxanes), which are synthesized via cyclooxygenase, and leukotrienes (hydroxy fatty acids and lipoxins), which are synthesized via lipoxygenase, as illustrated in Figure 31.4.

A proper balance of n-3 and n-6 fatty acids should be maintained in the diet and SL products. High concentrations of dietary 18:2n-6 may lead to increased production of immunosuppressive eicosanoids of the 2- and 4-series (prostaglandin E_2 (PGE_2), thromboxane A_2 (TXA_2), leukotriene B_4 (LTB_4)). However, diets high in 20:5n-3 inhibit eicosanoid production and reduce inflammation by increasing production of TXA_3, prostacyclin (PGI_3), and LTB_5. Diets including n-3 PUFAs also increase HDL-cholesterol and interleukin-2 (IL-2) levels. On the other hand, they inhibit or decrease the levels of IL-1, LDL-cholesterol, and very low density lipoprotein cholesterol (VLDL-cholesterol).

5. Omega-9 Fatty Acids

The n-9 fatty acids or monounsaturated fatty acids are found in vegetable oils such as canola, olive, peanut, and high-oleic sunflower as oleic acid (18:n-9). Oleic acid can be synthesized by the human body and is not considered an EFA. However, it plays a moderate role in reducing plasma cholesterol in the body [26]. Oleic acid is useful in SLs for fulfilling the LCT requirements of SLs, as given in Table 31.1.

6. Long-Chain Saturated Fatty Acids

Generally, saturated fatty acids are believed to increase plasma cholesterol levels, but it has been claimed that fatty acids with chains 4–10 carbon atoms do not raise cholesterol levels [37,38]. Stearic acid (18:0) has also been reported not to raise plasma cholesterol levels [39]. TAGs containing high amounts of LCFAs, particularly stearic acid, are poorly absorbed in man, partly because stearic acid has a melting point higher than body temperature; they exhibit poor emulsion formation and poor micellar solubilization [40]. The poor absorption of saturated LCTs [40] makes them potential substrates for low-calorie SL synthesis. Indeed, Nabisco Foods Group used this property of stearic acid to make the group of low-calorie SLs called Salatrim (now Benefat) (see Section II.B.2), which consist of short-chain aliphatic fatty acids and LCFAs, predominantly C18:0 [41]. Caprenin, an SL produced by Procter & Gamble, contains C22:0, which is also poorly absorbed. An SL containing two behenic acids and one oleic acid has been used in the food industry to prevent chocolate bloom and to enhance fine crystal formation of palm oil and lard products [42].

B. Synthesis of Structured Lipids

1. Chemical Synthesis

Chemical synthesis of SLs usually involves hydrolysis of a mixture of MCTs and LCTs and then reesterification after random mixing of the MCFAs and LCFAs has occurred, by a process called transesterification (ester interchange). The reaction is catalyzed by alkali metals or alkali metal alkylates. This process requires high temperature and anhydrous conditions. Chemical transesterification results in desired randomized TAG molecular species, known as SLs, and in a number of unwanted products, which can be difficult to remove. The SL product consists of one (MLL, LML) or two (LMM, MLM) MCFAs, in random order (Figure 31.2), and small quantities of pure unreacted MCTs and LCTs [19].

The starting molar ratios of the MCTs and LCTs, and the source or type of TAG, can be varied to produce new desired SL molecules. Coconut oil is a good source of MCTs, and soybean and safflower oils are excellent sources of (*n*-6)-containing fatty acids for SL synthesis. Isolation and purification of the products is tedious because of unwanted coproducts. SLs are also produced by physical blending of specific amounts of MCTs and LCTs, except there is no exchange or rearrangement of fatty acids within the same glycerol backbone. When consumed, the blend will retain the original absorption rates of the individual TAG. Positional specificity of fatty acids on the glycerol molecule is not achieved by chemical transesterification, and this is a key factor in the metabolism of SLs. A possible alternative is the use of enzymes—specifically lipases—as described later in this chapter (Section II.B.3).

2. Examples of Commercial Products

a. *Caprenin*

Caprenin is a common name for caprocaprylobehenin, an SL containing C8:0, C10:0, and C22:0 fatty acids esterified to glycerol moiety. It is manufactured from coconut, palm kernel, and rapeseed oils by a chemical transesterification process. The MCFAs are obtained from the coconut oil and the LCFAs from rapeseed oil. Caprenin's caloric density is 5 kcal/g compared with 9 kcal/g for a conventional TAG. Behenic acid is partially absorbed by the body and thus contributes few calories to the product. The MCFAs are metabolized quickly, like carbohydrates.

Procter & Gamble filed a Generally Recognized as Safe (GRAS) affirmation petition to the U.S. Food and Drug Administration (FDA) for use of Caprenin in soft candies such as candy bars, and in confectionery coatings for nuts, fruits, cookies, etc. Caprenin is made up of 95% TAGs, 2% DAGs, and 1% MAGs with C8:0 + C10:0 contributing 43%–45% and C22:0 40%–54% of the fatty acids. Caprenin has a bland taste, is liquid or semisolid at room temperature, and is fairly stable to heat. It can be used as a cocoa butter substitute. The structure of Caprenin is shown in Figure 31.6. Swift et al. [43] showed that Caprenin fed as an SL diet to male subjects for 6 days did not alter plasma cholesterol concentration but decreased HDL-cholesterol by 14%. However, the MCT diet raised plasma TAGs by 42% and reduced HDL-cholesterol by 15%.

b. *Benefat/salatrim*

Benefat contains C2:0–C4:0, and C18:0 esterified to glycerol moiety. Benefat is a brand name for Salatrim (short and long acyl triglyceride molecule), developed by Nabisco Foods Group (Parsippany, NJ), but now marketed as Benefat by Cultor Food Science (New York, NY). Benefat is produced by base-catalyzed interesterification of highly hydrogenated vegetable oils with TAGs of acetic, propionic, and butyric acids [44]. The product contains randomly distributed fatty acids attached to the glycerol molecule. Due to the random distribution of fatty acids, each preparation contains many molecular species. The ratio of SCFAs such as acetic, propionic, and butyric acids to LCFAs such as stearic acid can be varied to obtain SLs with physical and functional properties resembling those of conventional fats such as cocoa butter. The FDA accepted for filing in 1994 a GRAS affirmation petition by Nabisco.

Benefat is a low-calorie fat like Caprenin, with a caloric availability of 5 kcal/g. The caloric availability of C2:0, C3:0, C4:0, glycerol, and LCFA in the Benefat molecule are 3.5, 5.0, 6.0, 4.3,

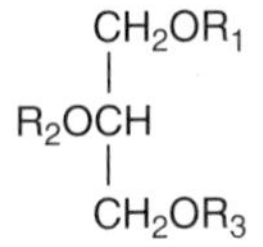

$$
\begin{array}{c}
CH_2OR_1 \\
| \\
R_2OCH \\
| \\
CH_2OR_3
\end{array}
$$

FIGURE 31.6 Structure of Caprenin (caprocaprylobehenin) with three randomized acyl groups: R_1, R_2, R_3, acyl part of capric acid, C10:0, caprylic acid, C8:0, and behenic acid, C22:0 in no particular order.

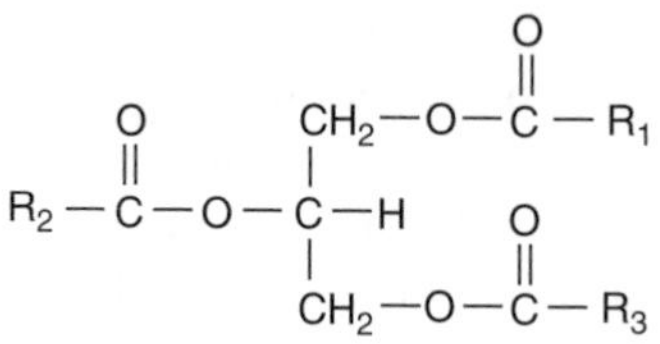

$$R_2-\overset{\overset{\displaystyle O}{\|}}{C}-O-\overset{\overset{\displaystyle \begin{array}{c} CH_2-O-\overset{\overset{\displaystyle O}{\|}}{C}-R_1 \\ | \end{array}}{\underset{\begin{array}{c} | \\ CH_2-O-\overset{\overset{\displaystyle O}{\|}}{C}-R_3 \end{array}}{C}}}{\underset{}{C}}-H$$

FIGURE 31.7 Structure of Benefat (brand name for Salatrim): R, alkyl part of C2:0–C4:0 and C18:0; must contain at least one short-chain C2:0 or C3:0 or C4:0 and one long chain (predominantly C18:0).

and 9.5 kcal/g, respectively. Stearic acid is poorly or only 50% absorbed [45], especially if it is esterified to the *sn*-1 and *sn*-3 positions of the glycerol. Acetyl and propionyl groups in Benefat are easily hydrolyzed by lipases in the stomach and upper intestine and readily converted to carbon dioxide [46]. Benefat is intended for use in baking chips, chocolate-flavored coatings, baked and dairy products, dressings, dips, and sauces, or as a cocoa butter substitute in foods. The consistency of Benefat varies from liquid to semisolid, depending on the fatty acid composition and the number of SCFAs attached to the glycerol molecule. The structure of Benefat is given in Figure 31.7.

c. Others

Other commercially available chemically synthesized SLs and lipid emulsions are listed in Table 31.3. These include Captex, Neobee, and Intralipid 20%. Typical fatty acid profiles of selected SL products and MCTs are given in Table 31.4. Applications of these products vary depending on the need of the patient or the function of the intended food product. Enzymes can be used to custom-produce SLs for specific applications. Unfortunately, many enzymatically synthesized SLs are not commercially available, although the potential is there. This technology needs to be commercialized.

3. Enzymatic Synthesis

a. Lipases in Fats and Oils Industry

TAG lipases, also known as TAG acylhydrolases (EC 3.1.1.3), are enzymes that hydrolyze TAGs to DAGs, MAGs, free fatty acids (FFAs), and glycerol. They can catalyze the hydrolysis of TAGs and the transesterification of TAGs with fatty acids (acidolysis) or direct esterification of FFAs with glycerol [47–49]. Annual sales of lipases account for over $20 million, which corresponds to less than 4% of the worldwide enzyme market estimated at $600 million [50]. Two main reasons for the apparent misconception of the economic significance of lipases are as follows: (1) lipases have been

TABLE 31.3

Commercial Sources of Structured Lipids and Lipid Emulsions

Product	Composition	Source
Caprenin	C8:0, C10:0, C22:0	Procter & Gamble Co., Cincinnati, Ohio
Benefat	C2:0–C4:0, C18:0	Cultor Food Science, New York, New York
Captex	C8:0, C10:0, C18:2	ABITEC Corp., Columbus, Ohio
Neobee	C8:0, C10:0, LCFA	Stepan Company, Maywood, New Jersey
Intralipid	20% soybean oil emulsion	KabiVitrum, Berkeley, California
		Pharmacia AB, Stockholm, Sweden
FE 73403	Fat emulsion of C8:0, C10:0, LCFA	Pharmacia AB, Stockholm, Sweden

LCFA, long-chain fatty acid (may vary from C16:0 to C18:3*n*-3); FE, fat emulsion.

TABLE 31.4

Fatty Acid Composition of Typical Lipid Emulsions and Medium-Chain Triacylglycerol

Fatty Acid	Composition (%)		
	FE Emulsion 73403	Intralipid 20%	MCT
8:0	27	—	65–75
10:0	10	—	25–35
12:0	—	—	1–2
16:0	7	13	—
18:0	3	4	—
18:1n-9	13	22	—
18:2n-6	33	52	—
18:3n-3	5	8	—
Other	2	1	1–2

MCT, Medium-chain triacylglycerol.

investigated extensively as a route to novel biotransformation and (2) diversity of the current and proposed industrial applications of lipases by far exceeds that of other enzymes such as proteases or carbohydrases [51].

Although enzymes have been used for several years to modify the structure and composition of foods, they have only recently become available for large-scale use in industry, mainly because of the high cost of enzymes. However, according to enzyme manufacturers, progress in genetics and in process technology may now enable the enzyme industry to offer products with improved properties and at reduced costs [51]. For lipases to be economically useful in industry, enzyme immobilization is necessary to enable enzyme reuse and to facilitate continuous processes. Immobilization of enzymes can simply be accomplished by mixing an aqueous solution of the enzyme with a suitable support material and removing the water at reduced pressure, after which small amounts of water are added to activate the enzyme. Suitable support materials for enzyme immobilization include glass beads, Duolite, acrylic resin, and Celite.

In spite of the obvious advantages of biological catalysis, the current level of commercial exploitation in the oleochemical industry is disappointing, probably because of the huge capital investments involved, and until recently, the high cost of lipase [51]. The introduction of cheap and thermostable enzymes should tip the economic balance in favor of lipase use for the commercial production of SL and lipid modifications.

b. Mode of Action of Lipases

TAG lipases are probably among the most frequently used enzymes in organic synthesis. This is in part because they do not require coenzymes and because they are stable enough in organic solvents at relatively high temperatures [52]. Lipases act at the oil–water interface of heterogeneous reaction systems. This property makes them well suited for reactions in hydrophobic media. Lipases differ from esterases in their involvement of a lipid–water interface in the catalytic process [53]. Some regions of the molecular structure responsible for the catalytic action of lipase are presumed to be different from those of ordinary enzymes that act on water-soluble substrates in a homogeneous medium [54]. Because lipases work at substrate–water interfaces, a large area of interface between the water-immiscible reaction phase and the aqueous phase that contains the catalyst is necessary to obtain reasonable rates of interesterification [55]. This is exemplified by the greater tendency for lipase to form off-flavors in homogenized milk than in unhomogenized milk.

Theoretical interpretations of the activation of lipase by interfaces can be divided into two groups: those assuming that the substrates can be activated by the presence of an oil–water interface and those assuming that the lipase undergoes a change to an activated form on contact with an oil–water interface. The first interpretation assumes higher concentrations of the substrate near the interface rather than in the bulk of the oil, and the second involves the existence of separate adsorption and catalytic sites for the lipase such that the lipase becomes catalytically active only after binding to the interface. More information on the action of microbial lipases is available in Chapter 29.

c. Enzymes in Organic Solvents

It is now commonly accepted that enzymes can function efficiently in anhydrous organic solvents. When enzymes are placed in an organic environment, they exhibit novel characteristics, such as altered chemo- and stereoselectivity, enhanced stability, and increased rigidity [56]. Lipases have also been shown to catalyze peptide synthesis, since they can catalyze the formation of amide links while lacking the ability to hydrolyze them [57]. Lipase can be used in several ways in the modification of TAGs [48]. In an aqueous medium, hydrolysis is the dominant reaction, but in organic media esterification and interesterification reactions are predominant. Lipases from different sources display hydrolytic positional specificity and some fatty acid specificity. The positional specificity is retained when lipases are used in organic media.

One application of lipases in organic solvents is their use as catalysts in the regio-specific interesterification of fats and oils for the production of TAGs with desired physical properties [58]. Lipases can also be used in the resolution of racemic alcohols and carboxylic acids by the asymmetric hydrolysis of the corresponding esters. An example of stereoselectivity of lipases is the esterification of menthol by *Candida cylindracea*. This enzyme was shown to esterify L-menthol while being catalytically inactive with the D-isomer [59,60]. Table 31.5 lists advantages of employing lipases in organic solvents for the modification of lipids as opposed to aqueous media [61].

d. Strategies for the Enzymatic Production of Structured Lipids

Various methods can be used for lipase-catalyzed production of SLs [4]. The method of choice depends to a large extent on the type of substrates available and the products desired.

Direct Esterification. Direct esterification can be used for the preparation of SLs by reacting FFAs with glycerol. The major problem is that the water molecules formed as a result of the esterification reaction must be removed as they are formed to prevent them from hydrolyzing back the product, leading to low product yield. Direct esterification, rarely used in SL synthesis (except in the synthesis of DAG oils; see Section VI.B), is presented in equation form as follows:

TABLE 31.5

Advantages of Lipase Modification of Lipids in Organic Solvents

Increased solubility of nonpolar lipid substrates in organic solvents such as hexane and isooctane.

Shift of thermodynamic equilibria to the right in favor of synthesis over hydrolysis.

Reduction in water-dependent side reactions, since very little water is required by lipases in synthetic reactions.

Enzyme recovery is made possible by simple filtration of the powdered or immobilized lipase.

If immobilization is desired, adsorption onto nonporous surfaces (e.g., glass beads) is satisfactory; enzymes are unable to desorb from these surfaces in nonaqueous media.

Ease of recovery of products from low boiling point solvents.

Enhanced thermal stability of enzymes in organic solvents.

Elimination of microbial contamination.

Potential of enzymes to be used directly within a chemical process.

Immobilized enzyme can be reused several times.

Source: Modified from Dordick, J.S., *Enzyme Microb. Technol.*, 11, 194, 1989.

$$\text{Glycerol} + \text{MCFA} + \text{LCFA} \xrightarrow{\text{Lipase}} \text{SL} + \text{Water,}$$

where

MCFA is medium-chain fatty acid
LCFA is long-chain fatty acid
SL is structured lipid moieties

Transesterification–Acidolysis. Acidolysis is a type of transesterification reaction involving the exchange of acyl groups or radicals between an ester and a free acid:

$$\text{MCT} + \text{LCFA} \xrightarrow{\text{Lipase}} \text{SL} + \text{MCFA,}$$

$$\text{LCT} + \text{MCFA} \xrightarrow{\text{Lipase}} \text{SL} + \text{LCFA,}$$

where

MCT is medium-chain triacylglycerol
LCT is long-chain triacylglycerol

Figure 31.8 shows an example of acidolysis reaction [62], in this case between caprylic acid and triolein. Shimada et al. [63] used acidolysis reaction catalyzed by immobilized *Rhizopus delemar* lipase to synthesize an SL containing 22:6*n*-3 (DHA) and caprylic acids. Product isolation is easy after acidolysis. FFAs are removed by distillation or by other appropriate techniques.

Transesterification–Ester Interchange. This reaction involves the exchange of acyl groups between one ester and another ester:

$$\text{MCT} + \text{LCT} \xrightarrow{\text{Lipase}} \text{SL,}$$

$$\text{LCT} + \text{MCFAEE} \xrightarrow{\text{Lipase}} \text{SL} + \text{LCFAEE,}$$

$$\text{MCT} + \text{LCFAEE} \xrightarrow{\text{Lipase}} \text{SL} + \text{MCFAEE,}$$

where

MCFAEE is medium-chain fatty acid ethyl ester
LCFAEE is long-chain fatty acid ethyl ester
This method is widely used in lipid modifications and in the synthesis of SLs [4,47,64,65]

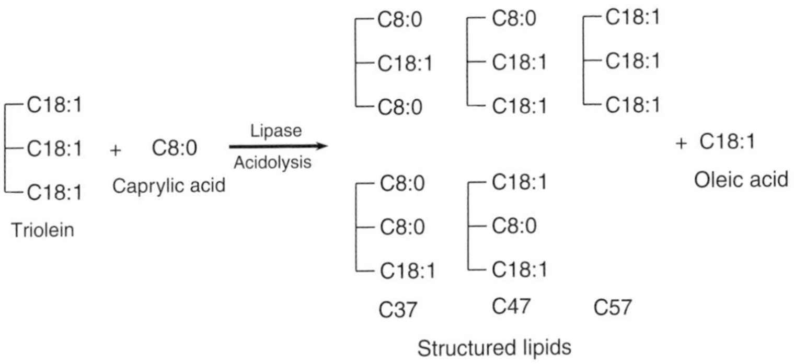

FIGURE 31.8 Reaction scheme showing acidolysis reaction in the synthesis of structured lipids from caprylic acid and triolein. (From Akoh, C.C. and Huang, K.H., *J. Food Lipids*, 2, 219, 1995.)

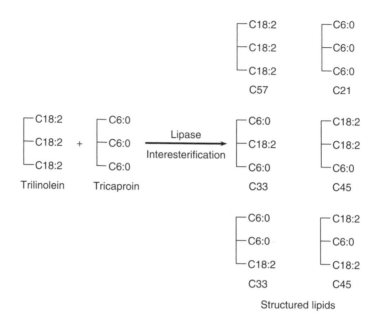

FIGURE 31.9 Ester interchange reaction between two triacylglycerols, trilinolein, and tricaproin in the enzymatic production of structured lipids.

In a transesterification reaction, generally, hydrolysis precedes esterification. In all the preceding examples, SCTs and SCFAs can replace MCTs and MCFAs, respectively, or can be used in combination. Figures 31.9 and 31.10 give examples of the suggested strategies involving interchange reactions between a TAG (trilinolein) and a TAG (tricaproin) ester and between EPA ethyl ester and tricaprin, respectively. We have successfully used enzymes to synthesize position-specific SLs containing n-3 PUFAs with ability to improve immune function and reduce serum cholesterol [31,32].

e. Factors That Affect Enzymatic Process and Product Yield

Water. It is generally accepted that water is essential for enzymatic catalysis. This status is attributed to the role water plays in all noncovalent interactions. Water is responsible for maintaining the

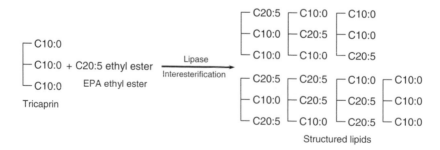

FIGURE 31.10 Ester interchange reaction in the production of structured lipids containing eicosapentaenoic acid (EPA) with tricaprin and EPA ethyl ester as substrates. An immobilized *Candida antarctica* lipase, SP 435, was the biocatalyst. Note EPA esterified to the *sn*-2 position.

active conformation of proteins, facilitating reagent diffusion, and maintaining enzyme dynamics [66]. Zaks and Klibanov [67] reported that for enzymes and solvents, tested enzymatic activity greatly increased with an increase in the water content of the solvent. The absolute amount of water required for catalysis for different enzymes varies significantly from one solvent to another [56]. Hydration levels corresponding to one monolayer of water can yield active enzymes [68]. Although many enzymes are active in a variety of organic solvents, the best nonaqueous reaction media for enzymatic reactions are hydrophobic, water-immiscible solvents [67,69,70]. Enzymes in these solvents tend to keep the layer of essential water, which allows them to maintain their native configuration, and therefore catalytic activity.

Solvent Type. The type of organic solvent employed can dramatically affect the reaction kinetics and catalytic efficiency of an enzyme. Therefore, the choice of solvent to be used in biocatalysis is critical. Two factors affecting this choice are the extent to which the solvent affects the activity or stability of the enzyme and the effect of the solvent on the equilibrium position of the desired reaction [71]. The equilibrium position in an organic phase is usually different from that in water due to differential solution of the reactants. For example, hydrolytic equilibrium is usually shifted in favor of the synthetic product because the product is less polar than the starting materials [71]. The nature of the solvent can also cause inhibition or inactivation of enzymes by directly interacting with the enzymes. Here the solvent alters the native conformation of the protein by disrupting hydrogen bonding and hydrophobic interactions, thereby leading to reduced activity and stability [72].

Lipases differ in their sensitivity to solvent type. An important solvent characteristic that determines the effect of solvent in enzymatic catalysis is the polarity of the solvent. Solvent polarity is measured by means of the partition coefficient (P) of a solvent between octanol and water [73], and this is taken as a quantitative measure of polarity, otherwise known as log P value [74]. The catalytic activity of enzymes in solvents with log $P < 2$ is usually lower than that of enzymes in solvents with log $P > 2$. This is because hydrophilic or polar solvents can penetrate into the hydrophilic core of the protein and alter the functional structure [75]. They also strip off the essential water of the enzyme [67]. Hydrophobic solvents are less able to remove or distort the enzyme-associated water and are less likely to cause inactivation of enzymes [61].

In choosing a solvent for a particular reaction, two important factors must be taken into consideration: the solubility of the reactants in the chosen solvent and the need for the chosen solvent to be inert to the reaction [61]. Other factors that must be taken into account in determining the most appropriate solvent for a given reaction include solvent density, viscosity, surface tension, toxicity, flammability, waste disposal, and cost [61]. A report by Akoh and Huang [62] on the effect of solvent polarity on the synthesis of SLs using Lipozyme RM IM (immobilized lipase from *Rhizomucor miehei*) showed that nonpolar solvents such as isooctane and hexane produced 40 mol% of disubstituted SL, whereas a more polar solvent such as acetone produced 1.4% of the same SL. Claon and Akoh [76] found that with SP 435 lipase from *C. antarctica*, a higher log P value does not necessarily sustain a higher enzyme activity. Some experimentation is therefore necessary in selecting solvents for enzymatic reactions.

pH. Enzymatic reactions are strongly pH dependent in aqueous solutions. Studies on the effect of pH on enzyme activity in organic solvents show that enzymes remember the pH of the last aqueous solution to which they were exposed [65,70]. That is, the optimum pH of the enzyme in an organic solvent coincides with the pH optimum of the last aqueous solution to which it was exposed. This phenomenon is called pH memory. A favorable pH range depends on the nature of the enzyme, the substrate concentration, the stability of the enzyme, the temperature, and the length of the reaction [77].

Thermostability. Temperature changes can affect parameters such as enzyme stability, affinity of enzyme for substrate, and preponderance of competing reactions [78]. Thermostability of enzymes is a major factor the industry considers before commercialization of any enzymatic process, mostly due to the potential for saving energy and minimizing thermal degradation. Thermostability of lipases varies considerably with enzyme origin: animal and plant lipases are usually less thermostable than microbial extracellular lipases [49].

Several processes that lead to the irreversible inactivation of enzymes involve water as a reactant [79]. This characteristic of enzymes makes them more thermostable in water-restricted environments such as organic solvents. Enzymes are usually inactivated in aqueous media at high temperatures. Several studies have been reported on the effect of temperature on lipase activity [64,76,80]. Zaks and Klibanov [80], who studied the effect of temperature on the activity of porcine pancreatic lipase, showed that in aqueous solution at 100°C, the lipase is completely inactivated within seconds, whereas in dry tributyrin containing heptanol, the lipase had a shelf life at 100°C of 12 h. These investigators concluded that in organic solvents, porcine pancreatic lipase remains rigid and cannot undergo partial unfolding, which causes inactivation. The heat stability of a lipase also depends on whether a substrate is present. This is because substrates remove excess water from the immediate vicinity of the enzyme, thus restricting its overall conformational mobility [81].

Most lipases in nonimmobilized form are optimally active between 30°C and 40°C [82]. Immobilization confers additional stability to the lipase compared with nonimmobilized lipase. Excellent reviews on the immobilization procedures and bioreactors for lipase catalysis were published recently [48,83,84]. The immobilization support must possess the following properties: high surface area to allow maximum contact with enzyme, high porosity to allow good flow properties, high physical strength, solvent resistance, high flow properties, and chemical and microbiological inertness [85,86].

Other Factors. Other factors that affect yield of products are substrate molar ratio; enzyme source, activity, and load; incubation time; specificity of the enzyme to substrate type and chain length; and regiospecificity.

f. Chemical versus Enzymatic Synthesis

The most useful property of lipases is their regio- and stereospecificity, which result in products with better defined and more predictable chemical composition and structure than those obtained by chemical catalysis. Potential advantages of using enzymes over chemical procedures may be found in the specificity of enzymes and the mild reaction conditions under which enzymes operate [87]. Enzymes form products that are more easily purified and produce less waste, and thus make it easier to meet environmental requirements [87]. Chemical catalysts randomize fatty acids in TAG mixtures and do not lead to the formation of specialty products with desired physicochemical characteristics [51]. The specificities of lipase have classically been divided into five major types: lipid class, positional, fatty acid, stereochemical, and combinations thereof [81]. Enzymes have high turnover numbers and are well suited for the production of chiral compounds important to the pharmaceutical industry.

Transesterification using *sn*-1,3-specific lipase results in SL products with fatty acids at the *sn*-2 position remaining almost intact. This is significant from a nutritional point of view because the 2-MAGs produced by pancreatic lipase digestion are the main carriers of fatty acids through the intestinal wall [88]. Fatty acids esterified at the *sn*-2 position are therefore more efficiently absorbed than those at the *sn*-1 and *sn*-3 positions. A TAG containing an EFA at the *sn*-2 position and SCFA or MCFA in the *sn*-1 and *sn*-3 positions has the advantage of efficiently providing an EFA and a quick energy source [89].

Some studies have shown that the rate of autoxidation and melting properties of TAGs can be affected by the position of unsaturated fatty acids on the glycerol molecule [90,91]. TAGs having unsaturated fatty acids at the *sn*-2 position of glycerol are more stable toward oxidation than those linked at the *sn*-1 and *sn*-3 positions.

The energy saved and minimizations of thermal degradation are probably among the greatest attractions in replacing the current chemical technology with enzyme biotechnology [51].

TABLE 31.6
Advantages of Enzymatic Approach to Structured Lipid Design

Position-specific SL (i.e., desirable fatty acids can be incorporated at specific positions of triacylglycerol).

Enzymes exhibit regioselectivity (discriminate based on bond to be cleaved), enantioselectivity (optical activity), chemoselectivity (based on functional group), and fatty acid chain length specificity.

Can design SL on case-by-case basis to target specific food or therapeutic use—custom synthesis.

Products with defined structure can be produced.

Novel products not possible by conventional plant breeding and genetic engineering can be obtained (e.g., by inserting specific fatty acid at the *sn*-2 position of glycerol molecule).

Mild reaction conditions.

Few or no unwanted side reactions or products.

Can control the overall process.

Ease of product recovery.

Add value to fats and oils.

Improve functionality and properties of fats.

Table 31.6 shows some of the potential advantages of the enzymatic approach to SL design. Potential food applications of SL are listed in Table 31.7.

4. Analysis of Structured Lipids

Figure 31.11 presents a purification and analysis scheme for enzymatically produced SLs. Method of analysis depends on whether the SL is synthesized by acidolysis or by interesterification reaction. The crude SL product can be analyzed with silica gel G or argentation $AgNO_3$ (based on unsaturation), thin-layer chromatography (TLC), gas–liquid chromatography (GLC) of the fatty acid methyl or ethyl esters for fatty acid profile, and by reversed phase high-performance liquid chromatography (RP-HPLC) of molecular species based on equivalent carbon number (ECN) or total carbon number (TCN). A typical HPLC chromatogram of SL products is shown in Figure 31.12.

Other methods of typical lipid analysis described in this book can be applied to studies of SLs. The choice of fractionation or purification technique depends on substrate or reactant types, products formed, overall cost, and whether a small-scale or large-scale synthesis was employed. The need for improved methodologies for the analysis of SCFA and MCFA components of SLs is emphasized here because of their volatility during extraction and GLC analysis.

TABLE 31.7
Potential Food Uses of Structured Lipids

Margarine, butter, spreads, shortening, dressings, dips, and sauces.

Improve melting properties of fats.

Cocoa butter substitute.

Confectioneries.

Soft candies.

As reduced- or low-calorie fats (e.g., Caprenin, Benefat).

Baking chips, baked goods.

Snack foods.

Dairy products.

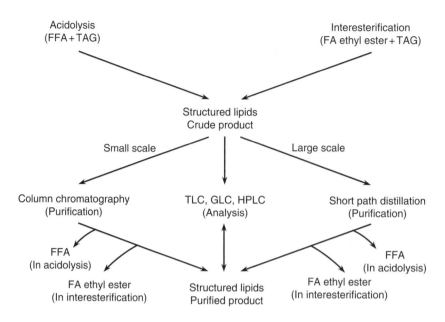

FIGURE 31.11 Purification and analysis scheme for enzymatically produced structured lipids.

a. Stereospecific Analysis

Figure 31.13 shows the stereochemical configuration of a TAG molecule, with *sn* notation indicating the stereochemical numbering system. The positional distribution of SFCA, MCFA, and LCFA on the glycerol moiety of SL is important in relation to the physical and functional properties of the SL and its metabolism. As indicated in Section III, the absorption and transport pathway of the SL depend somewhat on the fatty acid at the *sn*-2 position. In most vegetable oils, unsaturated fatty

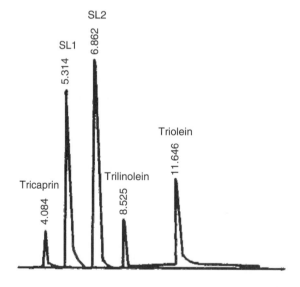

FIGURE 31.12 High-performance liquid chromatographic separation of structured lipid products from the reactants using a reversed-phase column: SL1, structured lipid containing two medium-chain fatty acids; SL2, structured lipid containing one medium-chain fatty acid. Trilinolein and tricaprin were the reactants, and triolein was the internal standard.

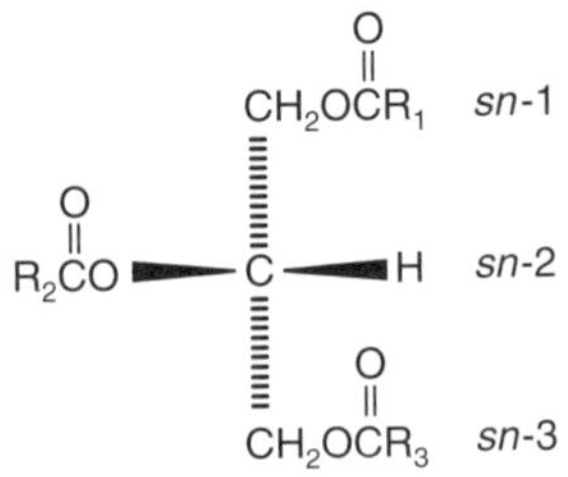

$$
\begin{array}{c}
\text{O} \\
\parallel \\
\text{CH}_2\text{OCR}_1 \quad sn\text{-}1 \\
\end{array}
$$

FIGURE 31.13 Stereochemical configuration of triacylglycerols or structured lipids with *sn* notation indicating stereochemical numbering of the carbon atoms of glycerol moiety. When the carbon in the 2-position is in the plane of the page and the 1- and 3-carbons behind the plane of the page, if the −OH on the 2-position of glycerol is drawn to the left, the top carbon becomes 1 and the bottom becomes 3. Thus, a structured lipid with octanoic acid on the 1-position, oleic acid on the 2-position, and decanoic acid on the 3-position is named *sn*-glycerol-1-octanoate-2-oleate-3-decanoate.

acids occupy the *sn*-2 position, and saturated fatty acids are located in the *sn*-1 and *sn*-3 positions [92,93,94,95]. The *sn*-2 position of TAG is determined by pancreatic lipase hydrolysis of the fatty acids at the *sn*-1 and *sn*-3 positions, followed by GLC analysis of the 2-MAG fatty acid methyl or ethyl ester. Detailed stereospecific analysis of the fatty acids at all three positions of the glycerol molecule was excellently reviewed by Small [93] and is not discussed in detail here. ^{13}C-NMR was used to determine acyl position of fatty acids on glycerol molecule [96].

Grignard reagent or Grignard degradation [97,98] is useful in obtaining the complete stereochemical structure of any TAG following pancreatic hydrolysis. In general, phospholipid derivatives (phosphatidylcholine, PC) of 1,2-DAG and 2,3-DAG are made by reacting with phospholipase A$_2$ (PLA$_2$). Since the *sn*-2 fatty acid is known, chemical analysis of the 2,3-diacyl-PC PLA$_2$ hydrolysis product gives the fatty acid at the *sn*-3 position. Similarly, chemical analysis of the 1,2-DAG hydrolysis product of PLA$_2$ gives the fatty acid at the *sn*-1 position. Alternatively, pancreatic hydrolysis of the 1,2-DAG followed by chemical analysis can give the fatty acid at the *sn*-1 position, since this enzyme is *sn*-1,3 specific.

III. ABSORPTION, TRANSPORT, AND METABOLISM OF STRUCTURED LIPIDS

The influence of TAG structure on lipid metabolism has been the subject of recent reviews and research efforts [92,93,99–101]. SLs may be targeted for either portal or lymphatic transport. In one widely accepted pathway, C2:0 to C12:0 fatty acids are transported via the portal system and C12:0 to C24:0 via the lymphatic system [2]. There is growing evidence that MCFAs may indeed be absorbed as 2-MAG, especially if they are esterified to the *sn*-2 position of the SL. The rate of hydrolysis at the *sn*-2 position of TAG is very slow, and as a result the fatty acid at this position remains intact as 2-MAG during digestion and absorption. Indeed, close to 75% of *sn*-2 fatty acids are conserved throughout the process of digestion and absorption [102].

LCTs are partially hydrolyzed by pancreatic lipase and absorbed slowly as partial acylglycerols in mixed micelles [93]. The resulting LCFAs are reesterified and incorporated into chylomicrons in the enterocyte, whereupon they enter the lymphatics to reach the general circulation through the thoracic duct. However, MCTs are nearly completely hydrolyzed and absorbed faster, mainly as FFAs and rarely as 2-MAGs. These MCFAs are then transported as FFAs bound to serum albumin in portal venous blood.

Figure 31.14 shows a proposed modified pathway for MCT, LCT, and SL metabolism. The metabolism of an SL is determined by the nature and position of the constituent fatty acids on the glycerol moiety. This may account for the differences in the pathway of absorption: lymphatics versus portal.

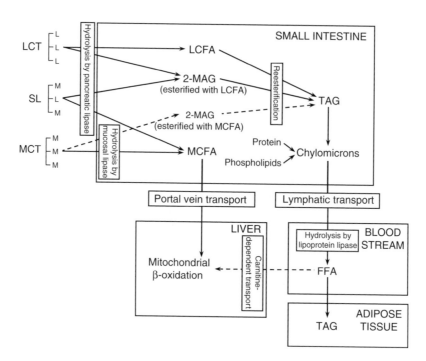

FIGURE 31.14 Proposed modified metabolic pathways for medium-chain and long-chain triacylglycerols and structured lipids.

Evidence for lymphatic absorption of MCFAs and storage in adipose tissue is accumulating [103–107]. Jensen et al. [107] observed the presence of more C10:0 than C8:0 fatty acids in the lymph of canine model fed an SL containing MCTs and fish oil versus its physical mixture, despite an overall ratio of C10:0 to C8:0 of 0.3 in their diets. Analysis of the SL molecular species revealed that MCFAs in lymph were present as mixed TAGs, suggesting that the MCFAs at the *sn*-2 position may account for the improved absorption. The 2-MAGs apparently were reesterified with endogenous or circulating LCFAs and subsequently absorbed through the lymphatic system. In addition, feeding high levels of MCTs can lead to lymphatic absorption and presence of MCFAs in the chylomicrons.

Enhanced absorption of 18:2*n*-6 was observed in cystic fibrosis patients fed SL containing LCFAs and MCFAs [89,108]. Rapid hydrolysis and absorption of an SL containing MCFAs at the *sn*-1 and *sn*-3 positions and an LCFA at the *sn*-2 position has been reported [89,109,110]. To improve the absorption of any fatty acid, its esterification to the *sn*-2 position of the glycerol moiety is suggested. Mok et al. [111] reported that the metabolism of an SL differs greatly from that of a similar physical mixture. The purported benefit of fish oil *n*-3 PUFAs can be attributed to their absorption as 2-MAGs. This factor is important in the construction of novel or designer SL molecules for food, therapeutic, and nutritional use.

IV. NUTRITIONAL AND MEDICAL APPLICATIONS

SLs can be synthesized to target specific metabolic effects or to improve physical characteristics of fats. An SL made from fish oil and MCTs was compared with conventional LCTs and found to decrease tumor protein synthesis, reduce tumor growth in Yoshida sarcoma-bearing rats, decrease body weight, and improve nitrogen maintenance [30]. In addition, the effects of fish/MCT on tumor growth were synergistic with tumor necrosis factor (TNF). A similar study by Mendez et al. [112] compared the effects of an SL (made from fish oil and MCFAs) with a physical mixture of fish oil and MCTs and found that the SL resulted in improved nitrogen balance in animals, probably because of the modified absorption rates of SL. Gollaher et al. [113] reported that the protein-sparing action associated with SL administration is not seen when the SLs provide 50% of protein calories and suggested that the protein-sparing action of SLs may be dependent on the ratio of MCTs to LCTs used to synthesize the SL.

Jandacek et al. [89] demonstrated that an SL containing caprylic acid at the *sn*-1 and *sn*-3 positions and an LCFA in the *sn*-2 position is more rapidly hydrolyzed and efficiently absorbed than a typical LCT. They proposed that the SL may be synthesized to provide the most desirable features of LCFAs and MCFAs for use as nutrients in cases of pancreatic insufficiency [89]. Metabolic infusion of an SL emulsion in healthy humans showed that the capacity of these subjects to hydrolyze SL is at least as high as that to hydrolyze LCT [114]. This finding is significant due to evidence of interaction and interference in the metabolism of LCT and MCT when both are present in a physical mixture [115,116]. An investigation into the in vivo fate of fat emulsions based on SL showed potential for use of SL as core material in fat emulsion-based drug delivery systems [117].

An SL made from safflower oil and MCFAs was fed to injured rats, and the animals receiving the SL were found to have greater gain in body weight, greater positive nitrogen balance, and higher serum albumin concentration than controls receiving a physical mixture [111]. Enhanced absorption of 18:2*n*-6 was observed in cystic fibrosis patients fed SLs containing LCFAs and MCFAs [25]. A mixed acid type of TAG composed of linoleic acid and MCFAs has been reported to improve immune functions [118], and evaluations in clinical nutrition are ongoing. However, a 3:1 admixture of MCT-LCT emulsions was reported to elevate plasma cholesterol concentrations compared with LCT emulsions in rats fed by intravenous infusion [119].

SL appears to preserve reticuloendothelial function while improving nitrogen balance as measured by the organ uptake of radiolabeled *Pseudomonas* in comparison with LCT [120]. Long-term feeding studies with an SL containing MCFAs and fish oil fatty acids showed that SL modified plasma fatty acid composition, reflecting dietary intake and induced systemic metabolic changes that persisted after the diet was discontinued [121]. An SL made by reacting tripalmitin with unsaturated fatty acids using an *sn*-1,3-specific lipase that closely mimicked the fatty acid distribution of human milk was commercially developed for application in infant formulas under the trade name Betapol [122]. HDL cholesterol decreased by 14% when a diet containing Caprenin as 40% of total calories was fed to healthy men, compared with no change in levels when an LCT diet was fed [43]. Table 31.8 lists the potential and other reported benefits of SL [1,25,32,89,106,107,111,120,123–129].

TABLE 31.8
Potential and Reported Benefits of Structured Lipids

Benefit	References
Superior nitrogen retention	[111]
Preservation of reticuloendothelial system (RES) function	[120]
Attenuation of protein catabolism and the hypermetabolic stress response to thermal injury	[123–125]
Enhanced absorption of the fatty acid at the *sn*-2 position (e.g., 18:2*n*-6 cystic fibrosis patients)	[25,126]
Reduction in serum TAG, LDL-cholesterol, and cholesterol	[32,106]
Improved immune function	[1,127]
Prevention of thrombosis	[1]
Lipid emulsion for enteral and parenteral feeding	[127,128]
Calorie reduction	[129]
Improved absorption of other fats	[89,106,107]

Source: Modified from Akoh, C.C., *Inform*, 6, 1055, 1995.

V. SAFETY AND REGULATORY ISSUES

The problem with consuming large doses of pure MCTs or their emulsions is the tendency to form ketone bodies (i.e., to induce metabolic acidosis). This outcome can be circumvented by using SLs or their emulsions. SL is safe and well tolerated in the body. Physiological and biochemical data suggest that SL emulsions, Intralipid 20%, and fat emulsion 73403 (Kabi Pharmacia AB, Stockholm, Sweden), when fed to postoperative patients, were rapidly cleared and metabolized [130]. The safety of Benefat was assessed, and no significant clinical effects were reported in subjects consuming up to 30 g/day [131]. Other studies also indicate that SLs are safe [132].

SLs that provide fewer calories (<9 kcal/g) than conventional TAGs (9 kcal/g) pose a great challenge to the FDA and other regulatory agencies around the world. These SLs include Caprenin, Benefat, and Captex. The issue is complicated by the labeling requirements for reduced fats. The big problem is how to establish uniform digestibility and absorbability coefficients for all available and soon-to-be-available SL molecules and other fat substitutes [133]. The current dietary guidelines recognize total fat and saturated fat, but not digestibility coefficients. The FDA needs to develop new guidelines for SLs and genetically engineered vegetable oils or to modify existing guidelines for TAGs to reflect the new generation of fats. FDA accepted for filing the GRAS petition for Benefat/Salatrim in 1994.

VI. DIACYLGLYCEROL OILS

A. WHAT ARE DAG OILS?

DAGs have been used as the principal emulsifiers in the mixture form with MAGs in the food industry, as covered in Chapters 3, 27, and 28. In addition to such traditional applications of DAGs based on their chemical properties, recently, the specialty SL products called DAG oils were developed for the purpose of obtaining positive effects on blood lipid profiles and obesity [134,135]. The physiological benefits of DAG oil will be described later (see Section VI.C).

DAG oils generally refer to the edible oils containing high concentrations of DAGs (>80%, w/w) and smaller amounts of TAGs (<20%, w/w) and MAGs (<3%, w/w) [135,136]. The predominant isoform of DAG present in the DAG oils is 1,3-DAGs (~70%, w/w) as compared with 1,2-DAGs (~30%, w/w) due to its enzymatic synthesis process using sn-1,3-specific lipase [135]. DAG oils are similar in taste, appearance, and fatty acid composition to conventional vegetable oils (TAG oils) [136].

DAG oils were first introduced to the Japanese market by Kao Corporation in February 1999 [136,137]. DAG oils produced by an enzymatic process are approved as food for specified health use (FOSHU) in Japan [138,139]. In the United States, DAGs have the status of GRAS as food ingredients according to the notice by FDA in 2000 [138,139]. DAG oil is also marketed as Enova oil by ADM Kao LLC, a joint venture of Archer Daniels Midland (ADM) Company and Kao Corporation [140].

B. ENZYMATIC SYNTHESIS OF DAG OILS

For the synthesis of DAG oils, enzymatic procedures using lipases are preferred to chemical procedures since the positional specificity of lipases results in oils rich in 1,3-DAGs, which cannot be obtained by chemical catalysts. DAGs can be synthesized by several enzymatic procedures via partial hydrolysis of TAGs in the presence of water [141], transesterification (glycerolysis) of TAGs with glycerols [142], and direct esterification of glycerol with FFAs [143–146].

Plou et al. [141] prepared mixtures of DAGs and MAGs in high yields (~79%) through partial hydrolysis of TAGs using sn-1,3-specific lipases from porcine pancreas, Rhizopus sp., and M. miehei and nonspecific lipases from C. rugosa. However, partial hydrolysis of TAGs is usually not a proper way to specifically synthesize 1,3-DAGs. Yamane et al. [142] obtained high yield

(~85%) of 1,3-DAGs from hydrogenated beef tallow and glycerol via enzymatic glycerolysis using a *Psesudomonas* sp. lipase. However, their procedure based on the selective crystallization of 1,3-DAGs is limited to the synthesis of DAGs of saturated fatty acids having relatively high melting points.

Direct esterification of glycerol with FFAs has been the most successful enzymatic procedure for the production of oils high in 1,3-DAGs [143–146]. Rosu et al. [143] performed the direct esterification of glycerol with several kinds of FFAs having relatively low melting points (<45°C), such as MCFAs and unsaturated fatty acids, in a solvent-free system. They obtained high yields (up to 85%) of 1,3-DAGs using Lipozyme RM IM (*sn*-1,3-specific lipase) as the biocatalyst. Watanabe et al. [144] synthesized 1,3-DAGs (maximal yield, ~75%) through Lipozyme RM IM-catalyzed direct esterification of glycerol with oleic and linoleic acids in a stirred batch reactor system. They removed water in the reaction mixture using a vacuum pump that helped shift the reaction equilibrium to the direction of esterification and thereby attained high yield (~75%) of 1,3-DAGs. Lo et al. [145] reported on the direct esterification of glycerol with fatty acids from soybean oil deodorizer distillates in a solvent-free system using several kinds of *sn*-1,3-specific lipases. They obtained maximal yield (47%) of 1,3-DAGs with Lipozyme RM IM. Yamada et al. [146] patented an enzymatic procedure for the industrial manufacture of DAG oils. Their procedure involves the partial hydrolysis of natural oils to obtain a partial hydrolysate with concentrated FFAs, followed by an *sn*-1,3-specific lipase-catalyzed direct esterification of glycerol with the FFAs.

However, in spite of using *sn*-1,3-specific lipase, a side reaction called acyl migration is known to occur during the esterification reaction. Acyl migration in the synthesis of 1,3-DAGs includes conversions of 1,3-DAGs to 1,2-DAGs or conversion of 1-MAGs to 2-MAGs. Therefore, acyl migration plays an important role in decreasing the yield of 1,3-DAGs in the enzymatic production of DAG oils. Recently, several studies have attempted to elucidate the reaction factors that influence acyl migration and to reduce it during laboratory-scale or pilot-scale lipase-catalyzed esterification reaction [147–149].

C. METABOLIC CHARACTERISTICS AND PHYSIOLOGICAL BENEFITS OF DAG OILS

Over the last few years, DAG oils have received much attention as one important class of SLs used as functional foods and nutraceuticals because it has unique physiological benefits compared with conventional TAG oils as follows: (1) suppressive effects on postprandial elevated blood TAG levels (known as hypertriglyceridemia) and (2) suppressive effects on body fat accumulation and obesity [134,135].

Taguchi et al. [150] compared the energy values of DAG oil containing 87% (w/w) DAGs and TAG oil with a similar fatty acid composition by measuring the combustion energies with a bomb calorimeter. According to their study, the energy values of DAG oil and TAG oil were 9.29 and 9.46 kcal/g, respectively, and the energy difference (<2%) was shown to be negligible in the total energy value of the practical diet. They also found that there was no difference between the absorption coefficients (weight percentage of ingested fat which was not excreted in the feces) of DAG oil and TAG oil in male Sprague–Dawley rats fed the diets containing DAG or TAG oils (20% of diet weight) [150]. Because the digestibility of fats is very similar for humans and rats, their findings might also be compatible in the case of humans [151,152]. Furthermore, the digestibilities of dietary DAG and TAG oils were proven to be nearly same to each other in another animal model, such as mice [153]. These results suggest that the physiological differences between DAG and TAG oils are due to their different metabolic characteristics after their digestion and absorption.

1. Hypotriglyceridemic Effect

Both dietary DAGs and TAGs are hydrolyzed to MAGs and fatty acids by pancreatic lipase in the small intestine. However, the 1-MAGs are produced during digestion of 1,3-DAGs, whereas, the main products of TAGs digestion are 2-MAGs [135]. The produced MAGs and fatty acids are absorbed by small intestinal epithelial cells (enterocytes) and are transported into the endoplasmic reticulum, where they are used to resynthesize TAGs for chylomicron formation and secretion [154]. The resynthesis of TAGs is predominantly catalyzed by two types of enzymes: MAG acyltransferase (MGAT) and DAG acyltransferase (DGAT). Sequentially, MGAT catalyzes the formation of DAGs from MAGs and fatty acyl-CoAs and DGAT catalyzes the production of TAGs from DAGs and fatty acyl-CoAs [154]. MGAT has a preference for 2-MAGs produced from dietary TAGs as the substrates, whereas 1-MAGs derived from 1,3-DAGs are not readily used for the resynthesis of DAGs [136]. In addition, 1,3-DAGs are little utilized by DGAT as substrates in the resynthesis of TAGs compared with 1,2-DAGs, which are the common substrates for DGAT [155]. Therefore, the lower postprandial blood TAG levels by dietary DAG oils rich in 1,3-DAGs may be due to their suppressive resynthesis of TAGs and retarded lymphatic transport of TAGs as chylomicrons compared with those of TAG oils [134].

2. Antiobesity Effect

Some experimental studies in humans and animals demonstrated that dietary DAG oils have antiobesity effects, such as reductions of body weight and body fat accumulation [153,156–160]. However, despite such observed beneficial effects of DAG oils, the evident antiobesity mechanisms are still not known in detail. As compared with TAG oils, the antiobesity effects of dietary DAG oils are shown to be related to the activated energy metabolism via the enhanced β-oxidation and energy expenditure phenomena found in some animal studies. That is, some researchers reported that dietary DAG increased hepatic fat oxidation [161], increased activities of enzymes involved in β-oxidation in the liver [153,158,162], and increased messenger RNA expression of some enzymes (e.g., acyl-CoA oxidase, medium-chain acyl-CoA dehydrogenase, acyl-CoA synthase) involved in β-oxidation and lipid metabolism in the small intestine and liver [153,158]. Furthermore, Kamphuis et al. [163] demonstrated that dietary DAG oil increases (hepatic) fat oxidation and decreases respiratory quotient (RQ) (i.e., carbon dioxide production divided by oxygen consumption) in humans. Their results suggest that the utilization of DAG oil as an energy source is greater than TAG oil even though a change in energy expenditure was not found. However, further studies are still required to determine the precise mechanism behind the antiobesity effects of DAG oils.

VII. PERSPECTIVES

This chapter discussed the currently available SLs, methods of synthesis, raw materials considerations, metabolism, and SL applications. An understanding of the functional properties and metabolic fate of the component fatty acids will aid in the synthesis of new SL molecules with beneficial end-use properties. The key to efficient absorption rests on the stereochemical structure of the SL. With this in mind, the outlook and potential for commercialization (Table 31.9) of the enzymatic process is bright.

Enzymes allow scientists to design SLs intended for various applications, which may include treatment of cystic fibrosis patients; individuals with pancreatic insufficiency; acquired immune deficiency syndrome (AIDS) patients, who need to boost their immune system by consuming SLs containing 20:5n-3 at the sn-2 position; stressed and septic and hospital patients; obese patients; and preterm infants. Potential nonmedical applications include foods and nutritional supports. SLs will continue to play a role in enteral and parenteral nutrition.

More research is needed on the effect of all lipid emulsions, especially SL emulsions, on the immune system. The notion that MCTs do not go via lymphatic transport is becoming less acceptable in the scientific community. The use of enzymes in constructing SLs destined for either portal or lymphatic transport will greatly enhance our knowledge on how SLs are metabolized. Chemical synthesis will lead to randomized SLs. Since the position and type of fatty acids in the

---

TABLE 31.9

Factors That Affect Outlook for Commercialization of the Enzymatic Process

Specialty needs or niche market as food ingredient, fine chemical use, nutritional supplement, enteral and parenteral feeding.

Cost of enzymatic versus chemical process and product yield.

Ease with which the enzymatic process can be scaled up.

Position specific with enzymes versus randomized products with chemical synthesis.

Consumer preference: natural versus synthetic products.

Cost–benefit assessment: investment and potential returns.

Catalyst reuse: immobilized enzyme can be reused several times without significant loss of activity.

Side or unwanted products of the reaction.

Processing costs to obtain products of high purity.

Regulation by the FDA (time-consuming process).

Competition with genetically engineered crops that produce structured lipids (e.g., high lauric acid canola oil).

TAG are key to their metabolism, the best alternative to chemical synthesis is the use of lipases. More applications of SLs in our regular diets is encouraged, meaning that food technologists need to explore this further. Genetic engineering of vegetable oil producing plants as covered in Chapter 32 will play a role in future commercial availability of SL.

REFERENCES

1. J.P. Kennedy. Structured lipids: fats of the future. *Food Technol.* 45(11):76–83 (1991).
2. V.K. Babayan. Medium chain triglycerides and structured lipids. *Lipids* 22:417–420 (1987).
3. V.K. Babayan. Medium chain triglycerides. In: *Dietary Fat Requirements in Health and Development* (J. Beare-Rogers, ed.). American Oil Chemists' Society, Champaign, IL, 1988, pp. 73–86.
4. C.C. Akoh. Structured lipids—enzymatic approach. *Inform* 6:1055–1061 (1995).
5. M.K. Schimdl. The role of lipids in medical and designer foods. In: *Food Lipids and Health* (R.E. McDonald and D.B. Min, eds.). Dekker, New York, NY, 1996, pp. 417–436.
6. M.J. Wolin. Fermentation in the rumen and human large intestine. *Science* 213:1463–1468 (1981).
7. W.C. Breckenridge and A. Kuksis. Molecular weight distributions of milk fat triglycerides from seven species. *J. Lipid Res.* 8:473–478 (1967).
8. G.A. Garton. The composition and biosynthesis of milk lipids. *J. Lipid Res.* 4:237–254 (1963).
9. J.C. Hawke and M.W. Taylor. Influence of nutritional factors on the yield, composition and physical properties of milk fat. In: *Developments in Dairy Chemistry*, Vol. 2 (P.F. Fox, ed.). Applied Science Publishers, New York, NY, 1983, pp. 37–81.
10. J. Bezard and M. Bugaut. Absorption of glycerides containing short, medium, and long chain fatty acids. In: *Fat Absorption*, Vol. 1 (A. Kuksis, ed.). CRC Press, Boca Raton, FL, 1986, pp. 119–158.
11. S.R. Elsden. The fermentation of carbohydrates in the rumen of the sheep. *J. Exp. Biol.* 22:51–62 (1945).
12. R.A. Leng. Formation and production of volatile fatty acids in the rumen. In: *Physiology of Digestion and Metabolism in the Ruminant* (A.T. Phillipson, ed.). Oriel Press, Newcastle Upon Tyne, England, 1970, pp. 406–421.
13. R.H. Dunlop and L. Bueno. Molasses neurotoxicity and higher volatile fatty acids in sheep. *Ann. Rech. Vet.* 10:462–464 (1979).
14. R.G. Jensen, R.M. Clark, F.A. Dejong, M. Hamosh, T.H. Liao, and N.R. Mehta. The lipolytic triad: human lingual, breast milk, and pancreatic lipases: physiological implications of their characteristics in digestion of dietary fats. *J. Pediatr. Gastroenterol. Nutr.* 1:243–255 (1982).
15. Y. Gargouri, G. Pieroni, C. Riviere, P.A. Lowe, J.F. Sauniere, L. Sarda, and R. Verger. Importance of human gastric lipase for intestinal lipolysis: an in vitro study. *Biochim. Biophys. Acta* 879:419–423 (1986).

16. F.H. Mattson and L.W. Beck. The specificity of pancreatic lipase for the primary hydroxyl groups of triglycerides. *J. Biol. Chem.* 219:735–740 (1956).

17. C.J. Megremis. Medium-chain triglycerides: a nonconventional fat. *Food Technol.* 45(2):108–114 (1991).

18. S.M. Kim and J.S. Rhee. Production of medium-chain glycerides by immobilized lipase in a solvent-free system. *J. Am. Oil Chem. Soc.* 68:499–503 (1991).

19. W.C. Heird, S.M. Grundy, and V.S. Hubbard. Structured lipids and their uses in clinical nutrition. *Am. J. Clin. Nutr.* 43:320–324 (1986).

20. A.C. Bach and V.K. Babayan. Medium chain triglycerides: an update. *Am. J. Clin. Nutr.* 36:950–962 (1982).

21. H. Kaunitz, C.A. Slanetz, R.E. Johnson, V.K. Babayan, and G. Barsky. Nutritional properties of the triglycerides of saturated fatty acids of medium chain-length. *J. Am. Oil Chem. Soc.* 35:10–13 (1958).

22. J.W. Stewart, K.D. Wiggers, N.L. Jacobson, and P.J. Berger. Effect of various triglycerides on blood and tissue cholesterol of calves. *J. Nutr.* 108:561–566 (1978).

23. N.B. Cater, H.J. Heller, and M.A. Denke. Comparison of the effects of medium-chain triacylglycerols, palm oil, and high oleic acid sunflower oil on plasma triacylglycerol fatty acids and lipid and lipoprotein concentrations in humans. *Am. J. Clin. Nutr.* 65:41–45 (1997).

24. A.C. Bach, Y. Ingenbleek, and A. Frey. The usefulness of dietary medium-chain triglycerides in body weight control: fact or fancy? *J. Lipid Res.* 37:708–726 (1996).

25. M.C. McKenna, V.S. Hubbard, and J.G. Bieri. Linoleic acid absorption from lipid supplements in patients with cystic fibrosis with pancreatic insufficiency and in control subjects. *J. Pediatr. Gastro-enterol. Nutr.* 4:45–51 (1985).

26. J.J. Gottenbos. Nutritional evaluation of n-6 and n-3 polyunsaturated fatty acids. In: *Dietary Fat Requirements in Health and Development* (J. Beare-Rogers, ed.). American Oil Chemists' Society, Champaign, IL, 1988, pp. 107–119.

27. G.O. Burr and M.M. Burr. A new deficiency disease produced by the rigid exclusion of fat from the diet. *Nutr. Rev.* 31:248–249 (1973).

28. P.V. Johnston. Essential fatty acids and the immune response. In: *Dietary Fat Requirements in Health and Development* (J. Beare-Rogers, ed.). American Oil Chemists' Society, Champaign, IL, 1988, pp. 151–162.

29. R.T. Holman, S.B. Johnson, and T.F. Hatch. A case of human linolenic acid deficiency involving neurological abnormalities. *Am. J. Clin. Nutr.* 35:617–623 (1982).

30. P.R. Ling, N.W. Istfan, S.M. Lopes, V.K. Babayan, G.L. Blackburn, and B.R. Bistrian. Structured lipid made from fish oil and medium-chain triglycerides alters tumor and host metabolism in Yoshida-sarcoma-bearing rats. *Am. J. Clin. Nutr.* 53:1177–1184 (1991).

31. K.T. Lee and C.C. Akoh. Immobilized lipase-catalyzed production of structured lipids with eicosapen-taenoic acid at specific positions. *J. Am. Oil Chem. Soc.* 73:611–615 (1996).

32. K.T. Lee, C.C. Akoh, and D.L. Dawe. Effects of structured lipid containing omega-3 and medium chain fatty acids on serum lipids and immunological variables in mice. *J. Food Biochem.* 23:197–208 (1999).

33. J. Dyerberg, H.O. Bang, E. Stoffersen, S. Moncada, and J.R. Vane. Eicosapentaenoic acid and prevention of thrombosis and atherosclerosis. *Lancet* 2:117–119 (1978).

34. S.E. Carlson, P.G. Rhodes, and M.G. Ferguson. Docosahexaenoic acid status of preterm infants at birth and following feeding with human milk or formula. *Am. J. Clin. Nutr.* 44:798–804 (1986).

35. W.T. Cave, Jr. Dietary n-3 (ω-3) polyunsaturated fatty acid effects on animal tumorigenesis. *FASEB J.* 5:2160–2166 (1991).

36. J.E. Kinsella. Sources of omega-3 fatty acids in human diets. In: *Omega-3 Fatty Acids in Health and Disease* (R.S. Lees and M. Karel, eds.). Dekker, New York, NY, 1990, pp. 157–200.

37. F. Grande. Dog serum lipid responses to dietary fats differing in the chain length of the saturated fatty acids. *J. Nutr.* 76:255–264 (1962).

38. S.A. Hashim, A. Arteaga, and T.B. Van Itallie. Effect of a saturated medium-chain triglyceride on serum-lipids in man. *Lancet* 1:1105–1108 (1960).

39. A. Bonanome and S.M. Grundy. Effect of dietary stearic acid on plasma cholesterol and lipoprotein levels. *N. Engl. J. Med.* 318:1244–1248 (1988).

40. S.A. Hashim and V.K. Babayan. Studies in man of partially absorbed dietary fats. *Am. J. Clin. Nutr.* 31: S273–S276 (1978).

41. J.W. Finley, L.P. Klemann, G.A. Leveille, M.S. Otterburn, and C.G. Walchak. Caloric availability of Salatrim in rats and humans. *J. Agric. Food Chem.* 42:495–499 (1994).

42. Y. Kawahara. Progress in fats, oils food technology. *Inform* 4:663–667 (1993).

43. L.L. Swift, J.O. Hill, J.C. Peters, and H.L. Greene. Plasma lipids and lipoproteins during 6 d of maintenance feeding with long chain, medium chain and mixed chain triglycerides. *Am. J. Clin. Nutr.* 56:881–886 (1992).

44. R.E. Smith, J.W. Finley, and G.A. Leveille. Overview of Salatrim, a family of low-calorie fats. *J. Agric. Food Chem.* 42:432–434 (1994).

45. L.P. Klemann, J.W. Finley, and G.A. Leveille. Estimation of the absorption coefficient of stearic acid in Salatrim fats. *J. Agric. Food Chem.* 42:484–488 (1994).

46. J.R. Hayes, J.W. Finley, and G.A. Leveille. In vivo metabolism of Salatrim fats in the rat. *J. Agric. Food Chem.* 42:500–514 (1994).

47. U.T. Bornscheuer, C. Bessler, R. Srinivas, and S.H. Krishna. Optimizing lipases and related enzymes for efficient application. *Trends Biotechnol.* 20:433–437 (2002).

48. C.C. Akoh. Enzymatic modification of lipids. In: *Food Lipids and Health* (R.E. McDonald and D.B. Min, eds.). Dekker, New York, NY, 1996, pp. 117–138.

49. T. Yamane. Enzyme technology for the lipids industry: an engineering overview. *J. Am. Oil Chem. Soc.* 64:1657–1662 (1987).

50. M.V. Arbige and W.H. Pitcher. Industrial enzymology: a look towards the future. *Trends Biotechnol.* 7:330–335 (1989).

51. E.N. Vulfson. Enzymatic synthesis of food ingredients in low-water media. *Trends Food Sci. Technol.* 4:209–215 (1993).

52. E. Santaniello, P. Ferraboschi, and P. Grisenti. Lipase-catalyzed transesterification in organic solvents: applications to the preparation of enantiomerically pure compounds. *Enzyme Microb. Technol.* 15:367–382 (1993).

53. H.L. Brockman. General features of lipolysis: reaction scheme, interfacial structure and experimental approaches. In: *Lipases* (B. Borgstrom and H.L. Brockman, eds.). Elsevier, Amsterdam, The Netherlands, 1984, pp. 3–46.

54. M. Iwai and Y. Tsujisaka. Fungal lipase. In: *Lipases* (B. Borgstrom and H.L. Brockman, eds.). Elsevier, Amsterdam, The Netherlands, 1984, pp. 443–469.

55. A.R. Macrae. Interesterification of fats and oils. In: *Biocatalysts in Organic Syntheses* (J. Tramper, H.C. Van der Plas, and P. Linko, eds.). Elsevier, Amsterdam, The Netherlands, 1985, pp. 195–208.

56. A. Zaks and A.J. Russell. Enzymes in organic solvents: properties and application. *J. Biotechnol.* 8:259–270 (1988).

57. A.L. Margolin, D.F. Tai, and A.M. Klibanov. Incorporation of d-amino acids into peptides via enzymatic condensation in organic solvents. *J. Am. Chem. Soc.* 109:7885–7887 (1987).

58. K. Yokozeki, S. Yamanaka, K. Takinami, Y. Hirose, A. Tanaka, K. Sonomoto, and S. Fukui. Application of immobilized lipase to regio-specific interesterification of triglyceride in organic solvent. *Eur. J. Appl. Microbiol. Biotechnol.* 14:1–5 (1982).

59. S. Koshiro, K. Sonomoto, A. Tanaka, and S. Fukui. Stereoselective esterification of dl-menthol by polyurethane-entrapped lipase in organic solvent. *J. Biotechnol.* 2:47–57 (1985).

60. W.H. Wu, C.C. Akoh, and R.S. Phillips. Lipase-catalyzed stereoselective esterification of dl-menthol in organic solvents using acid anhydrides as acylating agents. *Enzyme Microb. Technol.* 18:536–539 (1996).

61. J.S. Dordick. Enzymatic catalysis in monophasic organic solvents. *Enzyme Microb. Technol.* 11:194–211 (1989).

62. C.C. Akoh and K.H. Huang. Enzymatic synthesis of structured lipids: transesterification of triolein and caprylic acid. *J. Food Lipids* 2:219–230 (1995).

63. Y. Shimada, A. Sugihara, K. Maruyama, T. Nagao, S. Nakayama, H. Nakano, and Y. Tominaga. Production of structured lipid containing docosahexaenoic and caprylic acids using immobilized *Rhizopus delemar* lipase. *J. Ferment. Bioeng.* 81:299–303 (1996).

64. K.H. Huang and C.C. Akoh. Enzymatic synthesis of structured lipids: transesterification of triolein and caprylic acid ethyl ester. *J. Am. Oil Chem. Soc.* 73:245–250 (1996).

65. L.N. Yee, C.C. Akoh, and R.S. Phillips. Lipase *PS*-catalyzed transesterification of citronellyl butyrate and geranyl caproate: effect of reaction parameters. *J. Am. Oil Chem. Soc.* 74:255–260 (1997).

66. H. Hirata, K. Higuchi, and T. Yamashina. Lipase-catalyzed transesterification in organic solvent: effects of water and solvent, thermal stability and some applications. *J. Biotechnol.* 14:157–167 (1990).
67. A. Zaks and A.M. Klibanov. The effect of water on enzyme action in organic media. *J. Biol. Chem.* 263:8017–8021 (1988).
68. J.E. Schinkel, N.W. Downer, and J.A. Rupley. Hydrogen exchange of lysozyme powders. Hydration dependence of internal motions. *Biochemistry* 24:352–366 (1985).
69. M. Reslow, P. Adlercreutz, and B. Mattiasson. Organic solvents for bioorganic synthesis. 1. Optimization of parameters for a chymotrypsin catalyzed process. *Appl. Microbiol. Biotechnol.* 26:1–8 (1987).
70. A. Zaks and A.M. Klibanov. Enzymatic catalysis in nonaqueous solvents. *J. Biol. Chem.* 263:3194–3201 (1988).
71. P.J. Halling. Lipase-catalyzed reactions in low-water organic media: effects of water activity and chemical modification. *Biochem. Soc. Trans.* 17:1142–1145 (1989).
72. P. Cremonesi, G. Carrea, L. Ferrara, and E. Antonini. Enzymatic dehydrogenation of testosterone coupled to pyruvate reduction in a two-phase system. *Eur. J. Biochem.* 44:401–405 (1974).
73. A. Leo, C. Hansch, and D. Elkins. Partition coefficients and their uses. *Chem. Rev.* 71:525–616 (1971).
74. C. Laane, S. Boeren, K. Vos, and C. Veeger. Rules for optimization of biocatalysis in organic solvents. *Biotechnol. Bioeng.* 30:81–87 (1987).
75. M.V. Rodionova, A.B. Belova, V.V. Mozhaev, K. Martinek, and I.V. Berezin. Mechanism of denaturation of enzymes by organic solvents. *Chem. Abstr.* 106:171823 (1987).
76. P.A. Claon and C.C. Akoh. Effect of reaction parameters on SP435 lipase-catalyzed synthesis of citronellyl acetate in organic solvent. *Enzyme Microb. Technol.* 16:835–838 (1994).
77. G. Reed. Effect of temperature and pH. In: *Enzymes in Food Processing,* 2nd edn. (G. Reed, ed.). Academic Press, New York, NY, 1975, pp. 31–42.
78. J.R. Whitaker. *Principles of Enzymology for the Food Sciences.* Dekker, New York, NY, 1972, p. 320.
79. T.J. Ahern and A.M. Klibanov. The mechanism of irreversible enzyme inactivation at 100°C. *Science* 228:1280–1284 (1985).
80. A. Zaks and A.M. Klibanov. Enzymatic catalysis in organic media at 100°C. *Science* 224:1249–1251 (1984).
81. F.X. Malcata, H.R. Reyes, H.S. Garcia, C.G. Hill, Jr., and C.H. Amundson. Kinetics and mechanisms of reactions catalysed by immobilized lipases. *Enzyme Microb. Technol.* 14:426–446 (1992).
82. K.M. Shahani. Lipases and esterases. In: *Enzymes in Food Processing*, 2nd edn. (G. Reed, ed.). Academic Press, New York, NY, 1975, pp. 181–217.
83. V.M. Balcao, A.L. Paiva, and F.X. Malcata. Bioreactors with immobilized lipases: state of the art. *Enzyme Microb. Technol.* 18:392–416 (1996).
84. X. Xu. Enzyme bioreactors for lipid modifications. *Inform* 11:1004–1012 (2000).
85. S.I. West. Enzymes in the food processing industry. *Chem. Br.* 24:1220–1222 (1988).
86. P. Adlercreutz, R. Barros, and E. Wehtje. Immobilization of enzymes for use in organic media. *Enzyme Eng. XIII Ann. NY Acad. Sci.* 799:197–200 (1996).
87. F.D. Gunstone. Chemical reactions of oils and fats. *Biochem. Soc. Trans.* 17:1141–1142 (1989).
88. S. Ray and D.K. Bhattacharyya. Comparative nutritional study of enzymatically and chemically interesterified palm oil products. *J. Am. Oil Chem. Soc.* 72:327–330 (1995).
89. R.J. Jandacek, J.A. Whiteside, B.N. Holcombe, R.A. Volpenhein, and J.D. Taulbee. The rapid hydrolysis and efficient absorption of triglycerides with octanoic acid in the 1 and 3 positions and long-chain fatty acid in the 2 position. *Am. J. Clin. Nutr.* 45:940–945 (1987).
90. K.G. Raghuveer and E.G. Hammond. The influence of glyceride structure on the rate of autoxidation. *J. Am. Oil Chem. Soc.* 44:239–243 (1967).
91. S. Wada and C. Koizumi. Influence of the position of unsaturated fatty acid esterified glycerol on the oxidation rate of triglyceride. *J. Am. Oil Chem. Soc.* 60:1105–1109 (1983).
92. S. Kubow. The influence of positional distribution of fatty acids in native, interesterified and structure-specific lipids on lipoprotein metabolism and atherogenesis. *J. Nutr. Biochem.* 7:530–541 (1996).
93. D.M. Small. The effects of glyceride structure on absorption and metabolism. *Annu. Rev. Nutr.* 11:413–434 (1991).
94. K.T. Lee and T.A. Foglia. Synthesis, purification, and characterization of structured lipids produced from chicken fat. *J. Am. Oil Chem. Soc.* 77:1027–1034 (2000).

95. H. Brockerhoff and M. Yurkowski. Stereospecific analyses of several vegetable fats. *J. Lipid Res.* 7:62–64 (1966).

96. F.H. Mattson and R.A. Volpenhein. The specific distribution of fatty acids in the glycerides of vegetable fats. *J. Biol. Chem.* 236:1891–1894 (1961).

97. J.J. Myher and A. Kuksis. Stereospecific analysis of triacylglycerols via racemic phosphatidylcholines and phospholipase C. *Can. J. Biochem.* 57:117–124 (1979).

98. P.R. Redden, X. Lin, and D.F. Horrobin. Comparison of the Grignard deacylation TLC and HPLC methods and high resolution ^{13}C-NMR for the *sn*-2 positional analysis of triacylglycerols containing γ-linolenic acid. *Chem. Phys. Lipids* 79:9–19 (1996).

99. E.L. Lien. The role of fatty acid composition and positional distribution in fat absorption in infants. *J. Pediatr.* 125:S62–S68 (1994).

100. U. Bracco. Effect of triglyceride structure on fat absorption. *Am. J. Clin. Nutr.* 60:1002S–1009S (1994).

101. H. Sadou, C.L. Leger, B. Descomps, J.N. Barjon, L. Monnier, and A.C. De Paulet. Differential incorporation of fish-oil eicosapentaenoate and docosahexaenoate into lipids of lipoprotein fractions as related to their glyceryl esterification: a short-term (postprandial) and long-term study in healthy humans. *Am. J. Clin. Nutr.* 62:1193–1200 (1995).

102. F.H. Mattson and R.A. Volpenhein. The digestion and absorption of triglycerides. *J. Biol. Chem.* 239:2772–2777 (1964).

103. L.L. Swift, J.O. Hill, J.C. Peters, and H.L. Greene. Medium-chain fatty acids: evidence for incorporation into chylomicron triglycerides in humans. *Am. J. Clin. Nutr.* 52:834–836 (1990).

104. P. Sarda, G. Lepage, C.C. Roy, and P. Chessex. Storage of medium-chain triglycerides in adipose tissue of orally fed infants. *Am. J. Clin. Nutr.* 45:399–405 (1987).

105. M. Chanez, B. Bois-Joyeux, M.J. Arnaud, and J. Peret. Metabolic effects in rats of a diet with a moderate level of medium-chain triglycerides. *J. Nutr.* 121:585–594 (1991).

106. I. Ikeda, Y. Tomari, M. Sugano, S. Watanabe, and J. Nagata. Lymphatic absorption of structured glycerolipids containing medium-chain fatty acids and linoleic acid, and their effect on cholesterol absorption in rats. *Lipids* 26:369–373 (1991).

107. G.L. Jensen, N. McGarvey, R. Taraszewski, S.K. Wixson, D.L. Seidner, T. Pai, Y.Y. Yeh, T.W. Lee, and S.J. DeMichele. Lymphatic absorption of enterally fed structured triacylglycerol vs physical mix in a canine model. *Am. J. Clin. Nutr.* 60:518–524 (1994).

108. M.M. Jensen, M.S. Christensen, and C.E. Hoy. Intestinal absorption of octanoic, decanoic, and linoleic acids: effect of triglyceride structure. *Ann. Nutr. Metab.* 38:104–116 (1994).

109. M.S. Christensen, C.E. Hoy, C.C. Becker, and T.G. Redgrave. Intestinal absorption and lymphatic transport of eicosapentaenoic (EPA), docosahexaenoic (DHA), and decanoic acids: dependence on intramolecular triacylglycerol structure. *Am. J. Clin. Nutr.* 61:56–61 (1995).

110. M.S. Christensen, C.E. Hoy, and T.G. Redgrave. Lymphatic absorption of *n*-3 polyunsaturated fatty acids from marine oils with different intramolecular fatty acid distributions. *Biochim. Biophys. Acta* 1215:198–204 (1994).

111. K.T. Mok, A. Maiz, K. Yamazaki, J. Sobrado, V.K. Babayan, L.L. Moldawer, B.R. Bistrian, and G.L. Blackburn. Structured medium-chain and long-chain triglyceride emulsions are superior to physical mixtures in sparing body protein in the burned rat. *Metabolism* 33:910–915 (1984).

112. B. Mendez, P.R. Ling, N.W. Istfan, V.K. Babayan, and B.R. Bistrian. Effects of different lipid sources in total parenteral nutrition on whole body protein kinetics and tumor growth. *J. Parenter. Enteral Nutr.* 16:545–551 (1992).

113. C.J. Gollaher, E.S. Swenson, E.A. Mascioli, V.K. Babayan, G.L. Blackburn, and B.R. Bistrian. Dietary fat level as determinant of protein-sparing actions of structured triglycerides. *Nutrition* 8:348–353 (1992).

114. J. Nordenstrom, A. Thorne, and T. Olivercrona. Metabolic effects of infusion of a structured triglyceride emulsion in healthy subjects. *Nutrition* 11:269–274 (1995).

115. R.J. Deckelbaum, J.A. Hamilton, A. Moser, G. Bengtsson-Olivecrona, E. Butbul, Y.A. Carpentier, A. Gutman, and T. Olivecrona. Medium-chain versus long-chain triacylglycerol emulsion hydrolysis by lipoprotein lipase and hepatic lipase: implications for the mechanisms of lipase action. *Biochemistry* 29:1136–1142 (1990).

116. M. Adolph, J. Eckhart, C. Metges, G. Neeser, and G. Wolfram. Is there an influence of MCT on the LCT oxidation rate during total parenteral nutrition of trauma patients? *J. Parenter. Enteral Nutr.* 14:21S (1990).

117. H. Hedeman, H. Brondsted, A. Mullertz, and S. Frokjaer. Fat emulsions based on structured lipids (1,3-specific triglycerides): an investigation of the *in vivo* fate. *Pharm. Res.* 13:725–728 (1996).

118. J.M. Daly, M. Lieberman, J. Goldfine, J. Shou, F.N. Weintraub, E.F. Rosato, and P. Lavin. Enteral nutrition with supplemental arginine, RNA and omega-3 fatty acids: a prospective clinical trial. *J. Parenter. Enteral Nutr.* 15:19S (1991).

119. D.M. Ney, H. Yang, J. Rivera, and J.B. Lasekan. Total parenteral nutrition containing medium-vs. long-chain triglyceride emulsions elevates plasma cholesterol concentrations in rats. *J. Nutr.* 123:883–892 (1993).

120. J. Sobrado, L.L. Moldawer, J.J. Pomposelli, E.A. Mascioli, V.K. Babayan, B.R. Bistrian, and G.L. Blackburn. Lipid emulsions and reticuloendothelial system function in healthy and burned guinea pigs. *Am. J. Clin. Nutr.* 42:855–863 (1985).

121. E.S. Swenson, K.M. Selleck, V.K. Babayan, G.L. Blackburn, and B.R. Bistrian. Persistence of metabolic effects after long-term oral feeding of a structured triglyceride derived from medium-chain triglyceride and fish oil in burned and normal rats. *Metabolism* 40:484–490 (1991).

122. P. Quinlan and S. Moore. Modification of triglycerides by lipases: process technology and its application to the production of nutritionally improved fats. *Inform* 4:580–585 (1993).

123. S.J. DeMichele, M.D. Karlstad, B.R. Bistrian, N. Istfan, V.K. Babayan, and G.L. Blackburn. Enteral nutrition with structured lipid: effect on protein metabolism in thermal injury. *Am. J. Clin. Nutr.* 50:1295–1302 (1989).

124. T.C. Teo, S.J. DeMichele, K.M. Selleck, V.K. Babayan, G.L. Blackburn, and B.R. Bistrian. Administration of structured lipid composed of MCT and fish oil reduces net protein catabolism in enterally fed burned rats. *Ann. Surg.* 210:100–107 (1989).

125. T.C. Teo, K.M. Selleck, J.M.F. Wan, J.J. Pomposelli, V.K. Babayan, G.L. Blackburn, and B.R. Bistrian. Long-term feeding with structured lipid composed of medium-chain and *n*-3 fatty acids ameliorates endotoxic shock in guinea pigs. *Metabolism* 40:1152–1159 (1991).

126. V.S. Hubbard and M.C. McKenna. Absorption of safflower oil and structured lipid preparations in patients with cystic fibrosis. *Lipids* 22:424–428 (1987).

127. E.A. Mascioli, V.K. Babayan, B.R. Bistrian, and G.L. Blackburn. Novel triglycerides for special medical purposes. *J. Parenter. Enteral Nutr.* 12:127S–132S (1988).

128. G.L. Jensen and R.G. Jensen. Specialty lipids for infant nutrition. II. Concerns, new developments, and future applications. *J. Pediatr. Gastroenterol. Nutr.* 15:382–394 (1992).

129. S.A. Miller. Biochemistry and nutrition of low-calorie triglycerides. *Inform* 6:461 (1995).

130. R. Sandstrom, A. Hyltander, U. Korner, and K. Lundholm. Structured triglycerides to postoperative patients: a safety and tolerance study. *J. Parenter. Enteral Nutr.* 17:153–157 (1993).

131. J.W. Finley, G.A. Leveille, R.M. Dixon, C.G. Walchak, J.C. Sourby, R.E. Smith, K.D. Francis, and M.S. Otterburn. Clinical assessment of Salatrim, a reduced-calorie triacylglycerol. *J. Agric. Food Chem.* 42:581–596 (1994).

132. D.R. Webb, F.E. Wood, T.A. Bertram, and N.E. Fortier. A 91-day feeding study in rats with caprenin. *Food Chem. Toxicol.* 31:935–946 (1993).

133. C.C. Akoh. Lipid-based fat substitutes. *Crit. Rev. Food Sci. Nutr.* 35:405–430 (1995).

134. N. Matsuo and I. Tokimitsu. Metabolic characteristics of diacylglycerol: an edible oil that is less likely to become body fat. *Inform* 12:1098–1102 (2001).

135. T. Yasukawa and Y. Katsuragi. Diacylglycerols. In: *Diacylglycerol Oil* (Y. Katsuragi et al., eds.). AOCS Press, Champaign, IL, 2004, pp. 1–15.

136. B.D. Flickinger and N. Matsuo. Nutritional characteristics of DAG oil. *Lipids* 38:129–132 (2003).

137. Y. Shimada. Enzymatic modification of lipids for functional foods and nutraceuticals. In: *Handbook of Functional Lipids* (C.C. Akoh, ed.). CRC Press, Boca Raton, FL, 2005, pp. 437–455.

138. K. Yasunaga, W. Glinsmann, Y. Seo, Y. Katsuragi, S. Kobayashi, B. Flickinger, E. Kennepohl, T. Yasukawa, and J.F. Borzelleca. Safety aspects regarding the consumption of high-dose dietary diacylglycerol oil in men and women in a double-blind controlled trial in comparison with consumption of a triacylglycerol control oil. *Food Chem. Toxicol.* 42:1419–1429 (2004).

139. H. Takase, K. Shoji, T. Hase, and I. Tokimitsu. Effect of diacylglycerol on postprandial lipid metabolism in non-diabetic subjects with and without insulin resistance. *Atherosclerosis* 180:197–204 (2005).

140. C.C. Akoh. Structured and specialty lipids. In: *Healthful Lipids* (C.C. Akoh and O.M. Lai, eds.). AOCS Press, Champaign, IL, 2005, pp. 591–606.

141. F.J. Plou, M. Barandiaran, M.V. Calvo, A. Ballesteros, and E. Pastor. High-yield production of mono- and di-oleylglycerol by lipase-catalyzed hydrolysis of triolein. *Enzyme Microb. Technol.* 18:66–71 (1996).

142. T. Yamane, S.T. Kang, K. Kawahara, and Y. Koizumi. High-yield diacylglycerol formation by solid-phase enzymatic glycerolysis of hydrogenated beef tallow. *J. Am. Oil Chem. Soc.* 71:339–342 (1994).

143. R. Rosu, M. Yasui, Y. Iwasaki, and T. Yamane. Enzymatic synthesis of symmetrical 1,3-DAGs by direct esterification of glycerol in solvent-free system. *J. Am. Oil Chem. Soc.* 76:839–843 (1999).

144. T. Watanabe, M. Shimizu, M. Sugiura, M. Sato, J. Kohori, N. Yamada, and K. Nakanishi. Optimization of reaction conditions for the production of DAG using immobilized 1,3-regiospecific lipase lipozyme RM IM. *J. Am. Oil Chem. Soc.* 80:1201–1207 (2003).

145. S.K. Lo, B.S. Baharin, C.P. Tan, and O.M. Lai. Lipase-catalysed production and chemical composition of diacylglycerols from soybean oil deodoriser distillate. *Eur. J. Lipid Sci. Technol.* 106:218–224 (2004).

146. Y. Yamada, M. Shimizu, M. Sugiura, and N. Yamada. International patent WO 99/09119 (1999).

147. X. Xu, A.R.H. Skands, C.E. Høy, H. Mu, S. Balchen, and J. Alder-Nissen. Production of specific-structured lipids by enzymatic interesterification: elucidation of acyl migration by response surface design. *J. Am. Oil Chem. Soc.* 75:1179–1186 (1998).

148. X. Xu, H. Mu, A.R.H. Skands, C.E. Høy, and J. Alder-Nissen. Parameters affecting diacylglycerol formation during the production of specific-structured lipids by lipase-catalyzed interesterification. *J. Am. Oil Chem. Soc.* 76:175–181 (1999).

149. T. Yang, M.B. Fruekilde, and X. Xu. Suppression of acyl migration in enzymatic production of structured lipids through temperature programming. *Food Chem.* 92:101–107 (2005).

150. H. Taguchi, T. Nagao, H. Watanabe, K. Onizawa, N. Matsuo, I. Tokimitsu, and H. Itakura. Energy value and digestibility of dietary oil containing mainly 1,3-diacylglycerol are similar to those of triacylglycerol. *Lipids* 36:379–382 (2001).

151. K.E. Bach Knudsen, E. Wisker, M. Daniel, W. Feldheim, and B.O. Eggum. Digestibility of energy, protein, fat and non-starch polysaccharides in mixed diets: comparative studies between man and the rat. *Br. J. Nutr.* 71:471–487 (1994).

152. E. Wisker, K.E. Bach Knudsen, M. Daniel, W. Feldheim, and B.O. Eggum. Digestibilities of energy, protein, fat and nonstarch polysaccharides in a low fiber diet and diets containing coarse or fine whole meal rye are comparable in rats and humans. *J. Nutr.* 126:481–488 (1996).

153. T. Murase, T. Mizuno, T. Omachi, K. Onizawa, Y. Komine, H. Kondo, T. Hase, and I. Tokimitsu. Dietary diacylglycerol suppresses high fat and high sucrose diet-induced body fat accumulation in C57BL/6J Mice. *J. Lipid Res.* 42:372–378 (2001).

154. R.A. Coleman, T.M. Lewin, and D.M. Muoio. Physiological and nutritional regulation of enzymes of triacylglycerol synthesis. *Annu. Rev. Nutr.* 20:77–103 (2000).

155. H. Kondo, T. Hase, T. Murase, and I. Tokimitsu. Digestion and assimilation features of dietary DAG in the rat small intestine. *Lipids* 38:25–30 (2003).

156. T. Teramoto, H. Watanabe, K. Ito, Y. Omata, T. Furukawa, K. Shimoda, M. Hoshino, T. Nagao, and S. Naito. Significant effects of diacylglycerol on body fat and lipid metabolism in patients on hemodialysis. *Clin. Nutr.* 23:1122–1126 (2004).

157. X.H. Meng, D.Y. Zou, Z.P. Shi, Z.Y. Duan, and Z.G. Mao. Dietary diacylglycerol prevents high-fat diet-induced lipid accumulation in rat liver and abdominal adipose tissue. *Lipids* 39:37–41 (2004).

158. T. Murase, M. Aoki, T. Wakisaka, T. Hase, and I. Tokimitsu. Anti-obesity effect of dietary diacylglycerol in C57BL/6J Mice: dietary diacylglycerol stimulates intestinal lipid metabolism. *J. Lipid Res.* 43:1312–1319 (2002).

159. K.C. Maki, M.H. Davidson, R. Tsushima, N. Matsuo, I. Tokimitsu, D.M. Umporowicz, M.R. Dicklin, G.S. Foster, K.A. Ingram, B.D. Anderson, S.D. Frost, and M. Bell. Consumption of diacylglycerol oil as part of a reduced-energy diet enhances loss of body weight and fat in comparison with consumption of a triacylglycerol control oil. *Am. J. Clin. Nutr.* 76:1230–1236 (2002).

160. T. Nagao, H. Watanabe, N. Goto, K. Onizawa, H. Taguchi, N. Matsuo, T. Yasukawa, R. Tsushima, H. Shimasaki, and H. Itakura. Dietary diacylglycerol suppresses accumulation of body fat compared to triacylglycerol in men in a double-blind controlled trial. *J. Nutr.* 130:792–797 (2000).

161. H. Watanabe, K. Onizawa, H. Taguchi, M. Kobori, H. Chiba, S. Naito, N. Matsuo, T. Yasukawa, M. Hattori, and H. Shimasaki. Nutritional characterization of diacylglycerol in rats. *J. Jpn. Oil Chem. Soc.* 46:301–308 (1997).

162. M. Murata, T. Ide, and K. Hara. Reciprocal responses to dietary diacylglycerol of hepatic enzymes of fatty acid synthesis and oxidation in the rat. *Br. J. Nutr.* 77:107–121 (1997).

163. M.M.J.W. Kamphuis, D.J. Mela, and M.S. Westerterp-Plantenga. Diacylglycerols affect substrate oxidation and appetite in humans. *Am. J. Clin. Nutr.* 77:1133–1139 (2003).

32 Genetic Engineering of Crops That Produce Vegetable Oil

Vic C. Knauf and Anthony J. Del Vecchio

CONTENTS

I. INTRODUCTION

A new oilseed crop was introduced commercially in the southern United States in the fall of 1994. The crop looked no different from normal varieties of *Brassica napus* canola. The farmer cultivated and harvested the crop without departing from standard canola harvesting practice. The oilseed meal, after crushing, was essentially the same as regular canola meal and, indeed, was treated simply as standard canola meal for use in animal feeds. The resulting vegetable oil, however, was unique and different from any previously available for either food or industrial uses [1].

The new oilseed crop was derived from transgenic canola developed by a biotechnology company. The genetic engineering approach was targeted toward a lauric acid-rich oil in this case, but the technology is used to develop other new vegetable oils with novel structure and compositions. In this chapter, we discuss some of the oils that may result and also some of the factors that shape the feasibility of these and other potentially novel raw materials to be achieved by genetic engineering.

II. GENETIC ENGINEERING OF PLANTS

The chemical composition of vegetable oils is a highly heritable trait. For example, year to year, soybean oil is a reliably constant raw material for the food industry. Not only is the high lauric acid content of coconut oil characteristic of coconut oil, but also one does not find an occasional crop year in which soybean oil has lauric acid, or in which coconut oil is lacking lauric acid. These basic chemical compositions characteristic of various vegetable oils—which define the uses of specific vegetable oils in the food industry—are determined by the genes of each plant variety. The advent of genetic engineering technology in agriculture has enabled the directed modification of the gene set that determines oil composition in a given oilseed crop.

There is a long and fruitful history of plant breeders, who have deliberately selected for lines in which the seed oil is different in chemical composition. Examples include high-oleic sunflower, low-linolenic flax, and low-erucic rapeseed. These successes appear to be the cases in which a specific gene or genes become nonfunctional. Although genetic engineering techniques in yeasts and bacteria allow the specific targeting of genes to be "knocked out," this is not yet the case with plants. However, it is possible now to add genes to a plant, and some of these approaches can indeed be used to decrease the functional expression of genes resident in the host crop plant genome.

There is an ever-increasing knowledge base to explain the genetic bases that determine the chemical composition of seed oils. In summary [2], a number of lipids-related genes have been individually cloned in the laboratory, and most of these genes turn out to encode specific enzymes used by the plant to synthesize the triacylglycerols that make up vegetable oils. For example, cloning a gene from *Cuphea lanceolata* that encodes a specific enzyme in the seed with a unique activity on a capric acid precursor may help to explain why high levels of capric acid are found in *C. lanceolata* seed oil. Moreover, failure to find that enzyme in canola seed may at least partly explain why capric acid is not naturally found in canola oil. In its simplest manifestation, a genetic engineering approach would suggest taking the cloned *C. lanceolata* gene, transferring it into the chromosomes of a canola plant, and seeing whether the oil from the seeds of the resulting transgenic canola plant contains capric acid.

Currently, genetic engineering of plants allows the addition of genes. These genes are incorporated directly into plant chromosomes and basically behave in subsequent generations of progeny plant like other genes. That is, they are inherited in a Mendelian manner and are subjected to the same modifications to which the genes preexisting in the plant genome are subjected. The gene's specific "behavior," however, may be something entirely novel to the host plant. For example, it may encode for an enzyme that has never been found in that plant species. The ability to introduce such modifications is what genetic engineering adds to the plant breeder's tool kit. A breeder, of course, could not cross a coconut tree with a canola plant to get a canola plant with certain coconut tree properties. But a discrete number of coconut genes can be transferred into a canola plant to make some facet of a canola plant's metabolism more closely resemble that of a coconut tree.

As in the example of *C. lanceolata* and canola earlier, one can add a novel gene from a wide range of sources, including not only any plant but also animals, bacteria, and even genes encoding enzymes that were "designed" on a computer and synthesized in the laboratory. It is possible to clone gene from a specific plant species, engineer the gene so that it expresses the same enzyme at an unusually high level at the targeted stage of seed development, and add it back to the same plant species. By having more of an enzyme controlling a rate-limiting step in oil biosynthesis, one might achieve more oil and/or an oil with a different fatty acid composition. Finally, it is possible to add genes that interfere with the function of existing genes in the host plant genome, thus decreasing enzyme activities and enzymatic products that are not desired.

All these facets of adding genes can be used to redirect the molecular pathway by which oils are synthesized in plants; since these changes are determined by the genes, the changed oil in a transgenic plant will also be seen in the progeny of that plant. Thus, new crop species can be created that essentially look the same as the parent crop, yet produce a significantly different oil. A transgenic soybean plant looks like any other soybean plant and can be grown by the farmer in exactly the same way.

A review of the detailed methods used to genetically engineer oilseed crops is outside the scope of this chapter. The reader is referred to several reviews [1,3,4]. The intent of this summary is to describe some of the practical impacts on the food lipids area that will result from the introduction of a set of technologies that lead to new crop varieties, which in turn produce new raw materials for the food industry.

III. EXAMPLE OF LAURATE CANOLA

Coconut and oil palm kernel are the primary sources of lauric oils. Both are produced from trees grown in the tropical zone, and each contains about 50% lauric acid by weight. However, some temperate zone plants (e.g., *Cuphea glutinosa*, *Umbellularia californica*) accumulate even higher levels of lauric acid in seed oils. Therefore, a priori, it appeared in the early 1980s that lauric acid oils could conceivably be produced in temperate zone agricultural systems. For industries in the United Sates, a temperate zone crop source of lauric acid would (1) help address national import/export imbalances; (2) perhaps stabilize world price fluctuations for lauric acid oils; and (3) conceivably provide an oil with higher lauric acid content and the associated savings in processing costs.

When Calgene, Inc., of Davis, California, initiated a project in 1985 to engineer *B. napus* canola into a lauric oil producer, there was a long list of technical unknowns. There was no reliable system for putting genes into canola and getting back normal plants. The available data suggested that a gene from a monocotyledonous plant like coconut might not function correctly when transferred to a dicotyledonous plant like canola. The general opinion was that production of lauric acid in nonseed tissues might be detrimental or even lethal to a canola plant. Indeed, even as late as 1990, it was thought that lauric acid might "gum up" triacylglycerol synthesis in canola. Certainly in 1985, there was an insufficient understanding of how to limit foreign gene expression to just developing the seed of a transgenic canola plant. In addition, there was absolutely no experience, on which to base assumptions that a transgenic oil or simply the process of generating transgenic canola would not somehow compromise the agronomic productivity of canola. The perhaps most worrisome aspect of all was that no one knew, despite numerous tries, the mechanistic basis of laurate accumulation in coconut or oil palm kernel, let alone the wild species of *C. glutinosa* or *U. californica*.

This list of technical challenges provides a framework to understand the steps and areas of technical expertise required to generate a useful new crop type containing a genetically engineered oil.

In the example of lauric acid canola, scientists were able to demonstrate the existence of a unique enzyme comprising lauroyl-acyl carrier protein (ACP) thioesterase in embryos of seed from *U. californica* that could conceivably account for the production of lauric acid. Moreover, that specific enzyme activity appeared to be missing in similar extracts from canola seed. Encouraged by that finding, a team of biochemists purified the *U. californica* lauroyl-ACP thioesterase protein sufficiently to obtain, by means of an automated protein sequencer, a partial amino acid sequence. This amino acid sequence allowed the design of synthetic DNA primers, which were then used with templates made by molecular biologists from messenger RNA (mRNA) isolated from developing seed of *U. californica* to generate gene-specific DNA probes. The DNA probes were subsequently used to identify lauroyl-ACP thioesterase complementary DNA (cDNA) clones from a cDNA bank that was constructed using, again, mRNA from developing seed of *U. californica*.

Independently in the same laboratory, another group of molecular biologists isolated a canola gene specifically expressed at high levels in developing embryos during the normal period of canola seed development when storage lipids are formed. This natural *B. napus* gene was dissected down to the elements necessary to encode proper gene expression timing and tissue-specific local-ization within the canola plant. These "promoter" or genetic expression elements were then combined with the central portion of the *U. californica* lauroyl-ACP thioesterase cDNA close corresponding to the open reading frame encoded by the original naturally occurring mRNA. This synthetic gene was then combined with a second gene (the selectable marker gene) in a specific manner relative to other DNA signal sequences in a unique microbe known as a disarmed *Agrobacterium tumefaciens*.

Plant cell biologists then took the genetically engineered strain of *A. tumefaciens* and cultivated it for a brief time with sectioned hypocotyl tissues from germinated *B. napus* plantlets. This now-routine method of gene transfer into canola had to be developed while the biochemists were studying the lauroyl-ACP thioesterase enzyme in *U. californica* seed extracts and while molecular biologists were identifying the gene expression control elements from embryo-specific gene expres-sion in seed of *B. napus*. After the treatment of *B. napus* hypocotyl sections with *A. tumefaciens*, the bacteria were completely removed and the plant tissues cultured on a series of different growth media in different containers, to regenerate complete *B. napus* canola plants. During this process of regeneration, most of the plantlets were deliberately killed by the addition of an antibiotic. The selectable marker gene that was linked to the synthetic lauroyl-ACP thioesterase gene in the *A. tumefaciens* strain was a synthetic gene designed to detoxify the antibiotic used in these experiments. Thus, the only plants that survived the antibiotic treatment were transgenic plantlets that had received the kanamycin resistance gene (and almost always the lauroyl-ACP thioesterase gene as well). This is how the cell biologist can selectively produce 30–300 different transgenic plants without having to sort through thousands of plants that only might be transgenic. Each transgenic plant coming out of this process is potentially different and is generally regarded as a different event. As it turns out, each event tends to be unique in one or several features, most notably, in this case, with respect to how much lauric acid accumulates in the seed oil.

Progeny from each different event can be grown up and examined for the lauric acid content in the seed oil, and for other traits as well. As mentioned earlier, the transgenes show inheritance like regular canola genes, so gene segregation is possible, especially when a transgenic canola is crossed with another line.

The plant breeder's job to genetically fix the transgenes in a given canola line as well as to select for other traits such that the lauric acid content is reliably constant from generation to generation and agronomic characters such as seed yield are also preserved. Practically speaking, of course, this means multiple seasons of field testing and selections. In addition, just like other canolas, lines optimally adapted to different environments have to be tested and selected in different environ-ments. The importance of this phase to the eventual success of developing a new crop type with a novel oil composition should not be underestimated.

In summary, then, development of laurate canola required basic biochemistry research, applied protein purification, gene cloning molecular biology, embryo-specific gene expression molecular biology, cell biology, and breeding, with many of these tasks and much of the technology development carried out in parallel. As well as an agronomically acceptable lauric acid canola, there had to be development of an infrastructure at the farm level to produce and harvest a new canola without contamination by regular canola seed. And as we shall see, another set of expertises to recognize and take advantage of special properties that may be present in a new oil had also to be established. Clearly, the development of a new crop type by directed genetic modification of a seed oil is not a small undertaking.

IV. APPLICATIONS FOR FOOD LIPIDS

A. NATURAL LIMITATIONS

One of the first considerations in thinking of specific ways to modify seed storage lipids is feasibility. Certainly, there are some limits to what is practically possible. For example, seed storage lipids serve an important role in the life cycle of a crop plant. Germinating seeds presumably require the energy stored in the seed lipids and/or the young seedlings require access to those seed lipids in the cotyledons. If a lipid has been modified into particular structures that interfere with the ability of germinating seeds or young seedlings to tap that energy, or if the oil contains fatty acids that are difficult to metabolize, there may be a limit to the quantities of modified lipids of those kinds that one can achieve in a seed oil and still have a viable crop variety.

Moreover, the synthesis and incorporation of certain unusual fatty acids into storage lipids during seed development may also result in those fatty acids becoming incorporated into structural lipids that are essential for normal seed function. For example, the castor bean endosperm contains very high levels of ricinoleic fatty acid in the storage triacylglycerols, with very little ricinoleic acid in structural lipids. Clearly, as the castor bean lineage evolved synthetic mechanisms for ricinoleic fatty acid in seed oils, it also evolved mechanisms to either prevent incorporation of ricinoleic acid into structural lipids or to clear that fatty acid from structural lipid molecules. If one chooses to engineer soybean to produce ricinoleic acid in seeds, it may be found that soybean lacks the ability to maintain the integrity of the seed structural lipids and the seed is not fertile.

Lipid biosynthesis occurs in all plant cells and is essential to support growth. Whereas the triacylglycerols comprising vegetable oils found in seeds are neutral lipids used as a means to store energy and fixed carbon, polar lipids found in all cells—including those in seeds—serve important structural functions and are thus often referred to as structural lipids. Both structural lipids and seed triacylglycerols contain glycerol-bound fatty acids, and thus it is not surprising that their respective synthesis pathways share many common steps. Re-engineering triacylglycerol biosynthesis must not compromise the synthesis of the structural lipids necessary to support growth and viability.

If some desirable lipid compositions are indeed incompatible with viable seed, there may be a unique opportunity in engineering such oil types into either oil palm or avocado mesocarp tissues. These are oil-rich tissues whose natural fate is to rot away after fruit dehiscence. Since plant progeny do not directly depend on the viability of these tissues, they might be engineered to "self-destruct" by making, for example, very solid fats rich in long-chain saturated fatty acids. Engineered oil palm fruits could be harvested by standard means and the palm kernels (the oil palm seed) would be normal—assuming that the technology employed worked to ensure that the changed fatty acid composition was limited to just the mesocarp.

Another type of limitation that is of commercial interest is the ultimate amount of oil obtainable from a crop. While focused on traits like disease resistance and plant stature, much of the oilseed crop breeding community measures success by how much seed is harvested per acre. As crop yields go up, one expects the cost of food oils to come down. At a different level, breeders and genetic engineers alike are interested in increasing the oil content on a per-seed basis. Soybeans typically contain only 20%–25% oil by weight, whereas canola seed is typically 40%–45%. Peanuts can be 50% oil, whereas the cacao bean is 60% cocoa butter. It seems physiologically possible, therefore, to increase the seed oil content in soybean and canola with correspondingly dramatic effects on the cost of producing vegetable oils from these major crops.

B. PRACTICAL LIMITATIONS

Even though the scope of possible lipids modification projects must be very widely based on the natural variation in seed lipids found among plants from around the world, there are factors, other than natural ones, that practically limit what will be done in the near term.

Genetic engineering of crop plants requires a battery of expertises: plant physiology, enzymology, molecular biology, gene transfer cell biology, prototype evaluation, and plant breeding. Much of an engineering project entails sequential applications of each area of expertise to the eventual goal. Coupled to the obvious costs of having these expertises in place and having the time needed to go from concept to practice is a consideration of the technical risk entailed when one is embarking on a new project. Financial modeling and analyses of the eventual value and return versus development costs and risk may argue against ever starting certain projects.

Part of the eventual value may hinge on the production system that allows the value to be ensured and protected. Since genetically engineered genes behave like other genes once present in a plant, they are necessarily transmitted by pollen. Vegetable oils are often zygotic characters. That is, the chemical composition is determined by both parents. Thus if a transgenic canola crop is grown immediately adjacent to normal canola, an interchange of pollen will decrease the purity of both crops. So, novel oilseed crops may have to be grown in carefully managed districts. Of course, there are some precedents for producing "specialty" oils, including the previously mentioned low-linolenic flax, high-oleic sunflower, and erucic-containing industrial rapeseed. These identity-preserved production practices typically add costs. Moreover, seed from crops with different types of oil cannot be stored together or crushed together; this restriction adds more complexity to production. These special production practices need to be considered when the development of a new food oil is planned. In the case of canola, where the precedent of identity-preserved production of high erucic oils already exists, it appears that the incremental costs are likely <5¢/lb.

Regulatory and political issues should be anticipated as well. Since genetic engineering is a new technology, products are receiving additional scrutiny by the public and by regulatory agencies to ensure that environmental and food safety issues are as fully reviewed as feasible. The nature of the issues varies considerably according to the nature of the altered traits. For example, different issues are raised by herbicide resistance and the potential introduction of a new food oil rich in myristic acid. Some segments of society may be less willing to try foods containing ingredients based on biotechnology. Religious issues for some novel oils may need addressing (e.g., kosher and halal definitions). These market realities may affect what genetic engineering projects make sense.

Although plant lipid biosynthesis research has blossomed dramatically in the past 20 years, fueled by interest in transgenic plant applications, there is still much to be learned. Even though plant oils exist in nature that are >85% laurate, there may not be enough known to justify the laying out of a scientific strategy to convert soybean into a producer of an oil with 85% lauric acid. On the other hand, if a technical route to some financially attractive target appears to be straightforward, an organization should consider the competitive aspects carefully. Perhaps, a competitor elected to gamble at an earlier stage when the technical route was less clear and that organization may have gained an insurmountable lead to a commanding patent position.

Obviously, embarking on a long-term project with technical risk and complexity requires not only an evaluation of the utility and marketplace need of the final product, but also an extensive competitive analysis. The technology underlying the genetic engineering of oil composition may provide a clear basis for the patentability of the product of the research. Thus, since whoever is first may very well be in a position to block similar approaches by others, a consideration of which groups are after the same goal and where they are in the process is in order. In addition, for vegetable oils, there may be more than one means to the same end, and these alternatives should be anticipated as well as possible. For example, one could consider the genetic engineering of soybean with a *C. lanceolata* gene to produce an oil rich in capric acid. Alternatively, another group may choose more conventional means of domesticating *C. lanceolata* into an economically feasible crop. Either or neither approach may succeed; but if both succeed, the eventual marketplace value to each for the financial and time investment may be less.

In the case of structured oils and fats, there are of course chemically and enzymatically based synthetic methods that use glycerol and fatty acids sourced from animal fats as well as vegetable oils. Cost of raw materials and synthetic capacities are factors to consider when this approach is compared with the genetic engineering of crop plants. In the latter case, volumes are limited only to the number of acres that can be planted; in the former case, capital-intensive hard assets may be required in the form of manufacturing plants.

Of course, the possibility of competitive products becoming available and concerns about the profitability and financial return on investment versus risk are critical concepts for the private sector entities considering the development of new food lipids via the genetic engineering of crop plants. On the other hand, certain targets of lipids modification that will not pass these private sector hurdles for profitability versus risk may nonetheless have value and importance to the human community. For example, a cottonseed oil lacking malvalic acid may not be able to command much of a premium over regular cottonseed oil as long as cheap soybean oil (which naturally lacks malvalic acid) is also available. This might be a lipids modification project that could appropriately be undertaken by public sector research groups like the U.S. Department of Agriculture, since the benefit could be spread among a large number of independent cooperative ginning entities throughout the cotton-growing delta of the Mississippi River. Similarly, improving the nutritive value of lipids in rice may not be attractive to private sector companies for many reasons, including the long and perhaps risky research phases that would be necessary, the structure (or lack of structure) in how rice planting seed is sold, the prevalence of small (subsistence) family farms in specific growing regions of Asia, and the resistance of farmers to paying premiums. But clearly, if the technology can be used to improve traits like carotenoid or vitamin content even slightly in a basic commodity foodstuff, the benefits would be tremendous to society at large.

The important issue to address regarding public sector initiative research on tailored vegetable oils is the coordinated recognition of technical feasibility along with marketplace-defined needs (e.g., shelf life or processing traits like colors or free fatty acid content) and consumer common interest objectives (e.g., levels of atherogenic saturated fatty acids or essential dietary fatty acids). How this process sorts out common good objectives and provides for the necessary funding and coordination of properly qualified laboratories, plant breeding organizations, and the equally necessary product development and introduction functions is a challenge for everyone working in areas relating to foods and health.

C. TARGET CROPS

As of this writing, efficient systems exist to transfer genes into selective tissue types of rapeseed (including canola), soybean, cotton, and corn; it is then possible to regenerate transgenic plants with normal growth habits and harvest yields similar to parental types. Thus, transgenic crops are already in commercial production (or there are existing seed sales) for transgenic canola, cotton, corn, and soybean.

Success has been reported by a few groups for similar gene transfer and regeneration steps for sunflower, flax, and peanut. The procedures for these and perhaps other temperate zone annual oilseed crops like safflower and sesame will likely become routine relatively soon.

Because of the long generation times and the related lack of regeneration cell biology knowledge, current prospects for transgenic oil-producing tree crops like oil palm, coconut, olive, and cacao are less encouraging. Also of food interest is the modification of the fatty acids in oils of almond, walnut, and other nuts that tend to oxidize and produce off-flavors after storage. The time required with tree species to develop and apply methods, evaluate transgenically produced potential products, and scale-up for production may discourage any commercial projects. Like the projects described earlier that do not offer large profit margin opportunities to reward the risk takers, genetic engineering of the lipids in major oil crops like oil palm and coconut may be taken up primarily by public sector research units, which do not need to satisfy short-term investment return expectations.

Another consideration in the modification of seed oil content of specific crops may be the ability to properly express transgenes in specific tissues of the plant. That is, seed lipid biosynthesis tends to occur during a very specific phase of seed development. To be able to modify that biosynthetic pathway to alter the character of the vegetable oil, transgenes must exert their effects in the correct tissue at the correct time. Moreover, in some cases, it may be deleterious to have a transgene expressed in the wrong tissue or at the wrong time. Ricinoleic acid naturally appears only in the castor bean endosperm and not in leaves and other tissues, where it presumably has no function and might interfere with the synthesis of necessary structural molecules.

Technology to obtain the correct transgene expression for seed storage lipid modification has been well demonstrated in canola and soybean. One can expect the same principles to apply for peanut, sunflower, and other crops, although fine-tuning of the technology may be needed for maximum benefit in some cases. This potential requirement should be taken into account in considerations of transgenic approaches to oil modification in crops where there is less experience.

D. TARGET TRAITS

1. Short-Chain Saturates

The first genetically engineered vegetable oil, Calgene's Laurical, is already a commercial product. Researchers studying the seed embryos of *U. californica* (California bay tree) identified an enzyme known as a lauroyl-ACP thioesterase that appeared to account for the 60% laurate content in the California bay seed oil (Figure 32.1). When the cDNA corresponding to the mRNA for the bay tree lauroyl-ACP thioesterase was successfully cloned, adapted to canola gene expression controlling elements, and transferred into canola, the resulting oil from transgenic seeds contained lauric acid in amounts ranging from <1% to >45%. From these original plants, referred to as transgenic "events," lines were developed that produced genetically uniform seed that reliably contained an average 38%–42% lauric acid in the oil.

When the lauric acid-rich oil from transgenic canola seed was examined more closely, it was observed that very little of the lauric acid was in the second position of the triglyceride molecules. The practical implications of this result are discussed later. From a scientific point of view, it was assumed that the canola enzyme that converts lysophosphatidic acid to phosphatidic acid discriminated against lauroyl CoA as a substrate (Figure 32.2). Indeed, it had been reported that in canola, this enzyme, known as lysophosphatidic acid acyltransferase (LPAT) discriminates in vitro against saturated acyl CoAs as well as substrates with monounsaturated acyl groups exceeding 18 carbons in length.

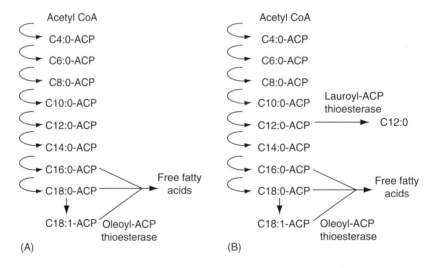

FIGURE 32.1 Biosynthesis of fatty acids in plant cells. (A) Synthesis of fatty acids in canola proplastids: fatty acids, or acyl groups, are built up in two-carbon increments while bound as a thioester to acyl carrier protein (ACP). As the acyl-ACP molecules lengthen into C16:0-ACP, C18:0-ACP, and (after a desaturation step) C18:1-ACP, an acyl-ACP thioesterase cleaves the thioester bond to ACP and the complete fatty acid is available for triacylglycerol synthesis after conversion to an acyl CoA in the cytoplasm (see Figure 32.2). (B) Engineered laurate biosynthesis in transgenic canola: addition of a lauroyl-ACP thioesterase from *Umbellularia californica* (or other lauric acid-producing plant) allows the thioester bond to ACP to be cleaved at the C12:0 stage, thus making free lauric acid available for triacylglycerol synthesis.

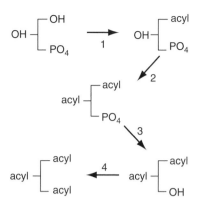

FIGURE 32.2 Biosynthesis of triacylglycerols in plants. Reactions 1, 2, and 4 each draw on acyl CoA pools available in the plant cell. (1) Glycerol-3-phosphate is acylated to form 1-acyl-*sn*-glycerol-3-phosphate, also known as lysophosphatidic acid. (2) 1-Acyl-*sn*-glycerol-3-phosphate is acylated to form phosphatidic acid by the enzyme lysophosphatidic acid acyltransferase (LPAT). (3) A phosphatase activity removes the phosphate group to generate diacylglycerol. (4) Diacylglycerol acyltransferase attaches the third acyl group to create triacylglycerol. LPAT in most plants is very selective for unsaturated 18-carbon acyl CoA, whereas the other enzymes in this pathway are typically nonselective. However, coconut LPAT in particular has high activity on shorter chain saturated fatty acid CoAs.

Subsequently, it was shown that coconut endosperm, which contains a triacylglycerol oil with high levels of laurate in the second position, also contains an LPAT enzyme with high activity for placing laurate into the second position. When the cDNA corresponding to the mRNA for the coconut endosperm LPAT was cloned, adapted to canola gene expression controlling elements, and transferred into canola along with the bay tree thioesterase gene construct, the resulting oil from transgenic seeds contained laurate in all three positions of the triacylglycerols. Thus, one type of lauric oil was obtained by engineering canola with the bay tree thioesterase enzyme, and a second type of oil was obtained by combining the bay tree lauroyl-ACP thioesterase with the coconut endosperm LPAT in seed of transgenic canola.

Creating novel oils in seeds of transgenic plants with increased levels of other saturated fatty acids of chain lengths <16 carbons is also possible. By using cDNA clones derived from mRNA from different species of New World genus *Cuphea* plants, seed oils enriched in C8 and C10 fatty acids have been obtained. Using cDNA clones derived from mRNA from developing seed of *Cuphea palustris* or from nutmeg, oils enriched in C14 fatty acids have been obtained. The latter myristate-type oils also tend to have higher (C16:0) palmitic levels. Again, the transgenic canola triacylglycerols tend to have these newly introduced, engineered fatty acids in the first and third positions on the glycerol backbone. Thus, these oils and fats may have special functional properties due to the underlying structure.

2. Naturally Solid Fats

The temperate zone oilseed crops tend to have storage lipids rich in unsaturated fatty acids, and thus liquid oils that are in themselves not suitable for products like shortenings and margarine. This shortcoming is typically addressed by a postharvest treatment of partial hydrogenation of vegetable oils. Partial hydrogenation does carry some incremental cost and generally results in the introduction of *trans*-unsaturated fatty acids, which may be undesirable in foods for health reasons. Of the tropical tree oils, only oil palm is produced on a scale large enough to provide a "fat" fraction rich in palmitic acid as a hardstock. Illipe, shea, and sal are tropical trees with seeds rich in the C18 fatty acid stearate, which contributes more solids to a fat; however, such tropical fats are limited in availability and generally too expensive to use as hardstock ingredients. Inspection of the pathway for biosynthesis of fatty acids reveals that both palmitic and stearic fatty acids are precursors to the unsaturated fatty acids making up the bulk of the oil found in seeds from temperate zone crops. Not surprisingly, then, it has been possible to engineer the pathway so that fewer unsaturated fatty acids are made and more of the precursor palmitic and stearic acids accumulate in the triacylglycerols.

Numerous thioesterase enzymes have been identified by cloning cDNAs from plant mRNAs that share DNA homology with the California bay tree thioesterase cDNA described earlier. Every plant so far examined appears to have an enzyme with high activity on oleoyl-ACP as a substrate, and this enzyme presumably is the basic thioesterase in the fatty acid biosynthesis pathway shown in Figure 32.1. The oleoyl-ACP thioesterase typically has less but significant activity on both palmitoyl-ACP and stearoyl-ACP substrates. Another class of thioesterase enzymes has been found in many plants, however, with enhanced activity on palmitoyl-ACP substrates; genetically engineered overexpression of such enzymes in seed of transgenic canola plants results in seed oils enriched in palmitic acid. On the other hand, thioesterase enzymes with low activity on palmitoyl-ACP substrates and relatively high activity on steroyl-ACP appear to be rare. When researchers looked at cDNA clones made from mRNA from developing seed of mangosteen fruit (mangosteen seed oil typically contains 40%–50% stearic acid), a clone was identified that corresponds to an enzyme with enhanced levels of stearoyl-ACP activity relative to other thioesterases. Genetically engineered overexpression of that mangosteen enzyme in seed of *trans*-genie canola plants results in a seed oil enriched in stearic acid, up to 30% in some seed.

Alternatively, one can consider engineering a seed to have less desaturase activity so that fewer molecules of the saturated precursor stearoyl-ACP are converted to oleoyl-ACP. This has been demonstrated in transgenic canola and soybean by suppressing levels of the stearoyl-ACP desaturase enzyme. Suppression of an enzyme can be achieved by genetic engineering methods of either antisense or cosuppression. Each allows for tissue-specific suppression. In this particular application, tissue-specific suppression is important because the stearoyl-ACP desaturase is an essential enzyme in leaves and other tissues for plant viability.

3. Yield

As noted earlier, yield is a very important trait in reducing the cost of vegetable oils. Theoretically, there are several arguments for the feasibility of raising seed oil content in most crops. The cacao bean has 60% oil by weight, so the physiological limit for crops like soybean and canola may be at least this much high. Individual seed of rapeseed can typically vary from 35% to 50%; so even within existing germplasm, the current average oil content from canola of 42% will likely be raised by straightforward breeding selections in the coming years. Historically, soybean protein content has been the most important component of the bean, with 20% of the seed that is oil a valuable by-product. Clearly, there is room to increase oil content in soybean by some means; however, increased oil will be balanced against maintaining value in the meal component for animal feed uses.

Despite the evident technical premise and motivation for increasing oil content, exact scientific strategies are still developing. Acetyl CoA carboxylase is an enzyme activity often considered to be a rate-limiting step for fatty acid biosynthesis. The data for this assumption seem clearest in animal cells and fairly convincing in the bacterium *Escherichia coli*, but somewhat less clear in higher plants. Transgenic modifications of levels of this enzyme activity have been achieved with slight and possibly significant increases in oil content; however, experiments continue. The development of other strategies to increase fatty acid biosynthesis at the expense of nondigestible fiber and other less desirable components of seed represent an area of increasing research.

4. Removing Negatives

The availability of gene suppression technologies allows one to think of decreasing certain constituents of vegetable oils. An early and far-reaching application is evident in the use of cosuppression in canola and soybean to dramatically reduce the levels of polyunsaturated fatty acids. Other potential project might be, for example, the suppression of enzymes directly responsible for the formation of cyclopropene fatty acids such as malvalic acid in cottonseed oil or achieving a decrease in the high palmitic acid content in palm oil by suppressing the level of palmitoyl-ACP thioesterase activity in the oil palm mesocarp so that more fatty acids can be elongated and desaturated to oleic acid.

5. Other Lipid Targets

For the most part, we have discussed modification of the kinds of fatty acid found in triacylglycerols. However, vegetable oils can also contain other important lipid components. These include the antioxidants of tocopherols (e.g., vitamin E) and of mixed carotenoids (including β-carotene and lutein), plant sterols such as sitosterol (which may have cholesterol-lowering properties in the human diet), and economically important by-products of vegetable oil processing such as lecithin. As the enzymatic bases for the synthesis of these compounds and the genetic bases for regulation of amounts of these compounds become better understood, there should be growing prospects for the genetic engineering of levels of these compounds as well as flavor components in vegetable oils.

E. NATURE OF INCREMENTAL PROGRESS

When low erucic acid lines of rapeseed were first commercialized in Canada, the oil content levels were typically <38% by weight. However, over years of breeding progress, oil content steadily increased to a current average of 42%–44%. The primary determinant of value, oil quality, was first selected for and developed; then further incremental improvements in yield, meal quality, and other traits were layered onto the original trait.

Genetically engineered plant oils will likely follow a similar path. That is, after the synthesis and accumulation of lauric acid have been established in a transgenic canola oil, it may become important to select (or genetically engineer) for lower levels of linoleic and linolenic fatty acids. Or as discussed earlier, a coconut LPAT enzyme may be engineered into the lauric canola to enable higher levels of laurate to be achieved. Such higher levels of lauric acid made possible by the presence of the medium-chain LPAT enzyme may actually be achieved by field selections (breeding) over a number of years of incremental improvements.

Finally, of course, once a valuable oil has been achieved in composition, one would like to keep selecting for varieties with better and better yields. Thus, it is clear that genetic engineering of oil composition, just like conventional breeding of oil composition, comprises one or a few big steps—transgenes in the case of genetic engineering—supplemented by a continuous process of fine-tuning the oil and the crop variety producing the oil.

V. IDENTIFICATION OF UTILITY AND VALUE

A. MARKET-DRIVEN PRODUCT DEVELOPMENT

1. Fatty Acid Strategies

Fats and oils are ubiquitous components of most compounded formulations, whether they are foods or industrial products. These applications may not involve the whole triglyceride, or the pure derivative fatty acids that comprise triglycerides, but may entail the splitting of the triglyceride into its component fatty acids (see Figure 32.3) and subsequent derivatization of those fatty acids into industrially important products that move into the oleochemical markets.

We shall not discuss these industrial applications at great length, but it is important to recognize their importance in the dynamics of the total fats and oils industry. Figure 32.4 is a typical process flow diagram showing the types of derivatives that are of current importance in the various industrial sectors. As can be seen from the end products synthesized and their uses, most fatty acids are converted to fatty alcohols and then to a variety of derivatives that use their surface activity.

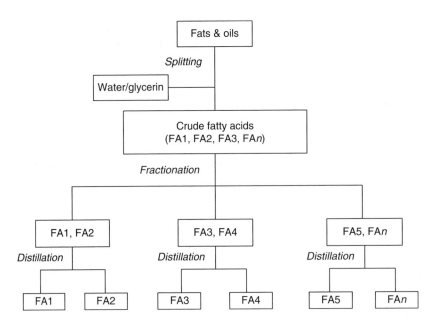

FIGURE 32.3 Basic oleochemical processing: steps for splitting of oils or fats into component fatty acids and glycerin.

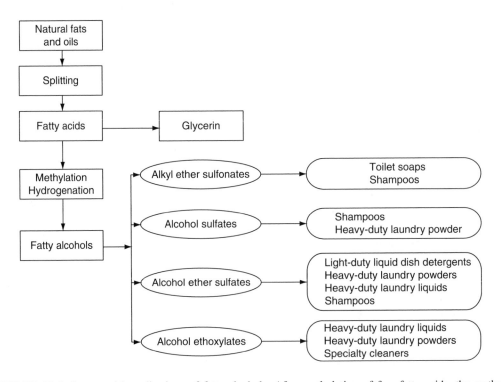

FIGURE 32.4 Commercial applications of fatty alcohols. After methylation of free fatty acids, the methyl esters are converted to fatty alcohols, which find their way into numerous consumer products.

TABLE 32.1
Volume and Value of World Fatty Acid Production

Country/Region	Volume (×10 lbs)			Value (1995)
	1988	1995	2000	
United States	1300	1499	1653	592
Western Europe	1973	2226	2424	879
Asia	1223	1455	1653	575
Other	419	496	573	196
Total	4915	5675	6303	2242

Although these compounds are, as a class, called surfactants, they perform a variety of functions in a variety of end products, including wetting agents, spreading agents, emulsifiers, foaming agents, and foam stabilizers. The majority of these surfactants are based on alcohols derived from lauric acid. Natural alcohols (from naturally occurring fats and oils containing C12:0 and/or fatty acids of interest >C12) compete with synthetic alcohols derived from petroleum. Synthetic alcohols historically dominated the market, primarily because of their low cost, but naturally derived alcohols are becoming increasingly important. Natural alcohol production is projected to equal that of the synthetics by the year 2000, as indicated by the data for fatty acid production in Table 32.1. No new synthetic alcohol capacity has come on-line in the last decade, and no new plants are projected for the near future. In contrast, six new natural alcohol facilities have been built since 1990.

The driver for this industry is lauric acid availability, since this product yields derivatives having the most functional end effects. Natural lauric acid is available on a commercial basis only from the splitting and fractionation of certain tropical oils: coconut oil and palm kernel oil.

The world's major coconut oil producer is the Philippines, where thousands of small farmers with small production acreage accounted for 43% of production and 65% of total exports in 1992–1993. Because the Philippine coconut industry has suffered from chronic underinvestment, production and exports have decreased and are expected to continue to do so for the foreseeable future. Estimates are that as many as one-third of the 300 million coconut trees in the Philippines will become nonproductive over the next 10 years. Significant coconut oil producers are shown in Table 32.2. Production in these secondary countries, spurred by government investment, is expected to increase. Coconut production in any given year is heavily influenced by weather factors, including rainfall amounts and the impact of typhoons.

The second major source of lauric oils is palm kernel. Malaysia is the largest producer of palm kernel oil accounting for 54% of the world's production in 1992–1993. Other large producers are shown in Table 32.2. Unlike coconut production in the Philippines, palm (along with palm kernel) is produced by large sophisticated plantations and enjoys major government and private investments that are allowing it to forward-integrate into end-use derivatives. As a result, palm kernel oil production is projected to increase significantly through the year 2002, with yields of 2175 metric tons as new plantations become productive.

In the world of foods, where, typically the whole triglyceride is used as one of the components, lauric oils can be replaced—to some degree—by the nonlauric oils, especially as specialty fat blends are developed that mimic the physical properties of these lauric fats through a series of processing steps including special hydrogenation and fractionation. In the industrial area, however, lauric acid-derived molecules invest unique properties into the surface activity of the resulting compound and formulation, and they allow for little or no substitution of the fatty acid moiety without a direct effect on performance. This is especially true in the soap, detergent, and personal care markets, where the C12 moiety provides unequaled detergency and mildness. Because the perennial nature of lauric oilseeds prevents quick response to changes in supply and demand, and much of the demand

TABLE 32.2
World Lauric Oils Production (Million Metric Tons); By Country, 1992–1993

Country	Coconut Oil Production	Coconut Oil Percent of World Total	Palm Kernel Oil Production	Palm Kernel Oil Percent of World Total
Cameroon	0	0	22	1
Colombia	0	0	31	2
Ecuador	0	0	11	1
Ghana	0	0	13	1
India	267	9	0	0
Indonesia	664	23	336	20
Ivory Coast	47	2	31	2
Malaysia	32	1	925	54
Mexico	106	4	0	0
Mozambique	39	1	0	0
Nigeria	0	0	176	10
Papua New Guinea	37	1	21	1
Philippines	1301	45	0	0
Sri Lanka	36	1	0	0
Thailand	40	1	27	2
Vietnam	107	4	0	0
Zaire	0	0	22	1
Others	212	7	89	5
Total	2888	100%	1704	100%

side is inelastic owing to the unique functionality of the oil, lauric oil prices are typically much more volatile than prices for nonlauric oils (Figure 32.5). Rising incomes and standards of living in Southeast Asia and Eastern Europe are expected to result in an increase in consumption of detergent and personal care products and, therefore, increased demand for lauric oils. Nonetheless, most experts project that increases in palm kernel oil production will compensate for increases in laurate demand and decreases in coconut oil production. Lauric oil price volatility is expected to decrease

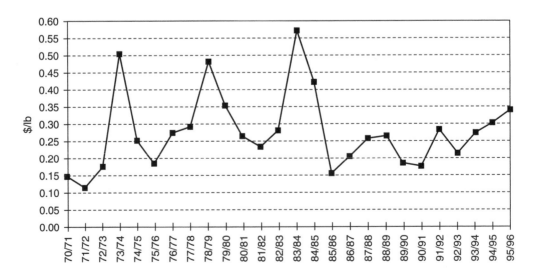

FIGURE 32.5 Coconut oil prices: average price by year, in U.S. dollars per pound as crude oil delivered.

from historic patterns, but prices are expected to remain much more volatile than those for the nonlauric products.

Under the circumstances just outlined, what better goal than to produce a temperate source of C12 fatty acids in a triglyceride using the techniques of genetic engineering coupled with traditional plant breeding? If high enough levels of C12 could be generated, several issues could be addressed:

1. Impact of wide fluctuations in commodity prices for world lauric oils due to weather, natural disaster, political unrest, and other uncontrollable factors could be mitigated by the development of an annual, temperate oilseed crop that could be used to offset the unpredictability of some of these tropical crops and to stabilize the pricing of the desired fatty acids.
2. Environmental concerns associated with the use of petrochemical products favor conversion to plant oil-based products that are biodegradable and renewable. Chemical processing plants devote a significant portion of their capital to postproduction cleanup systems to minimize the environment stress. Use of plant oil feedstocks would ease environmental concerns and, presumably, reduce manufacturing costs.

The real world, however, imposes the following constraints:

1. Basic oleochemical business is a coproduct-balanced, commodity operation. Profitability is dependent on tight control of manufacturing processes and costs, and on maximizing the value of every product produced as a result of the splitting process.
2. These industries are process-driven. The plants have been designed around and optimized for profitability based on a relatively narrow range of feedstock composition. In the case of lauric fats, coconut, and palm kernel, the balance of the fatty acid streams produced dictates the profitability of the total operation. Major shifts in fatty acid distribution in the feedstock disrupt this balance and require changes in the way the end products are treated as saleable items. To clarify this point, note the difference in fatty acid composition of laurate canola versus palm kernel oil and coconut oil as shown in Table 32.3.

As can be seen, laurate canola does not contain as much C12 as either coconut oil or palm kernel oil. In addition to decreased levels of this fatty acid, note also the total absence of the medium-chain fatty acids, C8:0 and C10:0, typically present in the tropical fats. Specific markets for these medium chains have been developed, and their sales contribution is important to the overall profitability of the splitting operation. The approach, then, to achieve utility within the existing framework of the massive oleochemical complex, would be to develop triglycerides having such massive doses of C12:0 that coproduct values would be obviated in standard feedstock sources. This would mean the

TABLE 32.3
Fatty Acid Composition (%) of Laurate Canola versus Coconut, Palm Kernel, and Canola Oils

Fatty Acids	Laurate Canola	Coconut Oil	Palm Kernel Oil	Canola
Lauric	38.0	49.0	47.0	0.0
Myristic	4.0	17.5	16.0	0.1
Oleic	31.0	5.0	16.5	61.5
Linoleic	11.0	1.8	2.5	20.0
Other	16.0	26.7	18.0	18.4
Total	100.0	100.0	100.0	100.0

development of triglycerides having C12:0 levels >85%. Is such a goal possible using this technology? The answers are not yet known, but the research still required is not trivial—as discussed earlier—and the impact of achieving such high levels of a specific fatty acid on the viability of the resulting plant is not known.

What has been learned from this exercise?

1. High expressions of specific fatty acids in plant triglycerides should be considered only in the context of alternative sources of such fatty acids, and the coproduct value of the rest of the components.
2. Supply limitations as drivers for specific fatty acids can be a viable foundation for product development if current sources of feedstock triglycerides do not necessarily offer cost offsets because of the value of coproducts.

In the context of developing increased levels of specific fatty acids in plant triglycerides, it is often fruitful to take a closer look at the actual coproducts derived from the feedstock splitting streams and to assess the value of those coproducts if they were not supply-limited (i.e., dependent on the levels present in the base feedstock oil). Again, coproduct value in the resultant genetically engineered oil must also be considered, so that value can be obtained using the whole product, and not depending on one or two of the fatty acids to carry the whole value of the product. Examples of potential products would include oils having very high levels of the medium-chain fatty acids, C8:0 and C10:0, or very high levels of myristic acid, C14:0. Again, although these particular fatty acids are limited based on their presence in tropical oil feedstock streams, an assessment needs to be made of their potential value in the marketplace if such level-rich sources become available through genetic engineering. It is not given that an increased availability of these fatty acids would lead to an automatic increase in their industrial or food uses. One must also be constantly aware of the need to balance coproduct volumes and values to achieve success in the marketplace.

2. Triglyceride Strategies

An example of a strategy developed for the application of transgenic oilseed products is the use of laurate canola as an alternative for coconut and palm kernel fats (the so-called tropical laurates) in food systems. Although these markets are smaller in volume than the industrial applications of laurics (see Figure. 32.6), the functional nature of the uses would allow the recapture of the research investment, assuming that at least equal functionality was shown in existing applications.

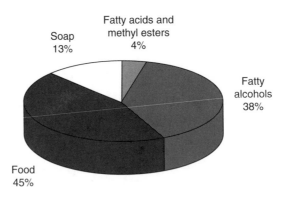

FIGURE 32.6 World consumption of lauric oils by end use in 1994. Although food uses account for much of the consumption of lauric oils, other oils with similar functionality could be used for cooking in most cases. That is, the low cost of lauric oils in developing countries is a primary factor for their use in food. However, soap and detergent uses take advantage of the unique functionality associated with the 12-carbon lauric acid.

The main function of lauric oils is to provide desirable structure and mouthfeel characteristics in a variety of foods. Both coconut and palm kernel provide relatively high solids at room temperature; but a steep melting curve at about body temperature gives food products based on these oils a good melt-away sensation in the mouth. The solids profile of a given fat is, typically, determined by dilatometric methods, although techniques using pulsed NMR spectrometry allow for similar determinations of the solid/liquid ratio in a given fat as a function of temperature. Usually, the solids fat index (SFI) is determined for food fats at a selective number of temperatures. The plots of solids versus temperatures have been related over the years to the actual performance of given fat systems in specific food formulas. The portion of the curve below 25°C (77°F) relates to the "hardness" of the fat, specifically the resistance to deformation it provides to the base food, while the solids present between the temperatures of 25°C and 30°C (86°F) represent the amount of solids typically found at ambient temperatures and are a measure of the degree of heat resistance provided to the food product by the fat system. Solids that occur above 37°C (or about body temperature, 98.6°F) are perceived as "waxiness" in the mouth, although low levels of solids are acceptable. A rapid rate of solids decreases between ambient temperatures and body temperature provides a desirable cooling sensation in the mouth and is deemed positive for confectionery fat systems. A major use of these fats is in confectionery coatings, where the lauric fats provide an effect similar to that of cocoa butter in chocolate, at a much reduced cost.

The major fat used in the manufacture of chocolate is cocoa butter. It is also one of the most expensive ingredients used in the formula, and because so much of it must be used to achieve the desired functional effects, it contributes the highest cost of goods for the total formulation. Aside from a significant contribution to the final flavor of the chocolate, cocoa butter has some physical properties that make it unique among the triglycerides found in nature. First of all, it is composed of a very few sets of highly structured triglycerides (see Table 32.4).

These triglycerides are essentially composed of only three fatty acids: palmitic (C16:0, P), stearic (C18:0, S), and oleic (C18:1, O). Practically, all the oleic acid occurs esterified at the *sn-2* position of the various triglycerides, with the *sn-1* and *sn-3* positions containing the esterified

TABLE 32.4
Triglyceride Composition
of Cocoa Butter (%)

Triglycerides	Percent
POSt	36.3–41.2
StOSt	23.7–28.8
POP	13.8–18.4
StOO	2.7–6.0
StLiP	2.4–6.0
PliS	2.4–4.3
POO	1.9–5.5
StOA	1.6–2.9
PliP	1.5–2.5
StLiSt	1.2–2.1
OOA	0.8–1.8
PPSt	8.0
PStSt	0.2–1.5
POLi	0.2–1.1
OOO	0.2–0.9

Abbreviations: A, Arachidate; Li, linoleate; O, oleate;
P, palmitate; St, stearate.

palmitic or stearic acids. This specificity results in the presence of three predominant structured triglycerides: POP, POS, and SOS. These three triglyceride types make up >80% of the triglycerides found in cocoa butter. These specific triglycerides, which resemble each other very closely, provide cocoa butter with its unique solids profile and desirable melting characteristics.

Going back to the basic issue, however, cocoa butter is expensive, and supply-limited. Over the years, various researchers have spent considerable amounts of time developing fats that could serve as alternatives to cocoa butter by providing properties similar to its melting profile and solids content at various temperatures. Several classes of alternatives have been developed, ranging from cocoa butter "equivalents" produced from selected blends or fractions of natural fats high in specific triglyceride contents that are miscible in all proportions to cocoa butter, to fats that contain totally different triglyceride distributions but mimic, in many ways, the melting behavior of cocoa butter. An example of the production of a cocoa butter "equivalent" would involve the purification and fractionation of a series of different naturally occurring fats to obtain the proper proportion of the desired triglycerides having the desired structure, at the appropriate levels. Commonly used sources of these specialty fats and their triglyceride distributions are given in Table 32.5. The majority of the POP portion required is obtained from palm midfraction.

The reconstruction of a cocoa butter equivalent from the isolation of specific triglycerides, through fractionation from a variety of source oils and subsequent blending, obviously is not an inexpensive process, especially since the oils used for feedstocks are sourced from tropical plants that are not grown and cultivated in the most efficient manner and in themselves are often supply-constrained. These "equivalents" are commercially viable only when the price of cocoa butter is high enough to justify their high costs. When cocoa butter is relatively inexpensive, they are not used.

Another approach to mimicking the functional properties of cocoa butter is through the use of special fractions of a combination of domestic and tropical (non-lauric) fats that have been selectively hydrogenated. The fats often used include soy, canola, cotton, and palm. The manufacture of chocolate-flavored coatings using these final blends of fat in place of cocoa butter is well known, and these coatings have a real place in the market in certain applications where rate of melting of the fat in the mouth and ultimate flavor release are not key factors in consumer acceptance of the product. They do have the drawback, however, of some amount of "waxiness" that can be detected in the mouth, since a fraction of their triglycerides does melt above body

TABLE 32.5

Triglyceride Composition of Sal, Kokum, Shea, and Illipe Fats

Triglyceride	Composition (wt%)			
	Sal	Kokum	Shea	Illipe
POO	3	—	2	—
POSt	11	5	5	35
StOSt	42	72	40	45
StOO	16	15	27	3
StOL	1	—	6	—
StOA	13	—	2	4
OOO	3	2	5	—
POP	1	0	0	7
Others	10	6	13	6

Abbreviations: A, arachidate; O, oleate; P, palmitate; St, stearate.

TABLE 32.6
Fatty Acid Composition (wt%) of Processed Palm Kernel and Coconut Oils

Oil	C12	C14	C16	C18	C18:1	C18:2
Palm kernel oil	47.0	17.0	8.5	3.0	11.0	2.0
Hydrogenated at 35°C	47.0	17.0	8.5	13.0	7.0	—
Hydrogenated at 40°C	47.0	17.0	8.5	19.0	1.0	—
Fractionated	55.0	21.0	8.5	2.0	7.0	1.0
Hydro/fractionated	55.0	21.0	8.5	10.0	—	—
Interesterified/hydrogenated	47.0	17.0	8.5	19.0	1.0	—
Coconut oil	47.5	17.5	8.5	2.5	7.0	1.5
Hydrogenated	47.5	17.5	8.5	10.0	1.0	—

temperature. This effect is minimized when chocolate-flavored coatings made with these cocoa butter replacers are used on baked goods, where the crumb structure of the substrate aids in the mastication of the coating and helps minimize the perception of waxiness in the mouth.

The most popular approach for the formulation of chocolate-flavored coatings is the use of cocoa butter substitutes (CBSs), based on lauric fats. These coatings are widely used in both the chocolate/confectionery and baking industries. The principal base fats used in their manufacture are palm kernel oil and coconut oil, with the former being the more widely used. The specific fatty acid composition of these fats and their various fractions after processing is shown in Table 32.6. These fats contain triglycerides that are high in the esters of lauric and myristic acid. Triglycerides of these compositions form crystals that are relatively small and uniform, and have a sharp melting point around body temperature. The set point of these fats is also, typically, very sharp, and they provide a coating base that has significant solids at room temperature. With their high levels of lauric and myristic acids, however, it is obvious that a significant percentage of these fatty acids will occur on all three of the available positions in a random manner. Because these triglycerides are of vastly different compositions and are nonstructured, coatings based on them are not compatible with cocoa butter; hence, the ultimate coatings can tolerate only minimal levels of cocoa butter containing ingredients, such as cocoa powder, and the flavor impact of these coatings is usually degraded by these restrictions. They do have better melting properties than the cocoa butter replacers discussed earlier, however, with the noticeable absence of any lingering waxy mouthfeel.

With the production of the first crop of laurate canola, we were faced with a totally new type of triglyceride: one that was high in lauric acid esters, though not as high as coconut or palm kernel oils, but one in which all the C12:0 and C14:0 occurred on the sn-1 and sn-3 positions, and the sn-2 position was occupied solely by C18:x fatty acids, where $x = 1, 2,$ and/or 3. Selective hydrogenation of this oil, then, would allow us to closely control the melting properties of the resulting triglyceride and to manipulate the solids profile of the final fat system. When these new fats were used in typical formulated food systems, it was noted that the systems offered flavor release that was far superior to that obtainable with the corresponding tropical laurics. Although this statement is based on anecdotal data, a great deal of effort is under way to quantify the flavor release properties of these new structured fats versus the random lauric fats typically used. It could be immediately demonstrated, however, that this new source of a lauric fat used in confectionery was highly compatible with cocoa butter—a trait not encountered with the tropical laurics—hence could be used with cocoa butter sources that lead to higher flavor in the finished product. These obvious differences in how the genetically engineered laurate canola performed with respect to existing laurics were explored at some length and are discussed later (Section V.B.2).

3. Investigations Driven by Purported Health Benefits to the Consumer

a. *Low Saturates*

Consumer concern over saturates has driven food marketers to look for ways to differentiate their products on the basis of saturated fat content. Food labeling regulations recently published by the U.S. Food and Drug Administration (FDA) allow products containing <3.4% total saturates to be labeled as containing no saturated fat. With the size of the current salad and frying oil markets as a target, the potential justified examining and developing a technical strategy to address this opportunity. Over 45% of the vegetable oil consumed in the United States is used in salad and cooking oils. More than 6.5 billion pounds of vegetable oils (valued at $1.7 billion, wholesale) was used in this segment in 1992. The U.S. salad and cooking oil market is characterized by slow but steady growth, driven largely by population increases. The principal oils used in the production of retail salad and cooking oils in the United States are soybean, canola, corn, sunflower, cottonseed, and peanut oils (see Figure 32.7).

Saturate levels in liquid oils have become a significant issue for consumers because of reports linking saturates to coronary heart disease. Saturates have been shown to lower the levels of high-density lipoprotein (HDL) cholesterol. The FDA now requires that food labels list saturate levels. These concerns have already had a major effect of the U.S. salad and cooking oil segment, driving canola oil (with lower saturates than other oils) consumption from 263 million pounds in 1987 to over 1.2 billion pounds in 1993. Consumption of saturated tropical fats decreased by 17.8% over the same period. Consumer concern over saturates is projected to continue to play a key role in this market segment, and food companies will continue to search for products that address this concern.

Since canola oil contains the lowest level of saturates of any of the commonly used food oils, it seemed a natural base from which to launch a variety of approaches that would lead, ultimately, to the desired product. As it turned out, the research was not trivial. A combination of mutagenesis and the use of two different genes led to a canola oil having saturates level between 4.0% and 4.5%, but with questionable agronomic performance. The presence of relatively high levels of polyunsaturated C18 fatty acids also contributed negative performance characteristics when the oil was used for frying purposes. Several different companies are continuing to work in this project area using conventional plant breeding to lower the saturates. In most cases, these companies are working to combine this attribute with low linolenic and/or high oleic traits to offer an oil that is low in

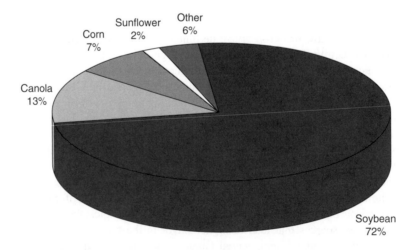

FIGURE 32.7 Vegetable oils used in salad and cooking oils for the United States in 1994.

saturates with improved stability under high temperature uses. None of these groups has yet developed an oil with <8% saturates.

b.   High Saturates

The presence of *trans*-fatty acids in the diet has become a major issue for United States and European consumers. In 1993, margarine sales began to decline as a number of studies indicated a potentially adverse impact of the *trans*-fatty acids found in partially hydrogenated vegetable oils used in the manufacture of most margarines and spreads. These studies (some of which received extensive publicity and are discussed in Ref. [5]) suggested that *trans*-fatty acids contributed to coronary heart disease by raising levels of low-density lipoprotein (LDL) cholesterol and lowering levels of HDL cholesterol. Although there is still considerable debate in the medical community about the true health impacts of *trans*-fatty acids, they have become a source of concern for many consumers and have affected buying habits. At the same time that U.S. consumers became concerned over the *trans* issue, butter manufacturers began to aggressively compete with margarine manufacturers for market share through price reductions and increased marketing and promotional activities. The net result is that margarine sales have declined since 1993, falling by >6% from 1993 to 1994.

An opportunity was seen to exist for the development of a suitable oil containing sufficient saturates in its natural makeup to eliminate the need for hydrogenation. Hydrogenation, normally used to increase the saturated fatty acid levels in a triglyceride system, is the primary producer of *trans*-fatty acids when polyunsaturated systems are reduced to monounsaturated systems. Thus elimination of hydrogenation would result in the elimination or minimizing of *trans*-fatty acid occurrence in food Systems. Key to assessing the opportunity is understanding the needs of the margarine and spreads industries.

Margarines and spreads are prepared by blending fats and oils with other ingredients, including water, milk, edible proteins, salt, flavorings, coloring. Spreads are differentiated from margarines by their lower fat content. By FDA regulation, a product must contain at least 80% fat to be labeled margarine; products with lower fat levels must be labeled as spreads. As consumers have become more concerned about overall fat consumption, spreads have become increasingly popular. The key to the formulation of both margarines and spreads is to provide a solid, spreadable fat with appropriate melting characteristics. This is achieved by using partially hydrogenated vegetable oils, which have had a portion of their unsaturated fatty acids converted to both saturated fatty acids and less saturated, *trans*-fatty acids. Saturated and *trans*-fatty acids have higher melting points than unsaturates and provide the requisite functional properties for the finished product. The specific oils used and the degree to which they are hydrogenated vary as a function of specific marketing objectives for the particular product.

In Europe, some margarines contain an oil portion consisting of a liquid vegetable oil blended with one that has been fully hydrogenated—a formulation ploy seen increasingly in the United States. Unlike partially hydrogenated oils, fully hydrogenated oils do not contain any *trans*-fatty acids. This allows for the manufacture of margarines that are *trans*-free, but also contain higher levels of saturates than those containing partially hydrogenated oils. European manufacturers also use more tropical oils in the manufacture of margarines and spreads than firms in the United States, since European consumers are not as concerned as U.S. consumers about tropical fats.

As of this writing, several companies are actively pursuing the development of seed oils that contain levels of saturated fatty acids high enough to permit the elimination or needs for hydrogenation, without the concomitant production of *trans*-fatty acids. As an integral part of this higher production of functional, saturated fatty acids, it is also necessary to have the desired fatty acids in a low polyunsaturated fatty acid background to assure the ultimate stability of the finished formulation. To compete in the large markets involved under these two sets of conditions of composition, one must also strive for a resulting plant with agronomic vigor and oil yields comparable with those of existing sources of commodity oils.

B. DISCOVERY OF NOVEL UTILITY

1. Conventional Wisdom versus Structured Triglycerides

Most of the body of knowledge that has been built up over the years on the functionality of lipids in food systems has been based on experiments designed around variations of naturally occurring fats and oils. The functional performance of these fats, from whatever source, was related back to specific analytical characteristics that still enjoy wide use in the industry: solid fat index, iodine value, and fatty acid composition. In addition to these characterizing values, a number of analytical tests were routinely performed on the fats that were indicators of their quality, or their ability to withstand the stresses of temperature and shelf-life requirements. These included free fatty acid content, peroxide value, color, and odor. None of these tests, however, related the functional performance of the fat to the presence (or absence) of any specific triglyceride having a specific structure (i.e., which fatty acid was on which carbon of the glycerol backbone). In the majority of cases, such knowledge would have been of academic interest only, since these structured fats were simply not available in any great proportion in a given fat system. Aside from cocoa butter and some of the other more exotic tropical fats, most fats used in foods consist of a random assortment of triglycerides driven by the types and levels of fatty acids in their composition, so that such knowledge would have no direct bearing on a formulator's capabilities.

Although many pioneering studies were conducted on synthesized and purified structured triglycerides to ascertain their physical chemical properties, especially those related to their melting characteristics and crystal forms, the quantities synthesized were not sufficient to be used in real food systems.

Recently, a great deal more effort has been expended to study structured triglycerides in foods, and these researches have led to the market introduction of synthesized species that have been almost exclusively targeted at the confectionery market for the replacement of cocoa butter, with the additional benefit of producing reduced-calorie products. These products take advantage of the effects of positional isomerism on the glycerin backbone to address the specific physical properties required in the final food product, and to use the differences in caloric contribution of the various fatty acids used to arrive at a lowered caloric intake. These novel ingredients, however, are costly to manufacture. Each one requires a series of synthetic steps along with requisite purification procedures. With a final price to the end-user that remains at several dollars per pound, the ultimate use of novel ingredients is restricted to specific niche markets in the food industry; a significant move toward their use in a wide array of food products cannot yet be projected.

The foregoing type of research is, however, needed to drive our understanding of the functionality of triglyceride structures in food systems. With the advent of the tools provided by genetic engineering, the opportunity to create new, structured triglycerides in the seed oil of an agricultural crop at costs much closer to a commodity seed oil base than to that of a synthesized product is very real. The goal, then, is to develop knowledge that relates structure to function so that a specific structure can be used for a specific end use. When this has been achieved, it is also likely that the total amount of fat required in any given food system will be reduced to levels well below those currently required using the various random systems as they occur in nature. The end result will very likely take us to the reduction in total fat intake that is so strongly recommended by health-care professionals.

2. Laurate Canola: A Case Study

The first genetically engineered oil approved for food use was developed by Calgene over a period of ~10 years, through the use of techniques discussed earlier. The product was ultimately brought to market using the common and usual name, laurate canola. The specific composition of this oil was given earlier (Table 32.3). Initial functional screening of this new oil was conducted to see if it would serve as a cocoa butter replacement in coating and confectionery products. Although laurate canola did not contain the same levels of laurate as the coconut and/or palm kernel products

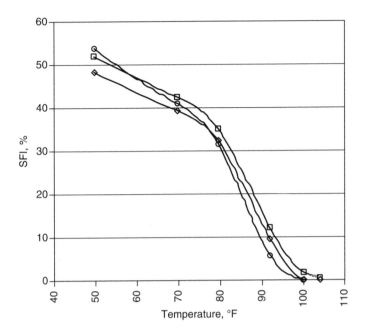

FIGURE 32.8 Solid fat index (SFI) of laurate canola as a function of C12 content (wt%): squares, 30/32%; diamonds, 36%; circles, 40%. All values for laurate canola hydrogenated to an iodine value (IV) of 37.

currently used for these applications, all the C12 and C14 fatty acids occurred exclusively at the *sn*-1 and *sn*-3 positions of the glycerin molecule, and this property encouraged us to look for novel functional effects. Since the *sn*-2 position contained only C18:*x* fatty acids, where $x = 1$, 2, or 3, selective hydrogenation would allow us to vary the SFI profiles of the resulting fats in a significant manner. Before commercializing a specific product, we were able to study a variety of plant oils that had increasing levels of the C12/C14 substitution, and to examine the effects of hydrogenation on the resultant SFI curves of the final fats. These data are shown in Figure 32.8.

The slope of the SFI curve clearly becomes steeper as the levels of C12 increase. It is also apparent that the tailing of the curve toward the higher melting end (the so-called waxy portion of the curve) is minimized as the C12 content increases. If we examine a specific laurate canola having ~38% C12 content, and vary the degree of hydrogenation (i.e., vary the content of C18:0 versus the C18-unsaturates that occupy the *sn*-2 position), we find the effects on the SFI curves illustrated in Figure 32.9.

As a reference, Figure 32.9 also provides the SFI for palm kernel stearine. The solids profile is bracketed by two laurate canola products: one with an IV of 17 and one with an IV of 37. Thus, we knew that we could effectively match the solids profile, at least in the melting range around body temperature, that most manufacturers required for their products. The next most important piece of information for the use of this new fat in confectionery was knowledge of its crystallization properties. Most lauric fats used in this application crystallize in the β-form, without the need to go through any elaborate tempering step during the processing of the coating mass. Through x-ray crystallography, we were able to show that laurate canola crystallizes predominantly into a β′-crystal. With this knowledge, we then evaluated the product in a standard confectionery coating formulation (Table 32.7) versus commercial lauric fats based on both PKO and coconut oil. The results of these experimental evaluations revealed some significant differences between the laurate canola coatings and those made using palm kernel oil- or coconut oil-based fat systems, including:

1. Significant increased flavor impact with laurate canola
2. Increased coating shelf life (decreased bloom) when laurate canola was used
3. Preferred mouthfeel over standard laurics

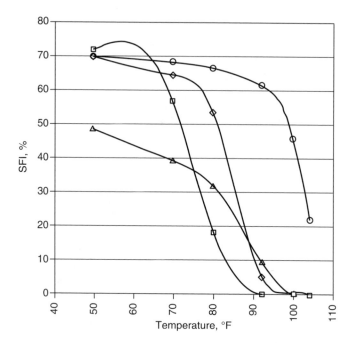

FIGURE 32.9 Solid fat index (SFI) of cocoa butter, palm kernel stearine, and laurical: 38% laurate canola as a function of IV (circles, IV 17; triangles, IV 37) compared with palm kernel stearine (diamonds) with an IV of 7 and cocoa butter (squares) with an IV of 34.

There were some negative findings also, including:

1. Decreased shrinkage on cooling (problems with demolding)
2. Less snap (coating not as hard as comparable controls)
3. Less gloss (reflectance of surface crystals below that of controls, indicating that crystal size was an issue)

All these negative issues are addressed through a combination of formulation techniques and procedures, and through variations in the structural makeup of the laurate canola itself. We believe that this fine-tuning process will be an integral part of any applications development effort for any of the structured triglycerides developed using this technology.

TABLE 32.7
Standard Confectionery Coating Formula

Ingredient	Weight (%)
Sugar 6	49.6
Cocoa powder, natural (10–12)	11.0
Nonfat milk powder	8.0
Laurical 25	28.0
Laurical 15	3.0
Lecithin	0.3
Vanillin	0.1

One of the most significant differences we found for the laurate canola versus the typical laurics used in confectionery coatings was the high degree of compatibility between laurate canola and cocoa butter. Typical lauric fats have limited compatibility with cocoa butter and tend to produce eutectic effects in admixtures that create softer fats than either of the two base fats alone. This result is primarily due to the interference of the crystallization path of the lauric fat on the crystallization dynamics of the cocoa butter. The addition of only a few percent of cocoa butter into a typical lauric fat will result in this softening effect, causing the resulting fat blend to be unsuitable for use in coating applications. With laurate canola, however, such negative interactions with cocoa butter did not occur until significant levels of cocoa butter (about 40% on an oil basis) were admixed. This means that sources of "chocolate" flavor, typically those high in cocoa butter, can now be used to impart more of the desired flavor to the finished goods when laurate canola is used as the base fat for the coating. The actual mechanism of this cocrystallization effect has not been determined, but the functional effects of such blends, as interpreted through SFI curves of the laurate canola–cocoa butter systems, are clear.

Again, finding advantages and relating them to the composition and structure of the base oil and then using breeding and selection to "grow" the optimal oil will be the ongoing efforts. However, once a baseline of information has been developed that relates structure to function, a predictive capability should be established that will significantly shorten the turnaround time required for new product development.

VI. PROSPECTS AND SUMMARY

We have seen that genetic engineering of dramatically new vegetable oil compositions is feasible, through a complex and protracted process leading to an economically viable transgenic crop that is then optimized by a continuing phase of improvements to that new crop variety. The fatty acid composition of vegetable oils is plastic; moreover, new lipid structures that take advantage of those fatty acid building blocks are also possible.

The complexity and cost of practicing this technology and the concomitant planning imply a careful examination of the value and utility of the resulting vegetable oils before a genetic engineering project is begun. However, the potentially most exciting applications are those that create truly novel oils. Such oils typically do not have any extensive history of use in the food industry, and thus there are only small knowledge bases available to help predict value and utility of hypothetical oils. Certainly in the case of laurate canola, the significant performance improvements based on the unique triacylglycerol structure could not have been anticipated because that structure had not been predicted. Industry laboratories cannot justify spending resources to develop application for raw material vegetable oils they cannot obtain. Conversely, biotechnology companies will have trouble in justifying the development of a novel oil if no industry group has demonstrated interest in buying that oil.

There is a long history of working with the established and conventional vegetable oils such as soybean or cottonseed, and therefore a certain familiarity with how they can be used. When truly novel vegetable oil and fatty acid compositions are made available, a careful reexamination of the basic assumptions for how commodity vegetable oils are currently used will be worth while. Important new applications may be possible due to inherently novel chemical composition and structures of vegetable oils created by the modification of crop plant genes. This, then, is a major challenge to lipid chemists in all fields: to understand the contributions of underlying composition and structure of oils to the eventual functionalities, and to use that understanding to intelligently predict the most advantageous uses of genetic engineering technology applied to vegetable oils.

REFERENCES

1. A.J. Del Vecchio. High-laurate canola. *INFORM* 7:230 (1996).
2. V.C. Knauf. Progress in the cloning of genes for plant storage lipid biosynthesis. In: *Genetic Engineering*, Vol. 15 (J.K. Setlow, ed.). Plenum Press, New York, 1993, pp. 149–164.
3. V.C. Knauf. The application of genetic engineering of oilseed crops. *Trends Biotechnol.* 5:40 (1987).
4. R. Töpfer, N. Martini, and J. Schell. Modification of plant lipid biosynthesis. *Science* 268:681 (1995).
5. W.C. Willett. Diet and health: what should we eat? *Science* 264:532 (1994).

Index

A

Aberrant crypt foci (ACF), 636–637
ACAT, *see* Acyl CoA-cholesterol acyltransferase
Accelerated solvent extraction (ASE), 131–132
Acetylated monoglycerides, 415
Acetyl CoA, 516
Acetyl CoA carboxylase, 882
Acetylenic fatty acids, 12–15
Acid/alkali hydrolysis, 128
Acid-catalyzed nucleophilic acyl substitution, 272
Acid degumming, 223
Acid digestion methods, 133
Acid oil, 224
Acidolysis, involving fatty acids, 269
Acinetobacter sp., 712
ACP, *see* Acyl carrier protein
Active oxygen method (AOM), 391
Acyl carrier protein, 516, 875
Acyl CoA-cholesterol acyltransferase, 114
Acylglycerols, 20–22
Adenomatous polyposis coli (APC) gene, 636
Adsorption activities, lipase, 781
Adsorption chromatography, 138–139
4a,5-epoxy-α-tocopherolquinone, formation, 443–444
Aerosol optical thickness (AOT), 778
Agrobacterium tumefaciens, 876
β−Alanyl-ʟ-histidine, *see* Carnosine
Alcoholysis, 269
Alkyl glycosides, 665
Alkylglycosides esters, preparation of, 730
γ−Allylic hydrogen, 333
Altered hepatic foci, 635–636
American Oil Chemists' Society (AOCS), 159
1-Amino-3-iminopropane, 397
Ammonolysis, 730–731
Anaerobic glycolysis, 325
Animal-derived products, CLA content in, 581
Animal fats and marine oils, recovery of, 221
Animal lipases, 779
p-Anisidine value (p-AnV), 395
9-Anthryl-diazomethane (ADAM), 140
Antiobesity effect, 864
Antioxidants
    chain-breaking, 410
    classification of, 410
    enzymes, 339
    in lipid oxidation, 467
    other natural sources of, 427
    in plants, 426
    synthetic and natural, 413–414
Antioxidants in oxidative reactions, mechanisms of
    free radicals inactivation, 476–478
        carotenoids, 484–485

        plant phenolics in, 484
        synthetic phenolic antioxidants, 480–483
        tocopherols, 479–480
        ubiquinone, 483–484
        water-soluble free radical inactivators, 485–486
    interactions, 491–492
    lipid oxidation breakdown products, alterations, 490
    lipid oxidation catalysts control
        prooxidant activity control, 486–487
        singlet oxygen and lipoxygenase activity
            inactivation, 488
    oxidation intermediates inactivation
        peroxides, 489–490
        photoactivated sensitizers, 490
        superoxide anion, 488–489
    surface-active antioxidants and physical effects,
        490–491
AOAC, *see* Association of Official Analytical Chemists
AOCS method, for trans fatty acid determination, 159–160
Apolipoprotein and functions, 553
Arachidonic acid (AA), 365
Arachidonic acid-containing single-cell oil (ARASCO), 711
Archer Daniels Midland (ADM) Company, 862
Argentation, 169
Arthrobacter oxydans, 773
Arthrobacter sp., 711
Ascorbic acid
    as antioxidant, 336
    in lipid oxidation, 467
Ascorbyl palmitate, 730
Aspergillus terreus, 710
Association colloids, 66
Association of Official Analytical Chemists, 4, 127
Atherosclerosis, 502, 546
Atherosclerotic plaque, 562
Atmospheric pressure chemical ionization (APCI), 402
ATR-FTIR infrared absorption spectra, 161–163
Attenuated total reflection (ATR), 160–163
Aureobasidium sp., 719
Autocatalytic free radical chain mechanism, 411
Autoxidation
    of lipids, 137
    mechanism of, 53
Autoxidizable materials, rate of oxidation of, 334
Avena sativa, 742
2,2′-Azobis(2,4-dimethyl-valeronitrile) (AMVN), 371
Azoxymethane (AOM), 636

B

Babcock and Gerber methods, *see* Acid digestion methods
Bacillus subtilis, 713
Bacterial esterases, 773